Instructor's Solutions Manual

to accompany

Functions Modeling Change: A Preparation for Calculus

Fourth Edition

Eric Connally
Harvard University Extension School

Deborah Hughes-Hallett
University of Arizona

Andrew M. Gleason
Harvard University

et al.

John Wiley& Sons, Inc.

This book was set in Times Roman by the Consortium using TeX Mathematica and the package AsTeX, which as written by Alex Kasman. It was printed and bound by G&H Soho. The process was managed by Elliot Marks. Cover photo ©Patrick Zephyr/Patrick Zephyr Nature Photography

Copyright © 2004, 2007, 2011 John Wiley & Sons, Inc. All rights reserved.
No part of this publication may be reproduced, stored in a retrieval system or transmitted in any form or by any means, electronic, mechanical, photocopying, recording, scanning or otherwise, except as permitted under Sections 107 or 108 of the 1976 United States Copyright Act, without either the prior written permission of the Publisher, or authorization through payment of the appropriate per-copy fee to the Copyright Clearance Center, Inc. 222 Rosewood Drive, Danvers, MA 01923, website www.copyright.com. Requests to the Publisher for permission should be addressed to the Permissions Department, John Wiley & Sons, Inc., 111 River Street, Hoboken, NJ 07030-5774, (201)748-6011, fax (201)748-6008, website http://www.wiley.com/go/permissions.

This material is based upon work supported by the National Science Foundation under Grant No. DUE-9352905. Opinions expressed are those of the authors and not necessarily those of the Foundation.

ISBN-13 978- 0-470-54736-6

Printed in the United States of America

10 9 8 7 6 5 4 3 2 1

CONTENTS

CHAPTER ONE

Solutions for Section 1.1

Skill Refresher

S1. Finding the common denominator we get $c + \frac{1}{2}c = \frac{2c+c}{2} = \frac{3c}{2} = \frac{3}{2}c$.

S2. $P(1 + 0.07 + 0.02) = 1.09P$.

S3. $2\pi r^2 + 2\pi r \cdot 2r = 2\pi r^2 + 4\pi r^2 = 6\pi r^2$.

S4. $\dfrac{12\pi - 2\pi}{6\pi} = \dfrac{10\pi}{6\pi} = \dfrac{5}{3}$.

S5. $\left(\frac{1}{2}\right) - 5(-5) = \frac{1}{2} + 25 = \frac{51}{2}$.

S6. $1 - 12(3) + (3)^2 = 1 - 36 + 9 = -26$.

S7. $\dfrac{3}{2-(-1)^3} = \dfrac{3}{2-(-1)} = \dfrac{3}{1+1} = \dfrac{3}{2}$.

S8. $\dfrac{4}{1+\frac{1}{-\frac{3}{4}}} = \dfrac{4}{1-\frac{4}{3}} = \dfrac{4}{-\frac{1}{3}} = -12$.

S9. The figure is a parallelogram, so $A = (-2, 8)$.

S10. The figure is a parallelogram, so $A = (3, 21)$.

Exercises

1. **(a)** On the graph, the high tides occur when the graph is at its highest points. On this particular day, there were two high tides.

(b) The low tides occur when the graph is at its lowest points. There were two low tides on this day.

(c) To find the amount of time elapsed between high tides, find the distance between the two highest points on the graph. It is about 12 hours.

2. $m = f(v)$.

3. $w = f(c)$.

4. Appropriate axes are shown in Figure 1.1.

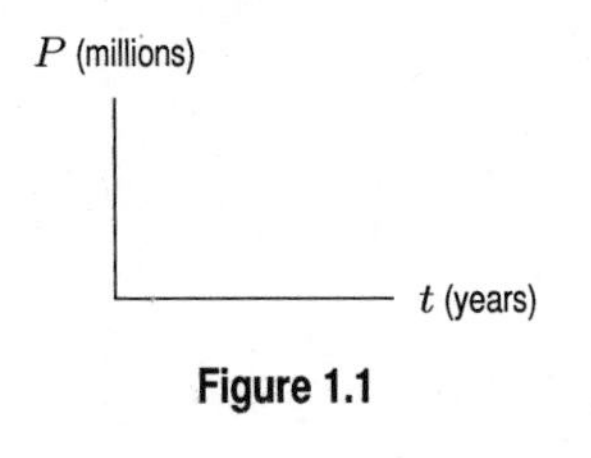

Figure 1.1

5. Appropriate axes are shown in Figure 1.2.

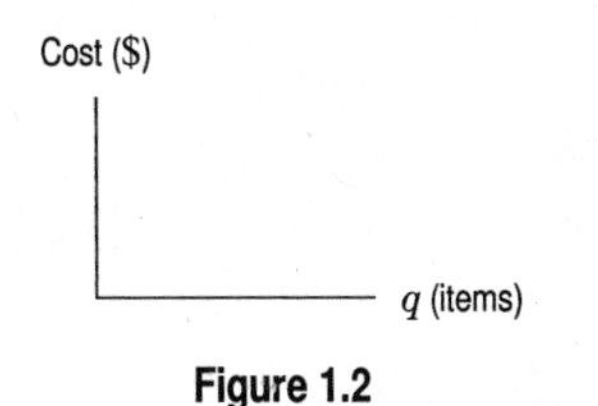

Figure 1.2

6. Appropriate axes are shown in Figure 1.3.

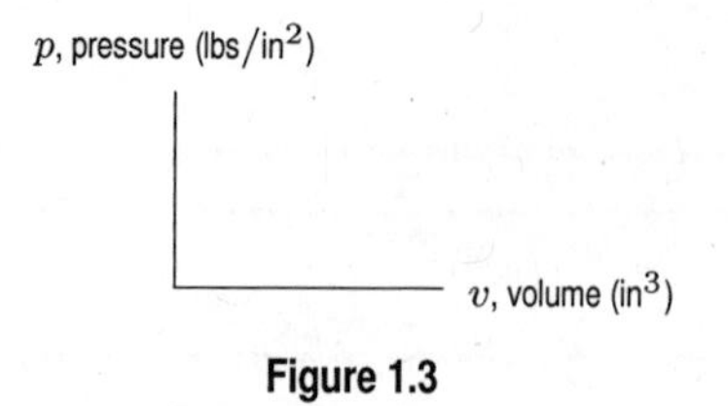

Figure 1.3

7. Appropriate axes are shown in Figure 1.4.

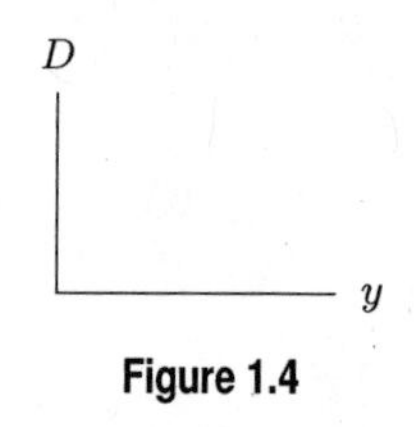

Figure 1.4

8. These data are plotted in Figure 1.5. The independent variable is A and the dependent variable is n.

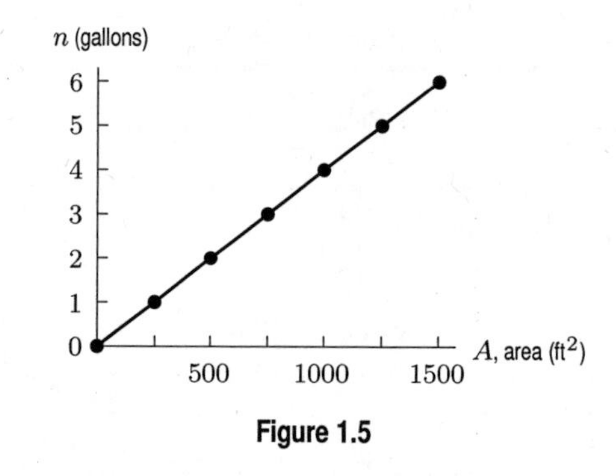

Figure 1.5

9. **(a)** Since $f(x)$ is 4 when $x = 0$, we have $f(0) = 4$.
(b) Since $x = 3$ when $f(x) = 0$, we have $f(3) = 0$.
(c) $f(1) = 2$
(d) There are two x values leading to $f(x) = 1$, namely $x = 2$ and $x = 4$. So $f(2) = 1$ and $f(4) = 1$.

10. **(a)** Since the vertical intercept is $(0, 40)$, we have $f(0) = 40$.
(b) Since the horizontal intercept is $(2, 0)$, we have $f(2) = 0$.

11. $f(6.9) = 2.9$.

12. $(2.2, 2.9)$; $(6.1, 4.9)$

13. Since $f(0) = f(4) = f(8) = 0$, the solutions are $x = 0, 4, 8$.

14. Since the graphs touch at $x = 2.2$ and $x = 6.1$, these are the solutions.

15. **(a)** w goes on the horizontal axis
(b) $(-4, 10)$
(c) $(6, 1)$

16. **(a)** No. Because there can be two different points sharing the same x-coordinate. For example, when $x = 0$, $y = 1$ or $y = -1$. So for each value of x, there is not a unique value of y.
(b) Yes. Because on the semi-circle above the x-axis there is only one point for each x-coordinate. Thus, each x-value corresponds to at most one y-value.

17. **(a)** Yes. For each value of s, there is exactly one area.
(b) No. Suppose $s = 4$ represents the length of the rectangle. The width could have any other value, say 7 or 1.5 or π or .... In this case, for one value of s, there are infinitely many possible values for A, so the area of a rectangle is not a function of the length of one of its sides.

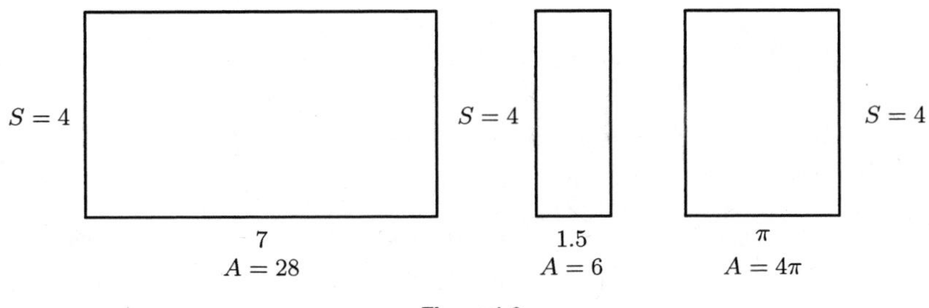

Figure 1.6

18. We apply the vertical-line test to each graph. As you can see, only in (a) do all vertical lines intersect only one point on the graph. So graph (a) defines the only function.

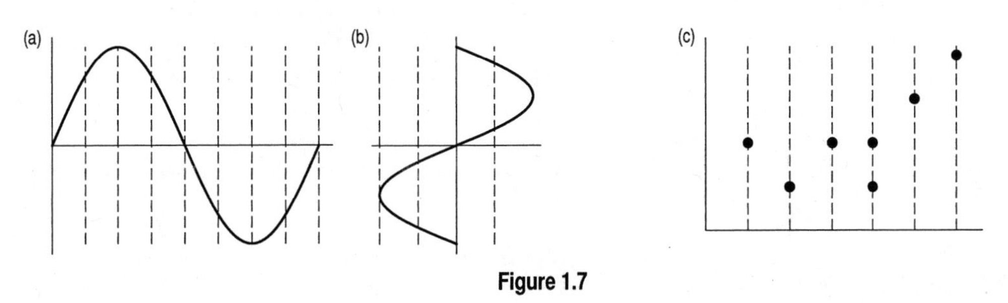

Figure 1.7

Problems

19. A possible graph is shown in Figure 1.8

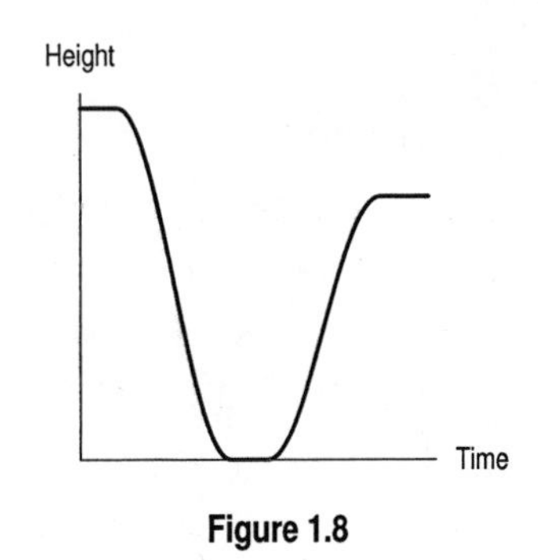

Figure 1.8

20. Figure 1.9 shows a possible graph of blood sugar level as a function of time over one day. Note that the actual curve is smooth, and does not have any sharp corners.

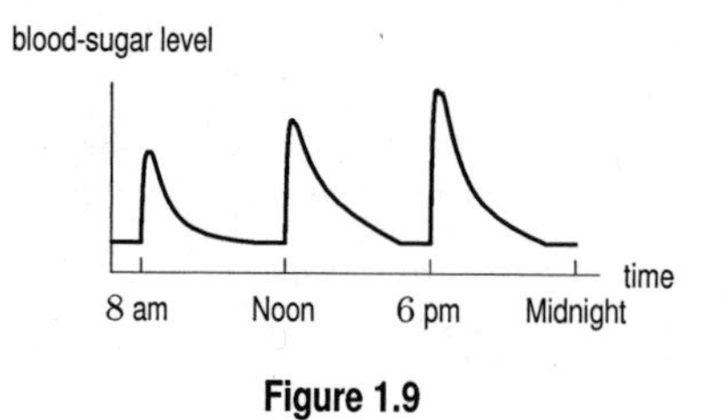

Figure 1.9

21. **(a)** The number of people who own cell phones in the year 2000 is 100,300,000.
(b) There are 20,000,000 people who own cell phones a years after 1990.
(c) There will be b million people who own cell phones in the year 2010.
(d) The number n is the number of people (in millions) who own cell phones t years after 1990.

22. The value of N is not necessarily a function of G, since each value of G does not need to have a unique value of N associated to it. For example, suppose we choose the value of G to be a B. There may be more than one student who received a B, so there may be more than one ID number corresponding to B.

The value of G must be a function of N, because each ID number (each student) receives exactly one grade. Therefore each value of N has a unique value of G associated with it. Writing $G = f(N)$ indicates that the ID number is the input which uniquely determines the grade, the output.

23. **(a)** In 1995, or year $t = 0$, the ranking for Hannah was 7, making it most popular, and the ranking for Madison was 29, making it least popular.
(b) In 2004, or year $t = 9$, the ranking for Madison was 3, making it the most popular, and the ranking for Alexis was 11, making it least popular.

24. **(a)** We have $r_m(0) - r_h(0) = 29 - 7 = 22$. This tells us that in 1995 (year $t = 0$), the name Hannah was ranked 22 places higher than Madison on the list of most popular names. (Recall that the lower the ranking, the higher a name's position on the list.)
(b) We have $r_m(9) - r_h(9) = 3 - 5 = -2$. This tells us that in 2004 (year $t = 9$), the name Hannah was ranked 2 places lower than Madison on the list of most popular names.
(c) We have $r_m(t) < r_a(t)$ for $t = 5$ to $t = 9$. This tells us that the name Madison was ranked higher than the name Alexis on the list of most popular names in the years 2000 to 2004.

25. **(a)** At 40 mph, fuel consumption is about 28 mpg, so the fuel used is $300/28 = 10.71$ gallons.
(b) At 60 mph, fuel consumption is about 29 mpg. At 70 mph, fuel consumption is about 28 mpg. Therefore, on a 200 mile trip

$$\text{Fuel saved} = \frac{200}{28} - \frac{200}{29} = 0.25 \text{ gallons.}$$

(c) The most fuel-efficient speed is where mpg is a maximum, which is about 55 mph.

26. **(a)** When there is no snow, it is equivalent to no rain. Thus, the vertical intercept is 0. Since every ten inches of snow is equivalent to one inch of rain, we can specify the slope as

$$\frac{\Delta\text{rain}}{\Delta\text{snow}} = \frac{1}{10} = 0.1$$

We have a vertical intercept of 0 and a slope of 0.1. Thus, the equation is: $r = f(s) = 0.1s$.
(b) By substituting 5 in for s, we get $f(5) = 0.1(5) = 0.5$ This tells us that five inches of snow is equivalent to approximately $1/2$ inch of rain.
(c) Substitute 5 inches for $r = f(s)$ in the equation: $5 = 0.1s$. Solving gives $s = 50$. Five inches of rain is equivalent to approximately 50 inches of snow.

27. Figure 1.10 shows the tank.

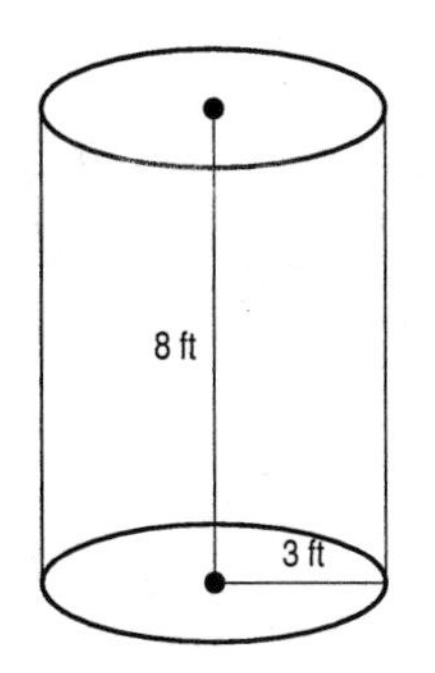

Figure 1.10: Cylindrical water tank

(a) The volume of a cylinder is equal to the area of the base times the height, where the area of the base is πr^2. Here, the radius of the base is $(1/2)(6) = 3$ ft, so the area is $\pi \cdot 3^2 = 9\pi$ ft^2. Therefore, the capacity of this tank is $(9\pi)8 = 72\pi$ ft^3.

(b) If the height of the water is 5 ft, the volume becomes $(9\pi)5 = 45\pi$ ft^3.

(c) In general, if the height of water is h ft, the volume of the water is $(9\pi)h$. If we let $V(h)$ be the volume of water in the tank as a function of its height, then

$$V(h) = 9\pi h.$$

Note that this function only makes sense for a non-negative value of h, which does not exceed 8 feet, the height of the tank.

28. **(a)** Since the person starts out 5 miles from home, the vertical intercept on the graph must be 5. Thus, (i) and (ii) are possibilities. However, since the person rides 5 mph away from home, after 1 hour the person is 10 miles from home. Thus, (ii) is the correct graph.

(b) Since this person also starts out 5 miles from home, (i) and (ii) are again possibilities. This time, however, the person is moving at 10 mph and so is 15 miles from home after 1 hour. Thus, (i) is correct.

(c) The person starts out 10 miles from home so the vertical intercept must be 10. The fact that the person reaches home after 1 hour means that the horizontal intercept is 1. Thus, (v) is correct.

(d) Starting out 10 miles from home means that the vertical intercept is 10. Being half way home after 1 hour means that the distance from home is 5 miles after 1 hour. Thus, (iv) is correct.

(e) We are looking for a graph with vertical intercept of 5 and where the distance is 10 after 1 hour. This is graph (ii). Notice that graph (iii), which depicts a bicyclist stopped 10 miles from home, does not match any of the stories.

29. **(a)** 69°F

(b) July 17th and 20th

(c) Yes. For each date, there is exactly one low temperature.

(d) No, it is not true that for each low temperature, there is exactly one date: for example, 73° corresponds to both the 17th and 20th.

30. **(a)** Figure 1.11 shows the plot of R versus t. R is a function of t because no vertical line intersects the graph in more than one place.

(b) Figure 1.12 shows the plot of F versus t. F is a function of t because no vertical line intersects the graph in more than one place.

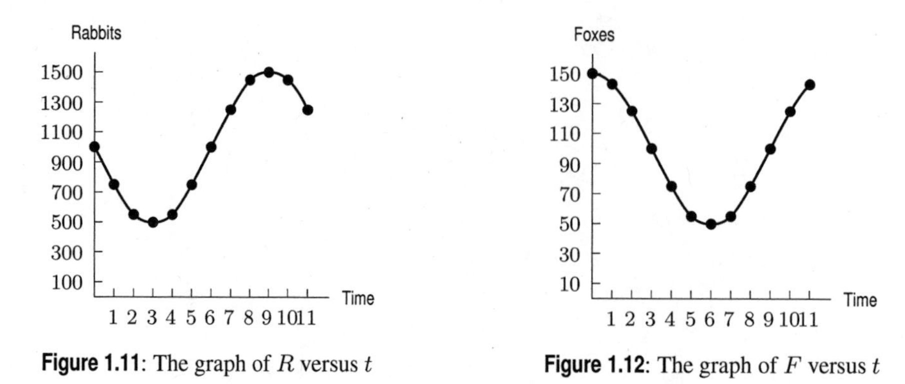

Figure 1.11: The graph of R versus t

Figure 1.12: The graph of F versus t

(c) Figure 1.13 shows the plot of F versus R. We have also drawn the vertical line corresponding to $R = 567$. This tells us that F is not a function of R because there is a vertical line that intersects the graph twice. In fact the lines $R = 567$, $R = 750$, $R = 1000$, $R = 1250$, and $R = 1433$ all intersect the graph twice. However, the existence of any one of them is enough to guarantee that F is not a function of R.

(d) Figure 1.14 shows the plot of R versus F. We have drawn the vertical line corresponding to $F = 57$. This tells us that R is not a function of F because there is a vertical line that intersects the graph twice. In fact the lines $F = 57$, $F = 75$, $F = 100$, $F = 125$, and $F = 143$ all intersect the graph twice. However, the existence of any one of them is enough to guarantee that R is not a function of F.

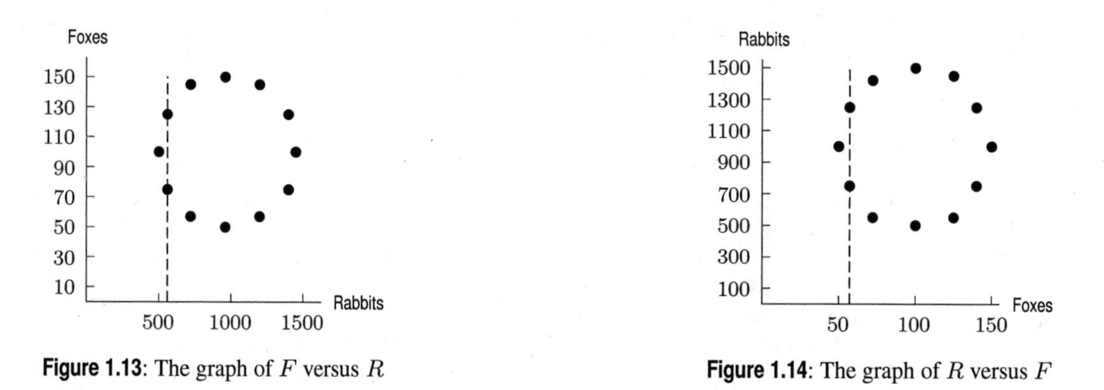

Figure 1.13: The graph of F versus R

Figure 1.14: The graph of R versus F

31. **(a)** No, in the year 1954 there were two world records; in the year 1981 there were three world records.
(b) Yes, each world record occurred in only one year.
(c) The world record of 3 minutes and 47.33 seconds was set in 1981.
(d) The statement $y(3{:}51.1) = 1967$ tells us that the world record of 3 minutes, 51.1 seconds was set in 1967.

32. **(a)** From the table, we see that $f(100) = 625$. This means that there is approximately \$625 billion worth of \$100 bills in circulation in the United States.
(b) To determine the number of \$5 bills, we divide 11 by 5. Thus, we have 2.2 billion \$5 bills in circulation. The number of \$1 bills is the same as the value, so there are 9.5 billion \$1 bills in circulation. There are more \$1 bills.

33. **(a)** Adding the male total to the female total gives $x + y$, the total number of applicants.
(b) Of the men who apply, 15% are accepted. So $0.15x$ male applicants are accepted. Likewise, 18% of the women are accepted so we have $0.18y$ women accepted. Summing the two tells us that $0.15x + 0.18y$ applicants are accepted.
(c) The number accepted divided by the number who applied times 100 gives the percentage accepted. This expression is

$$\frac{(0.15)x + (0.18)y}{x + y}(100), \quad \text{or} \quad \frac{15x + 18y}{x + y}.$$

34. Since the tax is $0.06P$, the total cost would be the price of the item plus the tax,or

$$C = P + 0.06P = 1.06P.$$

35. Let $A(r)$ be the area of a circle expressed as a function of its radius. Then

$$A(r) = \pi r^2.$$

If the radius is increased by 10%, then it is 110% of its original length. So we want to know the output when our input is $1.1r$:

$$A(1.1r) = \pi(1.1r)^2 = 1.21\pi r^2.$$

The new area is the old area multiplied by 1.21. So the new area is 121% of the old area. In other words, the area of a circle is increased by 21% when its radius is increased by 10%.

36. The original price is P. Inflation causes a 5% increase, giving

$$\text{Inflated price} = P + 0.05P = 1.05P.$$

Then there is a 10% decrease, giving

$$\begin{aligned} \text{Final price} &= 90\%(\text{Inflated price}) \\ &= 0.9(1.05P) \\ &= 0.945P. \end{aligned}$$

37. (a)

Table 1.1 *Relationship between cost, C, and number of liters produced, l*

l (millions of liters)	0	1	2	3	4	5
C (millions of dollars)	2.0	2.5	3.0	3.5	4.0	4.5

(b) The cost, C, consists of a fixed cost of \$2 million plus a variable cost of \$0.50 million per million liters produced. If l millions of liters are produced, the total variable costs are $(0.5)l$. Thus, the total cost C in millions of dollars is given by

$$C = \text{Fixed cost} + \text{Variable cost},$$

so

$$C = 2 + (0.5)l.$$

38. (a) Yes. If the person walks due west and then due north, the distance from home is represented by the hypotenuse of the right triangle that is formed (see Figure 1.15).

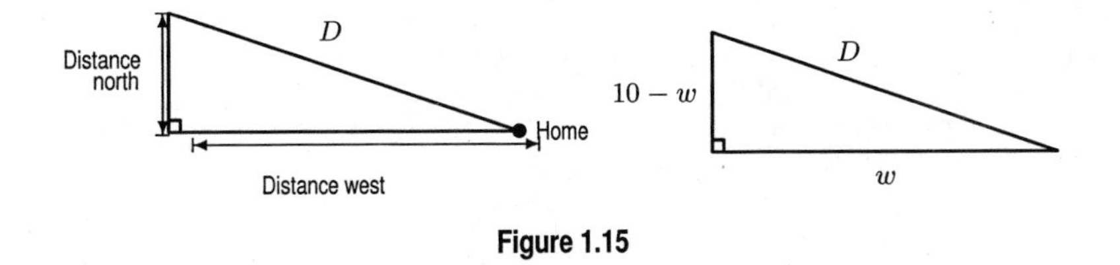

Figure 1.15

If the distance west is w miles and the total distance walked is 10 miles, then the distance north is $10 - w$ miles. We can use the Pythagorean Theorem to find that

$$D = \sqrt{w^2 + (10 - w)^2}.$$

So, for each value of w, there is a unique value of D given by this formula. Thus, the definition of a function is satisfied.

(b) No. Suppose she walks 10 miles, that is, $x = 10$. She might walk 1 mile west and 9 miles north, or 2 miles west and 8 miles north, or 3 miles west and 7 miles north, and so on. The right triangles in Fig 1.16 show three different routes she could take and still walk 10 miles.

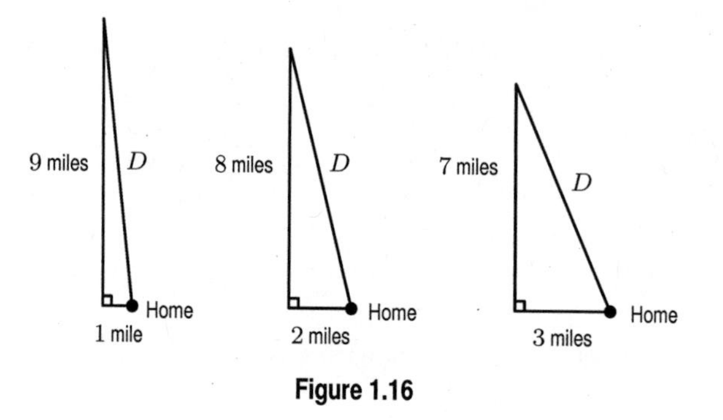

Figure 1.16

Each situation gives a different distance from home. The Pythagorean Theorem shows that the distances from home for these three examples are

$$D = \sqrt{1^2 + 9^2} = 9.06,$$
$$D = \sqrt{2^2 + 8^2} = 8.25,$$
$$D = \sqrt{3^2 + 7^2} = 7.62.$$

Thus, the distance from home cannot be determined from the distance walked.

Solutions for Section 1.2

Skill Refresher

S1. $\frac{4-6}{3-2} = \frac{-2}{1} = -2.$

S2. $\frac{1-3}{2^2-(-3)^2} = \frac{-2}{4-(+9)} = \frac{-2}{4-9} = \frac{-2}{-5} = \frac{2}{5}.$

S3. $\frac{-3-(-9)}{-1-2} = \frac{-3+9}{-3} = \frac{6}{-3} = -2.$

S4. $\frac{(1-3^2)-(1-4^2)}{3-4} = \frac{(1-9)-(1-16)}{-1} = \frac{-8+15}{-1} = -7.$

S5. $\frac{\frac{1}{2}-(-4)^2-(\frac{1}{2}-(5^2))}{-4-5} = \frac{\frac{1}{2}-16-\frac{1}{2}+25}{-9} = \frac{9}{-9} = -1.$

S6. $2(x+a) - 3(x-b) = 2x + 2a - 3x + 3b = -x + 2a + 2b.$

S7.

$$\begin{aligned} x^2 - (2x+a)^2 &= x^2 - (4x^2 + 4xa + a^2) \\ &= x^2 - 4x^2 - 4xa - a^2 \\ &= -3x^2 - 4ax - a^2. \end{aligned}$$

S8.

$$\begin{aligned} 4x^2 - (x-b)^2 &= 4x^2 - (x^2 - 2xb + b^2) \\ &= 4x^2 - x^2 + 2xb - b^2 \\ &= 3x^2 + 2bx - b^2. \end{aligned}$$

S9. $\frac{x^2-\frac{3}{4}-(y^2-\frac{3}{4})}{x-y} = \frac{x^2-\frac{3}{4}-y^2+\frac{3}{4}}{x-y} = \frac{x^2-y^2}{x-y} = \frac{(x-y)(x+y)}{x-y} = x+y.$

S10.

$$\begin{aligned}\frac{2(x+h)^2-2x^2}{(x+h)-x} &= \frac{2(x^2+2xh+h^2)-2x^2}{h}\\ &= \frac{2x^2+4xh+2h^2-2x^2}{h}\\ &= \frac{4xh+2h^2}{h}\\ &= 4x+2h.\end{aligned}$$

Exercises

1. We take the change in number, n, of CDs divided by the change in time, t years, to determine the average rate of change.

(a) $\frac{\Delta n}{\Delta t} = \frac{120-40}{2008-2005} = \frac{80}{3}$, so the average rate of change is $80/3$ CDs per year.

(b) $\frac{\Delta n}{\Delta t} = \frac{40-120}{2012-2008} = \frac{-80}{4} = -20$, so the average rate of change is -20 CDs per year. That is, you have on average 20 fewer CDs per year.

(c) $\frac{\Delta n}{\Delta t} = \frac{40-40}{2012-2005} = \frac{0}{7} = 0$, so the average rate of change is zero CDs per year.

2. **(a)** Let $s = V(t)$ be the sales (in millions) of VCRs in year t. Then

$$\begin{aligned}\text{Average rate of change of } s \text{ from } t=1998 \text{ to } t=2000 &= \frac{\Delta s}{\Delta t} = \frac{V(2000)-V(1998)}{2000-1998}\\ &= \frac{1869-2409}{2}\\ &= -270 \text{ million VCRs/year.}\end{aligned}$$

Let $q = D(t)$ be the sales (in millions) of DVD players in year t. Then

$$\begin{aligned}\text{Average rate of change of } q \text{ from } t=1998 \text{ to } t=2000 &= \frac{\Delta q}{\Delta t} = \frac{D(2000)-D(1998)}{2000-1998}\\ &= \frac{1717-421}{2}\\ &= 648 \text{ million DVD players/year.}\end{aligned}$$

(b) By the same argument

$$\begin{aligned}\text{Average rate of change of } s \text{ from } t=2000 \text{ to } t=2003 &= \frac{\Delta s}{\Delta t} = \frac{V(2003)-V(2000)}{2003-2000}\\ &= \frac{407-1869}{3}\\ &= -487.33 \text{ million VCRs/year.}\end{aligned}$$

$$\begin{aligned}\text{Average rate of change of } q \text{ from } t=2000 \text{ to } t=2003 &= \frac{\Delta q}{\Delta t} = \frac{D(2003)-D(2000)}{2003-2000}\\ &= \frac{3050-1717}{3}\\ &= 444.33 \text{ million DVD players/year.}\end{aligned}$$

(c) The fact that $\Delta s/\Delta t = -270$ tells us that VCR sales decreased at an average rate of 270 million VCRs/year between 1998 and 2000. The fact that the average rate of change is negative tells us that annual sales are decreasing.

The fact that $\Delta s/\Delta t = -487.33$ tells us that VCR sales decreased at an average rate of 487.33 million VCRs/year between 2000 and 2003.

The fact that $\Delta q/\Delta t = 648$ means that DVD player sales increased at an average rate of 648 million players/year between 1998 and 2000. The fact that $\Delta q/\Delta t = 444.33$ means that LP sales increased at an average rate of 444.33 million players/year between 2000 and 2003.

3. To decide if VCR sales are an increasing or decreasing function of DVD player sales, we must read the table in the direction in which DVD player sales increase. This means we read the table from right to left. As the number of DVD player sales increases, the number of VCR sales decrease. Thus, VCR sales are a decreasing function of DVD player sales.

4. **(a)** For 1990 to 2000, the rate of change of P_1 is

$$\frac{\Delta P_1}{\Delta t} = \frac{83-53}{2000-1990} = \frac{30}{10} = 3 \text{ hundred people per year,}$$

while for P_2 we have

$$\frac{\Delta P_2}{\Delta t} = \frac{70-85}{2000-1990} = \frac{-15}{10} = -1.5 \text{ hundred people per year.}$$

(b) For 1995 to 2007,

$$\frac{\Delta P_1}{\Delta t} = \frac{93-73}{2007-1995} = \frac{20}{12} = 1.67 \text{ hundred people per year,}$$

and

$$\frac{\Delta P_2}{\Delta t} = \frac{65-75}{2007-1995} = \frac{-10}{12} = -0.83 \text{ hundred people per year.}$$

(c) For 1990 to 2007,

$$\frac{\Delta P_1}{\Delta t} = \frac{93-53}{2007-1990} = 2.35 \text{ hundred people per year,}$$

and

$$\frac{\Delta P_2}{\Delta t} = \frac{65-85}{2007-1990} = -1.18 \text{ hundred people per year.}$$

5. We have

$$\begin{aligned}\frac{\Delta f}{\Delta x} &= \frac{f(6.1)-f(2.2)}{6.1-2.2} \\ &= \frac{4.9-2.9}{3.9} \\ &= 0.513.\end{aligned}$$

6. There are many such intervals. One way to find them is to look for points at which a horizontal line (a line of slope $m = 0$) intersects the graph. Here are some possible intervals: $0 \le x \le 4$, $0 \le x \le 8$, $4 \le x \le 8$, $2.2 \le x \le 5.2$, $2.2 \le x \le 6.9$, $5.2 \le x \le 6.9$.

7. Using the points on g

$$\text{Average rate of change} = \frac{g(6.1)-g(2.2)}{6.1-2.2} = \frac{4.9-2.9}{6.1-2.2} = 0.513.$$

8. They are equal; both are given by

$$\frac{4.9-2.9}{6.1-2.2}.$$

9. **(a)** Negative
(b) Positive

10. We have $G(3) - G(-1) > 0$.

11. We have $F(-2) > F(2)$.

12. (a) (i) After 2 hours 60 miles had been traveled. After 5 hours, 150 miles had been traveled. Thus on the interval from $t = 2$ to $t = 5$ the value of Δt is

$$\Delta t = 5 - 2 = 3$$

and the value of ΔD is

$$\Delta D = 150 - 60 = 90.$$

(ii) After 0.5 hours 15 miles had been traveled. After 2.5 hours, 75 miles had been traveled. Thus on the interval from $t = 0.5$ to $t = 2.5$ the value of Δt is

$$\Delta t = 2.5 - .5 = 2$$

and the value of ΔD is

$$\Delta D = 75 - 15 = 60.$$

(iii) After 1.5 hours 45 miles had been traveled. After 3 hours, 90 miles had been traveled. Thus on the interval from $t = 1.5$ to $t = 3$ the value of Δt is

$$\Delta t = 3 - 1.5 = 1.5$$

and the value of ΔD is

$$\Delta D = 90 - 45 = 45.$$

(b) For the interval from $t = 2$ to $t = 5$, we see

$$\text{Rate of change} = \frac{\Delta D}{\Delta t} = \frac{90}{3} = 30.$$

For the interval from $t = 0.5$ to $t = 2.5$, we see

$$\text{Rate of change} = \frac{\Delta D}{\Delta t} = \frac{60}{2} = 30.$$

For the interval from $t = 1.5$ to $t = 3$, we see

$$\text{Rate of change} = \frac{\Delta D}{\Delta t} = \frac{45}{1.5} = 30.$$

This suggests that the average speed is 30 miles per hour throughout the trip.

Problems

13. (a) The coordinates of point A are $(10, 30)$.
The coordinates of point B are $(30, 40)$.
The coordinates of point C are $(50, 90)$.
The coordinates of point D are $(60, 40)$.
The coordinates of point E are $(90, 40)$.

(b) From Figure 1.17, we see that F is on the graph but G is not.

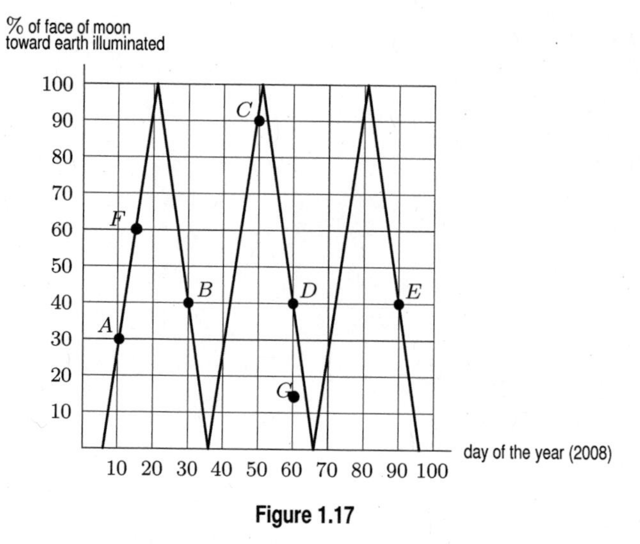

Figure 1.17

(c) The function is increasing from, approximately, days 6 through 21, 36 through 51, and 66 through 81.
(d) The function is decreasing from, approximately, days 22 through 35, 52 through 65, and 82 through 96.

14. A knowledge of when the record was established determines the world record time, so the world record time is a function of the time it was established. Also, when a world record is established it is smaller than the previous world record and occurs later in time. Thus, it is a decreasing function. Because a world record could be established twice in the same year, a knowledge of the year does not determine the world record time, so the world record time is not a function of the year it was established.

15. **(a)** Between $(-2,-7)$ and $(3,3)$,

$$\text{Average rate of change} = \frac{\Delta y}{\Delta x} = \frac{3-(-7)}{3-(-2)} = \frac{10}{5} = 2.$$

(b) The function is increasing over this interval, since the average rate of change is positive.
(c) As the x-values increase, so do the y-values. See Figure 1.18. Thus, this function is increasing everywhere.

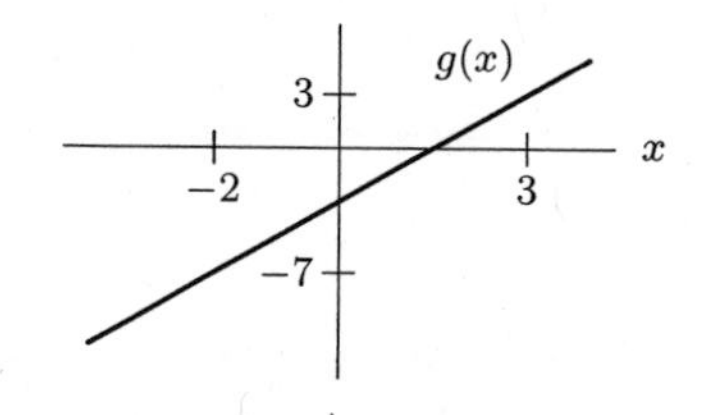

Figure 1.18

16. **(a)** (i) We have

$$\frac{f(2)-f(0)}{2-0} = \frac{16-2^2-(16-0)}{2} = -\frac{4}{2} = -2.$$

This means $f(x)$ decreases by an average of 2 units per unit change in x on the interval $0 \le x \le 2$.

(ii) We have

$$\frac{f(4)-f(2)}{4-2} = \frac{16-(4)^2-(16-2^2)}{2} = \frac{-16+4}{2} = -6.$$

This means $f(x)$ decreases by an average of 6 units per unit change in x on the interval $2 \le x \le 4$.

(iii) We have

$$\frac{f(4)-f(0)}{4-0}=\frac{16-(4)^2-(16-0)}{4}=-\frac{16}{4}=-4.$$

This means $f(x)$ decreases by an average of 4 units per unit change in x on the interval $0 \leq x \leq 4$.

(b) The graph of $f(x)$ is the solid curve in Figure 1.19. The secants corresponding to each rate of change are shown as dashed lines. The average rate of decrease is greatest on the interval $2 \leq x \leq 4$.

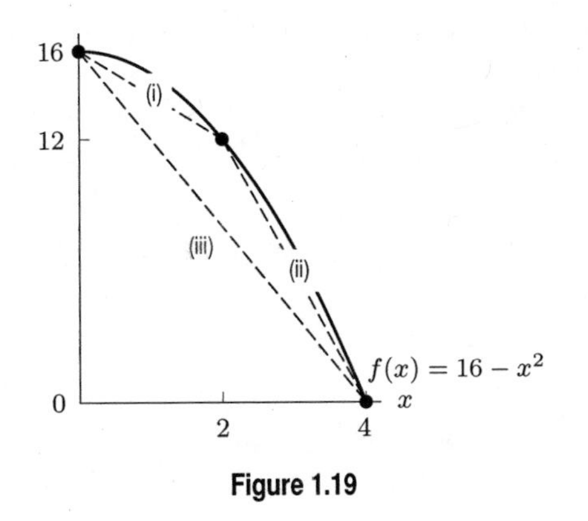

Figure 1.19

17. **(a)** We see from the figure that at time $t = 0$ the population of Town A is 5000 people, whereas the population of Town B is 10,000 people. Thus, Town B starts with more people than Town A.

(b) Town A grows from a population of 5000 at time $t = 0$ to a population of 25,000 at $t = 50$. Thus, it grows by $25{,}000 - 5{,}000 = 20{,}000$ people during these 50 years, and the

$$\begin{aligned}\begin{matrix}\text{average rate of change}\\ \text{of population A}\end{matrix} &= \frac{\text{change in value of } P}{\text{change in value of } t}\\ &= \frac{20{,}000 \text{ people}}{50 \text{ years}}\\ &= 400 \text{ people/year.}\end{aligned}$$

On the other hand, Town B grows from a population of 10,000 at time $t = 0$ to a population of 20,000 at time $t = 50$. Thus, it grows by $20{,}000 - 10{,}000 = 10{,}000$ people during these 50 years, and the

$$\begin{aligned}\begin{matrix}\text{average rate of change}\\ \text{of population B}\end{matrix} &= \frac{\text{change in value of } P}{\text{change in value of } t}\\ &= \frac{10{,}000 \text{ people}}{50 \text{ years}}\\ &= 200 \text{ people/year.}\end{aligned}$$

Thus, during the time interval shown, Town A is growing twice as fast as Town B.

18. You start with nothing. Each year, on average, your net worth increases by \$5000, so in forty years, you have 40 yrs $\cdot$ \$5000/yr = \$200,000.

19.

$$\begin{aligned}\text{Average rate of change} &= \frac{\text{Temp at 7:32 am} - \text{Temp at 7:30 am}}{7{:}32 - 7{:}30}\\ &= \frac{45-(-4)}{2} = 24.5 \text{ degrees/minute.}\end{aligned}$$

20. **(a)** The number of sunspots, s, is a function of the year, t, because knowing the year is enough to uniquely determine the number of sunspots. The graph passes the vertical line test.

(b) When read from left to right, the graph increases from approximately $t = 1964$ to $t = 1969$, from approximately 1971 to 1972, from approximately $t = 1976$ to $t = 1979$, from approximately 1986 to 1989, from approximately 1990 to 1991 and from approximately 1996 to 2000. Thus, s is an increasing function of t on the intervals $1964 < t < 1969$, $1971 < t < 1972$, $1976 < t < 1979$, $1986 < t < 1989$, and $1996 < t < 2000$. For each of these intervals, the average rate of change must be positive.

21. **(a)** According to the table, a 200-lb person uses 5.4 calories per minute while walking. Since a half hour is 30 minutes, a half-hour walk uses $(5.4)(30) = 162$ calories.

(b) A 120-lb swimmer uses 6.9 calories per minute. Thus, in one hour the swimmer uses $(6.9)(60) = 414$ calories. A 220-lb bicyclist uses 11.9 calories per minute. In a half-hour, the bicyclist uses $(11.9)(30) = 357$ calories. Thus, the swimmer uses more calories.

(c) Increases, since the numbers 2.7, 3.2, 4.0, 4.6, 5.4, 5.9 are increasing.

22. According to the table in the text, the tree has 139μg of carbon-14 after 3000 years from death and 123μg of carbon-14 after 4000 years from death. Because the function $L = g(t)$ is decreasing, the tree must have died between 3,000 and 4,000 years ago.

23. **(a)** Between $(1, 4)$ and $(2, 13)$,

$$\text{Average rate of change } = \frac{\Delta y}{\Delta x} = \frac{13-4}{2-1} = 9.$$

(b) Between (j, k) and (m, n),

$$\text{Average rate of change } = \frac{\Delta y}{\Delta x} = \frac{n-k}{m-j}.$$

(c) Between $(x, f(x))$ and $(x+h, f(x+h))$,

$$\begin{aligned}\text{Average rate of change } = \frac{\Delta y}{\Delta x} &= \frac{(3(x+h)^2+1)-(3x^2+1)}{(x+h)-x}\\ &= \frac{(3(x^2+2xh+h^2)+1)-(3x^2+1)}{h}\\ &= \frac{3x^2+6xh+3h^2+1-3x^2-1}{h}\\ &= \frac{6xh+3h^2}{h}\\ &= 6x+3h.\end{aligned}$$

24. **(a)** From the graph, we see that $g(4) \approx 2$ and $g(0) \approx 0$. Thus,

$$\frac{g(4)-g(0)}{4-0} \approx \frac{2-0}{4-0} = \frac{1}{2}.$$

(b) The line segment joining the points in part (a), as well as the line segment in part (d), is shown on the graph in Figure 1.20.

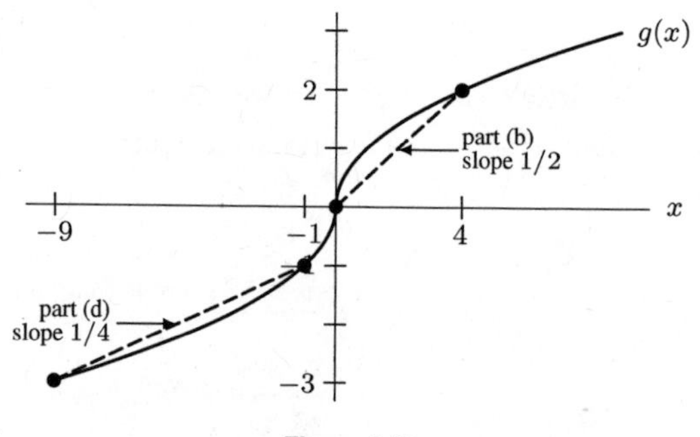

Figure 1.20

(c) From the graph, $g(-9) \approx -3$ and $g(-1) \approx -1$. Thus,

$$\frac{g(b)-g(a)}{b-a} \approx \frac{-1-(-3)}{-1-(-9)} = \frac{2}{8} = \frac{1}{4}.$$

(d) The line segment in part (c) with slope $(1/4)$ is shown in Figure 1.20.

25. **(a)** Since Δt refers to the change in the numbers of years, we calculate

$$\Delta t = 1970 - 1960 = 10, \qquad \Delta t = 1980 - 1970 = 10, \qquad \text{and so on until 2000.}$$

So from 1960 to 2000 $\Delta t = 10$ for all consecutive entries. From 2000 to 2007 $\Delta t = 7$, and from 2007 to 2008,$\Delta t =$ 1.

(b) Since ΔG is the change in the amount of garbage produced per year, for the period 1960-1970 we have

$$\Delta G = 121.1 - 88.1 = 30.$$

Continuing in this way gives the Table 1.2:

Table 1.2

Time period	1960–70	1970-80	1980–90	1990–2000	2000–2007	2007–2008
ΔG	30	30.5	53.6	33.9	115.5	−5

(c) Not all of the ΔG values are the same. We also know that not all the values of Δt are the same. Computing $\Delta G/\Delta t$, we see the average rate of change in the amount of garbage produced each year, is not constant. This tells us that the amount of garbage being produced each year is changing, but not at a constant rate.

(d) In 2007 the United States embarked on a recycling and composting program.

26. **(a)** Table 1.3 shows the average rate of change of distance, commonly called the average speed or average velocity.

Table 1.3 *Carl Lewis' times at 10 meter intervals*

Time (sec)	Distance (meters)	$\Delta d/\Delta t$ (meters/sec)
0.00 to 1.94	0 to 10	5.15
1.94 to 2.96	10 to 20	9.80
2.96 to 3.91	20 to 30	10.53
3.91 to 4.78	30 to 40	11.49
4.78 to 5.64	40 to 50	11.63
5.64 to 6.50	50 to 60	11.63
6.50 to 7.36	60 to 70	11.63
7.36 to 8.22	70 to 80	11.63
8.22 to 9.07	80 to 90	11.76
9.07 to 9.93	90 to 100	11.63

(b) He attained his maximum speed (11.76 meters/sec) between 80 and 90 meters. He does not appear to be running his fastest when he crossed the finish line.

Solutions for Section 1.3

Skill Refresher

S1. We have $f(0) = \frac{2}{3}(0) + 5 = 5$ and $f(3) = \frac{2}{3}(3) + 5 = 2 + 5 = 7$.

S2. We have $f(0) = 17 - 4(0) = 17$ and $f(3) = 17 - 4(3) = 17 - 12 = 5$.

S3. We have $f(2) - f(0) = 3 - (-2) = 5$.

S4. We have $f(2) - f(0) = (-1) - (2) = -3$.

S5. To find the y-intercept, we let $x = 0$,

$$\begin{aligned} y &= -4(0) + 3 \\ &= 3. \end{aligned}$$

To find the x-intercept, we let $y = 0$,

$$\begin{aligned} 0 &= -4x + 3 \\ 4x &= 3 \\ x &= \frac{3}{4}. \end{aligned}$$

S6. To find the y-intercept, we let $x = 0$,

$$\begin{aligned} 5(0) - 2y &= 4 \\ -2y &= 4 \\ y &= -2. \end{aligned}$$

To find the x-intercept, we let $y = 0$,

$$\begin{aligned} 5x - 2(0) &= 4 \\ 5x &= 4 \\ x &= \frac{5}{4} \end{aligned}$$

S7. Combining like terms we get

$$\frac{7}{2} - 2x.$$

Hence the constant term is $\frac{7}{2}$ and the coefficient of x is -2.

S8. Simplifying the expression we get

$$4 - 3x - 6 + 12x - 6 = 9x - 8.$$

Hence the constant term is -8 and the coefficient is 9.

S9. Combining like terms we get

$$(a - 3)x - ab + a + 3.$$

Hence the constant term is $-ab + a + 3$ and the coefficient is $a - 3$.

S10. Distributing 5 we get $5(x - 1) + 3 = 5x - 2$. So the constant term is -2 and the coefficient is 5.

Exercises

1. The function g is not linear even though $g(x)$ increases by $\Delta g(x) = 50$ each time. This is because the value of x does not increase by the same amount each time. The value of x increases from 0 to 100 to 300 to 600 taking steps that get larger each time.

2. The function h is not linear even though the value of x increases by $\Delta x = 10$ each time. This is because $h(x)$ does not increase by the same amount each time. The value of $h(x)$ increases from 20 to 40 to 50 to 55 taking smaller steps each time.

3. This table could not represent a linear function because the rate of change of $g(t)$ is not constant. We consider the first three points. Between $t = 1$ and $t = 2$, the value of $g(t)$ changes by $4 - 5 = -1$. Between $t = 2$ and $t = 3$, the value of $g(t)$ changes by $5 - 4 = 1$. Thus, the rate of change is not constant ($-1 \neq 1$), so the function is not linear.

4. The function f could be linear because the value of x increases by $\Delta x = 5$ each time and $f(x)$ increases by $\Delta f(x) = 10$ each time. Assuming that any values of f not shown by the table follow this same pattern, the function f is linear.

5. This table could represent a linear function because the rate of change of $p(\gamma)$ is constant. Between consecutive data points, $\Delta\gamma = -1$ and $\Delta p(\gamma) = 10$. Thus, the rate of change is $\Delta p(\gamma)/\Delta\gamma = -10$. Since this is constant, the function could be linear.

6. The function j could be linear if the pattern continues for values of x that are not shown, because we see that a one unit increase in x corresponds to a constant decrease of two units in $j(x)$.

7. **(a)** Since the slopes are 2 and -15, we see that $y = 7 + 2x$ has the greater slope.
   (b) Since the y-intercepts are 7 and 8, we see that $y = 8 - 15x$ has the greater y-intercept.

8. **(a)** Since the slopes are -2 and -3, we see that $y = 5 - 2x$ has the greater slope.
   (b) Since the y-intercepts are 5 and 7, we see that $y = 7 - 3x$ has the greater y-intercept.

9. The vertical intercept is 54.25, which tells us that in 1970 ($t = 0$) the population was $54,250$ (54.25 thousand) people. The slope is $-\frac{2}{7}$. Since

$$\text{Slope} = \frac{\Delta\text{population}}{\Delta\text{years}} = -\frac{2}{7},$$

we know that every seven years the population decreases by 2000 people. That is, the population decreases by 2/7 thousand per year.

10. The vertical intercept is 17.75, which tells us that the stalactite was 17.75 inches long when it was first measured. The slope is $\frac{1}{250}$. Since

$$\text{Slope} = \frac{\Delta\text{inches}}{\Delta\text{years}} = \frac{1}{250},$$

it means that, for every 250 years, the stalactite grows 1 inch. That is, the stalactite grows 1/250 inch per year.

11. The vertical intercept is -3000, which tells us that if no items are sold, the company loses \$3000. The slope is 0.98. Since

$$\text{Slope} = \frac{\Delta\text{profit}}{\Delta\text{number}} = \frac{0.98}{1},$$

this tells us that, for each item the company sells, their profit increases by \$0.98.

12. The vertical intercept is 29.99, which tells us that the company charges \$29.99 per month for the phone service, even if the person does not talk on the phone. The slope is 0.05. Since

$$\text{Slope} = \frac{\Delta\text{cost}}{\Delta\text{minutes}} = \frac{0.05}{1},$$

we see that, for each minute the phone is used, it costs an additional \$0.05.

Problems

13. **(a)** To find out if this function could be linear, we calculate rates of change between pairs of points and determine whether they are constant.

$$\frac{100 - 75}{65 - 60} = \frac{25}{5} = 5 \qquad \frac{125 - 100}{70 - 65} = \frac{25}{5} = 5$$

If we continue this process, we note that the rate of change is always 5, so the function could be linear.

(b) Using units while calculating the rate of change, we get

$$\frac{\$100 - \$75}{65 \text{ mph} - 60 \text{ mph}} = \frac{\$25}{5 \text{ mph}} = \frac{\$5}{1 \text{ mph}}.$$

This suggests that for each increase of 1 mile per hour of speed, the fine is increased by five dollars.

(c) See Figure 1.21.

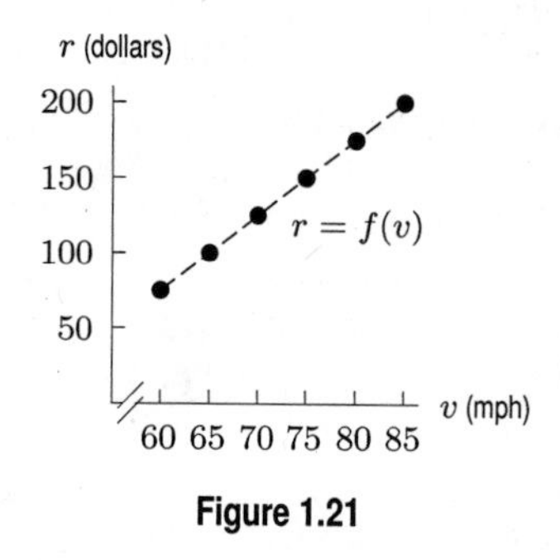

Figure 1.21

14. Table 1.4 shows the population of the town as a function of the number of years since 2006. So, a formula is $P = 18{,}310 + 58t$.

Table 1.4

t	P
0	18,310
1	$18{,}310 + 58$
2	$18{,}310 + 2 \cdot 58$
3	$18{,}310 + 3 \cdot 58$
4	$18{,}310 + 4 \cdot 58$
...	
t	$18{,}310 + t \cdot 58$

15. Since the depreciation can be modeled linearly, we can write the formula for the value of the car, V, in terms of its age, t, in years, by the following formula:

$$V = b + mt.$$

Since the initial value of the car is \$21,500, we know that $b = 21{,}500$.

Hence,

$$V = 21{,}500 + mt.$$

To find m, we know that $V = 11{,}900$ when $t = 3$, so

$$\begin{aligned} 11{,}900 &= 21{,}500 + m(3) \\ -9{,}600 &= 3m \\ \frac{-9{,}600}{3} &= m \\ -3200 &= m. \end{aligned}$$

So, $V = 21{,}500 - 3200t$.

16. The 78.9 tells us that there were approximately 79 cases on March 17. The 30.1 tells us that the number of cases increased by about 30 a day.

17. (a) If the relationship is linear we must show that the rate of change between any two points is the same. That is, for any two points (x_0, C_0) and (x_1, C_1), the quotient

$$\frac{C_1 - C_0}{x_1 - x_0}$$

is constant. From Table 1.25 we have taken the data $(0, 50)$, $(10, 52.50)$; $(5, 51.25)$, $(100, 75.00)$; and $(50, 62.50)$, $(200, 100.00)$.

$$\frac{52.50 - 50.00}{10 - 0} = \frac{2.50}{10} = 0.25$$
$$\frac{75.00 - 51.25}{100 - 5} = \frac{23.75}{95} = 0.25$$
$$\frac{100.00 - 62.50}{200 - 50} = \frac{37.50}{150} = 0.25$$

You can verify that choosing any one other pair of data points will give a slope of 0.25. The data are linear.

(b) The data from Table 1.25 are plotted below.

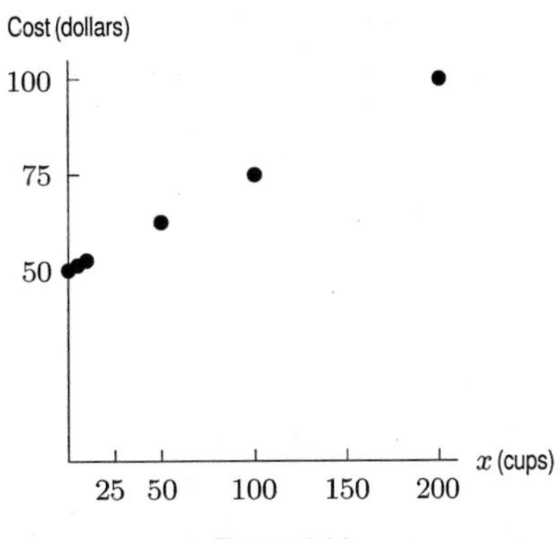

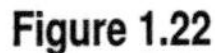
Figure 1.22

(c) Place a ruler on these points. You will see that they appear to lie on a straight line. The slope of the line equals the rate of change of the function, which is 0.25. Using units, we note that

$$\frac{\$52.50 - \$50.00}{10 \text{ cups} - 0 \text{ cups}} = \frac{\$2.50}{10 \text{ cups}} = \frac{\$0.25}{\text{cup}}.$$

In other words, the price for each additional cup of coffee is $\$0.25$.

(d) The vendor has fixed start-up costs for this venture, i.e. cart rental, insurance, salary, etc.

18. (a) Any line with a slope of 2.1, using appropriate scales on the axes. The horizontal axis should be labeled "days" and the vertical axis should be labeled "inches." See Figure 1.23.

(b) Any line with a slope of -1.3, using appropriate scales on the axes. The horizontal axis should be labeled "miles" and the vertical axis should be labeled "gallons." See Figure 1.24.

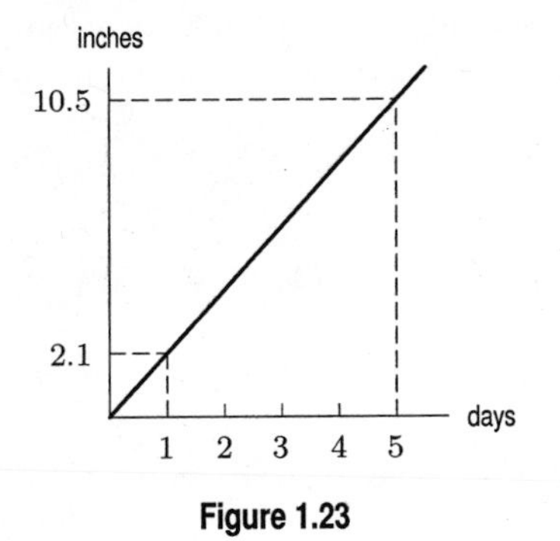

Figure 1.23

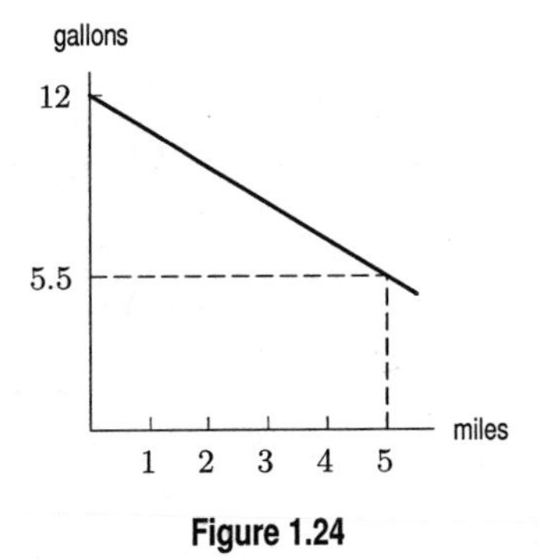

Figure 1.24

19. If zero passengers take the flight, there is no revenue, and the cost of operation is \$10,000, then the profit is $\pi = -10{,}000$, which is our π-intercept. Since each passenger pays \$127, the slope is 127 dollars per passenger, so

$$\pi(n) = -10{,}000 + 127n.$$

20. **(a)** The total amount of revenue for the café is equal to the price of a cup of coffee, \$0.95, times the number of cups sold, x:

$$\text{Revenue} = R = 0.95x.$$

The costs of the café are the fixed costs, \$200, plus the cost to make each cup of coffee, \$0.25:

$$\text{Cost} = C = 200 + 0.25x.$$

The profit is the revenue minus the costs:

$$\begin{aligned} P &= R - C \\ &= 0.95x - (200 + 0.25x) \\ &= 0.95x - 200 - 0.25x \\ &= -200 + 0.70x. \end{aligned}$$

(b) The graph is in Figure 1.25.

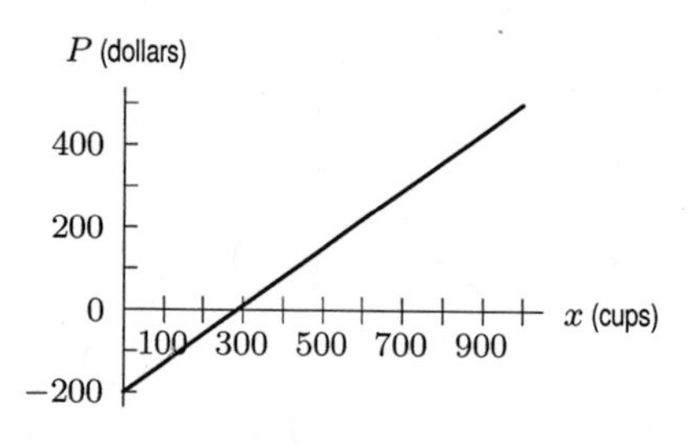

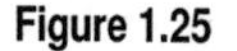

Figure 1.25

The line crosses the x-axis when $P = 0$.

$$\begin{aligned} P &= 0 \\ -200 + 0.70x &= 0 \\ 0.70x &= 200 \\ x &\approx 286 \end{aligned}$$

So, the line intersects the x-axis at $x \approx 286$. The graph of the profit function is below the x-axis for $x < 286$ and above it for $x > 286$. When the graph of the profit function is above the x-axis, the café's profits are positive, meaning that it is making money. When the graph of the profit function is below the x-axis, the café's profits are negative, meaning that it is losing money.

(c) The slope represents the increase in profit due to the sale of an additional single cup of coffee. From the equation for P, the slope is 0.70. This \$0.70 profit is equal to the price of a cup of coffee, \$0.95, minus the café's expense for the cup, \$0.25.

The y-intercept represents the profit if zero cups of coffee are sold. Since this is negative, we see that the café loses \$200 per day (its fixed costs) if no coffee is sold.

The x-intercept represents the number of cups of coffee that must be sold if the café is to break even, i.e., to make a profit equal to zero.

21. Each month, regardless of the amount of rocks mined from the quarry, the owners must pay 1000 dollars for maintenance and insurance, as well as 3000 dollars for monthly salaries. This totals to 4000 dollars in fixed costs. In addition, the cost for mining each ton of rocks is 80 dollars. The total cost incurred by the quarry's owners each month can be written:

$$\begin{aligned}\begin{pmatrix}\text{total}\\\text{cost}\end{pmatrix} &= \begin{pmatrix}\text{fixed}\\\text{costs}\end{pmatrix} + \begin{pmatrix}\text{mining cost}\\\text{per ton}\end{pmatrix} \cdot \begin{pmatrix}\text{tons of rocks}\\\text{mined}\end{pmatrix}\\ &= 4000 \text{ dollars} + (80 \text{ dollars/ton})(r \text{ tons})\\ c &= 4000 + 80r.\end{aligned}$$

22. **(a)** Looking at the data from Table 1.26 and calculating the rate of change of area versus side length between various points, we see that the function is not linear. For example, the rate of change between the points $(0,0)$ and $(1,1)$ is

$$\frac{\Delta\text{area}}{\Delta\text{length}} = \frac{1-0}{1-0} = \frac{1}{1} = 1$$

while the rate of change between the points $(1,1)$ and $(2,4)$ is

$$\frac{\Delta\text{area}}{\Delta\text{length}} = \frac{4-1}{2-1} = \frac{3}{1} = 3.$$

The rates of change are different. The relationship is not linear. On the other hand, when we view the data from Table 1.26, we see that the rate of change of perimeter versus side length between any two points is always constant. Thus, that function could be linear. For example, let's look at the pairs of points $(0,0)$, $(3,12)$; $(1,4)$, $(4,16)$ and $(2,8)$, $(5,20)$. For $(0,0)$, $(3,12)$ the rate of change is

$$\frac{\Delta\text{perimeter}}{\Delta\text{length}} = \frac{12-0}{3-0} = \frac{12}{3} = 4.$$

For $(1,4)$, $(4,16)$ the rate of change is

$$\frac{\Delta\text{perimeter}}{\Delta\text{length}} = \frac{16-4}{4-1} = \frac{12}{3} = 4.$$

For $(2,8)$, $(5,20)$ the rate of change is

$$\frac{\Delta\text{perimeter}}{\Delta\text{length}} = \frac{20-8}{5-2} = \frac{12}{3} = 4.$$

Check that using any two of the data points in Table 1.26 to calculate the rate of change gives a rate of change of 4.

(b) See Figures 1.26 and 1.27.

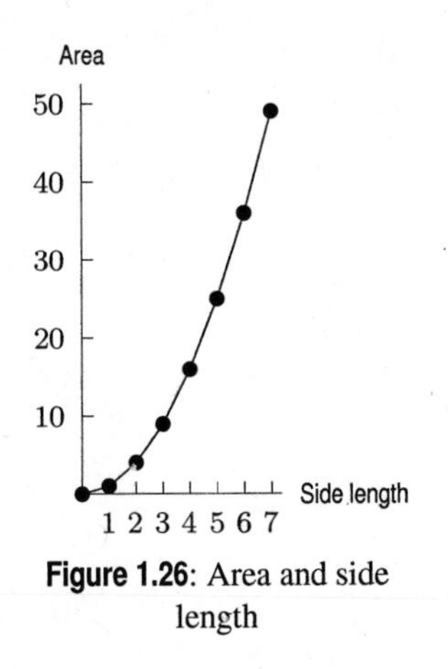

Figure 1.26: Area and side length

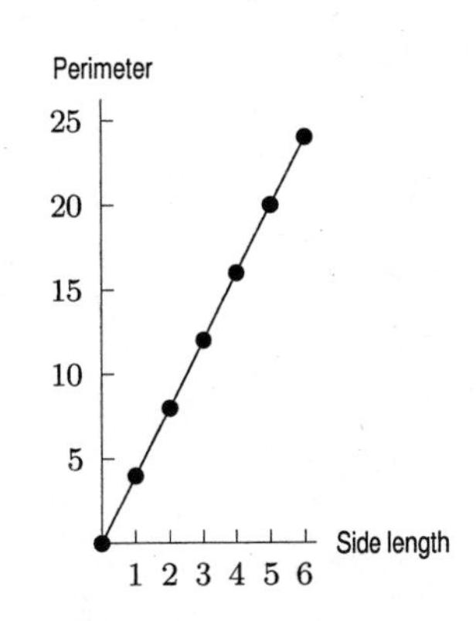

Figure 1.27: Perimeter and side length

(c) From part (a) we see that the rate of change of the function giving perimeter versus side length is 4. This tells us that for a given square, when we increase the length of each side by one unit, the length of the perimeter increases by four units.

23. We know that the area of a circle of radius r is

$$\text{Area } = \pi r^2$$

while its circumference is given by

$$\text{Circumference } = 2\pi r.$$

Thus, a table of values for area and circumference is

Table 1.5

Radius	0	1	2	3	4	5	6
Area	0	π	4π	9π	16π	25π	36π
Circumference	0	2π	4π	6π	8π	10π	12π

(a) In the area function we see that the rate of change between pairs of points does not remain constant and thus the function is not linear. For example, the rate of change between the points $(0,0)$ and $(2,4\pi)$ is not equal to the rate of change between the points $(3,9\pi)$ and $(6,36\pi)$. The rate of change between $(0,0)$ and $(2,4\pi)$ is

$$\frac{\Delta\text{area}}{\Delta\text{radius}} = \frac{4\pi - 0}{2-0} = \frac{4\pi}{2} = 2\pi$$

while the rate of change between $(3,9\pi)$ and $(6,36\pi)$ is

$$\frac{\Delta\text{area}}{\Delta\text{radius}} = \frac{36\pi - 9\pi}{6-3} = \frac{27\pi}{3} = 9\pi.$$

On the other hand, if we take only pairs of points from the circumference function, we see that the rate of change remains constant. For instance, for the pair $(0,0)$, $(1,2\pi)$ the rate of change is

$$\frac{\Delta\text{circumference}}{\Delta\text{radius}} = \frac{2\pi - 0}{1-0} = \frac{2\pi}{1} = 2\pi.$$

For the pair $(2,4\pi)$, $(4,8\pi)$ the rate of change is

$$\frac{\Delta\text{circumference}}{\Delta\text{radius}} = \frac{8\pi - 4\pi}{4-2} = \frac{4\pi}{2} = 2\pi.$$

For the pair $(1,2\pi)$, $(6,12\pi)$ the rate of change is

$$\frac{\Delta\text{circumference}}{\Delta\text{radius}} = \frac{12\pi - 2\pi}{6-1} = \frac{10\pi}{5} = 2\pi.$$

Picking any pair of data points would give a rate of change of 2π.

(b) The graphs for area and circumference as indicated in Table 1.5 are shown in Figure 1.28 and Figure 1.29.

Figure 1.28

Figure 1.29

(c) From part (a) we see that the rate of change of the circumference function is 2π. This tells us that for a given circle, when we increase the length of the radius by one unit, the length of the circumference would increase by 2π units. Equivalently, if we decreased the length of the radius by one unit, the length of the circumference would decrease by 2π.

24. **(a)** We see that the population of Country B grows at the constant rate of roughly 2.4 million every ten years. Thus Country B must be Sri Lanka. The population of country A did not change at a constant rate: In the ten years of 1970–1980 the population of Country A grew by 2.7 million while in the ten years of 1980–1990 its population dropped. Thus, Country A is Afghanistan.

(b) The rate of change of Country B is found by taking the population increase and dividing it by the corresponding time in which this increase occurred.Thus

$$\text{Rate of change of population} = \frac{9.9 - 7.5}{1960 - 1950} = \frac{2.4 \text{ million people}}{10 \text{ years}} = 0.24 \text{ million people/year.}$$

This rate of change tells us that on the average, the population of Sri Lanka increases by 0.24 million people every year. The rate of change for the other intervals is the same or nearly the same.

(c) In 1980 the population of Sri Lanka was 14.9 million. If the population grows by 0.24 million every year, then in the eight years from 1980 to 1988

$$\text{Population increase} = 8 \cdot 0.24 \text{ million} = 1.92 \text{ million.}$$

Thus in 1988

$$\text{Population of Sri Lanka} = 14.9 + 1.92 \text{ million} \approx 16.8 \text{ million.}$$

25. **(a)** No. The values of $f(d)$ first drop, then rise, so f is not linear.

(b) For $d \geq 150$, the graph looks linear. See Figure 1.30.

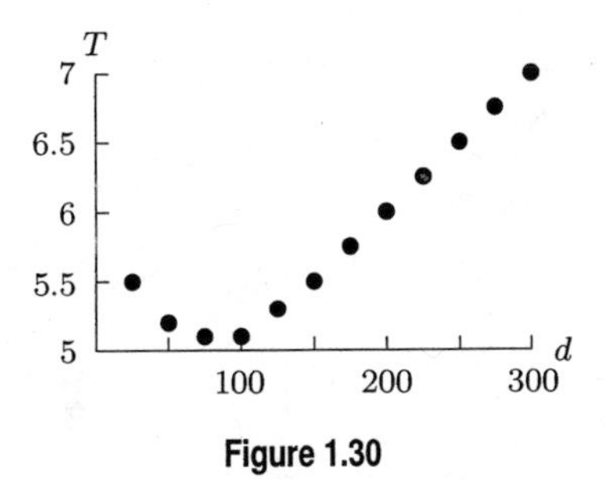

Figure 1.30

(c) For $d \geq 150$ the average rate of change appears to be constant. Each time the depth goes up by $\Delta d = 25$ meters, the temperature rises by $\Delta T = 0.25°$C, so the average rate of change is $\Delta T/\Delta d = 0.25/25 = 0.01°\text{C/meter}$. In other words, the temperature rises by 0.01°C for each extra meter in depth.

26. **(a)** The slope is $m = -56.57$, which tells us that the area of the ice cover goes down by 56.57 m^2/year. The A-intercept is $b = 1951$, which tells us that in the year 2000, the area was 1551 m^2. The equation that models this is therefore $A = 1951 - 56.57t$

(b) We have $f(11) = 1951 - 56.57(11) = 1328.73$, which tells us that in the year 2011, the area of the ice cover is predicted to be 1328.73 m^2.

(c) Solving $A = 1951 - 56.57t = 0$ we have $t = 34.49$, so in about 34.5 years or in the year 2034 the ice cover will disappear.

27. **(a)** Since C is 8, we have $T = 300 + 200C = 300 + 200(8) = 1900$. Thus, taking 8 credits costs \$1900.

(b) Here, the value of T is 1700 and we solve for C.

$$\begin{aligned} T &= 300 + 200C \\ 1700 &= 300 + 200C \\ 7 &= C \end{aligned}$$

Thus, \$1,700 is the cost of taking 7 credits.

(c) Table 1.6 is the table of costs.

Table 1.6

C	1	2	3	4	5	6	7	8	9	10	11	12
T	500	700	900	1100	1300	1500	1700	1900	2100	2300	2500	2700
$\frac{T}{C}$	500	350	300	275	260	250	243	238	233	230	227	225

(d) The largest value for C, that is, 12 credits, gives the smallest value of T/C. In general, the ratio of tuition cost to number of credits is getting smaller as C increases.

(e) This cost is independent of the number of credits taken; it might cover fixed fees such as registration, student activities, and so forth.

(f) The 200 represents the rate of change of cost with the number of credit hours. In other words, one additional credit hour costs an additional $200.

28. **(a)** Since for each additional $5000 spent the company will sell 20 more units, we have

$$m = \frac{\Delta y}{\Delta x} = \frac{20}{5000}.$$

Also, since 300 units will be sold even if no money is spent on advertising, the y–intercept, b, is 300. Our formula is

$$y = 300 + \frac{20}{5000}x = 300 + \frac{1}{250}x.$$

(b) If $x = \$25{,}000$, the number of units it sells will be

$$y = 300 + \frac{1}{250}(25000) = 300 + 100 = 400.$$

If $x = \$50{,}000$, the number of units it sells will be

$$y = 300 + \frac{1}{250}(50000) = 300 + 200 = 500.$$

(c) If $y = 700$, we need to solve for x:

$$\begin{aligned} 300 + \frac{1}{250}x &= 700 \\ \frac{1}{250}x &= 700 - 300 = 400 \\ x &= 250 \cdot 400 = 100{,}000. \end{aligned}$$

Thus, the firm would need to spend $100,000 to sell 700 units.

(d) The slope is the change in the value of y, the number of units sold, for a given change in x, the amount of money spent on ads. Thus, an interpretation of the slope is that for each additional $250 spent on ads, one additional unit is sold.

29. **(a)** Here we have

$$\begin{aligned} 2r &= 5 \\ r &= \frac{5}{2} \\ \text{and} \quad \sqrt{s} &= 4 \\ s &= 16 \\ \text{giving} \quad y &= \underbrace{2\left(\frac{5}{2}\right)}_{2r} + x\underbrace{\sqrt{16}}_{\sqrt{s}}. \end{aligned}$$

(b) Here we have

$$\begin{aligned} \frac{1}{k} &= 5 \\ k &= \frac{1}{5} = 0.2 \\ \text{and} \quad -(j-1) &= 4 \\ j &= -3 \end{aligned}$$

giving

$$y = \underbrace{\frac{1}{0.2}}_{1/k} - \underbrace{(-3-1)}_{j-1}x.$$

30. In the window $-10 \le x \le 10, -10 \le y \le 10$, the graph of this equation looks like a line. By choosing a larger viewing window, however, you can see that it is not a line. For example, using $-500 \le x \le 500, -500 \le y \le 500$ produces Figure 1.31.

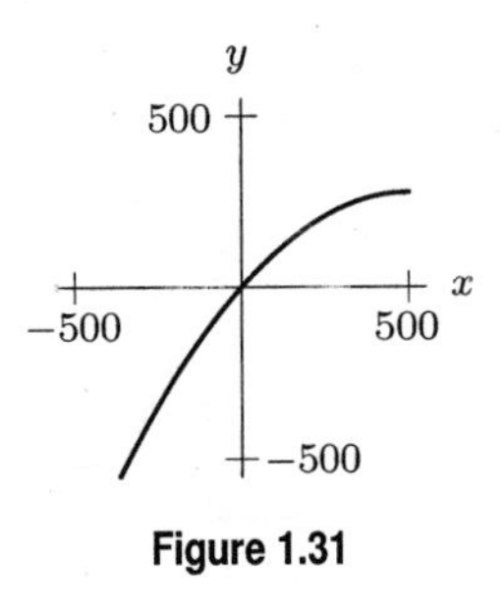

Figure 1.31

31. As Figure 1.32 shows, the graph of $y = 2x + 400$ does not appear in the window $-10 \le x \le 10, -10 \le y \le 10$. This is because all the corresponding y-values are between 380 and 420, which are outside this window. The graph can be seen by using a different viewing window: for example, $380 \le y \le 420$.

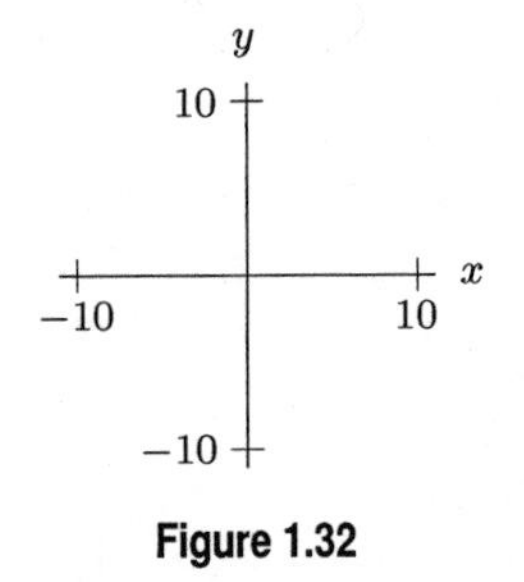

Figure 1.32

32. As Figure 1.33 shows, the graph is not visible in the window $10 \le x \le 10, -10 \le y \le 10$. The reason is that the graph of $y = 200x + 4$ is nearly vertical and almost coincides with the y-axis in this window. To see more clearly that the graph is not vertical, use a much larger y-range. For example, a window of $-10 \le x \le 10$, $-2000 \le y \le 2000$ gives a more informative graph. Alternatively, use a much smaller x-range; for example, try a window of $-0.1 \le x \le 0.1$, $-10 \le y \le 10$.

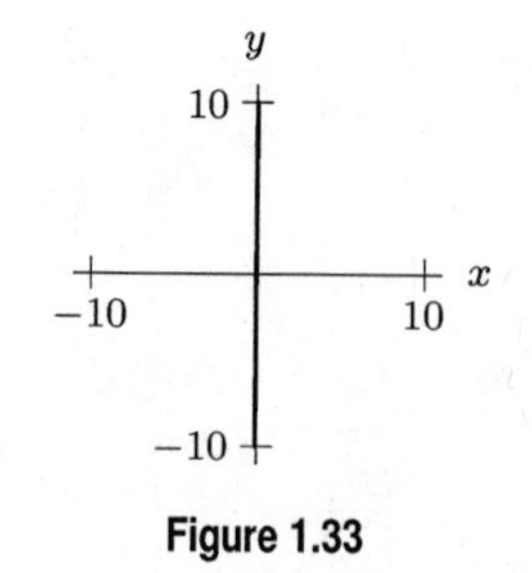

Figure 1.33

33. Most functions look linear if viewed in a small enough window. This function is not linear. We see this by graphing the function in the larger window $-100 \leq x \leq 100, -20 \leq y \leq 20$.

34. Since the radius is 10 miles, the longest ride will not be more than 20 miles. The maximum cost will therefore occur when $d = 20$, so the maximum cost is $C = 2.50 + 2d = 2.50 + 2(20) = 2.50 + 40 = 42.50$. Therefore, the window should be at least $0 \leq d \leq 20$ and $0 \leq C \leq 42.50$. See Figure 1.34.

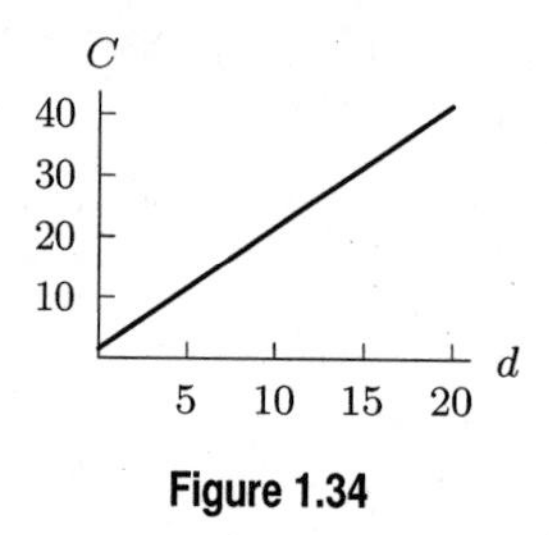

Figure 1.34

Solutions for Section 1.4

Skill Refresher

S1.

$$\begin{aligned} y - 5 &= 21 \\ y &= 26. \end{aligned}$$

S2.

$$\begin{aligned} 2x - 5 &= 13 \\ 2x &= 18 \\ \frac{2x}{2} &= \frac{18}{2} \\ x &= 9. \end{aligned}$$

S3.

$$\begin{aligned} 2x - 5 &= 4x - 9 \\ 2x &= 4x - 4 \\ -2x &= -4 \\ x &= 2. \end{aligned}$$

S4.

$$\begin{aligned} 17 - 28y &= 13y + 24 \\ -28y &= 13y + 7 \\ -41y &= 7 \\ y &= -\frac{7}{41}. \end{aligned}$$

S5. We first distribute $\frac{5}{3}(y+2)$ to obtain:

$$\begin{aligned} \frac{5}{3}(y+2) &= \frac{1}{2} - y \\ \frac{5}{3}y + \frac{10}{3} &= \frac{1}{2} - y \\ \frac{5}{3}y + y &= \frac{1}{2} - \frac{10}{3} \\ \frac{5}{3}y + \frac{3y}{3} &= \frac{3}{6} - \frac{20}{6} \\ \frac{8y}{3} &= -\frac{17}{6} \\ \left(\frac{3}{8}\right)\frac{8y}{3} &= \left(\frac{3}{8}\right)\left(-\frac{17}{6}\right) \\ y &= -\frac{17}{16}. \end{aligned}$$

S6. Dividing by rt gives

$$P = \frac{I}{rt}.$$

S7. We have

$$\begin{aligned} C &= \frac{5}{9}(F - 32) \\ \frac{9C}{5} &= F - 32 \\ F &= \frac{9}{5}C + 32. \end{aligned}$$

S8. Dividing by 2π gives

$$r = \frac{C}{2\pi}.$$

S9. We collect all terms involving x and then divide by $2a$:

$$\begin{aligned} ab + ax &= c - ax \\ 2ax &= c - ab \\ x &= \frac{c - ab}{2a}. \end{aligned}$$

S10. We collect all terms involving y and then factor out the y.

$$\begin{aligned} by - d &= ay + c \\ by - ay &= c + d \\ y(b - a) &= c + d \\ y &= \frac{c + d}{b - a}. \end{aligned}$$

Exercises

1. Rewriting in slope-intercept form:

$$\begin{aligned} 5(x+y) &= 4 \\ 5x+5y &= 4 \\ 5y &= 4-5x \\ \frac{5y}{5} &= \frac{4}{5}-\frac{5x}{5} \\ y &= \frac{4}{5}-x \end{aligned}$$

2. Rewriting in slope-intercept form:

$$\begin{aligned} 3x+5y &= 20 \\ 5y &= 20-3x \\ y &= \frac{20}{5}-\frac{3x}{5} \\ y &= 4-\frac{3}{5}x \end{aligned}$$

3. Rewriting in slope-intercept form:

$$\begin{aligned} 0.1y+x &= 18 \\ 0.1y &= 18-x \\ y &= \frac{18}{0.1}-\frac{x}{0.1} \\ y &= 180-10x \end{aligned}$$

4. Rewriting in slope-intercept form:

$$\begin{aligned} 5x-3y+2 &= 0 \\ -3y &= -2-5x \\ y &= \frac{-2}{-3}-\frac{5}{-3}x \\ y &= \frac{2}{3}+\frac{5}{3}x \end{aligned}$$

5. Rewriting in slope-intercept form:

$$\begin{aligned} y-0.7 &= 5(x-0.2) \\ y-0.7 &= 5x-1 \\ y &= 5x-1+0.7 \\ y &= 5x-0.3 \\ y &= -0.3+5x \end{aligned}$$

6. Writing $y = 5$ as $y = 5 + 0x$ shows that $y = 5$ is the form $y = b + mx$ with $b = 5$ and $m = 0$.

7. Rewriting in slope-intercept form:

$$\begin{aligned} 3x + 2y + 40 &= x - y \\ 2y + y &= x - 3x - 40 \\ 3y &= -40 - 2x \\ y &= -\frac{40}{3} - \frac{2}{3}x \end{aligned}$$

8. Not possible, the slope is not defined (vertical line).

9. Rewriting in slope-intercept form:

$$\begin{aligned} \frac{x+y}{7} &= 3 \\ x + y &= 21 \\ y &= 21 - x \end{aligned}$$

10. Yes. Write the function as

$$g(w) = -\frac{1-12w}{3} = -\left(\frac{1}{3} - \frac{12}{3}w\right) = -\frac{1}{3} + 4w,$$

so $g(w)$ is linear with $b = -1/3$ and $m = 4$.

11. Yes. Write the function as

$$F(P) = 13 - \frac{2^{-1}}{4}P = 13 - \frac{1}{8}P = 13 + \left(\frac{-1}{8}\right)P,$$

so $F(P)$ is linear with $b = 13$ and $m = -1/8$.

12. The function is not linear because the power of s is not 1.

13. Yes. Write the function as

$$C(r) = 0 + 2\pi r,$$

so $C(r)$ is linear with $b = 0$ and $m = 2\pi$.

14. The function $h(x)$ is not linear because the 3^x term has the variable in the exponent and is not the same as $3x$ which would be a linear term.

15. Yes. Write the function as $f(x) = n^2 + m^2x$. The constant term is $b = n^2$ and the coefficient of x is m^2.

16. We have the slope $m = -4$ so

$$y = b - 4x.$$

The line passes through $(7, 0)$ so

$$\begin{aligned} 0 &= b + (-4)(7) \\ 28 &= b \end{aligned}$$

and

$$y = 28 - 4x.$$

17. We can put the slope $m = 3$ and y-intercept $b = 8$ directly into the general equation $y = b + mx$ to get $y = 8 + 3x$.

18. We first use the points to find the slope m:

$$m = \frac{y_1 - y_0}{x_1 - x_0} = \frac{5 - (-1)}{-1 - 2} = \frac{6}{-3} = -2.$$

Next we use the equation:

$$y = b + mx.$$

Substituting -2 for m, we have

$$y = b + (-2)x.$$

Using the point $(-1, 5)$, we have:

$$\begin{aligned}5 &= b + (-2)(-1)\\ 5 &= b + 2\\ 3 &= b\end{aligned}$$

so,

$$y = 3 - 2x.$$

19. We can put $m = \frac{2}{3}$ and $(x_0, y_0) = (5, 7)$ into the equation

$$\begin{aligned}y &= b + mx\\ 7 &= b + \frac{2}{3}(5)\\ \frac{21}{3} &= b + \frac{10}{3}\\ \frac{11}{3} &= b\,.\end{aligned}$$

so

$$y = \frac{11}{3} + \frac{2}{3}x.$$

20. Since we know the x-intercept and y-intercepts are $(3, 0)$ and $(0, -5)$ respectively, we can find the slope:

$$\text{slope} = m = \frac{-5-0}{0-3} = \frac{-5}{-3} = \frac{5}{3}.$$

We can then put the slope and y-intercept into the general equation for a line.

$$y = -5 + \frac{5}{3}x.$$

21. Since the slope is $m = 0.1$, the equation is

$$y = b + 0.1x.$$

Substituting $x = -0.1$, $y = 0.02$ to find b gives

$$\begin{aligned}0.02 &= b + 0.1(-0.1)\\ 0.02 &= b - 0.01\\ b &= 0.03.\end{aligned}$$

The equation is $y = 0.03 + 0.1x$.

22. We have $f(0.3) = 0.8$ and $f(0.8) = -0.4$. This gives $y = b + mx$ where

$$m = \frac{f(0.8) - f(0.3)}{0.8 - 0.3} = \frac{-0.4 - 0.8}{0.5} = -2.4.$$

Solving for b, we have

$$\begin{aligned}f(0.3) &= b - 2.4(0.3)\\ b &= f(0.3) + 2.4(0.3) = 0.8 + 2.4(0.3) = 1.5,\end{aligned}$$

so $y = 1.5 - 2.4x$.

23. Two points on the graph of f are $(x_0, y_0) = (-2, 7)$ and $(x_1, y_1) = (3, -3)$. We can then find

$$m = \frac{y_1 - y_0}{x_1 - x_0} = \frac{-3-7}{3-(-2)} = \frac{-10}{5} = -2.$$

Therefore, $y = b - 2x$. To solve for b, we could use the point $(x_0, y_0) = (-2, 7)$:

$$7 = b - 2(-2) = b + 4$$
$$b = 3.$$

This gives $y = f(x) = 3 - 2x$. We can check this formula by plugging in $x = 3$:

$$f(3) = 3 - 2(3) = -3.$$

This is the y-value we expected.

24. We have a V intercept of 2000. Since the value is decreasing by \$500 per year, our slope is -500 dollars per year. So one possible equation is

$$V = 2000 - 500t.$$

25. Since the function is linear, we can choose any two points to find its formula. We use the form

$$q = b + mp$$

to get the number of bottles sold as a function of the price per bottle. We use the two points $(0.50, 1500)$ and $(1.00, 500)$. We begin by finding the slope, $\Delta q/\Delta p = (500 - 1500)/(1.00 - 0.50) = -2000$. Next, we substitute a point into our equation using our slope of -2000 bottles sold per dollar increase in price and solve to find b, the q-intercept. We use the point $(1.00, 500)$:

$$500 = b - 2000 \cdot 1.00$$
$$2500 = b.$$

Therefore,

$$q = 2500 - 2000p.$$

26. Since the function is linear, we can use any two points to find its formula. We use the form

$$y = b + mx$$

to get temperature in °C, y, as a function of temperature in °F, x. We use the two points, $(32, 0)$ and $(41, 5)$. We begin by finding the slope, $\Delta y/\Delta x = (5-0)/(41-32) = 5/9$. Next, we substitute a point into our equation using our slope of $5/9$°C per °F and solve to find b, the y-intercept. We use the point $(32, 0)$:

$$0 = b + \frac{5}{9} \cdot 32$$
$$-\frac{160}{9} = b.$$

Therefore,

$$y = -\frac{160}{9} + \frac{5}{9}x.$$

Traditionally, we give this formula as $y = (5/9)(x - 32)$, which is often easier to manipulate. You might want to check to see if the two are the same.

27. Since we are told that the function is linear, any two points will define the line for us. We will use the form

$$y = b + mx$$

to get temperature in ° Rankine, y, as a function of temperature in ° Fahrenheit, x. (Rankine is a rarely used absolute temperature scale.) We choose the two points, $(0, 459.7)$ and $(10, 469.7)$. We begin by finding the slope, $\Delta R/\Delta F = (469.7 - 459.7)/(10 - 0) = 10/10 = 1$. Next, we substitute a point into our equation using our slope of 1 °R per °F and solve to find b, the y-intercept. We use the point $(0, 459.7)$:

$$459.7 = b + 1 \cdot 0$$
$$459.7 = b.$$

Therefore,

$$y = 459.7 + 1x.$$

Note that each °R is the same as each °F, but the two systems choose different starting points (zero °R is absolute zero, while zero °F is an arbitrary point).

28. Since the function is linear, we can choose any two points (from the graph) to find its formula. We use the form

$$p = b + mh$$

to get the price of an apartment as a function of its height. We use the two points $(10, 175{,}000)$ and $(20, 225{,}000)$. We begin by finding the slope, $\Delta p/\Delta h = (225{,}000 - 175{,}000)/(20 - 10) = 5000$. Next, we substitute a point into our equation using our slope of 5000 dollars per meter of height and solve to find b, the p-intercept. We use the point $(10, 175{,}000)$:

$$175{,}000 = b + 5000 \cdot 10$$
$$125{,}000 = b.$$

Therefore,

$$p = 125{,}000 + 5000h.$$

29. Since the function is linear, we can use any two points (from the graph) to find its formula. We use the form

$$u = b + mn$$

to get the meters of shelf space used as a function of the number of different medicines stocked. We use the two points $(60, 5)$ and $(120, 10)$. We begin by finding the slope, $\Delta u/\Delta n = (10 - 5)/(120 - 60) = 1/12$. Next, we substitute a point into our equation using our slope of $1/12$ meters of shelf space per medicine and solve to find b, the u-intercept. We use the point $(60, 5)$:

$$5 = b + (1/12) \cdot 60$$
$$0 = b.$$

Therefore,

$$u = (1/12)n.$$

The fact that $b = 0$ is not surprising, since we would expect that, if no medicines are stocked, they should take up no shelf space.

30. Since the function is linear, we can use any two points (from the graph) to find its formula. We use the form

$$s = b + mq$$

to get the number of hours of sleep obtained as a function of the quantity of tea drunk. We use the two points $(4, 7)$ and $(12, 3)$. We begin by finding the slope, $\Delta s/\Delta q = (3-7)/(12-4) = -0.5$. Next, we substitute a point into our equation using our slope of -0.5 hours of sleep per cup of tea and solve to find b, the s-intercept. We use the point $(4, 7)$:

$$7 = b - 0.5 \cdot 4$$
$$9 = b.$$

Therefore,

$$s = 9 - 0.5q.$$

Problems

31. We have $f(-3) = -8$ and $f(5) = -20$. This gives $f(x) = b + mx$ where

$$m = \frac{f(5) - f(-3)}{5 - (-3)} = \frac{-20 - (-8)}{8} = -\frac{12}{8} = -1.5.$$

Solving for b, we have

$$\begin{aligned} f(-3) &= b - 1.5(-3) \\ b &= f(-3) + 1.5(-3) \\ &= -8 + 1.5(-3) = -12.5, \end{aligned}$$

so $f(x) = -12.5 - 1.5x$.

32. We have $g(100) = 2000$ and $g(400) = 3800$. This gives $g(x) = b + mx$ where

$$m = \frac{g(400) - g(100)}{400 - 100} = \frac{3800 - 2000}{300} = 6.$$

Solving for b, we have

$$\begin{aligned} g(100) &= b + 6(100) \\ b &= g(100) - 6(100) = 2000 - 6(100) = 1400, \end{aligned}$$

so $g(x) = 1400 + 6x$.

33. The starting value is $b = 12{,}000$, and the growth rate is $m = 225$, so $h(t) = 12{,}000 + 225t$.

34. The graph contains the points $(-2, (-2)^2) = (-2, 4)$ and $(3, 3^2) = (3, 9)$, so $h(-2) = 4$ and $h(3) = 9$. This gives $y = b + mx$ where

$$m = \frac{h(3) - h(-2)}{3 - (-2)} = \frac{9 - 4}{5} = 1.$$

Solving for b, we have

$$\begin{aligned} h(-2) &= b + 1(-2) \\ b &= h(-2) - 1(-2) = 4 - 1(-2) = 6, \end{aligned}$$

so $h(x) = 6 + x$.

35. **(a)** $C(175) = 11{,}375$, which means that it costs \$11,375 to produce 175 units of the good.

(b) $C(175) - C(150) = 125$, which means that the cost of producing 175 units is \$125 greater than the cost of producing 150 units. That is, the cost of producing the additional 25 units is an additional \$125.

(c) $\dfrac{C(175) - C(150)}{175 - 150} = \dfrac{125}{25} = 5$, which means that the average per-unit cost of increasing production to 175 units from 150 units is \$5.

36. We would like to find a table value that corresponds to $n = 0$. The pattern from the table, is that for each decrease of 25 in n, $C(n)$ goes down by 125. It takes four decreases of 25 to get from $n = 100$ to $n = 0$, and $C(100) = 11{,}000$, so we might estimate $C(0) = 11{,}000 - 4 \cdot 125 = 10{,}500$. This means that the fixed cost, before any goods are produced, is \$10,500.

37. We found in Problem 36 that the fixed cost of this good is \$10,500. We found the unit cost in Problem 35(c) to be \$5. (In that problem, we used $n = 150$ and $n = 175$, but since this is a linear total-cost function, any pair of values of n will give the same rate of change or cost per unit.) Thus,

$$C(n) = 10{,}500 + 5n.$$

38. **(a)** To find the formula for the line through the two points $(90, 1005)$ and $(140, 1205)$ we first find the slope of the line:

$$m = \frac{1205 - 1005}{140 - 90} = 4.$$

Now, to find b, we use the point-slope form, so $C - 1005 = 4(n - 90)$ or $C = 645 + 4n$.

(b) From the answer to part (a) the price per meal is \$4.00 and the membership fee is \$645.

(c) The cost of a meal plan is the membership fee plus n times the cost of a meal. Using our results from part (a), to find the cost of 120 meals we have:

$$C = 645 + 4 \cdot 120 = 1125.$$

So 120 meals cost \$1255

(d) Rewriting our expression for the cost of a meal plan:

$$\begin{aligned} 645 + 4 \cdot n &= C \\ 4 \cdot n &= C - 645 \\ n &= \frac{C - 645}{4}. \end{aligned}$$

(e) Given $C = \$1285$ you can buy:

$$n = \frac{C - 645}{4} = \frac{1285 - 645}{4} = 160.$$

39. **(a)** The bottle travels upward at first, and then begins to fall toward the ground. As it falls, it falls faster and faster. Negative values of v represent falling toward the ground.

(b) Notice that v decreases by 32 ft/sec each second. Since $m = \Delta v/\Delta t$,

$$m = \frac{-32}{1} = -32.$$

We have

$$v = b + mt = b - 32t,$$

and since at $t = 0$, $v = 40$, this gives

$$\begin{aligned} 40 &= b - 32 \cdot 0 \\ b &= 40 \end{aligned}$$

and so $v = 40 - 32t$.

(c) The slope is

$$\frac{\Delta v}{\Delta t} = \frac{-32 \text{ ft/sec}}{\text{sec}},$$

which tells us that the velocity of the bottle decreases by 32 ft/sec for each second elapsed.

(d) The v-axis intercept occurs when $t = 0$. If $t = 0$, then $v = 40$, which means that the bottle's initial velocity is $v = 40$ ft/sec. The t-axis intercept occurs when $v = 0$. If $v = 0$, then

$$\begin{aligned} 0 &= 40 - 32t \\ 32t &= 40 \\ t &= \frac{40}{32} = 1.25 \text{ seconds.} \end{aligned}$$

This means that at $t = 1.25$ seconds, the bottle's velocity is zero, meaning that it stopped rising and began to fall.

40. **(a)** A table of the allowable combinations of sesame and poppy seed rolls is shown below.

Table 1.7

s, sesame seed rolls	0	1	2	3	4	5	6	7	8	9	10	11	12
p, poppy seed rolls	12	11	10	9	8	7	6	5	4	3	2	1	0

(b) The sum of s and p is 12. So we can write $s + p = 12$, or $p = 12 - s$.

(c)

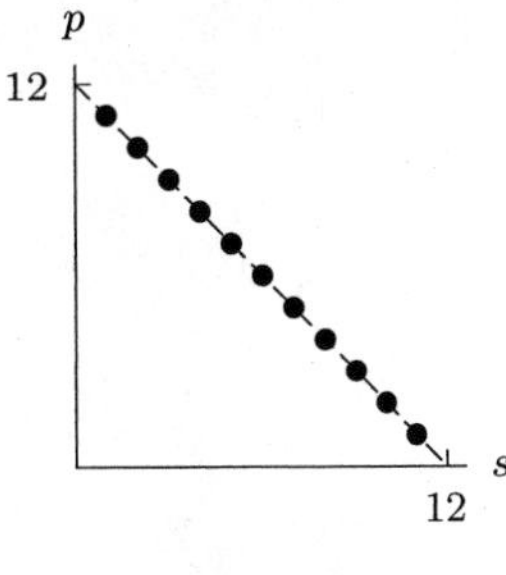

Figure 1.35

41. (a) Since q is linear, $q = b + mp$, where

$$m = \frac{\Delta q}{\Delta p} = \frac{65 - 45}{3.10 - 3.50} = \frac{20}{-0.40} = -50 \text{ gallons/dollar.}$$

Thus, $q = b - 50p$ and since $q = 65$ if $p = 3.10$,

$$\begin{aligned} 65 &= b - 50(3.10) \\ 65 &= b - 155 \\ b &= 65 + 155 = 210. \end{aligned}$$

So,

$$q = 210 - 50p.$$

(b) The slope is $m = -50$ gallons per dollar, which tells us that the quantity of gasoline demanded in one time period decreases by 50 gallons for each \$1 increase in price.

(c) If $p = 0$ then $q = 210$, which means that if the price of gas were \$0 per gallon, then the quantity demanded in one time period would be 210 gallons per month. This means if gas were free, a person would want 210 gallons. If $q = 0$ then $210 - 50p = 0$, so $210 = 50p$ and $p = 210/50 = 4.20$. This tells us that (according to the model), at a price of \$4.20 per gallon there will be no demand for gasoline. In the real world, this is not likely.

42. (a) We are looking at the amount of municipal solid waste, W, as a function of year, t, and the two points are $(1960, 88.1)$ and $(2000, 239.1)$. For the model, we assume that the quantity of solid waste is a linear function of year. The slope of the line is

$$m = \frac{239.1 - 88.1}{2000 - 1960} = \frac{151}{40} = 3.78 \frac{\text{millions of tons}}{\text{year}}.$$

This slope tells us that the amount of solid waste generated in the cities of the US has been going up at a rate of 3.78 million tons per year. To find the equation of the line, we must find the vertical intercept. We substitute the point $(1960, 88.1)$ and the slope $m = 3.78$ into the equation $W = b + mt$:

$$\begin{aligned} W &= b + mt \\ 88.1 &= b + (3.78)(1960) \\ 88.1 &= b + 7408.8 \\ -7320.7 &= b. \end{aligned}$$

The equation of the line is $W = -7320.7 + 3.78t$, where W is the amount of municipal solid waste in the US in millions of tons, and t is the year.

(b) How much solid waste does this model predict in the year 2020? We can graph the line and find the vertical coordinate when $t = 2020$, or we can substitute $t = 2020$ into the equation of the line, and solve for W:

$$\begin{aligned} W &= -7320.7 + 3.78t \\ W &= -7320.7 + (3.78)(2020) = 314.9. \end{aligned}$$

The model predicts that in the year 2020, the solid waste generated by cities in the US will be 314.9 million tons.

43. Point P is on the curve $y = x^2$ and so its coordinates are $(2, 2^2) = (2, 4)$. Since line l contains point P and has slope 4, its equation is

$$y = b + mx.$$

Using $P = (2, 4)$ and $m = 4$, we get

$$\begin{aligned} 4 &= b + 4(2) \\ 4 &= b + 8 \\ -4 &= b \end{aligned}$$

so,

$$y = -4 + 4x.$$

44. Using the point-slope form, we have $y = y_0 + m(x - x_0)$ where the slope m is given by the average rate of change of f on the interval $1 \leq x \leq 3$:

$$\begin{aligned} m &= \frac{\Delta f}{\Delta x} \\ &= \frac{f(3) - f(1)}{3 - 1} \\ &= \frac{1 - 5}{2} = -2. \end{aligned}$$

Letting $(x_0, y_0) = (1, f(1)) = (1, 5)$, we have

$$\begin{aligned} y &= 5 + (-2)(x - 1) \\ &= 7 - 2x. \end{aligned}$$

To verify that this line contains the other point, $(3, f(3))$, we let $x = 3$:

$$y = 7 - 2(3) = 1.$$

Since $f(3) = 1$, we see that the line $y = 7 - 2x$ is correct.

45. Both P and Q lie on $y = x^2 + 1$, so their coordinates must satisfy that equation. Point Q has x-coordinate 2, so $y = 2^2 + 1 = 5$. Point P has y-coordinate 8, so

$$\begin{aligned} 8 &= x^2 + 1 \\ x^2 &= 7, \end{aligned}$$

and $x = -\sqrt{7}$ because we know from the graph that $x < 0$.

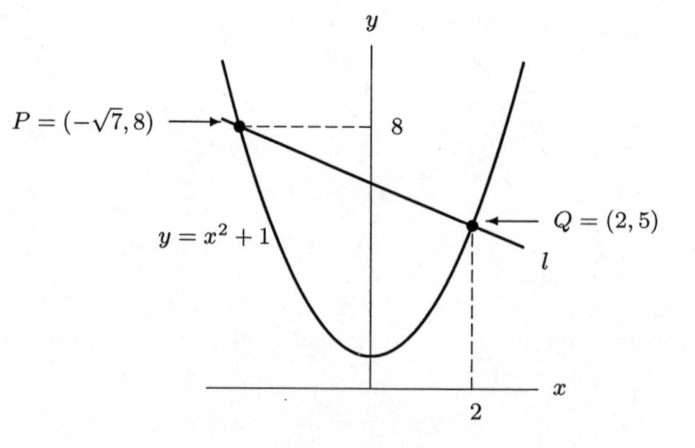

Figure 1.36

Using the coordinates of P and Q, we know that

$$m = \frac{\Delta y}{\Delta x} = \frac{5-8}{2-(-\sqrt{7})} = \frac{-3}{2+\sqrt{7}}.$$

Since $y = 5$ when $x = 2$, we have

$$\begin{aligned} 5 &= b - \frac{3}{2+\sqrt{7}} \cdot 2 \\ 5 &= b - \frac{6}{2+\sqrt{7}} \\ b &= 5 + \frac{6}{2+\sqrt{7}} = \frac{5(2+\sqrt{7})}{(2+\sqrt{7})} + \frac{6}{2+\sqrt{7}} \\ &= \frac{10+5\sqrt{7}+6}{2+\sqrt{7}} = \frac{16+5\sqrt{7}}{2+\sqrt{7}}. \end{aligned}$$

So the equation of the line is

$$y = \frac{16+5\sqrt{7}}{2+\sqrt{7}} - \frac{3}{2+\sqrt{7}}x.$$

Optionally, this can be simplified (by rationalizing denominators) to

$$y = 1 + 2\sqrt{7} + (2-\sqrt{7})x.$$

46.

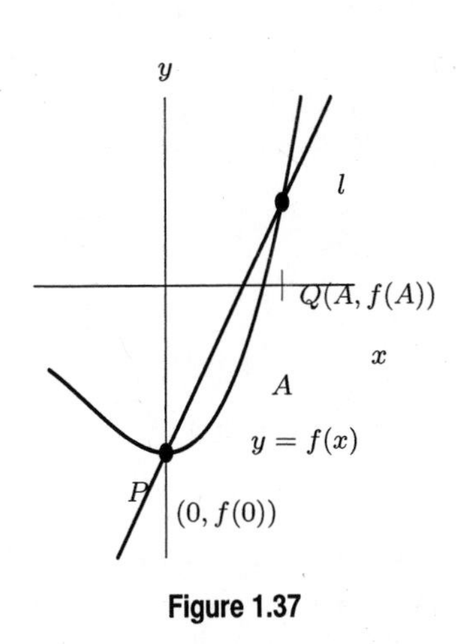

Figure 1.37

Using points $P = (0, f(0))$ and $Q = (A, f(A))$, we can find the slope of the line to be

$$m = \frac{\Delta y}{\Delta x} = \frac{f(A) - f(0)}{A - 0} = \frac{f(A) - f(0)}{A}.$$

Since the y-intercept of l is $b = f(0)$, we have

$$y = f(0) + \frac{f(A) - f(0)}{A}x.$$

47. **(a)** We know that the equation will be of the form

$$p = b + mt$$

where m is the slope and b is the p-intercept. Since there are 100 minutes in an hour and 40 minutes, two points on this line are $(100, 9)$ and $(50, 4)$. Solving for the slope we get

$$m = \frac{9-4}{100-50} = \frac{5}{50} = 0.1 \text{ pages/minute.}$$

Thus, we get

$$p = b + 0.1t.$$

Using the point $(50, 4)$ to solve for b, we get

$$\begin{aligned} 4 &= 50(0.1) + b \\ &= 5 + b. \end{aligned}$$

Thus

$$b = -1$$

and

$$p = -1 + 0.1t \quad \text{or} \quad p = 0.1t - 1.$$

Since p must be non-negative, we have $0.1t - 1 \geq 0$, or $t \geq 10$.

(b) In 2 hours there are 120 minutes. If $t = 120$ we get

$$p = 0.1(120) - 1 = 12 - 1 = 11.$$

Thus 11 pages can be typed in two hours.

(c) The slope of the function tells us that you type 0.1 pages per minute.

(d) Solving the equation for time in terms of pages we get

$$\begin{aligned} p &= 0.1t - 1 \\ 0.1t - 1 &= p \\ 0.1t &= p + 1 \\ t &= 10p + 10. \end{aligned}$$

(e) If $p = 15$ and we use the formula from part (d), we get

$$t = 10(15) + 10 = 150 + 10 = 160.$$

Thus it would take 160 minutes, or two hours and forty minutes to type a fifteen page paper.

(f) Answers vary. Sometimes we know the amount of time we have available to type and we could then use $p = f(t)$ to tell us how many pages can be typed in this time. On the other hand, $t = g(p)$ is useful when we know the number of pages we have and want to know how long it will take to type them.

48. Let us write the equation for the diameter $d(g)$ as follows:

$$d(g) = b + mg$$

where g is the gauge number (and in our case the independent variable), m is the slope of the function and b is the d-intercept. First find the slope, m, by using the data points $(2, 0.2656)$ and $(8, 0.1719)$:

$$\begin{aligned} m &= \frac{d(8) - d(2)}{8-2} \\ &= \frac{0.1719 - 0.2656}{8-2} = \frac{-.0937}{6} \approx -0.01562. \end{aligned}$$

We will use 5 decimal places. Thus $d(g) = b + (-0.01562)g$. Substituting the point $(2, 0.2656)$ in this equation and solving for b, gives

$$\begin{aligned} 0.2656 &= b + (-0.01562)(2) \\ 0.2656 &= b + (-0.03124) \\ \text{and} \quad b &= 0.29684. \end{aligned}$$

Thus,

$$d(g) = 0.29684 + (-0.01562)g.$$

So,

$$\begin{aligned} d(12.5) &= (-0.01562)(12.5) + 0.29684 \\ &= -0.19525 + 0.29684 = 0.10159 \end{aligned}$$

and

$$d(0) = (-0.01562)(0) + 0.29684 = 0.29684.$$

Thus, gauge 12.5 corresponds to a thickness of 0.1016 inches, while gauge 0 corresponds to a thickness of 0.2968 inches. We know that gauge numbers are no longer sensible when they correspond to a negative or zero thickness, thus we must solve

$$d(g) > 0.$$

Solving, we get

$$\begin{aligned} d(g) &> 0 \\ (-0.01562)g + 0.29684 &> 0 \\ (-0.01562)g &> -0.29684 \\ g &< \frac{-0.29684}{-0.01562} \approx 19 \quad \text{(since we divided by a negative number, we must flip the inequality sign).} \end{aligned}$$

Thus, the gauge number only makes sense for values less than 19.

49. (a) Since i is linear, we can write

$$i(x) = b + mx.$$

Since $i(10) = 25$ and $i(20) = 50$, we have

$$m = \frac{50 - 25}{20 - 10} = 2.5.$$

So,

$$i(x) = b + 2.5x.$$

Using $i(10) = 25$, we can solve for b:

$$\begin{aligned} i(10) &= b + 2.5(10) \\ 25 &= b + 25 \\ b &= 0. \end{aligned}$$

Our formula then is

$$i(x) = 2.5x.$$

(b) The increase in risk associated with *not* smoking is $i(0)$. Since there is no increase in risk for a non-smoker, we have $i(0) = 0$.

(c) The slope of $i(x)$ tells us that the risk increases by a factor of 2.5 with each additional cigarette a person smokes per day.

50. Here, s is the independent variable, so we write

$$\begin{aligned} v(s) &= \pi x^2 - 3xr - 4rs - s\sqrt{x} \\ &= \underbrace{\pi x^2 - 3xr}_{b} + \underbrace{(-4r - \sqrt{x})}_{m}\, s \\ \text{so}\quad b &= \pi x^2 - 3xr \\ m &= -4r - \sqrt{x}. \end{aligned}$$

51. Here, r is the independent variable, so we write

$$\begin{aligned} w(r) &= \pi x^2 - 3xr - 4rs - s\sqrt{x} \\ &= \pi x^2 - s\sqrt{x} - 3xr - 4rs \\ &= \underbrace{\pi x^2 - s\sqrt{x}}_{b} + \underbrace{(-3x - 4s)}_{m}\, r \\ \text{so}\quad b &= \pi x^2 - s\sqrt{x}. \\ m &= -3x - 4s. \end{aligned}$$

52. **(a)** The organism never matures at a development rate of $r = 0$. Solving for t, we have $b + kH_{\text{min}} = 0$, so $H_{\text{min}} = -b/k$. This is the t-intercept of the graph of r.

(b) We know that $r = 1/t$ and so $t = 1/r$. We have $S = (H - H_{\text{min}})t$, so

$$S = (H - H_{\text{min}})t = (H - H_{\text{min}})/r.$$

We know that $r = b + kH$ and, from part (a), that $H_{\text{min}} = -b/k$. We have

$$S = \frac{H - (-b/k)}{b + kH} = \frac{H + b/k}{b + kH} = \frac{k}{k} \cdot \frac{H + b/k}{b + kH} = \frac{1}{k}.$$

(c) We are told that $H_{\text{min}} = 15°\text{C}$ and that at a temperature of $H = 20°\text{C}$, development takes $t = 25$ days, so the required number of degree-days is given by

$$S = (H - H_{\text{min}})t = (20 - 15)25 = 125.$$

Since S is constant, we see that at a temperature of $H = 25°\text{C}$,

$$125 = (H - H_{\text{min}})t = (25 - 15)t = 10t,$$

so the organism requires $125/10 = 12.5$ days to develop.

(d) From Table 1.8, we see that after 1 day the total number of degree-days is 5. The total rises to 12 after 2 days, to 24 after 3 days, and so on. The total reaches 126 degree-days on the eleventh day, and since the organism requires only 125 degree-days, it has reached maturity by this point.

Table 1.8 *Accumulated number of degree-days over a twelve-day period*

Day	1	2	3	4	5	6	7	8	9	10	11	12
Degree-day	5	12	24	37	49	65	79	94	107	117	126	137

53. **(a)** We know that $r = 1/t$. Table 1.9 gives values of r. From the table, we see that $\Delta r/\Delta H \approx 0.01/2 = 0.005$, so $r = b + 0.005H$. Solving for b, we have

$$\begin{aligned} 0.070 &= b + 0.005 \cdot 20 \\ b &= 0.070 - 0.1 = -0.03. \end{aligned}$$

Thus, a formula for r is given by $r = 0.005H - 0.03$.

Table 1.9 *Development time t (in days) for an organism as a function of ambient temperature H (in °C)*

H, °C	20	22	24	26	28	30
r, rate	0.070	0.080	0.090	0.100	0.110	0.120

(b) From Problem 52, we know that if $r = b + kH$ then the number of degree-days is given by $S = 1/k$. From part (a) of this problem, we have $k = 0.005$, so $S = 1/0.005 = 200$.

Solutions for Section 1.5

Skill Refresher

S1. Plugging 5 in the first equation for y we get

$$x + 5 = 3$$
$$x = -2.$$

S2. Adding the two equations to eliminate y, we have

$$2x = 8$$
$$x = 4.$$

Using $x = 4$ in the first equation gives

$$4 + y = 3,$$

so

$$y = -1.$$

S3. Multiplying the first equation by -2 and adding the two equations we get

$$0 = 3.$$

So this system has no solution.

S4. Notice that multiplying the first equation by 2 and rewriting it, we get the second equation. Hence, this system has infinitely many solutions.

S5. Substituting the value of y from the first equation into the second equation, we obtain

$$x + 2(2x - 10) = 15$$
$$x + 4x - 20 = 15$$
$$5x = 35$$
$$x = 7.$$

Now we substitute $x = 7$ into the first equation, obtaining $2(7) - y = 10$, hence $y = 4$.

S6. From the first equation, we get $2x + 2y = 3$. From the second equation we get $-2x - y = -15$. So we have,

$$\begin{aligned} 2x + 2y &= 3 \\ -2x - y &= -15. \end{aligned}$$

Adding the two equations gives $y = -12$, and solving for x in either equation gives $x = 13.5$.

S7. We set the equations $y = x$ and $y = 3 - x$ equal to one another.

$$\begin{aligned} x &= 3 - x \\ 2x &= 3 \\ x &= \frac{3}{2} \quad \text{and} \quad y = \frac{3}{2}. \end{aligned}$$

So the point of intersection is $x = 3/2$, $y = 3/2$.

S8. The point of intersection lies on the two lines

$$y = 2x - 3.5 \qquad \text{and} \qquad y = -\frac{1}{2}x + 4.$$

To find the point, we solve this system of equations simultaneously. Setting these two equations equal to each other and solving for x, we have

$$\begin{aligned} 2x - 3.5 &= -\frac{1}{2}x + 4 \\ 2x + \frac{1}{2}x &= 3.5 + 4 = 7.5 = \frac{15}{2} \\ \frac{5}{2}x &= \frac{15}{2} \\ x &= \frac{15}{2} \cdot \frac{2}{5} = 3. \end{aligned}$$

Since $x = 3$, we have

$$y = 2x - 3.5 = 2(3) - 3.5 = 6 - 3.5 = 2.5.$$

Thus, the point of intersection is $(3, 2.5)$.

Exercises

1. **(a)** is (V), because slope is negative, vertical intercept is 0
(b) is (VI), because slope and vertical intercept are both positive
(c) is (I), because slope is negative, vertical intercept is positive
(d) is (IV), because slope is positive, vertical intercept is negative
(e) is (III), because slope and vertical intercept are both negative
(f) is (II), because slope is positive, vertical intercept is 0

2. **(a)** is (V), because slope is positive, vertical intercept is negative
(b) is (IV), because slope is negative, vertical intercept is positive
(c) is (I), because slope is 0, vertical intercept is positive
(d) is (VI), because slope and vertical intercept are both negative
(e) is (II), because slope and vertical intercept are both positive
(f) is (III), because slope is positive, vertical intercept is 0
(g) is (VII), because it is a vertical line with positive x-intercept.

3. The functions f and g have the same y-intercept, $b = 20$. u and v both have y-intercept $b = 60$. f and g are increasing functions, with slopes $m = 2$ and $m = 4$, respectively. u and v are decreasing functions, with slopes $m = -1$ and $m = -2$, respectively.

The figure shows that graphs A and B describe increasing functions with the same y-intercept. The functions f and g are good candidates since they are both linear functions with positive slope and their y-intercepts coincide. Since graph

A is steeper than graph B, the slope of A is greater than the slope of B. The slope of g is larger than the slope of f, so graph A corresponds to g and graph B corresponds to f.

Graphs D and E describe decreasing functions with the same y-intercept. u and v are good candidates since they both have negative slope and their y-intercepts coincide. Graph E is steeper than graph D. Thus, graph D corresponds to u, and graph E to v. Note that graphs D and E start at a higher point on the y-axis than A and B do. This corresponds to the fact that the y-intercept $b = 60$ of u and v is above the y-intercept $b = 20$ of f and g.

This leaves graph C and the function h. The y-intercept of h is -30, corresponding to the fact that graph C starts below the x-axis. The slope of h is 2, the same slope as f. Since graph C appears to climb at the same rate as graph B, it seems reasonable that f and h should have the same slope.

4. In Figure 4, we see that lines A and B both represent increasing functions with the same y-intercept. Thus, since f and h have positive slope and the same y-intercept, $b = 5$, lines A and B correspond to the functions f and h. Since line A is steeper than line B, its slope is greater. The slope of h is 3, while the slope of f is 2. Therefore, line A is $h(x) = 5 + 3x$ and line B is $f(x) = 5 + 2x$.

Line C also represents an increasing function. Furthermore, since it crosses the y-axis below the x-axis, it has a negative y-intercept. Since $g(x) = -5 + 2x$ is an increasing function with a negative y-intercept, it corresponds to line C.

Finally, lines D and E both represent decreasing functions, and so both have negative slopes. Since line E is steeper than line D, its slope is steeper—that is, more negative—than the slope of line D. Thus, line E represents $k(x) = 5 - 3x$ and line D represents $j(x) = 5 - 2x$.

5. **(a)** See Figures 1.38 and 1.39.

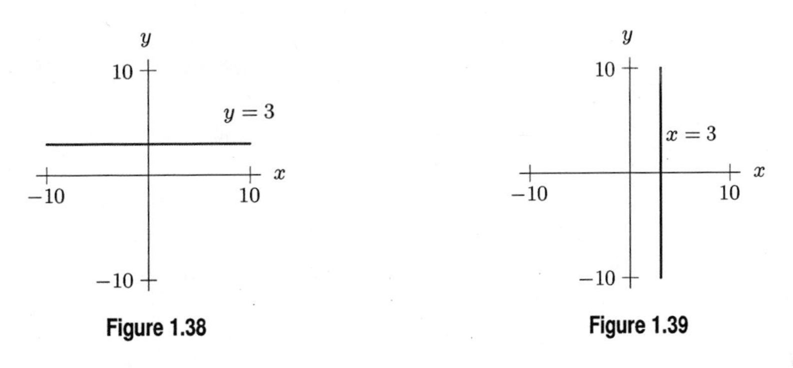

Figure 1.38

Figure 1.39

(b) Yes for $y = 3$: $y = 3 + 0x$. No for $x = 3$, since the slope is undefined, and there is no y-intercept.

6. These lines are parallel because they have the same slope, 5.

7. These lines are perpendicular because one slope, $-\frac{1}{4}$, is the negative reciprocal of the other, 4.

8. These line are parallel because they have the same slope, 2.

9. These lines are neither parallel nor perpendicular. They do not have the same slope, nor are their slopes negative reciprocals (if they were, one of the slopes would be negative).

10. These lines are neither parallel nor perpendicular. They do not have the same slopes, nor are their slopes negative reciprocals (if they were, one of the slopes would be negative).

11. Rewriting as $y = 8 - \frac{1}{2}x$ and $y = -2 - \frac{1}{2}x$ shows that these lines are parallel because their slopes are the same, $-\frac{1}{2}$.

Problems

12. This family of lines all have y–intercept equal to -2. Furthermore, the slopes of these lines are positive. A possible family is shown in Figure 1.40

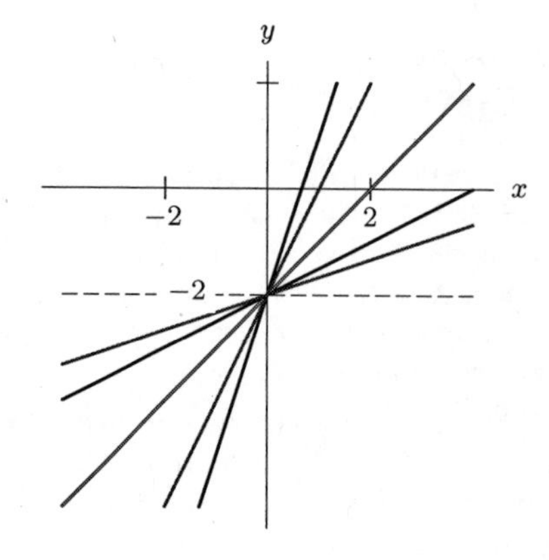

Figure 1.40

13. We can write the equation in slope-intercept form

$$\begin{aligned} 3x + 5y &= 6 \\ 5y &= 6 - 3x \\ y &= \frac{6}{5} - \frac{3}{5}x. \end{aligned}$$

The slope is $\frac{-3}{5}$. Lines parallel to this line all have slope $\frac{-3}{5}$. Since the line passes through $(0, 6)$, its y-intercept is equal to 6. So $y = 6 - \frac{3}{5}x$.

14. $y = 5x - 3$. Since the slope of this line is 5, we want a line with slope $-\frac{1}{5}$ passing through the point $(2, 1)$. The equation is $(y - 1) = -\frac{1}{5}(x - 2)$, or $y = -\frac{1}{5}x + \frac{7}{5}$.

15. The line $y + 4x = 7$ has slope -4. Therefore the parallel line has slope -4 and equation $y - 5 = -4(x - 1)$ or $y = -4x + 9$. The perpendicular line has slope $\frac{-1}{(-4)} = \frac{1}{4}$ and equation $y - 5 = \frac{1}{4}(x - 1)$ or $y = 0.25x + 4.75$.

16. Note that the x and y scales are different and the intercepts appear to be (0,3) and (7.5,0), giving

$$\text{Slope } = \frac{-3}{7.5} = -\frac{6}{15} = -\frac{2}{5}.$$

The y-intercept is at (0,3), so

$$y = -\frac{2}{5}x + 3$$

is a possible equation for the line (answers may vary).

17. Since P is the x-intercept, we know that point P has y-coordinate $= 0$, and if the x-coordinate is x_0, we can calculate the slope of line l using $P(x_0, 0)$ and the other given point $(0, -2)$.

$$m = \frac{-2 - 0}{0 - x_0} = \frac{-2}{-x_0} = \frac{2}{x_0}.$$

We know this equals 2, since l is parallel to $y = 2x + 1$ and therefore must have the same slope. Thus we have

$$\frac{2}{x_0} = 2.$$

So $x_0 = 1$ and the coordinates of P are $(1, 0)$.

18. We see in the figure from the problem that line l_2 is perpendicular to line l_1. We can find the slope of line l_1 because we are given the x-intercept $(3, 0)$ and the y-intercept $(0, 2)$.

$$m_1 = \frac{2 - 0}{0 - 3} = \frac{2}{-3} = \frac{-2}{3}$$

Therefore, we know that the slope of line l_2 is

$$m_2 = \frac{-1}{\frac{-2}{3}} = \frac{3}{2}$$

We also know that l_2 passes through the origin $(0,0)$ and therefore has a y-intercept of zero.

Hence, the equation of l_2 is

$$y = \frac{3}{2}x.$$

19. (a) After one year, the value of the Frigbox refrigerator is $\$950 - \$50 = \$900$; after two years, its value is $\$950 - 2 \cdot \$50 = \$850$; after t years, the value, V, of the Frigbox is given by

$$V = 950 - t \cdot 50 \quad \text{or} \quad V = 950 - 50t.$$

Similarly, after t years, the value of the ArcticAir refrigerator is

$$V = 1200 - 100t.$$

The two refrigerators have equal value when

$$\begin{aligned} 950 - 50t &= 1200 - 100t \\ -250 &= -50t \\ 5 &= t. \end{aligned}$$

In five years the two refrigerators have equal value.

(b) According to the formula, in 20 years time, the value of the Frigbox refrigerator will be

$$\begin{aligned} V &= 950 - 50(20) \\ &= 950 - 1000 = -50 \end{aligned}$$

This negative value is not realistic, so after some time, the linear model is no longer appropriate. Similarly, the value of the ArcticAir refrigerator is predicted to be $V = 1200 - 100(20) = 1200 - 2000 = -800$, which is also not realistic.

20. (a) The three formulas are linear with b being the fixed rate and m being the cost per mile. The formulas are,

$$\begin{aligned} \text{Company A} &= 20 + 0.2x \\ \text{Company B} &= 35 + 0.1x \\ \text{Company C} &= 70. \end{aligned}$$

(b)

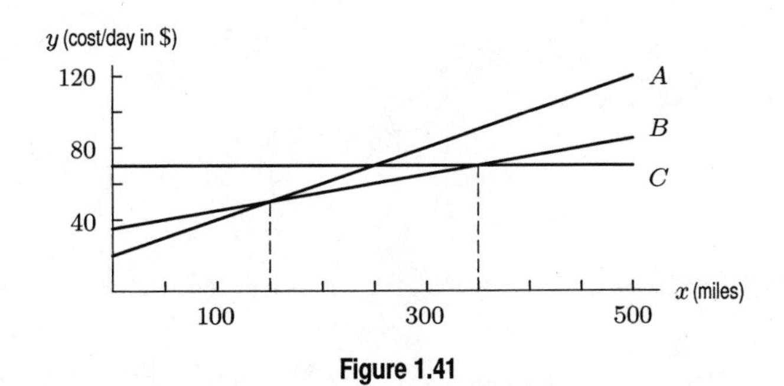

Figure 1.41

(c) The slope is the rate charged for each mile, and its units are dollars per mile. The vertical intercept is the fixed cost—what you pay for renting the car for a day, not considering mileage charges.

(d) By reading Figure 1.41 we see A is cheapest if you drive less than 150 miles; B is cheapest if you drive between 150 and 350 miles; C is cheapest if you drive more than 350 miles. We would expect A to be the cheapest for a small number of miles since it has the lowest fixed rate and C to be the cheapest for a large number of miles since it does not charge per mile.

21. **(a)** This line, being parallel to l, has the same slope. Since the slope of l is $-\frac{2}{3}$, the equation of this line is

$$y = b - \frac{2}{3}x.$$

To find b, we use the fact that $P = (6, 5)$ is on this line. This gives

$$\begin{aligned} 5 &= b - \frac{2}{3}(6) \\ 5 &= b - 4 \\ b &= 9. \end{aligned}$$

So the equation of the line is

$$y = 9 - \frac{2}{3}x.$$

(b) This line is perpendicular to line l, and so its slope is given by

$$m = \frac{-1}{-2/3} = \frac{3}{2}.$$

Therefore its equation is

$$y = b + \frac{3}{2}x.$$

We again use point P to find b:

$$\begin{aligned} 5 &= b + \frac{3}{2}(6) \\ 5 &= b + 9 \\ b &= -4. \end{aligned}$$

This gives

$$y = -4 + \frac{3}{2}x.$$

(c) Figure 1.42 gives a graph of line l together with point P and the two lines we have found.

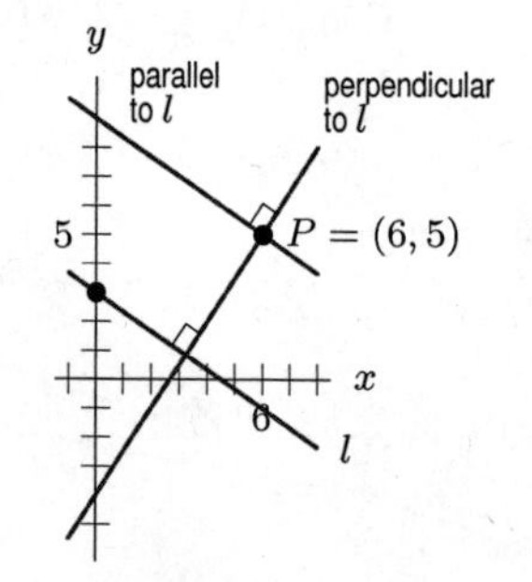

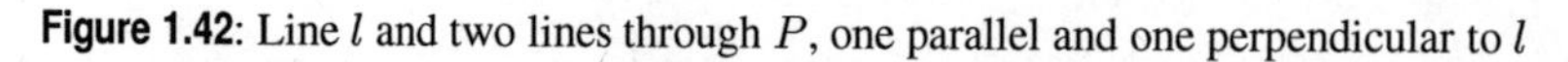

Figure 1.42: Line l and two lines through P, one parallel and one perpendicular to l

22. **(a)** Since $y = f(x)$, to show that $f(x)$ is linear, we can solve for y in terms of A, B, C, and x.

$$\begin{aligned} Ax + By &= C \\ By &= C - Ax, \text{ and, since } B \neq 0, \\ y &= \frac{C}{B} - \frac{A}{B}x \end{aligned}$$

Because C/B and $-A/B$ are constants, the formula for $f(x)$ is of the linear form:

$$f(x) = y = b + mx.$$

Thus, f is linear, with slope $m = -(A/B)$ and y-intercept $b = C/B$.
To find the x-intercept, we set $y = 0$ and solve for x:

$$\begin{aligned} Ax + B(0) &= C \\ Ax &= C, \text{ and, since } A \neq 0, \\ x &= \frac{C}{A}. \end{aligned}$$

Thus, the line crosses the x–axis at $x = C/A$.

(b) (i) Since $A > 0, B > 0, C > 0$, we know that C/A (the x-intercept) and C/B (the y-intercept) are both positive and we have Figure 1.43.

(ii) Since only $C < 0$, we know that C/A and C/B are both negative, and we obtain Figure 1.44.

(iii) Since $A > 0, B < 0, C > 0$, we know that C/A is positive and C/B is negative. Thus, we obtain Figure 1.45.

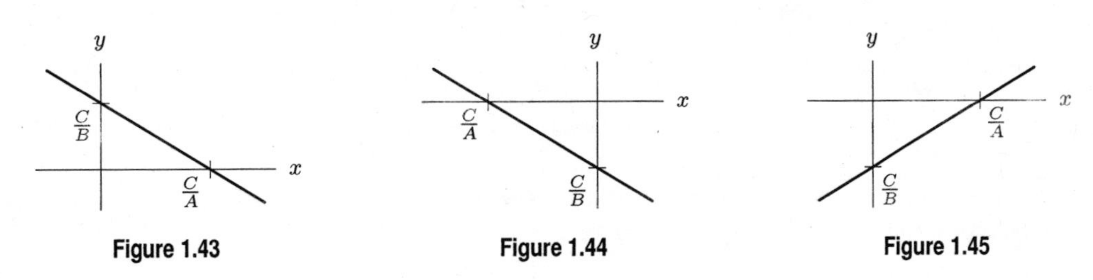

Figure 1.43 Figure 1.44 Figure 1.45

23. **(a)** $P = (a, 0)$
(b) $A = (0, b)$, $B = (-c, 0)$
$C = (a + c, b)$, $D = (a, 0)$

24. The graphs are shown in Figure 1.46.

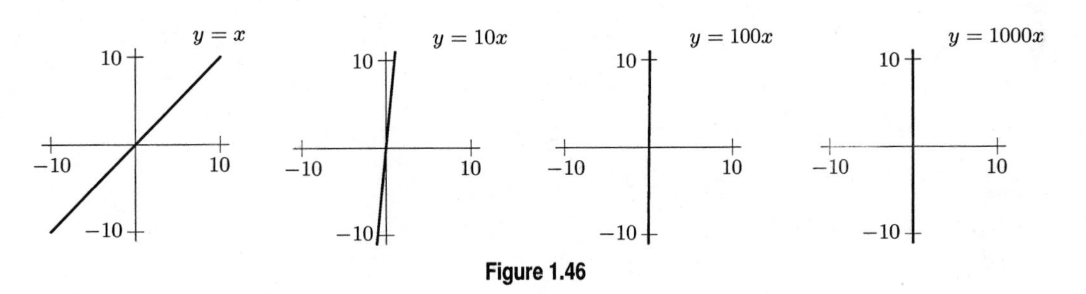

Figure 1.46

(a) As the slopes become larger, the lines become steeper, getting very close to the y-axis.

(b) We start with the equation of a line $y = b + mx$. Because the line passes through the origin, we want the graph to have a y-intercept of zero, so $b = 0$. Because the line is horizontal, we want a slope of zero, so $m = 0$. Thus, our equation is

$$y = 0.$$

25. The graphs are shown in Figure 1.47.

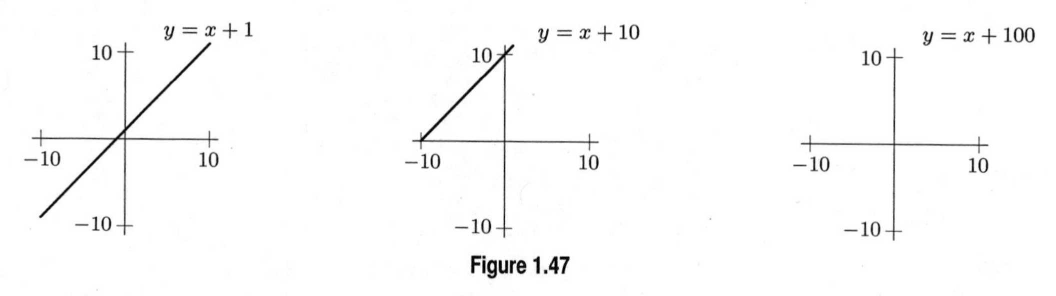

Figure 1.47

(a) As b becomes larger, the graph moves higher and higher up, until it disappears from the viewing rectangle.
(b) There are many correct answers, one of which is $y = x - 100$.

26.

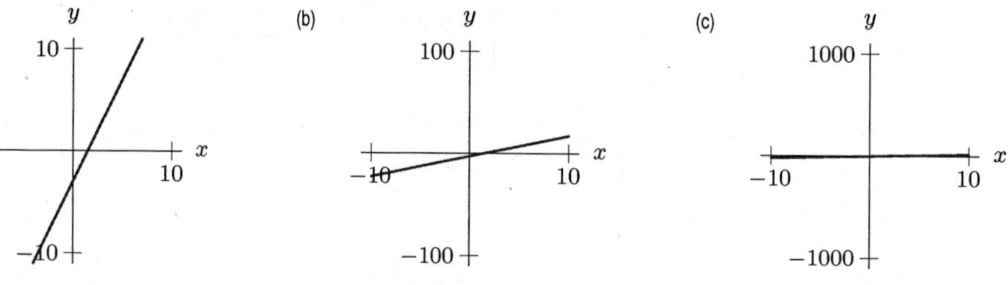

Figure 1.48

(d) If the width of the window remains constant and the height of the window increases, then the graph will appear less steep.

27. Writing this equation as

$$y = \underbrace{\frac{1}{\beta - 3}}_{m} \cdot x + \underbrace{\frac{1}{6 - \beta}}_{b},$$

we see that $m = 1/(\beta - 3)$ and $b = 1/(6 - \beta)$. For the slope to be positive, we require the denominator of m to be positive, so $\beta > 3$. For the y-intercept to be positive, we require the denominator of b to be positive, so $\beta < 6$. Putting together these two requirements gives $3 < \beta < 6$.

28. Writing this equation as

$$y = \underbrace{\frac{1}{\beta - 3}}_{m} \cdot x + \underbrace{\frac{1}{6 - \beta}}_{b},$$

we see that $m = 1/(\beta - 3)$ and $b = 1/(6 - \beta)$. The line $y = (\beta - 7)x - 3$ has slope $m_1 = \beta - 7$. For these two lines to be perpendicular, we require

$$\begin{aligned} m \cdot m_1 &= -1 \\ \frac{1}{\beta - 3} \cdot (\beta - 7) &= -1 \\ \beta - 7 &= -(\beta - 3) \\ \beta - 7 &= -\beta + 3 \\ 2\beta &= 10 \\ \beta &= 5. \end{aligned}$$

We can verify this by confirming that the resulting equations describe perpendicular lines:

$$y = \frac{1}{\beta - 3} \cdot x + \frac{1}{6 - \beta} \quad \text{First equation}$$

$$
\begin{aligned}
&= \frac{1}{5-3} \cdot x + \frac{1}{6-5} \\
&= \frac{1}{2} \cdot x + 1 \\
y &= (\beta - 7)x - 3 \qquad \text{Second equation} \\
&= (5-7)x - 3 \\
&= -2x - 3.
\end{aligned}
$$

Since the slopes are negative reciprocals, the lines are perpendicular, as required.

29. **(a)** Use the point-slope formula:

$$y - 0 = \frac{\sqrt{3}}{-1}(x - 0),$$

so $y = -\sqrt{3}x$.

(b) The slope of the tangent line is the negative reciprocal of $-\sqrt{3}$, so $m = \dfrac{1}{\sqrt{3}}$, and

$$y - \sqrt{3} = \frac{1}{\sqrt{3}}(x - (-1)),$$

or

$$y = \frac{1}{\sqrt{3}}x + \frac{4}{\sqrt{3}}.$$

30. The altitude is perpendicular to the side $\overline{BC}$. The slope of $\overline{BC}$ is

$$\frac{8-2}{9-(-3)} = \frac{6}{12} = \frac{1}{2}.$$

Therefore the slope of the altitude is the negative reciprocal of $1/2$, which is -2, and it passes through the point $(-4, 5)$. Using the point-slope formula, we find the equation $y - 5 = -2(x + 4)$, or $y = -2x - 3$.

31. When $x = 1$, $y = \sqrt{1} = 1$, and when $x = 4$, $y = \sqrt{4} = 2$, so the points of intersection are $(1, 1)$ and $(4, 2)$. See Figure 1.49.

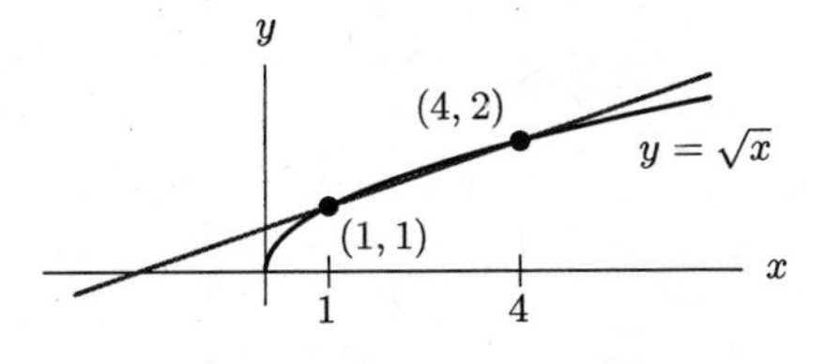

Figure 1.49

The line connecting $(1, 1)$ and $(4, 2)$ has slope $m = \frac{2-1}{4-1} = \frac{1}{3}$. To find the y-intercept, we can substitute one of the points, for example, $x = 1$, $y = 1$:

$$
\begin{aligned}
y &= \frac{1}{3}x + b \\
1 &= \frac{1}{3}(1) + b \\
b &= \frac{2}{3}
\end{aligned}
$$

The equation of the line is $y = \frac{1}{3}x + \frac{2}{3}$. Now we'll solve the system

$$
\begin{aligned}
y &= \sqrt{x} \\
y &= \frac{1}{3}x + \frac{2}{3}
\end{aligned}
$$

by setting the equations equal to each other:

$$\sqrt{x} = \frac{1}{3}x + \frac{2}{3}.$$

Squaring both sides gives

$$\begin{aligned} x &= \left(\frac{1}{3}x + \frac{2}{3}\right)^2 \\ x &= \frac{1}{9}x^2 + \frac{4}{9}x + \frac{4}{9} \\ \frac{1}{9}x^2 - \frac{5}{9}x + \frac{4}{9} &= 0 \\ x^2 - 5x + 4 &= 0 \quad \text{after multiplying both sides by 9} \\ (x-4)(x-1) &= 0 \\ x = 4 &\text{ or } x = 1 \end{aligned}$$

When $x = 4$, $y = \sqrt{4} = 2$, giving the point $(4, 2)$. When $x = 1$, $y = \sqrt{1} = 1$, giving the point $(1, 1)$. The results are consistent with the original problem

32. **(a)** To have no points in common the lines will have to be parallel and distinct. To be parallel their slopes must be the same, so $m_1 = m_2$. To be distinct we need $b_1 \neq b_2$.

(b) To have all points in common the lines will have to be parallel and the same. To be parallel their slopes must be the same, so $m_1 = m_2$. To be the same we need $b_1 = b_2$.

(c) To have exactly one point in common the lines will have to be nonparallel. To be nonparallel their slopes must be distinct, so $m_1 \neq m_2$.

(d) It is not possible for two lines to meet in just two points.

Solutions for Section 1.6

1. Although the points do not lie on a line, they are tending upward as x increases. So, there is a strong positive correlation; $r = 0.93$ is reasonable.

2. The points are scattered all over. There is no clear upward nor a downward trend, so there is probably no correlation between them. ($r = -0.15$.)

3. These points are very close to lying on a line with negative slope, so $r = 1$ is not reasonable. ($r = -0.9976$.)

4. Although the points do not lie on a line, they are tending upward as x increases. So, there is a positive correlation and so $r = 0.92$ is reasonable.

5. A scatter plot of the data is shown in Figure 1.50.

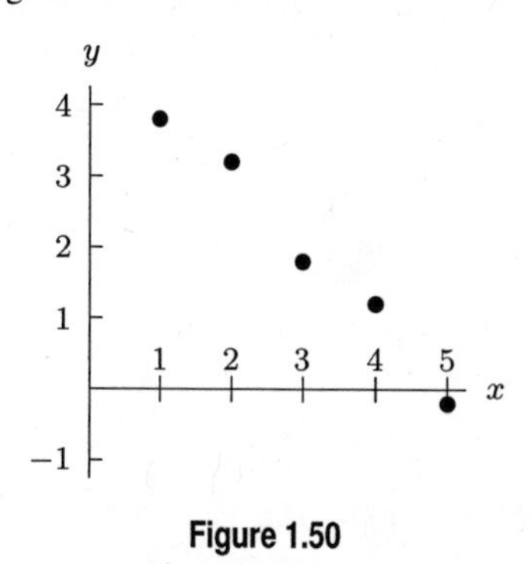

Figure 1.50

$r = 1$ is not possible. These points are very close to lying on a line with negative slope, so r will be negative. ($r = -0.98$.)

6. A scatter plot of the data is shown in Figure 1.51.

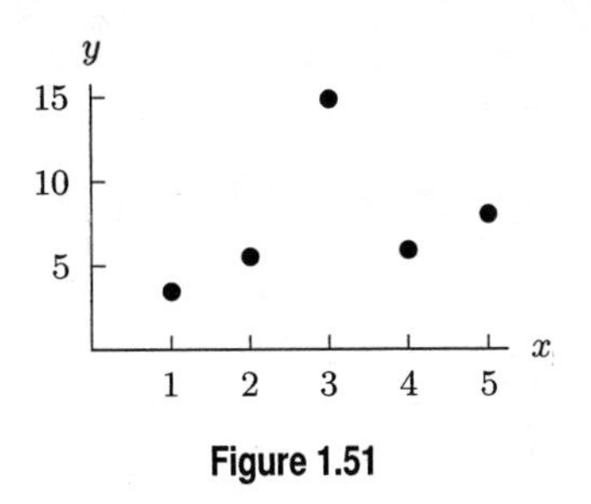

Figure 1.51

Although these points are quite scattered, there is an upward slope, so $r = 0.343$ is reasonable.

7. **(a)** Since the points lie on a line of positive slope, $r = 1$.
(b) Although the points do not lie on a line, they are tending upward as x increases. So, there is a positive correlation and a reasonable guess is $r = 0.7$.
(c) The points are scattered all over. There is neither an upward nor a downward trend, so there is probably no correlation between x and y, so $r = 0$.
(d) These points are very close to lying on a line with negative slope, so the best correlation coefficient is $r = -0.98$.
(e) Although these points are quite scattered, there is a downward slope, so $r = -0.25$ is probably a good answer.
(f) These points are less scattered than those in part (e). The best answer here is $r = -0.5$.

8. **(a)** The points are graphed in Figure 1.52.

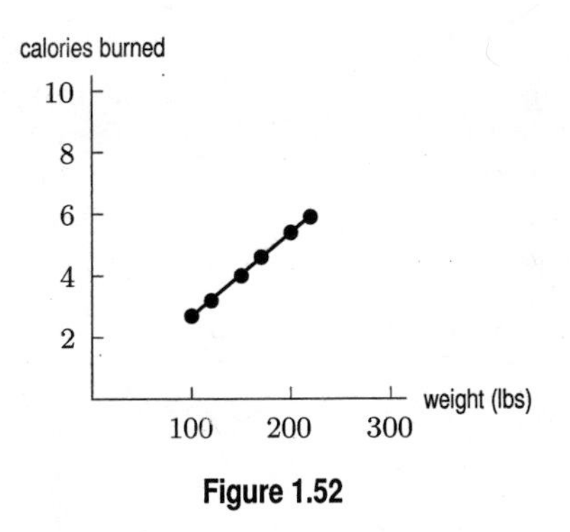

Figure 1.52

(b) See Figure 1.52.
(c) Since the points all seem to lie on a line, the correlation coefficient is close to one.

9. **(a)** See Figure 1.53.

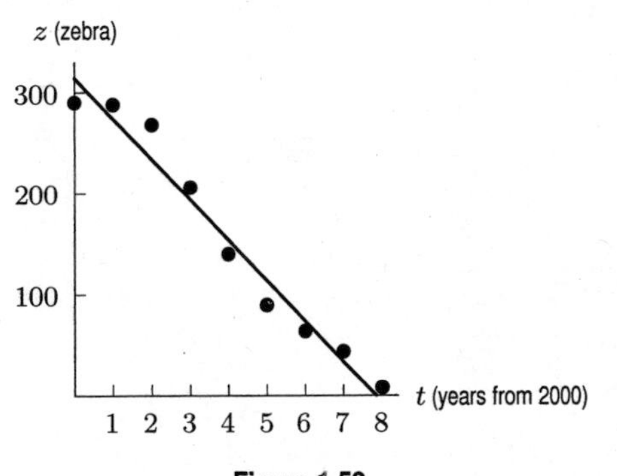

Figure 1.53

(b) Estimates vary, but should be roughly $z = -40t + 314$.
(c) A calculator gives $z = -40t + 314$.
(d) The slope of -40 tells us that, on average, 40 zebra die per year. The vertical intercept value is the initial population. The vertical intercept of the line is 314, which is close to the initial data value of 290. The horizontal intercept of 7.85 years is the number of years until all the zebra have died.
(e) There is a strong negative correlation, ($r = -0.983$).

10. (a) See Figure 1.54.

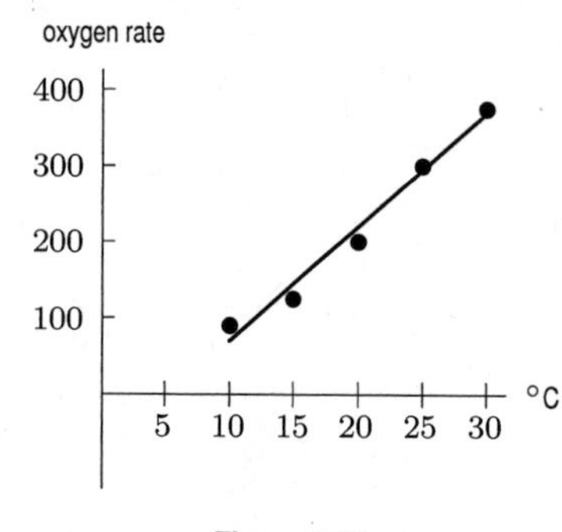

Figure 1.54

(b) Estimates vary, but should be roughly $y = 15x - 80$.
(c) A calculator gives $y = 15x - 80$. Without it, results will vary.
(d) The slope of 15 tells us that for every rise of one degree, the consumption rate increases by 15. The horizontal intercept value (5.33) tells us the temperature when the rate of consumption is 0 (the beetle stops breathing). The vertical intercept value would give us the oxygen rate at 0°C (freezing) but in this case a negative value ($-80°$) tells us that the model breaks down and is not valid for cold temperatures.
(e) There is a strong positive correlation ($r \approx 0.99$).

11. (a) See Figure 1.55.

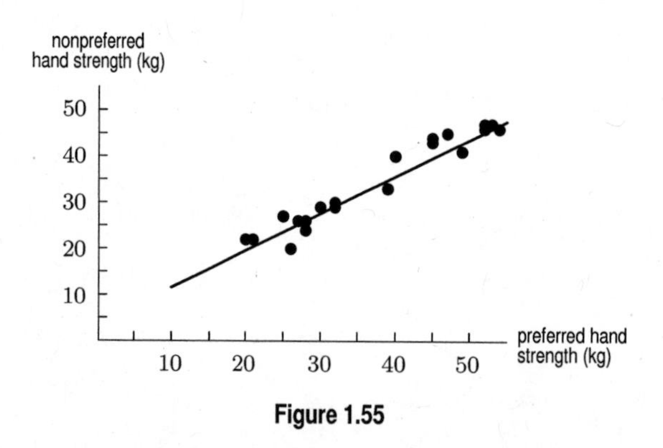

Figure 1.55

(b) Answers vary, but should be close to $y = 3.6 + 0.8x$.
(c) Answers may vary slightly. A possible equation is: $y = 3.623 + 0.825x$.
(d) The preferred hand strength is the independent quantity, so it is represented by x. Substituting $x = 37$ gives

$$y = 3.623 + 0.825(37) \approx 34.$$

So, the nonpreferred hand strength is about 34 kg.

(e) If we predict strength of the nonpreferred hand based on the strength of the preferred hand for values within the observed values of the preferred hand (such as 37), then we are interpolating. However, if we chose a value such as 10, which is below all the actual measurements, and use this to predict the nonpreferred hand strength, then we are extrapolating. Predicting from a value of 100 would be another example of extrapolation. In this case of hand strength, it seems safe to extrapolate; in other situations, extrapolation can be inaccurate.

(f) The correlation coefficient is positive because both hand strengths increase together, so the line has a positive slope. The value of r is close to 1 because the hand strengths lie close to a line of positive slope.

(g) The two clusters suggest that there are two distinct groups of students. These might be men and women, or perhaps students who are involved in college athletics (and therefore in excellent physical shape) and those who are not involved.

12. (a) We would expect a race of length 0 meters to take 0 seconds for both men and women, so we should insert a column of zeros at the start of the table.

(b) Figure 1.56 shows the men's and women's times against distance together with the regression lines $t = 0.587d - 9.069$ (men) and $t = 0.637d - 8.260$ (women). The slopes represent the average change in record time per meter increase in the length of the race. The women's line is steeper than the men's. This means that for a given increase in race length, the women's record time increases by more than the men's. The vertical intercepts are -9.069 and -8.260 seconds. We expect the vertical intercept to represent an estimate of the world record for swimming 0 meters. However, the practical interpretation of the fact that the intercepts are negative is that the model is not valid for $d = 0$.

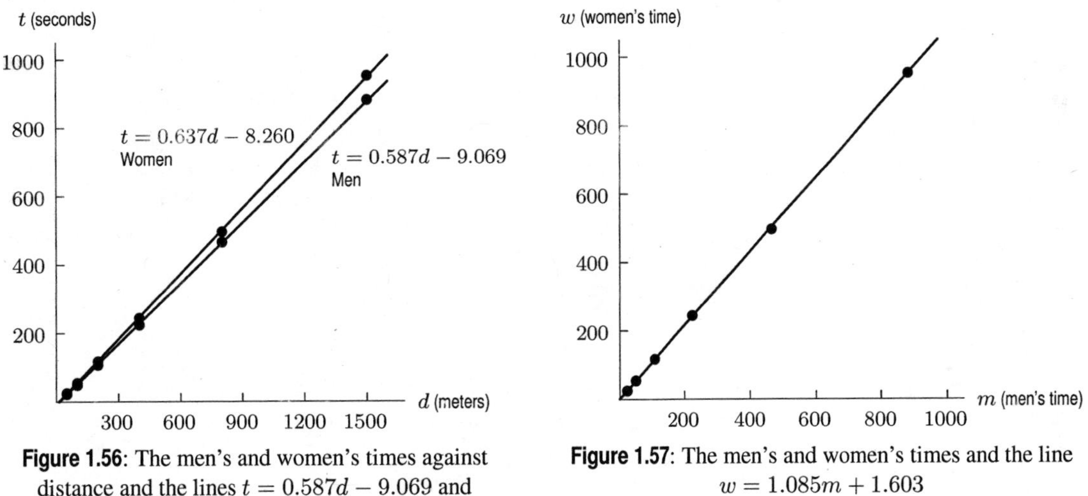

Figure 1.56: The men's and women's times against distance and the lines $t = 0.587d - 9.069$ and $t = 0.637d - 8.260$

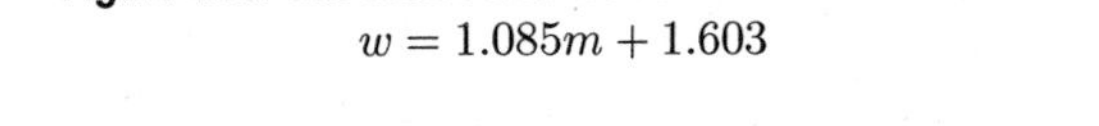

Figure 1.57: The men's and women's times and the line $w = 1.085m + 1.603$

(c) Suppose m = men's time and w = women's time. From part (b) we found $t = 0.587d - 9.069$ (men) and $t = 0.637d - 8.260$ (women), so that $m = 0.587d - 9.069$ and $w = 0.637d - 8.260$. We solve the first equation for d, obtaining $d = (m + 9.069)/0.587$, and substitute in the second to find $w = 0.637(m + 9.069)/0.587 - 8.260 = 1.085m + 1.603$. The slope is 1.085 and represents the average change in women's times per unit change in men's times. Figure 1.57 shows the men's and women's times and the line $w = 1.085m + 1.603$. Another way to say this is that the changes in women's times are 108.5% of changes in the men's. However this does not mean that women's times are 108.5% of men's times (only that *changes* are). For large times (that is, large enough to make the 1.603 negligible), women's times are 108.5% of men's times. Thus

$$952.10 \text{ is } 108.866\% \text{ of } 874.56$$

but

$$24.13 \text{ is } 111.506\% \text{ of } 21.64.$$

Thus the reporter's statement that the women's records are about 8% higher than the men's—that is, 108% of the men's—is approximately correct for large distances, but not for small.

The vertical intercept is 1.603 seconds. This would mean that 1.603 seconds would be the women's world record for swimming a race that took the men 0 seconds to swim. The practical interpretation is that the linear model is not valid for $m = 0$.

Solutions for Chapter 1 Review

Exercises

1. Any w value can give two z values. For example, if $w = 0$,

$$7 \cdot 0^2 + 5 = z^2$$
$$\pm\sqrt{5} = z,$$

so there are two z values (one positive and one negative) for $w = 0$. Thus, z is not a function of w.

A similar argument shows that w is not a function of z.

2. Here, y is a function of x, because any particular x value gives one and only one y value. For example, if we input the constant a as the value of x, we have $y = a^4 - 1$, which is one particular y value.

However, some values of y lead to more than one value of x. For example, if $y = 15$, then $15 = x^4 - 1$, so $x^4 = 16$, giving $x = \pm 2$. Thus, x is not a function of y.

3. Here, m is a function of t. For any t, there is only one possible value of m. In addition, for any m, there is only one possible value of t, given by $t = m^2$. Thus, t is a function of m.

4. Both of the relationships are functions because any quantity of gas determines the quantity of coffee that can be bought, and vice versa. For example, if you buy 30 gallons of gas, spending $60, you buy 4 pounds of coffee.

5. At the two points where the graph breaks (marked A and B in Figure 1.58), there are two y values for a single x value. The graph does not pass the vertical line test. Thus, y is not a function of x.

Similarly, x is not a function of y because there are many y values that give two x values (For example, $y = 0$.)

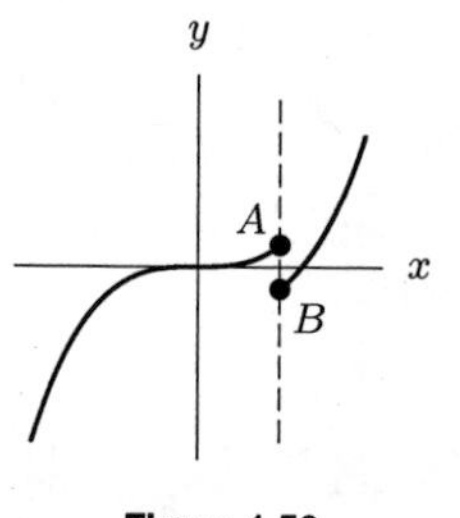

Figure 1.58

6. **(a)** The graphs in (I), (III), (IV), (V), (VII), and (VIII) are functions. The graphs in (II), (VI), and (IX) do not pass the vertical line test and so they cannot be the graphs of functions.

(b) (i) The graph of SAT Math score versus SAT Verbal score for a number of students will be a graph of a number of points. Graphs (V) and (VI) are of this type.

(ii) The graph of hours of daylight per day must be an oscillating function (since the number of hours of daylight fluctuates up and down throughout the year). Graph (VIII) represents this.

(c) If the train fare remains constant throughout the day, graph (III) describes the fare. If there are specific times of the day (rush hours, for example) when the train company raises its prices, then graph (IV) represents the train fare as a function of time of day.

7. (a) We have

$$f(0) = \frac{10}{1+0^2} = \frac{10}{1} = 10$$
$$f(1) = \frac{10}{1+1^2} = \frac{10}{2} = 5$$
$$f(2) = \frac{10}{1+2^2} = \frac{10}{5} = 2$$
$$f(3) = \frac{10}{1+3^2} = \frac{10}{10} = 1.$$

See Table 1.10.

Table 1.10

x	0	1	2	3
$f(x)$	10	5	2	1

(b) For $x = 0$, we have $f(0) = 10$. This value is largest because the x-value is smallest.

8. (a) (i) We find the average rate of change in the population as follows. For P_1 from 1990 to 2000,

$$\text{Rate of change} = \frac{\Delta P_1}{\Delta t} = \frac{62-42}{2000-1990} = 2 \text{ thousand people per year.}$$

Thus, P_1 is growing, on average, by two thousand people per year. For P_2 over the same period,

$$\text{Rate of change} = \frac{\Delta P_2}{\Delta t} = \frac{72-82}{2000-1990} = -1 \text{ thousand people per year.}$$

The negative sign tells us that P_2 is decreasing, on average, by one thousand people per year.

(ii) For 1990–2007, the average rate of change of P_1 is:

$$\text{Rate of change} = \frac{\Delta P_1}{\Delta t} = \frac{76-42}{2007-1990} = 2 \text{ thousand people per year.}$$

That is, the city is gaining 2 thousand people per year. The average rate of change of P_2 is:

$$\text{Rate of change} = \frac{\Delta P_2}{\Delta t} = \frac{65-82}{2007-1990} = -1 \text{ thousand people per year.}$$

That is, the city is losing a thousand people per year.

(iii) For 1995–2007, we have:

$$\frac{\Delta P_1}{\Delta t} = \frac{76-52}{2007-1995} = 2 \text{ thousand people per year.}$$

That is, the city is gaining 2 thousand people per year. The average rate of growth for the second population is:

$$\frac{\Delta P_2}{\Delta t} = \frac{65-77}{2007-1995} = -1 \text{ thousand people per year.}$$

That is, the city is losing a thousand people per year.

(b) The average rate of change of each population is the same on all three time intervals. Each population appears to be changing at a constant rate. The first population is growing, on average, by 2 thousand people per year in each time interval. The second population is dropping, on average, by 1 thousand people per year in each time interval.

9. (a) Machine #2 gives two different possible snacks for each button. Thus, S is not a function of N. It is a bad machine to use because you can't choose the snack you will get.

(b) Machines #1 and #3 give S as a function of N. This means that by choosing a button number, you can choose a snack.

(c) Machine #3. N is not a function of S because two different button numbers correspond to the same snack. For example, $N = 8$ and 9 both correspond to Snickers. This means that one snack corresponds to more than one button.

10. (a)

Table 1.11 *Average temperature in Albany, NY*

Month	1	2	3	4	5	6	7	8	9	10	11	12
Avg. Temp	22	23	33	45	57	67	71	70	62	51	39	26

(b) July is the warmest month, as it has the highest average temperature.
(c) The temperature is increasing from January to July. The temperature is decreasing from July to December.

11. (a) At the end of the race, Owens was running at 12 yards/sec and the horse was running at 20 yards/sec.
(b) We can find the time when they were both running the same speed by finding the point on the graph when the two lines intersect. This occurs at $t = 6$. So they were both running the same speed after 6 seconds.

12. This table could not represent a linear function, because the rate of change of $q(\lambda)$ is not constant. Consider the first three points in the table. Between $\lambda = 1$ and $\lambda = 2$, we have $\Delta\lambda = 1$ and $\Delta q(\lambda) = 2$, so the rate of change is $\Delta q(\lambda)/\Delta\lambda = 2$. Between $\lambda = 2$ and $\lambda = 3$, we have $\Delta\lambda = 1$ and $\Delta q(\lambda) = 4$, so the rate of change is $\Delta q(\lambda)/\Delta\lambda = 4$. Thus, the function could not be linear.

13. This table could represent a linear function, because, for the values shown, the rate of change of $a(t)$ is constant. For the given data points, between consecutive points, $\Delta t = 3$, and $\Delta a(t) = 2$. Thus, in each case, the rate of change is $\Delta a(t)/\Delta t = 2/3$. Since the rate of change is constant, the function could be linear.

14. The function g is linear, and its formula can be written as $y = b + mx$. We can find m using any pair of data points from the table. For example, letting $(x_0, y_0) = (200, 70)$ and $(x_1, y_1) = (230, 68.5)$, we have

$$\begin{aligned} m &= \frac{\Delta y}{\Delta x} = \frac{y_1 - y_0}{x_1 - x_0} \\ &= \frac{68.5 - 70}{230 - 200} = \frac{-1.5}{30} = -0.05. \end{aligned}$$

Alternatively, we could have picked any pair of data points such as $(x_0, y_0) = (400, 60)$ and $(x_1, y_1) = (300, 65)$. This gives

$$\begin{aligned} m &= \frac{\Delta y}{\Delta x} = \frac{y_1 - y_0}{x_1 - x_0} \\ &= \frac{65 - 60}{300 - 400} = \frac{5}{-100} = -0.05, \end{aligned}$$

the same answer as before.

Having found that the slope is $m = -0.05$, we know that the equation is of the form $y = b - 0.05x$. We still need to find b. The value of $g(0)$ is not given in the table but we don't need it. We can use any data point in the table to determine b. For example, we know that $y = 70$ if $x = 200$. Using our equation, we have

$$\begin{aligned} 70 &= b - 0.05 \cdot 200 \\ 70 &= b - 10 \\ b &= 80. \end{aligned}$$

You can check for yourself to see that any other data point in the table gives the same value of b. Thus, the formula for g is $g(x) = 80 - 0.05x$.

15. We know that the function $f(t)$ is linear, so knowing the coordinates of two data points from the table gives us sufficient information to determine the formula. Let's use the points $(1.2, 0.736)$ and $(1.4, 0.492)$. We know that the slope is given by

$$m = \frac{0.492 - 0.736}{1.4 - 1.2} = -1.22.$$

We must now solve for the y-intercept, b. We know that the function is of the form

$$f(t) = b - 1.22t.$$

Using the point $(1.4, 0.492)$, we get

$$0.492 = b - 1.22(1.4)$$
$$0.492 = b - 1.708$$

giving us

$$b = 2.2$$

and

$$f(t) = 2.2 - 1.22t.$$

16. We know that the function is linear so it is of the form $f(t) = b + mt$. We can choose any two points to find the slope. We use $(5.4, 49.2)$ and $(5.5, 37)$, so

$$m = \frac{37 - 49.2}{5.5 - 5.4} = -122.$$

Thus $f(t)$ is of the form $f(t) = b - 122t$. Substituting the coordinates of the point $(5.5, 37)$ we get

$$37 = b - 122 \cdot 5.5.$$

In other words,

$$b = 37 + 122 \cdot 5.5 = 708.$$

Thus

$$f(t) = 708 - 122t.$$

17. **(a)** $f(x)$ has a y-intercept of 1 and a positive slope. Thus, (ii) must be the graph of $f(x)$.
(b) $g(x)$ has a y-intercept of 1 and a negative slope. Thus, (iii) must be the graph of $g(x)$.
(c) $h(x)$ is a constant function with a y intercept of 1. Thus, (i) must be the graph of $h(x)$.

18. **(a)** Since the slopes are 2 and 3, we see that $y = -2 + 3x$ has the greater slope.
(b) Since the y-intercepts are -1 and -2, we see that $y = -1 + 2x$ has the greater y-intercept.

19. **(a)** Since the slopes are 4 and -2, we see that $y = 3 + 4x$ has the greater slope.
(b) Since the y-intercepts are 3 and 5, we see that $y = 5 - 2x$ has the greater y-intercept.

20. **(a)** Since the slopes are $\frac{1}{4}$ and -6, we see that $y = \frac{1}{4}x$ has the greater slope.
(b) Since the y-intercepts are 0 and 1, we see that $y = 1 - 6x$ has the greater y-intercept.

21. These lines are neither parallel nor perpendicular. They do not have the same slope, nor are their slopes negative reciprocals (if they were, one of the slopes would be negative).

22. These lines are perpendicular because one slope, $-\frac{1}{14}$, is the negative reciprocal of the other, 14.

23. These lines are perpendicular because one slope, $-\frac{1}{3}$, is the negative reciprocal of the other, 3.

24. Rewriting as $y = \frac{8}{7} + 3x$ and $y = \frac{77}{9} - \frac{1}{3}x$ shows that these lines are perpendicular. The slope of the first is 3, and the slope of the second is $-\frac{1}{3}$. These slopes are negative reciprocals of each other.

Problems

25. From the table, $r(300) = 120$, which tells us that at a height of 300 m the wind speed is 120 mph.

26. Judging from the table, $r(s) \geq 116$ for $200 \leq s \leq 1000$. This tells us that the wind speed is at least 116 mph between 200 m and 1000 m above the ground.

27. The wind reaches its greatest speed, $v = 122$ mph, at a height of $s = 500$ m.

28. **(a)** One, because otherwise it would automatically fail the vertical-line test using the y-axis as the vertical line.
(b) Yes, it can cross an infinite number of times. For example, the graph in Figure 1.59 oscillates an infinite number of times across the x-axis.

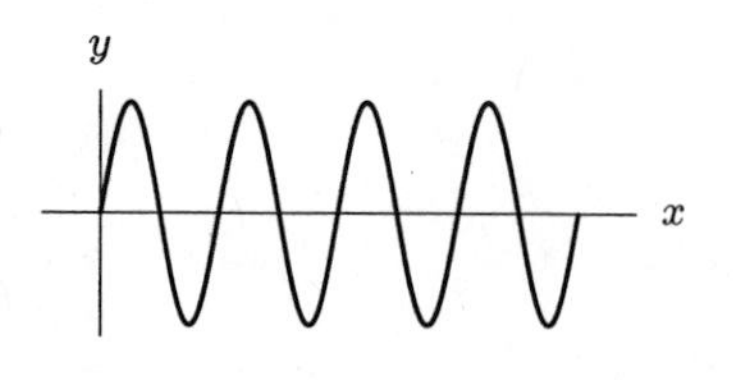

Figure 1.59

29. A possible graph is shown in Figure 1.60.

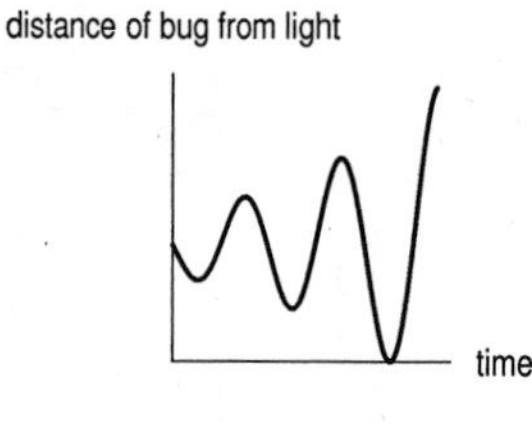

Figure 1.60

30. **(a)** Yes, if we know the Congress we can determine the number of female senators.
(b) No, there were 2 female senators in the 98th, 100th and 102nd Congresses.
(c) The number of female senators who served the 104th Congress is 8.
(d) The statement $S(110) = 16$ tells us that 17 female senators served the 110th Congress.

31. A possible graph is shown in Figure 1.61.

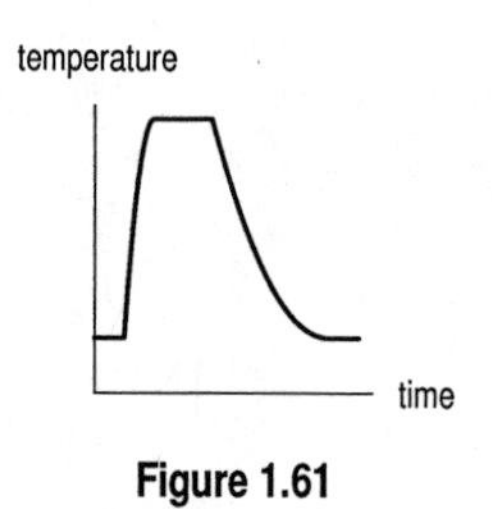

Figure 1.61

32. **(a)** It takes Charles Osgood 60 seconds to read 15 lines, so that means it takes him 4 seconds to read 1 line, 8 seconds for 2 lines, and so on. Table 1.12 shows this. From the table we see that it takes 36 seconds to read 9 lines.

Table 1.12 *The time it takes Charles Osgood to read*

Lines	0	1	2	3	4	5	6	7	8	9	10
Time	0	4	8	12	16	20	24	28	32	36	40

(b) Figure 1.62 shows the plot of the time in seconds versus the number of lines.
(c) In Figure 1.63 we have dashed in a line to see the trend. By drawing the vertical line at 9 lines, we see that this corresponds to approximately 36 seconds. By drawing a horizontal line at 30 seconds, we see that this corresponds to approximately 7.5 lines.

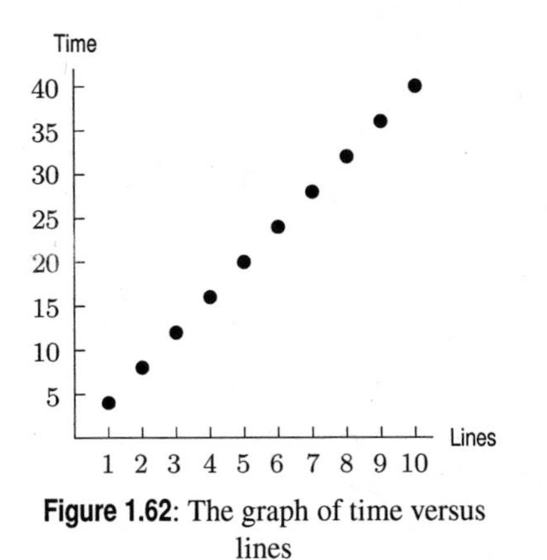

Figure 1.62: The graph of time versus lines

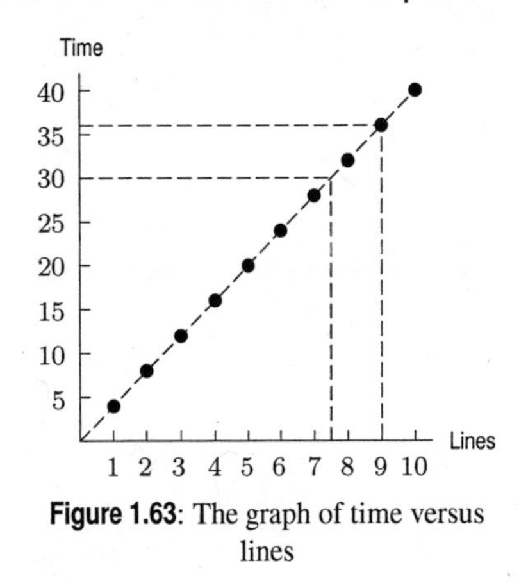

Figure 1.63: The graph of time versus lines

(d) If we let T be the time in seconds that it takes to read n lines, then $T = 4n$.

33. The diagram is shown in Figure 1.64.

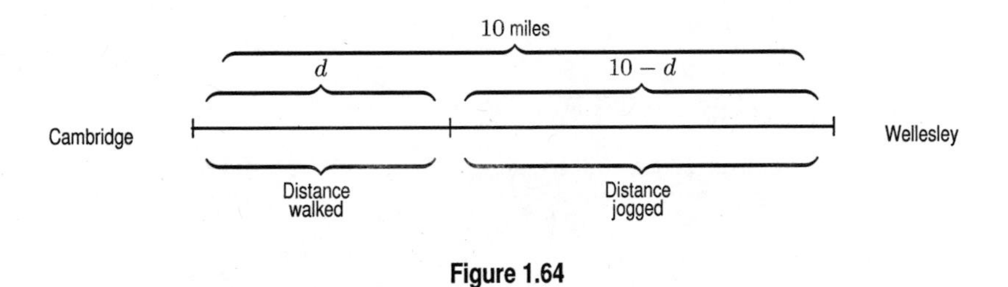

Figure 1.64

The total time the trip takes is given by the equation

$$\text{Total time} = \text{Time walked} + \text{Time jogged}.$$

The distance walked is d, and, since the total distance is 10, the remaining distance jogged is $(10 - d)$. See Figure 1.64. We know that time equals distance over speed, which means that

$$\text{Time walked} = \frac{d}{5} \quad \text{and} \quad \text{Time jogged} = \frac{10-d}{8}.$$

Thus, the total time is given by the equation

$$T(d) = \frac{d}{5} + \frac{10-d}{8}.$$

34. The area of each end of the can is πr^2. To find the surface area of the cylindrical side, imagine making vertical cut from top to bottom and unfolding the cylinder into a rectangle. See Figure 1.65.

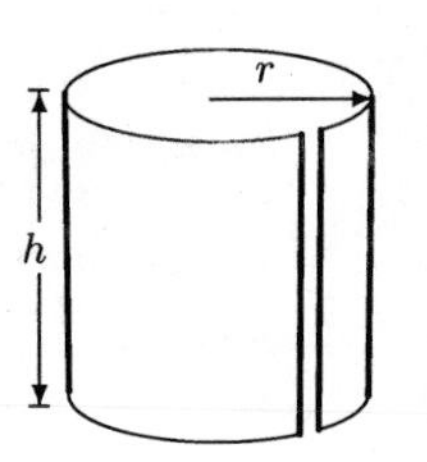

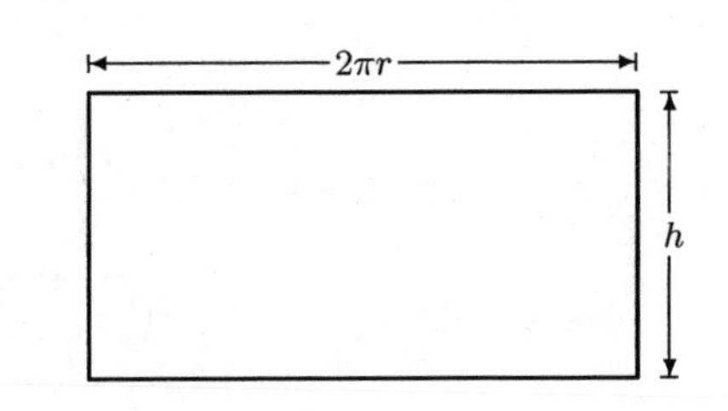

Figure 1.65

Thus, the surface area of the cylindrical side is $2\pi rh$.
The total surface area of the can is given by

$$S = 2(\text{Area of one end}) + \text{Area of cylindrical side}$$
$$S = 2(\pi r^2) + 2\pi rh.$$

Using the fact that height is twice radius, $h = 2r$, we get

$$S = 2\pi r^2 + 2\pi r(2r) = 6\pi r^2.$$

35. We know that $s + w = 1440$, so

$$s = 1440 - w.$$

36. **(a)** (i) Between $(-1, f(-1))$ and $(3, f(3))$

$$\text{Average rate of change} = \frac{f(3) - f(-1)}{3 - (-1)} = \frac{(15-4)-(-5-4)}{4} = \frac{11-(-9)}{4} = \frac{20}{4} = 5.$$

(ii) Between $(a, f(a))$ and $(b, f(b))$

$$\text{Average rate of change} = \frac{f(b) - f(a)}{b - a} = \frac{(5b-4)-(5a-4)}{b-a} = \frac{5b-4-5a+4}{b-a} = \frac{5b-5a}{b-a} = \frac{5(b-a)}{b-a} = 5.$$

(iii) Between $(x, f(x))$ and $(x + h, f(x + h))$

$$\begin{aligned}\text{Average rate of change} &= \frac{f(x+h) - f(x)}{(x+h) - x} = \frac{(5(x+h)-4)-(5x-4)}{(x+h)-x} \\ &= \frac{5x+5h-4-5x+4}{h} = \frac{5h}{h} = 5.\end{aligned}$$

(b) The average rate of change is always 5.

37. **(a)** (i) Between $(-1, f(-1))$ and $(3, f(3))$

$$\text{Average rate of change} = \frac{f(3) - f(-1)}{3 - (-1)} = \frac{\left(\frac{3}{2}+\frac{5}{2}\right)-\left(\frac{-1}{2}+\frac{5}{2}\right)}{4} = \frac{4-2}{4} = \frac{2}{4} = \frac{1}{2}.$$

(ii) Between $(a, f(a))$ and $(b, f(b))$

$$\text{Average rate of change} = \frac{f(b) - f(a)}{b - a} = \frac{\left(\frac{b}{2}+\frac{5}{2}\right)-\left(\frac{a}{2}+\frac{5}{2}\right)}{b-a} = \frac{\frac{b}{2}+\frac{5}{2}-\frac{a}{2}-\frac{5}{2}}{b-a} = \frac{\frac{b}{2}-\frac{a}{2}}{b-a} = \frac{\frac{1}{2}(b-a)}{b-a} = \frac{1}{2}.$$

(iii) Between $(x, f(x))$ and $(x + h, f(x + h))$

$$\begin{aligned}\text{Average rate of change} &= \frac{f(x+h) - f(x)}{(x+h) - x} = \frac{\left(\frac{x+h}{2}+\frac{5}{2}\right)-\left(\frac{x}{2}+\frac{5}{2}\right)}{(x+h)-x} \\ &= \frac{\frac{x+h}{2}+\frac{5}{2}-\frac{x}{2}-\frac{5}{2}}{x+h-x} = \frac{\frac{x+h-x}{2}}{h} = \frac{\frac{h}{2}}{h} = \frac{1}{2}.\end{aligned}$$

(b) The average rate of change is always $\frac{1}{2}$.

38. **(a)** (i) Between $(-1, f(-1))$ and $(3, f(3))$

$$\text{Average rate of change} = \frac{f(3) - f(-1)}{3 - (-1)} = \frac{(3^2+1)-((-1)^2+1)}{4} = \frac{10-2}{4} = \frac{8}{4} = 2.$$

(ii) Between $(a, f(a))$ and $(b, f(b))$

$$\begin{aligned}\text{Average rate of change} &= \frac{f(b)-f(a)}{b-a} = \frac{(b^2+1)-(a^2+1)}{b-a} \\ &= \frac{b^2+1-a^2-1}{b-a} = \frac{b^2-a^2}{b-a} = \frac{(b+a)(b-a)}{b-a} = b+a.\end{aligned}$$

(iii) Between $(x, f(x))$ and $(x+h, f(x+h))$

$$\begin{aligned}\text{Average rate of change} &= \frac{f(x+h)-f(x)}{(x+h)-x} = \frac{((x+h)^2+1)-(x^2+1)}{(x+h)-x} \\ &= \frac{x^2+2xh+h^2+1-x^2-1}{x+h-x} = \frac{2xh+h^2}{h} = \frac{h(2x+h)}{h} = 2x+h.\end{aligned}$$

(b) The average rate of change is different each time. However, it seems to be the sum of the two x-coordinates.

39. (a) We have

$$\frac{f(150)-f(25)}{150-25} = \frac{5.50-5.50}{125} = 0^\circ\text{C/meter}.$$

This tells us that on average the temperature changes by 0°C per meter of depth between 25 meters and 150 meters.

(b) We have

$$\frac{f(75)-f(25)}{75-25} = \frac{5.10-5.50}{50} = -0.008^\circ\text{C/meter}.$$

This tells us that on average the temperature drops by 0.008°C per meter of depth, or by 0.8°C per 100 meters of depth, on this interval.

(c) We have

$$\frac{f(200)-f(100)}{200-100} = \frac{6.00-5.10}{100} = 0.009^\circ\text{C/meter}.$$

This tells us that on average the temperature rises by 0.009°C per meter of depth, or by 0.9°C per 100 meters of depth, on this interval.

40. (a) See Table 1.13.

Table 1.13

t	0	10	20	30	40	50
$P(t)$	22	25	28	31	34	37

(b) See Figure 1.66.

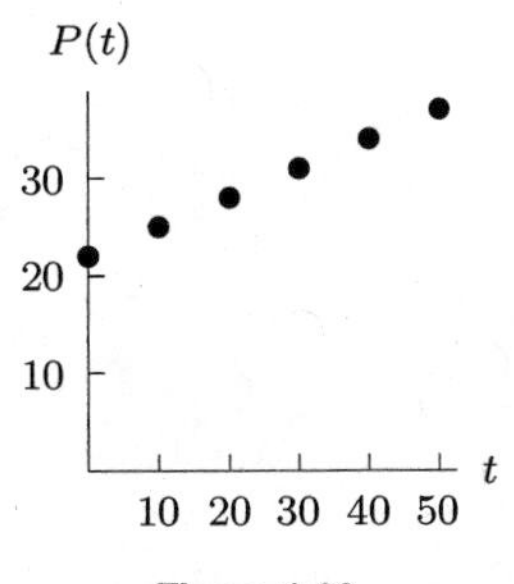

Figure 1.66

(c) From Table 1.13, we see that $P(0) = 22$, so the initial population is 22 million.

(d) The rate of change of the population is the difference in the number of people from one year to the next. The change from year 0 to year 10 is $P(10) - P(0) = 3$ million people, so the change from year 0 to year 1 is

$$\frac{P(10)-P(0)}{10-0} = 0.3 \text{ million people.}$$

The change from one year to the next will be the same no matter which year we choose since P is linear, so the rate of change is 0.3 million people/year.

41. **(a)** One horse costs the woodworker \$5000 in start-up costs plus \$350 for labor and materials, a total of \$5350. Thus, if $n = 1$, we have

$$C = \underbrace{5000}_{\text{Start-up costs}} + \underbrace{350}_{\text{Extra cost for 1 horse}} = 5350.$$

Similarly, for 2 horses

$$C = \underbrace{5000}_{\text{Start-up costs}} + \underbrace{350 \cdot 2}_{\text{Extra cost for 2 horses}} = 5700,$$

and for 5 horses

$$C = \underbrace{5000}_{\text{Start-up costs}} + \underbrace{350 \cdot 5}_{\text{Extra cost for 5 horses}} = 6750.$$

Similarly, for 10 horses, $C = 5000 + 350 \cdot 10 = 8500$, and for 20 horses, $C = 12{,}000$. See Table 1.14 and Figure 1.67.

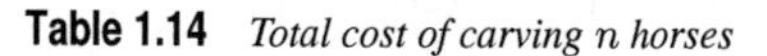

Table 1.14 *Total cost of carving n horses*

n, number of horses	C, total cost (\$)
0	5000
1	5350
2	5700
5	6750
10	8500
20	12,000

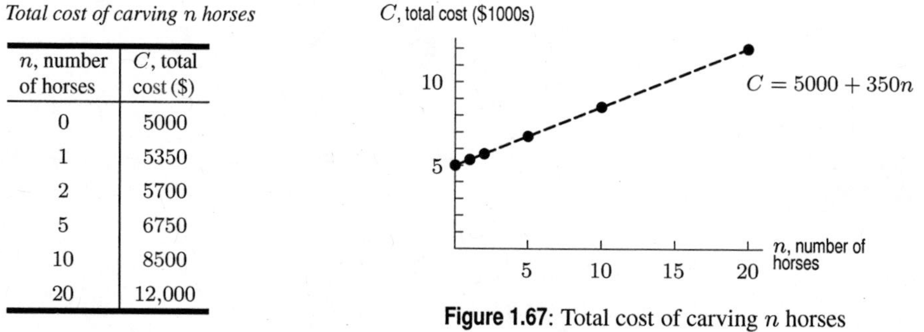

Figure 1.67: Total cost of carving n horses

Notice that it costs the woodworker \$5000 to carve 0 horses since he buys the tools, plans, and advertising even if he never carves a single horse.

(b) From part (a), a formula for C, as a function of n, is

$$C = \underbrace{5000}_{\text{Start-up costs}} + \underbrace{350 \cdot n}_{\text{Extra cost for } n \text{ horses}} = 5000 + 350n.$$

(c) The average rate of change of this function is

$$\text{Rate of change} = \frac{\Delta C}{\Delta n} = \frac{\text{Change in cost}}{\text{Change in number of horses carved}}.$$

Each additional horse costs an extra \$350, so

$$\Delta C = 350 \quad \text{if} \quad \Delta n = 1.$$

Thus, the rate of change is given by

$$\frac{\Delta C}{\Delta n} = \frac{\$350}{1 \text{ horse}} = \$350 \text{ per horse.}$$

The rate of change of C gives the additional cost to carve one additional horse. Since the total cost increases at a constant rate (\$350 per horse), the graph of C against n is a straight line sloping upward.

42. **(a)** $F = 2C + 30$

(b) Since we are finding the difference for a number of values, it would perhaps be easier to find a formula for the difference:

$$\begin{aligned}\text{Difference} &= \text{Approximate value} - \text{Actual value}\\ &= (2C+30) - \left(\frac{9}{5}C + 32\right) = \frac{1}{5}C - 2.\end{aligned}$$

If the Celsius temperature is $-5°$, $(1/5)C - 2 = (1/5)(-5) - 2 = -1 - 2 = -3$. This agrees with our results above.

Similarly, we see that when $C = 0$, the difference is $(1/5)(0) - 2 = -2$ or 2 degrees too low. When $C = 15$, the difference is $(1/5)(15) - 2 = 3 - 2 = 1$ or 1 degree too high. When $C = 30$, the difference is $(1/5)(30) - 2 = 6 - 2 = 4$ or 4 degrees too high.

(c) We are looking for a temperature C, for which the difference between the approximation and the actual formula is zero.

$$\begin{aligned}\frac{1}{5}C - 2 &= 0\\ \frac{1}{5}C &= 2\\ C &= 10\end{aligned}$$

Another way we can solve for a temperature C is to equate our approximation and the actual value.

$$\begin{aligned}\text{Approximation} &= \text{Actual value}\\ 2C + 30 &= 1.8C + 32,\\ 0.2C &= 2\\ C &= 10\end{aligned}$$

So the approximation agrees with the actual formula at 10° Celsius.

43. Using our formula for j, we have

$$\begin{aligned}h(-2) &= j(-2) = 30(0.2)^{-2} = 750\\ h(1) &= j(1) = 30(0.2)^{1} = 6.\end{aligned}$$

This means that $h(t) = b + mt$ where

$$m = \frac{h(1) - h(-2)}{1 - (-2)} = \frac{6 - 750}{3} = -248.$$

Solving for b, we have

$$\begin{aligned}h(-2) &= b - 248(-2)\\ b &= h(-2) + 248(-2) = 750 + 248(-2) = 254,\end{aligned}$$

so $h(t) = 254 - 248t$.

44. First we find the coordinates of the squares' corners on l. Because the area of the left square is 13, the length of a side is $\sqrt{13}$. Similarly for the right square with area equal to 8, the side length is $\sqrt{8}$.

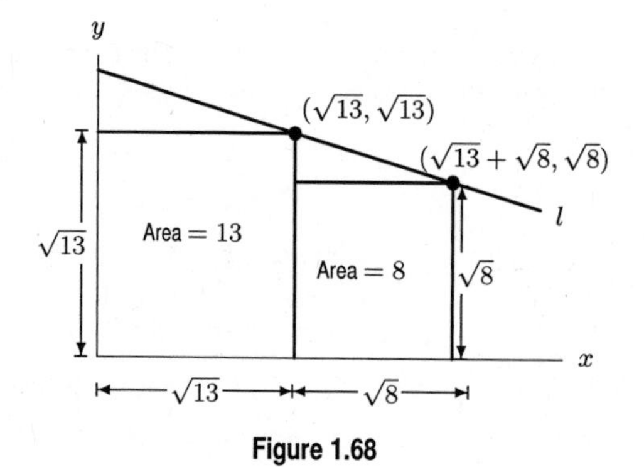

Figure 1.68

From Figure 1.68 we see that the coordinates of the corners on the line are $(\sqrt{13}+\sqrt{8}, \sqrt{8})$ and $(\sqrt{13}, \sqrt{13})$. We use this to find the slope of the line l:

$$m = \frac{\Delta y}{\Delta x} = \frac{\sqrt{8}-\sqrt{13}}{(\sqrt{13}+\sqrt{8})-\sqrt{13}} = \frac{\sqrt{8}-\sqrt{13}}{\sqrt{8}}.$$

So

$$y = b + \left(\frac{\sqrt{8}-\sqrt{13}}{\sqrt{8}}\right)x.$$

To find the y-intercept b, we put one of the known points into the equation and solve for b. Since $(\sqrt{13}, \sqrt{13})$ is on line l we have

$$\begin{aligned}
\sqrt{13} &= b + \left(\frac{\sqrt{8}-\sqrt{13}}{\sqrt{8}}\right)\sqrt{13}\\
\sqrt{13} &= b + \frac{\sqrt{8}\sqrt{13}-13}{\sqrt{8}}\\
b &= \sqrt{13} - \frac{\sqrt{8}\sqrt{13}-13}{\sqrt{8}}\\
&= \frac{\sqrt{13}\sqrt{8}}{\sqrt{8}} - \frac{\sqrt{8}\sqrt{13}-13}{\sqrt{8}}\\
&= \frac{\sqrt{13}\sqrt{8}-\sqrt{13}\sqrt{8}+13}{\sqrt{8}} = \frac{13}{\sqrt{8}}.
\end{aligned}$$

So the equation of the line is

$$y = \frac{13}{\sqrt{8}} + \left(\frac{\sqrt{8}-\sqrt{13}}{\sqrt{8}}\right)x.$$

Optionally, this can be simplified (including rationalizing denominators) to

$$y = \frac{13\sqrt{2}}{4} + \frac{4-\sqrt{26}}{4}x.$$

45. **(a)** The results are in Table 1.15.

Table 1.15

t	0	1	2	3	4
$v = f(t)$	1000	990.2	980.4	970.6	960.8

(b) The speed of the bullet is decreasing at a constant rate of 9.8 meters/sec every second. To confirm this, calculate the rate of change in velocity over every second. We get

$$\frac{\Delta v}{\Delta t} = \frac{990.2 - 1000}{1 - 0} = \frac{980.4 - 990.2}{2 - 1} = \frac{970.6 - 980.4}{3 - 2} = \frac{960.8 - 970.6}{4 - 3} = -9.8.$$

Since the value of $\Delta v / \Delta t$ comes out the same, -9.8, for every interval, we can say that the bullet is slowing down at a constant rate. This makes sense as the constant force of gravity acts to slow the upward moving bullet down at a constant rate.

(c) The slope, -9.8, is the rate at which the velocity is changing. The v-intercept of 1000 is the initial velocity of the bullet. The t-intercept of $1000/9.8 = 102.04$ is the time at which the bullet stops moving and starts to head back to Earth.

(d) Since Jupiter's gravitational field would exert a greater pull on the bullet, we would expect the bullet to slow down at a faster rate than a bullet shot from earth. On earth, the rate of change of the bullet is -9.8, meaning that the bullet is slowing down at the rate of 9.8 meters per second. On Jupiter, we expect that the coefficient of t, which represents the rate of change, to be a more negative number (less than -9.8). Similarly, since the gravitational pull near the surface of the moon is less, we expect that the bullet would slow down at a lesser rate than on earth. So, the coefficient of t should be a less negative number (greater than -9.8 but less than 0).

46. (a) The general equation for a line is

$$y = b + mx,$$

so we must find m, the slope, and b, the y-intercept of the line. Since we have two points on the line, we can find the slope. The coordinates of the two points are $(1324, 11328)$ and $(1529, 13275.50)$. The slope is then:

$$\frac{y_1 - y_0}{x_1 - x_0} = \frac{13275.50 - 11328}{1529 - 1324} = \frac{1947.50}{205} = 9.50.$$

We can put our value for the slope into the general equation to get:

$$y = b + (9.50)x.$$

To find b, we use either of the points and solve for b. Using the point $(1324, 11328)$, we get

$$\begin{aligned} (11328) &= b + (9.50)(1324) \\ b &= 11328 - (9.50)(1324) = -1250. \end{aligned}$$

We now have the slope and the intercept and can use them to get the equation of the line in part (a):

$$y = -1250 + (9.50)x.$$

(b) The y-intercept of the line is the profit the movie theater makes if zero patrons attend the theater during a week. Since this number is negative, we see that the theater loses \$1250 if nobody attends. The slope of the line represents the increase in profit the theater receives for each patron. The theater makes an additional \$9.50 in profit per patron.

(c) To find the break-even point, we find the number of patrons, x, that makes the profit, y, equal to 0, i.e. we set y equal to zero and solve for x.

$$\begin{aligned} 0 &= -1250 + (9.50)x \\ 1250 &= (9.50)x \\ x &= \frac{1250}{9.50} \approx 131.58 \end{aligned}$$

Therefore the theater needs 132 patrons per week to break even.

(d) The equation we found in part (a) gives the profit as a function of the number of patrons. To find the number of patrons as a function of profit, we solve this equation for x in terms of y.

$$\begin{aligned} y &= -1250 + (9.50)x \\ y + 1250 &= (9.50)x \\ 9.50x &= y + 1250 \\ x &= \frac{y}{9.50} + \frac{1250}{9.50} \end{aligned}$$

(e) Putting $y = 17759.50$ into the equation found in (d), we get

$$x = \frac{17759.50}{9.50} + \frac{1250}{9.50} = 2001.$$

So 2001 patrons attended the theater.

47. There are many possible answers. For example, when you buy something, the amount of sales tax depends on the sticker price of the item bought. Let's say Tax $= 0.05 \times$Price. This means that the sales tax rate is 5%.

48. (a) When the price of the product went from \$3 to \$4, the demand for the product went down by 200 units. Since we are assuming that this relationship is linear, we know that the demand will drop by another 200 units when the price increases another dollar, to \$5. When $p = 5$, $D = 300 - 200 = 100$. So, when the price for each unit is \$5, consumers will only buy 100 units a week.

(b) The slope, m, of a linear equation is given by

$$m = \frac{\text{change in dependent variable}}{\text{change in independent variable}} = \frac{\Delta D}{\Delta P}.$$

Since quantity demanded depends on price, quantity demanded is the dependent variable and price is the independent variable. We know that when the price changes by \$1, the quantity demand changes by -200 units. That is, the quantity demanded goes down by 200 units. Thus,

$$m = \frac{-200}{1}.$$

Since the relationship is linear, we know that its formula is of the form

$$D = b + mp.$$

We know that $m = -200$, so

$$D = b - 200p.$$

We can find b by using the fact that when $p = 3$ then $D = 500$ or by using the fact that if $p = 4$ then $D = 300$ (it does not matter which). Using $p = 3$ and $D = 500$, we get

$$\begin{aligned} D &= b - 200p \\ 500 &= b - 200(3) \\ 500 &= b - 600 \\ 1100 &= b. \end{aligned}$$

Thus, $D = 1100 - 200p$.

(c) We know that $D = 1100 - 200p$ and $D = 50$, so

$$\begin{aligned} 50 &= 1100 - 200p \\ -1050 &= -200p \\ 5.25 &= p. \end{aligned}$$

At a price of \$5.25, the demand would be only 50 units.

(d) The slope is -200, which means that the demand goes down by 200 units when the price goes up by \$1.

(e) The demand is 1100 when the price is 0. This means that even if you were giving this product away, people would only want 1100 units of it per week. When the price is \$5.50, the demand is zero. This means that at or above a unit price of \$5.50, the company cannot sell this product.

49. (a) Since the relationship is linear, the general formula for S in terms of p is

$$S = b + mp.$$

Since we know that the quantity supplied rises by 50 units when the rise in the price is \$0.50, we canwrite $\Delta S = 50$ units, when $\Delta p =$\$0.50. The slope is then:

$$m = \frac{\Delta S}{\Delta p} = \frac{50\text{ units}}{\$0.50} = 100\text{ units/dollar}.$$

Put this value of the slope into the formula for S and solve for b using $p = 2$ and $S = 100$:

$$\begin{aligned} S &= b + mp \\ 100 &= b + (100)(2) \\ 100 &= b + 200 \\ b &= -100. \end{aligned}$$

We now have the slope m and the S-intercept b. So, we know that

$$S = -100 + 100p.$$

(b) The slope in this problem is 100 units/dollar, which means that for every increase of \$1 in price, suppliers are willing to supply another 100 units.

(c) The price below which suppliers will not supply the good is represented by the point at which $S = 0$. Putting $S = 0$ into the equation found in (b) we get:

$$\begin{aligned} 0 &= -100 + 100p \\ 100 &= 100p \\ p &= 1. \end{aligned}$$

So when the price is \$1, or less, the suppliers will not want to produce anything.

(d) From Problem 48 we know that

$$D = 1100 - 200p.$$

To find when supply equals demand set the formulas for S and D equal and solve for p:

$$\begin{aligned} S &= D \\ -100 + 100p &= 1100 - 200p \\ 100p + 200p &= 1100 + 100 \\ 300p &= 1200 \\ p &= \frac{1200}{300} = 4. \end{aligned}$$

Therefore, the market clearing price is \$4.

50. (a) The easiest way to plot the lines is to use the two points given for each one. Let p be price and q be quantity demanded or supplied weekly.

For the demand line, we know that when $p = 3$, $q = 500$ and when $p = 4$, $q = 300$. With p on the vertical axis, this means the points $(500, 3)$ and $(300, 4)$ are on the line.

For the supply line, we know that when $p = 2$, $q = 100$ and when $p = 2.50$, $q = 150$. Thus the points $(100, 2)$ and $(150, 2.50)$ are on this line. See Figure 1.69.

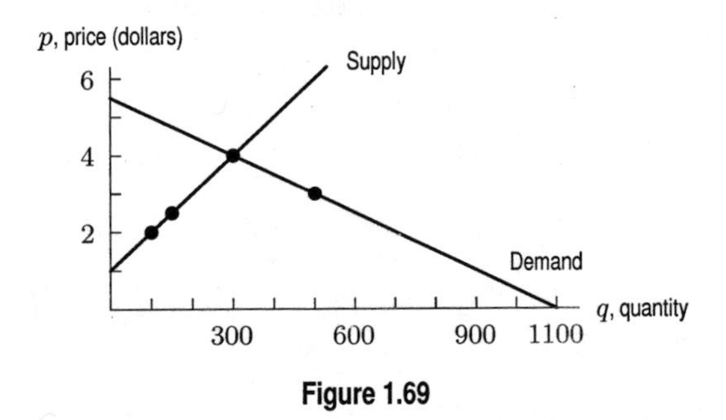

Figure 1.69

(b) The market clearing price is the price at which the lines cross, which is $(300, 4)$, so the market clearing price is \$4. This agrees with the answer in Problem 49(d).

51. We see that the graphs of f and g intersect at $x = -6, 2$. This means

$$\begin{aligned} g(2) = f(2) &= 12 - 0.5(2+4)^2 = -6 \\ g(-6) = f(-6) &= 12 - 0.5(-6+4)^2 = 10. \end{aligned}$$

This means

$$\begin{aligned} m &= \frac{g(2) - g(-6)}{2 - (-6)} \\ &= \frac{-6 - 10}{8} = -2. \end{aligned}$$

Using the point-slope form with $(x_0, y_0) = (2, -6)$, we have

$$\begin{aligned} g(x) &= y_0 + m(x - x_0) \\ &= -6 - 2(x - 2). \end{aligned}$$

Checking our answer, we see that

$$g(-6) = -6 - 2(-6-2) = -6 - 2(-8) = 10. \quad \text{as required}$$

We can also write our answer in slope-intercept form:

$$\begin{aligned} g(x) &= -6 - 2(x - 2) \\ &= -6 - 2x + 4 \\ &= -2 - 2x. \end{aligned}$$

52. Here, the independent variable is r. This means the other letters are constants, and we can write:

$$\begin{aligned} f(r) &= rx^3 + 3rx^2 + 2r + 4sx + 7s + 3 \\ &= \underbrace{(x^3 + 3x^2 + 2)}_{m}\, r + \underbrace{4sx + 7s + 3}_{b} \\ \text{so} \quad m &= x^3 + 3x^2 + 2 \\ b &= 4sx + 7s + 3. \end{aligned}$$

53. The line intersects f at $x = -2$ and $x = 5$. We see that

$$\begin{aligned} f(-2) &= 2 + \frac{3}{-2+5} = 2 + \frac{3}{3} = 3 \\ f(5) &= 2 + \frac{3}{5+5} = 2 + \frac{3}{10} = 2.3, \end{aligned}$$

so the line contains the points $(-2, 3)$ and $(5, 2.3)$. This means the slope is

$$m = \frac{2.3 - 3}{5 - (-2)} = \frac{-0.7}{7} = -0.1.$$

Using the point-slope formula with $(x_0, y_0) = (-2, 3)$, we have

$$\begin{aligned} y &= 3 - 0.1\,(x - (-2)) \\ &= 2.8 - 0.1x. \end{aligned}$$

Checking our answer, we see that

$$\begin{aligned} \text{At } x = -2: &\quad y = 2.8 - 0.1(-2) = 3 \\ \text{At } x = 5: &\quad y = 2.8 - 0.1(5) = 2.3, \end{aligned}$$

as required.

54. **(a)** If she holds no client meetings, she can hold 30 co-worker meetings. On the other hand, if she holds no co-worker meetings, she can hold 20 client meetings. A graph that describes the relationship is shown in Figure 1.70.

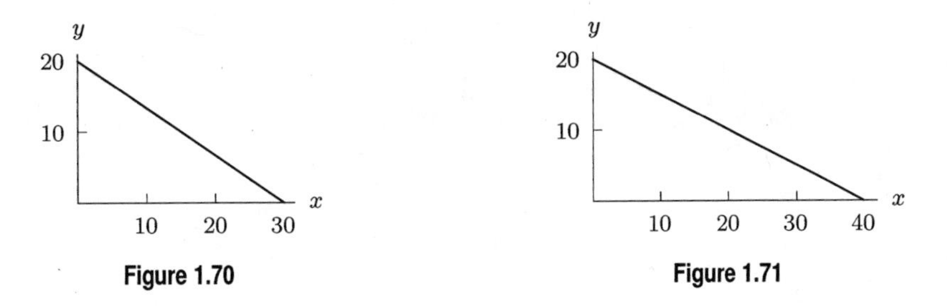

Figure 1.70

Figure 1.71

(b) Since $(0, 20)$ and $(30, 0)$ are on the line, $m = (20 - 0)/(0 - 30) = -(2/3)$. Using the slope intercept form of the line, we have $y = 20 - (2/3)x$.

(c) Since the slope is $-(2/3)$, we know that for every two additional client meetings she must sacrifice three co-worker meetings. Equivalently, for every two fewer client meetings, she gains time for three additional co-worker meetings. The x-intercept is 30. This means that she does not have time for any client meetings at all when she's scheduled 30 co-worker meetings. The y-intercept is 20. This means that she does not have time for any co-worker meetings at all when she's scheduled 20 client meetings.

(d) Instead of 2 hours, co-worker meetings now take $3/2$ hours. If all of her 60 hours are spent in co-worker meetings, she can have $60/(3/2) = 40$ co-worker meetings. The new graph is shown in Figure 1.71. The y-intercept remains at 20. However, the x-intercept is changed to 40. The slope changes, too, from $-(2/3)$ to $-(1/2)$. The new slope is still negative but is less steep because there is less of a decrease in the amount of time available for client meetings due to each extra co-worker meeting.

55. Since you are moving in a straight line away from Pittsburgh, your total distance is the initial distance, 60 miles, plus the additional miles covered. In each hour, you will travel fifty miles as shown in Figure 1.72.

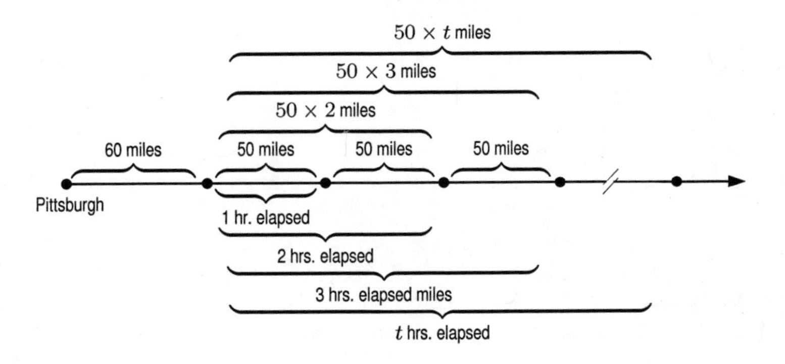

Figure 1.72

So, the total distance from Pittsburgh can be expressed as $d = 60 + 50t$.

56. Since the graph is parallel to the line $y = 20 - 4x$, the slope is the same, so $m = -4$. Using the point-slope formula with $(x_0, y_0) = (3, 12)$ we have

$$\begin{aligned} y &= 12 - 4(x - 3) \\ &= 12 - 4x + 12 \\ &= 24 - 4x. \end{aligned}$$

57. First we place $5x - 3y = 6$ into slope-intercept form:

$$\begin{aligned} 5x - 3y &= 6 \\ 3y &= -6 + 5x \\ y &= -2 + \frac{5}{3}x. \end{aligned}$$

The slope of this line is $5/3$. Since the graph of g is perpendicular to it, the slope of g is

$$m = -\frac{1}{5/3} = -\frac{3}{5}.$$

We see from the first equation that at $x = 15$,

$$y = \frac{5}{3} \cdot 15 - 2 = 25 - 2 = 23,$$

so the graph of the first equation contains the point $(15, 23)$. Since the graph of g also contains this point, we have

$$\begin{aligned} g(15) = b + m \cdot 15 &= 23 \\ b - \frac{3}{5}(15) &= 23 \\ b - 9 &= 23 \\ b &= 32, \end{aligned}$$

so $g(x) = 32 - (3/5)x$.

58. The sloping line has $m = 1$, so its equation is $y = x - 1$. The horizontal line is $y = 3$.
Solving simultaneously gives

$$3 = x - 1 \quad \text{so} \quad x = 4.$$

Thus, the point of intersection is $(4, 3)$.

59. **(a)** Company A charges \$0.37 per minute. So, the cost with company A is simply 0.37 times the number of minutes, or

$$Y_A = 0.37x.$$

Company B charges \$13.95 per month plus \$0.22 per minute. So, the cost for company B is

$$Y_B = 13.95 + 0.22x.$$

Company C charges a fixed rate of \$50 per month. So,

$$Y_C = 50.$$

(b) Using the fixed costs for each company – \$0 for company A, \$13.95 for company B, and \$50 for company C – we know that Y_A goes through the origin, that Y_B goes through the point $(0, 13.95)$ and that Y_C goes through $(0, 50)$. We also know that Y_A has a rate of change of 0.35, Y_B has a rate of change of 0.22 and Y_C is a constant function. So, Y_A has the steepest slope and Y_C has a slope of zero. So, we can label the graphs as follows:

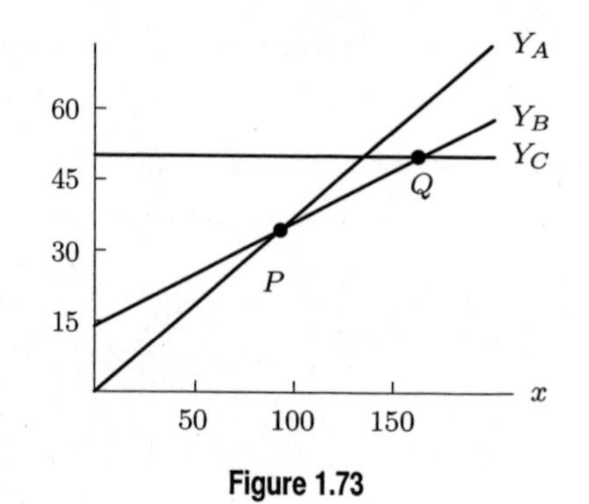

Figure 1.73

(c) From Figure 1.73 we see that between the points P and Q, the graph of Y_B is below the graphs of Y_A and Y_C. This means that company B is the cheapest in the interval between P and Q. We need to find the x-coordinates of P and Q.

To find the x-coordinate of point P, we note that the graphs of Y_A and Y_B intersect at P. So, we can set the formulas for Y_A and Y_B equal to each other and solve for x.

$$\begin{aligned} Y_A &= Y_B \\ 0.37x &= 13.95 + 0.22x \\ 0.15x &= 13.95 \\ x &= \frac{13.95}{0.15} = 93 \end{aligned}$$

Therefore, for $x > 93$ minutes, the graph of Y_B is below the graph of Y_A, meaning that company B is cheaper than company A.

To find the x-coordinate of point Q, we note that Q is the intersection of the graphs of Y_B and Y_C. By setting the formulas for Y_B and Y_C equal to each other, we can solve for x.

$$\begin{aligned} Y_B &= Y_C \\ 13.95 + 0.22x &= 50 \\ 0.22x &= 50 - 13.95 \\ 0.22x &= 36.05 \\ x &= \frac{36.05}{0.22} \approx 163.86 \end{aligned}$$

Thus, for $x \leq 163$ minutes, company B is cheaper than company C.

Putting these two results together, we conclude that company B is the cheapest for values of x between 93 and 163 minutes.

60. (a) Since the first option remains constant, an equation describing it is $y_1 = 100$. The second option has a base of 50 and increases at a rate of 10% for every dollar in sales. An equation is: $y_2 = 50 + 0.10x$. The break-even point is found by setting the two expressions equal to each other and solving:

$$100 = 50 + 0.10x,$$

so $x = 500$. The first option is better if the person has sales below \$500, and the second option is better if the person has sales above \$500.

(b) The first option has a base of 175 and increases at a rate of 7% for every dollar in sales. The second option also has a base of 175, but increases at a rate of 8% for every dollar in sales. Since both options have the same starting point, and the second option increases at a faster rate, it is better. Thus, choose the second option.

(c) The first option has a base of 145 and increases at a rate of 7% for every dollar in sales. The second option has a base of 165 and also increases at a rate of 7% for every dollar in sales. Since the two options have different starting points, but increase at the same rate, the one with the higher starting point is a higher salary. Thus, choose the second option.

(d) The first option has a base of \$225 and increases at a rate of 3% for every dollar in sales. An equation describing it is $y_1 = 225 + 0.03x$ The second option has a base of 180 and increases at a rate of 6% for every dollar in sales. An equation is: $y_2 = 180 + 0.06x$. The break-even point is found by setting the two equations equal to each other and solving:

$$225 + 0.03x = 180 + 0.06x,$$

so $x = 1500$. The first option is better if the person has sales below \$1500, and the second option is better if the person has sales above \$1500.

61. (a) See Figure 1.74.

(b) The scatterplot suggests that as IQ increases, the number of hours of TV viewing decreases. The points, though, are not close to being on a line, so a reasonable guess is $r \approx -0.5$.

(c) A calculator gives the regression equation $y = 27.5139 - 0.1674x$ with $r = -0.5389$.

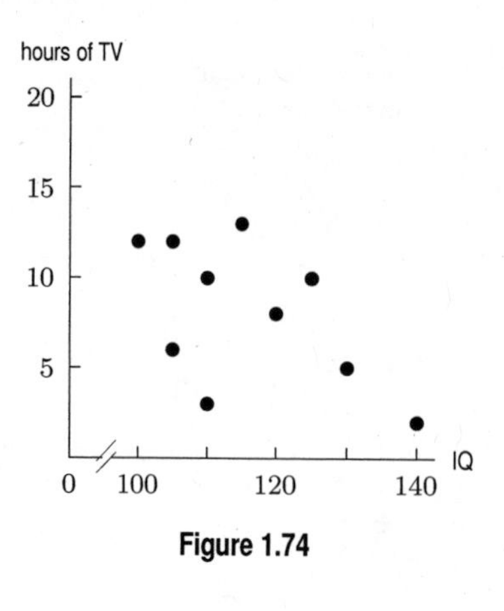

Figure 1.74

62. **(a)** Figure 1.75 shows the data.

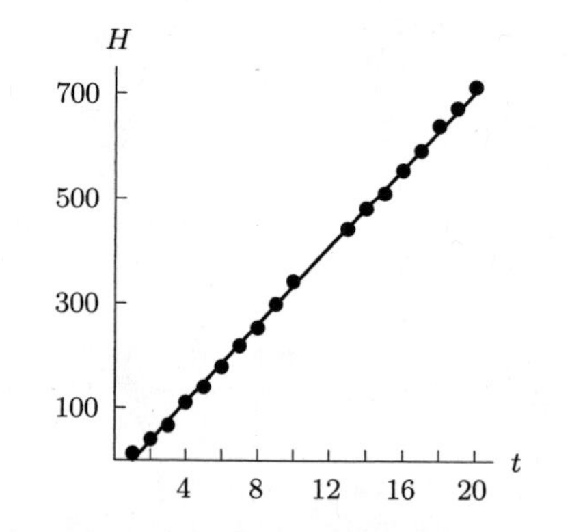

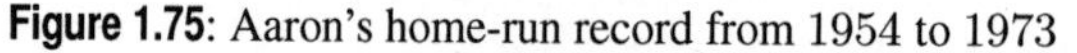

Figure 1.75: Aaron's home-run record from 1954 to 1973

(b) Estimates will vary but the equation $H = 37t - 37$ is typical.

(c) A calculator gives $H = 37.26t - 39.85$, with correlation coefficient $r = 0.9995$, which rounds to $r = 1$. The data set lies very close indeed to the regression line, which has a positive slope. In other words, Aaron's home-runs grew at a constant rate over his career.

(d) The slope gives the average number of home-runs per year, about 37.

(e) From the answer to part (d) we expect Henry Aaron to hit about 37 home-runs in each of the years 1974, 1975, 1976, and 1977. However, the knowledge that Aaron retired in 1976 means that he scored 0 home-runs in 1977. Also, people seldom retire at the peak of their abilities, so it is likely that Aaron's performance dropped off in the last few years. In fact he scored 20, 12, and 10 home-runs in the years 1974, 1975, and 1976, well below the average of 37.

63. **(a)** The inequalities place b to the right of a and $f(b)$ higher than $f(a)$. Since the function is linear, the graph is a line. See Figure 1.76.

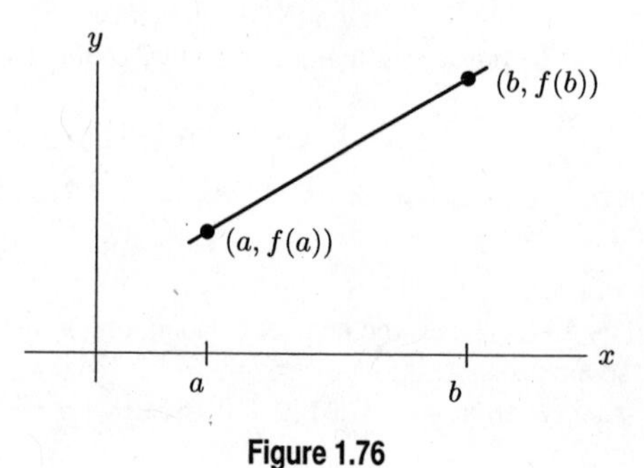

Figure 1.76

(b) The slope is

$$m = \frac{\Delta y}{\Delta x} = \frac{f(b) - f(a)}{b - a}.$$

64. (a) (i)

$$\frac{f(2) - f(1)}{2 - 1} = \frac{(0.003 - (1.246(2) + 0.37)) - (0.003 - (0.1246(1) + 0.37))}{1}$$
$$= -2.859 - (-1.613) = -1.246$$

(ii)

$$\frac{f(1) - f(2)}{1 - 2} = \frac{-1.613 - (-2.859)}{-1} = -1.246$$

(iii)

$$\frac{f(3) - f(3)}{3 - 4} = \frac{-4.105 - (-5.351)}{-1} = \frac{1.246}{-1} = -1.246$$

(b)

$$\begin{aligned} f(x) &= 0.003 - (1.246x + 0.37) \\ &= 0.003 - 0.37 - 1.246x \\ f(x) &= -0.367 - 1.246x \end{aligned}$$

65. Writing

$$\begin{aligned} y &= -3 - \frac{x}{2} \\ &= -3 + \left(-\frac{1}{2}\right) \cdot x, \end{aligned}$$

we see that in order for the coefficients of x to be the same, we have

$$\begin{aligned} -r^2 &= -\frac{1}{2} \\ r^2 &= \frac{1}{2} \\ r &= \sqrt{0.5}. \quad \text{because } r > 0 \end{aligned}$$

Likewise, in order for the constant terms to be the same, we have

$$\begin{aligned} \frac{p}{p-1} &= -3 \\ p &= -3(p - 1) \\ p &= -3p + 3 \\ 4p &= 3 \\ p &= 0.75, \\ \text{so we have } y &= \frac{0.75}{0.75 - 1} - \left(\sqrt{0.5}\right)^2 x. \end{aligned}$$

Checking our answer, we see that

$$\begin{aligned} y &= \frac{0.75}{0.75 - 1} - \left(\sqrt{0.5}\right)^2 x. \\ &= \frac{0.75}{-0.25} - 0.5x \\ &= -3 - \frac{x}{2}. \qquad \text{as required} \end{aligned}$$

66. Writing

$$\begin{aligned} y &= \frac{x+k}{z} \\ &= \frac{x}{z} + \frac{k}{z} \\ &= \frac{k}{z} + \frac{1}{z} \cdot x, \end{aligned}$$

and writing

$$\begin{aligned} y &= -3 - \frac{x}{2} \\ &= -3 + \left(-\frac{1}{2}\right) \cdot x, \end{aligned}$$

we see that in order for the coefficients of x to be the same, we have

$$\begin{aligned} \frac{1}{z} &= -\frac{1}{2} \\ z &= -2. \end{aligned}$$

Likewise, for the constant terms to be the same, we have

$$\begin{aligned} \frac{k}{z} &= -3 \\ \frac{k}{-2} &= -3 && \text{because } z = -2 \\ k &= 6, \\ \text{so we have} \quad y &= \frac{x+6}{-2} && \text{because } k = 6, z = -2. \end{aligned}$$

Checking our answer, we see that

$$\begin{aligned} y &= \frac{x+6}{-2} \\ &= \frac{x}{-2} + \frac{6}{-2} \\ &= -3 - \frac{x}{2}. && \text{as required.} \end{aligned}$$

67. (a) If you buy x apples and the cost is p dollars each, then the amount spent on apples is px. Similarly, if you buy y bananas and the cost is q dollars each, then the amount spent on bananas is qy. Since the total amount spent on the two goods is c dollars, we have the equation:

$$px + qy = c.$$

If $x = 0$, then

$$\begin{aligned} p \cdot (0) + qy &= c \\ qy &= c \\ y &= \frac{c}{q}. \end{aligned}$$

So the y-intercept is c/q. If $y = 0$, then

$$\begin{aligned} px + q \cdot (0) &= c \\ px &= c \\ x &= \frac{c}{p}. \end{aligned}$$

So c/p is the x-intercept. See Figure 1.77.

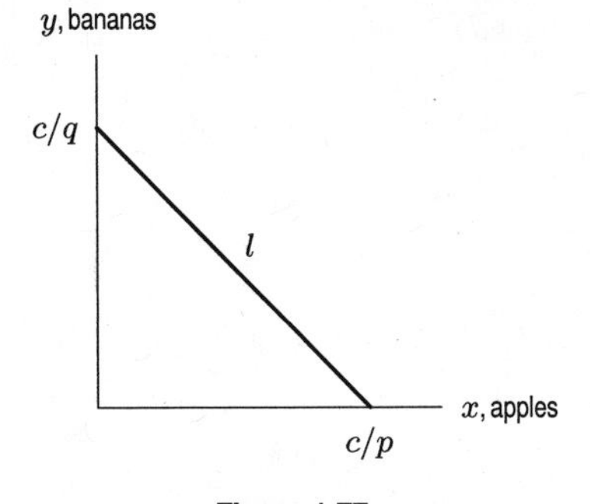

Figure 1.77

(b) Since $(0, c/q)$ and $(c/p, 0)$ are two points on this line, the slope is:

$$m = \frac{\frac{c}{q} - 0}{0 - \frac{c}{p}} = \frac{\frac{c}{q}}{-\frac{c}{p}}$$
$$= \frac{c}{q} \cdot -\frac{p}{c} = -\frac{p}{q}.$$

So, the graph of this line has a slope of $-p/q$.

68. In Figure 1.78 the decision to spend all c dollars of your money on apples is represented by the x-intercept; the decision to spend it all on bananas is represented the y-intercept. If we decide to spend all of our money on either all apples or all bananas, and bananas are cheaper, then we would be able to purchase more bananas than apples for our c dollars. So, we want the line for which the y-intercept is greater than the x-intercept. If we look at line l_1 we see that the y-intercept is greater than 10 and the x–intercept is less than 10. Thus, l_1 represents the case where we can buy more bananas than apples, so apples must be more expensive than bananas.

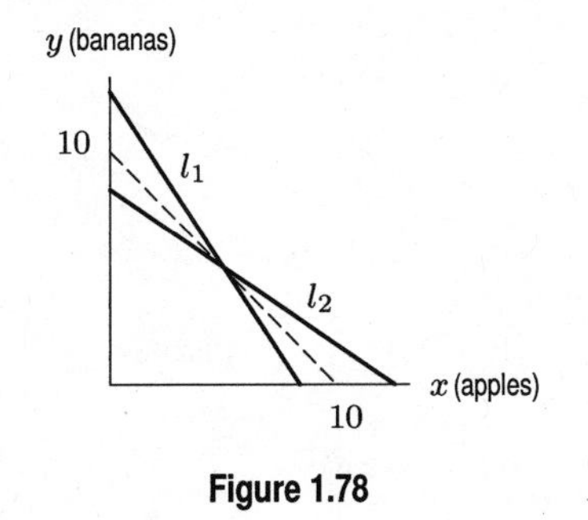

Figure 1.78

69. In Figure 1.79 the graph of the hair length is steepest just after each haircut, assumed to be at the beginning of each year. As the year progresses, the growth is slowed by split ends. By the end of the year, the hair is breaking off as fast as it is growing, so the graph has leveled off. At this time the hair is cut again. Once again it grows until slowed by the split ends. Then it is cut. This continues for five years when the longest hairs fall out because they have come to the end of their natural lifespan.

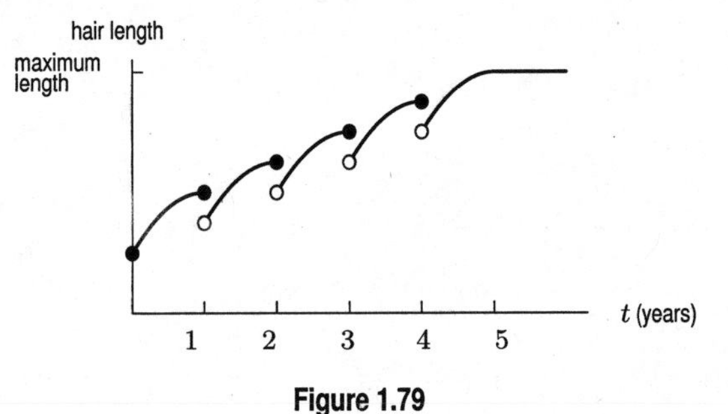

Figure 1.79

70. **(a)** We know that 75% of David Letterman's 7 million person audience belongs to the nation's work force. Thus

$$\left(\begin{array}{c}\text{Number of people from the}\\ \text{work force in Dave's audience}\end{array}\right) = 75\% \text{ of 7 million} = 0.75 \cdot (7 \text{ million}) = 5.25 \text{ million}.$$

Thus the percentage of the work force in Dave's audience is

$$\begin{aligned}\left(\begin{array}{c}\% \text{ of work force}\\ \text{in audience}\end{array}\right) &= \left(\frac{\text{People from work force in audience}}{\text{Total work force}}\right) \cdot 100\% \\ &= \left(\frac{5.25}{118}\right) \cdot 100\% = 4.45\%.\end{aligned}$$

(b) Since 4.45% of the work force belongs to Dave's audience, David Letterman's audience must contribute 4.45% of the GDP. Since the GDP is estimated at \$6.325 trillion,

$$\left(\begin{array}{c}\text{Dave's audience's contribution}\\ \text{to the G.D.P.}\end{array}\right) = (0.0445) \cdot (6.325 \text{ trillion}) \approx 281 \text{ billion dollars}.$$

(c) Of the contributions by Dave's audience, 10% is estimated to be lost. Since the audience's total contribution is \$281 billion, the "Letterman Loss" is given by

$$\text{Letterman loss} = 0.1 \cdot (281 \text{ billion dollars}) = \$28.1 \text{ billion}.$$

71. Let g be a linear function whose graph intersects the graph of f at $x = -2\pi$ and $x = 3\pi$. Judging from the figure, the graph of f contains the points $(-2\pi, 1)$ and $(3\pi, 7)$. Since g is linear, this means

$$\begin{aligned}g(-2\pi) &= f(-2\pi) = 1\\ g(3\pi) &= f(3\pi) = 7\\ m &= \frac{g(3\pi) - g(-2\pi)}{3\pi - (-2\pi)}\\ &= \frac{7-1}{5\pi}\\ &= \frac{6}{5\pi}.\end{aligned}$$

Using point-slope form with $(x_0, y_0) = (3\pi, 7)$, we have

$$\begin{aligned}g(x) &= y_0 + m(x - x_0)\\ &= 7 + \frac{6}{5\pi} \cdot (x - 3\pi).\end{aligned}$$

Checking our answer, we see that

$$\begin{aligned}g(-2\pi) &= 7 + \frac{6}{5\pi} \cdot (-2\pi - 3\pi)\\ &= 7 + \frac{6}{5\pi} \cdot (-5\pi)\\ &= 7 - 6 = 1,\end{aligned}$$

as required. We can also write our answer in slope-intercept form:

$$g(x) = 7 + \frac{6}{5\pi} \cdot (x - 3\pi)$$

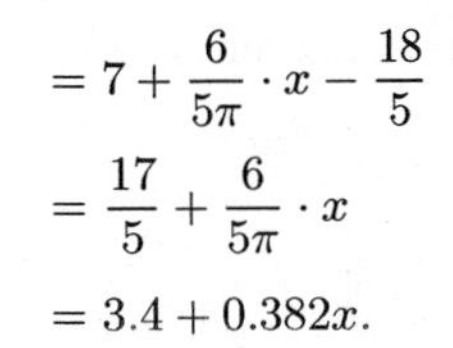

$$
\begin{aligned}
&= 7 + \frac{6}{5\pi}\cdot x - \frac{18}{5}\\
&= \frac{17}{5} + \frac{6}{5\pi}\cdot x\\
&= 3.4 + 0.382x.
\end{aligned}
$$

72. Looking up the table entry $n = 22$, we see that it is in column $c = 5$, row $r = 4$. Thus, $g(22) = 5^{1/4} = \sqrt[4]{5}$.

73. Extending the pattern in the table, we see that entry $n = 54$ is found in column $c = 8$, row $r = 5$. Thus, $g(54) = 8^{1/5} = \sqrt[5]{8}$.

74.
- We know that $\sqrt{3} = 3^{1/2}$, so looking in column $c = 3$, row $r = 2$, we see that $g(8) = 3^{1/2} = \sqrt{3}$. Thus, one possible solution is $n = 8$.
- We know that $\sqrt{3} = 3^{1/2} = 3^{2/4} = \left(3^2\right)^{1/4} = 9^{1/4}$. This means if we look in row $r = 4$, column $c = 9$, we will find another solution. (In fact, if you go to the trouble to write out the table, we find that this solution is $n = 78$.)

75. Each sheet is half the area of the next larger sheet, so

$$
\begin{aligned}
\text{Area of A1} &= \frac{1}{2}\cdot\text{Area of A0} = \frac{1}{2}\cdot 10{,}000 = 5000\\
\text{Area of A2} &= \frac{1}{2}\cdot\text{Area of A1} = \frac{1}{2}\cdot 5000 = 2500\\
\text{Area of A3} &= \frac{1}{2}\cdot\text{Area of A2} = \frac{1}{2}\cdot 2500 = 1250\\
\text{Area of A4} &= \frac{1}{2}\cdot\text{Area of A3} = \frac{1}{2}\cdot 1250 = 625\\
\text{Area of A5} &= \frac{1}{2}\cdot\text{Area of A4} = \frac{1}{2}\cdot 625 = 312.5.
\end{aligned}
$$

76. (a) Since the proportion (length over width) is the same for both paper sizes, we know that

$$\frac{\text{Length of A1}}{\text{Width of A1}} = \frac{\text{Length of A0}}{\text{Width of A0}}.$$

Since the length of A1 is the width of A0, and the width of A1 is half the length of A0,

$$\frac{\overbrace{\text{Width of A0}}^{\text{Length of A1}}}{\underbrace{\text{Half length of A0}}_{\text{Width of A1}}} = \frac{\text{Length of A0}}{\text{Width of A0}}.$$

Letting W and L stand for the width and length of A0, we have

$$
\begin{aligned}
\frac{W}{0.5L} &= \frac{L}{W}\\
\frac{2W}{L} &= \frac{L}{W}\\
2W^2 &= L^2\\
L &= W\sqrt{2}\\
&\approx 1.41W.
\end{aligned}
$$

This means that A0 paper is about 1.4 times as long as it is wide, Other sheets in the series have the same proportion. In comparison, US letter paper is $11/8.5 = 1.294$ times as long as it is wide, so A-series paper is slightly more "rectangular" and US paper is slightly more "squarish."

(b) We know from part (a) that $l = w\sqrt{2}$ where l and w are the length and width of A4 paper. Thus,

$$\begin{aligned} \text{Area} = \quad lw &= 625 \\ \underbrace{w\sqrt{2}}_{l}\cdot w &= 625 \\ w^2\sqrt{2} &= 625 \\ w^2 &= \frac{625}{\sqrt{2}} \\ \text{Width} = \quad w &= \sqrt{\frac{625}{\sqrt{2}}} = 21.02 \text{ cm} \\ \text{Length} = \quad l &= w\sqrt{2} = 29.73 \text{ cm.} \end{aligned}$$

Checking our answer, we see that

$$\begin{aligned} \text{Proportion} = \frac{l}{w} = \frac{29.73}{21.02} &= 1.414 \approx \sqrt{2} \\ \text{Area} = lw = 29.73(21.02) &= 624.92 \approx 625, \end{aligned}$$

as required.

CHECK YOUR UNDERSTANDING

1. False. $f(t)$ is functional notation, meaning that f is a function of the variable t.
2. False. A rule need not be a formula.
3. True. The $P = f(x)$ notation means that P is the dependent variable and x is the independent variable.
4. False. The independent variable is commonly denoted by the letter x or t, but any letter can be used.
5. True. The number of people who enter a store in a day and the total sales for the day are related, but neither quantity is uniquely determined by the other.
6. True. This is the definition of a function.
7. True. If one column of values has no repeated values then that column can be defined as the input values of a function and the other column as the output.
8. False. For example, if Q is a constant function, then P is not a function of Q.
9. True. A circle does not pass the vertical line test.
10. False. Just the reverse; the equation tells us that 100 angels can dance on a pin head of 10 square millimeters.
11. True. This is the definition of average speed.
12. False. The average rate of change of Q with respect to t is written $\Delta Q/\Delta t$.
13. True. This is the definition of an increasing function.
14. True. A decreasing function decreases on all intervals for which it is defined.
15. True. For a function $f(x)$ over the interval $a \le x \le b$ the slope is $\dfrac{f(b) - f(a)}{b - a}$. This is the slope of the line through the points $(a, f(a))$ and $(b, f(b))$ on the graph of f.
16. False. $\Delta y = (3 \cdot 6 - 4) - (3 \cdot 2 - 4) = 12$ and $\Delta x = 6 - 2 = 4$, so the average rate of change is $\Delta y/\Delta x = 3$.
17. False. Parentheses must be inserted. The correct ratio is $\dfrac{(10 - 2^2) - (10 - 1^2)}{2 - 1} = -3$.
18. False. The first slope is 5, the second slope is -5.
19. False. If $y = mx + b$ then m, the slope, is the rate of change over every interval.

20. True. A linear function has a constant rate of change.

21. False. Writing the equation as $y = (-3/2)x + 7/2$ shows that the slope is $-3/2$.

22. True. This ratio is the slope, which is constant for a linear function.

23. True. If a linear function is decreasing, then for all positive Δx, the change, Δy, is negative, so the slope, $\Delta y/\Delta x$, is negative.

24. True. The slope is $\Delta y/\Delta x$ and this quotient can only be negative when either $\Delta x > 0$ and $\Delta y < 0$ or $\Delta x < 0$ and $\Delta y > 0$.

25. True. A constant function has slope zero. Its graph is a horizontal line.

26. False. It can be written $y = 2x - 3$.

27. False. The slope is $-3/5$.

28. False. The line $y = 4x + 5$ has slope 4 but the given point is not on the line since $3 \neq 4 \cdot (-2) + 5$.

29. True. At $y = 0$, we have $4x = 52$, so $x = 13$. The x-intercept is $(13, 0)$.

30. True. Evaluate $f(2) = -2 \cdot 2 + 7 = 3$

31. False. Since the slope is $\Delta y/\Delta x = (-10 - 2)/(4 - 1) = -12/3 = -4$.

32. False. First simplify the right side to $y - 5 = 4x + 4$ then add 5 to both sides to get $y = 4x + 9$.

33. False. Substitute the point's coordinates in the equation: $-3 - 4 \neq -2(4 + 3)$.

34. True. The given equation is in the slope-intercept form so that the slope is the coefficient of x.

35. True. The given equation is in the slope-intercept form so that the y-intercept is the constant term.

36. True. The given equation is in the slope-intercept form for a line.

37. False. The first line does but the second, in slope-intercept form, is $y = (1/8)x + (1/2)$, so it crosses the y-axis at $y = 1/2$.

38. False. The graph is a horizontal line. The slope is zero.

39. False. The slopes are $-4/5$ and $4/5$. Thus the lines are not parallel because they have different slopes.

40. True. The slopes, 9 and $-1/9$, are negative reciprocals of one another.

41. True. The point $(1, 3)$ is on both lines because $3 = -2 \cdot 1 + 5$ and $3 = 6 \cdot 1 - 3$.

42. True. Parallel lines have equal slopes.

43. False. The line $y = -3/4$ is parallel to the x-axis.

44. True. A line parallel to the x-axis represents a constant function. Since $\Delta y = 0$ for any two points on the line, its slope is $\Delta y/\Delta x = 0$.

45. True. The slope, $\Delta y/\Delta x$ is undefined because Δx is zero for any two points on a vertical line.

46. True.

47. True. Interpolation estimates values within the range of data values. Extrapolation estimates values outside the range for which data is available.

48. False. An interpolation value is calculated between known data values so it is, in general, more reliable than extrapolation.

49. False. For example, in children there is a high correlation between height and reading ability, but it is clear that neither causes the other.

50. False. There can be a relationship, but if it is not linear then the correlation coefficient can be close to zero.

51. True. All values from -1.0 to 1.0 are possible values of a correlation coefficient.

52. True.

53. True. There is a perfect fit of the line to the data.

54. True. The least squares line refers to the criterion by which the regression line was chosen.

Solutions to Skills for Chapter 1

1.

$$\begin{aligned} 3x &= 15 \\ \frac{3x}{3} &= \frac{15}{3} \\ x &= 5 \end{aligned}$$

2.

$$\begin{aligned} -2y &= 12 \\ \frac{-2y}{-2} &= \frac{12}{-2} \\ y &= -6 \end{aligned}$$

3.

$$\begin{aligned} 4z &= 22 \\ \frac{4z}{4} &= \frac{22}{4} \\ z &= \frac{11}{2} \end{aligned}$$

4.

$$\begin{aligned} x + 3 &= 10 \\ x &= 7 \end{aligned}$$

5.

$$\begin{aligned} w - 23 &= -34 \\ w &= -11 \end{aligned}$$

6.

$$\begin{aligned} 7 - 3y &= -14 \\ -3y &= -21 \\ \frac{-3y}{-3} &= \frac{-21}{-3} \\ y &= 7 \end{aligned}$$

7.

$$\begin{aligned} 13t + 2 &= 47 \\ 13t &= 45 \\ \frac{13t}{13} &= \frac{45}{13} \\ t &= \frac{45}{13} \end{aligned}$$

8.

$$\begin{aligned} 0.5x - 3 &= 7 \\ 0.5x &= 10 \\ x &= 20. \end{aligned}$$

9. The common denominator for this fractional equation is 3. If we multiply both sides of the equation by 3, we obtain:

$$\begin{aligned} 3\left(3t - \frac{2(t-1)}{3}\right) &= 3(4) \\ 9t - 2(t-1) &= 12 \\ 9t - 2t + 2 &= 12 \\ 7t + 2 &= 12 \\ 7t &= 10 \\ t &= \frac{10}{7}. \end{aligned}$$

10.

$$\begin{aligned} 2(r+5) - 3 &= 3(r-8) + 21 \\ 2r + 10 - 3 &= 3r - 24 + 21 \\ 2r + 7 &= 3r - 3 \\ 2r &= 3r - 10 \\ -r &= -10 \\ r &= 10 \end{aligned}$$

11. Solving for B,

$$\begin{aligned} B - 4[B - 3(1 - B)] &= 42 \\ B - 4[B - 3 + 3B] &= 42 \\ B - 4[4B - 3] &= 42 \\ B - 16B + 12 &= 42 \\ -15B + 12 &= 42 \\ -15B &= 30 \\ B &= -2. \end{aligned}$$

12. Expanding yields

$$\begin{aligned} 1.06s - 0.01(248.4 - s) &= 22.67s \\ 1.06s - 2.484 + 0.01s &= 22.67s \\ -21.6s &= 2.484 \\ s &= -0.115. \end{aligned}$$

13. Dividing by w gives $l = A/w$.

14. Solving for w,

$$\begin{aligned} l &= l_0 + \frac{k}{2}w \\ l - l_0 &= \frac{k}{2}w \\ 2(l - l_0) &= kw \\ \frac{2}{k}(l - l_0) &= w. \end{aligned}$$

15. Putting $v_0 t$ on the other side of the equation:

$$h - v_0 t = \frac{1}{2}at^2$$
$$\frac{2}{t^2}(h - v_0 t) = a$$

16. We have

$$3xy + 1 = 2y - 5x$$
$$3xy - 2y = -5x - 1$$
$$y(3x - 2) = -5x - 1$$
$$y = \frac{-5x - 1}{3x - 2}.$$

17. We collect all terms involving v and then factor out the v.

$$u(v + 2) + w(v - 3) = z(v - 1)$$
$$uv + 2u + wv - 3w = zv - z$$
$$uv + wv - zv = 3w - 2u - z$$
$$v(u + w - z) = 3w - 2u - z$$
$$v = \frac{3w - 2u - z}{u + w - z}.$$

18. Multiplying by $(r - 1)$:

$$S(r - 1) = rL - a$$
$$Sr - S = rL - a$$
$$Sr - rL = S - a$$
$$r(S - L) = S - a$$
$$r = \frac{S - a}{S - L}.$$

19. Solving for x:

$$\frac{a - cx}{b + dx} + a = 0$$
$$\frac{a - cx}{b + dx} = -a$$
$$a - cx = -a(b + dx) = -ab - adx$$
$$adx - cx = -ab - a$$
$$(ad - c)x = -a(b + 1)$$
$$x = -\frac{a(b + 1)}{ad - c}.$$

20. Multiplying on both sides by $C - B(1 - 2t)$ gives

$$At - B = 3(C - B + 2Bt)$$
$$At - B = 3C - 3B + 6Bt$$

$$\begin{aligned} At - 6Bt &= 3C - 3B + B = 3C - 2B \\ t(A - 6B) &= 3C - 2B \\ t &= \frac{3C - 2B}{A - 6B}. \end{aligned}$$

21. Solving for y',

$$\begin{aligned} y'y^2 + 2xyy' &= 4y \\ y'(y^2 + 2xy) &= 4y \\ y' &= \frac{4y}{y^2 + 2xy} \\ y' &= \frac{4}{y + 2x} \text{ if } y \neq 0. \end{aligned}$$

Note that if $y = 0$, then y' could be any real number.

22. We collect all terms involving the variable y' and factor out the y'.

$$\begin{aligned} 2x - (xy' + yy') + 2yy' &= 0 \\ 2x - xy' - yy' + 2yy' &= 0 \\ 2x - xy' + yy' &= 0 \\ 2x - y'(x - y) &= 0 \\ -y'(x - y) &= -2x \\ y'(x - y) &= 2x \\ y' &= \frac{2x}{x - y}. \end{aligned}$$

23. Substituting the value of y from the second equation into the first, we obtain

$$\begin{aligned} 3x - 2(2x - 5) &= 6 \\ 3x - 4x + 10 &= 6 \\ -x &= -4 \\ x &= 4. \end{aligned}$$

From the second equation, we have

$$y = 2(4) - 5 = 3$$

so

$$y = 3.$$

24. Substituting the value of x from the first equation into the second equation, we obtain

$$\begin{aligned} 4(7y - 9) - 15y &= 26 \\ 28y - 36 - 15y &= 26 \\ 13y &= 62 \\ y &= \frac{62}{13}. \end{aligned}$$

From the first equation, we have

$$\begin{aligned} x &= 7\left(\frac{62}{13}\right) - 9 \\ &= \frac{434}{13} - \frac{117}{13} = \frac{317}{13}. \end{aligned}$$

25. We substitute the expression $-\frac{3}{5}x+6$ for y in the first equation.

$$\begin{aligned}
2x+3y &= 7\\
2x+3\left(-\frac{3}{5}x+6\right) &= 7\\
2x-\frac{9}{5}x+18 &= 7 \quad \text{or}\\
\frac{10}{5}x-\frac{9}{5}x+18 &= 7\\
\frac{1}{5}x+18 &= 7\\
\frac{1}{5}x &= -11\\
x &= -55\\
y &= -\frac{3}{5}(-55)+6\\
y &= 39
\end{aligned}$$

26. One way to solve this system is by substitution. Solve the first equation for y:

$$\begin{aligned}
3x-y &= 17\\
-y &= 17-3x\\
y &= 3x-17.
\end{aligned}$$

In the second equation, substitute the expression $3x-17$ for y:

$$\begin{aligned}
-2x-3y &= -4\\
-2x-3(3x-17) &= -4\\
-2x-9x+51 &= -4\\
-11x &= -4-51=-55\\
x &= \frac{-55}{-11}=5.
\end{aligned}$$

Since $x=5$ and $y=3x-17$, we have

$$y=3(5)-17=15-17=-2.$$

Thus, the solution to the system is $x=5$ and $y=-2$.

Check your results by substituting the values into the second equation:

$$\begin{aligned}
-2x-3y &= -4\\
\text{Substituting, we get } -2(5)-3(-2) &= -4\\
-10+6 &= -4\\
-4 &= -4.
\end{aligned}$$

27. We regard a as a constant. Multiplying the first equation by a and subtracting the second gives

$$\begin{aligned}
a^2x+ay &= 2a^2\\
x+ay &= 1+a^2
\end{aligned}$$

so, subtracting

$$(a^2-1)x=a^2-1.$$

Thus $x=1$ (provided $a\neq\pm1$). Solving for y in the first equation gives $y=2a-a(1)$, so $y=a$.

28. Substituting $y = x + 1$ into $2x + 3y = 12$ gives

$$\begin{aligned} 2x + 3(x+1) &= 12 \\ 5x + 3 &= 12 \\ 5x &= 9 \\ x &= \frac{9}{5} = 1.8. \end{aligned}$$

If $x = 1.8$, then $y = 1.8 + 1 = 2.8$. Thus, the point of intersection is $x = 1.8$, $y = 2.8$.

29. Substituting $y = 2x$ into $2x + y = 12$ gives

$$\begin{aligned} 2x + 2x &= 12 \\ 4x &= 12 \\ x &= 3. \end{aligned}$$

Thus, substituting $x = 3$ into $y = 2x$ gives $y = 6$, so the point of intersection is $x = 3$, $y = 6$.

30. The figure is a square, so $A = (17, 23)$; $B = (0, 40)$; $C = (-17, 23)$.

31. The figure is a parallelogram, so $A = (-4, 7)$.

32. The radius is 4, so $B = (7, 4)$, $A = (11, 0)$.

33. The radius is 8, so $A = (2, 9)$, $B = (10, 1)$.

34. Since A is a corner of the rectangle, $A = (2, 5)$. The radius of the circle is $(8 - 2)/2 = 3$, so $B = (5, 8)$.

35. The radius is 4 so $A = (-7, 8)$, $B = (-3, 4)$

CHAPTER TWO

Solutions for Section 2.1

Skill Refresher

S1. $5(x-3) = 5x - 15$.

S2. $a(2a+5) = 2a^2 + 5a$.

S3. $(m-5)(4(m-5)+2) = 4(m-5)^2 + 2(m-5) = 4(m^2 - 10m + 25) + 2m - 10 = 4m^2 - 40m + 100 + 2m - 10 = 4m^2 - 38m + 90$.

S4. $(x+2)(3x-8) = 3x^2 - 8x + 6x - 16 = 3x^2 - 2x - 16$.

S5.

$$\begin{aligned} 3\left(1 + \frac{1}{x}\right) &= 3\left(\frac{x+1}{x}\right) \\ &= \frac{3x+3}{x}. \end{aligned}$$

S6. $3 + 2\left(\frac{1}{x}\right)^2 - x = 3 + \frac{2}{x^2} - x = \frac{3x^2 + 2 - x^3}{x^2}$.

S7. $x^2 - 9 = (x-3)(x+3) = 0$. Hence $x = \pm 3$.

S8.

$$\begin{aligned} \sqrt{2x-1} + 3 &= 9 \\ \sqrt{2x-1} &= 6 \\ 2x - 1 &= 36 \\ 2x &= 37 \\ x &= \frac{37}{2}. \end{aligned}$$

S9.

$$\begin{aligned} \frac{21}{z-5} - \frac{13}{z^2 - 5z} &= 3 \\ \frac{21}{z-5} - \frac{13}{z(z-5)} &= 3 \\ \frac{21z - 13}{z(z-5)} &= 3 \\ 21z - 13 &= 3z(z-5) \\ 21z - 13 &= 3z^2 - 15z \\ 3z^2 - 36z + 13 &= 0 \end{aligned}$$

$$\begin{aligned} z &= \frac{-(-36) \pm \sqrt{(-36)^2 - 4(3)(13)}}{2(3)} \\ &= \frac{36 \pm \sqrt{1140}}{6} \\ &= \frac{36 \pm \sqrt{4 \cdot 285}}{6} \end{aligned}$$

$$= \frac{36 \pm 2\sqrt{285}}{6}$$
$$= \frac{18 \pm \sqrt{285}}{3}.$$

S10. Adding 1 to both sides we get $2x^{\frac{3}{2}} = 8$. Dividing both sides by 2, we have $x^{\frac{3}{2}} = 4$. To solve it for x we raise both sides to the $2/3$ power to get

$$x = 4^{\frac{2}{3}}.$$

Exercises

1. **(a)** Substituting $t = 0$ gives $f(0) = 0^2 - 4 = -4$.
(b) Setting $f(t) = 0$ and solving gives $t^2 - 4 = 0$, so $t^2 = 4$, so $t = \pm 2$.

2. **(a)** Substituting $x = 0$ gives $g(0) = 0^2 - 5(0) + 6 = 6$.
(b) Setting $g(x) = 0$ and solving gives $x^2 - 5x + 6 = 0$.
Factoring gives $(x-2)(x-3) = 0$, so $x = 2, 3$.

3. **(a)** Substituting $t = 0$ gives

$$g(0) = \frac{1}{0+2} - 1 = \frac{1}{2} - 1 = -\frac{1}{2}.$$

(b) Setting $g(t) = 0$ and solving gives

$$\begin{aligned} \frac{1}{t+2} - 1 &= 0 \\ \frac{1}{t+2} &= 1 \\ 1 &= t+2 \\ t &= -1. \end{aligned}$$

4. Substituting zero for x gives

$$h(0) = a \cdot 0^2 + b \cdot 0 + c = c.$$

5. Substituting -27 for x gives

$$g(-27) = -\frac{1}{2}(-27)^{1/3} = -\frac{1}{2}(-3) = \frac{3}{2}.$$

6. We need to solve for x in the equation:

$$0.3 = \frac{2x+1}{x+1}.$$

Multiplying both sides by $x+1$ gives:

$$\begin{aligned} 0.3(x+1) &= 2x+1 \\ 0.3x + 0.3 &= 2x + 1 \\ -1.7x &= 0.7 \\ x &= -0.412. \end{aligned}$$

7. To evaluate $p(7)$, we substitute 7 for each r in the formula:

$$p(7) = 7^2 + 5 = 54.$$

8. To evaluate $p(x) + p(8)$, we substitute x and 8 for each r in the formula and add the two expressions:

$$p(x) + p(8) = (x^2 + 5) + (8^2 + 5) = x^2 + 74.$$

9.

10.

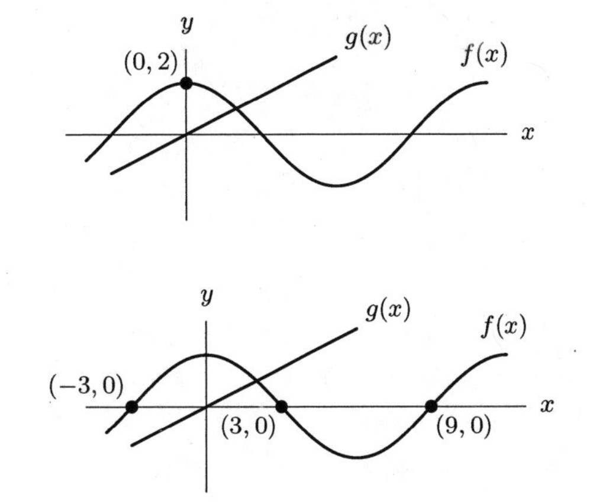

11. Curves cross at $x = 2$. See Figure 2.1. We do not know the y-coordinate of this point.

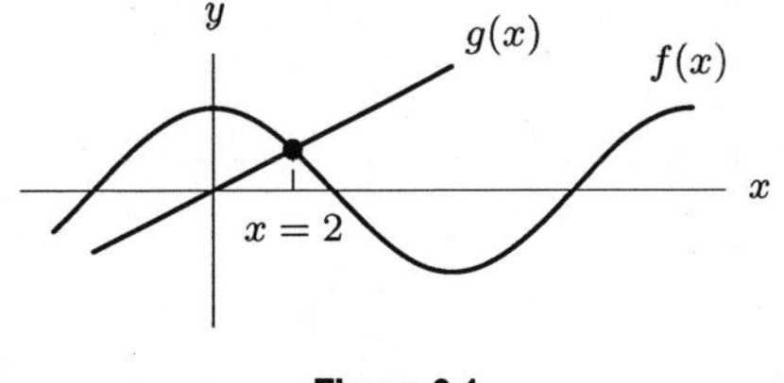

Figure 2.1

12. Graph of $g(x)$ is above graph of $f(x)$ for x to the right of 2. See Figure 2.2.

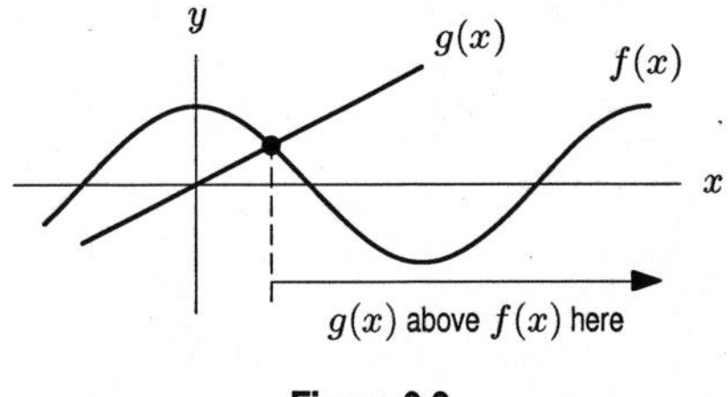

Figure 2.2

Problems

13. The input, t, is the number of months since January 1, and the output, F, is the number of foxes. The expression $g(9)$ represents the number of foxes in the park on October 1. Table 1.3 on page 5 of the text gives $F = 100$ when $t = 9$. Thus, $g(9) = 100$. On October 1, there were 100 foxes in the park.

14. The output $g(t)$ stands for a number of foxes. We want to know in what month there are 75 foxes. Table 1.3 on page 5 of the text tells us that this occurs when $t = 4$ and $t = 8$; that is, in May and in September.

15. Substituting $\frac{1}{3}$ for x gives

$$f\left(\frac{1}{3}\right) = 3 + 2\left(\frac{1}{3}\right)^2 = 3 + \frac{2}{9} = 3.222.$$

On the other hand

$$f(1) = 3 + 2(1)^2 = 5$$
$$f(3) = 3 + 2(3)^2 = 21.$$

So $\dfrac{f(1)}{f(3)} = \dfrac{5}{21} = 0.238$, and we see that

$$f\left(\frac{1}{3}\right) \neq \frac{f(1)}{f(3)}.$$

They are not equal.

16. **(a)** $g(x) = x^2 + x$
$g(-3x) = (-3x)^2 + (-3x)$
$g(-3x) = 9x^2 - 3x$

(b) $g(1-x) = (1-x)^2 + (1-x) = (1 - 2x + x^2) + (1-x) = x^2 - 3x + 2$

(c) $g(x+\pi) = (x+\pi)^2 + (x+\pi) = (x^2 + 2\pi x + \pi^2) + (x+\pi) = x^2 + (2\pi+1)x + \pi^2 + \pi$

(d) $g(\sqrt{x}) = (\sqrt{x})^2 + \sqrt{x} = x + \sqrt{x}$

(e) $g\left(\dfrac{1}{x+1}\right) = \left(\dfrac{1}{(x+1)^2}\right) + \dfrac{1}{x+1} = \dfrac{1}{(x+1)^2} + \dfrac{x+1}{(x+1)^2} = \dfrac{x+2}{(x+1)^2}$

(f) $g(x^2) = (x^2)^2 + x^2 = x^4 + x^2$

17. **(a)** (i) $\dfrac{\frac{1}{t}}{\frac{1}{t} - 1} = \dfrac{\frac{1}{t}}{\frac{1-t}{t}} = \dfrac{1}{t} \cdot \dfrac{t}{1-t} = \dfrac{1}{1-t}.$

(ii) $\dfrac{\frac{1}{t+1}}{\frac{1}{t+1} - 1} = \dfrac{1}{t+1} \cdot \dfrac{t+1}{1-t-1} = -\dfrac{1}{t}.$

(b) Solve $f(x) = \dfrac{x}{(x-1)} = 3$, so

$$x = 3x - 3$$
$$3 = 2x$$
$$x = \frac{3}{2}.$$

18. **(a)**

x	-2	-1	0	1	2	3
$h(x)$	0	9	8	3	0	6

(b) From the table, we see that $h(3) = 6$, while $h(1) = 3$. Thus, $h(3) - h(1) = 6 - 3 = 3$.

(c) From the table, we see that $h(2) = 0$, and $h(0) = 8$. Thus, $h(2) - h(0) = 0 - 8 = -8$.

(d) From the table, we see that $h(0) = 8$. Thus, $2h(0) = 2(8) = 16$.

(e) From the table, we see that $h(1) = 3$. Thus, $h(1) + 3 = 3 + 3 = 6$.

19. **(a)** Substituting into $h(t) = -16t^2 + 64t$, we get

$$h(1) = -16(1)^2 + 64(1) = 48$$
$$h(3) = -16(3)^2 + 64(3) = 48$$

Thus the height of the ball is 48 feet after 1 second and after 3 seconds.

(b) The graph of $h(t)$ is in Figure 2.3. The ball is on the ground when $h(t) = 0$. From the graph we see that this occurs at $t = 0$ and $t = 4$. The ball leaves the ground when $t = 0$ and hits the ground at $t = 4$ or after 4 seconds. From the graph we see that the maximum height is 64 ft.

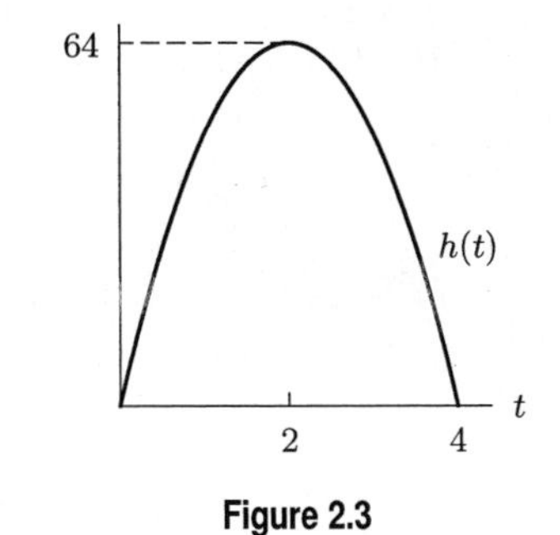

Figure 2.3

20. **(a)** Substituting $t = 0$ gives $v(0) = 0^2 - 2(0) = 0 - 0 = 0$.

(b) To find when the object has velocity equal to zero, we solve the equation

$$\begin{aligned} t^2 - 2t &= 0 \\ t(t-2) &= 0 \\ t &= 0 \quad \text{or} \quad t = 2. \end{aligned}$$

Thus the object has velocity zero at $t = 0$ and at $t = 2$.

(c) The quantity $v(3)$ represents the velocity of the object at time $t = 3$. Its units are ft/sec.

21. **(a)** The car's position after 2 hours is denoted by the expression $s(2)$. The position after 2 hours is

$$s(2) = 11(2)^2 + 2 + 100 = 44 + 2 + 100 = 146.$$

(b) This is the same as asking the following question: "For what t is $v(t) = 65$?"

(c) To find out when the car is going 67 mph, we set $v(t) = 67$. We have

$$\begin{aligned} 22t + 1 &= 67 \\ 22t &= 66 \\ t &= 3. \end{aligned}$$

The car is going 67 mph at $t = 3$, that is, 3 hours after starting. Thus, when $t = 3$, $S(3) = 11(3^2) + 3 + 100 = 202$, so the car's position when it is going 67 mph is 202 miles.

22. **(a)** From Figure 2.4, we see that $P = (b, a)$ and $Q = (d, e)$.

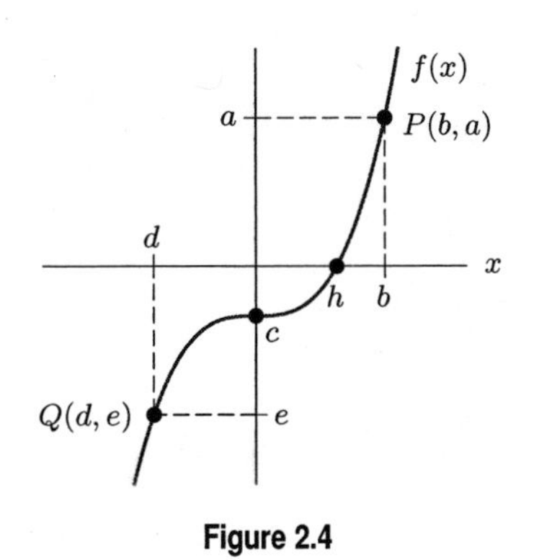

Figure 2.4

(b) To evaluate $f(b)$, we want to find the y-value when the x-value is b. Since (b, a) lies on this graph, we know that the y-value is a, so $f(b) = a$.

(c) To solve $f(x) = e$, we want to find the x-value for a y-value of e. Since (d, e) lies on this curve, $x = d$ is our solution.

(d) To solve $z = f(x)$, we need to first find a value for z; in other words, we need to first solve for $f(z) = c$. Since $(0, c)$ lies on this graph, we know that $z = 0$. Now we need to solve $0 = f(x)$ by finding the point whose y-value is 0. That point is $(h, 0)$, so $x = h$ is our solution.

(e) We know that $f(b) = a$ and $f(d) = e$. Thus, if $f(b) = -f(d)$, we know that $a = -e$.

23. **(a)** Her tax is \$973 on the first \$20,000 plus 6.85% of the remaining \$48,000:

$$\text{Tax owed} = \$973 + 0.0685(\$48{,}000) = \$973 + \$3288 = \$4261.$$

(b) Her taxable income, $T(x)$, is 80% of her total income, or 80% of x. So $T(x) = 0.8x$.

(c) Her tax owed is \$973 plus 6.85% of her taxable income over \$20,000. Since her taxable income is $0.8x$, her taxable income over \$20,000 is $0.8x - 20{,}000$. Therefore,

$$L(x) = 973 + 0.0685(0.8x - 20{,}000),$$

so multiplying out and simplifying, we obtain

$$L(x) = 0.0548x - 397.$$

(d) Evaluating for $x = \$85{,}000$, we have

$$\begin{aligned} L(85{,}000) &= 973 + 0.0685(0.8(85{,}000) - 20{,}000) \\ &= \$4261. \end{aligned}$$

The values are the same.

24. **(a)** We calculate the values of $f(x)$ and $g(x)$ using the formulas given in Table 2.1.

Table 2.1

x	-2	-1	0	1	2
$f(x)$	6	2	0	0	2
$g(x)$	6	2	0	0	2

The pattern is that $f(x) = g(x)$ for $x = -2, -1, 0, 1, 2$. Based on this, we might speculate that f and g are really the same function. This is, in fact, the case, as can be verified algebraically:

$$\begin{aligned} f(x) &= 2x(x-3) - x(x-5) \\ &= 2x^2 - 6x - x^2 + 5x \\ &= x^2 - x \\ &= g(x). \end{aligned}$$

Their graphs are the same, and are shown in Figure 2.5.

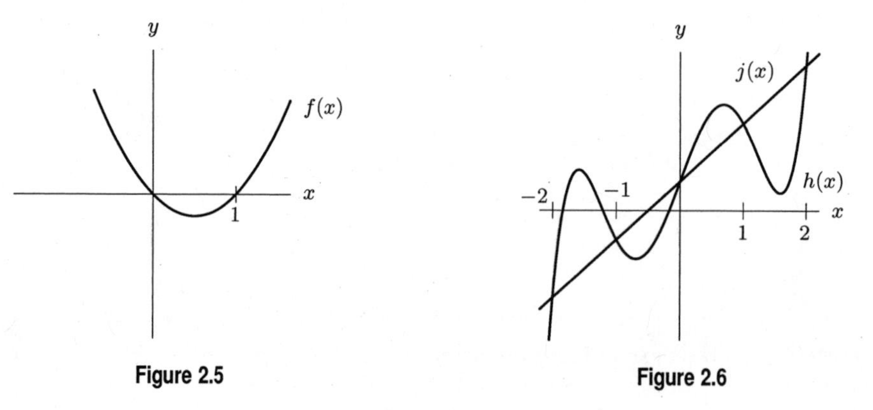

Figure 2.5

Figure 2.6

(b) Using the formulas for $h(x)$ and $j(x)$, we obtain Table 2.2.

Table 2.2

x	-2	-1	0	1	2
$h(x)$	-3	-1	1	3	5
$j(x)$	-3	-1	1	3	5

The pattern is that $h(x) = j(x)$ for $x = -2, -1, 0, 1, 2$. Based on this, we might speculate that h and j are really the same function. The graphs of these functions are shown in Figure 2.6. We see that the graphs share only the points of the table and are thus two different functions.

25. $r(0.5s_0)$ is the wind speed at a half the height above ground of maximum wind speed.

26. In $r(s) = 0.75v_0$, the variable s is the height (or heights) at which the wind speed is 75% of the maximum wind speed.

27. **(a)** $h(1) = (1)^2 + b(1)^2 + c = b + c + 1$

(b) Substituting $b+1$ for x in the formula for $h(x)$:

$$\begin{aligned} h(b+1) &= (b+1)^2 + b(b+1) + c \\ &= (b^2 + 2b + 1) + b^2 + b + c \\ &= 2b^2 + 3b + c + 1 \end{aligned}$$

28. **(a)** $g(100) = 100\sqrt{100} + 100 \cdot 100 = 100 \cdot 10 + 100 \cdot 100 = 11,000$

(b) $g(4/25) = 4/25 \cdot \sqrt{4/25} + 100 \cdot 4/25 = 4/25 \cdot 2/5 + 16 = 8/125 + 16 = 16.064$

(c) $g(1.21 \cdot 10^4) = g(12100) = (12100)\sqrt{12100} + 100 \cdot (12100) = 2,541,000$

29. $f(a) = \dfrac{a \cdot a}{a + a} = \dfrac{a^2}{2a} = \dfrac{a}{2}.$

30. $f(1-a) = \dfrac{a(1-a)}{a + (1-a)} = a(1-a) = a - a^2$

31. $f\left(\dfrac{1}{1-a}\right) = \dfrac{a\dfrac{1}{1-a}}{a + \dfrac{1}{1-a}} = \dfrac{\dfrac{a}{1-a}}{\dfrac{a(1-a)+1}{1-a}} = \dfrac{a}{1-a} \cdot \dfrac{1-a}{a - a^2 + 1} = \dfrac{a}{a - a^2 + 1}.$

32. **(a)** To evaluate $f(1)$, we need to find the value of f which corresponds to $x = 1$. Looking in the table, we see that that value is 2. So we can say $f(1) = 2$. Similarly, to find $g(3)$, we see in the table that the value of g which corresponds to $x = 3$ is 4. Thus, we know that $g(3) = 4$.

(b) The values of $f(x)$ increase by 3 as x increases by 1. For $x > 1$, the values of $g(x)$ are consecutive perfect squares. The entries for $g(x)$ are symmetric about $x = 1$. In other words, when $x < 1$ the values of $g(x)$ are the same as the values when $x > 1$, but the order is reversed.

(c) Since the values of $f(x)$ increase by 3 as x increases by 1 and $f(4) = 11$, we know that $f(5) = 11 + 3 = 14$. Similarly, $f(x)$ decreases by three as x goes down by one. Since $f(-1) = -4$, we conclude that $f(-2) = -4 - 3 = -7$.

The values of $g(x)$ are consecutive perfect squares. Since $g(4) = 9$, then $g(5)$ must be the next perfect square which is 16, so $g(5) = 16$. Since the values of $g(x)$ are symmetric about $x = 1$, the value of $g(-2)$ will equal $g(5)$ (since -2 and 4 are both a distance of 3 units from 1). Thus, $g(-2) = g(4) = 9$.

(d) To find a formula for $f(x)$, we begin by observing that $f(0) = -1$, so the value of $f(x)$ that corresponds to $x = 0$ is -1. We know that the value of $f(x)$ increases by 3 as x increases by 1, so

$$\begin{aligned} f(1) &= f(0) + 3 = -1 + 3 \\ f(2) &= f(1) + 3 = (-1 + 3) + 3 = -1 + 2 \cdot 3 \\ f(3) &= f(2) + 3 = (-1 + 2 \cdot 3) + 3 = -1 + 3 \cdot 3 \\ f(4) &= f(3) + 3 = (-1 + 3 \cdot 3) + 3 = -1 + 4 \cdot 3. \end{aligned}$$

The pattern is

$$f(x) = -1 + x \cdot 3 = -1 + 3x.$$

We can check this formula by choosing a value for x, such as $x = 4$, and use the formula to evaluate $f(4)$. We find that $f(4) = -1 + 3(4) = 11$, the same value we see in the table.

Since the values of $g(x)$ are all perfect squares, we expect the formula for $g(x)$ to have a square in it. We see that x^2 is not quite right since the table for such a function would look like Table 2.3.

Table 2.3

x	-1	0	1	2	3	4
x^2	1	0	1	4	9	16

But this table is very similar to the one that defines g. In order to make Table 2.3 look identical to the one given in the problem, we need to subtract 1 from each value of x so that $g(x) = (x-1)^2$. We can check our formula by choosing a value for x, such as $x = 2$. Using our formula to evaluate $g(2)$, we have $g(2) = (2-1)^2 = 1^2 = 1$. This result agrees with the value given in the problem.

33. **(a)** (i) From the table, $N(150) = 6$. When 150 students enroll, there are 6 sections.
(ii) Since $N(75) = 4$ and $N(100) = 5$, and 80 is between 75 and 100 students, we choose the higher value for $N(s)$. So $N(80) = 5$. When 80 students enroll, there are 5 sections.
(iii) The quantity $N(55.5)$ is not defined, since 55.5 is not a possible number of students.

(b) (i) The table gives $N(s) = 4$ sections for $s = 75$ and $s = 50$. For any integer between those in the table, the section number is the higher value. Therefore, for $50 \le s \le 75$, we have $N(s) = 4$ sections. We do not know what happens if $s < 50$.
(ii) First evaluate $N(125) = 5$. So we solve the equation $N(s) = 5$ for s. There are 5 sections when enrollment is between 76 and 125 students.

34. **(a)** See Figure 2.7.
(b) See Figure 2.7.

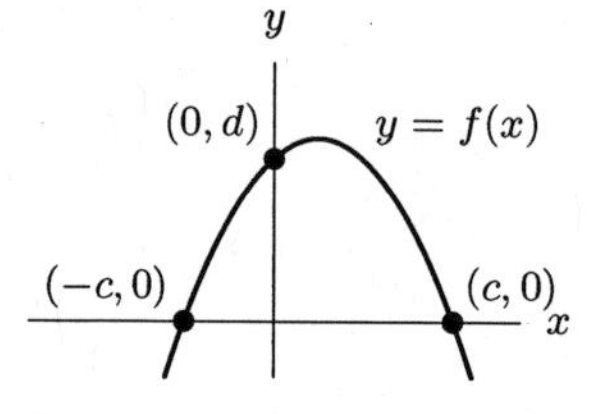

Figure 2.7

35. **(a)** To evaluate $f(2)$, we determine which value of I corresponds to $w = 2$. Looking at the graph, we see that $I \approx 7$ when $w = 2$. This means that ≈ 7000 people were infected two weeks after the epidemic began.

(b) The height of the epidemic occurred when the largest number of people were infected. To find this, we look on the graph to find the largest value of I, which seems to be approximately 8.5, or 8500 people. This seems to have occurred when $w = 4$, or four weeks after the epidemic began. We can say that at the height of the epidemic, at $w = 4$, $f(4) = 8.5$.

(c) To solve $f(x) = 4.5$, we must find the value of w for which $I = 4.5$, or 4500 people were infected. We see from the graph that there are actually two values of w at which $I = 4.5$, namely $w \approx 1$ and $w \approx 10$. This means that 4500 people were infected after the first week when the epidemic was on the rise, and that after the tenth week, when the epidemic was slowing, 4500 people remained infected.

(d) We are looking for all the values of w for which $f(w) \ge 6$. Looking at the graph, this seems to happen for all values of $w \ge 1.5$ and $w \le 8$. This means that more than 6000 people were infected starting in the middle of the second week and lasting until the end of the eighth week, after which time the number of infected people fell below 6000.

36. This represents the change in average hurricane intensity at average Caribbean Sea surface temperature after CO_2 levels rise to future projected levels.

37. This represents the change in average hurricane intensity at current CO_2 levels if sea surface temperature rises by 1°C.

Solutions for Section 2.2

Skill Refresher

S1. The function is undefined when the denominator is zero. Therefore, $x-3=0$ tells us the function is undefined for $x=3$.

S2. The function is undefined when the denominator is zero. Therefore, $x-3=0$ tells us the function is undefined for $x=3$ But it is also undefined at $x=0$.

S3. Because of the square root the function is undefined when $x-15$ is negative. That is, when $x<15$.

S4. Because of the square root the function is undefined when $15-x$ is negative. That is, when $x>15$.

S5. Adding 8 to both sides of the inequality we get $x>8$.

S6. First we subtract 5 from both sides to get $-x>-5$. Then multiplying both sides by -1 and changing the direction of the inequality we get $x<5$.

S7. Divide by -3 with a direction change, and then add 4.

$$\begin{aligned}
-3(n-4) &> 12 \\
\frac{-3(n-4)}{-3} &< \frac{12}{-3} \\
n-4 &< -4 \\
n-4+4 &< -4+4 \\
n &< 0.
\end{aligned}$$

S8. Subtract 24, then divide by -4 with a direction change:

$$\begin{aligned}
12 &\le 24-4a \\
12-24 &\le 24-4a-24 \\
-12 &\le -4a \\
\frac{-12}{-4} &\ge \frac{-4a}{-4} \\
3 &\ge a.
\end{aligned}$$

Or, equivalently, $a\le 3$.

S9. $x^2-25>0$ is true when $x>5$ or $x<-5$.

S10. We have $36-x^2\ge 0$ when $-6\le x\le 6$

Exercises

1. The graph of $f(x)=1/x$ for $-2\le x\le 2$ is shown in Figure 2.8. From the graph, we see that $f(x)=-(1/2)$ at $x=-2$. As we approach zero from the left, $f(x)$ gets more and more negative. On the other side of the y-axis, $f(x)=(1/2)$ at $x=2$. As x approaches zero from the right, $f(x)$ grows larger and larger. Thus, on the domain $-2\le x\le 2$, the range is $f(x)\le -(1/2)$ or $f(x)\ge (1/2)$.

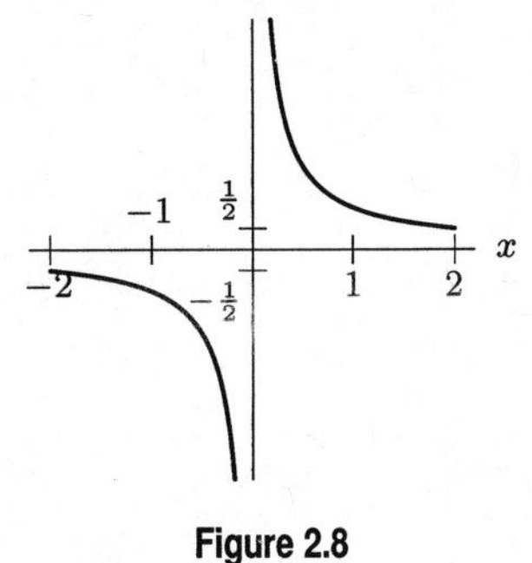

Figure 2.8

2. The graph of $f(x) = 1/x^2$ for $-1 \le x \le 1$ is shown in Figure 2.9. From the graph, we see that $f(x) = 1$ at $x = -1$ and $x = 1$. As we approach 0 from 1 or from -1, the graph increases without bound. The lower limit of the range is 1, while there is no upper limit. Thus, on the domain $-1 \le x \le 1$, the range is $f(x) \ge 1$.

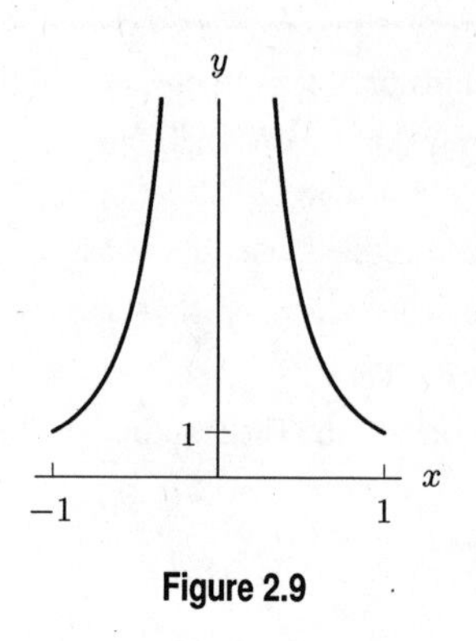

Figure 2.9

3. The graph of $f(x) = x^2 - 4$ for $-2 \le x \le 3$ is shown in Figure 2.10. From the graph, we see that $f(x) = 0$ at $x = -2$, that $f(x)$ decreases down to -4 at $x = 0$, and then increases to $f(x) = 3^2 - 4 = 5$ at $x = 3$. The minimum value of $f(x)$ is -4, while the maximum value is 5. Thus, on the domain $-2 \le x \le 3$, the range is $-4 \le f(x) \le 5$.

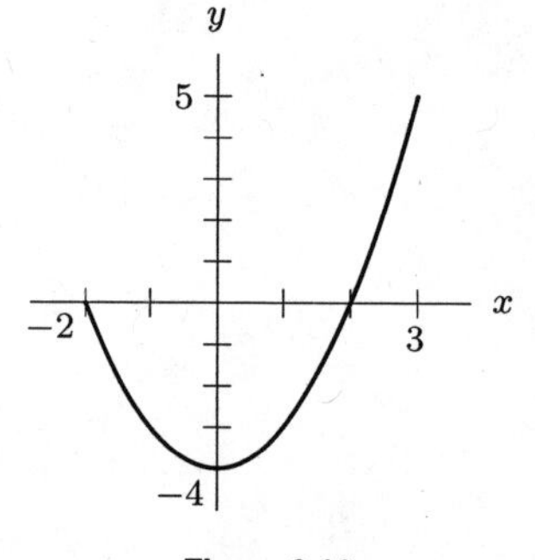

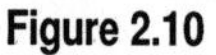

Figure 2.10

4. The graph of $f(x) = \sqrt{9 - x^2}$ for $-3 \le x \le 1$ is shown in Figure 2.11. From the graph, we see that $f(x) = 0$ at $x = -3$, and that $f(x)$ increases to a maximum value of 3 at $x = 0$, and then decreases to a value of $f(x) = \sqrt{9 - 1^2} \approx 2.83$ or $= 2\sqrt{2}$ at $x = 1$. Thus, on the domain $-3 \le x \le 1$, the range is $0 \le f(x) \le 3$.

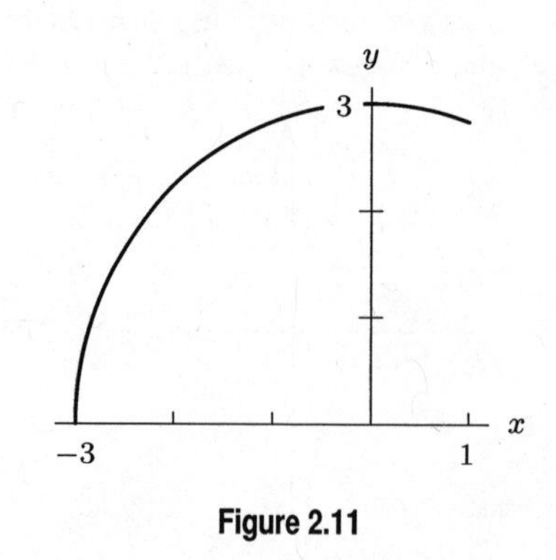

Figure 2.11

5. The domain is all real numbers except those which do not yield an output value. The expression $1/(x+3)$ is defined for any real number x except -3, since for $x = -3$ the denominator of $f(x)$, $x+3$, is 0 and division by 0 is undefined. Therefore, the domain of $f(x)$ is all real numbers $\neq -3$.

6. Since division by 0 is undefined, $p(t)$ is defined for any real number except for t, for which the denominator is equal to 0. Solving the equation $t^2 - 4 = 0$, we get $t = \pm 2$. So the domain is all real numbers $\neq \pm 2$.

7. $f(t)$ is not defined when $3t + 9 = 0$. So the domain is all real numbers $\neq -3$.

8. Since $q^4 + 2$ can not be equal to zero for any real number, the domain is all real numbers.

9. To evaluate $f(x)$, we must have $x - 4 > 0$. Thus

$$\text{Domain: } x > 4.$$

To find the range, we want to know all possible output values. We solve the equation $y = f(x)$ for x in terms of y. Since

$$y = \frac{1}{\sqrt{x-4}},$$

squaring gives

$$y^2 = \frac{1}{x-4},$$

and multiplying by $x - 4$ gives

$$\begin{aligned} y^2(x-4) &= 1 \\ y^2x - 4y^2 &= 1 \\ y^2x &= 1 + 4y^2 \\ x &= \frac{1+4y^2}{y^2}. \end{aligned}$$

This formula tells us how to find the x-value which corresponds to a given y-value. The formula works for any y except $y = 0$ (which puts a 0 in the denominator). We know that y must be positive, since $\sqrt{x-4}$ is positive, so we have

$$\text{Range: } y > 0.$$

10. $y(t)$ is not defined when $t^4 = 0$. So the domain is all real numbers $\neq 0$.

11. For $f(x)$ to be defined, the expression $x^2 - 4$, found inside the square root sign, must always be nonnegative. This happens when $x \geq 2$ or $x \leq -2$. So the domain is all real numbers x, such that $x \geq 2$ or $x \leq -2$.

12. We can take the cube root of any number, so the domain is all real numbers.

13. Any number can be squared, so the domain is all real numbers.

14. To evaluate $t(a)$, we must have $a - 2 \geq 0$. So the domain is all real numbers ≥ 2.

15. Since $m(q)$ is a linear function, the domain of $m(q)$ is all real numbers. For any value of $m(q)$ there is a corresponding value of q. So the range is also all real numbers.

16. The square root function is only defined for non-negative values, that is when $15 - 4x \geq 0$, thus we require that $x \leq 15/4$. The domain is $x \leq 15/4$. The values of the square root function are all non-negative, so the range is $f(x) \geq 0$.

17. $f(x)$ is not defined when $x = a$. So $a = 3$.

18. $p(t)$ is not defined when $2t - a = 0$ and $t + b = 0$. So $p(t)$ is not defined when $t = a/2$ and $t = -b$. A possible answer is $a/2 = 4$, namely $a = 8$ and $b = -5$.

19. $m(r)$ is defined when $r - a \geq 0$. So the domain of $m(r)$ is all real numbers $\geq a$. Hence $a = -3$.

20. $n(q)$ can be evaluated when $r^2 + a \geq 0$. Note that for all $a \geq 0$, $r^2 + a \geq 0$. Hence the domain of $n(q)$ is all real numbers for any $a \geq 0$.

Problems

21. The domain is $1 \leq x \leq 7$. The range is $2 \leq f(x) \leq 18$.

22. The domain is $2 \leq x \leq 6$. The range is $1 \leq f(x) \leq 3$.

23. One way to do this is to combine two operations, one of which forces x to be negative, the other of which forces x not to equal -5. One possibility is

$$y = \frac{1}{(x+5)\sqrt{-x}}.$$

The fraction's denominator must not equal 0, so x must not equal -5. The input, $-x$, of the square root function must not be negative, so $-x$ must be greater than or equal to zero, but being in the denominator, it cannot be zero.

24.
- A function such as $y = \sqrt{x+4}$ is undefined for $x < -4$, because the input of the square root operation is negative for these x-values.
- A function such as $y = 1/(x+2)$ is undefined for $x = -2$.
- Combining two functions such as these, for example by adding or multiplying them, yields a function with the required domain. Thus, possible formulas include

$$y = \frac{1}{x+2} + \sqrt{x+4} \quad \text{or} \quad y = \frac{\sqrt{x+4}}{x+2}.$$

25. Since the restaurant opens at 2 pm, $t = 0$, and closes at 2 am, $t = 12$, a reasonable domain is $0 \leq t \leq 12$.

Since there cannot be fewer than 0 clients in the restaurant and 200 can fit inside, the range is $0 \leq f(t) \leq 200$.

26. The domain is all possible input values, namely $t = 1, 2, 3, \ldots, 12$.

27. We know that the theater can hold anywhere from 0 to 200 people. Therefore the domain of the function is the integers, n, such that $0 \leq n \leq 200$.

We know that each person who enters the theater must pay \$4.00. Therefore, the theater makes $(0) \cdot (\$4.00) = 0$ dollars if there is no one in the theater, and $(200) \cdot (\$4.00) = \$800.00$ if the theater is completely filled. Thus the range of the function would be the integer multiples of 4 from 0 to 800. (That is, $0, 4, 8, \ldots$.)

The graph of this function is shown in Figure 2.12.

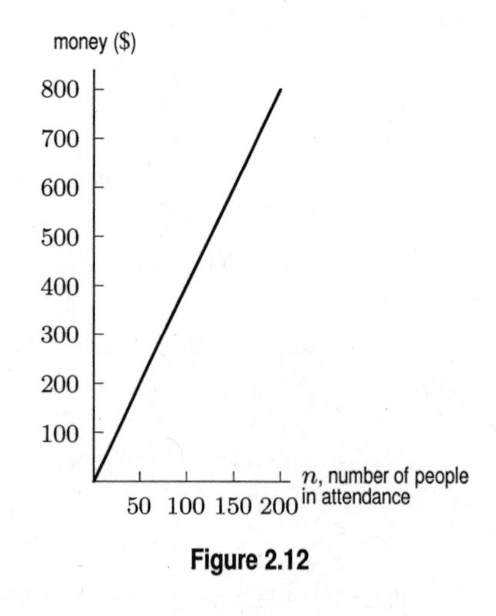

Figure 2.12

28. A possible graph of gas mileage (in miles per gallon, mpg) is shown in Figure 2.13. The function shown has a domain $0 \leq x \leq 120$ mph, as the car cannot have a negative speed and is not likely to go faster than 120 mph. The range of the function shown is $0 \leq y \leq 40$ mpg. A wide variety of other answers is possible.

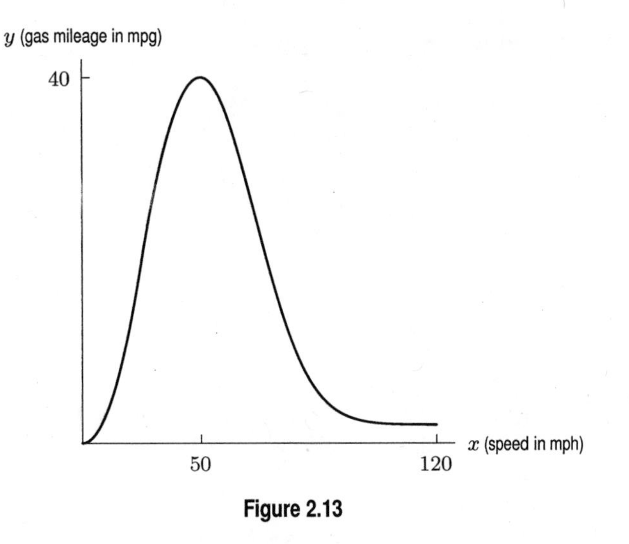

Figure 2.13

29. **(a)** From the table we find that a 200 lb person uses 5.4 calories per minute while walking. So a half-hour, or a 30 minute, walk burns $30(5.4) = 162$ calories.

(b) The number of calories used per minute is approximately proportional to the person's weight. The relationship is an approximately linear increasing function, where weight is the independent variable and number of calories burned is the dependent variable.

(c) (i) Since the function is approximately linear, its equation is $c = b + mw$, where c is the number of calories and w is weight. Using the first two values in the table, the slope is

$$m = \frac{3.2 - 2.7}{120 - 100} = \frac{0.5}{20} = 0.025 \text{ cal/lb.}$$

Using the point $(100, 2.7)$ we have

$$2.7 = b + 0.025(100)$$
$$b = 0.2.$$

So the equation is $c = 0.2 + 0.025w$. See Figure 2.14. All the values given lie on this line with the exception of the last two which are slightly above it.

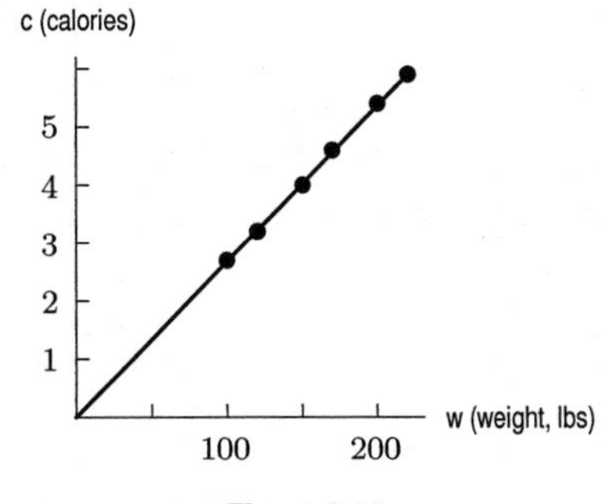

Figure 2.14

(ii) The intercept $(0, 0.2)$ is the number of calories burned by a weightless runner. Since 0.2 is a small number, most of the calories burned appear to be due to moving a person's weight. Other methods of finding the equation of the the line may give other values for the vertical intercept, but all values are close to 0.

(iii) Domain $0 < w$; range $0 < c$

(iv) Evaluating the function at $w = 135$,

$$\text{Calories} = 0.2 + 0.025(135) \approx 3.6.$$

30. We can substitute any positive number for x, which makes the domain $x > 0$. We note that as x approaches infinity $h(x)$, approaches zero, no matter the magnitude of a. If a is positive, then as x approaches zero, $h(x)$ approaches infinity. If a is negative, then as x approaches zero, $h(x)$ approaches negative infinity. Thus, the range is all positive numbers if $a > 0$ and all negative numbers if $a < 0$. Finally if $a = 0$, the range is just 0.

31. The function is defined for all values of x. Since $|x - b| \geq 0$, the range is all numbers greater than or equal to 6.

32. **(a)** We see that the 6th listing has a last digit of 9. Thus, $f(6) = 9$.

(b) The domain of the telephone directory function is $n = 1, 2, 3, \ldots, N$, where N is the total number of listings in the directory. We could find the value of N by counting the number of listings in the phone book.

(c) The range of this function is $d = 0, 1, 2, \ldots, 9$, because the last digit of any listing must be one of these numbers.

33. **(a)** Substituting $t = 0$ into the formula for $p(t)$ shows that $p(0) = 50$, meaning that there were 50 rabbits initially. Using a calculator, we see that $p(10) \approx 131$, which tells us there were about 131 rabbits after 10 months. Similarly, $p(50) \approx 911$ means there were about 911 rabbits after 50 months.

(b) The graph in Figure 2.15 tells us that the rabbit population grew quickly at first but then leveled off at about 1000 rabbits after around 75 months or so. It appears that the rabbit population increased until it reached the island's capacity.

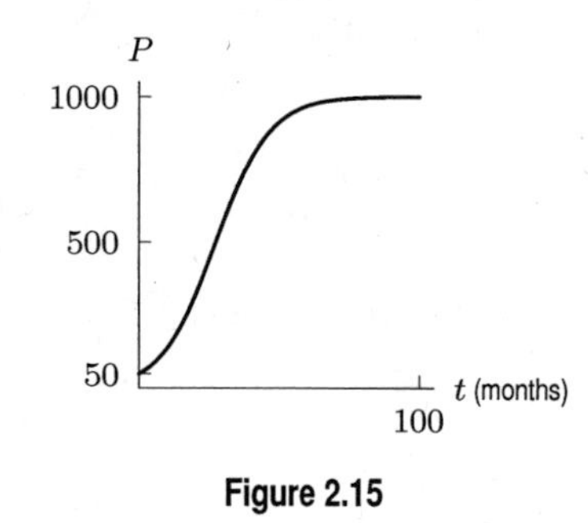

Figure 2.15

(c) From the graph in Figure 2.15, we see that the range is $50 \leq p(t) \leq 1000$. This tells us that (for $t \geq 0$) the number of rabbits is no less than 50 and no more than 1000.

(d) The smallest population occurred when $t = 0$. At that time, there were 50 rabbits. As t gets larger and larger, $(0.9)^t$ gets closer and closer to 0. Thus, as t increases, the denominator of

$$p(t) = \frac{1000}{1 + 19(0.9)^t}$$

decreases. As t increases, the denominator $1+19(0.9)^t$ gets close to 1 (try $t = 100$, for example). As the denominator gets closer to 1, the fraction gets closer to 1000. Thus, as t gets larger and larger, the population gets closer and closer to 1000. Thus, the range is $50 \leq p(t) < 1000$.

34. **(a)** We can add as much copper to our alloy as we like, so, since positive x-values represent quantities of added copper, x can be as big as we please. But, since the alloy starts off with only 3 kg of copper, we can remove no more than this. Therefore, the domain of f is $x \geq -3$.

For the range of f, note that the output of f is a percentage of copper. Since the alloy can contain no less than 0% copper (as would be the case if all 3 kg were removed), we see that $f(x)$ must be greater than (or equal to) 0%. On the other hand, no matter how much copper we add, the alloy will always contain 6 kg of tin. Thus, we can never obtain a pure, 100%-copper alloy. This means that if $y = f(x)$,

$$0\% \leq y < 100\%,$$

or

$$0 \leq y < 1.$$

(b) By definition, $f(x)$ is the percentage of copper in the bronze alloy after x kg of copper are added (or removed). We have

$$\begin{array}{c}\text{Percentage of copper}\\ \text{in the bronze alloy}\end{array} = \frac{\text{quantity of copper in the alloy}}{\text{total quantity of alloy}}.$$

Since x is the quantity of copper added or removed, this gives

$$f(x) = \frac{\text{initial quantity of copper} + x}{\text{initial quantity of alloy} + x},$$

and since the original 9 kg of alloy contained 3 kg of copper, we have

$$f(x) = \frac{3+x}{9+x}.$$

(c) If we think of the formula $f(x) = (3+x)/(9+x)$ as defining a function, but not as a model of an alloy of bronze, then the way we think about its domain and range changes. For example, we no longer need to ask, "Does this x-value make sense in the context of the model?" We need only ask "Is $f(x)$ algebraically defined for this value of x?" or "If we use this x-value for input, will there be a corresponding y-value as output?"

For the domain of f, we see that $y = (3+x)/(9+x)$ is defined for any x-value other than $x = -9$. Thus, the domain of f is any value of x such that $x \neq -9$.

To find the range of this function, we solve $y = f(x)$ for x in terms of y:

$$\begin{aligned}
y &= \frac{3+x}{9+x} \\
y(9+x) &= 3+x && \text{(multiply both sides by denominator)} \\
9y + xy &= 3+x && \text{(expand parentheses)} \\
xy - x &= 3 - 9y && \text{(collect all terms with } x \text{ at left)} \\
x(y-1) &= 3 - 9y && \text{(factor out } x\text{)} \\
x &= \frac{3-9y}{y-1} && \text{(divide by } y-1\text{).}
\end{aligned}$$

In solving for x, at the last step we had to divide by $y - 1$. This is valid if and only if $y \neq 1$, for otherwise we would be dividing by zero. There is no x-value resulting in a y-value of 1. Consequently, the range of f is any value of y such that $y \neq 1$.

Notice the difference between this situation and the situation where f is being used as a model for bronze.

35. **(a)** $r(0) = 800 - 40(0) = 800$ means water is entering the reservoir at 800 gallons per second at time $t = 0$. Since we don't know how much water was in the reservoir originally, this is not the amount of water in the reservoir.
$r(15) = 800 - 40(15) = 800 - 600 = 200$ means water is entering the reservoir at 200 gallons per second at time $t = 15$.
$r(25) = 800 - 40(25) = 800 - 1000 = -200$ means water is leaving the reservoir at 200 gallons per second at time $t = 25$.

(b) The intercepts occur at $(0, 800)$ and $(20, 0)$. The first tells us that the water is initially flowing in at the rate of 800 gallons per second. The other tells us that at 20 seconds, the flow has stopped.

The slope is $(800 - 0)/(0 - 20) = 800/-20 = -40$. This means that the rate at which water enters the reservoir decreases by 40 gallons per second each second. The water is flowing in at a decreasing rate.

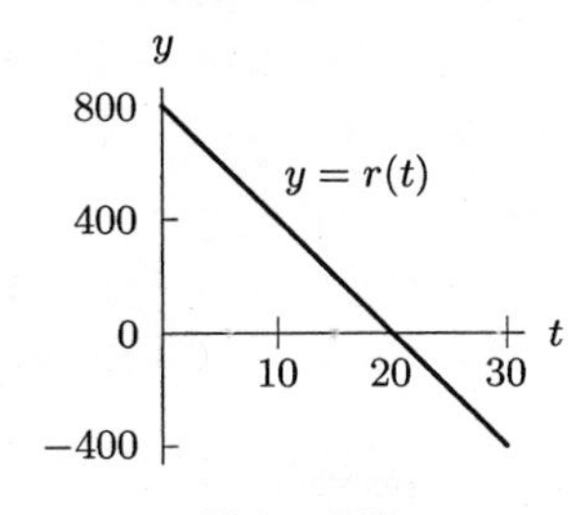

Figure 2.16

(c) The reservoir has more and more water when the rate is positive, because then water is being added. Water is being added until $t = 20$ when it starts flowing out. This means at $t = 20$, the most water is in the reservoir. The reservoir has water draining out between $t = 20$ and $t = 30$, but this amount is not as much as the water that entered between $t = 0$ and $t = 20$. Thus, the reservoir had the least amount of water at the beginning when $t = 0$. Remember the graph shows the rate of flow, not the amount of water in the reservoir.

(d) The domain is the number of seconds specified; $0 \le t \le 30$. The rate varies from 800 gallons per second at $t = 0$ to -400 gallons per second at $t = 30$, so the range is $-400 \le r(t) \le 800$.

36. (a) If $V = \pi r^2 h$ and $V = 355$, then $\pi r^2 h = 355$. So $h = (355)/(\pi r^2)$. Thus, since

$$A = 2\pi r^2 + 2\pi r h,$$

we have

$$A = 2\pi r^2 + 2\pi r\left(\frac{355}{\pi r^2}\right),$$

and

$$A = 2\pi r^2 + \frac{710}{r}.$$

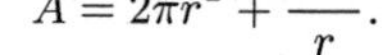

(b)

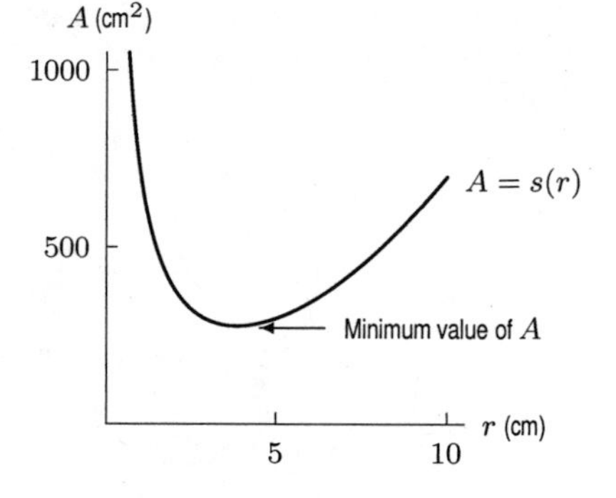

Figure 2.17: Graph of $A(r)$ for $0 < r \le 10$

(c) The domain is any positive value or $r > 0$, because (in practice) a cola can could have as large a radius as you wanted (it would just have to be very short to maintain its 12 oz size). From the graph in (b), the value of A is never less than about 277.5 cm^2. Thus, the range is $A > 277.5$ cm^2 (approximately).

(d) They need a little more than 277.5 cm^2 per can. The minimum A-value occurs (from graph) at $r \approx 3.83$ cm, and since $h = 355/\pi r^2$, $h \approx 7.7$ cm.

(e) Since the radius of a real cola can is less than the value required for the minimum value of A, it must use more aluminum than necessary. This is because the minimum value of A has $r \approx 3.83$ cm and $h \approx 7.7$ cm. Such a can has a diameter of $2r$ or 7.66 cm. This is roughly equal to its height—holding such a can would be difficult. Thus, real cans are made with slightly different dimensions.

Solutions for Section 2.3

Skill Refresher

S1. Since the point zero is not included, this graph represents $x > 0$.

S2. Since the point -3 is included, the graph represents $x \le -3$

S3. Since both end points of the interval are solid dots, this graph represents $x \ge 2$ and $x \le 3$. Combining these two inequalities we have $2 \le x \le 3$.

S4. Since the point zero is not included and the point 4 is included, this graph represents $x > 0$ and $x \le 4$. Combining these two we have $0 < x \le 4$.

S5. Since both end points of the interval are solid dots, this graph represents $x \le -1$ or $x \ge 2$.

S6. Since the point -1 is not included and the point 4 is included, this graph represents $x < -1$ or $x \ge 4$.

S7. Domain: $2 \le x < 6$ and Range: $3 \le x < 5$.

S8. Domain: $1 < x < 7$ and Range: $-8 < x < -1$.

S9. Domain: $-2 \le x \le 3$ and range: $-2 \le x \le 3$.

S10. Domain: $-2 \le x < 4$ and Range: $1 \le x \le 5$.

Exercises

1. $f(x) = \begin{cases} -1, & -1 \le x < 0 \\ 0, & 0 \le x < 1 \\ 1, & 1 \le x < 2 \end{cases}$ is shown in Figure 2.18.

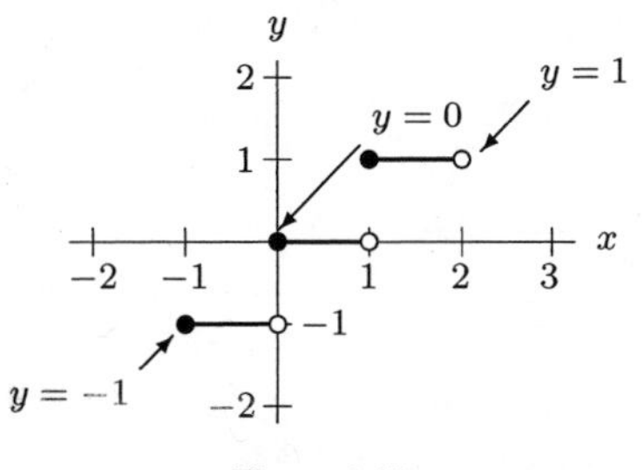

Figure 2.18

2. $f(x) = \begin{cases} x+1, & -2 \le x < 0 \\ x-1, & 0 \le x < 2 \\ x-3, & 2 \le x < 4 \end{cases}$ is shown in Figure 2.19.

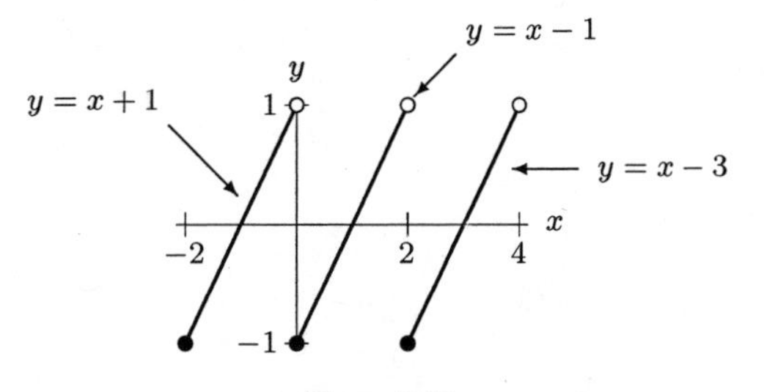

Figure 2.19

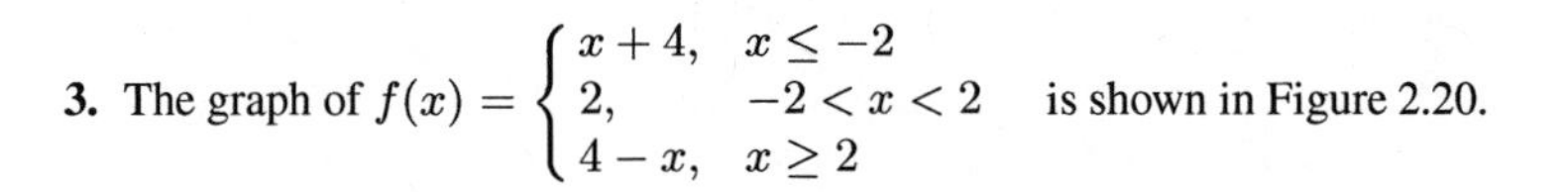

3. The graph of $f(x) = \begin{cases} x+4, & x \le -2 \\ 2, & -2 < x < 2 \\ 4-x, & x \ge 2 \end{cases}$ is shown in Figure 2.20.

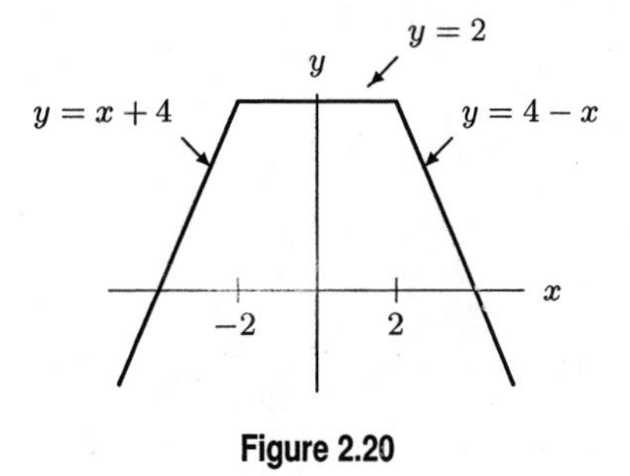

Figure 2.20

4. The graph of $f(x) = \begin{cases} x^2, & x \le 0 \\ \sqrt{x}, & 0 < x < 4 \\ x/2, & x \ge 4 \end{cases}$ is shown in Figure 2.21.

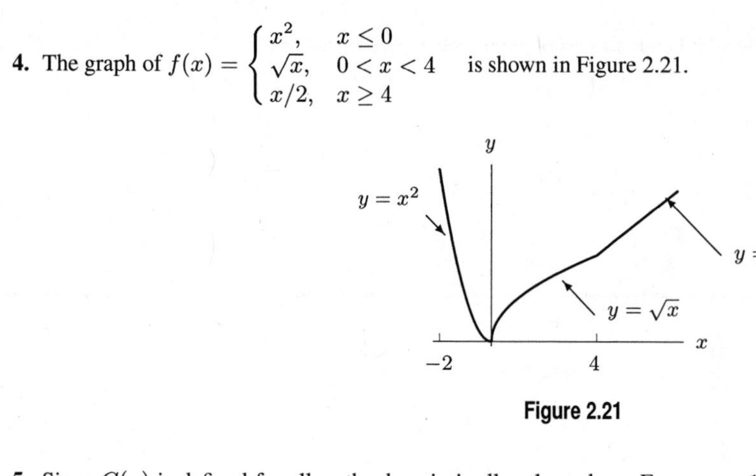

Figure 2.21

5. Since $G(x)$ is defined for all x, the domain is all real numbers. For $x < -1$ the values of the function are all negative numbers. For $-1 \ge x \ge 0$ the functions values are $4 \ge G(x) \ge 3$, while for $x > 0$ we see that $G(x) \ge 3$ and the values increase to infinity. The range is $G(x) < 0$ and $G(x) \ge 3$.

6. Since $F(x)$ is defined for all x, the domain is all real numbers. For $x \le 1$ the values of the function are $-\infty \le F(x) \le 1$, while for $x > 1$ we see that $0 < F(x) < 1$, so the range is $F(x) \le 1$.

7. We find the formulas for each of the lines. For the first, we use the two points we have, $(1, 4)$ and $(3, 2)$. We find the slope: $(2 - 4)/(3 - 1) = -1$. Using the slope of -1, we solve for the y-intercept:

$$\begin{aligned} 4 &= b - 1 \cdot 1 \\ 5 &= b. \end{aligned}$$

Thus, the first line is $y = 5 - x$, and it is for the part of the function where $x < 3$. Notice that we do not use this formula for the value $x = 3$.

We follow the same method for the second line, using the points $(3, \frac{1}{2})$ and $(5, \frac{3}{2})$. We find the slope: $(\frac{3}{2} - \frac{1}{2})/(5 - 3) = \frac{1}{2}$. Using the slope of $\frac{1}{2}$, we solve for the y-intercept:

$$\begin{aligned} \frac{1}{2} &= b + \frac{1}{2} \cdot 3 \\ -1 &= b. \end{aligned}$$

Thus, the second line is $y = -1 + \frac{1}{2}x$, and it is for the part of the function where $x \ge 3$.

Therefore, the function is:

$$y = \begin{cases} 5 - x & \text{for } x < 3 \\ -1 + \frac{1}{2}x & \text{for } x \ge 3. \end{cases}$$

8. We find the formulas for each of the lines. For the first, we use the two points we have, $(1, 6.5)$ and $(3, 5.5)$. We find the slope: $(5.5 - 6.5)/(3 - 1) = -\frac{1}{2}$. Using the slope of $-\frac{1}{2}$, we solve for the y-intercept:

$$\begin{aligned} 6.5 &= b - \frac{1}{2} \cdot 1 \\ 7 &= b. \end{aligned}$$

Thus, the first line is $y = 7 - \frac{1}{2}x$, and it is for the part of the function where $x \le 3$.

We follow the same method for the second line, using the points $(3, 2)$ and $(5, 2)$. Noting that the y values are the same, we know the slope is zero and that the line is $y = 2$ for the part of the function where $3 < x \le 5$.

We follow the same method for the third line, using the points $(5, 7)$ and $(7, 3)$. We find the slope: $(3 - 7)/(7 - 5) = -2$. Using the slope of -2, we solve for the y-intercept:

$$\begin{aligned} 3 &= b - 2 \cdot 7 \\ 17 &= b. \end{aligned}$$

Thus, the second line is $y = 17 - 2x$, and it is for the part of the function where $x > 5$.

Therefore, the function is:

$$y = \begin{cases} 7 - \frac{1}{2}x & \text{for } x \leq 3 \\ 2 & \text{for } 3 < x \leq 5 \\ 17 - 2x & \text{for } x > 5. \end{cases}$$

9. We find the formulas for each of the lines. For the first, we use the two points we have, $(1, 3.5)$ and $(3, 2.5)$. We find the slope: $(2.5 - 3.5)/(3 - 1) = -\frac{1}{2}$. Using the slope of $-\frac{1}{2}$, we solve for the y-intercept:

$$\begin{aligned} 3.5 &= b - \frac{1}{2} \cdot 1 \\ 4 &= b. \end{aligned}$$

Thus, the first line is $y = 4 - \frac{1}{2}x$, and it is for the part of the function where $1 \leq x \leq 3$.

We follow the same method for the second line, using the points $(5, 1)$ and $(8, 7)$. We find the slope: $(7-1)/(8-5) = 2$. Using the slope of 2, we solve for the y-intercept:

$$\begin{aligned} 1 &= b + 2 \cdot 5 \\ -9 &= b. \end{aligned}$$

Thus, the second line is $y = -9 + 2x$, and it is for the part of the function where $5 \leq x \leq 8$.

Therefore, the function is:

$$y = \begin{cases} 4 - \frac{1}{2}x & \text{for } 1 \leq x \leq 3 \\ -9 + 2x & \text{for } 5 \leq x \leq 8. \end{cases}$$

10. We find the formulas for each of the lines. For the first, we use the two points we have, $(1, 4)$ and $(3, 2)$. We find the slope: $(2 - 4)/(3 - 1) = -1$. Using the slope of -1, we solve for the y-intercept:

$$\begin{aligned} 4 &= b - 1 \cdot 1 \\ 5 &= b. \end{aligned}$$

Thus, the first line is $y = 5 - x$, and it is for the part of the function where $x \leq 3$.

We follow the same method for the second line, using the points $(3, 2)$ and $(5, 4)$. We find the slope: $(4-2)/(5-3) = 1$. Using the slope of 1, we solve for the y-intercept:

$$\begin{aligned} 4 &= b + 1 \cdot 5 \\ -1 &= b. \end{aligned}$$

Thus, the second line is $y = -1 + x$, and it is for the part of the function where $x \geq 3$.

Therefore, the function is:

$$y = \begin{cases} 5 - x & \text{for } x \leq 3 \\ -1 + x & \text{for } x \geq 3. \end{cases}$$

Notice that the value of $y = 2$ at $x = 3$ can be obtained from either formula.

Alternatively, this is the graph of the absolute value function,

$$y = |x - 3| + 2.$$

Problems

11. **(a)** Yes, because every value of x is associated with exactly one value of y.

(b) No, because some values of y are associated with more than one value of x.

(c) $y = 1, 2, 3, 4$.

12. **(a)** Figures 2.23 and 2.22 show the two functions x and $\sqrt{x^2}$. Because the two functions do not coincide for $x < 0$, they cannot be equal. The graph of $\sqrt{x^2}$ looks like the graph of $|x|$.

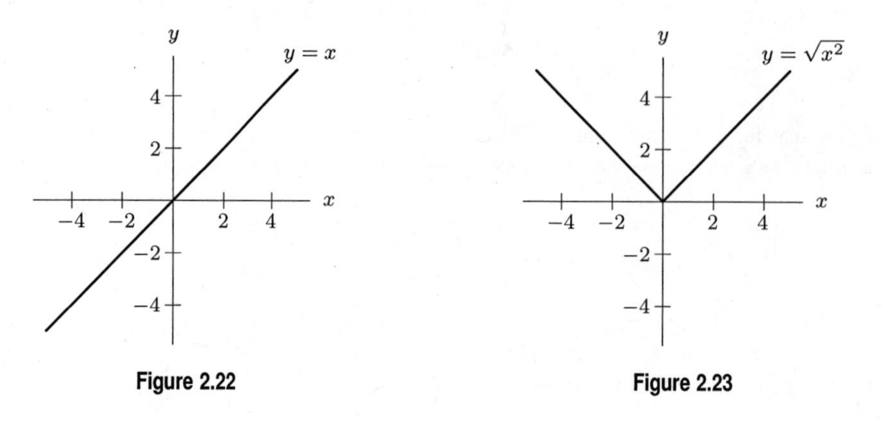

Figure 2.22

Figure 2.23

(b) Table 2.4 is the complete table. Because the two functions do not coincide for $x < 0$ they cannot be equal. The table for $\sqrt{x^2}$ looks like a table for of $|x|$.

Table 2.4

x	-5	-4	-3	-2	-1	0	1	2	3	4	5
$\sqrt{x^2}$	5	4	3	2	1	0	1	2	3	4	5

(c) If $x > 0$, then $\sqrt{x^2} = x$, whereas if $x < 0$ then $\sqrt{x^2} = -x$. This is the definition of $|x|$. Thus we have shown $\sqrt{x^2} = |x|$.

(d) We see nothing because $\sqrt{x^2} - |x| = 0$, and the graphing calculator or computer has drawn a horizontal line on top of the x-axis.

13. **(a)** Figure 2.24 shows the function $u(x)$. Some graphing calculators or computers may show a near vertical line close to the origin. The function seems to be -1 when $x < 0$ and 1 when $x > 0$.

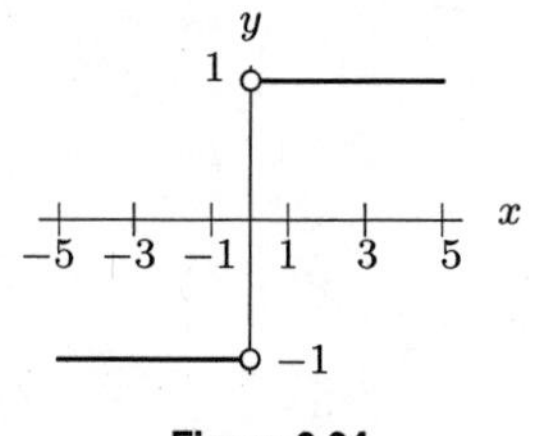

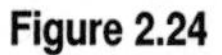
Figure 2.24

(b) Table 2.5 is the completed table. It agrees with what we found in part (a). The function is undefined at $x = 0$.

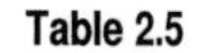
Table 2.5

x	-5	-4	-3	-2	-1	0	1	2	3	4	5
$\|x\|/x$	-1	-1	-1	-1	-1		1	1	1	1	1

(c) The domain is all x except $x = 0$. The range is -1 and 1.

(d) $u(0)$ is undefined, not 0. The claim is false.

14. (a) Upon entry, the cost is \$2.50. The tax surcharge of \$0.50 is added to the fare. So, the initial cost will be \$3.00. The cost for the first $1/5$ mile adds \$0.40, giving a fare of \$3.40. For a journey of $2/5$ mile, another \$0.40 is added for a fare of \$3.80. Each additional $1/5$ mile gives an another increment of \$0.40. See Table 2.6.

Table 2.6

Miles	0	0.2	0.4	0.6	0.8	1	1.2	1.4	1.6	1.8	2
Cost	3.00	3.40	3.80	4.20	4.60	5.00	5.40	5.80	6.20	6.60	7.00

(b) The table shows that the cost for a 1.2 mile trip is \$5.40.
(c) From the table, the maximum distance one can travel for 5.80 is 1.4 mile.
(d) See Figure 2.25

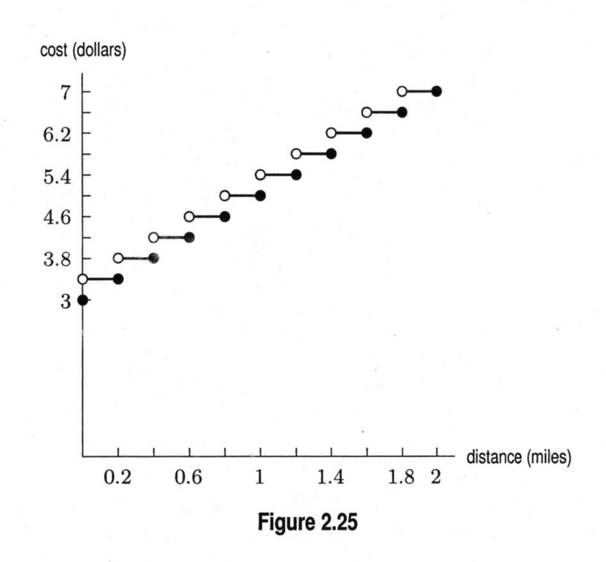

Figure 2.25

15. (a) The dots in Figure 2.26 represent the graph of the function.

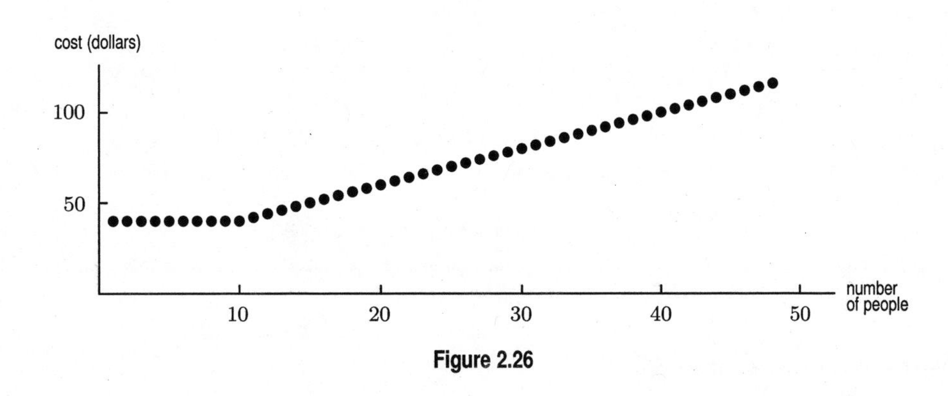

Figure 2.26

(b) Since admission is charged for whole numbers of people between 1 and 50, the domain is the integers from 1 to 50. The minimum cost is \$40. The maximum occurs for 50 people and is $\$40 + 40(\$2) = \$120$. Since the lowest cost is \$40, and each additional person costs \$2, the range only includes numbers which are multiples of 2. Thus, the range is all the even integers from 40 to 120.

16. **(a)** Let $y = f(x)$ be the cost of a stripping and refinishing job for a floor which is x square feet in area. When the area is less than or equal to 150 square feet, the price is \$1.83 times the number of square feet. Thus, for x-values up through 150, we have $f(x) = 1.83x$. However, if the area is more than 150 square feet, the extra cost of toxic waste disposal is added, giving $f(x) = 1.83x + 350$. The maximum total area for a job is 1000 square feet, so the formula is

$$f(x) = \begin{cases} 1.83x, & 0 \le x \le 150 \\ 1.83x + 350, & 150 < x \le 1000 \end{cases}$$

(b) The graph is in Figure 2.27. Note that when $x = 150$ sq ft, $y = 1.83(150) = \$274.5$. When x goes above 150 sq ft, the cost jumps by \$350 to \$624.5.

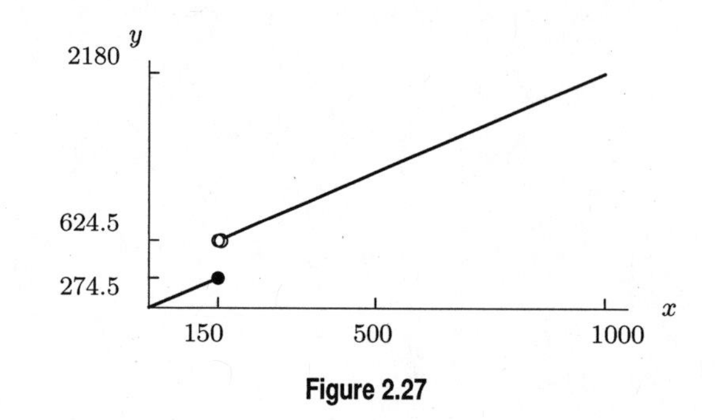

Figure 2.27

No floor has a negative area and the company will refinish any floor whose area is less than or equal to 1000 square feet, so

$$\text{Domain is} \quad 0 \le x \le 1000.$$

As the size of the floor gets bigger, the cost increases. The smallest value of the range occurs when $x = 0$ and the largest value occurs when $x = 1000$. So the smallest value is $f(0) = 0$ and the largest is $f(1000) = 2180$. There is a gap, though, in the values of the range. The value of $f(x)$ jumps from 274.5, when $x = 150$, to more than 624.5 when x is just slightly more than 150. Putting all these pieces together, we have

$$\text{Range is} \quad 0 \le y \le 274.5 \text{ or } 624.5 < y \le 2180.$$

17. **(a)** The smaller the difference, the smaller the refund. The smallest possible difference is \$0.01. This translates into a refund of $\$1.00 + \$0.01 = \$1.01$.

(b) Looking at the refund rules, we see that there are three separate cases to consider. The first case is when 10 times the difference is less than \$1. If the difference is more than 0 but less than 10¢, and you will receive \$1 plus the difference. The formula for this is:

$$y = 1 + x \quad \text{for} \quad 0 < x < 0.10.$$

In the second case, 10 times the difference is between \$1 and \$5. This will be true if the difference is between 10¢and 50¢. The formula for this is:

$$y = 10x + x \quad \text{for} \quad 0.10 \le x \le 0.50.$$

In the third case, 10 times the difference is more than \$5. If the difference is more than 50¢, then you receive \$5 plus the difference or:

$$y = 5 + x \quad \text{for} \quad x > 0.50.$$

Putting these cases together, we get:

$$y = \begin{cases} 1 + x & \text{for } 0 < x < 0.1 \\ 10x + x & \text{for } 0.1 \le x \le 0.5 \\ 5 + x & \text{for } x > 0.5. \end{cases}$$

(c) We want x such that $y = 9$. Since the highest possible value of y for the first case occurs when $x = 0.09$, and $y = 1 + 0.09 = \$1.09$, the range for this case does not go high enough. The highest possible value for the second case occurs when $x = 0.5$, and $y = 10(0.5) + 0.5 = \$5.50$. This range is also not high enough. So we look to the third case where $x > 0.5$ and $y = 5 + x$. Solving $5 + x = 9$ we find $x = 4$. So the price difference would have to be \$4.

(d) See Figure 2.28.

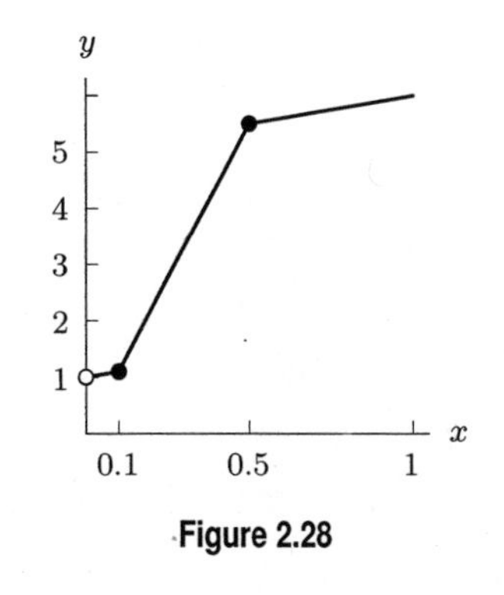

Figure 2.28

18. **(a)** See Table 2.7. Note that the cost is in fractions of a penny, so answers are rounded to the nearest penny.

Table 2.7

kWh	0	5	10	15	20	25	30	35	40
Cost in dollars	0.12	0.35	0.57	1.05	1.53	2.01	2.48	2.96	3.44

(b) In dollars, $C(n) = \begin{cases} 0.1157 + 0.0459n, & 0 \le n \le 10 \\ 0.5747 + 0.0955(n - 10), & n > 10 \end{cases}$

(c) The formula gives a cost of \$2.77 for 33 kWh usage.

(d) Since 33 kWh cost \$2.77, we use the formula for $n > 10$ and solve

$$3 = 0.5747 + 0.0955(n - 10),$$

giving $n = 35.396$.

19. **(a)** Figure 2.29 shows the rates for the first and last periods of the year. Figure 2.31 shows the rates for holiday periods (Dec 25-Jan 3, Jan16-18, Feb 3-21) and Figure 2.30 shows the rates for the other times.

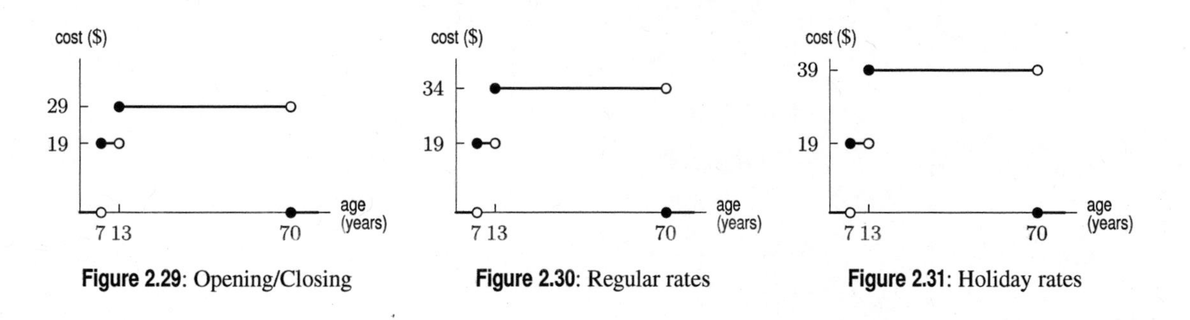

Figure 2.29: Opening/Closing

Figure 2.30: Regular rates

Figure 2.31: Holiday rates

(b) Ages 13-69.

(c) See Figure 2.32.

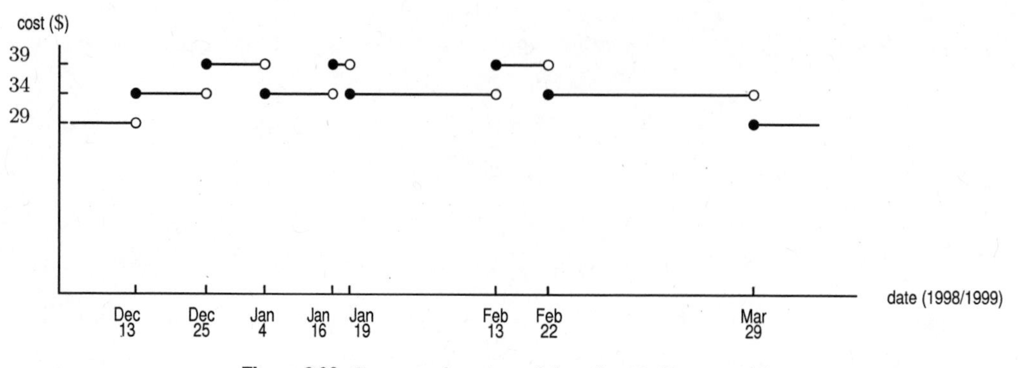

Figure 2.32: Cost as a function of date for 13-69 years old

(d) Rates through 12 December and after 19 March represent early-season and late-season rates, respectively; these are off-peak rates; it makes economic sense to cut rates when there are fewer skiers. Holiday rates took effect from 25 December through 3 January because of the Christmas/New Year's holiday; they took effect from 16 January through 18 January for Martin Luther King's holiday; they took effect from 13 February through 21 February for the Presidents' Week holiday; it makes economic sense to charge over peak rates during the holidays, as more skiers are available to use the facility. Other times represent rates during the heart of the winter skiing season; these are the regular rates.

20. **(a)** We have $f(x) = \begin{cases} x^2 - 4 & \text{for } x^2 - 4 \geq 0 \\ -(x^2 - 4) & \text{for } x^2 - 4 < 0 \end{cases}$

or

$$f(x) = \begin{cases} x^2 - 4 & \text{for } x \leq -2 \\ 4 - x^2 & \text{for } -2 < x < 2 \\ x^2 - 4 & \text{for } x \geq 2 \end{cases}.$$

(b) See Figure 2.33.

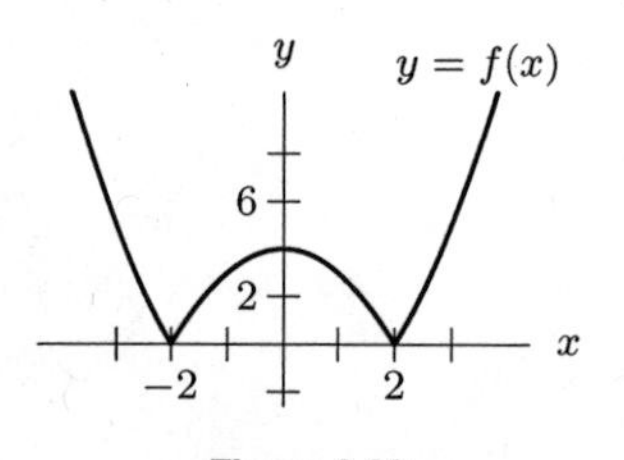

Figure 2.33

21. **(a)** We have $f(x) = \begin{cases} 2x - 6 & \text{for } 2x - 6 \geq 0 \\ -(2x - 6) & \text{for } 2x - 6 < 0 \end{cases}$

or

$$f(x) = \begin{cases} 2x - 6 & \text{for } x \geq 3 \\ 6 - 2x & \text{for } x < 3 \end{cases}.$$

(b) See Figure 2.34.

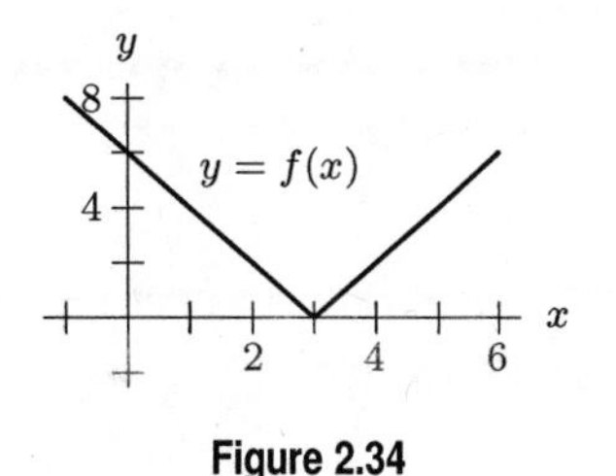

Figure 2.34

22. **(a)** We have $f(-2) = \dfrac{1}{-2} = -\dfrac{1}{2}$.

(b) We have $f(2) = \sqrt{2}$.

(c) For $x \geq -1$, $f(x) \geq 0$, because $\sqrt{x}$ and x^2 are always non-negative. For $x < -1$, $1/x$ is between -1 and 0, thus the range is $f(x) > -1$.

23. **(a)** Since zero lies in the interval $-1 \leq x \leq 1$, we find the function value from the formula $f(x) = 3x$. This gives $f(0) = 3 \cdot 0 = 0$. To find $f(3)$, we first note that $x = 3$ lies in the interval $1 < x \leq 5$, so we find the function value from the formula $f(x) = -x + 4$. The result is $f(3) = -3 + 4 = 1$.

(b) By graphing f we can see in Figure 2.35 that the combined domain is $-1 \leq x \leq 5$ and the range is $-3 \leq f(x) \leq 3$.

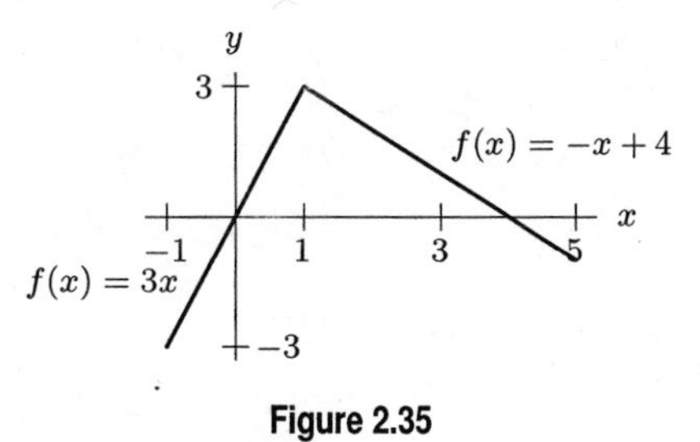

Figure 2.35

24. **(a)** Since $-2 < 0$, we choose the formula $g(x) = -1$, getting $g(-2) = -1$. Since $2 \geq 0$, we choose the formula $g(x) = x^3$, getting $g(2) = 2^3 = 8$. The value $x = 0$ is the place on the x-axis, where we switch from one formula to the other. But the equal sign is attached to the formula for $g(x) = x^3$, so $g(0) = 0^3 = 0$.

(b) By graphing g we can see in Figure 2.36 that the domain includes the entire x-axis. On the vertical axis the only negative value is -1, so the range is $g(x) \geq 0$ and $g(x) = -1$.

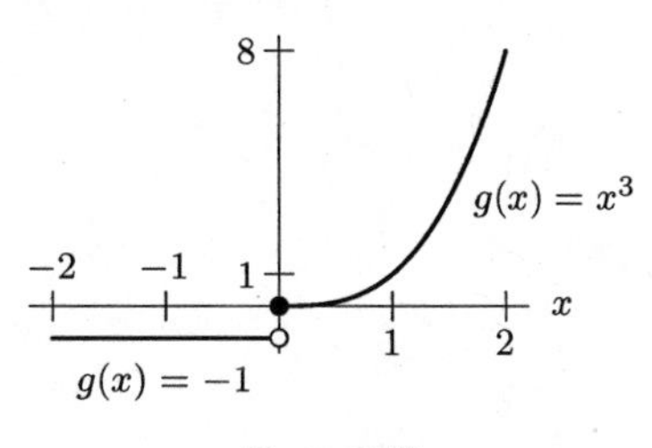

Figure 2.36

25. The statement

$$y = x^2 \text{ for } x \text{ less than zero, and } y = x - 1 \text{ for } x \text{ greater than or equal to zero}$$

can be condensed to $y = \begin{cases} x^2 & \text{for } x < 0 \\ x - 1 & \text{for } x \geq 0 \end{cases}$

Solutions for Section 2.4

Skill Refresher

S1. Adding 4 to both sides and dividing by 3 we get $y = \frac{x+4}{3}$.

S2. First we subtract $4x$ from both sides

$$\begin{aligned} 4x - 3y - 4x &= 7 - 4x \\ -3y &= 7 - 4x \\ y &= \frac{7 - 4x}{-3}. \end{aligned}$$

S3. First we multiply both sides by $y - 2$ and get $x(y - 2) = 2y + 1$.

$$\begin{aligned} xy - 2x &= 2y + 1 \\ xy - 2y &= 2x + 1 \\ y(x - 2) &= 2x + 1 \\ y &= \frac{2x + 1}{x - 2}. \end{aligned}$$

S4. Adding 2 to both sides we get $\sqrt{y} = x + 2$. Squaring both sides, we get

$$y = (x + 2)^2 = x^2 + 4x + 4.$$

S5. Add 4 to both sides and take the cube root to obtain $y = \sqrt[3]{x + 4}$.

S6. Add $(2y)^3$ to both sides and subtract x from both sides to get

$$\begin{aligned} (2y)^3 &= 5 - x \\ 2y &= \sqrt[3]{5 - x} \\ y &= \frac{\sqrt[3]{5 - x}}{2}. \end{aligned}$$

S7.

$$\begin{aligned} 5\left(\frac{1}{5}x - 1\right) + 5 &= x - 5 + 5 \\ &= x. \end{aligned}$$

S8.

$$4x^2 + 4x + 1 - 4 = 4x^2 + 4x - 3.$$

S9. We have

$$\begin{aligned} 3(y - 2)^2 - 7 &= 3(y^2 - 4y + 4) - 7 \\ &= 3y^2 - 12y + 12 - 7 \\ &= 3y^2 - 12y + 5. \end{aligned}$$

S10. We have $(1 - t)^2 - (1 - t) = 1 - 2t + t^2 - 1 + t = t^2 - t$.

Exercises

1. $A(f(t))$ is the area, in square centimeters, of the circle at time t minutes.

2. $R(f(p))$ is the revenue, in millions of dollars, when the price of oil is p dollars/barrel.

3. $C(A(d))$ is the price of a pizza with diameter d.

4. $f(g(0)) = f(1 - 0^2) = f(1) = 3 \cdot 1 - 1 = 2$.

5. $g(f(0)) = g(3 \cdot 0 - 1) = g(-1) = 1 - (-1)^2 = 0$.

6. $g(f(2)) = g(3 \cdot 2 - 1) = g(5) = 1 - (5)^2 = -24$.

7. $f(g(2)) = f(1 - 2^2) = f(-3) = 3(-3) - 1 = -10$.

8. $f(g(x)) = f(1 - x^2) = 3(1 - x^2) - 1 = 2 - 3x^2$.

9. $f(f(x)) = f(3x - 1) = 3(3x - 1) - 1 = 9x - 4$.

10. $g(g(x)) = g(1 - x^2) = 1 - (1 - x^2)^2 = 2x^2 - x^4$.

11. The inverse function, $f^{-1}(P)$, gives the year in which population is P million. Units of $f^{-1}(P)$ are years.

12. The inverse function, $f^{-1}(T)$, gives the temperature in °F needed if the cake is to bake in T minutes. Units of $f^{-1}(T)$ are °F.

13. The inverse function, $f^{-1}(N)$, is the number of days for N inches of snow to fall. Units of $f^{-1}(N)$ are days.

14. The inverse function, $C^{-1}(x)$, gives the weight in ounces of x calories of almonds.

15. The inverse function, $P^{-1}(c)$, gives the diameter in inches of a pizza costing c dollars.

16. Since $y = 2t + 3$, solving for t gives

$$\begin{aligned} 2t + 3 &= y \\ t &= \frac{y-3}{2} \\ f^{-1}(y) &= \frac{y-3}{2}. \end{aligned}$$

17. Since $Q = x^3 + 3$, solving for x gives

$$\begin{aligned} x^3 + 3 &= Q \\ x^3 &= Q - 3 \\ x &= (Q-3)^{1/3} \\ f^{-1}(Q) &= (Q-3)^{1/3} \end{aligned}$$

18. Since $A = \pi r^2$, solving for r gives

$$\begin{aligned} \frac{A}{\pi} &= r^2 \\ \sqrt{\frac{A}{\pi}} &= r \\ r = f^{-1}(A) &= \sqrt{\frac{A}{\pi}}. \end{aligned}$$

19. Since $y = 1 + \frac{1}{s}$, solving for s gives

$$y - 1 = \frac{1}{s}$$

$$\frac{1}{y-1} = s$$
$$g^{-1}(y) = \frac{1}{y-1}.$$

20. **(a)** Since the vertical intercept of the graph of f is $(0, 1.5)$, we have $f(0) = 1.5$.
(b) Since the horizontal intercept of the graph of f is $(2.2, 0)$, we have $f(2.2) = 0$.
(c) The function f^{-1} goes from y-values to x-values, so to evaluate $f^{-1}(0)$, we want the x-value corresponding to $y = 0$. This is $x = 2.2$, so $f^{-1}(0) = 2.2$.
(d) Solving $f^{-1}(?) = 0$ means finding the y-value corresponding to $x = 0$. This is $y = 1.5$, so $f^{-1}(1.5) = 0$.

21. **(a)** Since the vertical intercept of the graph of f is $(0, b)$, we have $f(0) = b$.
(b) Since the horizontal intercept of the graph of f is $(a, 0)$, we have $f(a) = 0$.
(c) The function f^{-1} goes from y-values to x-values, so to evaluate $f^{-1}(0)$, we want the x-value corresponding to $y = 0$. This is $x = a$, so $f^{-1}(0) = a$.
(d) Solving $f^{-1}(?) = 0$ means finding the y-value corresponding to $x = 0$. This is $y = b$, so $f^{-1}(b) = 0$.

Problems

22. Since
$$n = \frac{A}{250},$$
solving for A gives
$$A = 250n.$$
Thus, $A = f^{-1}(n) = 250n$.

23. Since $n = f(A)$, in $f(100)$ we have $A = 100$ ft^2. Evaluating $f(100)$ tells us how much paint is needed for 100 ft^2. Since
$$n = f(100) = \frac{100}{250} = 0.4,$$
it takes 0.4 gallon of paint to cover 100 ft^2.

In $f^{-1}(100)$, the 100 is the number of gallons, so $f^{-1}(100)$ represents the area which can be painted by 100 gallons:
$$A = f^{-1}(100) = 250 \cdot 100 = 25{,}000 \text{ ft}^2.$$

24. Since $f(A) = A/250$ and $f^{-1}(n) = 250n$, we have
$$f^{-1}(f(A)) = f^{-1}\left(\frac{A}{250}\right) = 250\frac{A}{250} = A.$$
$$f(f^{-1}(n)) = f(250n) = \frac{250n}{250} = n.$$
To interpret these results, we use the fact that $f(A)$ gives the number of gallons of paint needed to cover an area A, and $f^{-1}(n)$ gives the area covered by n gallons. Thus $f^{-1}(f(A))$ gives the area which can be covered by $f(A)$ gallons; that is, A square feet. Similarly, $f(f^{-1}(n))$ gives the number of gallons needed for an area of $f^{-1}(n)$; that is, n gallons.

25. **(a)** To find values of f, read the table from top to bottom, so
(i) $f(0) = 2$ (ii) $f(1) = 0$.
To find values of f^{-1}, read the table in the opposite direction (from bottom to top), so
(iii) $f^{-1}(0) = 1$ (iv) $f^{-1}(2) = 0$.
(b) Since the values $(0, 2)$ are paired in the table, we know $f(0) = 2$ and $f^{-1}(2) = 0$. Thus, knowing the answer to (i) (namely, $f(0) = 2$) tells us the answer to (iv). Similarly, the answer to (ii), namely $f(1) = 0$, tells us that the values $(1, 0)$ are paired in the table, so $f^{-1}(0) = 1$ too.

26. We solve the equation $C = g(x) = 600 + 45x$ for x. Subtract 600 from both sides and divide both sides by 45 to get

$$x = \frac{1}{45}(C - 600).$$

So

$$g^{-1}(C) = \frac{1}{45}(C - 600).$$

27. **(a)** The cost of producing 5000 loaves is \$653.
(b) $C^{-1}(80)$ is the number of loaves of bread that can be made for \$80, namely 0.62 thousand or 620.
(c) The solution is $q = 6.3$ thousand. It costs \$790 to make 6300 loaves.
(d) The solution is $x = 150$ dollars, so 1.2 thousand, or 1200, loaves can be made for \$150.

28. **(a)** $f(10) = 100 + 0.2 \cdot 10 = 102$ thousand dollars, the cost of producing 10 kg of the chemical.
(b) $f^{-1}(200)$ is the quantity of the chemical which can be produced for 200 thousand dollars. Since

$$\begin{aligned} 200 &= 100 + 0.2q \\ 0.2q &= 100 \\ q &= \frac{100}{0.2} = 500 \text{ kg}, \end{aligned}$$

we have $f^{-1}(200) = 500$.
(c) To find $f^{-1}(C)$, solve for q:

$$\begin{aligned} C &= 100 + 0.2q \\ 0.2q &= C - 100 \\ q &= \frac{C}{0.2} - \frac{100}{0.2} = 5C - 500 \\ f^{-1}(C) &= 5C - 500. \end{aligned}$$

29. **(a)** $f(3) = 4 \cdot 3 = 12$ is the perimeter of a square of side 3.
(b) $f^{-1}(20)$ is the side of a square of perimeter 20. If $20 = 4s$, then $s = 5$, so $f^{-1}(20) = 5$.
(c) To find $f^{-1}(P)$, solve for s:

$$\begin{aligned} P &= 4s \\ s &= \frac{P}{4} \\ f^{-1}(P) &= \frac{P}{4}. \end{aligned}$$

30. **(a)** $G(15)$ is the output corresponding to the input of $t = 15$. So $G(15)$ represents the GDP fifteen years after 1990. This tells us that, in 2005, the gross domestic product was \$12,638.4 billion.
(b) The input to the G^{-1} function is billions of dollars, so its output is a time in years after 1990. Thus, $G^{-1}(14, 441.4) = 18$ tells us that, eighteen years after 1990, the GDP was 14,441.4 billion dollars. Thus, the GDP was \$14,441.4 billion in 2008.

31. We can find the inverse function by solving for t in our equation:

$$\begin{aligned} H &= \frac{5}{9}(t - 32) \\ \frac{9}{5}H &= t - 32 \\ \frac{9}{5}H + 32 &= t. \end{aligned}$$

This function gives us the temperature in degrees Fahrenheit if we know the temperature in degrees Celsius.

32. **(a)** We substitute zero into the function, giving:

$$H = f(0) = \frac{5}{9}(0 - 32) = -\frac{160}{9} = -17.778.$$

This means that zero degrees Fahrenheit is about -18 degrees Celsius.

(b) In Exercise 31, we found the inverse function. Using it with $H = 0$, we have:

$$t = f^{-1}(0) = \frac{9}{5}0 + 32 = 32.$$

This means that zero degrees Celsius is equivalent to 32 degrees Fahrenheit (the temperature at which water freezes).

(c) We substitute 100 into the function, giving:

$$H = f(100) = \frac{5}{9}(100 - 32) = \frac{340}{9} = 37.778.$$

This means that 100 degrees Fahrenheit is about 38 degrees Celsius.

(d) In Exercise 31, we found the inverse function. Using it with $H = 100$, we have:

$$t = f^{-1}(100) = \frac{9}{5}100 + 32 = 212.$$

This means that 100 degrees Celsius is equivalent to 212 degrees Fahrenheit (the temperature at which water boils).

33. We have

$$H = f(g(n)) = f(68 + 10 \cdot 2^{-n}) = \frac{5}{9}\left(68 + 10 \cdot 2^{-n} - 32\right) = 20 + \frac{50}{9}2^{-n},$$

and $f(g(n))$ gives the temperature, H, in degrees Celsius after n hours.

34. Since

$$T = 2\pi\sqrt{\frac{l}{g}},$$

solving for l gives

$$T^2 = 4\pi^2\frac{l}{g}$$
$$l = \frac{gT^2}{4\pi^2}.$$

Thus,

$$f^{-1}(T) = \frac{gT^2}{4\pi^2}.$$

The function $f^{-1}(T)$ gives the length of a pendulum of period T.

35. **(a)** $A = f(r) = \pi r^2$

(b) $f(0) = 0$

(c) $f(r + 1) = \pi(r + 1)^2$. This is the area of a circle whose radius is 1 cm more than r.

(d) $f(r) + 1 = \pi r^2 + 1$. This is the area of a circle of radius r, plus 1 square centimeter more.

(e) Centimeters.

36. **(a)** Since the pizza is circular, $f(d) = \pi(d/2)^2$.

(b) Since a package costs \$2.99 and covers 250 in^2, we know

$$\text{Cost of pepperoni for 1 in}^2 = \frac{2.99}{250} = 0.01196 \text{ dollars.}$$

Thus

$$\text{Cost of pepperoni for } A \text{ in}^2 = 0.01196A \text{ dollars,}$$

so

$$C = g(A) = 0.01196A.$$

(c) Substituting for $A = f(d) = \pi(d/2)^2$ into g gives

$$C = g(f(d)) = 0.01196\pi(d/2)^2.$$

The function $g(f(d))$ gives the cost in dollars of adding pepperoni to a pizza of diameter d inches.

(d) The area of an 11 inch pizza is $f(11) = \pi(11/2)^2 = 95.033$ in^2. The cost of adding a pepperoni topping to an 11 inch pizza is $g(11) = 0.01196 \cdot 95.033 = 1.14$ dollars.

37. Since $V = \frac{4}{3}\pi r^3$ and $r = 50 - 2.5t$, substituting r into V gives

$$V = f(t) = \frac{4}{3}\pi(50 - 2.5t)^3.$$

38. **(a)** Since $f(2) = 3$, $f^{-1}(3) = 2$.
(b) Unknown
(c) Since $f^{-1}(5) = 4$, $f(4) = 5$.

39. Since the oil slick is circular, $A = \pi r^2$, so substituting $r = 2t - 0.1t^2$ into the formula for A gives

$$A = f(t) = \pi(2t - 0.1t^2)^2.$$

40. **(a)** Since $t = 0$ represents 2004, we see that $0 \leq t \leq 4$ is the domain. The corresponding outputs are the range, $375 \leq C(t) \leq 383$.
(b) $C(4)$ is the concentration of carbon dioxide in the earth's atmosphere when $t = 4$, which is the year 2008. We know that $C(4) = 383$, since in 2008, the concentration of carbon dioxide in the earth's atmosphere was 383 ppm.
(c) $C^{-1}(381)$ is the number of years after 2004 when the concentration was 381 ppm. From the data given in the problem, the actual number of years cannot be determined. (Note: For your information, the concentration was 381 ppm in the year 2007.)

41. **(a)** $f(2) = 2.80$ means that 2 pounds of apples cost \$2.80.
(b) $f(0.5) = 0.70$ means that 1/2 pound of apples cost \$0.70.
(c) $f^{-1}(0.35) = 0.25$ means that \$0.35 buys 1/4 pound of apples.
(d) $f^{-1}(7) = 5$ means that \$7 buys 5 pounds of apples.

42. We have

$$h\left(-\frac{1}{2}\right) = \frac{-1/2}{4} + 2 = -\frac{1}{8} + 2 = \frac{15}{8}.$$

If $8 = x/4 + 2$, then $x/4 = 6$, so $x = 24$. Thus, $h^{-1}(8) = 24$.

43. We have

$$g\left(\frac{3}{2}\right) = 2\left(\frac{3}{2}\right)^3 - 1 = \frac{23}{4}$$

If $-17 = 2x^3 - 1$, then $x^3 = -8$, so $x = -2$. Thus, $g^{-1}(-17) = -2$.

44. Since $y = 3x - 7$, solving for x gives

$$\begin{aligned} 3x - 7 &= y \\ x &= \frac{y+7}{3} \\ f^{-1}(y) &= \frac{y+7}{3}. \end{aligned}$$

45. Since $y = x^3 + 1$, solving for x gives

$$\begin{aligned} x^3 + 1 &= y \\ x^3 &= y - 1 \\ x &= (y-1)^{1/3} \\ f^{-1}(y) &= (y-1)^{1/3}. \end{aligned}$$

46. Since we can take cube root of any number, the domain of $t(a)$ is all real numbers. Since the range of $t(a)$ is the domain of its inverse function, we first compute the inverse of $t(a)$. Let $t(a) = y$. Solving for a gives

$$\begin{aligned} y &= \sqrt[3]{a+1} \\ y^3 &= a+1 \\ a &= y^3 - 1 \\ t^{-1}(y) &= y^3 - 1. \end{aligned}$$

Since $y^3 - 1$ is defined for any y, the domain of $t^{-1}(y)$ is all real numbers, hence the range of $t(a)$ is all real numbers.

47. Since we can cube any number, the domain of $n(r)$ is all real numbers. To find the range, we find the inverse function. Let $y = n(r)$. Solving for r, we get

$$\begin{aligned} y &= r^3 + 2 \\ y - 2 &= r^3 \\ r &= (y-2)^{1/3} \\ n^{-1}(y) &= (y-2)^{1/3}. \end{aligned}$$

Since the domain of $n^{-1}(y)$ is all real numbers, the range of $n(r)$ is all real numbers.

48. To evaluate $m(x)$, we must have $x - 2 > 0$. So the domain of $m(x)$ is all real numbers > 2. To find the range of $m(x)$, we find the inverse function of $m(x)$. Let $y = m(x)$. Solving for x, we get

$$\begin{aligned} y &= \frac{1}{\sqrt{x-2}} \\ \sqrt{x-2} &= \frac{1}{y} \\ x - 2 &= \frac{1}{y^2} \\ x &= \frac{1}{y^2} + 2 \\ m^{-1}(y) &= \frac{1}{y^2} + 2. \end{aligned}$$

The formula works for any y except $y = 0$. We know that y must be positive, since $\sqrt{x-2}$ is positive, so the range of $m(x)$ is all real numbers > 0.

49. To evaluate $p(x)$, we must have $3 - x > 0$. So the domain of $p(x)$ is all real numbers < 3. To find the range of $p(x)$, we find the inverse function of $p(x)$. Let $y = p(x)$. Solving for x, we get

$$\begin{aligned} y &= \frac{1}{\sqrt{3-x}} \\ \sqrt{3-x} &= \frac{1}{y} \\ 3 - x &= \frac{1}{y^2} \\ x &= 3 - \frac{1}{y^2} \\ p^{-1}(y) &= 3 - \frac{1}{y^2}. \end{aligned}$$

The formula works for any y except $y = 0$. We know that y must be positive, since $\sqrt{3-x}$ is positive, so the range of $p(x)$ is all real numbers > 0.

50. **(a)** $f(60) = 30$. A car traveling at 60 km/hr needs 30 meters to stop.
(b) $f(70)$ should be between $f(60) = 30$ and $f(80) = 50$, so we estimate 40 meters.
(c) $f^{-1}(70) = 100$ because $f(100) = 70$. A car that took 70 meters to stop was traveling at 100 km/hr.

Solutions for Section 2.5

Exercises

1. To determine concavity, we calculate the rate of change:

$$\frac{\Delta f(x)}{\Delta x} = \frac{1.3 - 1.0}{1 - 0} = 0.3$$

$$\frac{\Delta f(x)}{\Delta x} = \frac{1.7 - 1.3}{3 - 1} = 0.2$$

$$\frac{\Delta f(x)}{\Delta x} = \frac{2.2 - 1.7}{6 - 3} \approx 0.167.$$

The rates of change are decreasing, so we expect the graph of $f(x)$ to be concave down.

2. To determine concavity we calculate the rate of change:

$$\frac{\Delta f(t)}{\Delta t} = \frac{10 - 20}{1 - 0} = -10.$$

$$\frac{\Delta f(t)}{\Delta t} = \frac{6 - 10}{2 - 1} = -4.$$

$$\frac{\Delta f(t)}{\Delta t} = \frac{3 - 6}{3 - 2} = -3.$$

$$\frac{\Delta f(t)}{\Delta t} = \frac{1 - 3}{4 - 3} = -2.$$

It appears that a graph of this function would be concave up, because the average rate of change becomes less negative as t increases.

3. The graph appears to be concave up, as its slope becomes less negative as x increases.

4. The graph appears to be concave down, as its slope becomes less positive as x increases.

5. The slope of $y = x^2$ is always increasing, so its graph is concave up. See Figure 2.37.

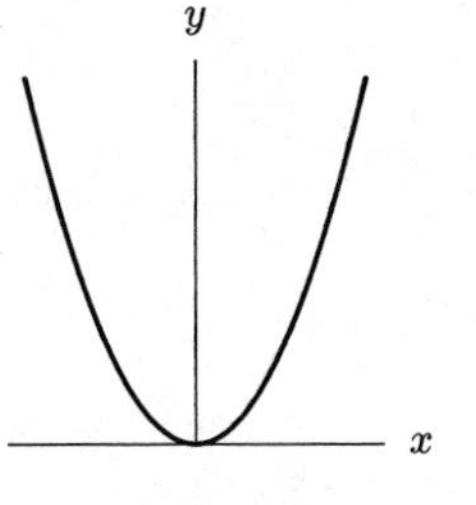

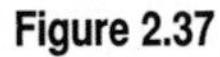
Figure 2.37

6. The slope of $y = -x^2$ is always decreasing, so its graph is concave down. See Figure 2.38.

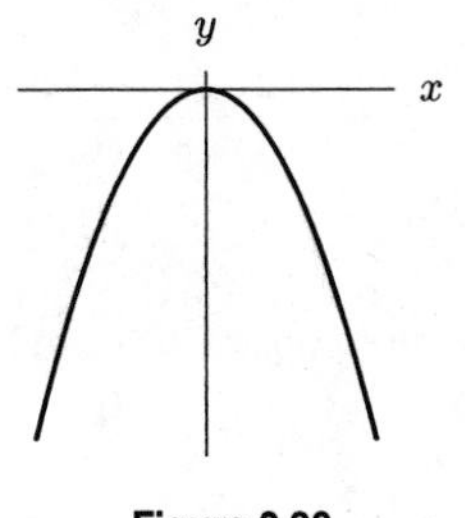

Figure 2.38

7. The slope of $y = x^3$ is always increasing on the interval $x > 0$, so its graph is concave up. See Figure 2.39.

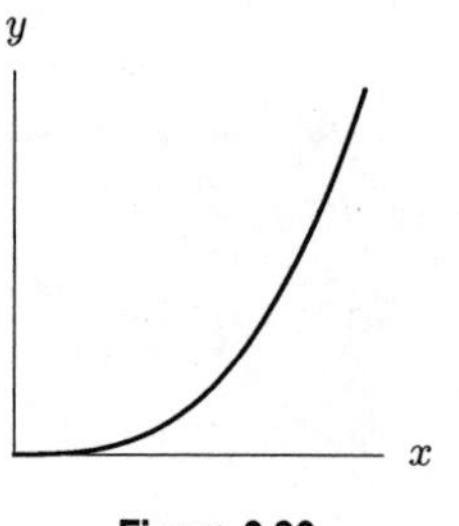

Figure 2.39

8. The slope of $y = x^3$ is always decreasing on the interval $x < 0$, so its graph is concave down. See Figure 2.40.

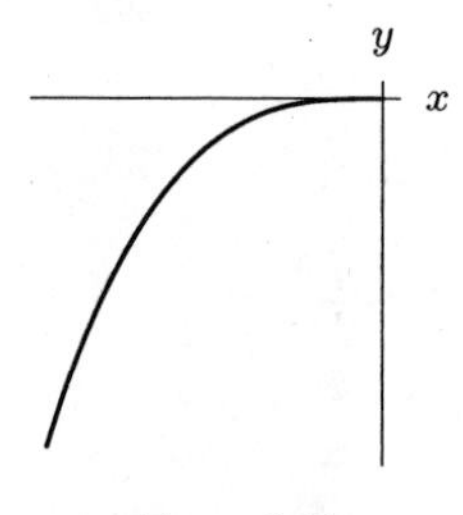

Figure 2.40

9. The rate of change between $t = 1.5$ and $t = 2.4$ is

$$\frac{\Delta R(t)}{\Delta t} = \frac{-3.1 - (-5.7)}{2.4 - 1.5} = 2.889.$$

Similarly, we have

$$\frac{\Delta R(t)}{\Delta t} = \frac{-1.4 - (-3.1)}{3.6 - 2.4} = 1.417$$

$$\frac{\Delta R(t)}{\Delta t} = \frac{0 - (-1.4)}{4.8 - 3.6} = 1.167.$$

The rate of change is decreasing, so we expect the graph to be concave down.

10. The rate of change between $x = 12$ and $x = 15$ is

$$\frac{\Delta H(x)}{\Delta x} = \frac{21.53 - 21.40}{15 - 12} \approx 0.043.$$

Similarly, we have

$$\frac{\Delta H(x)}{\Delta x} = \frac{21.75 - 21.53}{18 - 15} \approx 0.073$$

$$\frac{\Delta H(x)}{\Delta x} = \frac{22.02 - 21.75}{21 - 18} \approx 0.090.$$

The rate of change is increasing, so we expect the graph of $H(x)$ to be concave up.

11. A possible graph is in Figure 2.41.

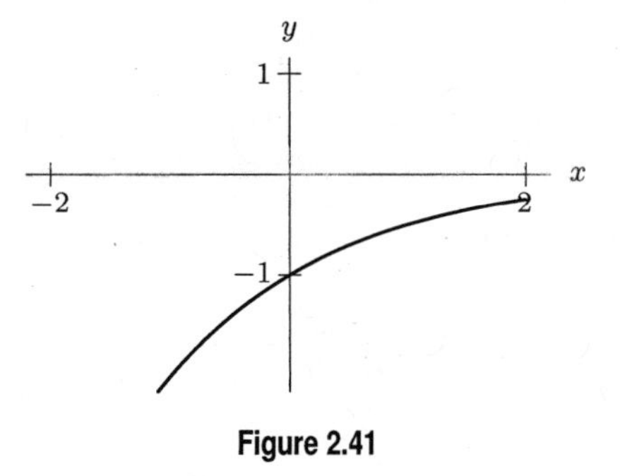

Figure 2.41

12. A possible graph is in Figure 2.42.

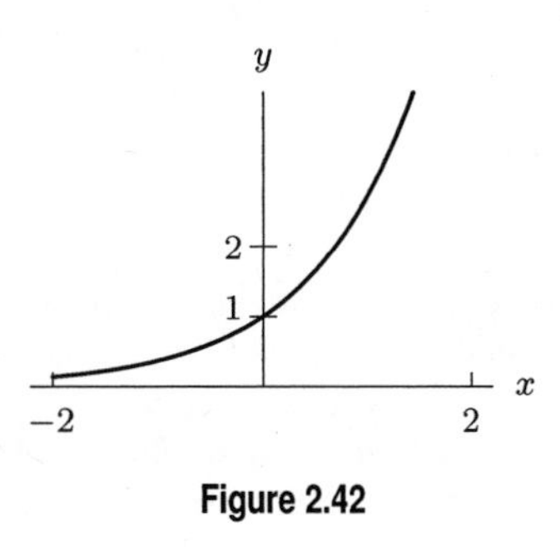

Figure 2.42

Problems

13. This function is increasing throughout and the rate of increase is increasing, so the graph is concave up.

14. This function is decreasing. As the coffee cools off, the temperature decreases at a slower rate. Since the rate of change is less negative, the graph is concave up.

15. The function is increasing throughout. At first, the graph is concave up. As more and more people hear the rumor, the rumor spreads more slowly, which means that the graph is then concave down.

16. Since more and more of the drug is being injected into the body, this is an increasing function. However, since the rate of increase of the drug is slowing down, the graph is concave down.

17. Since new people are always trying the product, it is an increasing function. At first, the graph is concave up. After many people start to use the product, the rate of increase slows down and the graph becomes concave down.

18. Many answers are possible. See Figure 2.43.

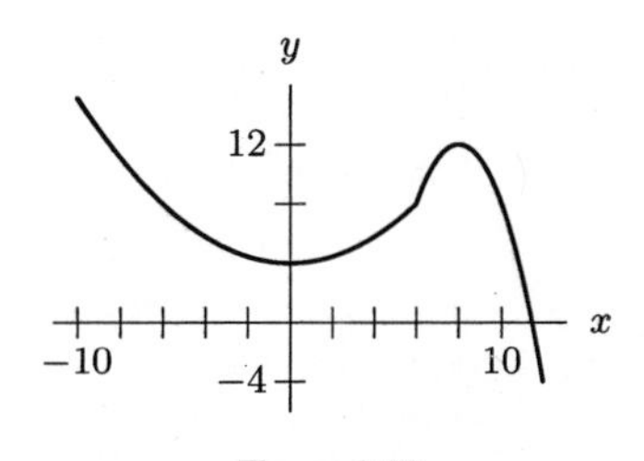

Figure 2.43

19. **(a)** This describes a situation in which y is increasing rapidly at first, then very slowly at the end. In Table (E), y increases dramatically at first (from 20 to 275) but is hardly growing at all by the end. In Graph (I), y is increasing at a constant rate, while in Graph (II), it is increasing faster at the end. Graph (III) increases rapidly at first, then slowly at the end. Thus, scenario (a) matches with Table (E) and Graph (III).

(b) Here, y is growing at a constant rate. In Table (G), y increases by 75 units for every 5-unit increase in x. A constant increase in y relative to x means a straight line, that is, a line with a constant slope. This is found in Graph (I).

(c) In this scenario, y is growing at a faster and faster rate as x gets larger. In Table (F), y starts out by growing by 16 units, then 30, then 54, and so on, so Table (F) refers to this case. In Graph (II), y is increasing faster and faster as x gets larger.

20. **(a)** This is a case in which the rate of decrease is constant, i.e., the change in y divided by the change in x is always the same. We see this in Table (F), where y decreases by 80 units for every decrease of 1 unit in x, and graphically in Graph (IV).

(b) Here, the change in y gets smaller and smaller relative to corresponding changes in x. In Table (G), y decreases by 216 units for a change of 1 unit in x initially, but only decreases by 6 units when x changes by 1 unit from 4 to 5. This is seen in Graph (I), where y is falling rapidly at first, but much more slowly for longer values of x.

(c) If y is the distance from the ground, we see in Table (E) that initially it is changing very slowly; by the end, however, the distance from the ground is changing rapidly. This is shown in Graph (II), where the decrease in y is larger and larger as x gets bigger.

(d) Here, y is decreasing quickly at first, then decreases only slightly for a while, then decreases rapidly again. This occurs in Table (H), where y decreases from 147 units, then 39, and finally by another 147 units. This corresponds to Graph (III).

21. The graphical representation of the data is misleading because in the graph the number of violent crimes is put on the horizontal axis which give the graph the appearance of leveling out. This can fool us into believing that crime is leveling out. Note that it took from 1998 to 2000, about 2 years, for the number of violent crimes to go from 500 to 1,000, but it took less than 1/2 a year for that number to go from 1,500 to 2,000, and even less time for it to go from 2,000 to 2,500. In actuality, this graph shows that crime is growing at an increasing rate. If we were to graph the number of crimes as a function of the year, the graph would be concave up.

22. **(a)** From O to A, the rate is zero, so no water is flowing into the reservoir, and the volume remains constant. From A to B, the rate is increasing, so the volume is going up more and more quickly. From B to C, the rate is holding steady, but water is still going into the reservoir—it's just going in at a constant rate. So volume is increasing on the interval from B to C. Similarly, it is increasing on the intervals from C to D and from D to E. Even on the interval from E to F, water is flowing into the reservoir; it is just going in more and more slowly (the *rate* of flow is decreasing, but the total amount of water is still increasing). So we can say that the volume of water increases throughout the interval from A to F.

(b) The volume of water is constant when the rate is zero, that is from O to A.

(c) According to the graph, the rate at which the water is entering the reservoir reaches its highest value at $t = D$ and stays at that high value until $t = E$. So the volume of water is increasing most rapidly from D to E. (Be careful. The rate itself is increasing most rapidly from C to D, but the volume of water is increasing fastest when the rate is at its highest points.)

(d) When the rate is negative, water is leaving the reservoir, so its volume is decreasing. Since the rate is negative from F to I, we know that the volume of water *decreases* on that interval.

23. **(a)** If l is the length of one salmon and its speed is u, then

$$u = 19.5\sqrt{l}.$$

Suppose the speed of the longer salmon is U and its length is $4l$. Then

$$U = 19.5\sqrt{4l} = 2 \cdot 19.5\sqrt{l} = 2u.$$

Thus, the larger one swims twice as fast as the smaller one.

(b) A typical graph is in Figure 2.44. Notice that the graphs are all of the shape of $y = \sqrt{x} = x^{1/2}$. All the graphs are increasing and concave down.

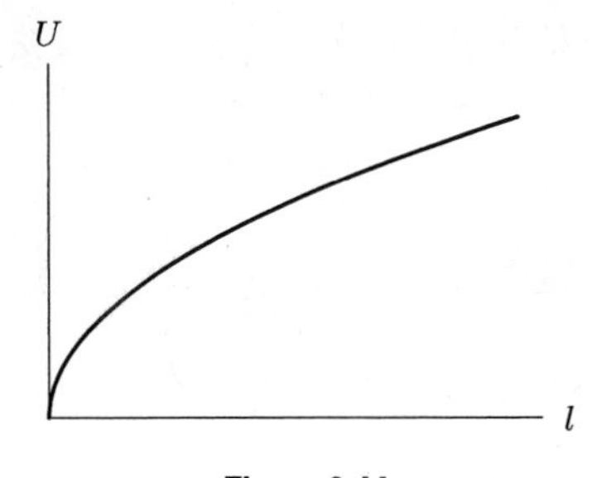

Figure 2.44

(c) The function $U = \sqrt{l}$ is an increasing function. Because $\sqrt{l}$ is an increasing function the equation predicts that larger salmon swim faster than smaller ones.

(d) The graph of $U = \sqrt{l}$ is concave down. Because the graph is concave down equal changes in l give smaller changes in U for the larger l. Thus, the difference in speed between the two smaller fish is greater than the difference in speed between the two larger fish.

24. Since f is concave down, the average rate of change is decreasing as x increases. Therefore,

$$\frac{f(3)-f(1)}{3-1} > \frac{f(5)-f(3)}{5-3}.$$

Solutions for Chapter 2 Review

Exercises

1. To evaluate when $x = -7$, we substitute -7 for x in the function, giving $f(-7) = -\frac{7}{2} - 1 = -\frac{9}{2}$.

2. To evaluate when $x = -7$, we substitute -7 for x in the function, giving $f(-7) = (-7)^2 - 3 = 49 - 3 = 46$.

3. We have

$$y = f(4) = \frac{6}{2-4^3} = \frac{6}{2-64} = \frac{6}{-62} = -\frac{3}{31}.$$

Solve for x:

$$\begin{aligned} \frac{6}{2-x^3} &= 6 \\ 6 &= 6(2-x^3) \\ 1 &= 2-x^3 \\ x^3 &= 1 \\ x &= 1. \end{aligned}$$

4. We have

$$y = f(4) = \sqrt{20+2\cdot 4^2} = \sqrt{52}.$$

Solve for x:

$$\begin{aligned} \sqrt{20+2x^2} &= 6 \\ 20+2x^2 &= 36 \\ 2x^2 &= 16 \\ x^2 &= 8 \\ x &= \pm\sqrt{8}. \end{aligned}$$

5. We have
$$y = f(4) = 4 \cdot 4^{3/2} = 4 \cdot 2^3 = 4 \cdot 8 = 32.$$
Solve for x:
$$\begin{aligned} 4x^{3/2} &= 6 \\ x^{3/2} &= 6/4 \\ x^3 &= 36/16 = 9/4 \\ x &= \sqrt[3]{9/4}. \end{aligned}$$

6. We find:
$$y = f(4) = (4)^{-3/4} - 2 = \frac{1}{(\sqrt[4]{4})^3} - 2 = \frac{1}{(\sqrt{2})^3} - 2 = \frac{1}{\sqrt{2}\sqrt{2}\sqrt{2}} - 2 = \frac{1}{2\sqrt{2}} - 2.$$
Solve for x:
$$\begin{aligned} x^{-3/4} - 2 &= 6 \\ x^{-3/4} &= 8 \\ x &= 8^{-4/3} = \frac{1}{(\sqrt[3]{8})^4} = \frac{1}{2^4} \\ &= \frac{1}{16}. \end{aligned}$$

7. **(a)** Substituting $x = 0$ gives $f(0) = 2(0) + 1 = 1$.
(b) Setting $f(x) = 0$ and solving gives $2x + 1 = 0$, so $x = -1/2$.

8. Substituting -2 for x gives
$$f(-2) = \frac{-2}{1-(-2)^2} = \frac{-2}{1-4} = \frac{2}{3}.$$

9. Substituting 4 for t gives
$$P(4) = 170 - 4 \cdot 4 = 154.$$
Similarly, with $t = 2$,
$$P(2) = 170 - 4 \cdot 2 = 162,$$
so
$$P(4) - P(2) = 154 - 162 = -8.$$

10. **(a)** Substituting, $h(x+3) = \dfrac{1}{x+3}$.
(b) Substituting and adding, $h(x) + h(3) = \dfrac{1}{x} + \dfrac{1}{3}$.

11. **(a)** Reading from the table, we have $f(1) = 2$, $f(-1) = 0$, and $-f(1) = -2$.
(b) When $x = -1$, $f(x) = 0$.

12. **(a)** $f(0)$ is the value of the function when $x = 0$, $f(0) = 3$.
(b) $f(x) = 0$ for $x = -1$ and $x = 3$.
(c) $f(x)$ is positive for $-1 < x < 3$.

13. The expression $x^2 - 9$, found inside the square root sign, must always be non-negative. This happens when $x \geq 3$ or $x \leq -3$, so our domain is $x \geq 3$ or $x \leq -3$.
For the range, the smallest value $\sqrt{x^2 - 9}$ can have is zero. There is no largest value, so the range is $q(x) \geq 0$.

14. To evaluate $m(r)$, we must have $r^2 - 1 > 0$. This happens when $r > 1$ or $r < -1$. So the domain is all real numbers r, such that $r > 1$ or $r < -1$. Since the square root of a number cannot be negative, the range is all positive real numbers, $m(r) > 0$.

15. Since for any value of x that you might choose you can find a corresponding value of $m(x)$, we can say that the domain of $m(x) = 9 - x$ is all real numbers.
For any value of $m(x)$ there is a corresponding value of x. So the range is also all real numbers.

16. Since you can choose any value of x and find an associated value for $n(x)$, we know that the domain of this function is all real numbers.
However, there are some restrictions on the range. Since x^4 is always positive for any value of x, $9 - x^4$ will have a largest value of 9 when $x = 0$. So the range is $n(x) \leq 9$.

17. Since $m(t)$ is a linear function, the domain of $m(t)$ is all real numbers. For any value of $m(t)$ there is a corresponding value of t. So the range is also all real numbers.

18. Since division by 0 is undefined, $s(q)$ is not defined when $5 - 4q = 0$. So the domain of $s(q)$ is all real numbers except $q = 5/4$. To find the range we find the inverse function. We solve the equation $y = s(q) = \frac{2q+3}{5-4q}$ for q.

$$\begin{aligned} y &= \frac{2q+3}{5-4q} \\ y(5-4q) &= 2q+3 \\ 5y - 4qy &= 2q+3 \\ 5y - 3 &= q(2+4y) \\ q &= \frac{5y-3}{2+4y} \\ s^{-1}(y) &= \frac{5y-3}{2+4y}. \end{aligned}$$

The domain of the inverse function is all real numbers except $-1/2$, so the range of $s(q)$ is all real numbers except $-1/2$.

19. **(a)** If $x = a$ is not in the domain of f there is no point on the graph with x-coordinate a. For example, there are no points on the graph of the function in Figure 2.45 with x-coordinates greater than 2. Therefore, $x = a$ is not in the domain of f for any $a > 2$.

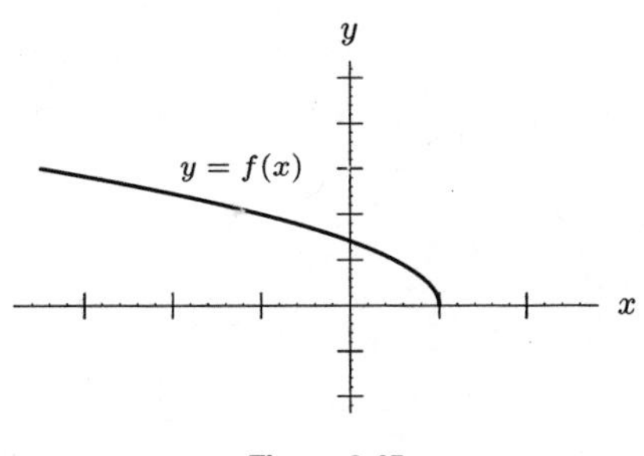

Figure 2.45

(b) If $x = a$ is not in the domain of f the formula is undefined for $x = a$. For example, if $f(x) = 1/(x-3)$, $f(3)$ is undefined, so 3 is not in the domain of f.

20. **(a)** $-3g(x) = -3(x^2 + x)$.
(b) $g(1) - x = (1^2 + 1) - x = 2 - x$.
(c) $g(x) + \pi = (x^2 + x) + \pi = x^2 + x + \pi$.
(d) $\sqrt{g(x)} = \sqrt{x^2 + x}$.
(e) $g(1)/(x+1) = (1^2 + 1)/(x+1) = 2/(x+1)$.
(f) $(g(x))^2 = (x^2 + x)^2$.

21. **(a)** $2f(x) = 2(1 - x)$.
(b) $f(x) + 1 = (1 - x) + 1 = 2 - x$.
(c) $f(1 - x) = 1 - (1 - x) = x$.
(d) $(f(x))^2 = (1 - x)^2$.
(e) $f(1)/x = (1 - 1)/x = 0$.
(f) $\sqrt{f(x)} = \sqrt{1 - x}$.

22. $f(g(x)) = f(x^3 + 1) = 3(x^3 + 1) - 7 = 3x^3 - 4$.

23. $g(f(x)) = g(3x - 7) = (3x - 7)^3 + 1$.

24. $a(g(w))$ is the acceleration in meters/sec^2 when the wind speed is w meters/second.

25. $P(f(t))$ is the period, in seconds, of the pendulum at time t minutes.

26. $f(g(0)) = f(2 \cdot 0 + 3) = f(3) = 3^2 + 1 = 10$.

27. $f(g(1)) = f(2 \cdot 1 + 3) = f(5) = 5^2 + 1 = 26$.

28. $g(f(0)) = g(0^2 + 1) = g(1) = 2 \cdot 1 + 3 = 5$.

29. $g(f(1)) = g(1^2 + 1) = g(2) = 2 \cdot 2 + 3 = 7$.

30. $f(g(x)) = f(2x + 3) = (2x + 3)^2 + 1 = 4x^2 + 12x + 10$.

31. $g(f(x)) = g(x^2 + 1) + 3 = 2(x^2 + 1) + 3 = 2x^2 + 5$.

32. $f(f(x)) = f(x^2 + 1) = (x^2 + 1)^2 + 1 = x^4 + 2x^2 + 2$.

33. $g(g(x)) = g(2x + 3) = 2(2x + 3) + 3 = 4x + 9$.

34. The inverse function, $f^{-1}(V)$, gives the time at which the speed is V. Units of $f^{-1}(V)$ are seconds.

35. The inverse function, $f^{-1}(I)$, gives the interest rate that gives $\$I$ in interest. Units of $f^{-1}(I)$ is percent per year.

36. We can put in any number for x except zero, which makes $1/x$ undefined. We note that as x approaches infinity or negative infinity, $1/x$ approaches zero, though it never arrives there, and that as x approaches zero, $1/x$ goes to negative or positive infinity. Thus, the range is all real numbers except a.

37. Since $(x - b)^{1/2} = \sqrt{x - b}$, we know that $x - b \geq 0$. Thus, $x \geq b$. If $x = b$, then $(x - b)^{1/2} = 0$, which is the minimum value of $\sqrt{x - b}$, since it can't be negative. Thus, the range is all real numbers greater than or equal to 6.

38. Since $y = \sqrt{t} + 1$, solving for t gives

$$\begin{aligned} \sqrt{t} + 1 &= y \\ \sqrt{t} &= y - 1 \\ t &= (y - 1)^2 \\ g^{-1}(y) &= (y - 1)^2. \end{aligned}$$

39. Since $P = 14q - 2$, solving for q gives

$$\begin{aligned} 14q - 2 &= P \\ q &= \frac{P + 2}{14} \\ f^{-1}(P) &= \frac{P + 2}{14}. \end{aligned}$$

40. The composition, $P = g(f(t))$ gives the daily electricity consumption in megawatts at time t.

41. If $P = f(t)$, then $t = f^{-1}(P)$, so $f^{-1}(P)$ gives the time in years at which the population is P million.

42. If $E = g(P)$, then $P = g^{-1}(E)$ so $g^{-1}(E)$ gives the population leading to a daily electricity consumption of E megawatts.

43. The rate of change between $t = 0.2$ and $t = 0.4$ is

$$\frac{\Delta p(t)}{\Delta t} = \frac{-2.32 - (-3.19)}{0.4 - 0.2} = 4.35.$$

Similarly, we have

$$\frac{\Delta p(t)}{\Delta t} = \frac{-1.50 - (-2.32)}{0.6 - 0.4} = 4.10$$

$$\frac{\Delta p(t)}{\Delta t} = \frac{-0.74 - (-1.50)}{0.8 - 0.6} = 3.80.$$

The rate of change is decreasing, so we expect the graph to be concave down.

44. We have

$$p(8) = \frac{12}{\sqrt{8}} = \frac{6}{\sqrt{2}}.$$

If $\sqrt{2} = 12/\sqrt{x}$, then $\sqrt{2x} = 12$, so $x = 72$. Thus, $p^{-1}(\sqrt{2}) = 72$.

45. We find $f(16) = 12 - \sqrt{16} = 8$. If $3 = 12 - \sqrt{x}$, then $\sqrt{x} = 9$, so $x = 81$. Thus, $f^{-1}(3) = 81$

46. The graph of $f(x) = \begin{cases} x^2 & \text{for } x \leq 1 \\ 2 - x & \text{for } x > 1 \end{cases}$ is shown in Figure 2.46.

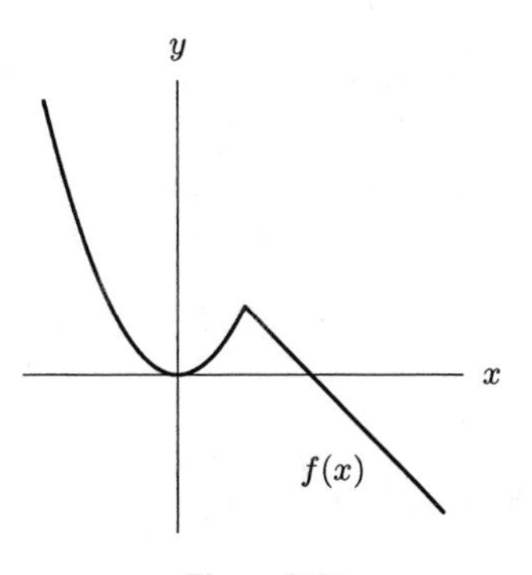

Figure 2.46

47. The graph of $g(x) = \begin{cases} x + 5 & \text{for } x < 0 \\ x^2 + 1 & \text{for } 0 \leq x \leq 2 \\ 3 & \text{for } x > 2 \end{cases}$ is shown in Figure 2.47.

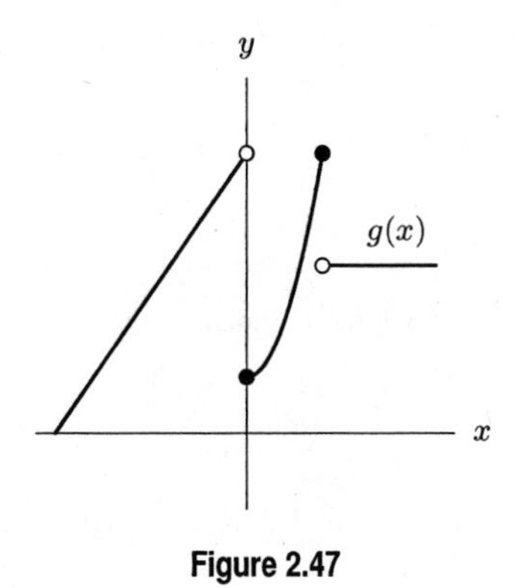

Figure 2.47

Problems

48. Substituting $r = 3$ and $h = 2$ gives

$$V = \frac{1}{3}\pi 3^2 \cdot 2 = 6\pi \text{ cubic inches.}$$

49. **(a)** Substituting, $q(5) = 3 - (5)^2 = -22$.
(b) Substituting, $q(a) = 3 - a^2$.
(c) Substituting, $q(a-5) = 3 - (a-5)^2 = 3 - (a^2 - 10a + 25) = -a^2 + 10a - 22$.
(d) Using the answer to part (b), $q(a) - 5 = 3 - a^2 - 5 = -a^2 - 2$.
(e) Using the answer to part (b) and (a), $q(a) - q(5) = (3 - a^2) - (-22) = -a^2 + 25$.

50. Substituting -1 gives $p(-1) = (-1)^2 + (-1) + 1 = 1$. Substituting 1 and taking the negative gives, $-p(1) = -((1)^2 + (1) + 1) = -3$. Thus, $p(-1) \neq -p(1)$. They are not equal.

51. **(a)** The table shows $f(6) = 3.7$, so $t = 6$. In a typical June, Chicago has 3.7 inches of rain.
(b) First evaluate $f(2) = 1.8$. Solving $f(t) = 1.8$ gives $t = 1$ or $t = 2$. Chicago has 1.8 inches of rain in January and in February.

52. **(a)** In order to find $f(0)$, we need to find the value which corresponds to $x = 0$. The point $(0, 24)$ seems to lie on the graph, so $f(0) = 24$.
(b) Since $(1, 10)$ seems to lie on this graph, we can say that $f(1) = 10$.
(c) The point that corresponds to $x = b$ seems to be about $(b, -7)$, so $f(b) = -7$.
(d) When $x = c$, we see that $y = 0$, so $f(c) = 0$.
(e) When your input is d, the output is about 20, so $f(d) = 20$.

53. **(a)** Substituting $x = 0$ gives $f(0) = \sqrt{0^2 + 16} - 5 = \sqrt{16} - 5 = 4 - 5 = -1$.
(b) We want to find x such that $f(x) = \sqrt{x^2 + 16} - 5 = 0$. Thus, we have

$$\begin{aligned}\sqrt{x^2 + 16} - 5 &= 0\\ \sqrt{x^2 + 16} &= 5\\ x^2 + 16 &= 25\\ x^2 &= 9\\ x &= \pm 3.\end{aligned}$$

Thus, $f(x) = 0$ for $x = 3$ or $x = -3$.
(c) In part (b), we saw that $f(3) = 0$. You can verify this by substituting $x = 3$ into the formula for $f(x)$:

$$f(3) = \sqrt{3^2 + 16} - 5 = \sqrt{25} - 5 = 5 - 5 = 0.$$

(d) The vertical intercept is the value of the function when $x = 0$. We found this to be -1 in part (a). Thus the vertical intercept is -1.
(e) The graph touches the x-axis when $f(x) = 0$. We saw in part (b) that this occurs at $x = 3$ and $x = -3$.

54. **(a)** Since the vertical intercept of the graph of f is $(0, 2)$, we have $f(0) = 2$.
(b) Since the horizontal intercept of the graph of f is $(-3, 0)$, we have $f(-3) = 0$.
(c) The function f^{-1} goes from y-values to x-values, so to evaluate $f^{-1}(0)$, we want the x-value corresponding to $y = 0$. This is $x = -3$, so $f^{-1}(0) = -3$.
(d) Solving $f^{-1}(?) = 0$ means finding the y-value corresponding to $x = 0$. This is $y = 2$, so $f^{-1}(2) = 0$.

55. The input values in the table of values of g^{-1} are the output values for g. See Table 2.8.

Table 2.8

y	7	12	13	19	22
$g^{-1}(y)$	1	2	3	4	5

56. We solve the equation $V = f(r) = \frac{4}{3}\pi r^3$ for r. Divide both sides by $\frac{4}{3}\pi$ and then take the cube root to get

$$r = \sqrt[3]{\frac{3V}{4\pi}}.$$

So

$$f^{-1}(V) = \sqrt[3]{\frac{3V}{4\pi}}.$$

57. **(a)** To write s as a function of A, we solve $A = 6s^2$ for s

$$s^2 = \frac{A}{6} \quad \text{so} \quad s = f(A) = +\sqrt{\frac{A}{6}} \quad \text{Because the length of a side of a cube is positive.}$$

The function f gives the side of a cube in terms of its area A.

(b) Substituting $s = f(A) = \sqrt{A/6}$ in the formula $V = g(s) = s^3$ gives the volume, V, as a function of surface area, A,

$$V = g(f(A)) = s^3 = \left(\sqrt{\frac{A}{6}}\right)^3.$$

58. **(a)** Since the deck is square, $f(s) = s^2$.

(b) Since a can costs \$29.50 and covers 200 ft^2, we know

$$\text{Cost of stain for 1 ft}^2 = \frac{29.50}{200} = 0.1475 \text{ dollars.}$$

Thus

$$\text{Cost of stain for } A \text{ ft}^2 = 0.1475A \text{ dollars,}$$

so

$$C = g(A) = 0.1475A.$$

(c) Substituting for $A = f(s) = s^2$ into g gives

$$C = g(f(s)) = 0.1475s^2.$$

The function $g(f(s))$ gives the cost in dollars of staining a square deck of side s feet.

(d) (i) $f(8) = 8^2 = 64$ square feet; the area of a deck of side 8 feet.

(ii) $g(80) = 0.1475 \cdot 80 = 11.80$ dollars; the cost of staining a deck of area 80 ft^2.

(iii) $g(f(10) = 0.1475 \cdot 10^2 = 14.75$ dollars; the cost of staining a deck of side 10 feet.

59. **(a)** This is the fare for a ride of 3.5 miles. $C(3.5) \approx \$6.25$.

(b) This is the number of miles you can travel for \$3.50. Between 1 and 2 miles the increase in cost is \$1.50. Setting up a proportion we have:

$$\frac{\text{1 additional mile}}{\text{\$1.50 additional fare}} = \frac{x \text{ additional miles}}{\text{\$3.50} - \text{\$2.50 additional fare}}$$

and $x = 0.67$ miles. Therefore

$$C^{-1}(\$3.5) \approx 1.67.$$

60. **(a)** $P = f(s) = 4s$.

(b) $f(s+4) = 4(s+4) = 4s+16$. This the perimeter of a square whose side is four meters larger than s.

(c) $f(s)+4 = 4s+4$. This is the perimeter of a square whose side is s, plus four meters.

(d) Meters.

61. **(a)** Using Pythagoras' Theorem, we see that the diagonal d is given in terms of s by

$$d^2 = 2s^2$$
$$s = \sqrt{\frac{d^2}{2}} = \frac{d}{\sqrt{2}}$$
$$s = f(d) = \frac{d}{\sqrt{2}}.$$

(b) $A = g(s) = s^2$.

(c) Substituting $s = d/\sqrt{2}$ in g gives

$$A = g(s) = \left(\frac{d}{\sqrt{2}}\right)^2 = \frac{d^2}{2}.$$

(d) The function h is the composition of f and g, with f as the inside function, that is $h(d) = g(f(d))$.

62. **(a)** $j(h(4)) = h^{-1}(h(4)) = 4$
(b) We don't know $j(4)$
(c) $h(j(4)) = h(h^{-1}(4)) = 4$
(d) $j(2) = 4$
(e) We don't know $h^{-1}(-3)$
(f) $j^{-1}(-3) = 5$, since $j(5) = -3$
(g) We don't know $h(5)$
(h) $(h(-3))^{-1} = \left(j^{-1}(-3)\right)^{-1} = 5^{-1} = 1/5$
(i) We don't know $(h(2))^{-1}$

63. **(a)** To find a point on the graph $k(x)$ with an x-coordinate of -2, we substitute -2 for x in the formula for $k(x)$. We obtain $k(-2) = 6 - (-2)^2 = 6 - 4 = 2$. Thus, we have the point $(-2, k(-2))$, or $(-2, 2)$.
(b) To find these points, we want to find all the values of x for which $k(x) = -2$. We have

$$\begin{aligned} 6 - x^2 &= -2 \\ -x^2 &= -8 \\ x^2 &= 8 \\ x &= \pm 2\sqrt{2}. \end{aligned}$$

Thus, the points $(2\sqrt{2}, -2)$ and $(-2\sqrt{2}, -2)$ both have a y-coordinate of -2.
(c) Figure 2.48 shows the desired graph. The point in part (a) is $(-2, 2)$. We have called this point A on the graph in Figure 2.48. There are two points in part (b): $(-2\sqrt{2}, -2)$ and $(2\sqrt{2}, -2)$. We have called these points B and C, respectively, on the graph in Figure 2.48.

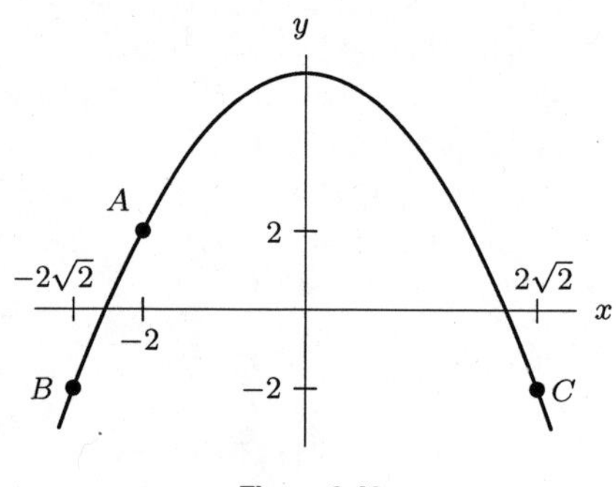

Figure 2.48

(d) For $p = 2$, $k(p) - k(p-1) = k(2) - k(1)$. Now $k(2) = 6 - 2^2 = 6 - 4 = 2$, while $k(1) = 6 - (1)^2 = 6 - 1 = 5$, thus, $k(2) - k(1) = 2 - 5 = -3$.

64. **(a)** To find a point on the graph of $h(x)$ whose x-coordinate is 5, we substitute 5 for x in the formula for $h(x)$. $h(5) = \sqrt{5+4} = \sqrt{9} = 3$. Thus, the point $(5, 3)$ is on the graph of $h(x)$.
(b) Here we want to find a value of x such that $h(x) = 5$. We set $h(x) = 5$ to obtain

$$\begin{aligned} \sqrt{x+4} &= 5 \\ x + 4 &= 25 \\ x &= 21. \end{aligned}$$

Thus, $h(21) = 5$, and the point $(21, 5)$ is on the graph of $h(x)$.
(c) Figure 2.49 shows the desired graph. The point in part (a) is $(5, h(5))$, or $(5, 3)$. This point is labeled A in Figure 2.49. The point in part (b) is $(21, 5)$. This point is labeled B in Figure 2.49.

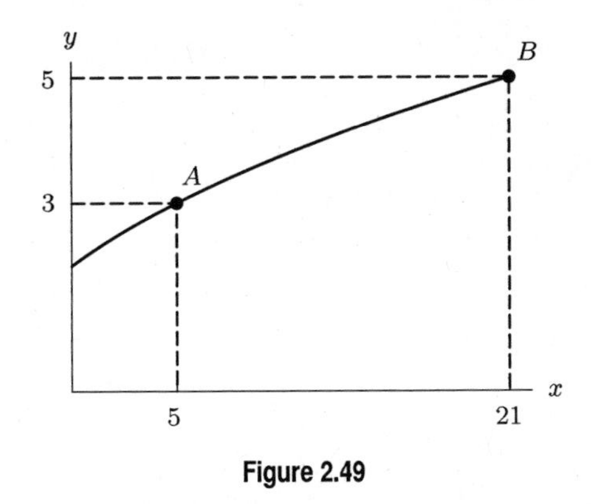

Figure 2.49

(d) If $p = 2$, then $h(p+1) - h(p) = h(2+1) - h(2) = h(3) - h(2)$. But $h(3) = \sqrt{3+4}$, while $h(2) = \sqrt{2+4}$, thus, $h(p+1) - h(p)$ for $p = 2$ equals $h(3) - h(2) = \sqrt{7} - \sqrt{6} \approx 0.1963$.

65. (a) $t(400) = 272$.

(b) (i) It takes 136 seconds to melt 1 gram of the compound at a temperature of 800°C.

(ii) It takes 68 seconds to melt 1 gram of the compound at a temperature of 1600°C.

(c) This means that $t(2x) = t(x)/2$, because if x is a temperature and $t(x)$ is a melting time, then $2x$ would be double this temperature and $t(x)/2$ would be half this melting time.

66. (a)

n	1	2	3	4	5	6	7	8	9	10	11	12
$f(n)$	1	1	2	3	5	8	13	21	34	55	89	144

(b) We note that for every value of n, we can find a unique value for $f(n)$ (by adding the two previous values of the function). This satisfies the definition of function, so $f(n)$ is a function.

(c) Using the pattern, we can figure out $f(0)$ from the fact that we must have

$$f(2) = f(1) + f(0).$$

Since $f(2) = f(1) = 1$, we have

$$1 = 1 + f(0),$$

so

$$f(0) = 0.$$

Likewise, using the fact that $f(1) = 1$ and $f(0) = 0$, we have

$$\begin{aligned} f(1) &= f(0) + f(-1) \\ 1 &= 0 + f(-1) \\ f(-1) &= 1. \end{aligned}$$

Similarly, using $f(0) = 0$ and $f(-1) = 1$ gives

$$\begin{aligned} f(0) &= f(-1) + f(-2) \\ 0 &= 1 + f(-2) \\ f(-2) &= -1. \end{aligned}$$

However, there is no obvious way to extend the definition of $f(n)$ to non-integers, such as $n = 0.5$. Thus we cannot easily evaluate $f(0.5)$, and we say that $f(0.5)$ is undefined.

67. **(a)** This tells us that a person who loses 30 minutes of sleep takes 5 minutes longer to complete the task than a person who loses no sleep.

(b) This tells us that a person who loses t_1 minutes of sleep takes twice as long to complete the task as a person who loses no sleep.

(c) This tells us that a person who loses $2t_1$ minutes of sleep takes 50% longer to complete the task as a person who loses t_1 minutes of sleep.

(d) This tells us that a person who loses $t_2 + 60$ minutes of sleep takes 10 minutes longer to complete the task than a person who loses only $t_2 + 30$ minutes of sleep.

68. One way to do this is to combine two operations, one of which forces x to be non-negative, the other of which forces x not to equal 3. One possibility is

$$y = \frac{1}{x-3} + \sqrt{x}.$$

The fraction's denominator must not equal 0, so x must not equal 3. Further, the input of the square root function must not be negative, so x must be greater than or equal to zero. Other possibilities include

$$y = \frac{\sqrt{x}}{x-3}.$$

69.

- A function such as $y = \sqrt{x-4}$ is undefined for $x < 4$, because the input of the square root operation is negative for these x-values.
- A function such as $y = 1/(x-8)$ is undefined for $x = 8$.
- Combining two functions such as these, for example by adding or multiplying them, yields a function with the required domain. Thus, possible formulas include

$$y = \frac{1}{x-8} + \sqrt{x-4} \quad \text{or} \quad y = \frac{\sqrt{x-4}}{x-8}.$$

70. **(a)** Each signature printed costs \$0.14, and in a book of p pages, there are at least $p/16$ signatures. In a book of 128 pages, there are

$$\frac{128}{16} = 8 \text{ signatures,}$$

$$\text{Cost for 128 pages} = 0.14(8) = \$1.12.$$

A book of 129 pages requires 9 signatures, although the ninth signature is used to print only 1 page. Therefore,

$$\text{Cost for 129 pages} = \$0.14(9) = \$1.26.$$

To find the cost of p pages, we first find the number of signatures. If p is divisible by 16, then the number of signatures is $p/16$ and the cost is

$$C(p) = 0.14\left(\frac{p}{16}\right).$$

If p is not divisible by 16, the number of signatures is $p/16$ rounded up to the next highest integer and the cost is 0.14 times that number. In this case, it is hard to write a formula for $C(p)$ without a symbol for "rounding up."

(b) The number of pages, p, is greater than zero. Although it is possible to have a page which is only half filled, we do not say that a book has 124 1/2 pages, so p must be an integer. Therefore, the domain of $C(p)$ is $p > 0$, p an integer. Because the cost of a book increases by multiples of \$0.14 (the cost of one signature), the range of $C(p)$ is $C > 0$, C an integer multiple of \$0.14,

(c) For 1 to 16 pages, the cost is \$0.14, because only 1 signature is required. For 17 to 32 pages, the cost is \$0.28, because 2 signatures are required. These data are continued in Table 2.9 for $0 \leq p \leq 128$, and they are plotted in Figure 2.50. A closed circle represents a point included on the graph, and an open circle indicates a point excluded from the graph. The unbroken lines in Figure 2.50 suggest, erroneously, that *fractions* of pages can be printed. It would be more accurate to draw each step as 16 separate dots instead of as an unbroken line.

Table 2.9 *The cost C for printing a book of p pages*

p, pages	C, dollars
1-16	0.14
17-32	0.28
33-48	0.42
49-64	0.56
65-80	0.70
81-96	0.84
97-112	0.98
113-128	1.12

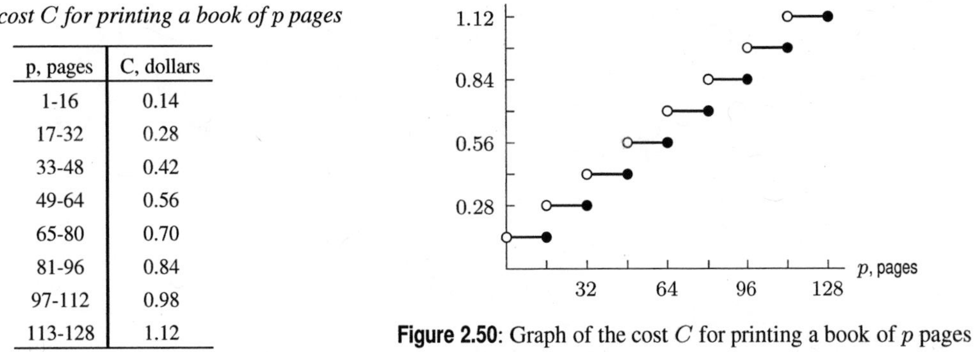

Figure 2.50: Graph of the cost C for printing a book of p pages

71. **(a)** Increasing until year 60 and then decreasing.

(b) The average rate of change of the population is given in Table 2.10.

Table 2.10 *The population of Ireland from 1780 to 1910*

Year (years)	Population (millions)	$\Delta P/\Delta t$ (millions/year)
0 to 20	4.0 to 5.2	0.060
20 to 40	5.2 to 6.7	0.075
40 to 60	6.7 to 8.3	0.080
60 to 70	8.3 to 6.9	−0.140
70 to 90	6.9 to 5.4	−0.075
90 to 110	5.4 to 4.7	−0.035
110 to 130	4.7 to 4.4	−0.015

(c) The average rate of change is increasing until between years 40 and 60. At year 60, the sign abruptly changes, but after 60, the rate of change is still increasing. Thus, the graph is concave up, although something strange is happening near year 60.

(d) The rate of change of the population was greatest between 40 and 60, that is 1820-1840. The rate of change of the population was least (most negative) between 60 and 70, that is 1840-1850. At this time the population was shrinking fastest.

Since the greatest rate of increase was directly followed by the greatest rate of decrease, something catastrophic must have happened to cause the population not only to stop growing, but to start shrinking.

(e) Figures 2.51 and 2.52 show the population of Ireland from 1780 to 1910 as a function of the time with two different curves dashed in—either of which could be correct. From the graphs we can see that the curve is increasing until about year 60, and then decreases. Also, it is concave up most of the time except, possibly, for a short time interval near year 60.

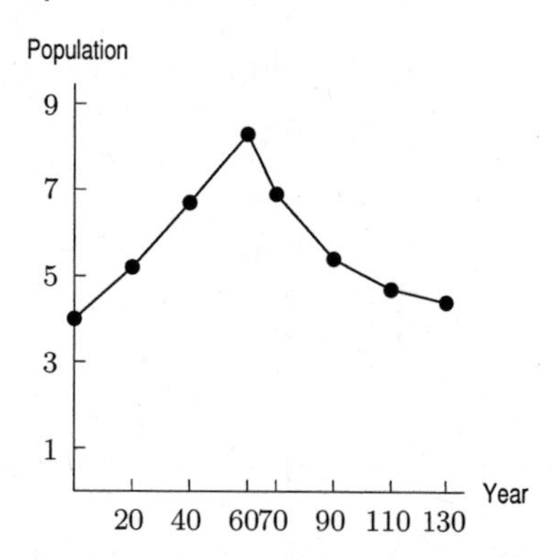

Figure 2.51: The population of Ireland from 1780 to 1910

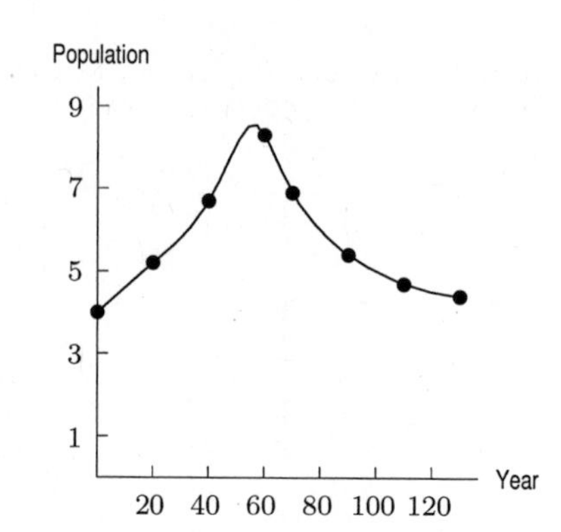

Figure 2.52: The population of Ireland from 1780 to 1910

(f) Something catastrophic happened in Ireland about year 60—that is, 1840. This is when the Irish potato famine took place.

CHECK YOUR UNDERSTANDING

1. False. $f(2) = 3 \cdot 2^2 - 4 = 8$.
2. True. Functions are evaluated by substituting a known value or variable, here b, for the independent variable, here x.
3. False. $f(x+h) = (x+h)^2 = x^2 + 2xh + h^2$.
4. True. If $q = 1/\sqrt{z^2+5} = 1/3$, then

$$\begin{aligned}\frac{1}{z^2+5} &= \frac{1}{9}\\ z^2+5 &= 9\\ z^2 &= 4\\ z &= \pm 2.\end{aligned}$$

5. False. $W = (8+4)/(8-4) = 3$.
6. True. $f(0) = 0^2 + 64 = 64$.
7. False. For example, if $f(x) = x - 3$, then $f(x) = 0$ for $x = 3$ but not for $x = 0$.
8. False. For example, $f(1) = 10$ but $f(-1) = 6$.
9. True. A fraction can only be zero if the numerator is zero.
10. False. $h(3) + h(4) = (-6 \cdot 3 + 9) + (-6 \cdot 4 + 9) = -9 + (-15) = -24$ but $h(7) = -6 \cdot 7 + 9 = -33$.
11. True. This is the definition of the domain.
12. True. This is the common practice.
13. False. The domain consists of all real numbers x, $x \neq 3$
14. False. The domain consists of all real numbers $x \leq 2$.
15. False. The range does not include zero, since $1/x$ does not equal zero for any x.
16. False. If $x < 0$, then $y > 4$.
17. True. Since f is an increasing function, the domain endpoints determine the range endpoints. We have $f(15) = 12$ and $f(20) = 14$.
18. True. The $x^2 + 1$ inside the square root is positive for all x so $f(x)$ is defined for all x.

19. True. It has a slope of -1 to the left of the origin, goes through the origin, and continues as an increasing function with slope 1 to the right of the origin.

20. True. It is defined for all x.

21. True. $|x| = |-x|$ for all x.

22. False. For example, if $x = 1$ then $f(1) = 1$ and $g(1) = -1$.

23. False. For example, if $x = -1$ then $y = -1$.

24. False. Since $0 \leq 3 < 4$, the middle formula must be used. So $f(3) = 3^2 = 9$.

25. True. If $x < 0$, then $f(x) = x < 0$, so $f(x) \neq 4$. If $x > 4$, then $f(x) = -x < 0$, so $f(x) \neq 4$. If $0 \leq x \leq 4$, then $f(x) = x^2 = 4$ only for $x = 2$. The only solution for the equation $f(x) = 4$ is $x = 2$.

26. False. If f is invertible, we know $f^{-1}(5) = 3$, but nothing else.

27. True. This is the definition of the inverse function.

28. False. Check to see if $f(0) = 8$, which it does not.

29. True. To find $f^{-1}(R)$, we solve $R = \frac{2}{3}S + 8$ for S by subtracting 8 from both sides and then multiplying both sides by $(3/2)$.

30. False. For example, if $f(x) = x + 1$ then $f^{-1}(x) = x - 1$ but $(f(x))^{-1} = \dfrac{1}{x+1}$.

31. True. Since $t^{-1} = 1/t$ this is a direct substitution for the independent variable x.

32. False. The output units of a function are the same as the input units of its inverse.

33. False. Since

$$f(g(x)) = 2\left(\frac{1}{2}x - 1\right) + 1 = x - 1 \neq x,$$

the functions do not undo each other.

34. True. The quantity of rice required is $f(x)$ tons. The cost of this quantity is $g(f(x))$ dollars. Thus, the cost to feed x million people for a year is $g(f(x))$ dollars.

35. True.

36. False. The composite function $g(f(t))$ gives the volume of the ball in meter^3 after t seconds. Thus, the units of $g(f(t))$ are meter^3.

37. True. Since the function is concave up, the average rate of change increases as we move right.

38. True. The rates of change are increasing:

$$\frac{f(0) - f(-2)}{0 - (-2)} = \frac{6 - 5}{2} = \frac{1}{2}.$$

$$\frac{f(2) - f(0)}{2 - 0} = \frac{8 - 6}{2} = 1.$$

$$\frac{f(4) - f(2)}{4 - 2} = \frac{12 - 8}{2} = 2.$$

39. True. The rates of change are decreasing:

$$\frac{g(1) - g(-1)}{1 - (-1)} = \frac{8 - 9}{2} = -\frac{1}{2}.$$

$$\frac{g(3) - g(1)}{3 - 1} = \frac{6 - 8}{2} = -1.$$

$$\frac{g(5) - g(3)}{5 - 3} = \frac{3 - 6}{2} = -\frac{3}{2}.$$

40. False. A straight line is neither concave up nor concave down.

41. True. For $x > 0$, the function $f(x) = -x^2$ is both decreasing and concave down.

42. False. For $x < 0$, the function $f(x) = x^2$ is both concave up and decreasing.

CHAPTER THREE

Solutions for Section 3.1

Skill Refresher

S1. In this example, we distribute the factors $50t$ and $2t$ across the two binomials t^2+1 and $25t^2+125$, respectively. Thus,

$$\begin{aligned}(t^2+1)(50t)-(25t^2+125)(2t) &= 50t^3+50t-(50t^3+250t)\\ &= 50t^3+50t-50t^3-250t=-200t.\end{aligned}$$

S2. Expanding $(A^2-B^2)^2=(A^2-B^2)(A^2-B^2)$, we get

$$A^4-2A^2B^2+B^4.$$

S3. $u^2-2u=u(u-2)$

S4. $x^2+3x+2=(x+2)(x+1)$

S5. $3x^2-x-4=(3x-4)(x+1)$

S6. Difference of squares: $(s+2t)^2-4p^2=(s+2t+2p)(s+2t-2p)$.

S7. $16x^2-1=(4x-1)(4x+1)$

S8. $y^3-y^2-12y=y(y^2-y-12)=y(y-4)(y+3)$

S9.

$$\begin{aligned}x^2+7x+6&=0\\ (x+6)(x+1)&=0\\ x+6=0 \quad &\text{or} \quad x+1=0\\ x=-6 \quad &\text{or} \quad x=-1\end{aligned}$$

S10.

$$\begin{aligned}2w^2+w-10&=0\\ (2w+5)(w-2)&=0\\ 2w+5=0 \quad &\text{or} \quad w-2=0\\ w=\frac{-5}{2} \quad &\text{or} \quad w=2\end{aligned}$$

Exercises

1. Yes. We rewrite the function giving

$$\begin{aligned}f(x)&=2(7-x)^2+1\\ &=2(49-14x+x^2)+1\\ &=98-28x+2x^2+1\\ &=2x^2-28x+99.\end{aligned}$$

So $f(x)$ is quadratic with $a=2$, $b=-28$ and $c=99$.

2. Yes.We rewrite the function giving

$$L(P) = (P+1)(1-P) = 1 - P^2 = (-1)P^2 + 0 \cdot P + 1.$$

So $L(P)$ is quadratic with $a = -1$, $b = 0$ and $c = 1$.

3. Yes.We rewrite the function giving

$$\begin{aligned} g(m) &= m(m^2 - 2m) + 3\left(14 - \frac{m^3}{3}\right) + \sqrt{3}m \\ &= m^3 - 2m^2 + 42 - m^3 + \sqrt{3}m \\ &= -2m^2 + \sqrt{3}m + 42. \end{aligned}$$

So $g(m)$ is quadratic with $a = -2$, $b = \sqrt{3}$ and $c = 42$.

4. Yes.We rewrite the function giving

$$h(t) = -16(t-3)(t+1) = -16(t^2 - 2t - 3) = -16t^2 + 32t + 48$$

So $h(t)$ is quadratic with $a = -16$, $b = 32$ and $c = 48$.

5. No. We rewrite the function giving

$$\begin{aligned} R(q) &= \frac{1}{q^2}(q^2+1)^2 \\ &= \frac{1}{q^2}(q^4 + 2q^2 + 1) \\ &= q^2 + 2 + \frac{1}{q^2} \\ &= q^2 + 2 + q^{-2}. \end{aligned}$$

So $R(q)$ is not quadratic since it contains a term with q to a negative power.

6. No. The function $K(x)$ is not quadratic since the term 13^x has the variable in the exponent.

7. Yes.We rewrite the function giving

$$T(n) = \sqrt{5} + \sqrt{3n^4} - \sqrt{\frac{n^4}{4}} = \sqrt{5} + \sqrt{3}n^2 - \frac{n^2}{2} = \left(\sqrt{3} - \frac{1}{2}\right)n^2 + \sqrt{5}$$

So $T(n)$ is quadratic with $a = \sqrt{3} - 1/2$, $b = 0$ and $c = \sqrt{5}$.

8. We solve for r in the equation by factoring

$$\begin{aligned} 2r^2 - 6r - 36 &= 0 \\ 2(r^2 - 3r - 18) &= 0 \\ 2(r-6)(r+3) &= 0. \end{aligned}$$

The solutions are $r = 6$ and $r = -3$.

9. We solve for x in the equation $5x - x^2 + 3 = 0$ using the quadratic formula with $a = -1$, $b = 5$ and $c = 3$.

$$\begin{aligned} x &= \frac{-5 \pm \sqrt{(-5)^2 - 4(-1)3}}{2(-1)} \\ x &= \frac{-5 \pm \sqrt{37}}{-2}. \end{aligned}$$

The solutions are $x \approx -0.541$ and $x \approx 5.541$.

10. **(a)** Rewriting $6x - \frac{1}{3} = 3x^2$ in the form $ax^2 + bx + c = 0$, we get $3x^2 - 6x + \frac{1}{3} = 0$, as shown in Figure 3.1. Applying the quadratic formula, we obtain

$$x = \frac{6 \pm \sqrt{6^2 - 4 \cdot 3 \cdot \frac{1}{3}}}{2 \cdot 3}$$
$$x = \frac{6 \pm \sqrt{36 - 4}}{6} = \frac{6 \pm \sqrt{32}}{6}$$
$$x = 1 \pm \frac{4\sqrt{2}}{6}$$
$$x \approx 0.057 \quad \text{or} \quad x \approx 1.943.$$

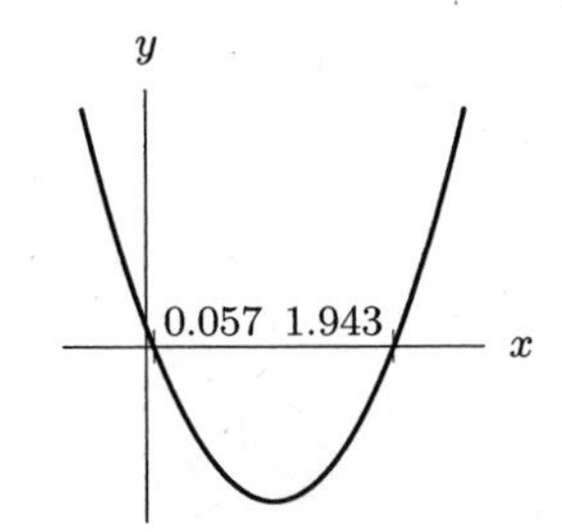

Figure 3.1: The graph of $y = 3x^2 - 6x + \frac{1}{3}$ crosses x-axis at $x \approx 0.057$, and $x \approx 1.943$

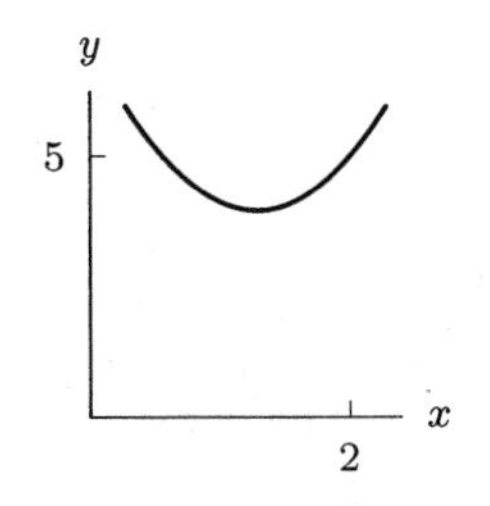

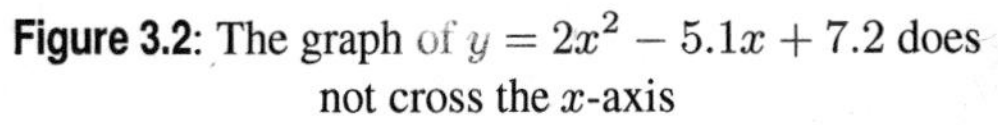

Figure 3.2: The graph of $y = 2x^2 - 5.1x + 7.2$ does not cross the x-axis

(b) Rewriting $2x^2 + 7.2 = 5.1x$ in the form $ax^2 + bx + c = 0$, we get $2x^2 - 5.1x + 7.2 = 0$. Applying the quadratic formula, we obtain

$$x = \frac{5.1 \pm \sqrt{5.1^2 - 4 \cdot 2 \cdot 7.2}}{2 \cdot 2}.$$

Notice that $5.1^2 - 4 \cdot 2 \cdot 7.2 = -31.59$, so the number under the square root sign is negative. Thus, there are no real solutions. (See Figure 3.2.)

11. Setting the factors equal to zero, we have

$$\begin{aligned} 2 - x &= 0 \\ x &= 2 \\ \text{and} \quad 3 - 2x &= 0 \\ x &= 3/2, \end{aligned}$$

so the zeros are $x = 2, 3/2$.

12. To find the zeros, we solve the equation

$$0 = 2x^2 + 5x + 2.$$

We see that this is factorable, as follows:

$$y = (2x + 1)(x + 2).$$

Therefore, the zeros occur where $x = -2$ and $x = -\frac{1}{2}$.

13. To find the zeros, we solve the equation

$$0 = 4x^2 - 4x - 8.$$

We see that this is factorable, as follows:

$$0 = 4(x^2 - x - 2)$$
$$0 = 4(x - 2)(x + 1).$$

Therefore, the zeros occur where $x = 2$ and $x = -1$.

14. To find the zeros, we solve the equation

$$0 = 7x^2 + 16x + 4.$$

We see that this is factorable, as follows:

$$0 = (7x + 2)(x + 2).$$

Therefore, the zeros occur where $x = -\frac{2}{7}$ and $x = -2$.

15. Using the quadratic formula, we have

$$\begin{aligned} x &= \frac{-2 \pm \sqrt{2^2 - 4(5)(-1)}}{2(5)} \\ &= \frac{-2 \pm \sqrt{24}}{10} \\ &= \frac{-1 \pm \sqrt{6}}{5}. \end{aligned}$$

16. To find the zeros, we solve the equation $0 = -17x^2 + 23x + 19$. This does not appear to be factorable. Thus, we use the quadratic formula with $a = -17$, $b = 23$, and $c = 19$:

$$\begin{aligned} x &= \frac{-23 \pm \sqrt{23^2 - 4(-17)(19)}}{2(-17)} \\ x &= \frac{-23 \pm \sqrt{1821}}{-34}. \end{aligned}$$

Therefore, the zeros occur where $x = (-23 \pm \sqrt{1821})/(-34) \approx 1.932$ or -0.579.

17. Letting $z = x^2$, we have $y = z^2 + 5z + 6$. This can be factored, giving

$$\begin{aligned} y &= (z + 2)(z + 3) \\ &= (x^2 + 2)(x^2 + 3). \end{aligned}$$

Setting the factors equal to zero, we have $x^2 + 2 = 0$, which has no solution, and $x^2 + 3 = 0$, which also has no solution, so this function has no real-valued zeros. Another way to see this is to notice that both x^4 and $5x^2$ are either positive or 0, so y can not be less than 6.

18. Letting $z = \sqrt{x}$, we have $z^2 = x$, which gives $y = z^2 - z - 12$. Factoring, we have

$$\begin{aligned} y &= (z - 4)(z + 3) \\ &= (\sqrt{x} - 4)(\sqrt{x} + 3). \end{aligned}$$

Setting the factors equal to zero, we have $\sqrt{x} = 4$, so $x = 16$, and $\sqrt{x} = -3$, so $x = 9$. Checking our answers, we have $16 - \sqrt{16} - 12 = 16 - 4 - 12 = 0$, so $x = 16$ is a zero of the original function. However, we also see that $9 - \sqrt{9} - 12 = 9 - 3 - 12 = -6$, so $x = 9$ is not a zero of the original function. Thus, $x = 16$ is the only zero.

Problems

19. The zeros are $x = 1, 4$, so $y = a(x - 1)(x - 4)$. Solving for a, we have

$$\begin{aligned} a(0 - 1)(0 - 4) &= 7 \\ 4a &= 7 \\ a &= \frac{7}{4}, \end{aligned}$$

so $y = (7/4)(x - 1)(x - 4)$.

20. There is one zero at $x = -2$, so by symmetry the vertex is $(-2, 0)$. We have $y = a(x+2)^2$. Solving for a, we have

$$\begin{aligned} a(0+2)^2 &= 7 \\ 4a &= 7 \\ a &= \frac{7}{4}, \end{aligned}$$

so $y = (7/4)(x+2)^2$.

21. We solve the equation $f(t) = -16t^2 + 47t + 3 = 0$ using the quadratic formula

$$\begin{aligned} -16t^2 + 47t + 3 &= 0 \\ t &= \frac{-47 \pm \sqrt{47^2 - 4(-16)3}}{2(-16)}. \end{aligned}$$

Evaluating gives $t = -1/16$ sec and $t = 3$ sec; the value $t = 3$ sec is the time we want. The baseball hits the ground 3 sec after it was hit.

22. No, there is not. The shape of a non-trivial quadratic function is a parabola, and a parabola cannot intersect the x-axis more than twice, whereas a function with zeros $x = 1$, $x = 2$, and $x = 3$ would intersect the x-axis three times.

23. The function $f(x) = (x-1)(x-2)$ has zeros $x = 1$ and $x = 2$. To get another function with the same zeros, we can multiply $f(x)$ by any constant: for example, let $g(x) = -7(x-1)(x-2)$. In general, any function of the form $y = a(x-1)9x-2)$ will do.

24. Between $x = -1$ and $x = 1$

$$\frac{\Delta f(x)}{\Delta x} = \frac{f(1) - f(-1)}{1-(-1)} = \frac{(4-1^2) - (4-(-1)^2)}{2} = 0.$$

Between $x = 1$ and $x = 3$

$$\frac{\Delta f(x)}{\Delta x} = \frac{f(3) - f(1)}{3-1} = \frac{(4-3^2) - (4-1^2)}{2} = -4.$$

Between $x = 3$ and $x = 5$

$$\frac{\Delta f(x)}{\Delta x} = \frac{f(5) - f(3)}{5-3} = \frac{(4-5^2) - (4-3^2)}{2} = -8.$$

Since rates of change are decreasing, the graph of $f(x)$ is concave down.

25. For example, we can use $y = (x+2)(x-3)$. See Figure 3.3.

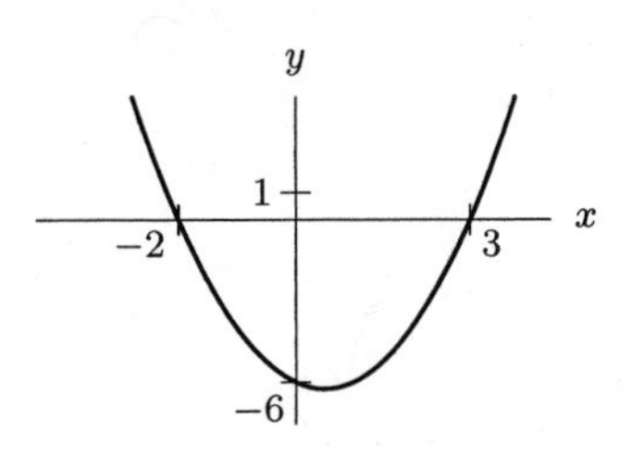

Figure 3.3

26. Factoring gives $y = 3x^2 - 16x - 12 = (3x+2)(x-6)$. So the zeros are $x = -\frac{2}{3}$ and $x = 6$, the axis of symmetry is halfway between the zeros at $x = \dfrac{-\frac{2}{3} + 6}{2} = \dfrac{8}{3}$. The y-coordinates of the vertex is

$$y = 3\left(\frac{8}{3}\right)^2 - 16\left(\frac{8}{3}\right) - 12 = -\frac{100}{3}$$

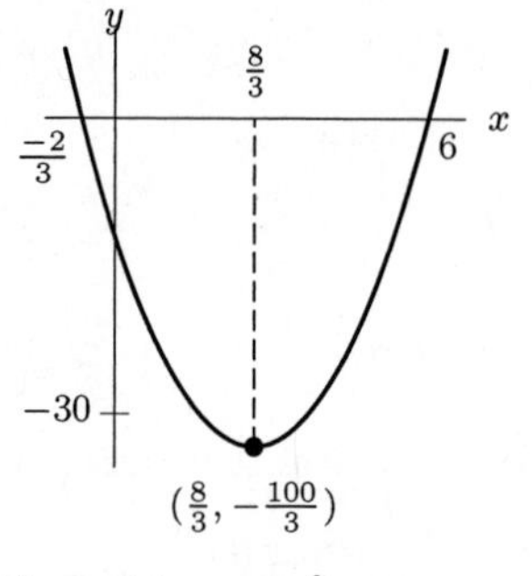

Figure 3.4: $y = 3x^2 - 16x - 12$

27. Factoring gives $y = -4cx + x^2 + 4c^2 = x^2 - 4ck + 4c^2 = (x - 2c)^2$. Since $c > 0$, this is the graph of $y = x^2$ shifted to the right $2c$ units. See Figure 3.5.

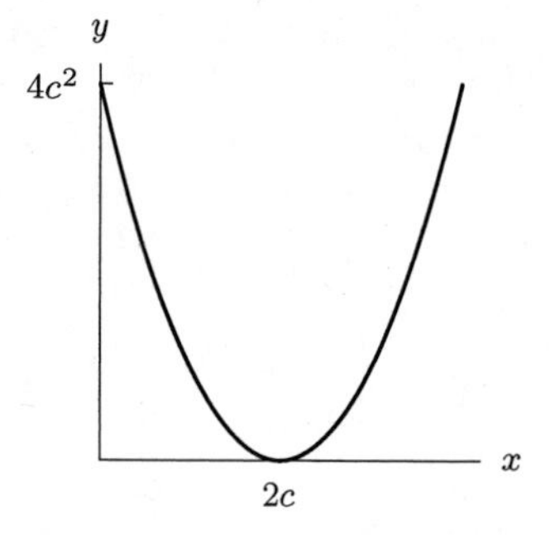

Figure 3.5: $y = -4cx + x^2 + 4c^2$ for $c > 0$

28. We know there are zeros at $x = -1$ and $x = 3$, so we use the factored form

$$y = a(x + 1)(x - 3).$$

We solve for a by substituting $x = 0$, $y = -1$ giving

$$-1 = a(1)(-3)$$
$$a = \frac{1}{3}.$$

Thus, the parabola is

$$y = \frac{1}{3}(x + 1)(x - 3)$$

or

$$y = \frac{1}{3}x^2 - \frac{2}{3}x - 1.$$

29. We know there are zeros at $x = -6$ and $x = 2$, so we use the factored form:

$$y = a(x + 6)(x - 2)$$

and solve for a. At $x = 0$, we have

$$5 = a(0 + 6)(0 - 2)$$
$$5 = -12a$$
$$-\frac{5}{12} = a.$$

Thus,

$$y = -\frac{5}{12}(x+6)(x-2)$$

or

$$y = -\frac{5}{12}x^2 - \frac{5}{3}x + 5.$$

30. Since the parabola has x-intercepts at $x = -1$ and $x = 5$, its formula is:

$$y = a(x+1)(x-5).$$

The coordinates $(-2, 6)$ must satisfy the equation, so

$$6 = a(-2+1)(-2-5).$$

Solving for a gives $a = \frac{6}{7}$. The formula is:

$$y = \frac{6}{7}(x+1)(x-5).$$

31. We have

$$\begin{aligned} y &= 3(0.5x-4)(4-20x) \\ &= 3(0.5)(x-8)(-20)(x-4/20) \\ &= -30(x-8)(x-0.2), \end{aligned}$$

so $k = -30, r = 8, s = 0.2$. Note that the order of r and s does not matter.

32. **(a)** $h(2) = 80(2) - 16(2)^2 = 160 - 64 = 96$. This means that after 2 seconds, the ball's height is 96 feet.

(b)

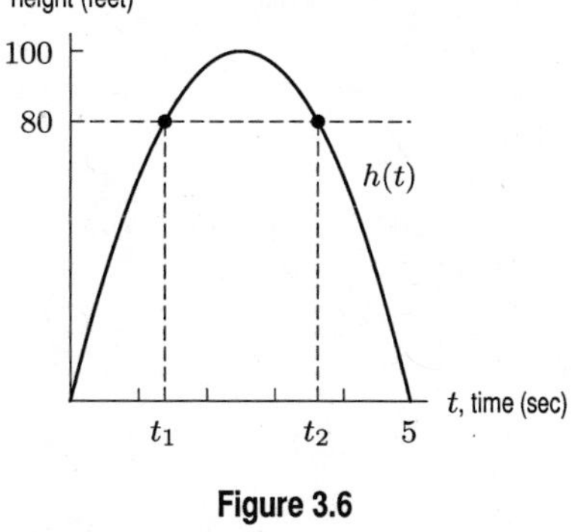

Figure 3.6

$h(t) = 80$ has 2 solutions, as you can see from Figure 3.6. One way to find these solutions is by using a graphing calculator. Another way is to solve

$$\begin{aligned} 80t - 16t^2 &= 80 \\ 16t^2 - 80t + 80 &= 0. \end{aligned}$$

Divide both sides of the equation by 16:

$$t^2 - 5t + 5 = 0.$$

Use the quadratic formula

$$t = \frac{5 \pm \sqrt{25 - 4 \cdot 5}}{2} = \frac{5 \pm \sqrt{5}}{2}.$$

The solutions are $t \approx 1.382$ and $t \approx 3.618$. This means that the ball reaches the height of 80 ft once on the way up, after approximately 1.382 seconds, and once on the way down, after 3.618 seconds.

33. **(a)** The initial velocity is the velocity when $t = 0$. So $v(0) = 0^2 - 4 \cdot 0 + 4 = 4$ meters per second.

(b) The object is not moving when its velocity is zero. This time is found by factoring $t^2 - 4t + 4 = (t-2)^2 = 0$. The solution, $t = 2$, tells us that the object is not moving at 2 seconds.

(c) From the graph of the velocity function in Figure 3.7, we can see that it is concave up.

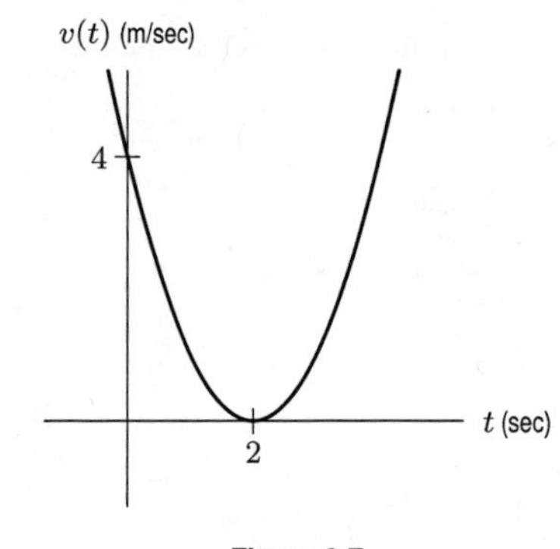

Figure 3.7

34. **(a)** At $t = 0$ the snowboarder is 5 meters below the edge of the half-pipe.

(b) We find the zeros of $y = -4.9t^2 + 14t - 5$ using the quadratic formula:

$$t = \frac{-14 \pm \sqrt{14^2 - 4(-4.9)(-5)}}{2(-4.9)}$$
$$t = 0.4184 \text{ or } 2.4387.$$

Thus snowboarder leaves the pipe and flies into the air before she returns to the pipe, so we choose the lower zero. She reaches the air after 0.4184 seconds.

She comes back to the pipe at the second zero, after 2.4387 seconds.

(c) She is in the air from the time she leaves the pipe until the time she returns, from 0.4184 seconds to 2.4387 seconds. Thus, she spends $2.4387 - 0.4184 = 2.0203$ seconds in the air.

35. To show that the data in the table is approximated by the formula $p(x) = -0.8x^2 + 8.8x + 7.2$, we substitute $x = 0, 1, 2, 3, 4$ (for years 1992-1996) into the formula:

$$p(0) = 7.2,\ p(1) = 15.2,\ p(2) = 21.6,\ p(3) = 26.4,\ p(4) = 29.6.$$

Our results approximate the table. In the year 2004, $x = 12$, and $p(12) = -2.4$, so the model predicts -2.4% of schools will have videodisc players in 2004. This is a reasonable model for the period 1992 to 1996, but not for the year 2004 since -2.4% does not make sense. Since the x^2 term in $p(x)$ has a negative coefficient, as x increases beyond 12, the values of $p(x)$ become more negative, and so are not reasonable predictions for the percentage of schools with a videodisc player. Thus, $p(x)$ is not a good model for predicting the future.

36. **(a)** In this window we see the expected parabolic shapes of $f(x)$ and $g(x)$. Both graphs open upward, so their shapes are similar and the end behaviors are the same. The differences in $f(x)$ and $g(x)$ are apparent at their vertices and intercepts. The graph of $f(x)$ has one intercept at $(0,0)$. The graph of $g(x)$ has x-intercepts at $x = -4$ and $x = 2$, and a y-intercept at $y = -8$. See Figure 3.8.

(b) As we extend the range to $y = 100$, the difference between the y-intercepts for $f(x)$ and $g(x)$ becomes less significant. See Figure 3.9.

Figure 3.8

Figure 3.9

(c) In the window $-20 \leq x \leq 20, -10 \leq y \leq 400$, the graphs are still distinguishable from one another, but all intercepts appear much closer. In the next window, the intercepts appear the same for $f(x)$ and $g(x)$. Only a thickening along the sides of the parabola gives the hint of two functions. In the last window, the graphs appear identical. See Figure 3.10.

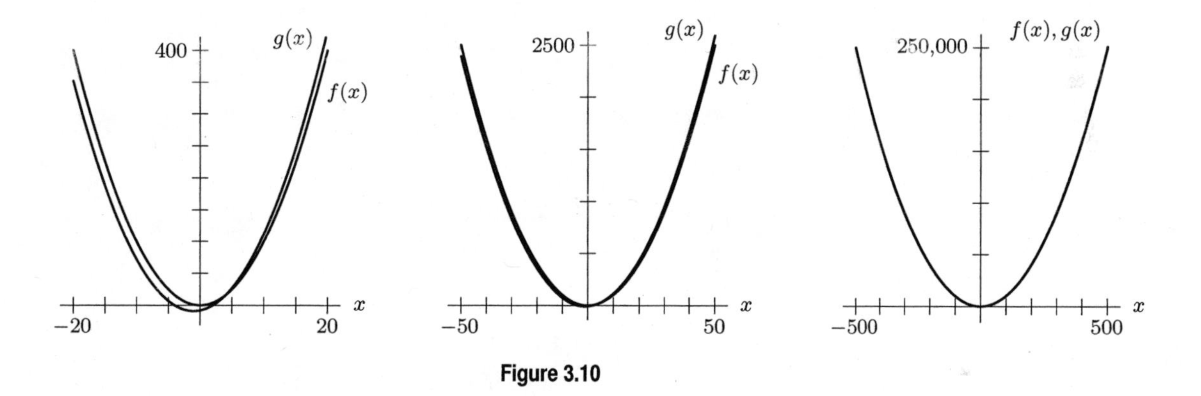

Figure 3.10

37. (a) According to the figure in the text, the package was dropped from a height of 5 km.

(b) When the package hits the ground, $h = 0$ and $d = 4430$. So, the package has moved 4430 meters forward when it lands.

(c) Since the maximum is at $d = 0$, the formula is of the form $h = ad^2 + b$ where a is negative and b is positive. Since $h = 5$ at $d = 0$, $5 = a(0)^2 + b = b$, so $b = 5$. We now know that $h = ad^2 + 5$. Since $h = 0$ when $d = 4430$, we have $0 = a(4430)^2 + 5$, giving $a = \dfrac{-5}{(4430)^2} \approx -0.000000255$. So $h \approx -0.000000255d^2 + 5$.

Solutions for Section 3.2

Skill Refresher

S1. $y^2 - 12y = y^2 - 12y + 36 - 36 = (y-6)^2 - 36$

S2. $s^2 + 6s - 8 = s^2 + 6s + 9 - 9 - 8 = (s+3)^2 - 17$

S3. We add and subtract the square of half the coefficient of the c-term, $(\frac{3}{2})^2 = \frac{9}{4}$, to get

$$\begin{aligned} c^2 + 3c - 7 &= c^2 + 3c + \frac{9}{4} - \frac{9}{4} - 7 \\ &= \left(c^2 + 3c + \frac{9}{4}\right) - \frac{9}{4} - 7 \\ &= \left(c + \frac{3}{2}\right)^2 - \frac{37}{4}. \end{aligned}$$

S4. Factoring out the coefficient of s^2 gives

$$4\left(s^2 + \frac{1}{4}s + \frac{1}{2}\right).$$

Inside the parenthesis, we add and subtract the square of half the coefficient of the s-term, $\left(\frac{1/4}{2}\right)^2 = \frac{1}{64}$, to get

$$\begin{aligned} 4\left(s^2 + \frac{1}{4}s + \frac{1}{2}\right) &= 4\left(s^2 + \frac{1}{4}s + \frac{1}{64} - \frac{1}{64} + \frac{1}{2}\right) \\ &= 4\left(\left(s + \frac{1}{8}\right)^2 - \frac{1}{64} + \frac{1}{2}\right) \\ &= 4\left(s + \frac{1}{8}\right)^2 + \frac{31}{16}. \end{aligned}$$

S5. Get the variables on the left side, the constants on the right side and complete the square using $(\frac{-6}{2})^2 = 9$.

$$\begin{aligned} r^2 - 6r &= -8 \\ r^2 - 6r + 9 &= 9 - 8 \\ (r-3)^2 &= 1. \end{aligned}$$

Take the square root of both sides and solve for r.

$$\begin{aligned} r - 3 &= \pm 1 \\ r &= 3 \pm 1. \end{aligned}$$

So, $r = 4$ or $r = 2$.

S6. Get the variables on the left side, the constants on the right side and complete the square using $\left(\frac{-3}{2}\right)^2 = \frac{9}{4}$.

$$\begin{aligned} n^2 - 3n &= 18 \\ n^2 - 3n + \frac{9}{4} &= 18 + \frac{9}{4} \\ \left(n - \frac{3}{2}\right)^2 &= 18 + \frac{9}{4} \\ \left(n - \frac{3}{2}\right)^2 &= \frac{81}{4}. \end{aligned}$$

Take the square root of both sides and solve for n.

$$\begin{aligned} n - \frac{3}{2} &= \pm\frac{9}{2} \\ n &= \frac{3}{2} \pm \frac{9}{2}. \end{aligned}$$

So, $n = 6$ or $n = -3$.

S7. Get the variables on the left side, the constants on the right side and complete the square using $(\frac{2/5}{2})^2 = \frac{1}{25}$.

$$\begin{aligned}
5q^2 - 2q &= 8 \\
5\left(q^2 - \frac{2}{5}q\right) &= 8 \\
5\left(q^2 - \frac{2}{5}q + \frac{1}{25}\right) &= 5\left(\frac{1}{25}\right) + 8 \\
5\left(q - \frac{1}{5}\right)^2 &= \frac{1}{5} + 8 \\
5\left(q - \frac{1}{5}\right)^2 &= \frac{41}{5}.
\end{aligned}$$

Divide by 5, take the square root of both sides and solve for q.

$$\begin{aligned}
\left(q - \frac{1}{5}\right)^2 &= \frac{41}{25} \\
q - \frac{1}{5} &= \pm\sqrt{\frac{41}{25}} \\
q - \frac{1}{5} &= \pm\frac{\sqrt{41}}{5} \\
q &= \frac{1}{5} \pm \frac{\sqrt{41}}{5}.
\end{aligned}$$

S8. Use the quadratic formula with $a = -3$, $b = 4$, and $c = 9$, to get

$$\begin{aligned}
t &= \frac{-4 \pm \sqrt{4^2 - 4\cdot(-3)\cdot 9}}{2\cdot(-3)} \\
&= \frac{-4 \pm \sqrt{16 + 108}}{-6} \\
&= \frac{-4 \pm \sqrt{124}}{-6} \\
&= \frac{-4 \pm 2\sqrt{31}}{-6} \\
&= \frac{2 \pm \sqrt{31}}{3}.
\end{aligned}$$

S9. Rewrite the equation to equal zero, and factor.

$$\begin{aligned}
n^2 + 4n - 5 &= 0 \\
(n+5)(n-1) &= 0.
\end{aligned}$$

So, $n + 5 = 0$ or $n - 1 = 0$, thus $n = -5$ or $n = 1$.

S10. Rewrite the equation as $2q^2 + 4q = 13$ and solve by completing the square.

$$\begin{aligned}
2(q^2 + 2q) &= 13 \\
2(q^2 + 2q + 1) &= 2\cdot 1 + 13 \\
2(q+1)^2 &= 2 + 13 \\
2(q+1)^2 &= 15.
\end{aligned}$$

Dividing by 2, taking the square root of both sides and solving for q, we get

$$(q+1)^2 = \frac{15}{2}$$
$$q+1 = \pm\sqrt{\frac{15}{2}}$$
$$q = -1 \pm \sqrt{\frac{15}{2}}.$$

Exercises

1. By comparing $f(x)$ to the vertex form, $y = a(x-h)^2 + k$, we see the vertex is $(h,k) = (1,2)$. The axis of symmetry is the vertical line through the vertex, so the equation is $x = 1$. The parabola opens upward because the value of a is positive 3.

2. To compare $g(x)$ with the vertex form, rewrite it as $g(x) = -1(x-(-3))^2 + (-4)$. We then see the vertex is $(h,k) = (-3,-4)$. The axis of symmetry is the vertical line through the vertex, so the equation is $x = -3$. The parabola opens downward because the value of a is negative 1.

3. To complete the square, we take $\frac{1}{2}$ of the coefficient of t and square the result. This gives $(\frac{1}{2}\cdot 11)^2 = (\frac{11}{2})^2 = \frac{121}{4}$. Using this number, we can rewrite the formula for $v(t)$:

$$v(t) = \underbrace{t^2 + 11t + \left(\frac{11}{2}\right)^2}_{\text{completing the square}} - \underbrace{\frac{121}{4}}_{\text{compensating term}} - 4$$
$$= \left(t + \frac{11}{2}\right)^2 - \frac{137}{4}.$$

Thus, the vertex of v is $(-\frac{11}{2}, -\frac{137}{4})$ and the axis of symmetry is $t = -\frac{11}{2}$.

4. Since the coefficient of x^2 is not 1, we first factor out the coefficient of x^2 from the formula. This gives

$$w(x) = -3\left(x^2 + 10x - \frac{31}{3}\right).$$

We next complete the square of the expression in parentheses. To do this, we add $(\frac{1}{2}\cdot 10)^2 = 25$ inside the parentheses:

$$w(x) = -3(\underbrace{x^2 + 10x + 25}_{\text{completing the square}} - \underbrace{25}_{\text{compensating term}} - 31/3).$$

Thus,

$$w(x) = -3((x+5)^2 - 106/3)$$
$$w(x) = -3(x+5)^2 + 106$$

so the vertex of the graph of this function is $(-5, 106)$, and the axis of symmetry is $x = -5$. Also, since $a = -3$ is negative, the graph is a downward opening parabola.

5. **(a)** See Figure 3.11. For g, we have $a = 1, b = 0$, and $c = 3$. Its vertex is at $(0,3)$, and its axis of symmetry is the y-axis, or the line $x = 0$. This function has no zeros.

   (b) See Figure 3.12. For f, we have $a = -2, b = 4$, and $c = 16$. The axis of symmetry is the line $x = 1$ and the vertex is at $(1,18)$. The zeros, or x-intercepts, are at $x = -2$ and $x = 4$. The y-intercept is at $y = 16$.

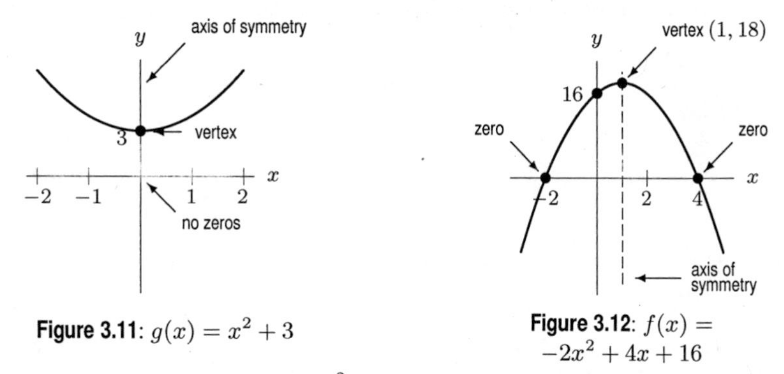

Figure 3.11: $g(x) = x^2 + 3$

Figure 3.12: $f(x) = -2x^2 + 4x + 16$

6. Factoring out negative one (to make the coefficient of x^2 equal 1) and completing the square gives

$$\begin{aligned} y &= -1 \cdot \left(x^2 - 7x + \left(-\frac{7}{2}\right)^2 - \left(-\frac{7}{2}\right)^2 + 13\right) \\ &= -\left(x - \frac{7}{2}\right)^2 + \frac{49}{4} - 13 \\ &= -\left(x - \frac{7}{2}\right)^2 - \frac{3}{4}. \end{aligned}$$

Thus, the graph of this function is a downward-opening parabola with a vertex below the x-axis. Since the graph is below the x-axis and opens down, it does not intersect the x-axis. We conclude that this function has no zeros which are real numbers.

To see this algebraically, notice that the equation $y = 0$ has no real-valued solution, because solving

$$-\left(x - \frac{7}{2}\right)^2 - \frac{3}{4} = 0$$

gives

$$x = \frac{7}{2} \pm \sqrt{-\frac{3}{4}}$$

and $\sqrt{-\frac{3}{4}}$ is not a real number.

7. The coordinates of the point $(6, 13)$ must satisfy the equation, so

$$13 = (6 - 3)^2 + k.$$

Solving for k gives $k = 4$. The formula is: $y = (x - 3)^2 + 4$.

8. Substituting the coordinates of the vertex gives

$$-2 = a(0)^2 + k.$$

Solving for k gives $k = -2$. The formula now is $y = ax^2 - 2$. Substituting the coordinates of the point now gives

$$4 = a(3)^2 - 2.$$

Solving for a gives $a = \frac{2}{3}$. The formula is:

$$y = \frac{2}{3}x^2 - 2.$$

9. Since the vertex is $(4, 7)$, we use the form $y = a(x-h)^2 + k$, with $h = 4$ and $k = 7$. We solve for a, substituting in the second point, $(0, 4)$.

$$\begin{aligned} y &= a(x-4)^2 + 7 \\ 4 &= a(0-4)^2 + 7 \\ -3 &= 16a \\ -\frac{3}{16} &= a. \end{aligned}$$

Thus, an equation for the parabola is

$$y = -\frac{3}{16}(x-4)^2 + 7.$$

10. Since the vertex is $(3, 3)$, we use the form $y = a(x-h)^2 + k$, with $h = 3$ and $k = 3$. We solve for a, substituting in the second point, $(5, 5)$.

$$\begin{aligned} y &= a(x-3)^2 + 3 \\ 5 &= a(5-3)^2 + 3 \\ 2 &= 4a \\ \frac{1}{2} &= a. \end{aligned}$$

Thus, an equation for the parabola is

$$y = \frac{1}{2}(x-3)^2 + 3.$$

11. The vertex is the point $(h, k) = (3, -5)$. Thus, a possible formula for this function is of the form

$$y = a(x-3)^2 - 5.$$

To find the value of a, we use the fact that the y-intercept of this function is $(0, 2)$. Thus, we have $x = 0, y = 2$, so

$$\begin{aligned} a(0-3)^2 - 5 &= 2 \\ 9a &= 7 \\ a &= \frac{7}{9}. \end{aligned}$$

The formula for this quadratic function is $y = \frac{7}{9}(x-3)^2 - 5$. Since $|a| < 1$, this graph is wider than the graph of $y = x^2$.

12. The function has zeros at $x = -4$ and $x = 5$, and appears quadratic, so it could be of the form $y = a(x+4)(x-5)$. Since $y = 36$ when $x = 2$, we know that $y = a(2+4)(2-5) = -18a = 36$, so $a = -2$. Therefore, $y = -2(x+4)(x-5)$ is a possible formula..

13. The square of half the coefficient of the x-term is $(\frac{8}{2})^2 = 16$. Adding and subtracting this number after the x-term gives

$$f(x) = x^2 + 8x + 16 - 16 + 3.$$

This can be simplified to $f(x) = (x+4)^2 - 13$. The vertex is $(-4, -13)$ and the axis of symmetry is $x = -4$.

14. Factoring out the coefficient of x^2 gives

$$g(x) = -2(x^2 - 6x - 2).$$

Inside the parenthesis, we add and subtract the square of half the coefficient of the x-term, $(-6/2)^2 = 9$, to get:

$$\begin{aligned} g(x) &= -2(x^2 - 6x + 9 - 9 - 2) \\ g(x) &= -2((x-3)^2 - 11) \\ g(x) &= -2(x-3)^2 + 22. \end{aligned}$$

The vertex is $(3, 22)$ and the axis of symmetry is $x = 3$.

15. Letting $y = p(t)$, we have:

$$\begin{aligned}
y &= 2t^2 - 0.12t + 0.1 \\
y - 0.1 &= 2t^2 - 0.12t && \text{subtract } 0.1 \\
0.5(y - 0.1) &= t^2 - 0.06t && \text{multiply by } 0.5 \\
0.5(y - 0.1) + (0.03)^2 &= t^2 - 0.06t + (0.03)^2 && \text{complete the square} \\
0.5(y - 0.1) + 0.0009 &= (t - 0.03)^2 && \text{factor right-hand side} \\
0.5(y - 0.1) &= (t - 0.03)^2 - 0.0009 && \text{subtract } 0.0009 \\
y - 0.1 &= 2(t - 0.03)^2 - 0.0018 && \text{multiply by } 2 \\
y &= 2(t - 0.03)^2 + 0.0982. && \text{add } 0.1
\end{aligned}$$

Thus, the vertex is $(0.03, 0.0982)$ and the axis of symmetry is $t = 0.03$.

16. Letting $y = -3z^2 + 9z - 2$, we have:

$$\begin{aligned}
y &= -3z^2 + 9z - 2 \\
y + 2 &= -3z^2 + 9z && \text{add } 2 \\
-\frac{1}{3}(y + 2) &= z^2 - 3z && \text{multiply by } -1/3 \\
-\frac{1}{3}(y + 2) + \left(\frac{3}{2}\right)^2 &= z^2 - 3x + \left(\frac{3}{2}\right)^2 && \text{complete the square} \\
-\frac{1}{3}(y + 2) + \frac{9}{4} &= \left(z - \frac{3}{2}\right)^2 && \text{factor right-hand side} \\
-\frac{1}{3}(y + 2) &= \left(z - \frac{3}{2}\right)^2 - \frac{9}{4} && \text{subtract } 9/4 \\
y + 2 &= -3\left(z - \frac{3}{2}\right)^2 + \frac{27}{4} && \text{multiply by } -3 \\
y &= -3\left(z - \frac{3}{2}\right)^2 + \frac{27}{4} - 2 && \text{subtract } 2 \\
&= -3\left(z - \frac{3}{2}\right)^2 + \frac{19}{4} && \text{simplify,}
\end{aligned}$$

so the vertex is $(3/2, 19/4)$ and the axis of symmetry is $z = 3/2$.

17. The standard form is obtained by writing the right-hand side as three terms and rearranging the terms:

$$y = \frac{1}{2}x^2 - \frac{1}{2}x - 6.$$

The vertex form can be obtained from the standard form by completing the square:

$$\begin{aligned}
y &= \frac{1}{2}x^2 - \frac{1}{2}x - 6 \\
&= \frac{1}{2}(x^2 - x - 12) \\
&= \frac{1}{2}(x^2 - x + \frac{1}{4} - \frac{1}{4} - 12) \\
&= \frac{1}{2}(x^2 - x + \frac{1}{4}) + \frac{1}{2}(-\frac{1}{4} - 12) \\
&= \frac{1}{2}(x - \frac{1}{2})^2 + \frac{1}{2}(-\frac{49}{4}) \\
&= \frac{1}{2}(x - \frac{1}{2})^2 - \frac{49}{8}.
\end{aligned}$$

The factored form can be obtained by factoring:

$$\begin{aligned}
y &= \frac{1}{2}(x^2 - x - 12) \\
&= \frac{1}{2}(x - 4)(x + 3).
\end{aligned}$$

18. The standard form is obtained by writing the right-hand side as two terms:

$$f(t) = \frac{5}{2}t^2 - 10.$$

(or $f(t) = \frac{5}{2}t^2 + 0t - 10$.) The vertex form can be obtained from the standard form by subtracting a zero in the square term:

$$f(t) = \frac{5}{2}t^2 - 10 = \frac{5}{2}(t-0)^2 - 10.$$

The factored form can be obtained as follows:

$$\begin{aligned} f(t) &= \frac{5t^2 - 20}{2} \\ &= \frac{5}{2}(t^2 - 4) \\ &= \frac{5}{2}(t-2)(t+2). \end{aligned}$$

19. The standard form is obtained by multiplying out the right-hand side and collecting like terms:

$$g(s) = (s-5)(2s+3) = 2s^2 + 3s - 10s - 15 = 2s^2 - 7s - 15.$$

The vertex form can be obtained from the standard form by completing the square:

$$\begin{aligned} g(s) &= 2s^2 - 7s - 15 \\ &= 2\left(s^2 - \frac{7}{2}s - \frac{15}{2}\right) \\ &= 2\left(s^2 - \frac{7}{2}s + \left(\frac{7}{4}\right)^2 - \left(\frac{7}{4}\right)^2 - \frac{15}{2}\right) \\ &= 2\left(s - \frac{7}{4}\right)^2 + 2\left(-\frac{49}{16} - \frac{120}{16}\right) \\ &= 2\left(s - \frac{7}{4}\right)^2 - \frac{169}{8}. \end{aligned}$$

The factored form is obtained by simply factoring the $(2s+3)$ term,

$$g(s) = (s-5)(2s+3) = 2(s-5)\left(s + \frac{3}{2}\right).$$

Problems

20. Using the vertex form $y = a(x-h)^2 + k$, where $(h, k) = (2, 5)$, we have

$$y = a(x-2)^2 + 5.$$

Since the parabola passes through $(1, 2)$, these coordinates must satisfy the equation, so

$$2 = a(1-2)^2 + 5.$$

Solving for a gives $a = -3$. The formula is:

$$y = -3(x-2)^2 + 5.$$

21. We have $(h, k) = (4, 2)$, so $y = a(x-4)^2 + 2$. Solving for a, we have

$$\begin{aligned} a(0-4)^2 + 2 &= 6 \\ 16a &= 4 \\ a &= \frac{1}{4}, \end{aligned}$$

so $y = (1/4)(x-4)^2 + 2$.

22. We have $(h, k) = (4, 2)$, so $y = a(x-4)^2 + 2$. Solving for a, we have

$$\begin{aligned} a(0-4)^2 + 2 &= -4 \\ 16a &= -6 \\ a &= -\frac{6}{16} = -\frac{3}{8}, \end{aligned}$$

so $y = (-3/8)(x-4)^2 + 2$.

23. We have $(h, k) = (4, 2)$, so $y = a(x-4)^2 + 2$. Solving for a, we have

$$\begin{aligned} a(11-4)^2 + 2 &= 0 \\ 49a &= -2 \\ a &= -\frac{2}{49}, \end{aligned}$$

so $y = (-2/49)(x-4)^2 + 2$. We can verify this formula using the other zero:

$$(-2/49)(-3-4)^2 + 2 = (-2/49)(-7)^2 + 2 = -2 + 2 = 0,$$

as required.

24. We have $(h, k) = (-7, -3)$, so $y = a(x+7)^2 - 3$. Solving for a, we have

$$\begin{aligned} a(-3+7)^2 - 3 &= -7 \\ 16a &= -4 \\ a &= -1/4, \end{aligned}$$

so $y = (-1/4)(x+7)^2 - 3$.

25. (a)

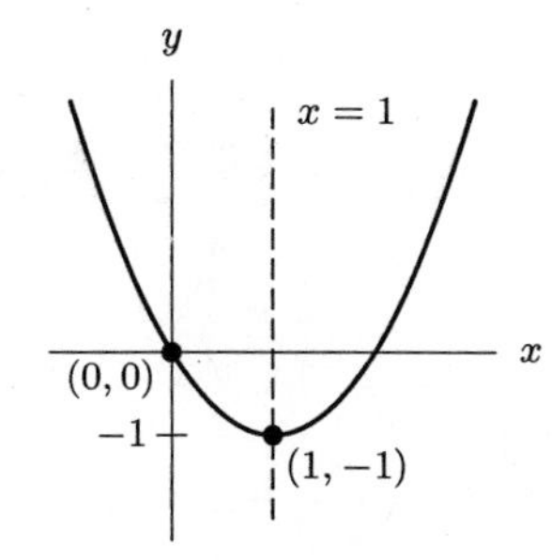

Figure 3.13

(b) Since the vertex is at $(1, -1)$, the parabola could be described by

$$f(x) = a(x-1)^2 - 1.$$

Since the parabola passes through $(0, 0)$

$$\begin{aligned} 0 &= a(0-1)^2 - 1 \\ 0 &= a - 1. \end{aligned}$$

Therefore

$$a = 1.$$

So, the equation is

$$f(x) = (x-1)^2 - 1$$

or

$$f(x) = x^2 - 2x.$$

(c) Since the vertex is at $(1, -1)$ and the parabola is concave up, the range of this function is all real numbers greater than or equal to -1.

(d) Since one zero is at $x = 0$, which is one unit to the left of the axis of symmetry at $x = 1$, the other zero will occur at one unit to the right of the axis of the symmetry at $x = 2$.

26. We have

$$\begin{aligned}
y &= 0.03x^2 + 1.8x + 2 \\
y - 2 &= 0.03x^2 + 1.8x \\
&= 0.03\left(x^2 + 60x\right) && \text{factor} \\
\frac{y-2}{0.03} &= x^2 + 60x \\
\frac{y-2}{0.03} + (30)^2 &= x^2 + 60x + (30)^2 && \text{complete the square} \\
\frac{y-2}{0.03} + 900 &= (x+30)^2 && \text{factor} \\
\frac{y-2}{0.03} &= (x+30)^2 - 900 \\
y - 2 &= 0.03(x+30)^2 - 0.03(900) \\
y &= 0.03\,(x - (-30))^2 - 25.
\end{aligned}$$

This is a quadratic function in vertex form with vertex $(h, k) = (-30, -25)$ and $a = 0.03$. From the original equation, we see that the y-intercept is $y = 2$. Since $a > 0$, the graph opens up, and since the vertex lies below the x-axis, there are two x-intercepts. Solving for $y = 0$ gives

$$\begin{aligned}
0.03(x+30)^2 - 25 &= 0 \\
0.03(x+30)^2 &= 25 \\
(x+30)^2 &= \frac{25}{0.03} \\
&= \frac{2500}{3} \\
x + 30 &= \pm\sqrt{\frac{2500}{3}} \\
x &= -30 \pm \sqrt{\frac{2500}{3}},
\end{aligned}$$

so the x-intercepts are $x = -58.868$ and $x = -1.133$. See Figure 3.14.

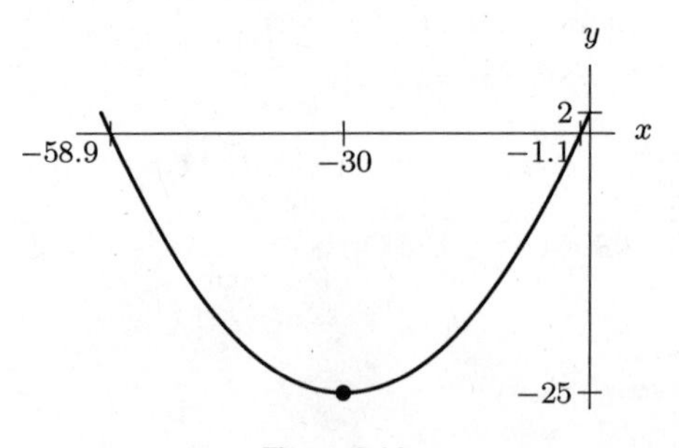

Figure 3.14

27. We have

$$\begin{aligned}
y &= 26 + 0.4x - 0.01x^2 \\
y - 26 &= 0.4x - 0.01x^2 \\
-100(y - 26) &= x^2 - 40x \\
-100(y - 26) + (-20)^2 &= x^2 - 40x + (-20)^2 \\
-100(y - 26) + 400 &= (x - 20)^2 \\
-100(y - 26) &= (x - 20)^2 - 400 \\
y - 26 &= -0.01(x - 20)^2 + 4 \\
y &= -0.01(x - 20)^2 + 30.
\end{aligned}$$

Thus, the graph is a downward-opening parabola with vertex $(h, k) = (20, 30)$. From the original equation, we see that the y-intercept is $y = 26$. We find the x-intercepts by solving $y = 0$:

$$\begin{aligned}
-0.01(x - 20)^2 + 30 &= 0 \\
-0.01(x - 20)^2 &= -30 \\
(x - 20)^2 &= 3000 \\
x - 20 &= \pm\sqrt{3000} \\
x &= 20 \pm \sqrt{3000},
\end{aligned}$$

so the x-intercepts are $x = 20 - \sqrt{3000} = -34.772$ and $x = 20 + \sqrt{3000} = 74.772$. See Figure 3.15.

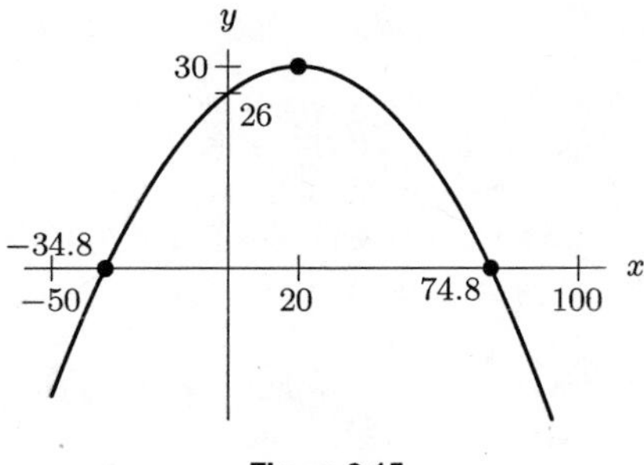

Figure 3.15

28. Yes, we can find the function. Because the vertex is $(1, 4)$, $f(x) = a(x - 1)^2 + 4$ for some a. To find a, we use the fact that $x = -1$ is a zero, that is, the fact that $f(-1) = 0$. We can write $f(-1) = a(-1 - 1)^2 + 4 = 0$, so $4a + 4 = 0$ and $a = -1$. Thus $f(x) = -(x - 1)^2 + 4$.

29. The distance around any rectangle with a height of h units and a base of b units is $2b + 2h$. See Figure 3.16. Since the string forming the rectangle is 50 cm long, we know that $2b + 2h = 50$ or $b + h = 25$. Therefore, $b = 25 - h$. The area, A, of such a rectangle is

$$\begin{aligned}
A &= bh \\
A &= (25 - h)(h).
\end{aligned}$$

The zeros of this quadratic function are $h = 0$ and $h = 25$, so the axis of symmetry, which is halfway between the zeros, is $h = 12.5$. Since the maximum value of A occurs on the axis of symmetry, the area will be the greatest when the height is 12.5 and the base is also 12.5 ($b = 25 - h = 25 - 12.5 = 12.5$).

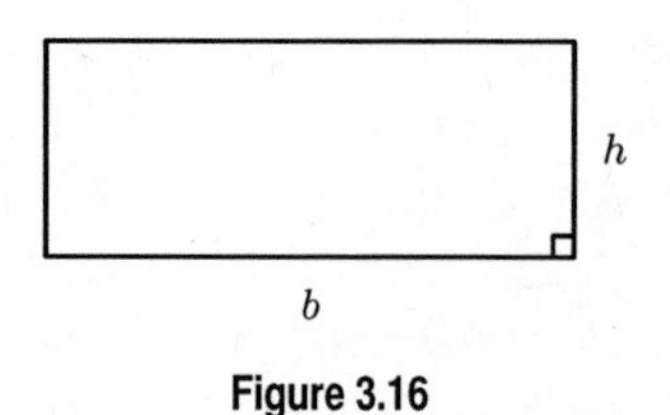

Figure 3.16

If the string were k cm long, $2b + 2h = k$ or $b + h = \frac{k}{2}$, so $b = \frac{k}{2} - h$. $A = bh = \left(\frac{k}{2} - h\right) h$. The zeros in this case are $h = 0$ and $h = \frac{k}{2}$, so the axis of symmetry is $h = \frac{k}{4}$. If $h = \frac{k}{4}$, then $b = \frac{k}{2} - h = \frac{k}{2} - \frac{k}{4} = \frac{k}{4}$. So the dimensions for maximum area are $\frac{k}{4}$ by $\frac{k}{4}$; in other words, the rectangle with the maximum area is a square whose side measures $\frac{1}{4}$ of the length of the string.

30. **(a)** See Figure 3.17

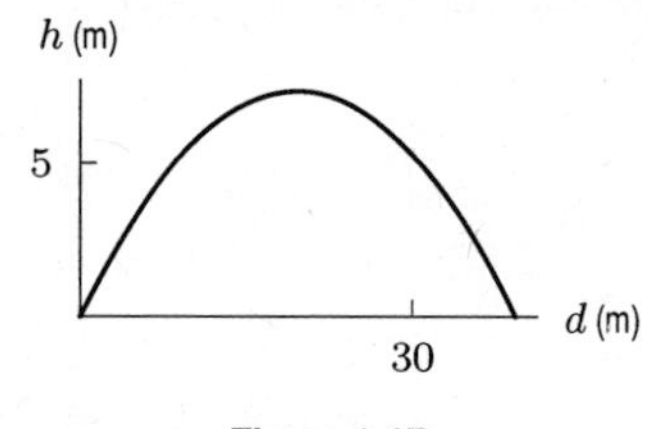

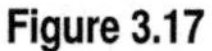

Figure 3.17

(b) When the ball hits the ground $h = 0$, so $h = 0.75d - 0.0192d^2 = d(0.75 - 0.0192d) = 0$ and we get $d = 0$ or $d \approx 39.063$ m. Since $d = 0$ is the position where the kicker is standing, the ball must hit the ground about 39.063 meters from the point where it is kicked.

(c) The path is parabolic and the maximum height occurs at the vertex, which lies on the axis of symmetry, mid-way between the zeros at $d \approx 19.531$ m. Since $h = 0.75(19.531) - 0.0192(19.531)^2 \approx 7.324$, we know that the ball reaches 7.324 meters above the ground before it begins to fall.

(d) From part (c), the horizontal distance traveled when the ball reaches its maximum height is ≈ 19.531 m.

31. **(a)** Factoring gives $h(t) = -16t^2 + 16Tt = 16t(T - t)$. Since $h(t) \geq 0$ only for $0 \leq t \leq T$, the model makes sense only for these values of t.

(b) The times $t = 0$ and $t = T$ give the start and end of the jump. The maximum height occurs halfway in between, at $t = T/2$.

(c) Since $h(t) = 16t(T - t)$, we have

$$h\left(\frac{T}{2}\right) = 16\left(\frac{T}{2}\right)\left(T - \frac{T}{2}\right) = 4T^2.$$

Solutions for Chapter 3 Review

Exercises

1. We have:

$$f(x) = (2x - 3)(5 - x) = -2x^2 + 13x - 15,$$

so $a = -2, b = 13, c = -15$.

2. We have:

$$g(t) = 3(t - 2)^2 + 7 = 3t^2 - 12t + 19,$$

so $a = 3, b = -12, c = 19$.

3. We have:

$$\begin{aligned} w(n) &= n(n-3)(n-2) - n^2(n-8) \\ &= n(n^2-5n+6) - (n^3-8n^2) \\ &= n^3 - 5n^2 + 6n - n^3 + 8n^2 \\ &= 3n^2 + 6n + 0, \end{aligned}$$

so $a = 3, b = 6, c = 0$.

4. We have:

$$\begin{aligned} r(v) &= \frac{v^2+\sqrt{2}}{3} + \frac{v-3}{5} + \pi v^2 \\ &= \frac{1}{3}v^2 + \frac{\sqrt{2}}{3} + \frac{v}{5} - \frac{3}{5} + \pi v^2 && \text{split numerators} \\ &= \frac{1}{3}v^2 + \pi v^2 + \frac{1}{5}v + \frac{\sqrt{2}}{3} - \frac{3}{5} && \text{gather like terms} \\ &= \left(\frac{1}{3}+\pi\right)v^2 + \frac{1}{5}v + \frac{5\sqrt{2}-9}{15}, && \text{factor and combine} \end{aligned}$$

so $a = 1/3 + \pi, b = 1/5, c = (5\sqrt{2} - 9)/15$.

5. To find the zeros, we solve the equation

$$0 = 9x^2 + 6x + 1.$$

We see that this is factorable, as follows:

$$\begin{aligned} y &= (3x+1)(3x+1) \\ y &= (3x+1)^2. \end{aligned}$$

Therefore, there is only one zero at $x = -\frac{1}{3}$.

6. This can be factored as follows:

$$\begin{aligned} 6x^2 - 17x + 12 &= 6x^2 \underbrace{-8x - 9x}_{-17x} + 12 \\ &= 2x(3x-4) - 3(3x-4) \\ &= (2x-3)(3x-4). \end{aligned}$$

Setting these factors equal to zero, we find that the zeros of this function are $x = 3/2, 4/3$.

7. To find the zeros, we solve the equation $0 = 89x^2 + 55x + 34$. This does not appear to be factorable. Thus, we use the quadratic formula with $a = 89$, $b = 55$, and $c = 34$:

$$\begin{aligned} x &= \frac{-55 \pm \sqrt{55^2 - 4(89)(34)}}{2(89)} \\ x &= \frac{-55 \pm \sqrt{-9079}}{178} \end{aligned}$$

Since $\sqrt{-9079}$ is undefined, there are no zeros.

8. Using the quadratic formula, we have

$$\begin{aligned} x &= \frac{-(-2) \pm \sqrt{(-2)^2 - 4(3)(6)}}{2(3)} \\ &= \frac{2 \pm \sqrt{-68}}{6}, \end{aligned}$$

so there are no real-valued zeros.

9. We solve for t in the equation by factoring

$$N(t) = t^2 - 7t + 10 = (t-2)(t-5) = 0.$$

The zeros of N(t) are $t = 2$ and $t = 5$.

10. We solve for r in the equation $Q(r) = 2r^2 - 6r - 36 = 0$ using the quadratic formula with $a = 2$, $b = -6$ and $c = -36$.

$$\begin{aligned} r &= \frac{-(-6) \pm \sqrt{(-6)^2 - 4(2)(-36)}}{2(2)} \\ r &= \frac{6 \pm \sqrt{36 + 288}}{4} \\ r &= \frac{6 \pm \sqrt{324}}{4} \\ r &= \frac{6 \pm 18}{4}. \end{aligned}$$

Therefore $r = (6+18)/4 = 6$ and $r = (6-18)/4 = -3$. The zeros of $Q(r)$ are $r = 6$ and $r = -3$.

11. The function will have real zeros if and only if $(b^2 - 4ac) \geq 0$. Since $(-1)^2 - 4(1)(41) = -163 < 0$, there are no real zeros.

12. To write the equation in vertex form we complete the square.

$$\begin{aligned} y &= 3x^2 - 6x + 5 \\ &= 3(x^2 - 2x) + 5 \\ &= 3(x^2 - 2x + 1) + 5 - 3 \\ &= 3(x-1)^2 + 2. \end{aligned}$$

Therefore, the vertex is $(1, 2)$ and the axis of symmetry is $x = 1$.

13. We have $y = a(x-1)^2 - 2$ and if $x = 0$, $y = -5$, so $-5 = a(-1)^2 - 2$. Therefore, $a = -3$, and we have $y = -3(x-1)^2 - 2$.

14. We have $y = a(x-4)^2 - 2$, and if $x = 0$, $y = -3$, so $-3 = a(16) - 2$. Therefore $a = -1/16$, and we have $y = (-1/16)(x-4)^2 - 2$.

15. We have $y = a(x-7)^2 + 3$. Since the point $(3, 7)$ is on the curve, we obtain $7 = a(-4)^2 + 3$, so $a = 1/4$. Therefore $(1/4)(x-7)^2 + 3$.

16. We have $y = a(x-1)^2 - 1$. Since $(0, 0)$ is on the curve, $0 = a(-1)^2 - 1$. Therefore, $a = 1$ and we have $y = (x-1)^2 - 1$.

17. We have $y = a(x+1)(x-2)$. Since $(-2, 16)$ is on the curve, $16 = a(-1)(-4)$. Therefore $a = 4$, so $y = 4(x+1)(x-2)$.

18. We have $y = a(x - 1/2)^2$. Since $(0, 3)$ is on the curve, $3 = a(-1/2)^2$ or $3 = a/4$. Therefore, $a = 12$. Thus, $y = 12(x - 1/2)^2$.

19. The function has zeros at $x = -1$ and $x = 3$, and appears quadratic, so it could be of the form $y = a(x+1)(x-3)$. Since $y = -3$ when $x = 0$, we know that $y = a(0+1)(0-3) = -3a = -3$, so $a = 1$. Thus $y = (x+1)(x-3)$ is a possible formula.

20. The vertex is the point $(h, k) = (-6, 9)$, so a possible equation is

$$y = a(x+6)^2 + 9.$$

Solving for a, we use the fact that the graph has an x-intercept of 15, so $y = 0$ when $x = 15$. This gives

$$\begin{aligned} a(-15+6)^2 + 9 &= 0 \\ 81a &= -9 \\ a &= -\frac{1}{9}, \end{aligned}$$

so $y = -\frac{1}{9}(x+6)^2 + 9$. Since $a < 0$, this graph is a flipped upside down and much wider than $y = x^2$.

21. The function appears quadratic with vertex at $(2,0)$, so it could be of the form $y = a(x-2)^2$. For $x = 0$, $y = -4$, so $y = a(0-2)^2 = 4a = -4$ and $a = -1$. Thus $y = -(x-2)^2$ is a possible formula.

22. Since the vertex is $(6,5)$, we use the form $y = a(x-h)^2 + k$, with $h = 6$ and $k = 5$. We solve for a, substituting in the second point, $(10,8)$.

$$\begin{aligned} y &= a(x-6)^2 + 5 \\ 8 &= a(10-6)^2 + 5 \\ 3 &= 16a \\ \frac{3}{16} &= a. \end{aligned}$$

Thus, an equation for the parabola is

$$y = \frac{3}{16}(x-6)^2 + 5.$$

Problems

23. By inspection the vertex is $(3/4, -2/3)$. The axis of symmetry is $x = 3/4$. The y-intercept occurs when $x = 0$, so $y = 2(3/4)^2 - 2/3 = 11/24$. Since the coefficient a is positive the curve is concave up.

24. By inspection the vertex is $(-6,-4)$. The axis of symmetry is $x = -6$. The y-intercept occurs when $x = 0$, so $y = -1/2(6)^2 - 4 = -22$. Since the a term is negative the curve is concave down.

25. By inspection the vertex is $(0.6, 0)$. The axis of symmetry is $x = 0.6$. The y-intercept occurs when $x = 0$, so $y = 0.36$. Since the coefficient a is positive the curve is concave up.

26. We can think of this parabola as $y = -0.3(x-0)^2 - 7$. By inspection the vertex is $(0,-7)$. The axis of symmetry is $x = 0$. The y-intercept occurs when $x = 0$, so $y = -7$. Since the coefficient a is negative the curve is concave down.

27. In factored form, we have

$$\begin{aligned} y &= 0.3x^2 - 0.6x - 7.2 \\ &= 0.3(x^2 - 2x - 24) \\ &= 0.3(x-6)(x+4). \end{aligned}$$

Therefore, the zeros are at $x = 6$ and $x = -4$. The axis of symmetry is midway between the zeros, so its equation is $x = 1$. The vertex occurs on the axis of symmetry, so substituting $x = 1$ into $y = 0.3(x-6)(x+4)$ gives $y = -7.5$. Hence the vertex is $(1, -7.5)$.

28. We have $y = 2x^2 - 4x - 2 = 2(x^2 - 2x) - 2$. Completing the square, we get

$$\begin{aligned} 2(x^2 - 2x) - 2 &= 2(x^2 - 2x + 1) - 2 - 2 \\ &= 2(x-1)^2 - 4. \end{aligned}$$

Therefore the vertex occurs at $(1,-4)$. The zeros occur when $y = 0$, so

$$\begin{aligned} 0 &= 2(x-1)^2 - 4 \\ 2 &= (x-1)^2 \\ x &= \pm\sqrt{2} + 1. \end{aligned}$$

Therefore the zeros occur at $x = \pm\sqrt{2} + 1$.

29. We have $y = -3x^2 + 24x - 36 = -3(x^2 - 8x + 12) = -3(x-6)(x-2)$. The zeros are at $x = 6$ and $x = 2$. The axis of symmetry occurs midway between the zeros, so it is $x = 4$. The vertex is on the axis of symmetry and can be found by setting $x = 4$ and solving for y. This means that $y = -3(-2)(2) = 12$. Therefore the vertex is $(4, 12)$.

30. Writing this in vertex form, we have:

$$
\begin{aligned}
y-1 &= 2x^2+\frac{7}{3}x && \text{subtract 1}\\
\frac{y-1}{2} &= x^2+\frac{7}{6}x && \text{divide by 2}\\
\frac{y-1}{2}+\left(\frac{7}{12}\right)^2 &= x^2+\frac{7}{6}x+\left(\frac{7}{12}\right)^2 && \text{complete square}\\
\frac{y-1}{2}+\left(\frac{7}{12}\right)^2 &= \left(x+\frac{7}{12}\right)^2 && \text{factor right-hand side}\\
\frac{y-1}{2} &= \left(x+\frac{7}{12}\right)^2-\left(\frac{7}{12}\right)^2 && \text{subtract } \left(\tfrac{7}{12}\right)^2\\
y-1 &= 2\left(x+\frac{7}{12}\right)^2-2\left(\frac{7}{12}\right)^2 && \text{multiply by 2}\\
y &= 2\left(x+\frac{7}{12}\right)^2-2\left(\frac{7}{12}\right)^2+1 && \text{add 1}\\
&= 2\left(x+\frac{7}{12}\right)^2+\frac{23}{72}. && \text{simplify}
\end{aligned}
$$

Therefore the vertex is $(-7/12, 23/72)$. We can check for zeros by setting $y=0$:

$$
\begin{aligned}
2\left(x+\frac{7}{12}\right)^2+\frac{23}{72} &= 0\\
2\left(x+\frac{7}{12}\right)^2 &= -\frac{23}{72}. && \text{no solution.}
\end{aligned}
$$

Thus, there are no zeros.

31. Between $x=-2$ and $x=0$

$$\frac{\Delta f(x)}{\Delta x}=\frac{f(0)-f(-2)}{0-(-2)}=\frac{(0-1)^2+2-((-2-1)^2+2)}{2}=-4.$$

Between $x=0$ and $x=2$

$$\frac{\Delta f(x)}{\Delta x}=\frac{f(2)-f(0)}{2-0}=\frac{(2-1)^2+2-((0-1)^2+2)}{2}=0.$$

Between $x=2$ and $x=4$

$$\frac{\Delta f(x)}{\Delta x}=\frac{f(4)-f(2)}{4-2}=\frac{(4-1)^2+2-((2-1)^2+2)}{2}=4.$$

Since rates of change are increasing, the graph of $f(x)$ is concave up.

32. **(a)**

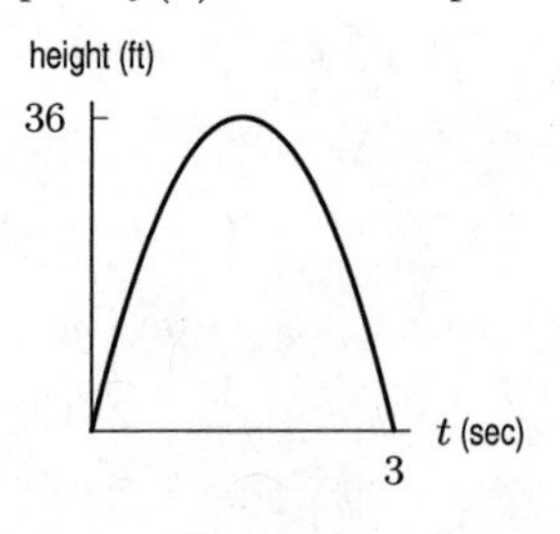

Figure 3.18

(b) To find t when $d(t)=0$, either use the graph or factor $-16t^2+48t$ and set it equal to zero. Factoring yields $-16t^2+48t=-16t(t-3)$, so $d(t)=0$ when $t=0$ or $t=3$. The first time $d(t)=0$ is at the moment the tomato is being thrown up into the air. The second time is when the tomato hits the ground.

(c) The maximum height occurs on the axis of symmetry, which is halfway between the zeros, at $t=1.5$. So, the tomato is highest 1.5 seconds after it is thrown.

(d) The maximum height reached is $d(1.5)=36$ feet.

33. Figure 3.19 shows a graph of the basketball player's trajectory for $T = 1$ second. Since this is the graph of a parabola, the maximum height occurs at the t-value which is halfway between the zeros, 0 and 1. Thus, the maximum occurs at $t = 1/2$ second. The maximum height is $h(1/2) = 4$ feet, and 75% of 4 is 3. Thus, when the basketball player is above 3 feet from the ground, he is in the top 25% of his trajectory. To find when he reaches a height of 3 feet, set $h(t) = 3$. Solving for t gives $t = 0.25$ or $t = 0.75$ seconds. Thus, from $t = 0.25$ to $t = 0.75$ seconds, the basketball player is in the top 25% of his jump, as indicated in Figure 3.19. We see that he spends half of the time at the top quarter of the height of this jump, giving the impression that he hangs in the air.

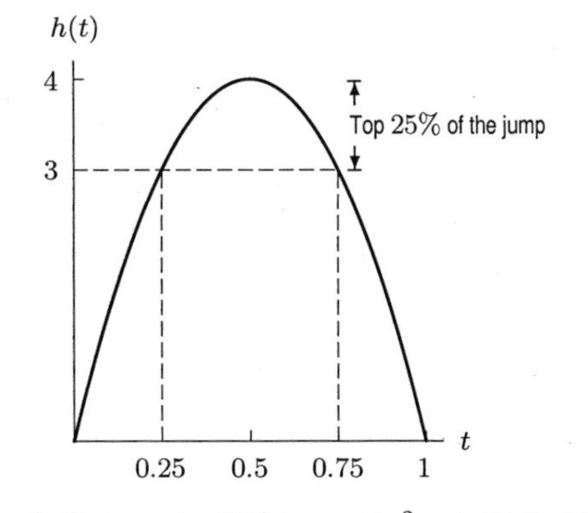

Figure 3.19: A graph of $h(t) = -16t^2 + 16Tt$ for $T = 1$

CHECK YOUR UNDERSTANDING

1. True. It is of the form $f(x) = a(x - r)(x - s)$ where $a = 1$, $r = 0$ and $s = -2$.
2. False. The zeros are -1 and -2.
3. False. It has a minimum but no maximum.
4. False. All quadratic equations have the form $y = ax^2 + bx + c$.
5. False. The time when the object hits the ground is when the height is zero, $s(t) = 0$. The value $s(0)$ gives the height when the object is launched at $t = 0$.
6. True. Solving $f(x) = 0$ gives the zeros of $f(x)$.
7. False. For example, $x^2 + 1 = 0$ has no solutions, and $x^2 = 0$ has one solution.
8. False. The functions $f(x) = x^2 - 4$ and $g(x) = -x^2 + 4$ both have zeros at $x = -2, x = 2$.
9. False. A quadratic function may have two, one, or no zeros.
10. True.
11. True. This is the definition of a vertex.
12. False. If a parabola is concave up its vertex is a minimum point.
13. False. The vertex is located at the point (h, k).
14. True. The vertex is (h, k), and the axis of symmetry is the vertical line $x = h$ through the vertex.
15. True. Transform $y = ax^2 + bx + c$ to the form $y = a(x - h)^2 + k$ where it can be seen that if $a < 0$, then the value of y has a maximum at the vertex (h, k), and the parabola opens downward.

Solutions to Skills for Factoring

1. $2(3x - 7) = 6x - 14$

2. $-4(y+6) = -4y - 24$

3. $12(x+y) = 12x + 12y$

4. $-7(5x - 8y) = -35x + 56y$

5. $x(2x+5) = 2x^2 + 5x$

6. $3z(2x - 9z) = 6xz - 27z^2$

7. $-10r(5r + 6rs) = -50r^2 - 60r^2s$

8. $x(3x-8) + 2(3x-8) = 3x^2 - 8x + 6x - 16 = 3x^2 - 2x - 16$

9. $5z(x-2) - 3(x-2) = 5xz - 10z - 3x + 6$

10. $(x+1)(x+3) = x^2 + 3x + x + 3 = x^2 + 4x + 3$

11. $(x-2)(x+6) = x^2 + 6x - 2x - 12 = x^2 + 4x - 12$

12. $(5x-1)(2x-3) = 10x^2 - 15x - 2x + 3 = 10x^2 - 17x + 3$

13. $(y+1)(z+3) = yz + 3y + z + 3$

14. $(12y - 5)(8w + 7) = 96wy + 84y - 40w - 35$

15. $(5z - 3)(x - 2) = 5xz - 10z - 3x + 6$

16. $-(x-3) - 2(5-x) = -x + 3 - 10 + 2x = x - 7.$

17. $(x-5)6 - 5(1-(2-x)) = 6x - 30 - 5(1 - 2 + x) = 6x - 30 + 5 - 5x = x - 25.$

18. First we multiply 4 by the terms $3x$ and $-2x^2$, and expand $(5+4x)(3x-4)$. Therefore,

$$\begin{aligned}\left(3x - 2x^2\right)(4) + (5+4x)(3x-4) &= 12x - 8x^2 + 15x - 20 + 12x^2 - 16x \\ &= 4x^2 + 11x - 20.\end{aligned}$$

19. The order of operations tell us to expand $(p - 3q)^2$ first and then multiply the result by p. Therefore,

$$\begin{aligned}P(p-3q)^2 &= P(p-3q)(p-3q) \\ &= P(p^2 - 3pq - 3pq + 9q^2) = P(p^2 - 6pq + 9q^2) \\ &= Pp^2 - 6Ppq + 9Pq^2.\end{aligned}$$

20. The order of operations tells us to expand $(x-3)^2$ first and then multiply the result by 4. Therefore,

$$\begin{aligned}4(x-3)^2 + 7 &= 4(x-3)(x-3) + 7 \\ &= 4(x^2 - 3x - 3x + 9) + 7 = 4(x^2 - 6x + 9) + 7 \\ &= 4x^2 - 24x + 36 + 7 = 4x^2 - 24x + 43.\end{aligned}$$

21. First we square $\sqrt{2x} + 1$ and then take the negative of this result. Therefore,

$$\begin{aligned}-\left(\sqrt{2x}+1\right)^2 &= -\left(\sqrt{2x}+1\right)\left(\sqrt{2x}+1\right) = -\left(2x + \sqrt{2x} + \sqrt{2x} + 1\right) \\ &= -(2x + 2\sqrt{2x} + 1) = -2x - 2\sqrt{2x} - 1.\end{aligned}$$

22. Multiplying from left to right we obtain:

$$\begin{aligned}u\left(u^{-1} + 2^u\right)2^u &= \left(u^0 + u \cdot 2^u\right)2^u = (1 + u \cdot 2^u)\,2^4 \\ &= 2^u + u \cdot 2^u \cdot 2^u = 2^u + u \cdot 2^{2u}.\end{aligned}$$

23. $2x + 6 = 2(x+3)$

24. $3y + 15 = 3(y+5)$

25. $5z - 30 = 5(z - 6)$

26. $4t - 6 = 2(2t - 3)$

27. $10w - 25 = 5(2w - 5)$

28. $3u^4 - 4u^3 = u^3(3u - 4)$

29. $3u^7 + 12u^2 = 3u^2(u^5 + 4)$

30. $12x^3y^2 - 18x = 6x(2x^2y^2 - 3)$

31. $14r^4s^2 - 21rst = 7rs(2r^3s - 3t)$

32. Can be factored no further.

33. $x^2 - 3x + 2 = (x - 2)(x - 1)$

34. Can be factored no further.

35. Can be factored no further.

36. $x^2 - 2x - 3 = (x - 3)(x + 1)$

37. Can be factored no further.

38. $x^2 + 2x - 3 = (x + 3)(x - 1)$

39. $2x^2 + 5x + 2 = (2x + 1)(x + 2)$

40. Since each term has a common factor of 2, we write:

$$\begin{aligned} 2x^2 - 10x + 12 &= 2\left(x^2 - 5x + 6\right) \\ &= 2(x - 3)(x - 2). \end{aligned}$$

41. $x^2 + 3x - 28 = (x + 7)(x - 4)$

42. $x^3 - 2x^2 - 3x = x(x^2 - 2x - 3) = x(x - 3)(x + 1)$

43. $x^3 + 2x^2 - 3x = x(x^2 + 2x - 3) = x(x + 3)(x - 1)$

44. $ac + ad + bc + bd = a(c + d) + b(c + d) = (c + d)(a + b).$

45. $x^2 + 2xy + 3xz + 6yz = x(x + 2y) + 3z(x + 2y) = (x + 2y)(x + 3z).$

46. $x^2 - 1.4x - 3.92 = (x + 1.4)(x - 2.8)$

47. $a^2x^2 - b^2 = (ax - b)(ax + b)$

48. The common factor is πr. Therefore,

$$\pi r^2 + 2\pi rh = \pi r(r + 2h).$$

49. We notice that the only factors of 24 whose sum is -10 are -6 and -4. Therefore,

$$B^2 - 10B + 24 = (B - 6)(B - 4).$$

50. $c^2 + x^2 - 2cx = (x - c)^2$

51. The expression $x^2 + y^2$ cannot be factored.

52. We factor and observe that $a^2 - 4$ is the difference of perfect squares. Thus,

$$\begin{aligned} a^4 - a^2 - 12 &= \left(a^2 - 4\right)\left(a^2 + 3\right) \\ &= (a - 2)(a + 2)\left(a^2 + 3\right). \end{aligned}$$

53. This example is factored as the difference of perfect squares. Thus,

$$\begin{aligned} (t + 3)^2 - 16 &= ((t + 3) - 4)\,((t + 3) + 4) \\ &= (t - 1)(t + 7). \end{aligned}$$

Alternatively, we could arrive at the same answer by multiplying the expression out and then factoring it.

54. $x^2+4x+4-y^2=(x+2)^2-(y)^2=(x+2+y)(x+2-y)$.

55. $a^3-2a^2+3a-6=a^2(a-2)+3(a-2)=(a-2)(a^2+3)$.

56.

$$\begin{aligned} b^3-3b^2-9b+27 &= b^2(b-3)-9(b-3) \\ &= (b-3)(b^2-9) \\ &= (b-3)(b-3)(b+3) \\ &= (b-3)^2(b+3). \end{aligned}$$

57.

$$\begin{aligned} c^2d^2-25c^2-9d^2+225 &= c^2(d^2-25)-9(d^2-25) \\ &= (d^2-25)(c^2-9) \\ &= (d+5)(d-5)(c+3)(c-3). \end{aligned}$$

58. By grouping the terms hx^2 and $-4hx$, we find a common factor of hx and for the terms 12 and $-3x$, we find a common factor of -3. Therefore,

$$\begin{aligned} hx^2+12-4hx-3x = hx^2-4hx+12-3x &= hx(x-4)-3(-4+x) \\ &= hx(x-4)-3(x-4)=(hx-3)(x-4). \end{aligned}$$

59. The idea here is to rewrite the second expression $-2(s-r)$ as $+2(r-s)$. This latter expression shares a common factor of $r-s$ with the first expression $r(r-s)$. Thus,

$$r(r-s)-2(s-r)=r(r-s)+2(r-s)=(r+2)(r-s).$$

60. Factor as:

$$y^2-3xy+2x^2=(y-2x)(y-x).$$

61. The common factor is xe^{-3x}. Therefore,

$$x^2e^{-3x}+2xe^{-3x}=xe^{-3x}(x+2).$$

62. $t^2e^{5t}+3te^{5t}+2e^{5t}=e^{5t}(t^2+3t+2)=e^{5t}(t+1)(t+2)$.

63. The two expressions $P(1+r)^2$ and $P(1+r)^2r$ share a common factor of $P(1+r)^2$. So,

$$P(1+r)^2+P(1+r)^2r=P(1+r)^2(1+r)=P(1+r)^3.$$

64. $x^2-6x+9-4z^2=(x-3)^2-(2z)^2=(x-3+2z)(x-3-2z)$.

65. $dk+2dm-3ek-6em=d(k+2m)-3e(k+2m)=(k+2m)(d-3e)$.

66. $\pi r^2-2\pi r+3r-6=\pi r(r-2)+3(r-2)=(r-2)(\pi r+3)$.

67. $8gs-12hs+10gm-15hm=4s(2g-3h)+5m(2g-3h)=(2g-3h)(4s+5m)$.

68.

$$\begin{aligned} y^2-5y-6 &= 0 \\ (y+1)(y-6) &= 0 \\ y+1=0 \quad &\text{or} \quad y-6=0 \\ y=-1 \quad &\text{or} \quad y=6 \end{aligned}$$

69.

$$x = \frac{-3 \pm \sqrt{3^2 - 4(4)(-15)}}{2(4)}$$
$$x = \frac{-3 \pm \sqrt{249}}{8}$$

70.

$$\frac{2}{x} + \frac{3}{2x} = 8$$
$$\frac{4+3}{2x} = 8$$
$$16x = 7$$
$$x = \frac{7}{16}$$

71.

$$\frac{3}{x-1} + 1 = 5$$
$$\frac{3}{x-1} = 4$$
$$4(x-1) = 3$$
$$4x - 4 = 3$$
$$4x = 7$$
$$x = \frac{7}{4}$$

72.

$$\sqrt{y-1} = 13$$
$$y - 1 = 169$$
$$y = 170$$

73.

$$-16t^2 + 96t + 12 = 60$$
$$-16t^2 + 96t - 48 = 0$$
$$t^2 - 6t + 3 = 0$$

$$t = \frac{-(-6) \pm \sqrt{(-6)^2 - 4(1)(3)}}{2(1)}$$
$$t = \frac{6 \pm \sqrt{24}}{2} = \frac{6 \pm 2\sqrt{6}}{2}$$
$$t = 3 \pm \sqrt{6}.$$

74. Rewrite the equation $g^3 - 4g = 3g^2 - 12$ with a zero on the right side and factor completely.

$$g^3 - 3g^2 - 4g + 12 = 0$$
$$g^2(g-3) - 4(g-3) = 0$$
$$(g-3)(g^2-4) = 0$$
$$(g-3)(g+2)(g-2) = 0.$$

So, $g - 3 = 0$, $g + 2 = 0$, or $g - 2 = 0$. Thus, $g = 3, -2$, or 2.

75. First multiply both sides by (-1):

$$-1(8+2x-3x^2) = (-1)(0).$$

$$\begin{aligned} 3x^2-2x-8 &= 0 \\ (3x+4)(x-2) &= 0 \\ 3x+4=0 \quad &\text{or} \quad x-2=0 \\ x=-\frac{4}{3} \quad &\text{or} \quad x=2. \end{aligned}$$

76. By grouping the first two and the last two terms, we obtain:

$$\begin{aligned} \left(2p^3+p^2\right)-18p-9 &= 0 \\ \left(2p^3+p^2\right)-(18p+9) &= 0 \\ p^2(2p+1)-9(2p+1) &= 0 \\ \left(p^2-9\right)(2p+1) &= 0 \\ (p-3)(p+3)(2p+1) &= 0 \\ p=3, \text{ or } p=-3, &\text{ or } p=-\frac{1}{2}. \end{aligned}$$

77.

$$\begin{aligned} N^2-2N-3 &= 2N(N-3) \\ N^2-2N-3 &= 2N^2-6N \\ N^2-4N+3 &= 0 \\ (N-3)(N-1) &= 0 \\ N=3 &\text{ or } N=1 \end{aligned}$$

78. Do not divide both sides by t, because you would lose the solution $t=0$ in that case. Instead, set one side $=0$ and factor.

$$\begin{aligned} \frac{1}{64}t^3 &= t \\ \frac{1}{64}t^3-t &= 0 \\ t(\frac{1}{64}t^2-1) &= 0 \\ t=0 \text{ or } &\frac{1}{64}t^2-1=0 \end{aligned}$$

The second equation still needs to be solved for t:

$$\begin{aligned} \frac{1}{64}t^2-1 &= 0 \\ \frac{1}{64}t^2 &= 1 \\ t^2 &= 64 \\ t &= \pm 8. \end{aligned}$$

So the final answer is $t=0$ or $t=8$ or $t=-8$.

79. We write $x^2 - 1 = 2x$ or $x^2 - 2x - 1 = 0$ which does not factor. Employing the quadratic formula, we have $a = 1$, $b = -2$, $c = -1$. Therefore

$$\begin{aligned} x &= \frac{-(-2) \pm \sqrt{(-2)^2 - 4(1)(-1)}}{2(1)} = \frac{2 \pm \sqrt{4+4}}{2} = \frac{2 \pm \sqrt{8}}{2} \\ &= \frac{2 \pm 2\sqrt{2}}{2} = 1 \pm \sqrt{2}. \end{aligned}$$

80.

$$\begin{aligned} 4x^2 - 13x - 12 &= 0 \\ (x-4)(4x+3) &= 0 \\ x = 4 \text{ or } x &= -\frac{3}{4} \end{aligned}$$

81. We rewrite the quadratic equation in standard form and use the quadratic formula. So

$$\begin{aligned} 60 &= -16t^2 + 96t + 12 \\ 16t^2 - 96t + 48 &= 0 \\ t^2 - 6t + 3 &= 0 \\ t &= \frac{-(-6) \pm \sqrt{(-6)^2 - 4(1)(3)}}{2} = \frac{6 \pm \sqrt{36 - 12}}{2} \\ &= \frac{6 \pm \sqrt{24}}{2} = \frac{6 \pm 2\sqrt{6}}{2} = 3 \pm \sqrt{6}. \end{aligned}$$

82. Rewrite the equation $n^5 + 80 = 5n^4 + 16n$ with a zero on the right side and factor completely.

$$\begin{aligned} n^5 - 5n^4 - 16n + 80 &= 0 \\ n^4(n-5) - 16(n-5) &= 0 \\ (n-5)(n^4 - 16) &= 0 \\ (n-5)(n^2-4)(n^2+4) &= 0 \\ (n-5)(n+2)(n-2)(n^2+4) &= 0. \end{aligned}$$

So, $n - 5 = 0$, $n + 2 = 0$, $n - 2 = 0$, or $n^2 + 4 = 0$. Note that $n^2 + 4 = 0$ has no real solutions, so, $n = 5, -2$, or 2.

83. Rewrite the equation $5a^3 + 50a^2 = 4a + 40$ with a zero on the right side and factor.

$$\begin{aligned} 5a^3 + 50a^2 - 4a - 40 &= 0 \\ 5a^2(a+10) - 4(a+10) &= 0 \\ (a+10)(5a^2 - 4) &= 0. \end{aligned}$$

So, $a + 10 = 0$ or $5a^2 - 4 = 0$. Thus, $a = -10$ or $a = \dfrac{\pm 2\sqrt{5}}{5}$.

84. Using the quadratic formula for

$$y^2 + 4y - 2 = 0, \quad a = 1,\ b = 4,\ c = -2,$$

we obtain,

$$\begin{aligned} y &= \frac{-4 \pm \sqrt{(4)^2 - 4(1)(-2)}}{2} = \frac{-4 \pm \sqrt{16+8}}{2} \\ &= \frac{-4 \pm \sqrt{24}}{2} = \frac{-4 \pm 2\sqrt{6}}{2} = -2 \pm \sqrt{6}. \end{aligned}$$

85. To find the common denominator, we factor the second denominator

$$\frac{2}{z-3}+\frac{7}{z^2-3z}=0$$
$$\frac{2}{z-3}+\frac{7}{z(z-3)}=0$$

which produces a common denominator of $z(z-3)$. Therefore:

$$\frac{2z}{z(z-3)}+\frac{7}{z(z-3)}=0$$
$$\frac{2z+7}{z(z-3)}=0$$
$$2z+7=0$$
$$z=-\frac{7}{2}.$$

86. First we combine like terms in the numerator.

$$\frac{x^2+1-2x^2}{(x^2+1)^2}=0$$
$$\frac{-x^2+1}{(x^2+1)^2}=0$$
$$-x^2+1=0$$
$$-x^2=-1$$
$$x^2=1$$
$$x=\pm 1$$

87.

$$4-\frac{1}{L^2}=0$$
$$4=\frac{1}{L^2}$$
$$4L^2=1$$
$$L^2=\frac{1}{4}$$
$$L=\pm\frac{1}{2}$$

88. The common denominator for this fractional equation is $(q+1)(q-1)$. If we multiply both sides of this equation by $(q+1)(q-1)$, we obtain:

$$2+\frac{1}{q+1}-\frac{1}{q-1}=0$$
$$2(q+1)(q-1)+1(q-1)-1(q+1)=0$$
$$2\left(q^2-1\right)+q-1-q-1=0$$
$$2q^2-2+q-1-q-1=0$$
$$2q^2-4=0$$
$$2q^2=4$$
$$q^2=2$$
$$q=\pm\sqrt{2}.$$

89. We can solve this equation by squaring both sides.

$$\begin{aligned}\sqrt{r^2+24} &= 7\\ r^2+24 &= 49\\ r^2 &= 25\\ r &= \pm 5\end{aligned}$$

90. We can solve this equation by cubing both sides of this equation.

$$\begin{aligned}\frac{1}{\sqrt[3]{x}} &= -2\\ \left(\frac{1}{\sqrt[3]{x}}\right)^3 &= (-2)^3\\ \frac{1}{x} &= -8\\ x &= -\frac{1}{8}\end{aligned}$$

91. We can solve this equation by squaring both sides.

$$\begin{aligned}3\sqrt{x} &= \frac{1}{2}x\\ 9x &= \frac{1}{4}x^2\\ \frac{1}{4}x^2 - 9x &= 0\\ x(\frac{1}{4}x - 9) &= 0\\ x = 0 &\text{ or } \frac{1}{4}x = 9\\ x = 0 &\text{ or } x = 36\end{aligned}$$

92. We can solve this equation by squaring both sides.

$$\begin{aligned}10 &= \sqrt{\frac{v}{7\pi}}\\ 100 &= \frac{v}{7\pi}\\ 700\pi &= v\end{aligned}$$

93. Multiply by $(x-5)(x-1)$ on both sides of the equation, giving

$$(3x+4)(x-2) = 0.$$

So, $3x+4=0$, or $x-2=0$, that is,

$$x = -\frac{4}{3}, \quad x = 2.$$

94. We begin by squaring both sides of the equation in order to eliminate the radical.

$$\begin{aligned} T &= 2\pi\sqrt{\frac{l}{g}} \\ T^2 &= 4\pi^2\left(\frac{l}{g}\right) \\ \frac{gT^2}{4\pi^2} &= l \end{aligned}$$

95. First solve for b^5, then take fifth root:

$$\begin{aligned} Ab^5 &= C \\ b^5 &= \frac{C}{A} \\ b &= \sqrt[5]{\frac{C}{A}}. \end{aligned}$$

96. We have $2x + 1 = 7$ or $2x + 1 = -7$, that is, $2x = 6$ or $2x = -8$. So,

$$x = 3, \quad x = -4.$$

97. For a fraction to equal zero, the numerator must equal zero. So, we solve

$$x^2 - 5mx + 4m^2 = 0.$$

Since $x^2 - 5mx + 4m^2 = (x - m)(x - 4m)$, we know that the numerator equals zero when $x = 4m$ and when $x = m$. But for $x = m$, the denominator will equal zero as well. So, the fraction is undefined at $x = m$, and the only solution is $x = 4m$.

98. We substitute -3 for y in the first equation.

$$\begin{aligned} y &= 2x - x^2 \\ -3 &= 2x - x^2 \\ x^2 - 2x - 3 &= 0 \\ (x - 3)(x + 1) &= 0 \\ x = 3 \quad &\text{and} \quad y = 2(3) - 3^2 = -3 \quad \text{or} \\ x = -1 \quad &\text{and} \quad y = 2(-1) - (-1)^2 = -3 \end{aligned}$$

99. We set the equations $y = 1/x$ and $y = 4x$ equal to one another.

$$\begin{aligned} \frac{1}{x} &= 4x \\ 4x^2 &= 1 \\ x^2 &= \frac{1}{4} \\ x &= \frac{1}{2} \quad \text{and} \quad y = \frac{1}{\frac{1}{2}} = 2 \quad \text{or} \\ x &= -\frac{1}{2} \quad \text{and} \quad y = \frac{1}{-\frac{1}{2}} = -2 \end{aligned}$$

100. Substituting the value of y from the second equation into the first equation, we obtain

$$x^2 + (x-3)^2 = 36$$
$$x^2 + x^2 - 6x + 9 = 36$$
$$2x^2 - 6x - 27 = 0,$$

$$\begin{aligned} x &= \frac{-(-6) \pm \sqrt{(-6)^2 - 4(2)(-27)}}{(2)(2)} \\ &= \frac{6 \pm \sqrt{252}}{4} \\ &= \frac{6 \pm \sqrt{4 \cdot 63}}{4} \\ &= \frac{6 \pm 2\sqrt{63}}{4} \\ &= \frac{3 \pm \sqrt{63}}{2}. \end{aligned}$$

Now we substitute the values of x into the second equation:

$$\begin{aligned} y &= \frac{3 \pm \sqrt{63}}{2} - 3 \\ &= \frac{-3 \pm \sqrt{63}}{2}. \end{aligned}$$

101. We substitute the expression $4 - x^2$ for y in the second equation.

$$\begin{aligned} y - 2x &= 1 \\ 4 - x^2 - 2x &= 1 \\ -x^2 - 2x + 3 &= 0 \\ x^2 + 2x - 3 &= 0 \\ (x+3)(x-1) &= 0 \end{aligned}$$

$$x = -3 \quad \text{and} \quad y = 4 - (-3)^2 = -5 \quad \text{or}$$
$$x = 1 \quad \text{and} \quad y = 4 - 1^2 = 3$$

102. These equations cannot be solved exactly. A calculator gives the solutions as

$$x = 2.081, \quad y = 8.013 \quad \text{and} \quad x = 4.504, \quad y = 90.348.$$

103. Using the point-slope formula for the equation of a line, we have

$$y - 0 = 3(x - 0)$$

or

$$y = 3x.$$

We need to find the points where this line intersects $y = x^2$. This means we want points such that

$$x^2 = 3x \quad \text{or} \quad x^2 - 3x = 0$$

$$x(x-3) = 0$$

$$x = 0 \quad \text{or} \quad x = 3.$$

So the points are $(0, 0)$ and $(3, 9)$.

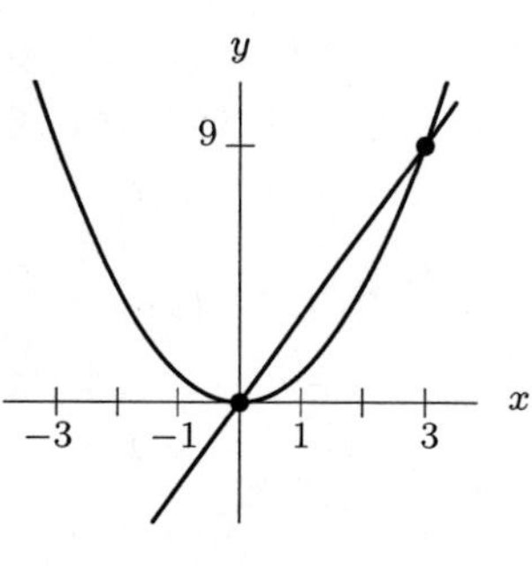

Figure 3.20

104. We substitute $y = x - 1$ in the equation $x^2 + y^2 = 25$.

$$\begin{aligned} x^2 + (x-1)^2 &= 25 \\ x^2 + x^2 - 2x + 1 &= 25 \\ 2x^2 - 2x - 24 &= 0 \\ x^2 - x - 12 &= 0 \\ (x-4)(x+3) &= 0 \\ x = 4 \quad \text{and} \quad y &= 4 - 1 = 3 \quad \text{or} \\ x = -3 \quad \text{and} \quad y &= -3 - 1 = -4 \end{aligned}$$

So the points of intersection are $(4, 3)$, $(-3, -4)$.

105. Solving $y = x^2$ and $y = 15 - 2x$ simultaneously, we have

$$\begin{aligned} x^2 &= 15 - 2x \\ x^2 + 2x - 15 &= 0 \\ (x+5)(x-3) &= 0 \\ x &= -5, 3. \end{aligned}$$

Thus, the points of intersection are $(-5, 25)$, $(3, 9)$.

Solutions to Skills for Completing the Square

1. $x^2 + 8x = x^2 + 8x + 16 - 16 = (x+4)^2 - 16$
2. $w^2 + 7w = w^2 + 7w + \left(\frac{7}{2}\right)^2 - \left(\frac{7}{2}\right)^2 = \left(w + \frac{7}{2}\right)^2 - \left(\frac{7}{2}\right)^2$
3. $2r^2 + 20r = 2(r^2 + 10r) = 2(r^2 + 10r + 25 - 25) = 2((r+5)^2 - 25) = 2(r+5)^2 - 50.$
4. $3t^2 + 24t - 13 = 3(t^2 + 8t) - 13 = 3(t^2 + 8t + 16) - 48 - 13 = 3(t+4)^2 - 61$
5. We add and subtract the square of half the coefficient of the a-term, $(\frac{-2}{2})^2 = 1$, to get

$$\begin{aligned} a^2 - 2a - 4 &= a^2 - 2a + 1 - 1 - 4 \\ &= (a^2 - 2a + 1) - 1 - 4 \\ &= (a-1)^2 - 5. \end{aligned}$$

6. We add and subtract the square of half the coefficient of the n-term, $(\frac{4}{2})^2 = 4$, to get

$$\begin{aligned} n^2 + 4n - 5 &= n^2 + 4n + 4 - 4 - 5 \\ &= (n^2 + 4n + 4) - 4 - 5 \\ &= (n + 2)^2 - 9. \end{aligned}$$

7. Factoring out the coefficient of r^2 gives

$$3r^2 + 9r - 4 = 3\left(r^2 + 3r - \frac{4}{3}\right).$$

Inside the parenthesis, we add and subtract the square of half the coefficient of the r-term, $\left(\frac{3}{2}\right)^2 = \frac{9}{4}$, to get

$$\begin{aligned} 3\left(r^2 + 3r - \frac{4}{3}\right) &= 3\left(r^2 + 3r + \frac{9}{4} - \frac{9}{4} - \frac{4}{3}\right) \\ &= 3\left(\left(r + \frac{3}{2}\right)^2 - \frac{9}{4} - \frac{4}{3}\right) \\ &= 3\left(r + \frac{3}{2}\right)^2 - \frac{43}{4}. \end{aligned}$$

8. Factoring out the coefficient of g^2 gives

$$12g^2 + 8g + 5 = 12\left(g^2 + \frac{2}{3}g + \frac{5}{12}\right).$$

Inside the parenthesis, we add and subtract the square of half the coefficient of the g-term, $\left(\frac{2/3}{2}\right)^2 = \frac{1}{9}$, to get

$$\begin{aligned} 12\left(g^2 + \frac{2}{3}g + \frac{5}{12}\right) &= 12\left(g^2 + \frac{2}{3}g + \frac{1}{9} - \frac{1}{9} + \frac{5}{12}\right) \\ &= 12\left(\left(g + \frac{1}{3}\right)^2 - \frac{1}{9} + \frac{5}{12}\right) \\ &= 12\left(g + \frac{1}{3}\right)^2 + \frac{11}{3}. \end{aligned}$$

9. Completing the square yields

$$x^2 - 2x - 3 = (x^2 - 2x + 1) - 1 - 3 = (x - 1)^2 - 4.$$

10. First we rewrite $10 - 6x + x^2$ as $x^2 - 6x + 10$ and then complete the square. So

$$10 - 6x + x^2 = x^2 - 6x + 10 = (x^2 - 6x + 9) - 9 + 10 = (x - 3)^2 + 1.$$

11. First we factor out -1. Then

$$\begin{aligned} -x^2 + 6x - 2 &= -(x^2 - 6x + 2) = -(x^2 - 6x + 9 - 9 + 2) \\ &= -(x^2 - 6x + 9 - 7) = -(x^2 - 6x + 9) + 7 \\ &= -(x - 3)^2 + 7. \end{aligned}$$

12. First we factor 3 from the first two terms. Then

$$\begin{aligned} 3x^2 - 12x + 13 &= 3(x^2 - 4x) + 13 = 3(x^2 - 4x + 4 - 4) + 13 \\ &= 3(x^2 - 4x + 4) - 12 + 13 = 3(x - 2)^2 + 1. \end{aligned}$$

13. Complete the square and write in vertex form.

$$\begin{aligned} y &= x^2 + 6x + 3 \\ &= x^2 + 6x + 9 - 9 + 3 \\ &= (x+3)^2 - 6. \end{aligned}$$

The vertex is $(-3, -6)$.

14. Complete the square and write in vertex form.

$$\begin{aligned} y &= x^2 - x + 4 \\ &= x^2 - x + \frac{1}{4} - \frac{1}{4} + 4 \\ &= \left(x - \frac{1}{2}\right)^2 - \frac{1}{4} + 4 \\ &= \left(x - \frac{1}{2}\right)^2 + \frac{15}{4}. \end{aligned}$$

The vertex is $(1/2, 15/4)$.

15. Complete the square and write in vertex form.

$$\begin{aligned} y &= -x^2 - 8x + 2 \\ &= -(x^2 + 8x - 2) \\ &= -(x^2 + 8x + 16 - 16 - 2) \\ &= -((x+4)^2 - 18) \\ &= -(x+4)^2 + 18. \end{aligned}$$

The vertex is $(-4, 18)$

16. Complete the square and write in vertex form.

$$\begin{aligned} y &= x^2 - 3x - 3 \\ &= x^2 - 3x + \frac{9}{4} - \frac{9}{4} - 3 \\ &= \left(x - \frac{3}{2}\right)^2 - \frac{9}{4} - 3 \\ &= \left(x - \frac{3}{2}\right)^2 - \frac{21}{4}. \end{aligned}$$

The vertex is $(3/2, -21/4)$.

17. Complete the square and write in vertex form.

$$\begin{aligned} y &= -x^2 + x - 6 \\ &= -(x^2 - x + 6) \\ &= -\left(x^2 - x + \frac{1}{4} - \frac{1}{4} + 6\right) \\ &= -\left(\left(x - \frac{1}{2}\right)^2 - \frac{1}{4} + 6\right) \\ &= -\left(x - \frac{1}{2}\right)^2 - \frac{23}{4}. \end{aligned}$$

The vertex is $(1/2, -23/4)$.

18. Complete the square and write in vertex form.

$$\begin{aligned} y &= 3x^2 + 12x \\ &= 3(x^2 + 4x) \\ &= 3(x^2 + 4x + 4 - 4) \\ &= 3((x+2)^2 - 4) \\ &= 3(x+2)^2 - 12. \end{aligned}$$

The vertex is $(-2, -12)$.

19. Complete the square and write in vertex form.

$$\begin{aligned} y &= -4x^2 + 8x - 6 \\ &= -4\left(x^2 - 2x + \frac{6}{4}\right) \\ &= -4\left(x^2 - 2x + 1 - 1 + \frac{6}{4}\right) \\ &= -4\left((x-1)^2 - 1 + \frac{6}{4}\right) \\ &= -4(x-1)^2 - 2. \end{aligned}$$

The vertex is $(1, -2)$.

20. Complete the square and write in vertex form.

$$\begin{aligned} y &= 5x^2 - 5x + 7 \\ &= 5\left(x^2 - x + \frac{7}{5}\right) \\ &= 5\left(x^2 - x + \left(\frac{1}{4} - \frac{1}{4} + \frac{7}{5}\right)\right) \\ &= 5\left(\left(x - \frac{1}{2}\right)^2 - \frac{1}{4} + \frac{7}{5}\right) \\ &= 5\left(x - \frac{1}{2}\right)^2 + \frac{23}{4}. \end{aligned}$$

The vertex is $(1/2, 23/4)$.

21. Complete the square and write in vertex form.

$$\begin{aligned} y &= 2x^2 - 7x + 3 \\ &= 2\left(x^2 - \frac{7}{2}x + \frac{3}{2}\right) \\ &= 2\left(x^2 - \frac{7}{2}x + \frac{49}{16} - \frac{49}{16} + \frac{3}{2}\right) \\ &= 2\left(\left(x - \frac{7}{4}\right)^2 - \frac{49}{16} + \frac{3}{2}\right) \\ &= 2\left(x - \frac{7}{4}\right)^2 - \frac{25}{8}. \end{aligned}$$

The vertex is $(7/4, -25/8)$.

22. Complete the square and write in vertex form.

$$\begin{aligned} -3x^2 - x - 2 &= -3\left(x^2 + \frac{1}{3}x + \frac{2}{3}\right) \\ &= -3\left(x^2 + \frac{1}{3}x + \frac{1}{36} - \frac{1}{36} + \frac{2}{3}\right) \end{aligned}$$

$$= -3\left(\left(x+\frac{1}{6}\right)^2 - \frac{1}{36} + \frac{2}{3}\right)$$
$$= -3\left(x+\frac{1}{6}\right)^2 - \frac{23}{12}.$$

The vertex is $(-1/6, -23/12)$.

23. Get the variables on the left side, the constants on the right side and complete the square using $(\frac{-2}{2})^2 = 1$.

$$\begin{aligned} g^2 - 2g &= 24 \\ g^2 - 2g + 1 &= 24 + 1 \\ (g-1)^2 &= 25. \end{aligned}$$

Take the square root of both sides and solve for g.

$$\begin{aligned} g - 1 &= \pm 5 \\ g &= 1 \pm 5. \end{aligned}$$

So, $g = 6$ or $g = -4$.

24. Complete the square using $(-2/2)^2 = 1$, take the square root of both sides and solve for p.

$$\begin{aligned} p^2 - 2p &= 6 \\ p^2 - 2p + 1 &= 6 + 1 \\ (p-1)^2 &= 7 \\ p - 1 &= \pm\sqrt{7} \\ p &= 1 \pm \sqrt{7}. \end{aligned}$$

25. Complete the square with $(\frac{1}{2})^2 = \frac{1}{4}$ and take the square root of both sides to solve for d.

$$\begin{aligned} d^2 - d + \frac{1}{4} &= \frac{1}{4} + 2 \\ \left(d - \frac{1}{2}\right)^2 &= \frac{9}{4} \\ d - \frac{1}{2} &= \pm\frac{3}{2} \\ d &= \frac{1}{2} \pm \frac{3}{2}. \end{aligned}$$

So $d = 2$ or $d = -1$.

26. Get the variables on the left side, the constants on the right side and complete the square using $(\frac{2}{2})^2 = 1$.

$$\begin{aligned} 2r^2 + 4r &= 5 \\ 2(r^2 + 2r) &= 5 \\ 2(r^2 + 2r + 1) &= 5 + 2 \cdot 1 \\ 2(r+1)^2 &= 7. \end{aligned}$$

Divide by 2, take the square root of both sides and solve for r.

$$\begin{aligned} (r+1)^2 &= \frac{7}{2} \\ r + 1 &= \pm\sqrt{\frac{7}{2}} \\ r &= -1 \pm \sqrt{\frac{7}{2}}. \end{aligned}$$

27. Get the variables on the left side, the constants on the right side and complete the square using $\left(\frac{5}{2}\right)^2 = \frac{25}{4}$.

$$\begin{aligned}
2s^2 + 10s &= 1 \\
2\left(s^2 + 5s\right) &= 1 \\
2\left(s^2 + 5s + \frac{25}{4}\right) &= 2\left(\frac{25}{4}\right) + 1 \\
2\left(s + \frac{5}{2}\right)^2 &= \frac{25}{2} + 1 \\
2\left(s + \frac{5}{2}\right)^2 &= \frac{27}{2}.
\end{aligned}$$

Divide by 2, take the square root of both sides and solve for s.

$$\begin{aligned}
\left(s + \frac{5}{2}\right)^2 &= \frac{27}{4} \\
s + \frac{5}{2} &= \pm\sqrt{\frac{27}{4}} \\
s + \frac{5}{2} &= \pm\frac{\sqrt{27}}{2} \\
s &= -\frac{5}{2} \pm \frac{\sqrt{27}}{2}.
\end{aligned}$$

28. Get the variables on the left side and the constants on the right side, and complete the square.

$$\begin{aligned}
7r^2 - 3r &= 6 \\
7\left(r^2 - \frac{3}{7}r\right) &= 6 \\
7\left(r^2 - \frac{3}{7}r + \left(\frac{3}{14}\right)^2\right) &= 7\left(\frac{3}{14}\right)^2 + 6 \\
7\left(r - \frac{3}{14}\right)^2 &= \frac{9}{28} + 6 \\
7\left(r - \frac{3}{14}\right)^2 &= \frac{177}{28}.
\end{aligned}$$

Divide by 7, take the square root of both sides and solve for r.

$$\begin{aligned}
\left(r - \frac{3}{14}\right)^2 &= \frac{177}{196} \\
r - \frac{3}{14} &= \pm\sqrt{\frac{177}{196}} \\
r - \frac{3}{14} &= \pm\frac{\sqrt{177}}{14} \\
r &= \frac{3}{14} \pm \frac{\sqrt{177}}{14}.
\end{aligned}$$

29. Complete the square on the left side.

$$\begin{aligned}
5\left(p^2 + \frac{9}{5}p\right) &= 1 \\
5\left(p^2 + \frac{9}{5}p + \frac{81}{100}\right) &= 5\left(\frac{81}{100}\right) + 1
\end{aligned}$$

$$5\left(p+\frac{9}{10}\right)^2 = \frac{81}{20}+1$$
$$5\left(p+\frac{9}{10}\right)^2 = \frac{101}{20}.$$

Divide by 5 and take the square root of both sides to solve for p.

$$\left(p+\frac{9}{10}\right)^2 = \frac{101}{100}$$
$$p+\frac{9}{10} = \pm\sqrt{\frac{101}{100}}$$
$$p+\frac{9}{10} = \pm\frac{\sqrt{101}}{10}$$
$$p = -\frac{9}{10} \pm \frac{\sqrt{101}}{10}.$$

30. With $a = 1$, $b = -4$, and $c = -12$, we use the quadratic formula,

$$\begin{aligned} n &= \frac{-b \pm \sqrt{b^2-4ac}}{2a} \\ &= \frac{4 \pm \sqrt{(-4)^2 - 4\cdot 1\cdot(-12)}}{2\cdot 1} \\ &= \frac{4 \pm \sqrt{16+48}}{2} \\ &= \frac{4 \pm \sqrt{64}}{2} \\ &= \frac{4 \pm 8}{2}. \end{aligned}$$

So, $n = 6$ or $n = -2$.

31. Rewrite the equation so the left side is zero and use the quadratic formula with $a = 2$, $b = 5$, and $c = 2$.

$$\begin{aligned} y &= \frac{-b \pm \sqrt{b^2-4ac}}{2a} \\ &= \frac{-5 \pm \sqrt{5^2 - 4\cdot 2\cdot 2}}{2\cdot 2} \\ &= \frac{-5 \pm \sqrt{25-16}}{4} \\ &= \frac{-5 \pm \sqrt{9}}{4} \\ &= \frac{-5 \pm 3}{4}. \end{aligned}$$

So, $y = -\frac{1}{2}$ or $y = -2$.

32. Set the equation equal to zero, $6k^2 + 11k + 3 = 0$. With $a = 6$, $b = 11$, and $c = 3$, we use the quadratic formula,

$$\begin{aligned} k &= \frac{-b \pm \sqrt{b^2-4ac}}{2a} \\ &= \frac{-11 \pm \sqrt{11^2 - 4\cdot 6\cdot 3}}{2\cdot 6} \\ &= \frac{-11 \pm \sqrt{121-72}}{12} \end{aligned}$$

$$= \frac{-11 \pm \sqrt{49}}{12}$$
$$= \frac{-11 \pm 7}{12}.$$

So, $k = -\frac{1}{3}$ or $k = -\frac{3}{2}$.

33. Set the equation equal to zero, $w^2 + w - 4 = 0$. With $a = 1$, $b = 1$, and $c = -4$, we use the quadratic formula,

$$\begin{aligned} w &= \frac{-b \pm \sqrt{b^2 - 4ac}}{2a} \\ &= \frac{-1 \pm \sqrt{1^2 - 4 \cdot 1 \cdot (-4)}}{2 \cdot 1} \\ &= \frac{-1 \pm \sqrt{1 + 16}}{2} \\ &= \frac{-1 \pm \sqrt{17}}{2}. \end{aligned}$$

34. Set the equation equal to zero, $z^2 + 4z - 6 = 0$. With $a = 1$, $b = 4$, and $c = -6$, we use the quadratic formula,

$$\begin{aligned} z &= \frac{-b \pm \sqrt{b^2 - 4ac}}{2a} \\ &= \frac{-4 \pm \sqrt{4^2 - 4 \cdot 1 \cdot (-6)}}{2 \cdot 1} \\ &= \frac{-4 \pm \sqrt{16 + 24}}{2} \\ &= \frac{-4 \pm \sqrt{40}}{2} \\ &= \frac{-4 \pm 2\sqrt{10}}{2} \\ &= -2 \pm \sqrt{10}. \end{aligned}$$

35. With $a = 2$, $b = 6$, and $c = -3$, we use the quadratic formula

$$\begin{aligned} q &= \frac{-b \pm \sqrt{b^2 - 4ac}}{2a} \\ &= \frac{-6 \pm \sqrt{6^2 - 4 \cdot 2 \cdot (-3)}}{2 \cdot 2} \\ &= \frac{-6 \pm \sqrt{36 + 24}}{4} \\ &= \frac{-6 \pm \sqrt{60}}{4} \\ &= \frac{-6 \pm 2\sqrt{15}}{4} \\ &= \frac{-3 \pm \sqrt{15}}{2} \end{aligned}$$

36. Rewrite the equation to equal zero, and factor.

$$\begin{aligned} r^2 - 2r - 8 &= 0 \\ (r - 4)(r + 2) &= 0. \end{aligned}$$

So, $r - 4 = 0$ or $r + 2 = 0$ and $r = 4$ or $r = -2$.

37. Solve by completing the square using $\left(\frac{3}{2}\right)^2 = \frac{9}{4}$.

$$\begin{aligned} s^2 + 3s + \frac{9}{4} &= 1 + \frac{9}{4} \\ \left(s + \frac{3}{2}\right)^2 &= 1 + \frac{9}{4} \\ \left(s + \frac{3}{2}\right)^2 &= \frac{13}{4}. \end{aligned}$$

Taking the square root of both sides and solving for s,

$$\begin{aligned} s + \frac{3}{2} &= \pm\sqrt{\frac{13}{4}} \\ s + \frac{3}{2} &= \pm\frac{\sqrt{13}}{2} \\ s &= -\frac{3}{2} \pm \frac{\sqrt{13}}{2} \\ s &= \frac{-3 \pm \sqrt{13}}{2}. \end{aligned}$$

38. Rewrite the equation to equal zero, and factor by grouping.

$$\begin{aligned} z^3 + 2z^2 - 3z - 6 &= 0 \\ z^2(z+2) - 3(z+2) &= 0 \\ (z+2)(z^2 - 3) &= 0. \end{aligned}$$

So, $z + 2 = 0$ or $z^2 - 3 = 0$, thus $z = -2$ or $z = \pm\sqrt{3}$.

39. Set the equation equal to zero, and use the quadratic formula with $a = 25$, $b = -30$, and $c = 4$.

$$\begin{aligned} u &= \frac{30 \pm \sqrt{(-30)^2 - 4 \cdot 25 \cdot 4}}{2 \cdot 25} \\ &= \frac{30 \pm \sqrt{900 - 400}}{50} \\ &= \frac{30 \pm \sqrt{500}}{50} \\ &= \frac{30 \pm 10\sqrt{5}}{50} \\ &= \frac{3 \pm \sqrt{5}}{5}. \end{aligned}$$

40. This equation can be solved by completing the square.

$$\begin{aligned} v^2 - 4v &= 9 \\ v^2 - 4v + 4 &= 9 + 4 \\ (v-2)^2 &= 13. \end{aligned}$$

Take the square root of both sides and solve for v to get $v = 2 \pm \sqrt{13}$.

41. Simplify by dividing by 3 and solve by completing the square.

$$\begin{aligned} y^2 &= 2y + 6 \\ y^2 - 2y &= 6 \\ y^2 - 2y + 1 &= 1 + 6 \\ (y-1)^2 &= 7. \end{aligned}$$

Take the square root of both sides and solve for y to get $y = 1 \pm \sqrt{7}$.

42. Set the equation equal to zero and use the quadratic formula with $a = 2$, $b = -14$, and $c = 23$.

$$\begin{aligned} p &= \frac{14 \pm \sqrt{(-14)^2 - 4 \cdot 2 \cdot 23}}{2 \cdot 2} \\ &= \frac{14 \pm \sqrt{196 - 184}}{4} \\ &= \frac{14 \pm \sqrt{12}}{4} \\ &= \frac{14 \pm 2\sqrt{3}}{4} \\ &= \frac{7 \pm \sqrt{3}}{2}. \end{aligned}$$

43. Set the equation equal to zero and use factoring.

$$\begin{aligned} 2w^3 - 6w^2 - 8w + 24 &= 0 \\ w^3 - 3w^2 - 4w + 12 &= 0 \\ w^2(w-3) - 4(w-3) &= 0 \\ (w-3)(w^2-4) &= 0 \\ (w-3)(w-2)(w+2) &= 0. \end{aligned}$$

So, $w - 3 = 0$ or $w - 2 = 0$ or $w + 2 = 0$ thus, $w = 3$ or $w = 2$ or $w = -2$.

44. Solve by completing the square.

$$\begin{aligned} 4x^2 + 16x &= 5 \\ 4(x^2 + 4x) &= 5 \\ 4(x^2 + 4x + 4) &= 4 \cdot 4 + 5 \\ 4(x+2)^2 &= 21 \\ (x+2)^2 &= \frac{21}{4}. \end{aligned}$$

Take the square root of both sides and solve for x.

$$\begin{aligned} x + 2 &= \pm\sqrt{\frac{21}{4}} \\ x + 2 &= \pm\frac{\sqrt{21}}{2} \\ x &= -2 \pm \frac{\sqrt{21}}{2}. \end{aligned}$$

45. Use the quadratic formula with $a = 49$, $b = 70$, $c = 22$, to solve this equation.

$$\begin{aligned} m &= \frac{-70 \pm \sqrt{70^2 - 4 \cdot 49 \cdot 22}}{2 \cdot 49} \\ &= \frac{-70 \pm \sqrt{4900 - 4312}}{98} \\ &= \frac{-70 \pm \sqrt{588}}{98} \\ &= \frac{-70 \pm 14\sqrt{3}}{98} \\ &= \frac{-5 \pm \sqrt{3}}{7}. \end{aligned}$$

46. Before completing the square, get all terms on the left, and divide by the coefficient of x^2:

$$8x^2 - 1 = 2x$$
$$8x^2 - 2x - 1 = 0$$
$$x^2 - \frac{1}{4}x - \frac{1}{8} = 0.$$

Now complete the square

$$x^2 - \frac{1}{4}x + \frac{1}{64} - \frac{1}{64} - \frac{1}{8} = 0$$
$$(x - \frac{1}{8})^2 - \frac{9}{64} = 0$$
$$x - \frac{1}{8} = \pm\sqrt{\frac{9}{64}}$$
$$x = \frac{1}{8} \pm \frac{3}{8}$$

So the solutions are $x = 1/2$ and $x = -1/4$.

CHAPTER FOUR

Solutions for Section 4.1

Skill Refresher

S1. We have $6\% = 0.06$.

S2. We have $0.6\% = 0.006$.

S3. We have $0.0012\% = 0.12\%$.

S4. We have $1.23\% = 123\%$.

Exercises

1. Yes. Writing the function as

$$g(w) = 2\left(2^{-w}\right) = 2\left(2^{-1}\right)^w = 2\left(\frac{1}{2}\right)^w,$$

we have $a = 2$ and $b = 1/2$.

2. Yes. Writing the function as

$$m(t) = (2 \cdot 3^t)^3 = 2^3 \cdot (3^t)^3 = 8 \cdot 3^{3t} = 8(3^3)^t = 8(27)^t,$$

we have $a = 8$ and $b = 27$.

3. Yes. Writing the function as

$$f(x) = \frac{3^{2x}}{4} = \frac{1}{4}(3^{2x}) = \frac{1}{4}(3^2)^x = \frac{1}{4}(9)^x,$$

we have $a = 1/4$ and $b = 9$.

4. No. The base must be a constant.

5. Yes. Writing the function as

$$q(r) = \frac{-4}{3^r} = -4\left(\frac{1}{3^r}\right) = -4\left(\frac{1^r}{3^r}\right) = -4\left(\frac{1}{3}\right)^r,$$

we have $a = -4$ and $b = 1/3$.

6. Yes. Writing the function as

$$j(x) = 2^x 3^x = (2 \cdot 3)^x = 6^x,$$

we have $a = 1$ and $b = 6$.

7. Yes. Writing the function as

$$Q(t) = 8^{t/3} = 8^{(1/3)t} = \left(8^{1/3}\right)^t = 2^t,$$

we have $a = 1$ and $b = 2$.

8. Yes. Writing the function as

$$K(x) = \frac{2^x}{3 \cdot 3^x} = \frac{1}{3}\left(\frac{2^x}{3^x}\right) = \frac{1}{3}\left(\frac{2}{3}\right)^x,$$

we have $a = 1/3$ and $b = 2/3$.

9. No. The two terms cannot be combined into the form b^r.

10. The annual growth factor is $1+$ the growth rate, so we have 1.03.

11. The decennial growth factor (growth factor per 10 years) is $1+$ the growth per decade: 1.28.

12. The daily growth factor is $1+$ the daily growth. Since the mine's resources are shrinking, the growth is -0.01, giving 0.99.

13. The growth factor per century is 1+ the growth per century. Since the forest is shrinking, the growth is negative, so we subtract 0.80, giving 0.20.

14. We have $a = 1750$, $b = 1.593$, and $r = b - 1 = 0.593 = 59.3\%$.

15. We have $a = 34.3$, $b = 0.788$, and $r = b - 1 = 0.788 - 1 = -0.212 = -21.2\%$.

16. Since $Q = 79.2(1.002)^t$, we have $a = 79.2$, $b = 1.002$, and $r = b - 1 = 0.002 = 0.2\%$.

17. We can rewrite this as

$$\begin{aligned} Q &= 0.0022(2.31^{-3})^t \\ &= 0.0022(0.0811)^t, \end{aligned}$$

so $a = 0.0022$, $b = 0.0811$, and $r = b - 1 = -0.9189 = -91.89\%$.

Problems

18. **(a)** The formula $f(t) = ab^t$ represents exponential growth if the base $b > 1$ and exponential decay if $0 < b < 1$. Towns (i), (ii), and (iv) are growing and towns (iii), (v), and (vi) are shrinking.

(b) Town (iv) is growing the fastest since its growth factor of 1.185 is the largest. Since $1.185 = 1 + 0.185$, it is growing at a rate of 18.5% per year.

(c) Town (v) is shrinking the fastest since its growth factor of 0.78 is the smallest. Since $0.78 = 1 - 0.22$, it is shrinking at a rate of 22% per year.

(d) In the exponential function $f(t) = ab^t$, the parameter a gives the value of the function when $t = 0$. We see that town (iii) has the largest initial population (2500) and town (ii) has the smallest initial population (600).

19. Since, after one year, 3% of the investment is added on to the original amount, we know that its value is 103% of what it had been a year earlier. Therefore, the growth factor is 1.03.

$$\begin{aligned} \text{So, after one year,} \quad V &= 100{,}000(1.03) \\ \text{After two years,} \quad V &= 100{,}000(1.03)(1.03) = 100{,}000(1.03)^2 \\ \text{After three years,} \quad V &= 100{,}000(1.03)(1.03)(1.03) = 100{,}000(1.03)^3 = \$109{,}272.70 \end{aligned}$$

20. If an investment decreases by 5% each year, we know that only 95% remains at the end of the first year. After 2 years there will be 95% of 95%, or 0.95^2 left. After 4 years, there will be $0.95^4 \approx 0.81451$ or 81.451% of the investment left; it therefore decreases by about 18.549% altogether.

21. To match formula and graph, we keep in mind the effect on the graph of the parameters a and b in $y = ab^t$.

If $a > 0$ and $b > 1$, then the function is positive and increasing.
If $a > 0$ and $0 < b < 1$, then the function is positive and decreasing.
If $a < 0$ and $b > 1$, then the function is negative and decreasing.
If $a < 0$ and $0 < b < 1$, then the function is negative and increasing.

(a) $y = 0.8^t$. So $a = 1$ and $b = 0.8$. Since $a > 0$ and $0 < b < 1$, we want a graph that is positive and decreasing. The graph in (ii) satisfies the conditions.

(b) $y = 5(3)^t$. So $a = 5$ and $b = 3$. The graph in (i) is both positive and increasing.

(c) $y = -6(1.03)^t$. So $a = -6$ and $b = 1.03$. Here, $a < 0$ and $b > 1$, so we need a graph which is negative and decreasing. The graph in (iv) satisfies these conditions.

(d) $y = 15(3)^{-t}$. Since $(3)^{-t} = (3)^{-1 \cdot t} = (3^{-1})^t = (\frac{1}{3})^t$, this formula can also be written $y = 15(\frac{1}{3})^t$. $a = 15$ and $b = \frac{1}{3}$. A graph that is both positive and decreasing is the one in (ii).

(e) $y = -4(0.98)^t$. So $a = -4$ and $b = 0.98$. Since $a < 0$ and $0 < b < 1$, we want a graph which is both negative and increasing. The graph in (iii) satisfies these conditions.

(f) $y = 82(0.8)^{-t}$. Since $(0.8)^{-t} = (\frac{8}{10})^{-t} = (\frac{8}{10})^{-1 \cdot t} = ((\frac{8}{10})^{-1})^t = (\frac{10}{8})^t = (1.25)^t$ this formula can also be written as $y = 82(1.25)^t$. So $a = 82$ and $b = 1.25$. A graph which is both positive and increasing is the one in (i).

22. **(a)** Since the initial amount is 2000 and the growth rate is 5% per year, the formula is $Q = 2000(1.05)^t$.

(b) At $t = 10$, we have $Q = 2000(1.05)^{10} = 3257.789$.

23. **(a)** Since the initial amount is 35 and the quantity is decreasing at rate of 8% per year, the formula is $Q = 35(1-0.08)^t = 35(0.92)^t$.
(b) At $t = 10$, we have $Q = 35(0.92)^{10} = 15.204$.

24. **(a)** Since the initial amount is 112.8 and the quantity is decreasing at a rate of 23.4% per year, the formula is $Q = 112.8(1-0.234)^t = 112.8(0.766)^t$.
(b) At $t = 10$, we have $Q = 112.8(0.766)^{10} = 7.845$.

25. **(a)** Since the initial amount is 5.35 and the growth rate is 0.8% per year, the formula is $Q = 5.35(1.008)^t$.
(b) At $t = 10$, we have $Q = 5.35(1.008)^{10} = 5.794$.

26. **(a)** Since the initial amount is 5 and the growth rate is 100% per year, the formula is $Q = 5(1+1)^t = 5(2)^t$.
(b) At $t = 10$, we have $Q = 5(2)^{10} = 5120$.

27. **(a)** Since the initial amount is 0.2 and the quantity is decreasing at a rate of 0.5% per year, the formula is $Q = 0.2(1-0.005)^t = 0.2(0.995)^t$.
(b) At $t = 10$, we have $Q = 0.2(0.995)^{10} = 0.190$.

28. Using the formula for slope, we have

$$\text{Slope} = \frac{f(5)-f(1)}{5-1} = \frac{4b^5-4b^1}{4} = \frac{4b(b^4-1)}{4} = b(b^4-1).$$

29. The population is growing at a rate of 1.9% per year. So, at the end of each year, the population is $100\%+1.9\% = 101.9\%$ of what it had been the previous year. The growth factor is 1.019. If P is the population of this country, in millions, and t is the number of years since 2010, then, after one year,

$$\begin{aligned}
& P = 70(1.019). \\
\text{After two years,} \quad & P = 70(1.019)(1.019) = 70(1.019)^2 \\
\text{After three years,} \quad & P = 70(1.019)(1.019)(1.019) = 70(1.019)^3 \\
\text{After } t \text{ years,} \quad & P = 70\underbrace{(1.019)(1.019)\ldots(1.019)}_{t \text{ times}} = 70(1.019)^t
\end{aligned}$$

30. If it decays at 5.626% per year, its growth factor is $1 - 0.05626 = 0.94374$. So, with t in years and an initial amount of 726 grams, we have:

$$Q = 726(0.94374)^t.$$

We graph it on a graphing calculator or with a computer to obtain the graph in Figure 4.1.

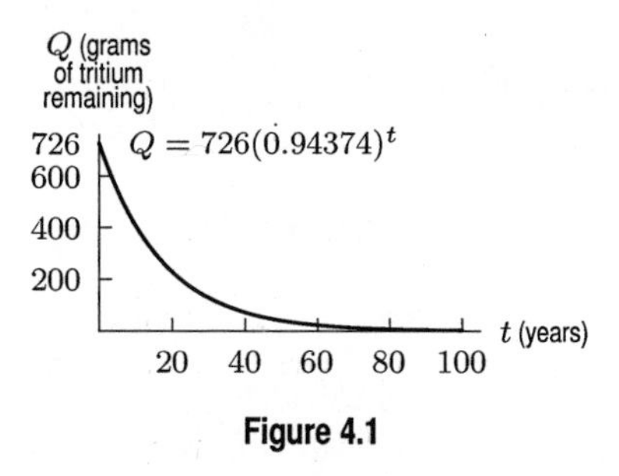

Figure 4.1

31. To find a formula for $f(n)$, we start with the number of people infected in 2010, namely P_0. In 2011, only 80% as many people, or $0.8P_0$, were infected. In 2012, again only 80% as many people were infected, which means that 80% of $0.8P_0$ people, or $0.8(0.8P_0)$ people, were infected. Continuing this line of reasoning, we can write

$$f(0) = P_0$$

$$f(1) = \underbrace{(0.80)}_{\substack{\text{one 20\%} \\ \text{reduction}}} P_0 = (0.8)^1 P_0$$

$$f(2) = \underbrace{(0.80)(0.80)}_{\substack{\text{two 20\%} \\ \text{reductions}}} P_0 = (0.8)^2 P_0$$

$$f(3) = \underbrace{(0.80)(0.80)(0.80)}_{\text{three 20\% reductions}} P_0 = (0.8)^3 P_0,$$

and so on, so that n years after 2010 we have

$$f(n) = \underbrace{(0.80)(0.80)\cdots(0.80)}_{n \text{ 20\% reductions}} P_0 = (0.8)^n P_0.$$

We see from its formula that $f(n)$ is an exponential function, because it is of the form $f(n) = ab^n$, with $a = P_0$ and $b = 0.8$. The graph of $y = f(n) = P_0(0.8)^n$, for $n \geq 0$, is given in Figure 4.2. Beginning at the P-axis, the curve decreases sharply at first toward the horizontal axis, but then levels off so that its descent is less rapid.

Figure 4.2 shows that the prevalence of the virus in the population drops quickly at first, and that it eventually levels off and approaches zero. The curve has this shape because in the early years of the vaccination program, there was a relatively large number of infected people. In later years, due to the success of the vaccine, the infection became increasingly rare. Thus, in the early years, a 20% drop in the infected population represented a larger number of people than a 20% drop in later years.

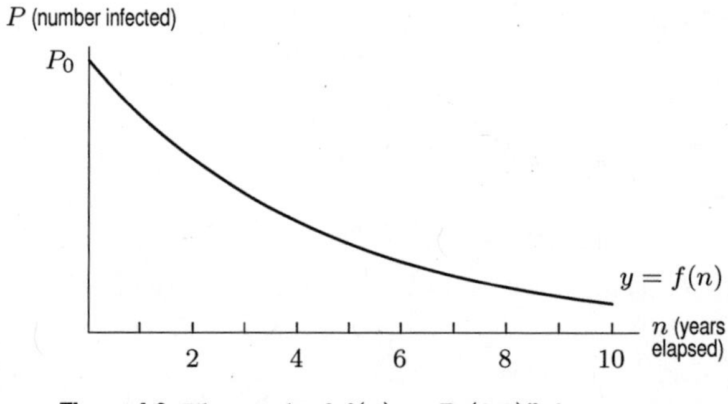

Figure 4.2: The graph of $f(n) = P_0(0.8)^n$ for $n \geq 0$

32. In year $t = 0$, there are one million organisms, which we take as our initial value. Our growth factor is 0.98, for a decay of 2% per year. Thus:

$$O = 1{,}000{,}000(0.98)^t.$$

33. **(a)** We assume that the price of a movie ticket increases at the rate of 3.5% per year. This means that the price is rising exponentially, so a formula for p is of the form $p = ab^t$. We have $a = 7.50$ and $b = 1 + r = 1.035$. Thus, a formula for p is

$$p = 7.50(1.035)^t.$$

(b) In 20 years ($t = 20$) we have $p = 7.50(1.035)^{20} \approx 14.92$. Thus, in 20 years, movie tickets will cost almost \$15 if the inflation rate remains at 3.5%.

34. We have $a = 2500$ and $b = 1.0325$, so $r = b - 1 = 0.0325 = 3.25\%$. Thus, the starting value is \$2500 and the percent growth rate is 3.25%.

35. **(a)** We have $C = C_0(1-r)^t = 100(1-0.16)^t = 100(0.84)^t$, so

$$C = 100(0.84)^t.$$

(b) At $t = 5$, we have $C = 100(0.84)^5 = 41.821$ mg

36. **(a)** Since the initial size is 142 and the size is shrinking at a rate of 4.4% per year, the formula is $S = 142(1-0.044)^t = 142(0.956)^t$.

(b) At $t = 8$, we have $S = 142(0.956)^8 = 99.072$. The glacier is predicted to be just under 100 acres by the year 2015, at about 99.072 acres.

(c) At $t = 3$, we have $S = 142(0.956)^3 = 124.069$. The glacier lost about $142 - 124 = 18$ acres of ice during this three-year period.

37. **(a)** At a time t years after 2005, the population P in millions is $P = 36.8(1.013)^t$.
In 2030, we have $t = 25$, so

$$P = 36.8(1.013)^{25} = 50.826 \text{ million.}$$

Between 2005 and 2030,

$$\text{Increase} = 50.826 - 36.8 = 14.026 \text{ million.}$$

In 2055, we have $t = 50$, so

$$P = 36.8(1.013)^{50} = 70.197 \text{ million.}$$

Between 2030 and 2055,

$$\text{Increase} = 70.197 - 50.826 = 19.371 \text{ million.}$$

(b) The increase between 2030 and 2055 is expected to be larger than the increase between 2005 and 2030 because the exponential function is concave up. Both increases are over 25 year periods, but since the graph of the function bends upward, the increase in the later time period is larger.

38. Let $D(t)$ be the difference between the oven's temperature and the yam's temperature, which is given by an exponential function $D(t) = ab^t$. The initial temperature difference is $300°\text{F} - 0°\text{F} = 300°\text{F}$, so $a = 300$. The temperature difference decreases by 3% per minute, so $b = 1 - 0.03 = 0.97$. Thus,

$$D(t) = 300(0.97)^t.$$

If the yam's temperature is represented by $Y(t)$, then the temperature difference is given by

$$D(t) = 300 - Y(t),$$

so, solving for $Y(t)$, we have

$$Y(t) = 300 - D(t),$$

giving

$$Y(t) = 300 - 300(0.97)^t.$$

39. **(a)** Since the initial population is 1.15 and the growth rate is 1.35% per year, the formula is $P = 1.15(1.0135)^t$.

(b) At $t = 5$, we have $P = 1.15(1.0135)^5 = 1.230$. At $t = 10$, we have $P = 1.15(1.0135)^{10} = 1.315$. This model predicts that the population of India will be 1.230 billion in the year 2015 and 1.315 billion in the year 2020.

(c) Since the population is 1.15 billion and is growing at a rate of 1.35% per year, we expect the population to grow by $1.15(0.0135) = 0.015525$ billion people during the year, or a rate of about 15.525 million people per year.

(d) Since there are $60 \cdot 24 \cdot 365 = 525{,}600$ minutes in a year, we see that 15.525 million people per year is an average increase of

$$\frac{15{,}525{,}000}{525{,}600} = 29.538 \text{ people per minute.}$$

40. Since each filter removes 85% of the remaining impurities, the rate of change of the impurity level is $r = -0.85$ per filter. Thus, the growth factor is $B = 1 + r = 1 - 0.85 = 0.15$. This means that each time the water is passed through a filter, the impurity level L is multiplied by a factor of 0.15. This makes sense, because if each filter removes 85% of the impurities, it will leave behind 15% of the impurities. We see that a formula for L is

$$L = 420(0.15)^n,$$

because after being passed through n filters, the impurity level will have been multiplied by a factor of 0.15 a total of n times.

41. **(a)** In 2012, we have

$$\text{World PV market installations } = 2826(1.62)^5 = 31{,}531.7 \text{ megawatts,}$$

and

$$\text{Japan PV market installations } = 230(0.77)^5 = 62.3 \text{ megawatts.}$$

(b) In 2007, the proportion of world market installations in Japan was $230/2826 = 0.081$, or about 8.1%. In 2012, we estimate that the proportion is $62.3/31{,}531.7 = 0.00197$, or less than 0.2%.

42. We have

$$\begin{aligned} y &= 5(0.5)^{t/3} \\ &= 5\left((0.5)^{\frac{1}{3}}\right)^t \\ &= 5\left(\sqrt[3]{0.5}\right)^t, \end{aligned}$$

so $a = 5$ and $b = (0.5)^(1/3) = \sqrt[3]{0.5}$.

43. We have

$$\begin{aligned} y &= 5(0.5)^{t/3} \\ &= 5\left(2^{-1}\right)^{\frac{1}{3}\cdot t} \\ &= 5\cdot 2^{-\frac{1}{3}\cdot t} \\ &= 5\left(4^{1/2}\right)^{-\frac{1}{3}\cdot t} \\ &= 5\cdot 4^{-\frac{1}{6}\cdot t}, \end{aligned}$$

so $a = 5, k = -1/6$.

44. We have

$$\begin{aligned} f(t) &= \frac{60}{5\cdot 2^{t/11.2}} \\ &= \frac{60}{5}\cdot\left(2^{\frac{1}{11.2}t}\right)^{-1} \\ &= 12\left(2^{-\frac{1}{11.2}}\right)^t \\ &= 12\,(0.940)^t, \\ \text{so} \quad a &= 12 \\ b &= 0.940. \end{aligned}$$

45. **(a)** Since N is growing by 5% per year, we know that N is an exponential function of t with growth factor $1 + 0.05 = 1.05$. Since $N = 13.4$ when $t = 0$, we have

$$N = 13.4(1.05)^t.$$

(b) In the year 2015, we have $t = 6$ and

$$N = 13.4(1.05)^6 \approx 17.957 \text{ million passengers.}$$

In the year 2005, we have $t = -4$ and

$$N = 13.4(1.05)^{-4} \approx 11.024 \text{ million passengers.}$$

46. **(a)** The initial dose equals the amount of drug in the body when $t = 0$. We have $A(0) = 25(0.85)^0 = 25(1) = 25$ mg.

(b) According to the formula,

$$\begin{aligned} A(0) &= 25(0.85)^0 = 25 \\ A(1) &= 25(0.85)^1 = 25(0.85) \\ A(2) &= 25(0.85)^2 = 25(0.85)(0.85) \end{aligned}$$

After each hour, the amount of the drug in the body is the amount at the end of the previous hour multiplied by 0.85. In other words, the amount remaining is 85% of what it had been an hour ago. So, 15% of the drug has left in that time.

(c) After 10 hours, $t = 10$. $A(10) = 4.922$ mg.

(d) Using trial and error, substitute integral values of t into $A(t) = 25(0.85)^t$ to determine the smallest value of t for which $A(t) < 1$. We find that $t = 20$ is the best choice. So, after 20 hours there will be less than one milligram in the body.

47. (a) $N(0)$ gives the number of teams remaining in the tournament after no rounds have been played. Thus, $N(0) = 64$. After 1 round, half of the original 64 teams remain in the competition, so

$$N(1) = 64(\frac{1}{2}).$$

After 2 rounds, half of these teams remain, so

$$N(2) = 64(\frac{1}{2})(\frac{1}{2}).$$

And, after r rounds, the original pool of 64 teams has been halved r times, so that

$$N(r) = 64 \underbrace{(\frac{1}{2})(\frac{1}{2})\cdots(\frac{1}{2})}_{\text{pool halved } r \text{ times}},$$

giving

$$N(r) = 64(\frac{1}{2})^r.$$

The graph of $y = N(r)$ is given in Figure 4.3. The domain of N is $0 \le r \le 6$, for r an integer. A curve has been dashed in to help you see the overall shape of the function.

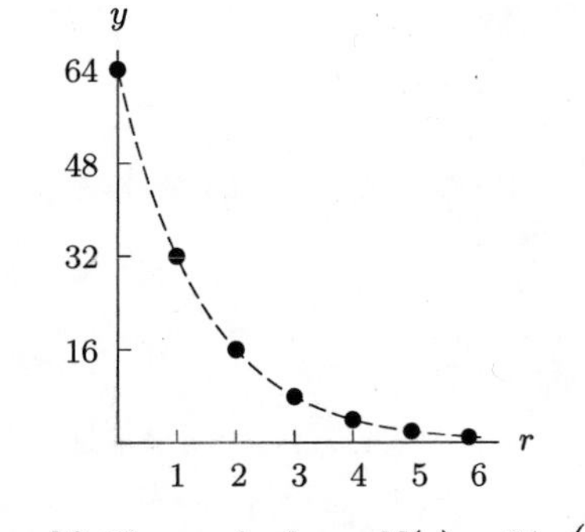

Figure 4.3: The graph of $y = N(r) = 64 \cdot \left(\frac{1}{2}\right)^r$

(b) There will be a winner when there is only one person left, North Carolina. So, $N(r) = 1$.

$$\begin{aligned} 64(\frac{1}{2})^r &= 1 \\ \left(\frac{1}{2}\right)^r &= \frac{1}{64} \\ \frac{1}{2^r} &= \frac{1}{64} \\ 2^r &= 64 \\ r &= 6 \end{aligned}$$

You can solve $2^r = 64$ either by taking successive powers of 2 until you get to 64 or by substituting values for r until you get the one that works.

48. **(a)** The amount of forest lost is 2.17% of 4077 million, or $0.0217 \cdot 4077 = 88.4709 \approx 88.471$ million hectares.

(b) The amount of world forest cover in 2000 is the amount in 1990 minus the amount lost, or $4077 - 88.4709 = 3988.5291 \approx 3988.529$ million hectares.

(c) Let $f(t)$ represent the number of million hectares of natural forest in the world t years after 1990. Since $f(0) = 4077$ and we are assuming exponential decay, we have $f(t) = 4077b^t$ for some base b. Since 2.17% decayed over a 10-year period, we know that $1 - 0.0217 = 0.9783$ was the growth factor for the 10-year period. We have

$$b^{10} = 0.9783$$
$$b = (0.9783)^{1/10}.$$

The formula is $f(t) = 4077\left((0.9783)^{1/10}\right)^t \approx 4077(0.9978)^t$.

(d) The annual growth factor is $0.9978 = 1 - 0.0022$ so the world forest cover decreased at a rate of 0.22% per year during the years 1990 – 2000.

(e) The annual growth factor is $1 - 0.0018 = 0.9982$. We calculated that there were approximately 3988.529 million hectares of forest in the year 2000. So, our formula is $f(t) = 3988.529(0.9982)^t$. When we substitute 5 for t we get $f(5) = 3988.529(0.9982)^5 \approx 3952.761$ million hectares.

49. Let r be the percentage by which the substance decays each year. Every year we multiply the amount of radioactive substance by $1 - r$ to determine the new amount. If a is the amount of the substance on hand originally, we know that after five years, there have been five yearly decreases, by a factor of $1 - r$. Since we know that there will be 60% of a, or $0.6a$, remaining after five years (because 40% of the original amount will have decayed), we know that

$$a \cdot \underbrace{(1-r)^5}_{\substack{\text{five annual decreases} \\ \text{by a factor of } 1-r}} = 0.6a.$$

Dividing both sides by a, we have $(1 - r)^5 = 0.6$, which means that

$$1 - r = (0.6)^{\frac{1}{5}} \approx 0.9029$$

so

$$r \approx 0.09712 = 9.712\%.$$

Each year the substance decays by 9.712%.

50. **(a)** Since

$$\begin{aligned} \text{New population} &= 1.134 \cdot (\text{old population}) \\ &= 113.4\% \text{ of old population} \\ &= 100\% \text{ of old population } + 13.4\% \text{ of old population,} \end{aligned}$$

so the town has increased in size by 13.4%.

(b) Let b be the annual growth factor. Then since 1.134 is the two-year growth factor,

$$b^2 = 1.134$$
$$b = \sqrt{1.134} \approx 1.06489.$$

With this result, we know that after one year the town is 106.489% of its size from the previous year. Thus, this town grew at an annual rate of 6.489%.

51. **(a)** The monthly payment on \$1000 each month at 4% for a loan period of 15 years is \$7.40. For \$60,000, the payment would be $\$7.40 \times 60 = \$444$ per month.

(b) The monthly payment on \$1000 each month at 4% for a loan period of 30 years is \$4.77. For \$60,000, the payment would be $\$4.77 \times 60 = \$286.20$ per month.

(c) The monthly payment on \$1000 each month at 6% for a loan period of 15 years is \$8.44. For \$60,000, the payment would be $\$8.44 \times 60 = \$506.40$ per month.

(d) As calculated in part (a), the monthly payment on a \$60,000 loan at 4% for 15 years would be \$444 per month. In part (c) we showed that the the monthly payment on a \$60,000 loan at 6% for 15 years would be \$506.40 per month. So taking the loan out at 4% rather that 6% would save the difference:

$$\text{Amount saved} = \$506.40 - \$444 = \$62.40 \text{ per month}$$

Since there are $15 \times 12 = 180$ months in 15 years,

$$\text{Total amount saved} = \$62.40 \text{ per month} \times 180 \text{ months} = \$11{,}232.$$

(e) In part (a) we found the monthly payment on an 4% mortgage of \$60,000 for 15 years to be \$444. The total amount paid over 15 years is then

$$\$444 \text{ per month} \times 180 \text{ months} = \$79,920.$$

In part (b) we found the monthly payment on an 4% mortgage of \$60,000 for 30 years to be \$286.20. The total amount paid over 30 years is then

$$286.20 \text{ per month} \times 360 \text{ months} = \$103,032.$$

The amount saved by taking the mortgage over a shorter period of time is the difference:

$$\$103,032 - \$79,920 = \$23,112.$$

52. **(a)**

Table 4.1

Month	Balance	Interest	Minimum payment
0	\$2000.00	\$30.00	\$50.00
1	\$1980.00	\$29.70	\$49.50
2	\$1960.20	\$29.40	\$49.01
3	\$1940.59	\$29.11	\$48.51
4	\$1921.19	\$28.82	\$48.03
5	\$1901.98	\$28.53	\$47.55
6	\$1882.96	\$28.24	\$47.07
7	\$1864.13	\$27.96	\$46.60
8	\$1845.49	\$27.68	\$46.14
9	\$1827.03	\$27.41	\$45.68
10	\$1808.76	\$27.13	\$45.22
11	\$1790.67	\$26.86	\$44.77
12	\$1772.76		

(b) After one year, your unpaid balance is \$1772.76. You have paid off $\$2000 - \$1772.76 = \$227.24$. The interest you have paid is the sum of the middle column: \$340.84

53. Each time we make a tri-fold, we triple the number of layers of paper, $N(x)$. So $N(x) = 3^x$, where x is the number of folds we make. After 20 folds, the letter would have 3^{20} (almost 3.5 billion!) layers. To find out how high our letter would be, we divide the number of layers by the number of sheets in one inch. So the height, h, is

$$h = \frac{3^{20}\text{sheets}}{150 \text{ sheets/inch}} \approx 23{,}245{,}229.34 \text{ inches.}$$

Since there are 12 inches in a foot and 5280 feet in a mile, this gives

$$\begin{aligned} h &\approx 23245229.34 \text{ in} \left(\frac{1 \text{ ft}}{12 \text{ in}}\right)\left(\frac{1 \text{ mile}}{5280 \text{ ft}}\right) \\ &\approx 366.875 \text{ miles.} \end{aligned}$$

54. We have

$$\begin{aligned} f(n) &= 1000 \cdot 2^{-\frac{1}{4}-\frac{n}{2}} \\ &= 1000 \cdot 2^{-\frac{1}{4}} 2^{-\frac{n}{2}} \\ &= \underbrace{\frac{1000}{2^{1/4}}}_{a} \underbrace{\left(2^{-1/2}\right)^n}_{b}, \end{aligned}$$

so

$$\begin{aligned} a &= \frac{1000}{2^{1/4}} = 840.896 \\ b &= 0.7071. \end{aligned}$$

The value of a tells us that a sheet of $A0$ paper is 840.896 m wide. The value of b tells us that the width decreases by a factor of 0.7071, or by 29.29%, for each higher-numbered sheet in the series.

55. We have

$$\begin{aligned} \frac{f(n+2)}{f(n)} &= \frac{1000 \cdot 2^{-\frac{1}{4}-\frac{n+2}{2}}}{1000 \cdot 2^{-\frac{1}{4}-\frac{n}{2}}} \\ &= \frac{1000}{1000} \cdot \frac{2^{-\frac{1}{4}}}{2^{-\frac{1}{4}}} \cdot \frac{2^{-\frac{n+2}{2}}}{2^{-\frac{n}{2}}} \\ &= 2^{-\frac{n+2}{2}} \cdot 2^{\frac{n}{2}} \\ &= 2^{-\frac{n}{2}-\frac{2}{2}+\frac{n}{2}} \\ &= 2^{-1} = 0.5. \end{aligned}$$

This means a sheet two numbers higher in the series is half as wide. For instance, a sheet of $A3$ is half as wide as a sheet of $A1$.

56. The vertical intercept of $a_1(b_1)^t$ is greater than that of $a_0(b_0)^t$, so $a_1 > a_0$.

57. The graph $a_0(b_0)^t$ climbs faster than that of $a_1(b_1)^t$, so $b_0 > b_1$.

58. The value of t_0 goes down. To see this, notice that as a_0 increases, the vertical intercept of $a_0(b_0)^t$ goes up, so the point of intersection moves to the left. In other words, as a_0 increases, the graph of $a_0(b_0)^t$ "catches up" to the graph of $a_1(b_1)^t$ earlier. If a_0 rises as high as a_1, the value of t_0 drops to 0, because the two graphs intersect at the vertical axis. If a_0 rises higher than a_1, the value of t_0 becomes negative, because the two graphs intersect to the left of the y-axis.

59. The value of t_0 decreases. To see that, notice that if b_1 is decreased, the graph of $a_1(b_1)^t$ climbs more slowly, and eventually (if b_1 falls below 1) begins to fall. Thus the point of intersection moves to the left, so t_0 goes down. However, since the graph of $a_1(b_1)^t$ intersects the vertical axis above the graph of $a_0(b_0)^t$, t_0 remains positive no matter how small b_1 becomes, so long as $b_1 > 0$. (Recall that the base of an exponential function must be positive.)

60.
- The y-intercept of g is above the y-intercept of f, so $a_1 > a_0$.
- The half-life of f is τ_0. For instance,

$$\begin{aligned} f(0) &= a_0 \\ f(\tau_0) &= a_0 \cdot 2^{-\tau_0/\tau_0} = a_0 \cdot 2^{-1} = 0.5a_0 \\ f(2\tau_0) &= a_0 \cdot 2^{-2\tau_0/\tau_0} = a_0 \cdot 2^{-2} = 0.25a_0, \end{aligned}$$

etc. We see that the value of $f(t)$ drops by a factor of 0.5 each time t increases by τ_0. By a similar argument, the half-life of g is τ_1. Since the graph of f lies above the graph of g at the left end of the range shown, but below the graph of g at the right end, this means f decreases more rapidly than g. We conclude that f has a shorter half-life (so that it decreases faster), which means $\tau_1 > \tau_0$.

61. **(a)** We have

$$\text{Total revenue} = \text{No. households} \times \text{Rate per household}$$
$$\text{so} \quad R = N \times r.$$

(b) We have

$$\text{Average revenue} = \frac{\text{Total revenue}}{\text{No. students}}$$
$$\text{so} \quad A = \frac{R}{P} = \frac{Nr}{P}.$$

(c) We have

$$N_{\text{new}} = N + (2\%)N = 1.02N$$
$$r_{\text{new}} = r + (3\%)r = 1.03r$$

(d) We have

$$R_{\text{new}} = N_{\text{new}} \times r_{\text{new}} = (1.02N)(1.03r)$$
$$= 1.0506Nr = 1.0506R.$$

Thus, R increased by 5.06%, or by just over 5%.

(e) We have

$$P_{\text{new}} = P + (8\%)P = 1.08P$$
$$A_{\text{new}} = \frac{R_{\text{new}}}{P_{\text{new}}}$$
$$= \frac{1.0506R}{1.08P} = (0.9728)\left(\frac{R}{P}\right) \approx (97.3\%)A,$$

so the average revenue fell by 2.7%, despite the fact that the tax rate and the tax base both grew.

Solutions for Section 4.2

Skill Refresher

S1. We have $b^4 \cdot b^6 = b^{4+6} = b^{10}$.

S2. We have $8g^3 \cdot (-4g)^2 = 8g^3 \cdot 16g^2 = 128g^{3+2} = 128g^5$.

S3. We have

$$\frac{18a^{10}b^6}{6a^3b^{-4}} = \frac{18}{3} \cdot \frac{a^{10}}{a^3} \cdot \frac{b^6}{b^{-4}}$$
$$= 6a^{10-3}b^{6-(-4)}$$
$$= 6a^7b^{10}.$$

S4. We have

$$\frac{(2a^3b^2)^3}{(4ab^{-4})^2} = \frac{2^3a^{3\cdot 3}b^{2\cdot 3)}}{4a^2b^{-4\cdot 2}}$$
$$= \frac{8a^9b^6}{4a^2b^{-8}}$$
$$= 2a^{9-2}b^{6-(-8)}$$
$$= 2a^7b^{14}.$$

S5. We have $f(0) = 5.6(1.043)^0 = 5.6$ and $f(3) = 5.6(1.043)^3 = 6.354$.

S6. We have $g(0) = 12{,}837(0.84)^0 = 12{,}837$ and $g(3) = 12{,}837(0.84)^3 = 7608.541$

S7. We have

$$\begin{aligned} 4x^3 &= 20 \\ x^3 &= 5 \\ x &= (5)^{1/3} = 1.710. \end{aligned}$$

S8. We have

$$\begin{aligned} \frac{5}{x^2} &= 125 \\ \frac{1}{x^2} &= 25 \\ x^2 &= 1/25 \\ x &= \pm(1/25)^{1/2} = \pm 1/5 = \pm 0.2. \end{aligned}$$

S9. We have

$$\begin{aligned} \frac{4}{3}x^5 &= 7 \\ x^5 &= 5.25 \\ x &= (5.25)^{1/5} = 1.393. \end{aligned}$$

S10. We have

$$\begin{aligned} \sqrt{4x^3} &= 5 \\ 2x^{3/2} &= 5 \\ x^{3/2} &= 2.5 \\ x &= (2.5)^{2/3} = 1.842. \end{aligned}$$

Exercises

1. **(a)** Since the rate of change is constant, the formula is the linear function $p = 2.50 + 0.03t$.
(b) Since the rate of change is constant, the formula is the linear function $p = 2.50 - 0.07t$.
(c) Since the percent rate of change is constant, the formula is the exponential function $p = 2.50(1.02)^t$.
(d) Since the percent rate of change is constant, the formula is the exponential function $p = 2.50(0.96)^t$.

2. **(a)** The population is decreasing linearly, with a slope of -100 people/year, so $P = 5000 - 100t$
(b) The population is decreasing exponentially with "growth" factor $1 - 0.08 = 0.92$, so $P = 5000(1 - 0.08)^t = 5000(0.92)^t$

3. The formula $P_A = 200 + 1.3t$ for City A shows that its population is growing linearly. In year $t = 0$, the city has 200,000 people and the population grows by 1.3 thousand people, or 1,300 people, each year.

The formulas for cities B, C, and D show that these populations are changing exponentially. Since $P_B = 270(1.021)^t$, City B starts with 270,000 people and grows at an annual rate of 2.1%. Similarly, City C starts with 150,000 people and grows at 4.5% annually.

Since $P_D = 600(0.978)^t$, City D starts with 600,000 people, but its population decreases at a rate of 2.2% per year. We find the annual percent rate by taking $b = 0.978 = 1 + r$, which gives $r = -0.022 = -2.2\%$. So City D starts out with more people than the other three but is shrinking.

Figure 4.4 gives the graphs of the three exponential populations. Notice that the P-intercepts of the graphs correspond to the initial populations (when $t = 0$) of the towns. Although the graph of P_C starts below the graph of P_B, it eventually catches up and rises above the graph of P_B, because City C is growing faster than City B.

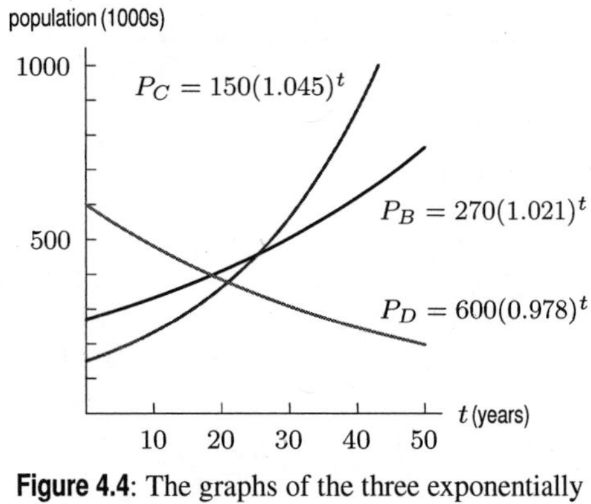

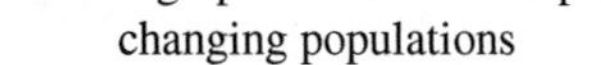
Figure 4.4: The graphs of the three exponentially changing populations

4. An exponential function should be used, because an exponential function grows by a constant percentage and a linear function grows by a constant absolute amount.

5. We have $g(10) = 50$ and $g(30) = 25$. Using the ratio method, we have

$$\begin{aligned}\frac{ab^{30}}{ab^{10}} &= \frac{g(30)}{g(10)}\\ b^{20} &= \frac{25}{50}\\ b &= \left(\frac{25}{50}\right)^{1/20} \approx 0.965936.\end{aligned}$$

Now we can solve for a:

$$\begin{aligned}a(0.965936)^{10} &= 50\\ a &= \frac{50}{(0.965936)^{10}} \approx 70.711.\end{aligned}$$

so $Q = 70.711(0.966)^t$.

6. We have $f(x) = ab^x$. Using our two points, we have

$$f(-8) = ab^{-8} = 200$$

and

$$f(30) = ab^{30} = 580.$$

Taking ratios, we have

$$\begin{aligned}\frac{580}{200} &= \frac{ab^{30}}{ab^{-8}} = b^{38}\\ \frac{580}{200} &= b^{38}.\end{aligned}$$

This gives

$$b = \left(\frac{580}{200}\right)^{1/38} \approx 1.0284.$$

We now solve for a. We know that $f(30) = 580$ and $f(x) = a(1.0284)^x$, so we have

$$\begin{aligned} f(30) &= a(1.0284)^{30} \\ 580 &= a(1.0284)^{30} \\ a &= \frac{580}{1.0284^{30}} \\ &\approx 250.4. \end{aligned}$$

Thus, $f(x) = 250.4(1.0284)^x$.

7. Since f is exponential, $f(x) = ab^x$. We know that

$$f(2) = ab^2 = \frac{2}{9}$$

and

$$f(-3) = ab^{-3} = 54,$$

so

$$\frac{2/9}{54} = \frac{ab^2}{ab^{-3}} = b^5.$$

$$\begin{aligned} b^5 &= \frac{1}{243} \\ b &= \left(\frac{1}{243}\right)^{1/5} = \frac{1}{3}. \end{aligned}$$

Thus, $f(x) = a\left(\frac{1}{3}\right)^x$. Since $f(2) = \frac{2}{9}$ and $f(2) = a(\frac{1}{3})^2$, we have

$$\begin{aligned} a\left(\frac{1}{3}\right)^2 &= \frac{2}{9} \\ \frac{a}{9} &= \frac{2}{9} \\ a &= 2. \end{aligned}$$

Thus, $f(x) = 2\left(\frac{1}{3}\right)^x$.

8. We know that $f(x) = ab^x$. Taking the ratio of $f(2)$ to $f(-1)$ we have

$$\begin{aligned} \frac{f(2)}{f(1)} &= \frac{1/27}{27} = \frac{ab^2}{ab^{-1}} \\ \frac{1}{(27)^2} &= b^3 \\ b^3 &= \frac{1}{27^2} \\ b &= (\frac{1}{27^2})^{\frac{1}{3}}. \end{aligned}$$

Thus, $b = \frac{1}{9}$. Therefore, $f(x) = a(\frac{1}{9})^x$.

Using the fact that $f(-1) = 27$, we have

$$f(-1) = a\left(\frac{1}{9}\right)^{-1} = a \cdot 9 = 27,$$

which means $a = 3$. Thus,

$$f(x) = 3\left(\frac{1}{9}\right)^x.$$

9. Let the equation of the exponential curve be $Q = ab^t$. Since this curve passes through the points $(-1, 2)$, $(1, 0.3)$, we have

$$\begin{aligned} 2 &= ab^{-1} \\ 0.3 &= ab^1 = ab \end{aligned}$$

So,

$$\frac{0.3}{2} = \frac{ab}{ab^{-1}} = b^2,$$

that is, $b^2 = 0.15$, thus $b = \sqrt{0.15} \approx 0.3873$ because b is positive. Since $2 = ab^{-1}$, we have $a = 2b = 2 \cdot 0.3873 = 0.7746$, and the equation of the exponential curve is

$$Q = 0.7746 \cdot (0.3873)^t.$$

10. We use $y = ab^x$. Since $y = 10$ when $x = 0$, we have $y = 10b^x$. We use the point $(3, 20)$ to find the base b:

$$\begin{aligned} y &= 10b^x \\ 20 &= 10b^3 \\ b^3 &= 2 \\ b &= 2^{1/3} = 1.260. \end{aligned}$$

The formula is

$$y = 10(1.260)^x.$$

11. We use $y = ab^x$. Since $y = 50$ when $x = 0$, we have $y = 50b^x$. We use the point $(5, 20)$ to find the base b:

$$\begin{aligned} y &= 50b^x \\ 20 &= 50b^5 \\ b^5 &= 0.4 \\ b &= 0.4^{1/5} = 0.833. \end{aligned}$$

The formula is

$$y = 50(0.833)^x.$$

12. Since the function is exponential, we know $y = ab^x$. We also know that $(0, 1/2)$ and $(3, 1/54)$ are on the graph of this function, so $1/2 = ab^0$ and $1/54 = ab^3$. The first equation implies that $a = 1/2$. Substituting this value in the second equation gives $1/54 = (1/2)b^3$ or $b^3 = 1/27$, or $b = 1/3$. Thus, $y = \frac{1}{2}\left(\frac{1}{3}\right)^x$.

13. Since this function is exponential, we know $y = ab^x$. We also know that $(-2, 8/9)$ and $(2, 9/2)$ are on the graph of this function, so

$$\frac{8}{9} = ab^{-2}$$

and

$$\frac{9}{2} = ab^2.$$

From these two equations, we can say that

$$\frac{\frac{9}{2}}{\frac{8}{9}} = \frac{ab^2}{ab^{-2}}.$$

Since $(9/2)/(8/9) = 9/2 \cdot 9/8 = 81/16$, we can re-write this equation to be

$$\frac{81}{16} = b^4.$$

Keeping in mind that $b > 0$, we get

$$b = \sqrt[4]{\frac{81}{16}} = \frac{\sqrt[4]{81}}{\sqrt[4]{16}} = \frac{3}{2}.$$

Substituting $b = 3/2$ in $ab^2 = 9/2$, we get

$$\frac{9}{2} = a(\frac{3}{2})^2 = \frac{9}{4}a$$
$$a = \frac{\frac{9}{2}}{\frac{9}{4}} = \frac{9}{2} \cdot \frac{4}{9} = \frac{4}{2} = 2.$$

Thus, $y = 2(3/2)^x$.

14. The formula for this function must be of the form $y = ab^x$. We know that $(-2, 400)$ and $(1, 0.4)$ are points on the graph of this function, so

$$400 = ab^{-2}$$

and

$$0.4 = ab^1.$$

This leads us to

$$\frac{0.4}{400} = \frac{ab^1}{ab^{-2}}$$
$$0.001 = b^3$$
$$b = 0.1.$$

Substituting this value into $0.4 = ab^1$, we get

$$0.4 = a(0.1)$$
$$a = 4.$$

So our formula for this function is $y = 4(0.1)^x$. Since $0.1 = 10^{-1}$ we can also write $y = 4(10^{-1})^x = 4(10)^{-x}$.

15. The graph contains the points $(40, 80)$ and $(120, 20)$. Using the ratio method, we have

$$\frac{ab^{120}}{ab^{40}} = \frac{20}{80}$$
$$b^{80} = \frac{20}{80}$$
$$b = \left(\frac{20}{80}\right)^{1/80} = 0.98282.$$

Now we can solve for a:

$$a(0.98282)^{40} = 80$$
$$a = \frac{80}{(0.98282)^{40}} = 160.$$

so $y = 160(0.983)^x$.

16. (a) See Table 4.2.

Table 4.2

t	0	1	2	3	4	5
$f(t)$	1000	1200	1440	1728	2073.6	2488.32

(b) We have

$$\begin{aligned}
\frac{f(1)}{f(0)} &= \frac{1200}{1000} = 1.2\\
\frac{f(2)}{f(1)} &= \frac{1440}{1200} = 1.2\\
\frac{f(3)}{f(2)} &= \frac{1728}{1440} = 1.2\\
\frac{f(4)}{f(3)} &= \frac{2073.6}{1728} = 1.2\\
\frac{f(5)}{f(4)} &= \frac{2488.32}{2073.6} = 1.2.
\end{aligned}$$

(c) All of the ratios of successive terms are 1.2. This makes sense because we have

$$\begin{aligned}
f(0) &= 1000\\
f(1) &= 1000 \cdot 1.2\\
f(2) &= 1000 \cdot 1.2 \cdot 1.2\\
f(3) &= 1000 \cdot 1.2 \cdot 1.2 \cdot 1.2
\end{aligned}$$

and so on. Each term is the previous term multiplied by 1.2. It follows that the ratio of successive terms will always be the growth factor, which is 1.2 in this case.

17. Let f be the function whose graph is shown. Were f exponential, it would increase by equal factors on equal intervals. However, we see that

$$\begin{aligned}
\frac{f(3)}{f(1)} &= \frac{11}{5} = 2.2\\
\frac{f(5)}{f(3)} &= \frac{30}{11} = 2.7.
\end{aligned}$$

On these equal intervals, the value of f does not increase by equal factors, so f is not exponential.

18. We see that the output doubles from 7 to 14, then from 14 to 28, each time the input rises by 2. So this function could be exponential. We have:

$$\begin{aligned}
\frac{f(3)}{f(1)} &= \frac{14}{7}\\
\frac{ab^3}{ab^1} &= 2\\
b^2 &= 2\\
b &= \sqrt{2}\\
&= 1.4142.
\end{aligned}$$

Solving for a, we have

$$\begin{aligned}
f(1) &= 7\\
a\left(\sqrt{2}\right)^1 &= 7\\
a &= \frac{7}{\sqrt{2}}\\
&= 4.9497.
\end{aligned}$$

Thus,

$$\begin{aligned}
y &= \frac{7}{\sqrt{2}}\left(\sqrt{2}\right)^x\\
&= 4.9497(1.4142)^x.
\end{aligned}$$

19. Since g is exponential, we have $g(t) = ab^t$. Taking ratios, we see that

$$\begin{aligned} \frac{g(3.5)}{g(2.3)} &= \frac{ab^{3.5}}{ab^{2.3}} = \frac{0.1}{0.4} \\ &= b^{1.2} = 0.25 \\ \text{so} \qquad b &= (0.25)^{1/1.2} \\ &= 0.315. \end{aligned}$$

Finding a, we see that

$$\begin{aligned} g(3.5) = a(0.315)^{3.5} &= 0.1 \\ \text{so} \qquad a &= \frac{0.1}{0.315^{3.5}} \\ &= 5.7. \end{aligned}$$

Thus, $g(t) = 5.7(0.315)^t$. Checking our answer, we see that

$$\begin{aligned} g(2.3) &= 5.7(0.315)^{2.3} \\ &= 0.4, \end{aligned}$$

as required.

20. Since f is exponential, we have $f(t) = ab^t$. Taking ratios, we see that

$$\begin{aligned} \frac{f(17)}{f(-5)} &= \frac{ab^{17}}{ab^{-5}} = \frac{46}{22} \\ &= b^{22} = 2.091 \\ \text{so} \qquad b &= 2.091^{1/22} \\ &= 1.034. \end{aligned}$$

Finding a, we see that

$$\begin{aligned} f(17) = a(1.034)^{17} &= 46 \\ \text{so} \qquad a &= \frac{46}{1.034^{17}} \\ &= 26.015. \end{aligned}$$

Thus, $f(t) = 26.015(1.034)^t$. Checking our answer, we see that

$$\begin{aligned} f(-5) &= 26.015(1.034)^{-5} \\ &= 22.01, \end{aligned}$$

which (allowing for rounding) is correct.

21. Note that in the table, the x-values are evenly spaced with $\Delta x = 5$.

- Taking ratios, we see that on equally spaced intervals, f appears to decrease by a constant factor:

$$\begin{aligned} \frac{f(5)}{f(0)} &= \frac{85.9}{95.4} = 0.90 \\ \frac{f(10)}{f(5)} &= \frac{77.3}{85.9} = 0.90 \\ \frac{f(15)}{f(10)} &= \frac{69.6}{77.3} = 0.90 \\ \frac{f(20)}{f(15)} &= \frac{62.6}{69.6} = 0.90. \end{aligned}$$

  This is the hallmark of an exponential function.

- Taking differences, we see that on equally spaced intervals, h appears to decrease by a constant amount:

$$\begin{aligned}h(5)-h(0)&=36.6-37.3=-0.7\\h(10)-h(5)&=35.9-36.6=-0.7\\h(15)-h(10)&=35.2-35.9=-0.7\\h(20)-h(15)&=34.5-35.2=-0.7.\end{aligned}$$

This is the hallmark of a linear function.

- Taking differences, we see that g does not change by a constant factor. For instance,

$$\begin{aligned}\frac{g(5)}{g(0)}&=\frac{40.9}{44.8}=0.91\\\frac{g(10)}{g(5)}&=\frac{36.8}{40.9}=0.90\\\frac{g(15)}{g(10)}&=\frac{32.5}{36.8}=0.88\\\frac{g(20)}{g(15)}&=\frac{28.0}{32.5}=0.86.\end{aligned}$$

Nor, however, does g change by a constant amount:

$$\begin{aligned}g(5)-g(0)&=40.9-44.8=-3.9\\g(10)-g(5)&=36.8-40.9=-4.1\\g(15)-g(10)&=32.5-36.8=-4.3\\g(20)-g(15)&=28-32.5\quad=-4.5.\end{aligned}$$

This means g is neither exponential nor linear.

22. **(a)** If a function is linear, then the differences in successive function values will be constant. If a function is exponential, the ratios of successive function values will remain constant. Now

$$f(1)-f(0)=13.75-12.5=1.25$$

while

$$f(2)-f(1)=15.125-13.75=1.375.$$

Thus, $f(x)$ is not linear. On the other hand,

$$\frac{f(1)}{f(0)}=\frac{13.75}{12.5}=1.1$$

and

$$\frac{f(2)}{f(1)}=\frac{15.25}{13.75}=1.1.$$

Checking the rest of the data, we see that the ratios of differences remains constant, so $f(x)$ is exponential.

(b) We know that f is exponential, so

$$f(x)=ab^x$$

for some constants a and b. We know that $f(0)=12.5$, so

$$\begin{aligned}12.5&=f(0)\\12.5&=ab^0\\12.5&=a(1).\end{aligned}$$

Thus,

$$a=12.5.$$

We also know

$$13.75 = f(1)$$
$$13.75 = 12.5b.$$

Thus,

$$b = \frac{13.75}{12.5} = 1.1.$$

As a result,

$$f(x) = 12.5(1.1)^x.$$

The graph of $f(x)$ is shown in Figure 4.5.

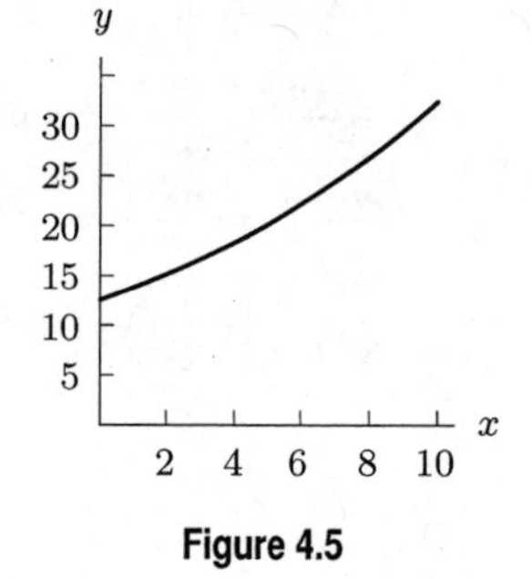

Figure 4.5

23. **(a)** If a function is linear, then the differences in successive function values will be constant. If a function is exponential, the ratios of successive function values will remain constant. Now

$$g(1) - g(0) = 2 - 0 = 2$$

and

$$g(2) - g(1) = 4 - 2 = 2.$$

Checking the rest of the data, we see that the differences remain constant, so $g(x)$ is linear.

(b) We know that $g(x)$ is linear, so it must be of the form

$$g(x) = b + mx$$

where m is the slope and b is the y-intercept. Since at $x = 0$, $g(0) = 0$, we know that the y-intercept is 0, so $b = 0$. Using the points $(0, 0)$ and $(1, 2)$, the slope is

$$m = \frac{2-0}{1-0} = 2.$$

Thus,

$$g(x) = 0 + 2x = 2x.$$

The graph of $y = g(x)$ is shown in Figure 4.6.

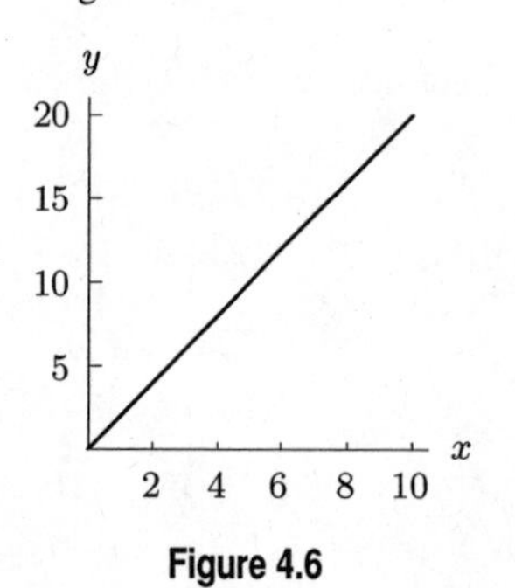

Figure 4.6

24. (a) If a function is linear, then the differences in successive function values will be constant. If a function is exponential, the ratios of successive function values will remain constant. Now

$$h(1) - h(0) = 12.6 - 14 = -1.4$$

while

$$h(2) - h(1) = 11.34 - 12.6 = -1.26.$$

Thus, $h(x)$ is not linear. On the other hand,

$$\frac{h(1)}{h(0)} = 0.9$$
$$\frac{h(2)}{h(1)} = \frac{11.34}{12.6} = 0.9.$$

Checking the rest of the data, we see that the ratio of differences remains constant, so $h(x)$ is exponential.

(b) We know that $h(x)$ is exponential, so

$$h(x) = ab^x,$$

for some constants a and b. We know that $h(0) = 14$, so

$$14 = h(0)$$
$$14 = ab^0$$
$$14 = a(1).$$

Thus, $a = 14$. Also

$$12.6 = h(1)$$
$$12.6 = 14b.$$

Thus,

$$b = \frac{12.6}{14} = 0.9.$$

So, we have $h(x) = 14(0.9)^x$. The graph of $h(x)$ is shown in Figure 4.7.

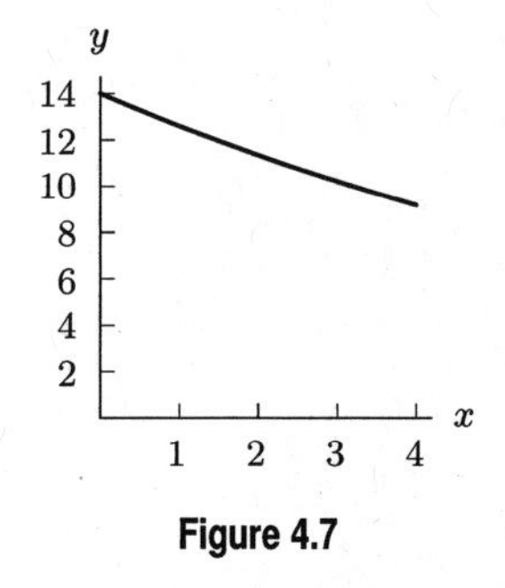

Figure 4.7

25. (a) If a function is linear, then the differences in successive function values will be constant. If a function is exponential, the ratios of successive function values will remain constant. Now

$$i(1) - i(0) = 14 - 18 = -4$$

and

$$i(2) - i(1) = 10 - 14 = -4.$$

Checking the rest of the data, we see that the differences remain constant, so $i(x)$ is linear.

(b) We know that $i(x)$ is linear, so it must be of the form

$$i(x) = b + mx,$$

where m is the slope and b is the y-intercept. Since at $x = 0$, $i(0) = 18$, we know that the y-intercept is 18, so $b = 18$. Also, we know that at $x = 1$, $i(1) = 14$, we have

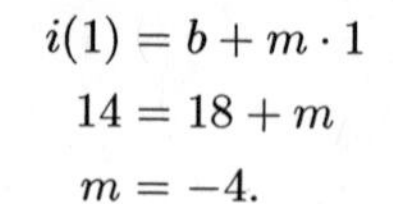

$$\begin{aligned} i(1) &= b + m \cdot 1 \\ 14 &= 18 + m \\ m &= -4. \end{aligned}$$

Thus, $i(x) = 18 - 4x$. The graph of $i(x)$ is shown in Figure 4.8.

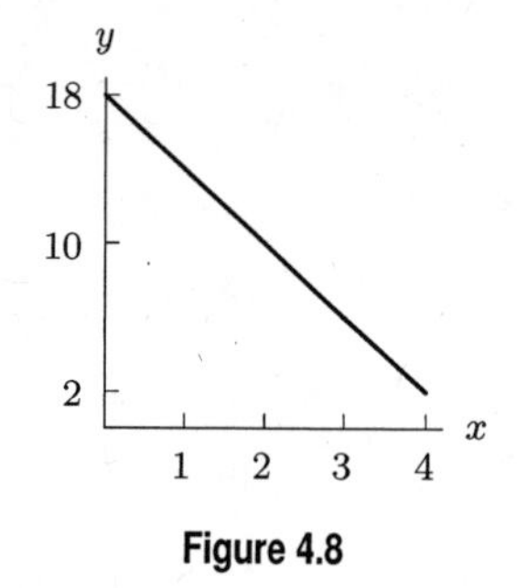

Figure 4.8

Problems

26. For the linear function, we first find the slope:

$$\text{Slope} = \frac{\Delta y}{\Delta x} = \frac{75 - 20}{5 - 0} = 11.$$

The vertical intercept is 20 so the linear function is $y = 20 + 11x$.

The vertical intercept is 20, so the exponential function is $y = 20(a)^x$. We substitute the point $(5, 75)$ to find a:

$$\begin{aligned} 75 &= 20(a)^5 \\ 3.75 &= a^5 \\ a &= (3.75)^{1/5} = 1.303. \end{aligned}$$

The exponential function is $y = 20(1.303)^x$.

27. One approach is to graph both functions and to see where the graph of $p(x)$ is below the graph of $q(x)$. From Figure 4.9, we see that $p(x)$ intersects $q(x)$ in two places; namely, at $x \approx -1.69$ and $x = 2$. We notice that $p(x)$ is above $q(x)$ between these two points and below $q(x)$ outside the segment defined by these two points. Hence $p(x) < q(x)$ for $x < -1.69$ and for $x > 2$.

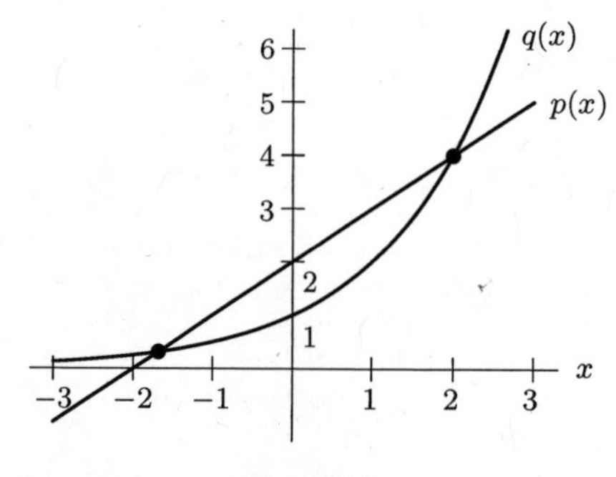

Figure 4.9

28. By graphing both functions in a window centered at the origin we get Figure 4.10 with graphs of f and g for $-1 \leq x \leq 1$ and $-1 \leq y \leq 2$. We see an intersection point to the left of the origin. So $g(x) < f(x)$ for $x > x_1$. Using a computer or graphing calculator, it can be found as $x_1 \approx -0.587$.

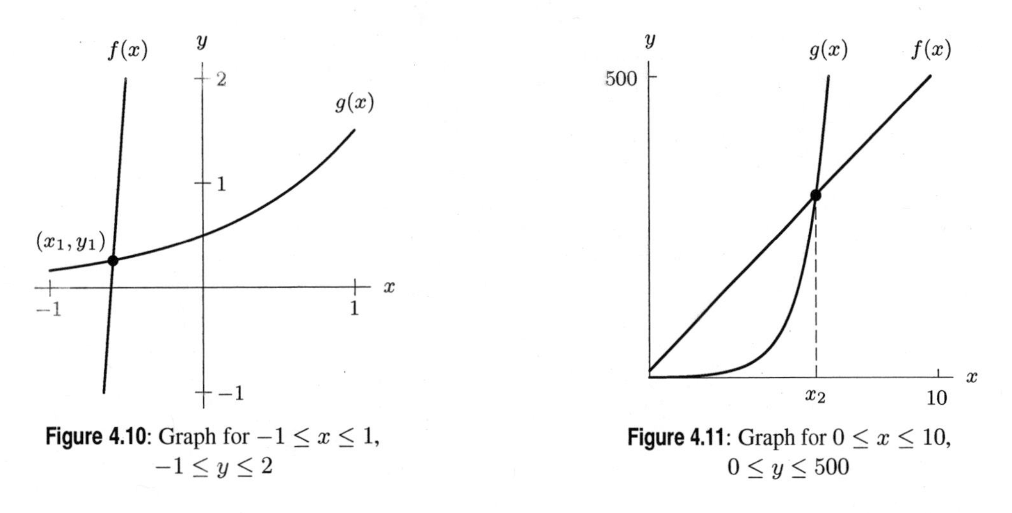

Figure 4.10: Graph for $-1 \leq x \leq 1$, $-1 \leq y \leq 2$

Figure 4.11: Graph for $0 \leq x \leq 10$, $0 \leq y \leq 500$

To the left of our viewing window there can be no more intersections because $f(x)$ will get more and more negative while $g(x)$ remains positive. So for $x < -1$, we must have $f(x) < 0 < g(x)$. Since an exponential function will eventually grow greater than any linear function, there must be another intersection point to the right of our first viewing window. See Figure 4.11. We find $g(x) < f(x)$ for $x < x_2$, where x_2 can be found, using a computer or a graphing calculator, to be $x_2 \approx 4.911$.

Thus, $g(x) < f(x)$ for $x_1 < x < x_2$, that is, the approximate interval

$$-0.587 < x < 4.911.$$

29. We have $p = f(x) = ab^x$. From the figure, we see that the starting value is $a = 20$ and that the graph contains the point $(10, 40)$. We have

$$\begin{aligned} f(10) &= 40 \\ 20b^{10} &= 40 \\ b^{10} &= 2 \\ b &= 2^{1/10} \\ &= 1.0718, \end{aligned}$$

so $p = 20(1.0718)^x$.

We have $q = g(x) = ab^x$, $g(10) = 40$, and $g(15) = 20$. Using the ratio method, we have

$$\begin{aligned} \frac{ab^{15}}{ab^{10}} &= \frac{g(15)}{g(10)} \\ b^5 &= \frac{20}{40} \\ b &= \left(\frac{20}{40}\right)^{1/5} \\ &= (0.5)^{1/5} = 0.8706. \end{aligned}$$

Now we can solve for a:

$$a((0.5)^{1/5})^{10} = 40$$

$$\begin{aligned} a &= \frac{40}{(0.5)^2} \\ &= 160. \end{aligned}$$

so $q = 160(0.8706)^x$.

30. If a function is linear, then the rate of change is constant. For $Q(t)$,

$$\frac{8.70 - 7.51}{10 - 3} = 0.17.$$

and

$$\frac{9.39 - 8.7}{14 - 10} = 0.17.$$

So this function appears to be very close to linear. Thus, $Q(t) = b + mt$ where $m = 0.17$ as shown above. We solve for b by using the point $(3, 7.51)$.

$$\begin{aligned} Q(t) &= b + 0.17t \\ 7.51 &= b + 0.17(3) \\ 7.51 - 0.51 &= b \\ b &= 7. \end{aligned}$$

Therefore, $Q(t) = 7 + 0.17t$.

31. Testing the rates of change for $R(t)$, we find that

$$\frac{2.61 - 2.32}{9 - 5} = 0.0725$$

and

$$\frac{3.12 - 2.61}{15 - 9} = 0.085,$$

so we know that $R(t)$ is not linear. If $R(t)$ is exponential, then $R(t) = ab^t$, and

$$R(5) = a(b)^5 = 2.32$$

and

$$R(9) = a(b)^9 = 2.61.$$

So

$$\begin{aligned} \frac{R(9)}{R(5)} &= \frac{ab^9}{ab^5} = \frac{2.61}{2.32} \\ \frac{b^9}{b^5} &= \frac{2.61}{2.32} \\ b^4 &= \frac{2.61}{2.32} \\ b &= (\frac{2.61}{2.32})^{\frac{1}{4}} \approx 1.030. \end{aligned}$$

Since

$$\begin{aligned} R(15) &= a(b)^{15} = 3.12 \\ \frac{R(15)}{R(9)} &= \frac{ab^{15}}{ab^9} = \frac{3.12}{2.61} \\ b^6 &= \frac{3.12}{2.61} \\ b &= \left(\frac{3.12}{2.61}\right)^{\frac{1}{6}} \approx 1.030. \end{aligned}$$

Since the growth factor, b, is constant, we know that $R(t)$ could be an exponential function and that $R(t) = ab^t$. Taking the ratios of $R(5)$ and $R(9)$, we have

$$\begin{aligned}\frac{R(9)}{R(5)} = \frac{ab^9}{ab^5} &= \frac{2.61}{2.32}\\ b^4 &= 1.125\\ b &= 1.030.\end{aligned}$$

So $R(t) = a(1.030)^t$. We now solve for a by using $R(5) = 2.32$,

$$\begin{aligned}R(5) &= a(1.030)^5\\ 2.32 &= a(1.030)^5\\ a &= \frac{2.32}{1.030^5} \approx 2.001.\end{aligned}$$

Thus, $R(t) = 2.001(1.030)^t$.

32. Testing rates of change for $S(t)$, we find that

$$\frac{6.72 - 4.35}{12 - 5} = 0.339$$

and

$$\frac{10.02 - 6.72}{16 - 12} = 0.825.$$

Since the rates of change are not the same we know that $S(t)$ is not linear.

Testing for a possible constant growth factor we see that

$$\begin{aligned}\frac{S(12)}{S(5)} = \frac{ab^{12}}{ab^5} &= \frac{6.72}{4.35}\\ b^7 &= \frac{6.72}{4.35}\\ b &\approx 1.064\end{aligned}$$

and

$$\begin{aligned}\frac{S(16)}{S(12)} = \frac{ab^{16}}{ab^{12}} &= \frac{10.02}{6.72}\\ b^4 &= \frac{10.02}{6.72}\\ b &\approx 1.105.\end{aligned}$$

Since the growth factors are different, $S(t)$ is not an exponential function.

33. **(a)** Since this function is exponential, its formula is of the form $f(t) = ab^t$, so

$$\begin{aligned}f(3) &= ab^3\\ f(8) &= ab^8.\end{aligned}$$

From the graph, we know that

$$\begin{aligned}f(3) &= 2000\\ f(8) &= 5000.\end{aligned}$$

So

$$\begin{aligned}\frac{f(8)}{f(3)} = \frac{ab^8}{ab^3} &= \frac{5000}{2000}\\ b^5 &= \frac{5}{2} = 2.5\\ (b^5)^{1/5} &= (2.5)^{1/5}\\ b &= 1.20112.\end{aligned}$$

We now know that $f(t) = a(1.20112)^t$. Using either of the pairs of values on the graph, we can find a. In this case, we use $f(3) = 2000$. According to the formula,

$$f(3) = a(1.20112)^3$$
$$2000 = a(1.20112)^3$$
$$a = \frac{2000}{(1.20112)^3} \approx 1154.160.$$

The formula we want is $f(t) = 1154.160(1.20112)^t$ or $P = 1154.160(1.20112)^t$.

(b) The initial value of the account occurs when $t = 0$.

$$f(0) = 1154.160(1.20112)^0 = 1154.160(1) = \$1154.16.$$

(c) The value of b, the growth factor, is related to the growth rate, r, by

$$b = 1 + r.$$

We know that $b = 1.20112$, so

$$1.20112 = 1 + r$$
$$0.20112 = r$$

Thus, in percentage terms, the annual interest rate is 20.112%.

34. (a) Under Penalty A, the total fine is \$1 million for August 2 and \$10 million for each day after August 2 . By August 31, the fine had been increasing for 29 days so the total fine would be $1 + 10(29) = \$291$ million.

Under the Penalty B, the penalty on August 2 is 1 cent. On August 3, it is $1(2)$ cents; on August 4, it is $1(2)(2)$ cents; on August 5, it is $1(2)(2)(2)$ cents. By August 31, the fine has doubled 29 times, so the total fine is $(1) \cdot (2)^{29}$ cents, which is 536,870,912 cents or \$5,368,709.12 or, approximately, \$5.37 million.

(b) If t represents the number of days after August 2, then the total fine under Penalty A would be \$1 million plus the number of days after August 2 times \$10 million, or $A(t) = 1 + 10 \cdot t$ million dollars $= (1 + 10t)10^6$ dollars. The total fine under Penalty B would be 1 cent doubled each day after August 2, so $B(t) = 1\underbrace{(2)(2)(2)\ldots(2)}_{t \text{ times}}$ cents or $B(t) = 1 \cdot (2)^t$ cents, or $B(t) = (0.01)2^t$ dollars.

(c) We plot $A(t) = (1 + 10t)10^6$ and $B(t) = (0.01)2^t$ on the same set of axes and observe that they intersect at $t \approx 35.032$ days. Another possible approach is to find values of $A(t)$ and $B(t)$ for different values of t, narrowing in on the value for which they are most nearly equal.

35. We use an exponential function of the form $P = ab^t$. Since $P = 1046$ when $t = 0$, we use $P = 1046b^t$ for some base b. Since $P = 338$ when $t = 5$, we have

$$P = 1046b^t$$
$$338 = 1046b^5$$
$$b^5 = \frac{338}{1046} = 0.3231$$
$$b = (0.3231)^{1/5} = 0.798.$$

An exponential formula for global production of CFCs as a function of t, the number of years since 1989 is

$$P = 1046(0.798)^t.$$

Since $0.798 = 1 - 0.202$, CFC production was decreasing at a rate of 20.2% per year during this time period.

36. (a) Since, for $t < 0$, we know that the voltage is a constant 80 volts, $V(t) = 80$ on that interval.

For $t \geq 0$, we know that $v(t)$ is an exponential function, so $V(t) = ab^t$. According to this formula, $V(0) = ab^0 = a(1) = a$. According to the graph, $V(0) = 80$. From these two facts, we know that $a = 80$, so $V(t) = 80b^t$. If $V(10) = 80b^{10}$ and $V(10) = 15$ (from the graph), then

$$\begin{aligned} 80b^{10} &= 15 \\ b^{10} &= \frac{15}{80} \\ (b^{10})^{\frac{1}{10}} &= (\frac{15}{80})^{\frac{1}{10}} \\ b &\approx 0.8459 \end{aligned}$$

so that $V(t) = 80(0.8459)^t$ on this interval. Combining the two pieces, we have

$$V(t) = \begin{cases} 80 & \text{for } t < 0 \\ 80(0.8459)^t & \text{for } t \geq 0. \end{cases}$$

(b) Using a computer or graphing calculator, we can find the intersection of the line $y = 0.1$ with $y = 80(0.8459)^t$. We find $t \approx 39.933$ seconds.

37. We let W represent the winning time and t represent the number of years since 1994.

(a) To find the linear function, we first find the slope:

$$\text{Slope} = \frac{\Delta W}{\Delta t} = \frac{41.94 - 43.45}{12 - 0} = -0.126.$$

The vertical intercept is 43.45 so the linear function is $W = 43.45 - 0.126t$. The predicted winning time in 2018 is $W = 43.45 - 0.126(24) = 40.43$ seconds.

(b) The time at $t = 0$ is 43.45, so the exponential function is $W = 43.45(a)^t$. We use the fact that $W = 41.94$ when $t = 12$ to find a:

$$\begin{aligned} 41.94 &= 43.45(a)^{12} \\ 0.965247 &= a^{12} \\ a &= (0.965247)^{1/12} = 0.997057. \end{aligned}$$

The exponential function is $W = 43.45(0.997057)^t$. The predicted winning time in 2018 is $W = 43.45(0.997057)^{24} = 40.48$ seconds.

38. (a) We use $N = b + mt$. Since $N = 178.8$ when $t = 0$, we have $b = 178.8$. We find the slope:

$$m = \frac{\Delta N}{\Delta t} = \frac{187.2 - 178.8}{10 - 0} = 0.84.$$

The formula is

$$N = 178.8 + 0.84t.$$

(b) We use $N = ab^t$. Since $N = 178.8$ when $t = 0$, we have $a = 178.8$. We find the base b using the fact that $N = 187.2$ when $t = 10$:

$$\begin{aligned} N &= 178.8b^t \\ 187.2 &= 178.8b^{10} \\ b^{10} &= \frac{187.2}{178.8} = 1.047 \\ b &= (1.047)^{1/10} = 1.0046. \end{aligned}$$

The formula is

$$N = 178.8(1.0046)^t.$$

39. **(a)** We have

$$\text{Slope} = \frac{\Delta P}{\Delta t} = \frac{1700 - 3500}{10 - 0} = -180.$$

The vertical intercept is 3500 so the function is $P = 3500 - 180t$. The rate of change is -180 fish per year. The population has been shrinking by about 180 fish each year.

(b) The vertical intercept is 3500, so the exponential function is $P = 3500(a)^t$. We substitute the point at $t = 10$ to find a:

$$\begin{aligned} 1700 &= 3500(a)^{10} \\ \frac{1700}{3500} &= a^{10} \\ a &= \left(\frac{1700}{3500}\right)^{1/10} = 0.930. \end{aligned}$$

The function is $P = 3500(0.930)^t$. The population has been shrinking by about 7% per year.

(c) See Figure 4.12.

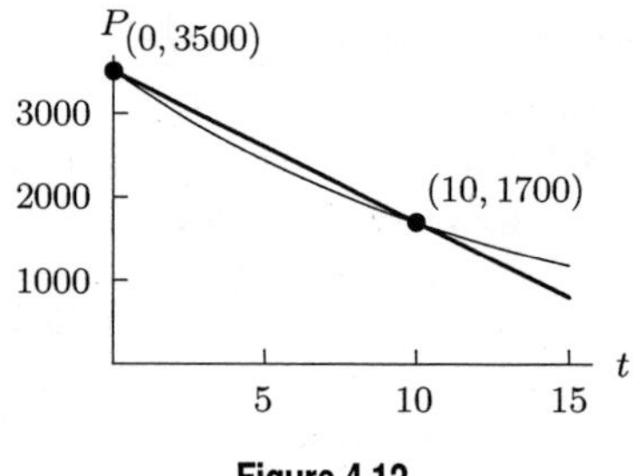

Figure 4.12

40. We let t represent the number of years since 2000. To see if world cocoa production is linear, we look at differences between successive values. Since $3.875 - 3.1 = 0.775$ and $4.844 - 3.875 = 0.969$, we see that world cocoa production is not linear. To see if it is exponential, we look at ratios of successive terms:

$$\frac{3.875}{3.1} = 1.25, \quad \frac{4.844}{3.875} = 1.250, \quad \frac{6.055}{4.844} = 1.25, \quad \frac{7.568}{6.055} = 1.250.$$

Since ratios of successive terms are constant (up to round-off error), we see that world cocoa production is exponential with base 1.25. Production at $t = 0$ is 3.1 million tons, so we have

$$\text{World cocoa production} = 3.1(1.25)^t.$$

To see if production in the Ivory Coast is linear, we look at successive differences:

$$1.34 - 1.3 = 0.04, \quad 1.38 - 1.34 = 0.04; \quad 1.42 - 1.38 = 0.04, \quad 1.46 - 1.42 = 0.04.$$

Since successive differences are constant, production is linear with slope 0.04. Production at $t = 0$ is 1.3 million tons, so we have

$$\text{Ivory Coast cocoa production} = 1.3 + 0.04t.$$

41. **(a)** Since the rate of change is constant, the increase is linear.

(b) Life expectancy is increasing at a constant rate of 3 months, or 0.25 years, each year. The slope is 0.25. When $t = 9$ we have $L = 78.1$. We use the point-slope form to find the linear function:

$$\begin{aligned} L - 78.1 &= 0.25(t - 9) \\ L - 78.1 &= 0.25t - 2.25 \\ L &= 0.25t + 75.85. \end{aligned}$$

(c) When $t = 50$, we have $L = 0.25(50) + 75.85 = 88.35$. It the rate of increase continues, babies born in 2050 will have a life expectancy of 88.35 years.

42. **(a)** For the population to decrease linearly, it must change by the same amount over each one year period. If P_0 represents the original population, then it will be reduced by $10\%P_0$ each year. So after one year the remaining population will be $P = P_0 - 0.1P_0 = 0.9P_0$. After two years the remaining population will be $P = 0.9P_0 - 0.1P_0 = 0.8P_0$. At the end of ten years there will be no population remaining. See Figure 4.13.

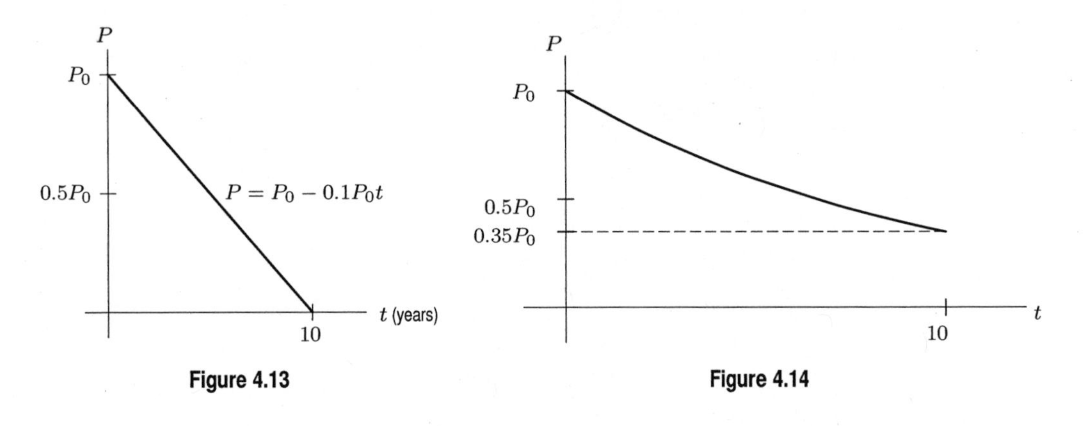

Figure 4.13 **Figure 4.14**

(b) In this case the population also decreases each year but not by the same amount. After one year the population remaining is the same as in the linear case:

$$P = P_0 - 0.1P_0 = 0.9P_0.$$

However, after two years, the population remaining is

$$P = 0.9P_0 - 0.1(0.9P_0) = 0.9P_0(1 - 0.1) = P_0(0.9)^2.$$

In general, after t years the remaining population will be given by

$$P = P_0(0.9)^t$$

and after 10 years there will be a population remaining of

$$P = P_0(0.9)^{10} = 0.35P_0 = 35\%P_0.$$

In other words, an exponential decrease of 10% a year will leave 35% of the original population after 10 years. See Figure 4.14.

43. **(a)** We want $N = f(t)$ so we have

$$\text{Slope} = \frac{\Delta N}{\Delta t} = \frac{300 - 84}{2009 - 1990} = \frac{216}{19} = 11.3684.$$

Since $N = 84$ when $t = 0$, the vertical intercept is 84 and the linear formula is

$$N = 84 + 11.3684t.$$

The slope is 11.3684. The number of asthma sufferers has increased, on average, by 11.3684 million people per year during this period.

(b) Since $N = 84$ when $t = 0$, we have the exponential function $N = 84b^t$ for some base b. Since $N = 300$ when $t = 19$, we have

$$\begin{aligned} N &= 84b^t \\ 300 &= 84b^{19} \\ b^{19} &= \frac{300}{84} = 3.5714 \\ b &= (3.5714)^{1/19} = 1.0693. \end{aligned}$$

The exponential formula is

$$N = 84(1.0693)^t.$$

The growth factor is 1.0693. The number of asthma sufferers has increased, on average, by 6.93% per year during this period.

(c) In the year 2020, we have $t = 30$. Using the linear formula, the predicted number in 2020 is

$$N = 84 + 11.3684(30) = 425.0520 \text{ million asthma sufferers.}$$

Using the exponential formula, the predicted number in 2020 is

$$N = 84(1.0693)^{30} = 626.9982 \text{ million asthma sufferers.}$$

44. (a) For a linear model, we assume that the population increases by the same amount every year. Since it grew by 4.14% in the first year, the town had a population increase of $0.0414(20{,}000) = 828$ people in one year. According to a linear model, the population in 2010 would be $20{,}000 + 10 \cdot 828 = 28{,}280$. Using an exponential model, we assume that the population increases by the same percent every year, so the population in 2010 would be $20{,}000 \cdot (1.0414)^{10} = 30{,}006$. Clearly the exponential model is a better fit.

(b) Assuming exponential growth at 4.14% a year, the formula for the population is

$$P(t) = 20{,}000(1.0414)^t.$$

45. (a) Assuming linear growth at 250 per year, the population in 2010 would be

$$18{,}500 + 250 \cdot 10 = 21{,}000.$$

Using the population after one year, we find that the percent rate would be $250/18{,}500 \approx 0.013514 = 1.351\%$ per year, so after 10 years the population would be

$$18{,}500(1.013514)^{10} \approx 21{,}158.$$

The town's growth is poorly modeled by both linear and exponential functions.

(b) We do not have enough information to make even an educated guess about a formula.

Solutions for Section 4.3

Exercises

1. (a) See Table 4.3.

Table 4.3

x	-3	-2	-1	0	1	2	3
$f(x)$	1/8	1/4	1/2	1	2	4	8

(b) For large negative values of x, $f(x)$ is close to the x-axis. But for large positive values of x, $f(x)$ climbs rapidly away from the x-axis. As x gets larger, y grows more and more rapidly. See Figure 4.15.

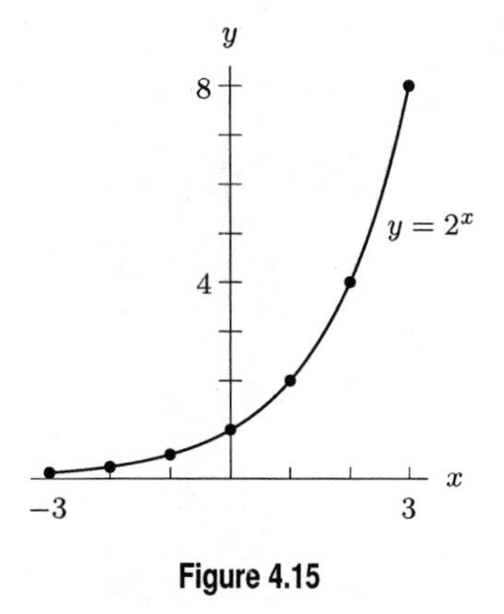

Figure 4.15

2. **(a)** See Table 4.4.

Table 4.4

x	-3	-2	-1	0	1	2	3
$f(x)$	8	4	2	1	1/2	1/4	1/8

(b) For large positive values of x, $f(x)$ is close to the x-axis. But for large negative values of x, $f(x)$ climbs rapidly away from the x-axis. See Figure 4.16.

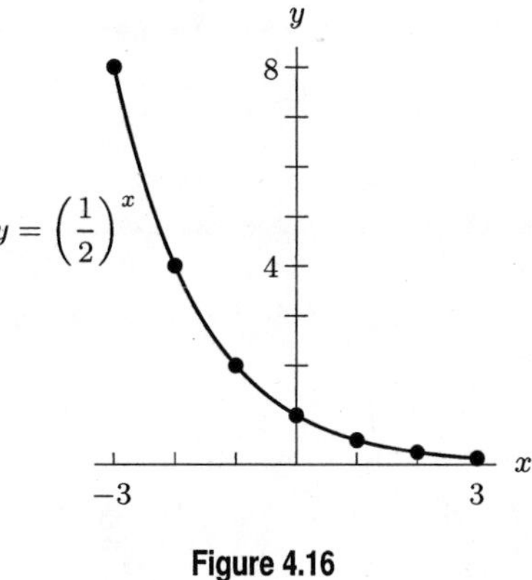

Figure 4.16

3. Let $f(x) = (1.1)^x$, $g(x) = (1.2)^x$, and $h(x) = (1.25)^x$. We note that for $x = 0$,

$$f(x) = g(x) = h(x) = 1;$$

so all three graphs have the same y-intercept. On the other hand, for $x = 1$,

$$f(1) = 1.1, \quad g(1) = 1.2, \quad \text{and} \quad h(1) = 1.25,$$

so $0 < f(1) < g(1) < h(1)$. For $x = 2$,

$$f(2) = 1.21, \quad g(2) = 1.44, \quad \text{and} \quad h(2) = 1.5625,$$

so $0 < f(2) < g(2) < h(2)$. In general, for $x > 0$,

$$0 < f(x) < g(x) < h(x).$$

This suggests that the graph of $f(x)$ lies below the graph of $g(x)$, which in turn lies below the graph of $h(x)$, and that all lie above the x-axis. Alternately, you can consider 1.1, 1.2, and 1.25 as growth factors to conclude $h(x) = (1.25)^x$ is the top function, and $g(x) = (1.2)^x$ is in the middle, $f(x)$ is at the bottom.

4. Let $f(x) = (0.7)^x$, $g(x) = (0.8)^x$, and $h(x) = (0.85)^x$. We note that for $x = 0$,

$$f(x) = g(x) = h(x) = 1.$$

On the other hand, $f(1) = 0.7$, $g(1) = 0.8$, and $h(1) = 0.85$, while $f(2) = 0.49$, $g(2) = 0.64$, and $h(2) = 0.7225$; so

$$0 < f(x) < g(x) < h(x).$$

So the graph of $f(x)$ lies below the graph of $g(x)$, which in turn lies below the graph of $h(x)$.

Alternately, you can consider 0.7, 0.8, and 0.85 as growth factors (decaying). The $f(x) = (0.7)^x$ will be the lowest graph because it is decaying the fastest. The $h(x) = (0.85)^x$ will be the top graph because it decays the least.

5. Yes, the graphs will cross. The graph of $g(x)$ has a smaller y-intercept but increases faster and will eventually overtake the graph of $f(x)$.

6. No, the graphs will not cross. The graph of $g(x)$ has a larger y-intercept and increases faster so it will always be higher than the graph of $f(x)$.

7. No, the graphs will not cross. The graph of $f(x)$ has a larger y-intercept and is increasing while the graph of $g(x)$ starts lower and is decreasing.

8. Yes, the graphs will cross. The graph of $f(x)$ has a larger y-intercept and is decreasing while the graph of $g(x)$ starts lower and is increasing.

9. No, the graphs will not cross. Both functions are decreasing but the graph of $f(x)$ has a larger y-intercept and is decreasing at a slower rate than $g(x)$ so it will always be above the graph of $g(x)$.

10. Yes, the graphs will cross. Both functions are decreasing. The graph of $f(x)$ has a larger y-intercept and is decreasing at a faster rate than $g(x)$ so it will eventually cross the graph of $g(x)$.

11. Since $y = a$ when $t = 0$ in $y = ab^t$, a is the y-intercept. Thus, the function with the greatest y-intercept, D, has the largest a.

12. Since $y = a$ when $t = 0$ in $y = ab^t$, a is the y-intercept. Thus, the two functions with the same y-intercept, A and B, have the same a.

13. The function with the smallest b should be the one that is decreasing the fastest. We note that D approaches zero faster than the others, so D has the smallest b.

14. The function with the largest b should be the one that is increasing the fastest. We note that A increases faster than the others, so A has the largest b.

15. **(a)** It appears that the volume is about 13 ft^3 at $t = 5$.
(b) It appears that when the volume is 20 ft^3, we have $t = 3.2$ weeks.

16. Graphing $y = 46(1.1)^x$ and tracing along the graph on a calculator gives us an answer of $x = 7.158$. See Figure 4.17.

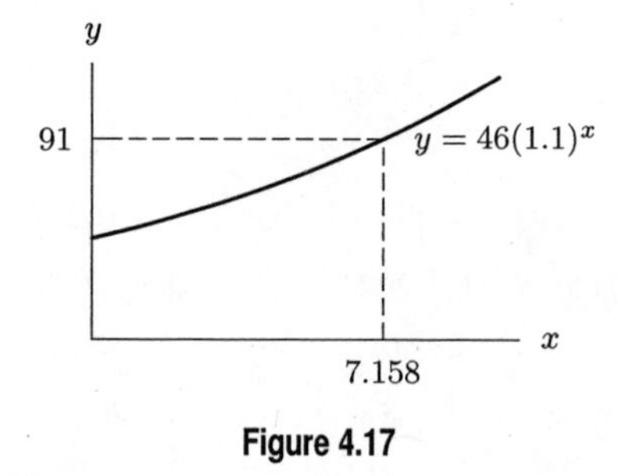

Figure 4.17

17. Graphing $p = 22(0.87)^q$ and tracing along the graph on a calculator gives us an answer of $q = 5.662$. See Figure 4.18.

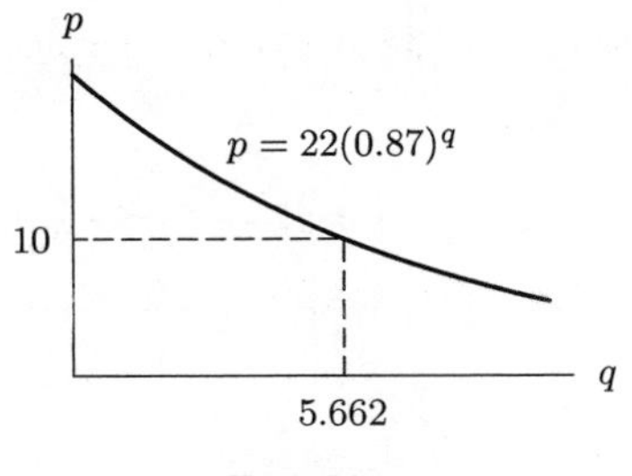

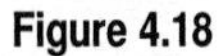
Figure 4.18

18. We solve for m to see $m = 4.25(2.3)^w$. Graphing $m = 4.25(2.3)^w$ and tracing along the graph on a calculator gives us an answer of $w = 1.246$. See Figure 4.19.

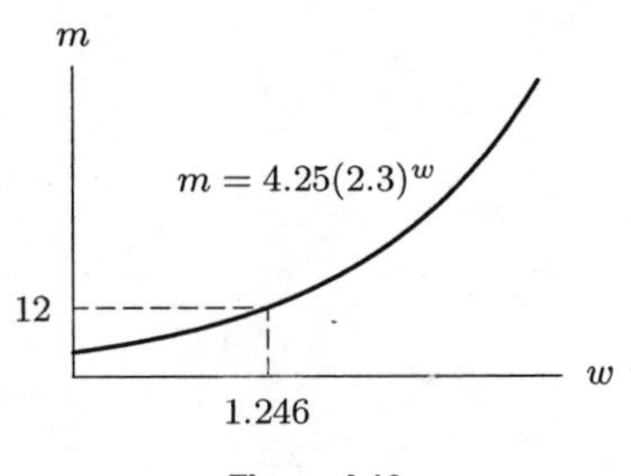

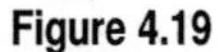
Figure 4.19

19. Solve for P to obtain $P = 7(0.6)^t$. Graphing $P = 7(0.6)^t$ and tracing along the graph on a calculator gives us an answer of $t = 2.452$. See Figure 4.20.

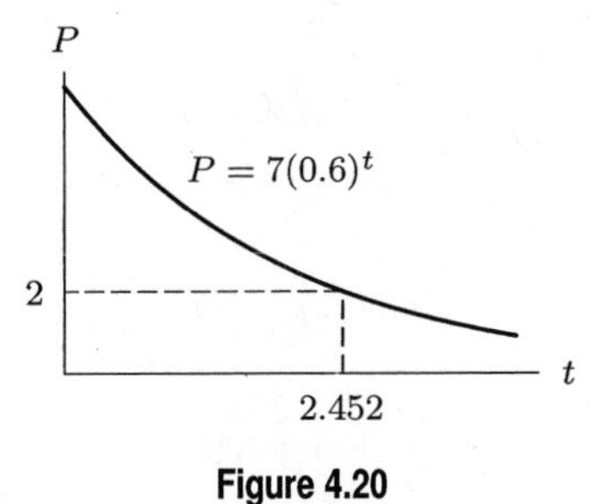

Figure 4.20

20. As t approaches $-\infty$, the value of ab^t approaches zero for any a, so the horizontal asymptote is $y = 0$ (the x-axis).

21. As t approaches ∞, the value of ab^t approaches zero for any a, so the horizontal asymptote is $y = 0$ (the x-axis).

22. The value of b is less than 1, so $0 < b < 1$.

Problems

23. A possible graph is shown in Figure 4.21.

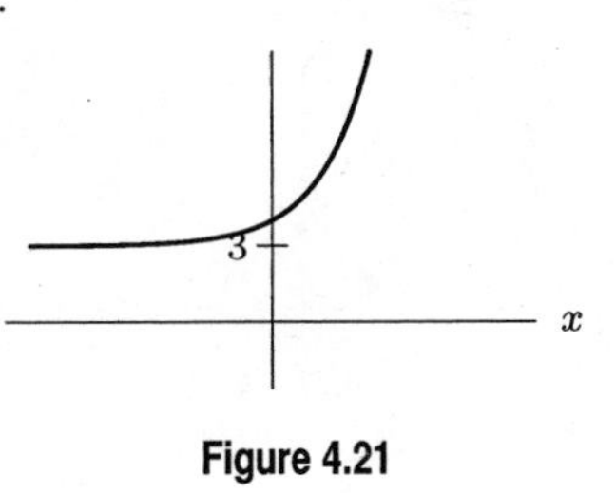

Figure 4.21

24. A possible graph is shown in Figure 4.22.

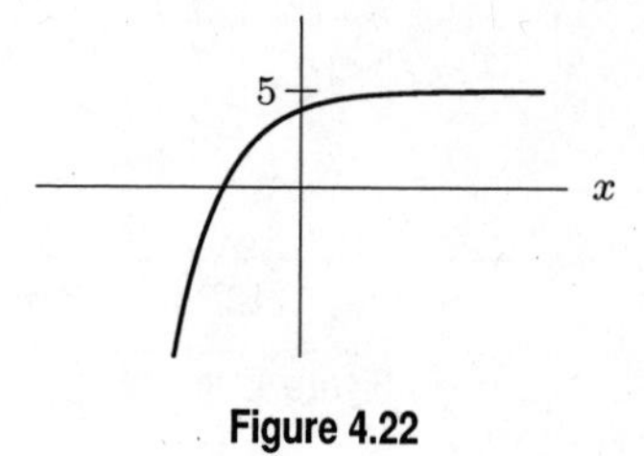

Figure 4.22

25. A possible graph is shown in Figure 4.23.

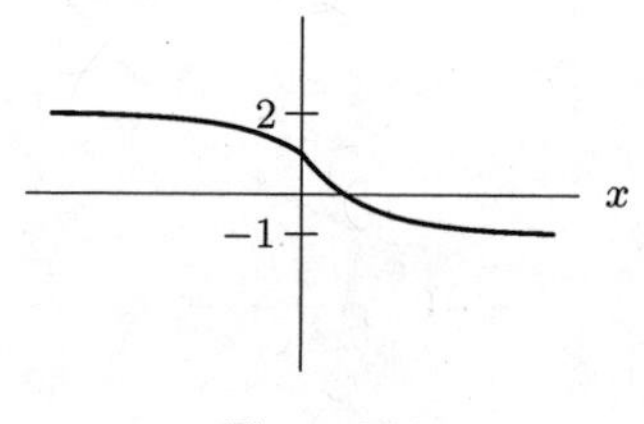

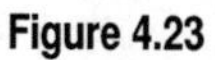
Figure 4.23

26. A possible graph is shown in Figure 4.24.

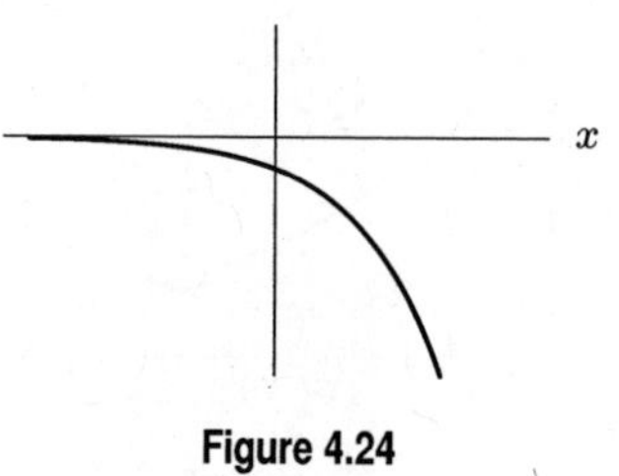

Figure 4.24

27. A possible graph is shown in Figure 4.25. There are many possible answers.

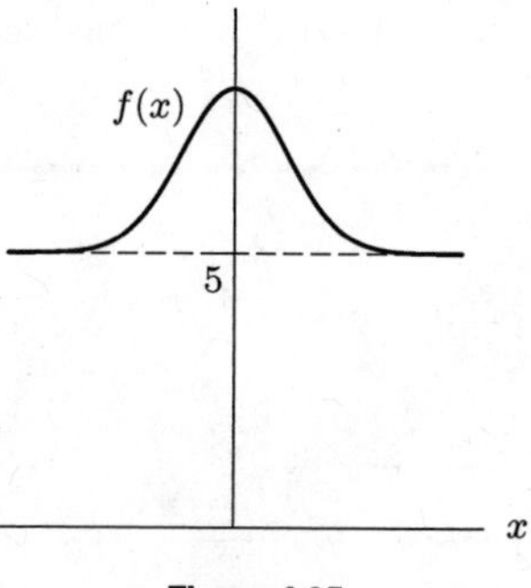

Figure 4.25

28. It appears in the graph that

(a) $\lim_{x\to-\infty} f(x) = 5$

(b) $\lim_{x\to\infty} f(x) = -3.$

Of course, we need to be sure that we are seeing all the important features of the graph in order to have confidence in these estimates.

29. It appears in the graph that

(a) $\lim_{x\to-\infty} f(x) = -\infty$

(b) $\lim_{x\to\infty} f(x) = -\infty.$

Of course, we need to be sure that we are seeing all the important features of the graph in order to have confidence in these estimates.

30. As $x \to \infty$, we know that $3(0.9)^x \to 0$ since it exponentially decays to zero. Therefore, as $x \to \infty$, we have $f(x) = 5 + 3(0.9)^x \to 5 + 0 = 5$. This function has a horizontal asymptote at 5. We can also see this graphically in Figure 4.26.

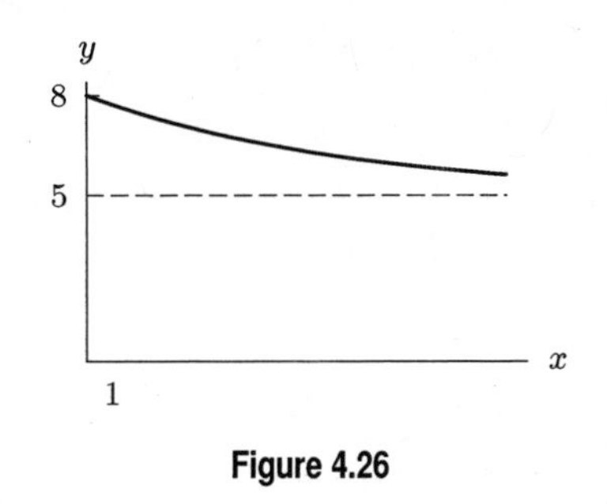

Figure 4.26

31. (a) All constants are positive.

(b) The constant b is definitely between 0 and 1, because $y = a \cdot b^x$ represents a decreasing function.

(c) In addition to b, the constants a, c, p could be between 0 and 1.

(d) Since the curves $y = a \cdot b^x$ and $y = c \cdot d^x$ cross on the y-axis, we must have $a = c$.

(e) The values of a and p are not equal as curves cross y axis at different points. The values of b and d and, likewise, b and q cannot be equal because in each case, one graph climbs while the other falls. However d and q could be equal.

32. (a) Note that all the graphs in Figure 4.27 are increasing and concave up. As the value of a increases, the graphs become steeper, but they are all going in the same general direction.

(b) Note that, in this case, while most of the graphs in Figure 4.28 are concave up, some are increasing (when $a > 1$), some are decreasing (when $0 < a < 1$), and one is a constant function (when $a = 1$).

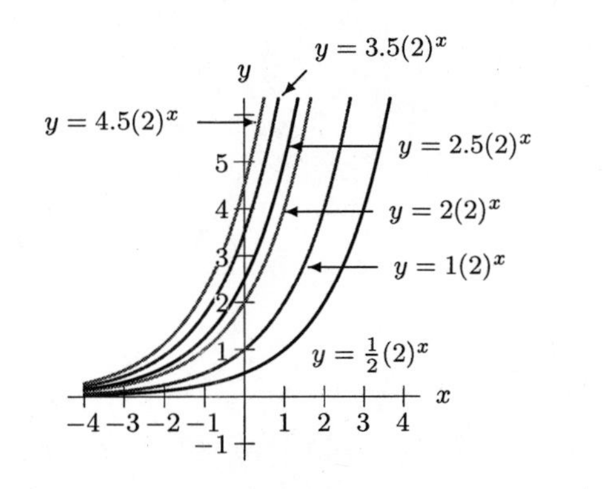

Figure 4.27

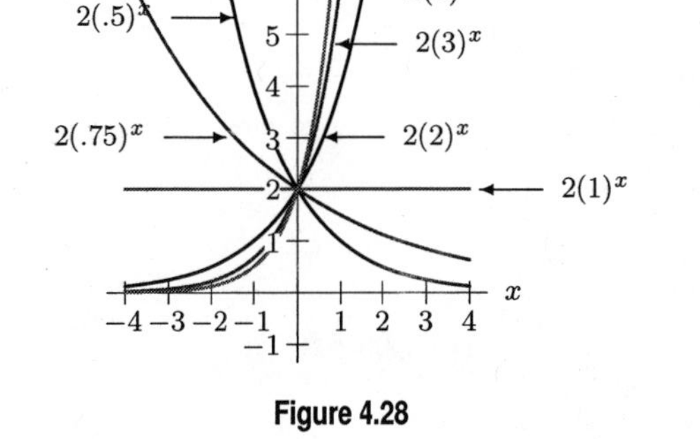

Figure 4.28

33. Increasing: $b > 1$, $a > 0$ or $0 < b < 1$, $a < 0$;
Decreasing: $0 < b < 1$, $a > 0$ or $b > 1$, $a < 0$;
The function is concave up for $a > 0$, $0 < b < 1$ or $b > 1$.

34. The domain is all possible t-values, so

$$\text{Domain: all } t\text{-values.}$$

The range is all possible Q-values. Since Q must be positive,

$$\text{Range: all } Q > 0.$$

35. As r increases, the graph of $y = a(1+r)^t$ rises more steeply, so the point of intersection moves to the left and down. However, no matter how steep the graph becomes, the point of intersection remains above and to the right of the y-intercept of the second curve, or the point $(0, b)$. Thus, the value of y_0 decreases but does not reach b.

36. As a increases, the y-intercept of the curve rises, and the point of intersection shifts down and to the left. Thus y_0 decreases. If a becomes larger than b, the point of intersection shifts to the left side of the y-axis, and the value of y_0 continues to decrease. However, y_0 will not decrease to 0, as the point of intersection will always fall above the x-axis.

37. **(a)** The growth factor is $1 - 0.0075 = 0.9925$ and the initial value is 651, so we have

$$P = 651(0.9925)^t.$$

(b) Using $t = 10$, we have $P = 651(0.9925)^{10} = 603.790$. If the current trend continues, the population of Baltimore is predicted to be 603,790 in the year 2010.

(c) See Figure 4.29. We see that $t = 22.39$ when $P = 550$. The population is expected to be 550 thousand in approximately the year 2023.

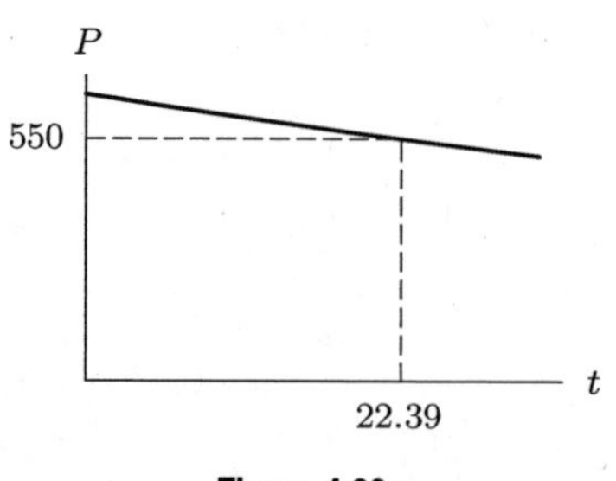

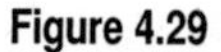

Figure 4.29

38. **(a)** $f(0) = 1000(1.04)^0 = 1000$, which means there are 1000 people in year 0.
$f(10) = 1000(1.04)^{10} \approx 1480.244$, which means there are 1480.244 people in year 10.

(b) For the first 10 years, use $0 \le t \le 10$, $0 \le P \le 1500$. See Figure 4.30. For the first 50 years, use $0 \le t \le 50$, $0 \le P \le 8000$. See Figure 4.31.

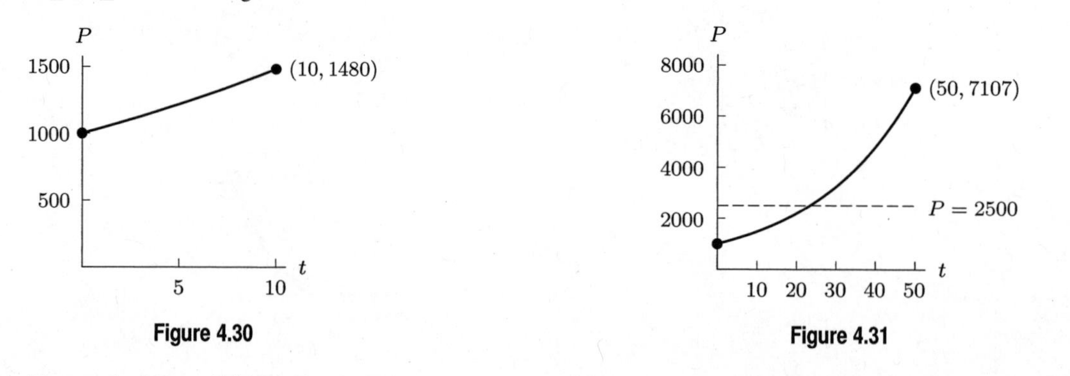

Figure 4.30

Figure 4.31

(c) The graph of $P = f(t)$ and $P = 2500$ intersect at $t \approx 23.362$. Thus, about 23.362 years after $t = 0$, the population will be 2500.

39. The function, when entered as $y = 1.04\hat{}5x$ is interpreted as $y = (1.04^5)x = 1.217x$. This function's graph is a straight line in all windows. Parentheses must be used to ensure that x is in the exponent.

40. **(a)** Since the number of cases is reduced by 10% each year, there are 90% as many cases in one year as in the previous one. So, after one year there are 90% of 10000 or 10000(0.90) cases, while after two years, there are $10000(0.90)(0.90) = 10000(0.90)^2$ cases. In general, the number of cases after t years is $y = (10000)(0.9)^t$.

(b) Setting $t = 5$, we obtain the number of cases 5 years from now

$$y = (10000) \cdot (0.9)^5 = 5904.9 \approx 5905 \text{ cases.}$$

(c) Plotting $y = (10000) \cdot (0.9)^t$ and approximating the value of t for which $y = 1000$, we obtain $t \approx 21.854$ years.

41. **(a)** For each kilometer above sea level, the atmospheric pressure is $86\%(= 100\% - 14\%)$ of the pressure one kilometer lower. If P represents the number of millibars of pressure and h represents the number of kilometers above sea level. Table 4.5 leads to the formula $P = 1013(0.86)^h$. So, at 50 km, $P = 1013(0.86)^{50} \approx 0.538$ millibars.

Table 4.5

h	P
0	1013
1	$1013(0.86) = 871.18$
2	$871.18(0.86) = 1013(0.86)(0.86) = 1013(0.86)^2$
3	$1013(0.86)^2 \cdot (0.86) = 1013(0.86)^3$
4	$1013(0.86)^4$
...	...
h	$1013(0.86)^h$

(b) If we graph the function $P = 1013(0.86)^h$, we can find the value of h for which $P = 900$. One approach is to see where it intersects the line $P = 900$. Doing so, you will see that at an altitude of $h \approx 0.784$ km, the atmospheric pressure will have dropped to 900 millibars.

42. **(a)** Since the population grows exponentially, it can be described by $P = ab^t$, where P is the number of rabbits and t is the number of years which have passed. We know that a represents the initial number of rabbits, so $a = 10$ and $P = 10b^t$. After 5 years, there are 340 rabbits so

$$\begin{aligned} 340 &= 10b^5 \\ 34 &= b^5 \\ (b^5)^{1/5} &= 34^{1/5} \\ b &\approx 2.024. \end{aligned}$$

From this, we know that $P = 10(2.024)^t$.

(b) We want to find t when $P = 1000$. Using a graph of $P = 10(2.024)^t$, we see that $P = 1000$ when $t = 6.53$ years.

43. **(a)** See Figure 4.32

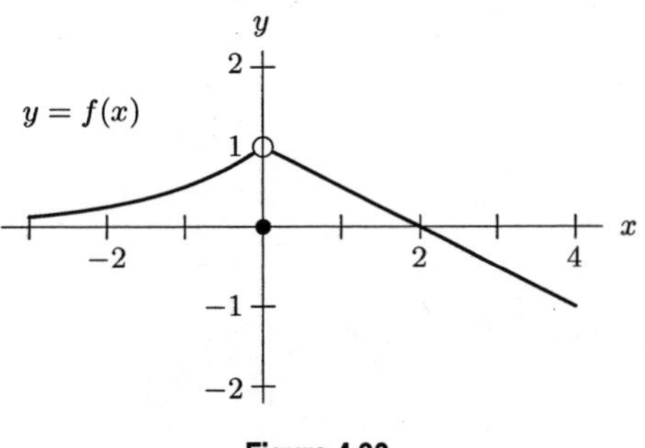

Figure 4.32

(b) The range of this function is all real numbers less than one — i.e. $f(x) < 1$.

(c) The y-intercept occurs at $(0, 0)$. This point is also an x-intercept. To solve for other x-intercepts we must attempt to solve $f(x) = 0$ for each of the two remaining parts of f. In the first case, we know that the function $f(x) = 2^x$ has no x-intercepts, as there is no value of x for which 2^x is equal to zero. In the last case, for $x > 0$, we set $f(x) = 0$ and solve for x:

$$\begin{aligned} 0 &= 1 - \frac{1}{2}x \\ \frac{1}{2}x &= 1 \\ x &= 2. \end{aligned}$$

Hence $x = 2$ is another x-intercept of f.

(d) As x gets large, the function is defined by $f(x) = 1 - 1/2x$. To determine what happens to f as $x \to +\infty$, find values of f for very large values of x. For example,

$$f(100) = 1 - \frac{1}{2}(100) = -49, \quad f(10000) = 1 - \frac{1}{2}(10000) = -4999$$

$$\text{and} \quad f(1{,}000{,}000) = 1 - \frac{1}{2}(1{,}000{,}000) = -499{,}999.$$

As x becomes larger, $f(x)$ becomes more and more negative. A way to write this is:

$$\text{As } x \to +\infty, \ f(x) \to -\infty.$$

As x gets very negative, the function is defined by $f(x) = 2^x$.

Choosing very negative values of x, we get $f(-100) = 2^{-100} = 1/2^{100}$, and $f(-1000) = 2^{-1000} = 1/2^{1000}$. As x becomes more negative the function values get closer to zero. We write

$$\text{As } x \to -\infty, \ f(x) \to 0.$$

(e) Increasing for $x < 0$, decreasing for $x > 0$.

44. (a) After the first hour all C values are measured at a common 2 hour interval, so we can estimate b by looking at the ratio of successive concentrations after the $t = 0$ concentration, namely,

$$\frac{7}{10} = 0.7, \quad \frac{5}{7} \approx 0.714, \quad \frac{3.5}{5} = 0.7, \quad \frac{2.5}{3.5} \approx 0.714.$$

These are nearly equal, the average being approximately 0.707, so $b^2 \approx 0.707$ and $b \approx 0.841$. Using the data point $(0, 12)$ we estimate $a = 12$. This gives $C = 12\,(0.841)^t$. Figure 4.33 shows these data plotted against time with this exponential function, which seems in good agreement.

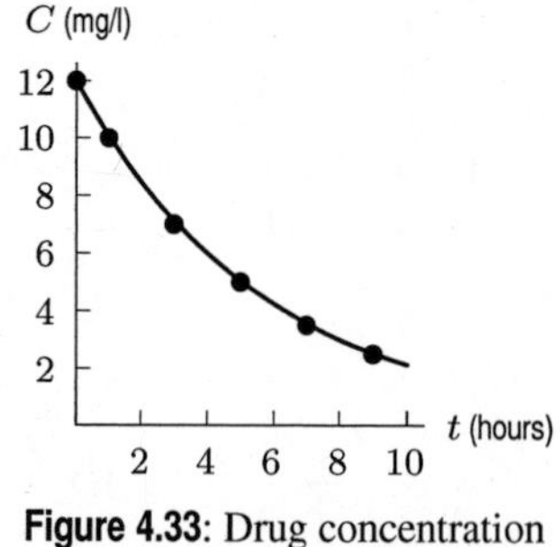

Figure 4.33: Drug concentration versus time with exponential fit

(b) One algorithm used by a calculator or computer gives the exponential regression function as

$$C = 11.914(0.840)^t$$

Other algorithms may give different formulas. This function is very similar to the answer to part (a).

45. **(a)** See Figure 4.34. The points appear to represent a function that is increasing and concave up so it makes sense to model these data with an exponential function.

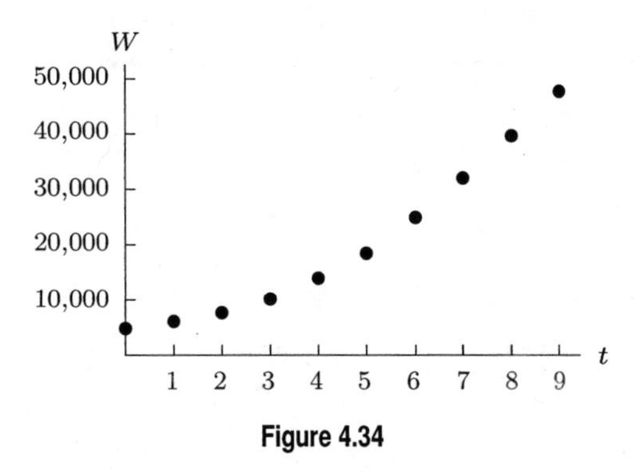

Figure 4.34

(b) One algorithm used by a calculator or computer gives the exponential regression function as

$$W = 4710(1.306)^t$$

Other algorithms may give different formulas.

(c) Since the base of this exponential function is 1.306, the global wind energy generating capacity was increasing at a rate of about 30.6% per year during this period.

46. **(a)** Figure 4.35 shows the three populations. From this graph, the three models seem to be in good agreement. Models 1 and 3 are indistinguishable; model 2 appears to rise a little faster. However notice that we cannot see the behavior beyond 50 months because our function values go beyond the top of the viewing window.

(b) Figure 4.36 shows the population differences. The graph of $y = f_2(x) - f_1(x) = 3(1.21)^x - 3(1.2)^x$ grows very rapidly, especially after 40 months. The graph of $y = f_3(x) - f_1(x) = 3.01(1.2)^x - 3(1.2)^x$ is hardly visible on this scale.

(c) Models 1 and 3 are in good agreement, but model 2 predicts a much larger mussel population than does model 1 after only 50 months. We can come to at least two conclusions. First, even small differences in the base of an exponential function can be highly significant, while differences in initial values are not as significant. Second, although two exponential curves can look very similar, they can actually be making very different predictions as time increases.

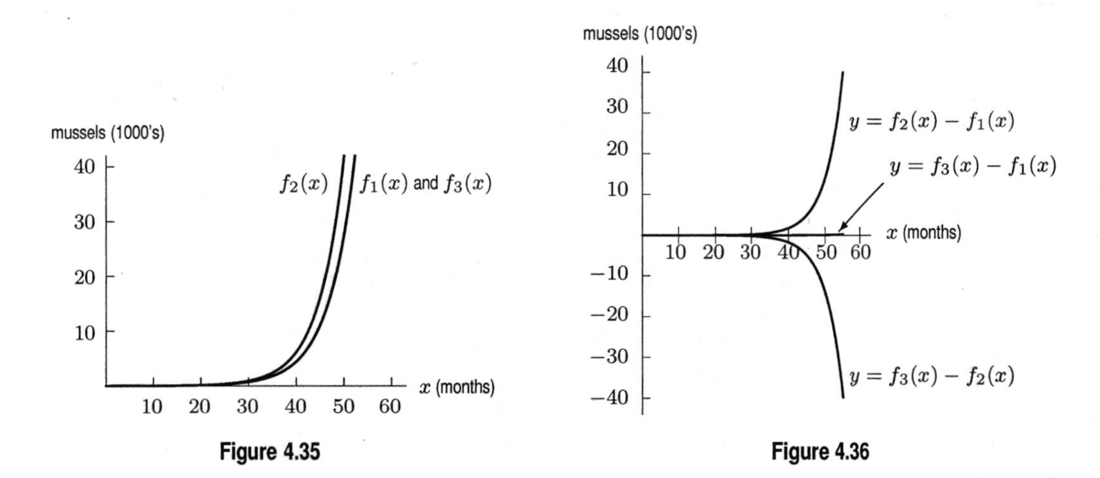

Figure 4.35

Figure 4.36

Solutions for Section 4.4

Exercises

1. **(a)** Suppose $1 is put in the account. The interest rate per month is $0.08/12$. At the end of a year,

$$\text{Balance} = \left(1 + \frac{0.08}{12}\right)^{12} = \$1.08300,$$

which is 108.3% of the original amount. So the effective annual yield is 8.300%.

(b) With weekly compounding, the interest rate per week is $0.08/52$. At the end of a year,

$$\text{Balance} = \left(1 + \frac{0.08}{52}\right)^{52} = \$1.08322,$$

which is 108.322% of the original amount. So the effective annual yield is 8.322%.

(c) Assuming it is not a leap year, the interest rate per day is $0.08/365$. At the end of a year

$$\text{Balance} = \left(1 + \frac{0.08}{365}\right)^{365} = \$1.08328,$$

which is 108.328% of the original amount. So the effective annual yield is 8.328%.

2. The amount in the account at time t is given by $1000b^t$. We set $1000b^{15}$ equal to 3500 and solve for b:

$$\begin{aligned} 1000b^{15} &= 3500 \\ b^{15} &= 3.5 \\ b &= (3.5)^{1/15} = 1.0871. \end{aligned}$$

The effective annual yield over the 15-year period was 8.71% per year.

3. If P is the initial amount, the amount after 20 years is $P(1.05)^{20} = P(2.653)$. Since $2.653 = 1 + 1.653$, the investment has increased by 165.3% over the 20-year period.

4. If P is the initial amount, the amount after 8 years is $0.5P$. To find the effective annual yield, we set Pb^8 equal to $0.5P$ and solve for b:

$$\begin{aligned} b^8 &= 0.5 \\ b &= (0.5)^{1/8} = 0.917. \end{aligned}$$

Since $0.917 = 1 - 0.083$, the investment has decreased by an effective annual rate of -8.3% per year.

5. **(a)** The nominal interest rate is 8%, so the interest rate per month is $0.08/12$. Therefore, at the end of 3 years, or 36 months,

$$\text{Balance} = \$1000\left(1 + \frac{0.08}{12}\right)^{36} = \$1270.24.$$

(b) There are 52 weeks in a year, so the interest rate per week is $0.08/52$. At the end of $52 \times 3 = 156$ weeks,

$$\text{Balance} = \$1000\left(1 + \frac{0.08}{52}\right)^{156} = \$1271.01.$$

(c) Assuming no leap years, the interest rate per day is $0.08/365$. At the end of 3×365 days

$$\text{Balance} = \$1000\left(1 + \frac{0.08}{365}\right)^{3\cdot 365} = \$1271.22.$$

6. **(a)** $B = B_0(1.013)^1 = B_0(1.013)$, so the effective annual rate is 1.3%.

(b) $B = B_0\left(1 + \frac{.013}{12}\right)^{12} \approx B_0(1.0131)$, so the effective annual rate is approximately 1.31%.

7. **(a)** If the interest is compounded annually, there will be $\$500 \cdot 1.01 = \$505$ after one year.
(b) If the interest is compounded weekly, after one year, there will be $\$500 \cdot (1 + 0.01/52)^{52} = \$505.02$.
(c) If the interest is compounded every minute, after one year, there will be $\$500 \cdot (1+0.01/525{,}600)^{525{,}600} = \$505.03$.

8. **(a)** If the interest is compounded annually, there will be $\$500 \cdot 1.03 = \$515$ after one year.
(b) If the interest is compounded weekly, there will be $500 \cdot (1 + 0.03/52)^{52} = \$515.22$ after one year.
(c) If the interest is compounded every minute, there will be $500 \cdot (1 + 0.03/525{,}600)^{525{,}600} = \$515.23$ after one year.

9. **(a)** If the interest is compounded annually, there will be $\$500 \cdot 1.05 = \$525$ after one year.
(b) If the interest is compounded weekly, there will be $500 \cdot (1 + 0.05/52)^{52} = \$525.62$ after one year.
(c) If the interest is compounded every minute, there will be $500 \cdot (1 + 0.05/525{,}600)^{525{,}600} = \$525.64$ after one year.

10. **(a)** If the interest is compounded annually, there will be $\$500 \cdot 1.08 = \$540$ after one year.
(b) If the interest is compounded weekly, there will be $500 \cdot (1 + 0.08/52)^{52} = \$541.61$ after one year.
(c) If the interest is compounded every minute, there will be $500 \cdot (1 + 0.08/525{,}600)^{525{,}600} = \$541.64$ after one year.

11. **(a)** The nominal rate is the stated annual interest without compounding, thus 1%.
The effective annual rate for an account paying 1% compounded annually is 1%.
(b) The nominal rate is the stated annual interest without compounding, thus 1%.
With quarterly compounding, there are four interest payments per year, each of which is $1/4 = 0.25\%$. Over the course of the year, this occurs four times, giving an effective annual rate of $1.0025^4 = 1.01004$, which is 1.004%.
(c) The nominal rate is the stated annual interest without compounding, thus 1%.
With daily compounding, there are 365 interest payments per year, each of which is $(1/365)\%$. Over the course of the year, this occurs 365 times, giving an effective annual rate of $(1+0.01/365)^{365} = 1.01005$, which is 1.005%.

12. **(a)** The nominal rate is the stated annual interest without compounding, thus 100%.
The effective annual rate for an account paying 1% compounded annually is 100%.
(b) The nominal rate is the stated annual interest without compounding, thus 100%.
With quarterly compounding, there are four interest payments per year, each of which is $100/4 = 25\%$. Over the course of the year, this occurs four times, giving an effective annual rate of $1.25^4 = 2.44141$, which is 144.141%.
(c) The nominal rate is the stated annual interest without compounding, thus 100%.
With daily compounding, there are 365 interest payments per year, each of which is $(100/365)\%$. Over the course of the year, this occurs 365 times, giving an effective annual rate of $(1 + 1/365)^{365} = 2.71457$, which is 171.457%.

13. **(a)** The nominal rate is the stated annual interest without compounding, thus 3%.
The effective annual rate for an account paying 1% compounded annually is 3%.
(b) The nominal rate is the stated annual interest without compounding, thus 3%.
With quarterly compounding, there are four interest payments per year, each of which is $3/4 = 0.75\%$. Over the course of the year, this occurs four times, giving an effective annual rate of $1.0075^4 = 1.03034$, which is 3.034%.
(c) The nominal rate is the stated annual interest without compounding, thus 3%.
With daily compounding, there are 365 interest payments per year, each of which is $(3/365)\%$. Over the course of the year, this occurs 365 times, giving an effective annual rate of $(1+0.03/365)^{365} = 1.03045$, which is 3.045%.

14. **(a)** The nominal rate is the stated annual interest without compounding, thus 6%.
The effective annual rate for an account paying 1% compounded annually is 6%.
(b) The nominal rate is the stated annual interest without compounding, thus 6%.
With quarterly compounding, there are four interest payments per year, each of which is $6/4 = 1.5\%$. Over the course of the year, this occurs four times, giving an effective annual rate of $1.015^4 = 1.06136$, which is 6.136%.
(c) The nominal rate is the stated annual interest without compounding, thus 6%.
With daily compounding, there are 365 interest payments per year, each of which is $(6/365)\%$. Over the course of the year, this occurs 365 times, giving an effective annual rate of $(1+0.06/365)^{365} = 1.06183$, which is 6.183%.

Problems

15. If the investment is growing by 3% per year, we know that, at the end of one year, the investment will be worth 103% of what it had been the previous year. At the end of two years, it will be 103% of $103\% = (1.03)^2$ as large. At the end of 10 years, it will have grown by a factor of $(1.03)^{10}$, or 1.34392. The investment will be 134.392% of what it had been, so we know that it will have increased by 34.392%. Since $(1.03)^{10} \approx 1.34392$, it increases by 34.392%.

16. If the annual growth factor is b, then we know that, at the end of 5 years, the investment will have grown by a factor of b^5. But we are told that it has grown by 30%, so it is 130% of its original size. So

$$\begin{aligned} b^5 &= 1.30 \\ b &= 1.30^{\frac{1}{5}} \approx 1.05387. \end{aligned}$$

Since the investment is 105.387% as large as it had been the previous year, we know that it is growing by about 5.387% each year.

17. Let b represent the growth factor, since the investment decreases, $b < 1$. If we start with an investment of P_0, then after 12 years, there will be $P_0 b^{12}$ left. But we know that since the investment has decreased by 60% there will be 40% remaining after 12 years. Therefore,

$$\begin{aligned} P_0 b^{12} &= P_0 0.40 \\ b^{12} &= 0.40 \\ b &= (0.40)^{1/12} = 0.92648. \end{aligned}$$

This tells us that the value of the investment will be 92.648% of its value the previous year, or that the value of the investment decreases by approximately 7.352% each year, assuming a constant percent decay rate.

18. **(a)** Let x be the amount of money you will need. Then, at 5% annual interest, compounded annually, after 6 years you will have the following dollar amount:

$$x(1+0.05)^6 = x(1.05)^6.$$

If this needs to equal \$25,000, then we have

$$\begin{aligned} x(1.05)^6 &= 25{,}000 \\ x &= \frac{25{,}000}{(1.05)^6} \approx \$18{,}655.38. \end{aligned}$$

(b) At 5% annual interest, compounded monthly, after 6 years, or $6 \cdot 12 = 72$ months, you will have the following dollar amount:

$$x\left(1+\frac{0.05}{12}\right)^{72}.$$

If this needs to equal \$25,000, then we have

$$\begin{aligned} x\left(1+\frac{0.05}{12}\right)^{72} &= 25{,}000 \\ x &= \frac{25{,}000}{(1+\frac{0.05}{12})^{72}} \approx \$18{,}532.00. \end{aligned}$$

(c) At 5% annual interest, compounded daily, after 6 years, or $6 \cdot 365 = 2190$ days, you will have the following dollar amount:

$$x\left(1+\frac{0.05}{365}\right)^{2190} = x(1.000136986)^{2190}.$$

If this needs to equal \$25,000, then we have

$$\begin{aligned} x(1.000136986)^{2190} &= 25{,}000 \\ x &= \frac{25{,}000}{(1.000136986)^{2190}} \approx \$18{,}520.84. \end{aligned}$$

(d) The effective yield on an account increases with the number of times of compounding. So, as the number of times increases, the amount of money you need to begin with in order to end up with 25,000 in 6 years decreases.

19. Let r represent the nominal annual rate. Since the interest is compounded quarterly, the investment earns $\frac{r}{4}$ each quarter. So, at the end of the first quarter, the investment is $850\left(1+\frac{r}{4}\right)$, and at the end of the second quarter is $850\left(1+\frac{r}{4}\right)^2$. By the end of 40 quarters (which is 10 years), it is $850\left(1+\frac{r}{4}\right)^{40}$. But we are told that the value after 10 years is \$1,000, so

$$\begin{aligned}
1000 &= 850\left(1+\frac{r}{4}\right)^{40} \\
\frac{1000}{850} &= \left(1+\frac{r}{4}\right)^{40} \\
\frac{20}{17} &= \left(1+\frac{r}{4}\right)^{40} \\
\left(\frac{20}{17}\right)^{\frac{1}{40}} &= 1+\frac{r}{4} \\
1.00407 &\approx 1+\frac{r}{4} \\
0.00407 &\approx \frac{r}{4} \\
r &\approx 0.01628.
\end{aligned}$$

We see that the nominal interest rate is 1.628%.

20. **(a)** The effective annual rate is the rate at which the account is actually increasing in one year. According to the formula, $M = M_0(1.07763)^t$, at the end of one year you have $M = 1.07763M_0$, or 1.07763 times what you had the previous year. The account is 107.763% larger than it had been previously; that is, it increased by 7.763%. Thus the effective rate being paid on this account each year is about 7.763%.

(b) Since the money is being compounded each month, one way to find the nominal annual rate is to determine the rate being paid each month. In t years there are $12t$ months, and so, if b is the monthly growth factor, our formula becomes

$$M = M_0 b^{12t} = M_0(b^{12})^t.$$

Thus, equating the two expressions for M, we see that

$$M_0(b^{12})^t = M_0(1.07763)^t.$$

Dividing both sides by M_0 yields

$$(b^{12})^t = (1.07763)^t.$$

Taking the tth root of both sides, we have

$$b^{12} = 1.07763$$

which means that

$$b = (1.07763)^{1/12} \approx 1.00625.$$

Thus, this account earns 0.625% interest every month, which amounts to a nominal interest rate of about $12(0.625\%) =$ 7.5%.

21. (i) Equation (b). Since the growth factor is 1.12, or 112%, the annual interest rate is 12%.

(ii) Equation (a). An account earning at least 1% monthly will have a monthly growth factor of at least 1.01, which means that the annual (12-month) growth factor will be at least

$$(1.01)^{12} \approx 1.1268.$$

Thus, an account earning at least 1% monthly will earn at least 12.68% yearly. The only account that earns this much interest is account (a).

(iii) Equation (c). An account earning 12% annually compounded semi-annually will earn 6% twice yearly. In t years, there are $2t$ half-years.

(iv) Equations (b), (c) and (d). An account that earns 3% each quarter ends up with a yearly growth factor of $(1.03)^4 \approx$ 1.1255. This corresponds to an annual percentage rate of 12.55%. Accounts (b), (c) and (d) earn less than this. Check this by determining the growth factor in each case.

(v) Equations (a) and (e). An account that earns 6% every 6 months will have a growth factor, after 1 year, of $(1 + 0.06)^2 = 1.1236$, which is equivalent to a 12.36% annual interest rate, compounded annually. Account (a), earning 20% each year, clearly earns more than 6% twice each year, or 12.36% annually. Account (e), which earns 3% each quarter, earns $(1.03)^2 = 1.0609$, or 6.09% every 6 months, which is greater than 6%.

22. **(a)** The investment is initially worth \$1500, and it grows in value by 7.7% each year.
(b) The investment is initially worth \$9500, but it loses value by 5.5% each year.
(c) The investment is initially worth \$1000, and it triples in value once every five years.
(d) The investment is initially worth \$500, and it earns 4% annual interest, compounded monthly.

Solutions for Section 4.5

Skill Refresher

S1. We have $e^{0.07} = 1.073$.

S2. We have $10e^{-0.14} = 8.694$.

S3. We have $\dfrac{2}{\sqrt[3]{e}} = 1.433$.

S4. We have $e^{3e} = 3480.202$.

S5. We have $f(0) = 2.3e^{0.3(0)} = 2.3$ and $f(4) = 2.3e^{0.3(4)} = 7.636$.

S6. We have $g(0) = 4.2e^{-0.12(0)} = 4.2$ and $g(4) = 4.2e^{-0.12(4)} = 2.599$.

S7. We have $h(0) = 153 + 8.6e^{0.43(0)} = 153 + 8.6 = 161.6$ and $h(4) = 153 + 8.6e^{0.43(4)} = 153 + 48.027 = 202.027$.

S8. We have $k(0) = 289 - 4.7e^{-0.0018(0)} = 289 - 4.7 = 284.3$ and $k(4) = 289 - 4.7e^{-0.0018(4)} = 289 - 4.666 = 284.334$.

S9. Writing the function as

$$f(t) = \left(3e^{0.04t}\right)^3 = 3^3e^{0.04t\cdot 3} = 27e^{0.12t},$$

we have $a = 27$ and $k = 0.12$.

S10. Writing the function as

$$g(z) = 5e^{7z}\cdot e^{4z}\cdot 3e^{z} = (5\cdot 3)e^{7z+4z+z} = 15e^{12z},$$

we have $a = 15$ and $k = 12$.

S11. To convert to the form $Q = ae^{kt}$, we use the property that $e^{m+n} = e^m\cdot e^n$. Thus, we have $Q = e^7\cdot e^{-3t} = 1{,}096.633e^{-3t}$.

S12. To convert to the form $Q = ae^{kt}$, we first use the property that $(e^m)^n = e^{mn}$,

$$\begin{aligned} Q &= \sqrt{e^{3+6t}} \\ &= \left(e^{3+6t}\right)^{1/2} \\ &= e^{\frac{1}{2}(3+6t)} \\ &= e^{\frac{3}{2}+3t}. \end{aligned}$$

Next we use the property that $e^{m+n} = e^m\cdot e^n$. Thus, we have $Q = e^{\frac{3}{2}+3t} = e^{3/2}\cdot e^{3t} = 4.482e^{3t}$.

S13. Writing the function as

$$m(x) = \frac{7e^{0.2x}}{\sqrt{3e^x}} = \frac{7}{\sqrt{3}}e^{0.2x}\cdot e^{-0.5x} = \frac{7}{\sqrt{3}}e^{-0.3x},$$

we have $a = \frac{7}{\sqrt{3}}$ and $k = -0.3$.

S14. Writing the function as

$$P(t) = \left(2\sqrt[3]{e^{5t}}\right)^4 = 2^4\left(e^{\frac{5}{3}t}\right)^4 = 16e^{(4\cdot\frac{5}{3}t)} = 16e^{\frac{20}{3}t},$$

we have $a = 16$ and $k = 20/3$.

S15. Writing the function as

$$H(r) = \frac{\left(e^{0.4r}\right)^2}{6e^{0.15r}} = \frac{1}{6}\frac{e^{(2\cdot 0.4r)}}{e^{0.15r}} = \frac{1}{6}e^{(0.8r-0.15r)} = \frac{1}{6}e^{0.65r},$$

we have $a = 1/6$ and $k = 0.65$.

Exercises

1. Using the formula $y = ab^x$, each of the functions has the same value for b, but different values for a and thus different y-intercepts.
When $x = 0$, the y-intercept for $y = e^x$ is 1 since $e^0 = 1$.
When $x = 0$, the y-intercept for $y = 2e^x$ is 2 since $e^0 = 1$ and $2(1) = 2$.
When $x = 0$, the y-intercept for $y = 3e^x$ is 3 since $e^0 = 1$ and $3(1) = 3$.
Therefore, $y = e^x$ is the bottom graph, above it is $y = 2e^x$ and the top graph is $y = 3e^x$.

2. We know that $e \approx 2.71828$, so $2 < e < 3$. Since e lies between 2 and 3, the graph of $y = e^x$ lies between the graphs of $y = 2^x$ and $y = 3^x$. Since 3^x increases faster than 2^x, the correct matching is shown in Figure 4.37.

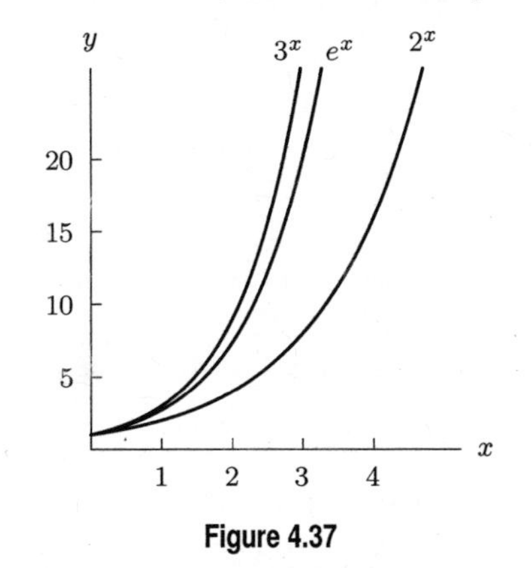

Figure 4.37

3. We know that $e^{0.3t}$ grows faster than $e^{0.25t}$ (because a 30% continuous growth rate is faster than a 25% continuous growth rate). Similarly $(1.25)^t$ grows faster than $(1.2)^t$ (because a 25% annual growth rate is faster than a 20% annual growth rate). In addition, a 25% continuous growth rate is faster than a 25% annual growth rate.
Thus $(1.2)^t$=(c) is (IV); $(1.25)^t$=(b) is (III); $e^{0.25t}$=(a) is (II); $e^{0.3t}$=(d) is (I).

4. Calculating the equivalent continuous rates, we find $e^{0.45} = 1.568$, $e^{0.47} = 1.600$, $e^{0.5} = 1.649$. Thus the functions to be matched are

(a) $(1.5)^x$ **(b)** $(1.568)^x$ **(c)** $(1.6)^x$ **(d)** $(1.649)^x$

So (a) is (IV), (b) is (III), (c) is (II), (d) is (I).

5. $y = e^x$ is an increasing exponential function, since $e > 1$. Therefore, it rises when read from left to right. It matches $g(x)$.
If we rewrite the function $y = e^{-x}$ as $y = (e^{-1})^x$, we can see that in the formula $y = ab^x$, we have $a = 1$ and $b = e^{-1}$. Since $0 < e^{-1} < 1$, this graph has a positive y-intercept and falls when read from left to right. Thus its graph is $f(x)$.
In the function $y = -e^x$, we have $a = -1$. Thus, the vertical intercept is $y = -1$. The graph of $h(x)$ has a negative y-intercept.

6. The functions given in (a) and (c) represent exponential decay while the functions given in (b) and (d) represent exponential growth. Thus, (a) and (c) correspond to (III) and (IV) while (b) and (d) correspond to (I) and (II). The function in (d) grows by 20% per time unit while the function in (b) grows by 5% per time unit. Since (d) is growing faster, formula (d) must correspond to graph (I) while formula (b) corresponds to graph (II). Graphs (III) and (IV) correspond to the exponential decay formulas, with graph (IV) decaying at a more rapid rate. Thus formula (a) corresponds to graph (III) and formula (c) corresponds to graph (IV). We have:

(a) (III)

(b) (II)
(c) (IV)
(d) (I)

7. Since $y = e^x$ is the only increasing function in the list, (a) corresponds to (I).
The other three functions all decrease as x increases, and $y = e^{-x} = 1/e^x$ decreases slowest while $y = e^{-3x} = 1/e^{3x}$ decreases fastest. Thus (b) corresponds to (II), and (c) corresponds to (III), and (d) corresponds to (IV).

Problems

8. **(a)** Since $\sqrt{2} = 1.414\ldots$ and $e = 2.718\ldots$, we have

$$\sqrt{2} < e < 3,$$

so

$$(\sqrt{2})^{2.2} < e^{2.2} < 3^{2.2}.$$

(b) Note $3^{-2.2} = 1/3^{2.2}$. Thus, since

$$e^{2.2} < 3^{2.2},$$

we know that

$$\frac{1}{e^{2.2}} > \frac{1}{3^{2.2}},$$

which gives

$$3^{-2.2} < e^{-2.2}.$$

9. $\lim_{x\to\infty} e^{-3x} = 0.$

10. $\lim_{t\to-\infty} 5e^{0.07t} = 0.$

11. $\lim_{t\to\infty} (2 - 3e^{-0.2t}) = 2 - 3 \cdot 0 = 2.$

12. $\lim_{t\to-\infty} 2e^{-0.1t+6} = \infty.$

13. The values of a and k are both positive.

14. **(a)** We see that $Q_0 = 25$
(b) Since the exponent is positive, the quantity is increasing.
(c) The growth rate is 3.2% per unit time.
(d) Yes, the growth rate is continuous.

15. **(a)** We see that $Q_0 = 2.7$
(b) Since the base is less than one, the quantity is decreasing.
(c) Since the base is $0.12 = 1 - 0.88$, the decay rate is 88% per unit time.
(d) No, the growth rate is not continuous.

16. **(a)** We see that $Q_0 = 158$
(b) Since the base is greater than one, the quantity is increasing.
(c) Since the base is $1.137 = 1 + 0.137$, the growth rate is 13.7% per unit time.
(d) No, the growth rate is not continuous.

17. **(a)** We see that $Q_0 = 0.01$
(b) Since the exponent is negative, the quantity is decreasing.
(c) The decay rate is 20% per unit time.
(d) Yes, the growth rate is continuous.

18. **(a)** We see that $Q_0 = 50$
(b) Since the exponent is positive, the quantity is increasing.
(c) The growth rate is 105% per unit time, so the quantity is more than doubling every time unit.
(d) Yes, the growth rate is continuous.

19. **(a)** We see that $Q_0 = 1$
(b) Since the base is greater than one, the quantity is increasing.
(c) Since the base is $2 = 1 + 1$, the growth rate is 100% per unit time. The quantity is doubling every time unit.
(d) No, the growth rate is not continuous.

20. **(a)** Since the growth rate is not continuous, we have $Q = 100(1.05)^t$. At $t = 10$ we have $Q = 100(1.05)^{10} = 162.889$.
(b) Since the growth rate is continuous,we have $Q = 100e^{0.05t}$. At $t = 10$ we have $Q = 100e^{0.05(10)} = 164.872$. As we expect, the results are similar for continuous and not continuous assumptions, but slightly larger if we assume a continuous growth rate.

21. **(a)** Since the growth rate is not continuous, we have $Q = 8(1.12)^t$. At $t = 10$ we have $Q = 8(1.12)^{10} = 24.847$.
(b) Since the growth rate is continuous,we have $Q = 8e^{0.12t}$. At $t = 10$ we have $Q = 8e^{0.12(10)} = 26.561$. As we expect, the results are similar for continuous and not continuous assumptions, but slightly larger if we assume a continuous growth rate.

22. **(a)** Since the decay rate is not continuous, we have $Q = 500(0.93)^t$. At $t = 10$ we have $Q = 500(0.93)^{10} = 241.991$.
(b) Since the decay rate is continuous,we have $Q = 500e^{-0.07t}$. At $t = 10$ we have $Q = 500e^{-0.07(10)} = 248.293$. As we expect, the results are similar for continuous and not continuous assumptions, but slightly larger if we assume a continuous decay rate.

23. **(a)** (i) The population, P, in millions, is given by $P = 3.2(1.02)^t$, so a century later

$$P = 3.2(1.02)^{100} = 23.183 \text{ million.}$$

(ii) The population, P, in millions, is given by $P = 3.2e^{0.02t}$, so a century later

$$P = 3.2e^{0.02(100)} = 23.645 \text{ million.}$$

(b) Since $e^{0.02} = 1.0202\ldots$ the growth factor in part (ii) is larger than the growth factor of 1.02 in part (i). Thus we expect the answer to part (ii) to be larger.

24. **(a)** The population starts at 200 and grows by 2.8% per year.
(b) The population starts at 50 and shrinks at a continuous rate of 17% per year.
(c) The population starts at 1000 and shrinks by 11% per year.
(d) The population starts at 600 and grows at a continuous rate of 20% per year.
(e) The population starts at 2000 and shrinks by 300 animals per year.
(f) The population starts at 600 and grows by 50 animals per year.

25. If the population is growing or shrinking at a constant rate of m people per year, the formula is linear. Since the vertical intercept is 3000, we have $P = 3000 + mt$.
If the population is growing or shrinking at a constant percent rate of r percent per year, the formula is exponential in the form $P = a(1 + r)^t$. Since the vertical intercept is 3000, we have $P = 3000(1 + r)^t$.
If the population is growing or shrinking at a constant continuous percent rate of k percent per year, the formula is exponential in the form $P = ae^{kt}$. Since the vertical intercept is 3000, we have $P = 3000e^{kt}$.
We have:

(a) $P = 3000 + 200t$.
(b) $P = 3000(1.06)^t$.
(c) $P = 3000e^{0.06t}$.
(d) $P = 3000 - 50t$.
(e) $P = 3000(0.96)^t$.
(f) $P = 3000e^{-0.04t}$.

26. **(a)** Using $P = P_0e^{kt}$ where $P_0 = 25{,}000$ and $k = 7.5\%$, we have

$$P(t) = 25{,}000e^{0.075t}.$$

(b) We first need to find the growth factor so will rewrite

$$P = 25{,}000e^{0.075t} = 25{,}000(e^{0.075})^t \approx 25{,}000(1.07788)^t.$$

At the end of a year, the population is 107.788% of what it had been at the end of the previous year. This corresponds to an increase of approximately 7.788%. This is greater than 7.5% because the rate of 7.5% per year is being applied to larger and larger amounts. In one instant, the population is growing at a rate of 7.5% per year. In the next instant, it grows again at a rate of 7.5% a year, but 7.5% of a slightly larger number. The fact that the population is increasing in tiny increments continuously results in an actual increase greater than the 7.5% increase that would result from one, single jump of 7.5% at the end of the year.

27. **(a)** Since $P(t)$ has continuous growth, its formula will be $P(t) = P_0e^{kt}$. Since P_0 is the initial population, which is 22,000, and k represents the continuous growth rate of 7.1%, our formula is

$$P(t) = 22{,}000e^{0.071t}.$$

(b) While, at any given instant, the population is growing at a rate of 7.1% a year, the effect of compounding is to give us an actual increase of more than 7.1%. To find that increase, we first need to find the growth factor, or b. Rewriting $P(t) = 22{,}000e^{0.071t}$ in the form $P = 22000b^t$ will help us accomplish this. Thus, $P(t) = 22{,}000(e^{0.071})^t \approx 22{,}000(1.07358)^t$. Alternatively, we can equate the two formulas and solve for b:

$$\begin{aligned} 22{,}000e^{0.071t} &= 22{,}000b^t \\ e^{0.071t} &= b^t \quad \text{(dividing both sides by 22,000)} \\ e^{0.071} &= b \quad \text{(taking the } t^{\text{th}} \text{ root of both sides).} \end{aligned}$$

Using your calculator, you can find that $b \approx 1.07358$. Either way, we see that at the end of the year, the population is 107.358% of what it had been at the end of the previous year, and so the population increases by approximately 7.358% each year.

28. A formula for the investment's value is $V = 7000e^{0.052t}$. After 7 years, we have

$$\begin{aligned} V &= 7000e^{0.052\cdot 7} \\ &= \$10{,}073.52. \end{aligned}$$

29.
- All three investments begin (in year $t = 0$) with \$1000.
- The investment $V = 1000e^{0.115t}$ earns interest at a continuous annual rate of 11.5%.
- The investment $V = 1000 \cdot 2^{t/6}$ doubles in value every 6 years.
- The investment $V = 1000(1.122)^t$ grows by 12.2% every year.

30. **(a)** For an annual interest rate of 5%, the balance B after 15 years is

$$B = 2000(1.05)^{15} = 4157.86 \text{ dollars.}$$

(b) For a continuous interest rate of 5% per year, the balance B after 15 years is

$$B = 2000e^{0.05\cdot 15} = 4234.00 \text{ dollars.}$$

31. Tracing along a graph of $V = 1000e^{0.02t}$ until $V = 3000$ gives $t \approx 54.931$ years. See Figure 4.38.

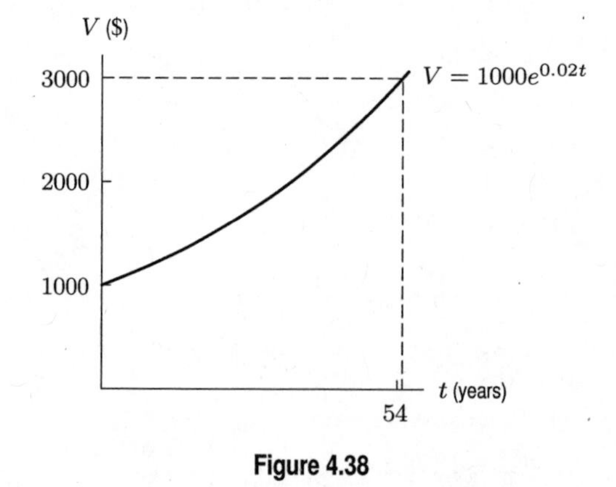

Figure 4.38

32. We want to know when $V = 1074$. Tracing along the graph of $V = 537e^{0.015t}$ gives $t \approx 46.210$ years. See Figure 4.39.

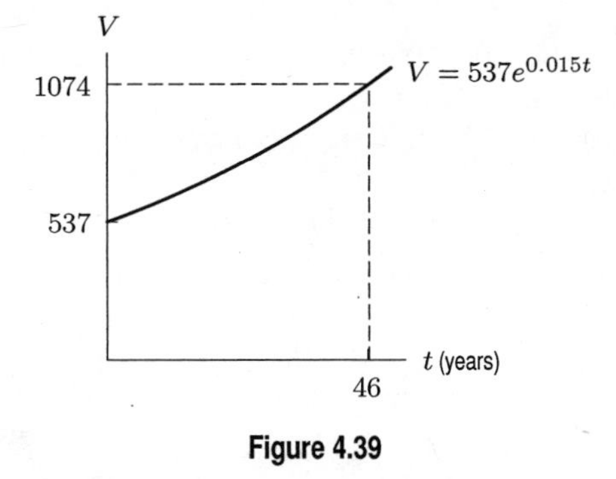

Figure 4.39

33. **(a)** At the end of 100 years,
$$B = 1200e^{0.03(100)} = 24{,}102.64 \text{ dollars}.$$
(b) Tracing along a graph of $B = 1200e^{0.03t}$ until $B = 50000$ gives $t \approx 124.323$ years.

34. With continuous compounding, the interest earns interest during the year, so the balance grows faster with continuous compounding than with annual compounding. Curve A corresponds to continuous compounding and curve B corresponds to annual compounding. The initial amount in both cases is the vertical intercept, \$500.

35. The continuous growth rate is $k = 0.19 = 19\%$ per year. To calculate the effective annual yield, we rewrite the function in the form $Q = 5500e^{0.19t} = 5500b^t$. Thus,
$$b = e^{0.19} = 1.20925,$$
Writing
$$Q = 5500(1.209250)^t$$
indicates that the effective annual yield is 20.925% per year.

36. We let B represent the balance in the account after t years.

(a) If interest is compounded annually, we have $B = 5000(1.04)^t$. After ten years, the amount in the account is $5000(1.04)^{10} = \$7401.22$.

(b) If interest is compounded continuously, we have $B = 5000e^{0.04t}$. After ten years, the amount in the account is $5000e^{0.04(10)} = \$7459.12$. As expected, the account contains more money if the interest is compounded continuously.

37. The value of the deposit is given by
$$V = 1000e^{0.05t}.$$
To find the effective annual rate, we use the fact that $e^{0.05t} = (e^{0.05})^t$ to rewrite the function as
$$V = 1000(e^{0.05})^t.$$
Since $e^{0.05} = 1.05127$, we have
$$V = 1000(1.05127)^t.$$
This tells us that the effective annual rate is 5.127%.

38. We have
$$\begin{aligned} V &= 500e^{0.0675t} \qquad \text{continuous rate is 6.75\%} \\ &= 500\left(e^{0.0675}\right)^t \\ &= 500(1.06983)^t, \\ \text{so} \quad a &= 500 \\ b &= 1.06983 \\ r &= 6.983\%. \end{aligned}$$

39. **(a)** (i) $B = B_0\left(1+\frac{.06}{4}\right)^4 \approx B_0(1.0614)$, so the APR is approximately 6.14%.

(ii) $B = B_0\left(1+\frac{.06}{12}\right)^{12} \approx B_0(1.0617)$, so the APR is approximately 6.17%.

(iii) $B = B_0\left(1+\frac{.06}{52}\right)^{52} \approx B_0(1.0618)$, so the APR is approximately 6.18%.

(iv) $B = B_0\left(1+\frac{.06}{365}\right)^{365} \approx B_0(1.0618)$, so the APR is approximately 6.18%.

(b) $e^{0.06} \approx 1.0618$. No matter how often we compound interest, we'll never get more than $\approx 6.18\%$ APR.

40. **(a)** For investment A, we have

$$P = 875(1 + \frac{0.135}{365})^{365(2)} = \$1146.16.$$

For investment B,

$$P = 1000(e^{0.067(2)}) = \$1143.39.$$

For investment C,

$$P = 1050(1 + \frac{0.045}{12})^{12(2)} = \$1148.69.$$

(b) A comparison of final balances does not reflect the fact that the initial investment amounts are different. One way to take initial amount into consideration is to look at the overall growth in the account. Comparing final balance to initial deposit for each account we find

$$\text{Investment A: } \frac{1146.16}{875} \approx 1.31$$

$$\text{Investment B: } \frac{1143.39}{1000} \approx 1.143$$

$$\text{Investment C: } \frac{1148.69}{1050} \approx 1.093.$$

Thus, in the two year period Investment A has grown by approximately 31%, followed by Investment B (14.3%) and finally Investment C (9.3%). From best to worst, we have A, B, C.

[Note: Comparing the effective annual rates for each account would be a more efficient way to solve the problem and would give the same result.]

41. To see which investment is best after 1 year, we compute the effective annual yield:
For Bank A, $P = P_0(1 + \frac{0.07}{365})^{365(1)} \approx 1.0725P_0$
For Bank B, $P = P_0(1 + \frac{0.071}{12})^{12(1)} \approx 1.0734P_0$
For Bank C, $P = P_0(e^{0.0705(1)}) \approx 1.0730P_0$
Therefore, the best investment is with Bank B, followed by Bank C and then Bank A.

42. Since $e^{0.053} = 1.0544$, the effective annual yield of the account paying 5.3% interest compounded continuously is 5.44%. Since this is less than the effective annual yield of 5.5% from the 5.5% compounded annually, we see that the account paying 5.5% interest compounded annually is slightly better.

43. **(a)** We have $G = 145.8e^{0.051t}$.

(b) Since $e^{0.051(1)} = 1.0523$, we see that the GDP increases by 5.23% each year.

(c) We have $G = 145.8(1.0523)^t$.

(d) The two functions $G = 145.8e^{0.051t}$ and $G = 145.8(1.0523)^t$ are shown in Figure 4.40. One lies exactly on top of the other because the two formulas have the same graph. They are two different representations of the same function.

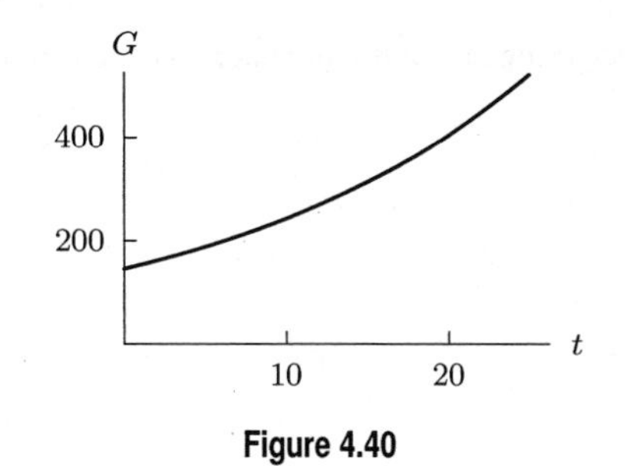

Figure 4.40

44. After 5 years, the first investment is worth:

$$1000(1+\frac{0.05}{12})^{12(5)} = \$1283.36.$$

After 5 years, the second investment is worth:

$$1100e^{0.04(5)} = \$1343.54.$$

After 5 years, the second investment is worth more.

After 10 years, the first investment is worth:

$$1000(1+\frac{0.05}{12})^{12(10)} = \$1647.01.$$

After 10 years, the second investment is worth:

$$1100e^{0.04(10)} = \$1641.01.$$

After 10 years, the first investment is worth slightly more.

45. The balance in the first bank is $10{,}000(1.05)^8 = \$14{,}774.55$. The balance in the second bank is $10{,}000e^{0.05(8)} = \$14{,}918.25$. The bank with continuously compounded interest has a balance \$143.70 higher.

46. **(a)** The \$1000 investment is represented by graph A and the \$1500 investment is represented by graph B.

(b) Figure 4.41 shows a graph of the two functions $A = 1000e^{0.08t}$ and $B = 1500(1.06)^t$. We use a graphing calculator to estimate the point of intersection at $t = 18.66$. The two investments will be equal after about 18.66 years.

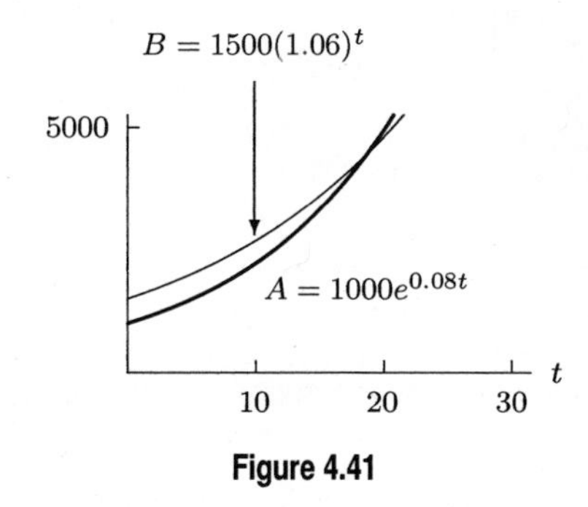

Figure 4.41

47. Since the graphs of ae^{kx} and be^{lx} have the same vertical intercept, we know $a = b$. Since their common intercept is above the vertical intercept of e^x, we know $a = b > 1$.

Since ae^{kx} increases as x increases, we know $k > 0$. But ae^{kx} increases more slowly than e^x, so $0 < k < 1$.

Since be^{lx} decreases as x increases, we know $l < 0$.

48. **(a)** Since poultry production is increasing at a constant continuous percent rate, we use the exponential formula $P = ae^{kt}$. Since $P = 94.7$ when $t = 0$, we have $a = 94.7$. Since $k = 0.011$, we have

$$P = 94.7e^{0.011t}.$$

(b) When $t = 6$, we have $P = 94.7e^{0.011(6)} = 101.16$. In the year 2015, the formula predicts that world poultry production will be about 101 million tons.

(c) A graph of $P = 94.7e^{0.011t}$ is given in Figure 4.42. We see that when $P = 110$ we have $t = 13.6$. We expect production to be 110 million tons near the middle of the year 2022.

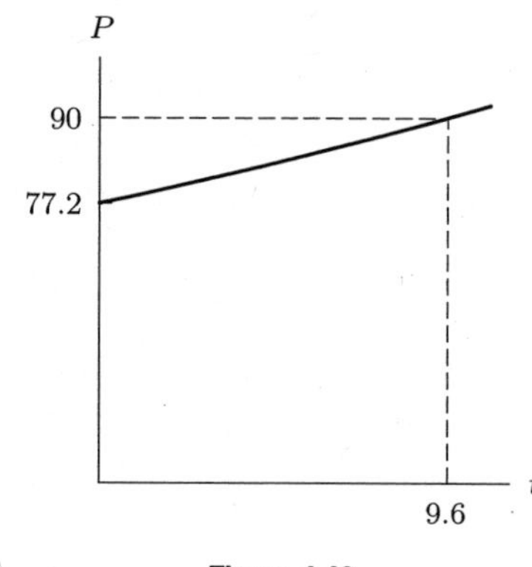

Figure 4.42

49. **(a)** The substance decays according to the formula

$$A = 50e^{-0.14t}.$$

(b) At $t = 10$, we have $A = 50e^{-0.14(10)} = 12.330$ mg.

(c) We see in Figure 4.43 that $A = 5$ at approximately $t = 16.45$, which corresponds to the year 2025.

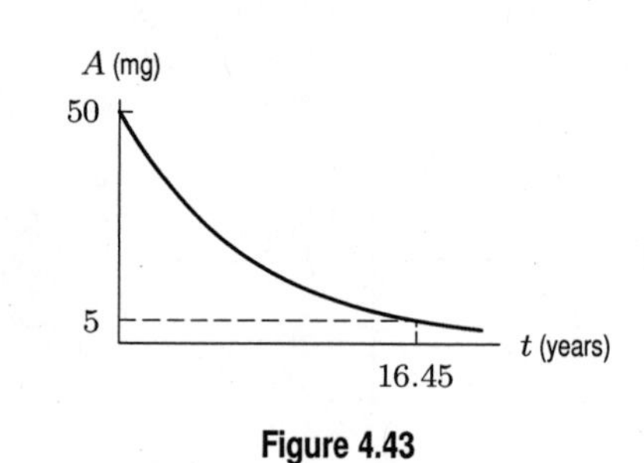

Figure 4.43

50. **(a)** Let p_0 be the price of an item at the beginning of 2000. At the beginning of 2001, its price will be 103.4% of that initial price or $1.034p_0$. At the beginning of 2002, its price will be 102.8% of the price from the year before, that is:

$$\text{Price beginning 2002} = (1.028)(1.034p_0).$$

By the beginning of 2003, the price will be 101.6% of its price the previous year.

$$\begin{aligned}\text{Price beginning 2003} &= 1.016(\text{price beginning 2002})\\ &= 1.016(1.028)(1.034p_0).\end{aligned}$$

Continuing this process,

$$\begin{aligned}(\text{Price beginning 2005}) &= (1.027)(1.023)(1.016)(1.028)(1.034)p_0\\ &\approx 1.135p_0.\end{aligned}$$

So, the cost at the beginning of 2005 is 113.5% of the cost at the beginning of 2000 and the total percent increase is 13.5%.

(b) If r is the average inflation rate for this time period, then $b = 1 + r$ is the factor by which the population on the average grows each year. Using this average growth factor, if the price of an item is initially p_0, at the end of a year its value would be p_0b, at the end of two years it would be $(p_0b)b = p_0b^2$, and at the end of five years p_0b^5. According to the answer in part (a), the price at the end of five years is $1.135p_0$. So

$$\begin{aligned}p_0b^5 &= 1.135p_0\\ b^5 &= 1.135\\ b &= (1.135)^{1/5} \approx 1.026.\end{aligned}$$

If $b = 1.026$, then $r = 0.026$ or 2.6%, the average annual inflation rate.

(c) We assume that the average rate of 2.6% inflation for 2000 through 2004 holds through the beginning of 2010. So, on average, the price of the shower curtain is 102.6% of what it was the previous year for ten years. Then the price of the shower curtain would be $20(1.026)^{10} \approx \$25.85$.

51. To find the fee for six hours, we convert 6 hours to years: (6)·(1 year/365 days)·(1 day/24 hours)= $\frac{6}{(365)(24)}$ years.

Since the interest is being compounded continuously, the total amount of money is given by $P = P_0e^{kt}$, where, in this case, $k = 0.20$ is the continuous annual rate and t is the number of years. So

$$P = 200{,}000{,}000e^{0.20(\frac{6}{365\cdot 24})} = 200{,}027{,}399.14$$

The value of the money at the end of the six hours was \$200,027,399.14, so the fee for that time was \$27,399.14.

52. **(a)** A calculator or computer gives the values in Table 4.6. We see that the values of $(1 + 1/n)^n$ increase as n increases.

Table 4.6

n	$(1+1/n)^n$
1000	2.7169239
10,000	2.7181459
100,000	2.7182682
1,000,000	2.7182805

(b) Extending Table 4.6 gives Table 4.7. Since, correct to 6 decimal places, $e = 2.718282$, we need approximately $n = 10^7$ to achieve an estimate for e that is correct to 6 decimal places.

Table 4.7

n	$(1+1/n)^n$	Correct to 6 decimal places
10^5	2.71826824	2.718268
10^6	2.71828047	2.718280
10^7	2.71828169	2.718282
10^8	2.71828179	2.718282

(c) Using most calculators, when $n = 10^{16}$ the computed value of $(1 + 1/n)^n$ is 1. The reason is that calculators use only a limited number of decimal places, so the calculator finds that $1 + 1/10^{16} = 1$.

53. **(a)** The sum is 2.708333333.

(b) The sum of $1+\frac{1}{1}+\frac{1}{1\cdot 2}+\frac{1}{1\cdot 2\cdot 3}+\frac{1}{1\cdot 2\cdot 3\cdot 4}+\frac{1}{1\cdot 2\cdot 3\cdot 4\cdot 5}+\frac{1}{1\cdot 2\cdot 3\cdot 4\cdot 5\cdot 6}$ is 2.718055556.

(c) 2.718281828 is the calculator's internal value for e. The sum of the first five terms has two digits correct, while the sum of the first seven terms has four digits correct.

(d) One approach to finding the number of terms needed to approximate e is to keep a running sum. We already have the total for seven terms displayed, so we can add the eighth term, $\frac{1}{1\cdot 2\cdot 3\cdot 4\cdot 5\cdot 6\cdot 7}$, and compare the result with 2.718281828. Repeat this process until you get the required degree of accuracy. Using this process, we discover that 13 terms are required.

Solutions for Chapter 4 Review

Exercises

1. For a 10% increase, we multiply by 1.10 to obtain $500 \cdot 1.10 = 550$.

2. For a 100% increase, we multiply by $1 + 1.00 = 2$ to obtain $500 \cdot 2 = 1000$.

3. For a 1% decrease, we multiply by $1 - 0.01 = 0.99$ to obtain $500 \cdot 0.99 = 495$.

4. For a 42% decrease, we multiply by $1 - 0.42 = 0.58$ to obtain $500 \cdot 0.58 = 290$.

5. For a 42% increase, we multiply by 1.42 to obtain $500 \cdot 1.42 = 710$. For a 42% decrease, we multiply by $1 - 0.42 = 0.58$ to obtain $710 \cdot 0.58 = 411.8$.

6. For a 42% decrease, we multiply by $1 - 0.42 = 0.58$ to obtain $500 \cdot 0.58 = 290$. For a 42% increase, we multiply by 1.42 to obtain $290 \cdot 1.42 = 411.8$.

7. The starting value is $a = 2200$. The growth rate is $r = -3.2\% = -0.032$, so $b = 1 + r = 0.968$. We have $P = 2200(0.968)^t$.

8. In the exponential formula $f(t) = ab^t$, the parameter a represents the vertical intercept. All the formulas given have graphs that intersect the vertical axis at 10, 20, or 30. We see that formulas (a) and (b) intersect the vertical axis at 10 and correspond (in some order) to graphs I and III.

Formula (c) intersects the vertical axis at 20 and must be graph II. Formulas (d), (e), and (f) intersect the vertical axis at 30, and correspond (in some order) to graphs IV, V, and VI. The parameter b in the exponential formula $f(t) = ab^t$ gives the growth factor. Since the growth factor for formula (b), 1.5, is greater than the growth factor for formula (a), 1.2, formula (a) must correspond to graph III while formula (b) corresponds to graph I.

Formula (f) represents exponential growth (and thus must correspond to graph IV), while formulas (e) and (f) represent exponential decay. Since formula (d) decays at a more rapid rate (15% per unit time compared to 5% per unit time) than formula (e), formula (d) corresponds to graph VI and formula (e) corresponds to graph V. We have:

(a) III
(b) I
(c) II
(d) VI
(e) V
(f) IV

9. The percent of change is given by

$$\text{Percent of change} = \frac{\text{Amount of change}}{\text{Old amount}} \cdot 100\%.$$

So in these two cases,

$$\text{Percent of change from 10 to 12} = \frac{12-10}{10} \cdot 100\% = 20\%$$

$$\text{Percent of change from 100 to 102} = \frac{102-100}{100} \cdot 100\% = 2\%$$

10. To use the ratio method we must have the y-values given at equally spaced x-values, which they are not. However, some of them are spaced 1 apart, namely, 1 and 2; 4 and 5; and 8 and 9. Thus, we can use these values, and consider

$$\frac{f(2)}{f(1)}, \frac{f(5)}{f(4)}, \text{ and } \frac{f(9)}{f(8)}.$$

We find

$$\frac{f(2)}{f(1)} = \frac{f(5)}{f(4)} = \frac{f(9)}{f(8)} = \frac{1}{4}.$$

With $f(x) = ab^x$ we also have

$$\frac{f(2)}{f(1)} = \frac{f(5)}{f(4)} = \frac{f(9)}{f(8)} = b,$$

so $b = \frac{1}{4}$. Using $f(1) = 4096$ we find $4096 = ab = a\left(\frac{1}{4}\right)$, so $a = 16{,}384$. Thus, $f(x) = 16{,}384\left(\frac{1}{4}\right)^x$.

11. The average rate of change of this function appears to be constant, and thus it could be linear. Taking any pair of data points, $\Delta p/\Delta r = 3$. So the slope of this linear function should be 3. Using the form $p(r) = b + mr$, we solve for b, substituting in the point $(1, 13)$ (any point will work):

$$\begin{aligned} p(r) &= b + 3r \\ 13 &= b + 3 \cdot 1 \\ b &= 10. \end{aligned}$$

Therefore, $p(r) = 10 + 3r$.

12. The table could represent an exponential function. For every change in x of 3, there is a 10% increase in $q(x)$. We note that there is a 21% increase in $q(x)$ when $\Delta x = 6$, which is the same $(1.1^2 = 1.21)$. Using the ratio method, we find b in the form $q(x) = ab^x$.

$$\frac{q(9)}{q(6)} = \frac{110}{100} = 1.1$$

and

$$\frac{q(9)}{q(6)} = \frac{ab^9}{ab^6} = b^3.$$

Thus, $b^3 = 1.1$, and $b = \sqrt[3]{1.1} \approx 1.03228$. We solve for a by substituting $x = 6$ and $q(x) = 100$ into the equation $q(x) = a(1.03228)^x$:

$$\begin{aligned} a \cdot 1.03228^6 &= 100 \\ a &= \frac{100}{1.03228^6} \approx 82.6446. \end{aligned}$$

Thus, $q(x) = 82.6446 \cdot 1.03228^x$.

13. This cannot be linear, since $\Delta f(x)/\Delta x$ is not constant, nor can it be exponential, since between $x = 15$ and $x = 12$, we see that $f(x)$ doubles while $\Delta x = 3$. Between $x = 15$ and $x = 16$, we see that $f(x)$ doubles while $\Delta x = 1$, so the percentage increase is not constant. Thus, the function is neither.

14. This table could represent an exponential function, since for every Δt of 1, the value of $g(t)$ halves. This means that b in the form $g(t) = ab^t$ must be $\frac{1}{2} = 0.5$. We can solve for a by substituting in $(1, 512)$ (or any other point):

$$\begin{aligned} 512 &= a \cdot 0.5^1 \\ 512 &= a \cdot 0.5 \\ 1024 &= a. \end{aligned}$$

Thus, a possible formula to describe the data in the table is $g(t) = 1024 \cdot 0.5^t$.

Problems

15. In 2011, the cost of tickets will be 1.07 times their cost in 2010, (i.e. 7% greater). Thus, the price in 2011 is $1.07 \cdot \$95 = \$101.65$. The price in 2012 is then $1.07 \cdot \$101.65 = \$108.77$, and so forth. See Table 4.8.

Table 4.8

Year	2010	2011	2012	2013	2014
Cost ($)	95	101.65	108.77	116.38	124.53

16. **(a)** If gallium-67 decays at the rate of 1.48% each hour, then 98.52% remains at the end of each hour. The growth factor is 0.9852. Since the initial quantity is 100, we have $f(t) = 100(0.9852)^t$, where $f(t)$ represents the number of milligrams of gallium-67 remaining after t hours.

(b) After 24 hours, we have $t = 24$ and

$$f(24) = 100(0.9852)^{24} = 69.92 \text{ mg gallium-67 remaining.}$$

After 1 week, or $7 \cdot 24 = 168$ hours, we have

$$f(168) = 100(0.9852)^{168} = 8.17 \text{ mg gallium-67 remaining.}$$

17. **(a)** $B = B_0(1.042)^1 = B_0(1.042)$, so the effective annual rate is 4.2%.

(b) $B = B_0\left(1 + \dfrac{.042}{12}\right)^{12} \approx B_0(1.0428)$, so the effective annual rate is approximately 4.28%.

(c) $B = B_0 e^{0.042(1)} \approx B_0(1.0429)$, so the effective annual rate is approximately 4.29%.

18. If a function is linear and the x-values are equally spaced, you get from one y-value to the next by adding (or subtracting) the same amount each time. On the other hand, if the function is exponential and the x-values are evenly spaced, you get from one y-value to the next by multiplying by the same factor each time.

19. Since $h(x) = ab^x$, $h(0) = ab^0 = a(1) = a$. We are given $h(0) = 3$, so $a = 3$. If $h(x) = 3b^x$, then $h(1) = 3b^1 = 3b$. But we are told that $h(1) = 15$, so $3b = 15$ and $b = 5$. Therefore $h(x) = 3(5)^x$.

20. Since $f(x) = ab^x$, $f(3) = ab^3$ and $f(-2) = ab^{-2}$. Since we know that $f(3) = -\frac{3}{8}$ and $f(-2) = -12$, we can say

$$ab^3 = -\frac{3}{8}$$

and

$$ab^{-2} = -12.$$

Forming ratios, we have

$$\begin{aligned} \frac{ab^3}{ab^{-2}} &= \frac{-\frac{3}{8}}{-12} \\ b^5 &= -\frac{3}{8} \times -\frac{1}{12} = \frac{1}{32}. \end{aligned}$$

Since $32 = 2^5$, $\frac{1}{32} = \frac{1}{2^5} = (\frac{1}{2})^5$. This tells us that

$$b = \frac{1}{2}.$$

Thus, our formula is $f(x) = a(\frac{1}{2})^x$. Use $f(3) = a(\frac{1}{2})^3$ and $f(3) = -\frac{3}{8}$ to get

$$\begin{aligned} a(\frac{1}{2})^3 &= -\frac{3}{8} \\ a(\frac{1}{8}) &= -\frac{3}{8} \\ \frac{a}{8} &= -\frac{3}{8} \\ a &= -3. \end{aligned}$$

Therefore $f(x) = -3(\frac{1}{2})^x$.

21. Since $g(x) = ab^x$, we can say that $g(\frac{1}{2}) = ab^{1/2}$ and $g(\frac{1}{4}) = ab^{1/4}$. Since we know that $g(\frac{1}{2}) = 4$ and $g(\frac{1}{4}) = 2\sqrt{2}$, we can conclude that

$$ab^{1/2} = 4 = 2^2$$

and

$$ab^{1/4} = 2\sqrt{2} = 2 \cdot 2^{1/2} = 2^{3/2}.$$

Forming ratios, we have

$$\begin{aligned}
\frac{ab^{1/2}}{ab^{1/4}} &= \frac{2^2}{2^{3/2}} \\
b^{1/4} &= 2^{1/2} \\
(b^{1/4})^4 &= (2^{1/2})^4 \\
b &= 2^2 = 4.
\end{aligned}$$

Now we know that $g(x) = a(4)^x$, so $g(\frac{1}{2}) = a(4)^{1/2} = 2a$. Since we also know that $g(\frac{1}{2}) = 4$, we can say

$$\begin{aligned}
2a &= 4 \\
a &= 2.
\end{aligned}$$

Therefore $g(x) = 2(4)^x$.

22. Since $g(x) = ab^x$ and $g(0) = 5$, we have $a = 5$, so

$$g(x) = 5b^x.$$

Now $g(-2) = 10$ means that

$$5b^{-2} = 10.$$

Solving for b gives

$$\begin{aligned}
\frac{5}{b^2} &= 10 \\
\frac{1}{2} &= b^2 \\
b &= \frac{1}{\sqrt{2}} = 0.707.
\end{aligned}$$

So $g(x) = 5(0.707)^x$.

23. If g is exponential, then $g(x) = ab^x$, so

$$g(1.7) = ab^{1.7} = 6$$

and

$$g(2.5) = ab^{2.5} = 4.$$

We use ratios to see

$$\begin{aligned}
\frac{ab^{2.5}}{ab^{1.7}} &= \frac{4}{6} = \frac{g(2.5)}{g(1.7)} \\
b^{0.8} &= \frac{4}{6} = \frac{2}{3} \\
b &= (\frac{2}{3})^{\frac{1}{0.8}} = 0.6024.
\end{aligned}$$

Thus, our formula becomes

$$g(x) = a(0.6024)^x.$$

We can use one of our data points to solve for a. For example,

$$\begin{aligned} g(1.7) &= a(0.6024)^{1.7} = 6 \\ a &= \frac{6}{0.6024^{1.7}} \\ &\approx 14.20. \end{aligned}$$

Thus, $g(x) = 14.20(0.6024)^x$.

24. We use the exponential formula $f(x) = ab^x$. Since $f(1) = 4$ and $f(3) = d$, we have

$$ab^1 = 4 \quad \text{and} \quad ab^3 = d.$$

Dividing these two equations, we have

$$\frac{ab^3}{ab^1} = \frac{d}{4}.$$

Now we cancel and solve for b in terms of d.

$$\begin{aligned} b^2 &= \frac{d}{4} \\ b &= \frac{d^{0.5}}{2}. \end{aligned}$$

To find a in terms of d, we use the fact that $f(1) = 4$:

$$ab^1 = 4.$$

Substituting for b gives

$$\begin{aligned} a\left(\frac{d^{0.5}}{2}\right) &= 4 \\ a &= \frac{8}{d^{0.5}}. \end{aligned}$$

Thus, we have

$$\begin{aligned} f(x) &= ab^x \\ &= \left(\frac{8}{d^{0.5}}\right)\left(\frac{d^{0.5}}{2}\right)^x \\ &= \frac{8}{2^x} \cdot \frac{d^{0.5x}}{d^{0.5}} \\ &= 2^{3-x} \cdot d^{0.5(x-1)}. \end{aligned}$$

25. **(a)** If f is linear, then $f(x) = b + mx$, where m, the slope, is given by:

$$m = \frac{\Delta y}{\Delta x} = \frac{f(2) - f(-3)}{(2) - (-3)} = \frac{20 - \frac{5}{8}}{5} = \frac{\frac{155}{8}}{5} = \frac{31}{8}.$$

Using the fact that $f(2) = 20$, and substituting the known values for m, we write

$$\begin{aligned} 20 &= b + m(2) \\ 20 &= b + \left(\frac{31}{8}\right)(2) \\ 20 &= b + \frac{31}{4} \end{aligned}$$

which gives

$$b = 20 - \frac{31}{4} = \frac{49}{4}.$$

So, $f(x) = \frac{31}{8}x + \frac{49}{4}$.

(b) If f is exponential, then $f(x) = ab^x$. We know that $f(2) = ab^2$ and $f(2) = 20$. We also know that $f(-3) = ab^{-3}$ and $f(-3) = \frac{5}{8}$. So

$$\frac{f(2)}{f(-3)} = \frac{ab^2}{ab^{-3}} = \frac{20}{\frac{5}{8}}$$

$$b^5 = 20 \times \frac{8}{5} = 32$$

$$b = 2.$$

Thus, $f(x) = a(2)^x$. Solve for a by using $f(2) = 20$ and (with $b = 2$), $f(2) = a(2)^2$.

$$20 = a(2)^2$$
$$20 = 4a$$
$$a = 5.$$

Thus, $f(x) = 5(2)^x$.

26. If the function is exponential, its formula is of the form $y = ab^x$. Since $(0, 1)$ is on the graph

$$y = ab^x$$
$$1 = ab^0$$

Since $b^0 = 1$,

$$1 = a(1)$$
$$a = 1.$$

Since $(2, 100)$ is on the graph and $a = 1$,

$$y = ab^x$$
$$100 = (1)b^2$$
$$b^2 = 100$$
$$b = 10 \text{ or } b = -10$$

$b = -10$ is excluded, since b must be greater than zero. Therefore, $y = 1(10)^x$ or $y = 10^x$ is a possible formula for this function.

27. The formula for an exponential function is of the form $y = ab^x$. Since $(0, 1)$ is on the graph,

$$y = ab^x$$
$$1 = ab^0.$$

Since $b^0 = 1$,

$$1 = a(1)$$
$$a = 1.$$

Since $(4, 1/16)$ is on the graph and $a = 1$,

$$y = ab^x$$
$$\frac{1}{16} = 1(b)^4$$
$$b^4 = \frac{1}{16}.$$

Since $2 \cdot 2 \cdot 2 \cdot 2 = 16$, we know that $(1/2) \cdot (1/2) \cdot (1/2) \cdot (1/2) = 1/16$, so

$$b = \frac{1}{2}.$$

(Although $b = -1/2$ is also a solution, it is rejected since b must be greater than zero.) Therefore $y = (1/2)^x$ is a possible formula for this function.

28. Since the function is exponential, we know that $y = ab^x$. Since $(0, 1.2)$ is on the graph, we know $1.2 = ab^0$, and that $a = 1.2$. To find b, we use point $(2, 4.8)$ which gives

$$\begin{aligned}4.8 &= 1.2(b)^2\\ 4 &= b^2\\ b &= 2, \text{ since } b > 0.\end{aligned}$$

Thus, $y = 1.2(2)^x$ is a possible formula for this function.

29. The formula is of the form $y = ab^x$. Since the points $(-1, 1/15)$ and $(2, 9/5)$ are on the graph, so

$$\begin{aligned}\frac{1}{15} &= ab^{-1}\\ \frac{9}{5} &= ab^2.\end{aligned}$$

Taking the ratio of the second equation to the first we obtain

$$\begin{aligned}\frac{9/5}{1/15} &= \frac{ab^2}{ab^{-1}}\\ 27 &= b^3\\ b &= 3.\end{aligned}$$

Substituting this value of b into $\frac{1}{15} = ab^{-1}$ gives

$$\begin{aligned}\frac{1}{15} &= a(3)^{-1}\\ \frac{1}{15} &= \frac{1}{3}a\\ a &= \frac{1}{15}\cdot 3\\ a &= \frac{1}{5}.\end{aligned}$$

Therefore $y = \frac{1}{5}(3)^x$ is a possible formula for this function.

30. Since the function is exponential, we know that $y = ab^x$. The points $(-2, 45/4)$ and $(1, 10/3)$ are on the graph so,

$$\begin{aligned}\frac{45}{4} &= ab^{-2}\\ \frac{10}{3} &= ab^1\end{aligned}$$

Taking the ratio of the second equation to the first one we have

$$\frac{10/3}{45/4} = \frac{ab^1}{ab^{-2}}.$$

Since $\frac{10}{3}/\frac{45}{4} = \frac{10}{3}\cdot\frac{4}{45} = \frac{8}{27}$,

$$\frac{8}{27} = b^3.$$

Since $8 = 2^3$ and $27 = 3^3$, we know that $\frac{8}{27} = \frac{2^3}{3^3} = (\frac{2}{3})^3$, so

$$\begin{aligned}(\tfrac{2}{3})^3 &= b^3\\ b &= \frac{2}{3}.\end{aligned}$$

Substituting this value of b into the second equation gives

$$\begin{aligned}\frac{10}{3} &= a(\frac{2}{3})^1\\ \frac{2}{3}a &= \frac{10}{3}\\ a &= 5.\end{aligned}$$

Thus, $y = 5\left(\frac{2}{3}\right)^x$.

31. Since the function is exponential, we know that $y = ab^x$. Since the points $(-1, 2.5)$ and $(1, 1.6)$ are on the graph, we know that

$$\begin{aligned}2.5 &= a(b^{-1})\\ 1.6 &= a(b^1)\end{aligned}$$

Dividing the second equation by the first and canceling a gives

$$\begin{aligned}\frac{1.6}{2.5} &= \frac{a \cdot b^1}{a \cdot b^{-1}}\\ \frac{1.6}{2.5} &= b^{1-(-1)} = b^2.\end{aligned}$$

Solving for b and using the fact that $b > 0$ gives

$$b = \sqrt{\frac{1.6}{2.5}} = 0.8.$$

Substituting $b = 0.8$ in the equation $1.6 = a(b^1) = a(0.8)$ and solving for a gives

$$a = \frac{1.6}{0.8} = 2.$$

Thus, $y = 2(0.8)^x$ is a possible formula for this function.

32. Assuming f is linear, we have $f(t) = b + mt$ where $f(5) = 22$ and $f(25) = 6$. This gives

$$m = \frac{f(25) - f(5)}{25 - 5} = \frac{6 - 22}{20} = -0.8.$$

Solving for b, we have

$$\begin{aligned}f(5) &= b - (0.8)5\\ b &= f(5) + (0.8)5 = 22 + (0.8)5 = 26,\end{aligned}$$

so $f(t) = 26 - 0.8t$.

Assuming g is exponential, we have $g(t) = ab^t$, where $g(5) = 22$ and $g(25) = 6$. Using the ratio method, we have

$$\begin{aligned}\frac{ab^{25}}{ab^5} &= \frac{g(25)}{g(5)}\\ b^{20} &= \frac{6}{22}\\ b &= \left(\frac{6}{22}\right)^{1/20} = 0.9371.\end{aligned}$$

Now solve for a:

$$\begin{aligned}a(0.9371)^5 &= 22\\ a &= \frac{22}{(0.9371)^5} = 30.443.\end{aligned}$$

so $g(t) = 30.443(0.9371)^t$.

33. **(a)** If P is linear, then $P(t) = b + mt$ and

$$m = \frac{\Delta P}{\Delta t} = \frac{P(13) - P(7)}{13 - 7} = \frac{3.75 - 3.21}{13 - 7} = \frac{0.54}{6} = 0.09.$$

So $P(t) = b + 0.09t$ and $P(7) = b + 0.09(7)$. We can use this and the fact that $P(7) = 3.21$ to say that

$$\begin{aligned} 3.21 &= b + 0.09(7) \\ 3.21 &= b + 0.63 \\ 2.58 &= b. \end{aligned}$$

So $P(t) = 2.58 + 0.09t$. The slope is 0.09 million people per year. This tells us that, if its growth is linear, the country grows by $0.09(1{,}000{,}000) = 90{,}000$ people every year.

(b) If P is exponential, $P(t) = ab^t$. So

$$P(7) = ab^7 = 3.21$$

and

$$P(13) = ab^{13} = 3.75.$$

We can say that

$$\begin{aligned} \frac{P(13)}{P(7)} &= \frac{ab^{13}}{ab^7} = \frac{3.75}{3.21} \\ b^6 &= \frac{3.75}{3.21} \\ (b^6)^{1/6} &= \left(\frac{3.75}{3.21}\right)^{1/6} \\ b &= 1.026. \end{aligned}$$

Thus, $P(t) = a(1.026)^t$. To find a, note that

$$\begin{aligned} P(7) &= a(1.026)^7 = 3.21 \\ a &= \frac{3.21}{(1.026)^7} = 2.68. \end{aligned}$$

We have $P(t) = 2.68(1.026)^t$. Since $b = 1.026$ is the growth factor, the country's population grows by about 2.6% per year, assuming exponential growth.

34. **(a)** $P = 100 + 10t$.

(b) $P = 100(1.10)^t$.

(c) See Figure 4.44.

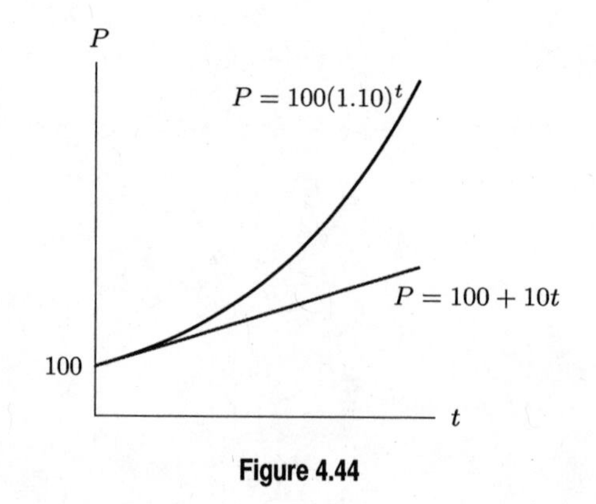

Figure 4.44

35. We have $\lim_{x\to\infty} 257(0.93)^x = 0$.

36. We have $\lim_{t\to\infty} 5.3e^{-0.12t} = 0$.

37. We have $\lim_{x\to-\infty} (15 - 5e^{3x}) = 15 - 5 \cdot 0 = 15$.

38. We have $\lim_{t\to-\infty} (21(1.2)^t + 5.1) = 21(0) + 5.1 = 5.1$.

39. We have $\lim_{x\to\infty} (7.2 - 2e^{3x}) = -\infty$.

40. We have $\lim_{x\to-\infty} (5e^{-7x} + 1.5) = \infty$.

41. Since $N = 10$ when $t = 0$, we use $N = 10b^t$ for some base b. Since $N = 20000$ when $t = 62$, we have

$$\begin{aligned} N &= 10b^t \\ 20000 &= 10b^{62} \\ b^{62} &= \frac{20000}{10} = 2000 \\ b &= (2000)^{1/62} = 1.13. \end{aligned}$$

An exponential formula for the brown tree snake population is

$$N = 10(1.13)^t.$$

The population has been growing by about 13% per year.

42. **(a)** Take 2010 to be the time $t = 0$ where t is measured in years. If V is the value, and V and t are related exponentially, then

$$V = ab^t.$$

If $V =$ 64,680 at time $t = 0$, then $a =$ 64,680, so

$$V = (64{,}680)b^t.$$

We find b by calculating another point that would be on the graph of V. If the car depreciates 42% during its first 5 years, then its value when $t = 5$ is 58% of the initial price. This is $(0.58)(\$64{,}680) = \$37{,}514.40$. So we have the data point $(5, 37514.40)$. To find b:

$$\begin{aligned} 37{,}514.4 &= (64{,}680)b^5 \\ 0.58 &= b^5 \\ b &= (0.58)^{1/5}. \end{aligned}$$

So the exponential formula relating price and time is:

$$V = (64{,}680)(0.58)^{1/5})^t \approx (64{,}680)(0.897)^t.$$

(b) If the depreciation is linear, then the value of the car at time t is

$$V = b + mt$$

where b is the value at time $t = 0$ (the year 2010). So $b =$ 64,680. We already calculated the value of the car after 5 years to be $(0.54)(\$64{,}680) = \$37{,}514.40$. Since $V =$ 37,514.4 when $t = 5$, and $b =$ 64,680, we have

$$\begin{aligned} 37{,}514.40 &= 64{,}680 + 5m, \\ -27{,}165.6 &= 5m \\ -5433.12 &= m. \end{aligned}$$

So $V = 64{,}680 - 5433.12t$.

(c) Using the exponential model, the value of the car after 4 years would be:

$$V = (64{,}680)((0.58)^{1/5})^4 \approx \$41{,}832.36.$$

Using the linear model, the value would be:

$$V = 64{,}680 - (5433.12)(4) = \$42{,}947.52.$$

So the linear model would result in a higher resale price and would therefore be preferable.

43. **(a)** To see if an exponential function fits the data well, we can look at ratios of successive terms. Giving each ratio to two decimal places, we have

$$\frac{140.8}{128.4} = 1.10, \quad \frac{158.7}{140.8} = 1.13, \quad \frac{182.1}{158.7} = 1.15, \quad \frac{207.9}{182.1} = 1.14, \quad \frac{233}{207.9} = 1.12.$$

Since the ratios are all similar, an exponential function approximates this data using a growth factor (or base) of 1.13. Since $S = 128.4$ when $t = 0$, an exponential function to model these data is $S = 128.4(1.13)^t$.

(b) The number of cell phone subscribers worldwide was growing at a rate of approximately 13% per year during this period.

(c) Using the model $S = 128.4(1.13)^t$ with $t = 7$ for 2008, we find $S \approx 302.1$. The model does not fit the 2008 data; growth has slowed.

44. For the following, let Q be the quantity after t years, and Q_0 be the initial amount.

(a) If Q doubles in size every 7 years, we have

$$\begin{aligned} 2Q_0 &= Q_0(b)^7 \\ b^7 &= 2 \\ b &= 2^{\frac{1}{7}} \approx 1.10409 \end{aligned}$$

and so Q grows by 10.409% per year.

(b) If Q triples in size every 11 years, we have

$$\begin{aligned} 3Q_0 &= Q_0 b^{11} \\ b^{11} &= 3 \\ b &= (3^{1/11}) \approx 1.10503 \end{aligned}$$

and so Q grows by 10.503% per year.

(c) If Q grows by 3% per month, we have

$$\begin{aligned} Q &= Q_0(1.03)^{12t} \quad \text{(because } 12t \text{ is number of months)} \\ &= Q_0(1.03^{12})^t \approx Q_0(1.42576)^t, \end{aligned}$$

and so the quantity grows by 42.576% per year.

(d) In t years there are $12t$ months. Thus, the number of 5-month periods in $12t$ months is $\frac{12}{5}t$. So, if Q grows by 18% every 5 months, we have

$$\begin{aligned} Q &= Q_0(1.18)^{\frac{12}{5}t} \\ &= Q_0(1.18^{12/5})^t \approx Q_0(1.48770)^t. \end{aligned}$$

Thus, Q grows by 48.770% per year.

45. A possible graph is shown in Figure 4.45.

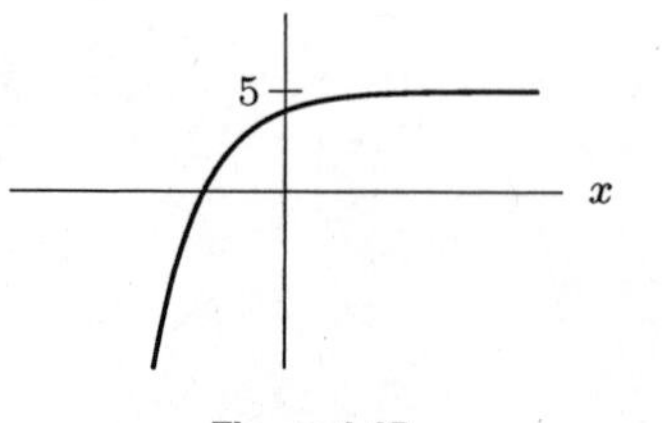

Figure 4.45

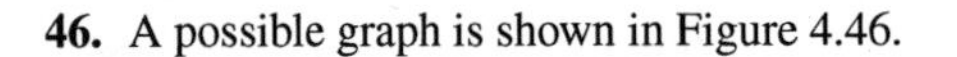

46. A possible graph is shown in Figure 4.46.

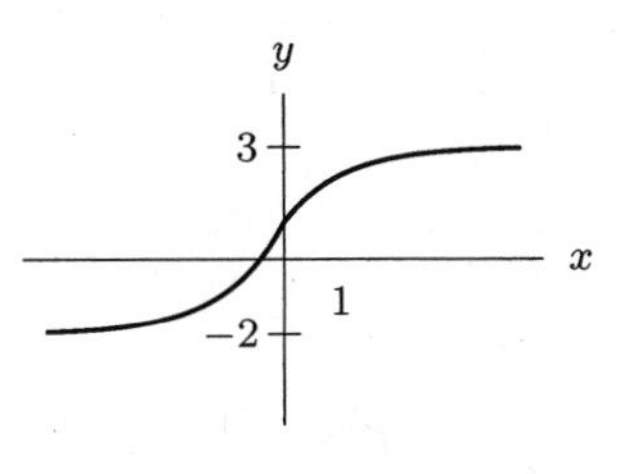

Figure 4.46

47. A possible graph is shown in Figure 4.47.

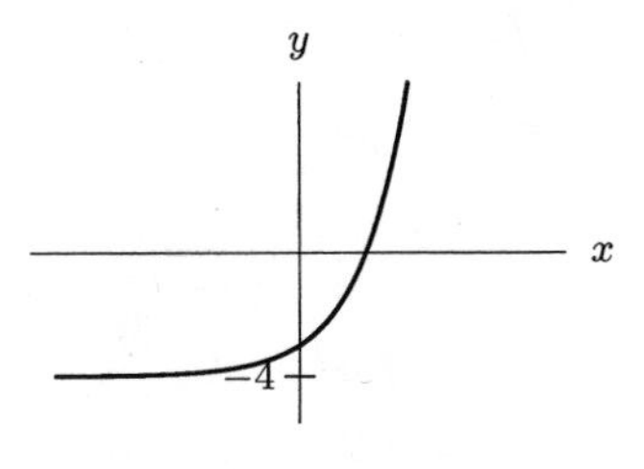

Figure 4.47

48. A possible graph is shown in Figure 4.48.

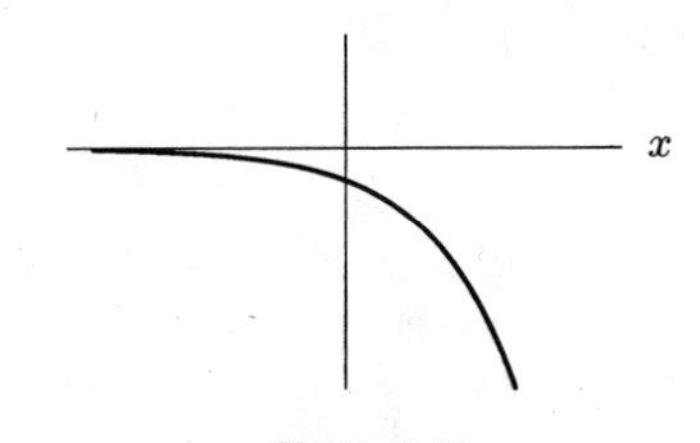

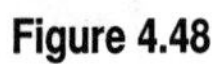

Figure 4.48

49. According to Figure 4.49, f seems to approach its horizontal asymptote, $y = 0$, faster. To convince yourself, compare values of f and g for very large values of x.

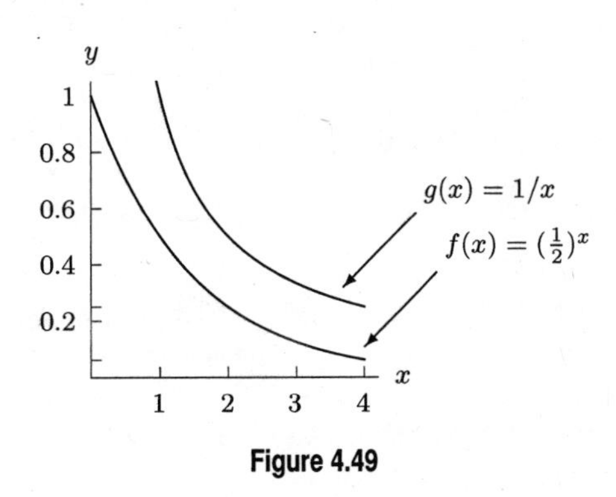

Figure 4.49

50. We see in Figure 4.50 that this function has a horizontal asymptote of $y = 8$.

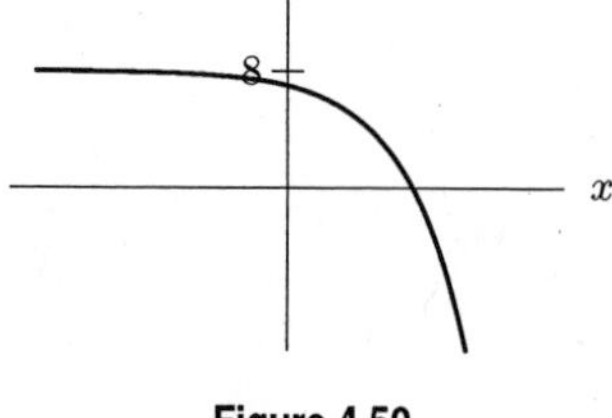

Figure 4.50

51. We see in Figure 4.51 that this function has a horizontal asymptote of $y = 2$.

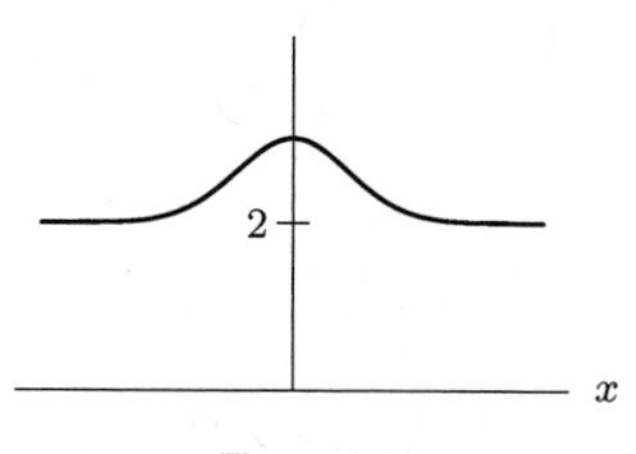

Figure 4.51

52. **(a)** The investment is initially worth \$2500 and decreases in value at a 4.34% annual rate, compounded continuously.
(b) The investment is initially worth \$4000 and grows by 0.5% twelve times a year. Notice that this means the investment earns 6% annual interest, compounded monthly,
(c) The investment is initially worth \$8000 and loses half (50%) of its value every 14 years.
(d) The investment is worth \$5000 in year $t = 10$ and grows by \$250 every year.

53. **(a)** This population initially numbers 5200, and it grows in size by 11.8% every year.
(b) This population initially numbers 4600. There are $12t$ months in t years, which means the population grows by a factor 1.01 twelve times each year, or once every month. In other words, this population grows by 1% every month.
(c) This population initially numbers 3800. It decreases by one-half (50%) every twelve years.
(d) This population initially numbers 8000. It grows at a continuous annual rate of 7.78%.
(e) Note that unlike the other functions, this is a linear function in point-slope form. It tells us that this population numbers 1675 in year $t = 30$, and the it falls by 25 members every year.

54. To match formula and graph, we keep in mind the effect on the graph of the parameters a and b in $y = ab^t$.
If $a > 0$ and $b > 1$, then the function is positive and increasing.
If $a > 0$ and $0 < b < 1$, then the function is positive and decreasing.
If $a < 0$ and $b > 1$, then the function is negative and decreasing.
If $a < 0$ and $0 < b < 1$, then the function is negative and increasing.

(a) $y = 8.3e^{-t}$, so $a = 8.3$ and $b = e^{-1}$. Since $a > 0$ and $0 < b < 1$, we want a graph which is positive and decreasing. The graph in (ii) satisfies this condition.
(b) $y = 2.5e^{t}$, so $a = 2.5$ and $b = e$. Since $a > 0$ and $b > 1$, we want a graph which is positive and increasing, such as (i).
(c) $y = -4e^{-t}$, so $a = -4$ and $b = e^{-1}$. Since $a < 0$ and $0 < b < 1$, we want a graph which is negative and increasing, such as (iii).

55. **(a)** In this account, the initial balance in the account is \$1100 and the effective yield is 5 percent each year.
(b) In this account, the initial balance in the account is \$1500 and the effective yield is approximately 5.13%, because $e^{0.05} \approx 1.0513$.

56. **(a)** Accion's interest rate $= (1150 - 1000)/1000 = 0.15 = 15\%$.
(b) Payment to loan shark $= 1000 + 22\% \cdot 1000 = \$1220$.
(c) The one from Accion, since the interest rate is lower.

57. Writing the function as

$$p(x) = \frac{7e^{6x} \cdot \sqrt{e} \cdot (2e^x)^{-1}}{10e^{4x}} = \frac{7 \cdot \sqrt{e} \cdot 2^{-1}}{10} e^{(6x-x-4x)} = \frac{7\sqrt{e}}{20} e^x,$$

we have $a = \frac{7\sqrt{e}}{20}$ and $k = 1$.

58. Here, v, the independent variable, is not in the exponent, so this can't be ab exponential function—it must be linear. We will write it in the form $r(v) = b + mv$:

$$\begin{aligned} r(v) &= vj^w - 4tj^w + kvj^w \\ &= \underbrace{-4tj^w}_{b} + \underbrace{(j^w + kj^w)}_{m} v, \end{aligned}$$

so $b = -4tj^w, m = j^w + kj^w$.

59. Here, w, the independent variable, is in the exponent, so this can't be a linear function—it must be exponential. We will write it in the form $s(w) = ab^w$:

$$\begin{aligned} s(w) &= vj^w - 4tj^w + kvj^w \\ &= \underbrace{(v - 4t + kv)}_{a} \cdot \underbrace{j^w}_{b^w}, \end{aligned}$$

so $a = v - 4t + kv, b = j$.

60. We have

$$\begin{aligned} g(1) &= \sqrt{f(1) \cdot f(1-1)} \\ &= \sqrt{f(1) \cdot f(0)} \\ \text{where} \quad f(1) &= 1000 \cdot 2^{-\frac{1}{4}-\frac{1}{2}} \\ &= 1000 \cdot 2^{-3/4} \\ &= 594.604 \\ \text{and} \quad f(0) &= 1000 \cdot 2^{-\frac{1}{4}-\frac{0}{2}} \\ &= 1000 \cdot 2^{-1/4} \\ &= 840.896 \\ \text{so} \quad g(1) &= \sqrt{594.605(840.896)} \\ &= 707.107. \end{aligned}$$

This tells us that B_1 paper is 707.107 mm wide.

61. We have $g(n) = \sqrt{f(n) \cdot f(n-1)}$ where

$$\begin{aligned} f(n) &= 1000 \cdot 2^{-\frac{1}{4}-\frac{n}{2}} \\ &= 1000 \cdot 2^{-\frac{1}{4}} \cdot 2^{-\frac{n}{2}} \\ \text{and} \quad f(n-1) &= 1000 \cdot 2^{-\frac{1}{4}-\frac{n-1}{2}} \\ &= 1000 \cdot 2^{-\frac{1}{4}} \cdot 2^{-\frac{n-1}{2}} \\ &= 1000 \cdot 2^{-\frac{1}{4}} \cdot 2^{\frac{1-n}{2}} \\ &= 1000 \cdot 2^{-\frac{1}{4}} \cdot 2^{\frac{1}{2}-\frac{n}{2}} \\ &= 1000 \cdot 2^{-\frac{1}{4}} \cdot 2^{\frac{1}{2}} \cdot 2^{-\frac{n}{2}} \\ &= 1000 \cdot 2^{-\frac{1}{4}+\frac{1}{2}} \cdot 2^{-\frac{n}{2}} \\ &= 1000 \cdot 2^{\frac{1}{4}} \cdot 2^{-\frac{n}{2}} \\ \text{so} \quad f(n) \cdot f(n-1) &= \left(1000 \cdot 2^{-\frac{1}{4}} 2^{-\frac{n}{2}}\right) \left(1000 \cdot 2^{\frac{1}{4}} \cdot 2^{-\frac{n}{2}}\right) \end{aligned}$$

$$= 1000 \cdot 1000 \cdot 2^{-\frac{1}{4}} \cdot 2^{\frac{1}{4}} \cdot 2^{-\frac{n}{2}} \cdot 2^{-\frac{n}{2}}$$
$$= 1{,}000{,}000 \cdot 2^{-n}.$$

This means

$$\begin{aligned} g(n) &= \sqrt{f(n) \cdot f(n-1)} \\ &= \sqrt{1{,}000{,}000 \cdot 2^{-n}} \\ &= \sqrt{1{,}000{,}000}\sqrt{\cdot 2^{-n}} \\ &= 1000\left(2^{-n}\right)^{0.5} \\ &= 1000 \cdot 2^{-0.5n} \\ &= 1000\left(2^{-0.5}\right)^{n} \\ &= 1000(0.7071)^{n}. \end{aligned}$$

62. Much as two points determine a straight line, two points determine a single exponential function. To see why, imagine letting (x_0, y_0) be the first point and (x_1, y_1) the second. We see that on the interval from x_0 to x_1, the growth factor is y_1/y_0. In order for the functions to both be exponential, they would grow by this same factor on equally spaced intervals from x_1 to x_2, x_2 to x_3, and so on. In short, a pair of points establishes a fixed starting position and growth factor, much as a pair of points establishes a fixed starting position and growth rate for a linear function.

In this particular case, we see that both f and g double from $y = 20$ to $y = 40$ for a growth factor of 2 on the interval from $x = 4$ to $x = 8$. It follows that, in order for both functions to be exponential, they must double from $y = 40$ to $y = 80$ on the interval from $x = 8$ to $x = 12$. However, we see from the figure that the graphs pass through two different points at $x = 12$, so at most one of them can be exponential, provided it passes through the point $(12, 80)$.

63. Answers will vary, but they should mention that $f(x)$ is increasing and $g(x)$ is decreasing, that they have the same domain, range, and horizontal asymptote. Some may see that $g(x)$ is a reflection of $f(x)$ about the y-axis whenever $b = 1/a$. Graphs might resemble the following:

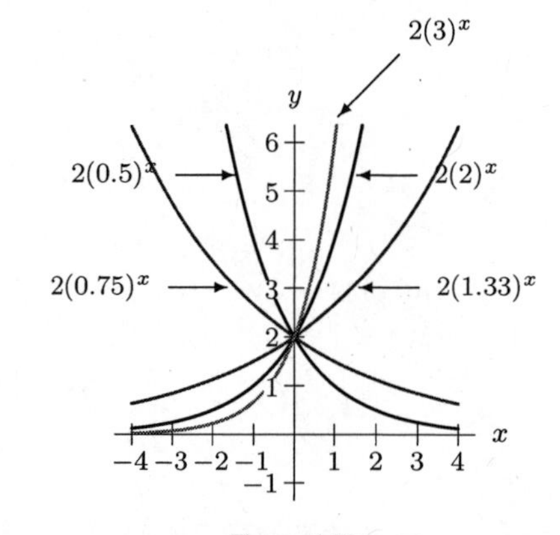

Figure 4.52

64. The y-intercept of f is greater than that of g, so $a > c$.

65. In (a), we see that the graph of g starts out below the graph of f. In (c), we see that at some point, the lower graph rises to intersect the higher graph, which tells us that g grows faster than f. This means that $d > b$.

66. In (a), we can see the y-intercept of f, which starts above g. The graph of g leaves the window in (a) to the right, not at the top, so the values of g are less than the values of f throughout the interval $0 \leq x \leq x_1$. Thus, the point of intersection is to the right of x_1, so x_1 is the smallest of these three values. The x-value of the point of intersection of the graphs in (b) is closer to x_3 than the x-value of the point of intersection in (c) is to x_2. Thus $x_3 < x_2$, so we have $x_1 < x_3 < x_2$.

67.
- f matches (ii) and (iv).
- g matches (i) and (iii)

68. **(a)** Since the human population is growing by a certain percent each year, it can be described by the formula $P = ab^t$. If t is the number of years since 1953, then a represents the population in 1953. If the growth rate is 6%, then each year the population is multiplied by the growth factor 1.06, so $b = 1.06$. Thus,

$$P = a(1.06)^t.$$

We know that in 2009 ($t = 56$) the population was 16 million, so

$$\begin{aligned} 16{,}000{,}000 &= a(1.06)^{56} \\ a &= \frac{16{,}000{,}000}{1.06^{56}} \approx 612{,}338. \end{aligned}$$

Therefore in 1953, the population of humans in Florida was about 600,000 people.

(b) In 1953 ($t = 0$), the bear population was 11,000, so $a = 11,000$. The population has been decreasing at a rate of 6% a year, so the growth rate is $100\% - 6\% = 94\%$ or 0.94. Thus, the growth function for black bears is

$$P = (11{,}000)(0.94)^t.$$

In 2009, $t = 56$, so

$$P = (11{,}000)(0.94)^{56} \approx 344.$$

(c) To find the year t when the bear population would be 100, we set P equal to 100 in the equation found in part (b) and get an equation involving t:

$$\begin{aligned} P &= (11{,}000)(0.94)^t \\ 100 &= (11000)(0.94)^t \\ \frac{100}{11000} &= (0.94)^t \\ 0.00909 &\approx (0.94)^t. \end{aligned}$$

By looking at the intersection of the graphs $P = 0.00909$ and $P = (0.94)^t$, or by trial and error, we find that $t \approx 75.967$ years. Our model predicts that in 76 years from 1953, which is the year 2029, the population of black bears would fall below 100.

69. **(a)**

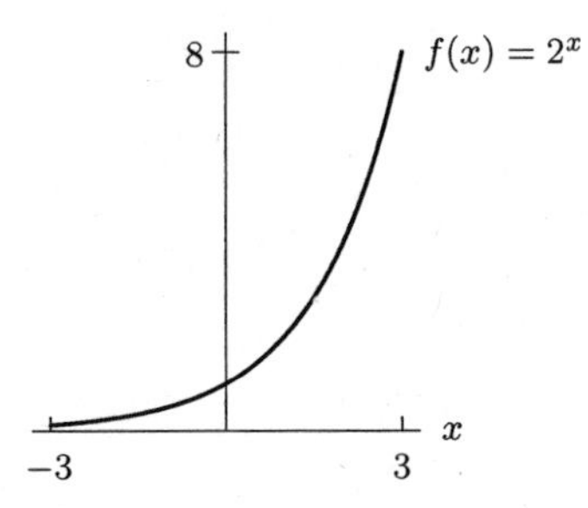

(b) The point $(0, 1)$ is on the graph. So is $(0.01, 1.00696)$. Taking $\dfrac{y_2 - y_1}{x_2 - x_1}$, we get an estimate for the slope of 0.696. We may zoom in still further to find that $(0.001, 1.000693)$ is on the graph. Using this and the point $(0, 1)$ we would get a slope of 0.693. Zooming in still further we find that the slope stabilizes at around 0.693; so, to two digits of accuracy, the slope is 0.69.

(c) Using the same method as in part (b), we find that the slope is ≈ 1.10.

(d) We might suppose that the slope of the tangent line at $x = 0$ increases as b increases. Trying a few values, we see that this is the case. Then we can find the correct b by trial and error: $b = 2.5$ has slope around 0.916, $b = 3$ has slope around 1.1, so $2.5 < b < 3$. Trying $b = 2.75$ we get a slope of 1.011, just a little too high. $b = 2.7$ gives a slope of 0.993, just a little too low. $b = 2.72$ gives a slope of 1.0006, which is as good as we can do by giving b to two decimal places. Thus $b \approx 2.72$.

In fact, the slope is exactly 1 when $b = e = 2.718\ldots$.

70. **(a)** One algorithm used by a calculator or computer gives the exponential regression function as

$$S = 16.6(1.423)^t$$

Other algorithms may give different formulas.

(b) See Figure 4.53. The exponential function appears to fit the points reasonably well.

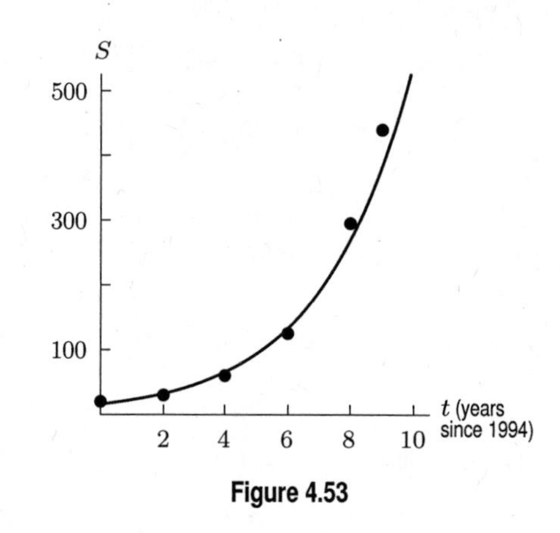

Figure 4.53

(c) Since the base of this exponential function is 1.423, sales have been increasing at a rate of about 42.3% per year during this period.

(d) Using $t = 16$, we have $S = 16.6(1.423)^{16} = 4692$ million sales.

71. We have $V = ae^{kt}$ where $a = 12{,}000$ and $k = 0.042$, so $V = 12{,}000e^{0.042t}$.

72. We have $p(20) = 300$ and $p(50) = 40$. Using the ratio method, we have

$$\begin{aligned} \frac{ab^{50}}{ab^{20}} &= \frac{p(50)}{p(20)} \\ b^{30} &= \frac{40}{300} \\ b &= \left(\frac{40}{300}\right)^{1/30} \approx 0.935. \end{aligned}$$

Now we can solve for a:

$$\begin{aligned} a\left(\left(\frac{40}{300}\right)^{1/30}\right)^{20} &= 300 \\ a &= \frac{300}{((40/300)^{1/30})^{20}} \\ &= 1149.4641, \end{aligned}$$

so $Q = 1149.4641(0.935)^t$. We can also write this in the form $Q = ae^{kt}$ where

$$k = \ln b = -0.06716.$$

73. At $x = 50$,

$$y = 5000e^{-50/40} = 1432.5240.$$

At $x = 150$,

$$y = 5000e^{-150/40} = 117.5887.$$

We have $q(50) = 1432.524$ and $q(150) = 117.5887$. This gives $y = b + mx$ where

$$m = \frac{q(150) - q(50)}{150 - 50} = \frac{117.5887 - 1432.524}{100} = -13.1.$$

Solving for b, we have

$$\begin{aligned} q(50) &= b - 13.1(50) \\ b &= q(50) + 13.1(50) \\ &= 1432.524 + 13.1(50) \\ &= 2090, \end{aligned}$$

so $y = -13.1x + 2090$.

74. The starting value is $a = 2500$, and the continuous growth rate is $k = 0.042$, so $V = 2500e^{0.042t}$.

75. We have $a = 12{,}000$. This tells us that in year $t = 0$ the population begins with 12,000 members. The constant $k = -0.122 = -12.2\%$. This tells us that the population is decreasing at a continuous annual rate of 12.2%. We have $b = e^k = e^{-0.122} = 0.8851$. This is the annual growth factor; since it is less than 1, we know the population is decreasing. The constant $r = b - 1 = -0.1149 = -11.49\%$. This tells us that the population decreases by 11.49% each year.

76. **(a)** See Figure 4.54.

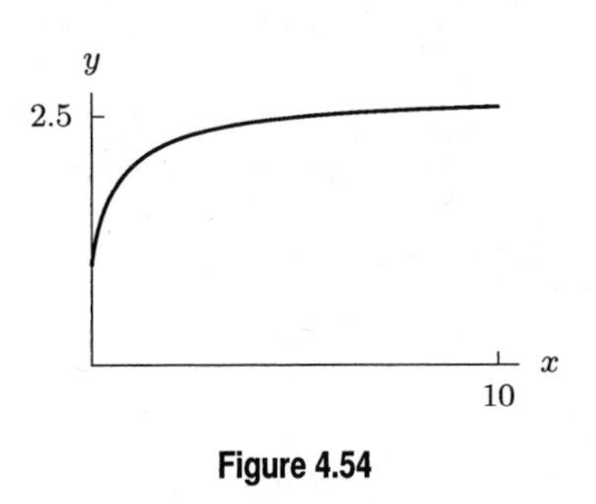

Figure 4.54

(b) The y values are increasing.
(c) The values of $(1 + 1/x)^x$ appear to approach a limiting value as x gets larger.
(d) Figures 4.55 and 4.56 show $y = (1 + 1/x)^x$ for $1 \leq x \leq 100$ and $1 \leq x \leq 1000$, respectively. We see y appears to approach a limiting value of slightly above 2.5

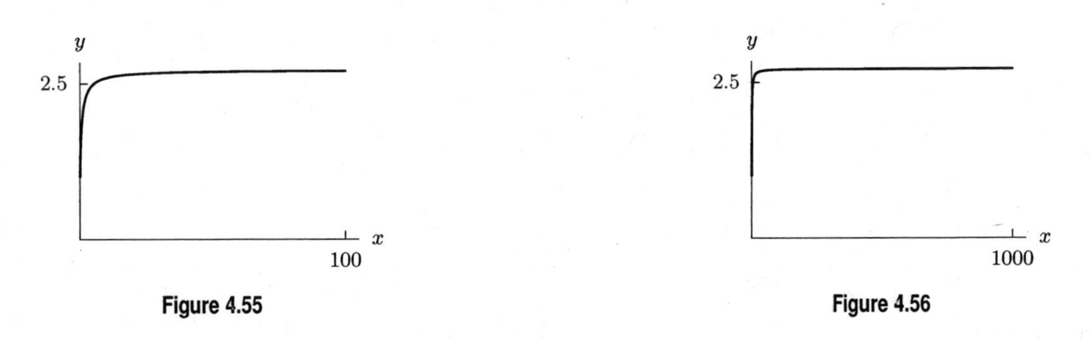

Figure 4.55 Figure 4.56

(e) The graphs of $y = (1+1/x)^x$ and $y = e$ are indistinguishable in Figure 4.57, suggesting that $(1+1/x)^x$ approaches e as x gets larger. However, a graph cannot tell us that $(1 + 1/x)^x$ approaches e exactly as x gets larger—only that $(1 + 1/x)^x$ gets very close to e.

(f) Table 4.9 shows that the value of $(1 + 1/x)^x$ agrees with $e = 2.718281828 \approx 2.7183$ for $x =$ 50,000 and above.

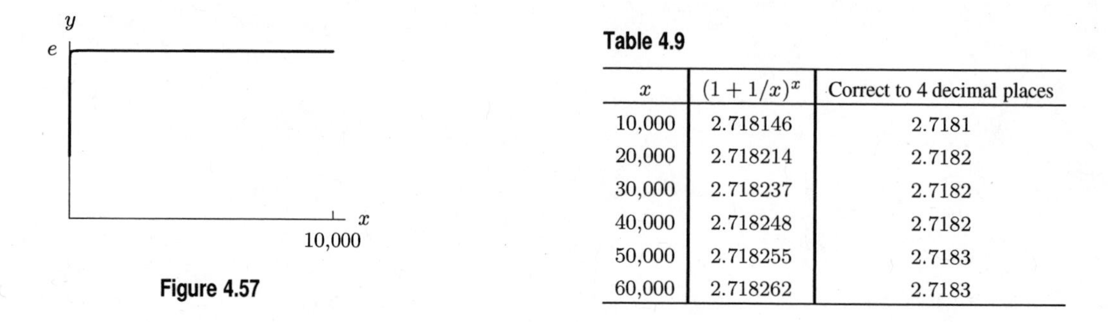

Figure 4.57

Table 4.9

x	$(1+1/x)^x$	Correct to 4 decimal places
10,000	2.718146	2.7181
20,000	2.718214	2.7182
30,000	2.718237	2.7182
40,000	2.718248	2.7182
50,000	2.718255	2.7183
60,000	2.718262	2.7183

77. **(a)** The data points are approximately as shown in Table 4.10. This results in $a \approx 15.269$ and $b \approx 1.122$, so $E(t) = 15.269(1.122)^t$.

Table 4.10

t (years)	0	1	2	3	4	5	6	7	8	9	10	11	12
$E(t)$ (thousands)	22	18	20	20	22	22	19	30	45	42	62	60	65

(b) In 1997 we have $t = 17$ so $E(17) = 15.269(1.122)^{17} \approx 108{,}066$.

(c) The model is probably not a good predictor of emigration in the year 2010 because Hong Kong was transferred to Chinese rule in 1997. Thus, conditions which affect emigration in 2010 may be markedly different than they were in the period from 1989 to 1992, for which data is given. In 2000, emigration was about 12,000.

78. **(a)** Because the time intervals are equally spaced at $t = 1$ units apart, we can estimate b by looking at the ratio of successive populations, namely,

$$\frac{0.901}{0.755} \approx 1.193, \quad \frac{1.078}{0.901} \approx 1.196, \quad \frac{1.285}{1.078} \approx 1.192.$$

These are nearly equal, the average being approximately 1.194, so $b \approx 1.194$. Using the data point $(0, 0.755)$ we estimate $a = 0.755$. Figure 4.58 shows the population data as well as the function $P = 0.755\,(1.194)^t$, where t is the number of 5-year intervals since 1975.

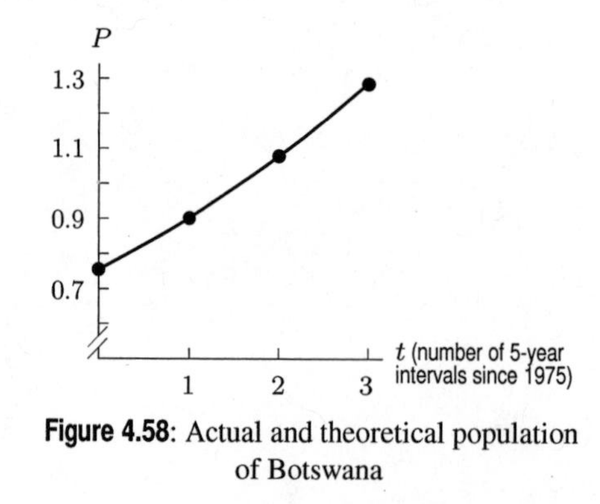

Figure 4.58: Actual and theoretical population of Botswana

(b) To find when the population doubles, we need to find the time t when $P = 2 \cdot 0.755$, that is, when

$$2 \cdot 0.755 = 0.755\,(1.194)^t,$$

$$2 = (1.194)^t.$$

By using a calculator to compute $(1.194)^t$ for $t = 1, 2, 3$, and so on, we find

$$(1.194)^4 \approx 2.03.$$

(Recall that t is measured in 5-year intervals.) This means the population doubles about every $5 \cdot 4 = 20$ years. Continuing in this way we see that

$$0.755\,(1.194)^{32} \approx 219.8.$$

Thus, according to this model, about $5 \cdot 32 = 160$ years from 1975, or 2135, the population of Botswana will exceed the 1975 population of the United States.

79. Figure 4.59 shows three different values of b, labeled b_1, b_2, b_3, and the corresponding values of t, labeled t_1, t_2, t_3. As you can see from the figure, as b is decreased, the point of intersection shifts to the left, so the t-coordinate decreases. (Note that if b is decreased to 0 or to a negative number, there is no point of intersection, and the value of t_0 is undefined.)

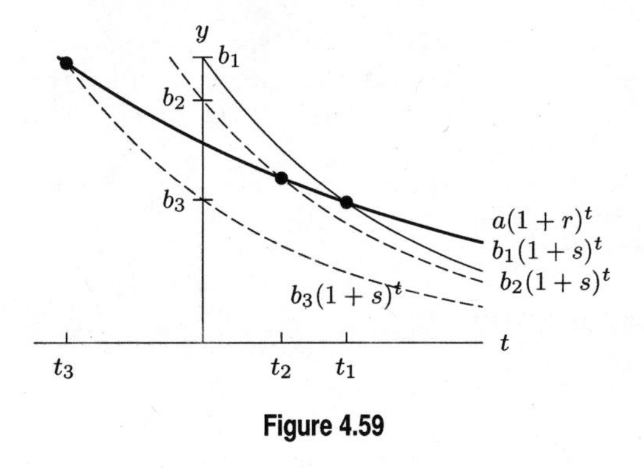

Figure 4.59

80. Figure 4.60 shows three different values of r, labeled r_1, r_2, r_3, and the corresponding values of t, labeled t_1, t_2, t_3. As you can see from the figure, as r is increased, the point of intersection shifts to the left, so the t-coordinate decreases.

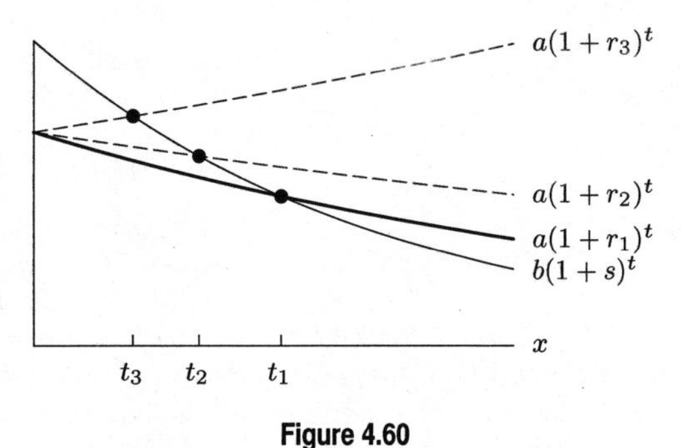

Figure 4.60

81. (a) See Figure 4.61.

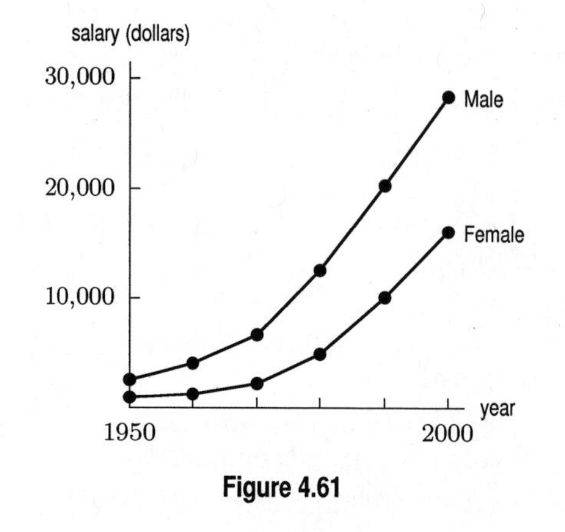

Figure 4.61

(b) For females, $W(1950) = ae^{b(0)} = a = 953$. Using trial and error, we find a value for $b = 0.062$ which approximates values in the table. So a possible formula for the median income of women is $W_F(t) = 953e^{0.062(t-1950)}$.
For males, $W(1950) = ae^{b(0)} = a = 2570$. A possible value for b is 0.051. These values give us the formula for median income of men of $W_M(t) = 2570e^{0.051(t-1950)}$.

(c) Through the year 2000, women's incomes trail behind those of men. See Figure 4.62. However Figure 4.63 shows women's incomes eventually overtake men's.

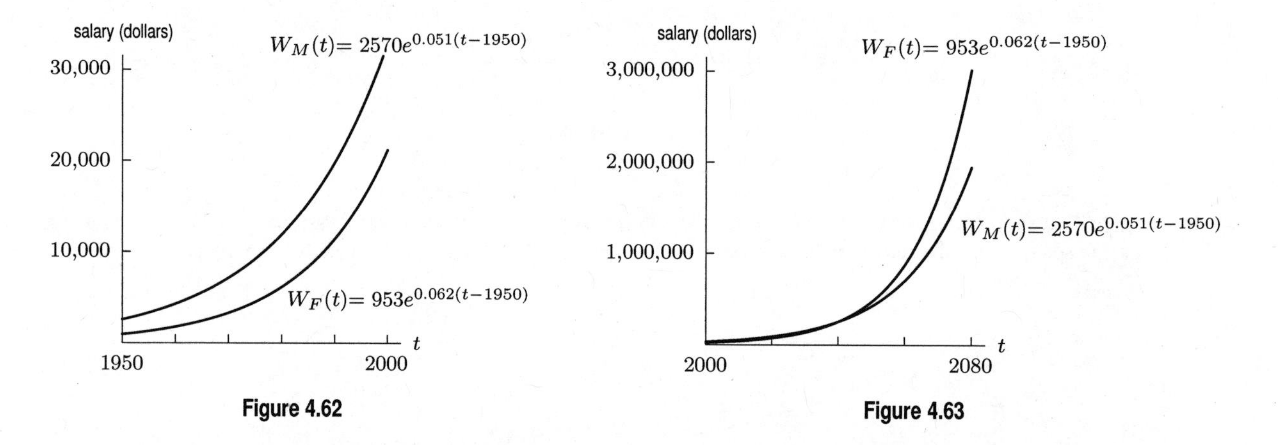

Figure 4.62 **Figure 4.63**

(d) The graph over a larger interval predicts a more promising outlook for equality in incomes. If we look carefully at each formula, $W_F(t) = 953e^{0.062(t-1950)}$ and $W_M(t) = 2570e^{0.051(t-1950)}$, we observe that women should ultimately catch up with, and even surpass, the income of men. The justification for this is the fact that the exponent in W_F is larger than the exponent in W_M, so W_F increases more quickly than W_M. See Figure 4.63.
The graph suggests that women will earn approximately the same amount as men in about the year 2060.

(e) The trends observed may not continue into the future, so the model may not apply. Thus these predictions are not reliable.

82. (a) Since f is an exponential function, we can write $f(x) = ab^x$, where a and b are constants. Since the blood alcohol level of a non-drinker is zero, we know that $f(0) = p_0$. Since $f(x) = ab^x$,

$$f(0) = ab^0 = a \cdot 1 = a,$$

and so $a = p_0$ and $f(x) = p_0 b^x$. With a BAC of 0.15, the probability of an accident is $25p_0$, so

$$f(0.15) = 25p_0.$$

From the formula we know that $f(0.15) = p_0 b^{0.15}$, so

$$\begin{aligned} p_0 b^{0.15} &= 25p_0 \\ b^{0.15} &= 25 \\ (b^{0.15})^{1/0.15} &= 25^{1/0.15} \\ b &\approx 2{,}087{,}372{,}982. \end{aligned}$$

Thus, $f(x) = p_0(2{,}087{,}372{,}982)^x$.

(b) Since

$$f(x) = p_0(2{,}087{,}372{,}982)^x,$$

using our formula, we see that $f(0.1) = p_0(2{,}087{,}372{,}982)^{0.1} \approx 8.55p_0$. This means that a legally intoxicated person is about 8.55 times as likely as a nondrinker to be involved in a single-car accident.

(c) If the probability of an accident is only three times the probability for a non-drinker, then we need to find the value of x for which

$$f(x) = 3p_0.$$

Since

$$f(x) = p_0(2{,}087{,}372{,}982)^x,$$

we have

$$p_0(2{,}087{,}372{,}982)^x = 3p_0$$

and

$$2{,}087{,}372{,}982^x = 3.$$

Using a calculator or computer, we find that $x \approx 0.051$. This is about half the BAC currently used in the legal definition.

83. The wavelength of red light is $\lambda = 650$ nm, and its absorption coefficient is $\mu(650) = 0.0034$. We have

$$\begin{aligned} T_\lambda(200) &= e^{-\mu(650)\cdot 200} \\ &= e^{-0.0034(200)} \\ &= 0.5066, \end{aligned}$$

so about 50.7% of red light is transmitted.

84. The wavelength of blue light is $\lambda = 475$ nm, and its absorption coefficient is $\mu(475) = 0.000114$. We have

$$\begin{aligned} T_\lambda(200) &= e^{-\mu(475)\cdot 200} \\ &= e^{-0.000114(200)} \\ &= 0.97746, \end{aligned}$$

so about 97.7% of blue light is transmitted.

CHECK YOUR UNDERSTANDING

1. True. If the constant rate is r then the formula is $f(t) = a \cdot (1+r)^t$. The function decreases when $0 < 1+r < 1$ and increases when $1 + r > 1$.

2. True. The exponential function formula $f(t) = a \cdot b^t$ shows that the independent variable, t, is in the exponent.

3. True. This fits the general exponential form $y = a \cdot b^t$ with $a = 40$ and $b = 1.05$.

4. False. When x increases from 1 to 2, the value of y doubles, but when x increase from 4 to 5, the value of y does not double.

5. False. The annual growth factor would be 1.04, so $S = S_0(1.04)^t$.

6. False. Evaluate $f(2) = 4(2)^2 = 16$.

7. True. The growth factor is $(2/5)$, which is less than 1, so the function values are decreasing.

8. True. Evaluate $Q = f(3) = 1000(0.5)^3 = 1000/8 = 125$. Since f is decreasing, there is no second value of t where $Q = f(t) = 3$.

9. True. The initial value means the value of Q when $t = 0$, so $Q = f(0) = a \cdot b^0 = a \cdot 1 = a$.

10. True. Using two data points, the parameters a and b can be found for the general linear function $y = b + ax$. In addition, new a and b can be calculated from the same data points for the exponential function $y = ab^x$.

11. False. The correct formula is $P = 1000(1 - 0.10)^t$ or $P = 1000(0.90)^t$.

12. True. The exponential function increases over intervals of length 1 by multiplication by a constant greater than 1, so it increases at an increasing rate. A linear function grows at a constant rate.

13. False. This is the formula of a linear function.

14. False. Suppose the initial population is 1 million, then $P = f(t) = 1(1.5)^t$ and $f(2) = 2.25$ so the population has increased from 1 million to 2.25 million, which is more than 100%.

15. True. The graph crosses the Q-axis when $t = 0$, and there $Q = ab^0 = a$.

16. False. For example, the graph of $Q = 2(0.5)^t$ falls as we read from left to right because $b = 0.5$ and $0 < b < 1$.

17. True. The irrational number $e = 2.71828\cdots$ has this as a good approximation.

18. True. As the x values get large the values of $f(x)$ gets close to k.

19. False. The graph of $y = a \cdot b^x$ is concave up when $a > 0$ but is concave down when $a < 0$.

20. False. The quantity is given by $Q = 110(1 - 0.03)^t = 110(0.97)^t$ grams.

21. True. The initial value is 200 and the growth factor is 1.04.

22. False. Since $0.2 = 20\%$, we see the given formula is for 20% continuous growth.

23. False. Since $-0.90 = -90\%$, we see the given formula is for 90% continuous decay.

24. False. Since for $t = 5$, we have $Q = 3e^{0.2 \cdot 5} = 3e = 8.15$.

25. True. Since k is the continuous growth rate and negative, Q is decreasing.

26. True. You will earn interest on the interest of each previous month.

27. False. The formula is nearly correct but 6% should be in its decimal form of 0.06:

$$B = 500\left(1 + \frac{0.06}{4}\right)^{3 \cdot 4}.$$

28. True. The formula is correct for the continuous compounding calculation.

29. True. The interest from any quarter is compounded in subsequent quarters.

30. False. The rate makes a difference. It is better to invest at 10% annually than 5% continuously.

31. False. For a 5% nominal rate, no matter how many times the interest is compounded, the earnings can never exceed the continuous rate of 5%. In twenty years, the investment cannot grow in value above $\$10{,}000e^{(0.05)20} = \$27{,}183$.

32. False. The rule of 70 estimates that it takes $70/5.5$ or about 13 years. You can also find that in 18 years your investment would be worth $\$1000e^{(0.055)18} = \$2691$.

Solutions to Skills for Chapter 4

1. $(-5)^2 = (-5)(-5) = 25$

2. $11^2 = 11 \cdot 11 = 121$

3. $10^4 = 10 \cdot 10 \cdot 10 \cdot 10 = 10{,}000$

4. $(-1)^{13} = \underbrace{(-1)(-1)\cdots(-1)}_{13 \text{ factors}} = -1$

5. $\frac{5^3}{5^2} = 5^{3-2} = 5^1 = 5$

6. $\frac{10^8}{10^5} = 10^{8-5} = 10^3 = 10 \cdot 10 \cdot 10 = 1{,}000$

7. $\frac{6^4}{6^4} = 6^{4-4} = 6^0 = 1$

8. $\sqrt{4} = 2$

9. $\sqrt{4^2} = 4$

10. $\sqrt{4^4} = \sqrt{256} = 16$

11. $\sqrt{(-4)^2} = \sqrt{16} = 4$

12. Since $\dfrac{1}{7^{-2}}$ is the same as 7^2, we obtain $7 \cdot 7$ or 49.

13. Since the base of 2 is the same in both numerator and denominator, we have $\dfrac{2^7}{2^3} = 2^{7-3} = 2^4$ or $2 \cdot 2 \cdot 2 \cdot 2$ or 16.

14. In this example, a negative base is raised to an odd power. The answer will thus be negative. Therefore $(-1)^{445} = -1$.

15. The order of operations tells us we have to square 11 first (giving 121), then take the negative. Thus $-11^2 = -\left(11^2\right) = -121$.

16. First we see that $5^0 = 1$. Then $\left(5^0\right)^3 = 1^3 = 1$.

17. The order of operations tells us to find 10^3 and then multiply by 2.1. Therefore $(2.1)\left(10^3\right) = (2.1)(1,000) = 2,100$.

18. $16^{1/2} = (2^4)^{1/2} = 2^2 = 4$

19. $16^{1/4} = (2^4)^{1/4} = 2^1 = 2$

20. $16^{3/4} = (2^4)^{3/4} = 2^3 = 8$

21. $16^{5/4} = (2^4)^{5/4} = 2^5 = 32$

22. $16^{5/2} = (2^4)^{5/2} = 2^{10} = 1024$

23. $100^{5/2} = (\sqrt{100})^5 = 10^5 = 100{,}000$

24. First we see within the radical that $(-4)^2 = 16$. Therefore $\sqrt{(-4)^2} = \sqrt{16} = 4$.

25. Exponentiation is done first, with the result that $(-1)^3 = -1$. Therefore $(-1)^3\sqrt{36} = (-1)\sqrt{36} = (-1)(6) = -6$.

26. Since the exponent is $\dfrac{1}{2}$, we can write $(0.04)^{1/2} = \sqrt{0.04} = 0.2$.

27. We can obtain the answer to $(-8)^{2/3}$ in two different ways: either by finding the cube root of $(-8)^2$ yielding $\sqrt[3]{(-8)^2} = \sqrt[3]{64} = 4$, or by finding the square of $\sqrt[3]{-8}$ yielding $\left(\sqrt[3]{-8}\right)^2 = (-2)^2 = 4$.

28. $3^{-1} = \frac{1}{3}^1 = \frac{1}{3}$

29. $3^{-3/2} = \frac{1}{3^{3/2}} = \frac{1}{(3^3)^{1/2}} = \frac{1}{(27)^{1/2}} = \frac{1}{(9 \cdot 3)^{1/2}} = \frac{1}{9^{1/2} \cdot 3^{1/2}} = \frac{1}{3\sqrt{3}}$

30. $25^{-1} = \frac{1}{25}$

31. $25^{-2} = \frac{1}{25}^2 = \frac{1}{625}$

32. For this example, we have $\left(\dfrac{1}{27}\right)^{-1/3} = (27)^{1/3} = 3$. This is because $\left(\dfrac{1}{27}\right)^{-1/3} = \left(\left(\dfrac{1}{27}\right)^{-1}\right)^{1/3} = \left(\dfrac{27}{1}\right)^{1/3} = 3$.

33. The cube root of 0.125 is 0.5. Therefore $(0.125)^{1/3} = \sqrt[3]{0.125} = 0.5$.

34. $\sqrt{x^4} = (x^4)^{1/2} = x^{4/2} = x^2$

35. $\sqrt{y^8} = (y^8)^{1/2} = y^{8/2} = y^4$

36. $\sqrt{w^8 z^4} = (w^8 z^4)^{1/2} = (w^8)^{1/2} \cdot (z^4)^{1/2} = w^{8/2} \cdot z^{4/2} = w^4 z^2$

37. $\sqrt{x^5 y^4} = (x^5 \cdot y^4)^{1/2} = x^{5/2} \cdot y^{4/2} = x^{5/2} y^2$

38. $\sqrt{49w^9} = (49w^9)^{1/2} = 49^{1/2} \cdot w^{9/2} = 7w^{9/2}$

39. $\sqrt{25x^3z^4} = (25x^3z^4)^{1/2} = 25^{1/2} \cdot x^{3/2} \cdot z^{4/2} = 5x^{3/2}z^2$

40. $\sqrt{r^2} = (r^2)^{1/2} = |r^1| = |r|$

41. $\sqrt{r^3} = (r^3)^{1/2} = r^{3/2}$

42. $\sqrt{r^4} = (r^4)^{1/2} = r^{4/2} = r^2$

43. $\sqrt{64s^7} = (64s^7)^{1/2} = 64^{1/2} \cdot s^{7/2} = 8s^{7/2}$

44.

$$\begin{aligned}\sqrt{50x^4y^6} &= 50^{1/2} \cdot (x^4)^{1/2} \cdot (y^6)^{1/2} \\ &= 50^{1/2}x^2y^3 \\ &= (25 \cdot 2)^{1/2}x^2y^3 \\ &= 25^{1/2} \cdot 2^{1/2} \cdot x^2 \cdot y^3 \\ &= 5\sqrt{2}x^2y^3\end{aligned}$$

45.

$$\begin{aligned}\sqrt{48u^{10}v^{12}y^5} &= (48)^{1/2} \cdot (u^{10})^{1/2} \cdot (v^{12})^{1/2} \cdot (y^5)^{1/2} \\ &= (16 \cdot 3)^{1/2}u^5v^6y^{5/2} \\ &= 16^{1/2} \cdot 3^{1/2} \cdot u^5v^6y^{5/2} \\ &= 4\sqrt{3}u^5v^6y^{5/2}\end{aligned}$$

46.

$$\begin{aligned}\sqrt{6s^2t^3v^5}\sqrt{6st^5v^3} &= \sqrt{36s^3t^8v^8} \\ &= (36)^{1/2} \cdot (s^3)^{1/2} \cdot (t^8)^{1/2} \cdot (v^8)^{1/2} \\ &= 6s^{3/2}t^4v^4\end{aligned}$$

47. $\left(S\sqrt{16xt^2}\right)^2 = S^2(\sqrt{16xt^2})^2 = S^2 \cdot 16xt^2 = 16S^2xt^2$

48. $\sqrt{e^{2x}} = (e^{2x})^{\frac{1}{2}} = e^{2x \cdot \frac{1}{2}} = e^x$

49.

$$(3AB)^{-1}\left(A^2B^{-1}\right)^2 = \left(3^{-1} \cdot A^{-1} \cdot B^{-1}\right)\left(A^4 \cdot B^{-2}\right) = \frac{A^4}{3^1 \cdot A^1 \cdot B^1 \cdot B^2} = \frac{A^3}{3B^3}.$$

50. Since we are multiplying numbers with the same base, e, we need only add the exponents. Thus, $e^{kt} \cdot e^3 \cdot e^1 = e^{kt+4}$.

51. First we write the radical exponentially. Therefore, $\sqrt{m+2}(2+m)^{3/2} = (m+2)^{1/2}(2+m)^{3/2}$ or $(m+2)^{1/2} \cdot (m+2)^{3/2}$. Then since the base is the same and we are multiplying, we simply add the exponents, or $(m+2)^{1/2}(m+2)^{3/2} = (m+2)^2$.

52. $\left(y^{-2}e^y\right)^2 = y^{-4} \cdot e^{2y} = \dfrac{e^{2y}}{y^4}$

53. $\dfrac{a^{n+1}3^{n+1}}{a^n3^n} = a^{n+1-n}3^{n+1-n} = a^1 \cdot 3^1 = 3a$

54. $\left(a^{-1} + b^{-1}\right)^{-1} = \dfrac{1}{\frac{1}{a} + \frac{1}{b}} = \dfrac{1}{\frac{b+a}{ab}} = \dfrac{ab}{b+a}.$

55. First we divide within the larger parentheses. Therefore,

$$\left(\frac{35(2b+1)^9}{7(2b+1)^{-1}}\right)^2 = \left(5(2b+1)^{9-(-1)}\right)^2 = \left(5(2b+1)^{10}\right)^2.$$

Then we expand to obtain

$$25(2b+1)^{20}.$$

56. $(-32)^{3/5} = (\sqrt[5]{-32})^3 = (-2)^3 = -8$

57. $-32^{3/5} = -(\sqrt[5]{32})^3 = -(2)^3 = -8$

58. $-625^{3/4} = -(\sqrt[4]{625})^3 = -(5)^3 = -125$

59. $(-625)^{3/4} = (\sqrt[4]{-625})^3$. Since $\sqrt[4]{-625}$ is not a real number, $(-625)^{3/4}$ is undefined.

60. $(-1728)^{4/3} = (\sqrt[3]{-1728})^4 = (-12)^4 = 20,736$

61. $64^{-3/2} = (\sqrt{64})^{-3} = (8)^{-3} = \left(\frac{1}{8}\right)^3 = \frac{1}{512}$

62. $-64^{3/2} = -(\sqrt{64})^3 = -(8)^3 = -512$

63. $(-64)^{3/2} = (\sqrt{-64})^3$. Since $\sqrt{-64}$ is not a real number, $(-64)^{3/2}$ is undefined.

64. $81^{5/4} = (\sqrt[4]{81})^5 = 3^5 = 243$.

65. We have

$$
\begin{aligned}
7x^4 &= 20x^2 \\
\frac{x^4}{x^2} &= \frac{20}{7} \\
x^2 &= 20/7 \\
x &= \pm(20/7)^{1/2} = \pm 1.690.
\end{aligned}
$$

66. We have

$$
\begin{aligned}
2(x+2)^3 &= 100 \\
(x+2)^3 &= 50 \\
x+2 &= (50)^{1/3} = 3.684. \\
x &= 1.684.
\end{aligned}
$$

67. The point of intersection occurs where the curves have the same x and y values. We set the two formulas equal and solve:

$$
\begin{aligned}
0.8x^4 &= 5x^2 \\
\frac{x^4}{x^2} &= \frac{5}{0.8} \\
x^2 &= 6.25 \\
x &= (6.25)^{1/2} = 2.5.
\end{aligned}
$$

The x coordinate of the point of intersection is 2.5. We use either formula to find the y-coordinate:

$$y = 5(2.5)^2 = 31.25,$$

or

$$y = 0.8(2.5)^4 = 31.25.$$

The coordinates of the point of intersection are $(2.5, 31.25)$.

68. The point of intersection occurs where the curves have the same x and y values. We set the two formulas equal and solve:

$$
\begin{aligned}
2x^3 &= 100\sqrt{x} \\
\frac{x^3}{x^{1/2}} &= \frac{100}{2} \\
x^{5/2} &= 50 \\
x &= (50)^{2/5} = 4.78176.
\end{aligned}
$$

The x coordinate of the point of intersection is about 4.782. We use either formula to find the y-coordinate:

$$y = 100\sqrt{4.78176} = 218.672,$$

or

$$y = 2(4.78176)^3 = 218.672.$$

The coordinates of the point of intersection are $(4.782, 218.672)$.

69. False

70. False

71. True

72. True

73. True

74. False

75. We have

$$\begin{aligned} 2^x &= 35 \\ &= 5 \cdot 7 \\ &= 2^r \cdot 2^s \\ &= 2^{r+s}, \end{aligned}$$

so $x = r + s$.

76. We have

$$\begin{aligned} 2^x &= 140 \\ &= 5 \cdot 7 \cdot 4 \\ &= 2^r \cdot 2^s \cdot 2^2 \\ &= 2^{r+s+2}, \end{aligned}$$

so $x = r + s + 2$.

77. We have

$$\begin{aligned} 5^x &= 32 \\ (2^a)^x &= 2^5 \\ 2^{ax} &= 2^5 \\ ax &= 5 \\ x &= \frac{5}{a}. \end{aligned}$$

78. We have

$$\begin{aligned} 7^x &= \frac{1}{8} \\ \left(2^b\right)^x &= \frac{1}{2^3} \\ 2^{bx} &= 2^{-3} \\ bx &= -3 \\ x &= \frac{-3}{b}. \end{aligned}$$

79. We have

$$\begin{aligned}
25^x &= 64 \\
\left(5^2\right)^x &= 64 \\
5^{2x} &= 64 \\
(2^a)^{2x} &= 64 \\
2^{2ax} &= 2^6 \\
2ax &= 6 \\
x &= \frac{3}{a}.
\end{aligned}$$

80. We have

$$\begin{aligned}
14^x &= 16 \\
(2\cdot 7)^x &= 16 \\
\left(2\cdot 2^b\right)^x &= 16 \\
\left(2^{b+1}\right)^x &= 16 \\
2^{(b+1)x} &= 2^4 \\
(b+1)x &= 4 \\
x &= \frac{4}{b+1}.
\end{aligned}$$

81. We have

$$\begin{aligned}
5^x &= 7 \\
(2^a)^x &= 2^b \\
2^{ax} &= 2^b \\
ax &= b \\
x &= \frac{b}{a}.
\end{aligned}$$

82.

$$\begin{aligned}
0.4^x &= 49 \\
\left(\frac{2}{5}\right)^x &= 7^2 \\
\left(2\cdot 5^{-1}\right)^x &= 7^2 \\
\left(2\left(2^a\right)^{-1}\right)^x &= 7^2 \\
\left(2\left(2^{-a}\right)\right)^x &= 7^2 \\
\left(2^{-a+1}\right)^x &= \left(2^b\right)^2 \\
2^{(1-a)x} &= 2^{2b} \\
(1-a)x &= 2b \\
x &= \frac{2b}{1-a}.
\end{aligned}$$

CHAPTER FIVE

Solutions for Section 5.1

Skill Refresher

S1. Since 1,000,000 $= 10^6$, we have $x = 6$.

S2. Since $0.01 = 10^{-2}$, we have $t = -2$.

S3. Since $\sqrt{e^3} = \left(e^3\right)^{1/2} = e^{3/2}$, we have $z = \dfrac{3}{2}$.

S4. Since $10^0 = 1$, we have $x = 0$. Similarly it follows for any constant b, if $b^x = 1$, then $x = 0$.

S5. Since e^w can never equal zero, the equation $e^w = 0$ has no solution. It similarly follows for any constant b, b^x can never equal zero.

S6. Since $\dfrac{1}{e^5} = e^{-5}$, we have $e^{3x} = e^{-5}$. Solving $3x = -5$, we have $x = \dfrac{-5}{3}$.

S7. Since $\sqrt{e^{9t}} = e^{\frac{9}{2}t}$, we have $e^{\frac{9}{2}t} = e^7$. Solving the equation $\dfrac{9}{2}t = 7$, we have $t = \dfrac{14}{9}$.

S8. We have $10^{-x} = \left(10^{-1}\right)^x = \left(\dfrac{1}{10}\right)^x$. For any constant $b > 0$, $b^x > 0$ for all x. Since $\dfrac{1}{10} > 0$, it follows $10^{-x} = \left(\dfrac{1}{10}\right)^x = -100$ has no solution.

S9. Since $\sqrt[4]{0.1} = 0.1^{1/4} = \left(10^{-1}\right)^{1/4} = 10^{-1/4}$, we have $2t = -\dfrac{1}{4}$. Solving for t we therefore have $t = -\dfrac{1}{8}$.

S10.

$$\begin{aligned}\frac{e^{3x}}{\sqrt[3]{e}} &= \sqrt[3]{e^5}\\ e^{3x} &= e^{5/3}\\ 3x &= 5/3\\ x &= 5/9.\end{aligned}$$

Exercises

1. The statement is equivalent to $19 = 10^{1.279}$.
2. The statement is equivalent to $4 = 10^{0.602}$.
3. The statement is equivalent to $26 = e^{3.258}$.
4. The statement is equivalent to $0.646 = e^{-0.437}$.
5. The statement is equivalent to $P = 10^t$.
6. The statement is equivalent to $q = e^z$.
7. The statement is equivalent to $8 = \log 100{,}000{,}000$.
8. The statement is equivalent to $-4 = \ln(0.0183)$.
9. The statement is equivalent to $v = \log \alpha$.
10. The statement is equivalent to $a = \ln b$.

11. To do these problems, keep in mind that we are looking for a power of 10. For example, log 10,000 is asking for the power of 10 which will give 10,000. Since $10^4 = 10{,}000$, we know that $\log 10{,}000 = 4$.

(a) $\log 1000 = \log 10^3 = 3 \log 10 = 3$
(b) $\log \sqrt{1000} = \log 1000^{1/2} = \frac{1}{2} \log 1000 = \frac{1}{2} \cdot 3 = 1.5$
(c) In this problem, we can use the identity $\log 10^N = N$. So $\log 10^0 = 0$. We can check this by observing that $10^0 = 1$, similar to what we saw in (a), that $\log 1 = 0$.
(d) To find the $\log \sqrt{10}$ we need to recall that $\sqrt{10} = 10^{1/2}$. Now we can use our identity and say $\log \sqrt{10} = \log 10^{1/2} = \frac{1}{2}$.
(e) Using the identity, we get $\log 10^5 = 5$.
(f) Using the identity, we get $\log 10^2 = 2$.
(g) $\log \frac{1}{\sqrt{10}} = \log 10^{-1/2} = -\frac{1}{2}$

For the last three problems, we'll use the identity $10^{\log N} = N$.

(h) $10^{\log 100} = 100$
(i) $10^{\log 1} = 1$
(j) $10^{\log 0.01} = 0.01$

12. **(a)** Since $1 = e^0$, $\ln 1 = 0$.
(b) Using the identity $\ln e^N = N$, we get $\ln e^0 = 0$. Or we could notice that $e^0 = 1$, so using part (a), $\ln e^0 = \ln 1 = 0$.
(c) Using the identity $\ln e^N = N$, we get $\ln e^5 = 5$.
(d) Recall that $\sqrt{e} = e^{1/2}$. Using the identity $\ln e^N = N$, we get $\ln \sqrt{e} = \ln e^{1/2} = \frac{1}{2}$.
(e) Using the identity $e^{\ln N} = N$, we get $e^{\ln 2} = 2$.
(f) Since $\frac{1}{\sqrt{e}} = e^{-1/2}$, $\ln \frac{1}{\sqrt{e}} = \ln e^{-1/2} = -\frac{1}{2}$

13. We are solving for an exponent, so we use logarithms. We can use either the common logarithm or the natural logarithm. Since $2^3 = 8$ and $2^4 = 16$, we know that x must be between 3 and 4. Using the log rules, we have

$$\begin{aligned} 2^x &= 11 \\ \log(2^x) &= \log(11) \\ x \log(2) &= \log(11) \\ x &= \frac{\log(11)}{\log(2)} = 3.459. \end{aligned}$$

If we had used the natural logarithm, we would have

$$x = \frac{\ln(11)}{\ln(2)} = 3.459.$$

14. We are solving for an exponent, so we use logarithms. We can use either the common logarithm or the natural logarithm. Using the log rules, we have

$$\begin{aligned} 1.45^x &= 25 \\ \log(1.45^x) &= \log(25) \\ x \log(1.45) &= \log(25) \\ x &= \frac{\log(25)}{\log(1.45)} = 8.663. \end{aligned}$$

If we had used the natural logarithm, we would have

$$x = \frac{\ln(25)}{\ln(1.45)} = 8.663.$$

15. We are solving for an exponent, so we use logarithms. Since the base is the number e, it makes the most sense to use the natural logarithm. Using the log rules, we have

$$\begin{aligned} e^{0.12x} &= 100 \\ \ln(e^{0.12x}) &= \ln(100) \\ 0.12x &= \ln(100) \\ x &= \frac{\ln(100)}{0.12} = 38.376. \end{aligned}$$

16. We begin by dividing both sides by 22 to isolate the exponent:

$$\frac{10}{22} = (0.87)^q.$$

We then take the log of both sides and use the rules of logs to solve for q:

$$\begin{aligned} \log\frac{10}{22} &= \log(0.87)^q \\ \log\frac{10}{22} &= q\log(0.87) \\ q &= \frac{\log\frac{10}{22}}{\log(0.87)} = 5.662. \end{aligned}$$

17. We begin by dividing both sides by 17 to isolate the exponent:

$$\frac{48}{17} = (2.3)^w.$$

We then take the log of both sides and use the rules of logs to solve for w:

$$\begin{aligned} \log\frac{48}{17} &= \log(2.3)^w \\ \log\frac{48}{17} &= w\log(2.3) \\ w &= \frac{\log\frac{48}{17}}{\log(2.3)} = 1.246. \end{aligned}$$

18. We take the log of both sides and use the rules of logs to solve for t:

$$\begin{aligned} \log\frac{2}{7} &= \log(0.6)^{2t} \\ \log\frac{2}{7} &= 2t\log(0.6) \\ \frac{\log\frac{2}{7}}{\log(0.6)} &= 2t \\ t &= \frac{\frac{\log\frac{2}{7}}{\log(0.6)}}{2} = 1.226. \end{aligned}$$

Problems

19. (a)

$$\begin{aligned} \log 100^x &= \log(10^2)^x \\ &= \log 10^{2x}. \end{aligned}$$

Since $\log 10^N = N$, then

$$\log 10^{2x} = 2x.$$

(b)

$$\begin{aligned}1000^{\log x} &= (10^3)^{\log x}\\ &= (10^{\log x})^3\end{aligned}$$

Since $10^{\log x} = x$ we know that

$$(10^{\log x})^3 = (x)^3 = x^3.$$

(c)

$$\begin{aligned}\log 0.001^x &= \log\left(\frac{1}{1000}\right)^x\\ &= \log(10^{-3})^x\\ &= \log 10^{-3x}\\ &= -3x.\end{aligned}$$

20. **(a)** Using the identity $\ln e^N = N$, we get $\ln e^{2x} = 2x$.
(b) Using the identity $e^{\ln N} = N$, we get $e^{\ln(3x+2)} = 3x+2$.
(c) Since $\frac{1}{e^{5x}} = e^{-5x}$, we get $\ln\left(\frac{1}{e^{5x}}\right) = \ln e^{-5x} = -5x$.
(d) Since $\sqrt{e^x} = (e^x)^{1/2} = e^{\frac{1}{2}x}$, we have $\ln\sqrt{e^x} = \ln e^{\frac{1}{2}x} = \frac{1}{2}x$.

21. **(a)** $\log(10\cdot 100) = \log 1000 = 3$
$\log 10 + \log 100 = 1 + 2 = 3$
(b) $\log(100\cdot 1000) = \log 100{,}000 = 5$
$\log 100 + \log 1000 = 2 + 3 = 5$
(c) $\log\frac{10}{100} = \log\frac{1}{10} = \log 10^{-1} = -1$
$\log 10 - \log 100 = 1 - 2 = -1$
(d) $\log\frac{100}{1000} = \log\frac{1}{10} = \log 10^{-1} = -1$
$\log 100 - \log 1000 = 2 - 3 = -1$
(e) $\log 10^2 = 2$
$2\log 10 = 2(1) = 2$
(f) $\log 10^3 = 3$
$3\log 10 = 3(1) = 3$
In each case, both answers are equal. This reflects the properties of logarithms.

22. **(a)** Patterns:

$$\begin{aligned}\log(A\cdot B) &= \log A + \log B\\ \log\frac{A}{B} &= \log A - \log B\\ \log A^B &= B\log A\end{aligned}$$

(b) Using these formulas, we rewrite the expression as follows,

$$\log\left(\frac{AB}{C}\right)^p = p\log\left(\frac{AB}{C}\right) = p(\log(AB) - \log C) = p(\log A + \log B - \log C).$$

23. **(a)** True.
(b) False. $\frac{\log A}{\log B}$ cannot be rewritten.
(c) False. $\log A \log B = \log A\cdot\log B$, not $\log A + \log B$.
(d) True.
(e) True. $\sqrt{x} = x^{1/2}$ and $\log x^{1/2} = \frac{1}{2}\log x$.
(f) False. $\sqrt{\log x} = (\log x)^{1/2}$.

24. Using properties of logs, we have

$$\begin{aligned} \log(3 \cdot 2^x) &= 8 \\ \log 3 + x \log 2 &= 8 \\ x \log 2 &= 8 - \log 3 \\ x &= \frac{8 - \log 3}{\log 2} = 24.990. \end{aligned}$$

25. Using properties of logs, we have

$$\begin{aligned} \ln(25(1.05)^x) &= 6 \\ \ln(25) + x \ln(1.05) &= 6 \\ x \ln(1.05) &= 6 - \ln(25) \\ x &= \frac{6 - \ln(25)}{\ln(1.05)} = 57.002. \end{aligned}$$

26. Using properties of logs, we have

$$\begin{aligned} \ln(ab^x) &= M \\ \ln a + x \ln b &= M \\ x \ln b &= M - \ln a \\ x &= \frac{M - \ln a}{\ln b}. \end{aligned}$$

27. Using properties of logs, we have

$$\begin{aligned} \log(MN^x) &= a \\ \log M + x \log N &= a \\ x \log N &= a - \log M \\ x &= \frac{a - \log M}{\log N}. \end{aligned}$$

28. Using properties of logs, we have

$$\begin{aligned} \ln(3x^2) &= 8 \\ \ln 3 + 2 \ln x &= 8 \\ 2 \ln x &= 8 - \ln 3 \\ \ln x &= \frac{8 - \ln 3}{2} \\ x &= e^{(8 - \ln 3)/2} = 31.522. \end{aligned}$$

Notice that to solve for x, we had to convert from an equation involving logs to an equation involving exponents in the last step.

An alternate way to solve the original equation is to begin by converting from an equation involving logs to an equation involving exponents:

$$\begin{aligned} \ln(3x^2) &= 8 \\ 3x^2 &= e^8 \\ x^2 &= \frac{e^8}{3} \\ x &= \sqrt{e^8/3} = 31.522. \end{aligned}$$

Of course, we get the same answer with both methods.

29. Using properties of logs, we have

$$\begin{aligned}
\log(5x^3) &= 2\\
\log 5 + 3\log x &= 2\\
3\log x &= 2 - \log 5\\
\log x &= \frac{2-\log 5}{3}\\
x &= 10^{(2-\log 5)/3} = 2.714.
\end{aligned}$$

Notice that to solve for x, we had to convert from an equation involving logs to an equation involving exponents in the last step.

An alternate way to solve the original equation is to begin by converting from an equation involving logs to an equation involving exponents:

$$\begin{aligned}
\log(5x^3) &= 2\\
5x^3 &= 10^2 = 100\\
x^3 &= \frac{100}{5} = 20\\
x &= (20)^{1/3} = 2.714.
\end{aligned}$$

Of course, we get the same answer with both methods.

30. **(a)** The initial value of P is 25. The growth rate is 7.5% per time unit.

(b) We see on the graph that $P = 100$ at approximately $t = 19$. We can use graphing technology to estimate t as accurately as we like.

(c) We substitute $P = 100$ and use logs to solve for t:

$$\begin{aligned}
P &= 25(1.075)^t\\
100 &= 25(1.075)^t\\
4 &= (1.075)^t\\
\log(4) &= \log(1.075^t)\\
t\log(1.075) &= \log(4)\\
t &= \frac{\log(4)}{\log(1.075)} = 19.169.
\end{aligned}$$

31. **(a)** The initial value of Q is 10. The quantity is decaying at a continuous rate of 15% per time unit.

(b) We see on the graph that $Q = 2$ at approximately $t = 10.5$. We can use graphing technology to estimate t as accurately as we like.

(c) We substitute $Q = 2$ and use the natural logarithm to solve for t:

$$\begin{aligned}
Q &= 10e^{-0.15t}\\
2 &= 10e^{-0.15t}\\
0.2 &= e^{-0.15t}\\
\ln(0.2) &= -0.15t\\
t &= \frac{\ln(0.2)}{-0.15} = 10.730.
\end{aligned}$$

32. (a)

$$\begin{aligned}
\log 6 &= \log(2\cdot 3)\\
&= \log 2 + \log 3\\
&= u + v.
\end{aligned}$$

(b)

$$\begin{aligned}\log 0.08 &= \log \frac{8}{100} \\ &= \log 8 - \log 100 \\ &= \log 2^3 - 2 \\ &= 3\log 2 - 2 \\ &= 3u - 2.\end{aligned}$$

(c)

$$\begin{aligned}\log \sqrt{\frac{3}{2}} &= \log \left(\frac{3}{2}\right)^{1/2} \\ &= \frac{1}{2}\log \frac{3}{2} \\ &= \frac{1}{2}(\log 3 - \log 2) \\ &= \frac{1}{2}(v - u).\end{aligned}$$

(d)

$$\begin{aligned}\log 5 &= \log \frac{10}{2} \\ &= \log 10 - \log 2 \\ &= 1 - u.\end{aligned}$$

33. **(a)** $\log 3 = \log \frac{15}{5} = \log 15 - \log 5$
(b) $\log 25 = \log 5^2 = 2\log 5$
(c) $\log 75 = \log(15 \cdot 5) = \log 15 + \log 5$

34. To find a formula for S, we find the points labeled (x_0, y_1) and (x_1, y_0) in Figure 5.1. We see that $x_0 = 4$ and that $y_1 = 27$. From the graph of R, we see that

$$y_0 = R(4) = 5.1403(1.1169)^4 = 8.$$

To find x_1 we use the fact that $R(x_1) = 27$:

$$\begin{aligned}5.1403(1.1169)^{x_1} &= 27 \\ 1.1169^{x_1} &= \frac{27}{5.1403} \\ x_1 &= \frac{\log(27/5.1403)}{\log 1.1169} \\ &= 15.\end{aligned}$$

We have $S(4) = 27$ and $S(15) = 8$. Using the ratio method, we have

$$\begin{aligned}\frac{ab^{15}}{ab^4} &= \frac{S(15)}{S(4)} \\ b^{11} &= \frac{8}{27} \\ b &= \left(\frac{8}{27}\right)^{1/11} \approx 0.8953.\end{aligned}$$

Now we can solve for a:

$$\begin{aligned} a(0.8953)^4 &= 27 \\ a &= \frac{27}{(0.8953)^4} \\ &\approx 42.0207. \end{aligned}$$

so $S(x) = 42.0207(0.8953)^x$.

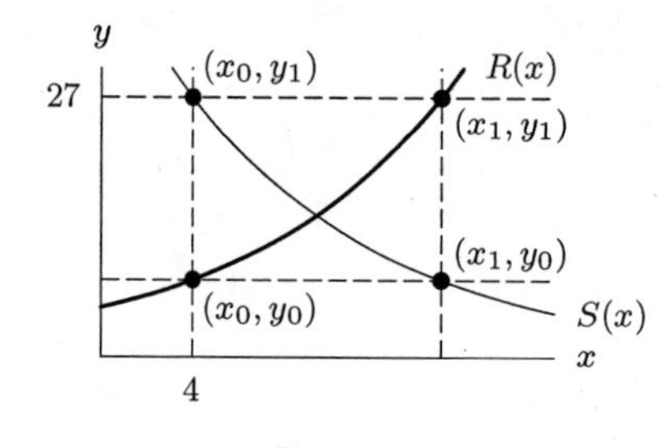

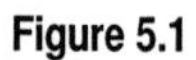
Figure 5.1

35. We begin by dividing both sides by 46 to isolate the exponent:

$$\frac{91}{46} = (1.1)^x.$$

We then take the log of both sides and use the rules of logs to solve for x:

$$\begin{aligned} \log \frac{91}{46} &= \log(1.1)^x \\ \log \frac{91}{46} &= x \log(1.1) \\ x &= \frac{\log \frac{91}{46}}{\log(1.1)}. \end{aligned}$$

36. Using the log rules

$$\begin{aligned} 84(0.74)^t &= 38 \\ 0.74^t &= \frac{38}{84} \\ \log(0.74)^t &= \log \frac{38}{84} \\ t \log 0.74 &= \log \frac{38}{84} \\ t &= \frac{\log(\frac{38}{84})}{\log 0.74} \approx 2.63 \end{aligned}$$

37.

$$\begin{aligned} e^{0.044t} &= 6 \\ \ln e^{0.044t} &= \ln 6 \\ 0.044t &= \ln 6 \\ t &= \frac{\ln 6}{0.044}. \end{aligned}$$

38.

$$
\begin{aligned}
200 \cdot 2^{t/5} &= 355 \\
2^{t/5} &= \frac{355}{200} \\
\ln 2^{t/5} &= \ln \frac{355}{200} \\
\frac{t}{5} \cdot \ln 2 &= \ln \frac{355}{200} \\
t &= \frac{5}{\ln 2} \cdot \ln \frac{355}{200} \\
&= 4.1391.
\end{aligned}
$$

Checking our answer, we see that

$$200 \cdot 2^{4.1391/5} = 355,$$

as required.

39.

$$
\begin{aligned}
e^{x+4} &= 10 \\
\ln e^{x+4} &= \ln 10 \\
x + 4 &= \ln 10 \\
x &= \ln 10 - 4
\end{aligned}
$$

40.

$$
\begin{aligned}
e^{x+5} &= 7 \cdot 2^x \\
\ln e^{x+5} &= \ln(7 \cdot 2^x) \\
x + 5 &= \ln 7 + \ln 2^x \\
x + 5 &= \ln 7 + x \ln 2 \\
x - x \ln 2 &= \ln 7 - 5 \\
x(1 - \ln 2) &= \ln 7 - 5 \\
x &= \frac{\ln 7 - 5}{1 - \ln 2}
\end{aligned}
$$

41.

$$
\begin{aligned}
0.4(\frac{1}{3})^{3x} &= 7 \cdot 2^{-x} \\
0.4(\frac{1}{3})^{3x} \cdot 2^x &= 7 \cdot 2^{-x} \cdot 2^x = 7 \\
0.4\left((\frac{1}{3})^3\right)^x \cdot 2^x &= 7 \\
\left((\frac{1}{3})^3 \cdot 2\right)^x &= \frac{7}{0.4} = 7\left(\frac{5}{2}\right) = \frac{35}{2} \\
\log(\frac{2}{27})^x &= \log \frac{35}{2} \\
x \log(\frac{2}{27}) &= \log \frac{35}{2} \\
x &= \frac{\log(\frac{35}{2})}{\log(\frac{2}{27})}.
\end{aligned}
$$

42.

$$\begin{aligned}\log_3 3^{5x+1} &= 2\\ 3^{5x+1} &= 3^2\\ 5x+1 &= 2\\ 5x &= 1\\ x &= \frac{1}{5}\end{aligned}$$

43. We have

$$\begin{aligned}400e^{0.1t} &= 500e^{0.08t}\\ \frac{e^{0.1t}}{e^{0.08t}} &= 500/400\\ e^{0.1t-0.08t} = e^{0.02t} &= 500/400\\ 0.02t &= \ln(500/400)\\ t &= \ln(500/400)/0.02.\end{aligned}$$

44. We have

$$\begin{aligned}6000\left(\frac{1}{2}\right)^{t/15} &= 1000\\ \left(\frac{1}{2}\right)^{t/15} &= \frac{1}{6}\\ \ln\left(\frac{1}{2}\right)^{t/15} &= \ln\left(\frac{1}{6}\right)\\ \frac{t}{15}\cdot\ln\left(\frac{1}{2}\right) &= -\ln 6\\ t &= -\frac{15\ln 6}{\ln 0.5}.\end{aligned}$$

45. Taking natural logs, we get

$$\begin{aligned}e^{x+4} &= 10\\ \ln e^{x+4} &= \ln 10\\ x+4 &= \ln 10\\ x &= \ln 10 - 4\end{aligned}$$

46. Taking logs and using the log rules:

$$\begin{aligned}\log(ab^x) &= \log c\\ \log a + \log b^x &= \log c\\ \log a + x\log b &= \log c\\ x\log b &= \log c - \log a\\ x &= \frac{\log c - \log a}{\log b}.\end{aligned}$$

47. Take natural logs and use the log rules

$$\begin{aligned}\ln(Pe^{kx}) &= \ln Q\\ \ln P + \ln(e^{kx}) &= \ln Q\end{aligned}$$

$$\begin{aligned}\ln P + kx &= \ln Q\\ kx &= \ln Q - \ln P\\ x &= \frac{\ln Q - \ln P}{k}.\end{aligned}$$

48.

$$\begin{aligned}58e^{4t+1} &= 30\\ e^{4t+1} &= \frac{30}{58}\\ \ln e^{4t+1} &= \ln(\frac{30}{58})\\ 4t+1 &= \ln(\frac{30}{58})\\ t &= \frac{1}{4}\left(\ln(\frac{30}{58}) - 1\right).\end{aligned}$$

49. $\log(2x+5)\cdot\log(9x^2) = 0$
In order for this product to equal zero, we know that one or both terms must be equal to zero. Thus, we will set each of the factors equal to zero to determine the values of x for which the factors will equal zero. We have

$$\begin{aligned}\log(2x+5) &= 0\\ 2x+5 &= 1\\ 2x &= -4\\ x &= -2\end{aligned}$$

or

$$\begin{aligned}\log(9x^2) &= 0\\ 9x^2 &= 1\\ x^2 &= \frac{1}{9}\\ x &= \frac{1}{3} \text{ or } x = -\frac{1}{3}.\end{aligned}$$

Thus our solutions are $x = -2, \frac{1}{3}$, or $-\frac{1}{3}$.

50. Using $\log a - \log b = \log\left(\frac{a}{b}\right)$ we can rewrite the left side of the equation to read

$$\log\left(\frac{1-x}{1+x}\right) = 2.$$

This logarithmic equation can be rewritten as

$$10^2 = \frac{1-x}{1+x},$$

since if $\log a = b$ then $10^b = a$. Multiplying both sides of the equation by $(1+x)$ yields

$$\begin{aligned}10^2(1+x) &= 1-x\\ 100+100x &= 1-x\\ 101x &= -99\\ x &= -\frac{99}{101}\end{aligned}$$

Check your answer:

$$\begin{aligned}\log\left(1-\frac{-99}{101}\right) - \log\left(1+\frac{-99}{101}\right) &= \log\left(\frac{101+99}{101}\right) - \log\left(\frac{101-99}{101}\right)\\ &= \log\left(\frac{200}{101}\right) - \log\left(\frac{2}{101}\right)\\ &= \log\left(\frac{200}{101}/\frac{2}{101}\right)\\ &= \log 100 = 2\end{aligned}$$

51. We have $\log(2x+5)\cdot\log(9x^2)=0$.
In order for this product to equal zero, we know that one or both terms must be equal to zero. Thus, we will set each of the factors equal to zero to determine the values of x for which the factors will equal zero. We have

$$\begin{aligned}\log(2x+5)&=0\\2x+5&=1\\2x&=-4\\x&=-2\end{aligned}\qquad\text{or}\qquad\begin{aligned}\log(9x^2)&=0\\9x^2&=1\\x^2&=\frac{1}{9}\\x&=\frac{1}{3}\text{ or }x=-\frac{1}{3}.\end{aligned}$$

Checking and substituting back into the original equation, we see that the three solutions work. Thus our solutions are $x=-2,\ \frac{1}{3}$, or $-\frac{1}{3}$.

52. **(a)** We combine like terms and then use properties of logs.

$$\begin{aligned}e^{2x}+e^{2x}&=1\\2e^{2x}&=1\\e^{2x}&=0.5\\2x&=\ln(0.5)\\x&=\frac{\ln(0.5)}{2}=-0.347.\end{aligned}$$

(b) We combine like terms and then use properties of logs.

$$\begin{aligned}2e^{3x}+e^{3x}&=b\\3e^{3x}&=b\\e^{3x}&=\frac{b}{3}\\3x&=\ln(b/3)\\x&=\frac{\ln(b/3)}{3}.\end{aligned}$$

53. Doubling n, we have

$$\log(2n)=\log 2+\log n=0.3010+\log n.$$

Thus, doubling a quantity increases its log by 0.3010.

54. Since $Q=r\cdot s^t$ and $q=\ln Q$, we see that

$$\begin{aligned}q&=\ln\underbrace{\left(r\cdot s^t\right)}_{Q}\\&=\ln r+\ln\left(s^t\right)\\&=\underbrace{\ln r}_{b}+\underbrace{(\ln s)}_{m}t,\end{aligned}$$

so $b=\ln r$ and $m=\ln s$.

55. We see that

$$\begin{aligned}\begin{matrix}\text{Geometric mean}\\\text{of } v \text{ and } w\end{matrix}&=\sqrt{vw}\\\log\begin{pmatrix}\text{Geometric mean}\\\text{of } v \text{ and } w\end{pmatrix}&=\log\sqrt{vw}\\&=\log\left((vw)^{1/2}\right)\end{aligned}$$

$$
\begin{aligned}
&= \frac{1}{2} \cdot \log vw \\
&= \frac{1}{2} \cdot (\log v + \log w) \\
&= \frac{\overbrace{\log v}^{p} + \overbrace{\log w}^{q}}{2} \\
&= \frac{p+q}{2} \\
&= \text{Arithmetic mean of } p \text{ and } q.
\end{aligned}
$$

Thus, letting $p = \log v$ and $q = \log w$, we see that the log of the geometric mean of v and w is the arithmetic mean of p and q.

56. We have

$$
\begin{aligned}
\ln A &= \ln 2^{2^{2^{83}}} \\
&= 2^{2^{83}} \cdot \ln 2 \\
\text{so} \quad \ln(\ln A) &= \ln\left(2^{2^{83}} \cdot \ln 2\right) \\
&= \ln\left(2^{2^{83}}\right) + \ln(\ln 2) \\
&= 2^{83} \cdot \ln 2 + \ln(\ln 2) \\
&= 6.704 \times 10^{24} \qquad \text{with a calculator} \\
\text{and} \quad \ln B &= \ln 3^{3^{3^{52}}} \\
&= 3^{3^{52}} \cdot \ln 3 \\
\text{so} \quad \ln(\ln B) &= \ln\left(3^{3^{52}} \cdot \ln 3\right) \\
&= \ln\left(3^{3^{52}}\right) + \ln(\ln 3) \\
&= 3^{52} \cdot \ln 3 + \ln(\ln 3) \\
&= 7.098 \times 10^{24}. \qquad \text{with a calculator}
\end{aligned}
$$

Thus, since $\ln(\ln B)$ is larger than $\ln(\ln A)$, we infer that $B > A$.

57. A standard graphing calculator will evaluate both of these expressions to 1. However, by taking logs, we see that

$$
\begin{aligned}
\ln A &= \ln\left(5^{3^{-47}}\right) \\
&= 3^{-47} \ln 5 \\
&= 6.0531 \times 10^{-23} \quad \text{with calculator} \\
\ln B &= \ln\left(7^{5^{-32}}\right) \\
&= 5^{-32} \ln 7 \\
&= 8.3576 \times 10^{-23} \quad \text{with calculator.}
\end{aligned}
$$

Since $\ln B > \ln A$, we see that $B > A$.

Solutions for Section 5.2

Skill Refresher

S1. Rewrite as $10^{-\log 5x} = 10^{\log(5x)^{-1}} = (5x)^{-1}$.

S2. Rewrite as $e^{-3\ln t} = e^{\ln t^{-3}} = t^{-3}$.

S3. Rewrite as $t \ln e^{t/2} = t(t/2) = t^2/2$.

S4. Rewrite as $10^{2+\log x} = 10^2 \cdot 10^{\log x} = 100x$.

S5. Taking logs of both sides we get

$$\log(4^x) = \log 9.$$

This gives

$$x \log 4 = \log 9$$

or in other words

$$x = \frac{\log 9}{\log 4} = 1.585.$$

S6. Taking natural logs of both sides we get

$$\ln(e^x) = \ln 8.$$

This gives

$$x = \ln 8 = 2.079.$$

S7. Dividing both sides by 2 gives

$$e^x = \frac{13}{2}.$$

Taking natural logs of both sides we get

$$\ln(e^x) = \ln\left(\frac{13}{2}\right).$$

This gives

$$x = \ln(13/2) = 1.872.$$

S8. Taking natural logs of both sides we get

$$\ln(e^{7x}) = \ln(5e^{3x}).$$

Since $\ln(MN) = \ln M + \ln N$, we then get

$$\begin{aligned} 7x &= \ln 5 + \ln e^{3x} \\ 7x &= \ln 5 + 3x \\ 4x &= \ln 5 \\ x &= \frac{\ln 5}{4} = 0.402. \end{aligned}$$

S9. We begin by converting to exponential form:

$$\begin{aligned} \log(2x+7) &= 2 \\ 10^{\log(2x+7)} &= 10^2 \\ 2x+7 &= 100 \\ 2x &= 93 \\ x &= \frac{93}{2}. \end{aligned}$$

S10. We first convert to exponential form and then use the properties of exponents:

$$\begin{aligned} \log(2x) &= \log(x+10) \\ 10^{\log(2x)} &= 10^{\log(x+10)} \\ 2x &= x+10 \\ x &= 10. \end{aligned}$$

Exercises

1. We want $25e^{0.053t} = 25(e^{0.053})^t = ab^t$, so we choose $a = 25$ and $b = e^{0.053} = 1.0544$. The given exponential function is equivalent to the exponential function $y = 25(1.0544)^t$. The annual percent growth rate is 5.44% and the continuous percent growth rate per year is 5.3% per year.

2. We want $100e^{-0.07t} = 100(e^{-0.07})^t = ab^t$, so we choose $a = 100$ and $b = e^{-0.07} = 0.9324$. The given exponential function is equivalent to the exponential function $y = 100(0.9324)^t$. Since $1 - 0.9324 = 0.0676$, the annual percent decay rate is 6.76% and the continuous percent decay rate per year is 7% per year.

3. We want $6000(0.85)^t = ae^{kt} = a(e^k)^t$ so we choose $a = 6000$ and we find k so that $e^k = 0.85$. Taking logs of both sides, we have $k = \ln(0.85) = -0.1625$. The given exponential function is equivalent to the exponential function $y = 6000e^{-0.1625t}$. The annual percent decay rate is 15% and the continuous percent decay rate per year is 16.25% per year.

4. We want $5(1.12)^t = ae^{kt} = a(e^k)^t$ so we choose $a = 5$ and we find k so that $e^k = 1.12$. Taking logs of both sides, we have $k = \ln(1.12) = 0.1133$. The given exponential function is equivalent to the exponential function $y = 5e^{0.1133t}$. The annual percent growth rate is 12% and the continuous percent growth rate per year is 11.33% per year.

5. The continuous percent growth rate is the value of k in the equation $Q = ae^{kt}$, which is 7.

To convert to the form $Q = ab^t$, we first say that the right sides of the two equations equal each other (since each equals Q), and then we solve for a and b. Thus, we have $ab^t = 4e^{7t}$. At $t = 0$, we can solve for a:

$$\begin{aligned} ab^0 &= 4e^{7\cdot 0} \\ a\cdot 1 &= 4\cdot 1 \\ a &= 4. \end{aligned}$$

Thus, we have $4b^t = 4e^{7t}$, and we solve for b:

$$\begin{aligned} 4b^t &= 4e^{7t} \\ b^t &= e^{7t} \\ b^t &= \left(e^7\right)^t \\ b &= e^7 \approx 1096.633. \end{aligned}$$

Therefore, the equation is $Q = 4\cdot 1096.633^t$.

6. The continuous percent growth rate is the value of k in the equation $Q = ae^{kt}$, which is 0.7.

To convert to the form $Q = ab^t$, we first say that the right sides of the two equations equal each other (since each equals Q), and then we solve for a and b. Thus, we have $ab^t = 0.3e^{0.7t}$. At $t = 0$, we can solve for a:

$$\begin{aligned} ab^0 &= 0.3e^{0.7\cdot 0} \\ a\cdot 1 &= 0.3\cdot 1 \\ a &= 0.3. \end{aligned}$$

Thus, we have $0.3b^t = 0.3e^{0.7t}$, and we solve for b:

$$\begin{aligned} 0.3b^t &= 0.3e^{0.7t} \\ b^t &= e^{0.7t} \\ b^t &= \left(e^{0.7}\right)^t \\ b &= e^{0.7} \approx 2.014. \end{aligned}$$

Therefore, the equation is $Q = 0.3\cdot 2.014^t$.

7. The continuous percent growth rate is the value of k in the equation $Q = ae^{kt}$, which is 0.03.

To convert to the form $Q = ab^t$, we first say that the right sides of the two equations equal each other (since each equals Q), and then we solve for a and b. Thus, we have $ab^t = \frac{14}{5}e^{0.03t}$. At $t = 0$, we can solve for a:

$$ab^0 = \frac{14}{5}e^{0.03\cdot 0}$$

$$\begin{aligned} a \cdot 1 &= \frac{14}{5} \cdot 1 \\ a &= \frac{14}{5}. \end{aligned}$$

Thus, we have $\frac{14}{5} b^t = \frac{14}{5} e^{0.03t}$, and we solve for b:

$$\begin{aligned} \frac{14}{5} b^t &= \frac{14}{5} e^{0.03t} \\ b^t &= e^{0.03t} \\ b^t &= \left(e^{0.03}\right)^t \\ b &= e^{0.03} \approx 1.030. \end{aligned}$$

Therefore, the equation is $Q = \frac{14}{5} \cdot 1.030^t$.

8. The continuous percent growth rate is the value of k in the equation $Q = ae^{kt}$, which is -0.02.

To convert to the form $Q = ab^t$, we first say that the right sides of the two equations equal each other (since each equals Q), and then we solve for a and b. Thus, we have $ab^t = e^{-0.02t}$. At $t = 0$, we can solve for a:

$$\begin{aligned} ab^0 &= e^{-0.02 \cdot 0} \\ a \cdot 1 &= 1 \\ a &= 1. \end{aligned}$$

Thus, we have $1b^t = e^{-0.02t}$, and we solve for b:

$$\begin{aligned} 1b^t &= e^{-0.02t} \\ b^t &= e^{-0.02t} \\ b^t &= \left(e^{-0.02}\right)^t \\ b &= e^{-0.02} \approx 0.980. \end{aligned}$$

Therefore, the equation is $Q = 1(0.980)^t$.

9. To convert to the form $Q = ae^{kt}$, we first say that the right sides of the two equations equal each other (since each equals Q), and then we solve for a and k. Thus, we have $ae^{kt} = 12(0.9)^t$. At $t = 0$, we can solve for a:

$$\begin{aligned} ae^{k \cdot 0} &= 12(0.9)^0 \\ a \cdot 1 &= 12 \cdot 1 \\ a &= 12. \end{aligned}$$

Thus, we have $12e^{kt} = 12(0.9)^t$, and we solve for k:

$$\begin{aligned} 12e^{kt} &= 12(0.9)^t \\ e^{kt} &= (0.9)^t \\ \left(e^k\right)^t &= (0.9)^t \\ e^k &= 0.9 \\ \ln e^k &= \ln 0.9 \\ k &= \ln 0.9 \approx -0.105. \end{aligned}$$

Therefore, the equation is $Q = 12e^{-0.105t}$.

10. To convert to the form $Q = ae^{kt}$, we first say that the right sides of the two equations equal each other (since each equals Q), and then we solve for a and k. Thus, we have $ae^{kt} = 16(0.487)^t$. At $t = 0$, we can solve for a:

$$\begin{aligned} ae^{k \cdot 0} &= 16(0.487)^0 \\ a \cdot 1 &= 16 \cdot 1 \\ a &= 16. \end{aligned}$$

Thus, we have $16e^{kt} = 16(0.487)^t$, and we solve for k:

$$\begin{aligned} 16e^{kt} &= 16(0.487)^t \\ e^{kt} &= (0.487)^t \\ \left(e^k\right)^t &= (0.487)^t \\ e^k &= 0.487 \\ \ln e^k &= \ln 0.487 \\ k &= \ln 0.487 \approx -0.719. \end{aligned}$$

Therefore, the equation is $Q = 16e^{-0.719t}$.

11. To convert to the form $Q = ae^{kt}$, we first say that the right sides of the two equations equal each other (since each equals Q), and then we solve for a and k. Thus, we have $ae^{kt} = 14(0.862)^{1.4t}$. At $t = 0$, we can solve for a:

$$\begin{aligned} ae^{k\cdot 0} &= 14(0.862)^0 \\ a \cdot 1 &= 14 \cdot 1 \\ a &= 14. \end{aligned}$$

Thus, we have $14e^{kt} = 14(0.862)^{1.4t}$, and we solve for k:

$$\begin{aligned} 14e^{kt} &= 14(0.862)^{1.4t} \\ e^{kt} &= \left(0.862^{1.4}\right)^t \\ \left(e^k\right)^t &= (0.812)^t \\ e^k &= 0.812 \\ \ln e^k &= \ln 0.812 \\ k &= -0.208. \end{aligned}$$

Therefore, the equation is $Q = 14e^{-0.208t}$.

12. To convert to the form $Q = ae^{kt}$, we first say that the right sides of the two equations equal each other (since each equals Q), and then we solve for a and k. Thus, we have $ae^{kt} = 721(0.98)^{0.7t}$. At $t = 0$, we can solve for a:

$$\begin{aligned} ae^{k\cdot 0} &= 721(0.98)^0 \\ a \cdot 1 &= 721 \cdot 1 \\ a &= 721. \end{aligned}$$

Thus, we have $721e^{kt} = 721(0.98)^{0.7t}$, and we solve for k:

$$\begin{aligned} 721e^{kt} &= 721(0.98)^{0.7t} \\ e^{kt} &= \left(0.98^{0.7}\right)^t \\ \left(e^k\right)^t &= (0.986)^t \\ e^k &= 0.986 \\ \ln e^k &= \ln 0.986 \\ k &= -0.0141. \end{aligned}$$

Therefore, the equation is $Q = 721e^{-0.0141t}$.

13. We have $a = 230$, $b = 1.182$, $r = b - 1 = 18.2\%$, and $k = \ln b = 0.1672 = 16.72\%$.

14. We have $a = 0.181$, $b = e^{0.775} = 2.1706$, $r = b - 1 = 1.1706 = 117.06\%$, and $k = 0.775 = 77.5\%$.

15. We have $a = 0.81$, $b = 2$, $r = b - 1 = 1 = 100\%$, and $k = \ln 2 = 0.6931 = 69.31\%$.

16. Writing this as $Q = 5 \cdot (2^{\frac{1}{8}})^t$, we have $a = 5$, $b = 2^{\frac{1}{8}} = 1.0905$, $r = b - 1 = 0.0905 = 9.05\%$, and $k = \ln b = 0.0866 = 8.66\%$.

17. Writing this as $Q = 12.1(10^{-0.11})^t$, we have $a = 12.1$, $b = 10^{-0.11} = 0.7762$, $r = b - 1 = -22.38\%$, and $k = \ln b = -25.32\%$.

18. We can rewrite this as

$$\begin{aligned} Q &= 40e^{t/12-5/12} \\ &= 40e^{-5/12}\left(e^{1/12}\right)^t. \end{aligned}$$

We have $a = 40e^{-5/12} = 26.3696$, $b = e^{1/12} = 1.0869$, $r = b - 1 = 8.69\%$, and $k = 1/12 = 8.333\%$.

19. We can use exponent rules to place this in the form ae^{kt}:

$$\begin{aligned} 2e^{(1-3t/4)} &= 2e^1e^{-3t/4} \\ &= (2e)e^{-\frac{3}{4}t}, \end{aligned}$$

so $a = 2e = 5.4366$ and $k = -3/4$. To find b and r, we have

$$\begin{aligned} b &= e^k = e^{-3/4} = 0.4724 \\ r &= b - 1 = -0.5276 = -52.76\%. \end{aligned}$$

20. We can use exponent rules to place this in the form ab^t:

$$\begin{aligned} 2^{-(t-5)/3} &= 2^{-(1/3)(t-5)} \\ &= 2^{5/3-(1/3)t} \\ &= 2^{5/3} \cdot 2^{-(1/3)t} \\ &= 2^{5/3} \cdot \left(2^{-1/3}\right)^t, \end{aligned}$$

so $a = 2^{5/3} = 3.1748$ and $b = 2^{(-1/3)} = 0.7937$. To find k and r, we use the fact that:

$$\begin{aligned} r &= b - 1 = -0.2063 = -20.63\% \\ k &= \ln b = -0.2310 = -23.10\%. \end{aligned}$$

Problems

21. Let t be the doubling time, then the population is $2P_0$ at time t, so

$$\begin{aligned} 2P_0 &= P_0e^{0.2t} \\ 2 &= e^{0.2t} \\ 0.2t &= \ln 2 \\ t &= \frac{\ln 2}{0.2} \approx 3.466. \end{aligned}$$

22. Since the growth factor is $1.26 = 1 + 0.26$, the formula for the city's population, with an initial population of a and time t in years, is

$$P = a(1.26)^t.$$

The population doubles for the first time when $P = 2a$. Thus, we solve for t after setting P equal to $2a$ to give us the doubling time:

$$\begin{aligned}
2a &= a(1.26)^t \\
2 &= (1.26)^t \\
\log 2 &= \log(1.26)^t \\
\log 2 &= t\log(1.26) \\
t &= \frac{\log 2}{\log(1.26)} = 2.999.
\end{aligned}$$

So the doubling time is about 3 years.

23. Since the growth factor is $1.027 = 1 + 0.027$, the formula for the bank account balance, with an initial balance of a and time t in years, is

$$B = a(1.027)^t.$$

The balance doubles for the first time when $B = 2a$. Thus, we solve for t after putting B equal to $2a$ to give us the doubling time:

$$\begin{aligned}
2a &= a(1.027)^t \\
2 &= (1.027)^t \\
\log 2 &= \log(1.027)^t \\
\log 2 &= t\log(1.027) \\
t &= \frac{\log 2}{\log(1.027)} = 26.017.
\end{aligned}$$

So the doubling time is about 26 years.

24. Since the growth factor is 1.12, the formula for the company's profits, Π, with an initial annual profit of a and time t in years, is

$$\Pi = a(1.12)^t.$$

The annual profit doubles for the first time when $\Pi = 2a$. Thus, we solve for t after putting Π equal to $2a$ to give us the doubling time:

$$\begin{aligned}
2a &= a(1.12)^t \\
2 &= (1.12)^t \\
\log 2 &= \log(1.12)^t \\
\log 2 &= t\log(1.12) \\
t &= \frac{\log 2}{\log(1.12)} = 6.116.
\end{aligned}$$

So the doubling time is about 6.1 years.

25. The growth factor for Tritium should be $1 - 0.05471 = 0.94529$, since it is decaying by 5.471% per year. Therefore, the decay equation starting with a quantity of a should be:

$$Q = a(0.94529)^t,$$

where Q is quantity remaining and t is time in years. The half life will be the value of t for which Q is $a/2$, or half of the initial quantity a. Thus, we solve the equation for $Q = a/2$:

$$\begin{aligned}
\frac{a}{2} &= a(0.94529)^t \\
\frac{1}{2} &= (0.94529)^t \\
\log(1/2) &= \log(0.94529)^t
\end{aligned}$$

$$\log(1/2) = t\log(0.94529)$$
$$t = \frac{\log(1/2)}{\log(0.94529)} = 12.320.$$

So the half-life is about 12.3 years.

26. The growth factor for Einsteinium-253 should be $1 - 0.03406 = 0.96594$, since it is decaying by 3.406% per day. Therefore, the decay equation starting with a quantity of a should be:

$$Q = a(0.96594)^t,$$

where Q is quantity remaining and t is time in days. The half life will be the value of t for which Q is $a/2$, or half of the initial quantity a. Thus, we solve the equation for $Q = a/2$:

$$\begin{aligned}
\frac{a}{2} &= a(0.96594)^t \\
\frac{1}{2} &= (0.96594)^t \\
\log(1/2) &= \log(0.96594)^t \\
\log(1/2) &= t\log(0.96594) \\
t &= \frac{\log(1/2)}{\log(0.96594)} = 20.002.
\end{aligned}$$

So the half-life is about 20 days.

27. Let $Q(t)$ be the mass of the substance at time t, and Q_0 be the initial mass of the substance. Since the substance is decaying at a continuous rate, we know that $Q(t) = Q_0e^{kt}$ where $k = -0.11$ (This is an 11% decay). So $Q(t) = Q_0e^{-0.11t}$. We want to know when $Q(t) = \frac{1}{2}Q_0$.

$$\begin{aligned}
Q_0e^{-0.11t} &= \frac{1}{2}Q_0 \\
e^{-0.11t} &= \frac{1}{2} \\
\ln e^{-0.11t} &= \ln\left(\frac{1}{2}\right) \\
-0.11t &= \ln\left(\frac{1}{2}\right) \\
t &= \frac{\ln\frac{1}{2}}{-0.11} \approx 6.301
\end{aligned}$$

So the half-life is 6.301 minutes.

28. Since the formula for finding the value, $P(t)$, of an \$800 investment after t years at 4% interest compounded annually is $P(t) = 800(1.04)^t$ and we want to find the value of t when $P(t) = 2{,}000$, we must solve:

$$\begin{aligned}
800(1.04)^t &= 2000 \\
1.04^t &= \frac{2000}{800} = \frac{20}{8} = \frac{5}{2} \\
\log 1.04^t &= \log\frac{5}{2} \\
t\log 1.04 &= \log\frac{5}{2} \\
t &= \frac{\log(5/2)}{\log 1.04} \approx 23.362 \text{ years.}
\end{aligned}$$

So it will take about 23.362 years for the \$800 to grow to \$2,000.

29. **(a)** We find k so that $ae^{kt} = a(1.08)^t$. We want k so that $e^k = 1.08$. Taking logs of both sides, we have $k = \ln(1.08) = 0.0770$. An interest rate of 7.70%, compounded continuously, is equivalent to an interest rate of 8%, compounded annually.

(b) We find b so that $ab^t = ae^{0.06t}$. We take $b = e^{0.06} = 1.0618$. An interest rate of 6.18%, compounded annually, is equivalent to an interest rate of 6%, compounded continuously.

30. **(a)** To find the annual growth rate, we need to find a formula which describes the population, $P(t)$, in terms of the initial population, a, and the annual growth factor, b. In this case, we know that $a = 11{,}000$ and $P(3) = 13{,}000$. But $P(3) = ab^3 = 11{,}000b^3$, so

$$\begin{aligned} 13000 &= 11000b^3 \\ b^3 &= \frac{13000}{11000} \\ b &= \left(\frac{13}{11}\right)^{\frac{1}{3}} \approx 1.05726. \end{aligned}$$

Since b is the growth factor, we know that, each year, the population is about 105.726% of what it had been the previous year, so it is growing at the rate of 5.726% each year.

(b) To find the continuous growth rate, we need a formula of the form $P(t) = ae^{kt}$ where $P(t)$ is the population after t years, a is the initial population, and k is the rate we are trying to determine. We know that $a = 11{,}000$ and, inthis case, that $P(3) = 11{,}000e^{3k} = 13{,}000$. Therefore,

$$\begin{aligned} e^{3k} &= \frac{13000}{11000} \\ \ln e^{3k} &= \ln\left(\frac{13}{11}\right) \\ 3k &= \ln\left(\frac{13}{11}\right) \qquad \text{(because } \ln e^{3k} = 3k\text{)} \\ k &= \frac{1}{3}\ln\left(\frac{13}{11}\right) \approx 0.05568. \end{aligned}$$

So our continuous annual growth rate is 5.568%.

(c) The annual growth rate, 5.726%, describes the actual percent increase in one year. The continuous annual growth rate, 5.568%, describes the percentage increase of the population at any given instant, and so should be a smaller number.

31. We have

$$\begin{aligned} \text{First investment} &= 5000(1.072)^t \\ \text{Second investment} &= 8000(1.054)^t \\ \text{so we solve}\quad 5000(1.072)^t &= 8000(1.054)^t \\ \frac{1.072^t}{1.054^t} &= \frac{8000}{5000} && \text{divide} \\ \left(\frac{1.072}{1.054}\right)^t &= 1.6 && \text{exponent rule} \\ t\ln\left(\frac{1.072}{1.054}\right) &= \ln 1.6 && \text{take logs} \\ t &= \frac{\ln 1.6}{\ln\left(\frac{1.072}{1.054}\right)} \\ &= 27.756, \end{aligned}$$

so it will take almost 28 years.

32.

$$\begin{aligned} \text{Value of first investment} &= 9000e^{0.056t} && a = 9000,\ k = 5.6\% \\ \text{Value of second investment} &= 4000e^{0.083t} && a = 4000,\ k = 8.3\% \end{aligned}$$

$$\begin{aligned}
\text{so}\quad 4000e^{0.083t} &= 9000e^{0.056t} && \text{set values equal}\\
\frac{e^{0.083t}}{e^{0.056t}} &= \frac{9000}{4000} && \text{divide}\\
e^{0.083t-0.056t} &= 2.25 && \text{simplify}\\
e^{0.027t} &= 2.25\\
0.027t &= \ln 2.25 && \text{take logs}\\
t &= \frac{\ln 2.25}{0.027}\\
&= 30.034,
\end{aligned}$$

so it will take almost exactly 30 years. To check our answer, we see that

$$\begin{aligned}
9000e^{0.056(30.034)} &= 48{,}383.26\\
4000e^{0.083(30.034)} &= 48{,}383.26. \quad \text{values are equal}
\end{aligned}$$

33. **(a)** Let $P(t) = P_0 b^t$ describe our population at the end of t years. Since P_0 is the initial population, and the population doubles every 15 years, we know that, at the end of 15 years, our population will be $2P_0$. But at the end of 15 years, our population is $P(15) = P_0 b^{15}$. Thus

$$\begin{aligned}
P_0 b^{15} &= 2P_0\\
b^{15} &= 2\\
b &= 2^{\frac{1}{15}} \approx 1.04729
\end{aligned}$$

Since b is our growth factor, the population is, yearly, 104.729% of what it had been the previous year. Thus it is growing by 4.729% per year.

(b) Writing our formula as $P(t) = P_0 e^{kt}$, we have $P(15) = P_0 e^{15k}$. But we already know that $P(15) = 2P_0$. Therefore,

$$\begin{aligned}
P_0 e^{15k} &= 2P_0\\
e^{15k} &= 2\\
\ln e^{15k} &= \ln 2\\
15k \ln e &= \ln 2\\
15k &= \ln 2\\
k &= \frac{\ln 2}{15} \approx 0.04621.
\end{aligned}$$

This tells us that we have a continuous annual growth rate of 4.621%.

34. We have $P = ab^t$ where $a = 5.2$ and $b = 1.031$. We want to find k such that

$$P = 5.2e^{kt} = 5.2(1.031)^t,$$

so

$$e^k = 1.031.$$

Thus, the continuous growth rate is $k = \ln 1.031 \approx 0.03053$, or 3.053% per year.

35. We let N represent the amount of nicotine in the body t hours after it was ingested, and we let N_0 represent the initial amount of nicotine. The half-life tells us that at $t = 2$ the quantity of nicotine is $0.5N_0$. We substitute this point to find k:

$$\begin{aligned}
N &= N_0 e^{kt}\\
0.5N_0 &= N_0 e^{k(2)}\\
0.5 &= e^{2k}\\
\ln(0.5) &= 2k\\
k &= \frac{\ln(0.5)}{2} = -0.347.
\end{aligned}$$

The continuous decay rate of nicotine is -34.7% per hour.

36. If 17% of the substance decays, then 83% of the original amount of substance, P_0, remains after 5 hours. So $P(5) = 0.83P_0$. If we use the formula $P(t) = P_0e^{kt}$, then

$$P(5) = P_0e^{5t}.$$

But $P(5) = 0.83P_0$, so

$$\begin{aligned} 0.83P_0 &= P_0e^{k(5)} \\ 0.83 &= e^{5k} \\ \ln 0.83 &= \ln e^{5k} \\ \ln 0.83 &= 5k \\ k &= \frac{\ln 0.83}{5} \approx -0.037266. \end{aligned}$$

Having a formula $P(t) = P_0e^{-0.037266t}$, we can find its half-life. This is the value of t for which $P(t) = \frac{1}{2}P_0$. To do this, we will solve

$$\begin{aligned} P_0e^{-0.037266t} &= \frac{1}{2}P_0 \\ e^{-0.037266t} &= \frac{1}{2} \\ \ln e^{-0.037266t} &= \ln\frac{1}{2} \\ -0.037266t &= \ln\frac{1}{2} \\ t &= \frac{\ln\frac{1}{2}}{-0.037266} \approx 18.583. \end{aligned}$$

So the half-life of this substance is about 18.6 hours

37. Since $W = 90$ when $t = 0$, we have $W = 90e^{kt}$. We use the doubling time to find k:

$$\begin{aligned} W &= 90e^{kt} \\ 2(90) &= 90e^{k(3)} \\ 2 &= e^{3k} \\ \ln 2 &= 3k \\ k &= \frac{\ln 2}{3} = 0.231. \end{aligned}$$

World wind energy generating capacity is growing at a continuous rate of 23.1% per year. We have $W = 90e^{0.231t}$.

38. We know the y-intercept is 0.8 and that the y-value doubles every 12 units. We can make a quick table and then plot points. See Table 5.1 and Figure 5.2.

Table 5.1

t	y
0	0.8
12	1.6
24	3.2
36	6.4
48	12.8
60	25.6

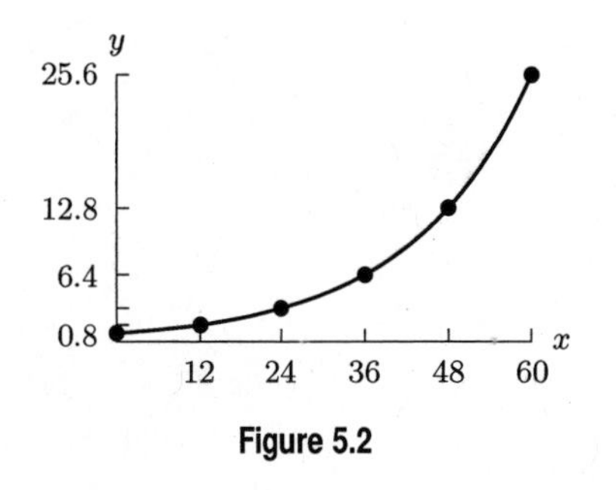

Figure 5.2

39. **(a)** Here, a formula for the population is $P = 5000 + 500t$. Solving $P = 10{,}000$ for t gives

$$\begin{aligned} 5000 + 500t &= 10{,}000 \\ 500t &= 5000 \\ t &= 10. \end{aligned}$$

Thus, it takes 10 years for the town to double to 10,000 people. But, as you can check for yourself, the town doubles in size a second time in year $t = 30$ and doubles a third time in year $t = 70$. Thus, it takes longer for the town to double each time: 10 years the first time, 30 years the second time, and 70 years the third time.

(b) Here, a formula for the population is $P = 5000(1.05)^t$. Solving $P = 10{,}000$ for t gives

$$\begin{aligned} 5000(1.05)^t &= 10{,}000 \\ 1.05^t &= 2 \\ \log(1.05)^t &= \log 2 \\ t \cdot \log 1.05 &= \log 2 \\ t = \frac{\log 2}{\log 1.05} &= 14.207. \end{aligned}$$

As you can check for yourself, it takes about 28.413 years for the town to double a second time, and about 42.620 years for it to double a third time. Thus, the town doubles every 14.207 years or so.

40. **(a)** We see that the initial value of the function is 50 and that the value has doubled to 100 at $t = 12$ so the doubling time is 12 days. Notice that 12 days later, at $t = 24$, the value of the function has doubled again, to 200. No matter where we start, the value will double 12 days later.

(b) We use $P = 50e^{kt}$ and the fact that the function increases from 50 to 100 in 12 days. Solving for k, we have:

$$\begin{aligned} 50e^{k(12)} &= 100 \\ e^{12k} &= 2 \\ 12k &= \ln 2 \\ k = \frac{\ln 2}{12} &= 0.0578. \end{aligned}$$

The continuous percent growth rate is 5.78% per day. The formula is $P = 50e^{0.0578t}$.

41. **(a)** We see that the initial value of Q is 150 mg and the quantity of caffeine has dropped to half that at $t = 4$. The half-life of caffeine is about 4 hours. Notice that no matter where we start on the graph, the quantity will be halved four hours later.

(b) We use $Q = 150e^{kt}$ and the fact that the quantity decays from 150 to 75 in 4 hours. Solving for k, we have:

$$\begin{aligned} 150e^{k(4)} &= 75 \\ e^{4k} &= 0.5 \\ 4k &= \ln(0.5) \\ k = \frac{\ln(0.5)}{4} &= -0.173. \end{aligned}$$

The continuous percent decay rate is -17.3% per hour. The formula is $Q = 150e^{-0.173t}$.

42. **(a)** At time $t = 0$ we see that the temperature is given by

$$\begin{aligned} H &= 70 + 120(1/4)^0 \\ &= 70 + 120(1) \\ &= 190. \end{aligned}$$

At time $t = 1$, we see that the temperature is given by

$$H = 70 + 120(1/4)^1$$

$$
\begin{aligned}
&= 70 + 120(1/4) \\
&= 70 + 30 \\
&= 100.
\end{aligned}
$$

At $t = 2$, we see that the temperature is given by

$$
\begin{aligned}
H &= 70 + 120(1/4)^2 \\
&= 70 + 120(1/16) \\
&= 70 + 7.5 \\
&= 77.5.
\end{aligned}
$$

(b) We solve for t to find when the temperature reaches $H = 90°$F:

$$
\begin{aligned}
70 + 120(1/4)^t &= 90 \\
120(1/4)^t &= 20 \qquad \text{subtracting} \\
(1/4)^t &= 20/120 \qquad \text{dividing} \\
\log(1/4)^t &= \log(1/6) \qquad \text{taking logs} \\
t\log(1/4) &= \log(1/6) \qquad \text{using a log property} \\
t &= \frac{\log(1/6)}{\log(1/4)} \qquad \text{dividing} \\
&= 1.292,
\end{aligned}
$$

so the coffee temperature reaches $90°$F after about 1.292 hours. Similar calculations show that the temperature reaches $75°$F after about 2.292 hours.

43. We have $P = ab^t$ and $2P = ab^{t+d}$. Using the algebra rules of exponents, we have

$$2P = ab^{t+d} = ab^t \cdot b^d = Pb^d.$$

Since P is nonzero, we can divide through by P, and we have

$$2 = b^d.$$

Notice that the time it takes an exponential growth function to double does not depend on the initial quantity a and does not depend on the time t. It depends only on the growth factor b.

44. (a) If prices rise at 3% per year, then each year they are 103% of what they had been the year before. After 5 years, they will be $(103\%)^5 = (1.03)^5 \approx 1.15927$, or 115.927% of what they had been initially. In other words, they have increased by 15.927% during that time.

(b) If it takes t years for prices to rise 25%, then

$$
\begin{aligned}
1.03^t &= 1.25 \\
\log 1.03^t &= \log 1.25 \\
t\log 1.03 &= \log 1.25 \\
t &= \frac{\log 1.25}{\log 1.03} \approx 7.549.
\end{aligned}
$$

With an annual inflation rate of 3%, it takes approximately 7.5 years for prices to increase by 25%.

45. (a) Initially, the population is $P = 300 \cdot 2^{0/20} = 300 \cdot 2^0 = 300$. After 20 years, the population reaches $P = 300 \cdot 2^{20/20} = 300 \cdot 2^1 = 600$.

(b) To find when the population reaches $P = 1000$, we solve the equation:

$$
\begin{aligned}
300 \cdot 2^{t/20} &= 1000 \\
2^{t/20} &= \frac{1000}{300} = \frac{10}{3} \qquad \text{dividing by 300}
\end{aligned}
$$

$$\log\left(2^{t/20}\right) = \log\left(\frac{10}{3}\right) \quad \text{taking logs}$$
$$\left(\frac{t}{20}\right)\cdot\log 2 = \log\left(\frac{10}{3}\right) \quad \text{using a log property}$$
$$t = \frac{20\log(10/3)}{\log 2} = 34.739,$$

and so it will take the population a bit less than 35 years to reach 1000.

46. Taking ratios, we have

$$\begin{aligned}
\frac{f(40)}{f(-20)} &= \frac{30}{5}\\
\frac{ab^{40}}{ab^{-20}} &= 6\\
b^{60} &= 6\\
b &= 6^{1/60}\\
&= 1.0303.
\end{aligned}$$

Solving for a gives

$$\begin{aligned}
f(40) &= ab^{40}\\
30 &= a\left(6^{1/60}\right)^{40}\\
a &= \frac{30}{6^{40/60}}\\
&= 9.0856.
\end{aligned}$$

To verify our answer, we see that

$$f(-20) = ab^{-20} = 9.0856(1.0303)^{-20} = 5.001,$$

which is correct within rounding.

Having found a and b, we now find k, using the fact that

$$\begin{aligned}
k &= \ln b\\
&= \ln\left(6^{1/60}\right)\\
&= \frac{\ln 6}{60}\\
&= 0.02986.
\end{aligned}$$

To check our answer, we see that

$$e^{0.02986} = 1.0303 = b,$$

as required.

Finally, to find s we use the fact that

$$\begin{aligned}
a\cdot 2^{t/s} &= ae^{kt} &&\\
\left(2^{1/s}\right)^t &= \left(e^k\right)^t && \text{exponent rule}\\
2^{1/s} &= e^k &&\\
\ln\left(2^{1/s}\right) &= k && \text{take logs}\\
\frac{1}{s}\cdot\ln 2 &= k && \text{log property}\\
\frac{1}{s} &= \frac{k}{\ln 2} && \text{divide}\\
s &= \frac{\ln 2}{k} && \text{reciprocate both sides}
\end{aligned}$$

$$\begin{aligned} &= \frac{\ln 2}{0.02986} \\ &= 23.213. \end{aligned}$$

To check our answer, we see that

$$a \cdot 2^{t/23.213} = a\left(2^{1/23.213}\right)^t = 1.0303 = b,$$

as required.

47. **(a)** If $P(t)$ is the investment's value after t years, we have $P(t) = P_0 e^{0.04t}$. We want to find t such that $P(t)$ is three times its initial value, P_0. Therefore, we need to solve:

$$\begin{aligned} P(t) &= 3P_0 \\ P_0 e^{0.04t} &= 3P_0 \\ e^{0.04t} &= 3 \\ \ln e^{0.04t} &= \ln 3 \\ 0.04t &= \ln 3 \\ t &= (\ln 3)/0.04 \approx 27.465 \text{ years.} \end{aligned}$$

With continuous compounding, the investment should triple in about $27\frac{1}{2}$ years.

(b) If the interest is compounded only once a year, the formula we will use is $P(t) = P_0 b^t$ where b is the percent value of what the investment had been one year earlier. If it is earning 4% interest compounded once a year, it is 104% of what it had been the previous year, so our formula is $P(t) = P_0(1.04)^t$. Using this new formula, we will now solve

$$\begin{aligned} P(t) &= 3P_0 \\ P_0(1.04)^t &= 3P_0 \\ (1.04)^t &= 3 \\ \log(1.04)^t &= \log 3 \\ t\log 1.04 &= \log 3 \\ t &= \frac{\log 3}{\log 1.04} \approx 28.011 \text{ years.} \end{aligned}$$

So, compounding once a year, it will take a little more than 28 years for the investment to triple.

48. **(a)** We see that the function $P = f(t)$, where t is years since 2005, is approximately exponential by looking at ratios of successive terms:

$$\begin{aligned} \frac{f(1)}{f(0)} &= \frac{92.175}{89.144} = 1.034 \\ \frac{f(2)}{f(1)} &= \frac{95.309}{92.175} = 1.034 \\ \frac{f(3)}{f(2)} &= \frac{98.550}{95.309} = 1.034 \\ \frac{f(4)}{f(3)} &= \frac{101.901}{98.550} = 1.034. \end{aligned}$$

Since $P = 89.144$ when $t = 0$, the formula is $P = 89.144(1.034)^t$. Alternately, we could use exponential regression to find the formula. The world population age 80 or older is growing at an annual rate of 3.4% per year.

(b) We find k so that $e^k = 1.034$, giving $k = \ln(1.034) = 0.0334$. The continuous percent growth rate is 3.34% per year. The corresponding formula is $P = 89.144e^{0.0334t}$.

(c) We can use either formula to find the doubling time (and the results will differ slightly due to round off error.) Using the continuous version, we have

$$2(89.144) = 89.144e^{0.0334t}$$

$$\begin{aligned} 2 &= e^{0.0334t} \\ \ln 2 &= 0.0334t \\ t &= \frac{\ln 2}{0.0334} = 20.753. \end{aligned}$$

The number of people in the world age 80 or older is doubling approximately every 21 years.

49. **(a)** We want a function of the form $R(t) = A\,B^t$. We know that when $t = 0$, $R(t) = 200$ and when $t = 6$, $R(t) = 100$. Therefore, $100 = 200B^6$ and $B \approx 0.8909$. So $R(t) = 200(0.8909)^t$.

(b) Setting $R(t) = 120$ we have $120 = 200(0.8909)^t$ and so $t = \dfrac{\ln 0.6}{\ln 0.8909} \approx 4.422$.

(c) Between $t = 0$ and $t = 2$

$$\frac{\Delta R(t)}{\Delta t} = \frac{200(0.8909)^2 - 200(0.8909)^0}{2} = -20.630$$

Between $t = 2$ and $t = 4$

$$\frac{\Delta R(t)}{\Delta t} = \frac{200(0.8909)^4 - 200(0.8909)^2}{2} = -16.374$$

Between $x = 4$ and $x = 6$

$$\frac{\Delta R(t)}{\Delta t} = \frac{200(0.8909)^6 - 200(0.8909)^4}{2} = -12.996$$

Since rates of change are increasing, the graph of $R(t)$ is concave up.

50. **(a)** For a function of the form $N(t) = ae^{kt}$, the value of a is the population at time $t = 0$ and k is the continuous growth rate. So the continuous growth rate is $0.013 = 1.3\%$.

(b) In year $t = 0$, the population is $N(0) = a = 5.4$ million.

(c) We want to find t such that the population of 5.4 million triples to 16.2 million. So, for what value of t does $N(t) = 5.4e^{0.013t} = 16.2$?

$$\begin{aligned} 5.4e^{0.013t} &= 16.2 \\ e^{0.013t} &= 3 \\ \ln e^{0.013t} &= \ln 3 \\ 0.013t &= \ln 3 \\ t &= \frac{\ln 3}{0.013} \approx 84.509 \end{aligned}$$

So the population will triple in approximately 84.5 years.

(d) Since $N(t)$ is in millions, we want to find t such that $N(t) = 0.000001$.

$$\begin{aligned} 5.4e^{0.013t} &= 0.000001 \\ e^{0.013t} &= \frac{0.000001}{5.4} \approx 0.000000185 \\ \ln e^{0.013t} &\approx \ln(0.000000185) \\ 0.013t &\approx \ln(0.000000185) \\ t &\approx \frac{\ln(0.000000185)}{0.013} \approx -1192.455 \end{aligned}$$

According to this model, the population of Washington State was 1 person 1192.455 years ago. It is unreasonable to suppose the formula extends so far into the past.

51. Since the half-life of carbon-14 is 5,728 years, and just a little more than 50% of it remained, we know that the man died nearly 5,700 years ago. To obtain a more precise date, we need to find a formula to describe the amount of carbon-14 left in the man's body after t years. Since the decay is continuous and exponential, it can be described by $Q(t) = Q_0e^{kt}$. We first find k. After 5,728 years, only one-half is left, so

$$Q(5{,}728) = \frac{1}{2}Q_0.$$

Therefore,

$$\begin{aligned}
Q(5{,}728) = Q_0 e^{5728k} &= \frac{1}{2}Q_0 \\
e^{5728k} &= \frac{1}{2} \\
\ln e^{5728k} &= \ln\frac{1}{2} \\
5728k = \ln\frac{1}{2} &= \ln 0.5 \\
k &= \frac{\ln 0.5}{5728}
\end{aligned}$$

So, $Q(t) = Q_0 e^{\frac{\ln 0.5}{5728}t}$.

If 46% of the carbon-14 has decayed, then 54% remains, so that $Q(t) = 0.54Q_0$.

$$\begin{aligned}
Q_0 e^{\left(\frac{\ln 0.5}{5728}\right)t} &= 0.54Q_0 \\
e^{\left(\frac{\ln 0.5}{5728}\right)t} &= 0.54 \\
\ln e^{\left(\frac{\ln 0.5}{5728}\right)t} &= \ln 0.54 \\
\frac{\ln 0.5}{5728}t &= \ln 0.54 \\
t &= \frac{(\ln 0.54)\cdot(5728)}{\ln 0.5} = 5092.013
\end{aligned}$$

So the man died about 5092 years ago.

52. If $P(t)$ describes the number of people in the store t minutes after it opens, we need to find a formula for $P(t)$. Perhaps the easiest way to develop this formula is to first find a formula for $P(k)$ where k is the number of 40-minute intervals since the store opened. After the first such interval there are $500(2) = 1{,}000$ people; after the second interval, there are $1{,}000(2) = 2{,}000$ people. Table 5.2 describes this progression:

Table 5.2

k	$P(k)$
0	500
1	$500(2)$
2	$500(2)(2) = 500(2)^2$
3	$500(2)(2)(2) = 500(2)^3$
4	$500(2)^4$
⋮	⋮
k	$500(2)^k$

From this, we conclude that $P(k) = 500(2)^k$. We now need to see how k and t compare. If $t = 120$ minutes, then we know that $k = \frac{120}{40} = 3$ intervals of 40 minutes; if $t = 187$ minutes, then $k = \frac{187}{40}$ intervals of 40 minutes. In general, $k = \frac{t}{40}$. Substituting $k = \frac{t}{40}$ into our equation for $P(k)$, we get an equation for the number of people in the store t minutes after the store opens:

$$P(t) = 500(2)^{\frac{t}{40}}.$$

To find the time when we'll need to post security guards, we need to find the value of t for which $P(t) = 10{,}000$.

$$\begin{aligned}
500(2)^{t/40} &= 10{,}000 \\
2^{t/40} &= 20 \\
\log\left(2^{t/40}\right) &= \log 20
\end{aligned}$$

$$\begin{aligned}\frac{t}{40}\log(2) &= \log 20\\ t(\log 2) &= 40\log 20\\ t &= \frac{40\log 20}{\log 2} \approx 172.877\end{aligned}$$

The guards should be commissioned about 173 minutes after the store is opened, or 12:53 pm.

53. **(a)** Since $f(x)$ is exponential, its formula will be $f(x) = ab^x$. Since $f(0) = 0.5$,

$$f(0) = ab^0 = 0.5.$$

But $b^0 = 1$, so

$$\begin{aligned}a(1) &= 0.5\\ a &= 0.5.\end{aligned}$$

We now know that $f(x) = 0.5b^x$. Since $f(1) = 2$, we have

$$\begin{aligned}f(1) = 0.5b^1 &= 2\\ 0.5b &= 2\\ b &= 4\end{aligned}$$

So $f(x) = 0.5(4)^x$.

We will find a formula for $g(x)$ the same way.

$$g(x) = ab^x.$$

Since $g(0) = 4$,

$$\begin{aligned}g(0) = ab^0 &= 4\\ a &= 4.\end{aligned}$$

Therefore,

$$g(x) = 4b^x.$$

We'll use $g(2) = \frac{4}{9}$ to get

$$\begin{aligned}g(2) = 4b^2 &= \frac{4}{9}\\ b^2 &= \frac{1}{9}\\ b &= \pm\frac{1}{3}.\end{aligned}$$

Since $b > 0$,

$$g(x) = 4\left(\frac{1}{3}\right)^x.$$

Since $h(x)$ is linear, its formula will be

$$h(x) = b + mx.$$

We know that b is the y-intercept, which is 2, according to the graph. Since the points $(a, a+2)$ and $(0, 2)$ lie on the graph, we know that the slope, m, is

$$\frac{(a+2)-2}{a-0} = \frac{a}{a} = 1,$$

so the formula is

$$h(x) = 2 + x.$$

(b) We begin with

$$f(x) = g(x)$$
$$\frac{1}{2}(4)^x = 4\left(\frac{1}{3}\right)^x.$$

Since the variable is an exponent, we need to use logs, so

$$\log\left(\frac{1}{2}\cdot 4^x\right) = \log\left(4\cdot\left(\frac{1}{3}\right)^x\right)$$
$$\log\frac{1}{2} + \log(4)^x = \log 4 + \log\left(\frac{1}{3}\right)^x$$
$$\log\frac{1}{2} + x\log 4 = \log 4 + x\log\frac{1}{3}.$$

Now we will move all expressions containing the variable to one side of the equation:

$$x\log 4 - x\log\frac{1}{3} = \log 4 - \log\frac{1}{2}.$$

Factoring out x, we get

$$x(\log 4 - \log\frac{1}{3}) = \log 4 - \log\frac{1}{2}$$
$$x\log\left(\frac{4}{1/3}\right) = \log\left(\frac{4}{1/2}\right)$$
$$x\log 12 = \log 8$$
$$x = \frac{\log 8}{\log 12}.$$

This is the exact value of x. Note that $\frac{\log 8}{\log 12} \approx 0.837$, so $f(x) = g(x)$ when x is exactly $\frac{\log 8}{\log 12}$ or about 0.837.

(c) Since $f(x) = h(x)$, we want to solve

$$\frac{1}{2}(4)^x = x + 2.$$

The variable does not occur only as an exponent, so logs cannot help us solve this equation. Instead, we need to graph the two functions and note where they intersect. The points occur when $x \approx 1.378$ or $x \approx -1.967$.

54. At an annual growth rate of 1%, the Rule of 70 tells us this investment doubles in $70/1 = 70$ years. At a 2% rate, the doubling time should be about $70/2 = 35$ years. The doubling times for the other rates are, according to the Rule of 70,

$$\frac{70}{5} = 14 \text{ years}, \qquad \frac{70}{7} = 10 \text{ years}, \qquad \text{and} \qquad \frac{70}{10} = 7 \text{ years}.$$

To check these predictions, we use logs to calculate the actual doubling times. If V is the dollar value of the investment in year t, then at a 1% rate, $V = 1000(1.01)^t$. To find the doubling time, we set $V = 2000$ and solve for t:

$$1000(1.01)^t = 2000$$
$$1.01^t = 2$$
$$\log(1.01^t) = \log 2$$
$$t\log 1.01 = \log 2$$
$$t = \frac{\log 2}{\log 1.01} \approx 69.661.$$

This agrees well with the prediction of 70 years. Doubling times for the other rates have been calculated and recorded in Table 5.3 together with the doubling times predicted by the Rule of 70.

Table 5.3 *Doubling times predicted by the Rule of 70 and actual values*

Rate (%)	1	2	5	7	10
Predicted doubling time (years)	70	35	14	10	7
Actual doubling time (years)	69.661	35.003	14.207	10.245	7.273

The Rule of 70 works reasonably well when the growth rate is small. The Rule of 70 does not give good estimates for growth rates much higher than 10%. For example, at an annual rate of 35%, the Rule of 70 predicts that the doubling time is $70/35 = 2$ years. But in 2 years at 35% growth rate, the \$1000 investment from the last example would be not worth \$2000, but only

$$1000(1.35)^2 = \$1822.50.$$

55. Let $Q = ae^{kt}$ be an increasing exponential function, so that k is positive. To find the doubling time, we find how long it takes Q to double from its initial value a to the value $2a$:

$$\begin{aligned} ae^{kt} &= 2a \\ e^{kt} &= 2 \qquad \text{(dividing by } a\text{)} \\ \ln e^{kt} &= \ln 2 \\ kt &= \ln 2 \qquad \text{(because } \ln e^x = x \text{ for all } x\text{)} \\ t &= \frac{\ln 2}{k}. \end{aligned}$$

Using a calculator, we find $\ln 2 = 0.693 \approx 0.70$. This is where the 70 comes from.

If, for example, the continuous growth rate is $k = 0.07 = 7\%$, then

$$\text{Doubling time} = \frac{\ln 2}{0.07} = \frac{0.693}{0.07} \approx \frac{0.70}{0.07} = \frac{70}{7} = 10.$$

If the growth rate is $r\%$, then $k = r/100$. Therefore

$$\text{Doubling time} = \frac{\ln 2}{k} = \frac{0.693}{k} \approx \frac{0.70}{r/100} = \frac{70}{r}.$$

56. We have

$$Q = 0.1e^{-(1/2.5)t},$$

and need to find t such that $Q = 0.04$. This gives

$$\begin{aligned} 0.1e^{-\frac{t}{2.5}} &= 0.04 \\ e^{-\frac{t}{2.5}} &= 0.4 \\ \ln e^{-\frac{t}{2.5}} &= \ln 0.4 \\ -\frac{t}{2.5} &= \ln 0.4 \\ t &= -2.5 \ln 0.4 \approx 2.291. \end{aligned}$$

It takes about 2.3 hours for their BAC to drop to 0.04.

57. **(a)** Applying the given formula,

$$\text{Number toads in year 0 is } P = \frac{1000}{1+49(1/2)^0} = 20$$

$$\text{Number toads in year 5 is } P = \frac{1000}{1+49(1/2)^5} = 395$$

$$\text{Number toads in year 10 is } P = \frac{1000}{1+49(1/2)^{10}} = 954.$$

(b) We set up and solve the equation $P = 500$:

$$\begin{aligned}
\frac{1000}{1+49(1/2)^t} &= 500 \\
500\left(1+49(1/2)^t\right) &= 1000 \quad \text{multiplying by denominator} \\
1+49(1/2)^t &= 2 \quad \text{dividing by 500} \\
49(1/2)^t &= 1 \\
(1/2)^t &= 1/49 \\
\log(1/2)^t &= \log(1/49) \quad \text{taking logs} \\
t\log(1/2) &= \log(1/49) \quad \text{using a log rule} \\
t &= \frac{\log(1/49)}{\log(1/2)} \\
&= 5.615,
\end{aligned}$$

and so it takes about 5.6 years for the population to reach 500. A similar calculation shows that it takes about 7.2 years for the population to reach 750.

(c) The graph in Figure 5.3 suggests that the population levels off at about 1000 toads. We can see this algebraically by using the fact that $(1/2)^t \to 0$ as $t \to \infty$. Thus,

$$P = \frac{1000}{1+49(1/2)^t} \to \frac{1000}{1+0} = 1000 \text{ toads} \qquad \text{as} \quad t \to \infty.$$

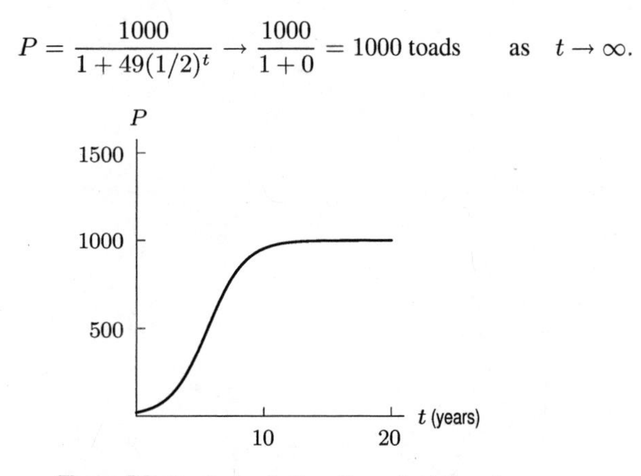

Figure 5.3: Toad population, P, against time, t

58. We have

$$\begin{aligned}
e^{k(t-t_0)} &= e^{kt-kt_0} \\
&= e^{kt}\cdot e^{-kt_0} \\
&= \underbrace{e^{-kt_0}}_{a}\underbrace{\left(e^k\right)^t}_{b^t}
\end{aligned}$$

$$\begin{aligned}
\text{so} \quad \left(e^k\right)^t &= b^t \\
e^k &= b \\
k &= \ln b \\
\text{and} \quad e^{-kt_0} &= a \\
-kt_0 &= \ln a \\
t_0 &= -\frac{\ln a}{k} \\
&= -\frac{\ln a}{\ln b} \quad \text{because } k = \ln b.
\end{aligned}$$

59. We have

$$\begin{aligned} 6e^{-0.5e^{-0.1t}} &= 3 & \\ e^{-0.5e^{-0.1t}} &= 0.5 & \text{divide} \\ -0.5e^{-0.1t} &= \ln 0.5 & \text{take logs} \\ e^{-0.1t} &= -2\ln 0.5 & \text{multiply} \\ -0.1t &= \ln(-2\ln 0.5) & \text{take logs} \\ t &= -10\ln(-2\ln 0.5) & \text{multiply} \\ &= -3.266. & \end{aligned}$$

Checking our answer, we see that

$$\begin{aligned} f(-3.266) &= 6e^{-0.5e^{-0.1(-3.266)}} \\ &= 6e^{-0.5e^{0.3266}} \\ &= 6e^{-0.5(1.3862)} \\ &= 6e^{-0.6931} = 3, \end{aligned}$$

as required.

60. **(a)** We have

$$\begin{aligned} 23\,(1.36)^t &= 85 \\ e^{\ln 23}\left(e^{\ln 1.36}\right)^t &= e^{\ln 85} \\ e^{\ln 23}\cdot e^{t\ln 1.36} &= e^{\ln 85} \\ e^{\overbrace{\ln 23}^{k}+t\overbrace{\ln 1.36}^{r}} &= e^{\overbrace{\ln 85}^{s}}, \end{aligned}$$

so $k = \ln 23, r = \ln 1.36, s = \ln 85$.

(b) We see that

$$\begin{aligned} e^{k+rt} &= e^s & \\ k + rt &= s & \text{same base, so exponents are equal} \\ rt &= s - k & \\ t &= \frac{s-k}{r}. & \end{aligned}$$

A numerical approximation is given by

$$\begin{aligned} t &= \frac{\ln 85 - \ln 23}{\ln 1.36} \\ &= 4.251. \end{aligned}$$

Checking our answer, we see that $23\,(1.36)^{4.251} = 84.997$, which is correct within rounding.

61. **(a)** We have

$$\begin{aligned} 1.12^t &= 6.3 \\ \left(10^{\log 1.12}\right)^t &= 10^{\log 6.3} \\ 10^{t\cdot\overbrace{\log 1.12}^{v}} &= 10^{\overbrace{\log 6.3}^{w}}, \end{aligned}$$

so $v = \log 1.12$ and $w = \log 6.3$.

(b) We see that

$$\begin{aligned} 10^{vt} &= 10^{w} \\ vt &= w \quad \text{same base, so exponents are equal} \\ t &= \frac{w}{v}. \end{aligned}$$

A numerical approximation is given by

$$\begin{aligned} t &= \frac{\log 1.12}{\log 6.3} \\ &= 16.241. \end{aligned}$$

Checking out answer, we see that $1.12^{16.241} = 6.300$, as required.

Solutions for Section 5.3

Skill Refresher

S1. $\log 0.0001 = \log 10^{-4} = -4 \log 10 = -4.$

S2. $\dfrac{\log 100^6}{\log 100^2} = \dfrac{6 \log 100}{2 \log 100} = \dfrac{6}{2} = 3.$

S3. The equation $10^5 = 100{,}000$ is equivalent to $\log 100{,}000 = 5$.

S4. The equation $e^2 = 7.389$ is equivalent to $\ln 7.389 = 2$.

S5. The equation $-\ln x = 12$ can be expressed as $\ln x = -12$, which is equivalent to $x = e^{-12}$.

S6. The equation $\log(x + 3) = 2$ is equivalent to $10^2 = x + 3$.

S7. Rewrite the expression as a sum and then use the power property,

$$\ln(x(7 - x)^3) = \ln x + \ln(7 - x)^3 = \ln x + 3\ln(7 - x).$$

S8. Using the properties of logarithms we have

$$\begin{aligned} \ln\left(\frac{xy^2}{z}\right) &= \ln xy^2 - \ln z \\ &= \ln x + \ln y^2 - \ln z \\ &= \ln x + 2\ln y - \ln z. \end{aligned}$$

S9. Rewrite the sum as $\ln x^3 + \ln x^2 = \ln(x^3 \cdot x^2) = \ln x^5$.

S10. Rewrite with powers and combine,

$$\begin{aligned} \frac{1}{3}\log 8 - \frac{1}{2}\log 25 &= \log 8^{1/3} - \log 25^{1/2} \\ &= \log 2 - \log 5 \\ &= \log \frac{2}{5}. \end{aligned}$$

Exercises

1. The graphs of $y = 10^x$ and $y = 2^x$ both have horizontal asymptotes, $y = 0$. The graph of $y = \log x$ has a vertical asymptote, $x = 0$.
2. The graphs of both $y = e^x$ and $y = e^{-x}$ have the same horizontal asymptote. Their asymptote is the x-axis, whose equation is $y = 0$. The graph of $y = \ln x$ is asymptotic to the y-axis, hence the equation of its asymptote is $x = 0$.
3. A is $y = 10^x$, B is $y = e^x$, C is $y = \ln x$, D is $y = \log x$.
4. A is $y = 3^x$, B is $y = 2^x$, C is $y = \ln x$, D is $y = \log x$, E is $y = e^{-x}$.
5. See Figure 5.4. The graph of $y = 2 \cdot 3^x + 1$ is the graph of $y = 3^x$ stretched vertically by a factor of 2 and shifted up by 1 unit.

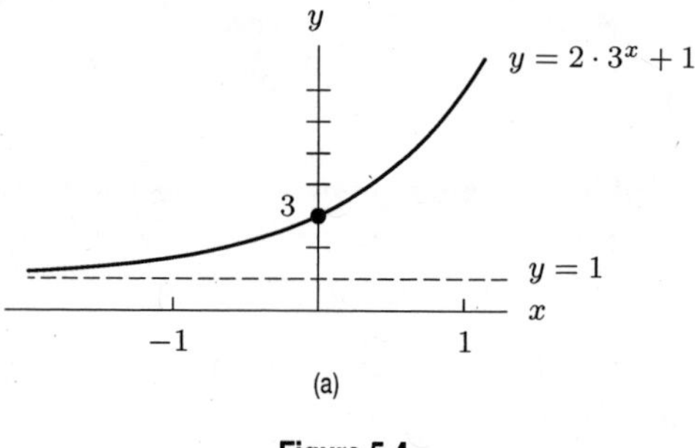

(a)

Figure 5.4

6. See Figure 5.5. The graph of $y = -e^{-x}$ is the graph of $y = e^x$ flipped over the y-axis and then over the x-axis.

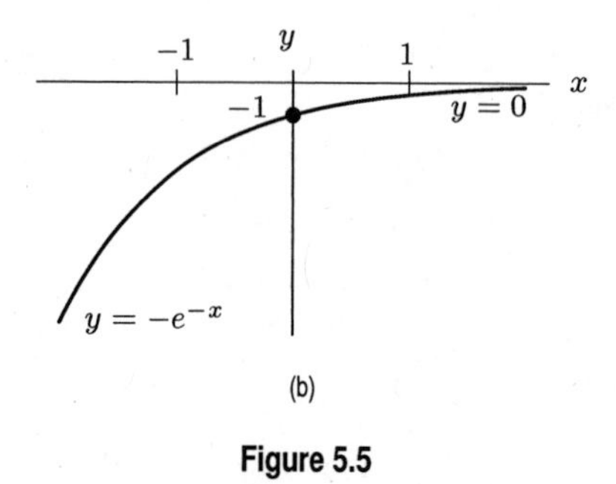

(b)

Figure 5.5

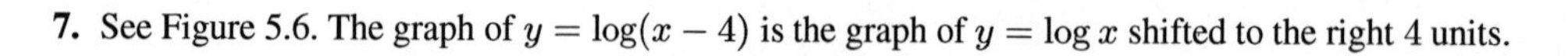

7. See Figure 5.6. The graph of $y = \log(x - 4)$ is the graph of $y = \log x$ shifted to the right 4 units.

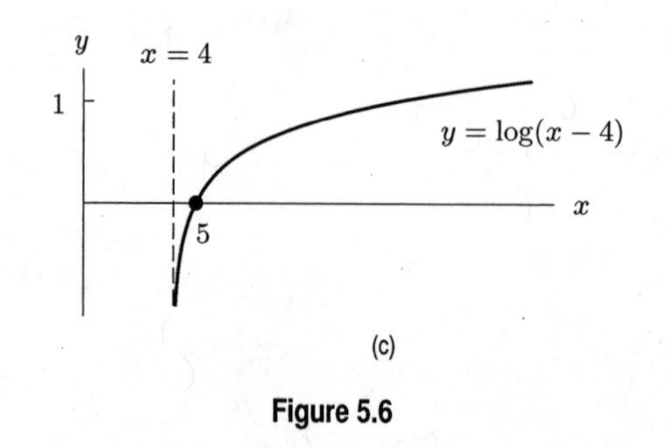

(c)

Figure 5.6

8. See Figure 5.7. The vertical asymptote is $x = -1$; there is no horizontal asymptote.

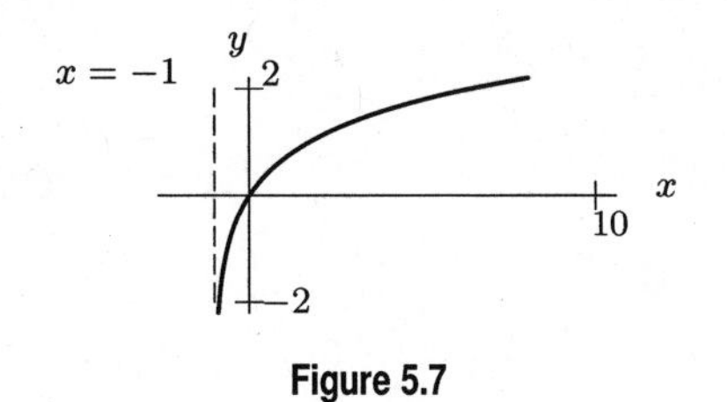

Figure 5.7

9. A graph of this function is shown in Figure 5.8. We see that the function has a vertical asymptote at $x = 3$. The domain is $(3, \infty)$.

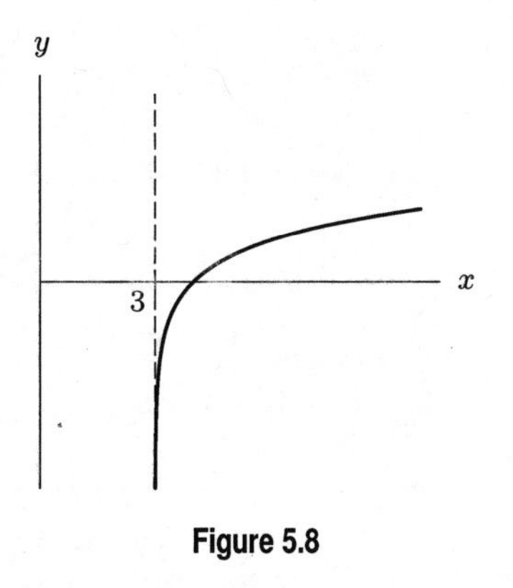

Figure 5.8

10. A graph of this function is shown in Figure 5.9. We see that the function has a vertical asymptote at $x = 2$. The domain is $(-\infty, 2)$.

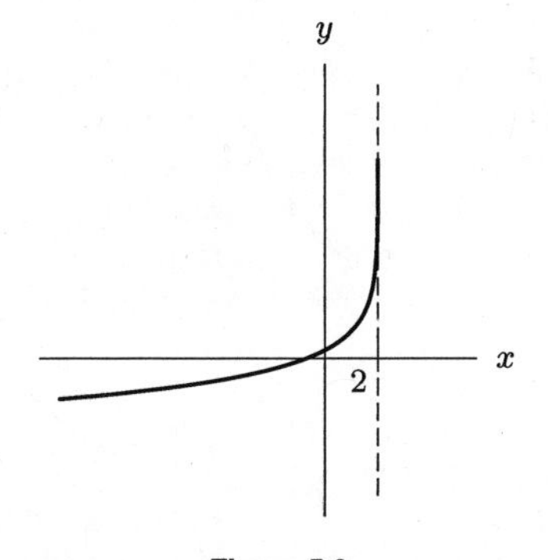

Figure 5.9

11. **(a)** $10^{-x} \to 0$ as $x \to \infty$.
(b) The values of $\log x$ get more and more negative as $x \to 0^+$, so

$$\log x \to -\infty.$$

12. **(a)** $e^x \to 0$ as $x \to -\infty$.

(b) The values of $\ln x$ get more and more negative as $x \to 0^+$, so

$$\ln x \to -\infty.$$

13. **(a)** Since $\log x$ becomes more and more negative as x decreases to 0 from above,

$$\lim_{x \to 0^+} \log x = -\infty.$$

(b) Since $-x$ is positive if x is negative and $-x$ decreases to 0 as x increases to 0 from below,

$$\lim_{x \to 0^-} \ln(-x) = -\infty.$$

14. We know, by the definition of pH, that $13 = -\log[H^+]$. Therefore, $-13 = \log[H^+]$, and $10^{-13} = [H^+]$. Thus, the hydrogen ion concentration is 10^{-13} moles per liter.

15. We know, by the definition of pH, that $1 = -\log[H^+]$. Therefore, $-1 = \log[H^+]$, and $10^{-1} = [H^+]$. Thus, the hydrogen ion concentration is $10^{-1} = 0.1$ moles per liter.

16. We know, by the definition of pH, that $8.3 = -\log[H^+]$. Therefore, $-8.3 = \log[H^+]$, and $10^{-8.3} = [H^+]$. Thus, the hydrogen ion concentration is $10^{-8.3} = 5.012 \times 10^{-9}$ moles per liter.

17. We know, by the definition of pH, that $4.5 = -\log[H^+]$. Therefore, $-4.5 = \log[H^+]$, and $10^{-4.5} = [H^+]$. Thus, the hydrogen ion concentration is $10^{-4.5} = 3.162 \times 10^{-5}$ moles per liter.

18. We know, by the definition of pH, that $0 = -\log[H^+]$. Therefore, $-0 = \log[H^+]$, and $10^{-0} = [H^+]$. Thus, the hydrogen ion concentration is $10^{-0} = 10^0 = 1$ mole per liter.

Problems

19. The log function is increasing but is concave down and so is increasing at a decreasing rate. It is not a compliment—growing exponentially would have been better. However, it is most likely realistic because after you are proficient at something, any increase in proficiency takes longer and longer to achieve.

20. **(a)** The graph in (III) has a vertical asymptote at $x = 0$ and $f(x) \to -\infty$ as $x \to 0^+$.

(b) The graph in (IV) goes through the origin, so $f(x) \to 0$ as $x \to 0^-$.

(c) The graph in (I) goes upward without bound, that is $f(x) \to \infty$, as $x \to \infty$.

(d) The graphs in (I) and (II) tend toward the x-axis, that is $f(x) \to 0$, as $x \to -\infty$.

21. **(a)** Let the functions graphed in (a), (b), and (c) be called $f(x)$, $g(x)$, and $h(x)$ respectively. Looking at the graph of $f(x)$, we see that $f(10) = 3$. In the table for $r(x)$ we note that $r(10) = 1.699$ so $f(x) \neq r(x)$. Similarly, $s(10) = 0.699$, so $f(x) \neq s(x)$. The values describing $t(x)$ do seem to satisfy the graph of $f(x)$, however. In the graph, we note that when $0 < x < 1$, then y must be negative. The data point $(0.1, -3)$ satisfies this. When $1 < x < 10$, then $0 < y < 3$. In the table for $t(x)$, we see that the point $(2, 0.903)$ satisfies this condition. Finally, when $x > 10$ we see that $y > 3$. The values $(100, 6)$ satisfy this. Therefore, $f(x)$ and $t(x)$ could represent the same function.

(b) For $g(x)$, we note that

$$\begin{cases} \text{when } 0 < x < 0.2, & \text{then } y < 0; \\ \text{when } 0.2 < x < 1, & \text{then } 0 < y < 0.699; \\ \text{when } x > 1, & \text{then } y > 0.699. \end{cases}$$

All the values of x in the table for $r(x)$ are greater than 1 and all the corresponding values of y are greater than 0.699, so $g(x)$ could equal $r(x)$. We see that, in $s(x)$, the values $(0.5, -0.060)$ do not satisfy the second condition so $g(x) \neq s(x)$. Since we already know that $t(x)$ corresponds to $f(x)$, we conclude that $g(x)$ and $r(x)$ correspond.

(c) By elimination, $h(x)$ must correspond to $s(x)$. We see that in $h(x)$,

$$\begin{cases} \text{when } x < 2, & \text{then } y < 0; \\ \text{when } 2 < x < 20, & \text{then } 0 < y < 1; \\ \text{when } x > 20, & \text{then } y > 1. \end{cases}$$

Since the values in $s(x)$ satisfy these conditions, it is reasonable to say that $h(x)$ and $s(x)$ correspond.

22. We use the formula $N = 10 \cdot \log\left(\frac{I}{I_0}\right)$, where N denotes the noise level in decibels and I and I_0 are sound's intensities in watts/cm^2. The intensity of a standard benchmark sound is $I_0 = 10^{-16}$ watts/cm^2 and the noise level of a whisper is $N = 30$ dB. Thus, we have $30 = 10 \cdot \log\left(\frac{I}{10^{-16}}\right)$. Solving for I we have

$$\begin{aligned}
10\log\left(\frac{I}{10^{-16}}\right) &= 30 \\
\log\left(\frac{I}{10^{-16}}\right) &= 3 && \text{Dividing by 10} \\
\frac{I}{10^{-16}} &= 10^3 && \text{Raising 10 to the power of both sides} \\
I &= 10^3 \cdot 10^{-16} && \text{Multiplying both sides by } 10^{-16} \\
I &= 10^{-13} \text{ watts/cm}^2.
\end{aligned}$$

23. We use the formula $N = 10 \cdot \log\left(\frac{I}{I_0}\right)$, where N denotes the noise level in decibels and I and I_0 are sound's intensities in watts/cm^2. The intensity of a standard benchmark sound is $I_0 = 10^{-16}$ watts/cm^2, and the noise level when loss of hearing tissue occurs is $N = 180$ dB. Thus, we have $180 = 10 \cdot \log\left(\frac{I}{10^{-16}}\right)$. Solving for I we have

$$\begin{aligned}
10\log\left(\frac{I}{10^{-16}}\right) &= 180 \\
\log\left(\frac{I}{10^{-16}}\right) &= 18 && \text{Dividing by 10} \\
\frac{I}{10^{-16}} &= 10^{18} && \text{Raising 10 to the power of both sides} \\
I &= 10^{18} \cdot 10^{-16} && \text{Multiplying both sides by } 10^{-16} \\
I &= 100 \text{ watts/cm}^2.
\end{aligned}$$

24. If I_B is the sound intensity of Broncos fans' roar, then

$$10\log\left(\frac{I_B}{I_0}\right) = 128.7 \text{ dB.}$$

Similarly, if I_S is the sound intensity of the Irish soccer fans, then

$$10\log\left(\frac{I_S}{I_0}\right) = 125.4 \text{ dB.}$$

Computing the difference of the decibel ratings gives

$$10\log\left(\frac{I_B}{I_0}\right) - 10\log\left(\frac{I_S}{I_0}\right) = 128.7 - 125.4 = 3.3.$$

Dividing by 10 gives

$$\begin{aligned}
\log\left(\frac{I_B}{I_0}\right) - \log\left(\frac{I_S}{I_0}\right) &= 0.33 \\
\log\left(\frac{I_B/I_0}{I_s/I_0}\right) &= 0.33 && \text{Using the property } \log b - \log a = \log(b/a) \\
\log\left(\frac{I_B}{I_S}\right) &= 0.33 && \text{Canceling } I_0 \\
\frac{I_B}{I_S} &= 10^{0.33} && \text{Raising 10 to the power of both sides}
\end{aligned}$$

So $I_B = 10^{0.33} I_S$, which means that crowd of Denver Bronco fans was $10^{0.33} \approx 2$ times as intense as the Irish soccer fans.

Notice that although the difference in decibels between the Broncos fans' roar and the Irish soccer fans' roar is only 3.3 dB, the sound intensity is twice as large.

25. Let I_A and I_B be the intensities of sound A and sound B, respectively. We know that $I_B = 5I_A$ and, since sound A measures 30 dB, we know that $10\log(I_A/I_0) = 30$. We have:

$$\begin{aligned}
\text{Decibel rating of B } &= 10\log\left(\frac{I_B}{I_0}\right) \\
&= 10\log\left(\frac{5I_A}{I_0}\right) \\
&= 10\log 5 + 10\log\left(\frac{I_A}{I_0}\right) \\
&= 10\log 5 + 30 \\
&= 10(0.699) + 30 \\
&\approx 37.
\end{aligned}$$

Notice that although sound B is 5 times as loud as sound A, the decibel rating only goes from 30 to 37.

26. **(a)** We know that $D_1 = 10\log\left(\frac{I_1}{I_0}\right)$ and $D_2 = 10\log\left(\frac{I_2}{I_0}\right)$. Thus

$$\begin{aligned}
D_2 - D_1 &= 10\log\left(\frac{I_2}{I_0}\right) - 10\log\left(\frac{I_1}{I_0}\right) \\
&= 10\left(\log\left(\frac{I_2}{I_0}\right) - \log\left(\frac{I_1}{I_0}\right)\right) && \text{factoring} \\
&= 10\log\left(\frac{I_2/I_0}{I_1/I_0}\right) && \text{using a log property}
\end{aligned}$$

so

$$D_2 - D_1 = 10\log\left(\frac{I_2}{I_1}\right).$$

(b) Suppose the sound's initial intensity is I_1 and that its new intensity is I_2. Then here we have $I_2 = 2I_1$. If D_1 is the original decibel rating and D_2 is the new rating then

$$\begin{aligned}
\text{Increase in decibels } &= D_2 - D_1 \\
&= 10\log\left(\frac{I_2}{I_1}\right) && \text{using formula from part (a)} \\
&= 10\log\left(\frac{2I_1}{I_1}\right) \\
&= 10\log 2 \\
&\approx 3.01.
\end{aligned}$$

Thus, the sound increases by 3 decibels when it doubles in intensity.

27. We set $M = 7.9$ and solve for the value of the ratio W/W_0.

$$\begin{aligned}
7.9 &= \log\left(\frac{W}{W_0}\right) \\
10^{7.9} &= \frac{W}{W_0} \\
&= 79{,}432{,}823.
\end{aligned}$$

Thus, the Sichuan earthquake had seismic waves that were 79,432,823 times more powerful than W_0.

28. We set $M = 3.5$ and solve for the ratio W/W_0.

$$\begin{aligned}
3.4 &= \log\left(\frac{W}{W_0}\right) \\
10^{3.5} &= \frac{W}{W_0} \\
&= 3{,}162.
\end{aligned}$$

Thus, the Chernobyl nuclear explosion had seismic waves that were 3,162 times more powerful than W_0.

29. We know $M_1 = \log\left(\frac{W_1}{W_0}\right)$ and $M_2 = \log\left(\frac{W_2}{W_0}\right)$. Thus,

$$\begin{aligned} M_2 - M_1 &= \log\left(\frac{W_2}{W_0}\right) - \log\left(\frac{W_1}{W_0}\right) \\ &= \log\left(\frac{W_2}{W_1}\right). \end{aligned}$$

30. Let $M_2 = 8.7$ and $M_1 = 7.1$, so

$$M_2 - M_1 = \log\left(\frac{W_2}{W_1}\right)$$

becomes

$$\begin{aligned} 8.7 - 7.1 &= \log\left(\frac{W_2}{W_1}\right) \\ 1.6 &= \log\left(\frac{W_2}{W_1}\right) \end{aligned}$$

so

$$\frac{W_2}{W_1} = 10^{1.6} \approx 40.$$

Thus, the seismic waves of the 2005 Sumatran earthquake were about 40 times as large as those of the 1989 California earthquake.

31. **(a)** (i) $\text{pH} = -\log x = 2$ so $\log x = -2$ so $x = 10^{-2}$
(ii) $\text{pH} = -\log x = 4$ so $\log x = -4$ so $x = 10^{-4}$
(iii) $\text{pH} = -\log x = 7$ so $\log x = -7$ so $x = 10^{-7}$

(b) Solutions with high pHs have low concentrations and so are less acidic.

32. **(a)** We are given the number of H^+ ions in 12 oz of coffee, and we need to find the number of moles of ions in 1 liter of coffee. So we need to convert numbers of ions to moles of ions, and ounces of coffee to liters of coffee. Finding the number of moles of H^+, we have:

$$2.41 \cdot 10^{18} \text{ ions} \cdot \frac{1 \text{ mole of ions}}{6.02 \cdot 10^{23} \text{ ions}} = 4 \cdot 10^{-6} \text{ ions}.$$

Finding the number of liters of coffee, we have:

$$12 \text{ oz} \cdot \frac{1 \text{ liter}}{30.3 \text{ oz}} = 0.396 \text{ liters}.$$

Thus, the concentration, $[H^+]$, in the coffee is given by

$$\begin{aligned} [H^+] &= \frac{\text{Number of moles } H^+ \text{ in solution}}{\text{Number of liters solution}} \\ &= \frac{4 \cdot 10^{-6}}{0.396} \\ &= 1.01 \cdot 10^{-5} \text{ moles/liter}. \end{aligned}$$

(b) We have

$$\begin{aligned} \text{pH} &= -\log[H^+] \\ &= -\log\left(1.01 \cdot 10^{-5}\right) \\ &= -(-4.9957) \\ &\approx 5. \end{aligned}$$

Thus, the pH is about 5. Since this is less than 7, it means that coffee is acidic.

33. **(a)** The pH is 2.3, which, according to our formula for pH, means that

$$-\log\left[H^+\right] = 2.3.$$

This means that

$$\log\left[H^+\right] = -2.3.$$

This tells us that the exponent of 10 that gives $[H^+]$ is -2.3, so

$$\begin{aligned}[H^+] &= 10^{-2.3} \quad \text{because } -2.3 \text{ is exponent of 10}\\ &= 0.005 \text{ moles/liter.}\end{aligned}$$

(b) From part (a) we know that 1 liter of lemon juice contains 0.005 moles of H^+ ions. To find out how many H^+ ions our lemon juice has, we need to convert ounces of juice to liters of juice and moles of ions to numbers of ions. We have

$$2 \text{ oz} \times \frac{1 \text{ liter}}{30.3 \text{ oz}} = 0.066 \text{ liters.}$$

We see that

$$0.066 \text{ liters juice} \times \frac{0.005 \text{ moles } H^+ \text{ ions}}{1 \text{ liter}} = 3.3 \times 10^{-4} \text{ moles } H^+ \text{ ions.}$$

There are 6.02×10^{23} ions in one mole, and so

$$3.3 \times 10^{-4} \text{ moles} \times \frac{6.02 \times 10^{23} \text{ ions}}{\text{mole}} = 1.987 \times 10^{20} \text{ ions.}$$

34. A possible formula is $y = \log x$.

35. This graph could represent exponential decay, so a possible formula is $y = b^x$ with $0 < b < 1$.

36. This graph could represent exponential growth, with a y-intercept of 2. A possible formula is $y = 2b^x$ with $b > 1$.

37. A possible formula is $y = \ln x$.

38. This graph could represent exponential decay, with a y-intercept of 0.1. A possible formula is $y = 0.1b^x$ with $0 < b < 1$.

39. This graph could represent exponential "growth", with a y-intercept of -1. A possible formula is $y = (-1)b^x = -b^x$ for $b > 1$.

40. We see that:

- In order for the square root to be defined,

$$\begin{aligned}7 - e^{2t} &\geq 0\\ e^{2t} &\leq 7\\ 2t &\leq \ln 7\\ t &\leq 0.5\ln 7.\end{aligned}$$

- In order for the denominator not to equal zero,

$$\begin{aligned}2 - \sqrt{7 - e^{2t}} &\neq 0\\ 2 &\neq \sqrt{7 - e^{2t}}\\ 4 &\neq 7 - e^{2t}\\ e^{2t} &\neq 3\\ 2t &\neq \ln 3\\ t &\neq 0.5\ln 3.\end{aligned}$$

Putting these together, we see that $t \leq 0.5\ln 7$ and $t \neq 0.5\ln 3$.

Solutions for Section 5.4

Skill Refresher

S1. 1.455×10^6

S2. 4.23×10^{11}

S3. 6.47×10^4

S4. 1.231×10^7

S5. 3.6×10^{-4}

S6. 4.71×10^{-3}

S7. Since $10{,}000 < 12{,}500 < 100{,}000$, $10^4 < 12{,}500 < 10^5$.

S8. Since $0.0001 < 0.000881 < 0.001$, $10^{-4} < 0.000881 < 10^{-3}$.

S9. Since $0.1 < \frac{1}{3} < 1$, $10^{-1} < \frac{1}{3} < 1 = 10^0$.

S10. Since

$$\begin{aligned} 3{,}850 \cdot 10^8 &= \left(3.85 \cdot 10^3\right) 10^8 \\ &= 3.85\left(10^3 \cdot 10^8\right) \\ &= 3.85 \cdot 10^{3+8} \\ &= 3.85 \cdot 10^{11}, \end{aligned}$$

we have $10^{11} < 3.85 \cdot 10^{11} < 10^{12}$.

Exercises

1. Using a linear scale, the wealth of everyone with less than a million dollars would be indistinguishable because all of them are less than one one-thousandth of the wealth of the average billionaire. A log scale is more useful.

2. In all cases, the average number of diamonds owned is probably less than 100 (probably less than 20). Therefore, the data will fit neatly into a linear scale.

3. As nobody eats fewer than zero times in a restaurant per week, and since it's unlikely that anyone would eat more than 50 times per week in a restaurant, a linear scale should work fine.

4. This should be graphed on a log scale. Someone who has never been exposed presumably has zero bacteria. Someone who has been slightly exposed has perhaps one thousand bacteria. Someone with a mild case may have ten thousand bacteria, and someone dying of tuberculosis may have hundreds of thousands or millions of bacteria. Using a linear scale, the data points of all the people not dying of the disease would be too close to be readable.

5. (a)

Table 5.4

n	1	2	3	4	5	6	7	8	9
$\log n$	0	0.3010	0.4771	0.6021	0.6990	0.7782	0.8451	0.9031	0.9542

Table 5.5

n	10	20	30	40	50	60	70	80	90
$\log n$	1	1.3010	1.4771	1.6021	1.6990	1.7782	1.8451	1.9031	1.9542

(b) The first tick mark is at $10^0 = 1$. The dot for the number 2 is placed $\log 2 = 0.3010$ of the distance from 1 to 10. The number 3 is placed at $\log 3 = 0.4771$ units from 1, and so on. The number 30 is placed 1.4771 units from 1, the number 50 is placed 1.6989 units from 1, and so on.

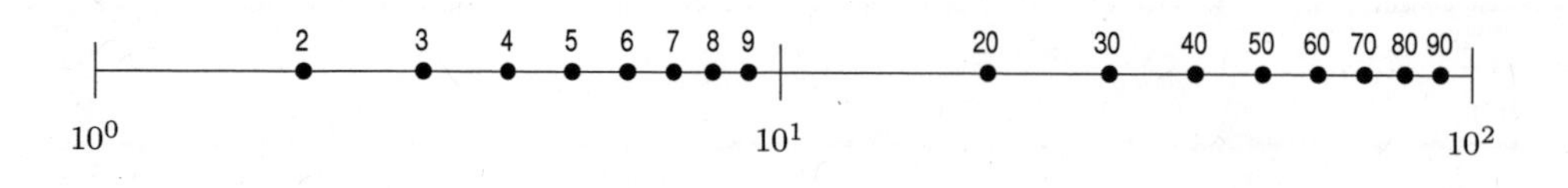

Figure 5.10

6. **(a)** Using linear regression we find that the linear function $y = 48.097 + 0.803x$ gives a correlation coefficient of $r = 0.9996$. We see from the sketch of the graph of the data that the estimated regression line provides an excellent fit. See Figure 5.11.

(b) To check the fit of an exponential we make a table of x and $\ln y$:

x	30	85	122	157	255	312
$\ln y$	4.248	4.787	4.977	5.165	5.521	5.704

Using linear regression, we find $\ln y = 4.295 + 0.0048x$. We see from the sketch of the graph of the data that the estimated regression line fits the data well, but not as well as part (a). See Figure 5.12. Solving for y to put this into exponential form gives

$$\begin{aligned} e^{\ln y} &= e^{4.295+0.0048x} \\ y &= e^{4.295}e^{0.0048x} \\ y &= 73.332e^{0.0048x}. \end{aligned}$$

This gives us a correlation coefficient of $r \approx 0.9728$. Note that since $e^{0.0048} = 1.0048$, we could have written $y = 73.332(1.0048)^x$.

Figure 5.11

Figure 5.12

(c) Both fits are good. The linear equation gives a slightly better fit.

7. **(a)** Run a linear regression on the data. The resulting function is $y = -3582.145 + 236.314x$, with $r \approx 0.7946$. We see from the sketch of the graph of the data that the estimated regression line provides a reasonable but not excellent fit. See Figure 5.13.

(b) If, instead, we compare x and $\ln y$ we get

$$\ln y = 1.568 + 0.200x.$$

We see from the sketch of the graph of the data that the estimated regression line provides an excellent fit with $r \approx 0.9998$. See Figure 5.14. Solving for y, we have

$$\begin{aligned} e^{\ln y} &= e^{1.568+0.200x} \\ y &= e^{1.568}e^{0.200x} \\ y &= 4.797e^{0.200x} \\ \text{or} \quad y &= 4.797(e^{0.200})^x \approx 4.797(1.221)^x. \end{aligned}$$

Figure 5.13

Figure 5.14

(c) The linear equation is a poor fit, and the exponential equation is a better fit.

8. (a) Using linear regression on x and y, we find $y = -169.331 + 57.781x$, with $r \approx 0.9707$. We see from the sketch of the graph of the data that the estimated regression line provides a good fit. See Figure 5.15.
   (b) Using linear regression on x and $\ln y$, we find $\ln y = 2.258 + 0.463x$, with $r \approx 0.9773$. We see from the sketch of the graph of the data that the estimated regression line provides a good fit. See Figure 5.16. Solving as in the previous problem, we get $y = 9.566(1.589)^x$.

Figure 5.15

Figure 5.16

(c) The exponential function $y = 9.566(1.589)^x$ gives a good fit, and so does the linear function $y = -169.331 + 57.781x$.

Problems

9. The Declaration of Independence was signed in 1776, about 225 years ago. We can write this number as

$$\frac{225}{1{,}000{,}000} = 0.000225 \text{ million years ago.}$$

This number is between $10^{-4} = 0.0001$ and $10^{-3} = 0.001$. Using a calculator, we have

$$\log 0.000225 \approx -3.65,$$

which, as expected, lies between -3 and -4 on the log scale. Thus, the Declaration of Independence is placed at

$$10^{-3.65} \approx 0.000224 \text{ million years ago} = 224 \text{ years ago.}$$

10. (a) An appropriate scale is from 0 to 70 at intervals of 10. (Other answers are possible.) See Figure 5.17. The points get more and more spread out as the exponent increases.

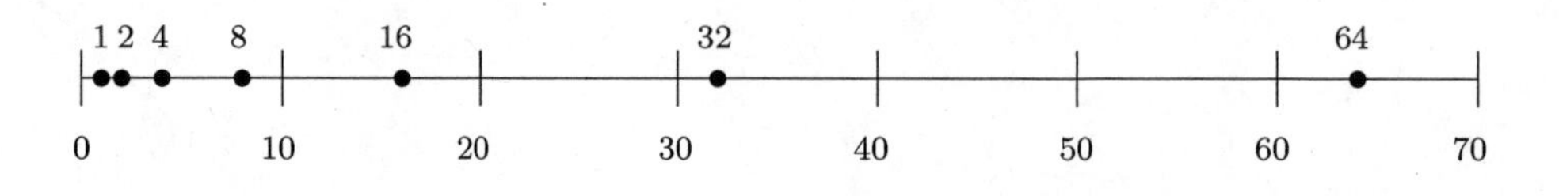

Figure 5.17

(b) If we want to locate 2 on a logarithmic scale, since $2 = 10^{0.3}$, we find $10^{0.3}$. Similarly, $8 = 10^{0.9}$ and $32 = 10^{1.5}$, so 8 is at $10^{0.9}$ and 32 is at $10^{1.5}$. Since the values of the logs go from 0 to 1.8, an appropriate scale is from 0 to 2 at intervals of 0.2. See Figure 5.18. The points are spaced at equal intervals.

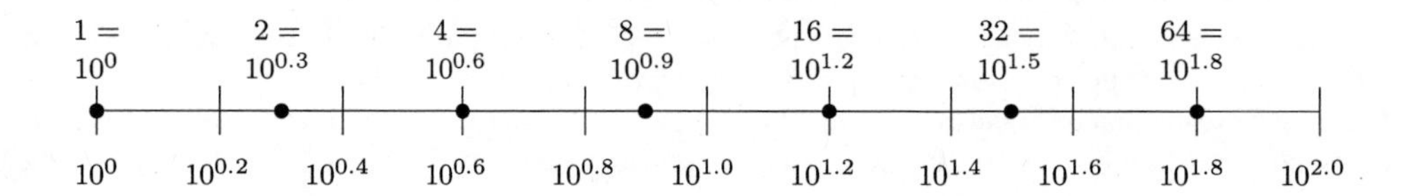

Figure 5.18

11. Point A appears to be at the value 0.2 so the log of the price of A is 0.2. Thus, we have

$$\text{Price of A} \approx 10^{0.2} = \$1.58.$$

Similarly, estimating the values of the other points, we have We have

$$\begin{aligned}
\text{Price of B} &\approx 10^{0.8} = \$6.31,\\
\text{Price of C} &\approx 10^{1.5} = \$31.62,\\
\text{Price of D} &\approx 10^{2.8} = \$630.96,\\
\text{Price of E} &\approx 10^{4.0} = \$10{,}000.00,\\
\text{Price of F} &\approx 10^{5.1} = \$125{,}892.54,\\
\text{Price of G} &\approx 10^{6.8} = \$6{,}309{,}573.45.
\end{aligned}$$

A log scale was necessary because the prices of the items range from less than 2 dollars to more than 6 million dollars.

12. **(a)** See Figure 5.19.

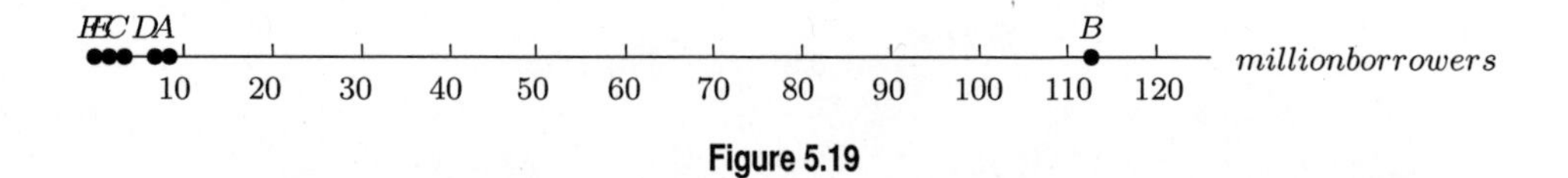

Figure 5.19

(b) We compute the logs of the values:

$$\begin{aligned}
\text{A:}\quad & \log(8.4) = 0.92\\
\text{B:}\quad & \log(112.7) = 2.05\\
\text{C:}\quad & \log(3.4) = 0.53\\
\text{D:}\quad & \log(6.8) = 0.83\\
\text{E:}\quad & \log(1.7) = 0.23\\
\text{F:}\quad & \log(0.05) = -1.30.
\end{aligned}$$

See Figure 5.20.

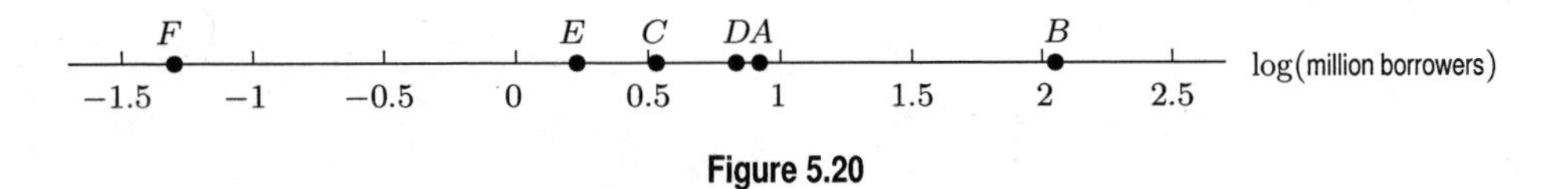

Figure 5.20

(c) The log scale is more appropriate because the numbers are of significantly different magnitude.

13. (a) Figure 5.21 shows the track events plotted on a linear scale.

100 400
0 200 800 1500 3000 5000 10000

Figure 5.21:

(b) Figure 5.22 shows the track events plotted on a logarithmic scale.

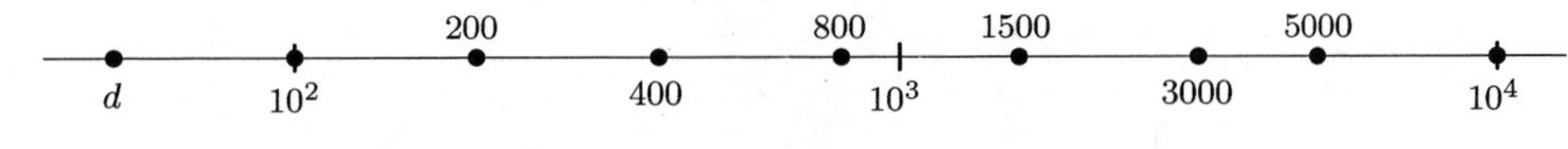

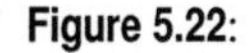

Figure 5.22:

(c) Figure 5.21 gives a runner better information about pacing for the distance.

(d) On Figure 5.21 the point 50 is $\frac{1}{2}$ the distance from 0 to 100. On Figure 5.22 the point 50 is the same distance to the left of 100 as 200 is to the right. This is shown as point *d*.

14. (a) A log scale is necessary because the numbers are of such different magnitudes. If we used a small scale (such as 0, 10, 20,...) we could see the small numbers but would never get large enough for the big numbers. If we used a large scale (such as counting by 100, 000s), we would not be able to differentiate between the small numbers. In order to see all of the values, we need to use a log scale.

(b) See Table 5.6.

Table 5.6 *Deaths due to various causes in the US in 2002*

Cause	Log of number of deaths
Scarlet fever	0.30
Whooping cough	0.95
Asthma	3.56
HIV	4.08
Kidney Diseases	4.66
Accidents	5.08
Malignant neoplasms	5.75
Cardiovascular Disease	5.92
All causes	6.38

(c) See Figure 5.23.

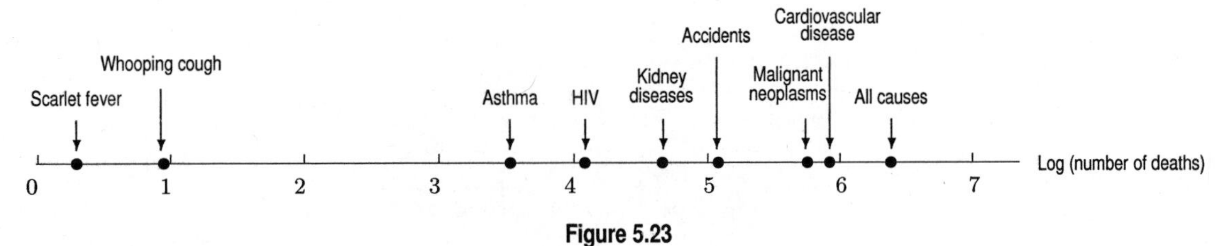

Figure 5.23

15. For the pack of gum, $\log(0.50) = -0.30$, so the pack of gum is plotted at -0.3. For the movie ticket, $\log(9) = 0.95$, so the ticket is plotted at 0.95, and so on. See Figure 5.24.

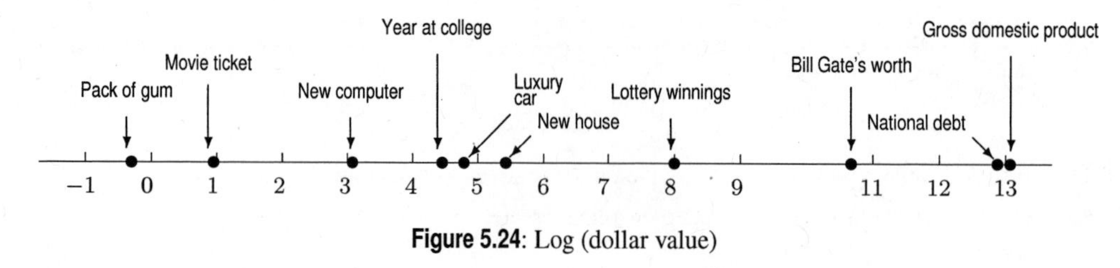

Figure 5.24: Log (dollar value)

16. Table 5.7 gives the logs of the sizes of the various organisms. The log values from Table 5.7 have been plotted in Figure 5.25.

Table 5.7 *Size (in cm) and* log(*size*) *of various organisms*

Animal	Size	log(size)	Animal	Size	log(size)
Virus	0.0000005	−6.3	Cat	60	1.8
Bacterium	0.0002	−3.7	Wolf	200	2.3
Human cell	0.002	−2.7	Shark	600	2.8
Ant	0.8	−0.1	Squid	2200	3.3
Hummingbird	12	1.1	Sequoia	7500	3.9

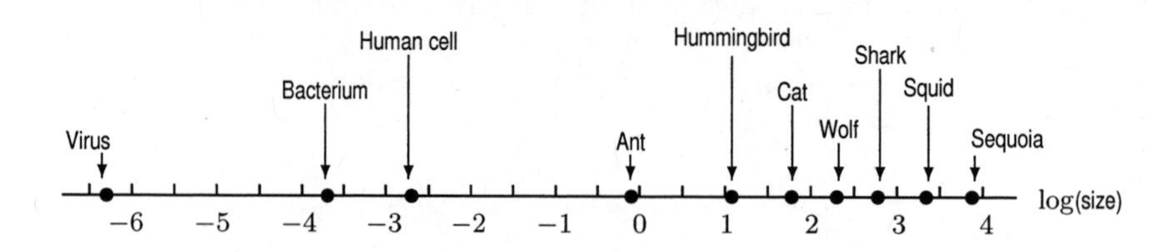

Figure 5.25: The log(sizes) of various organisms (sizes in cm)

17. **(a)**

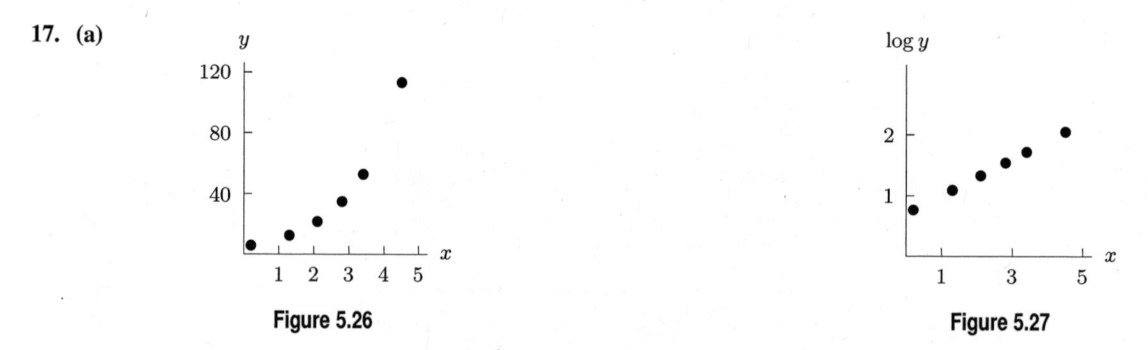

Figure 5.26

Figure 5.27

(b) The data appear to be exponential.
(c) See Figure 5.27. The data appear to be linear.

Table 5.8

x	0.2	1.3	2.1	2.8	3.4	4.5
$\log y$	.76	1.09	1.33	1.54	1.72	2.05

18. (a)

Table 5.9

x	0	1	2	3	4	5
$y = 3^x$	1	3	9	27	81	243

(b)

Table 5.10

x	0	1	2	3	4	5
$y = \log(3^x)$	0	0.477	0.954	1.431	1.908	2.386

The differences between successive terms are constant(≈ 0.477), so the function is linear.

(c)

Table 5.11

x	0	1	2	3	4	5
$f(x)$	2	10	50	250	1250	6250

Table 5.12

x	0	1	2	3	4	5
$g(x)$	0.301	1	1.699	2.398	3.097	3.796

We see that $f(x)$ is an exponential function (note that it is increasing by a constant growth factor of 5), while $g(x)$ is a linear function with a constant rate of change of 0.699.

(d) The resulting function is linear. If $f(x) = a \cdot b^x$ and $g(x) = \log(a \cdot b^x)$ then

$$\begin{aligned} g(x) &= \log(ab^x) \\ &= \log a + \log b^x \\ &= \log a + x \log b \\ &= k + m \cdot x, \end{aligned}$$

where the y intercept $k = \log a$ and $m = \log b$. Thus, g will be linear.

19.

Table 5.13

x	0	1	2	3	4	5
$y = \ln(3^x)$	0	1.0986	2.1972	3.2958	4.3944	5.4931

Table 5.14

x	0	1	2	3	4	5
$g(x) = \ln(2 \cdot 5^x)$	0.6931	2.3026	3.9120	5.5215	7.1309	8.7403

Yes, the results are linear.

20. (a) Using linear regression we get $y = 14.227 - 0.233x$ as an approximation for the percent share x years after 1950. Table 5.15 gives $\ln y$:

Table 5.15

Year	x	$\ln y$
1950	0	2.773
1960	10	2.380
1970	20	2.079
1980	30	1.902
1990	40	1.758
1992	42	1.609

(b) Using linear regression on the values in Table 5.15 we get $\ln y = 2.682 - 0.0253x$.

(c) Taking e to the power of both sides we get $y = e^{2.682-0.0253x} = e^{2.682}(e^{-0.0253x}) \approx 14.614e^{-.0253x}$ as an exponential approximation for the percent share.

21. **(a)** Find the values of $\ln t$ in the table, use linear regression on a calculator or computer with $x = \ln t$ and $y = P$. The line has slope -7.787 and P-intercept 86.283 ($P = -7.787 \ln t + 86.283$). Thus $a = -7.787$ and $b = 86.283$.

(b) Figure 5.28 shows the data points plotted with P against $\ln t$. The model seems to fit well.

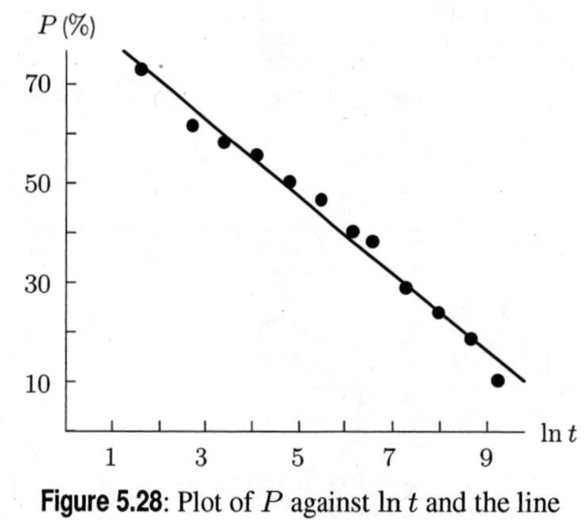

Figure 5.28: Plot of P against $\ln t$ and the line with slope -7.787 and intercept 86.283

(c) The subjects will recognize no words when $P = 0$, that is, when $-7.787 \ln t + 86.283 = 0$. Solving for t:

$$-7.787 \ln t = -86.283$$
$$\ln t = \frac{86.283}{7.787}$$

Taking both sides to the e power,

$$e^{\ln t} = e^{\frac{86.283}{7.787}}$$
$$t \approx 64{,}918.342,$$

so $t \approx 45$ days.

The subject recognized all the words when $P = 100$, that is, when $-7.787 \ln t + 86.283 = 100$. Solving for t:

$$-7.787 \ln t = 13.717$$
$$\ln t = \frac{13.717}{-7.787}$$
$$t \approx 0.172,$$

so $t \approx 0.172$ minutes (≈ 10 seconds) from the start of the experiment.

(d)

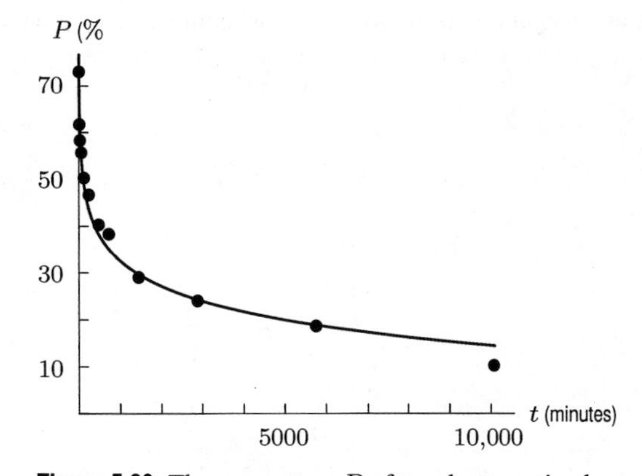

Figure 5.29: The percentage P of words recognized as a function of t, the time elapsed and the function $P = -7.787 \ln t + 86.283$

22. **(a)** We obtain the linear approximation $y = 47316.214 + 38101.643x$ using linear regression.
(b) Table 5.16 gives the natural log of the cost of imports, rather than the cost of imports itself.

Table 5.16

Year	x	$\ln y$
2002	0	11.7376
2003	1	11.9345
2004	2	12.1893
2005	3	12.4027
2006	4	12.5699
2007	5	12.6806
2008	6	12.7301

Using linear regression we get $\ln y = 11.801 + 0.1732x$ as an approximation.
(c) To find a formula for the cost and not for the natural log of the cost, we need to solve

$$\begin{aligned}
\ln y &= 11.801 + 0.1732x \quad \text{for } y. \\
e^{\ln y} &= e^{11.801+0.1732x} \\
y &= e^{11.801} e^{0.1732x} \\
y &= 133{,}385.672 e^{0.1732x}
\end{aligned}$$

23. **(a)** Table 5.17 gives values of $L = \ln \ell$ and $W = \ln w$. The data in Table 5.17 have been plotted in Figure 5.30, and a line of best fit has been drawn in. See part (b).
(b) The formula for the line of best fit is $W = 3.06L - 4.54$, as determined using a spreadsheet. However, you could also obtain comparable results by fitting a line by eye.
(c) We have

$$\begin{aligned}
W &= 3.06L - 4.54 \\
\ln w &= 3.06 \ln \ell - 4.54 \\
\ln w &= \ln \ell^{3.06} - 4.54 \\
w &= e^{\ln \ell^{3.06} - 4.54} \\
&= \ell^{3.06} e^{-4.54} \approx 0.011\ell^{3.06}.
\end{aligned}$$

(d) Weight tends to be directly proportional to volume, and in many cases volume tends to be proportional to the cube of a linear dimension (e.g., length). Here we see that w is in fact very nearly proportional to the cube of ℓ.

Table 5.17 *$L = \ln \ell$ and $W = \ln w$ for 16 different fish*

Type	1	2	3	4	5	6	7	8
L	2.092	2.208	2.322	2.477	2.501	2.625	2.695	2.754
W	1.841	2.262	2.451	2.918	3.266	3.586	3.691	3.857
Type	9	10	11	12	13	14	15	16
L	2.809	2.874	2.929	2.944	3.025	3.086	3.131	3.157
W	4.184	4.240	4.336	4.413	4.669	4.786	5.131	5.155

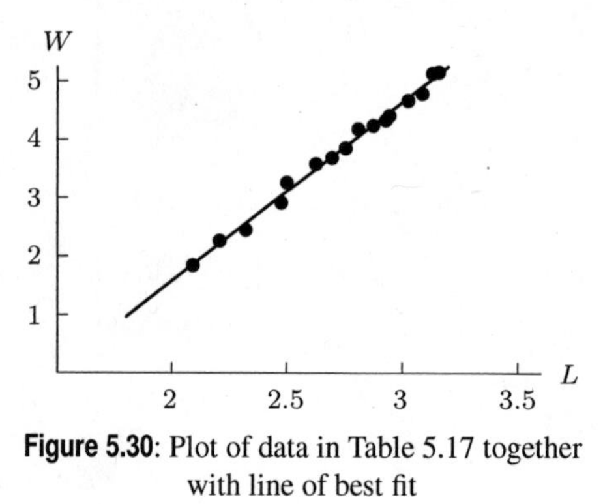

Figure 5.30: Plot of data in Table 5.17 together with line of best fit

24. (a) After converting the I values to $\ln I$, we use linear regression on a computer or calculator with $x = \ln I$ and $y = F$. We find $a \approx 4.26$ and $b \approx 8.95$ so that $F = 4.26 \ln I + 8.95$. Figure 5.31 shows a plot of F against $\ln I$ and the line with slope 4.26 and intercept 8.95.

(b) See Figure 5.31.

(c) Figure 5.32 shows a plot of $F = 4.26 \ln I + 8.95$ and the data set in Table 5.21. The model seems to fit well.

(d) Imagine the units of I were changed by a factor of $\alpha > 0$ so that $I_{\text{old}} = \alpha I_{\text{new}}$.

Then

$$\begin{aligned} F &= a_{\text{old}} \ln I_{\text{old}} + b_{\text{old}} \\ &= a_{\text{old}} \ln(\alpha I_{\text{new}}) + b_{\text{old}} \\ &= a_{\text{old}}(\ln \alpha + \ln I_{\text{new}}) + b_{\text{old}} \\ &= a_{\text{old}} \ln \alpha + a_{\text{old}} \ln I_{\text{new}} + b_{\text{old}}. \end{aligned}$$

Rearranging and matching terms, we see:

$$F = \underbrace{a_{\text{old}} \ln I_{\text{new}}}_{a_{\text{new}} \ln I_{\text{new}}} + \underbrace{a_{\text{old}} \ln \alpha + b_{\text{old}}}_{b_{\text{new}}}$$

so

$$a_{\text{new}} = a_{\text{old}} \quad \text{and} \quad b_{\text{new}} = b_{\text{old}} + a_{\text{old}} \ln \alpha.$$

We can also see that if $\alpha > 1$ then $\ln \alpha > 0$ so the term $a_{\text{old}} \ln \alpha$ is positive and $b_{\text{new}} > b_{\text{old}}$. If $\alpha < 1$ then $\ln \alpha < 0$ so the term $a_{\text{old}} \ln \alpha$ is negative and $b_{\text{new}} < b_{\text{old}}$.

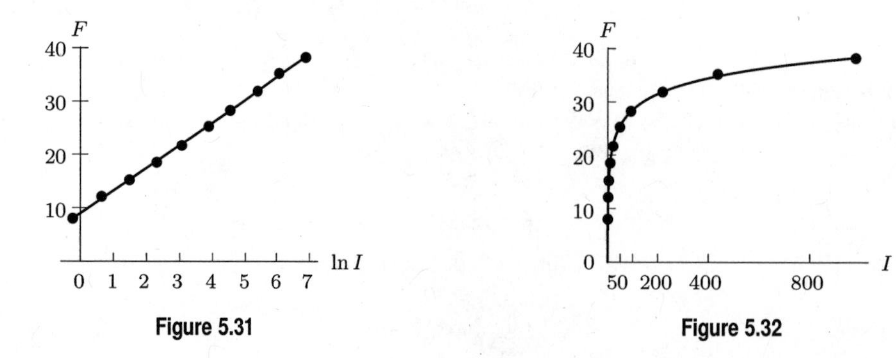

Figure 5.31 **Figure 5.32**

Solutions for Chapter 5 Review

Exercises

1. The continuous percent growth rate is the value of k in the equation $Q = ae^{kt}$, which is -10.

To convert to the form $Q = ab^t$, we first say that the right sides of the two equations equal each other (since each equals Q), and then we solve for a and b. Thus, we have $ab^t = 7e^{-10t}$. At $t = 0$, we can solve for a:

$$\begin{aligned} ab^0 &= 7e^{-10\cdot 0} \\ a\cdot 1 &= 7\cdot 1 \\ a &= 7. \end{aligned}$$

Thus, we have $7b^t = 7e^{-10t}$, and we solve for b:

$$\begin{aligned} 7b^t &= e^{-10t} \\ b^t &= e^{-10t} \\ b^t &= \left(e^{-10}\right)^t \\ b &= e^{-10} \approx 0.0000454. \end{aligned}$$

Therefore, the equation is $Q = 7(0.0000454)^t$.

2. The continuous percent growth rate is the value of k in the equation $Q = ae^{kt}$, which is 1 (since $t \cdot 1 = 1$).

To convert to the form $Q = ab^t$, we first say that the right sides of the two equations equal each other (since each equals Q), and then we solve for a and b. Thus, we have $ab^t = 5e^t$. At $t = 0$, we can solve for a:

$$\begin{aligned} ab^0 &= 5e^0 \\ a\cdot 1 &= 5\cdot 1 \\ a &= 5. \end{aligned}$$

Thus, we have $5b^t = 5e^t$, and we solve for b:

$$\begin{aligned} 5b^t &= 5e^t \\ b^t &= e^t \\ b &= e \approx 2.718. \end{aligned}$$

Therefore, the equation is $Q = 5e^t$ or $Q = 5 \cdot 2.718^t$.

3. To convert to the form $Q = ae^{kt}$, we first say that the right sides of the two equations equal each other (since each equals Q), and then we solve for a and k. Thus, we have $ae^{kt} = 4 \cdot 7^t$. At $t = 0$, we can solve for a:

$$\begin{aligned} ae^{k\cdot 0} &= 4\cdot 7^0 \\ a\cdot 1 &= 4\cdot 1 \\ a &= 4. \end{aligned}$$

Thus, we have $4e^{kt} = 4 \cdot 7^t$, and we solve for k:

$$\begin{aligned} 4e^{kt} &= 4\cdot 7^t \\ e^{kt} &= 7^t \\ \left(e^k\right)^t &= 7^t \\ e^k &= 7 \\ \ln e^k &= \ln 7 \\ k &= \ln 7 \approx 1.946. \end{aligned}$$

Therefore, the equation is $Q = 4e^{1.946t}$.

4. To convert to the form $Q = ae^{kt}$, we first say that the right sides of the two equations equal each other (since each equals Q), and then we solve for a and k. Thus, we have $ae^{kt} = 2 \cdot 3^t$. At $t = 0$, we can solve for a:

$$\begin{aligned} ae^{k \cdot 0} &= 2 \cdot 3^0 \\ a \cdot 1 &= 2 \cdot 1 \\ a &= 2. \end{aligned}$$

Thus, we have $2e^{kt} = 2 \cdot 3^t$, and we solve for k:

$$\begin{aligned} 2e^{kt} &= 2 \cdot 3^t \\ e^{kt} &= 3^t \\ \left(e^k\right)^t &= 3^t \\ e^k &= 3 \\ \ln e^k &= \ln 3 \\ k &= \ln 3 \approx 1.099. \end{aligned}$$

Therefore, the equation is $Q = 2e^{1.099t}$.

5. To convert to the form $Q = ae^{kt}$, we first say that the right sides of the two equations equal each other (since each equals Q), and then we solve for a and k. Thus, we have $ae^{kt} = 4 \cdot 8^{1.3t}$. At $t = 0$, we can solve for a:

$$\begin{aligned} ae^{k \cdot 0} &= 4 \cdot 8^0 \\ a \cdot 1 &= 4 \cdot 1 \\ a &= 4. \end{aligned}$$

Thus, we have $4e^{kt} = 4 \cdot 8^{1.3t}$, and we solve for k:

$$\begin{aligned} 4e^{kt} &= 4 \cdot 8^{1.3t} \\ e^{kt} &= \left(8^{1.3}\right)^t \\ \left(e^k\right)^t &= 14.929^t \\ e^k &= 14.929 \\ \ln e^k &= \ln 14.929 \\ k &= 2.703. \end{aligned}$$

Therefore, the equation is $Q = 4e^{2.703t}$.

6. To convert to the form $Q = ae^{kt}$, we first say that the right sides of the two equations equal each other (since each equals Q), and then we solve for a and k. Thus, we have $ae^{kt} = 973 \cdot 6^{2.1t}$. At $t = 0$, we can solve for a:

$$\begin{aligned} ae^{k \cdot 0} &= 973 \cdot 6^0 \\ a \cdot 1 &= 973 \cdot 1 \\ a &= 973. \end{aligned}$$

Thus, we have $973e^{kt} = 973 \cdot 6^{2.1t}$, and we solve for k:

$$\begin{aligned} 973e^{kt} &= 973 \cdot 6^{2.1t} \\ e^{kt} &= \left(6^{2.1}\right)^t \\ \left(e^k\right)^t &= 43.064^t \\ e^k &= 43.064 \\ \ln e^k &= \ln 43.064 \\ k &= 3.763. \end{aligned}$$

Therefore, the equation is $Q = 973e^{3.763t}$.

7. Using the log rules,

$$\begin{aligned}
1.04^t &= 3\\
\log(1.04)^t &= \log 3\\
t\log 1.04 &= \log 3\\
t &= \frac{\log 3}{\log 1.04}.
\end{aligned}$$

Using your calculator, you will find that $\frac{\log 3}{\log 1.04} \approx 28$. You can check your answer: $1.04^{28} \approx 3$.

8. We are solving for an exponent, so we use logarithms. Using the log rules, we have

$$\begin{aligned}
e^{0.15t} &= 25\\
\ln(e^{0.15t}) &= \ln(25)\\
0.15t &= \ln(25)\\
t &= \frac{\ln(25)}{0.15} = 21.459.
\end{aligned}$$

9. Since the goal is to get t by itself as much as possible, first divide both sides by 3, and then use logs.

$$\begin{aligned}
3(1.081)^t &= 14\\
1.081^t &= \frac{14}{3}\\
\log(1.081)^t &= \log(\frac{14}{3})\\
t\log 1.081 &= \log(\frac{14}{3})\\
t &= \frac{\log(\frac{14}{3})}{\log 1.081}.
\end{aligned}$$

10. We are solving for an exponent, so we use logarithms. We first divide both sides by 40 and then use logs:

$$\begin{aligned}
40e^{-0.2t} &= 12\\
e^{-0.2t} &= 0.3\\
\ln(e^{-0.2t}) &= \ln(0.3)\\
-0.2t &= \ln(0.3)\\
t &= \frac{\ln(0.3)}{-0.2} = 6.020.
\end{aligned}$$

11. Using the log rules

$$\begin{aligned}
5(1.014)^{3t} &= 12\\
1.014^{3t} &= \frac{12}{5}\\
\log(1.014)^{3t} &= \log(\frac{12}{5}) = \log 2.4\\
3t\log 1.014 &= \log 2.4\\
3t &= \frac{\log 2.4}{\log 1.014}\\
t &= \frac{\log 2.4}{3\log 1.014}.
\end{aligned}$$

12. Get all expressions containing t on one side of the equation and everything else on the other side. So we will divide both sides of the equation by 5 and by $(1.07)^t$.

$$\begin{aligned}
5(1.15)^t &= 8(1.07)^t \\
\frac{1.15^t}{1.07^t} &= \frac{8}{5} \\
(\frac{1.15}{1.07})^t &= \frac{8}{5} \\
\log(\frac{1.15}{1.07})^t &= \log\frac{8}{5} \\
t\log(\frac{1.15}{1.07}) &= \log\frac{8}{5} \\
t &= \frac{\log(\frac{8}{5})}{\log(\frac{1.15}{1.07})}.
\end{aligned}$$

13.

$$\begin{aligned}
5(1.031)^x &= 8 \\
1.031^x &= \frac{8}{5} \\
\log(1.031)^x &= \log\frac{8}{5} \\
x\log 1.031 &= \log\frac{8}{5} = \log 1.6 \\
x &= \frac{\log 1.6}{\log 1.031} = 15.395.
\end{aligned}$$

Check your answer: $5(1.031)^{15.395} approx 8$.

14.

$$\begin{aligned}
4(1.171)^x &= 7(1.088)^x \\
\frac{(1.171)^x}{(1.088)^x} &= \frac{7}{4} \\
\left(\frac{1.171}{1.088}\right)^x &= \frac{7}{4} \\
\log\left(\frac{1.171}{1.088}\right)^x &= \log\left(\frac{7}{4}\right) \\
x\log\left(\frac{1.171}{1.088}\right) &= \log\left(\frac{7}{4}\right) \\
x &= \frac{\log\left(\frac{7}{4}\right)}{\log\left(\frac{1.171}{1.088}\right)} \approx 7.612.
\end{aligned}$$

Checking your answer, you will see that

$$4(1.171)^{7.612} \approx 13.302 \qquad 7(1.088)^{7.612} \approx 13.302.$$

15.

$$3\log(2x+6) = 6$$

Dividing both sides by 3, we get:

$$\log(2x+6) = 2$$

Rewriting in exponential form gives

$$\begin{aligned}
2x + 6 &= 10^2 \\
2x + 6 &= 100 \\
2x &= 94 \\
x &= 47.
\end{aligned}$$

Check and get:

$$3\log(2 \cdot 47 + 6) = 3\log(100).$$

Since $\log 100 = 2$,

$$3\log(2 \cdot 47 + 6) = 3(2) = 6,$$

which is the result we want.

16.

$$\begin{aligned}
\frac{(2.1)^{3x}}{(4.5)^x} &= \frac{2}{1.7} \\
\left(\frac{(2.1)^3}{4.5}\right)^x &= \frac{2}{1.7} \\
\log\left(\frac{(2.1)^3}{4.5}\right)^x &= \log\left(\frac{2}{1.7}\right) \\
x\log\left(\frac{(2.1)^3}{4.5}\right) &= \log\left(\frac{2}{1.7}\right) \\
x &= \frac{\log(\frac{2}{1.7})}{\log\frac{(2.1)^3}{4.5}} \approx 0.225178
\end{aligned}$$

17.

$$\begin{aligned}
3^{(4\log x)} &= 5 \\
\log 3^{(4\log x)} &= \log 5 \\
(4\log x)\log 3 &= \log 5 \\
4\log x &= \frac{\log 5}{\log 3} \\
\log x &= \frac{\log 5}{4\log 3} \\
x &= 10^{\frac{\log 5}{4\log 3}} \approx 2.324
\end{aligned}$$

18. Rewriting each side to the base 10 gives

$$\begin{aligned}
(10^2)^{2x+3} &= \left(10^4\right)^{1/3} \\
10^{4x+6} &= 10^{4/3}.
\end{aligned}$$

Since the base of each side is the same, we can equate the exponents:

$$\begin{aligned}
4x + 6 &= \frac{4}{3} \\
12x + 18 &= 4 \\
12x &= -14 \\
x &= -\frac{14}{12} = -\frac{7}{6}.
\end{aligned}$$

19. Dividing by 13 and 25 before taking logs gives

$$
\begin{aligned}
13e^{0.081t} &= 25e^{0.032t} \\
\frac{e^{0.081t}}{e^{0.032t}} &= \frac{25}{13} \\
e^{0.081t-0.032t} &= \frac{25}{13} \\
\ln e^{0.049t} &= \ln\left(\frac{25}{13}\right) \\
0.049t &= \ln\left(\frac{25}{13}\right) \\
t &= \frac{1}{0.049}\ln\left(\frac{25}{13}\right) \approx 13.345.
\end{aligned}
$$

20. This equation cannot be solved analytically. Graphing $y = 87e^{0.066t}$ and $y = 3t + 7$ it is clear that these graphs will not intersect, which means $87e^{0.66t} = 3t + 7$ has no solution. The concavity of the graphs ensures that they will not intersect beyond the portions of the graphs shown in Figure 5.33.

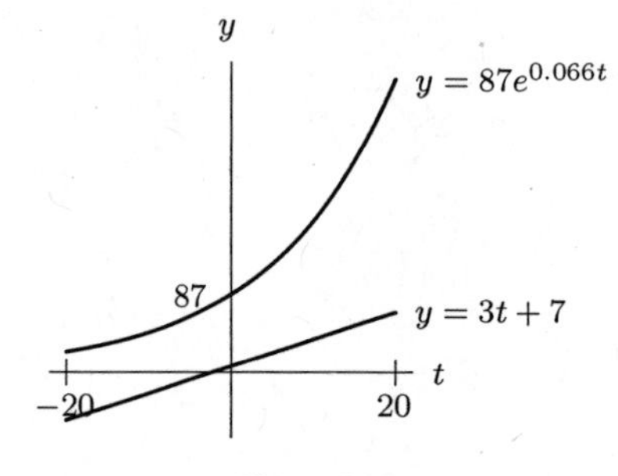

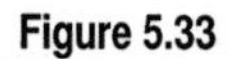
Figure 5.33

21.

$$
\begin{aligned}
\frac{\log x^2 + \log x^3}{\log(100x)} &= 3 \\
\log x^2 + \log x^3 &= 3\log(100x) \\
2\log x + 3\log x &= 3(\log 100 + \log x) \\
5\log x &= 3(2 + \log x) \\
5\log x &= 6 + 3\log x \\
2\log x &= 6 \\
\log x &= 3 \\
x &= 10^3 = 1000.
\end{aligned}
$$

To check, we see that

$$
\begin{aligned}
\frac{\log x^2 + \log x^3}{\log(100x)} &= \frac{\log(1000^2) + \log(1000^3)}{\log(100 \cdot 1000)} \\
&= \frac{\log(1{,}000{,}000) + \log(1{,}000{,}000{,}000)}{\log(100{,}000)} \\
&= \frac{6+9}{5} \\
&= 3,
\end{aligned}
$$

as required.

22.

$$\begin{aligned}\log x + \log(x-1) &= \log 2\\ \log(x(x-1)) &= \log 2\\ x(x-1) &= 2\\ x^2 - x - 2 &= 0\\ (x-2)(x+1) &= 0\\ x &= 2 \text{ or } -1\end{aligned}$$

but $x \neq -1$ since $\log x$ is undefined at $x = -1$. Thus $x = 2$.

23. $\log\left(100^{x+1}\right) = \log\left((10^2)^{x+1}\right) = 2(x+1)$.

24. $\ln\left(e \cdot e^{2+M}\right) = \ln\left(e^{3+M}\right) = 3 + M$.

25. Using the fact that $A^{-1} = 1/A$ and the log rules:

$$\begin{aligned}\ln(A+B) - \ln(A^{-1}+B^{-1}) &= \ln(A+B) - \ln\left(\frac{1}{A}+\frac{1}{B}\right)\\ &= \ln(A+B) - \ln\frac{A+B}{AB}\\ &= \ln\left((A+B)\cdot\frac{AB}{A+B}\right)\\ &= \ln(AB).\end{aligned}$$

26.
- The natural logarithm is defined only for positive inputs, so the domain of this function is given by

$$\begin{aligned}x + 8 &> 0\\ x &> -8.\end{aligned}$$

- The graph of $y = \ln x$ has a vertical asymptote at $x = 0$, that is, where the input is zero. So the graph of $y = \ln(x+8)$ has a vertical asymptote where its input is zero, at $x = -8$.

27.
- The common logarithm is defined only for positive inputs, so the domain of this function is given by

$$\begin{aligned}x - 20 &> 0\\ x &> 20.\end{aligned}$$

- The graph of $y = \log x$ has an asymptote at $x = 0$, that is, where the input is zero. So the graph of $y = \log(x-20)$ has a vertical asymptote where its input is zero, at $x = 20$.

28.
- The common logarithm is defined only for positive inputs, so the domain of this function is given by

$$\begin{aligned}12 - x &> 0\\ x &< 12.\end{aligned}$$

- The graph of $y = \log x$ has a vertical asymptote at $x = 0$, that is, where the input is zero. So the graph of $y = \log(12-x)$ has a vertical asymptote where its input is zero, at $x = 12$.

29.
- The natural logarithm is defined only for positive inputs, so the domain of this function is given by

$$\begin{aligned}300 - x &> 0\\ x &< 300.\end{aligned}$$

- The graph of $y = \ln x$ has a vertical asymptote at $x = 0$, that is, where the input is zero. So the graph of $y = \ln(300-x)$ has a vertical asymptote where its input is zero, at $x = 300$.

30.
- The natural logarithm is defined only for positive inputs, so the domain of this function is given by

$$\begin{aligned} x - e^2 &> 0 \\ x &> e^2. \end{aligned}$$

- The graph of $y = \ln x$ has a vertical asymptote at $x = 0$, that is, where the input is zero. So the graph of $y = \ln\left(x - e^2\right)$ has a vertical asymptote where its input is zero, at $x = e^2$.

31.
- The common logarithm is defined only for positive inputs, so the domain of this function is given by

$$\begin{aligned} x + 15 &> 0 \\ x &> -15. \end{aligned}$$

- The graph of $y = \log x$ has an asymptote at $x = 0$, that is, where the input is zero. So the graph of $y = \log(x + 15)$ has a vertical asymptote where its input is zero, at $x = -15$.

32. We have $\log 80{,}000 = 4.903$, so this lifespan would be marked at 4.9 inches.

33. We have $\log 4838 = 3.685$, so this lifespan would be marked at 3.7 inches.

34. We have $\log 1550 = 3.190$, so this lifespan would be marked at 3.2 inches.

35. We have $\log 160 = 2.204$, so this lifespan would be marked at 2.2 inches.

36. We have $\log 29 = 1.462$, so this lifespan would be marked at 1.5 inches.

37. We have $\log 4 = 0.602$, so this lifespan would be marked at 0.6 inches.

Problems

38. **(a)** $\log AB = \log A + \log B = x + y$

(b) $\log(A^3 \cdot \sqrt{B}) = \log A^3 + \log \sqrt{B} = 3\log A + \log B^{\frac{1}{2}} = 3\log A + \frac{1}{2}\log B = 3x + \frac{1}{2}y$

(c) $\log(A - B) = \log(10^x - 10^y)$ because $A = 10^{\log A} = 10^x$ and $B = 10^{\log B} = 10^y$, and this can't be further simplified.

(d) $\dfrac{\log A}{\log B} = \dfrac{x}{y}$

(e) $\log\left(\dfrac{A}{B}\right) = \log A - \log B = x - y$

(f) $AB = 10^x \cdot 10^y = 10^{x+y}$

39. **(a)** We have $\ln(nm^4) = \ln n + 4\ln m = q + 4p$.

(b) We have $\ln(1/n) = \ln 1 - \ln n = 0 - \ln n = -q$.

(c) We have $(\ln m)/(\ln n) = p/q$.

(d) We have $\ln(n^3) = 3\ln n = 3q$.

40. **(a)**

$$\begin{aligned} \log xy &= \log(10^U \cdot 10^V) \\ &= \log 10^{U+V} \\ &= U + V \end{aligned}$$

(b)

$$\begin{aligned} \log \frac{x}{y} &= \log \frac{10^U}{10^V} \\ &= \log 10^{U-V} \\ &= U - V \end{aligned}$$

(c)

$$\begin{aligned} \log x^3 &= \log(10^U)^3 \\ &= \log 10^{3U} \\ &= 3U \end{aligned}$$

(d)

$$\begin{aligned}\log\frac{1}{y} &= \log\frac{1}{10^V}\\ &= \log 10^{-V}\\ &= -V\end{aligned}$$

41. (a)

$$\begin{aligned}e^{x+3} &= 8\\ \ln e^{x+3} &= \ln 8\\ x+3 &= \ln 8\\ x &= \ln 8 - 3 \approx -0.9206\end{aligned}$$

(b)

$$\begin{aligned}4(1.12^x) &= 5\\ 1.12^x &= \frac{5}{4} = 1.25\\ \log 1.12^x &= \log 1.25\\ x\log 1.12 &= \log 1.25\\ x &= \frac{\log 1.25}{\log 1.12} \approx 1.9690\end{aligned}$$

(c)

$$\begin{aligned}e^{-0.13x} &= 4\\ \ln e^{-0.13x} &= \ln 4\\ -0.13x &= \ln 4\\ x &= \frac{\ln 4}{-0.13} \approx -10.6638\end{aligned}$$

(d)

$$\begin{aligned}\log(x-5) &= 2\\ x-5 &= 10^2\\ x &= 10^2 + 5 = 105\end{aligned}$$

(e)

$$\begin{aligned}2\ln(3x)+5 &= 8\\ 2\ln(3x) &= 3\\ \ln(3x) &= \frac{3}{2}\\ 3x &= e^{\frac{3}{2}}\\ x &= \frac{e^{\frac{3}{2}}}{3} \approx 1.4939\end{aligned}$$

(f)

$$\begin{aligned}\ln x - \ln(x-1) &= \frac{1}{2}\\ \ln\left(\frac{x}{x-1}\right) &= \frac{1}{2}\end{aligned}$$

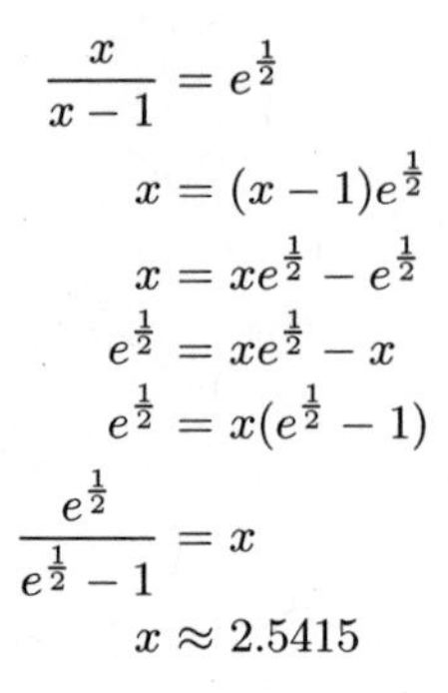

$$
\begin{aligned}
\frac{x}{x-1} &= e^{\frac{1}{2}} \\
x &= (x-1)e^{\frac{1}{2}} \\
x &= xe^{\frac{1}{2}} - e^{\frac{1}{2}} \\
e^{\frac{1}{2}} &= xe^{\frac{1}{2}} - x \\
e^{\frac{1}{2}} &= x(e^{\frac{1}{2}} - 1) \\
\frac{e^{\frac{1}{2}}}{e^{\frac{1}{2}} - 1} &= x \\
x &\approx 2.5415
\end{aligned}
$$

Note: (g) (h) and (i) can not be solved analytically, so we use graphs to approximate the solutions.

(g) From Figure 5.34 we can see that $y = e^x$ and $y = 3x + 5$ intersect at $(2.534, 12.601)$ and $(-1.599, 0.202)$, so the values of x which satisfy $e^x = 3x+5$ are $x = 2.534$ or $x = -1.599$. We also see that $y_1 \approx 12.601$ and $y_2 \approx 0.202$.

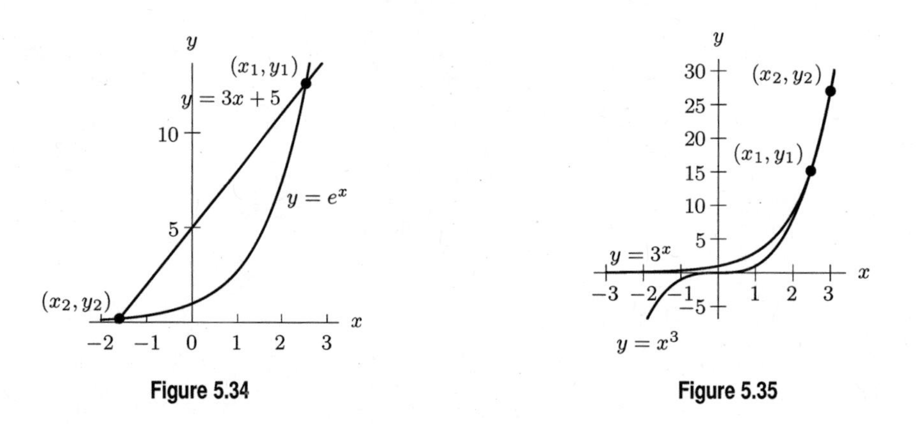

Figure 5.34

Figure 5.35

(h) The graphs of $y = 3^x$ and $y = x^3$ are seen in Figure 5.35. It is very hard to see the points of intersection, though $(3, 27)$ would be an immediately obvious choice (substitute 3 for x in each of the formulas). Using technology, we can find a second point of intersection, $(2.478, 15.216)$. So the solutions for $3^x = x^3$ are $x = 3$ or $x = 2.478$.

Since the points of intersection are very close, it is difficult to see these intersections even by zooming in. So, alternatively, we can find where $y = 3^x - x^3$ crosses the x-axis. See Figure 5.36.

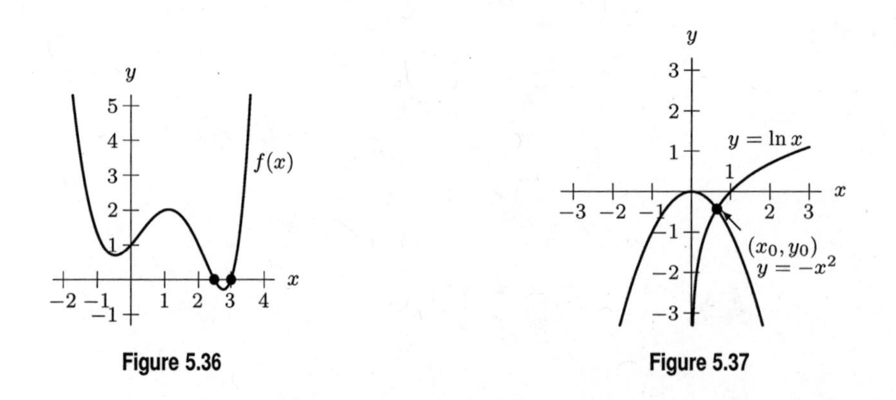

Figure 5.36

Figure 5.37

(i) From the graph in Figure 5.37, we see that $y = \ln x$ and $y = -x^2$ intersect at $(0.6529, -0.4263)$, so $x = 0.6529$ is the solution to $\ln x = -x^2$.

42. **(a)** Solving for x exactly:

$$\begin{aligned}
\frac{3^x}{5^{(x-1)}} &= 2^{(x-1)} \\
3^x &= 5^{x-1}\cdot 2^{x-1} \\
3^x &= (5\cdot 2)^{x-1} \\
3^x &= 10^{x-1} \\
\log 3^x &= \log 10^{x-1} \\
x\log 3 &= (x-1)\log 10 = (x-1)(1) \\
x\log 3 &= x-1 \\
x\log 3 - x &= -1 \\
x(\log 3 - 1) &= -1 \\
x &= \frac{-1}{\log 3 - 1} = \frac{1}{1-\log 3}
\end{aligned}$$

(b)

$$\begin{aligned}
-3+e^{x+1} &= 2+e^{x-2} \\
e^{x+1}-e^{x-2} &= 2+3 \\
e^x e^1 - e^x e^{-2} &= 5 \\
e^x(e^1-e^{-2}) &= 5 \\
e^x &= \frac{5}{e-e^{-2}} \\
\ln e^x &= \ln\left(\frac{5}{e-e^{-2}}\right) \\
x &= \ln\left(\frac{5}{e-e^{-2}}\right)
\end{aligned}$$

(c)

$$\begin{aligned}
\ln(2x-2)-\ln(x-1) &= \ln x \\
\ln\left(\frac{2x-2}{x-1}\right) &= \ln x \\
\frac{2x-2}{x-1} &= x \\
\frac{2(x-1)}{(x-1)} &= x \\
2 &= x
\end{aligned}$$

(d) Let $z=3^x$, then $z^2=(3^x)^2=9^x$, and so we have

$$\begin{aligned}
z^2-7z+6 &= 0 \\
(z-6)(z-1) &= 0 \\
z &= 6 \quad \text{or} \quad z=1.
\end{aligned}$$

Thus, $3^x=1$ or $3^x=6$, and so $x=0$ or $x=\ln 6/\ln 3$.

(e)

$$\begin{aligned}
\ln\left(\frac{e^{4x}+3}{e}\right) &= 1 \\
e^1 &= \frac{e^{4x}+3}{e}
\end{aligned}$$

$$\begin{aligned}
e^2 &= e^{4x} + 3 \\
e^2 - 3 &= e^{4x} \\
\ln(e^2 - 3) &= \ln\left(e^{4x}\right) \\
\ln(e^2 - 3) &= 4x \\
\frac{\ln(e^2 - 3)}{4} &= x
\end{aligned}$$

(f)

$$\begin{aligned}
\frac{\ln(8x) - 2\ln(2x)}{\ln x} &= 1 \\
\ln(8x) - 2\ln(2x) &= \ln x \\
\ln(8x) - \ln\left((2x)^2\right) &= \ln x \\
\ln\left(\frac{8x}{(2x)^2}\right) &= \ln x \\
\ln\left(\frac{8x}{4x^2}\right) &= \ln x \\
\frac{8x}{4x^2} &= x \\
8x &= 4x^3 \\
4x^3 - 8x &= 0 \\
4x(x^2 - 2) &= 0 \\
x &= 0, \sqrt{2}, -\sqrt{2}
\end{aligned}$$

Only $\sqrt{2}$ is a valid solution, because when $-\sqrt{2}$ and 0 are substituted into the original equation we are taking the logarithm of negative numbers and 0, which is undefined.

43. We have

$$\begin{aligned}
M_2 - M_1 &= \log\left(\frac{W_2}{W_1}\right) \\
6.4 - 4.2 &= \log\left(\frac{W_2}{W_1}\right) \\
2.2 &= \log\left(\frac{W_2}{W_1}\right) \\
\frac{W_2}{W_1} &= 10^{2.2} = 158.489 \\
W_2 &= 158.489 \cdot W_1.
\end{aligned}$$

The seismic waves of the second earthquake are about 158.5 times larger.

44. We have

$$\begin{aligned}
M_2 - M_1 &= \log\left(\frac{W_2}{W_1}\right) \\
5.8 - 5.3 &= \log\left(\frac{W_2}{W_1}\right) \\
0.5 &= \log\left(\frac{W_2}{W_1}\right) \\
\frac{W_2}{W_1} &= 10^{0.5} = 3.162 \\
W_2 &= 3.162 \cdot W_1.
\end{aligned}$$

The seismic waves of the second earthquake are about 3.2 times larger.

45. We have

$$\begin{aligned}
M_2 - M_1 &= \log\left(\frac{W_2}{W_1}\right)\\
5.6 - 4.4 &= \log\left(\frac{W_2}{W_1}\right)\\
1.2 &= \log\left(\frac{W_2}{W_1}\right)\\
\frac{W_2}{W_1} &= 10^{1.2} = 15.849\\
W_2 &= 15.849 \cdot W_1.
\end{aligned}$$

The seismic waves of the second earthquake are about 15.85 times larger.

46. We have

$$\begin{aligned}
M_2 - M_1 &= \log\left(\frac{W_2}{W_1}\right)\\
8.1 - 5.7 &= \log\left(\frac{W_2}{W_1}\right)\\
2.4 &= \log\left(\frac{W_2}{W_1}\right)\\
\frac{W_2}{W_1} &= 10^{2.4} = 251.189\\
W_2 &= 251.189 \cdot W_1.
\end{aligned}$$

The seismic waves of the second earthquake are about 251 times larger.

47. **(a)** For f, the initial balance is $\$1100$ and the effective rate is 5 percent each year.

(b) For g, the initial balance is $\$1500$ and the effective rate is 5.127%, because $e^{0.05} = 1.05127$.

(c) To find the continuous interest rate we must have $e^k = 1.05$. Therefore

$$\begin{aligned}
\ln e^k &= \ln 1.05\\
k &= \ln 1.05 = 0.04879.
\end{aligned}$$

To earn an effective annual rate of 5%, the bank would need to pay a continuous annual rate of 4.879%.

48. **(a)**

$$\begin{aligned}
\text{If } B = 5000(1.06)^t &= 5000e^{kt},\\
1.06^t &= (e^k)^t\\
\text{we have } e^k &= 1.06.
\end{aligned}$$

Use the natural log to solve for k,

$$k = \ln(1.06) \approx 0.0583.$$

This means that at a continuous growth rate of 5.83%/year, the account has an effective annual yield of 6%.

(b)

$$\begin{aligned}
7500e^{0.072t} &= 7500b^t\\
e^{0.072t} &= b^t\\
e^{0.072} &= b\\
b &\approx 1.0747
\end{aligned}$$

This means that an account earning 7.2% continuous annual interest has an effective yield of 7.47%.

49. **(a)** The number of bacteria present after 1/2 hour is

$$N = 1000e^{0.69(1/2)} \approx 1412.$$

If you notice that $0.69 \approx \ln 2$, you could also say

$$N = 1000e^{0.69/2} \approx 1000e^{\frac{1}{2}\ln 2} = 1000e^{\ln 2^{1/2}} = 1000e^{\ln\sqrt{2}} = 1000\sqrt{2} \approx 1412.$$

(b) We solve for t in the equation

$$\begin{aligned} 1{,}000{,}000 &= 1000e^{0.69t} \\ e^{0.69t} &= 1000 \\ 0.69t &= \ln 1000 \\ t &= \left(\frac{\ln 1000}{0.69}\right) \approx 10.011 \text{ hours.} \end{aligned}$$

(c) The doubling time is the time t such that $N = 2000$, so

$$\begin{aligned} 2000 &= 1000e^{0.69t} \\ e^{0.69t} &= 2 \\ 0.69t &= \ln 2 \\ t &= \left(\frac{\ln 2}{0.69}\right) \approx 1.005 \text{ hours.} \end{aligned}$$

If you notice that $0.69 \approx \ln 2$, you see why the half-life turns out to be 1 hour:

$$\begin{aligned} e^{0.69t} &= 2 \\ e^{t\ln 2} &\approx 2 \\ e^{\ln 2^t} &\approx 2 \\ 2^t &\approx 2 \\ t &\approx 1 \end{aligned}$$

50. **(a)** If t represents the number of years since 2010, let $W(t) =$ population of Erehwon at time t, in millions of people, and let $C(t) =$ population of Ecalpon at time t, in millions of people. Since the population of both Erehwon and Ecalpon are increasing by a constant percent, we know that they are both exponential functions. In Erehwon, the growth factor is 1.029. Since its population in 2010 (when $t = 0$) is 50 million people, we know that

$$W(t) = 50(1.029)^t.$$

In Ecalpon, the growth factor is 1.032, and the population starts at 45 million, so

$$C(t) = 45(1.032)^t.$$

(b) The two countries will have the same population when $W(t) = C(t)$. We therefore need to solve:

$$\begin{aligned} 50(1.029)^t &= 45(1.032)^t \\ \frac{1.032^t}{1.029^t} = \left(\frac{1.032}{1.029}\right)^t &= \frac{50}{45} = \frac{10}{9} \\ \log\left(\frac{1.032}{1.029}\right)^t &= \log\left(\frac{10}{9}\right) \\ t\log\left(\frac{1.032}{1.029}\right) &= \log\left(\frac{10}{9}\right) \\ t &= \frac{\log(10/9)}{\log(1.032/1.029)} = 36.191. \end{aligned}$$

So the populations are equal after about 36.191 years.

(c) The population of Ecalpon is double the population of Erehwon when

$$C(t) = 2W(t)$$

that is, when

$$45(1.032)^t = 2 \cdot 50(1.029)^t.$$

We use logs to solve the equation.

$$\begin{aligned}
45(1.032)^t &= 100(1.029)^t \\
\frac{(1.032)^t}{(1.029)^t} &= \frac{100}{45} = \frac{20}{9} \\
\left(\frac{1.032}{1.029}\right)^t &= \frac{20}{9} \\
\log\left(\frac{1.032}{1.029}\right)^t &= \log\left(\frac{20}{9}\right) \\
t\log\left(\frac{1.032}{1.029}\right) &= \log\left(\frac{20}{9}\right) \\
t &= \frac{\log(20/9)}{\log(1.032/1.029)} = 274.287 \text{ years.}
\end{aligned}$$

So it will take about 274 years for the population of Ecalpon to be twice that of Erehwon.

51. (a) The length of time it will take for the price to double is suggested by the formula. In the formula, 2 is raised to the power $\frac{t}{7}$. If $t = 7$ then $P(7) = 5(2)^{\frac{7}{7}} = 5(2)^1 = 10$. Since the original price is $5, we see that the price doubles to $10 in seven years.

(b) To find the annual inflation rate, we need to find the annual growth factor. One way to find this is to rewrite $P(t) = 5(2)^{\frac{t}{7}}$ in the form $P(t) = 5b^t$, where b is the annual growth factor:

$$P(t) = 5(2)^{\frac{t}{7}} = 5(2^{1/7})^t \approx 5(1.104)^t.$$

Since the price each year is 110.4% of the price the previous year, we know that the annual inflation rate is about 10.4%.

52. (a) The town population is 15,000 when $t = 0$ and grows by 4% each year.

(b) Since the two formulas represent the same amount, set the expressions equal to each other:

$$\begin{aligned}
15(b)^{12t} &= 15(1.04)^t \\
(b^{12})^t &= 1.04^t.
\end{aligned}$$

Take the tth root of both sides:

$$b^{12} = 1.04.$$

Now take the 12th root of both sides

$$b = 1.04^{1/12} \approx 1.003.$$

If t represents the number of years, then $12t$ represents the number of months in that time. If we are calculating $(b)^{12t}$ then b represents the monthly growth factor. Since $b = 1.04^{1/12}$ which is approximately equal to 1.003, we know that each month the population is approximately 100.3% larger than the population the previous month. The growth rate, then, is approximately 0.3% per month.

(c) Once again, the two formulas represent the same thing, so we will set them equal to one another.

$$\begin{aligned}
15(1.04)^t &= 15(2)^{\frac{t}{c}} \\
(1.04)^t &= (2)^{\frac{t}{c}} \\
\log(1.04)^t &= \log 2^{\frac{t}{c}} \\
t\log(1.04) &= \frac{t}{c}\log 2
\end{aligned}$$

$$\begin{aligned} ct\log(1.04) &= t\log 2 \\ c &= \frac{t\log 2}{t\log(1.04)} = \frac{\log 2}{\log 1.04} \\ c &\approx 17.67 \end{aligned}$$

To determine the meaning of c, we note that it is part of the exponent with 2 as a growth factor. If $t = 17.67$, then $P = 15(2)^{\frac{t}{c}} = 15(2)^{\frac{17.67}{17.67}} = 15(2)^1 = 30$. Since the population was 15,000, we see that it doubled in 17.67 years. If we add another 17.67 years onto t, we will have $P = 15(2)^{\frac{35.34}{17.67}} = 15(2)^2 = 60$. The population will have doubled again after another 17.67 years. This tells us that c represents the number of years it takes for the population to double.

53. We have

$$\begin{aligned} v(t) &= 30 \\ 20e^{0.2t} &= 30 \\ e^{0.2t} &= \frac{30}{20} \\ 0.2t &= \ln\left(\frac{30}{20}\right) \\ t &= \frac{\ln\left(\frac{30}{20}\right)}{0.2} = 2.027. \end{aligned}$$

54. We have

$$\begin{aligned} 3v(2t) &= 2w(3t) \\ 3\cdot 20e^{0.2(2t)} &= 2\cdot 12e^{0.22(3t)} \\ 60e^{0.4t} &= 24e^{0.66t} \\ \frac{e^{0.4t}}{e^{0.66t}} &= \frac{24}{60} = 0.4 \\ e^{0.4t-0.66t} = e^{-0.26t} &= 0.4 \\ -0.26t &= \ln(0.4) \\ t &= \frac{\ln(0.4)}{-0.26} = 3.524. \end{aligned}$$

55. The starting value of w is 12, so to find the doubling time, we can solve:

$$\begin{aligned} w(t) &= 24 \\ 12e^{0.22t} &= 24 \\ e^{0.22t} &= 2 \\ 0.22t &= \ln 2 \\ t &= \frac{\ln 2}{0.22} = 3.151. \end{aligned}$$

56. Another way to say $5 \approx 10^{0.7}$ is

$$\log 5 \approx 0.7.$$

Using this we can compute $\log 25$,

$$\log 25 = \log 5^2 = 2\log 5 \approx 2(0.7) = 1.4.$$

57. **(a)** For $f(x) = 10^x$,

$$\text{Domain of } f(x) \text{ is all } x$$
$$\text{Range of } f(x) \text{ is all } y > 0.$$

There is one asymptote, the horizontal line $y = 0$.

(b) Since $g(x) = \log x$ is the inverse function of $f(x) = 10^x$, the domain of $g(x)$ corresponds to range of $f(x)$ and range of $g(x)$ corresponds to domain of $g(x)$.

$$\text{Domain of } g(x) \text{ is all } x > 0$$

$$\text{Range of } g(x) \text{ is all } y.$$

The asymptote of $f(x)$ becomes the asymptote of $g(x)$ under reflection across the line $y = x$. Thus, $g(x)$ has one asymptote, the line $x = 0$.

58. The quadratic $y = x^2 - x - 6 = (x-3)(x+2)$ has zeros at $x = -2, 3$. It is positive outside of this interval and negative within this interval. Therefore, the function $y = \ln(x^2 - x - 6)$ is undefined on the interval $-2 \le x \le 3$, so the domain is all x not in this interval.

59. **(a)** Based on Figure 5.38, a log function seems as though it might give a good fit to the data in the table.

(b)

z	-1.56	-0.60	0.27	1.17	1.64	2.52
y	-11	-2	6.5	16	20.5	29

(c)

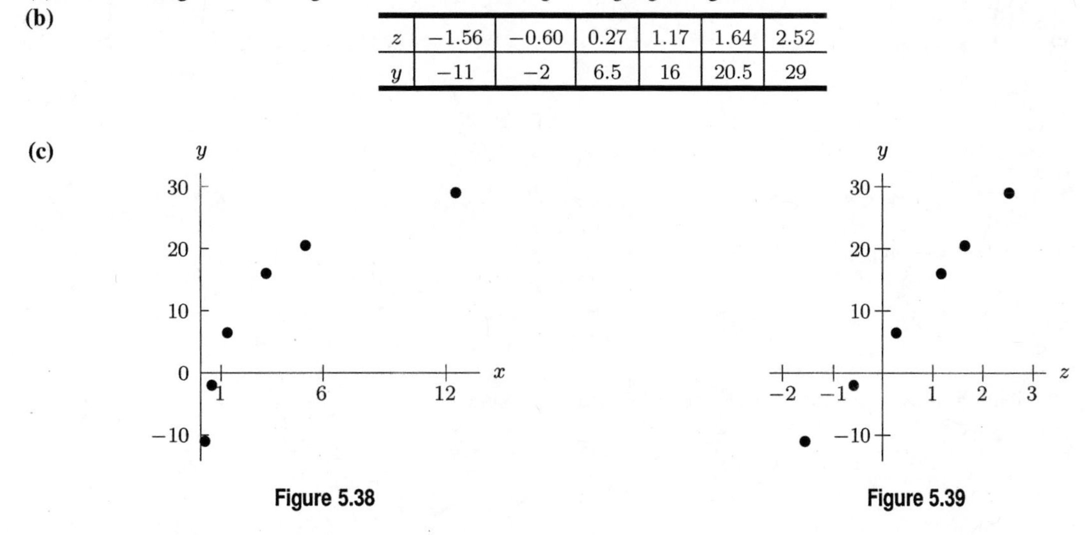

Figure 5.38 **Figure 5.39**

As you can see from Figure 5.39, the transformed data falls close to a line. Using linear regression, we see that $y = 4 + 9.9z$ gives an excellent fit to the data.

(d) Since $z = \ln x$, we see that the logarithmic function $y = 4 + 9.9\ln x$ gives an excellent fit to the data.

(e) Solving $y = 4 + 9.9\ln x$ for x, we have

$$\begin{aligned} y - 4 &= 9.9\ln x \\ \ln x &= \frac{y}{9.9} - \frac{4}{9.9} \\ e^{\ln x} &= e^{\frac{y}{9.9} - \frac{4}{9.9}} \\ x &= (e^{y/9.9})(e^{-4/9.9}). \end{aligned}$$

Since $e^{-\frac{4}{9.9}} \approx 0.67$ and $1/9.9 \approx 0.1$, we have

$$x \approx 0.67e^{0.1y}.$$

Thus, x is an exponential function of y.

60. For what value of t will $Q(t) = 0.23Q_0$?

$$\begin{aligned} 0.23Q_0 &= Q_0 e^{-0.000121t} \\ 0.23 &= e^{-0.000121t} \\ \ln 0.23 &= \ln e^{-0.000121t} \\ \ln 0.23 &= -0.000121t \\ t &= \frac{\ln 0.23}{-0.000121} = 12146.082. \end{aligned}$$

So the skull is about 12,146 years old.

61. **(a)** Since the drug is being metabolized continuously, the formula for describing the amount left in the bloodstream is $Q(t) = Q_0 e^{kt}$. We know that we start with 2 mg, so $Q_0 = 2$, and the rate of decay is 4%, so $k = -0.04$. (Why is k negative?) Thus $Q(t) = 2e^{-0.04t}$.

(b) To find the percent decrease in one hour, we need to rewrite our equation in the form $Q = Q_0 b^t$, where b gives us the percent left after one hour:

$$Q(t) = 2e^{-0.04t} = 2(e^{-0.04})^t \approx 2(0.96079)^t.$$

We see that $b \approx 0.96079 = 96.079\%$, which is the percent we have left after one hour. Thus, the drug level decreases by about 3.921% each hour.

(c) We want to find out when the drug level reaches 0.25 mg. We therefore ask when $Q(t)$ will equal 0.25.

$$\begin{aligned} 2e^{-0.04t} &= 0.25 \\ e^{-0.04t} &= 0.125 \\ -0.04t &= \ln 0.125 \\ t &= \frac{\ln 0.125}{-0.04} \approx 51.986. \end{aligned}$$

Thus, the second injection is required after about 52 hours.

(d) After the second injection, the drug level is 2.25 mg, which means that Q_0, the initial amount, is now 2.25. The decrease is still 4% per hour, so when will the level reach 0.25 again? We need to solve the equation

$$2.25e^{-0.04t} = 0.25,$$

where t is now the number of hours since the second injection.

$$\begin{aligned} e^{-0.04t} &= \frac{0.25}{2.25} = \frac{1}{9} \\ -0.04t &= \ln(1/9) \\ t &= \frac{\ln(1/9)}{-0.04} \approx 54.931. \end{aligned}$$

Thus the third injection is required about 55 hours after the second injection, or about $52 + 55 = 107$ hours after the first injection.

62. **(a)** Table 5.18 describes the height of the ball after n bounces:

Table 5.18

n	$h(n)$
0	6
1	90% of $6 = 6(0.9) = 5.4$
2	90% of $5.4 = 5.4(0.9) = 6(0.9)(0.9) = 6(0.9)^2$
3	90% of $6(0.9)^2 = 6(0.9)^2 \cdot (0.9) = 6(0.9)^3$
4	$6(0.9)^3 \cdot (0.9) = 6(0.9)^4$
5	$6(0.9)^5$
$\vdots$	$\vdots$
n	$6(0.9)^n$

so $h(n) = 6(0.9)^n$.

(b) We want to find the height when $n = 12$, so we will evaluate $h(12)$:

$$h(12) = 6(0.9)^{12} \approx 1.695 \text{ feet (about 1 ft 8.3 inches).}$$

(c) We are looking for the values of n for which $h(n) \leq 1$ inch $= \frac{1}{12}$ foot. So

$$\begin{aligned} h(n) &\leq \frac{1}{12} \\ 6(0.9)^n &\leq \frac{1}{12} \\ (0.9)^n &\leq \frac{1}{72} \\ \log(0.9)^n &\leq \log\frac{1}{72} \\ n\log(0.9) &\leq \log\frac{1}{72} \end{aligned}$$

Using your calculator, you will notice that $\log(0.9)$ is negative. This tells us that when we divide both sides by $\log(0.9)$, we must reverse the inequality. We now have

$$n \geq \frac{\log\frac{1}{72}}{\log(0.9)} \approx 40.591$$

So, the ball will rise less than 1 inch by the 41st bounce.

63. **(a)** If $Q(t) = Q_0 b^t$ describes the number of gallons left in the tank after t hours, then Q_0, the amount we started with, is 250, and b, the percent left in the tank after 1 hour, is 96%. Thus $Q(t) = 250(0.96)^t$. After 10 hours, there are $Q(10) = 250(0.96)^{10} \approx 166.208$ gallons left in the tank. This $\frac{166.208}{250} = 0.66483 = 66.483\%$ of what had initially been in the tank. Therefore approximately 33.517% has leaked out. It is less than 40% because the loss is 4% of 250 only during the first hour; for each hour after that it is 4% of whatever quantity is left.

(b) Since $Q_0 = 250$, $Q(t) = 250e^{kt}$. But we can also define $Q(t) = 250(0.96)^t$, so

$$\begin{aligned} 250e^{kt} &= 250(0.96)^t \\ e^{kt} &= 0.96^t \\ e^k &= 0.96 \\ \ln e^k &= \ln 0.96 \\ k\ln e &= \ln 0.96 \\ k &= \ln 0.96 \approx -0.04082. \end{aligned}$$

Since k is negative, we know that the value of $Q(t)$ is decreasing by 4.082% per hour. Therefore, k is the continuous hourly decay rate.

64. **(a)** $10\log(2) = \log(2^{10}) \approx \log(1000) = \log(10^3) = 3$. If $10\log 2 \approx 3$, then $\log 2 \approx \frac{3}{10} = 0.3$.

(b) $2\log(7) = \log(7^2) = \log 49 \approx \log 50 = \log\left(\frac{10^2}{2}\right) = \log(10^2) - \log(2) = 2 - \log(2)$. Since $2\log 7 \approx 2 - \log 2$ then $\log 7 \approx \frac{1}{2}(2 - \log 2) \approx \frac{1}{2}(2 - 0.30) = 0.85$.

65. **(a)** We have

$$\begin{aligned} \sqrt{\log(\text{googol})} &= \sqrt{\log(10^{100})} \\ &= \sqrt{100} \\ &= 10. \end{aligned}$$

(b) We have

$$\log\sqrt{\text{googol}} = \log\sqrt{10^{100}}$$

$$\begin{aligned}&= \log\left(\left(10^{100}\right)^{0.5}\right)\\&= \log\left(10^{50}\right)\\&= 50.\end{aligned}$$

(c) We have

$$\begin{aligned}\sqrt{\log(\text{googolplex})} &= \sqrt{\log\left(10^{\text{googol}}\right)}\\&= \sqrt{\text{googol}}\\&= \sqrt{10^{100}}\\&= \left(10^{100}\right)^{0.5}\\&= 10^{50}.\end{aligned}$$

66. **(a)** We will divide this into two parts and first show that $1 < \ln 3$. Since

$$e < 3$$

and ln is an increasing function, we can say that

$$\ln e < \ln 3.$$

But $\ln e = 1$, so

$$1 < \ln 3.$$

To show that $\ln 3 < 2$, we will use two facts:

$$3 < 4 = 2^2 \qquad \text{and} \qquad 2^2 < e^2.$$

Combining these two statements, we have $3 < 2^2 < e^2$, which tells us that $3 < e^2$. Then, using the fact that $\ln e^2 = 2$, we have

$$\ln 3 < \ln e^2 = 2$$

Therefore, we have

$$1 < \ln 3 < 2.$$

(b) To show that $1 < \ln 4$, we use our results from part (a), that $1 < \ln 3$. Since

$$3 < 4,$$

we have

$$\ln 3 < \ln 4.$$

Combining these two statements, we have

$$1 < \ln 3 < \ln 4, \qquad \text{so} \qquad 1 < \ln 4.$$

To show that $\ln 4 < 2$ we again use the fact that $4 = 2^2 < e^2$. Since $\ln e^2 = 2$, and ln is an increasing function, we have

$$\ln 4 < \ln e^2 = 2.$$

67. We have

$$\begin{aligned}\sqrt{1000^{\frac{1}{12}\cdot\log k}} &= \left(\left(10^3\right)^{\frac{1}{12}\cdot\log k}\right)^{0.5}\\&= \left(10^{3\cdot\frac{1}{12}\cdot\log k}\right)^{0.5}\\&= \left(10^{\frac{1}{4}\cdot\log k}\right)^{0.5}\\&= \left(\left(10^{\log k}\right)^{1/4}\right)^{1/2}\\&= k^{\frac{1}{4}\cdot\frac{1}{2}}\\&= \sqrt[8]{k}.\end{aligned}$$

CHECK YOUR UNDERSTANDING

1. False. Since the $\log 1000 = \log 10^3 = 3$ we know $\log 2000 > 3$. Or use a calculator to find that $\log 2000$ is about 3.3.
2. True. $\ln e^x = x$.
3. True. Check directly to see $2^{10} = 1024$. Or solve for x by taking the log of both sides of the equation and simplifying.
4. False. It grew to four times its original amount by doubling twice. So its doubling time is half of 8, or 4 hours.
5. True. Comparing the equation, we see $b = e^k$, so $k = \ln b$.
6. True. If x is a positive number, $\log x$ is defined and $10^{\log x} = x$.
7. True. This is the definition of a logarithm.
8. False. Since $10^{-k} = 1/10^k$ we see the value is positive.
9. True. The log function outputs the power of 10 which in this case is n.
10. True. The value of $\log n$ is the exponent to which 10 is raised to get n.
11. False. If a and b are positive, $\log\left(\frac{a}{b}\right) = \log a - \log b$.
12. False. If a and b are positive, $\ln a + \ln b = \ln(ab)$. There is no simple formula for $\ln(a + b)$.
13. False. For example, $\log 10 = 1$, but $\ln 10 \approx 2.3$.
14. True. The natural log function and the e^x function are inverses.
15. False. The log function has a vertical asymptote at $x = 0$.
16. True. As x increases, $\log x$ increases but at a slower and slower rate.
17. True. The two functions are inverses of one another.
18. True. Since $y = \log \sqrt{x} = \log(x^{1/2}) = \frac{1}{2}\log x$.
19. False. Consider $b = 10$ and $t = 2$, then $\log(10^2) = 2$, but $(\log 10)^2 = 1^2 = 1$. For $b > 0$, the correct formula is $\log(b^t) = t \log b$.
20. True. Think of these as $\ln e^1 = 1$ and $\log 10^1 = 1$.
21. False. Taking the natural log of both sides we see $t = \ln 7.32$.
22. True. Divide both sides of the first equation by 50. Then take the log of both sides and finally divide by $\log 0.345$ to solve for t.
23. True. Divide both sides of the first equation by a. Then take the log of both sides and finally divide by $\log b$ to solve for t.
24. False. It is the time it takes for the Q-value to double.
25. True. This is the definition of half-life.
26. False. Since $\frac{1}{4} = \frac{1}{2} \cdot \frac{1}{2}$, it takes only two half- life periods. That is 10 hours.
27. True. Replace the base 3 in the first equation with $3 = e^{\ln 3}$.
28. False. Since for $2P = Pe^{20r}$, we have $r = \frac{\ln 2}{20} = 0.035$ or 3.5%.
29. True. Solve for t by dividing both sides by Q_0, taking the ln of both sides and then dividing by k.
30. True. For example, astronomical distances.
31. True. Since $8000 \approx 10000 = 10^4$ and $\log 10^4 = 4$, we see that it would be just before 4 on a log scale.
32. False. Since $0.0000005 = 5 \cdot 10^{-7}$, we see that it would be between -6 and -7 on a log scale. (Closer to -6.)
33. False. Since $26,395,630,000,000 \approx 2.6 \cdot 10^{13}$, we see that it would be between 13 and 14 on a log scale.
34. True. Both scales are calibrated with powers of 10, which is a log scale.
35. False. An order of magnitude is a power of 10. They differ by a multiple of 1000 or three orders of magnitude.
36. False. There is no simple relation between the values of A and B and the data set.
37. False. The fit will not be as good as $y = x^3$ but an exponential function can be found.

Solutions to Skills for Chapter 5

1. $\log(\log 10) = \log(1) = 0.$

2. $\ln(\ln e) = \ln(1) = 0.$

3. $2\ln e^4 = 2(4\ln e) = 2(4) = 8.$

4. $\ln\left(\frac{1}{e^5}\right) = \ln 1 - \ln e^5 = \ln 1 - 5\ln e = 0 - 5 = -5.$

5. $\dfrac{\log 1}{\log 10^5} = \dfrac{0}{5\log 10} = \dfrac{0}{5} = 0.$

6. $e^{\ln 3} - \ln e = 3 - 1 = 2.$

7. $\sqrt{\log 10{,}000} = \sqrt{\log 10^4} = \sqrt{4\log 10} = \sqrt{4} = 2.$

8. By definition, $10^{\log 7} = 7.$

9. The equation $10^{-4} = 0.0001$ is equivalent to $\log 0.0001 = -4.$

10. The equation $10^{0.477} = 3$ is equivalent to $\log 3 = 0.477.$

11. The equation $e^{-2} = 0.135$ is equivalent to $\ln 0.135 = -2.$

12. The equation $e^{2x} = 7$ is equivalent to $2x = \ln 7.$

13. The equation $\log 0.01 = -2$ is equivalent to $10^{-2} = 0.01.$

14. The equation $\ln x = -1$ is equivalent to $e^{-1} = x.$

15. The equation $\ln 4 = x^2$ is equivalent to $e^{x^2} = 4.$

16. Rewrite the logarithm of the product as a sum, $\log 2x = \log 2 + \log x.$

17. The expression is not the logarithm of a quotient, so it cannot be rewritten using the properties of logarithms.

18. The logarithm of a quotient rule applies, so $\log\left(\frac{x}{5}\right) = \log x - \log 5.$

19. The logarithm of a quotient and the power property apply, so

$$\begin{aligned}\log\left(\frac{x^2+1}{x^3}\right) &= \log(x^2+1) - \log x^3\\ &= \log(x^2+1) - 3\log x.\end{aligned}$$

20. Rewrite the power,

$$\ln\sqrt{\frac{x-1}{x+1}} = \ln\frac{(x-1)^{1/2}}{(x+1)^{1/2}}.$$

Use the quotient and power properties,

$$\ln\frac{(x-1)^{1/2}}{(x+1)^{1/2}} = \ln(x-1)^{1/2} - \ln(x+1)^{1/2} = (1/2)\ln(x-1) - (1/2)\ln(x+1).$$

21. There is no rule for the logarithm of a sum, it cannot be rewritten.

22. In general the logarithm of a difference cannot be simplified. In this case we rewrite the expression so that it is the logarithm of a product.

$$\log(x^2 - y^2) = \log((x+y)(x-y)) = \log(x+y) + \log(x-y).$$

23. The expression is a product of logarithms, not a logarithm of a product, so it cannot be simplified.

24. The expression is a quotient of logarithms, not a logarithm of a quotient, but we can use the properties of logarithms to rewrite it without the x^2 term:

$$\frac{\ln x^2}{\ln(x+2)} = \frac{2\ln x}{\ln(x+2)}.$$

25. Rewrite the sum as $\log 12 + \log x = \log 12x$.

26. Rewrite the difference as

$$\ln x^2 - \ln(x+10) = \ln\frac{x^2}{x+10}.$$

27. Rewrite with powers and combine,

$$\frac{1}{2}\log x + 4\log y = \log\sqrt{x} + \log y^4 = \log(\sqrt{x}y^4).$$

28. Rewrite with powers and combine,

$$\log 3 + 2\log\sqrt{x} = \log 3 + \log(\sqrt{x})^2 = \log 3 + \log x = \log 3x.$$

29. Rewrite with powers and combine,

$$\begin{aligned} 3\left(\log(x+1) + \frac{2}{3}\log(x+4)\right) &= 3\log(x+1) + 2\log(x+4) \\ &= \log(x+1)^3 + \log(x+4)^2 \\ &= \log\left((x+1)^3(x+4)^2\right) \end{aligned}$$

30. Rewrite as

$$\ln x + \ln\left(\frac{y}{2}(x+4)\right) + \ln z^{-1} = \ln x + \ln\left(\frac{xy+4y}{2}\right) - \ln z = \ln\left(\frac{(x^2)y+4xy}{2}\right) - \ln z = \ln\left(\frac{(x^2)y+4xy}{2z}\right).$$

31. Rewrite with powers and combine,

$$\begin{aligned} 2\log(9-x^2) - (\log(3+x) + \log(3-x)) &= \log(9-x^2)^2 - (\log(3+x)(3-x)) \\ &= \log(9-x^2)^2 - \log(9-x^2) \\ &= \log\frac{(9-x^2)^2}{(9-x^2)} \\ &= \log(9-x^2). \end{aligned}$$

32. Rewrite as $2\ln e^{\sqrt{x}} = 2\sqrt{x}$.

33. The logarithm of a sum cannot be simplified.

34. Rewrite as $\log(10x) - \log x = \log(10x/x) = \log 10 = 1$.

35. Rewrite as $2\ln x^{-2} + \ln x^4 = 2(-2)\ln x + 4\ln x = 0$.

36. Rewrite as $\ln\sqrt{x^2+16} = \ln(x^2+16)^{1/2} = \frac{1}{2}\ln(x^2+16)$.

37. Rewrite as $\log 100^{2z} = 2z\log 100 = 2z(2) = 4z$.

38. Rewrite as $\dfrac{\ln e}{\ln e^2} = \dfrac{\ln e}{2\ln e} = \dfrac{1}{2}$.

39. Rewrite as $\ln \dfrac{1}{e^x+1} = \ln 1 - \ln(e^x+1) = -\ln(e^x+1)$.

40. Taking logs of both sides we get

$$\log(12^x) = \log 7.$$

This gives

$$x \log 12 = \log 7$$

or in other words

$$x = \frac{\log 7}{\log 12} \approx 0.783.$$

41. We divide both sides by 3 to get

$$5^x = 3.$$

Taking logs of both sides we get

$$\log(5^x) = \log 3.$$

This gives

$$x \log 5 = \log 3$$

or in other words

$$x = \frac{\log 3}{\log 5} \approx 0.683.$$

42. We divide both sides by 4 to get

$$13^{3x} = \frac{17}{4}.$$

Taking logs of both sides we get

$$\log(13^{3x}) = \log\left(\frac{17}{4}\right).$$

This gives

$$3x \log 13 = \log\left(\frac{17}{4}\right)$$

or in other words

$$x = \frac{\log(17/4)}{3\log 13} \approx 0.188.$$

43. Taking natural logs of both sides we get

$$\ln(e^{-5x}) = \ln 9.$$

This gives

$$\begin{aligned} -5x &= \ln 9 \\ x &= -\frac{\ln 9}{5} \approx -0.439. \end{aligned}$$

44. Taking logs of both sides we get

$$\log 12^{5x} = \log(3 \cdot 15^{2x}).$$

This gives

$$\begin{aligned} 5x\log 12 &= \log 3 + \log 15^{2x} \\ 5x \log 12 &= \log 3 + 2x \log 15 \\ 5x\log 12 - 2x\log 15 &= \log 3 \\ x(5\log 12 - 2\log 15) &= \log 3 \\ x &= \frac{\log 3}{5\log 12 - 2\log 15} \approx 0.157. \end{aligned}$$

45. Taking logs of both sides we get

$$\log 19^{6x} = \log(77 \cdot 7^{4x}).$$

This gives

$$\begin{aligned} 6x \log 19 &= \log 77 + \log 7^{4x} \\ 6x \log 19 &= \log 77 + 4x \log 7 \\ 6x \log 19 - 4x \log 7 &= \log 77 \\ x(6 \log 19 - 4 \log 7) &= \log 77 \\ x &= \frac{\log 77}{6 \log 19 - 4 \log 7} \approx 0.440. \end{aligned}$$

46. We first re-arrange the equation so that the log is alone on one side, and we then convert to exponential form:

$$\begin{aligned} 3\log(4x+9) - 6 &= 2 \\ 3\log(4x+9) &= 8 \\ \log(4x+9) &= \frac{8}{3} \\ 10^{\log(4x+9)} &= 10^{8/3} \\ 4x + 9 &= 10^{8/3} \\ 4x &= 10^{8/3} - 9 \\ x &= \frac{10^{8/3} - 9}{4} \approx 113.790. \end{aligned}$$

47. We first re-arrange the equation so that the log is alone on one side, and we then convert to exponential form:

$$\begin{aligned} 4\log(9x+17) - 5 &= 1 \\ 4\log(9x+17) &= 6 \\ \log(9x+17) &= \frac{3}{2} \\ 10^{\log(9x+17)} &= 10^{3/2} \\ 9x + 17 &= 10^{3/2} \\ 9x &= 10^{3/2} - 17 \\ x &= \frac{10^{3/2} - 17}{9} \approx 1.625. \end{aligned}$$

48. We begin by converting to exponential form:

$$\begin{aligned} \ln(3x+4) &= 5 \\ e^{\ln(3x+4)} &= e^5 \\ 3x + 4 &= e^5 \\ 3x &= e^5 - 4 \\ x &= \frac{e^5 - 4}{3} \approx 48.138. \end{aligned}$$

49. We first re-arrange the equation so that the natural log is alone on one side, and we then convert to exponential form:

$$\begin{aligned} 2\ln(6x-1) + 5 &= 7 \\ 2\ln(6x-1) &= 2 \\ \ln(6x-1) &= 1 \end{aligned}$$

$$\begin{aligned}
e^{\ln(6x-1)} &= e^1 \\
6x-1 &= e \\
6x &= e+1 \\
x &= \frac{e+1}{6} \approx 0.620.
\end{aligned}$$

CHAPTER SIX

Solutions for Section 6.1

Skill Refresher

S1. Substituting $x = 4$ into $f(x)$, we have $f(4) = \sqrt{4} = 2$.

S2. Substituting $x = 4$ into $g(x)$, we have $g(4) = \sqrt{4} + 6 = 8$. Notice that $g(x)$ is a vertical shift of $f(x) = \sqrt{x}$ up 6 units, and thus the point $(4, 2)$ on the graph of $f(x)$ has been shifted to the point $(4, 8)$ on the graph of $g(x)$.

S3. Substituting $x = 4$ into $h(x)$, we have $h(4) = \sqrt{4} - 3 = -1$. Notice that $h(x)$ is a vertical shift of $f(x) = \sqrt{x}$ down 3 units, and thus the point $(4, 2)$ on the graph of $f(x)$ has been shifted to the point $(4, -1)$ on the graph of $h(x)$.

S4. Substituting $x = 4$ into $k(x)$, we have $k(4) = \sqrt{4 + 5} = \sqrt{9} = 3$. Notice that $k(x)$ is a horizontal shift of $f(x) = \sqrt{x}$ to the left by 5 units.

S5. We know $e^x = 1$ has only one solution, $x = 0$.

S6. We first apply the natural log to both sides of the equation, $\ln e^{x-5} = \ln 1$, and we have $x - 5 = 0$. Thus $x = 5$ is the only solution. Notice $y = e^{x-5}$ is a horizontal shift of $y = e^x$ to the right by 5 units. The point $(0, 1)$ on the graph of $y = e^x$ we found in the previous exercise has been moved to the right 5 units to $(5, 1)$ on the graph of $y = e^{x-5}$.

S7. We first apply the natural log to both sides of the equation, $\ln e^{x+8} = \ln 1$, and we have $x + 8 = 0$. Thus $x = -8$ is the only solution. Notice $y = e^{x+8}$ is a horizontal shift of $y = e^x$ to the left by 8 units. The point $(0, 1)$ on the graph of $y = e^x$ we previously found has been moved to the left 8 units to $(-8, 1)$ on the graph of $y = e^{x+8}$.

S8. We first add 3 to both sides in order to isolate the exponent before applying the natural log to both sides of the equation, and we have $e^x = 4$. Next we apply the natural log to both sides, $\ln e^x = x = \ln 4$. Thus $x = \ln 4$ is the only solution. Notice $y = e^x - 3$ is a vertical shift of $y = e^x$ down by 3 units.

S9. **(a)** The translation $y = f(x - 4) = \ln(x - 4)$ shifts the graph to the right 4 units.
(b) The translation $y = f(x) - 7 = \ln(x) - 7$ shifts the graph down 7 units.
(c) The translation $y = f(x + \sqrt{2}) = \ln(x + \sqrt{2})$ shifts the graph to the left $\sqrt{2}$ units.
(d) The translation $y = f(x - 3) + 5 = \ln(x - 3) + 5$ shifts the graph to the right 3 units, and shifts the graph up 5 units.

S10. **(a)** The translation $y = f(x + 8) = \dfrac{1}{x + 8}$ shifts the graph to the left 8 units.
(b) The translation $y = f(x) + 3 = \dfrac{1}{x} + 3$ shifts the graph up 3 units.
(c) The translation $y = f(x) - \dfrac{1}{5} = \dfrac{1}{x} - \dfrac{1}{5}$ shifts the graph down $1/5$ units.
(d) The translation $y = f(x - 2) - \ln 7 = \dfrac{1}{x - 2} - \ln 7$ shifts the graph to the right 2 units, and shifts the graph down $\ln 7$ units.

Exercises

1. **(a)**

x	-1	0	1	2	3
$g(x)$	-3	0	2	1	-1

The graph of $g(x)$ is shifted one unit to the right of $f(x)$.

(b)

x	-3	-2	-1	0	1
$h(x)$	-3	0	2	1	-1

The graph of $h(x)$ is shifted one unit to the left of $f(x)$.

(c)

x	-2	-1	0	1	2
$k(x)$	0	3	5	4	2

The graph $k(x)$ is shifted up three units from $f(x)$.

(d)

x	-1	0	1	2	3
$m(x)$	0	3	5	4	2

The graph $m(x)$ is shifted one unit to the right and three units up from $f(x)$.

2. **(a)** This is the graph of the function $y = |x|$ shifted both up and to the right. Thus the formula is (vi).
(b) This is the graph of the function $y = |x|$ shifted to the right. Thus the formula is (iii).
(c) This is the graph of the function $y = |x|$ shifted down. Thus formula is (ii).
(d) This is the graph of the function $y = |x|$ shifted to the left. Thus the formula is (v).
(e) This is the graph of the function $y = |x|$. Thus the formula is (i).
(f) This is the graph of the function $y = |x|$ shifted up. Thus formula is (iv).

3. **(a)** The translation $f(x) + 5$ moves the graph up 5 units. The x-coordinate is not changed, but the y-coordinate is $-4 + 5 = 1$. The new point is $(3, 1)$.
(b) The translation $f(x + 5)$ shifts the graph to the left 5 units. The y-coordinate is not changed, but the x-coordinate is $3 - 5 = -2$. The new point is $(-2, -4)$.
(c) This translation shifts both the x and y coordinates; 3 units right and 2 units down resulting in $(6, -6)$.

4. See Figure 6.1.

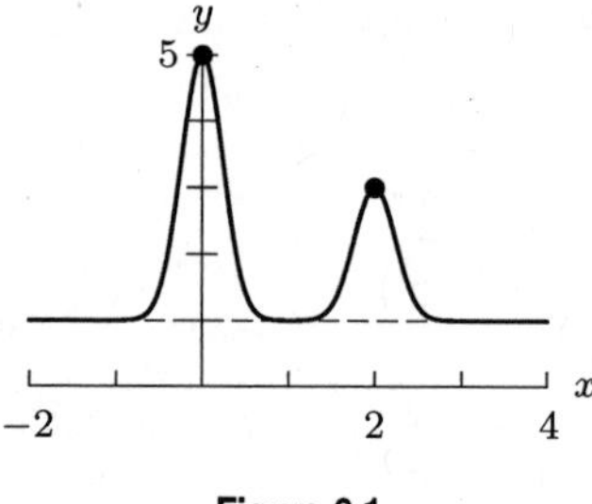

Figure 6.1

5. See Figure 6.2.

Figure 6.2

6. See Figure 6.3.

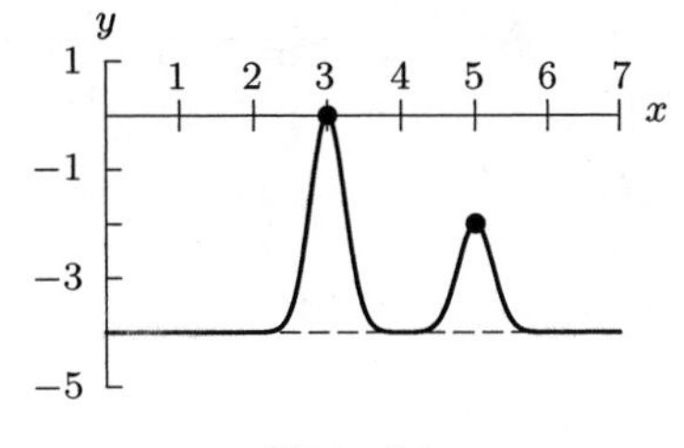

Figure 6.3

7. See Figure 6.4.

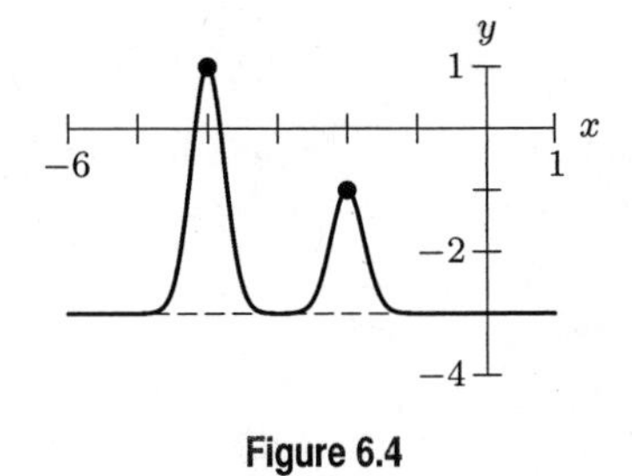

Figure 6.4

8. For all x, the graph of $g(x)$ is two units higher than the graph of $f(x)$, while the graph of $h(x)$ is 3 units lower than the graph of $f(x)$. See Figure 6.5.

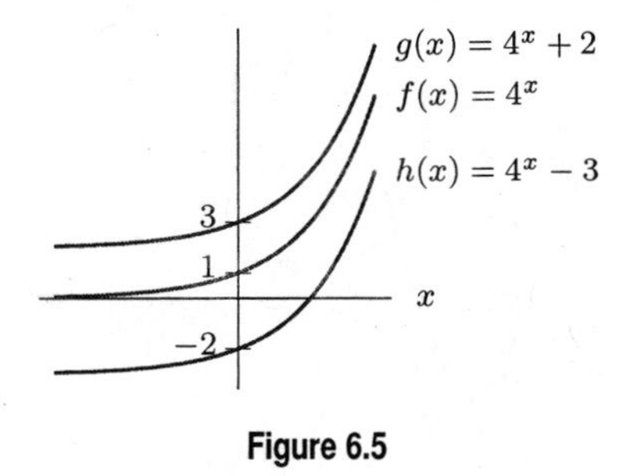

Figure 6.5

9. The graph of $g(x)$ is shifted four units to the left of $f(x)$, and the graph of $h(x)$ is shifted two units to the right of $f(x)$.

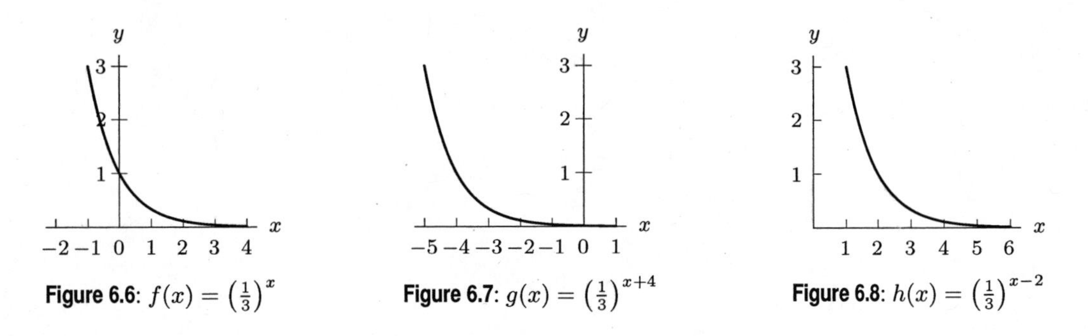

Figure 6.6: $f(x) = \left(\frac{1}{3}\right)^x$ Figure 6.7: $g(x) = \left(\frac{1}{3}\right)^{x+4}$ Figure 6.8: $h(x) = \left(\frac{1}{3}\right)^{x-2}$

10. The translation shifts the graph to the right 2 units, so the new domain is $0 < x < 9$.

11. The range shifts the graph down 150 units, so the new range is $-50 \leq R(s) - 150 \leq 50$.

12. The graph of $y = f(x+3) + 3$ is the graph of f shifted left by 3 units and up by 3 units.

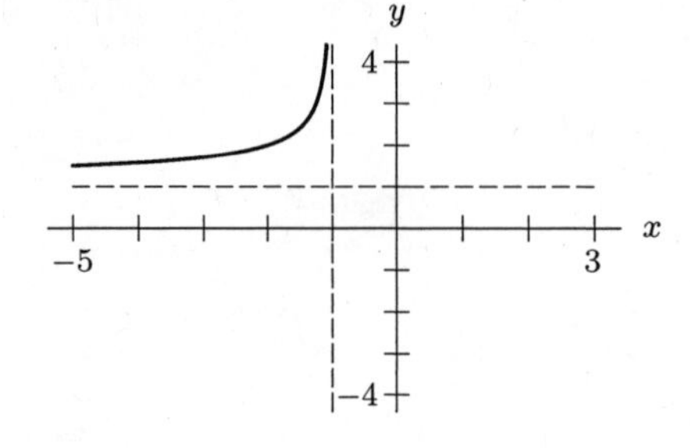

13. $m(n) + 1 = \dfrac{1}{2}n^2 + 1$

To graph this function, shift the graph of $m(n) = \frac{1}{2}n^2$ one unit up. See Figure 6.9.

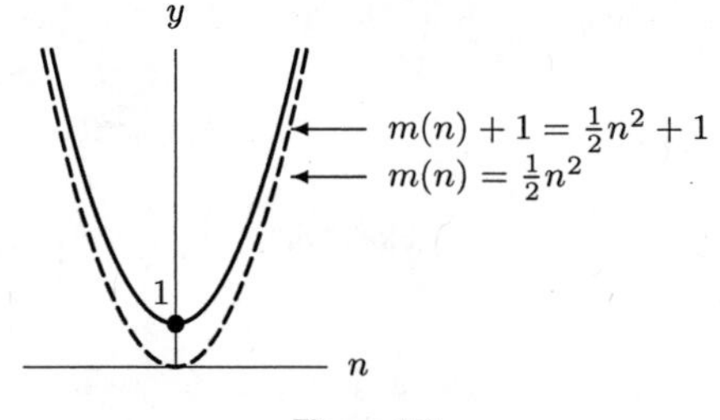

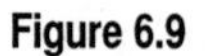
Figure 6.9

14. $m(n+1) = \dfrac{1}{2}(n+1)^2 = \dfrac{1}{2}n^2 + n + \dfrac{1}{2}$

To sketch, shift the graph of $m(n) = \frac{1}{2}n^2$ one unit to the left, as in Figure 6.10.

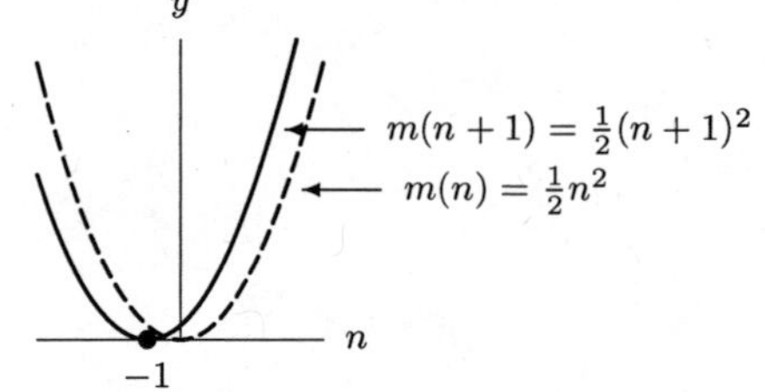

Figure 6.10

15. $m(n) - 3.7 = \dfrac{1}{2}n^2 - 3.7$

Sketch by shifting the graph of $m(n) = \frac{1}{2}n^2$ down by 3.7 units, as in Figure 6.11.

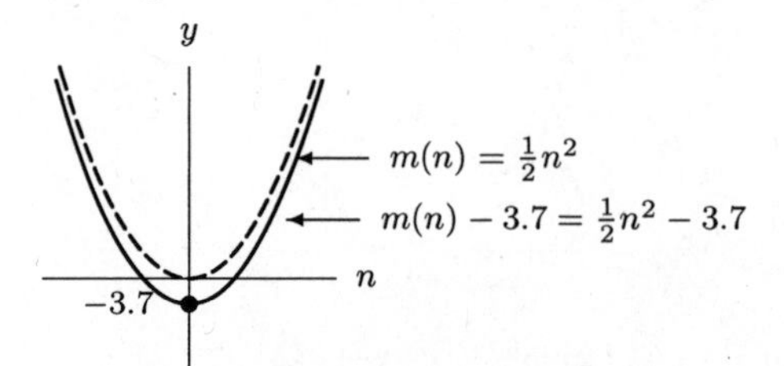

Figure 6.11

16. $m(n-3.7) = \dfrac{1}{2}(n-3.7)^2 = \dfrac{1}{2}n^2 - 3.7n + 6.845$

To sketch, shift the graph of $m(n) = \frac{1}{2}n^2$ to the right by 3.7 units, as in Figure 6.12.

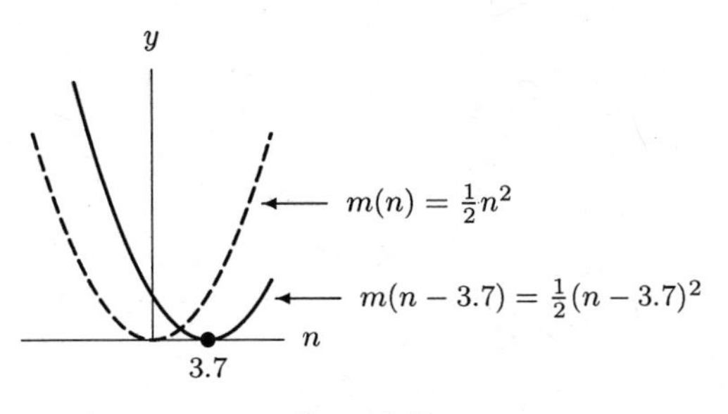

Figure 6.12

17. $m(n) + \sqrt{13} = \dfrac{1}{2}n^2 + \sqrt{13}$

To sketch, shift the graph of $m(n) = \frac{1}{2}n^2$ up by $\sqrt{13}$ units, as in Figure 6.13.

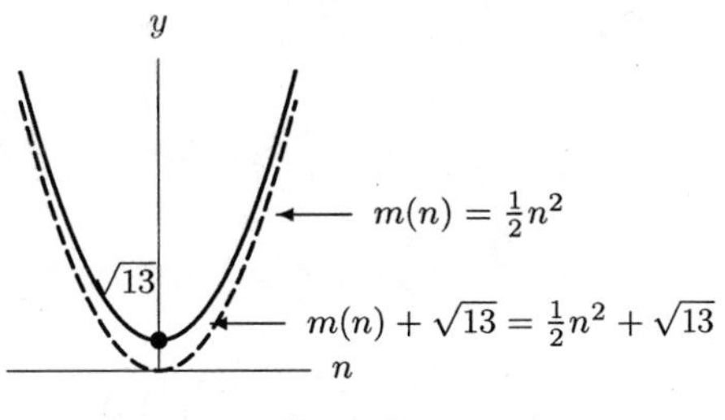

Figure 6.13

18. $m(n+2\sqrt{2}) = \dfrac{1}{2}(n+2\sqrt{2})^2 = \dfrac{1}{2}n^2 + 2\sqrt{2}n + 4$

To sketch, shift the graph of $m(n) = \frac{1}{2}n^2$ by $2\sqrt{2}$ units to the left, as in Figure 6.14.

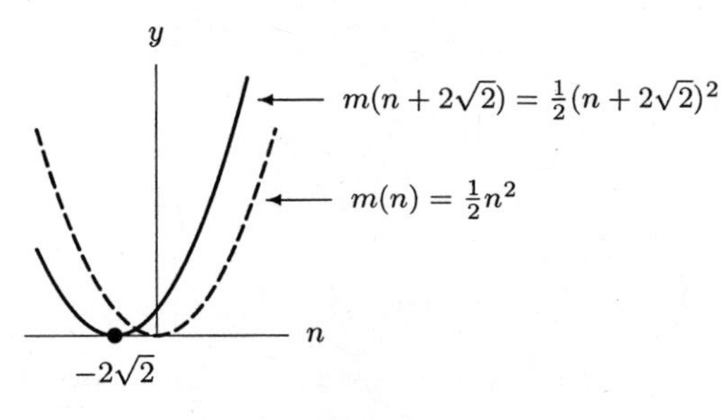

Figure 6.14

19. $m(n+3) + 7 = \dfrac{1}{2}(n+3)^2 + 7 = \left(\dfrac{1}{2}n^2 + 3n + \dfrac{9}{2}\right) + 7 = \dfrac{1}{2}n^2 + 3n + \dfrac{23}{2}.$

To sketch, shift the graph of $m(n) = \frac{1}{2}n^2$ by 3 units to the left and 7 units up, as in Figure 6.15.

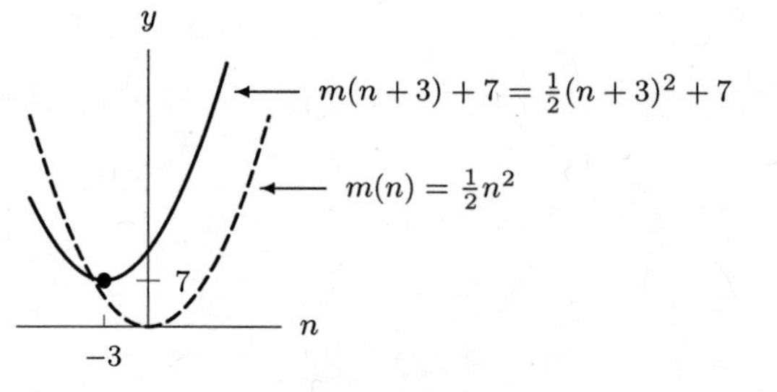

Figure 6.15

20. $m(n-17)-159 = \frac{1}{2}(n-17)^2 - 159 = \left(\frac{1}{2}n^2 - 17n + \frac{289}{2}\right) - 159 = \frac{1}{2}n^2 - 17n - \frac{29}{2}$
To sketch, shift the graph of $m(n) = \frac{1}{2}n^2$ by 17 units to the right and 159 units down, as in Figure 6.16.

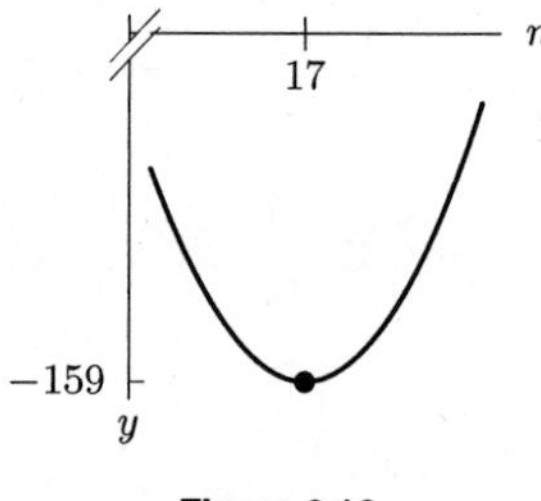

Figure 6.16

21. $k(w) - 3 = 3^w - 3$
To sketch, shift the graph of $k(w) = 3^w$ down 3 units, as in Figure 6.17.

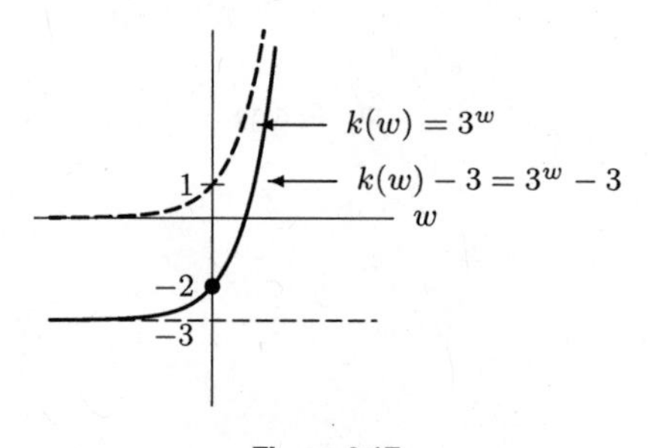

Figure 6.17

22. $k(w-3) = 3^{w-3}$
To sketch, shift the graph of $k(w) = 3^w$ to the right by 3 units, as in Figure 6.18.

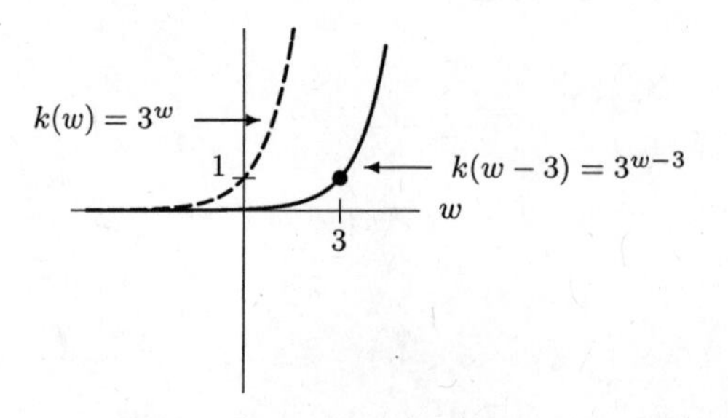

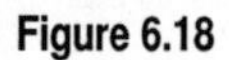
Figure 6.18

23. $k(w) + 1.8 = 3^w + 1.8$
To sketch, shift the graph of $k(w) = 3^w$ up by 1.8 units, as in Figure 6.19.

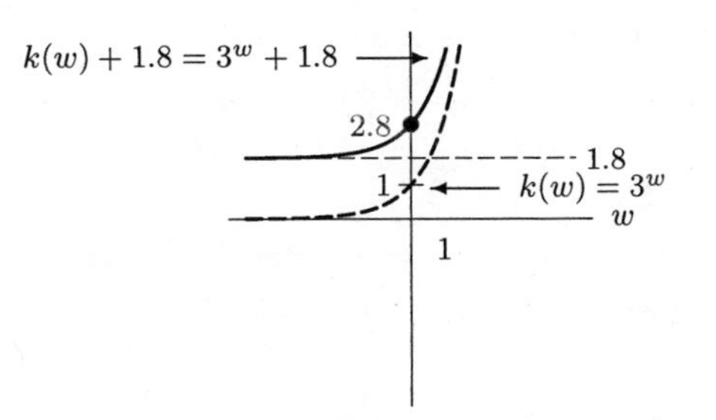

Figure 6.19

24. $k(w + \sqrt{5}) = 3^{w+\sqrt{5}}$
To sketch, shift the graph of $k(w) = 3^w$ to the left by $\sqrt{5}$ units, as in Figure 6.20.

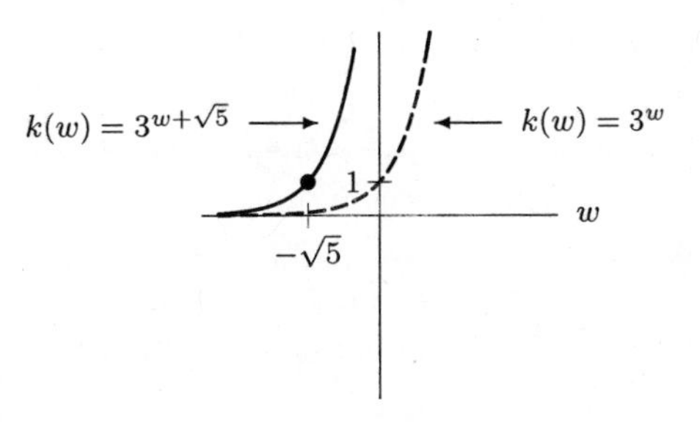

Figure 6.20

25. $k(w + 2.1) - 1.3 = 3^{w+2.1} - 1.3$
To sketch, shift the graph of $k(w) = 3^w$ to the left by 2.1 units and down 1.3 units, as in Figure 6.21.

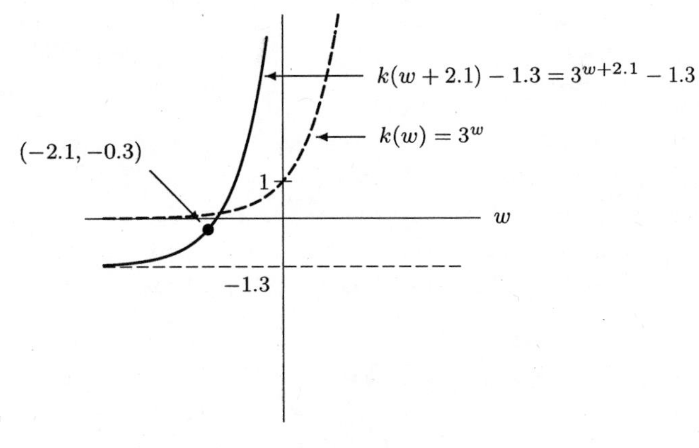

Figure 6.21

26. $k(w-1.5)-0.9 = 3^{w-1.5}-0.9$

To sketch, shift the graph of $k(w)=3^w$ to the right by 1.5 units and down by 0.9 units, as in Figure 6.22.

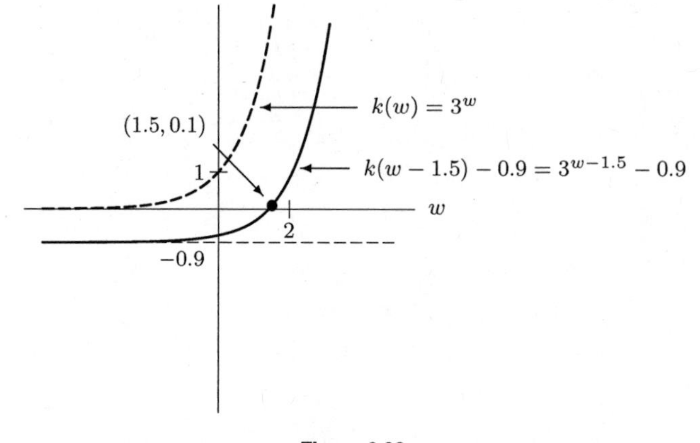

Figure 6.22

Problems

27. **(a)** (i) To evaluate $f(x)$ for $x=6$, we find from the table the value of $f(x)$ corresponding to an x-value of 6. In this case, the corresponding value is 248. Thus, $f(x)$ at $x=6$ is 248.

(ii) $f(5)$ equals the value of $f(x)$ corresponding to $x=5$, or 145. $f(5)-3 = 145-3 = 142$.

(iii) $f(5-3)$ is the same thing as $f(2)$, which is the value of $f(x)$ corresponding to $x=2$. Since $f(5-3)=f(2)$, and $f(2)=4$, $f(5-3)=4$.

(iv) $g(x)+6$ for $x=2$ equals $g(2)+6$. $g(2)$ is the value of $g(x)$ corresponding to an x-value of 2, thus $g(2)=6$. $g(2)+6 = 6+6 = 12$.

(v) $g(x+6)$ for $x=2$ equals $g(2+6)=g(8)$. Looking at the table in the problem, we see that $g(8)=378$. Thus, $g(x+6)$ for $x=2$ equals 378.

(vi) $g(x)$ for $x=0$ equals $g(0)=-6$. $3\cdot(g(0)) = 3\cdot(-6) = -18$.

(vii) $f(3x)$ for $x=2$ equals $f(3\cdot 2)=f(6)$. From part (a), we know that $f(6)=248$; thus, $f(3x)$ for $x=2$ equals 248.

(viii) $f(x)-f(2)$ for $x=8$ equals $f(8)-f(2)$. $f(8)=574$ and $f(2)=4$, so $f(8)-f(2)=574-4=570$.

(ix) $g(x+1)-g(x)$ for $x=1$ equals $g(1+1)-g(1)=g(2)-g(1)$. $g(2)=6$ and $g(1)=-7$, so $g(2)-g(1) = 6-(-7) = 6+7 = 13$.

(b) (i) To find x such that $g(x)=6$, we look for the entry in the table at which $g(x)=6$ and then see what the corresponding x-value is. In this case, it is 2. Thus, $g(x)=6$ for $x=2$.

(ii) We use the same principle as that in part (i): $f(x)=574$ when $x=8$.

(iii) Again, this is just like part (i): $g(x)=281$ when $x=7$.

(c) Solving $x^3+x^2+x-10 = 7x^2-8x-6$ involves finding those values of x for which both sides of the equation are equal, or where $f(x)=g(x)$. Looking at the table, we see that $f(x)=g(x)=-7$ for $x=1$, and $f(x)=g(x)=74$ for $x=4$.

28. **(a)** This graph is the graph of $m(r)$ shifted upwards by two units. Thus, the formula for $n(r)$ is

$$n(r) = m(r)+2.$$

(b) This graph is the graph of $m(r)$ shifted to the right by one unit. Thus, the formula for $p(r)$ is

$$p(r) = m(r-1).$$

(c) This graph is the graph of $m(r)$ shifted to the left by 1.5 units. Thus, the formula for $k(r)$ is

$$k(r) = m(r + 1.5).$$

(d) This graph is the graph of $m(r)$ shifted to the right by 0.5 units and downwards by 2.5 units. Thus, the formula for $w(r)$ is

$$w(r) = m(r - 0.5) - 2.5.$$

29. The graph in Figure 6.10 is a result of shifting the function in Figure 6.9 right 2 units and down 6 units. Thus $y = f(x-2) - 6$ is a formula for the function in Figure 6.10.

30. The graph of g is the graph of f shifted down by 1 unit and to the left by 2 units. Thus, $g(x) = f(x+2) - 1$.

31. **(a)** The translation should leave the x-coordinate unchanged, and shift the y-coordinate up 3; so $y = g(x) + 3$.
(b) The translation should leave the y-coordinate unchanged, and shift the x-coordinate right by 2; so $y = g(x-2)$.

32. **(a)** $f(-6) = ((-6)/2)^3 + 2 = (-3)^3 + 2 = -27 + 2 = -25$
(b) We are trying to find x so that $f(x) = 6$. Setting $f(x) = -6$, we have

$$\begin{aligned} -6 &= \left(\frac{x}{2}\right)^3 + 2 \\ -8 &= \left(\frac{x}{2}\right)^3 \\ -2 &= \left(\frac{x}{2}\right) \\ -4 &= x. \end{aligned}$$

Thus, $f(x) = -6$ for $x = -4$.
(c) In part (a), we found that $f(-6) = -25$. This means that the point $(-6, f(-6))$, or $(-6, -25)$ is on the graph of $f(x)$. We call this point A in Figure 6.23. In part (b), we found that $f(x) = -6$ at $x = -4$. This means the point $(-4, -6)$ is also on the graph of $f(x)$. We call this point B in Figure 6.23.

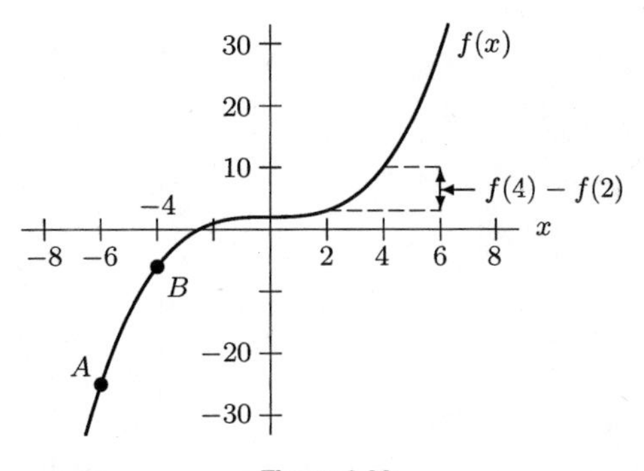

Figure 6.23

(d) We have $f(4) = (4/2)^3 + 2 = 8 + 2 = 10$ and $f(2) = (2/2)^3 + 2 = 1^3 + 2 = 3$. Thus $f(4) - f(2) = 10 - 3 = 7$. This is shown in Fig 6.23.
(e) If $a = -2$, we have $f(a+4) = f(-2+4) = f(2) = 3$. Thus, $f(a+4) = 3$ for $a = -2$. $f(-2) + 4 = (-2/2)^3 + 2 + 4 = -1 + 2 + 4 = 5$. Thus, $f(a) + 4 = 5$ for $a = -2$.
(f) $f(a+4) = f(-2+4) = f(2)$. Thus, an x-value of 2 corresponds to $f(a+4)$ for $a = -2$. $f(a)+4 = f(-2)+4 = 5$ for $a = -2$. To find an x-value which corresponds to $f(a) + 4$, we need to find the value of x for which $f(x) = 5$. Setting $f(x) = 5$,

$$\begin{aligned} \left(\frac{x}{2}\right)^3 + 2 &= 5 \\ \frac{x^3}{8} + 2 &= 5 \end{aligned}$$

$$\begin{aligned}\frac{x^3}{8} &= 3\\ x^3 &= 24\\ x &= \sqrt[3]{24} = 2\sqrt[3]{3}\\ &\approx 2.884.\end{aligned}$$

33. **(a)** $P(t) + 100$ describes a population that is always 100 people larger than the original population.
(b) $P(t + 100)$ describes a population that has the same number of people as the original population, but the number occurs 100 years earlier.

34. At $t = 3$ months, Jonah's weight is

$$V = s(3) + 2.$$

Since $s(3)$ is the average weight of a 3-month old boy, we see that at 3 months, Jonah weighs 2 pounds more than average. Similarly, at $t = 6$ months we have

$$V = s(6) + 2,$$

which means that, at 6 months, Jonah weighs 2 pounds more than average. In general, Jonah weighs 2 pounds more than average for babies of his age.

35. Since $W = s(t + 4)$, at age $t = 3$ months Ben's weight is given by

$$W = s(3 + 4) = s(7).$$

We defined $s(7)$ to be the average weight of a 7-month old baby. At age 3 months, Ben's weight is the same as the average weight of 7-month old babies. Since, on average, a baby's weight increases as the baby grows, this means that Ben is heavier than the average for a 3-month old. Similarly, at age $t = 6$, Ben's weight is given by

$$W = s(6 + 4) = s(10).$$

Thus, at 6 months, Ben's weight is the same as the average weight of 10-month old babies. In both cases, we see that Ben is above average in weight.

36. **(a)** If g is horizontal shift of $y = x^2$, then each point (a, b) on the graph of $y = x^2$ is shifted to the point $(a - k, b)$ on g. Since the point $(4, 16)$ is on the graph of $y = x^2$, we need to shift the point to the left one unit in order for the point $(3, 16)$ to be on the graph of g. Thus we have $g(x) = (x + 1)^2$.
(b) If g is vertical shift of $y = x^2$, then each point (a, b) on the graph of $y = x^2$ is shifted to the point $(a, b + k)$ on g. Since the point $(3, 9)$ is on the graph of $y = x^2$, we need to shift the point to up 7 units in order for the point $(3, 16)$ to be on the graph of g. Thus we have $g(x) = x^2 + 7$.
(c) If g is the result of applying a shift right 2 units followed by a vertical shift, then each point (a, b) on the graph of $y = x^2$ is shifted to the point $(a + 2, b + k)$. After being shifted to the right by 2 units, the point $(1, 1)$ on the graph of $y = x^2$ is moved to $(3, 1)$. If $g(3) = 16$, then we need to follow the shift right 2 units by a vertical shift up 15 units. Thus we have $g(x) = (x - 2)^2 + 15$.

37. To compensate for the down shift, we shift up 1. To compensate for the left shift by 3, we shift right by 3.

38. The graph is in Figure 6.24. Notice that the domain of $f(x)$ is all real numbers except 3 and the domain of $g(x)$ is all real numbers except 0. The vertical asymptotes of $f(x)$ is $x = 3$ and the vertical asymptote of $g(x)$ is $x = 0$.

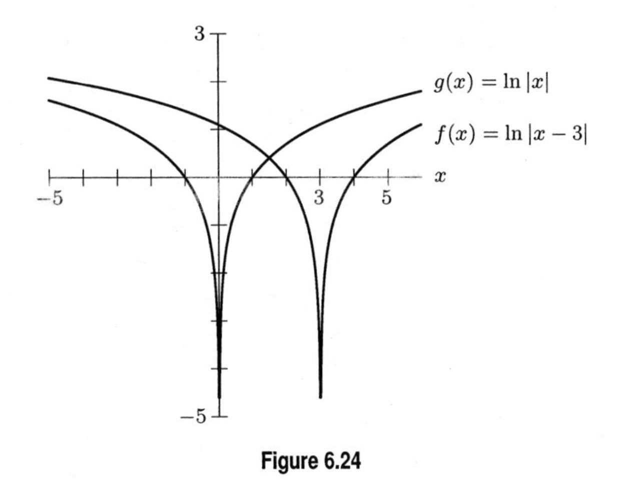

Figure 6.24

39. The graphs in Figure 6.25 appear to be vertical shifts of each other. The explanation for this relies on the property of logs which says that $\log(ab) = \log a + \log b$. Since

$$y = \log(10x) = \log 10 + \log x = 1 + \log x,$$

the graph of $y = \log(10x)$ is the graph of $y = \log x$ shifted up 1 unit. Similarly,

$$y = \log(100x) = \log 100 + \log x = 2 + \log x,$$

the graph of $y = \log(100x)$ is the graph of $y = \log x$ shifted up 2 units. Thus, we see that the graphs are indeed vertical shifts of one another.

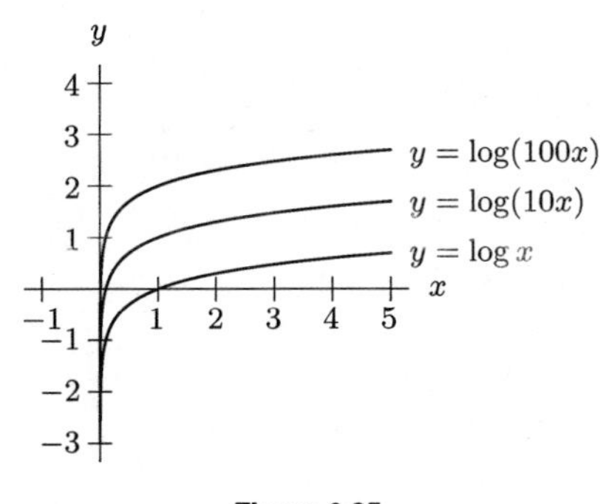

Figure 6.25

40. **(a)** Note that each value of $a(t)$ is 0.5 greater than the value of $g(t)$ for the same t. For example, $a(0) = g(0) + 0.5$, and in general

$$a(t) = g(t) + 0.5$$

(b) Observe that the values for $g(t)$ have been shifted to the left. For example, $b(-1) = g(0)$ and $b(0) = g(1)$. Thus,

$$b(t) = g(t+1).$$

(c) In this case, it is easier first to compare $c(t)$ and $b(t)$. For each t, the value of $c(t)$ is 0.3 less than the value of $b(t)$, so a possible formula is $c(t) = b(t) - 0.3$. Since $b(t) = g(t+1)$, we can say that

$$c(t) = g(t+1) - 0.3.$$

(d) The function $d(t)$ has the same values as $g(t)$ except they are shifted to the right by 0.5. For example, $d(0) = g(-0.5) = g(0-0.5)$ and $d(1) = g(0.5) = g(1-0.5)$. In each case

$$d(t) = g(t - 0.5)$$

.

(e) Compare values of $e(t)$ and $d(t)$. For any value of t, $e(t)$ is 1.2 more than $d(t)$. Thus, $e(t) = d(t) + 1.2$. But, $d(t) = g(t - 0.5)$, so

$$e(t) = g(t - 0.5) + 1.2$$

.

41. The graph of $y = x^2 - 10x + 25$ appears to be the graph of $y = x^2$ moved to the right by 5 units. See Figure 6.26. If this were so, then its formula would be $y = (x-5)^2$. Since $(x-5)^2 = x^2 - 10x + 25$, $y = x^2 - 10x + 25$ is, indeed, a horizontal shift of $y = x^2$.

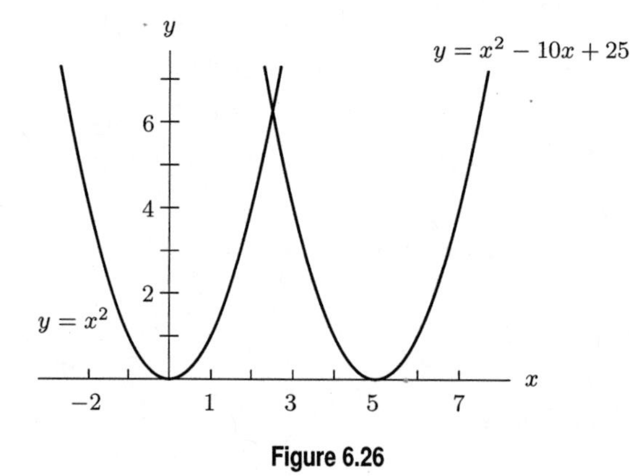

Figure 6.26

42. (a) The graph of g can be found by shifting the graph of f to the right 3 units and then up 2 units; $g(x) = f(x-3) + 2$.

(b) Yes, g is a quadratic function. To see this, notice that

$$\begin{aligned} g(x) &= (x-3)^2 + 2 \\ &= x^2 - 6x + 11. \end{aligned}$$

Thus, g is a quadratic function with parameters $a = 1$, $b = -6$, and $c = 11$.

(c) Figure 6.27 gives graphs of $f(x) = x^2$ and $g(x) = (x-3)^2 + 2$. Notice that g's axis of symmetry can be found by shifting f's axis of symmetry to the right 3 units. The vertex of g can be found by shifting f's vertex to the right 3 units and then up 2 units.

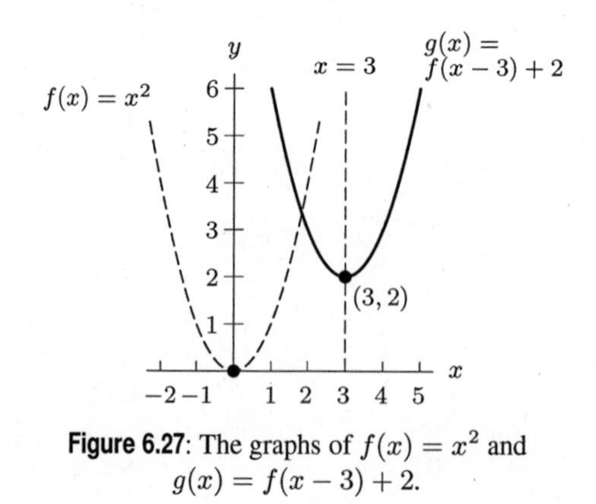

Figure 6.27: The graphs of $f(x) = x^2$ and $g(x) = f(x-3) + 2$.

43. Since the $+3$ is an outside change, this transformation shifts the entire graph of $q(z)$ up by 3 units. That is, for every z, the value of $q(z)+3$ is three units greater than $q(z)$.

44. Since the $-a$ is an outside change, this transformation shifts the entire graph of $q(z)$ down by a units. That is, for every z, the value of $q(z)-a$ is a units less than $q(z)$.

45. Since this is an inside change, the graph is four units to the left of $q(z)$. That is, for any given z value, the value of $q(z+4)$ is the same as the value of the function q evaluated four units to the right of z (at $z+4$).

46. Since this is an inside change, the graph is a units to the right of $q(z)$. That is, for any given z value, the value of $q(z-a)$ is the same as the value of the function q evaluated a units to the left of z (at $z-a$).

47. From the inside change, we know that the graph is shifted b units to the left. From the outside change, we know that it is shifted a units down. So, for any given z value, the graph of $q(z+b)-a$ is b units to the left and a units below the graph of $q(z)$.

48. From the inside change, we know that the graph is shifted $2b$ units to the right. From the outside change, we know that it is shifted ab units up. So, for any given z value, the graph of $q(z-2b)+ab$ is $2b$ units to the right and ab units above the graph of $q(z)$.

49. **(a)** On day d, high tide in Tacoma, $T(d)$, is 1 foot higher than high tide in Seattle, $S(d)$. Thus, $T(d)=S(d)+1$.
(b) On day d, height of the high tide in Portland equals high tide of the previous day, i.e. $d-1$, in Seattle. Thus, $P(d)=S(d-1)$.

50. **(a)** Notice that the value of $h(x)$ at every value of x is 2 less than the value of $f(x)$ at the same x value. Thus

$$h(x)=f(x)-2.$$

(b) Observe that $g(0)=f(1)$, $g(1)=f(2)$, and so on.In general,

$$g(x)=f(x+1).$$

(c) The values of $i(x)$ are two less than the values of $g(x)$at the same x value. Thus

$$i(x)=f(x+1)-2.$$

51. Notice that w reaches its maximum value of $y=13$ at $x=5$, whereas v reaches its maximum value of $y=20$ at $x=0$. Since we know that w is a transformation of v, this suggests that the graph of w is the graph of v shifted to the right by 5 units and down by 7 units:

$$w(x)=v(x-5)-7.$$

Thus, $h=5$ and $k=-7$. Checking our answer for $x=3$ and $x=4$, we see that

$$w(3)=v(3-5)-7=\underbrace{v(-2)}_{11}-7=4$$
$$w(4)=v(4-5)-7=\underbrace{v(-1)}_{17}-7=10,$$

and so on, as required.

52. **(a)** If

$$\begin{aligned} H(t)&=68+93(0.91)^t\\ \text{then}\quad H(t+15)&=68+93(0.91)^{(t+15)}=68+93(0.91)^{t+15}\\ \text{and}\quad H(t)+15&=(68+93(.91)^t)+15=83+93(0.91)^t \end{aligned}$$

(b)

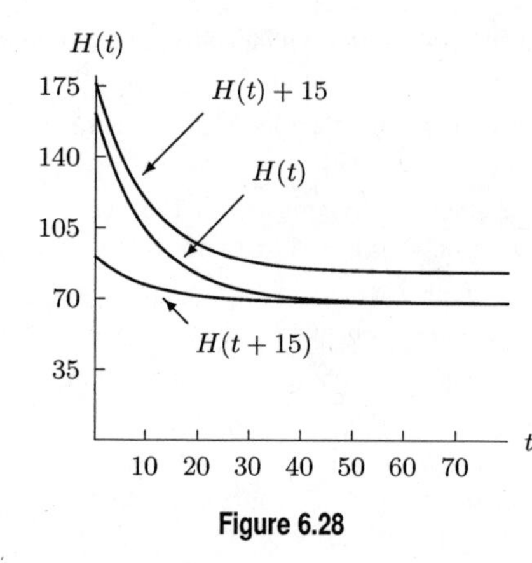

Figure 6.28

(c) $H(t+15)$ is the function $H(t)$ shifted 15 units to the left. This function could describe the temperature of the cup of coffee if it had been brought to class fifteen minutes earlier. $H(t)+15$ is the function $H(t)$ shifted upward 15 units, or $15^\circ F$. This function could describe the temperature of the coffee if it had been brought into a warmer classroom.

(d) As t gets very large, both $H(t+15)$ and $H(t)$ approach a final temperature of 68°F. In contrast, $H(t)+15$ approaches $68^\circ\text{F} + 15^\circ\text{F} = 83^\circ\text{F}$.

53. (a) If each drink costs \$7 then x drinks cost $\$7x$. Adding this to the \$20 cover charge gives $20+7x$. So

$$t(x) = 20 + 7x, \quad \text{for } x \geq 0.$$

(b) The cover charge is now \$25, so we have

$$\begin{aligned} n(x) &= 25 + 7x \\ &= 5 + \underbrace{20 + 7x} \\ &= 5 + t(x). \end{aligned}$$

Alternatively, notice that for any number of drinks the new cost, $n(x)$, is \$5 more than the old cost, $t(x)$. So

$$n(x) = t(x) + 5.$$

Thus $n(x)$ is the vertical shift of $t(x)$ up 5 units.

(c) Since 2 drinks are free, a customer who orders x drinks pays for only $(x-2)$ drinks at \$7/drink if $x \geq 2$. Thus

$$p(x) = 30 + 7(x-2), \quad \text{if } x \geq 2.$$

The formula for $p(x)$ if $x \geq 2$ can be written in terms of $t(x)$ as follows:

$$\begin{aligned} p(x) &= 10 + \underbrace{20 + 7(x-2)}_{t(x-2)} \\ &= 10 + t(x-2) \text{if } x \geq 2. \end{aligned}$$

Another way to think of this is to subtract two from your total number of drinks, x. Use $t(x-2)$ to determine the cost of two fewer drinks with the initial cover charge. Then add this 10 dollar increase in the cover charge to the result, so $p(x) = t(x-2) + 10$. This shows that the cover charge is \$10 more but you are charged for 2 fewer drinks.

54. Since the difference in temperatures decays exponentially, first we find a formula describing that difference over time. Let $D(t)$ represent the difference between the temperature of the brick and the temperature of the room.

When the brick comes out of the kiln, the difference between its temperature and room temperature is $350^\circ - 70^\circ = 280^\circ$. This difference will decay at the constant rate of 3% per minute. Therefore, a formula for $D(t)$ is

$$D(t) = 280(0.97)^t.$$

Since $D(t)$ is the difference between the brick's temperature, $H(t)$, and room temperature, $70°$, we have

$$D(t) = H(t) - 70.$$

Add 70 to both sides of the equation so that

$$H(t) = D(t) + 70,$$

Since $D(t) = 280(0.97)^t$,

$$H(t) = 280(0.97)^t + 70.$$

This function, $H(t)$, is *not* exponential because it is not of the form $y = ab^x$. However, since $D(t) = 280(0.97)^t$ *is* exponential, and since

$$H(t) = D(t) + 70,$$

$H(t)$ is a transformation of an exponential function. The graph of $H(t)$ is the graph of $D(t)$ shifted upward by 70. Figures 6.29 and 6.30 give the graphs of $D(t)$ and $H(t)$ for the first 4 hours, or 240 minutes, after the brick is removed from the kiln—that is, for $0 \leq t \leq 240$. As you can see, the brick cools off rapidly at first, and then levels off toward $70°$, or room temperature, where the graph of $H(t)$ has a horizontal asymptote. Notice that, by shifting the graph of $D(t)$upward by 70, the horizontal asymptote is also shifted, resulting in the asymptote at $T = 70$ for $H(t)$.

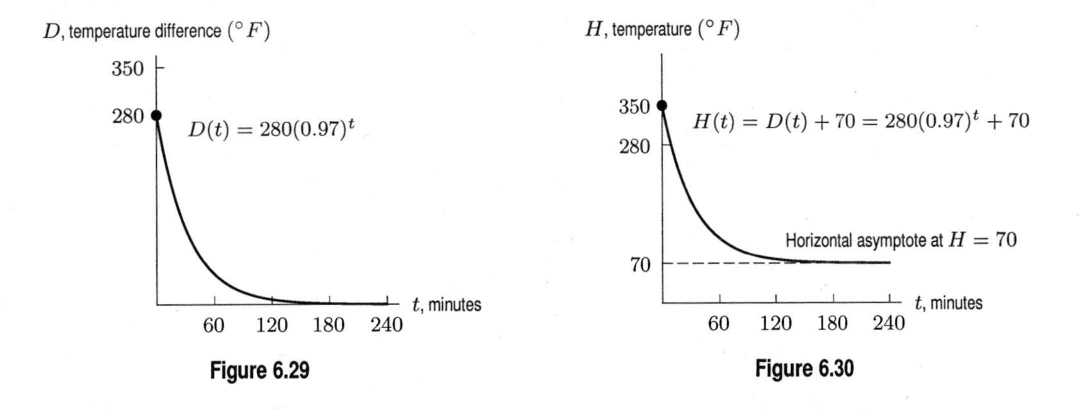

Figure 6.29 **Figure 6.30**

55. (a) There are many possible graphs, but all should show seasonally-related cycles of temperature increases and decreases, as in Figure 6.31.

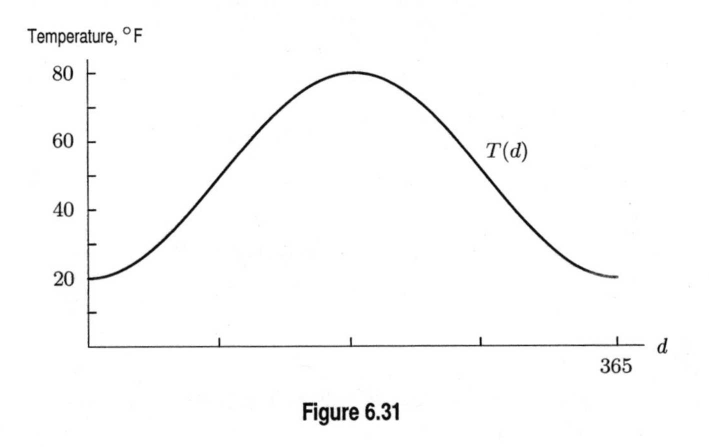

Figure 6.31

(b) While there are a wide variety of correct answers, the value of $T(6)$ is a temperature for a day in early January, $T(100)$ for a day in mid-April, and $T(215)$ for a day in early August. The value for $T(371) = T(365 + 6)$ should be close to that of $T(6)$.

(c) Since there are usually 365 days in a year, $T(d)$ and $T(d+365)$ represent average temperatures on days which are a year apart.

(d) $T(d+365)$ is the average temperature on the same day of the year a year earlier. They should be about the same value. Therefore, the graph of $T(d+365)$ should be about the same as that of $T(d)$.

(e) The graph of $T(d)+365$ is a shift upward of $T(d)$, by 365 units. It has no significance in practical terms, other than to represent a temperature that is 365° hotter than the average temperature on day d.

56. **(a)** Day t of the trip is day $t+400$ of the car's life. On the morning of that day, the odometer shows $f(t+400)$ miles. Therefore

$$h(t) = f(t+400).$$

(b) To find the number of miles driven during the first t days of the trip, find the difference between the odometer reading after t days and the reading at the start of the trip. That difference is

$$k(t) = h(t) - 3000 = f(t+400) - 3000.$$

57. Subtract the normal temperature from the thermometer reading. The fever is

$$f(t) = H(t) - 37 \text{ degrees}.$$

58. Since $g(x) = 5e^x$ and $g(x) = f(x-h) = e^{x-h}$, we have

$$5e^x = e^{x-h}.$$

Solve for h by taking the natural log of both sides

$$\begin{aligned} \ln(5e^x) &= \ln(e^{x-h}) \\ \ln 5 + x &= x - h \\ h &= -\ln 5. \end{aligned}$$

Solutions for Section 6.2

Skill Refresher

S1. Evaluating $f(x)$ at $x=3$, we have $f(3) = e^3 = 20.086$.

S2. Evaluating $g(x)$ at $x=3$, we have $g(3) = -e^3 = -20.086$.

S3. Evaluating $h(x)$ at $x=3$, we have $h(3) = e^{-3} = 0.050$.

S4. Evaluating $k(x)$ at $x=3$, we have $k(3) = -e^{-3} = -0.050$.

S5. **(a)** $f(-x) = 2(-x)^2 = 2x^2$

(b) $-f(x) = -\left(2x^2\right) = -2x^2$

Notice since $f(-x) = f(x) = 2x^2$, we see that $f(x)$ is an even function.

S6. **(a)** $f(-x) = \dfrac{1}{-x} = -\dfrac{1}{x}$

(b) $-f(x) = -\left(\dfrac{1}{x}\right) = -\dfrac{1}{x}$

Notice since $f(-x) = -f(x) = -\dfrac{1}{x}$, we see that $f(x) = 1/x$ is an odd function.

S7. **(a)** $f(-x) = 2(-x)^3 - 3 = -2x^3 - 3$

(b) $-f(x) = -\left(2x^3 - 3\right) = -2x^3 + 3$

Notice since $f(-x) \neq -f(x)$ and $f(-x) \neq -f(x)$, we see that $f(x) = 2x^3 - 3$ is neither an even nor an odd function.

S8. **(a)** $f(-x) = 4(-x)^3 + 9(-x) = -4x^3 - 9x$
(b) $-f(x) = -\left(4x^3 + 9x\right) = -4x^3 - 9x$
Notice since $f(-x) = -f(x)$, we see that $f(x) = 4x^3 + 9x$ is an odd function.

S9. **(a)** $f(-x) = 3(-x)^4 - 2(-x) = 3x^4 + 2x$
(b) $-f(x) = -\left(3x^4 - 2x\right) = -3x^4 + 2x$
Notice since $f(-x) \neq -f(x)$ and $f(-x) \neq -f(x)$, we see that $f(x) = 3x^4 - 2x$ is neither an even nor an odd function.

S10. **(a)** $f(-x) = \dfrac{3(-x)^3}{(-x)^2 - 1} = \dfrac{-3x^3}{x^2 - 1} = -\dfrac{3x^3}{x^2 - 1}$
(b) $-f(x) = -\left(\dfrac{3x^3}{x^2 - 1}\right) = -\dfrac{3x^3}{x^2 - 1}$
Notice since $f(-x) = -f(x)$, we see that $f(x) = \dfrac{3x^3}{x^2 - 1}$ is an odd function.

Exercises

1. **(a)** The y-coordinate is unchanged, but the x-coordinate is the same distance to the left of the y-axis, so the point is $(-2, -3)$.
   (b) The x-coordinate is unchanged, but the y-coordinate is the same distance above the x-axis, so the point is $(2, 3)$.
2. **(a)** Even symmetry means that the function is symmetric about the P-axis, so the P-coordinate is unchanged, but the t-coordinate is the same distance to the right of the P-axis, so the point is $(1, -5)$.
   (b) The negative sign causes a reflection across the t-axis giving $(-1, 5)$.
3. Since $H(x)$ is symmetric about the origin, $H(-x) = -H(x)$. So $H(3) = -H(-3) = -7$.
4. The negative sign reflects the graph about the x-axis, so the highest point on the original graph becomes the lowest on the new graph, similarly, the lowest becomes the highest, giving a new range of $-12 \leq -Q(x) \leq 2$.
5. The negative sign reflects the graph of $Q(t)$ horizontally about the y-axis, so the domain of $y = Q(-t)$ is $t < 0$. A horizontal reflection of the graph of $Q(t)$ about the y-axis will not change the range. The range of $Q(-t)$ is therefore the same as the range of $Q(t)$, $-4 \leq Q(-t) \leq 7$.
6. The negative sign reflects the graph of $Q(t)$ vertically about the t-axis, so the highest point on the original graph becomes the lowest on the new graph, similarly, the lowest becomes the highest, giving a new range of $-7 \leq -Q(t) \leq 4$. Since a vertical reflection of the graph of $Q(t)$ will not change the domain, the domain of $-Q(t)$ is therefore $t > 0$.
7. The two negative signs reflect the graph of $Q(t)$ vertically about the t-axis and horizontally about the y-axis. The domain of $-Q(-t)$ is therefore $t < 0$, and the range is $-7 \leq -Q(-t) \leq 4$.
8. The graph of $-Q(t - 4)$ is both a horizontal shift right 4 units and a vertical reflection about the t-axis of the original graph of $Q(t)$. The domain is therefore $t > 4$, and the range is $-7 \leq -Q(t - 4) \leq 4$.
9. To reflect about the x-axis, we make all the y-values negative, getting $y = -e^x$ as the formula.
10. To reflect about the y-axis, we substitute $-x$ for x in the formula getting $y = e^{-x}$.
11.

Table 6.1

p	-3	-2	-1	0	1	2	3
$f(p)$	0	-3	-4	-3	0	5	12

Table 6.2

p	-3	-2	-1	0	1	2	3
$g(p)$	12	5	0	-3	-4	-3	0

Table 6.3

p	-3	-2	-1	0	1	2	3
$h(p)$	0	3	4	3	0	-5	-12

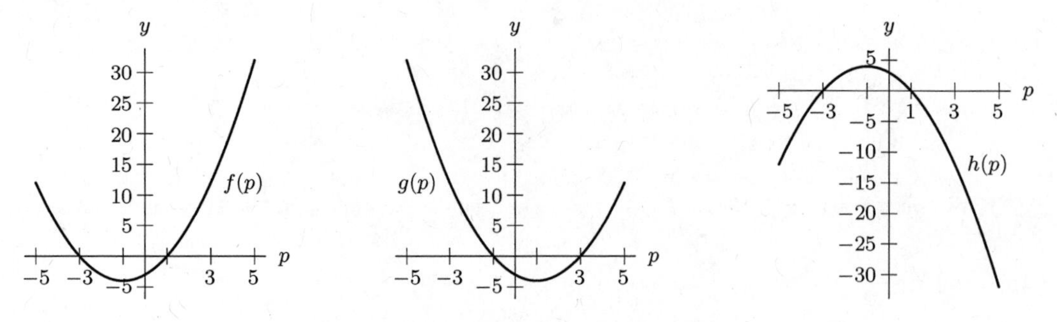

Figure 6.32: Graphs of $f(p)$, $g(p)$, and $h(p)$

Since $g(p) = f(-p)$, the graph of g is a horizontal reflection of the graph of f across the y-axis. Since $h(p) = -f(p)$, the graph of h is a reflection of the graph of f across the p-axis.

12. See Figure 6.33. The graph of $y = f(-x)$ is the graph of $y = f(x)$ reflected across the y-axis. The formula for $y = f(-x)$ is $y = 4^{-x}$.

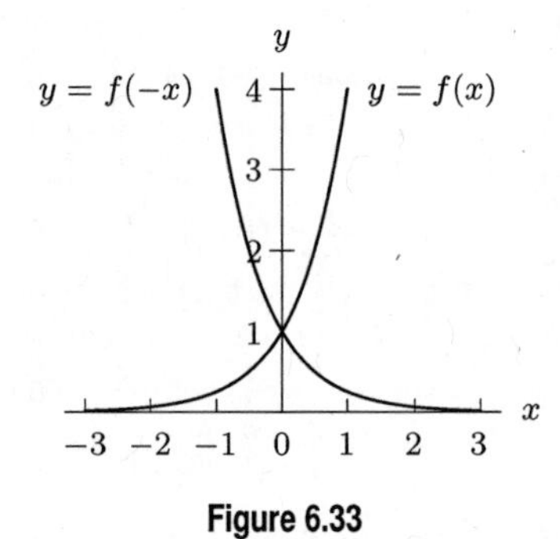

Figure 6.33

13. The graph of $y = -g(x) = -(1/3)^x$ is the graph of $y = g(x)$ reflected across the x-axis. See Figure 6.34.

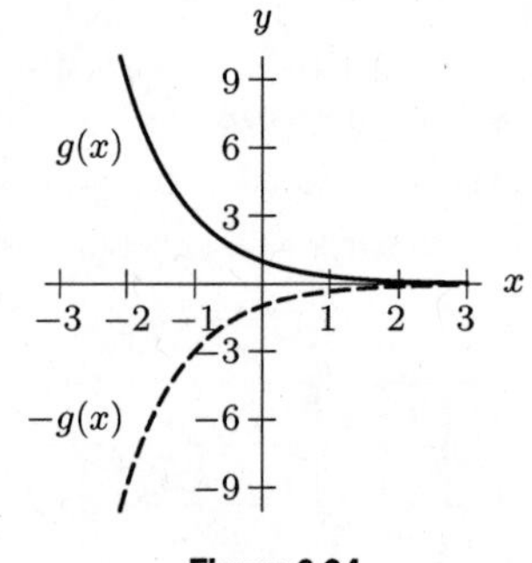

Figure 6.34

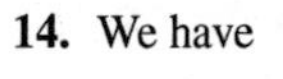

14. We have

$$\begin{aligned} y = m(-n) &= (-n)^2 - 4(-n) + 5 \\ &= n^2 + 4n + 5 \end{aligned}$$

To graph this function, reflect the graph of m across the y-axis. See Figure 6.35.

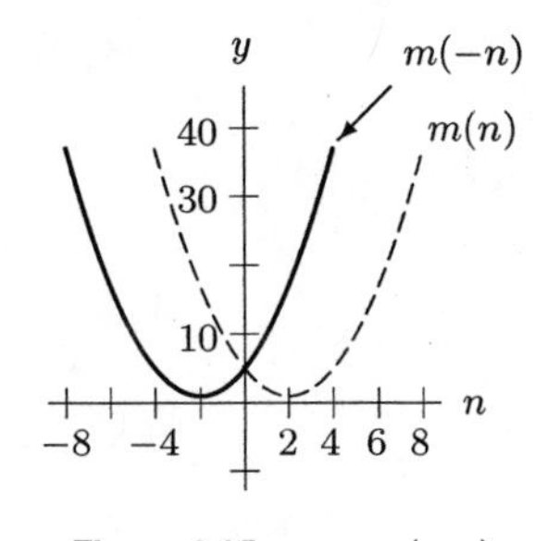

Figure 6.35: $y = m(-n)$

15. We have

$$y = -m(n) = -(n)^2 + 4n - 5$$

To graph this function, reflect the graph of m across the n-axis. See Figure 6.36.

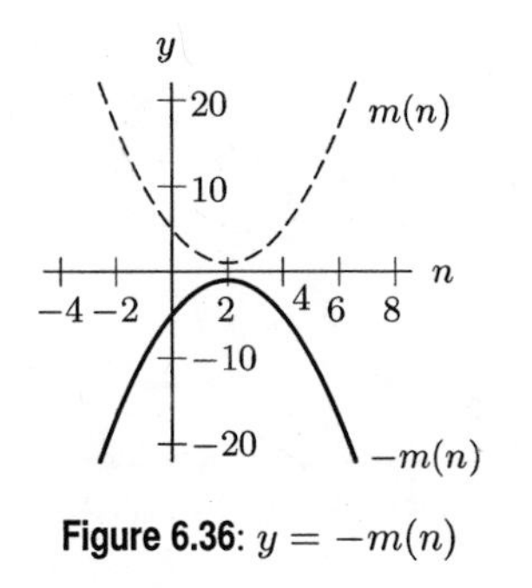

Figure 6.36: $y = -m(n)$

16. We have

$$y = -m(-n) = -(-n)^2 + 4(-n) - 5 = -n^2 - 4n - 5$$

To graph this function, first reflect the graph of m across the y-axis, then reflect it again across the n-axis. See Figure 6.37.

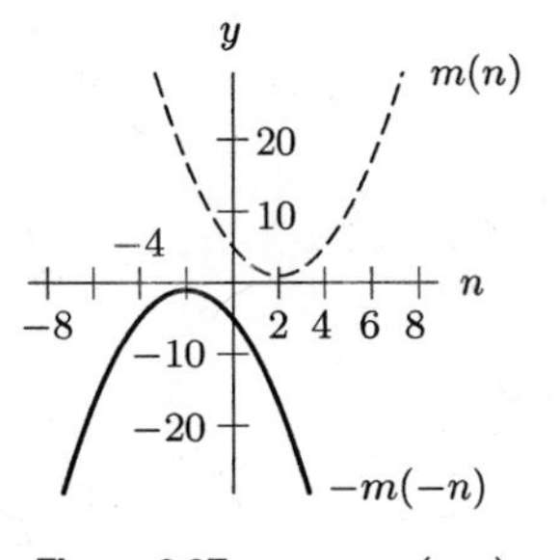

Figure 6.37: $y = -m(-n)$

17. We have

$$\begin{aligned} y &= m(-n) + 3 = (-n)^2 - 4(-n) + 5 + 3 \\ &= n^2 + 4n + 8. \end{aligned}$$

To graph this function, first reflect the graph of m across the y-axis, and then shift it up by 3 units. See Figure 6.38.

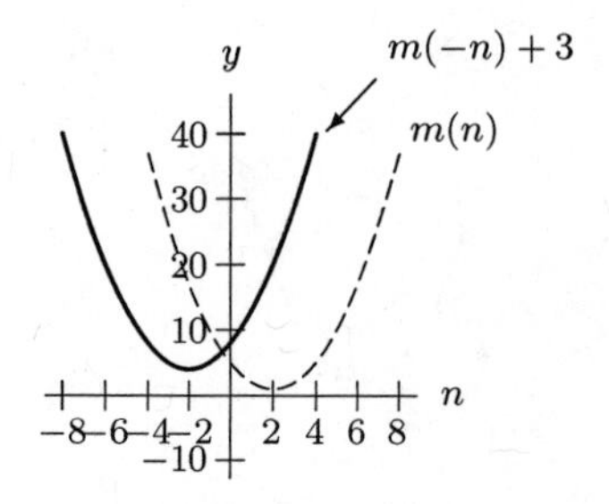

Figure 6.38: $y = m(-n) + 3$

18. We have

$$y = k(-w) = 3^{-w}$$

To graph this function, reflect the graph of k across the y-axis. See Figure 6.39.

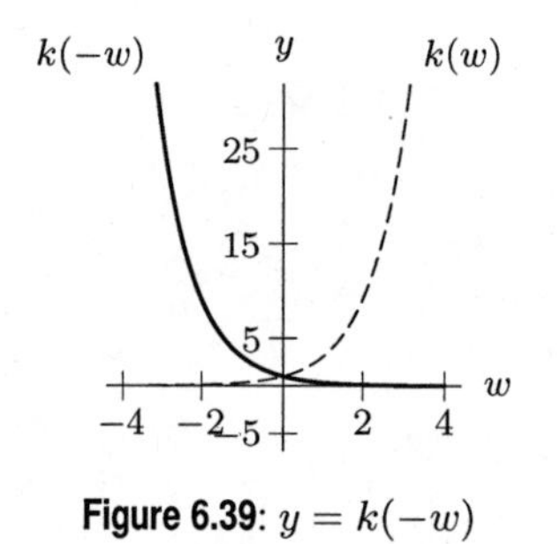

Figure 6.39: $y = k(-w)$

19. We have

$$y = -k(w) = -3^{w}$$

To graph this function, reflect the graph of k across the w-axis. See Figure 6.40.

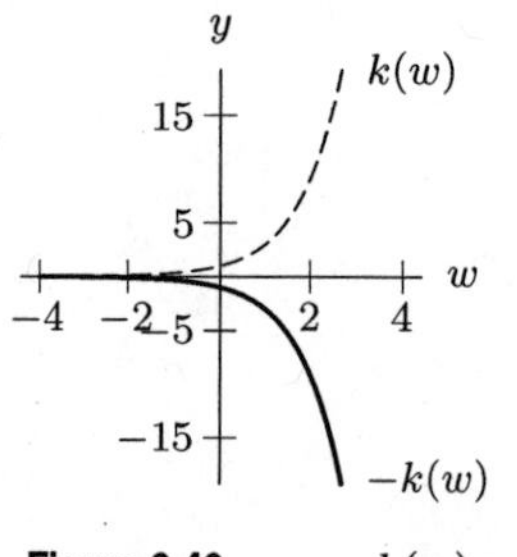

Figure 6.40: $y = -k(w)$

20. We have

$$y = -k(-w) = -3^{-w}$$

To graph this function, first reflect the graph of k across the y-axis, then reflect it again across the w-axis. Figure 6.41.

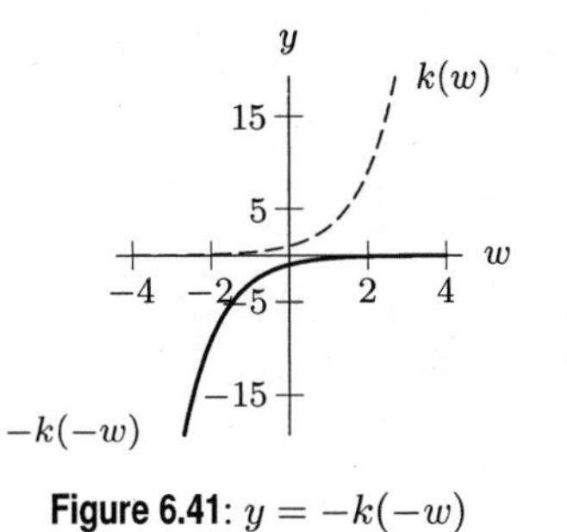

Figure 6.41: $y = -k(-w)$

21. We have

$$y = -k(w-2) = -3^{w-2}$$

To graph this function, first reflect the graph of k across the w-axis, then shift it to the right by 2 units. See Figure 6.42.

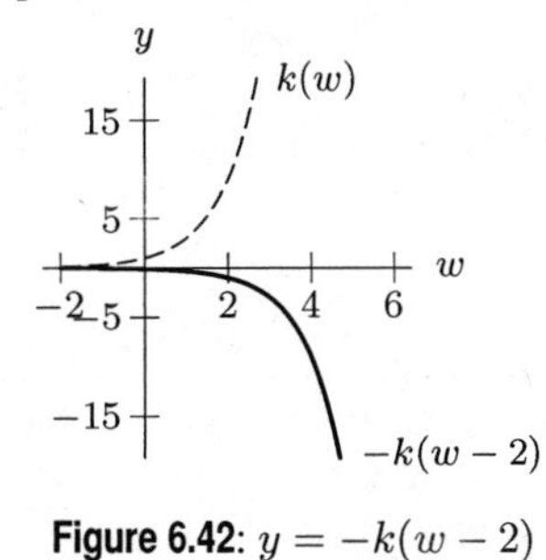

Figure 6.42: $y = -k(w-2)$

22. Since $f(-x) = 7(-x)^2 - 2(-x) + 1 = 7x^2 + 2x + 1$ is equal to neither $f(x)$ or $-f(x)$, the function is neither even nor odd.

23. The definition of an odd function is that $f(-x) = -f(x)$. Since $f(-x) = 4(-x)^7 - 3(-x)^5 = -4x^7 + 3x^5$, we see that $f(-x) = -f(x)$, so the function is odd.

24. The definition of an even function is that $f(-x) = f(x)$. Since $f(-x) = 8(-x)^6 + 12(-x)^2 = 8x^6 + 12x^2$, we see that $f(-x) = f(x)$, so the function is even.

25. Since $f(-x) = (-x)^5 + 3(-x)^3 - 2 = -x^5 - 3x^3 - 2$ is equal to neither $f(x)$ or $-f(x)$, the function is neither even nor odd.

Problems

26. **(a)** See Figure 6.43.
(b) See Figure 6.44.

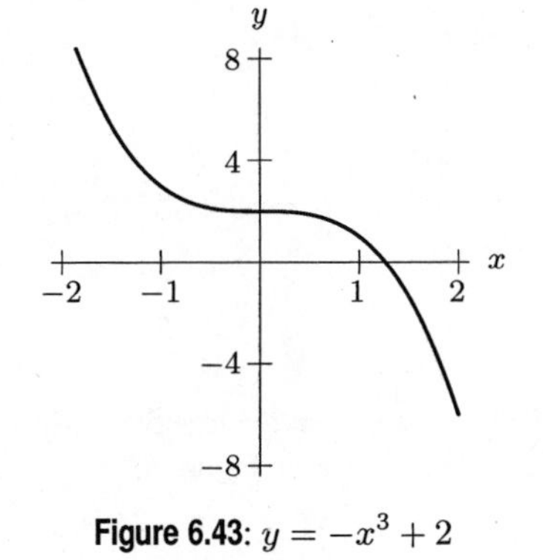

Figure 6.43: $y = -x^3 + 2$

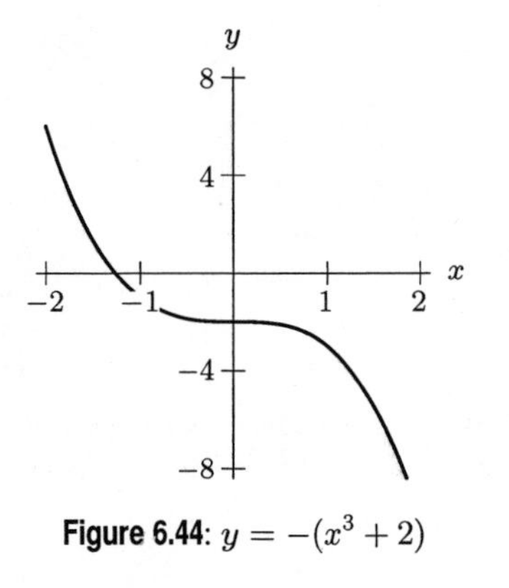

Figure 6.44: $y = -(x^3 + 2)$

(c) The two functions are not the same.

27. **(a)** See Figure 6.45.
(b) See Figure 6.45.

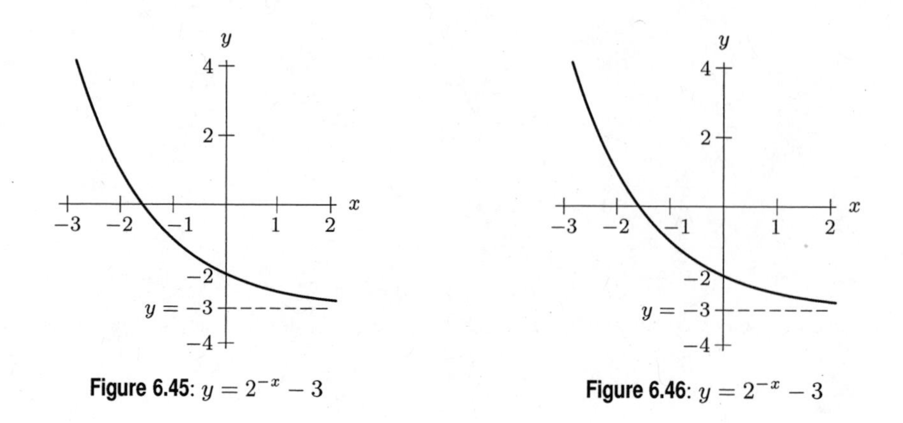

Figure 6.45: $y = 2^{-x} - 3$ **Figure 6.46**: $y = 2^{-x} - 3$

(c) The two functions are the same in this case. Note that you will not always obtain the same result if you change the order of the transformations.

28. **(a)** $f(-x)$ for $x = -4$ equals $f(-(-4)) = f(4)$. To evaluate $f(4)$ we find the y-value of the point on the graph when $x = 4$. In this case the point $(4, -10)$ is on the graph of $y = f(x)$; thus, $f(-x)$ for $x = -4$ equals -10.
(b) $-f(x)$ for $x = -6$ equals $-f(-6)$. To evaluate $f(-6)$ we find the y-value of the point on the graph when $x = -6$. In this case the point $(-6, 25)$ is on the graph of $y = f(x)$; thus, $-f(x)$ for $x = -6$ equals -25.
(c) $-f(-x)$ for $x = -4$ equals $-f(-(-4)) = -f(4)$. From part (a), we know that $f(4) = -10$. Thus, $-f(-x)$ for $x = -4$ is 10.
(d) $-f(x+2)$ for $x = 0$ equals $-f(0+2) = -f(2)$. Using the graph we see that the y-value of the point at $x = 2$ is approximately -3. Thus, $-f(x+2)$ for $x = 0$ is approximately 3
(e) $f(-x) + 4$ for $x = -6$ equals $f(-(-6)) + 4 = f(6) + 4$. Using the graph we see that the y-value of the point at $x = 6$ is -30. Thus we have $f(6) + 4 = -30 + 4 = -26$, so $f(-x) + 4$ for $x = -6$ is -26.

29. **(a)** $g(-x) = \sqrt[3]{-x} = -\sqrt[3]{x} = -g(x)$.
(b) From part (a), we see that $g(-x) = -g(x)$, so the graphs of these two functions coincide in Figure 6.47.

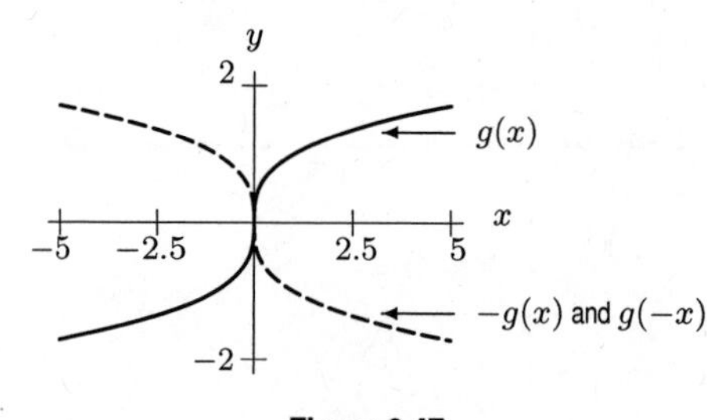

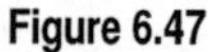

Figure 6.47

(c) The function is odd, since $g(-x) = -g(x)$.

30. The equation of the reflected line is

$$y = b + m(-x) = b - mx.$$

The reflected line has the same y-intercept as the original; that is b. Its slope is $-m$, the negative of the original slope, and its x-intercept is b/m, the negative of the original x-intercept. A possible graph is in Figure 6.48.

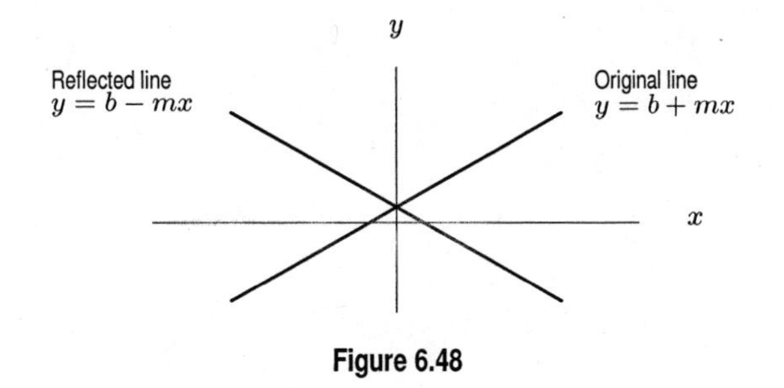

Figure 6.48

31. The graphs in Figure 6.49 are reflections of each other across the x-axis. To see this algebraically, note that

$$y = \log\left(\frac{1}{x}\right) = \log 1 - \log x = -\log x.$$

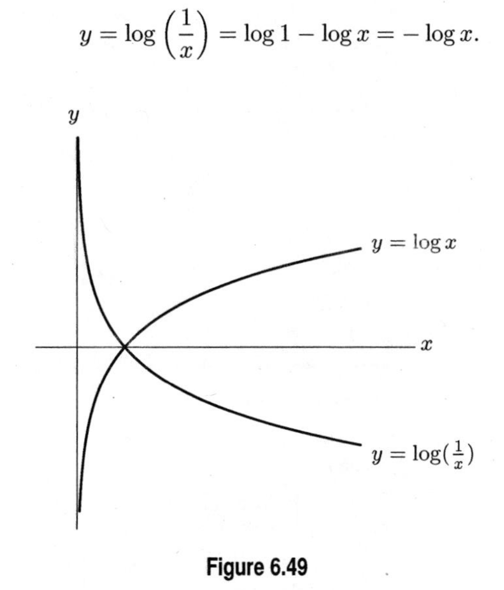

Figure 6.49

32. **(a)** See Figure 6.50.

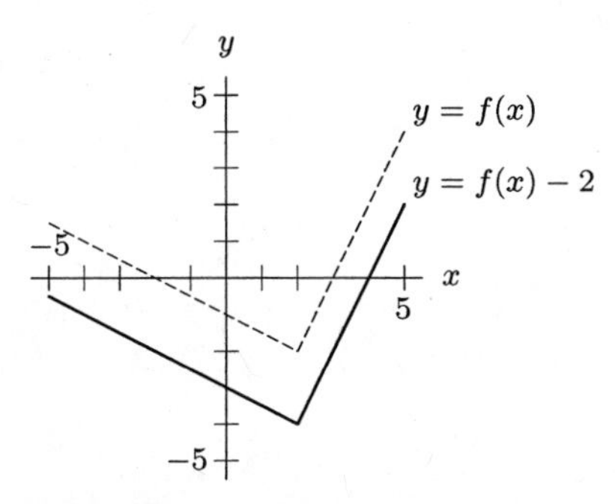

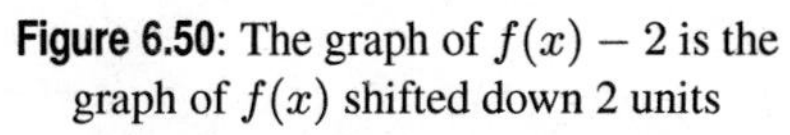
Figure 6.50: The graph of $f(x) - 2$ is the graph of $f(x)$ shifted down 2 units

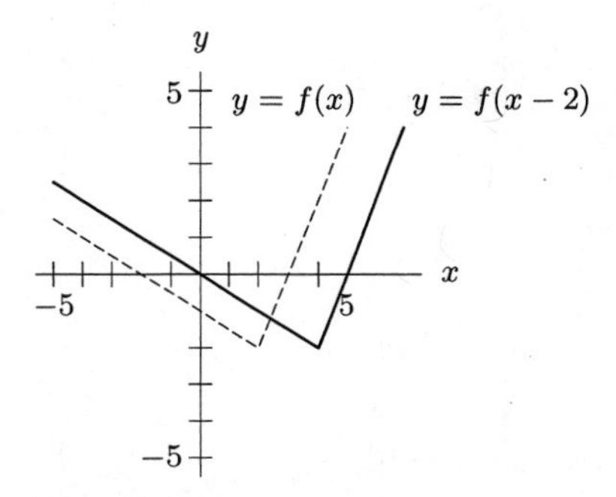

Figure 6.51: The graph of $f(x - 2)$ is the graph of $f(x)$ shifted right 2 units

(b) See Figure 6.51.

(c) See Figure 6.52.

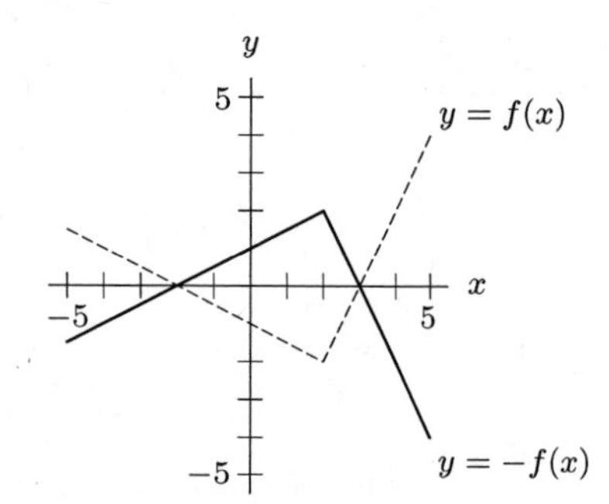

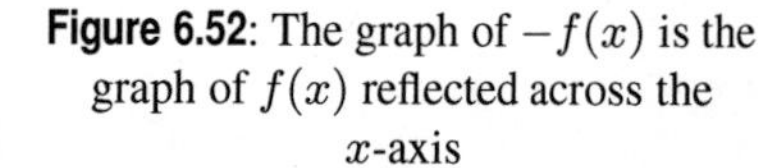

Figure 6.52: The graph of $-f(x)$ is the graph of $f(x)$ reflected across the x-axis

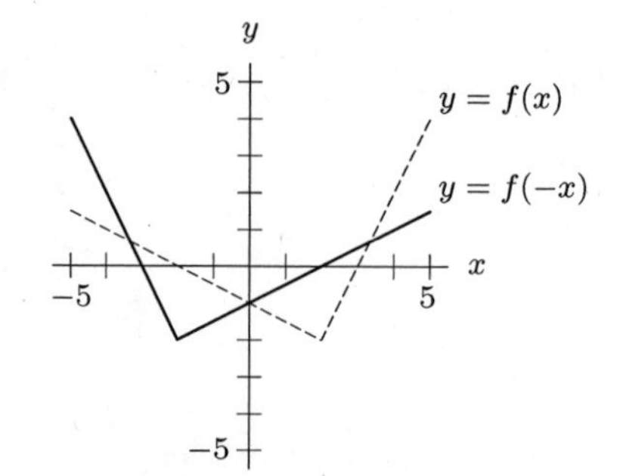

Figure 6.53: The graph of $f(-x)$ is the graph of $f(x)$ reflected across the y-axis

(d) See Figure 6.53.

33. The answers are

(i) b (ii) c (iii) d
(iv) e (v) a

34. **(a)**

Table 6.4

x	−3	−2	−1	0	1	2	3
y	5	−8	−4	?	−4	−8	5

(b)

Table 6.5

x	−3	−2	−1	0	1	2	3
y	5	8	−4	0	4	−8	−5

35. **(a)** Figure 6.54 shows the graph of a function f that is symmetric across the y-axis.
(b) Figure 6.55 shows the graph of function f that is symmetric across the origin.

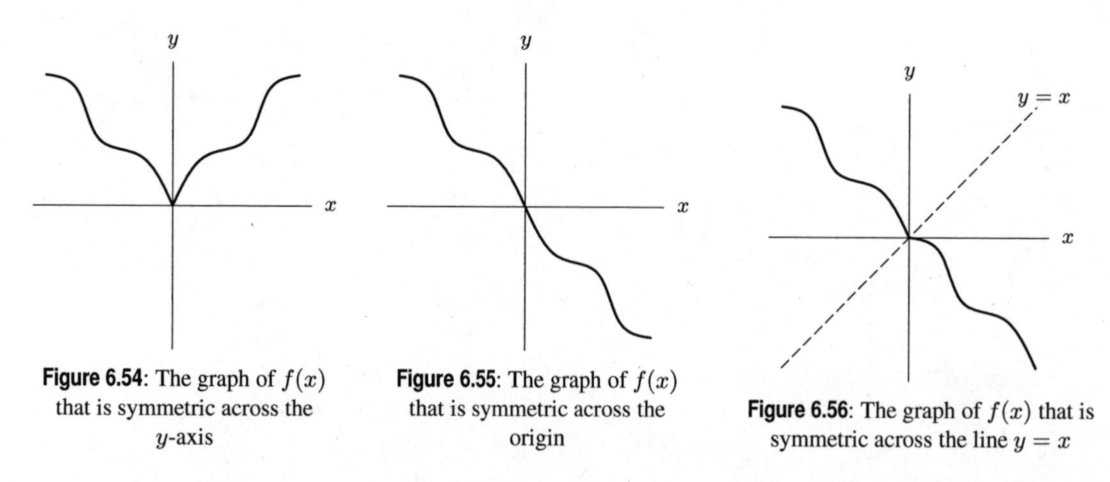

Figure 6.54: The graph of $f(x)$ that is symmetric across the y-axis

Figure 6.55: The graph of $f(x)$ that is symmetric across the origin

Figure 6.56: The graph of $f(x)$ that is symmetric across the line $y = x$

(c) Figure 6.56 shows the graph of function f that is symmetric across the line $y = x$.

36. **(a)** Since the values of $f(x)$ and $f(-x)$ are the same, $f(x)$ appears to be symmetric across the y-axis. Thus, $f(x)$ could be an even function.

(b) Since the value of $g(-x)$ is the opposite of $g(x)$, we know that $g(x)$ could be symmetric about the origin. Thus, $g(x)$ could be an odd function.

(c) Let $h(x) = f(x) + g(x)$. The value of $h(-x) = f(-x) + g(-x)$ is not the same as either $h(x) = f(x) + g(x)$ or $-h(x) = -(f(x) + g(x))$, so $h(x)$ is not symmetric.

(d) Let $j(x) = f(x+1)$. Note that $j(1) = f(1+1) = f(2) = 1$ and $j(-1) = f(-1+1) = f(0) = -3$. Thus, $j(-x)$ does not equal either $j(x)$ or $-j(x)$, so $j(x)$ is not symmetric.

37. Since $f(x)$ is an even function we know $f(-x) = f(x)$. Similarly since $g(x)$ is an odd function we know $g(-x) = -g(x)$. Using both of these identities, we evaluate at $-x$.

(a)

$$\begin{aligned} h(-x) &= f(-x)g(-x) \\ &= f(x)(-g(x)) \\ &= -f(x)g(x) \\ &= -h(x). \end{aligned}$$

Since $h(-x) = -h(x)$, the function $h(x) = f(x)g(x)$ must be odd.

(b)

$$\begin{aligned} k(-x) &= f(-x) + g(-x) \\ &= f(x) + (-g(x)) \\ &= f(x) - g(x). \end{aligned}$$

Since $k(-x) = f(x) - g(x)$ and $k(x) = f(x) + g(x)$, it follows $k(-x) = k(x)$ only if $g(x) = 0$, and $k(-x) = -k(x)$ only if $f(x) = 0$. Therefore unless $f(x) = 0$ or $g(x) = 0$, $k(x)$ will be neither even nor odd.

(c)

$$\begin{aligned} m(-x) &= g(f(-x)) \\ &= g(f(x)) \\ &= m(x). \end{aligned}$$

Since $m(-x) = m(x)$, the function $m(x) = g(f(x))$ must be even.

38. Since both $f(x)$ and $g(x)$ are odd functions, we know $f(-x) = -f(x)$ and $g(-x) = -g(x)$. Using both of these identities, we evaluate at $-x$.

(a)

$$\begin{aligned} h(-x) &= f(-x)g(-x) \\ &= (-f(x))(-g(x)) \\ &= f(x)g(x) \\ &= h(x). \end{aligned}$$

Since $h(-x) = h(x)$, the function $h(x) = f(x)g(x)$ must be even.

(b)

$$\begin{aligned} k(-x) &= f(-x) - g(-x) \\ &= -f(x) - (-g(x)) \\ &= -f(x) + g(x) \\ &= -(f(x) - g(x)) \\ &= -k(x). \end{aligned}$$

Since $k(-x) = -k(x)$, the function $h(x) = f(x)g(x)$ must be odd.

(c)

$$\begin{aligned} m(-x) &= f(g(-x)) \\ &= f(-g(x)) \\ &= -f(g(x)) \\ &= -m(x). \end{aligned}$$

Since $m(-x) = -m(x)$, the function $m(x) = f(g(x))$ must be odd.

39. See Figure 6.57.

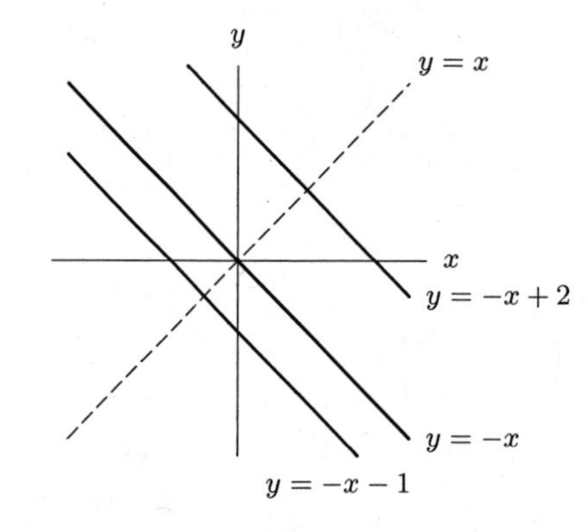

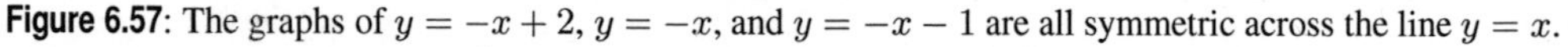

Figure 6.57: The graphs of $y = -x + 2$, $y = -x$, and $y = -x - 1$ are all symmetric across the line $y = x$.

Any straight line perpendicular to $y = x$ is symmetric across $y = x$. Its slope must be -1, so $y = -x + b$, for an arbitrary constant b, is symmetric across $y = x$.

Also, the line $y = x$ is symmetric about itself.

40. One way to do this is to sketch a graph of $y = h(x)$ to see that it appears to be symmetric across the origin. In other words, we can visually check to see that flipping the graph of $y = h(x)$ about the y-axis and then the x-axis (or vice-versa) does not change the appearance of the function's graph. See Figure 6.58.

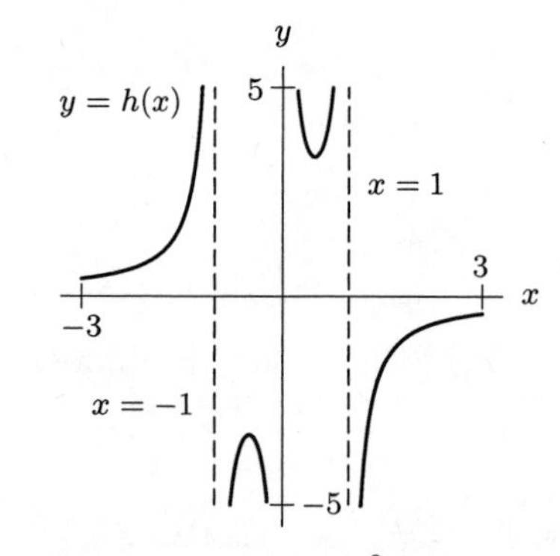

Figure 6.58: The graph of $y = h(x) = \frac{1+x^2}{x-x^3}$ is symmetric across the origin

To confirm that $h(x)$ is symmetric across the origin, we use algebra. We need to show that $h(-x) = -h(x)$ for any x. Finding the formula for $h(-x)$, we have

$$\begin{aligned} h(-x) &= \frac{1+(-x)^2}{(-x)-(-x)^3} = \frac{1+x^2}{-x+x^3} = \frac{1+x^2}{-(x-x^3)} \\ &= \frac{1+x^2}{x-x^3} = -h(x). \end{aligned}$$

Thus, the formula for $h(-x)$ is the same as the formula for $-h(x)$, and so the graph of $y = h(x)$ is symmetric across the origin.

41. The argument that $f(x)$ is not odd is correct. However, the statement "something is either even or odd" is false. This function is neither an odd function nor an even function.

42. If $f(x)$ is always increasing and concave down, its graph must have the same shape as the graph in Figure 6.59.

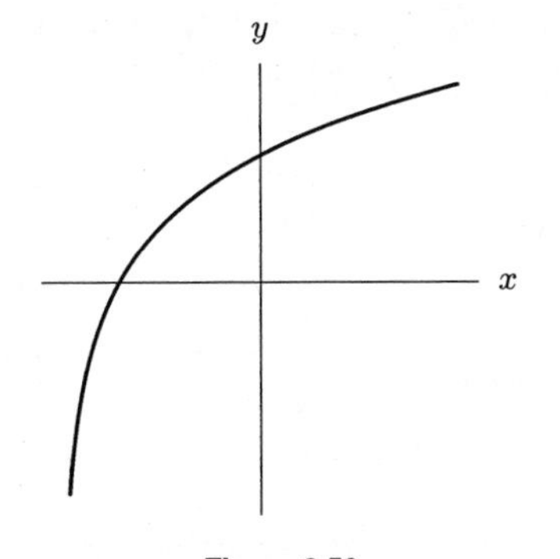

Figure 6.59

(a) The function $f(-x)$ is a horizontal reflection of the graph of $f(x)$ about the y-axis. Thus the graph of $f(-x)$ must have the same shape as the graph in Figure 6.60. We can see in Figure 6.60 the graph of $f(-x)$ is always decreasing and concave down.

(b) The function $-f(x)$ is a vertical reflection of the graph of $f(x)$ about the x-axis. Thus the graph of $-f(x)$ must have the same shape as the graph in Figure 6.61. We can see in Figure 6.61 the graph of $-f(x)$ is always decreasing and concave up.

(c) The graph of $-f(-x)$ is obtained by reflecting the graph of $f(x)$ horizontally about the y-axis and vertically about the x-axis. Thus the graph of $-f(-x)$ must have the same shape as the graph Figure 6.62. We can see in Figure 6.62 the graph of $-f(-x)$ is always increasing and concave up.

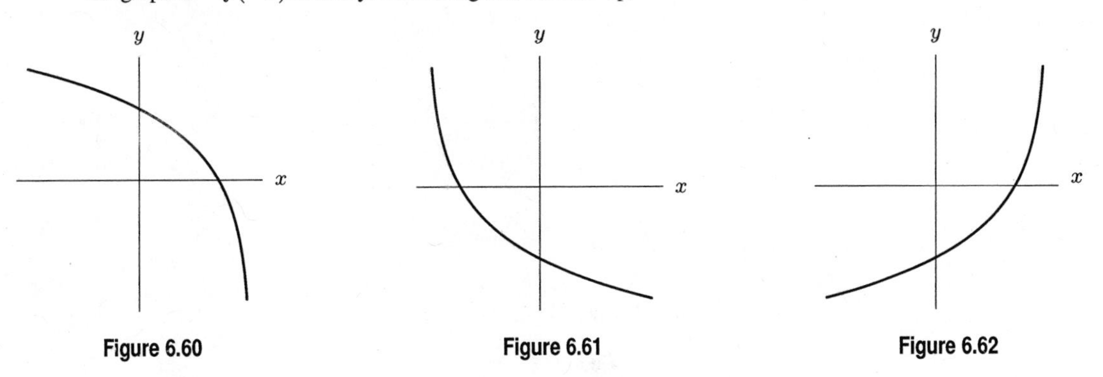

Figure 6.60 Figure 6.61 Figure 6.62

43. No, it is not possible for an odd function to be strictly concave up. If it were concave up in the first or second quadrants, then the fact that it is odd would mean it would have to be symmetric across the origin, and so would be concave down in the third or fourth quadrants.

44. (a) In order for $f(x)$ to be even,

$$\begin{aligned} f(-x) &= f(x) \\ m(-x) + b &= mx + b \\ -mx + b &= mx + b \text{ for all } x. \end{aligned}$$

This is true if and only if $-m = m$, which is true if and only if $m = 0$, so $f(x) = b$. Thus, a linear function is even only when it is a constant; its graph is a horizontal line.

(b) In order for $f(x)$ to be odd,

$$\begin{aligned} f(-x) &= -f(x) \\ m(-x)+b &= -(mx+b) \\ -mx+b &= -mx-b \text{ for all } x. \end{aligned}$$

This is true if and only if $-b=b$, which is true if and only if $b=0$, so $f(x)=mx$.

(c) If $f(x)$ is both even and odd, then both (a) and (b) are true, which means $m=0$ and $b=0$, so $f(x)=0$. The function $f(x)=0$ is both even and odd, and its graph is the line $y=0$, or the x-axis.

45. Because $f(x)$ is an odd function, $f(x)=-f(-x)$. Setting $x=0$ gives $f(0)=-f(0)$, so $f(0)=0$. Since $c(0)=1$, $c(x)$ is not odd. Since $d(0)=1$, $d(x)$ is not odd.

46. Because $f(x)$ is an even function, it is symmetric across the y-axis. Thus, in the second quadrant it must be decreasing and concave down.

47. To show that $f(x)=x^{1/3}$ is an odd function, we must show that $f(x)=-f(-x)$:

$$-f(-x)=-(-x)^{1/3}=x^{1/3}=f(x).$$

However, not all power functions are odd. The function $f(x)=x^2$ is an even function because $f(x)=f(-x)$ for all x. Another counter-example is $f(x)=\sqrt{x}=x^{1/2}$. This function is not odd because it is not defined for negative values of x.

48. Figure 6.63 shows the graphs of $s(x)$, $c(x)$, and $n(x)$. Based on the graphs, it appears that $s(x)$ is an even function (symmetric across the y-axis), $c(x)$ is an odd function (symmetric across the origin), and $n(x)$ is neither.

$s(-x)=2^{-x}+(\frac{1}{2})^{-x}=(\frac{1}{2})^x+2^x=2^x+(\frac{1}{2})^x=s(x)$, so $s(x)$ is an even function.

$c(-x)=2^{-x}-(\frac{1}{2})^{-x}=(\frac{1}{2})^x-2^x=-2^x+(\frac{1}{2})^x=-c(x)$, so $c(x)$ is an odd function.

$n(-x)=2^{-x}-(\frac{1}{2})^{-x-1}=(\frac{1}{2})^x+2^{x+1}=2\cdot 2^x+(\frac{1}{2})^x$. Since $n(-x)\neq n(x)$ and $n(-x)\neq -n(x)$, $n(x)$ is neither even nor odd.

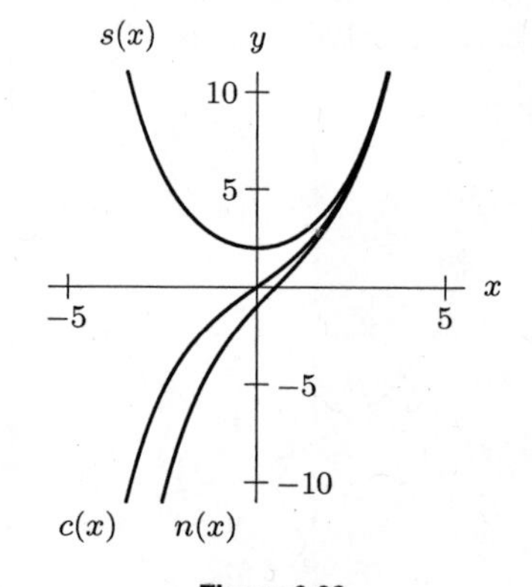

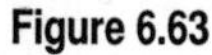
Figure 6.63

49. Suppose $f(x)$ is both even and odd. If $f(x)$ is even, then

$$f(-x)=f(x).$$

If $f(x)$ is odd, then

$$f(-x)=-f(x).$$

Since $f(-x)$ equals both $f(x)$ and $-f(x)$, we have

$$f(x)=-f(x).$$

Add $f(x)$ to both sides of the equation to get

$$2f(x) = 0$$

or

$$f(x) = 0.$$

Thus, the function $f(x) = 0$ is the only function which is both even and odd. There are no *nontrivial* functions that have both symmetries.

50. There is only one such function, and a rather unexciting one at that. Any function with symmetry about the x-axis would look unchanged if you flipped its graph about the x-axis. For any function f, the graph of $y = -f(x)$ is the graph of $y = f(x)$ flipped about the x-axis. Assuming this does not change the appearance of its graph, we have the equation

$$f(x) = -f(x).$$

Adding $f(x)$ to both sides gives

$$2f(x) = 0,$$

or simply

$$f(x) = 0.$$

Thus the only function that is symmetrical about the x-axis is the x-axis itself – that is, the line $y = 0$. If you think about it, you will see that any other curve that is symmetrical about the x-axis would necessarily fail the vertical line test, and would thus not represent the graph of a function.

Solutions for Section 6.3

Skill Refresher

S1. **(a)** In order to evaluate $2f(6)$, we first evaluate $f(6) = 6^2 = 36$. Then we multiply by 2, and thus we have $2f(6) = 2 \cdot 36 = 72$.
(b) Since $f(6) = 36$, we have $-\frac{1}{2}f(6) = -\frac{1}{2} \cdot 36 = -18$.
(c) Since $f(6) = 36$, we have $5f(6) - 3 = 5(36) - 3 = 177$.
(d) In order to evaluate $\frac{1}{4}f(x-1)$ at $x = 6$, we first evaluate $f(6-1) = f(5) = 5^2 = 25$. Next we divide by 4, and thus we have $\frac{1}{4}f(6-1) = \frac{1}{4} \cdot 25 = \frac{25}{4}$.

S2. **(a)** In order to evaluate $-4g(-2)$, we first evaluate $g(-2) = \frac{1}{-2}$. Then we multiply by -4, and thus we have $-4g(-2) = -4 \cdot -\frac{1}{2} = 2$.
(b) Since $g(-2) = -\frac{1}{2}$, we have $\frac{2}{3}g(-2) = \frac{2}{3} \cdot -\frac{1}{2} = -\frac{1}{3}$.
(c) In order to evaluate $2g(x+5)$ at $x = -2$, we first evaluate $g(-2+5) = g(3) = \frac{1}{3}$. Then we multiply by 2, and thus we have $2g(-2+5) = 2 \cdot \frac{1}{3} = \frac{2}{3}$.
(d) In order to evaluate $-\frac{1}{5}g(-x)$ at $x = -2$, we first evaluate $g\big(-(-2)\big) = g(2) = \frac{1}{2}$. Thus, we have

$$\begin{aligned} -\frac{1}{5}g\big(-(-2)\big) &= -\frac{1}{5} \cdot \frac{1}{2} \\ &= -\frac{1}{10}. \end{aligned}$$

S3. **(a)** $-\frac{1}{3}f(x) = -\frac{1}{3}\sqrt{x}$
(b) We have $f(-x) = \sqrt{-x}$, and thus $5f(-x) = 5\sqrt{-x}$.
(c) We have $f(x-8) = \sqrt{x-8}$, and thus $6f(x-8) = 6\sqrt{x-8}$.
(d) We have $f(2-x) = \sqrt{2-x}$, and thus $\frac{1}{4}f(2-x) = \frac{1}{4}\sqrt{2-x}$.

S4. **(a)** $5p(x) = 5\left(3x^2 - 6\right) = 15x^2 - 30.$

(b)

$$\begin{aligned} -\frac{1}{3}p(-x) &= -\frac{1}{3}\left(3(-x)^2 - 6\right) \\ &= -\frac{1}{3}(3x^2 - 6) \\ &= -x^2 + 2 \end{aligned}$$

(c)

$$\begin{aligned} -2p(x+3) &= -2\left(3(x+3)^2 - 6\right) \\ &= -2\left(3(x^2 + 6x + 9) - 6\right) \\ &= -2(3x^2 + 18x + 27 - 6) \\ &= -2(3x^2 + 18x + 21) \\ &= -6x^2 - 36x - 42 \end{aligned}$$

(d)

$$\begin{aligned} \frac{5}{3}p(x-1) &= \frac{5}{3}\left(3(x-1)^2 - 6\right) \\ &= \frac{5}{3}\left(3(x^2 - 2x + 1) - 6\right) \\ &= \frac{5}{3}(3x^2 - 6x + 3 - 6) \\ &= \frac{5}{3}(3x^2 - 6x - 3) \\ &= 5x^2 - 10x - 5 \end{aligned}$$

Exercises

1. To increase by a factor of 10, multiply by 10. The right shift of 2 is made by substituting $x - 2$ for x in the function formula. Together they give $y = 10f(x - 2)$.

2. The multiplication by 3 on the outside stretches $f(x)$ vertically by a factor of 3 and affects only the y-coordinate, multiplying it by 3. Thus, the point $(5, \frac{1}{3})$ on the graph of g is first moved to the point $(5, 3 \cdot \frac{1}{3}) = (5, 1)$ on the graph of $y = 3g(x)$.

The inside transformation $x + 1$ then shifts the graph of $y = 3g(x)$ to the left by 1 unit, and affects only the x-coordinate. The point $(5, 1)$ on the graph of $y = 3g(x)$ is then shifted to the point $(5 - 1, 1) = (4, 1)$ on the graph of $y = 3g(x + 1)$. Thus, combining these transformations, the point $(5, \frac{1}{3})$ on the graph of g is moved to the point $(4, 1)$ on the graph of $y = 3g(x + 1)$.

3. All of the output values are smaller by a factor of 0.25, so the compressed range is $-0.25 \leq 0.25C(x) \leq 0.25$.

4. Since the domain of $Q(n)$ is the same as the domain of $P(n)$, no horizontal transformations have been applied. The minimum value of $Q(n)$ is -2, which is one-third of the minimum value of $P(n)$. Similarly the maximum value that $P(n)$ approaches is 4, which is also one-third of the maximum value the function $P(n)$ approaches. Thus, $P(n)$ has been compressed vertically by a factor of $1/3$, and

$$Q(n) = \frac{1}{3}P(n)$$

.

5. Since the domain of $R(n)$ is the same as the domain of $P(n)$, no horizontal transformations have been applied. Since the maximum value of $R(n)$ is -5 times the minimum value of $P(n)$, and the minimum value $R(n)$ approaches is -5 times the maximum value $P(n)$ approaches, $P(n)$ has been stretched vertically by a factor of 5 and reflected about the x-axis. Thus, we have

$$R(n) = -5P(n)$$

.

6. The domain of $S(n)$ is obtained by reflecting the domain of $P(n)$ about the y-axis. In addition, the range of $S(n)$ is a result of shifting the range of $P(n)$ up by 8 units. Thus, we have

$$S(n) = P(-n) + 8$$

.

7. The domain of $T(n)$ is 7 units to the left of the original domain of $P(n)$. In addition, the range of $P(n)$ has been compressed vertically by a factor of $1/4$. Thus, $P(n)$ has been shifted to the left by 7 units and compressed vertically by a factor of $1/4$, and we have

$$T(n) = \frac{1}{4}P(n+7)$$

.

8. See Figure 6.64.

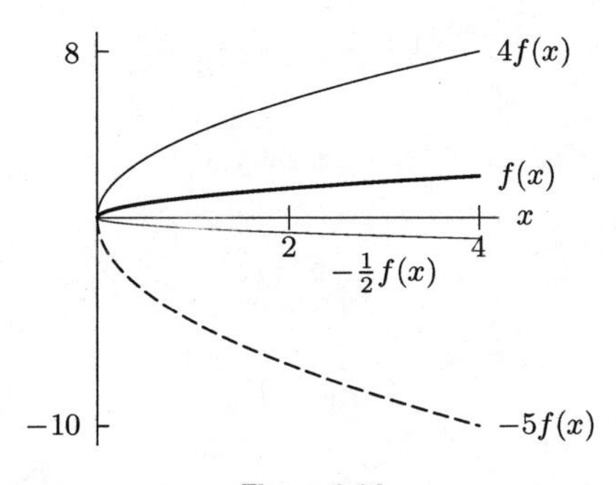

Figure 6.64

9. See Figure 6.65.

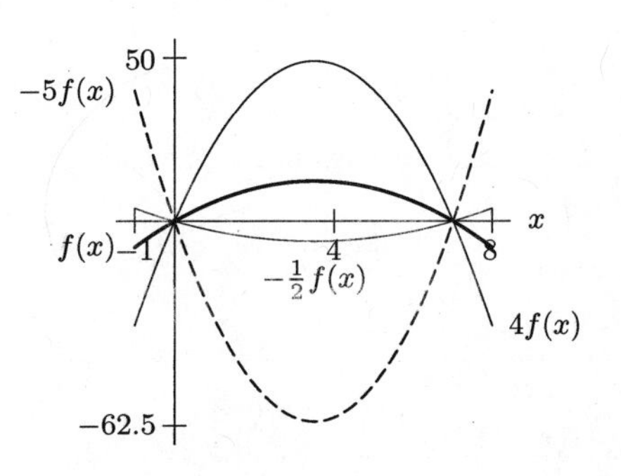

Figure 6.65

10. See Figure 6.66.

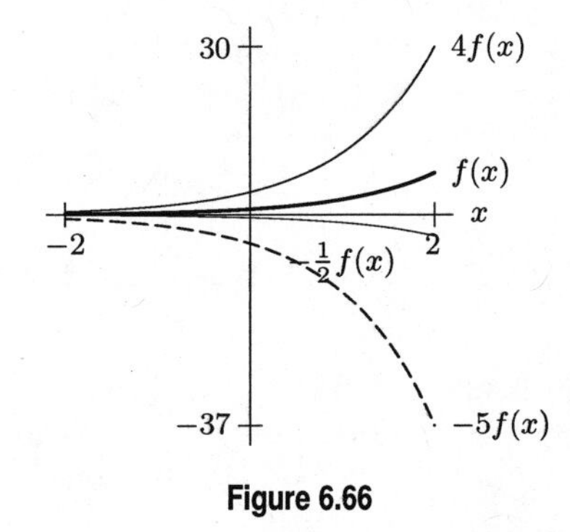

Figure 6.66

11. See Figure 6.67.

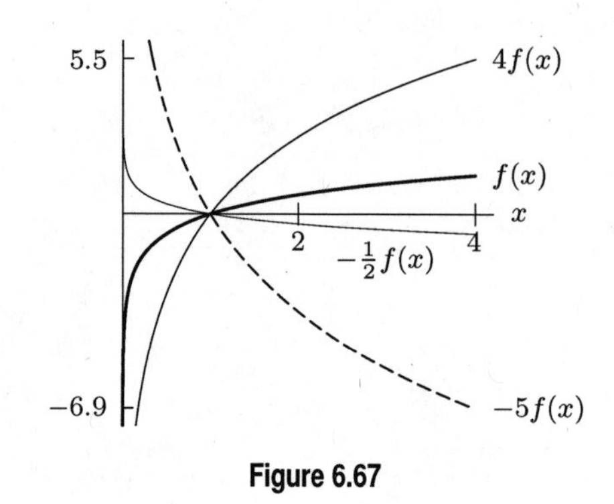

Figure 6.67

12. **(a)** To get the table for $f(x)/2$, you need to divide each entry for $f(x)$ by 2. See Table 6.6.

Table 6.6

x	−3	−2	−1	0	1	2	3
$f(x)/2$	1	1.5	3.5	−0.5	−1.5	2	4

(b) In order to get the table for $-2f(x+1)$, first get the table for $f(x+1)$. To do this, note that, if $x = 0$, then $f(x+1) = f(0+1) = f(1) = -3$ and if $x = -4$, then $f(x+1) = f(-4+1) = f(-3) = 2$. Since $f(x)$ is defined for $-3 \leq x \leq 3$, where x is an integer, then $f(x+1)$ is defined for $-4 \leq x \leq 2$.

Table 6.7

x	−4	−3	−2	−1	0	1	2
$f(x+1)$	2	3	7	−1	−3	4	8

Next, multiply each value of $f(x+1)$ entry by -2.

Table 6.8

x	−4	−3	−2	−1	0	1	2
$-2f(x+1)$	−4	−6	−14	2	6	−8	−16

(c) To get the table for $f(x) + 5$, you need to add 5 to each entry for $f(x)$ in in the table given in the problem.

Table 6.9

x	−3	−2	−1	0	1	2	3
$f(x)+5$	7	8	12	4	2	9	13

(d) If $x = 3$, then $f(x-2) = f(3-2) = f(1) = -3$. Similarly if $x = 2$ then $f(x-2) = f(0) = -1$, since $f(x)$ is defined for integral values of x from -3 to 3, $f(x-2)$ is defined for integral values of x, which are two units higher, that is from -1 to 5.

Table 6.10

x	−1	0	1	2	3	4	5
$f(x-2)$	2	3	7	−1	−3	4	8

(e) If $x = 3$, then $f(-x) = f(-3) = 2$, whereas if $x = -3$, then $f(-x) = f(3) = 8$. So, to complete the table for $f(-x)$, flip the values of $f(x)$ given in the problem about the origin.

Table 6.11

x	-3	-2	-1	0	1	2	3
$f(-x)$	8	4	-3	-1	7	3	2

(f) To get the table for $-f(x)$, take the negative of each value of $f(x)$ from the table given in the problem.

Table 6.12

x	-3	-2	-1	0	1	2	3
$-f(x)$	-2	-3	-7	1	3	-4	-8

13. **(a)**

Table 6.13

x	-4	-3	-2	-1	0	1	2	3	4
$f(-x)$	13	6	1	-2	-3	-2	1	6	13

(b)

Table 6.14

x	-4	-3	-2	-1	0	1	2	3	4
$-f(x)$	-13	-6	-1	2	3	2	-1	-6	-13

(c)

Table 6.15

x	-4	-3	-2	-1	0	1	2	3	4
$3f(x)$	39	18	3	-6	-9	-6	3	18	39

(d) All three functions are even.

14. The function is $y = f(x+3)$. Since $f(x) = |x|$, we want $y = |x+3|$. The transformation shifts the graph of $f(x)$ by 3 units to the left. See Figure 6.68.

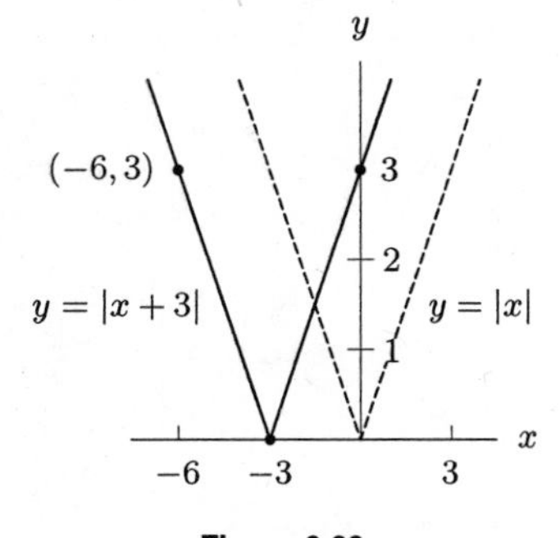

Figure 6.68

15. Again, $f(x) = |x|$. Therefore, $y = f(x) + 3$ means that we would shift the graph of $y = |x|$ upward 3 units. See Figure 6.69.

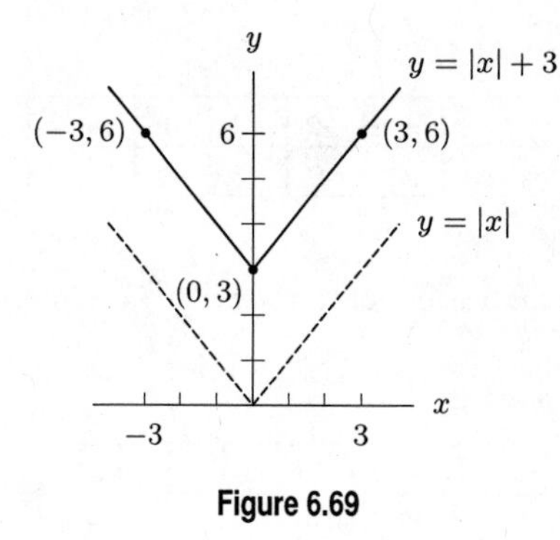

Figure 6.69

16. Since $g(x) = x^2$, $-g(x) = -x^2$. The graph of $g(x)$ is flipped over the x-axis. See Figure 6.70.

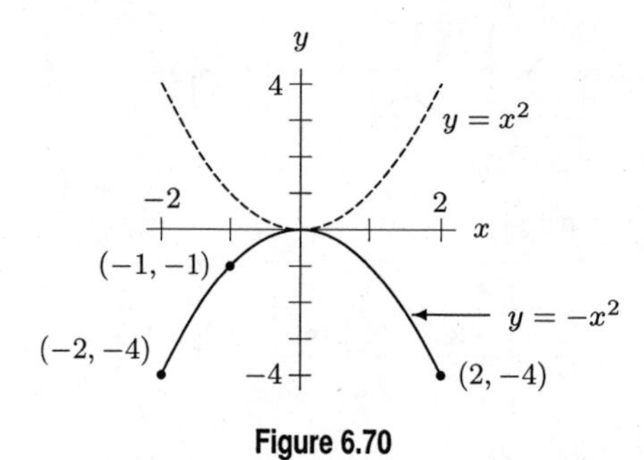

Figure 6.70

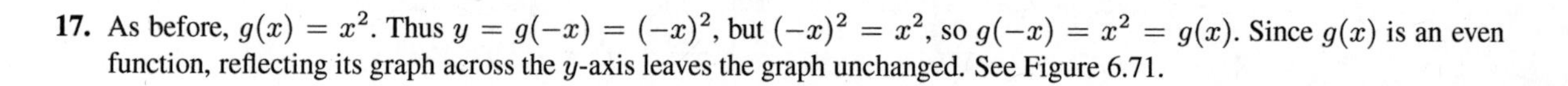
17. As before, $g(x) = x^2$. Thus $y = g(-x) = (-x)^2$, but $(-x)^2 = x^2$, so $g(-x) = x^2 = g(x)$. Since $g(x)$ is an even function, reflecting its graph across the y-axis leaves the graph unchanged. See Figure 6.71.

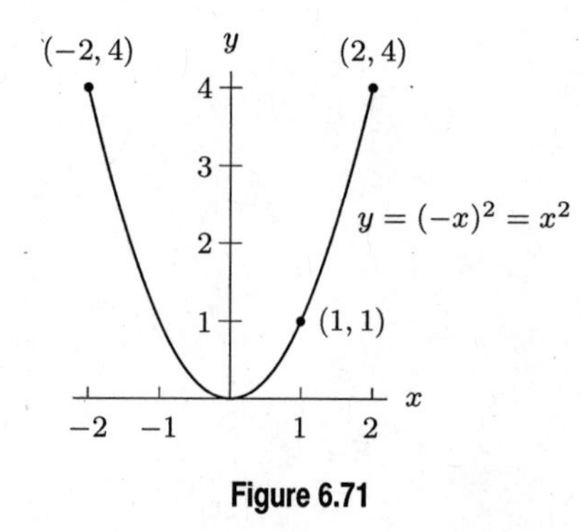

Figure 6.71

18. Since $h(x) = 2^x$, $3h(x) = 3 \cdot 2^x$. The graph of $h(x)$ is stretched vertically by a factor of 3. See Figure 6.72.

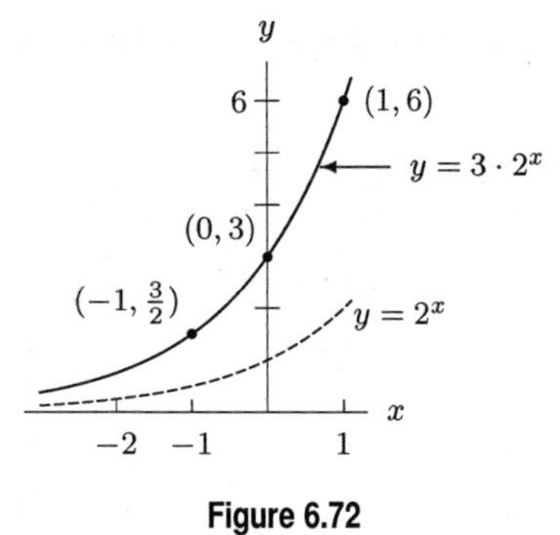

Figure 6.72

19. Since $h(x) = 2^x$, $0.5h(x) = 0.5 \cdot 2^x$. The graph of $h(x)$ is compressed vertically by a factor of $1/2$. See Figure 6.73.

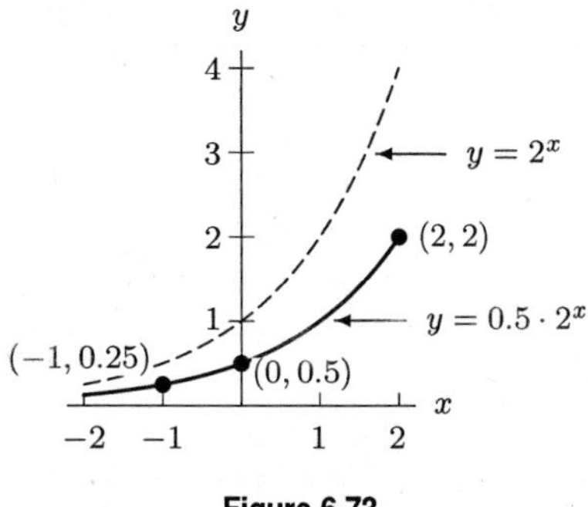

Figure 6.73

20. (i) i: The graph of $y = f(x)$ has been stretched vertically by a factor of 2.
(ii) c: The graph of $y = f(x)$ has been compressed vertically by a factor of $1/3$.
(iii) b: The graph of $y = f(x)$ has been reflected over the x-axis and shifted left by 1.
(iv) g: The graph of $y = f(x)$ has been shifted left by 2, and raised by 1.
(v) d: The graph of $y = f(x)$ has been reflected over the y-axis.

Problems

21. The factor of 2 doubles the resulting y-values (stretching the graph vertically). Then the input $(x + 1)$ moves the graph 1 unit to the left.

22. The new average rate of change is:

$$\frac{\frac{1}{2}s(4) - \frac{1}{2}s(0)}{4 - 0} = \frac{1}{2} \cdot \frac{s(4) - s(0)}{4 - 0} = \frac{1}{2}(70) = 35 \text{ mph},$$

which is half the original average rate of change.

23. The graph of $f(t) = 1/(1 + x^2)$ resembles a bell-shaped curve. (It is not, however, a true "bell curve.") See Figure 6.74.

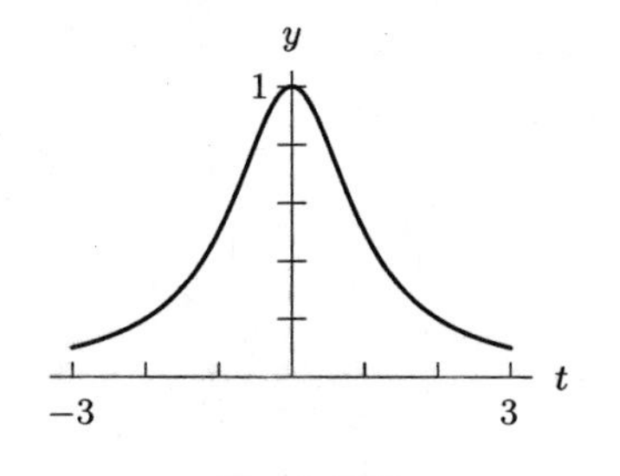

Figure 6.74

24. The graph of $f(t-3)$ is the graph of $f(t)$ shifted to the right by 3 units. See Figure 6.75.

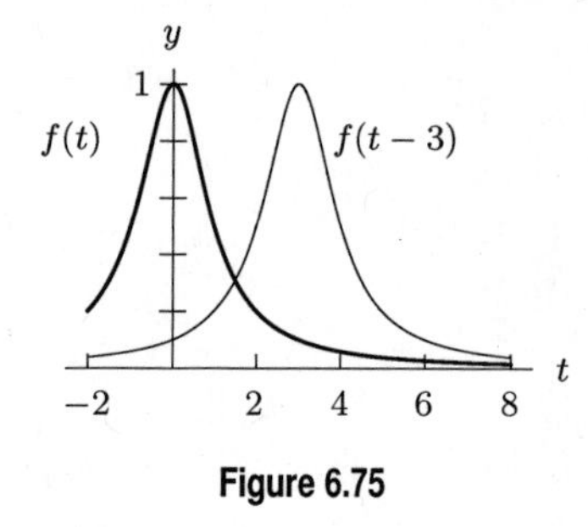

Figure 6.75

25. The graph of $0.5f(t)$ resembles a flattened version of the graph of $f(t)$. Each point on the new graph is half as far from the t-axis as the same point on the graph of f. See Figure 6.76.

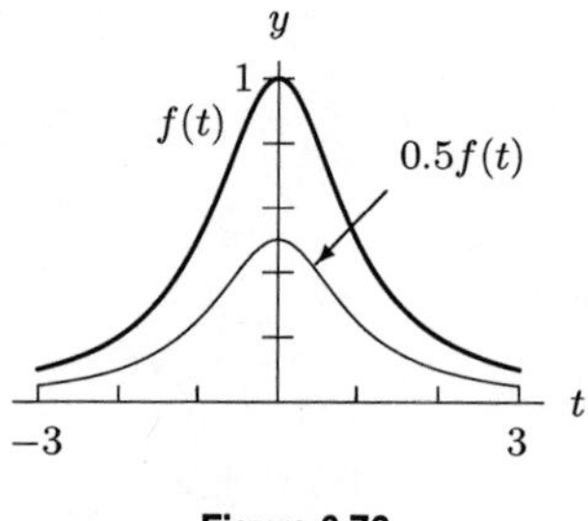

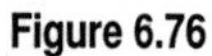

Figure 6.76

26. The graph of $-f(t)$ is a vertically flipped version of the graph of $f(t)$. See Figure 6.77.

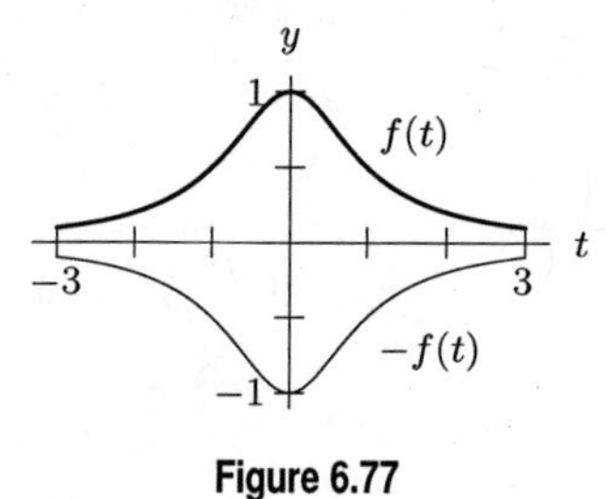

Figure 6.77

27. The graph of $f(t+5)-5$ is the graph of $f(t)$ shifted to the left by 5 units and then down by 5 units. See Figure 6.78.

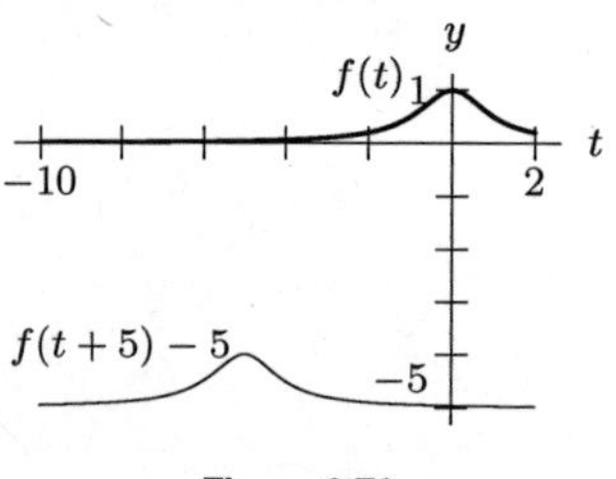

Figure 6.78

28. **(a)** (iii) The number of gallons needed to cover the house is $f(A)$; two more gallons will be $f(A) + 2$.
(b) (i) To cover the house twice, you need $f(A) + f(A) = 2f(A)$.
(c) (ii) The sign is an extra 2 ft^2 so we need to cover the area $A + 2$. Since $f(A)$ is the number of gallons needed to cover A square feet, $f(A + 2)$ is the number of gallons needed to cover $A + 2$ square feet.

29. I is (b)
II is (d)
III is (c)
IV is (h)

30. **(a)** If t represents the number of the months, then $t + 1$ represents one month later than month t. So $P(t + 1)$ represents the number of rabbits one month later.
(b) $2P(t)$ stands for twice the number of rabbits in the park in month t.

31. Since 1 euro is equivalent to 1.3 dollars, the cost of the first t days in dollars is

$$h(t) = 1.3C(t).$$

32. Since 1 year is 0.01 centuries, we have $g(x) = 0.01f(x)$.

33. See Figure 6.79. The graph is shifted to the right by 3 units.

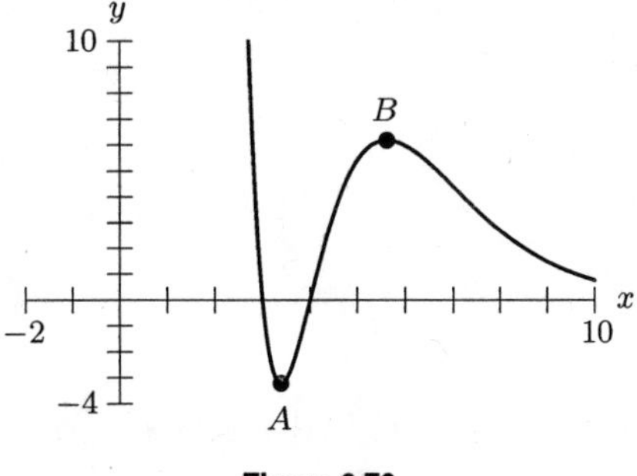

Figure 6.79

34. See Figure 6.80. The graph is shifted down by 3 units.

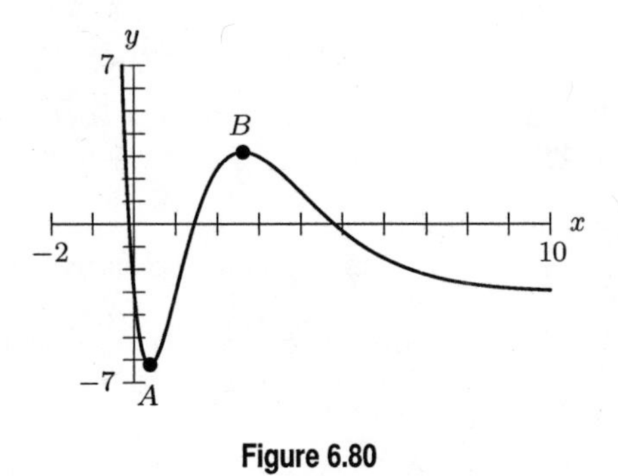

Figure 6.80

35. See Figure 6.81. The graph is horizontally flipped and vertically compressed by a factor of $1/3$.

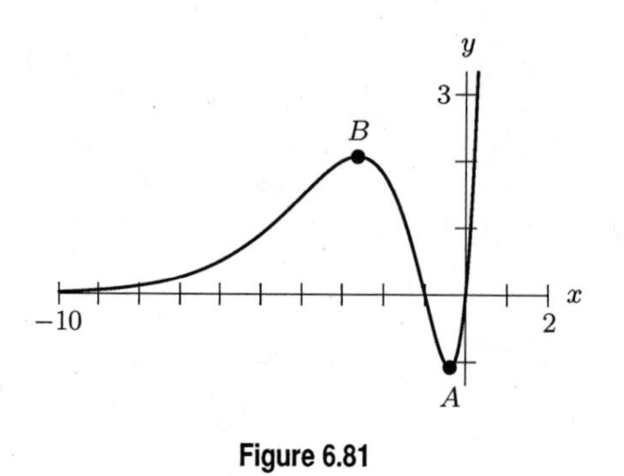

Figure 6.81

36. See Figure 6.82. The graph is vertically flipped and vertically stretched by a factor of 2.

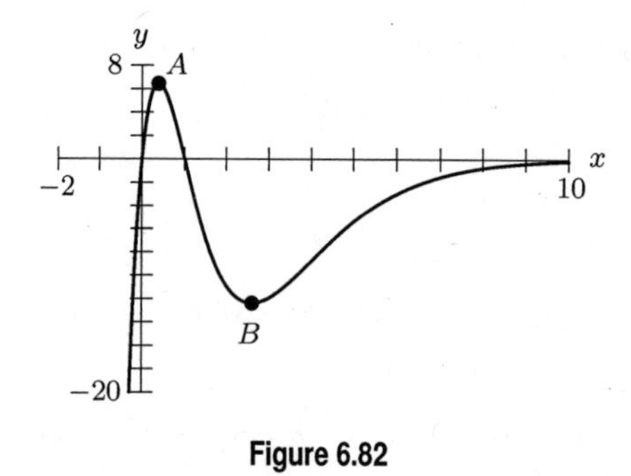

Figure 6.82

37. **(a)** Notice that the value of $h(x)$ at every value of x is one-half the value of $f(x)$ at the same x value. Thus, $f(x)$ has been compressed vertically by a factor of $1/2$, and

$$h(x) = \frac{1}{2}f(x).$$

(b) Observe that $k(-6) = f(6)$, $k(-4) = f(4)$, and so on. Thus, we have

$$k(x) = f(-x).$$

(c) The values of $m(x)$ are 4 less than the values of $f(x)$at the same x value. Thus, we have

$$m(x) = f(x) - 4.$$

38. **(a)** This figure is the graph of $f(t)$ shifted upward by two units. Thus its formula is $y = f(t) + 2$. Since on the graph of $f(t)$ the asymptote occurs at $y = 5$ on this graph the asymptote must occur at $y = 7$.

(b) This figure is the graph of $f(t)$ shifted to the left by one unit. Thus its formula is $y = f(t + 1)$. Since on the graph of $f(t)$ the asymptote occurs at $y = 5$, on this graph the asymptote also occurs at $y = 5$. Note that a horizontal shift does not affect the horizontal asymptotes.

(c) This figure is the graph of $f(t)$ shifted downward by three units and to the right by two units. Thus its formula is $y = f(t - 2) - 3$. Since on the graph of $f(t)$ the asymptote occurs at $y = 5$, on this graph the asymptote must occur at $y = 2$. Again, the horizontal shift does not affect the horizontal asymptote. However, outside changes (vertical shifts) do change the horizontal asymptote.

39. **(a)** $y = -2f(x)$. The function has been reflected over the x-axis, and stretched vertically by a factor of 2.

(b) $y = f(x) + 2$. This function has been shifted upward 2 units.

(c) $y = 3f(x - 2)$. This function has been shifted 2 units to the right and stretched vertically by a factor of 3.

40.

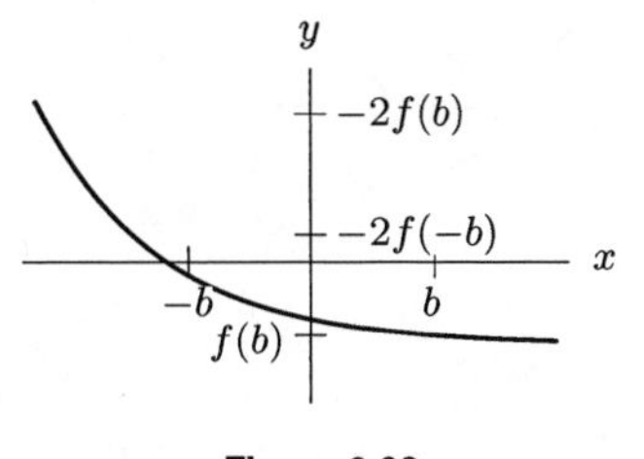

Figure 6.83

41. Figure 6.84 gives a graph of a function $y = f(x)$ together with graphs of $y = \frac{1}{2}f(x)$ and $y = 2f(x)$. All three graphs cross the x-axis at $x = -2, x = -1$, and $x = 1$. Likewise, all three functions are increasing and decreasing on the same intervals. Specifically, all three functions are increasing for $x < -1.55$ and for $x > 0.21$ and decreasing for $-1.55 < x < 0.21$.

Even though the stretched and compressed versions of f shown by Figure 6.84 are increasing and decreasing on the same intervals, they are doing so at different rates. You can see this by noticing that, on every interval of x, the graph of $y = \frac{1}{2}f(x)$ is less steep than the graph of $y = f(x)$. Similarly, the graph of $y = 2f(x)$ is steeper than the graph of $y = f(x)$. This indicates that the magnitude of the average rate of change of $y = \frac{1}{2}f(x)$ is less than that of $y = f(x)$, and that the magnitude of the average rate of change of $y = 2f(x)$ is greater than that of $y = f(x)$.

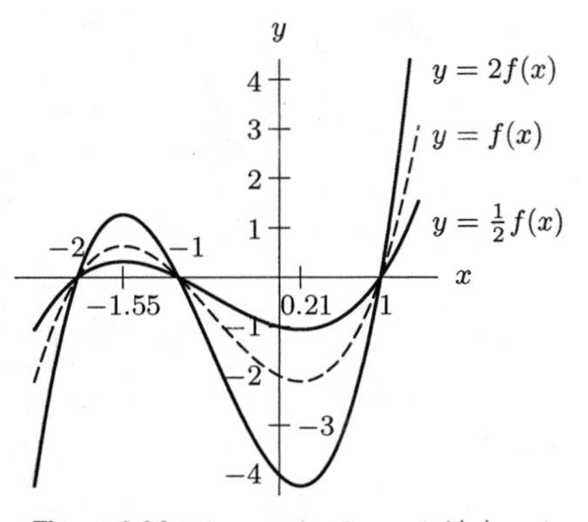

Figure 6.84: The graph of $y = 2f(x)$ and $y = \frac{1}{2}f(x)$ compared to the graph of $f(x)$

42.

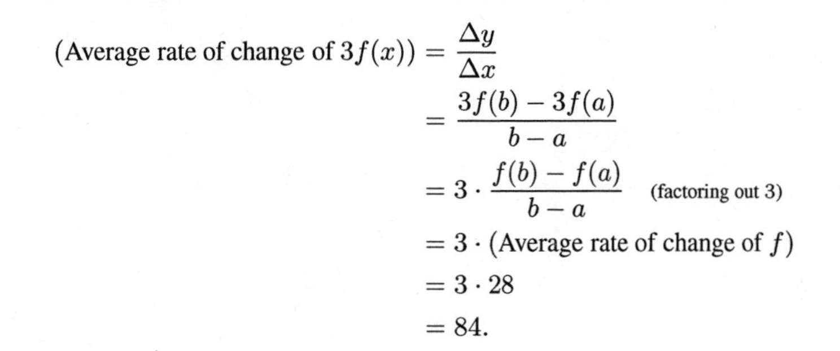

$$\begin{aligned}
\text{(Average rate of change of } 3f(x)) &= \frac{\Delta y}{\Delta x} \\
&= \frac{3f(b) - 3f(a)}{b-a} \\
&= 3 \cdot \frac{f(b) - f(a)}{b-a} \quad \text{(factoring out 3)} \\
&= 3 \cdot \text{(Average rate of change of } f) \\
&= 3 \cdot 28 \\
&= 84.
\end{aligned}$$

Recall in general that if $g(x) = k \cdot f(x)$, then on any interval,

$$\text{Average rate of change of } g = k \cdot \text{(Average rate of change of } f).$$

43.

$$\begin{aligned}
\left(\text{Average rate of change of } -\tfrac{1}{4}f(x)\right) &= \frac{\Delta y}{\Delta x}\\
&= \frac{-1/4f(b)-(-1/4f(a))}{b-a}\\
&= -\frac{1}{4}\cdot\frac{f(b)-f(a)}{b-a} \quad \text{(factoring out } -1/4\text{)}\\
&= -\frac{1}{4}\cdot(\text{Average rate of change of } f)\\
&= -\frac{1}{4}\cdot 28\\
&= -7.
\end{aligned}$$

Recall in general that if $g(x) = k\cdot f(x)$, then on any interval,

$$\text{Average rate of change of } g = k\cdot(\text{Average rate of change of } f)\,.$$

44.

$$\begin{aligned}
(\text{Average rate of change of } f(x)+3) &= \frac{\Delta y}{\Delta x}\\
&= \frac{(f(b)+3)-(f(a)+3)}{b-a}\\
&= \frac{f(b)+3-f(a)-3}{b-a} \quad \text{(distributing terms)}\\
&= \frac{f(b)-f(a)}{b-a} \quad \text{(canceling the 3's)}\\
&= \text{Average rate of change of } f\\
&= 28.
\end{aligned}$$

Notice vertical shifts do not affect the average rate of change.

Solutions for Section 6.4

Skill Refresher

S1. We have $f(2x) = (2x)^3 - 5 = 8x^3 - 5$.

S2. We have $2f(x) = 2\left(x^3 - 5\right) = 2x^3 - 10$.

S3. $f\left(-\frac{1}{3}x\right) = \left(\frac{-x}{3}\right)^3 - 5 = -\frac{x^3}{27} - 5$.

S4.

$$\begin{aligned}
\frac{1}{5}f(3x) &= \frac{1}{5}\left((3x)^3 - 5\right)\\
&= \frac{1}{5}\left(27x^3 - 5\right)\\
&= \frac{27}{5}x^3 - 1.
\end{aligned}$$

S5. $Q\left(\frac{1}{3}t\right) = 4e^{6\left(\frac{1}{3}t\right)} = 4e^{2t}$.

S6. $\frac{1}{3}Q(t) = \frac{1}{3}\left(4e^{6t}\right) = \frac{4}{3}e^{6t}$.

S7. $Q(2t) + 11 = 4e^{6(2t)} + 11 = 4e^{12t} + 11$.

S8. $7Q(t-3) = 7\left(4e^{6(t-3)}\right) = 28e^{6t-18}$.

Exercises

1. The graph is compressed horizontally by a factor of $1/2$ so the transformed function gives an output of 3 when the input is 1. Thus the point $(1, 3)$ lies on the graph of $g(2x)$.

2. The inside change stretches the graph horizontally by a factor of 10, while the outside change stretches it vertically by a factor of 10.

3. If $x = -2$, then $f(\frac{1}{2}x) = f(\frac{1}{2}(-2)) = f(-1) = 7$, and if $x = 6$, then $f(\frac{1}{2}x) = f(\frac{1}{2} \cdot 6) = f(3) = 8$. In general, $f(\frac{1}{2}x)$ is defined for values of x which are twice the values for which $f(x)$ is defined.

Table 6.16

x	-6	-4	-2	0	2	4	6
$f(\frac{1}{2}x)$	2	3	7	-1	-3	4	8

4.

Table 6.17

x	-3	-2	-1	0	1	2	3
$f(x)$	-4	-1	2	3	0	-3	-6
$f(\frac{1}{2}x)$	–	2	–	3	–	0	–
$f(2x)$	–	–	-1	3	-3	–	–

5. The graph in Figure 6.85 of $n(x) = e^{2x}$ is a horizontal compression of the graph of $m(x) = e^x$. The graph of $p(x) = 2e^x$ is a vertical stretch of the graph of $m(x) = e^x$. All three graphs have a horizontal asymptote at $y = 0$. The y-intercept of $n(x) = e^{2x}$ is the same as for $m(x)$, but the graph of $p(x) = 2e^x$ has a y-intercept of $(0, 2)$.

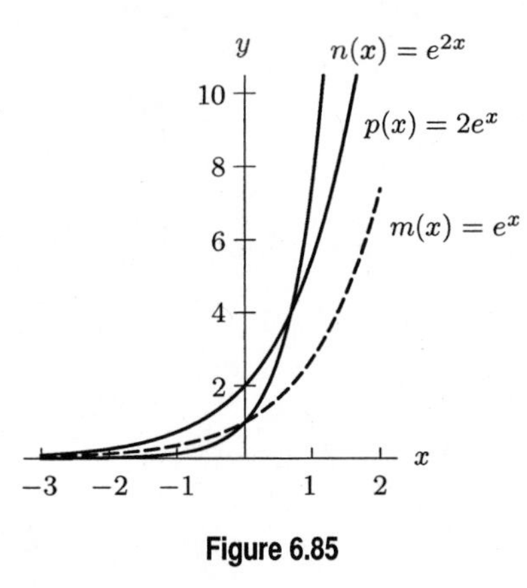

Figure 6.85

6. Since $h(x) = 2^x$, we know that $h(3x) = 2^{(3x)}$. Since we are multiplying x by a factor of 3, the graph of $2^{(3x)}$ is going to be 1/3 as wide as the graph of 2^x.

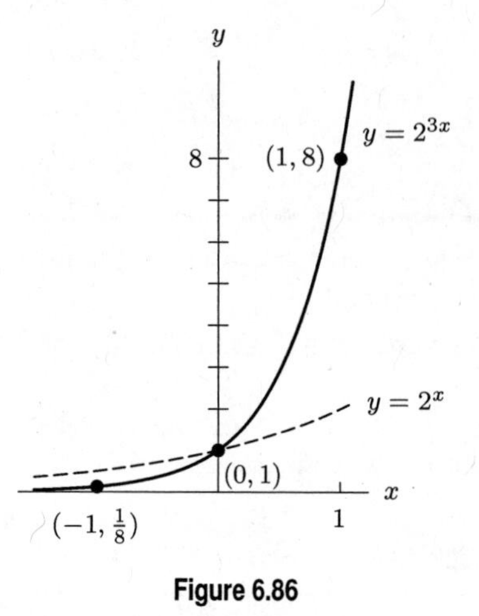

Figure 6.86

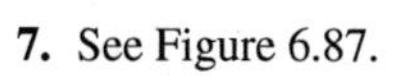
7. See Figure 6.87.

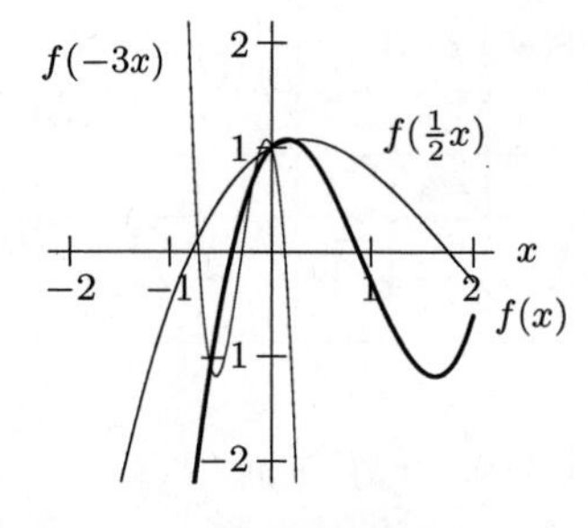

Figure 6.87

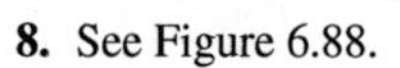
8. See Figure 6.88.

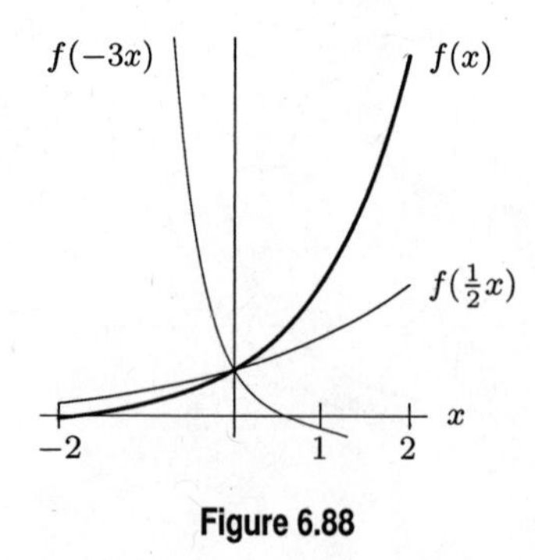

Figure 6.88

9. See Figure 6.89.

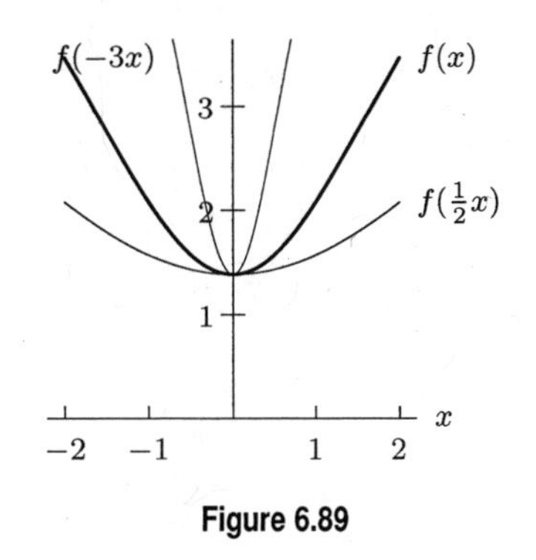

Figure 6.89

10. (i) e: The graph of $y = f(x)$ has been compressed horizontally by a factor of 1/2.
(ii) i: The graph of $y = f(x)$ has been compressed horizontally by a factor of 1/2, and stretched vertically by a factor of 2.
(iii) No match. None of these figures show $y = f(x)$ stretched horizontally by a factor of 2.

11. Since horizontal transformations do not affect features such as horizontal asymptotes or y-intercepts, each of the graphs in (a)-(c) will still have a horizontal asymptote at $y = 2$ and a y-intercept at $(0, -2)$.

(a) The graph of f has been compressed horizontally by a factor of $1/3$. The x-intercepts $(-1, 0)$ and $(3, 0)$ on the graph of f are moved to $(-\frac{1}{3}, 0)$ and $(1, 0)$ respectively on the graph of $y = f(3x)$ in Figure 6.90.

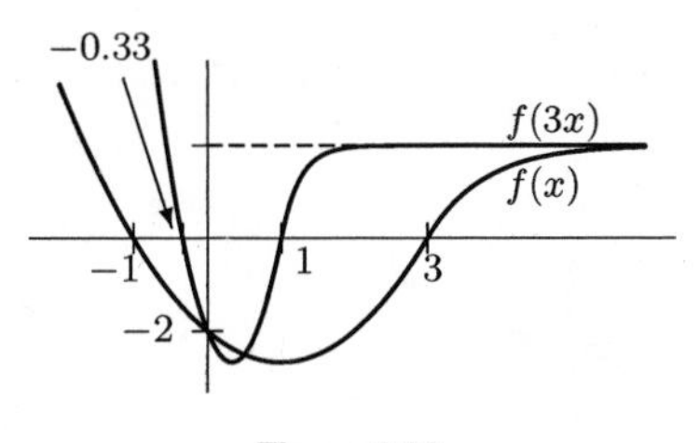

Figure 6.90

(b) The graph of f has been compressed horizontally by a factor of $1/2$ and reflected about the y-axis. The x-intercepts $(-1, 0)$ and $(3, 0)$ on the graph of f are moved to $(\frac{1}{2}, 0)$ and $(-\frac{3}{2}, 0)$ respectively on the graph of $y = f(-2x)$ in Figure 6.91.

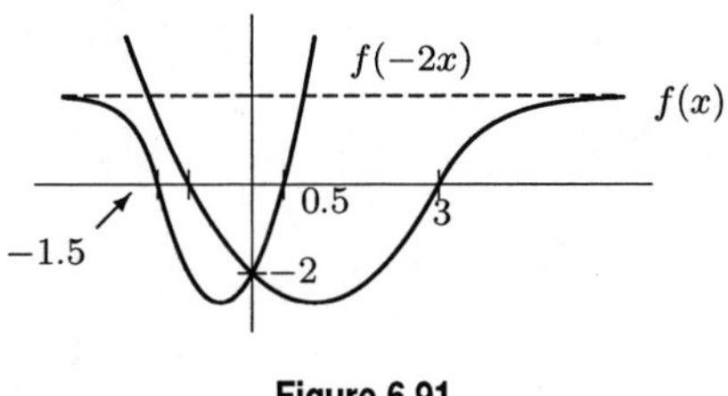

Figure 6.91

(c) The graph of f has been stretched horizontally by a factor of 2. The x-intercepts $(-1, 0)$ and $(3, 0)$ on the graph of f are moved to $(-2, 0)$ and $(6, 0)$ respectively on the graph of $y = f(\frac{1}{2}x)$ in Figure 6.92.

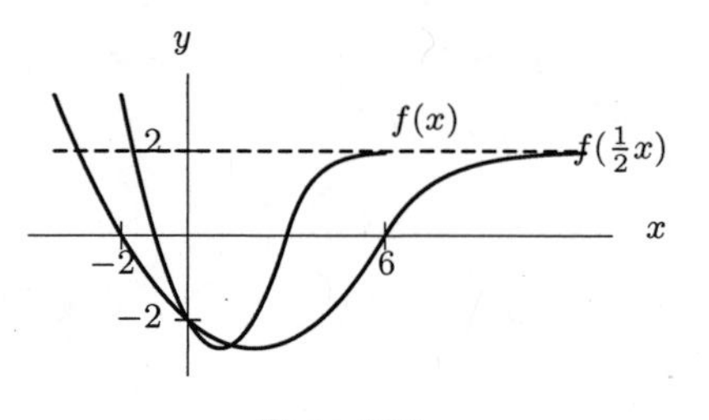

Figure 6.92

Problems

12. To get an output of 4 in the transformed function, we need the input to f to be 2. So $3p = 2$ and thus $p = \frac{2}{3}$.

13. **(a)** The graph is compressed horizontally by a factor of $\frac{1}{2}$, so the new domain is $-6 \le x \le 6$. There is no outside change so the range is still $0 \le l(2x) \le 3$.

(b) The graph is stretched horizontally by a factor of 2, so the new domain is $-24 \le x \le 24$. The change is only an inside change, so the range is still $0 \le l(\frac{1}{2}x) \le 3$.

14. The stretch away from the y-axis by a factor of d requires multiplying the x-coordinate by $1/d$, and the upward translation requires adding c to the y-coordinate. This gives $(a/d, b + c)$.

15. This graph is the graph of f shifted to the right by one unit, flipped vertically, and stretched vertically by a factor of 2. See Figure 6.93.

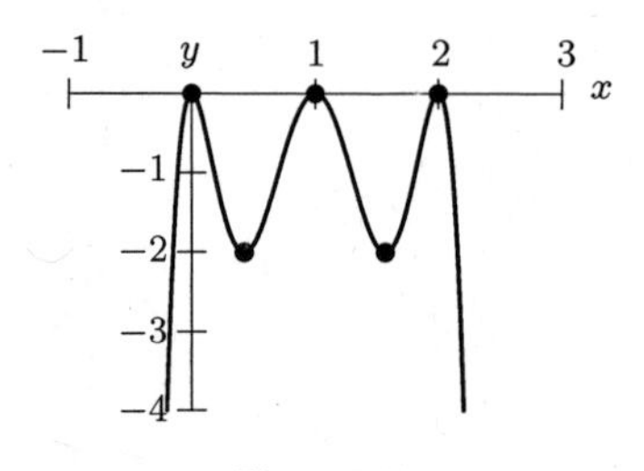

Figure 6.93

16. This graph is the graph of f stretched horizontally by a factor of 2 and then shifted down by 1 unit. See Figure 6.94.

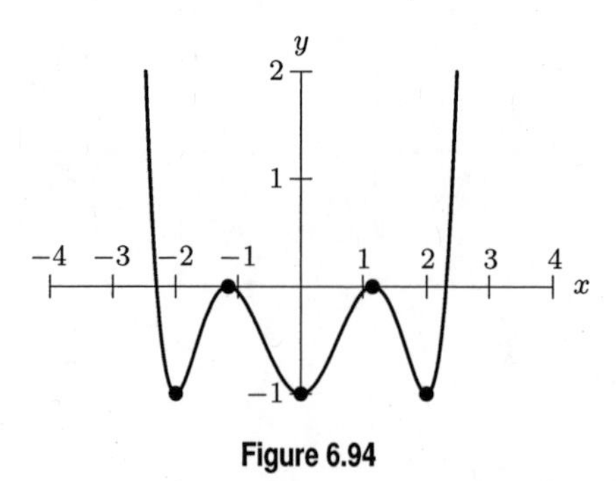

Figure 6.94

17. Since x km is $1000x$ meters, the temperature at a depth of x km is

$$p(x) = T(1000x).$$

18. Since n years is $t = 12n$ months, the man's height at n years, or $t = 12n$ months, is $f(t) = f(12n)$ meters. To find his height in centimeters, multiply by 100, so

$$g(n) = 100f(12n).$$

19. **(a)** Since s seconds is $s/60$ minutes, the altitude after s seconds is

$$f(s) = A(s/60) \text{ meters}$$

(b) Since h hours is $60h$ minutes, the altitude after h hours is

$$g(h) = A(60h).$$

20. **(a)** (ii) The \$5 tip is added to the fare $f(x)$, so the total is $f(x) + 5$.
(b) (iv) There were 5 extra miles so the trip was $x + 5$. I paid $f(x + 5)$.
(c) (i) Each trip cost $f(x)$ and I paid for 5 of them, or $5f(x)$.
(d) (iii) The miles were 5 times the usual so $5x$ is the distance, and the cost is $f(5x)$.

21. If profits are $r(t) = 0.5P(t)$ instead of $P(t)$, then profits are half the dollar level expected. If profits are $s(t) = P(0.5t)$ instead of $P(t)$, then profits are accruing half as fast as the projected rate.

22. **(a)** The formula is $A = f(r) = \pi r^2$.
(b) If the radius is increased by 10%, then the new radius is $r + (10\%)r = (110\%)r = 1.1r$. We want to know the output when our input is $1.1r$, so the appropriate expression is $f(1.1r)$.
(c) Since $f(1.1r) = \pi(1.1r)^2 = 1.21\pi r^2$, the new area is the old area multiplied by 1.21, or 121% of the old area. In other words, the area of a circle is increased by 21% when its radius is increased by 10%.

23. **(a)** Since III is horizontally stretched compared to one graph and compressed compared to another, it should be $f(x)$.
(b) The most horizontally compressed of the graphs are II and IV, so they should be $f(-2x)$ and $f(2x)$. Since II appears to be III reflected over the y-axis and compressed, it should be $f(-2x)$.
(c) The most horizontally stretched of the graphs should be $f(-\frac{1}{2}x)$, which is I.
(d) The most horizontally compressed of the graphs are II and IV, so they should be $f(-2x)$ and $f(2x)$. Since IV appears to be III compressed, it should be $f(2x)$.

24. **(a)** Since I is horizontally stretched compared to one graph and compressed compared to another, it should be $f(x)$.
(b) The most horizontally compressed of the graphs are III and II, so they should be $f(-2x)$ and $f(2x)$. Since III appears to be a compressed version of I reflected across the y-axis, it should be $f(-2x)$.
(c) The most horizontally stretched of the graphs should be $f(-\frac{1}{2}x)$, which is IV.
(d) The most horizontally compressed of the graphs are III and II, so they should be $f(-2x)$ and $f(2x)$. Since II appears to be a compressed version of I, it should be $f(2x)$.

25. The function f has been reflected over the x-axis and the y-axis and stretched horizontally by a factor of 2. Thus, $y = -f(-\frac{1}{2}x)$.

26. **(a)** See Figure 6.95. The zeros of $f(x)$ are at $x = \pm 2$.

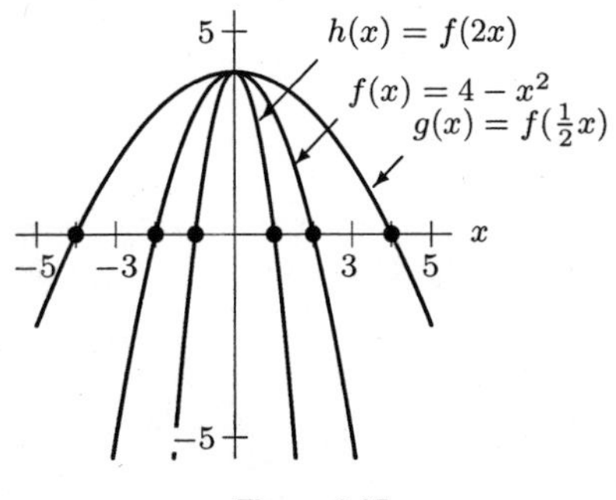

Figure 6.95

(b) $g(x) = f(0.5x) = 4 - (0.5x)^2 = 4 - (0.25x)^2$. From the graph we see the zeros at ± 4.
(c) $h(x) = f(2x) = 4 - (2x)^2 = 4 - 4x^2$. From the graph we see the zeros at ± 1.
(d) The zeros are compressed toward the origin by a factor of $1/10$, so the zeros are ± 0.2.

27. **(a)** To find $g(c)$, locate the point on $y = g(x)$ whose x-coordinate is c. The corresponding y-coordinate is $g(c)$.
(b) On the y-axis, go up twice the length of $g(c)$ to locate $2g(c)$.
(c) To find $g(2c)$, you must first find the location of $2c$ on the x-axis. This occurs twice as far from the origin as c. Then find the point on $y = g(x)$ whose x-coordinate is $2c$. The corresponding y-coordinate is $g(2c)$. See Figure 6.96.

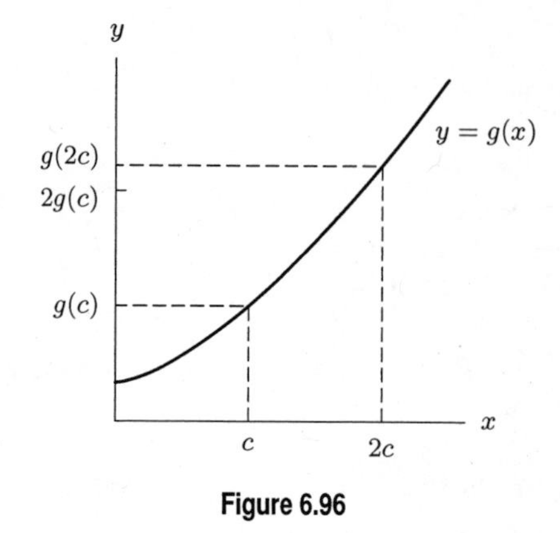

Figure 6.96

28. **(a)** Since $f(x)$ has been compressed horizontally by a factor of $1/2$, the domain of $f(x)$ is compressed by a factor of $1/2$. Thus, the domain of $g(x) = f(2x)$ is $-3 \leq x \leq 1$.
(b) Since the average rate of change of $f(x)$ over $-6 \leq x \leq 2$ is given as -3, we have

$$\frac{f(2) - f(-6)}{2 - (-6)} = \text{Average rate of change of } f(x)$$
$$\frac{f(2) - f(-6)}{8} = -3$$
$$f(2) - f(-6) = 8(-3) = -24.$$

We now use the fact that $f(2) - f(-6) = -24$ to calculate the average rate of change of $g(x)$ over its domain $-3 \leq x \leq 1$.

$$\begin{aligned} \text{Average rate of change of } g(x) &= \frac{\Delta y}{\Delta x} = \frac{g(1) - g(-3)}{1 - (-3)} \\ &= \frac{f(2(1)) - f(2(-3))}{4} \quad \text{(since } g(x) = f(2x)) \\ &= \frac{f(2) - f(-6)}{4} \\ &= \frac{-24}{4} \quad \text{(since } f(2) - f(-6) = -24) \\ &= -6. \end{aligned}$$

Notice the average rate of change of $g(x) = f(2x)$ over the compressed interval $-3 \leq x \leq 1$ is equal to 2 times the average rate of change of $f(x)$ over the original interval $-6 \leq x \leq 2$ since Δy has not changed while Δx has been compressed by a factor of $1/2$.

29. **(a)** Since $f(x)$ has been stretched horizontally by a factor of 4, the domain of $h(x)$ is $-24 \leq x \leq 8$.
(b) Since the average rate of change of $f(x)$ over $-6 \leq x \leq 2$ is given as -3, we have

$$\frac{f(2) - f(-6)}{2 - (-6)} = \text{Average rate of change of } f(x)$$
$$\frac{f(2) - f(-6)}{8} = -3$$
$$f(2) - f(-6) = 8(-3) = -24.$$

We now use the fact that $f(2) - f(-6) = -24$ to calculate the average rate of change of $h(x)$ over its domain $-24 \le x \le 8$.

$$\begin{aligned}
\text{Average rate of change of } h(x) &= \frac{\Delta y}{\Delta x} = \frac{h(8) - h(-24)}{8 - (-24)} \\
&= \frac{f\left(\frac{1}{4}(8)\right) - f\left(\frac{1}{4}(-24)\right)}{32} \quad \text{(since } h(x) = f(1/4 \cdot x)\text{)} \\
&= \frac{f(2) - f(-6)}{32} \\
&= \frac{-24}{32} \quad \text{(since } f(2) - f(-6) = -24\text{)} \\
&= -\frac{3}{4}.
\end{aligned}$$

Notice the average rate of change of $h(x) = f\left(\frac{1}{4}x\right)$ over the stretched interval $-24 \le x \le 8$ is one-fourth of the average rate of change of $f(x)$ over the original interval $-6 \le x \le 2$ since Δy has not changed while Δx has been stretched by a factor of 4.

30. **(a)** Since $f(x)$ has been shifted left by 2 units, the entire domain of $f(x)$ is shifted to the left by 2. Thus, the domain of $k(x) = f(x+2)$ is $-8 \le x \le 0$.

(b) Since the average rate of change of $f(x)$ over $-6 \le x \le 2$ is given as -3, we have

$$\begin{aligned}
\frac{f(2) - f(-6)}{2 - (-6)} &= \text{Average rate of change of } f(x) \\
\frac{f(2) - f(-6)}{8} &= -3 \\
f(2) - f(-6) &= 8(-3) = -24.
\end{aligned}$$

We now use the fact that $f(2) - f(-6) = -24$ to calculate the average rate of change of $k(x)$ over its domain $-8 \le x \le 0$.

$$\begin{aligned}
\text{Average rate of change of } k(x) &= \frac{\Delta y}{\Delta x} = \frac{k(0) - k(-8)}{0 - (-8)} \\
&= \frac{f(0+2) - f(-8+2)}{8} \quad \text{(since } k(x) = f(x+2)\text{)} \\
&= \frac{f(2) - f(-6)}{8} \\
&= \frac{-24}{8} \quad \text{(since } f(2) - f(-6) = -24\text{)} \\
&= -3.
\end{aligned}$$

31. **(a)** Since $f(x)$ has been reflected across the y-axis, the domain of $m(x)$ is $-2 \le x \le 6$.

(b) Since the average rate of change of $f(x)$ over $-6 \le x \le 2$ is given as -3, we have

$$\begin{aligned}
\frac{f(2) - f(-6)}{2 - (-6)} &= \text{Average rate of change of } f(x) \\
\frac{f(2) - f(-6)}{8} &= -3 \\
f(2) - f(-6) &= 8(-3) = -24.
\end{aligned}$$

We now use the fact that $f(2) - f(-6) = -24$ to calculate the average rate of change of $m(x)$ over its domain $-2 \le x \le 6$.

$$\text{Average rate of change of } m(x) = \frac{\Delta y}{\Delta x} = \frac{m(6) - m(-2)}{6 - (-2)}$$

$$
\begin{aligned}
&= \frac{f(-(6)) - f(-(-2))}{8} \quad \text{(since } m(x) = f(-x))\\
&= \frac{f(-6) - f(2)}{8}\\
&= \frac{-(f(2) - f(-6))}{8}\\
&= \frac{24}{8} \quad \text{(since } f(2) - f(-6) = -24)\\
&= 3.
\end{aligned}
$$

32. If $f(x)$ is a horizontal stretch of $g(x)$, then $f(x) = g(kx)$. Setting $x = 0$ we see that $f(0) = g(0)$, that is, the y-intercepts must be the same. In Figure 6.54 the y-intercepts are different, so $f(x)$ is a not horizontal stretch of $g(x)$.

Solutions for Section 6.5

Skill Refresher

S1. Factoring 4 out from the left side of the equation, we have

$$4(x + 3) = 4(x - h).$$

Thus, we see $h = -3$.

S2. Factoring $1/5$ out from the left side of the equation, we have

$$\frac{1}{5}(t - 50) = \frac{1}{5}(t - h).$$

Thus, we see $h = 50$.

S3. Factoring -3 out from the left side of the equation, we have

$$-3(z - \frac{10}{3}) = -3(z - h).$$

Thus, we see $h = \frac{10}{3}$.

S4. Factoring $-1/9$ out from the left side of the equation, we have

$$-\frac{1}{9}(x + 36) = -\frac{1}{9}(x - h).$$

Thus, we see $h = -36$.

S5. **(a)**

$$f(8(1)) - 3 = f(8) - 3 = \sqrt[3]{8} - 3 = 2 - 3 = -1$$

(b)

$$8f(1 - 3) = 8f(-2) = 8\sqrt[3]{-2}$$

(c)

$$8f(1) - 3 = 8\sqrt[3]{1} - 3 = 8(1) - 3 = 5$$

(d)

$$8\,(f(1) - 3) = 8(\sqrt[3]{1} - 3) = 8(-2) = -16$$

(e)

$$f(8(1 - 3)) = f(8(-2)) = f(-16) = \sqrt[3]{-16} = -2\sqrt[3]{2}$$

(f)

$$f(8(1) - 3) = f(5) = \sqrt[3]{5}$$

S6. **(a)**

$$y = -7Q(t-4) = -7e^{t-4}$$

(b)

$$y = -7Q(t) - 4 = -7e^{t} - 4$$

(c)

$$y = -7\left(Q(t) - 4\right) = -7\left(e^{t} - 4\right) = -7e^{t} + 28$$

(d)

$$y = Q(-7t) - 4 = e^{-7t} - 4$$

(e)

$$y = Q(-7(t-4)) = e^{-7(t-4)} = e^{-7t+28}$$

(f)

$$y = \frac{1}{4}Q(-7t+2) - 4 = \frac{1}{4}e^{-7t+2} - 4$$

S7. The function is already in the appropriate form since

$$y = f(-2x) + 9 = 1 \cdot f(-2(x-0)) + 9.$$

We see from the form above that $A = 1$, $B = -2$, $h = 0$, and $k = 9$.

S8. We must write the expression $2x - 6$ inside the function in the form $B(x-h)$. Factoring out the 2, we have $2x - 6 = 2(x-3)$. Thus, we have

$$y = -f(2x-6) + 9 = -f(2(x-3)) + 9.$$

We see from the form above that $A = -1$, $B = 2$, $h = 3$, and $k = 9$.

S9. We must write the expression $-1/3x - 9$ inside the function in the form $B(x-h)$. Factoring out the $-1/3$, we have $-1/3x - 9 = -1/3(x+27)$. Thus, we have

$$y = 6f\left(-\frac{1}{3}x - 9\right) = 6f\left(-\frac{1}{3}(x+27)\right).$$

We see from the form above that $A = 6$, $B = -1/3$, $h = -27$, and $k = 0$.

S10. We must write the expression $-x - 7$ inside the function in the form $B(x-h)$. Factoring out the -1, we have $-x - 7 = -(x+7)$. Thus, we have

$$y = -5\left(f(-x-7) + 2\right) = -5\left(f(-(x+7)) + 2\right).$$

Lastly, we distribute the -5 on the outside, and we have

$$y = -5\left(f(-(x+7)) + 2\right) = -5f(-(x+7)) - 10.$$

We see from the form above that $A = -5$, $B = -1$, $h = -7$, and $k = -10$.

Exercises

1. We transform the function $f(x)$ horizontally since all of the operations occur inside the function $f(x)$. To identify the transformations, we first factor out 3 in the expression $3x - 2$:

$$y = f(3x-2) = f(3(x - \frac{2}{3})).$$

So there are two inside transformations applied to f: First a horizontal compression by a factor of $1/3$ and then a horizontal shift right by $2/3$ units.

2. We transform the function $g(x)$ vertically since all of the operations occur outside the function $g(x)$. To identify the transformations, we distribute 5 in the expression $5(g(x) - 8)$:

$$y = 5(g(x) - 8) = 5g(x) - 40.$$

So there are two outside transformations applied to g: First a vertical stretch by a factor of 5 and then a vertical shift down by 40 units.

3. (a) The function $g(2x) - 5$ is already in the form $Ag(B(x - h)) + k$, with $A = 1$, $B = 2$, $h = 0$, and $k = -5$. Thus, $g(x)$ has been compressed horizontally by a factor of $1/2$, and then shifted down by 5 units. As a result of these transformations, the point $(6, -9)$ is first horizontally compressed to the point $(\frac{1}{2} \cdot 6, -9) = (3, -9)$, and then vertically shifted to the point $(3, -9 - 5) = (3, -14)$. The point on the new graph is $(3, -14)$.

(b) The function $3g(x) + 1$ is already in the form $Ag(B(x - h)) + k$, with $A = 3$, $B = 1$, $h = 0$, and $k = 1$. Thus, $g(x)$ has been stretched vertically by a factor of 3, and then shifted up by 1 unit. As a result of these transformations, the point $(6, -9)$ is first vertically stretched to the point $(6, 3 \cdot (-9)) = (6, -27)$, and then vertically shifted to the point $(6, -27 + 1) = (6, -26)$. The point on the new graph is $(6, -26)$.

(c) We first must express the function $-\left(g(\frac{1}{3}(x + 4)) - 8\right)$ in the form $Ag(B(x - h)) + k$. Distributing the -1 on the outside, we have

$$-\left(g(\frac{1}{3}(x + 4)) - 8\right) = -g(\frac{1}{3}(x + 4)) + 8,$$

and thus $A = -1$, $B = 1/3$, $h = -4$, and $k = 8$. So $g(x)$ has been reflected about the x-axis, then stretched horizontally by a factor of 3, next shifted left by 4 units, and finally shifted up by 8 units. As a result of this sequence of transformations, the point $(6, -9)$ is first moved to $(6, 9)$, then to $(18, 9)$, next to $(14, 9)$, and finally to the new point at $(18, 17)$.

(d) We first must express the function $\frac{1}{2}g(-5x - 15) - 8$ in the form $Ag(B(x - h)) + k$. Factoring the -5 inside the function, we have

$$\frac{1}{2}g(-5x - 15) - 8 = \frac{1}{2}g(-5(x + 3)) - 8,$$

and thus $A = 1/2$, $B = -5$, $h = -3$, and $k = -8$. So $g(x)$ has been compressed vertically by a factor of $1/2$, then compressed horizontally by a factor of $1/5$, then reflected about the y-axis, then shifted left 3 units, and finally shifted down 8 units. As a result of this sequence of transformations, the point $(6, -9)$ is first moved to $(6, -\frac{9}{2})$, then to $(30, -\frac{9}{2})$, then to $(-30, -\frac{9}{2})$, then to $(-33, -\frac{9}{2})$, and finally to the new point at $(-33, -\frac{25}{2})$.

4.

x	-2	-1	0	1	2
$f(x)$	-3	-4	2	0	5
$2f(x) + 3$	-3	-5	7	3	13
$f(x - 1) + 1$	–	-2	-3	3	1
$f(x + 2) - 1$	1	-1	4	–	–
$3f(2x + 2) - 1$	-10	5	14	–	–

5. We first write the function $w(t) = 40 - 2v(-0.5t)$ in the form $w(t) = Av(b(t - h)) + k$, and we have

$$w(t) = 40 - 2v(-0.5t) = -2v(-0.5t) + 40.$$

We have vertically stretched $v(t)$ by a factor of 2 and reflected about the t-axis, then horizontally stretched by a factor of 2 and reflected about the v-axis, and finally vertically shifted up by 40 units.

Notice a result of the horizontal transformations (stretch and reflection), the new domain of $w(t)$ is $t = -8, -6, -4, -2$, and 0. Evaluating $w(t)$ at these values, we have

$$\begin{aligned} w(-8) &= -2v\left(-0.5(-8)\right) + 40 \\ &= -2v(4) + 40 \\ &= -2(23) + 40 \\ &= -6, \end{aligned}$$

$$\begin{aligned} w(-6) &= -2v\left(-0.5(-6)\right) + 40 \\ &= -2v(3) + 40 \\ &= -2(19) + 40 \end{aligned}$$

$$= 2,$$

and so. Table 6.18 gives the rest of the values of $w(t)$.

Table 6.18

t	-8	-6	-4	-2	0
$w(t)$	-6	2	8	6	0

6. (a) This is a vertical shift of one unit upward. See Figure 6.97.

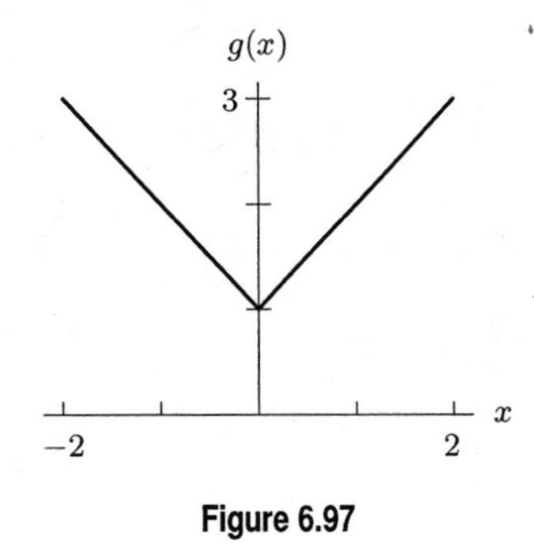

Figure 6.97

(b) Writing $h(x) = |x+1|$ as $h(x) = |x-(-1)|$, we see that this is a horizontal shift of one unit to the left. See Figure 6.98.

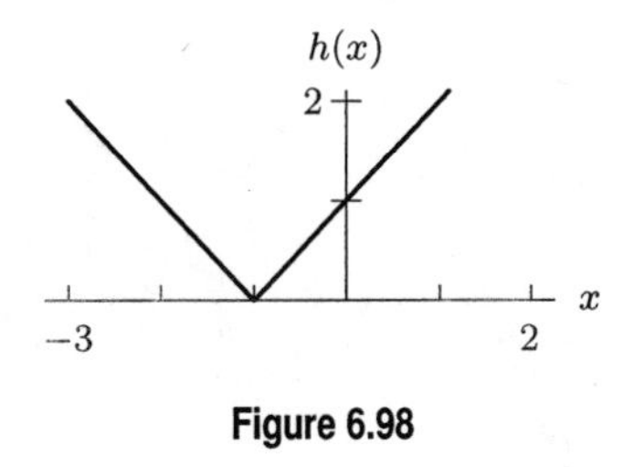

Figure 6.98

(c) Writing $j(x) = |2x+1| - 3$ as $j(x) = |2(x+\frac{1}{2})| - 3$, we see that this is a horizontal compression by a factor of $1/2$, then a horizontal shift of $1/2$ unit to the left, and finally a vertical shift down 3 units. See Figure 6.99.

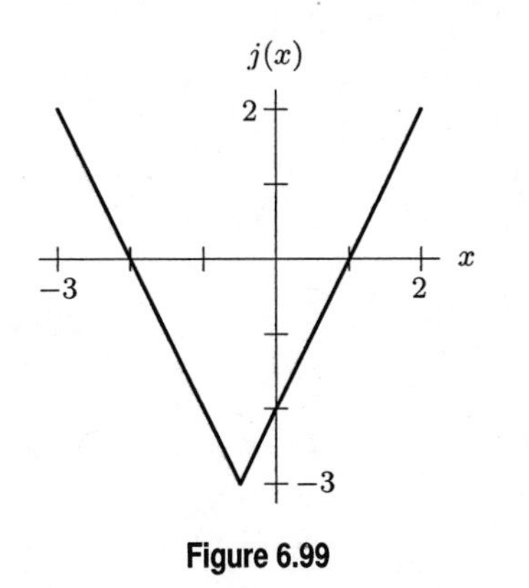

Figure 6.99

(d) Writing $k(x) = \frac{1}{2}|2x - 4| + 1$ as $k(x) = \frac{1}{2}|2(x - 2)| + 1$, we see that this is a vertical compression by a factor of $1/2$, then a horizontal compression by a factor of $1/2$, then a horizontal shift right 2 units, and finally a vertical shift up 1 unit. See Figure 6.100.

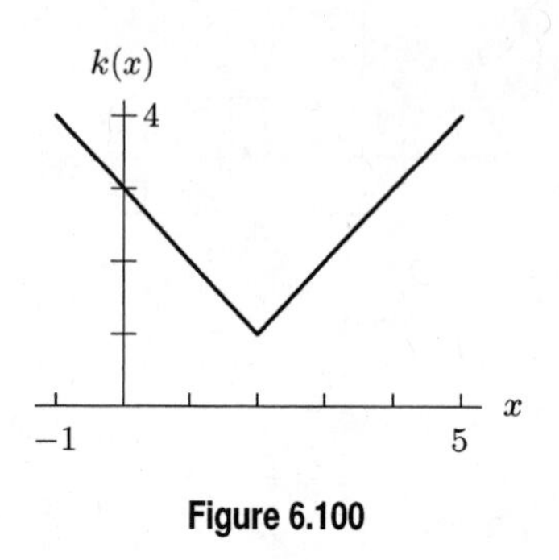

Figure 6.100

(e) Writing $m(x) = -\frac{1}{2}|4x + 12| - 3$ as $m(x) = -\frac{1}{2}|4(x + 3)| - 3$, we see that this is a vertical compression by a factor of $1/2$, then a reflection about the x-axis, then a horizontal compression by a factor of $1/4$, then a horizontal shift left 3 units, and finally a vertical shift down 3 units. See Figure 6.101.

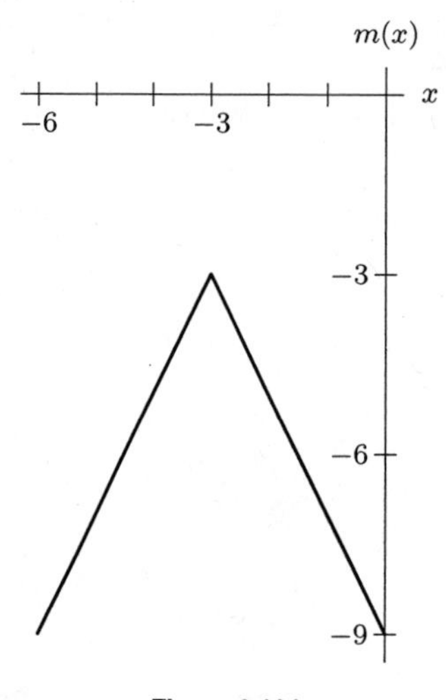

Figure 6.101

7. (a) Writing $f(3x + 6) = f(3(x + 2)) = 2^{3(x+2)} = 2^{3(x-(-2))}$, we see that this is a horizontal compression by a factor of $1/3$ followed by a horizontal shift left 2 units. See Figure 6.102.

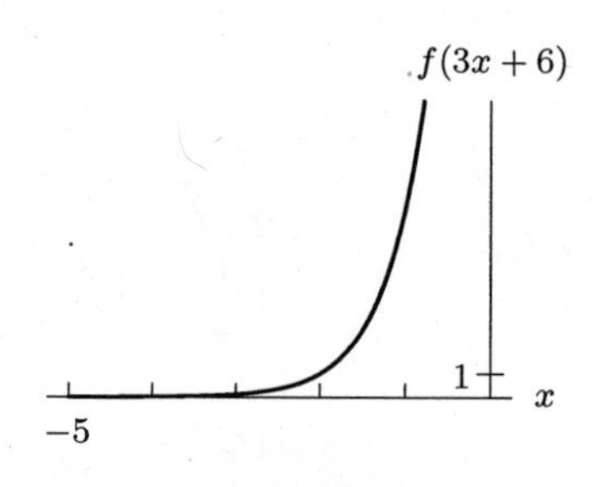

Figure 6.102

(b) Writing $f(-\frac{1}{2}x-1)=f(-\frac{1}{2}(x+2))=2^{(-1/2)(x+2)}$, we see that this is a horizontal stretch by a factor of 2, then a reflection about the y-axis, and finally a horizontal shift left 2 units. See Figure 6.103.

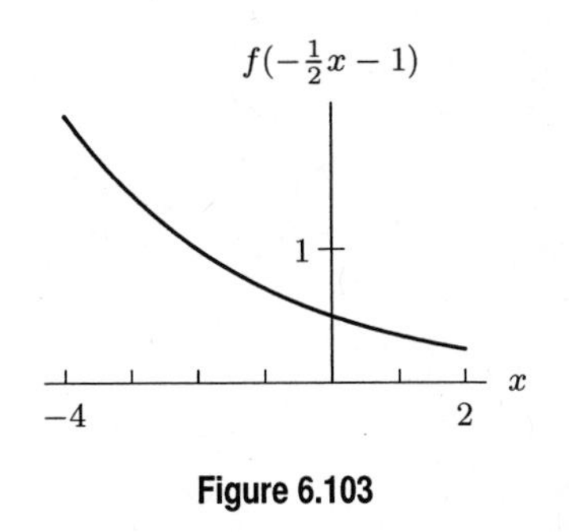

Figure 6.103

(c) Writing $0.4f(-x+1)-2=0.4f(-(x-1))-2=(0.4)2^{-(x-1)}-2$, we see that this is a vertical compression by a factor of 0.4, then a reflection about the y-axis, then a horizontal shift right 1 unit, and finally a vertical shift down 2 units. See Figure 6.104.

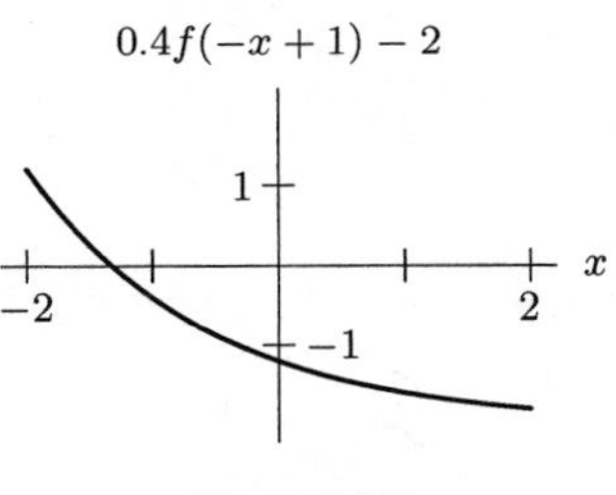

Figure 6.104

(d) Writing $-f(\frac{1}{2}x+4)+1=-2^{(1/2)(x-(-8))}+1$, we see that this is a reflection across the x-axis, then a horizontal stretch by a factor of 2, then a horizontal shift left 8 units, and finally a vertical shift up 1 unit. See Figure 6.105.

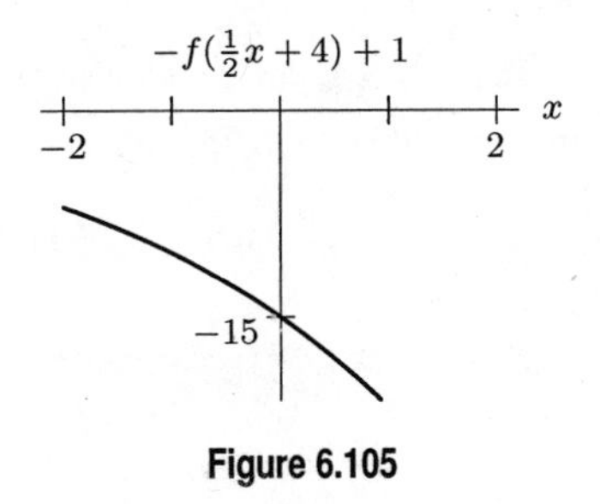

Figure 6.105

8. (a) Since $y=-f(x)+2$, we first need to reflect the graph of $y=f(x)$ over the x-axis and then shift it upward two units. See Figure 6.106.

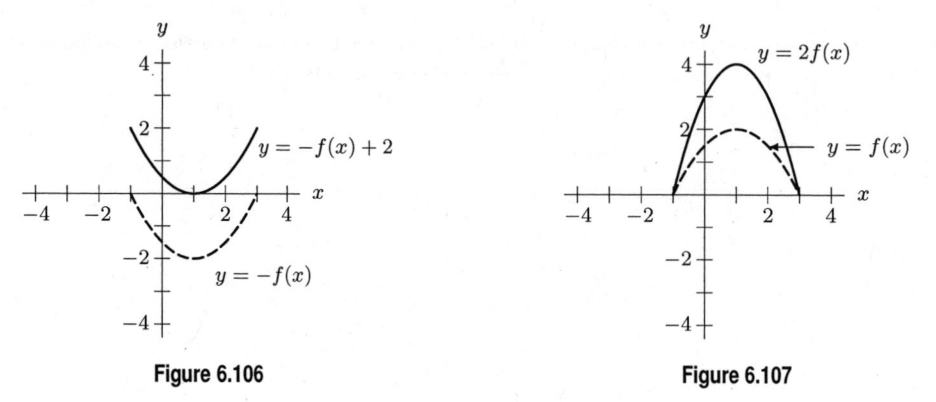

Figure 6.106

Figure 6.107

(b) We need to stretch the graph of $y = f(x)$ vertically by a factor of 2 in order to get the graph of $y = 2f(x)$. See Figure 6.107.

(c) In order to get the graph of $y = f(x-3)$, we will move the graph of $y = f(x)$ to the right by 3 units. See Figure 6.108.

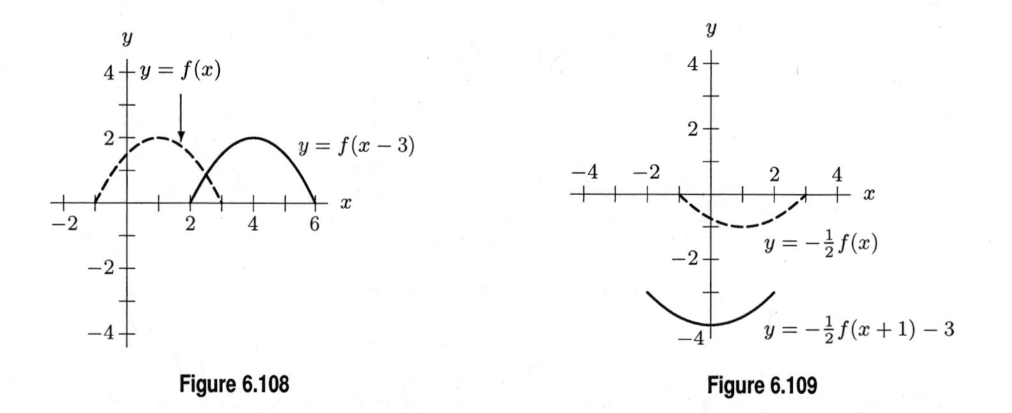

Figure 6.108

Figure 6.109

(d) To get the graph of $y = -\frac{1}{2}f(x+1) - 3$, first vertically compress the graph of $y = f(x)$ by a factor of $1/2$, then reflect about the x-axis, then horizontally shift left 1 unit, and finally vertically shift down 3 units. See Figure 6.109.

9. The graph is stretched vertically by a factor of 2, then stretched horizontally by a factor of 2, an finally shifted vertically by 20 units. See Figure 6.110.

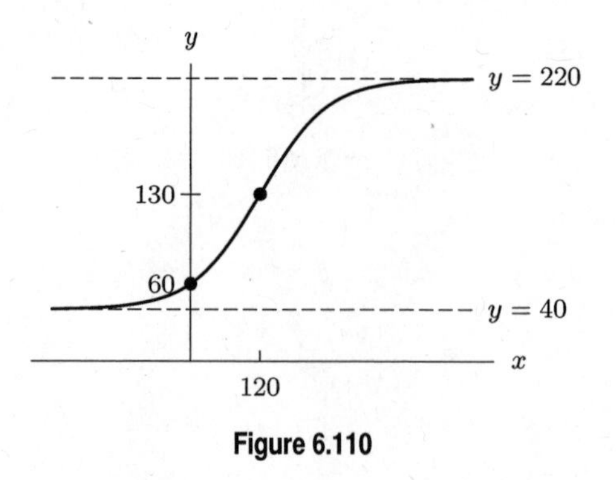

Figure 6.110

10. The graph is stretched vertically by a factor of 2, then flipped vertically across the x-axis, then stretched horizontally by a factor of 2, and finally flipped horizontally across the y-axis. See Figure 6.111.

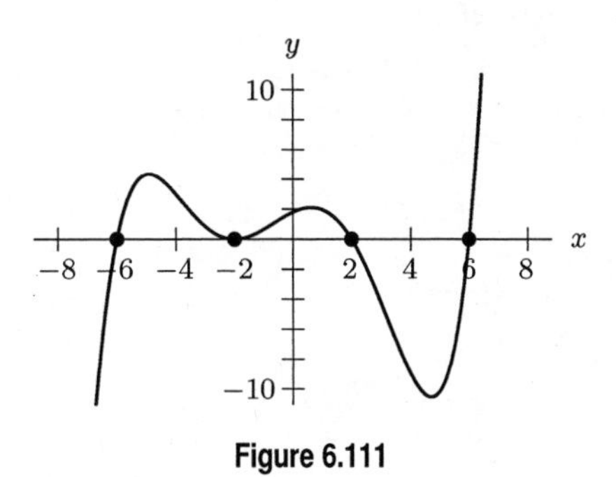

Figure 6.111

11. We have $h(x) = f(2x-1) = f(2(x-\frac{1}{2}))$. Thus, the graph of h is the graph of f first compressed horizontally by a factor of $1/2$, and then shifted horizontally to the right by $1/2$. See Figure 6.112.

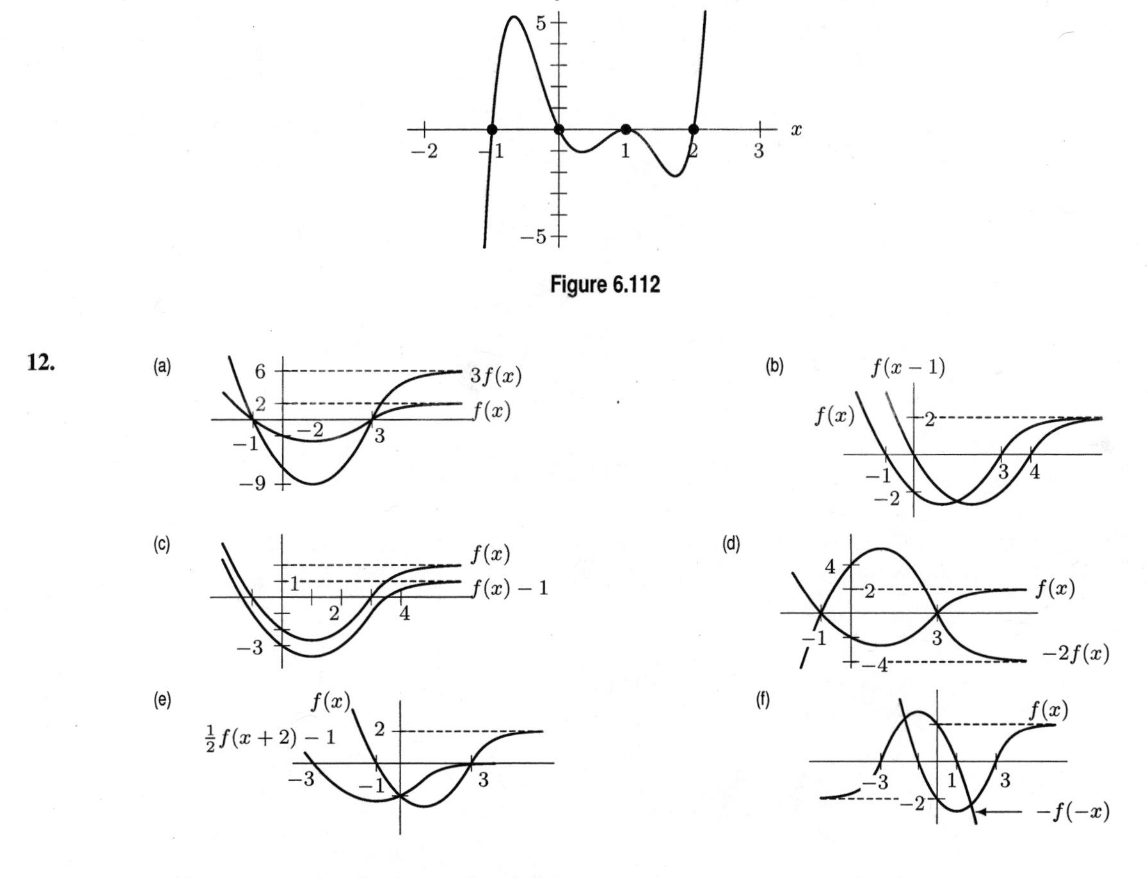

13. We have $w(t) = 3 - 0.5v(-2t) = -0.5v(-2t) + 3$.

The inside changes only affect the graph's x-coordinates, and so they do not affect its horizontal asymptote, since horizontally shifting or stretching a horizontal line has no effect.

- First, compress the graph horizontally by a factor of $1/2$. This moves the vertical asymptote from $t = 5$ to $t = 2.5$.

- Next, flip the graph horizontally across the y-axis. This moves the vertical asymptote from $t = 2.5$ to $t = -2.5$.

The outside changes only affect the graph's y-coordinates, and so they do not affect its vertical asymptote, since vertically shifting or stretching a vertical line has no effect.

- First, compress vertically by a factor of $1/2$. This moves the horizontal asymptote from $y = -4$ to $y = -2$.
- Next, flip the graph vertically across the t-axis. This moves the horizontal asymptote from $y = -2$ to $y = 2$.
- Finally, shift the graph vertically by 3 units. This moves the horizontal asymptote from $y = 2$ to $y = 5$.

Thus, the graph of w has vertical asymptote $t = -2.5$ and horizontal asymptote $y = 5$.

14. **(a)** Since we have only inside transformations, the y-coordinates will not change, and the range will remain $-40 < p \leq 160$. In order to determine the new domain, we first write $p(\frac{1}{5}t + 4)$ in the form $p(B(t - h))$,

$$p(\frac{1}{5}t + 4) = p(\frac{1}{5}(t + 20)).$$

Thus, $p(s)$ has been stretched horizontally by a factor of 5, and then shifted to the left 20 units. The domain is therefore first stretched to $5 \leq t \leq 60$, and then shifted to $-15 \leq t \leq 40$.

(b) Notice the function $-\frac{1}{10}p(2t) + 50$ is already written in the form $Ap(B(t - h)) + k$, where $A = -1/10$, $B = 2$, $h = 0$, and $k = 50$. Thus, $p(t)$ is first compressed vertically by a factor of $1/10$. The domain is not affected by the vertical compression, and the range becomes $-4 < p \leq 16$. Next we reflect about the p-axis, and the domain is still not affected while the range becomes $-16 \leq p < 4$. Then we compress horizontally by a factor of $1/2$. The range remains $-16 \leq p < 4$, and the domain now becomes $\frac{1}{2} \leq t \leq 6$. Finally we shift up 50 units. Thus, the new range is $34 \leq p < 54$, and the new domain is $\frac{1}{2} \leq t \leq 6$.

15. We have $h(x) = f(2(x + 3))$. Thus, we have first horizontally compressed f by a factor of $1/2$ and then horizontally shifted to the left by 3 units. See Figure 6.113.

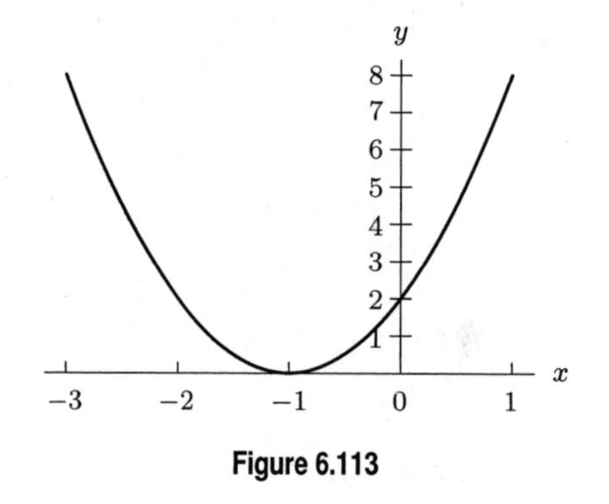

Figure 6.113

16. Since $q(x) = f(2(x + 6)$, we have first horizontally compressed f by a factor of $1/2$ and then horizontally shifted to the left by 6 units. See Figure 6.114.

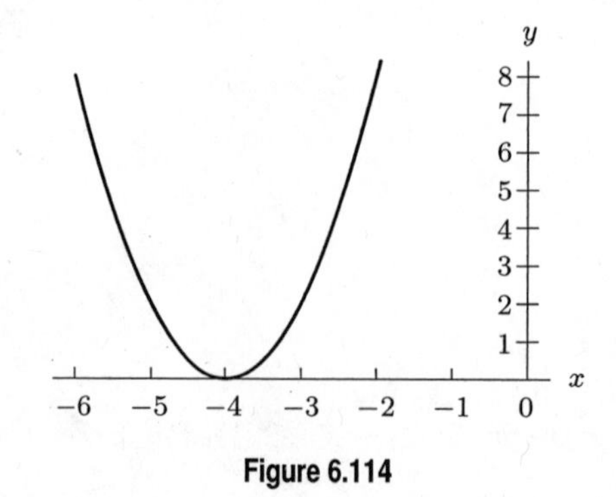

Figure 6.114

Problems

17. **(a)**

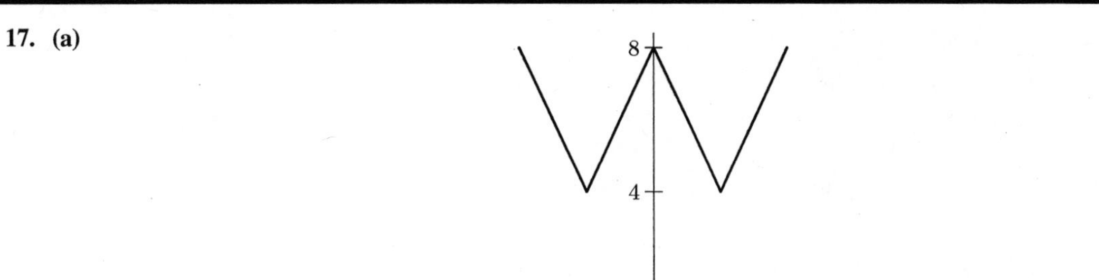

Figure 6.115

(b)

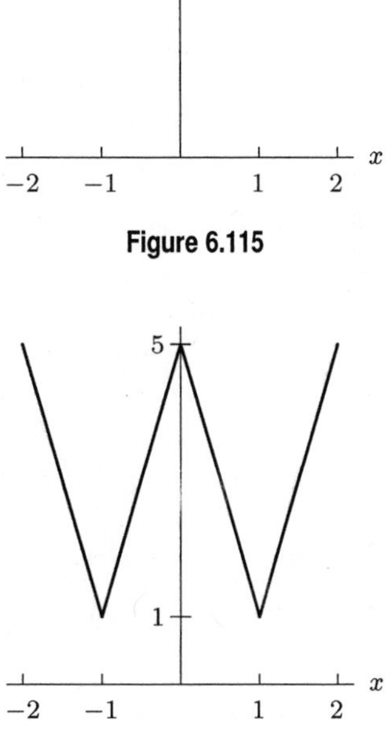

Figure 6.116

(c) Switching the order of transformations in parts (a) and (b) leads to two different graphs. Notice in (a), Figure 6.115 has maximum value $y = 8$ and minimum value $y = 4$. In (b), Figure 6.116 has maximum value $y = 5$ and minimum value $y = 1$. In both cases however, the difference between the maximum and minimum values is 4.

18. Applying these transformations in this order, we have

$$y = f(2x) \quad \text{horizontal compression by 1/2}$$
$$y = f(2(x-6)) = f(2x-12). \quad \text{horizontal shift right 6}$$

Thus, this sequence of transformations does not result in the function $y = f(2x - 6)$.

19. Applying these transformations in this order, we have

$$y = 5g(x) \quad \text{vertical stretch by 5}$$
$$y = 5g(x) - 2. \quad \text{vertical shift down 2}$$

Thus, this sequence of transformations does not result in the function $y = 5(g(x) - 2)$. We see by first distributing the 5, the function $y = 5g(x) - 10$ is obtained by a vertical stretch by a factor of 5 followed by a vertical shift down 10 units.

20. Before we graph the transformations of the Heaviside step function, we graph $H(x)$ itself. The graph of the function is shown in Figure 6.117.

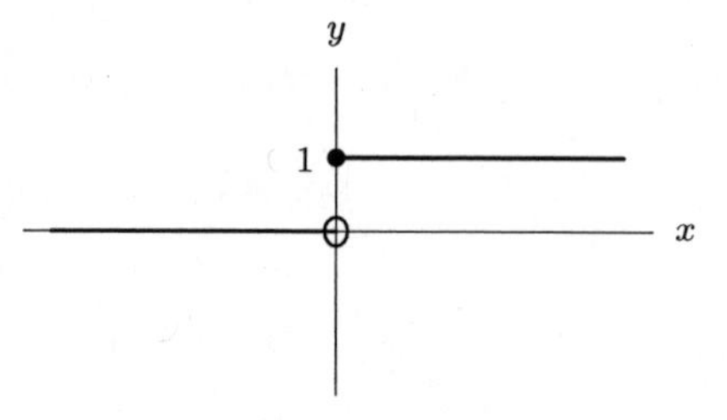

Figure 6.117: The Heaviside step function $H(x)$

(a) The graph of $y = H(x) - 2$ is the graph of $y = H(x)$ shifted down by 2 units.

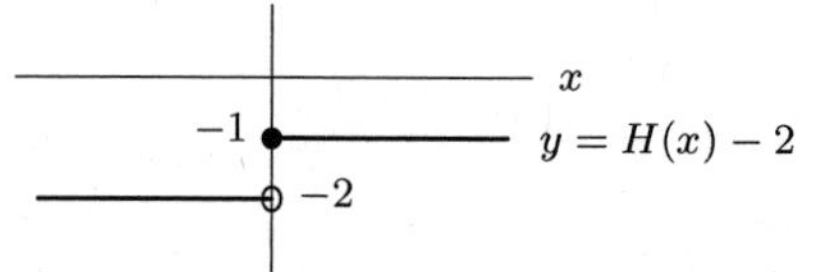

Figure 6.118: Graph of $y = H(x) - 2$

(b) The graph of $y = H(x + 2)$ is the graph of $y = H(x)$ shifted 2 units to the left.

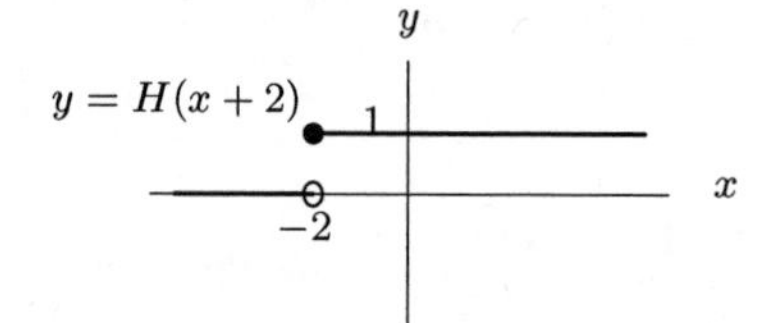

Figure 6.119: Graph of $y = H(x + 2)$

(c) We have $y = -3H(-x) + 4$, and thus we have first vertically stretched H by a factor of 3, then vertically reflected about the x-axis, then horizontally reflected about the y-axis, and finally vertically shifted up by 4 units.

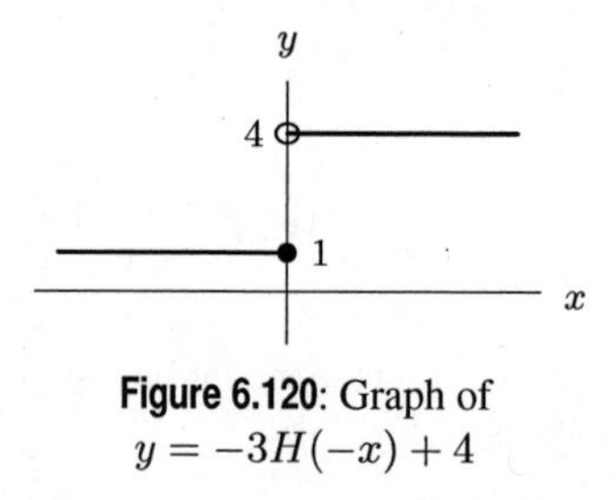

Figure 6.120: Graph of $y = -3H(-x) + 4$

21. We have

$$
\begin{aligned}
y &= f(x+4) && \text{shift left 4 units} \\
y &= -f(x+4) && \text{then flip vertically} \\
y &= -f(x+4) + 2 && \text{then shift up 2 units} \\
y &= 3\left(-f(x+4) + 2\right) && \text{then stretch by 3} \\
\text{so } g(x) &= -3f(x+4) + 6.
\end{aligned}
$$

22. We have

$$\begin{array}{lll} \text{Step 1:} & y = f(x-3) & \text{shift right 3 units} \\ \text{Step 2:} & y = -f(x-3) & \text{flip vertically} \\ \text{Step 3:} & y = -f(x-3)-2 & \text{shift down 2 units} \\ \text{Step 4:} & y = 2\left(-f(x-3)-2\right) & \text{stretch vertically.} \end{array}$$

Thus,

$$\begin{aligned} g(x) &= 2\left(-f(x-3)-2\right) \\ &= -2f(x-3)-4. \end{aligned}$$

23. We have

$$\begin{aligned} g(x) &= f(x+3)+2 & \text{shift } f \text{ left by 3 then up by 2} \\ h(x) &= -2g(-x) & \\ &= -2\underbrace{\left(f(-x+3)+2\right)}_{g(-x)} & \\ &= -2f(-x+3)-4. & \end{aligned}$$

24. Writing this as $g(x) = -2f(-3x)+10$, we see that to find the graph of g from the graph of f, we perform the following steps:

$$\begin{array}{rl} \text{First, stretch vertically by a factor of 2:} & (-12,20) \to (-12,40) \\ & (0,6) \to (0,12) \\ & (36,-2) \to (36,-4) \\ \text{Next, reflect about the } x\text{-axis:} & (-12,40) \to (-12,-40) \\ & (0,12) \to (0,-12) \\ & (36,-4) \to (36,4) \\ \text{Next, compress horizontally by a factor of } 1/3\text{:} & (-12,-40) \to (-4,-40) \\ & (0,-12) \to (0,-12) \\ & (36,4) \to (12,4) \\ \text{Next, reflect about the } y\text{-axis:} & (-4,-40) \to (4,-40) \\ & (0,-12) \to (0,-12) \\ & (12,4) \to (-12,4) \\ \text{Finally, shift vertically up by 10:} & (4,-40) \to (4,-30) \\ & (0,-12) \to (0,-2) \\ & (-12,4) \to (-12,14). \end{array}$$

Thus, three points on the graph of g are $(4,-30), (0,-2), (-12,14)$. To check our answer, we can verify that $g(4) = -30$:

$$g(4) = 10 - 2f(-3\cdot 4) = 10 - 2\underbrace{f(-12)}_{20} = 10 - 2\cdot 20 = -30,$$

as expected. Likewise, we can verify that $g(0) = -2$ and $g(-12) = 14$:

$$\begin{aligned} g(0) &= 10 - 2f(-3\cdot 0) &&= 10 - 2\underbrace{f(0)}_{6} &&= 10 - 2\cdot 6 &&= -2 \\ g(-12) &= 10 - 2f(-3(-12)) &&= 10 - 2\underbrace{f(36)}_{-2} &&= 10 - 2(-2) &&= 14, \end{aligned}$$

as expected.

25. We see that

$$\begin{aligned} y &= h(x+3) && \text{shift left 3 units} \\ y &= 2h(x+3) && \text{stretch vertically by 2} \\ y &= 2h(x+3)+6 && \text{shift vertically by 6} \\ \text{so} \quad f(x) &= 2h(x+3)+6. \end{aligned}$$

Since $(-12, 20)$ is on the graph of f, this means

$$\begin{aligned} f(-12) = 2h(-12+3)+6 &= 20 \\ 2h(-9)+6 &= 20 \\ 2h(-9) &= 14 \\ h(-9) &= 7. \end{aligned}$$

Thus, $(-9, 7)$ is a point on the graph of h. Likewise, since $(0, 6)$ and $(36, -2)$ are points on the graph of f, we have

$$\begin{aligned} f(0) = 2h(0+3)+6 &= 6 \\ 2h(3)+6 &= 6 \\ 2h(3) &= 0 \\ h(3) &= 0 \quad \text{so } (3,0) \text{ is on graph of } h \\ f(36) = 2h(36+3)+6 &= -2 \\ 2h(39)+6 &= -2 \\ 2h(39) &= -8 \\ h(39) &= -4. \quad \text{so } (39,-4) \text{ is on graph of } h \end{aligned}$$

Thus, points on the graph of h include $(-9, 7), (3, 0), (39, -4)$. To check our answer, if $(-9, 7)$ is on the graph of h and we shift it to the left 3 units, then $(-12, 7)$ is the new point. If we stretch this vertically by a factor of 2, we get the point $(-12, 14)$. Finally, if we shift it up by 6, we get the point $(-12, 20)$, which is in fact on the graph of f, as required. We can perform a similar check with the other two points.

26. **(a)** Notice $y = (x+3)^2 - 4$ is in the form $y = Af(x-h)+k$ for $f(x) = x^2$. We have horizontally shifted the parabola $f(x) = x^2$ to the left by 3 units and then vertically shifted down by 4 units. The vertex at $(0, 0)$ on the graph of $f(x) = x^2$ is shifted to the new vertex at $(-3, -4)$. Writing in standard form, we have $y = (x+3)^2 - 4 = x^2 + 6x + 5$. See Figure 6.121.

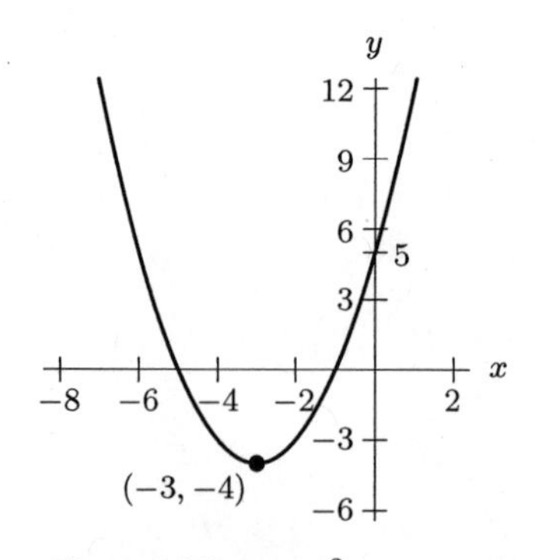

Figure 6.121: $y = x^2 + 6x + 5$

(b) Notice $y = -2(x+1)^2 + 3$ is in the form $y = Af(x-h)+k$ for $f(x) = x^2$. We have vertically stretched $f(x) = x^2$ by a factor of 2, then vertically reflected about the x-axis, then horizontally shifted to the left by 1 unit, and finally vertically shifted up by 3 units. The vertex at $(0,0)$ on the graph of $f(x) = x^2$ is moved to the new vertex at $(-1,3)$. Writing in standard form, we have $y = -2(x+1)^2 + 3 = -2x^2 - 4x + 1$. See Figure 6.122.

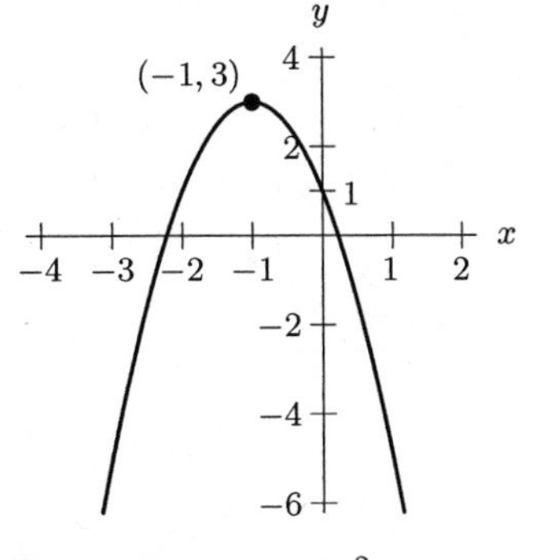

Figure 6.122: $y = -2x^2 - 4x + 1$

27. We see that f is an exponential function with y-intercept $a = 20$. The value of $f(x)$ doubles each time x increases by 1. In the transformation $y = 160 - 4f(-x/10) = -4f(-x/10) + 160$, the graph of f is moved in the following order:

- stretched vertically by a factor of 4;
- flipped vertically;
- stretched horizontally by a factor of 10, so that the doubling time is now 10 instead of 1;
- flipped horizontally; and finally
- shifted vertically by 160 units, so that the horizontal asymptote is now $y = 160$ instead of $y = 0$.

See Figure 6.123. Note that another way to think about this problem is to write

$$\begin{aligned} y &= -4f(-x/10) + 160 \\ &= \underbrace{-80 \cdot 2^{-x/10}}_{\text{exponential}} + 160. \end{aligned}$$

The function $y = -80 \cdot 2^{-x/10}$ is an exponential function with y-intercept $a = -80$ whose value is halved each time x increases by 10, or, equivalently, whose value doubles each time x decreases by 10.

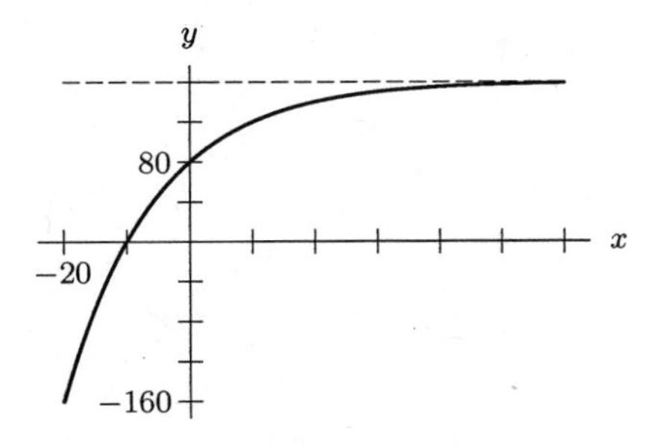

Figure 6.123

28. (a) We first shift $h(t)$ to the right 6 units, and we have

$$y = h(t-6) = (t-6)^2.$$

We next compress horizontally by a factor of $1/2$, and we have

$$y = h(2t-6)$$

$$\begin{aligned} &= (2t-6)^2 \\ &= (2(t-3))^2 \quad \text{(factoring out 2)} \\ &= 2^2(t-3)^2 \quad \text{(since } (ab)^2 = a^2b^2\text{)} \\ &= 4(t-3)^2. \end{aligned}$$

(b) We first compress $h(t)$ horizontally by a factor of $1/2$, and we have

$$y = h(2t) = (2t)^2.$$

We next shift to the right 6 units, and we have

$$\begin{aligned} y &= h(2(t-6)) \\ &= (2(t-6))^2 \\ &= 2^2(t-6)^2 \quad \text{(since } (ab)^2 = a^2b^2\text{)} \\ &= 4(t-6)^2. \end{aligned}$$

(c) Notice from part (a) the function $y = 4(t-3)^2 = h(2(t-3))$ is equivalent to first applying a horizontal compression by a factor of $1/2$ to $h(t)$, and then shifting the right by 3 units.

29. All four transformations are equivalent.

(a) We have:

$$\begin{aligned} \text{First step:} \quad & y = f(x) + 3 && \text{shift up} \\ \text{Second step:} \quad & y = 2\left(f(x) + 3\right) && \text{stretch} \\ \text{Third step:} \quad & y = -2\left(f(x) + 3\right). && \text{flip} \end{aligned}$$

Multiplying out, this gives $y = -2f(x) - 6$.

(b) We have:

$$\begin{aligned} \text{First step:} \quad & y = -f(x) && \text{flip} \\ \text{Second step:} \quad & y = -f(x) - 3 && \text{shift down} \\ \text{Third step:} \quad & y = 2(-f(x) - 3). && \text{stretch} \end{aligned}$$

Multiplying out, this gives $y = -2f(x) - 6$, which is the same as part (a).

(c) We have:

$$\begin{aligned} \text{First step:} \quad & y = 2f(x) && \text{stretch} \\ \text{Second step:} \quad & y = 2f(x) + 6 && \text{shift up} \\ \text{Third step:} \quad & y = -\left(2f(x) + 6\right). && \text{flip} \end{aligned}$$

Multiplying out, this gives $y = -2f(x) - 6$, which is the same as parts (a) and (b).

(d) We have:

$$\begin{aligned} \text{First step:} \quad & y = -f(x) && \text{flip} \\ \text{Second step:} \quad & y = -2f(x) && \text{stretch} \\ \text{Third step:} \quad & y = -2f(x) - 6. && \text{shift down} \end{aligned}$$

This is the same as parts (a)–(c).

30. Since $g(x) = 5e^{x-2}$ and $g(x) = kf(x) = ke^x$, we have

$$5e^{x-2} = ke^x.$$

Divide both sides by e^x to isolate k,

$$5\frac{e^{x-2}}{e^x} = k,$$

and simplify to find $k = \dfrac{5}{e^2}$.

31. **(a)** Graphing f and g shows that there is a vertical shift up 1 unit. See Figure 6.124.

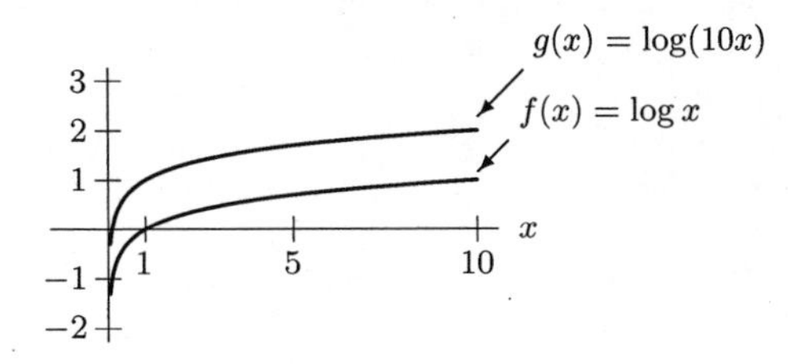

Figure 6.124: A vertical shift of $+1$

(b) Using the property that $\log(ab) = \log a + \log b$, we have

$$g(x) = \log(10x) = \log 10 + \log x = 1 + f(x).$$

Thus, $g(x)$ is $f(x)$ shifted vertically upward by 1.

(c) Using the same property of logarithms

$$\log(ax) = \log a + \log x \qquad \text{so} \qquad k = \log a.$$

32. A horizontal stretch of f by a factor of k gives the formula

$$\begin{aligned} y &= f\left(\frac{1}{k}x\right) \\ &= \ln\left(\frac{1}{k}x\right) \\ &= \ln\left(\frac{1}{k}\right) + \ln x \quad \text{(since } \ln(ab) = \ln a + \ln b\text{)} \\ &= f(x) + \ln\left(\frac{1}{k}\right) \\ &= f(x) + \ln\left(k^{-1}\right) \quad \text{(since } 1/a = a^{-1}\text{)} \\ &= f(x) - \ln k. \quad \text{(since } \ln(b^t) = t \cdot \ln b\text{)} \end{aligned}$$

Since $k > 1$, we know $\ln k > 0$. Thus a horizontal stretch of f by a factor of k is equivalent to a shifting f down $\ln k$ units.

33. A horizontal shift of g to the right k units gives

$$\begin{aligned} y &= g(x - k) \\ &= e^{(x-k)} \\ &= e^x \cdot e^{-k} \quad \text{(since } e^{ab} = e^a \cdot e^b\text{)} \\ &= e^{-k} \cdot g(x). \end{aligned}$$

If $k > 0$, then $0 < e^{-k} < 1$. Thus we see shifting g right k units is equivalent to vertically compressing of g by a factor of e^{-k}.

34. **(a)** Since the point $(-2, 5)$ is on the graph of r, we know $r(-2) = 5$. All the transformations in $y = 3(r(x) + 2)$ are outside r, so the x-coordinate remains unchanged. The new y-coordinate is given by $y = 3(r(-2)+2) = 3(5+2) = 21$. The transformed point is $(-2, 21)$.

(b) The function $r(x)$ is transformed vertically since all the operations occur outside the function $r(x)$. To identify the transformations, we first distribute 3 in the expression $3(r(x) + 2)$:

$$y = 3(r(x) + 2) = 3r(x) + 6.$$

So there are two outside transformations applied to $r(x)$: First a vertical stretch by a factor of 3 and then a vertical shift upward 6 units.

35. Notice that we are making only outside changes to f. Thus, any shifts and stretches are vertical, not horizontal.

(a) First, stretch the graph of f vertically by a factor of 2, giving the graph of the intermediate function $w(x) = 2f(x)$. Next, shift this graph up by 8 units, obtaining the graph of

$$\begin{aligned} y &= w(x) + 8 && \text{shift intermediate graph up by 8} \\ &= 2f(x) + 8. && \text{since } w(x) = 2f(x) \end{aligned}$$

(b) First, shift the graph of f vertically by k units, giving the graph of the intermediate function $v(x) = f(x) + k$. To determine the value of k, we next stretch the intermediate graph vertically by a factor of 2, giving the graph of

$$\begin{aligned} y &= 2v(x) && \text{stretch intermediate graph by 2} \\ &= 2\left(f(x) + k\right) && \text{since } v(x) = f(x) + k \\ &= 2f(x) + 2k. \end{aligned}$$

Since we want $2f(x) + 2k = 2f(x) + 8$, solving for k gives $k = 4$. We conclude that shifting f up 4 then stretching by 2 gives the same result as stretching f vertically by 2 then shifting up 8.

36. Notice that we are making only inside changes to f. Thus, any shifts and compressions are horizontal, not vertical.

(a) First stretch the graph of f horizontally by a factor of 3, giving the graph of the intermediate function $w(x) = f\left(\frac{1}{3}x\right)$. Next, shift this graph left by h units. To determine the value of h, we write

$$\begin{aligned} y &= w(x + h) && \text{shift intermediate graph left by } h \\ &= f\left(\frac{1}{3}(x + h)\right) && \text{since } w(x) = f\left(\tfrac{1}{3}x\right) \\ &= f\left(\frac{1}{3}x + \frac{1}{3}h\right). \end{aligned}$$

Since we want this to equal $f\left(\frac{1}{3}x + 4\right)$ we let $\frac{1}{3}h = 4$, so $h = 12$. Thus, to find the graph of $y = f\left(\frac{1}{3}x + 4\right)$, first stretch the graph of f horizontally by 3, then shift it left 12.

(b) First shift the graph of f left by 4, giving the graph of the intermediate function $v(x) = f(x + 4)$. Next, stretch this graph horizontally by a factor of 3, giving the graph of

$$\begin{aligned} y &= v\left(\frac{1}{3}x\right) && \text{stretch intermediate graph by 3} \\ &= f\left(\frac{1}{3}x + 4\right). && \text{since } v(x) = f(x + 4) \end{aligned}$$

Thus, another way to find the graph of $y = f\left(\frac{1}{3}x + 4\right)$ is first to shift the graph of f left by 4, then stretch the resulting graph horizontally by 3.

37. **(a)** The building is kept at 60° F until 5 am when the heat is turned up. The building heats up at a constant rate until 7 am when it is 68° F. It stays at that temperature until 3 pm when the heat is turned down. The building cools at a constant rate until 5 pm. At that time, the temperature is 60° F and it stays that level through the end of the day.

(b) Since $c(t) = 142 - d(t) = -d(t) + 142$, the graph of $c(t)$ will look like the graph of $d(t)$ that has been first vertically reflected across the t-axis and then vertically shifted up 142 units.

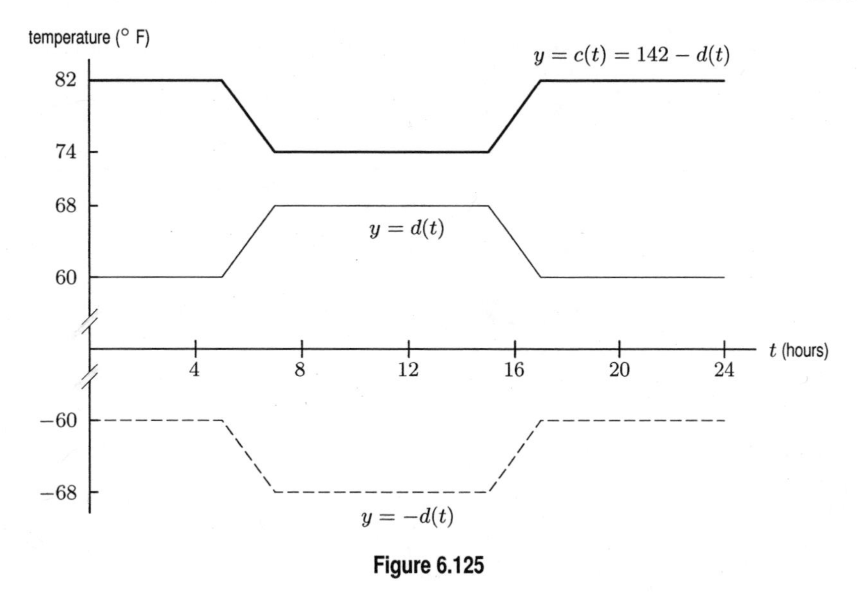

Figure 6.125

(c) This could describe the cooling schedule in the summer months when the temperature is kept at 82° F at night and cooled down to 74° during the day.

38. (a) Since 1 meter is equivalent to 0.001 meters, the depth of the tide in kilometers is

$$f(t) = 0.001D(t).$$

(b) Since 1 hour is equivalent to 60 minutes, we have $g(60) = f(1)$, $g(120) = f(2)$, and so on. Thus, the depth of the tide in kilometers is as a function of x minutes from noon is

$$g(x) = f\left(\frac{1}{60}x\right) = 0.001D\left(\frac{1}{60}x\right).$$

(c) Since we are now measuring time from a benchmark of 2PM instead of noon, we must shift our answer in part (c), $g(x)$, to the left by 120 minutes. Thus, the depth of the tide in kilometers as a function of x minutes from 2 pm is

$$j(x) = g(x + 120) = 0.001D\left(\frac{1}{60}(x + 120)\right).$$

The inside transformation $\frac{1}{60}(x + 120)$ means we first shift the benchmark time by 120 minutes and then convert from minutes to hours.

We also have written a formula for $j(x)$ in the equivalent form

$$j(x) = 0.001D\left(\frac{1}{60}(x + 120)\right) = 0.001D\left(\frac{1}{60}x + 2\right),$$

which is equivalent to first converting time from minutes to hours and then shifting by 2 hours.

39. (a) If f is shifted vertically, the resulting function is given by

$$y = f(x) + 3.$$

To flip this graph across the x-axis, we multiply by -1. Thus,

$$y = -(f(x) + 3).$$

The resulting graph is shown in Figure 6.126.

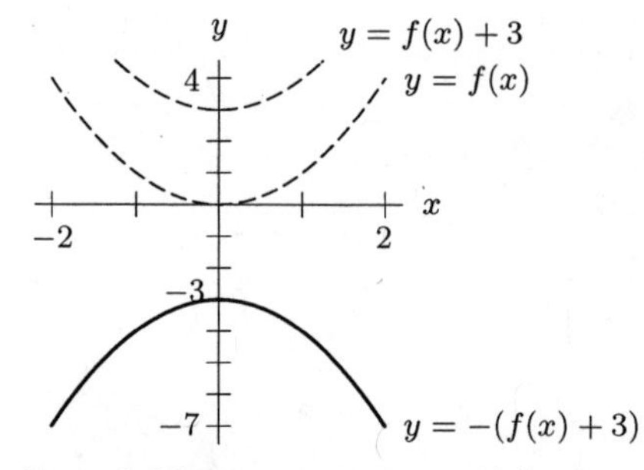

Figure 6.126: The graph of $y = f(x)$ shifted upward 3 units and then flipped across the x-axis

(b) If f is flipped across the x-axis, the resulting function is given by

$$y = -f(x).$$

To shift this graph upward by 3 units, we add 3:

$$y = -f(x) + 3.$$

The resulting graph is shown in Figure 6.127.

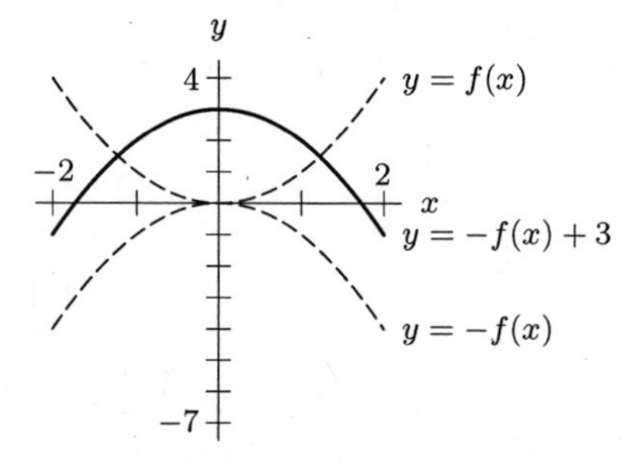

Figure 6.127: The graph of $y = f(x)$ flipped across the x-axis and then shifted upward 3 units

40. **(a)** See Figure 6.128.

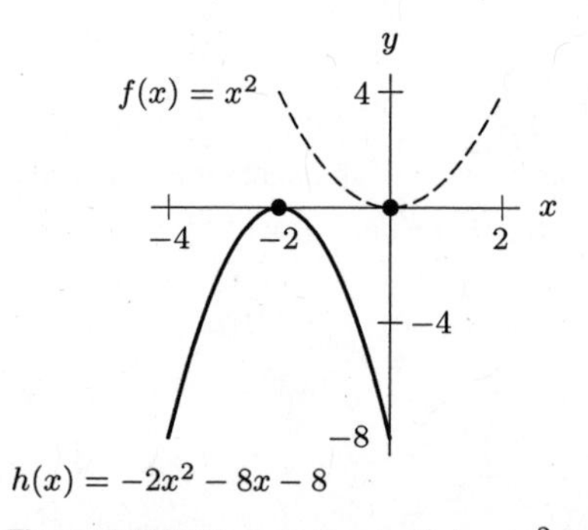

Figure 6.128: The graphs of $f(x) = x^2$ and $h(x) = -2x^2 - 8x - 8$

(b) From Figure 6.128, it appears as though the graph of h might be found by flipping the graph of $f(x) = x^2$ over the x-axis and then shifting it to the left. In other words, we might guess that the graph of h is given by $y = -f(x+2)$. This graph is shown in Figure 6.129. Notice, though, that $y = -f(x+2)$ is not as steep as the graph of $y = h(x)$. However, we can look at the y–intercepts of both graphs and notice that -4 must be stretched to -8. Thus, we can try applying a stretch factor of 2 to our guess. The resulting graph, given by

$$y = -2f(x+2),$$

does indeed match h.

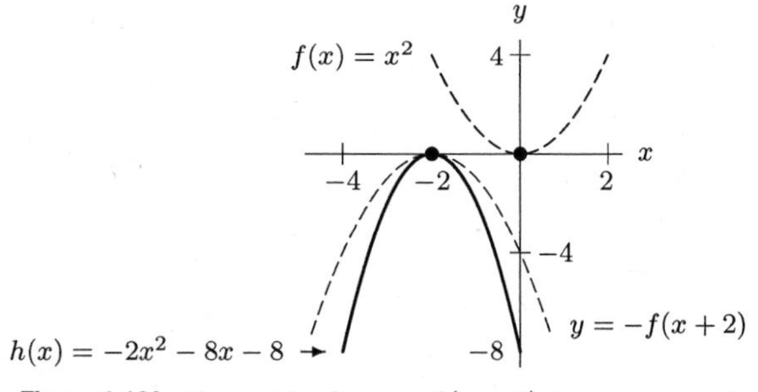

Figure 6.129: The graph of $y = -f(x+2)$ is a compressed version of the graph of h. The graph of $y = -2f(x+2)$ is the same as $y = h(x)$

We can verify the last result algebraically. If the graph of h is given by $y = -2f(x+2)$, then

$$\begin{aligned} y &= -2f(x+2) \\ &= -2(x+2)^2 \qquad \text{(because } f(x) = x^2\text{)} \\ &= -2(x^2 + 4x + 4) \\ &= -2x^2 - 8x - 8. \end{aligned}$$

This is the formula given for h.

41. **(a)** Vertically stretching the graph of $f(x) = x$ by a factor of $m > 1$ generates the family of functions $y = mx$, that is, straight lines with slope $m > 1$ through the origin. Since m is greater than 1, this family is all lines through the origin with slope greater than 1.

(b) Vertical stretching and horizontally shifting the graph of $f(x) = x$ generates the family of functions $y = m(x-h) = mx - mh$. If we write $-mh = b$, then we have $y = mx + b$, that is, the family of straight lines with slope $m > 1$ and y-intercept b. Since m is greater than 1 and b can have any value, this is the entire family of increasing linear functions with slope greater than 1. The family in part (a) is contained in the family found in part (b).

42. **(a)** Vertically stretching the graph of $f(x) = 2^x$ generates the family of functions $y = a \cdot 2^x$, where $a > 1$. Graphs of this family have y-intercepts of $(0, a)$ and a horizontal asymptote of $y = 0$.

(b) Horizontally shifting the graph of $f(x) = 2^x$ generates the family $y = 2^{x+h} = 2^h \cdot 2^x$. Since $2^h > 1$ for $h > 0$, shifting $f(x) = 2^x$ to the left by h units is equivalent to vertically stretching $f(x) = 2^x$ by a factor of 2^h. Setting $2^h = a$, we have the same family of function in part (a).

If $h < 0$, then $0 < 2^h < 1$. Thus, shifting $f(x) = 2^x$ right h units is equivalent to compressing $f(x) = 2^x$ by a factor of $2^h = a$. These functions are different than the family of functions we found in part (a). The family in part (a) is contained in the family in part (b).

43. Yes, g is also linear. We see that

$$f(sx) = b + m \cdot (sx) \qquad \text{because } f(x) = b + mx$$
$$\text{which means} \quad g(x) = r \cdot f(sx) + j$$

$$
\begin{aligned}
&= r(b+msx)+j \\
&= \underbrace{rb+j}_{B} + \underbrace{rms}_{M}\cdot x, \\
\text{so}\quad B &= rb+j \\
M &= rms.
\end{aligned}
$$

44. No, g is not also exponential. We see that:

$$
\begin{aligned}
f(sx) &= ae^{k\cdot(sx)} && \text{because } f(x)=ae^{kx} \\
\text{which means}\quad g(x) &= r\cdot f(sx)+j \\
&= r\cdot ae^{ksx}+j.
\end{aligned}
$$

This can not be written in the form $g(x) = Ae^{KX}$.

45. Yes, g is also quadratic. We see that:

$$
\begin{aligned}
f(sx) &= a(sx-h)^2+k && \text{because } f(x)=a(x-h)^2+k \\
&= a\left(s(x-h/s)\right)^2+k && \text{factor out } s \\
&= as^2(x-h/s)^2+k \\
\text{which means}\quad g(x) &= r\cdot f(sx)+j \\
&= r\left(as^2(x-h/s)^2+k\right)+j \\
&= \underbrace{ras^2}_{A}\cdot\left(x-\underbrace{h/s}_{H}\right)^2+\underbrace{rk+j}_{K} \\
\text{so}\quad A &= ras^2 \\
H &= h/s \\
K &= rk+j.
\end{aligned}
$$

Solutions for Chapter 6 Review

Exercises

1. **(a)** The input is $2x = 2\cdot 2 = 4$.
(b) The input is $\frac{1}{2}x = \frac{1}{2}\cdot 2 = 1$.
(c) The input is $x+3 = 2+3 = 5$.
(d) The input is $-x = -2$.

2. **(a)** The input is $2x$, and so $2x = 2$, which means $x = 1$.
(b) The input is $\frac{1}{2}x$, and so $\frac{1}{2}x = 2$, which means $x = 4$.
(c) The input is $x+3$, and so $x+3 = 2$, which means $x = -1$.
(d) The input is $-x$, and so $-x = 2$, which means $x = -2$.

3. **(a)** $(6,5)$ **(b)** $(2,1)$ **(c)** $(\frac{1}{2},5)$ **(d)** $(2,20)$

4. **(a)** $(-9,4)$ **(b)** $(-3,\frac{4}{3})$ **(c)** $(1,4)$ **(d)** $(-1,-4)$

5. A function is odd if $a(-x) = -a(x)$.

$$a(x) = \frac{1}{x}$$
$$a(-x) = \frac{1}{-x} = -\frac{1}{x}$$
$$-a(x) = -\frac{1}{x}$$

Since $a(-x) = -a(x)$, we know that $a(x)$ is an odd function.

6. $m(-x) = \dfrac{1}{(-x)^2} = \dfrac{1}{x^2} = m(x)$, so $m(x)$ is an even function.

7. In this case, $e(x) = x + 3$, $e(-x) = -x + 3$ and $-e(x) = -(x + 3) = -x - 3$. Since $e(-x)$ equals neither $e(x)$ nor $-e(x)$, the function $e(x)$ is neither even nor odd.

8. $p(-x) = (-x)^2 + 2(-x) = x^2 - 2x$, and $-p(x) = -x^2 - 2x$. Since $p(-x) \neq p(x)$ and $p(-x) \neq -p(x)$, the function p is neither even nor odd.

9. A function is even if $b(-x) = b(x)$.

$$b(x) = |x|$$
$$b(-x) = |-x| = |x|$$

Since $b(-x) = b(x)$, we know that $b(x)$ is an even function.

10. $q(-x) = 2^{-x+1}$, and $-q(x) = -2^{x+1}$. Since $q(-x) \neq q(x)$ and $q(-x) \neq -q(x)$, the function q is neither even nor odd.

11. **(a)** $f(2x) = 1 - (2x) = 1 - 2x$
(b) $f(x + 1) = 1 - (x + 1) = -x$
(c) $f(1 - x) = 1 - (1 - x) = x$
(d) $f(x^2) = 1 - x^2$
(e) $f(1/x) = 1 - (1/x) = (x/x) - (1/x) = (x - 1)/x$
(f) $f(\sqrt{x}) = 1 - \sqrt{x}$

12.

Table 6.19

x	-3	-2	-1	0	1	2	3
$f(x)$	-4	-1	2	3	0	-3	-6
$f(-x)$	-6	-3	0	3	2	-1	-4
$-f(x)$	4	1	-2	-3	0	3	6
$f(x) - 2$	-6	-3	0	1	-2	-5	-8
$f(x - 2)$	–	–	-4	-1	2	3	0
$f(x) + 2$	-2	1	4	5	2	-1	-4
$f(x + 2)$	2	3	0	-3	-6	–	–
$2f(x)$	-8	-2	4	6	0	-6	-12
$-f(x)/3$	4/3	1/3	$-2/3$	-1	0	1	2

Problems

13. The graph is the graph of f shifted to the left by 2 and up by 2. See Figure 6.130.

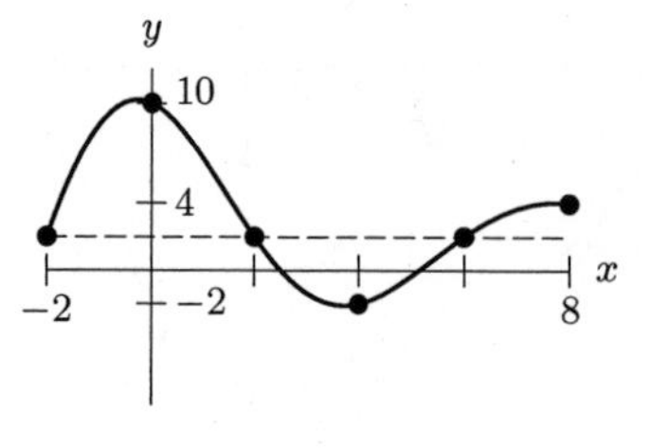

Figure 6.130

14. The graph is the graph of f flipped across the y-axis, stretched vertically by a factor of 2, and flipped across the x-axis. See Figure 6.131.

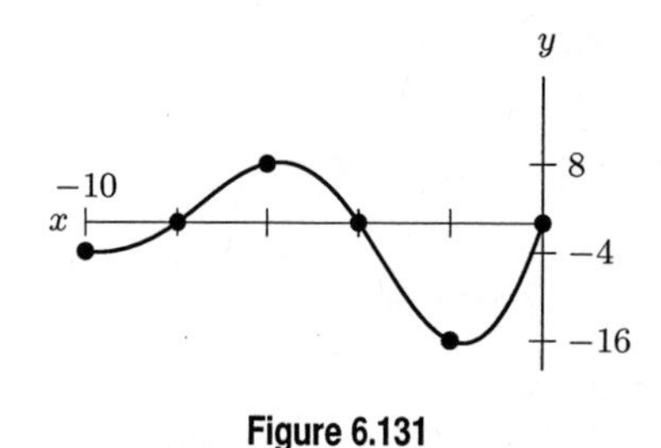

Figure 6.131

15. The graph appears to be the graph of f shifted to the left by 4 and down by 8, so $y = f(t+4) - 8$.

16. The graph appears to be the graph of f flattened vertically by a factor of 0.5, flipped across the x-axis, and shifted vertically by 4 units, so $y = -0.5f(x) + 4$.

17. **(a)** Since the x-coordinate of the point $(-3, 1)$ on the graph of $f(x)$ has been multiplied by -1 in order to obtain the point $(3, 1)$ on the graph of $g(x)$, $g(x)$ must be obtained by reflecting the graph of $f(x)$ horizontally about the y-axis.
(b) Since the y-coordinate remains constant and only the x-coordinate is moved, $f(x)$ must be shifted horizontally. The x-coordinate of the point $(3, 1)$ on the graph of $g(x)$ has been shifted to the right by 6 units from the point $(-3, 1)$ on the graph of $f(x)$.

18. **(a)** $D(225)$ represents the number of iced cappuccinos sold at a price of $2.25.
(b) $D(p)$ is likely to be a decreasing function. The coffeehouse will probably sell fewer iced cappuccinos if they charge a higher price for them.
(c) p is the price the coffeehouse should charge if they want to sell 180 iced cappuccinos per week.
(d) $D(1.5t)$ represents the number of iced cappuccinos the coffeehouse will sell if they charge one and a half times the average price. $1.5D(t)$ represents 1.5 times the number of cappuccinos sold at the average price. $D(t + 50)$ is the number of iced cappuccinos they will sell if they charge 50 cents more than the average price. $D(t) + 50$ represents 50 more cappuccinos than the number they can sell at the average price.

19. **(a)** (VI) We know that $y = e^x$ is an increasing function with a y-intercept of 1. When $x = 1$, $y = e$, or, a little less than 3.
(b) (V) The graph of $y = e^{5x}$ is similar to that of $y = e^x$, but it is compressed horizontally by a factor of $1/5$. It is a more rapidly increasing function with a y-intercept of 1.
(c) (III) The graph of $y = 5e^x$ is a vertical stretch, by a factor of 5, of $y = e^x$. It is an increasing function with a y-intercept of 5.
(d) (IV) The graph of $y = e^{x+5}$ is the graph of $y = e^x$ shifted to the left by 5 units.
(e) (I) The graph of $y = e^{-x}$ is the graph of $y = e^x$ transposed across the y-axis. It is a decreasing function with a y-intercept of 1.
(f) (II) The graph of $y = e^x + 5$ is the graph of $y = e^x$ shifted vertically upward by 5 units. The y-intercept will be $1 + 5 = 6$.

20. In the southern hemisphere the seasons are reversed, that is, they come a half year earlier (or later) than in the northern hemisphere. So we need to shift the graph horizontally by half a year. Whether we shift the graph left or right makes no difference. Therefore, possible formulas for the shifted curve include $L(d + \frac{365}{2})$ and $L(d - \frac{365}{2})$. In Figure 6.132, we have shifted $L(d)$ one-half year forward (to the left) giving $L(d + \frac{365}{2})$.

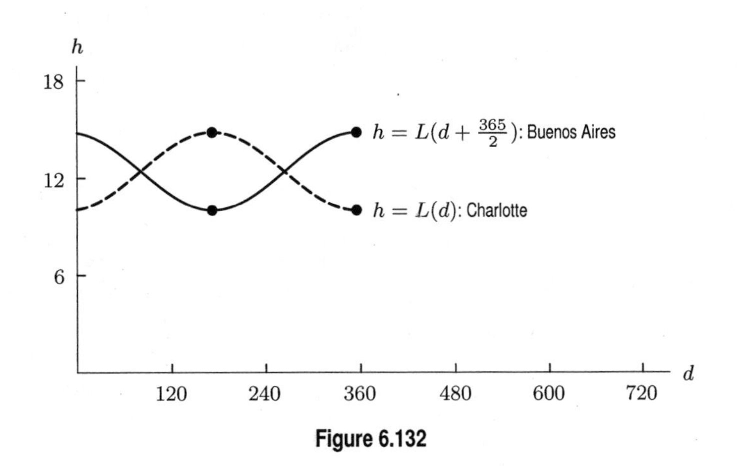

Figure 6.132

21. **(a)** Using the formula for $d(t)$, we have

$$\begin{aligned} d(t) - 15 &= (-16t^2 + 38) - 15 \\ &= -16t^2 + 23. \end{aligned}$$

$$\begin{aligned} d(t - 1.5) &= -16(t - 1.5)^2 + 38 \\ &= -16(t^2 - 3t + 2.25) + 38 \\ &= -16t^2 + 48t + 2. \end{aligned}$$

(b)

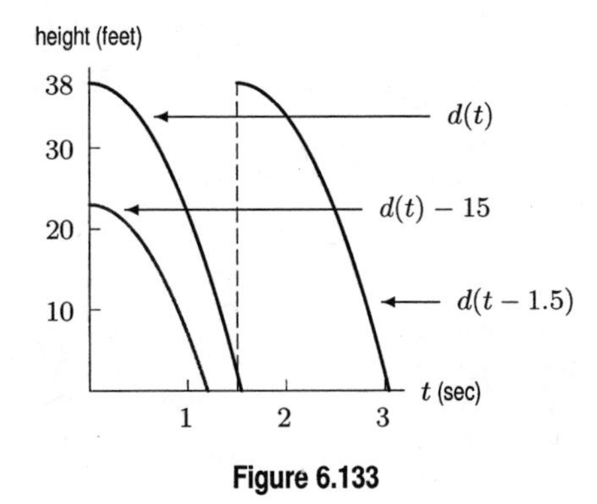

Figure 6.133

(c) $d(t) - 15$ represents the height of a brick which falls from $38 - 15 = 23$ feet above the ground. On the other hand, $d(t-1.5)$ represents the height of a brick which began to fall from 38 feet above the ground at one and a half seconds after noon.

(d) (i) The brick hits the ground when its height is 0. Thus, if we represent the brick's height above the ground by $d(t)$, we get

$$0 = d(t)$$

$$\begin{aligned}0 &= -16t^2 + 38\\ -38 &= -16t^2\\ t^2 &= \frac{38}{16}\\ t^2 &= 2.375\\ t &= \pm\sqrt{2.375} \approx \pm 1.541.\end{aligned}$$

We are only interested in positive values of t, so the brick must hit the ground 1.541 seconds after noon.

(ii) If we represent the brick's height above the ground by $d(t) - 15$ we get

$$\begin{aligned}0 &= d(t) - 15\\ 0 &= -16t^2 + 23\\ -23 &= -16t^2\\ t^2 &= \frac{23}{16}\\ t^2 &= 1.4375\\ t &= \pm\sqrt{1.4375} \approx \pm 1.199.\end{aligned}$$

Again, we are only interested in positive values of t, so the the brick hits the ground 1.199 seconds after noon.

(e) Since the brick, whose height is $d(t-1.5)$, begins falling 1.5 seconds after the brick whose height is $d(t)$, we expect the brick whose height is $d(t-1.5)$ to hit the ground 1.5 seconds after the brick whose height is $d(t)$. Thus, the brick should hit the ground $1.5 + 1.541 = 3.041$ seconds after noon.

22. The graph could be $y = |x|$ shifted right by one unit and up two units. Thus, let

$$y = |x - 1| + 2.$$

23. The graph is the cubic function that has been flipped about the x-axis and shifted left and up one unit. Thus, we could try

$$y = -(x+1)^3 + 1.$$

24. The graph appears to have been compressed horizontally by a factor of 1/2, flipped vertically, and shifted vertically by 2 units, so

$$y = -h(2x) + 2.$$

25. The graph appears to have been shifted to the left 6 units, compressed vertically by a factor of 2, and shifted vertically by 1 unit, so

$$y = \frac{1}{2}h(x+6) + 1.$$

26. **(a)** $f(10) = 6000$. The total cost for a carpenter to build 10 wooden chairs is \$6000.

(b) $f(30) = 7450$. The total cost for a carpenter to build 30 wooden chairs is \$7450.

(c) $z = 40$. A carpenter can build 40 wooden chairs for \$8000.

(d) $f(0) = 5000$. This is the fixed cost of production, or how much it costs the carpenter to set up before building any chairs.

27. Assuming f is linear between 10 and 20, we get

$$\begin{aligned}\frac{6400 - f(10)}{p - 10} &= \frac{f(20) - f(10)}{20 - 10},\\ \text{or} \quad \frac{400}{p - 10} &= \frac{800}{10}.\end{aligned}$$

Solving for p yields $p = 15$. Assuming f is linear between 20 and 30, we get

$$\frac{q - f(20)}{26 - 20} = \frac{f(30) - f(20)}{30 - 20},$$

or

$$\frac{q - 6800}{6} = \frac{650}{10}.$$

Solving for q yields $q = 7190$.

28. **(a)** We have

$$d_1 = f(30) - f(20) = 650$$
$$d_2 = f(40) - f(30) = 550$$
$$d_3 = f(50) - f(40) = 500$$

(b) d_1, d_2, and d_3 tell us how much building an additional 10 chairs will cost if the carpenter has already built 20, 30, and 40 chairs respectively.

29.

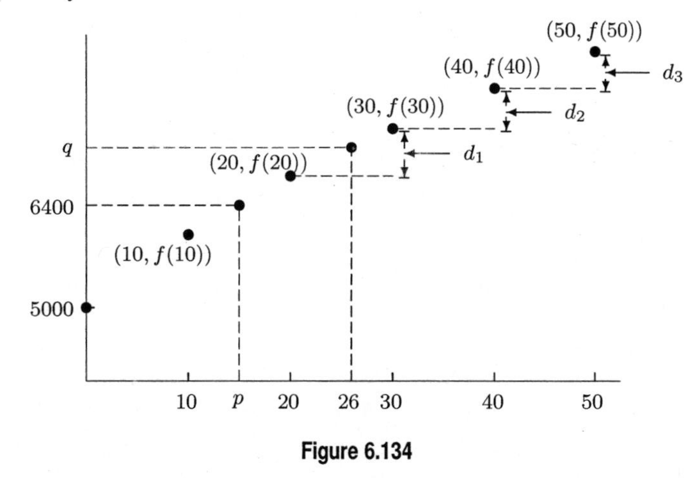

Figure 6.134

30. **(a)** (i) $f(k + 10)$ is how much it costs to produce 10 more than the normal weekly number of chairs.
(ii) $f(k) + 10$ is 10 dollars more than the cost of a normal week's production.
(iii) $f(2k)$ is the normal cost of two week's production.
(iv) $2f(k)$ is twice the normal cost of one week's production (which may be greater than $f(2k)$ since the fixed costs are included twice in $2f(k)$).

(b) The total amount the carpenter gets will be $1.8f(k)$ plus a five percent sales tax: that is, $1.05(1.8f(k)) = 1.89f(k)$.

31. See Figure 6.135.

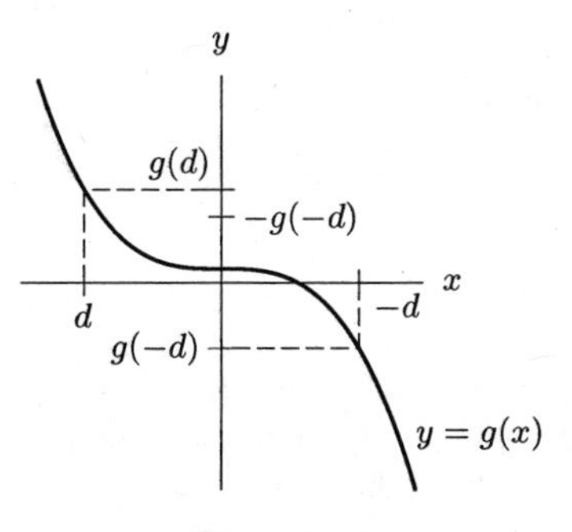

Figure 6.135

32. See Figure 6.136.

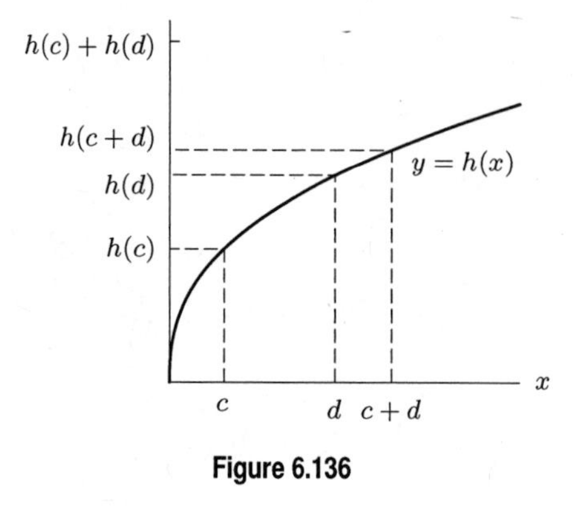

Figure 6.136

33. There is a vertical stretch of 3 so

$$y = 3h(x).$$

34. We have a reflection through the x-axis and a horizontal shift to the right by 1.

$$y = -h(x-1)$$

35. There is a reflection through the y-axis, a horizontal compression by a factor of $1/2$, a horizontal shift to the right by 1 unit, and a reflection through the x-axis. Combining these transformations we get

$$y = -h(-2(x-1)) \quad \text{or} \quad y = -h(2-2x).$$

36. Since we have $y = 2 - f(x-2) = -f(x-2) + 2$, the graph of f has been vertically reflected about the x-axis, then horizontally shifted to the right by 2 units, and finally vertically shifted up by 2 units. See Figure 6.137.

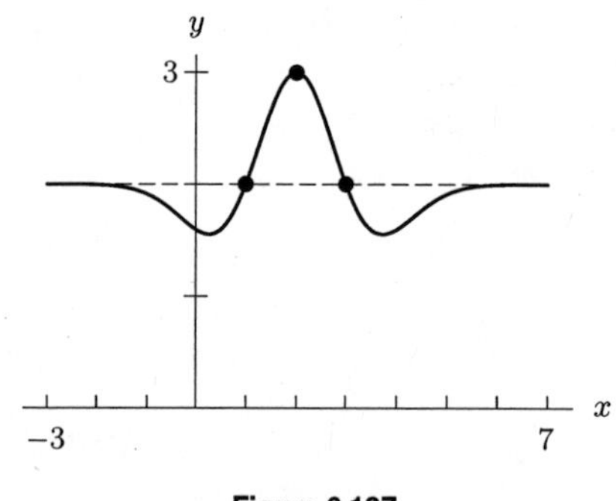

Figure 6.137

37.
- The vertical gap between the horizontal asymptote and the lowest point (at the y-intercept) is 1 in the original graph but 2 in the new graph. Thus, the graph appears to have been vertically stretched by a factor of 2.
- The two points where the graph crosses the horizontal asymptote are separated by a distance of 2 in the old graph but by a distance of 4 in the new graph. So, the graph appears to have been horizontally stretched by a factor of 2.
- The horizontal asymptote is at $y = 0$ in the old graph but $y = 3$ in the new graph, so the graph appears to have been vertically shifted by 3 units.

Putting all this together gives $y = 2f(x/2) + 3$.

38. We will reverse Gwendolyn's actions. First, we can shift the parabola back two units to the right by replacing x in $y = (x-1)^2 + 3$ with $(x-2)$. This gives

$$\begin{aligned} y &= ((x-2)-1)^2 + 3 \\ &= (x-3)^2 + 3. \end{aligned}$$

We subtract 3 from this function to move the parabola down three units, so

$$\begin{aligned} y &= (x-3)^2 + 3 - 3 \\ &= (x-3)^2. \end{aligned}$$

Finally, to flip the parabola back across the horizontal axis, we multiply the function by -1. Thus, Gwendolyn's original equation was

$$y = -(x-3)^2.$$

39. **(a)** $j(25)$ is the average amount of water (in gallons) required daily by a 25-foot oak. However, $j^{-1}(25)$ is the height of an oak requiring an average of 25 gallons of water per day.

(b) $j(v) = 50$ means that an oak of height v requires 50 gallons of water daily. This statement can be rewritten $j^{-1}(50) = v$.

(c) This statement can be written $j(z) = p$, or as $j^{-1}(p) = z$.

(d)
- $j(2z)$ is the amount of water required by a tree that is twice average height.
- $2j(z)$ is enough water for two oak trees of average height. This expression equals $2p$.
- $j(z+10)$ is enough water for an oak tree ten feet taller than average.
- $j(z) + 10$ is the amount of water required by an oak of average height, plus 10 gallons. Thus, this expression equals $p + 10$.
- $j^{-1}(2p)$ is the height of an oak requiring $2p$ gallons of water.
- $j^{-1}(p+10)$ is the height of an oak requiring $p+10$ gallons of water.
- $j^{-1}(p) + 10$ is the height of an oak that is 10 feet taller than average. Thus, this expression equals $z + 10$.

40. Temperatures in this borehole are 3°C lower than at the same depth in the Belleterre borehole. See Table 6.20.

Table 6.20

d	25	50	75	100	125	150	175	200
$g(d)$	2.5	2.2	2.1	2.1	2.3	2.5	2.75	3

41. Temperatures in this borehole are the same as temperatures 5 meters deeper in the Belleterre borehole. See Table 6.21.

Table 6.21

d	20	45	70	95	120	145	170	195
$h(d)$	5.5	5.2	5.1	5.1	5.3	5.5	5.75	6

42. Temperatures in this borehole are the same as temperatures 10 meters less deep in the Belleterre borehole. See Table 6.22.

Table 6.22

d	35	60	85	110	135	160	185	210
$m(d)$	5.5	5.2	5.1	5.1	5.3	5.5	5.75	6

43. Temperatures in this borehole are 50% higher than temperatures at the same depth in the Belleterre borehole. See Table 6.23.

Table 6.23

d	25	50	75	100	125	150	175	200
$n(d)$	8.25	7.8	7.65	7.65	7.95	8.25	8.63	9

44. Temperatures in this borehole are the same as temperatures 20% less deep in the Belleterre borehole. See Table 6.24.

Table 6.24

d	31.25	62.5	93.75	125	156.25	187.5	218.75	250
$p(d)$	5.5	5.2	5.1	5.1	5.3	5.5	5.75	6

45. Temperatures in this borehole are 2°C warmer than temperatures 50% higher than temperatures at the same depth in the Belleterre borehole. See Table 6.25.

Table 6.25

d	25	50	75	100	125	150	175	200
$q(d)$	10.25	9.8	9.65	9.65	9.95	10.25	10.63	11

CHECK YOUR UNDERSTANDING

1. True. The graph of $g(x)$ is a copy of the graph of f shifted vertically up by three units.

2. False. The horizontal shift is two units to the right.

3. True. The graph is shifted down by $|k|$ units.

4. True.

5. True. The reflection across the x-axis of $y = f(x)$ is $y = -f(x)$.

6. False. For an odd function $f(x) = -f(-x)$.

7. False. The graphs of odd functions are symmetric about the origin.

8. True. Any point (x, y) on the graph of $y = f(x)$ reflects across the x-axis to the point $(x, -y)$, which lies on the graph of $y = -f(x)$.

9. True. Any point (x, y) on the graph of $y = f(x)$ reflects across the y-axis to the point $(-x, y)$, which lies on the graph of $y = f(-x)$.

10. True. Symmetry about the y-axis means that if any point (x, y) is on the graph of the function then $(-x, y)$ must also be on the graph of the function. If (x, y) is on the graph, then $y = f(x)$. Thus, if the graph is symmetric, we also know that $y = f(-x)$, so $f(x) = f(-x)$.

11. False. In the figure in the problem, it appears that $g(x) = f(x - 2) + 1$ because the graph is two units to the right and one unit up from the graph of f.

12. False. Substituting $(x - 2)$ in to the formula for g gives $g(x - 2) = (x - 2)^2 + 4 = x^2 - 4x + 4 + 4 = x^2 - 4x + 8$.

13. False. If $f(x) = x^2$, then $f(x + 1) = x^2 + 2x + 1 \neq x^2 + 1 = f(x) + 1$.

14. True. This looks like the absolute value function shifted right 1 unit and down 2 units.

15. True. The reflection across the x-axis of the graph of $f(x)$ has equation $y = -f(x)$, and a four unit upward shift of that graph has equation $y = -f(x) + 4 = -3^x + 4$.

16. False. If $q(p) = p^2 + 2p + 4$ then $-q(-p) = -((-p)^2 + 2(-p) + 4) = -(p^2 - 2p + 4) = -p^2 + 2p - 4$.

17. True.

18. True. Since

$$\text{Rate of change} = \frac{f(b) - f(a)}{b - a},$$

multiplying f by k multiplies the rate of change by k.

19. False. The graph of g has the same shape as f so it has not been stretched. It appears that $g(x) = -f(x+1) + 3$.

20. False. From the table, we have $g(-2) = -\frac{1}{2}f(-2+1) - 3 = -\frac{1}{2}f(-1) - 3 = -\frac{1}{2}(4) - 3 = -5$.

21. False. Consider $f(x) = x^2$. Shifting up first and then compressing vertically gives the graph of $g(x) = \frac{1}{2}(x^2 + 1) = \frac{1}{2}x^2 + \frac{1}{2}$. Compressing first and then shifting gives the graph of $h(x) = \frac{1}{2}x^2 + 1$.

22. False. In the graph, it appears that $g(x) = 3f(2x)$.

23. True. For $3f(2x) + 1$, at $x = -2$, we have $3f(2(-2)) + 1 = 3f(-4) + 1 = 3(1) + 1 = 4$. For $-2f(\frac{1}{2}x)$, at $x = 2$, we have $-2f(\frac{1}{2}(-2)) = -2f(-1) = -2(-2) = 4$.

CHAPTER SEVEN

Solutions for Section 7.1

Exercises

1. Graphs (I), (II), and (IV) appear to decribe period functions.

(I) This function appears periodic. The rapid variation overlays a slower variation that appears to repeat every 8 units. (It almost appears to repeat every 4 units, but there is subtle difference between consecutive 4-second intervals. Do you see it?)

(II) This function also appears period, again with a period of about 4 units. For instance, the x-intercepts appear to be evenly spaced, at approximately $-11, -7, -3, 1, 5, 9$, and the peaks are also evenly spaced, at $-9, -5, -1, 3, 7, 11$.

(III) This function does not appear periodic. For instance, the x-intercepts grow increasingly close together (when read from left to right).

(IV) At first glance this function might appear to vary unpredictably. But on closer inspection we see that the graph repeats the same pattern on the interval $-12 \leq x \leq 0$ and $0 \leq x \leq 12$.

(V) This function does not appear periodic. The peaks of the graph appear to rise slowly (when read from left to right), and the troughs appear to fall slowly.

(VI) This function does not appear to be periodic. The peaks and troughs of its graph seem to vary unpredictably, although they are more or less evenly spaced.

2. In the 12 o'clock position the person is at the top of the wheel, or 165 m above the ground.

3. In the 3 o'clock position, the person is midway between the top and bottom of the wheel. Since the diameter is 150 m, the radius is 75 m, so the person is 75 m below the top, or $165 - 75 = 90$ m above the ground.

4. In the 6 o'clock position, the person is at the bottom of the wheel. The diameter is 150 m, so the person is 150 m below the top, or $165 - 150 = 15$ m above the ground.

5. In the 9 o'clock position, the person is midway between the top and bottom of the wheel. Since the diameter is 150 m, the radius is 75 m, so the person is 75 m below the top, or $165 - 75 = 90$ m above the ground.

6. The period is approximately 4.

7. The graph appears to have a period of b. Every change in the x value of b brings us back to the same y value.

8. The period appears to be 3.

9. The period appears to be $41 - 1 = 40$.

Problems

10. After 9.25 minutes, the person is one fourth of the way through one rotation. Since the wheel is turning clockwise, this means she is in the 9 o'clock position, midway between the top and bottom of the wheel. Since the diameter is 150 m, the radius is 75 m, so the person is 75 m below the top, or $165 - 75 = 90$ m above the ground.

11. After 18.5 minutes, the person is halfway through one rotation. This means she is in the 12 o'clock position or 165 m above the ground.

12. After 27.75 minutes, the person is three-fourths of the way through one rotation. Since the wheel is turning clockwise, this means she is in the 3 o'clock position, midway between the top and bottom of the wheel. Since the diameter is 150 m, the radius is 75 m, so the person is 75 m below the top, or $165 - 75 = 90$ m above the ground.

13. After 37 minutes, the person has completed one rotation. This means she is in the 6 o'clock position, at the bottom of the wheel. The diameter is 150 m, so the person is 150 m below the top, or $165 - 150 = 15$ m above the ground.

14. The wheel will complete two full revolutions after 20 minutes, so the function is graphed on the interval $0 \leq t \leq 20$. See Figure 7.1.

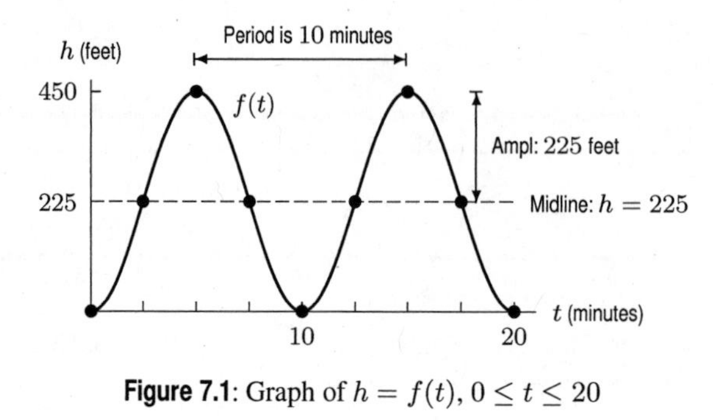

Figure 7.1: Graph of $h = f(t)$, $0 \leq t \leq 20$

15. The wheel will complete two full revolutions after 20 minutes, and the height ranges from $h = 0$ to $h = 600$. So the function is graphed on the interval $0 \leq t \leq 20$. See Figure 7.2.

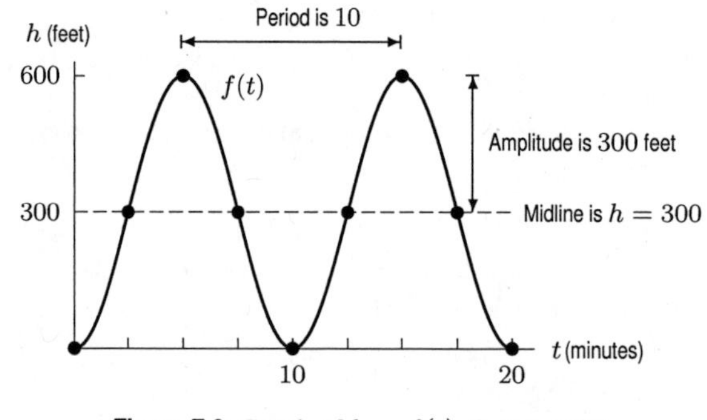

Figure 7.2: Graph of $h = f(t)$, $0 \leq t \leq 20$

16. The wheel will complete two full revolutions after 10 minutes. See Figure 7.3.

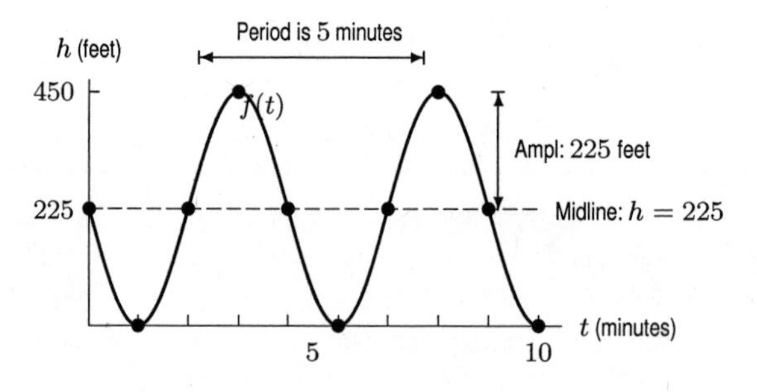

Figure 7.3: Graph of $h = f(t)$, $0 \leq t \leq 10$

17. See Figure 7.4.

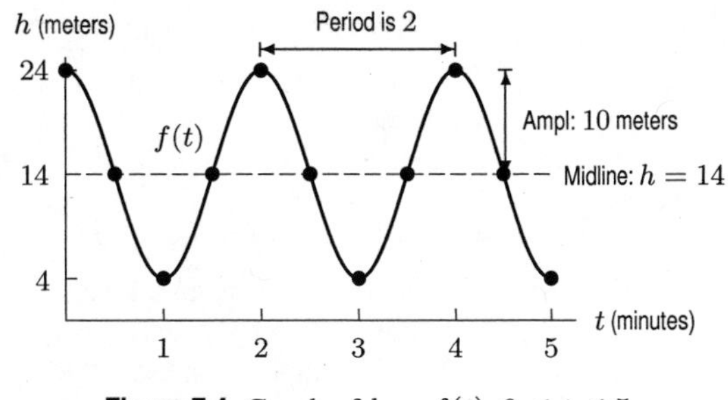

Figure 7.4: Graph of $h = f(t), 0 \leq t \leq 5$

18. See Figure 7.5.

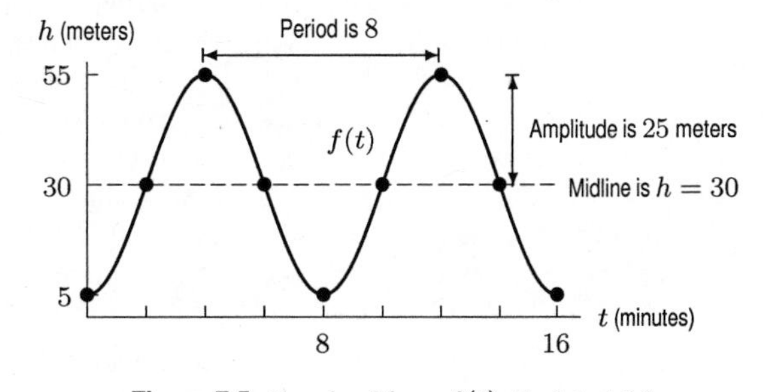

Figure 7.5: Graph of $h = f(t), 0 \leq t \leq 16$

19. See Figure 7.6.

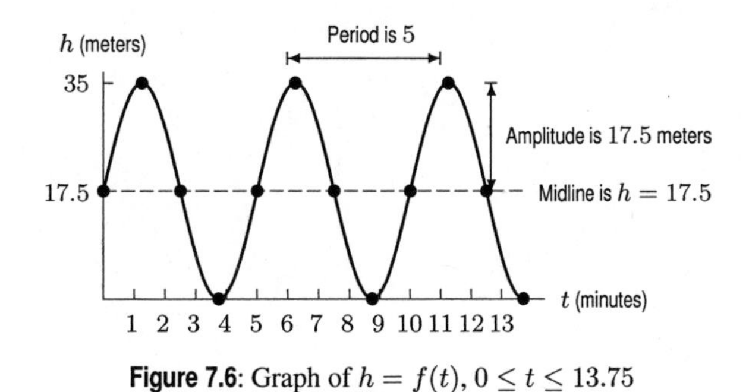

Figure 7.6: Graph of $h = f(t), 0 \leq t \leq 13.75$

20. At $t = 0$, we see $h = 20$, so you are level with the center of the wheel. Your initial position is at three o'clock (or nine o'clock) and initially you are rising. On the interval $0 \leq t \leq 7$ the wheel completes seven fourths of a revolution. Therefore, if p is the period, we know that

$$\frac{7}{4}p = 7$$

which gives $p = 4$. This means that the Ferris wheel takes 4 minutes to complete one full revolution. The minimum value of the function is $h = 5$, which means that you get on and get off of the wheel from a 5 meter platform. The maximum height above the midline is 15 meters, so the wheel's diameter is 30 meters. Notice that the wheel completes a total 2.75 cycles. Since each period is 4 minutes long, you ride the wheel for $4(2.75) = 11$ minutes.

21. Your initial position is twelve o'clock, since at $t = 0$, the value of h is at its maximum of 35. The period is 4 because the wheel completes one cycle in 4 minutes. The diameter is 30 meters and the boarding platform is 5 meters above ground. Because you go through 2.5 cycles, the length of time spent on the wheel is 10 minutes.

22. At $t = 0$, we see $h = 20$ m, and you are at the midline, so your initial height is level with the center of the wheel. Your initial position is at the three o'clock (or nine o'clock) position, and you are moving upward at $t = 0$. The amplitude of this function is 20, which means that the wheel's diameter is 40 meters. The minimum value of the function is $h = 0$, which means you board and get off the wheel at ground level. The period of this function is 5, which means that it takes 5 minutes for the wheel to complete one full revolution. Notice that the function completes 2.25 periods. Since each period is 5 minutes long, this means you ride the wheel for $5(2.25) = 11.25$ minutes.

23. At $t = 0$, we see $h = 20$ m, and you are at the midline, so your initial height is level with the center of the wheel. Your initial position is at the three o'clock (or the nine o'clock) position and at first your height is decreasing, so you are descending. The amplitude of this function is 20, which means that the wheel's diameter is 40 meters. The minimum value of the function is $h = 0$, which means you board and get off the wheel at ground level. The period of this function is 5, which means that it takes 5 minutes for the wheel to complete one full revolution. Notice that the function completes 2.25 periods. Since each period is 5 minutes long, this means you ride the wheel for $5(2.25) = 11.25$ minutes.

24. The midline of f is $d = 10$. The period of f is 1, the amplitude 4 cm, and its minimum and maximum values are 6 cm and 14 cm, respectively. The fact that $f(t)$ is wave-shaped means that the spring is bobbing up and down, or *oscillating*. The fact that the period of f is 1 means that it takes the weight one second to complete one oscillation and return to its original position. Studying the graph, we see that it takes the weight 0.25 seconds to move from its initial position at the midline to its maximum at $d = 14$, where it is farthest from the ceiling (and the spring is at its maximum extension). It takes another 0.25 seconds to return to its initial position at $d = 10$ cm. It takes another 0.25 seconds to rise up to its closest distance from the ceiling at $d = 6$ (the minimum extension of the spring). In 0.25 seconds more it moves back down to its initial position at $d = 10$. (This sequence of motions by the weight, completed in one second, represents one full oscillation.) Since Figure 7.10 of the text gives 3 full periods of $f(t)$, it represents the 3 complete oscillations made by the weight in 3 seconds.

25. The amplitude, period, and midline are the same for Figures 7.10 and 7.11 in the text. In Figure 7.11, the weight is initially moving upward toward the ceiling, since *d*, the distance from the ceiling, begins to decrease at $t = 0$, whereas in Figure 7.10, d begins to increase at $t = 0$. Thus, the motion described in Figure 7.11 must have resulted from pulling the weight away from the ceiling at $t = -0.25$, whereas the motion described by Figure 7.10 must have resulted from pushing the weight toward the ceiling at $t = -0.25$.

26.

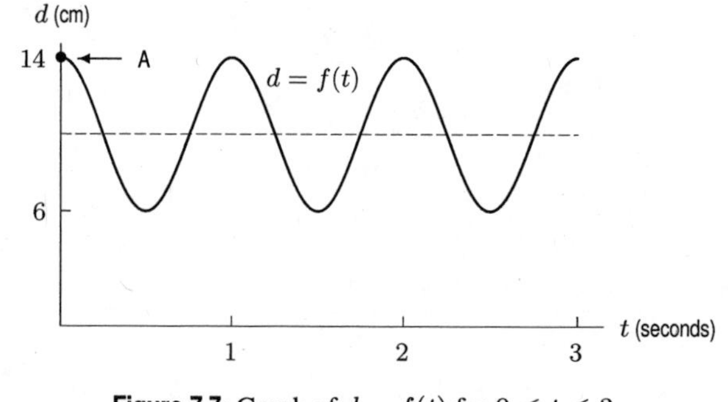

Figure 7.7: Graph of $d = f(t)$ for $0 \leq t \leq 3$

Since the weight is released at $d = 14$ cm when $t = 0$, it is initially at the point in Figure 7.7 labeled A. The weight will begin to oscillate in the same fashion as described by Figures 7.10 and 7.11. Thus, the period, amplitude, and midline for Figure 7.7 are the same as for Figures 7.10 and 7.11 in the text.

27. **(a)** Weight B, because the midline is $d = 10$, compared to $d = 20$ for weight A. This means that when the spring is not oscillating, weight B is 10 cm from the ceiling, while weight A is 20 cm from the ceiling.

(b) Weight A, because its amplitude is 10 cm, compared to the amplitude of 5 cm for weight B.

(c) Weight A, because its period is 0.5, compared to the period of 2 for weight B. This means that it takes weight A only half a second to complete one oscillation, whereas weight B completes one oscillation in 2 seconds.

28. **(a)** Two possible answers are shown in Figures 7.8 and 7.9.

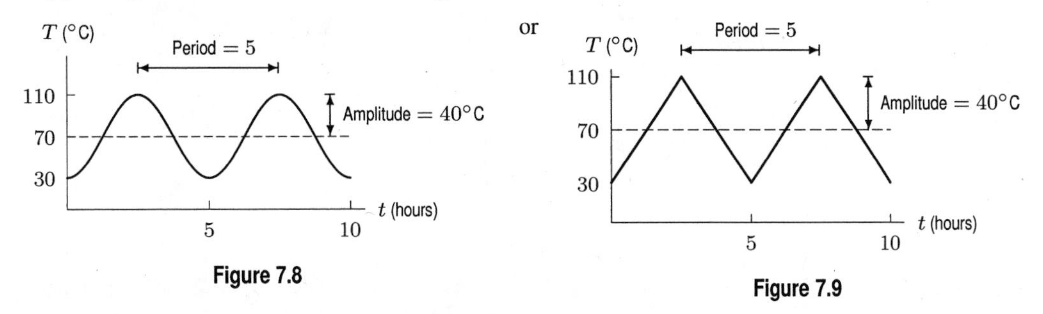

Figure 7.8

Figure 7.9

(b) The period is 5 hours. This is the time required for the temperature to cycle from 30° to 110° and back to 30°. The midline, or average temperature, is $T = (110 + 30)/2 = 70°$. The amplitude is 40° since this is the amount of temperature variation (up or down) from the average.

29. Notice that the function is only approximately periodic. See Figure 7.10.

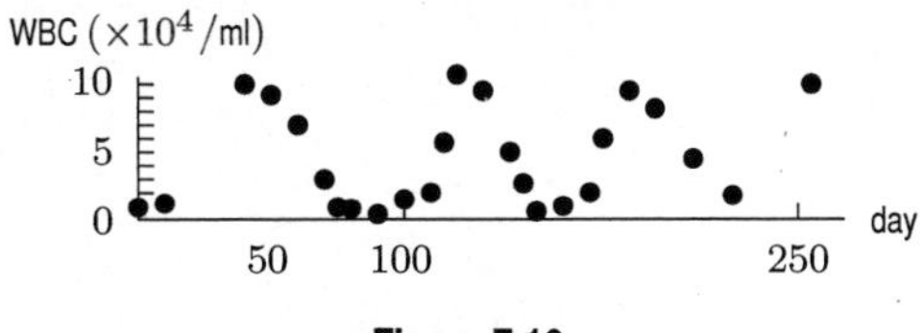

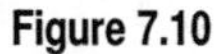

Figure 7.10

The midline is half way between the maximum and minimum WBC values.

$$y = \frac{(10.7 + 0.4)}{2} = 5.55.$$

The amplitude is the difference between the maximum and midline, so $A = 5.15$. The period is the length of time from peak to peak. Measuring between successive peaks gives $p_1 = 120 - 40 = 80$ days; $p_2 = 185 - 120 = 65$ days; $p_3 = 255 - 185 = 70$ days. Using the average of the three periods we get $p \approx 72$ days.

Solutions for Section 7.2

Exercises

1.

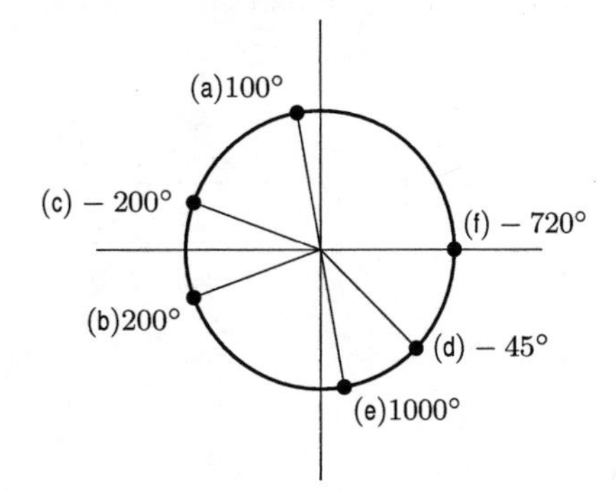

Figure 7.11

(a) $(\cos 100°, \sin 100°) = (-0.174, 0.985)$

(b) $(\cos 200°, \sin 200°) = (-0.940, -0.342)$
(c) $(\cos(-200°), \sin(-200°)) = (-0.940, 0.342)$
(d) $(\cos(-45°), \sin(-45°)) = (0.707, -0.707)$
(e) $(\cos 1000°, \sin 1000°) = (0.174, -0.985)$
(f) $(\cos 720°, \sin 720°) = (1, 0)$

2. If we go around four times, we make four full circles, which is $360° \cdot 4 = 1440$ degrees.

3. If we go around two times, we make two full circles, which is $360° \cdot 2 = 720$ degrees. Since we're going around in the negative (clockwise) direction, we have -720 degrees.

4. If we go around 16.4 times, we make 16.4 full circles, which is $360° \cdot 16.4 = 5904$ degrees.

5. To locate the points D, E, and F, we mark off their respective angles, $-90°$, $-135°$, and $-225°$, by measuring these angles from the positive x-axis in the clockwise direction. See Figure 7.12.

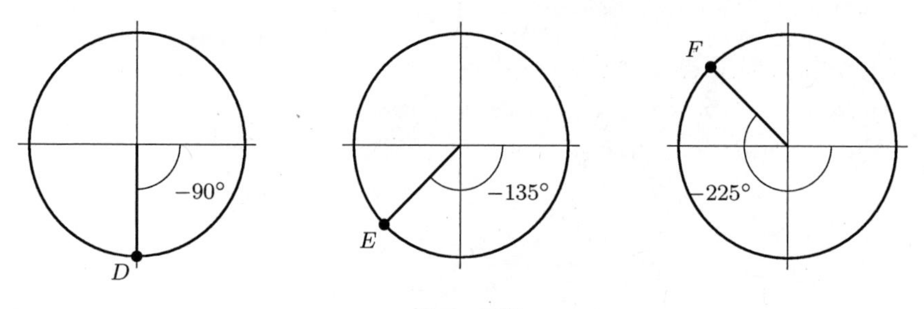

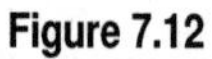

Figure 7.12

$D = (0, -1)$, $E = (-0.707, -0.707)$, $F = (-0.707, 0.707)$

6. To locate the points P, Q, and R, we mark off their respective angles, $540°$, $-180°$, and $450°$, by measuring these angles from the positive x-axis in the counterclockwise direction if the angle is positive and in the clockwise direction if the angle is negative. See Figure 7.13.

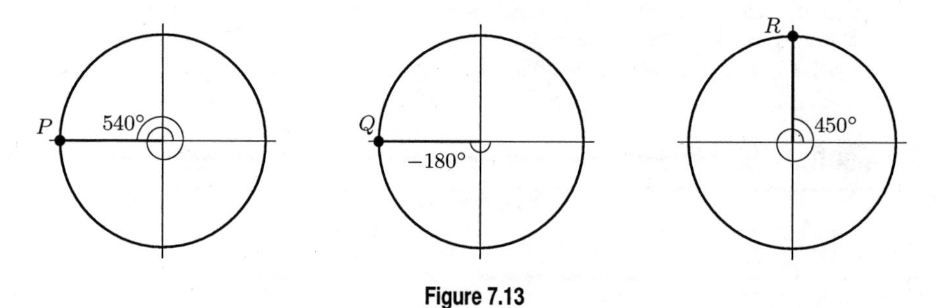

Figure 7.13

$P = (-1, 0)$, $Q = (-1, 0)$, $R = (0, 1)$

7. **(a)** We have
$$\sin\theta = \frac{\text{Side opposite}}{\text{Hypotenuse}} = \frac{24}{26} = 0.923.$$

(b) We have
$$\cos\theta = \frac{\text{Side adjacent}}{\text{Hypotenuse}} = \frac{10}{26} = 0.385.$$

8. **(a)** We have
$$\sin\theta = \frac{\text{Side opposite}}{\text{Hypotenuse}} = \frac{6}{\sqrt{117}} = 0.555.$$

(b) We have
$$\cos\theta = \frac{\text{Side adjacent}}{\text{Hypotenuse}} = \frac{9}{\sqrt{117}} = 0.832.$$

9. We use the Pythagorean theorem to find the length of the hypotenuse:

$$\begin{aligned}\text{Hypotenuse}^2 &= (0.1)^2 + (0.2)^2 = 0.01 + 0.04 = 0.05\\ \text{Hypotenuse} &= \sqrt{0.05}.\end{aligned}$$

(a) We have

$$\sin\theta = \frac{\text{Side opposite}}{\text{Hypotenuse}} = \frac{0.1}{\sqrt{0.05}} = 0.447.$$

(b) We have

$$\cos\theta = \frac{\text{Side adjacent}}{\text{Hypotenuse}} = \frac{0.2}{\sqrt{0.05}} = 0.894.$$

10. We use the Pythagorean theorem to find the length of the opposite side:

$$\begin{aligned}12^2 + (\text{Opposite side})^2 &= 32^2\\ 144 + (\text{Opposite side})^2 &= 1024\\ (\text{Opposite side})^2 &= 880\\ \text{Opposite side} &= \sqrt{880}.\end{aligned}$$

(a) We have

$$\sin\theta = \frac{\text{Side opposite}}{\text{Hypotenuse}} = \frac{\sqrt{880}}{32} = 0.927.$$

(b) We have

$$\cos\theta = \frac{\text{Side adjacent}}{\text{Hypotenuse}} = \frac{12}{32} = 0.375.$$

Problems

11. The car on the Ferris wheel starts at the 3 o'clock position. Let's suppose that you see the wheel rotating counterclockwise. (If not, move to the other side of the wheel.)

The angle $\phi = 420°$ indicates a counterclockwise rotation of the Ferris wheel from the 3 o'clock position all the way around once ($360°$), and then two-thirds of the way back up to the top (an additional $60°$). This leaves you in the 1 o'clock position, or at the angle $60°$.

A negative angle represents a rotation in the opposite direction, that is clockwise. The angle $\theta = -150°$ indicates a rotation from the 3 o'clock position in the clockwise direction, past the 6 o'clock position and two-thirds of the way up to the 9 o'clock position. This leaves you in the 8 o'clock position, or at the angle $210°$. (See Figure 7.14.)

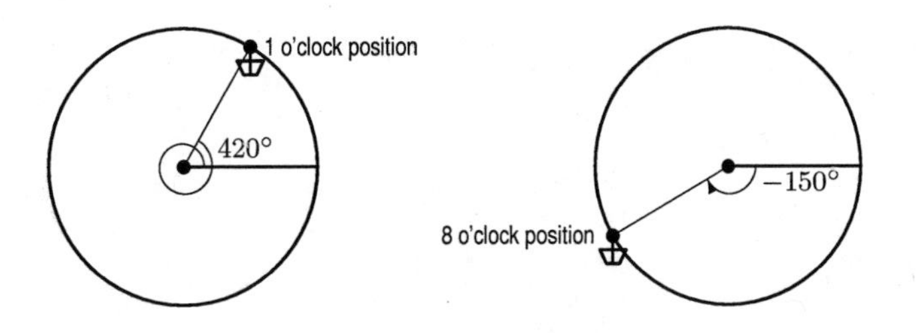

Figure 7.14: The positions and displacements on the Ferris wheel described by $420°$ and $-150°$

12. See Figure 7.15.

(a) $\cos 240°$ is negative, so we need an angle in the second quadrant with the same x-coordinate since $240° = 180° + 60°$. This angle is $180° - 60° = 120°$.

(b) $\sin 240°$ is negative, so we need an angle in the fourth quadrant with the same y-coordinate. This angle is $360° - 60° = 300°$.

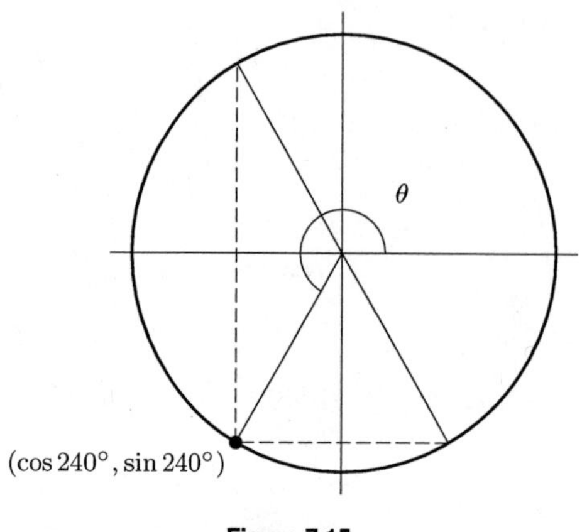

Figure 7.15

13. See Figure 7.16.

(a) $\cos 53°$ is positive, so we need an angle in the fourth quadrant with the same x-coordinate. This angle is $360° - 53° = 307°$.

(b) $\sin 53°$ is positive, so we need an angle in the second quadrant with the same y-coordinate. This angle is $180° - 53° = 127°$.

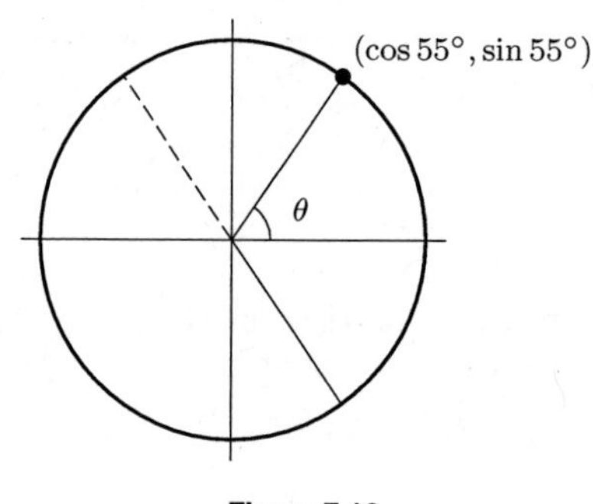

Figure 7.16

14. (a) As we see from Figure 7.17, the angle 135° specifies a point P' on the unit circle directly across the y-axis from the point P. Thus, P' has the same y-coordinate as P, but its x-coordinate is opposite in sign to the x-coordinate of P. Therefore, $\sin 135° = 0.707$, and $\cos 135° = -0.707$.

(b) As we see from Figure 7.18, the angle 285° specifies a point Q' on the unit circle directly across the x-axis from the point Q. Thus, Q' has the same x-coordinate as Q, but its y-coordinate is opposite in sign to the y-coordinate of Q. Therefore, $\sin 285° = -0.966$, and $\cos 285° = 0.259$.

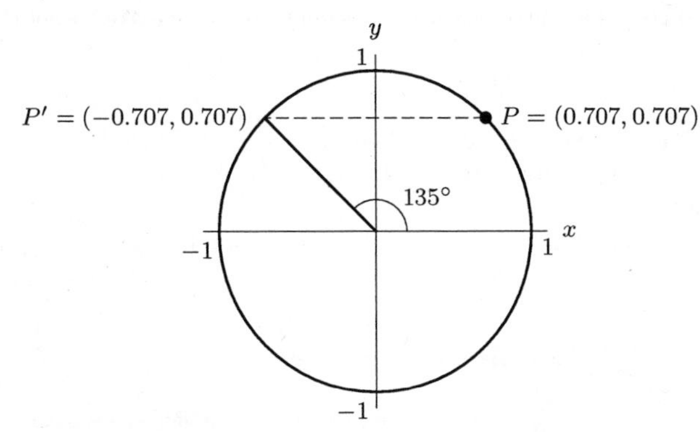

Figure 7.17: The sine and cosine of 135° can be found by referring to the sine and cosine of 45°

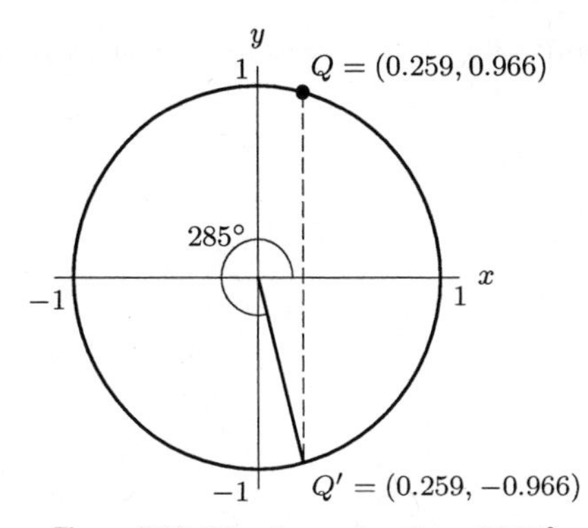

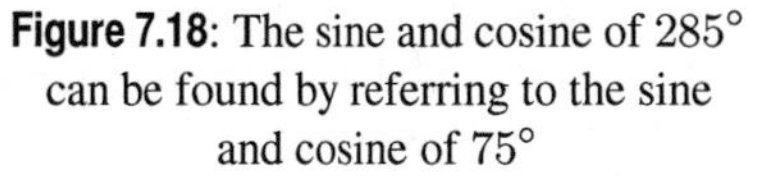

Figure 7.18: The sine and cosine of 285° can be found by referring to the sine and cosine of 75°

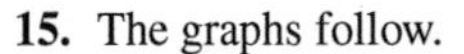

15. The graphs follow.

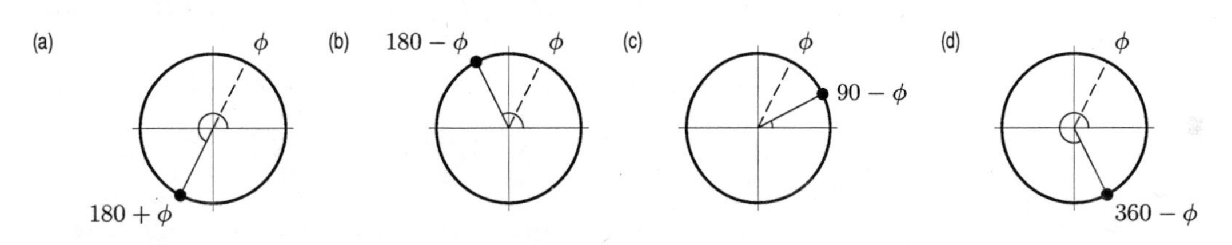

16. **(a)** $\sin(\theta + 360^\circ) = \sin\theta = a$, since the sine function is periodic with a period of 360°.

(b) $\sin(\theta + 180^\circ) = -a$. (A point on the unit circle given by the angle $\theta + 180^\circ$ diametrically opposite the point given by the angle θ. So the y-coordinates of these two points are opposite in sign, but equal in magnitude.)

(c) $\cos(90^\circ - \theta) = \sin\theta = a$. This is most easily seen from the right triangles in Figure 7.19.

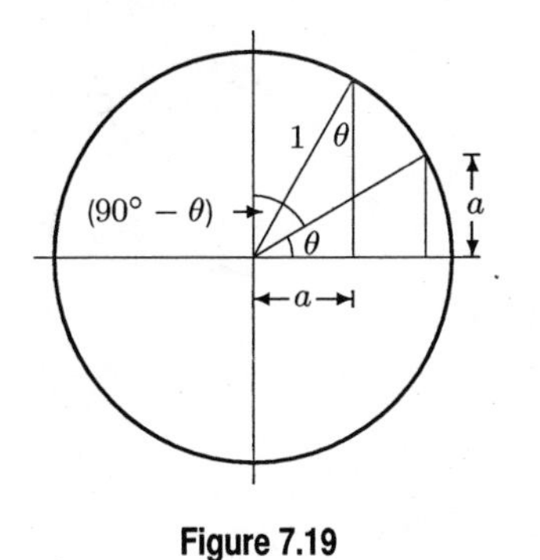

Figure 7.19

(d) $\sin(180^\circ - \theta) = a$. (A point on the unit circle given by the angle $180^\circ - \theta$ has a y-coordinate equal to the y-coordinate of the point on the unit circle given by θ.)

(e) $\sin(360^\circ - \theta) = -a$. (A point on the unit circle given the the angle $360^\circ - \theta$ has a y-coordinate of the same magnitude as the y-coordinate of the point on the unit circle given by θ, but is of opposite sign.)

(f) $\cos(270^\circ - \theta) = -\sin\theta = -a$.

17. Given the angle θ, draw a line l through the origin making an angle θ with the x-axis. Go counterclockwise if $\theta > 0$ and clockwise if $\theta < 0$, wrapping around the unit circle more than once if necessary. Let $P = (x, y)$ be the point where l intercepts the unit circle. Then the definition of sine is that $\sin\theta = y$.

18. **(a)** Since the four panels divide a full rotation or 360° into four equal spaces, the angle between two adjacent panels is

$$\frac{360^\circ}{4} = 90^\circ.$$

(b) The angle created by rotating a panel from B to A is equal to the angle between each panel, or 90°.
(c) Point B is directly across from point D. So the angle between the two is 180°.
(d) If the door moves from B to D, the angle of rotation is 180°.
(e) Each person, whether entering or leaving, must rotate the door by 180°. Thus the total rotation is $(3+5)(180^\circ) = 8(180^\circ) = 1440^\circ$. Since $1440^\circ = 4(360^\circ)$ the rotation is equivalent to 0°. Thus, the panel at point A ends up at point A.

19. **(a)** The five panels split the circle into five equal parts, so the angle between each panel is $360^\circ/5 = 72^\circ$.
(b) Point B is directly across the circle from D, so 180°.
(c) The angle from A to D is the same as the angle from B to C, and the BC angle is the angle between panels, which is 72°. So moving the panel between A and D gives an angle of $(72^\circ)/2 = 36^\circ$. The panel then goes from point D to point B spanning another 180°. Thus in total the panel traveled $36^\circ + 180^\circ = 216^\circ$.

20. Since 45° is half 90°, the point P in Figure 7.20 lies on the line $y = x$. Substituting $y = x$ into the equation of the circle, $x^2 + y^2 = 1$, gives $x^2 + x^2 = 1$. Solving for x, we get

$$\begin{aligned} 2x^2 &= 1 \\ x^2 &= \frac{1}{2} \\ x &= \pm\sqrt{\frac{1}{2}} = \pm\frac{1}{\sqrt{2}}. \end{aligned}$$

Since P is in the first quadrant, x and y are positive, so

$$x = \cos 45^\circ = \frac{1}{\sqrt{2}} \quad \text{and} \quad y = \sin 45^\circ = \frac{1}{\sqrt{2}}.$$

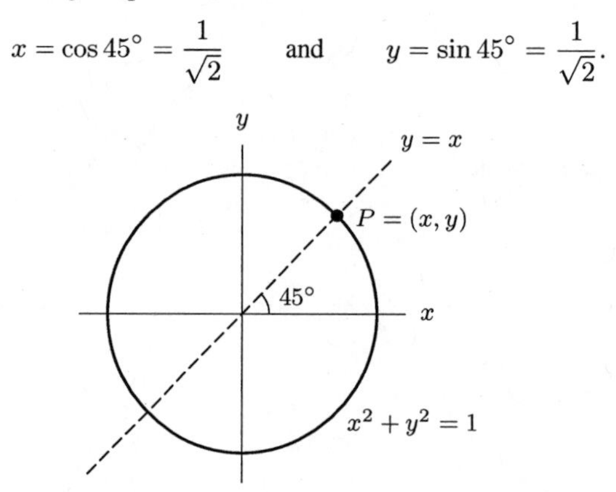

Figure 7.20: Calculating $\cos 45^\circ$ and $\sin 45^\circ$

21. **(a)** All sides have length 1, since triangle ΔKOL is an equilateral triangle. This is because all three angles are 60°.
(b) Since triangles ΔOPK and ΔOPL are congruent, the length from K to P must be half the length of KL. Thus the length of KP is $\frac{1}{2}(1) = 1/2$.
(c) Using the Pythagorean theorem we find that

$$\begin{aligned} \text{Distance from } O \text{ to } P &= \sqrt{(\text{Length of hypotenuse})^2 - (\text{Distance from } K \text{ to } P)^2} \\ &= \sqrt{1^2 - (1/2)^2} \\ &= \sqrt{\frac{3}{4}} \\ &= \frac{\sqrt{3}}{2}. \end{aligned}$$

(d) Since $OP = \sqrt{3}/2$ and $KP = 1/2$, the coordinates of K are $(\sqrt{3}/2, 1/2)$.
(e) It follows from part (d) that the cosine of 30° is $\sqrt{3}/2$ while the sine of 30° is $1/2$.
(f) In triangle KOP, we have $KP = 1/2$ and $OP = \sqrt{3}/2$. So

$$\sin 60^\circ = \frac{\text{Opposite}}{\text{Hypotenuse}} = \frac{\sqrt{3}/2}{1} = \frac{\sqrt{3}}{2}$$
$$\cos 60^\circ = \frac{\text{Adjacent}}{\text{Hypotenuse}} = \frac{1/2}{1} = \frac{1}{2}.$$

22.

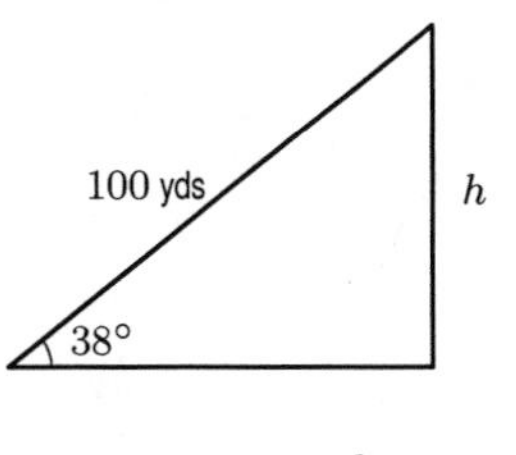

$$\sin(38^\circ) = \frac{h}{100},$$

which implies that

$$h = 100\sin(38^\circ) \approx 61.566 \text{ yards} \quad \text{or} \quad 184.698 \text{ feet.}$$

23. Let d be the distance from the base of the ladder to the wall; see Figure 7.21. Then, $d/3 = \cos\alpha$, so $d = 3\cos\alpha$ meters.

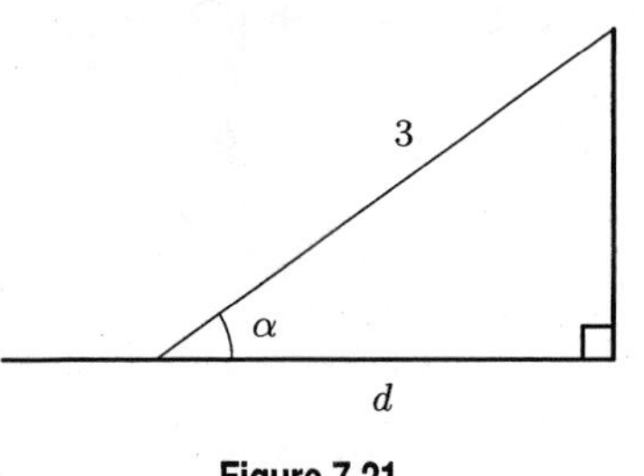

Figure 7.21

24. (a) Since $\sin 45^\circ = h/125$, we have $h = 125\sin 45^\circ \approx 88.388$ feet.
(b) Since $\sin 30^\circ = h/125$, we have $h = 125\sin 30^\circ = 62.5$ feet.
(c) Since $\cos 45^\circ = c/125$, we have $c = 125\cos(45^\circ) \approx 88.388$ feet.
Since $\cos 30^\circ = d/125$, we have $d \approx 108.253$ feet.

Solutions for Section 7.3

Exercises

1. Since the maximum value of the function is 3 and the minimum is 1, its midline is $y = 2$, and its amplitude is 1.
2. Since the maximum value of the function is 1 and the midline appears to be $y = -2$, the amplitude is 3.
3. Since the minimum value of the function is -10 and the midline appears to be $y = -3$, the amplitude is 7.
4. Since the middle of the clock's face is at 185 cm, the midline is at $h(t) = 185$ cm, and since the hand is 15 cm long, the amplitude is 15 cm.

5. Since the middle of the clock's face is at 223 cm, the midline is at $i(t) = 223$ cm, and since the hand is 20 cm long, the amplitude is 20 cm.

6.

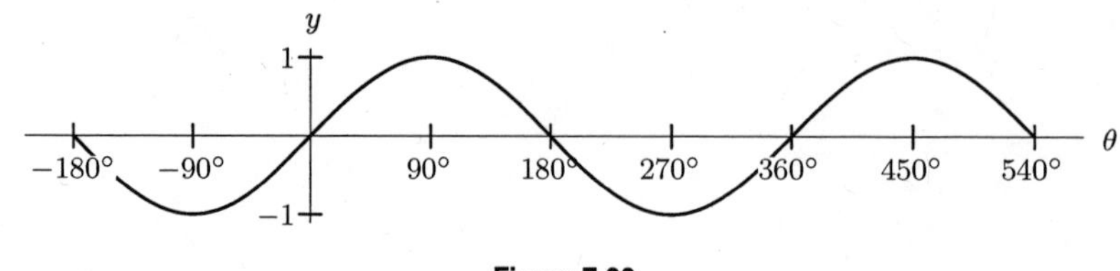

Figure 7.22

(a) (i) For $0 < \theta < 180°$ and $360° < \theta < 540°$ the function $\sin\theta$ is positive.
(ii) It is increasing for $-90° < \theta < 90°$ and $270° < \theta < 450°$.
(iii) For $-180° < \theta < 0$ and $180° < \theta < 360°$ it is concave up.
(b) The function appears to have the maximum rate of increase at $\theta = 0°, 360°$.

7. Since the x-coordinate is $r\cos\theta$ and the y-coordinate is $r\sin\theta$ and $r = 3.8$ and $\theta = 90°$, the point is $(3.8\cos 90, 3.8\sin 90) = (0, 3.8)$.

8. Since the x-coordinate is $r\cos\theta$ and the y-coordinate is $r\sin\theta$ and $r = 3.8$ and $\theta = 180°$, the point is $(3.8\cos 180, 3.8\sin 180) = (-3.8, 0)$.

9. Since the x-coordinate is $r\cos\theta$ and the y-coordinate is $r\sin\theta$ and $r = 3.8$ and $\theta = -180°$, the point is $(3.8\cos(-180), 3.8\sin(-180)) = (-3.8, 0)$.

10. Since the x-coordinate is $r\cos\theta$ and the y-coordinate is $r\sin\theta$ and $r = 3.8$ and $\theta = -90°$, the point is $(3.8\cos(-90), 3.8\sin(-90)) = (0, -3.8)$.

11. Since the x-coordinate is $r\cos\theta$ and the y-coordinate is $r\sin\theta$ and $r = 3.8$ and $\theta = -270°$, the point is $(3.8\cos(-270), 3.8\sin(-270)) = (0, 3.8)$.

12. Since the x-coordinate is $r\cos\theta$ and the y-coordinate is $r\sin\theta$ and $r = 3.8$ and $\theta = -540°$, the point is $(3.8\cos(-540), 3.8\sin(-540)) = (-3.8, 0)$.

13. Since the x-coordinate is $r\cos\theta$ and the y-coordinate is $r\sin\theta$ and $r = 3.8$ and $\theta = 1426°$, the point is $(3.8\cos 1426, 3.8\sin 1426) = (3.687, -0.919)$.

14. Since the x-coordinate is $r\cos\theta$ and the y-coordinate is $r\sin\theta$ and $r = 3.8$ and $\theta = 1786°$, the point is $(3.8\cos 1786, 3.8\sin 1786) = (3.687, -0.919)$.

15. Since the x-coordinate is $r\cos\theta$ and the y-coordinate is $r\sin\theta$ and $r = 3.8$, the point is $(3.8\cos 45, 3.8\sin 45) = (3.8\sqrt{2}/2, 3.8\sqrt{2}/2) = (2.687, 2.687)$.

16. Since the x-coordinate is $r\cos\theta$ and the y-coordinate is $r\sin\theta$ and $r = 3.8$, the point is $(3.8\cos 135, 3.8\sin 135) = (-3.8\sqrt{2}/2, 3.8\sqrt{2}/2) = (-2.687, 2.687)$.

17. Since the x-coordinate is $r\cos\theta$ and the y-coordinate is $r\sin\theta$ and $r = 3.8$, the point is $(3.8\cos 225, 3.8\sin 225) = (-3.8\sqrt{2}/2, -3.8\sqrt{2}/2) = (-2.687, -2.687)$.

18. Since the x-coordinate is $r\cos\theta$ and the y-coordinate is $r\sin\theta$ and $r = 3.8$, the point is $(3.8\cos 315, 3.8\sin 315) = (3.8\sqrt{2}/2, -3.8\sqrt{2}/2) = (2.687, -2.687)$.

19. Since the x-coordinate is $r\cos\theta$ and the y-coordinate is $r\sin\theta$ and $r = 3.8$, the point is $(3.8\cos(-10), 3.8\sin(-10)) = (3.742, -0.660)$.

20. Since the x-coordinate is $r\cos\theta$ and the y-coordinate is $r\sin\theta$ and $r = 3.8$, the point is $(3.8\cos(-20), 3.8\sin(-20)) = (3.571, -1.300)$.

21.

$$x = r\cos\theta = 10\cos 210° = 10(-\sqrt{3}/2) = -5\sqrt{3}$$

and

$$y = r\sin\theta = 10\sin 210° = 10(-1/2) = -5,$$

so the coordinates of W are $(-5\sqrt{3}, -5)$.

Problems

22. Judging from the figure:

- The curve looks the same from $t = 0$ to $t = 8$ as from $t = 8$ to $t = 16$, so it repeats with a period of 8.
- The midline is the dashed horizontal line $y = 30$.
- The vertical distance from the first peak to the midline is 20, so the amplitude is 20.

23. Judging from the figure:

- The curve looks the same from $t = 0$ to $t = 50$ as from $t = 50$ to $t = 100$, so it repeats with a period of 50.
- The midline is the dashed horizontal line $y = 12$.
- The vertical distance from the first peak to the midline is 5, so the amplitude is 5.

24. Judging from the figure:

- The curve looks the same from $t = 0$ to $t = 0.7$ as from $t = 0.7$ to $t = 1.4$, so it repeats with a period of 0.7.
- The midline is the dashed horizontal line $y = 0.05$.
- The vertical distance from the first peak to the midline is 0.03, so the amplitude is 0.03.

25. Judging from the figure:

- The curve looks the same from $t = -24$ to $t = 0$ as from $t = 0$ to $t = 24$ and as from $t = 24$ to $t = 48$, so it repeats with a period of 24.
- The midline is the dashed horizontal line $y = -500$.
- The vertical distance from the first peak to the midline is 2000, so the amplitude is 2000.

26. Judging from the figure:

- The curve looks the same from $t = 0$ to $t = 0.5$ as from $t = 0.5$ to $t = 1$, so it repeats with a period of 0.5.
- The midline is the dashed horizontal line $y = 0.5$.
- The vertical distance from the first peak to the midline is 0.5, so the amplitude is 0.5.

27. Judging from the figure:

- The curve looks the same from $t = -21$ to $t = 4$ as from $t = 4$ to $t = 29$ and as from $t = 29$ to $t = 54$, so it repeats with a period of 25.
- The midline is the dashed horizontal line $y = 30$.
- The vertical distance from the first peak to the midline is 25, so the amplitude is 25.

28. $f(x) = (\sin x) + 1$
$g(x) = (\sin x) - 1$

29. $g(x) = \cos x$, $a = 90°$ and $b = 1$.

30. Since $\sin\theta$ is the y-coordinate of a point on the unit circle, its height above the x-axis can never be greater than 1. Otherwise the point would be outside the circle. See Figure 7.23.

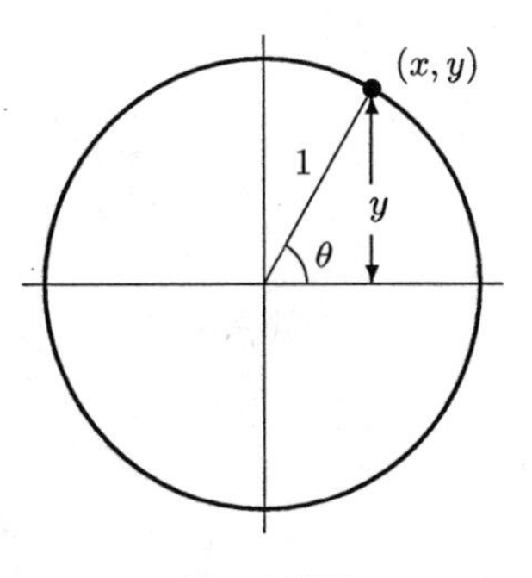

Figure 7.23

31. We can sketch these graphs using a calculator or computer. Figure 7.24 gives a graph of $y = \sin\theta$, together with the graphs of $y = 0.5\sin\theta$ and $y = -2\sin\theta$, where θ is in degrees and $0 \le \theta \le 360°$.

These graphs are similar but not the same. The amplitude of $y = 0.5\sin\theta$ is 0.5 and the amplitude of $y = -2\sin\theta$ is 2. The graph of $y = -2\sin\theta$ is vertically reflected relative to the other two graphs. These observations are consistent with the fact that the constant A in the equation

$$y = A\sin\theta$$

may result in a vertical stretching or shrinking and/or a reflection over the x-axis. Note that all three graphs have a period of 360°.

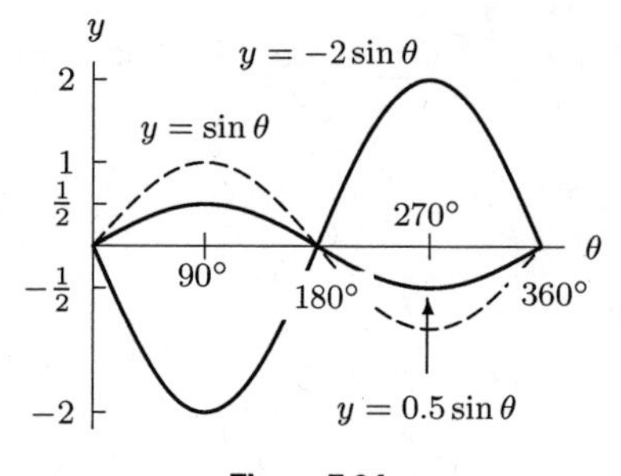

Figure 7.24

32. If the circle were centered at $(0, 0)$ we would see immediately that $x = 5\cos\theta$. The shift up 7 units has no effect on the x-value but the shift 6 units left means $x = 5\cos\theta - 6$. Thus $f(\theta) = 5\cos\theta - 6$. We can check this by plugging in convenient θ-values. For instance, $f(90°) = -6$ makes sense because it is the 12 o'clock position on the circle.

33. $f(x) = \sin(x + 90°)$
$g(x) = \sin(x - 90°)$

34. The radius is 10 meters. So when the seat height is 15 meters, the seat will be 5 meters above the horizontal line through the center of the wheel. This produces an angle whose sine is $5/10 = 1/2$, which we know is the angle 30°, or the 2 o'clock position. This situation is shown in Figure 7.25:

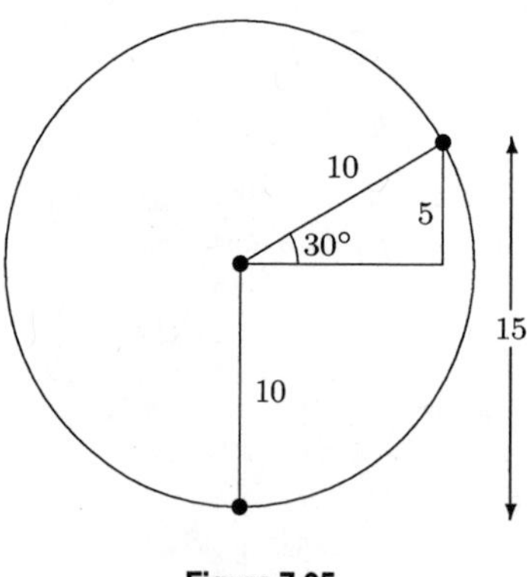

Figure 7.25

The seat is above 15 meters when it is between the 2 o'clock and 10 o'clock positions. This happens $4/12 = 1/3$ of the time, or for $4/3$ minutes each revolution.

35. See Figure 7.26. Since the diameter is 120 mm, the radius is 60 mm. The coordinates of the outer edge point, A, on the x-axis is $(60, 0)$. Similarly the inner edge at point B has coordinates $(7.5, 0)$.

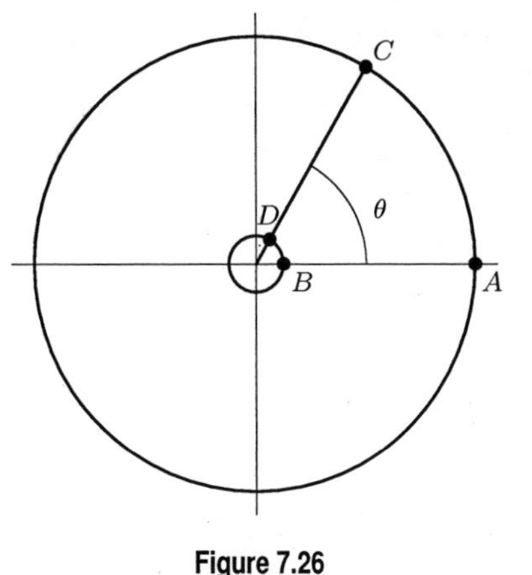

Figure 7.26

Points C and D are at an angle θ from the x-axis and have coordinates of the form $(r\cos\theta, r\sin\theta)$. For the outer edge, $r = 60$ so $C = (60\cos\theta, 60\sin\theta)$. The inner edge has $r = 7.5$, so $D = (7.5\cos\theta, 7.5\sin\theta)$.

36. **(a)** The midline of the function is 50 meters and the amplitude is 45 meters. So the highest point on the ride occurs when the person on the Ferris wheel is in the 12 o'clock position, where $\theta = 90^\circ$. Since $\sin 90^\circ = 1$, $f(90) = 50 + 45\sin 90^\circ = 95$ m. Similarly, the lowest point occurs when $\theta = 270^\circ$, which is in the 6 o'clock position: $f(270) = 50 + 45\sin 270^\circ = 5$ m, since $\sin 270^\circ = -1$.

(b) Since the amplitude of the height function is 45 meters, this is also the radius of the Ferris wheel. The midline of the function is 50 meters, so the center of the wheel is 50 meters off the ground. This means that as the Ferris wheel rotates, the lowest point on the wheel will be 5 meters above the ground.

37. Since the wheel has diameter 4.5 meters, it has radius 2.25 meters. Half of the bucket dips below the water surface in its lowest position, so the center of the bucket is $2.25 + 0.25 = 2.5$ meters away from the center of the water wheel. The lowest height of the center of the bucket, and thus the lowest function value, will be 0 when $\theta = 270^\circ$. The highest position of the center of the bucket will be 5 m above the water when $\theta = 90^\circ$. Therefore the height of the center of the water bucket above the river is given by a sinusoidal function with midline 2.5 and amplitude 2.5: $h(\theta) = 2.5 + 2.5\sin\theta$.

38. **(a)** The tip of blade 1 is 60 m above the ground when $\theta = 0^\circ$. When $\theta = 90^\circ$ it is at its highest point which is $60 + 30 = 90$ m. At $\theta = 270^\circ$ it is at its lowest point $60 - 30 = 30$ m. At $\theta = 360^\circ$ it is back at the starting position. So the amplitude of the function $h(\theta)$ is 30 meters, the midline is 60 meters and the period is 360°. Therefore, a function describing the height of the tip of blade 1 above the ground is $h(\theta) = 60 + 30\sin\theta$.

(b) Since the first bug landed at the tip of the blade and we already found a function for the height of the tip of the blade in part (a) of this problem, the same function works here. For the second bug, the only difference is its position along the blade. Its highest point is 75 meters and its lowest point is 45 meters. Therefore the only difference for the function describing the height of the second bug is the amplitude of the function. The height of the second bug is given by $h_2(\theta) = 60 + 15\sin\theta$.

Solutions for Section 7.4

Exercises

1. $\sin 0^\circ = 0$, $\cos 0^\circ = 1$, $\tan 0^\circ = \sin 0^\circ/\cos 0^\circ = 0/1 = 0$.

2. **(a)** $\cos 90^\circ = 0$

(b) $\tan 90^\circ$ is undefined, because at $\theta = 90^\circ$, the x-coordinate is 0. So in order to evaluate $\tan 90^\circ$, we would have to divide by 0.

(c) $\cos 540^\circ = \cos(360^\circ + 180^\circ) = \cos 180^\circ = -1$.

(d) $\tan 540^\circ = \tan 180^\circ = \dfrac{\sin 180^\circ}{\cos 180^\circ} = \dfrac{0}{-1} = 0$.

3. By the Pythagorean theorem, the hypotenuse has length $\sqrt{1^2+2^2}=\sqrt{5}$.

(a) $\tan\theta = \dfrac{\text{opposite}}{\text{adjacent}} = \dfrac{2}{1} = 2.$

(b) $\sin\theta = \dfrac{\text{opposite}}{\text{hypotenuse}} = \dfrac{2}{\sqrt{5}}.$

(c) $\cos\theta = \dfrac{\text{adjacent}}{\text{hypotenuse}} = \dfrac{1}{\sqrt{5}}.$

4. **(a)** $\dfrac{5}{\sqrt{125}}$

(b) $\dfrac{10}{\sqrt{125}}$

(c) $\dfrac{10}{\sqrt{125}}$

(d) $\dfrac{5}{\sqrt{125}}$

(e) $1/2$

(f) 2

5. By the Pythagorean Theorem, we know that the third side must be $\sqrt{7^2-2^2}=\sqrt{45}$.

(a) Since $\sin\theta$ is opposite side over hypotenuse, we have $\sin\theta = \sqrt{45}/7$.
(b) Since $\cos\theta$ is adjacent side over hypotenuse, we have $\cos\theta = 2/7$.
(c) Since $\tan\theta$ is opposite side over adjacent side, we have $\tan\theta = \sqrt{45}/2$.

6. By the Pythagorean Theorem, we know that the third side must be $\sqrt{9^2+5^2}=\sqrt{106}$.

(a) Since $\sin\theta$ is opposite side over hypotenuse, we have $\sin\theta = 5/\sqrt{106}$.
(b) Since $\cos\theta$ is adjacent side over hypotenuse, we have $\cos\theta = 9/\sqrt{106}$.
(c) Since $\tan\theta$ is opposite side over adjacent side, we have $\tan\theta = 5/9$.

7. By the Pythagorean Theorem, we know that the third side must be $\sqrt{12^2-8^2}=\sqrt{80}$.

(a) Since $\sin\theta$ is opposite side over hypotenuse, we have $\sin\theta = 8/12$.
(b) Since $\cos\theta$ is adjacent side over hypotenuse, we have $\cos\theta = \sqrt{80}/12$.
(c) Since $\tan\theta$ is opposite side over adjacent side, we have $\tan\theta = 8/\sqrt{80}$.

8. Because the two angles are the same, we know that the two missing sides must be equal. Therefore, by the Pythagorean Theorem, we know that the missing sides are (which we call s) are given by:

$$\begin{aligned} 17^2 &= s^2 + s^2 \\ 289 &= 2s^2 \\ \sqrt{\frac{289}{2}} &= s. \end{aligned}$$

(a) Since $\sin\theta$ is opposite side over hypotenuse, we have $\sin\theta = \sqrt{289/2}/17 = 1/\sqrt{2}$.
(b) Since $\cos\theta$ is adjacent side over hypotenuse, we have $\cos\theta = \sqrt{289/2}/17 = 1/\sqrt{2}$.
(c) Since $\tan\theta$ is opposite side over adjacent side, we have $\tan\theta = 1$.

9. By the Pythagorean Theorem, we know that the third side must be $\sqrt{11^2-2^2}=\sqrt{117}$.

(a) Since $\sin\theta$ is opposite side over hypotenuse, we have $\sin\theta = \sqrt{117}/11$.
(b) Since $\cos\theta$ is adjacent side over hypotenuse, we have $\cos\theta = 2/11$.
(c) Since $\tan\theta$ is opposite side over adjacent side, we have $\tan\theta = \sqrt{117}/2$.

10. By the Pythagorean Theorem, we know that the third side must be $\sqrt{a^2+b^2}$.

(a) Since $\sin\theta$ is opposite side over hypotenuse, we have $\sin\theta = a/\sqrt{a^2+b^2}$.
(b) Since $\cos\theta$ is adjacent side over hypotenuse, we have $\cos\theta = b/\sqrt{a^2+b^2}$.
(c) Since $\tan\theta$ is opposite side over adjacent side, we have $\tan\theta = a/b$.

11. Since $\sin 17^\circ = r/7$, we have $r = 7\sin 17^\circ$. Similarly, since $\cos 17^\circ = q/7$, we have $q = 7\cos 17^\circ$.

12. Since $\sin 12^\circ = 4/r$, we have $r = 4/\sin 12^\circ$. Similarly, since $\tan 12^\circ = 4/q$, we have $q = 4/\tan 12^\circ$.

13. Since $\cos 37^\circ = 6/r$, we have $r = 6/\cos 37^\circ$. Similarly, since $\tan 37^\circ = q/6$, we have $q = 6\tan 37^\circ$.

14. Since $\sin 40^\circ = r/15$, we have $r = 15\sin 40^\circ$. Similarly, since $\cos 40^\circ = q/15$, we have $q = 15\cos 40^\circ$.

15. Since $\tan 77^\circ = 9/r$, we have $r = 9/\tan 77^\circ$. Similarly, since $\sin 77^\circ = 9/q$, we have $q = 9/\sin 77^\circ$.

16. Since $\sin 22^\circ = \lambda/r$, we have $r = \lambda/\sin 22^\circ$. Similarly, since $\tan 22^\circ = \lambda/q$, we have $q = \lambda/\tan 22^\circ$.

17. $\cos 90^\circ = 0$

18. $\sin 90^\circ = 1$

19. $\tan 90^\circ$ is undefined, because at $\theta = 90^\circ$, the x-coordinate is 0. So in order to evaluate $\tan 90^\circ$, we would have to divide by 0.

20. $\sin 270^\circ = -1$

21. Since 225° is in the third quadrant,

$$\tan 225^\circ = \tan 45^\circ = 1.$$

22. Since 135° is in the second quadrant,

$$\tan 135^\circ = -\tan 45^\circ = -1.$$

23. $\tan 540^\circ = \tan 180^\circ = \dfrac{\sin 180^\circ}{\cos 180^\circ} = \dfrac{0}{-1} = 0.$

Problems

24. Using $\tan 13^\circ = \dfrac{\text{height}}{200}$ to find the height we get

$$\text{height} = 200\tan 13^\circ \approx 46.174\text{feet}.$$

Using $\cos 13^\circ = \dfrac{200}{\text{incline}}$ to find the incline we get

$$\text{incline} = 200/\cos 13^\circ \approx 205.261\text{feet}.$$

25. Figure 7.27 illustrates this situation.

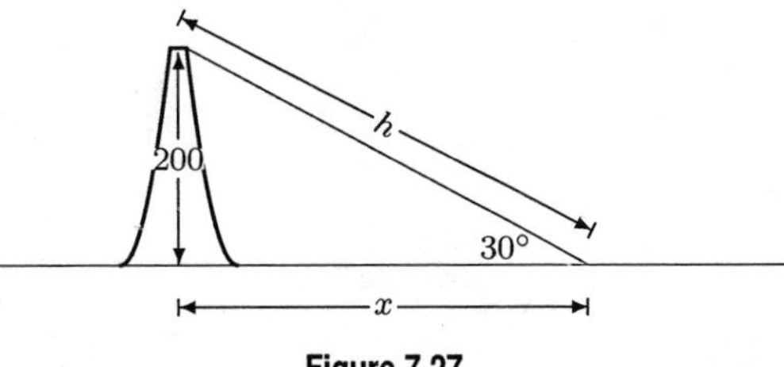

Figure 7.27

We have a right triangle with legs x and 200 and hypotenuse h. Thus,

$$\sin 30^\circ = \frac{200}{h}$$
$$h = \frac{200}{\sin 30^\circ} = \frac{200}{0.5} = 400 \text{ feet.}$$

To find the distance x, we can relate the angle and its opposite and adjacent legs by writing

$$\tan 30^\circ = \frac{200}{x}$$
$$x = \frac{200}{\tan 30^\circ} \approx 346.410 \text{ feet.}$$

We could also write the equation $x^2 + 200^2 = h^2$ and substitute $h = 400$ ft to solve for x.

26. If the horizontal distance is d, then

$$\frac{20}{d} = \tan 15^\circ,$$

so

$$d = \frac{20}{\tan 15^\circ} \approx 74.641 \text{ feet.}$$

27. Let d be the horizontal distance from the airplane to the arch. See Figure 7.28. Then, $\tan\theta = 35000/d$, or $d = 35000/\tan\theta$ feet.

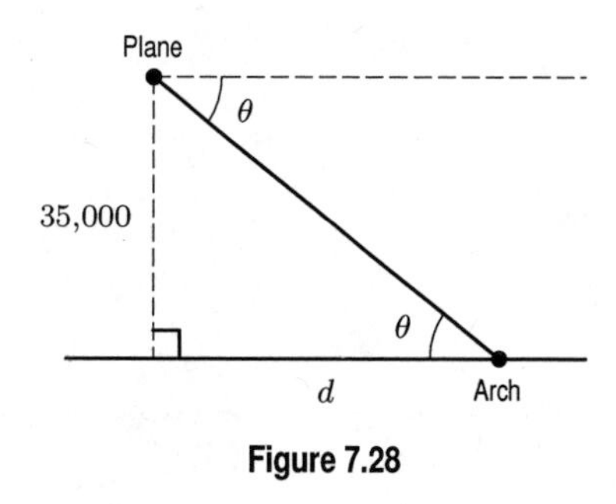

Figure 7.28

28. Let d be the distance from Hampton to the point where the beam strikes the shore. Then, $\tan\phi = d/3$, so $d = 3\tan\phi$ miles.

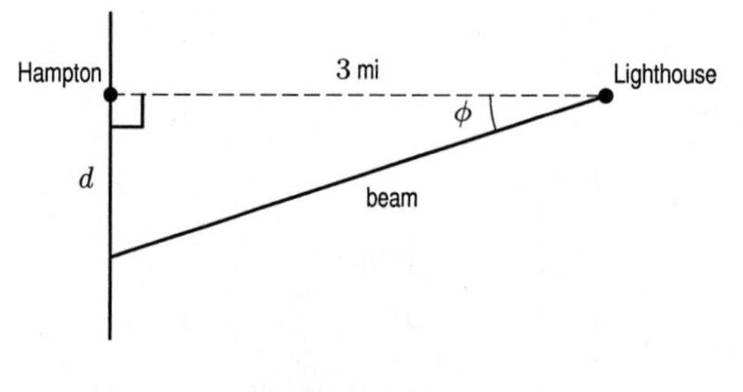

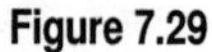

Figure 7.29

29. Since the distance from P to A is $\dfrac{50}{\tan 42^\circ}$ and the distance from P to B is $\dfrac{50}{\tan 35^\circ}$,

$$d = \frac{50}{\tan 35^\circ} - \frac{50}{\tan 42^\circ} \approx 15.877 \text{ feet.}$$

30. The point-slope formula for a line is $y = y_0 + m(x - x_0)$, where m is the slope and (x_0, y_0) is a point on the line. Here the slope of line l is $(\sin\theta)/(\cos\theta) = \tan\theta$. Thus, $y = y_0 + (\tan\theta)(x - x_0)$, where (x_0, y_0) is a point on the line.

31. **(a)** The side opposite of angle ϕ has length b and the side adjacent to angle ϕ has length a. Therefore,

$$\sin\phi = \frac{\text{side opposite}}{\text{hypotenuse}} = \frac{b}{c}$$
$$\cos\phi = \frac{\text{side adjacent}}{\text{hypotenuse}} = \frac{a}{c}$$
$$\tan\phi = \frac{\text{side opposite}}{\text{side adjacent}} = \frac{b}{a}.$$

(b)

$$\sin\phi = \frac{\text{side opposite }\phi}{\text{hypotenuse}} = \frac{b}{c},$$
$$\cos\theta = \frac{\text{side adjacent to }\theta}{\text{hypotenuse}} = \frac{b}{c}.$$

Thus $\sin\phi = \cos\theta$. Reversing the roles of ϕ and θ one can show $\cos\phi = \sin\theta$ in exactly the same way.

32. To solve for the distance x, we use $\tan 53^\circ = \frac{954}{x}$ and solve for x:

$$x = 954/\tan 53^\circ = 718.891 \text{ ft}.$$

To solve for the height of the Sea First Tower, we can use $\tan 37^\circ = y/x$ and solve for y:

$$y = 718.891 \tan 37^\circ = 541.723 \text{ ft}.$$

(The actual height of the Seafirst Tower is 543 ft.)

Solutions for Section 7.5

Exercises

1. Since we are looking for the angle θ, we have $\sin\theta = 0.876$, so $\theta = \sin^{-1} 0.876 = 61.164^\circ$.

2. Since we are looking for the angle θ, we have $\cos\theta = 0.016$, so $\theta = \cos^{-1} 0.016 = 89.083^\circ$.

3. Since we are looking for the angle θ, we have $\tan\theta = 0.123$, so $\theta = \tan^{-1} 0.123 = 7.012^\circ$.

4. Since the output of the sine function is $0 < \sin\theta < 1$, there is no angle whose sine is 1.342.

5. Since the output of the cosine function is $0 < \cos\theta < 1$, there is no angle whose cosine is 2.614.

6. Since we are looking for the angle θ, we have $\tan\theta = 54.169$, so $\theta = \tan^{-1} 54.169 = 88.942^\circ$.

7. Since we are looking for the angle θ, we have $\sin\theta = 0.999$, so $\theta = \sin^{-1} 0.999 = 89.190^\circ$.

8. Since we are looking for the angle θ, we have $\cos\theta = 0.999$, so $\theta = \cos^{-1} 0.999 = 2.563^\circ$.

9. We know that $\sin 60^\circ = \frac{\sqrt{3}}{2}$. Therefore, $\theta = 60^\circ$.

10. We know that $\tan 60^\circ = \sqrt{3}$. Therefore, $\theta = 60^\circ$.

11. We know that $\cos\ 60^\circ = \frac{1}{2}$. Therefore, $\theta = 60^\circ$.

12. We know that $\tan\ 45^\circ = 1$. Therefore, $\theta = 45^\circ$.

13. We know that $\sin 45^\circ = \frac{\sqrt{2}}{2}$. Therefore, $\theta = 45^\circ$.

14. We know that $\tan 30^\circ = \frac{\sqrt{3}}{3}$. Therefore, $\theta = 30^\circ$.

15. We have

$$\begin{aligned} c &= \sqrt{a^2+b^2} = \sqrt{1184} \approx 34.409 \\ \sin A &= \frac{a}{c} \\ A &= \sin^{-1}\frac{a}{c} = 35.538^\circ \\ B &= 90^\circ - A = 54.462^\circ. \end{aligned}$$

16. We have

$$\begin{aligned} b &= \sqrt{c^2-a^2} = \sqrt{384} \approx 19.596 \\ \sin A &= \frac{a}{c} \\ A &= \sin^{-1}\frac{a}{c} = 45.585^\circ \\ B &= 90^\circ - A = 44.415^\circ. \end{aligned}$$

17. We have

$$\begin{aligned} B &= 90^\circ - A = 62^\circ \\ a &= c \cdot \sin A = 20\sin 28^\circ = 9.389 \\ b &= c \cdot \sin B = 20\sin 62^\circ = 17.659. \end{aligned}$$

18. We have

$$\begin{aligned} A &= 90^\circ - B = 62^\circ \\ a &= c \cdot \sin A \\ c &= \frac{20}{\sin 62^\circ} = 22.651 \\ \tan B &= \frac{b}{20} \\ b &= 20\tan 62^\circ = 10.634. \end{aligned}$$

19. The letter k represents the angle and the letter a represents the value of the function.

20. Since $\cos^{-1} a = z$ means $\cos z = a$, the angle is z and the value is a.

21. Since $(\tan c)^{-1} = \dfrac{1}{\tan c} = d$, we have $\tan c = \dfrac{1}{d}$. Therefore the angle is c and the value is $\dfrac{1}{d}$.

22. Since $m = \arcsin y$ means the $y = \sin m$, the angle is m and the value is y.

23. The angle is n; the value is p.

24. Since $\tan^{-1}\left(\dfrac{1}{y}\right)$ means $\tan z = \dfrac{1}{y}$, the angle is z and the value is $\dfrac{1}{y}$.

Problems

25. **(a)** We are looking for the value of the sine of $\left(\dfrac{1}{2}\right)^\circ$. Using a calculator or computer, we have $\sin\left(\dfrac{1}{2}\right)^\circ = 0.009$.

(b) We are looking for the angle in a right triangle whose sine is $\dfrac{1}{2}$. Therefore, we have $\sin^{-1}\left(\dfrac{1}{2}\right) = 30^\circ$.

(c) We are looking for the reciprocal of the sine of $\left(\dfrac{1}{2}\right)^\circ$. We have

$$(\sin x)^{-1} = \left(\sin\frac{1}{2}\right)^{-1}$$

$$\begin{aligned} &= \frac{1}{\sin \frac{1}{2}} \\ &= 114.593. \end{aligned}$$

26. **(a)** We are looking for the value of the tangent of 10°. Using a calculator or computer, we have $\tan 10^\circ = 0.176$.

(b) We are looking for the angle in a right triangle whose tangent is 10. Using a calculator or computer, we have $\tan^{-1}(10) = 84.289^\circ$

(c) We are looking for the reciprocal of $\tan 10^\circ$. We have

$$\begin{aligned} (\tan x)^{-1} &= (\tan 10^\circ)^{-1} \\ &= \frac{1}{\tan 10^\circ} \\ &= 5.671. \end{aligned}$$

27. **(a)** We have $\sin 45^\circ + \cos 45^\circ + \tan 45^\circ = \frac{\sqrt{2}}{2} + \frac{\sqrt{2}}{2} + 1 = \sqrt{2} + 1$.

(b) We have $(\sin 45^\circ)^{-1} + (\cos 45^\circ)^{-1} + (\tan 45^\circ)^{-1} = \frac{2}{\sqrt{2}} + \frac{2}{\sqrt{2}} + 1 = 2\sqrt{2} + 1$.

(c) We have $\sin^{-1}(0.45) + \cos^{-1}(0.45) + \tan^{-1}(0.45) = 26.744^\circ + 63.256^\circ + 0.008^\circ = 90.008^\circ$.

28. We have

$$\begin{aligned} 4 \sin \theta &= 1 \\ \sin \theta &= \frac{1}{4} \\ \theta &= \sin^{-1}\left(\frac{1}{4}\right) \\ \theta &= 14.478^\circ. \end{aligned}$$

29. We have

$$\begin{aligned} 6 \cos \theta - 2 &= 3 \\ 6 \cos \theta &= 5 \\ \cos \theta &= \left(\frac{5}{6}\right) \\ \theta &= \cos^{-1}\left(\frac{5}{6}\right) \\ \theta &= 33.557^\circ. \end{aligned}$$

30. We have

$$\begin{aligned} 10 \tan \theta - 5 &= 15 \\ 10 \tan \theta &= 20 \\ \tan \theta &= 2 \\ \theta &= \tan^{-1}(2) \\ \theta &= 63.435^\circ. \end{aligned}$$

31. We have

$$\begin{aligned} 2 \cos \theta + 6 &= 9 \\ 2 \cos \theta &= 3 \\ \cos \theta &= \frac{3}{2}. \end{aligned}$$

Since the range of $\cos \theta$ is $0 < \cos \theta < 1$, there is no angle whose cosine is 1.5.

32. We have

$$\begin{aligned} 5\sin(3\theta) &= 4 \\ \sin(3\theta) &= \frac{4}{5} \\ (3\theta) &= \sin^{-1}\left(\frac{4}{5}\right) \\ 3\theta &= 53.130^\circ \\ \theta &= 17.710^\circ. \end{aligned}$$

33. We have

$$\begin{aligned} 9\tan(5\theta) + 1 &= 10 \\ 9\tan(5\theta) &= 9 \\ \tan(5\theta) &= 1 \\ 5\theta &= \tan^{-1}(1) \\ 5\theta &= 45^\circ \\ \theta &= 9^\circ. \end{aligned}$$

34. We have

$$\begin{aligned} 2\sqrt{3}\tan(2\theta) + 1 &= 3 \\ 2\sqrt{3}\tan(2\theta) &= 2 \\ \tan(2\theta) &= \frac{2}{2\sqrt{3}} \\ \tan(2\theta) &= \frac{1}{\sqrt{3}} \\ 2\theta &= \tan^{-1}\left(\frac{1}{\sqrt{3}}\right) \\ 2\theta &= 30^\circ \\ \theta &= 15^\circ. \end{aligned}$$

35. We have

$$\begin{aligned} 3\sin\theta + 3 &= 5\sin\theta + 2 \\ 1 &= 2\sin\theta \\ \frac{1}{2} &= \sin\theta \\ \sin^{-1}\left(\frac{1}{2}\right) &= \theta \\ 30^\circ &= \theta. \end{aligned}$$

36. We have

$$\begin{aligned} 6\cos(3\theta) + 3 &= 4\cos(3\theta) + 4 \\ 2\cos(3\theta) &= 1 \\ \cos(3\theta) &= \frac{1}{2} \\ 3\theta &= \cos^{-1}\left(\frac{1}{2}\right) \\ 3\theta &= 60^\circ \\ \theta &= 20^\circ. \end{aligned}$$

37. We have

$$
\begin{aligned}
5\tan(4\theta)+4 &= 2(\tan(4\theta)+5)\\
5\tan(4\theta)+4 &= 2\tan(4\theta)+10\\
3\tan(4\theta) &= 6\\
\tan(4\theta) &= 2\\
4\theta &= \tan^{-1}(2)\\
4\theta &= 63.435^\circ\\
\theta &= 15.859^\circ.
\end{aligned}
$$

38. **(a)** Since the grade of the ramp is $7\% = 7/100$, this means that a 7 foot height difference occurs over a horizontal distance of 100 feet. So we have $\tan\theta = \frac{7}{100}$. Using the $\tan^{-1}$ button on the calculator we get

$$\theta = \tan^{-1}\left(\frac{7}{100}\right) = 4.004^\circ.$$

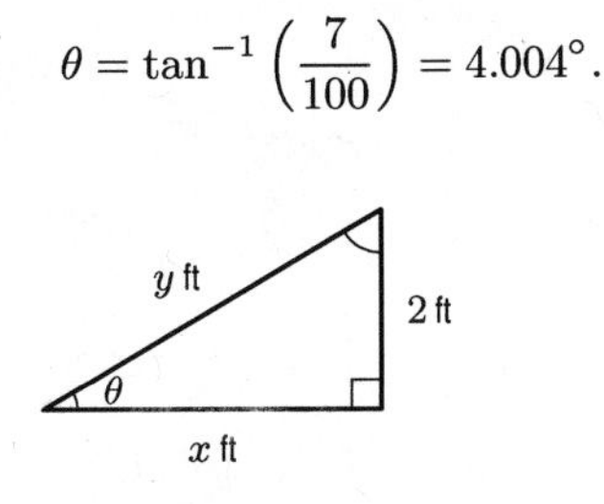

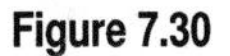
Figure 7.30

(b) From the right triangle representing the ramp we see that one leg represents the height difference between the driveway and the front door, which is 2 ft. The other leg represents the driveway, of which we would like to find the length x. Then $2/x = \tan 4.004^\circ$. Solving this equation for x we have

$$x = \frac{2}{\tan 4.004^\circ} = 28.57.$$

So the driveway has to be 28.57 feet long. (We can also use similar triangles: $2/7 = x/100$.)

(c) The ramp is represented by the hypotenuse y of the right triangle. Using the Pythagorean Theorem we have $y = \sqrt{28.57^2 + 2^2} = 28.64$ feet. (We can also use $\sin 4.004^\circ = 2/y$.)

39. Since the rise, the run and the roof form a right triangle with a horizontal leg of 12 and vertical leg of 10, we have $\tan\theta = \frac{10}{12}$. Using the $\tan^{-1}$ button on the calculator, we have $\theta = \tan^{-1}(\frac{10}{12}) = 39.806^\circ$.

40.

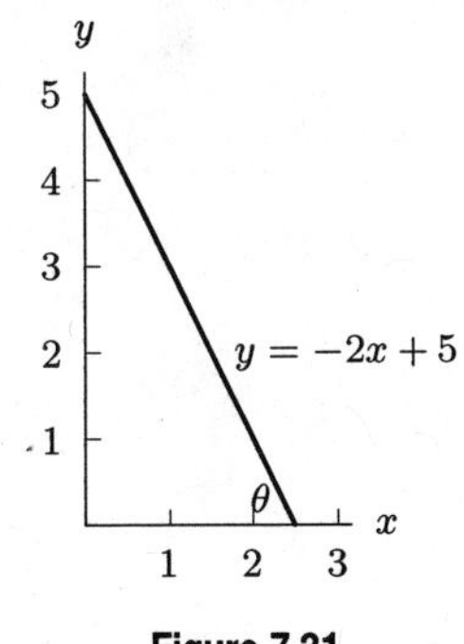

Figure 7.31

The y-intercept of this line is 5; the x-intercept is 2.5. These are the lengths of the legs of the right triangle in Figure 7.31 so $\tan\theta = \dfrac{5}{2.5} = 2$, or $\theta = \tan^{-1}(2) \approx 63.435^\circ$.

41. See Figure 7.32. The angle θ is the sun's angle of elevation. Here, $\tan\theta = \frac{50}{60} = \frac{5}{6}$. So, $\theta = \tan^{-1}\left(\frac{5}{6}\right) \approx 39.806°$.

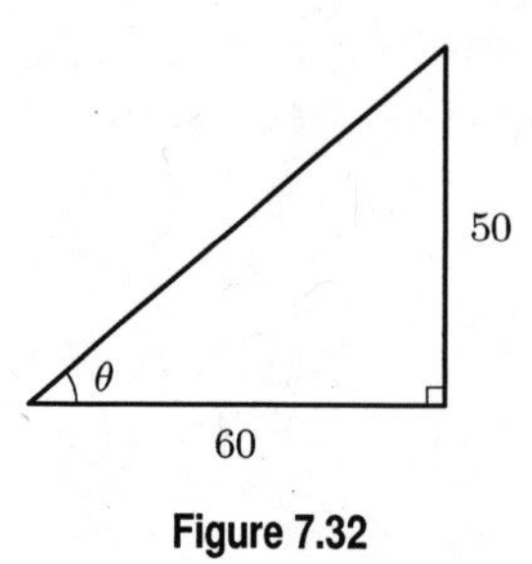

Figure 7.32

42. Draw a picture as in Figure 7.33. The angle that we want is labeled θ in this picture. We see that $\tan\theta = \frac{17.3}{10} = 1.73$. Evaluating $\tan^{-1}(1.73)$ on a calculator, we get $\theta \approx 59.971°$.

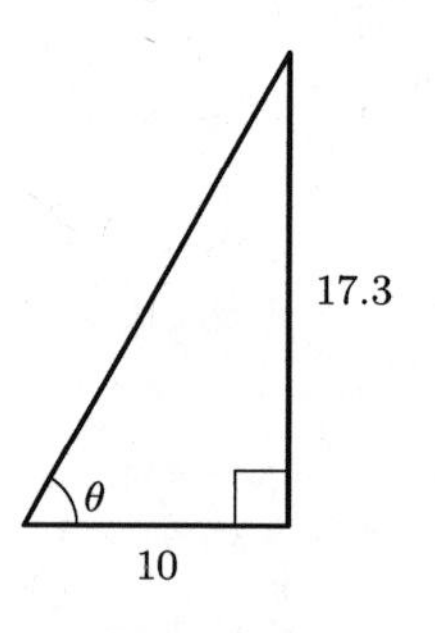

Figure 7.33

43. **(a)** First assume $A = 30°$ and $b = 2\sqrt{3}$. Since this is a right triangle, we know that $B = 90° - 30° = 60°$. Now we can determine a by writing

$$\cos A = \frac{\text{adjacent}}{\text{hypotenuse}} = \frac{2\sqrt{3}}{a}.$$

We also know that $\cos A = \frac{\sqrt{3}}{2}$, because A is $30°$. So $\frac{\sqrt{3}}{2} = \frac{2\sqrt{3}}{a}$, which means that $a = 4$. It follows from the Pythagorean theorem that c is $\sqrt{16-12} = 2$.

(b) Now assume that $a = 25$ and $c = 24$. The Pythagorean theorem, $a^2 + b^2 = c^2$, implies

$$b = \sqrt{25^2 - 24^2} = 7.$$

To determine angles, we use $\sin A = \frac{\text{opposite}}{\text{hypotenuse}} = \frac{c}{a} = \frac{24}{25}$. So evaluating $\sin^{-1}\left(\frac{24}{25}\right)$ on the calculator, we find that $A = 73.740°$. Therefore, $B = 90° - 73.740° = 16.260°$. (Be sure that your calculator is in degree mode when using the $\sin^{-1}$.)

44.

$$\tan\alpha = 0.2, \quad \text{so} \quad \alpha = \tan^{-1}(0.2) \approx 11.310°$$
$$\tan\beta = 0.3, \quad \text{so} \quad \beta = \tan^{-1}(0.3) \approx 16.699°$$

Solutions for Section 7.6

Exercises

1. By the Law of Sines, we have

$$\frac{x}{\sin 100^\circ} = \frac{6}{\sin 18^\circ}$$
$$x = 6\left(\frac{\sin 100^\circ}{\sin 18^\circ}\right) \approx 19.121.$$

2.

By the Law of Cosines, we have

$$x^2 = 3^2 + 5^2 - 2(3)(5)\cos(21^\circ)$$
$$x \approx 2.448.$$

3. In Figure 7.34, we have

$$\beta = 180^\circ - 90^\circ - 38^\circ$$
$$\beta = 52^\circ.$$

$$\sin 38^\circ = \frac{4}{c}$$
$$c = \frac{4}{\sin 38^\circ} \approx 6.497.$$

$$c^2 = 4^2 + b^2$$
$$b = \sqrt{c^2 - 16} \approx 5.120.$$

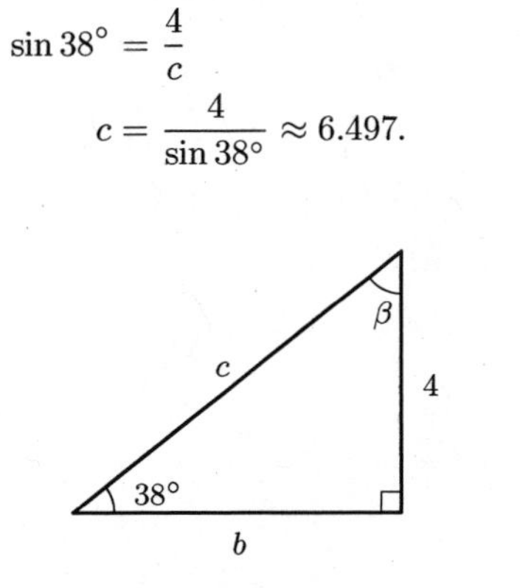

Figure 7.34

4. In Figure 7.35, we have

$$\tan\theta = \frac{7}{2} = 3.5$$
$$\theta = \tan^{-1}(3.5)$$
$$\theta \approx 74.055^\circ.$$

$$\tan\psi = \frac{2}{7}$$
$$\psi = \tan^{-1}(\tfrac{2}{7})$$
$$\psi \approx 15.945^\circ.$$

$$c^2 = 2^2 + 7^2 = 53$$
$$c = \sqrt{53} \approx 7.280.$$

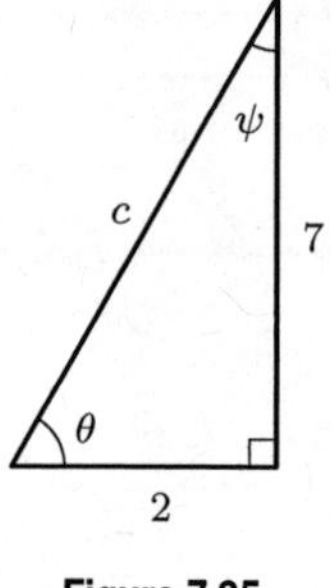

Figure 7.35

5. In Figure 7.36, we have

$$\theta = 180^\circ - 90^\circ - 10^\circ$$
$$\theta = 80^\circ.$$

$$a = 12\cos 10^\circ$$
$$a \approx 12(0.985)$$
$$a \approx 11.818.$$

$$b = 12\sin 10^\circ$$
$$b \approx 12(0.174)$$
$$b \approx 2.084.$$

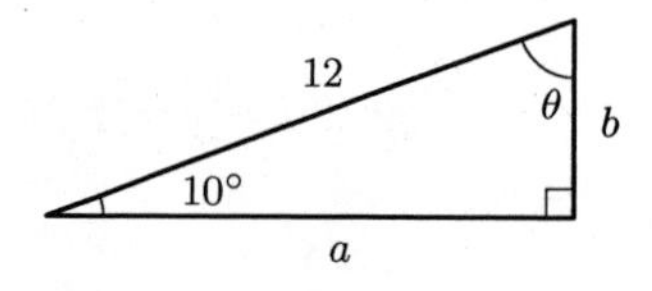

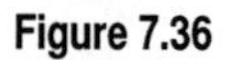
Figure 7.36

6. Using the Law of Cosines in Figure 7.37, we have

$$b^2 = 11^2 + 8^2 - 2 \cdot 11 \cdot 8 \cdot \cos 32^\circ$$
$$b \approx 5.979.$$

Using the Law of Cosines again to find θ,

$$11^2 = 8^2 + 5.979^2 - 2 \cdot 8 \cdot 5.979 \cos\theta$$
$$\cos\theta = -0.22199$$
$$\theta \approx 102.826^\circ$$
$$\psi = 180^\circ - 32^\circ - 102.826^\circ \approx 45.174^\circ.$$

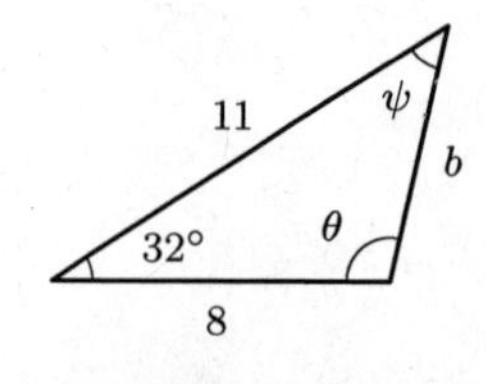

Figure 7.37

7. Using the Law of Sines in Figure 7.38, we have

$$\frac{\sin\theta}{5} = \frac{\sin 20^\circ}{6}$$
$$\theta = \sin^{-1}\left(\frac{5\sin 20^\circ}{6}\right) = 16.560^\circ$$

This is correct since $\theta < 90^\circ$ in the triangle. We expect $\theta < 20^\circ$ because θ is opposite a side which is shorter than 6. Therefore

$$\psi = 180^\circ - 16.560^\circ - 20^\circ = 143.440^\circ.$$

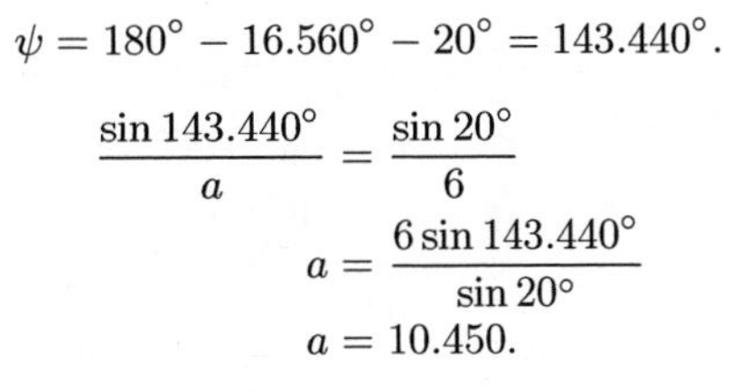

$$\frac{\sin 143.440^\circ}{a} = \frac{\sin 20^\circ}{6}$$
$$a = \frac{6\sin 143.440^\circ}{\sin 20^\circ}$$
$$a = 10.450.$$

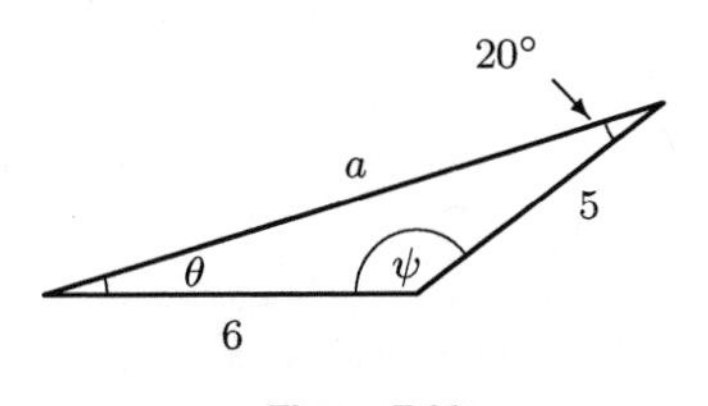

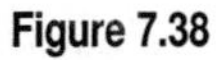
Figure 7.38

8. In Figure 7.39, we first determine α using the Law of Cosines:

$$7^2 = 10^2 + 12^2 - 2\cdot 10\cdot 12\cdot\cos\alpha$$
$$240\cos\alpha = 195$$
$$\cos\alpha = \frac{195}{240},$$
$$\alpha = \cos^{-1}\left(\frac{195}{240}\right) \approx 35.659^\circ.$$

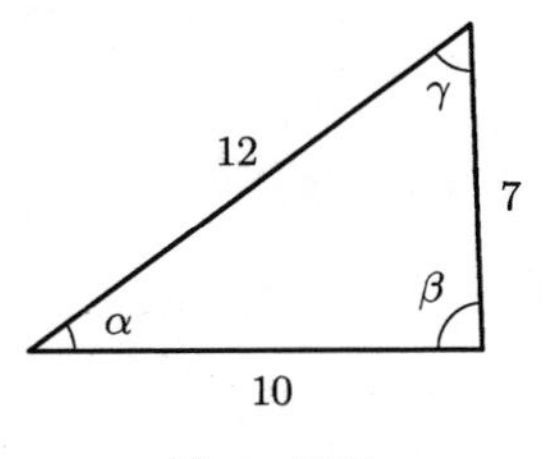

Figure 7.39

Then, we determine β using the Law of Cosines:

$$12^2 = 10^2 + 7^2 - 2\cdot 10\cdot 7\cos\beta$$
$$\cos\beta = \frac{5}{140}$$
$$\beta = \cos^{-1}\left(\frac{5}{140}\right) \approx 87.953^\circ.$$

Finally, $\gamma = 180^\circ - \alpha - \beta \approx 56.388^\circ$.

9. Using the Law of Cosines, we have

$$\begin{aligned}
41^2 &= 20^2 + 28^2 - 2 \cdot 20 \cdot 28 \cos C \\
497 &= -1120 \cos C \\
C &= \cos^{-1}\left(-\frac{497}{1120}\right) \\
C &= 116.343^\circ.
\end{aligned}$$

Using the Law of Cosines again (though we could use the Law of Sines), we have

$$\begin{aligned}
20^2 &= 41^2 + 28^2 - 2 \cdot 41 \cdot 28 \cos A \\
-2065 &= -2296 \cos A \\
A &= \cos^{-1}\left(\frac{2065}{2296}\right) \\
A &= 25.922^\circ.
\end{aligned}$$

Thus, $B = 180 - A - C = 37.735^\circ$.

10. The Law of Cosines gives

$$\begin{aligned}
c^2 &= 14^2 + 12^2 - 2 \cdot 14 \cdot 12 \cos 23^\circ = 30.7104 \\
c &= 5.5417
\end{aligned}$$

From the Law of Sines, we can solve for B, which must be an acute angle since it is across from b, which is shorter than a, so $B < A$, and there can be no more than one obtuse angle in a triangle.

$$\begin{aligned}
\frac{\sin B}{12} &= \frac{\sin 23^\circ}{5.5417} \\
\sin B &= \frac{12\sin(23^\circ)}{5.5417} \\
B &= \sin^{-1}\left(\frac{12\sin(23^\circ)}{5.5417}\right) \\
&= 57.789^\circ,
\end{aligned}$$

so $A = 180^\circ - B - C = 99.211^\circ$.

11. Using the Law of Cosines, we have

$$\begin{aligned}
b^2 &= 20^2 + 28^2 - 2 \cdot 20 \cdot 28 \cos 81^\circ \\
b^2 &= 1008.793 \\
b &= 31.762.
\end{aligned}$$

Using the Law of Sines, we have

$$\begin{aligned}
\frac{\sin A}{20} &= \frac{\sin 81^\circ}{31.762} \\
\sin A &= 0.622 \\
A &= \sin^{-1}(0.622) \text{ or } 180 - \sin^{-1}(0.622) \\
A &= 38.458^\circ \text{ or } 141.542^\circ.
\end{aligned}$$

We can discard the larger answer, as it does not fit in a triangle with $B = 81^\circ$. Thus we have $C = 180^\circ - A - B = 60.542^\circ$.

12. Using the Law of Cosines, we have

$$\begin{aligned} c^2 &= 20^2 + 28^2 - 2 \cdot 20 \cdot 28 \cos 12^\circ \\ &= 88.4747 \\ c &= 9.4061. \end{aligned}$$

Using the Law of Sines, we have

$$\begin{aligned} \frac{\sin A}{20} &= \frac{\sin 12^\circ}{9.4061} \\ \sin A &= 20 \cdot \frac{\sin 12^\circ}{9.4061} \\ A &= \sin^{-1}\left(20 \cdot \frac{\sin 12^\circ}{9.4061}\right) \text{ or } 180^\circ - \sin^{-1}\left(20 \cdot \frac{\sin 12^\circ}{9.4061}\right) \\ A &= 26.237^\circ \text{ or } 153.763^\circ. \end{aligned}$$

Since B is a larger angle than A (since $b > a$), we can discard the larger answer, as it does not fit in a triangle. Thus we have $B = 180^\circ - A - C = 141.763^\circ$.

13. We begin by using the law of cosines to find side c:

$$\begin{aligned} c^2 &= 9^2 + 8^2 - 2 \cdot 9 \cdot 8 \cos 80^\circ \\ c^2 &= 119.995 \\ c &= 10.954. \end{aligned}$$

We can now use the law of sines to find the other two angles.

$$\begin{aligned} \frac{\sin B}{8} &= \frac{\sin 80^\circ}{10.954} \\ \sin B &= 8\frac{\sin 80^\circ}{10.954} \\ B &= \sin^{-1} 0.719 \\ B &= 45.990^\circ. \end{aligned}$$

Therefore, $A = 180^\circ - 80^\circ - 45.990 = 54.010^\circ$.

14. By the Law of Cosines,

$$\begin{aligned} c^2 &= 8^2 + 11^2 - 2 \cdot 8 \cdot 11 \cos 114^\circ \\ c^2 &= 256.5856 \\ c &= 16.0183. \end{aligned}$$

By the Law of Sines,

$$\begin{aligned} \frac{\sin A}{8} &= \frac{\sin 114^\circ}{16.0183} \\ A &= 27.145^\circ. \end{aligned}$$

That means $B = 180^\circ - A - C = 38.855^\circ$.

15. By the Law of Cosines,

$$\begin{aligned} c^2 &= 5^2 + 11^2 - 2 \cdot 5 \cdot 11 \cos 32^\circ \\ c &= 7.2605. \end{aligned}$$

Angle A is acute (since $a < b$ and thus $A < B$), so we solve for it using the Law of Sines. We have

$$\begin{aligned}\frac{\sin A}{a} &= \frac{\sin C}{c}\\ \frac{\sin A}{5} &= \frac{\sin 32^\circ}{7.2605}\\ \sin A &= \frac{5\sin 32^\circ}{7.2605}\\ A &= \sin^{-1}\left(\frac{5\sin 32^\circ}{7.2605}\right)\\ &= 21.403^\circ.\end{aligned}$$

This gives $B = 180 - A - C = 126.597^\circ$.

16. We begin by finding the angle C, which is $180^\circ - 13^\circ - 25^\circ = 142^\circ$.
We can now use the law of sines to find the other two sides.

$$\begin{aligned}\frac{b}{\sin 25^\circ} &= \frac{4}{\sin 142^\circ}\\ b &= \sin 25^\circ \cdot \frac{4}{\sin 142^\circ}\\ b &= 2.746.\end{aligned}$$

Similarly,

$$\begin{aligned}\frac{a}{\sin 13^\circ} &= \frac{4}{\sin 142^\circ}\\ a &= \sin 13^\circ \cdot \frac{4}{\sin 142^\circ}\\ a &= 1.462.\end{aligned}$$

17. We begin by finding the angle C, which is $180^\circ - 105^\circ - 9^\circ = 66^\circ$.
We can now use the law of sines to find the other two sides.

$$\begin{aligned}\frac{b}{\sin 9^\circ} &= \frac{15}{\sin 66^\circ}\\ b &= \sin 9^\circ \cdot \frac{15}{\sin 66^\circ}\\ b &= 2.569.\end{aligned}$$

Similarly,

$$\begin{aligned}\frac{a}{\sin 105^\circ} &= \frac{15}{\sin 66^\circ}\\ a &= \sin 105^\circ \cdot \frac{15}{\sin 66^\circ}\\ a &= 15.860.\end{aligned}$$

18. We begin by finding the angle C, which is $180^\circ - 95^\circ - 22^\circ = 63^\circ$.
We can now use the law of sines to find the other two sides.

$$\begin{aligned}\frac{b}{\sin 22^\circ} &= \frac{7}{\sin 63^\circ}\\ b &= \sin 22^\circ \cdot \frac{7}{\sin 63^\circ}\\ b &= 2.943.\end{aligned}$$

Similarly,

$$\begin{aligned}\frac{a}{\sin 95°} &= \frac{7}{\sin 63°}\\ a &= \sin 95° \cdot \frac{7}{\sin 63°}\\ a &= 7.826.\end{aligned}$$

19. We begin by finding the angle C, which is $180° - 77° - 42° = 61°$.
We can now use the law of sines to find the other two sides.

$$\begin{aligned}\frac{b}{\sin 42°} &= \frac{9}{\sin 61°}\\ b &= \sin 42° \cdot \frac{9}{\sin 61°}\\ b &= 6.885.\end{aligned}$$

Similarly,

$$\begin{aligned}\frac{a}{\sin 77°} &= \frac{9}{\sin 61°}\\ a &= \sin 77° \cdot \frac{9}{\sin 61°}\\ a &= 10.026.\end{aligned}$$

20. From the Law of Sines, we have

$$\begin{aligned}\frac{\sin A}{a} &= \frac{\sin C}{c}\\ \frac{\sin A}{8} &= \frac{\sin 98°}{17}\\ A &= \sin^{-1}\left(\frac{8\sin 98°}{17}\right)\\ &= 27.7755°.\end{aligned}$$

Since C is obtuse, A and B are acute, so this is the correct value of A. (Had A been obtuse, we would have had to subtract this value from 180° to compensate.) Now we have $B = 180° - A - C = 54.2245°$. Once again, we can use the Law of Sines to solve for b:

$$\begin{aligned}\frac{b}{\sin B} &= \frac{c}{\sin C}\\ \frac{b}{\sin 54.2245} &= \frac{17}{\sin 98°}\\ b &= \frac{17\sin 54.2245}{\sin 98°}\\ &= 13.9279.\end{aligned}$$

21. We begin by finding the angle B, which is $180° - 150° - 12° = 18°$.
We can now use the law of sines to find the other two sides.

$$\begin{aligned}\frac{a}{\sin 12°} &= \frac{5}{\sin 150°}\\ a &= \sin 12° \cdot \frac{5}{\sin 150°}\\ a &= 2.079.\end{aligned}$$

Similarly,

$$\frac{b}{\sin 18^\circ} = \frac{5}{\sin 150^\circ}$$
$$b = \sin 18^\circ \cdot \frac{5}{\sin 150^\circ}$$
$$b = 3.090.$$

22. We begin by finding the angle B, which is $180^\circ - 35^\circ - 92^\circ = 53^\circ$.
We can now use the law of sines to find the other two sides.

$$\frac{a}{\sin 92^\circ} = \frac{9}{\sin 35^\circ}$$
$$a = \sin 92^\circ \cdot \frac{9}{\sin 35^\circ}$$
$$a = 15.681.$$

Similarly,

$$\frac{b}{\sin 53^\circ} = \frac{9}{\sin 35^\circ}$$
$$b = \sin 53^\circ \cdot \frac{9}{\sin 35^\circ}$$
$$b = 12.531.$$

23. We begin by finding the angle B, which is $180^\circ - 9^\circ - 5^\circ = 166^\circ$.
We can now use the law of sines to find the other two sides.

$$\frac{a}{\sin 5^\circ} = \frac{3}{\sin 9^\circ}$$
$$a = \sin 5^\circ \cdot \frac{3}{\sin 9^\circ}$$
$$a = 1.671.$$

Similarly,

$$\frac{b}{\sin 166^\circ} = \frac{3}{\sin 9^\circ}$$
$$b = \sin 166^\circ \cdot \frac{3}{\sin 9^\circ}$$
$$b = 4.639.$$

24. First, we recognize that it is possible that there are two triangles, since we may have the ambiguous case. However, since 95° is greater than 90°, there are no other obtuse angles possible, so there is but one possible triangle.
We begin by finding the angle C using the law of sines:

$$\frac{\sin C}{10} = \frac{\sin 95^\circ}{5}$$
$$\sin C = 10 \cdot \frac{\sin 95^\circ}{5}$$
$$C = \sin^{-1} 1.992.$$

Since there is no arcsine of 1.992, we notice that there is a problem. There are no solutions. We could have seen this before because the longest side is always across from the largest angle, and since C cannot be greater than 95°, the side across from it (10) cannot be longer than the side across from 95°. Since it is bigger, no triangle fulfills the conditions given.

25. First, we recognize that it is possible that there are two triangles, since we may have the ambiguous case. However, we know that angle C must be less than 72°, since the side across from it is shorter than 13. Thus, we begin by finding the angle C using the law of sines:

$$\begin{aligned}\frac{\sin C}{4} &= \frac{\sin 72^\circ}{13}\\ \sin C &= 4\cdot\frac{\sin 72^\circ}{13}\\ C &= \sin^{-1} 0.293\\ C &= 17.016^\circ.\end{aligned}$$

We can now solve for A, which is $180^\circ - 72^\circ - 17.016^\circ = 90.984^\circ$.

Using the law of sines, we can solve for side a:

$$\begin{aligned}\frac{a}{\sin 90.984} &= \frac{13}{\sin 72^\circ}\\ a &= \sin 90.984\cdot\frac{13}{\sin 72^\circ}\\ a &= 13.667.\end{aligned}$$

26. First, we recognize that it is possible that there are two triangles, since we may have the ambiguous case. However, we know that angle C must be less than 75°, since the side across from it is shorter than 7. Thus, we begin by finding the angle C using the law of sines:

$$\begin{aligned}\frac{\sin C}{2} &= \frac{\sin 75^\circ}{7}\\ \sin C &= 2\cdot\frac{\sin 75^\circ}{7}\\ C &= \sin^{-1} 0.276\\ C &= 16.020^\circ.\end{aligned}$$

We can now solve for A, which is $180^\circ - 75^\circ - 16.020^\circ = 88.980^\circ$.

Using the law of sines, we can solve for side a:

$$\begin{aligned}\frac{a}{\sin 88.980} &= \frac{7}{\sin 75^\circ}\\ a &= \sin 88.980\cdot\frac{7}{\sin 75^\circ}\\ a &= 7.246.\end{aligned}$$

27. First, we recognize that it is possible that there are two triangles, since we may have the ambiguous case. Therefore, we know that angle C might be obtuse or acute. We take this into account while using the law of sines to find C:

$$\begin{aligned}\frac{\sin C}{8} &= \frac{\sin 17^\circ}{5}\\ \sin C &= 8\cdot\frac{\sin 17^\circ}{5}\\ C &= \sin^{-1} 0.468\\ C &= 27.891^\circ.\end{aligned}$$

However, C could also equal $180^\circ - \sin^{-1} 0.468 = 152.109^\circ$.

Thus, A can be either $180^\circ - 27.891^\circ - 17^\circ = 135.109^\circ$ or $180^\circ - 152.109^\circ - 17^\circ = 10.891^\circ$.

Using the law of sines, we can solve for side a in either case:

$$\frac{a}{\sin 135.109^\circ} = \frac{5}{\sin 17^\circ}$$
$$a = \sin 135.109^\circ \cdot \frac{5}{\sin 17^\circ}$$
$$a = 12.070,$$

or

$$\frac{a}{\sin 10.891^\circ} = \frac{5}{\sin 17^\circ}$$
$$a = \sin 10.891^\circ \cdot \frac{5}{\sin 17^\circ}$$
$$a = 3.231.$$

28. We use the Law of Cosines in Figure 7.40 to find the length of side b, getting

$$b^2 = 18.7^2 + 21^2 - 2(21)(18.7)\cos 22^\circ \approx 62.480.$$

So $b \approx 7.904$ cm. To find γ we use the Law of Cosines:

$$21^2 = 18.7^2 + 7.904^2 - 2 \cdot 18.7 \cdot 7.904 \cos\gamma$$
$$\cos\gamma = \frac{-28.830}{294.625}$$
$$\gamma = \cos^{-1}\left(\frac{-28.830}{294.625}\right) = 95.597^\circ$$

Then $\alpha = 180^\circ - 22^\circ - 95.597^\circ = 62.403^\circ$.

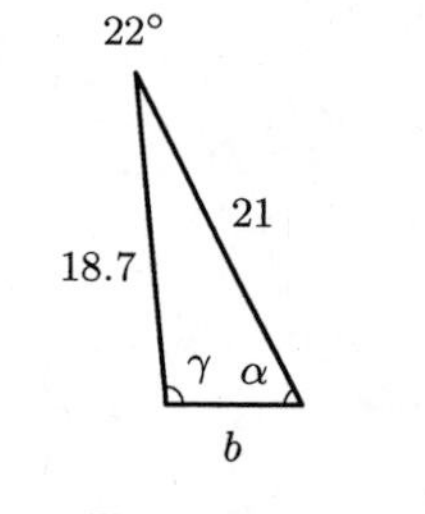

Figure 7.40

29.

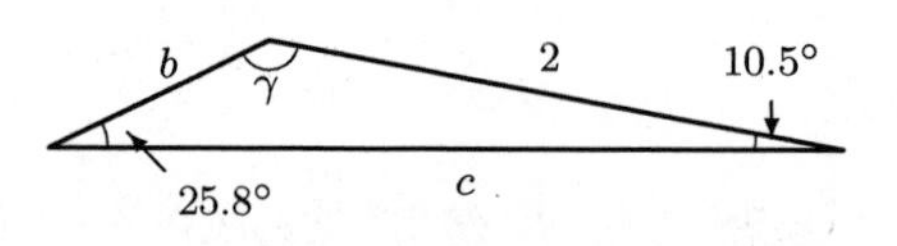

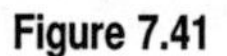
Figure 7.41

The Law of Sines tells us that $\frac{b}{\sin 10.5^\circ} = \frac{2}{\sin 25.8^\circ}$, so $b = 0.837$ m. We have $\gamma = 180^\circ - 10.5^\circ - 25.8^\circ = 143.7^\circ$. We use this value to find side length c. We have $c^2 = 2^2 + (.837)^2 - 2(2)(.837)\cos 143.7^\circ \approx 7.401$, or $c = 2.720$ m.

30. In Figure 7.42, use the Law of Sines:

$$\frac{\sin(30^\circ)}{259} = \frac{\sin\beta}{510}$$

to obtain $\sin\beta \approx 0.9846$ and use $\sin^{-1}$ to find $\beta_1 \approx 79.917$ or $\beta_2 \approx 100.083$. We then know $\alpha_1 = 180^\circ - 30^\circ - 79.917^\circ \approx 70.083^\circ$, or $\alpha_2 = 180^\circ - 30^\circ - 100.083^\circ \approx 49.917^\circ$. We can use the value of α and the Law of Sines to find the length of side a:

$$\frac{a_1}{\sin(70.083^\circ)} = \frac{259}{\sin 30^\circ}, \quad \text{or} \quad \frac{a_2}{\sin(49.917^\circ)} = \frac{259}{\sin 30^\circ}.$$
$$a_1 \approx 487.016 \text{ ft} \qquad a_2 \approx 396.330 \text{ ft}$$

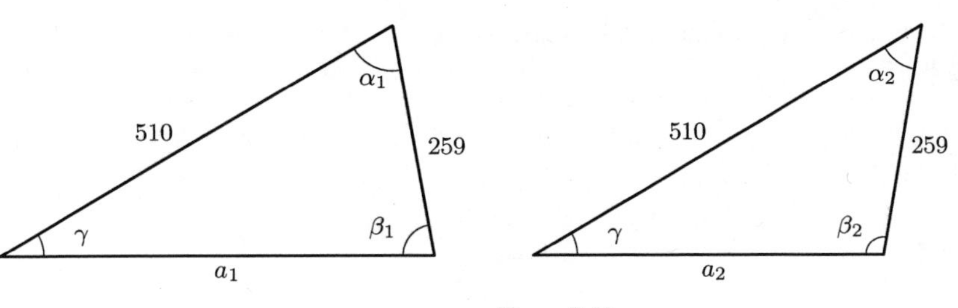

Figure 7.42

31. We use the Law of Cosines in Figure 7.43 to find α and β:

$$16^2 = 20^2 + 24^2 - 2(20)(24)\cos\alpha$$
$$\cos\alpha = 0.75$$
$$\alpha \approx 41.410^\circ.$$

$$24^2 = 20^2 + 16^2 - 2(20)(16)\cos(\beta)$$
$$\cos\beta = 0.125$$
$$\beta \approx 82.819^\circ.$$

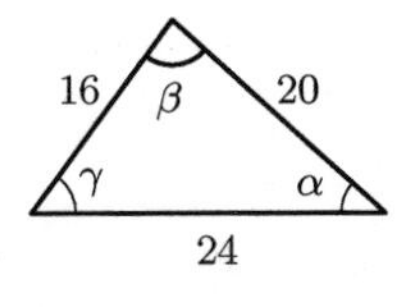

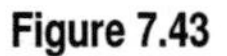
Figure 7.43

Finally $\gamma = 180^\circ - 41.410^\circ - 82.819^\circ = 55.771^\circ$.

Problems

32. **(a)** In a right triangle $\sin\theta = \text{Opp}/\text{Hyp}$. Thus $\sin\theta = 3/7$.
To find $\sin\phi$ we use the Law of Sines:

$$\frac{\sin\phi}{15} = \frac{\sin(20^\circ)}{8}.$$

This implies that

$$\sin\phi = \frac{15\sin(20^\circ)}{8}.$$

(b) Since $\sin\theta = \frac{3}{7}$,

$$\theta = \sin^{-1}\left(\frac{3}{7}\right) \approx 25.377^\circ.$$

This makes sense, as we expect $0^\circ < \theta < 90^\circ$.

For $\sin\phi = \dfrac{15\sin(20^\circ)}{8}$, there are two solutions

$$\phi = \sin^{-1}\left(\frac{15\sin(20^\circ)}{8}\right) \approx 39.888^\circ \quad \text{and} \quad \phi = 180^\circ - \sin^{-1}\left(\frac{15\sin(20^\circ)}{8}\right) \approx 140.112^\circ.$$

Assuming that $\phi > 90^\circ$, as the figure suggests, we choose the second solution $\phi \approx 140.112^\circ$.

33. **(a)** By the Law of Sines, we have

$$\frac{\sin\theta}{3} = \frac{\sin 110^\circ}{10}$$
$$\sin\theta = \left(\frac{3}{10}\right)\sin 110^\circ \approx 0.282.$$

(b) If $\sin\theta = 0.282$, then $\theta \approx 16.374°$ (as found on a calculator) or $\theta \approx 180° - 16.374° \approx 163.626°$. Since the triangle already has a $110°$ angle, $\theta \approx 16.374°$. (The $163.626°$ angle would be too large.)

(c) The height of the triangle is $10\sin\theta = 10 \cdot 0.282 = 2.819$ cm. Since the sum of the angles of a triangle is $180°$, and we know two of the angles, $\theta = 16.374°$ and $110°$, so the third angle is $180° - 16.374° - 110° = 53.626°$. By the Law of of Sines, we have

$$\frac{\sin 110°}{10} = \frac{\sin 53.626°}{\text{Base}}$$
$$\text{Base} = \frac{10\sin 53.626°}{\sin 110°} \approx 8.568 \text{ cm.}$$

Thus, the triangle has

$$\text{Area} = \frac{1}{2}\text{Base} \cdot \text{Height} = \frac{1}{2} \cdot 8.568 \cdot 2.819 = 12.077 \text{ cm}^2.$$

34.

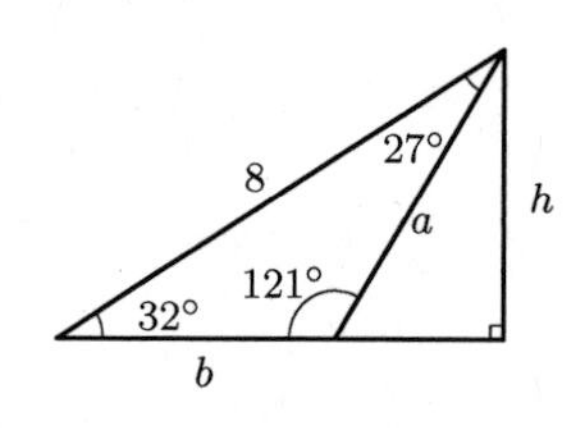

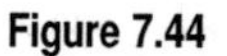
Figure 7.44

(a) $\dfrac{\sin 121°}{8} = \dfrac{\sin 32°}{a}$, so $a = \dfrac{8\sin 32°}{\sin 121°} \approx 4.946$. Similarly, $b = \dfrac{8\sin 27°}{\sin 121°} \approx 4.2371$.

(b) Construct an altitude h as in Figure 7.44. We have $\sin 32° = \dfrac{h}{8}$, so $h = 8\sin 32° \approx 4.2394$. Then area of the triangle is $\frac{1}{2}(4.2371)(4.2394) = 8.981$.

35. In Figure 7.45, the fire stations are at A and B and the forest fire is at C. The angle at C is $180° - 54° - 58° = 68°$. Solving for a and b using the Law of Sines, we get

$$\frac{56.7}{\sin 68°} = \frac{a}{\sin 54°} \qquad \frac{56.7}{\sin 68°} = \frac{b}{\sin 58°}$$

$$a = \frac{56.7\sin 54°}{\sin 68°} \qquad b = \frac{56.7\sin 58°}{\sin 68°}$$

$$a = 49.4738 \qquad b = 51.8606.$$

The fire station at point B is closer by $51.8606 - 49.4738 = 2.387$ miles.

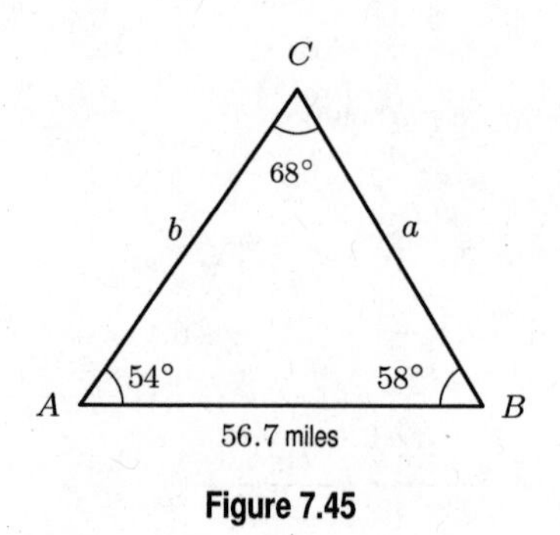

Figure 7.45

36. Figure 7.46 shows the Eiffel Tower, CD, and the two angles of elevation, $\angle CAD = 60°$ and $\angle CBD = 70°$. In triangle ABC, we find $\angle ABC = 180° - 70° = 110°$ and $\angle ACB = 180° - 110° - 60° = 10°$. We use the Law of Sines to solve for x:

$$\frac{210}{\sin 10°} = \frac{x}{\sin 60°}$$
$$x = \frac{210 \sin 60°}{\sin 10°} = 1047.321$$

Using the value of x in the right triangle CBD to find y, we get

$$\sin 70° = \frac{y}{1047.321}$$
$$y = \sin 70°(1047.321) \approx 984.160 \text{ feet.}$$

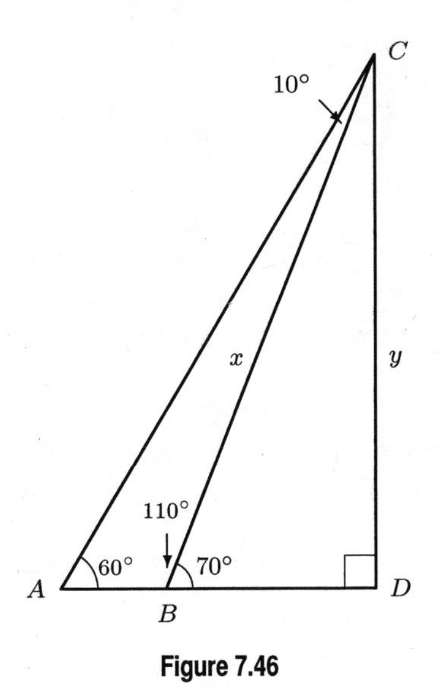

Figure 7.46

37. After half an hour, the DC-10 has traveled $509 \cdot (1/2) = 254.5$ miles and the L-1011 has traveled $503 \cdot (1/2) = 251.5$ miles. See Figure 7.47. Using the Law of Cosines:

$$x^2 = 254.5^2 + 251.5^2 - 2(254.5)(251.5)\cos 103°$$
$$x = 396.004 \text{ miles.}$$

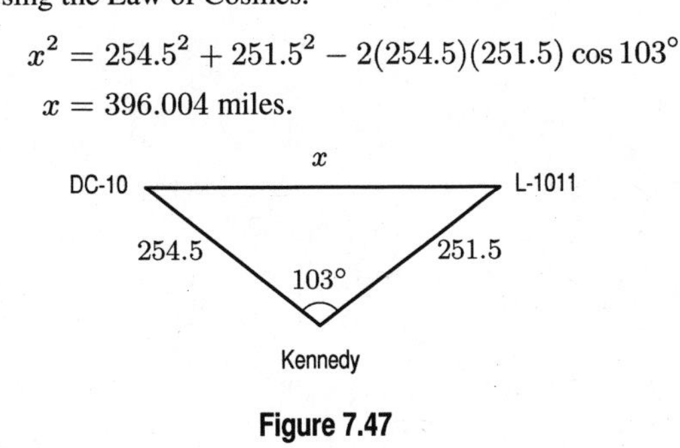

Figure 7.47

38. Figure 7.48 shows the triangle; we want to find x. The other two angles are $(180 - 39)°/2 = 70.5°$. Using the Law of Sines

$$\frac{425}{\sin 39°} = \frac{x}{\sin 70.5°}$$
$$x = \frac{425 \sin 70.5°}{\sin 39°} = 636.596 \text{ feet.}$$

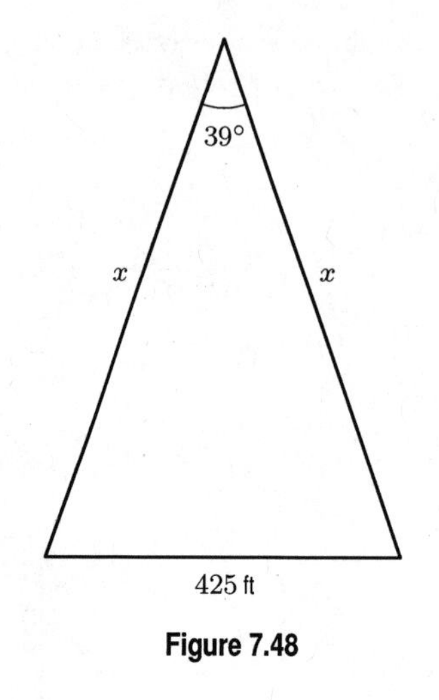

Figure 7.48

39. The horizontal and vertical displacements of the image are given by

$$\Delta x = 12\cos 25^\circ = 10.876$$
$$\Delta y = 12\sin 25^\circ = 5.071.$$

Thus, the new coordinates are $(x, y) = (8 + \Delta x, 5 + \Delta y) = (18.876, 10.071)$.

40. Initially, the angle θ made by the image with respect to the origin is given by

$$\tan\theta = \frac{3}{5}$$
$$\theta = \tan^{-1}(3/5) \approx 30.964^\circ,$$

and the distance from the origin is given by

$$r = \sqrt{5^2 + 3^2} = \sqrt{34}.$$

After the rotation, the new angle is $\phi = \theta + 42^\circ = 72.964^\circ$, but the distance does not change. The x- and y-coordinates are

$$x = r\cos\phi = \sqrt{34}\cos(72.964^\circ) = 1.708$$
$$y = r\sin\phi = \sqrt{34}\sin(72.964^\circ) = 5.575,$$

so the new position is $(1.708, 5.575)$.

41. If we look at Figure 7.49, we see that there are two triangles: the original triangle with angles A, B, C and the right triangle with hypotenuse b.

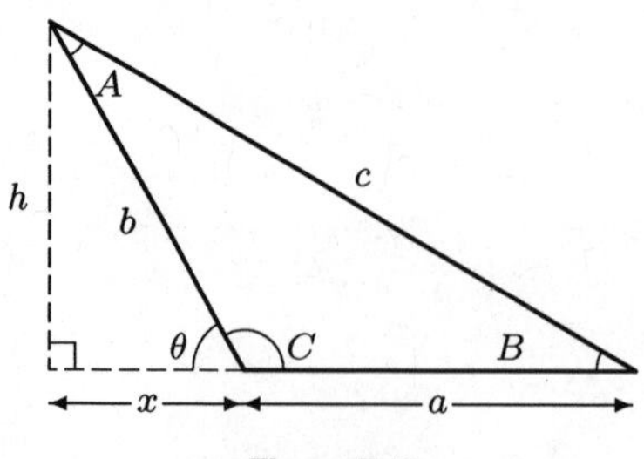

Figure 7.49

The Pythagorean theorem gives

$$x^2 + h^2 = b^2,$$

or

$$h^2 = b^2 - x^2.$$

If we apply the Pythagorean theorem to the right triangle with legs h and $x + a$ we obtain

$$(x + a)^2 + h^2 = c^2.$$

Substituting $h^2 = b^2 - x^2$ into this equation gives

$$x^2 + 2ax + a^2 + \underbrace{b^2 - x^2}_{h^2} = c^2.$$

This, in turn, reduces to

$$a^2 + b^2 + 2ax = c^2.$$

We now determine x. Since C is obtuse, $\cos C$ will be negative. We have

$$\cos C = -\cos\theta = -\frac{x}{b},$$

or

$$x = -b\cos C.$$

Substituting this expression for x into our equation gives the Law of Cosines:

$$a^2 + b^2 - 2a\underbrace{b\cos C}_{x} = c^2.$$

42. From the figure, we see that $\sin A = \frac{h}{b}$, which gives $h = b\sin A$. We also have $\sin B = \frac{h}{a}$, which gives $h = a\sin B$. Thus, $b\sin A = a\sin B$, which gives the Law of Sines:

$$\frac{\sin A}{a} = \frac{\sin B}{b}.$$

43. **(a)** See Figure 7.50.

Let x be the distance from the pitcher's mound to first base. Then, by the Law of Cosines

$$x^2 = 60.5^2 + 90^2 - 2(60.5)(90)\cos 45^\circ$$
$$x = 63.717.$$

To find the distance from the pitcher's mound to second base, let y be the distance from home plate to second base. Then

$$y^2 = 90^2 + 90^2$$
$$y = 127.279.$$

Then we find the distance from the pitcher's mound to second:

$$\text{Distance} = 127.279 - 60.5 = 66.779.$$

From the pitcher's mound to first base is closer by $66.779 - 63.717 = 3.062$ feet.

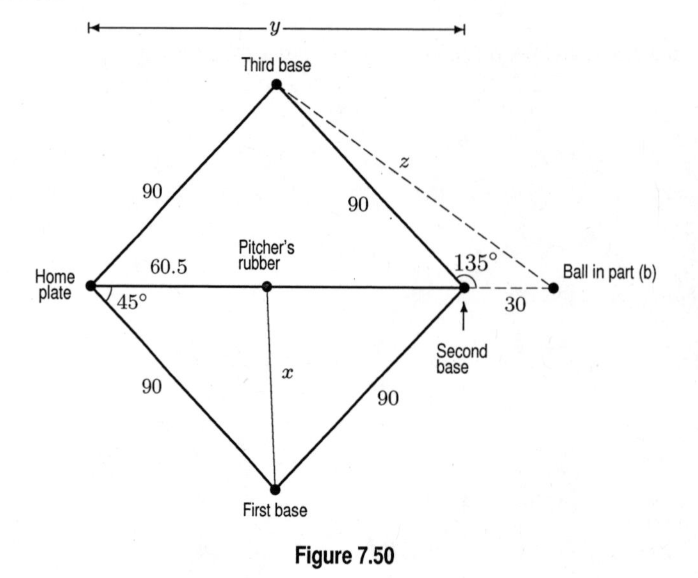

Figure 7.50

(b) Using a result from part (a), the distance from 30 feet past second base to home plate is given by

$$\text{Distance} = 30 + 127.279 = 157.279 \text{ feet.}$$

Let z be the distance from 30 feet past second base to third base:

$$z^2 = 30^2 + 90^2 - 2(30)(90)\cos 135^\circ$$
$$z = 113.218 \text{ feet.}$$

44. From Figure 7.51, we see $\angle ABC = 180^\circ - 93^\circ - 49^\circ = 38^\circ$. Using the Law of Sines, we have

$$\frac{102}{\sin 38^\circ} = \frac{c}{\sin 49^\circ}$$
$$c = \frac{102 \sin 49^\circ}{\sin 38^\circ} = 125.037 \text{ feet.}$$

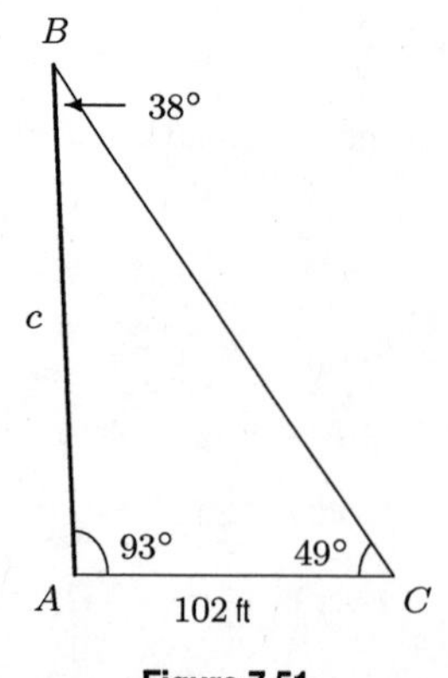

Figure 7.51

45. The three stakes are at A, B, C in Figure 7.52. Using the Law of Cosines, we have

$$x^2 = 82^2 + 97^2 - 2(82)(97)\cos 125^\circ$$
$$x \approx 158.926.$$

The mound is approximately 158.926 feet wide.

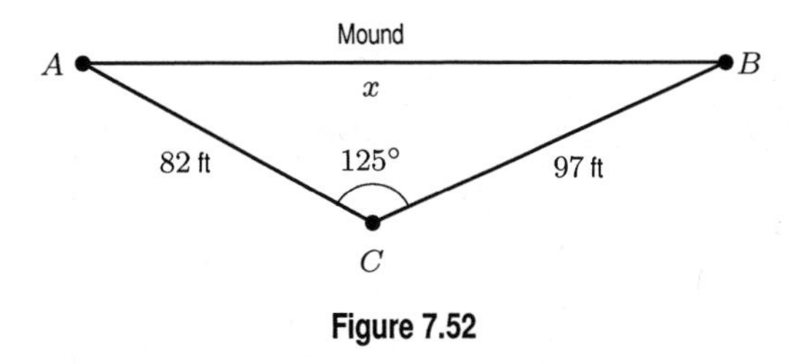

Figure 7.52

46. One way to organize this situation is to use the abbreviations from high school geometry. The six possibilities are { SSS, SAS, SSA, ASA, AAS, AAA }.

SSS Knowing all three sides allows us to find the angles by using the Law of Cosines.

SAS Knowing two sides and the included angle allows us to find the third side length by using the Law of Cosines. We can then use the SSS procedure.

SSA Knowing two sides but not the included angle is called the ambiguous case, because there could be two different solutions. Use the Law of Sines to find one of the missing angles, which, because we use the arcsine, may give two values. Or, use the Law of Cosines, which produces a quadratic equation that may also give two values. Treating these cases separately we can continue to find all sides and angles using the SAS procedure.

ASA Knowing two angles allows us to easily find the third angle. Use the Law of Sines to find each side.

AAS Find the third angle and then use the Law of Sines to find each side.

AAA This has an infinite number of solutions because of similarity of triangles. Once one side is known, then the ASA or AAS procedure can be followed.

47. Using the Law of Sines on triangle DEF in Figure 7.53, we have

$$\frac{105.2}{\sin 29^\circ} = \frac{ED}{\sin 68^\circ}$$
$$ED = \frac{105.2\sin 68^\circ}{\sin 29^\circ} = 201.192 \text{ feet.}$$

$$\text{Total amount of wire needed} = 201.192 + 145.3 + 23.5 + 20 = 389.992 \text{ feet.}$$

Since wire is sold in 100 feet rolls, 4 rolls of wire are needed.

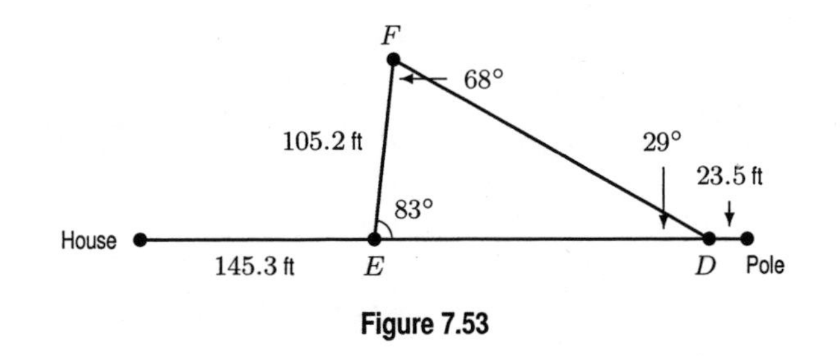

Figure 7.53

48. **(a)** By the Law of Sines,

$$\frac{\sin 82.6}{435} = \frac{\sin \psi}{112}.$$

Solving for ψ gives

$$\begin{aligned} 112\frac{\sin 82.6}{435} &= \sin \psi \\ \psi &= \sin^{-1}\left(112\frac{\sin 82.6}{435}\right) \approx 14.8^\circ \end{aligned}$$

Thus $\theta \approx 180 - 82.6 - 14.8 = 82.6^\circ$.

Knowing the angle θ now allows us to solve for the distance LT. By Law of Cosines:

$$\begin{aligned} LT &= \sqrt{112^2 + 435^2 - 2(112)(435)\cos 82.6} \\ &\approx 435 \text{ ft.} \end{aligned}$$

This answer can be also seen by figuring out that TOL is an isosceles triangle and therefore $LT = TO = 435$ ft.

(b) One way is to use the fact that it is an isosceles triangle and calculate the height from point T to the base OL. Then

$$\text{Area } = \frac{1}{2}(112)(56\tan 82.6^\circ) \approx 24{,}145.86 \text{ ft}^2.$$

This is about half an acre.

Solutions for Chapter 7 Review

Exercises

1. This function appears to be periodic because it repeats regularly.
2. This appears to be a periodic function, because it repeats regularly.
3. This does not appear to be a periodic function, because, while it rises and falls, it does not rise and fall to the same level regularly.
4. This appears to be a periodic function, because it repeats regularly.
5. This function does not appear to be periodic. Though it does rise and fall, it does not do so regularly.
6. This is not the graph of a periodic function. Although the y-values repeat, they do not repeat at regular intervals. The intervals get progressively shorter.
7. This could be a periodic function with a period of 3. The values of $f(t)$ repeat each time t increases by 3.
8. This function does not appear to be periodic. Though it does rise and fall, it seems to do so irregularly.
9.

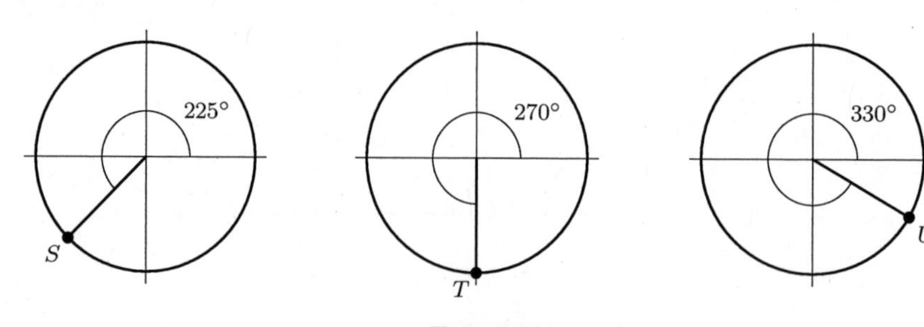

Figure 7.54

$S = (-0.707, -0.707), T = (0, -1), U = (0.866, -0.5)$

10. Point A is at 390°. The angle 390° is located by first wrapping around the circle, which accounts for 360°, and then continuing for an additional 30°. Similarly, B, which is at 495°, is located by first wrapping around the circle, and then continuing for an additional 495° − 360° = 135°. Finally, C, which as at 690°, is located by first wrapping around the circle, giving 360°, and then continuing for an additional 690° − 360° = 330°.

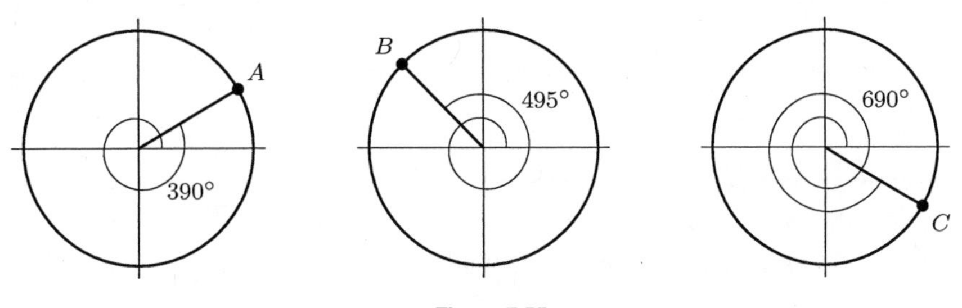

Figure 7.55

$A = (0.866, 0.5)$, $B = (-0.707, 0.707)$, $C = (0.866, -0.5)$

11. Since $x = r\cos\theta$ and $y = r\sin\theta$ we have

$$\begin{aligned} S &= (5\cos 225°, 5\sin 225°) = (-3.536, -3.536)\\ T &= (5\cos 270°, 5\sin 270°) = (0, -5)\\ U &= (5\cos 330°, 5\sin 330°) = (4.330, -2.5)\end{aligned}$$

12. Since $x = r\cos\theta$ and $y = r\sin\theta$ we have

$$\begin{aligned} A &= (3\cos 390°, 3\sin 390°) = (2.598, 1.5)\\ B &= (3\cos 495°, 3\sin 495°) = (-2.121, 2.121)\\ C &= (3\cos 690°, 3\sin 690°) = (2.598, -1.5)\end{aligned}$$

13.

$$x = r\cos\theta = 16\cos(-72°) \approx 4.944$$

and

$$y = r\sin\theta = 16\sin(-72°) \approx -15.217,$$

so the approximate coordinates of Z are $(4.944, -15.217)$.

14. **(a)** $\sin\theta$ and $\cos\theta$ are both positive in quadrant I only.
(b) $\tan\theta > 0$ in quadrants I and III.
(c) $\tan\theta < 0$ in quadrants II and IV.
(d) $\sin\theta < 0$ in quadrants III and IV. $\cos\theta > 0$ in quadrants I and IV. Thus both are true only in quadrant IV.
(e) $\cos\theta < 0$ in quadrants II and III, and $\tan\theta > 0$ in quadrants I and III. Both are true only in quadrant III.

15. Since we are looking for the angle θ, we have $\tan\theta = 0.999$, so $\theta = \tan^{-1} 0.999 = 44.971°$.

16. Since we are looking for the angle θ, we have $\sin\theta = \frac{3}{5}$, so $\theta = \sin^{-1}\left(\frac{3}{5}\right) = 36.870°$.

17. Since we are looking for the angle θ, we have $\tan\theta = \frac{5}{3}$, so $\theta = \tan^{-1}\frac{5}{3} = 59.036°$.

18. Since the output of the cosine function is $0 < \theta < 1$, there is no angle whose cosine is $\frac{5}{3}$.

19. We know that $\cos 30° = \frac{\sqrt{3}}{2}$. Therefore, $\theta = 30°$.

20. We know that $\sin 30° = \frac{1}{2}$. Therefore, $\theta = 30°$.

21. We know that $\cos 45° = \frac{\sqrt{2}}{2}$. Therefore, $\theta = 45°$.

22. When $\theta = 45°$, both the sine and cosine functions have the same value. Therefore, $\theta = 45°$.

23. Since $\cos^{-1} x = y$ means $\cos y = x$, the angle is y and the value is x.

24. Since $(\sin w)^{-1} = \dfrac{1}{\sin w} = g$, we have $\sin w = \dfrac{1}{g}$. Therefore the angle is w and the value is $\dfrac{1}{g}$.

25. Since $\tan^{-1}\left(c^{-1}\right) = d$ means $\tan d = \dfrac{1}{c}$, the angle is d and the value is $\dfrac{1}{c}$.

26. Since $(\cos t)^{-1} = p^{-1}$ means $\dfrac{1}{\cos t} = \dfrac{1}{p}$, we have $\cos t = p$. Therefore, the angle is t and the value is p.

Problems

27. A clock's face shows twelve hours, so each hour spans $360°/12 = 30°$.

(a) Two o'clock is $30°$ away from 3 o'clock in the counter-clockwise direction, so it can be specified by the angle $\theta = 30°$.

(b) Four o'clock is $30°$ away from 3 o'clock in the clockwise direction, so it can be specified by the angle $\theta = -30°$. Alternatively, it is $11 \times 30° = 330°$ away in the counter-clockwise direction, so it can be specified by the angle $\theta = 330°$.

(c) Ten o'clock is five hours away from 3 o'clock in the counter-clockwise direction, so it can be specified by the angle $\theta = 150°$.

See Figure 7.56.

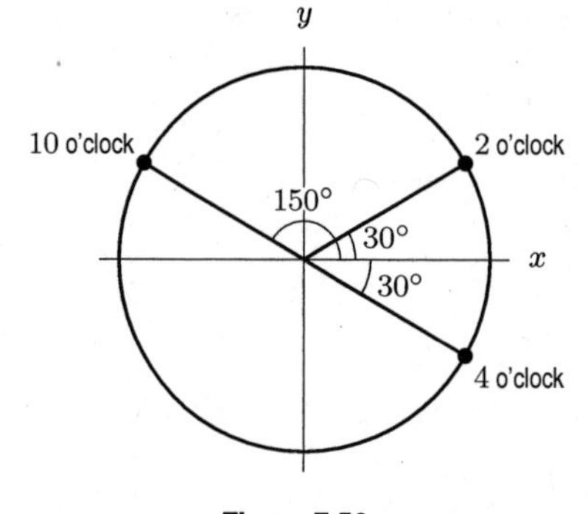

Figure 7.56

28. Let $P = (\cos 65°, \sin 65°) = (0.4226, 0.9063)$ be the point on the unit circle given by the angle $65°$. In Figure 7.57, we see that $-65°$ gives a point labeled Q that is the reflection of P across the x-axis. Thus, the y-coordinate of Q is the negative of the y-coordinate of P, so $Q = (0.4226, -0.9063)$. This means that

$$\sin(-65°) = -0.9063 \quad \text{and} \quad \cos(-65°) = 0.4226.$$

In Figure 7.58, we see that $245° = 180° + 65°$ gives point R that is diametrically opposite the point P. The coordinates of R are the negatives of the coordinates of P, so $R = (-0.4226, -0.9063)$. Thus,

$$\sin 245° = -0.9063 \quad \text{and} \quad \cos 245° = -0.4226.$$

Finally in Figure 7.59, we see that $785° = 720° + 65°$, so this angle specifies the same point as $65°$. This means that

$$\sin 785° = 0.9063 \quad \text{and} \quad \cos 785° = 0.4226.$$

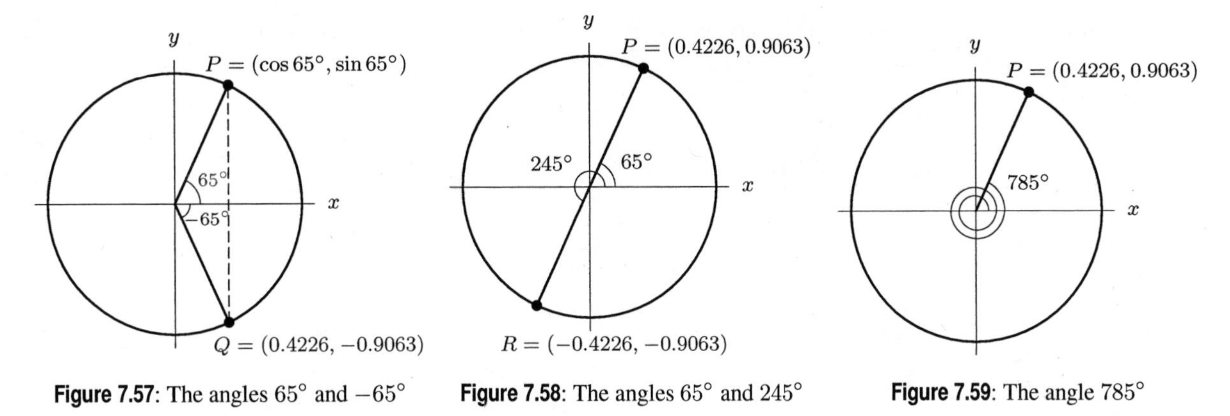

Figure 7.57: The angles 65° and -65° **Figure 7.58**: The angles 65° and 245° **Figure 7.59**: The angle 785°

29. In graph A, the average speed is relatively high with little variation, which corresponds to (iii). In graph B, the average speed is lower and there are significant speed-ups and slow-downs, which corresponds to (i). In graph C, the average speed is low and frequently drops to 0, which this corresponds to (ii).

30. $g(x) = 2\sin x$, $a = 180^\circ$ and $b = 2$.

31. The function completes one and a half oscillations in 9 units of t, so the period is $9/1.5 = 6$, the amplitude is 5, and the midline is $y = 0$.

32. The period is 4, the amplitude 3, and the midline $y = -3$.

33. By plotting the data in Figure 7.60, we can see that the midline is at $h = 2$ (approximately). Since the maximum value is 3 and the minimum value is 1, we have

$$\text{Amplitude} = 2 - 1 = 1.$$

Finally, we can see from the graph that one cycle has been completed from time $t = 0$ to time $t = 1$, so the period is 1 second.

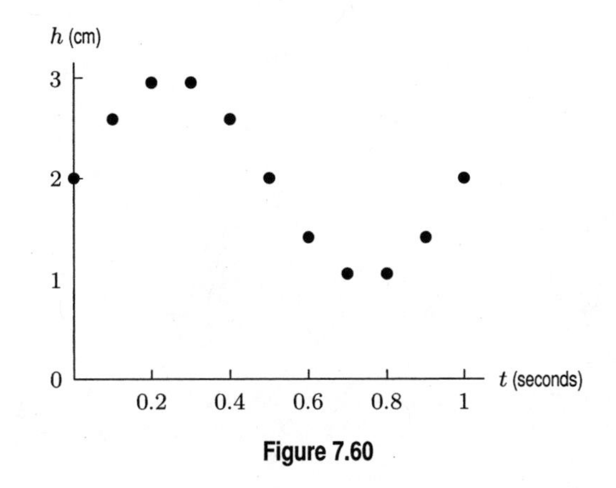

Figure 7.60

34. By plotting the data in Figure 7.61, we can see that the midline is at $h = 4$ (approximately). Since the maximum value is 6.5 and the minimum value is 1.5, we have

$$\text{Amplitude} = 4 - 1.5 = 6.5 - 4 = 2.5.$$

Finally, we can see from the graph that one cycle has been completed from time $t = 0$ to $t = 12$, so the period is 12 seconds.

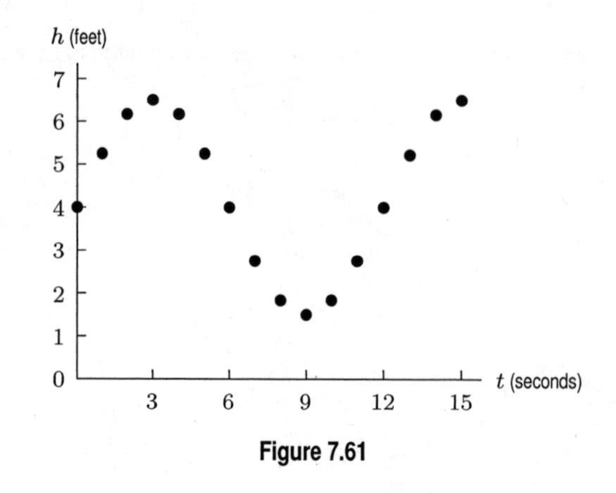

Figure 7.61

35. The period of the Ferris wheel is 30 minutes, so in 17.5 minutes you will travel $17.5/30$, or $7/12$, of a complete revolution, which is $7/12(360°) = 210°$ from your starting position (the 6 o'clock position). This location is $120°$ from the horizontal line through the center of the wheel. Thus, your height is $225 + 225 \sin 120° = 225 + 225(0.866) \approx 419.856$ feet above the ground.

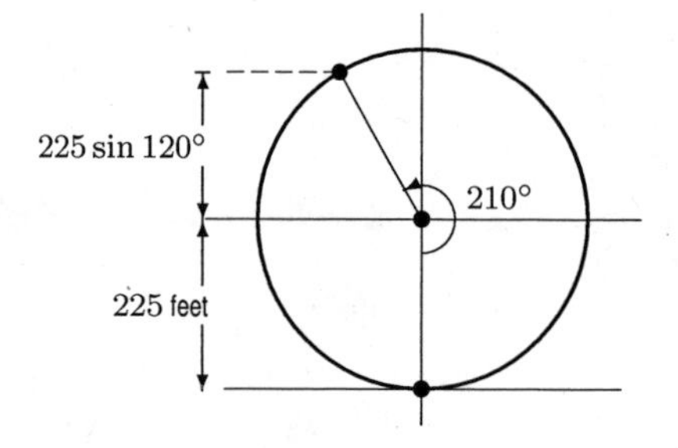

Figure 7.62

36. **(a)**

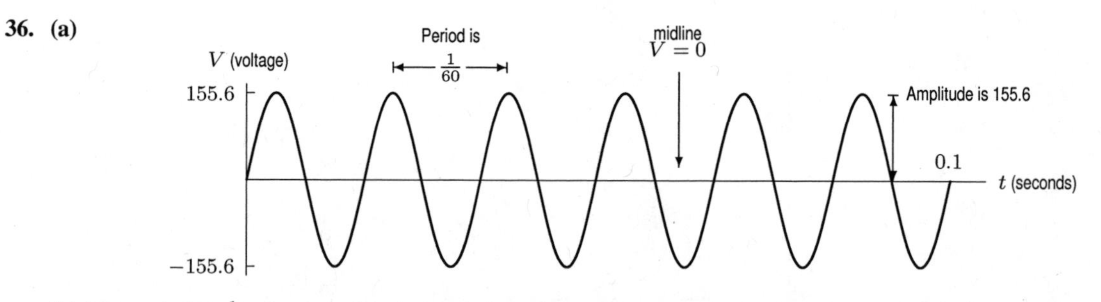

(b) The period is $\frac{1}{60}$ of a second as there are 60 cycles each second. The midline value, 0 volts, is the average voltage over one whole period. The amplitude, 155.6 volts, is the maximum amount by which the voltage can vary, either above or below the midline.

37. We know all three sides of this triangle, but only one of its angles. We find the value of $\sin\theta$ and $\sin\phi$ in this right triangle:

$$\sin\theta = \frac{\text{opposite}}{\text{hypotenuse}} = \frac{3}{5} = 0.6$$

and

$$\sin\phi = \frac{\text{opposite}}{\text{hypotenuse}} = \frac{4}{5} = 0.8.$$

Using inverse sines, we know that if $\sin\phi = 0.8$, then $\phi = \sin^{-1}(0.8) \approx 53.130°$. Similarly $\sin\theta = 0.6$ means $\theta = \sin^{-1}(0.6) \approx 36.870°$. Notice $\phi + \theta = 90°$, which has to be true in a right triangle.

38. The other angle must be $\theta = 90° - 59° = 31°$ (See Figure 7.63). By definition of the tangent,

$$\tan 59° = \frac{5}{x}$$
$$x = \frac{5}{\tan 59°} \approx 3.004.$$

By definition of the sine,

$$\sin 59° = \frac{5}{y}$$
$$y = \frac{5}{\sin 59°} \approx 5.833.$$

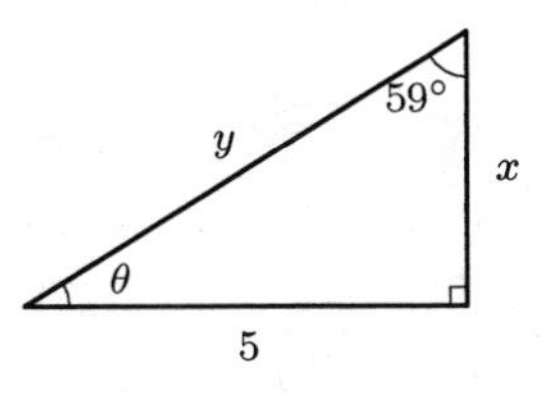

Figure 7.63

39. See Figure 7.64. By the Pythagorean theorem, $x = \sqrt{13^2 - 12^2} = 5$.

$$\sin\theta = \frac{12}{13}$$
$$\theta = \sin^{-1}\left(\frac{12}{13}\right)$$
$$\theta \approx 67.380°.$$

Thus $\varphi = 90° - \theta \approx 22.620°$.

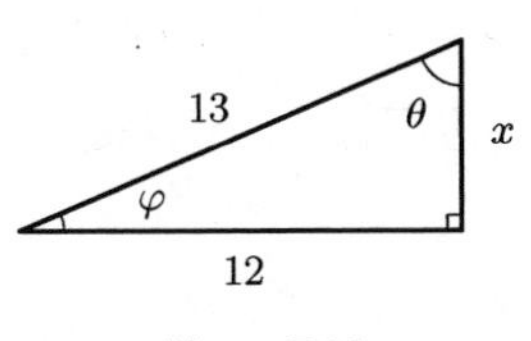

Figure 7.64

40. The other angle must be $\theta = 180° - 33° - 42° = 105°$ (See Figure 7.65). By the Law of Sines,

$$\frac{y}{\sin 42°} = \frac{8}{\sin 33°}$$
$$y = 8\left(\frac{\sin 42°}{\sin 33°}\right) \approx 9.829.$$

Again using the Law of Sines,

$$\frac{x}{\sin 105°} = \frac{8}{\sin 33°}$$
$$x = 8\left(\frac{\sin 105°}{\sin 33°}\right) \approx 14.188.$$

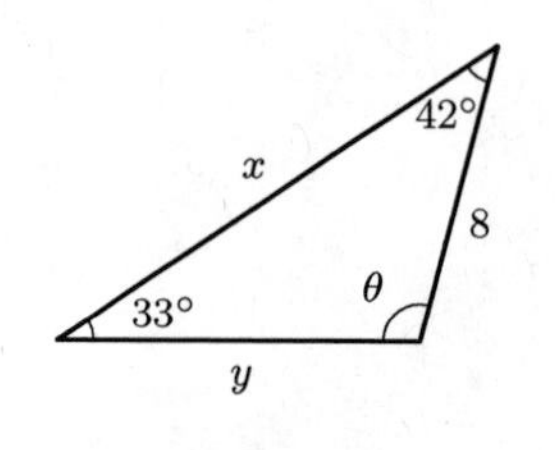

Figure 7.65

41. First check to see if there is a right triangle. It is not because $25^2 + 52^2 \neq 63^2$. So we must use the Law of Cosines:

$$\begin{aligned} 25^2 &= 63^2 + 52^2 - 2(63)(52)\cos\theta \\ \frac{25^2 - 63^2 - 52^2}{-2(63)(52)} &= \cos\theta \\ 0.9231 &\approx \cos\theta \\ \arccos(0.9231) &\approx \theta \\ 22.620^\circ &\approx \theta \end{aligned}$$

42. We have

$$\tan 28^\circ = \frac{\overbrace{\text{Leg opposite}}^{x}}{\underbrace{\text{Leg adjacent}}_{4}} = \frac{x}{4}$$

so

$$\begin{aligned} x &= 4\tan 28^\circ \\ &= 4(0.5317) \quad \text{using a calculator} \\ &= 2.1268. \end{aligned}$$

43. See Figure 7.91. Since the tangent is the length of the opposite side divided by the length of the adjacent side,

$$\tan 58^\circ = \frac{d}{50}$$

$$d = 50\tan 58^\circ \approx 80.017$$

The width of the river is about 80 meters.

44. Using right triangle trigonometry, we have

$$\begin{aligned} \tan\theta &= \frac{30}{16} \\ \theta &= (\tan)^{-1}\left(\frac{30}{16}\right) \\ &= 61.928^\circ. \end{aligned}$$

45. The angle of observation is labeled θ in Figure 7.92. The distance 200 meters forms the adjacent side for this angle and the height of the balloon, h, is the opposite side. Thus,

$$\tan\theta = \frac{h}{200},$$

so

$$h = 200\tan\theta.$$

46. Consider lines OP_1 and OP_2 in Figure 7.66. Let A and B be the lengths of OP_1 and OP_2 respectively. The right angle formed by OP_1 and P_3P_1 gives $200/A = \sin 25°$, so $A = 200/\sin 25° \approx 473.240$. Also, $400/B = \sin 50°$, so $B = 400/\sin 50° \approx 522.163$. In triangle OP_1P_2, the angle at O is $50 - 25 = 25°$. Applying the Law of Cosines to the triangle OP_1P_2 gives $A^2 + B^2 - 2AB\cos 25° = d^2$. So $d \approx 220.676$ m.

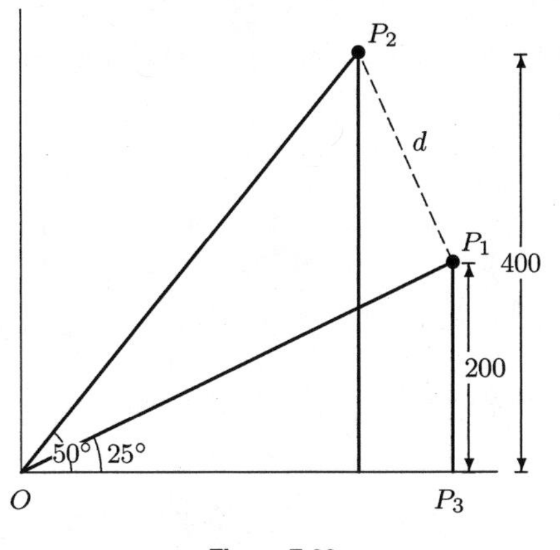

Figure 7.66

CHECK YOUR UNDERSTANDING

1. True. The point $(1, 0)$ on the unit circle is the starting point to measure angles so $\theta = 0°$.
2. True. This is the definition of $\sin\theta$.
3. True. Since $\cos\theta = x/r$ we have $x = r\cos\theta$.
4. False. The point is $(2\cos 240°, 2\sin 240°) = (-1, -\sqrt{3})$.
5. True. Because P and Q have the same x-coordinates.
6. True. Because P and R have the same y-coordinates.
7. True. Since the angle is greater than 180° but less than 270°, the point is in the third quadrant.
8. False. Since both angles are negative they are measured in the clockwise direction. Going clockwise beyond 180° takes the point into the second quadrant.
9. False. The angle 315° is in the fourth quadrant, so the cosine value is positive. The correct value is $\sqrt{2}/2$.
10. True. This is the definition of the period of f.
11. False. The amplitude is half the difference between its maximum and minimum values.
12. True.
13. False. A unit circle must have a radius of 1.
14. False. A parabola is a quadratic function and can never repeat its values more than twice.
15. True. This is the definition of a periodic function.
16. True. The graph of a periodic function is obtained by starting with the graph on one period and horizontally shifting it by all multiples of the period. The entire graph is determined by the graph of a single period. Since the graphs of the periodic functions $f(t)$ and $g(t)$ are the same on the period $0 \le t < A$, their graphs are the same everywhere.
17. False. Both acute angles are 45 degrees, and $\sin 45° = \sqrt{2}/2$.
18. False. The ladder is the hypotenuse of a right triangle. Using the 60° angle we know $\sqrt{3}/2 = \sin 60° = w/16$, where w is the height above ground. Solving this equation we get $w = 8\sqrt{3}$.
19. False. It can be applied to any triangle.

20. True. For example, to find angle C, substitute the values a, b and c, into $c^2 = a^2 + b^2 - 2ab\cos C$. Solve for the angle C by using algebra and the $\cos^{-1}$ function.

21. True. By the Law of Cosines, we have $p^2 = n^2 + r^2 - 2nr\cos P$, so $\cos P = (n^2 + r^2 - p^2)/(2nr)$.

22. True. If $C = 90°$, then $\cos C = 0$, so the Law of Cosines, $c^2 = a^2 + b^2 - 2ab\cos C$, becomes $c^2 = a^2 + b^2$.

23. False. The Law of Cosines requires that the length of two sides be known.

24. True. Use the Law of Sines and multiply by the common denominator.

25. True. Identify the opposite angles as B and L and use the Law of Sines to obtain $\dfrac{LA}{\sin B} = \dfrac{BA}{\sin L}$. Thus $\dfrac{LA}{BA} = \dfrac{\sin B}{\sin L}$.

26. True. Use the Law of Cosines to find the square of the third side, then take the square root to find the length.

27. True. Given two angles, the third can be determined since the sum of the angles in a triangle is 180°. Use the Law of Sines to create a ratio involving the unknown and known side lengths. Solve for the unknown side length.

28. False. Using the Law of Sines we find $\dfrac{\sin 60°}{8} = \dfrac{\sin A}{12}$. However, this has no solution for A because it evaluates to $\sin A = \dfrac{3\sqrt{3}}{4} > 1$. There is no solution for A.

Solutions to Skills for Chapter 7

Exercises

1. We have $\sin 30° = 1/2$.

2. We have $\sin 150° = \sin 30° = 1/2$, using a reference angle of 30°. See Figure 7.67.

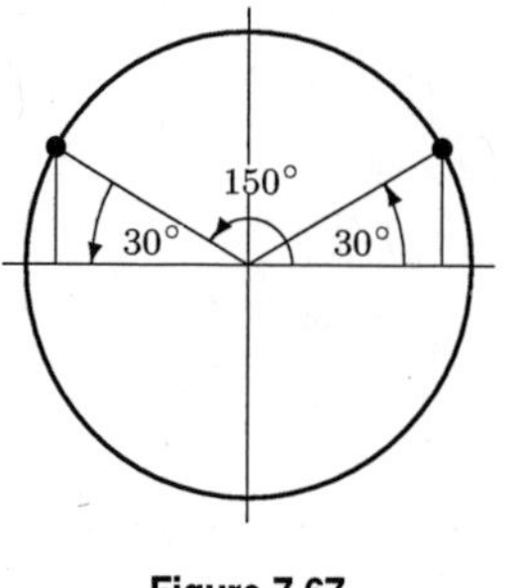

Figure 7.67

3. Since 300° is in the fourth quadrant,

$$\sin 300° = -\sin 60° = -\frac{\sqrt{3}}{2}.$$

4. Since we know that the y-coordinate on the unit circle at $-60°$ is the negative of the y-coordinate at 60°, we know that $\sin(-60°) = -\sin 60° = -\sqrt{3}/2$.

5. Since we know that the x-coordinate on the unit circle at $-60°$ is the same as the x-coordinate at 60°, we know that $\cos(-60°) = \cos 60° = 1/2$.

6. Since we know that the y-coordinate on the unit circle at 120° is the same as the y-coordinate at 60°, we know that $\sin 120° = \sin 60° = \sqrt{3}/2$.

7. Since we know that the x-coordinate on the unit circle at $-30°$ is the same as the x-coordinate at 30°, we know that $\cos(-30°) = \cos 30° = \sqrt{3}/2$.

8. Since we know that the y-coordinate on the unit circle at $210°$ is the negative of the y-coordinate at $30°$, we know that $\sin 210° = -\sin 30° = -1/2$.

9. Since $135°$ is in the second quadrant,
$$\sin 135° = \sin 45° = \frac{1}{\sqrt{2}}.$$

10. Since $300°$ is in the fourth quadrant,
$$\cos 300° = \cos 60° = \frac{1}{2}.$$

11. Since $210°$ is in the third quadrant,
$$\cos 210° = -\cos 30° = -\frac{\sqrt{3}}{2}.$$

12. Since $330°$ is in the fourth quadrant,
$$\sin 330° = -\sin 30° = -\frac{1}{2}.$$

13. Since $405°$ is in the first quadrant,
$$\sin 405° = \sin 45° = \frac{1}{\sqrt{2}}.$$

14. The reference angle for $120°$ is $180° - 120° = 60°$, so $\cos 120° = -\cos 60° = -1/2$.

15. Since $135°$ is in the second quadrant,
$$\cos 135° = -\cos 45° = -\frac{1}{\sqrt{2}}.$$

16. The reference angle for $225°$ is $225° - 180° = 45°$, so $\cos 225° = -\cos 45° = -\sqrt{2}/2$.

17. The reference angle for $300°$ is $360° - 300° = 60°$, so $\sin 300° = -\sin 60° = -\sqrt{3}/2$.

Problems

18. The calculator gives the value 0.707107... for both expressions $\sqrt{\frac{1}{2}}$ and $\frac{\sqrt{2}}{2}$. In fact, $\sqrt{\frac{1}{2}} = \frac{\sqrt{2}}{2}$. This is because
$$\sqrt{\frac{1}{2}} = \frac{\sqrt{1}}{\sqrt{2}} = \frac{1}{\sqrt{2}}\frac{\sqrt{2}}{\sqrt{2}} = \frac{\sqrt{2}}{2}.$$
The value 0.7071 is a good approximation of $\sqrt{2}/2$.

19. The calculator gives the value 0.866025... for both $\sqrt{\frac{3}{4}}$ and $\frac{\sqrt{3}}{2}$. In fact, $\sqrt{\frac{3}{4}} = \frac{\sqrt{3}}{\sqrt{4}} = \frac{\sqrt{3}}{2}$. The value 0.8660 is a good approximation of $\frac{\sqrt{3}}{2}$.

20. Since $\cos 45° = x/30$, we have $x = 30(\sqrt{2}/2) = 15\sqrt{2}$.

21. Since $\sin 30° = x/10$, we have $x = 10(1/2) = 5$.

22. Since $\cos 30° = x/10$, we have $x = 10(\sqrt{3}/2) = 5\sqrt{3}$.

23. Since $\cos 30° = 10/x$, we have $x = 10/\cos 30° = 10/(\sqrt{3}/2) = 20/\sqrt{3}$.

24. Since $\cos 45° = 10/x$, we have $x = 10/\cos 45° = 10/(\sqrt{2}/2) = 20/\sqrt{2} = 10\sqrt{2}$.

25. In a $45°$-$45°$-$90°$ triangle, the two legs are equal and the hypotenuse is $\sqrt{2}$ times the length of a leg. So the sides are 5 and 5 and $5\sqrt{2}$.

26. In a $30°$-$60°$-$90°$ triangle, the length of the longer leg is $\sqrt{3}$ times the length of the shorter leg. So the shorter leg is $4/\sqrt{3}$ and the hypotenuse is twice the length of the shorter leg, so it is $8/\sqrt{3}$.

27. In a $45°$-$45°$-$90°$ triangle, the two legs are equal and the hypotenuse is $\sqrt{2}$ times the length of a leg. So each leg is $7/\sqrt{2}$.

28. In a $30°$-$60°$-$90°$ triangle, the length of the longer leg is $\sqrt{3}$ times the length of the shorter leg, so the longer leg is $3\sqrt{3}$. The hypotenuse is twice the shorter leg, so it is 6.

29. A right triangle with two equal sides is a $45°$-$45°$-$90°$ triangle. In such triangles the length of the hypotenuse side is $\sqrt{2}$ times the length of a leg, so the third side is $4\sqrt{2}$.

30. A right triangle whose hypotenuse is twice the length of a leg is a $30°$-$60°$-$90°$ triangle. In such triangles the length of the longer leg is $\sqrt{3}$ times the length of the shorter leg, so the third side is $7\sqrt{3}$.

31. Point B is in the fourth quadrant, where $\cos 315°$ is positive and $\sin 315°$ is negative. The reference angle for $315°$ is $45°$, so $\cos 315° = \cos 45°$ and $\sin 315° = -\sin 45°$.

The coordinates of point B are given by

$$\begin{aligned} x &= r\cos\theta \\ &= 6\cos 315° \\ &= 6\left(\frac{\sqrt{2}}{2}\right) = 3\sqrt{2} \end{aligned} \qquad \text{and} \qquad \begin{aligned} y &= r\sin\theta \\ &= 6\sin 315° \\ &= 6\left(\frac{-\sqrt{2}}{2}\right) = -3\sqrt{2} \end{aligned}$$

Thus, the coordinates of B are $(3\sqrt{2}, -3\sqrt{2})$.

32. The simplest solution is to apply the Pythagorean theorem to the right triangle whose legs are panels and whose hypotenuse is d. Then

$$d = \sqrt{1^2 + 1^2} = \sqrt{2}.$$

An alternate solution uses trigonometry, an approach that works even if the panels do not have a right angle between them. If we consider the circle in Figure 7.99 as describing a unit circle centered at the origin, we can give the coordinates of the points B and C. We do this by first noting that C makes a $45°$ angle with the positive x-axis, and that B makes a $135°$ angle with the positive x-axis. Then the coordinates of point C are $(\cos 45°, \sin 45°) = (\sqrt{2}/2, \sqrt{2}/2)$. The coordinates of point B are $(\cos 135°, \sin 135°) = (-\sqrt{2}/2, \sqrt{2}/2)$. The difference between the x-values of these coordinates equals d, so

$$d = \frac{\sqrt{2}}{2} - \left(-\frac{\sqrt{2}}{2}\right) = \sqrt{2} \text{ meters} = 1.414 \text{ meters}.$$

CHAPTER EIGHT

Solutions for Section 8.1

Exercises

1. To convert 60° to radians, multiply by $\pi/180^\circ$:

$$60^\circ\left(\frac{\pi}{180^\circ}\right)=\left(\frac{60^\circ}{180^\circ}\right)\pi=\frac{\pi}{3}.$$

We say that the radian measure of a 60° angle is $\pi/3$.

2. To convert 45° to radians, multiply by $\pi/180^\circ$:

$$45^\circ\left(\frac{\pi}{180^\circ}\right)=\left(\frac{45^\circ}{180^\circ}\right)\pi=\frac{\pi}{4}.$$

Thus we say that the radian measure of a 45° angle is $\pi/4$.

3. If ϕ is the radian measure of 100°, then

$$\phi=\left(\frac{\pi}{180^\circ}\right)100^\circ\approx 1.7453 \text{ radians}.$$

4. If θ is the radian measure of 17°, then

$$\theta=\left(\frac{\pi}{180^\circ}\right)17^\circ\approx 0.297 \text{ radians}.$$

5. In order to change from degrees to radians, we multiply the number of degrees by $\pi/180$, so we have $150\cdot\pi/180$, giving $\frac{5}{6}\pi$ radians.

6. In order to change from degrees to radians, we multiply the number of degrees by $\pi/180$, so we have $120\cdot\pi/180$, giving $\frac{2}{3}\pi$ radians.

7. In order to change from degrees to radians, we multiply the number of degrees by $\pi/180$, so we have $-270\cdot\pi/180$, giving $-\frac{3}{2}\pi$ radians.

8. In order to change from degrees to radians, we multiply the number of degrees by $\pi/180$, so we have $\pi\cdot\pi/180$, giving $\pi^2/180\approx 0.0548$ radians.

9. In order to change from radians to degrees, we multiply the number of radians by $180/\pi$, so we have $\frac{7}{2}\pi\cdot 180/\pi$, giving 630 degrees.

10. In order to change from radians to degrees, we multiply the number of radians by $180/\pi$, so we have $5\pi\cdot 180/\pi$, giving 900 degrees.

11. In order to change from radians to degrees, we multiply the number of radians by $180/\pi$, so we have $90\cdot 180/\pi$, giving $16{,}200/\pi\approx 5156.620$ degrees.

12. In order to change from radians to degrees, we multiply the number of radians by $180/\pi$, so we have $2\cdot 180/\pi$, giving $360/\pi\approx 114.592$ degrees.

13. In order to change from radians to degrees, we multiply the number of radians by $180/\pi$, so we have $45\cdot 180/\pi$, giving $8100/\pi\approx 2578.310$ degrees.

14. **(a)** $30 \cdot \dfrac{\pi}{180} = \dfrac{\pi}{6}$ or 0.52

(b) $120 \cdot \dfrac{\pi}{180} = \dfrac{2\pi}{3}$ or 2.09

(c) $200 \cdot \dfrac{\pi}{180} = \dfrac{10\pi}{9}$ or 3.49

(d) $315 \cdot \dfrac{\pi}{180} = \dfrac{7\pi}{4}$ or 5.50

15. **(a)** I
(b) II
(c) II
(d) III
(e) IV
(f) IV
(g) I
(h) II
(i) II
(j) III

16. If we go around once, we make one full circle, which is 2π radians.

17. If we go around twice, we make two full circles, which is $2\pi \cdot 2 = 4\pi$ radians. Since we're going around in the negative direction, we have -4π radians.

18. If we go around 0.75 times, we make three-fourths of a full circle, which is $2\pi \cdot \frac{3}{4} = 3\pi/2$ radians.

19. If we go around 4.27 times, we make 4.27 full circles, which is $2\pi \cdot 4.27 = 8.54\pi$ radians.

20. The arc length, s, corresponding to an angle of θ radians in a circle of radius r is $s = r\theta$. In order to change from degrees to radians, we multiply the number of degrees by $\pi/180$, so we have $-180 \cdot \pi/180$, giving $-\pi$ radians. The negative sign indicates rotation in a clockwise, rather than counterclockwise, direction. Since length cannot be negative, we find the arc length corresponding to π radians. Thus, our arc length is $6.2\pi \approx 19.478$.

21. The arc length, s, corresponding to an angle of θ radians in a circle of radius r is $s = r\theta$. In order to change from degrees to radians, we multiply the number of degrees by $\pi/180$, so we have $45 \cdot \pi/180$, giving $\frac{\pi}{4}$ radians. Thus, our arc length is $6.2\pi/4 \approx 4.869$.

22. The arc length, s, corresponding to an angle of θ radians in a circle of radius r is $s = r\theta$. In order to change from degrees to radians, we multiply the number of degrees by $\pi/180$, so we have $180/\pi \cdot \pi/180$, giving 1 radian. Thus, our arc length is $6.2 \cdot 1 = 6.2$.

23. The arc length, s, corresponding to an angle of θ radians in a circle of radius r is $s = r\theta$. In order to change from degrees to radians, we multiply the number of degrees by $\pi/180$, so we have $a \cdot \pi/180$ radians. Thus, our arc length is $6.2a\pi/180$.

24. Figure 8.1 gives the coordinates of the points on the unit circle specified by $0, 3\pi/2$, and 2π. We use these coordinates to evaluate the sines and cosines of these angles.
The definition of the cosine function tells us that $\cos 0$ is the x-coordinate of the point on the unit circle specified by the angle 0. Since the x-coordinate of this point is $x = 1$, we have

$$\cos 0 = 1.$$

Similarly, the y-coordinate of the point $(1, 0)$ is $y = 0$ which gives

$$\sin 0 = 0.$$

From Figure 8.1 we see that the coordinates of the remaining points are $(0, -1)$, and $(1, 0)$, so

$$\cos \frac{3\pi}{2} = 0, \qquad \sin \frac{3\pi}{2} = -1, \qquad \text{and} \qquad \cos 2\pi = 1, \qquad \sin 2\pi = 0.$$

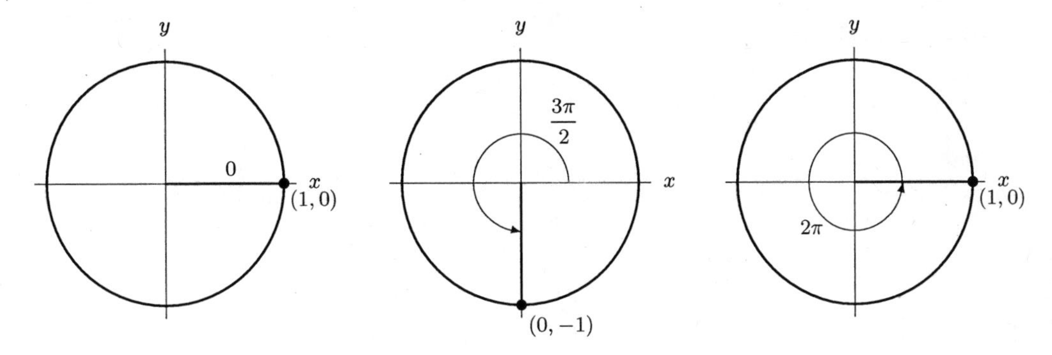

Figure 8.1: The coordinates of the points on the unit circle specified by the angles $0, 3\pi/2$, and 2π

25. **(a)** The reference angle for $2\pi/3$ is $\pi - 2\pi/3 = \pi/3$, so $\sin(2\pi/3) = \sin(\pi/3) = \sqrt{3}/2$.
(b) The reference angle for $3\pi/4$ is $\pi - 3\pi/4 = \pi/4$, so $\cos(3\pi/4) = -\cos(\pi/4) = -1/\sqrt{2}$.
(c) The reference angle for $-3\pi/4$ is $\pi - 3\pi/4 = \pi/4$, so $\tan(-3\pi/4) = \tan(\pi/4) = 1$.
(d) The reference angle for $11\pi/6$ is $2\pi - 11\pi/6 = \pi/6$, so $\cos(11\pi/6) = \cos(\pi/6) = \sqrt{3}/2$.

Problems

26. Using $s = r\theta$ gives $30 = 3r$. Solving for r we have $r = 10$ cm.

27. First $225°$ has to be converted to radian measure:

$$225 \cdot \frac{\pi}{180} = \frac{5\pi}{4}.$$

Using $s = r\theta$ gives

$$s = 4 \cdot \frac{5\pi}{4} = 5\pi \text{ feet}.$$

28. Using $s = r\theta$, we have $s = 8(2) = 16$ inches.

29.

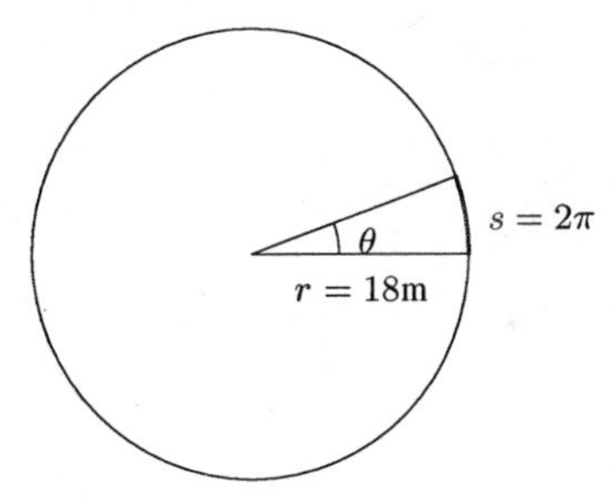

Figure 8.2

In Figure 8.2, we have $s = 2\pi$ and $r = 18$. Therefore,

$$\theta = \frac{s}{r} = \frac{2\pi}{18} = \frac{\pi}{9}.$$

Now,

$$\frac{\pi}{9} \text{ radians} = \frac{\pi}{9}\left(\frac{180°}{\pi}\right) = 20°.$$

Therefore, an arc of length 2π m on a circle of radius 18 m determines an angle of $\pi/9$ radians or $20°$.

30. We have $s = 3$ and $r = 5$. Using the formula $s = r\theta$, we have $\theta = 3/5$ radians or

$$\theta = \frac{3}{5} \cdot \frac{180^\circ}{\pi} = 34.3775^\circ.$$

The coordinates of P are $(r\cos\theta, r\sin\theta) = (4.1267, 2.8232)$.

31. From the figure, we see that $P = (7,4)$. This means $r = \sqrt{7^2+4^2} = \sqrt{65}$. We know that $\sin\theta = 4/\sqrt{65}$, and we can use a graphing calculator to estimate that $\theta = 0.5191$ radians or 29.7449°. This gives $s = 0.5191\sqrt{65} = 4.185$.

32. We have $\theta = 22^\circ$ or, in radians,

$$22^\circ \cdot \frac{\pi}{180^\circ} = 0.3840.$$

We also know that $r = 0.05$, so

$$s = 0.05(0.3840) = 0.01920.$$

The coordinates of P are $(r\cos\theta, r\sin\theta) = (0.0464, 0.0187)$.

33. We have $\theta = 1.3$ rad or, in degrees,

$$1.3\left(\frac{180^\circ}{\pi}\right) = 74.4845^\circ.$$

We also have $r = 12$, so

$$s = 12(1.3) = 15.6,$$

and $P = (r\cos\theta, r\sin\theta) = (3.2100, 11.5627)$.

34. We have $\theta = 3\pi/7$ or, in degrees,

$$\theta = \frac{3\pi}{7} \cdot \frac{180^\circ}{\pi} = 77.1429^\circ.$$

We also have $r = 80$, so

$$s = 80 \cdot \frac{3\pi}{7} = \frac{240\pi}{7} = 107.7117,$$

and $P = (r\cos\theta, r\sin\theta) = (17.8017, 77.9942)$.

35. We do not know the value of r or s, but we know that $s = r\theta$, so

$$\theta = \frac{s}{r} = 0.4,$$

or, in degrees,

$$\theta = 0.4\left(\frac{180^\circ}{\pi}\right) = 22.918^\circ.$$

This means that $P = (r\cos\theta, r\sin\theta) = (0.9211r, 0.3894r)$.

36. **(a)** Yes, in both it seems to be roughly 60°.
(b) Just over 6 arcs fit into the circumference, since the circumference is $2\pi r = 2\pi(2) = 12.566$.

37. **(a)** Negative
(b) Negative
(c) Positive
(d) Positive

38. **(a)** $-2\pi/3 < 2/3 < 2\pi/3 < 2.3$
(b) $\cos 2.3 < \cos(-2\pi/3) = \cos(2\pi/3) < \cos(2/3)$

39. $\sin\theta = 0.6$, $\cos\theta = -0.8$.

40. $\sin\theta = 0.8$, $\cos\theta = -0.6$.

41. Since the ant traveled three units on the unit circle, the traversed arc must be spanned by an angle of three radians. Thus the ant's coordinates must be

$$(\cos 3, \sin 3) \approx (-0.99, 0.14).$$

42. The graphs are found below.

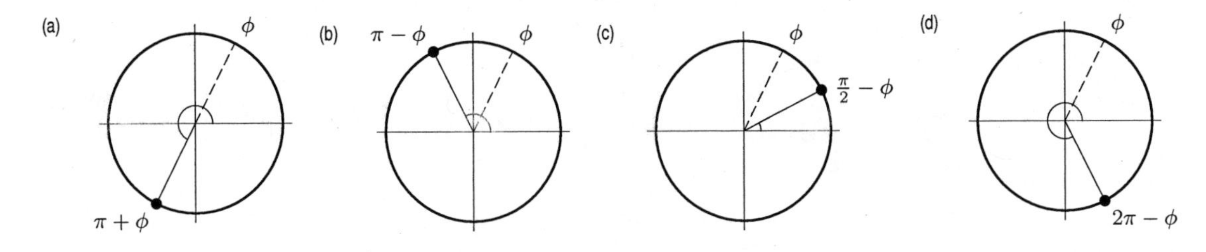

43. The angle spanned by the arc shown is $\theta = s/r = 4/5$ radians, so $m = r\cos\theta = 5\cos(4/5)$ and $n = r\sin\theta = 5\sin(4/5)$. By the Pythagorean theorem,

$$\begin{aligned} p^2 &= n^2 + (5-m)^2 \\ &= n^2 + m^2 - 10m + 25 \\ &= 25\sin^2(4/5) + 25\cos^2(4/5) - 10m + 25 \\ &= 50 - 10m \end{aligned}$$

so

$$\begin{aligned} p &= \sqrt{50 - 10m} \\ &= \sqrt{50 - 50\cos(4/5)} \\ &= 5\sqrt{2(1-\cos(4/5))}. \end{aligned}$$

44. The angle spanned by the arc shown is $\theta = s/r = 10/5 = 2$ radians, so $v = r\sin\theta = 5\sin 2$. Since u is positive as it is a length, $u = -r\cos\theta = -5\cos 2$, because $\cos 2$ is negative. By the Pythagorean theorem,

$$\begin{aligned} w^2 &= v^2 + (5+u)^2 \\ &= v^2 + u^2 + 10u + 25 \\ &= 25\sin^2 2 + 25\cos^2 2 + 10u + 25 \\ &= 50 + 10u, \end{aligned}$$

and

$$w = \sqrt{50 + 10u} = \sqrt{50 - 50\cos 2} = 5\sqrt{2(1-\cos 2)}.$$

45. **(a)** 1 radian is $180/\pi$ degrees so 30 radians is

$$30 \cdot \frac{180^\circ}{\pi} \approx 1718.873^\circ.$$

To check this answer, divide 1718.873° by 360° to find this is roughly 5 revolutions. A revolution in radians has a measure of $2\pi \approx 6$, so $5 \cdot 6 = 30$ radians makes sense.

(b) 1 degree is $\pi/180$ radians, so $\pi/6$ degrees is

$$\frac{\pi}{6} \cdot \frac{\pi}{180} = \frac{\pi^2}{6 \cdot 180} \approx 0.00914 \text{ radians}.$$

This makes sense because $\pi/6$ is about $1/2$, and $1/2$ a degree is very small. One radian is about 60° so $\frac{1}{2}^\circ$ is a very small part of a radian.

46. **(a)** Using the formula $s = \theta r$ with $r = 38/2 = 19$ cm we find

$$s = \text{ Arc length } = (19)(3.83) = 72.77 \text{ cm}.$$

(b) Using $s = \varphi r$ with s and r known, we have $3.83 = \varphi(19)$. Thus $\varphi = 3.83/19 \approx 0.202$ radians.

47. A complete revolution is an angle of 2π radians and this takes 60 minutes. In 35 minutes, the angle of movement in radians is $35/60 \cdot 2\pi = 7\pi/6$. The arc length is equal to the radius times the radian measure, which is $6(7\pi/6) = 7\pi \approx 21.991$ inches.

48. Converting the angle into radians, we get

$$1.4333^\circ = 1.4333\left(\frac{2\pi}{360}\right) \approx 0.0250 \text{ radians.}$$

Now we know that

$$\text{Arc length} = \text{Radius} \cdot \text{Angle in radians.}$$

Thus the radius is

$$\text{Radius} = \frac{\text{Arc length}}{\text{Angle in radians}} \approx \frac{100 \text{ miles}}{1.4333(2\pi/360)} = 3998.310 \text{ miles.}$$

49. The value of t is bigger than the value of $\sin t$ on $0 < t < \pi/2$. On a unit circle, the vertical segment, $\sin t$, is shorter than the arc, t. See Figure 8.3.

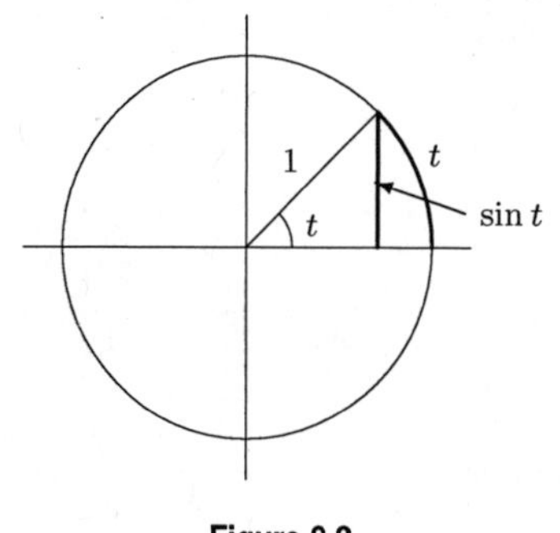

Figure 8.3

50. Make a table, such as Table 8.1, using your calculator to see that $\cos t$ is decreasing and the values of t are increasing.

Table 8.1

t	0	0.1	0.2	0.3	0.4	0.5	0.6	0.7	0.8	0.9
$\cos t$	1	0.995	0.980	0.953	0.921	0.878	0.825	0.765	0.697	0.622

Use a more refined table to see $t \approx 0.74$. (See Table 8.2.) Further refinements lead to $t \approx 0.739$.

Table 8.2

t	0.70	0.71	0.72	0.73	0.74	0.75	0.76
$\cos t$	0.765	0.758	0.752	0.745	0.738	0.732	0.725

Alternatively, consider the graphs of $y = t$ and $y = \cos t$ in Figure 8.4. They intersect at a point in the first quadrant, so for the t-coordinate of this point, $t = \cos t$. Trace with a calculator to find $t \approx 0.739$.

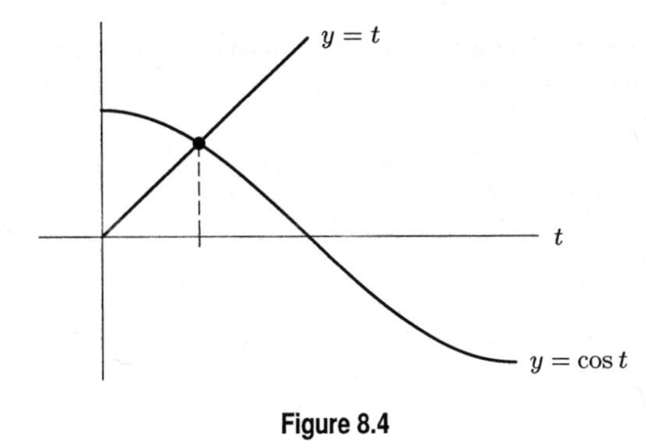

Figure 8.4

Solutions for Section 8.2

Exercises

1. The midline is $y = 0$. The amplitude is 6. The period is 2π.
2. The midline is $y = -8$. The amplitude is 7. The period is $2\pi/4 = \pi/2$.
3. We first divide both sides by 2, giving
$$y = \frac{1}{2}\cos(8(t-6)) + 1.$$
The midline is $y = 1$. The amplitude is $\frac{1}{2}$. The period is $2\pi/8 = \pi/4$.
4. The midline is $y = -1$. The amplitude is π. The period is $2\pi/2 = \pi$.
5. We see that the phase shift is -4, since the function is in a form that shows it. To find the horizontal shift, we factor out a 3 within the cosine function, giving us
$$y = 2\cos\left(3\left(t + \frac{4}{3}\right)\right) - 5.$$
Thus, the horizontal shift is $-4/3$.
6. We see that the phase shift is -13, since the function is in a form that shows it. To find the horizontal shift, we factor out a 7 within the cosine function, giving us
$$y = -4\cos\left(7\left(t + \frac{13}{7}\right)\right) - 5.$$
Thus, the horizontal shift is $-13/7$.
7. Both f and g have periods of 1, amplitudes of 1, and midlines $y = 0$.
8. **(a)** The function $y = \sin(-t)$ is periodic, and its period is 2π. The function begins repeating every 2π units, as is clear from its graph. Recall that $f(-x)$ is a reflection about the y-axis of the graph of $f(x)$, so the periods for $\sin(t)$ and $\sin(-t)$ are the same. See Figure 8.5.

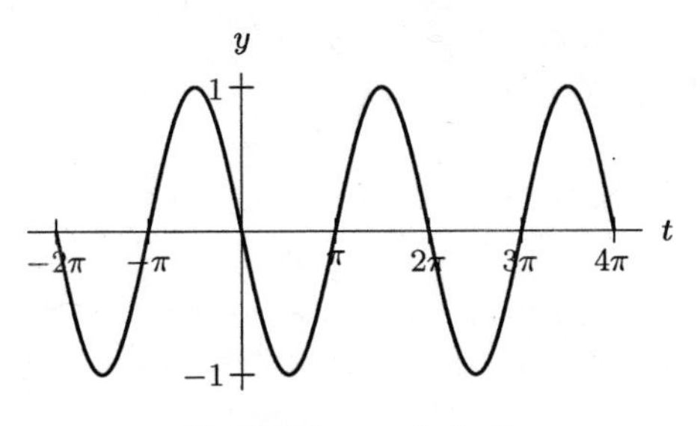

Figure 8.5: $y = \sin(-t)$

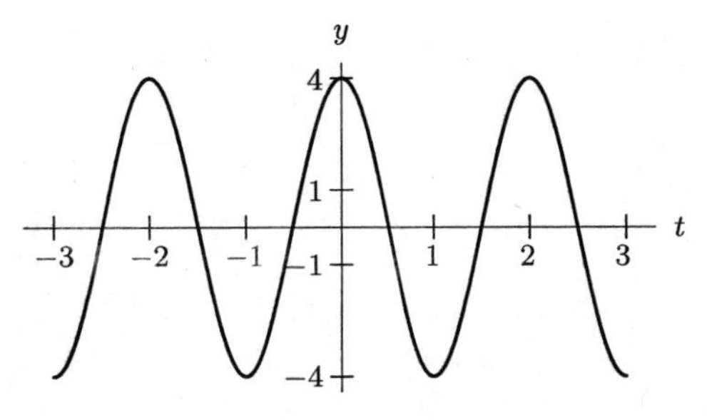

Figure 8.6: $y = 4\cos(\pi t)$

(b) The function $y = 4\cos(\pi t)$ is periodic, and its period is 2. This is because when $0 \le t \le 2$, we have $0 \le \pi t \le 2\pi$ and the cosine function has period 2π. Note the amplitude of $4\cos(\pi t)$ is 4, but changing the amplitude does not affect the period. See Figure 8.6.

(c) The function $y = \sin(t) + t$ is not periodic, because as t gets large, $\sin(t) + t$ gets large as well. In fact, since $\sin(t)$ varies from -1 to 1, y is always between $t - 1$ and $t + 1$. So the values of y cannot repeat. See Figure 8.7.

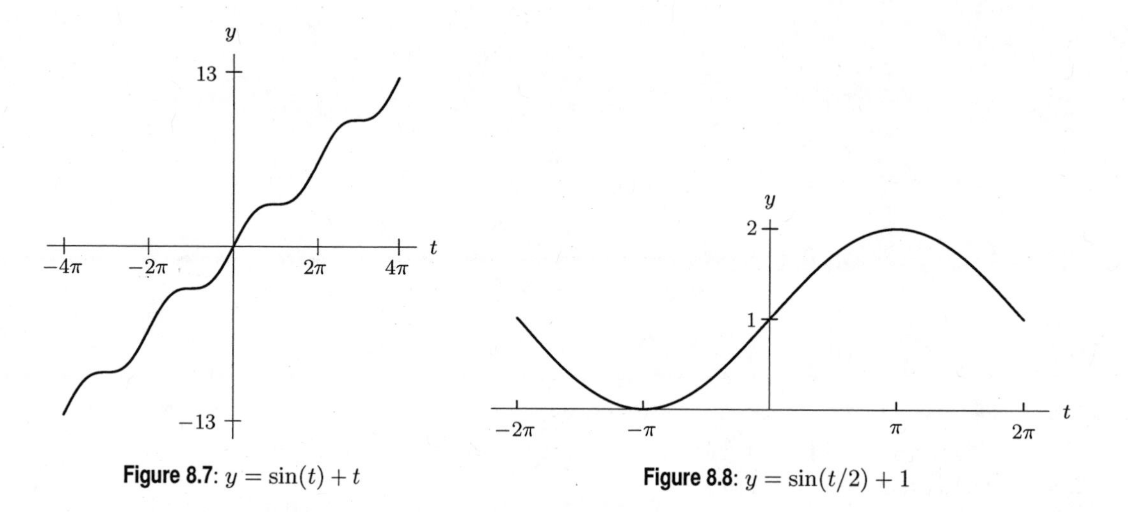

Figure 8.7: $y = \sin(t) + t$

Figure 8.8: $y = \sin(t/2) + 1$

(d) In general $f(x)$ and $f(x) + c$ will have the same period if they are periodic. The function $y = \sin(\frac{t}{2}) + 1$ is periodic, because $\sin(\frac{t}{2})$ is periodic. Since $\sin(t/2)$ completes one cycle for $0 \le t/2 \le 2\pi$, or $0 \le t \le 4\pi$, we see the period of $y = \sin(t/2) + 1$ is 4π. See Figure 8.8.

9. This function resembles a sine curve in that it passes through the origin and then proceeds to grow from there. We know that the smallest value it attains is -4 and the largest it attains is 4, thus its amplitude is 4. It has a period of 1. Thus in the equation

$$g(t) = A\sin(Bt)$$

we know that $A =$ and

$$1 = \text{period} = \frac{2\pi}{B}.$$

So $B = 2\pi$, and then

$$h(t) = 4\sin(2\pi t).$$

10. This function resembles a cosine curve in that it attains its maximum value when $t = 0$. We know that the smallest value it attains is -3 and that its midline is $y = 0$. Thus its amplitude is 3. It has a period of 4. Thus in the equation

$$f(t) = A\cos(Bt)$$

we know that $A = 3$ and

$$4 = \text{period} = \frac{2\pi}{B}.$$

So $B = \pi/2$, and then

$$f(t) = 3\cos\left(\frac{\pi}{2}t\right).$$

11. This function resembles an inverted cosine curve in that it attains its minimum value when $t = 0$. We know that the smallest value it attains is 0 and that its midline is $y = 2$. Thus its amplitude is 2 and it is shifted upward by two units. It has a period of 4π. Thus in the equation

$$g(t) = -A\cos(Bt) + D$$

we know that $A = -2$, $D = 2$, and

$$4\pi = \text{period} = \frac{2\pi}{B}.$$

So $B = 1/2$, and then

$$g(t) = -2\cos\left(\frac{t}{2}\right) + 2.$$

12. The graph is a horizontally and vertically compressed sine function. The midline is $y = 0$. The amplitude is 0.8. We see that $\pi/7 =$ two periods, so the period is $\pi/14$. Hence $B = 2\pi/(\text{period}) = 28$, and so

$$y = 0.8\sin(28\theta).$$

13. The midline is $y = 4000$. The amplitude is $8000 - 4000 = 4000$. The period is 60, so the angular frequency is $2\pi/60$. The graph at $x = 0$ rises from its midline, so we use the sine. Thus,

$$y = 4000 + 4000\sin\left(\frac{2\pi}{60}x\right).$$

14. The midline is $y = 20$. The amplitude is $30 - 20 = 10$. The period is 12, so the angular frequency is $2\pi/12$. The graph at $x = 0$ decreases from its maximum, like the cosine function. Thus,

$$y = 10\cos\left(\frac{2\pi}{12}x\right) + 20.$$

15. The graph resembles a sine function that is vertically reflected, horizontally and vertically stretched, and vertically shifted. There is no horizontal shift since the function hits its midline at $\theta = 0$. The midline is halfway between 0 and 4, so it has the equation $y = 2$. The amplitude is 2. Since we see 9 is $\frac{3}{4}$ of the length of a cycle, the period is 12. Hence $B = 2\pi/(\text{period}) = \pi/6$, and so

$$y = -2\sin\left(\frac{\pi}{6}\theta\right) + 2.$$

16. We see the interval from 0 to 2 is half a period, so the period $P = 4$. Hence $B = 2\pi/P = \pi/2$. The midline is shown at $y = 3$, so $D = 3$. We see the amplitude $|A| = 3$. Since g has a minimum at $\theta = 0$ like $-\cos\theta$, A is negative. Hence $A = -3$. Thus,

$$g(\theta) = -3\cos\left(\frac{\pi}{2}\theta\right) + 3.$$

Problems

17. See Figure 8.9.

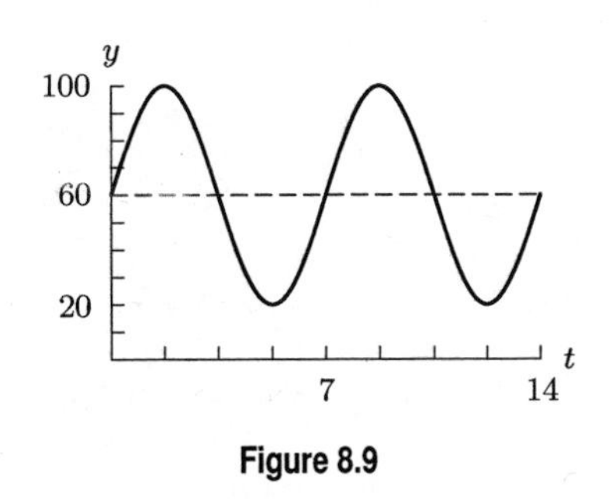

Figure 8.9

18. See Figure 8.10.

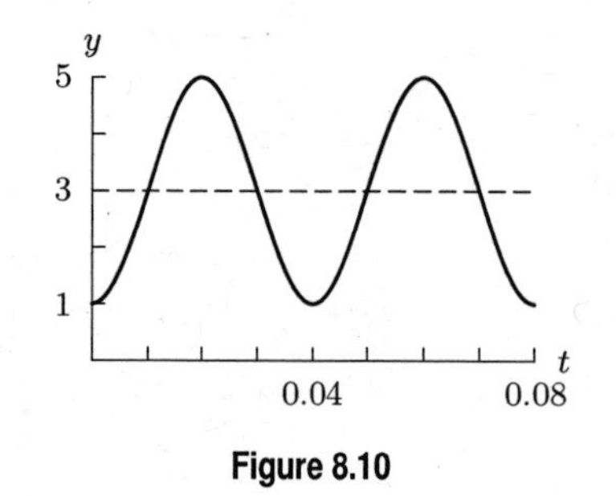

Figure 8.10

19. See Figure 8.11.

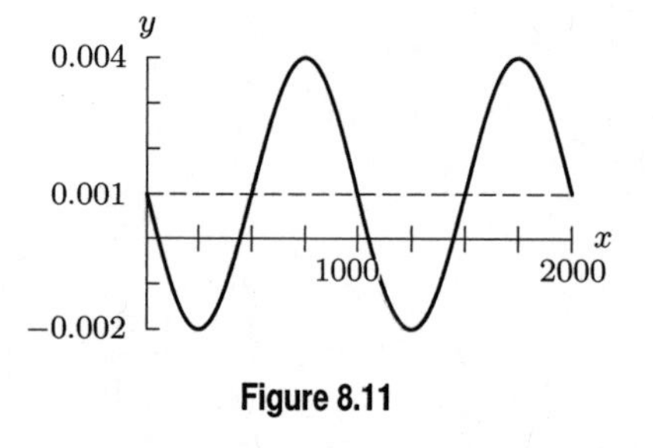

Figure 8.11

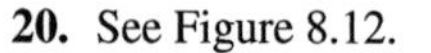

20. See Figure 8.12.

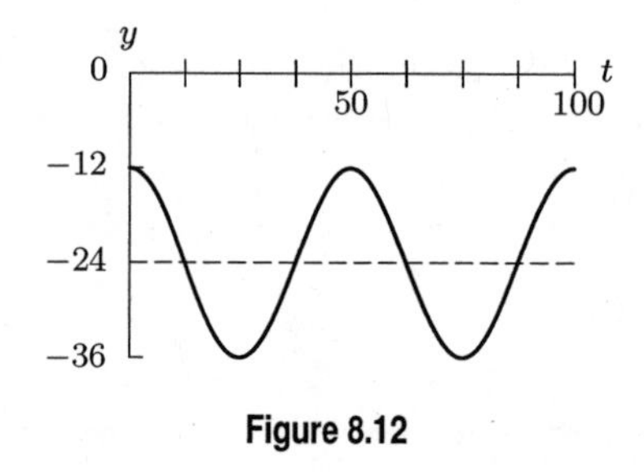

Figure 8.12

21. Because the period of $\sin x$ is 2π, and the period of $\sin 2x$ is π, so from the figure in the problem we see that

$$f(x) = \sin x.$$

The points on the graph are $a = \pi/2$, $b = \pi$, $c = 3\pi/2$, $d = 2\pi$, and $e = 1$.

22. This graph of the function $y = \cos(5t + \pi/4)$ is the graph of $y = \cos 5t$ shifted horizontally to the left. (In general, the graph of $f(x + k)$ is the graph of $f(x)$ shifted left if k is positive.) Since

$$5t + \frac{\pi}{4} = 5\left(t + \frac{\pi}{20}\right),$$

the graph shifts left a distance of $\pi/20$ units. The phase shift is $\pi/4$. Since $\pi/4$ is one-eighth of the period of $\cos t$, the graph of this function is $y = \cos 5t$ shifted left by one-eighth of its period.

23. From Figure 8.13, we see the amplitude A of this function is 20. We see 4 cycles in 3 seconds so the period is $\frac{3}{4}$.

The amplitude tells us the maximum amount by which the blood pressure can vary from its average value of 100 mm Hg. The period of $\frac{3}{4}$ seconds tells us the duration of one cycle of blood pressure change.

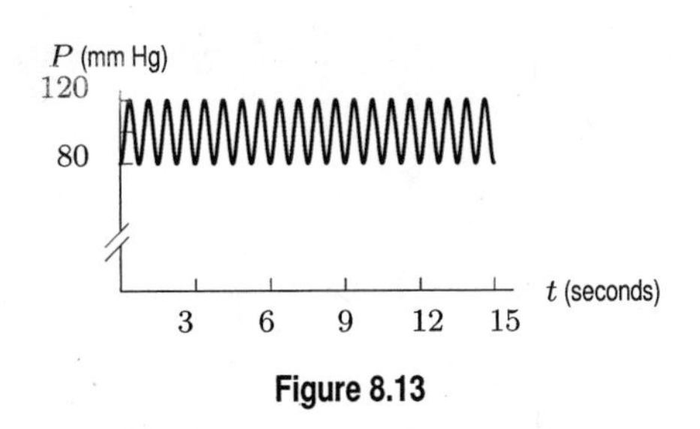

Figure 8.13

24. First, we note that the graph indicates that f completes one cycle in 8 units. Therefore, the period of f equals 8, and we can also see that by shifting the graph of f 2 units to the right, we obtain the graph of g. Thus, f must be shifted by $1/4$ of a period to the right to obtain g, meaning that the phase shift of g is $2\pi \cdot (1/4) = \pi/2$. Therefore, we have $g(x) = 3\sin((\pi/4)x - \pi/2)$.

25. First, we note that the graph indicates that f completes one cycle in 10 units. Therefore, the period of f equals 10, and we can also see that by shifting the graph of f 3 units to the right, we obtain the graph of g. Thus, f must be shifted by $3/10$ of a period to the right to obtain g, meaning that the phase shift of g is $2\pi \cdot (3/10) = 3\pi/5$. Therefore, we have $g(x) = 10\sin((\pi/5)x - 3\pi/5)$.

26. $f(t) = 17.5 + 17.5\sin\left(\frac{2\pi}{5}t\right)$

27. $f(t) = 14 + 10\sin\left(\pi t + \frac{\pi}{2}\right)$

28. The amplitude and midline are 20 and the period is 5. The graph is a sine curve shifted half a period to the right (or left), so the phase shift is $-\pi$. Thus

$$h = 20 + 20\sin\left(\frac{2\pi}{5}t - \pi\right).$$

29. $f(t) = 20 + 15\sin\left(\frac{\pi}{2}t + \frac{\pi}{2}\right)$

30. From Figure 8.2, we see that the midline and amplitude are both 225 and the period is 30. Since the graph looks like a cosine reflected about the midline, we take $k = 225$ and $A = -225$. In addition

$$\frac{2\pi}{B} = 30 \quad \text{so} \quad B = \frac{\pi}{15}.$$

The cosine function does not need to be shifted horizontally, so

$$f(t) = -225\cos\left(\frac{\pi}{15}t\right) + 225.$$

31. **(a)** The Ferris wheel makes one full revolution in 30 minutes. Since one revolution is 360°, the wheel turns

$$\frac{360}{30} = 12^\circ \text{ per minute.}$$

(b) The angle representing your position, measured from the 6 o'clock position, is $12t^\circ$. However, the angle shown in Figure 8.14 is measured from the 3 o'clock position, so

$$\theta = (12t - 90)^\circ.$$

(c) With y as shown in Figure 8.14, we have

$$\text{Height} = 225 + y = 225 + 225\sin\theta,$$

so

$$f(t) = 225 + 225\sin(12t - 90)^\circ.$$

Note that the expression $\sin(12t - 90)^\circ$ means $\sin\left((12t - 90)^\circ\right)$.

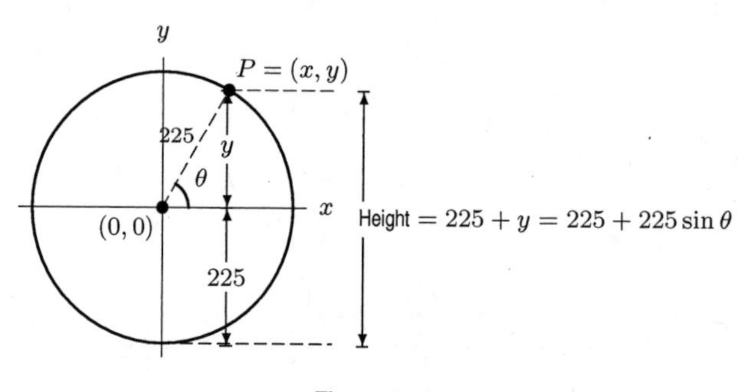

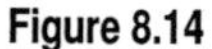
Figure 8.14

(d) Using a calculator, we obtain the graph of $h = f(t) = 225 + 225\sin(12t - 90)^\circ$ in Figure 8.15. The period is 30 minutes, the midline and amplitude are 225 feet.

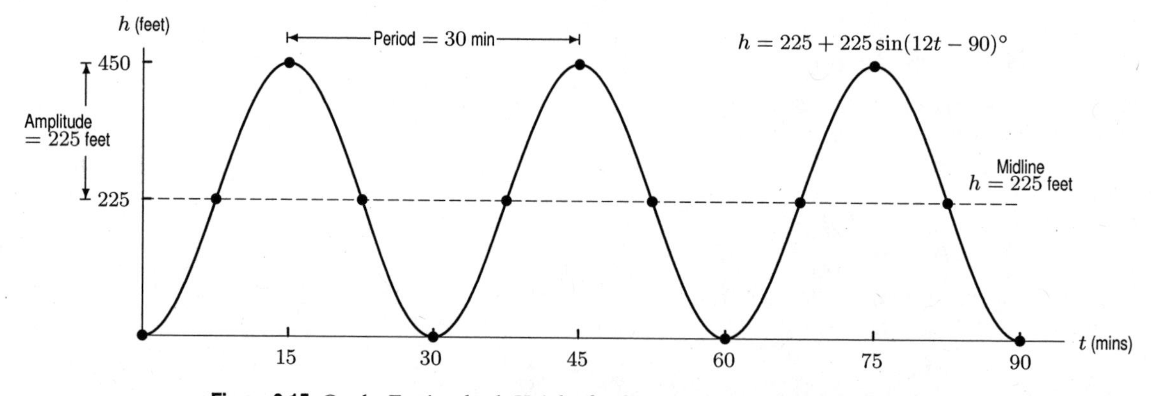

Figure 8.15: On the Ferris wheel: Height, h, above ground as function of time, t

32. (a) The population has initial value 1500 and grows at a constant rate of 200 animals per year.

(b) The population has initial value 2700 and decreases at a constant rate of 80 animals per year.

(c) The population has initial value 1800 and increases at the constant percent rate of 3% per year.

(d) The population has initial value 800 and decreases at the *continuous* percent rate of 4% per year.

(e) The population has initial value 3800, climbs to $3800 + 230 = 4030$, drops to $3800 - 230 = 3570$, and climbs back to 3800 over a 7 year period. This pattern keeps repeating itself.

33. (a) The midline is at $P = (2200 + 1300)/2 = 1750$. The amplitude is $|A| = 2200 - 1750 = 450$. The population starts at its minimum so it is modeled by vertically reflected cosine curve. This means $A = -450$, and that there is no phase shift. Since the period is 12, we have $B = 2\pi/12 = \pi/6$. This means the formula is

$$P = f(t) = -450\cos\left(\frac{\pi}{6}t\right) + 1750.$$

(b) The midline, $P = 1750$, is the average population value over one year. The period is 12 months (or 1 year), which means the cycle repeats annually. The amplitude is the amount that the population varies above and below the average annual population.

(c) Figure 8.16 is a graph of $f(t) = -450\cos(\frac{\pi}{6}t) + 1750$ and $P = 1500$.

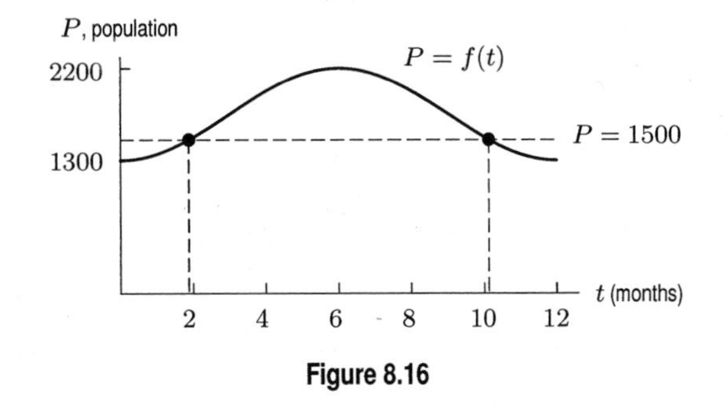

Figure 8.16

From a graph we get approximations of $t_1 \approx 1.9$ and $t_2 \approx 10.1$. This means that the population is 1500 sometime in late February and again sometime in early November.

34. The data given describe a trigonometric function shifted vertically because all the $g(x)$ values are greater than 1. Since the maximum is approximately 3 and the minimum approximately 1, the midline value is 2. We choose the sine function over the cosine function because the data tell us that at $x = 0$ the function takes on its midline value, and then increases. Thus our function will be of the form

$$g(x) = A\sin(Bx) + k.$$

We know that A represents the amplitude, k represents the vertical shift, and the period is $2\pi/B$.

We've already noted the midline value is $k \approx 2$. This means $A = \max - k = 1$. We also note that the function completes a full cycle after 1 unit. Thus

$$1 = \frac{2\pi}{B}$$

so

$$B = 2\pi.$$

Thus

$$g(x) = \sin(2\pi x) + 2.$$

35. This function has an amplitude of 3 and a period 1, and resembles a sine graph. Thus $y = 3f(x)$.

36. This function has an amplitude of 2 and a period of 3, and resembles vertically reflected cosine graph. Thus $y = -2g(x/3)$.

37. This function has an amplitude of 1 and a period of 0.5, and resembles an inverted sine graph. Thus $y = -f(2x)$.

38. This function has an amplitude of 2 and a period of 1 and a midline of $y = -3$, and resembles a cosine graph. Thus $y = 2g(x) - 3$.

39.

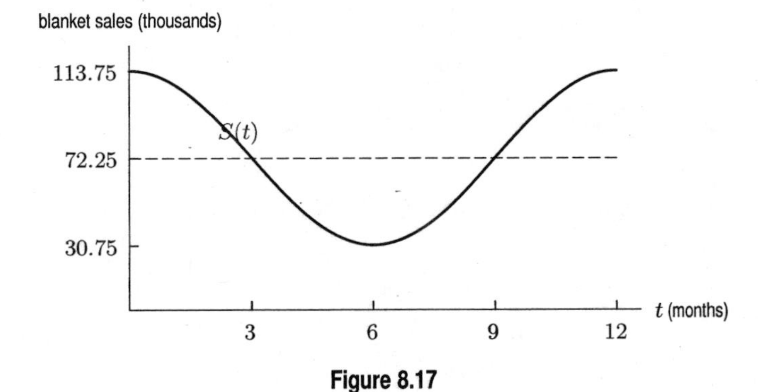

Figure 8.17

The amplitude of this graph is 41.5. The period is $P = 2\pi/B = (2\pi)/(\pi/6) = 12$ months. The amplitude of 41.5 tells us that during winter months sales of electric blankets are 41,500 above the average. Similarly, sales reach a minimum of 41,500 below average in the summer months. The period of one year indicates that this seasonal sales pattern repeats annually.

40. **(a)**

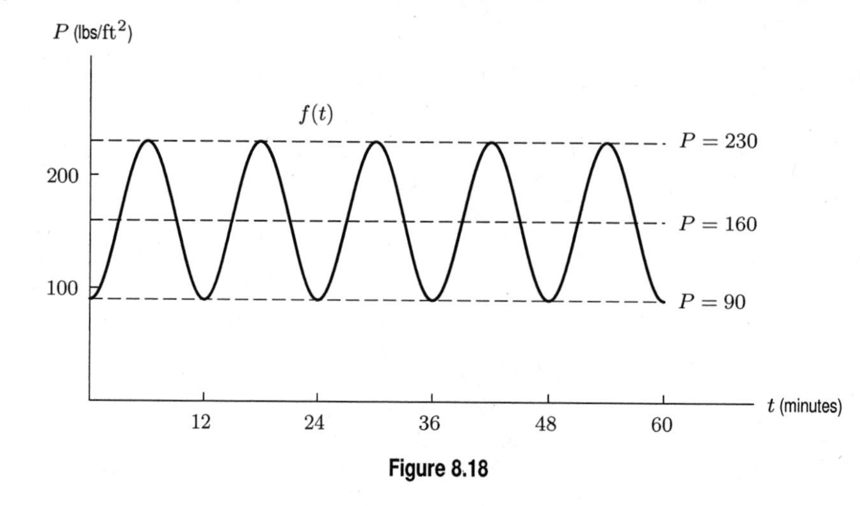

Figure 8.18

This function is a vertically reflected cosine function which has been vertically shifted. Thus the function for this equation will be of the form

$$P = f(t) = -A\cos(Bt) + k.$$

(b) The midline value is $k = (90 + 230)/2 = 160$.
The amplitude is $|A| = 230 - 160 = 70$.
A complete oscillation is made each 12 minutes, so the period is 12. This means $B = 2\pi/12 = \pi/6$. Thus $P = f(t) = -70\cos(\pi t/6) + 160$.

(c) Graphing $P = f(t)$ on a calculator for $0 \le t \le 2$ and $90 \le P \le 230$, we see that $P = f(t)$ first equals 115 when $t \approx 1.67$ minutes.

41. Figure 8.19 highlights the two parts of the graph. In the first hour, the plane is approaching Boston. In the second hour, the plane is circling Boston.

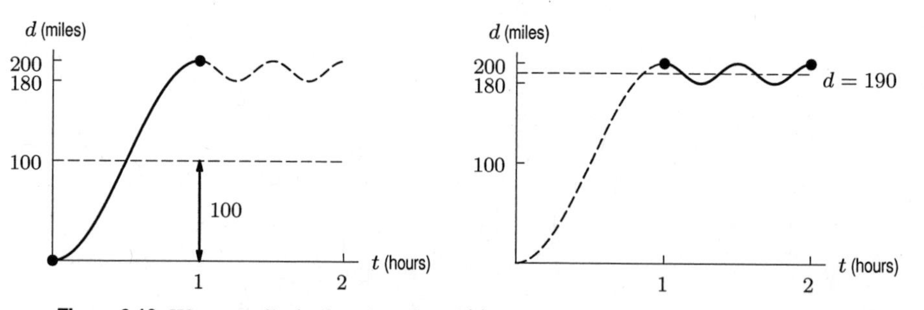

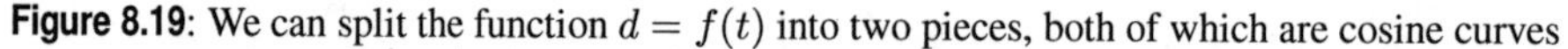

Figure 8.19: We can split the function $d = f(t)$ into two pieces, both of which are cosine curves

From Figure 8.19, we see that both parts of $f(t)$ look like cosine curves. The first part has the equation

$$f(t) = -100\cos(\pi t) + 100, \quad \text{for } 0 \le t \le 1.$$

In the second part, the period is 1/2, the midline is 190, and the amplitude is $200 - 190 = 10$, so

$$f(t) = 10\cos(4\pi t) + 190, \quad \text{for } 1 \le t \le 2.$$

Thus, a piecewise formula for $f(t)$ could be

$$f(t) = \begin{cases} -100\cos(\pi t) + 100 & \text{for } 0 \le t \le 1 \\ 10\cos(4\pi t) + 190 & \text{for } 1 < t \le 2. \end{cases}$$

42. **(a)** See Figure 8.20.

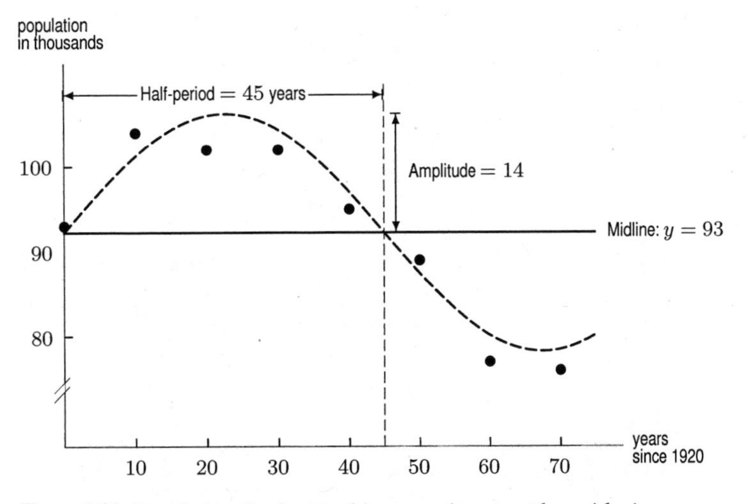

Figure 8.20: Population (in thousands) versus time, together with sine curve.

(b) The variation is possibly sinusoidal, but not necessarily. The population rises first, then falls. From the graph, it appears it could soon rise again, but this need not be the case.

(c) See Figure 8.20. There are many possible answers.

(d) For the graph drawn, the amplitude of the population function is $107-93 = 14$. The average value of P from the data is 93. The graph of $P = f(t)$ behaves as the graph of $\sin t$, for $3/4$ of a period. Therefore, we look for a reasonable approximation to the data of the form $P = f(t) = 14\sin(Bt) + 93$. To determine B, we assume that 45 years is half the period of f. Thus, the period equals 90 years and so $B = 2\pi/90 = \pi/45$. Hence an approximation to the data is

$$P = f(t) = 14\sin\left(\frac{\pi}{45}t\right) + 93.$$

(e) $P = f(-10) \approx 84.001$, which means that our formula predicts a population of about 84,000. This is not too far off the mark, but not all that close, either.

43. **(a)** See Figure 8.21, where January is represented by $t = 0$.

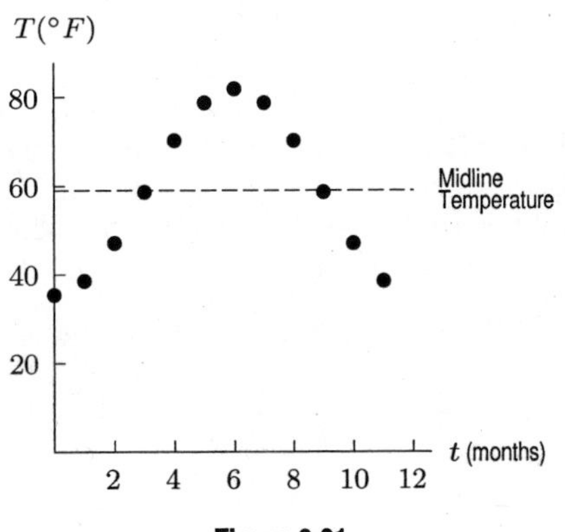

Figure 8.21

(b) The midline temperature is approximately $(81.8 + 35.4)/2 = 58.6$ degrees. The amplitude of the temperature function is then $81.8 - 58.6 = 23.2$ degrees. The period equals 12 months.

(c) We choose the approximating function $T = f(t) = -A\cos(Bt) + D$. Since the graph resembles an inverted cosine curve, we know $A = 23.2$ and $D = 58.6$. Since the period is 12, $B = 2\pi/12 = \pi/6$. Thus

$$T = f(t) = -23.2\cos\left(\frac{\pi}{6}t\right) + 58.6$$

is a good approximation, though it does not exactly agree with all the data.

(d) In October, $T = f(9) = -23.2\cos((\pi/6)9) + 58.6 \approx 58.6$ degrees, while the table shows an October value of 62.5.

44. The petroleum import data is graphed in Figure 8.22.

We start with a sine function of the form

$$f(t) = A\sin(B(t-h)) + k.$$

Since the maximum here is 18 and the minimum is 12, the midline value is $k = (18 + 12)/2 = 15$. The amplitude is then $A = 18 - 15 = 3$. The period, measured peak to peak, is 12. So $B = 2\pi/12 = \pi/6$. Lastly, we calculate h, the horizontal shift to the right. Our data are close to the midline value for $t \approx 74$, whereas $\sin t$ is at its midline value for $t = 0$. So $h = 74$ and the equation is

$$f(t) = 3\sin\frac{\pi}{6}(t - 74) + 15.$$

We can check our formula by graphing it and seeing how close it comes to the data points. See Figure 8.23.

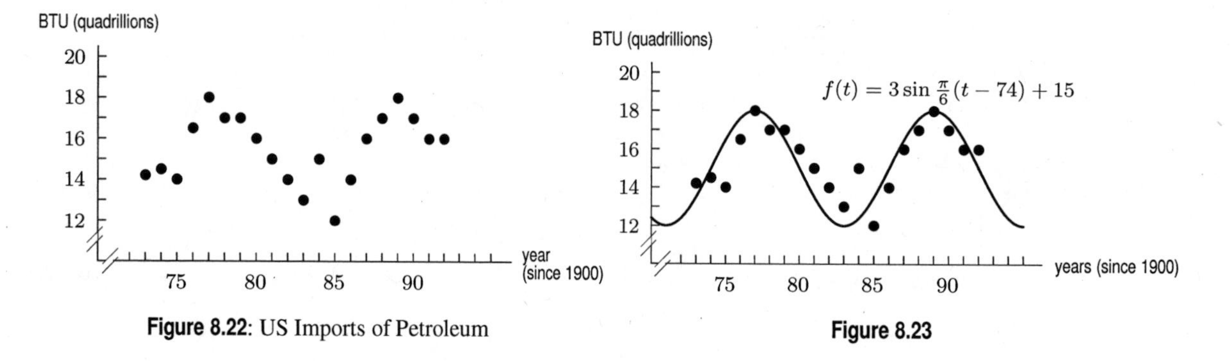

Figure 8.22: US Imports of Petroleum

Figure 8.23

45. **(a)** Although the graph has a rough wavelike pattern, the wave is not perfectly regular in each 7-day interval. A true periodic function has a graph which is absolutely regular, with values that repeat exactly every period.

(b) Usage spikes every 7 days or so, usually about midweek (8/7, 8/14, 8/21, etc.). It drops to a low point every 7 days or so, usually on Saturday or Sunday (8/10, 8/17, 8/25, etc.). This indicates that scientists use the site less frequently on weekends and more frequently during the week.

(c) See Figure 8.24 for one possible approximation. The function shown here is given by $n = a\cos(B(t-h)) + k$ where t is the number of days from Monday, August 5, and $a = 45{,}000$, $B = 2\pi/7$, $h = 2$, and $k = 100{,}000$. The midline $k = 100{,}000$ tells us that usage rises and falls around an approximate average of 100,000 connections per day. The amplitude $a = 45{,}000$ tells us that usage tends to rise or fall by about 45,000 from the average over the course of the week. The period is 7 days, or one week, giving $B = 2\pi/7$, and the curve resembles a cosine function shifted to the right by about $h = 2$ days. Thus,

$$n = 45{,}000\cos\left(\frac{2\pi}{7}(t-2)\right) + 100{,}000.$$

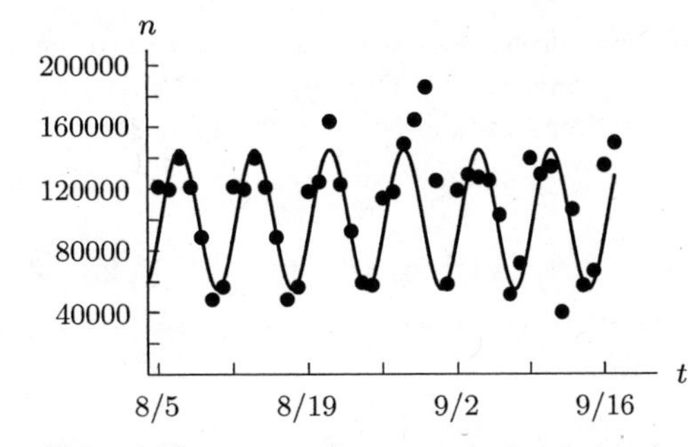

Figure 8.24: Fitting a trigonometric function to the arXiv.org usage data

46. **(a)** Yes. Since $\sin t$ repeats periodically, the input for f will repeat periodically, and thus so will the output of f, which makes $f(\sin t)$ periodic. In symbols, if we have $h(t) = f(\sin t)$, then

$$\begin{aligned} h(t+2\pi) &= f(\sin(t+2\pi)) \\ &= f(\sin t) \\ &= h(t). \end{aligned}$$

So h is periodic.

(b) No. For example, $y = \sin(t^2)$ is not periodic, as is clear from Figure 8.25. Although the graph does oscillate up and down, the time from peak to peak is shorter as time increases. Thus there is no constant period.

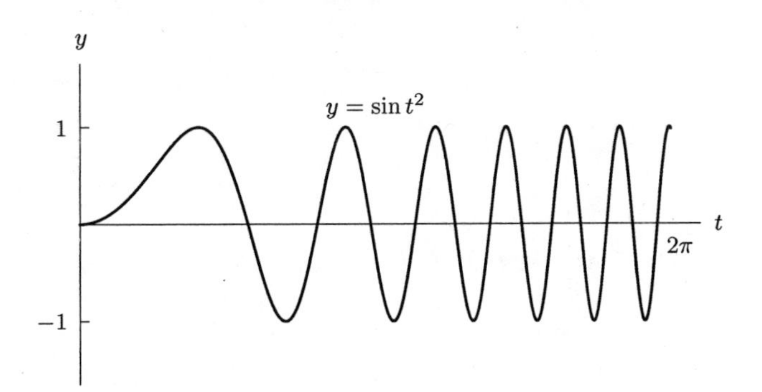

Figure 8.25: The function $y = \sin(t^2)$ is not periodic.

Solutions for Section 8.3

Exercises

1. 1

2. $\sqrt{3}$

3. $-\sqrt{3}$

4. 1

5. $\dfrac{-1}{\sqrt{3}}$

6. Since $\csc(5\pi/4) = 1/\sin(5\pi/4)$, we know that $\csc(5\pi/4) = 1/(-1/\sqrt{2}) = -\sqrt{2}$.

7. Since $\cot(5\pi/3) = 1/(\tan(5\pi/3) = 1/(\sin(5\pi/3)/\cos(5\pi/3)) = \cos(5\pi/3)/\sin(5\pi/3)$, we know that $\cot(5\pi/3) = (1/2)/(-\sqrt{3}/2) = -1/\sqrt{3}$.

8. Since $\sec(11\pi/6) = 1/\cos(11\pi/6)$, we know that $\sec(11\pi/6) = 1/(\sqrt{3}/2) = 2/\sqrt{3}$.

9. Since $\sec(-\pi/6) = 1/\cos(-\pi/6)$, we know that $\sec(-\pi/6) = 1/(\sqrt{3}/2) = 2/\sqrt{3}$.

10. Writing $\tan x = \sin x/\cos x$, we have

$$\tan x \cos x = \frac{\sin x}{\cos x} \cdot \cos x = \sin x.$$

11. Let $x = 2\theta$. Then $(\cos(2\theta))^2 + (\sin(2\theta))^2 = (\cos x)^2 + (\sin x)^2 = 1$, using the Pythagorean identity.

12. We have:

$$\begin{aligned} 3\sin t - 2(1 - 2\sin t) &= 3\sin t - 2 + 4\sin t \\ &= 7\sin t - 2. \end{aligned}$$

13. Writing $\sec t = 1/\cos t$, we have

$$\sec t \cos t = 1.$$

14. Writing $\tan\theta = \sin\theta/\cos\theta$, we have

$$\frac{\sin\theta}{\tan\theta} = \frac{\sin\theta}{\sin\theta/\cos\theta} = \cos\theta.$$

15. Writing $1 - \sin^2 t = \cos^2 t$ from the Pythagorean identity, we have

$$\frac{1 - \sin^2 t}{\cos t} = \frac{\cos^2 t}{\cos t} = \cos t.$$

16. We have:

$$\begin{aligned} 5\cos t - (2\cos t - 3\sin t) &= 5\cos t - 2\cos t + 3\sin t \\ &= 3\cos t + 3\sin t. \end{aligned}$$

17. Expanding the square and combining terms gives

$$(\sin x - \cos x)^2 + 2\sin x\cos x = \sin^2 x - 2\sin x\cos x + \cos^2 + 2\sin x\cos x = \sin^2 x + \cos^2 x = 1.$$

Problems

18. **(a)** Angle 2π is a complete revolution so the point Q at angle $t + 2\pi$ is the same as the point P at angle t. Thus $Q = P$. We have $Q = (x, y)$.

(b) By definition of cosine, $\cos t$ is the first coordinate of the point P so $\cos t = x$, and $\cos(t+2\pi)$ is the first coordinate of Q so $\cos(t + 2\pi) = x$. Thus $\cos(t + 2\pi) = \cos t$.

(c) By definition of sine, $\sin t$ is the second coordinate of the point P so $\sin t = y$, and $\sin(t + 2\pi)$ is the second coordinate of Q so $\sin(t + 2\pi) = y$. Thus $\sin(t + 2\pi) = \sin t$.

19. **(a)** Angle π is a half revolution so the point Q at angle $t+\pi$ is the point on the unit circle at the opposite end from P of the diameter that passes through P. We have $Q = (-x, -y)$.

(b) By definition of cosine, $\cos t$ is the first coordinate of the point P so $\cos t = x$, and $\cos(t+\pi)$ is the first coordinate of Q so $\cos(t+\pi) = -x$. Thus $\cos(t+\pi) = -\cos t$.

(c) By definition of sine, $\sin t$ is the second coordinate of the point P so $\sin t = y$, and $\sin(t+\pi)$ is the second coordinate of Q so $\sin(t+\pi) = -y$. Thus $\sin(t+\pi) = -\sin t$.

(d) By definition of tangent, $\tan t$ is the ratio of the second coordinate to the first coordinate of the point P so $\tan t = y/x$, and $\sin(t+\pi)$ is the ratio of the coordinates of Q so $\tan(t+\pi) = -y/(-x) = y/x$. Thus $\tan(t+\pi) = \tan t$.

20. **(a)** The point Q at angle $-t$ is the reflection across the horizontal axis of the point P at angle t. Since $P = (x, y)$ we have $Q = (x, -y)$.

(b) By definition of cosine, $\cos t$ is the first coordinate of the point P so $\cos t = x$, and $\cos(-t)$ is the first coordinate of Q so $\cos(-t) = x$. Thus $\cos(-t) = \cos t$.

(c) By definition of sine, $\sin t$ is the second coordinate of the point P so $\sin t = y$, and $\sin(-t)$ is the second coordinate of Q so $\sin(-t) = -y$. Thus $\sin(-t) = -\sin t$.

(d) By definition of tangent, $\tan t$ is the ratio of the second coordinate to the first coordinate of the point P so $\tan t = y/x$, and $\tan(-t)$ is the ratio of the coordinates of Q so $\tan(-t) = -y/x$. Thus $\tan(-t) = -\tan t$.

21. Since $\sec\theta = 1/\cos\theta$, we have $\sec\theta = 1/(1/2) = 2$. Since $1 + \tan^2\theta = \sec^2\theta$,

$$\begin{aligned} 1 + \tan^2\theta &= 2^2 \\ \tan^2\theta &= 4 - 1 \\ \tan\theta &= \pm\sqrt{3}. \end{aligned}$$

Since $0 \le \theta \le \pi/2$, we know that $\tan\theta \ge 0$, so $\tan\theta = \sqrt{3}$.

22. Since $\csc\theta = 1/\sin\theta$, we begin by using the Pythagorean Identity to find $\sin\theta$. Since $\cos^2\theta + \sin^2\theta = 1$, we have

$$\begin{aligned} \left(\frac{1}{2}\right)^2 + \sin^2\theta &= 1 \\ \sin^2\theta &= 1 - \frac{1}{4} \\ \sin\theta &= \pm\sqrt{\frac{3}{4}}. \end{aligned}$$

Since $0 \le \theta \le \pi/2$, $\sin\theta \ge 0$, so $\sin\theta = \sqrt{\frac{3}{4}}$. Thus, $\csc\theta = 1/\sqrt{3/4} = \sqrt{4/3}$.

Since $\cot\theta = \cos\theta/\sin\theta$, we have $\cot\theta = (1/2)/\sqrt{3/4} = (1/2)(2\sqrt{3}) = 1/\sqrt{3}$.

23. Since $\sec\theta = 1/\cos\theta$, we begin with the Pythagorean Identity, $\cos^2\theta + \sin^2\theta = 1$. We have

$$\begin{aligned} \cos^2\theta + \left(\frac{1}{3}\right)^2 &= 1 \\ \cos^2\theta &= 1 - \frac{1}{9} \\ \cos\theta &= \pm\sqrt{\frac{8}{9}} = \pm\frac{\sqrt{8}}{3}. \end{aligned}$$

Since $0 \le \theta \le \pi/2$, we know that $\cos\theta \ge 0$, so $\cos\theta = \sqrt{8}/3$. Therefore, $\sec\theta = 1/(\sqrt{8}/3) = 3/\sqrt{8}$.

Since $\tan\theta = \sin\theta/\cos\theta$, we have $\tan\theta = (1/3)/(\sqrt{8}/3) = 1/\sqrt{8}$.

24. Since $\cos\theta = 1/\sec\theta$, $\cos\theta = 1/17$. Using the Pythagorean Identity, $\cos^2\theta + \sin^2\theta = 1$, we have

$$\begin{aligned} \sin^2\theta + \left(\frac{1}{17}\right)^2 &= 1 \\ \sin^2\theta &= 1 - \frac{1}{17^2} \\ \sin\theta &= \pm\sqrt{\frac{288}{289}} = \pm\frac{\sqrt{288}}{17}. \end{aligned}$$

Since $0 \le \theta \le \pi/2$, we know that $\sin\theta \ge 0$, so $\sin\theta = \sqrt{288}/17$.

Using the identity $\tan\theta = \sin\theta/\cos\theta$, we see that $\tan\theta = (\sqrt{288}/17)/(1/17) = \sqrt{288}$.

25. This looks like a tangent graph. At $\pi/4$, $\tan\theta = 1$. Since on this graph, $f(\pi/4) = 1/2$, and since it appears to have the same period as $\tan\theta$ without a horizontal or vertical shift, a possible formula is $f(\theta) = \frac{1}{2}\tan\theta$.

26. This looks like a tangent graph. At $\pi/4$, $\tan\theta = 1$. Since on this graph, $f(\pi/2) = 1$, and since it appears to have the same period as $\tan\theta$ without a vertical shift, but shifted $\pi/4$ to the right, a possible formula is $f(\theta) = \tan(\theta - \pi/4)$.

27. Since $y = \sin\theta$, we can construct the following triangle:

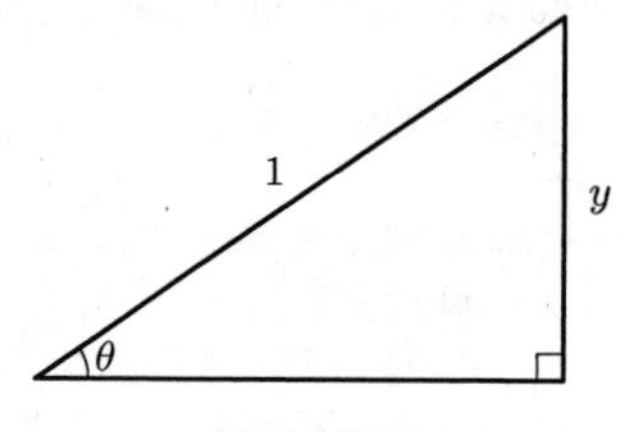

Figure 8.26

The adjacent side, using the Pythagorean theorem, has length $\sqrt{1-y^2}$. So, $\cos\theta = \frac{\text{adj}}{\text{hyp}} = \frac{\sqrt{1-y^2}}{1} = \sqrt{1-y^2}$.

28. **(a)** $\sin^2\alpha = 1 - \cos^2\alpha = 1 - \left(\frac{-\sqrt{3}}{5}\right)^2 = 1 - \frac{3}{25} = \frac{22}{25}$. Since α is in the third quadrant,

$$\sin\alpha = -\sqrt{\frac{22}{25}} = \frac{-\sqrt{22}}{5}.$$

$$\tan\alpha = \frac{\sin\alpha}{\cos\alpha} = \frac{-\sqrt{22}/5}{-\sqrt{3}/5} = \sqrt{\frac{22}{3}}.$$

(b) $\tan\beta = \frac{\sin\beta}{\cos\beta} = \frac{4}{3}$, so $\sin\beta = \frac{4}{3}\cos\beta$. Then

$$1 = \sin^2\beta + \cos^2\beta = \left(\frac{4}{3}\cos\beta\right)^2 + \cos^2\beta = \frac{25}{9}\cos^2\beta.$$

Since β is in the third quadrant, $\cos\beta = -\sqrt{\frac{9}{25}} = \frac{-3}{5}$ and $\sin\beta = \frac{4}{3}\left(\frac{-3}{5}\right) = \frac{-4}{5}$.

29. **(a)** $\sin^2\phi = 1 - \cos^2\phi = 1 - (0.4626)^2$ and $\sin\phi$ is negative, so $\sin\phi = -\sqrt{1-(0.4626)^2} = -0.8866$. Thus $\tan\phi = (\sin\phi)/(\cos\phi) = (-0.8866)/(0.4626) = -1.9166$.

(b) $\cos^2\theta = 1 - \sin^2\theta = 1 - (-0.5917)^2$ and $\cos\theta$ is negative, so $\cos\theta = -\sqrt{1-(-0.5917)^2} = -0.8062$. Thus $\tan\theta = (\sin\theta)/(\cos\theta) = (-0.5917)/(-0.8062) = 0.7339$.

30. For all t where the functions are defined

$$\tan(-t) = \frac{\sin(-t)}{\cos(-t)} = \frac{-\sin t}{\cos t} = -\tan t.$$

This shows that the tangent function satisfies the definition of an odd function.

31.

$$\cos^2\theta = 1 - \sin^2\theta = 1 - \left(\frac{x}{3}\right)^2 = 1 - \frac{x^2}{9} = \frac{9-x^2}{9},$$

so

$$\cos\theta = \sqrt{\frac{9-x^2}{9}} = \frac{\sqrt{9-x^2}}{3}.$$

$$\tan\theta = \frac{\sin\theta}{\cos\theta} = \frac{x}{3}\cdot\frac{3}{\sqrt{9-x^2}} = \frac{x}{\sqrt{9-x^2}}$$

32. $\sin^2\theta = 1-\cos^2\theta = 1-(4/x)^2 = 1-16/x^2 = (x^2-16)/x^2$, so $\sin\theta = \sqrt{(x^2-16)/x^2} = \sqrt{(x^2-16)}/x$. Thus $\tan\theta = \sin\theta/\cos\theta = \sqrt{x^2-16}/x \cdot x/4 = \sqrt{x^2-16}/4$.

33. First notice that $\cos\theta = \frac{x}{2}$, then $\sin^2\theta = 1-\cos^2\theta = 1-(x/2)^2 = 1-x^2/4 = (4-x^2)/4$, so $\sin\theta = \sqrt{(4-x^2)/4} = \sqrt{4-x^2}/2$. Thus $\tan\theta = \sin\theta/\cos\theta = \sqrt{4-x^2}/2 \cdot 2/x = \sqrt{4-x^2}/x$.

34. First notice that $\tan\theta = \frac{x}{9}$ so $\tan\theta = \sin\theta/\cos\theta = x/9$, so $\sin\theta = x/9\cdot\cos\theta$. Now to find $\cos\theta$ by using $1 = \sin^2\theta+\cos^2\theta = (x^2/81)\cos^2\theta+\cos^2\theta = \cos^2\theta(x^2/81+1)$, so $\cos^2\theta = 81/(x^2+81)$ and $\cos\theta = 9/\sqrt{x^2+81}$. Thus, $\sin\theta = (x/9)\cdot(9/\sqrt{x^2+81}) = x/\sqrt{x^2+81}$.

35. **(a)** (i) This is an identity. The graph of $y=\sin t$ is the graph of $y=\cos t$ shifted to the right by $\pi/2$. This means that value of $\cos\left(t-\frac{\pi}{2}\right)$ equals the value of $\sin t$ for all values of t.

(ii) This is not an identity. In Figure 8.27, we see that the graph of $y=\sin 2t$ has a period of π and an amplitude of 1, whereas the graph of $y=2\sin t$ has a period of 2π and an amplitude of 2. Instead of being true for all values of t, this equation is true only where the graphs of these two functions intersect.

(b) From the Figure 8.27, we see there are three solutions to the equation $\sin 2t = 2\sin t$ on the interval $0\le t\le 2\pi$, at $t=0, \pi$, and 2π.

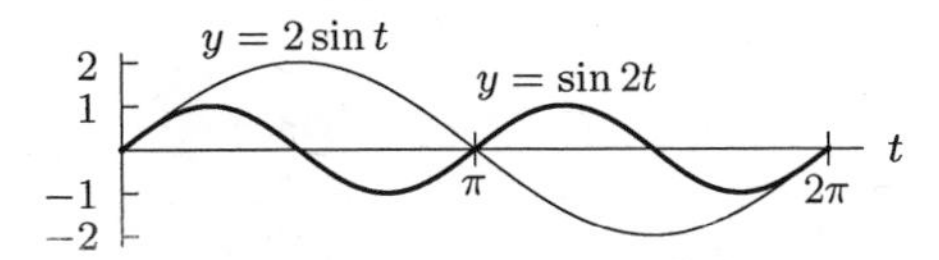

Figure 8.27: The graphs of $y=\sin 2t$ and $y=2\sin t$ are different, so $\sin 2t = 2\sin t$ is not an identity. It is an equation with three solutions on the interval shown

36. **(a)** Figure 8.28 shows these two graphs on the interval $-2\pi < t < 2\pi$. On this scale, the two functions appear identical.

(b) On the larger scale $-100\pi \le t \le 100\pi$, we see the function $y=\sin t+\sin(1.01t)$ is not a sinusoidal function at all.

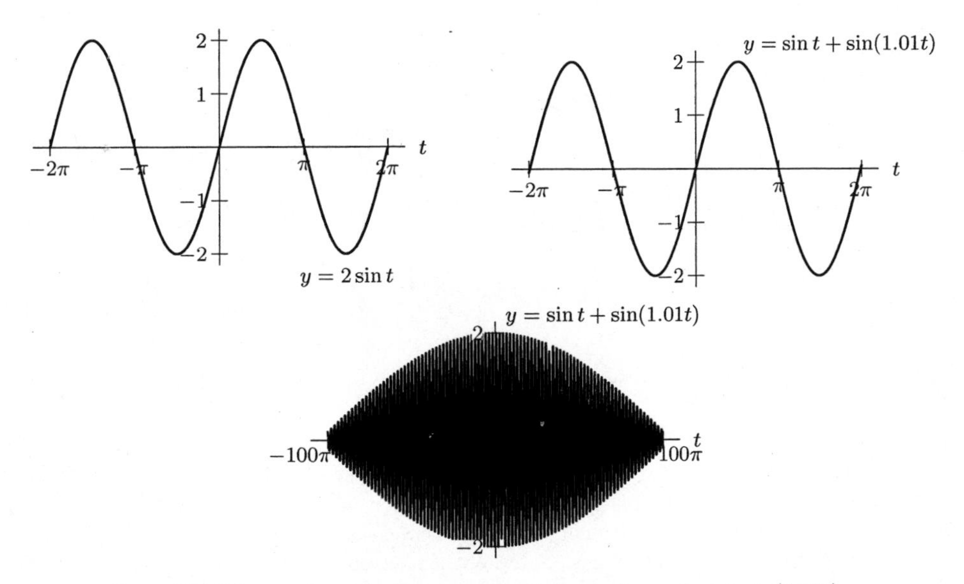

Figure 8.28: The graphs of the functions $y=2\sin t$ and $y=\sin t+\sin(1.01t)$

37. **(a)** The graph of $y = \tan t$ has vertical asymptotes at odd multiples of $\pi/2$, that is, at $\pi/2$, $3\pi/2$, $5\pi/2$, etc., and their negatives. The graph of $y = \cos t$ has t-intercepts at the same values.

(b) The graph of $y = \tan t$ has t-intercepts at multiples of π, that is, at 0, $\pm\pi$, $\pm 2\pi$, $\pm 3\pi$, etc. The graph of $y = \sin t$ has t-intercepts at the same values.

38.

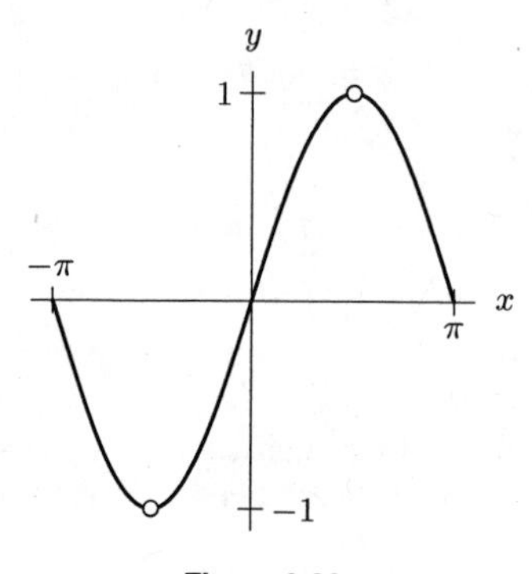

Figure 8.29

Though the function $y = f(x) = \cos x \cdot \tan x$ can be simplified by

$$\cos x \cdot \tan x = \cos x \cdot \frac{\sin x}{\cos x} = \sin x,$$

it is important to notice that $f(x)$ is not defined at the points where $\cos x = 0$. There would be division by zero at such points. Note the holes in the graph which denote undefined values of the function.

39. The functions $\sec x$ and $\csc x$, shown in Figures 8.30 and 8.31, have period 2π, while the function $\cot x$, shown in Figure 8.32, has period π. Each of these functions tends toward infinity where the reciprocal function approaches zero. The function $y = \sec x = 1/\cos x$ is positive on the intervals where $\cos x$ is positive and negative where $\cos x$ is negative. Similarly, the function $y = \csc x = 1/\sin x$ is positive on the intervals where $\sin x$ is positive and negative where $\sin x$ is negative. The function $y = \cot x = 1/\tan x$ is positive on the intervals where $\tan x$ is positive and negative where $\tan x$ is negative. The function $\cot x$ has zeros where $\tan x$ values approach infinity.

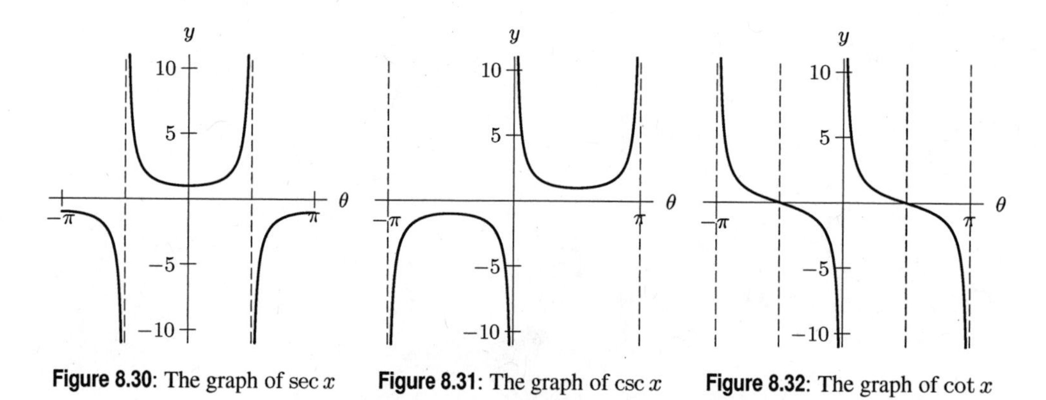

Figure 8.30: The graph of $\sec x$ **Figure 8.31**: The graph of $\csc x$ **Figure 8.32**: The graph of $\cot x$

The functions $\sec x$ and $\csc x$, shown in Figures 8.30 and 8.31, have period 2π, while the function $\cot x$, shown in Figure 8.32, has period π. Each of these functions tends toward infinity where the reciprocal function approaches zero. The function $y = \sec x = 1/\cos x$ is positive on the intervals where $\cos x$ is positive and negative where $\cos x$ is negative. Similarly, the function $y = \csc x = 1/\sin x$ is positive on the intervals where $\sin x$ is positive and negative where $\sin x$ is negative. The function $y = \cot x = 1/\tan x$ is positive on the intervals where $\tan x$ is positive and negative where $\tan x$ is negative. The function $\cot x$ has zeros where $\tan x$ values approach infinity.

Solutions for Section 8.4

Exercises

1. We use the inverse tangent function on a calculator to get $\theta = 1.570$.

2. We use the inverse sine function on a calculator to get $\theta = 0.608$.

3. We divide both sides by 3 to get $\cos\theta = 0.238$. We use the inverse cosine function on a calculator to get $\theta = 1.330$.

4. Since the cosine function always has a value between -1 and 1, there are no solutions.

5. We use the inverse tangent function on a calculator to get $5\theta + 7 = -0.236$. Solving for θ, we get $\theta = -1.447$.

6. We divide both sides by 2, giving us $\sin(4\theta) = 0.3335$. We then use the inverse sine function on a calculator to get $4\theta = 0.340$, so $\theta = 0.085$.

7. Graph $y = \sin\theta$ on $0 \le \theta \le 2\pi$ and locate the two points with y-coordinate 0.65. The θ-coordinates of these points are approximately $\theta = 0.708$ and $\theta = 2.434$. See Figure 8.33.

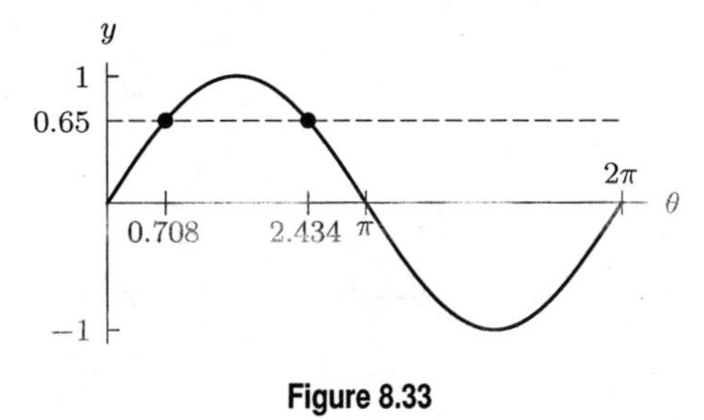

Figure 8.33

8. Graph $y = \tan x$ on $0 \le x \le 2\pi$ and locate the two points with y-coordinate 2.8. The x-coordinates of these points are approximately $x = 1.228$ and $x = 4.369$. See Figure 8.34.

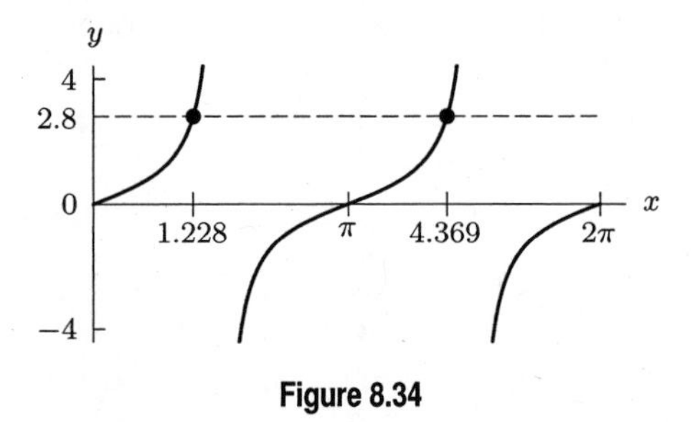

Figure 8.34

9. Graph $y = \cos t$ on $0 \le t \le 2\pi$ and locate the two points with y-coordinate -0.24. The t-coordinates of these points are approximately $t = 1.813$ and $t = 4.473$. See Figure 8.35.

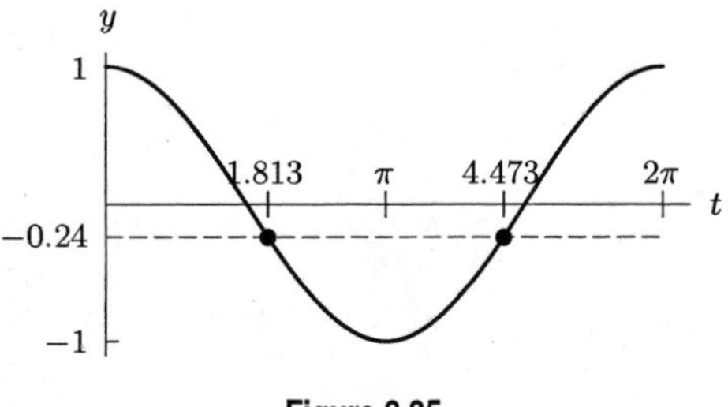

Figure 8.35

10. We draw a graph of $y = \cos t$ for $-\pi \le t \le 3\pi$ and trace along it on a calculator to find points at which $y = 0.4$. We read off the t-values at the points t_0, t_1, t_2, t_3 in Figure 8.36. If t is in radians, we find $t_0 = -1.159$, $t_1 = 1.159$, $t_2 = 5.124$, $t_3 = 7.442$. We can check these values by evaluating:

$$\cos(-1.159) = 0.40, \quad \cos(1.159) = 0.40, \quad \cos(5.124) = 0.40, \quad \cos(7.442) = 0.40.$$

Notice that because the cosine function is periodic, the equation $\cos t = 0.4$ has infinitely many solutions. The symmetry of the graph suggests that the solutions are related.

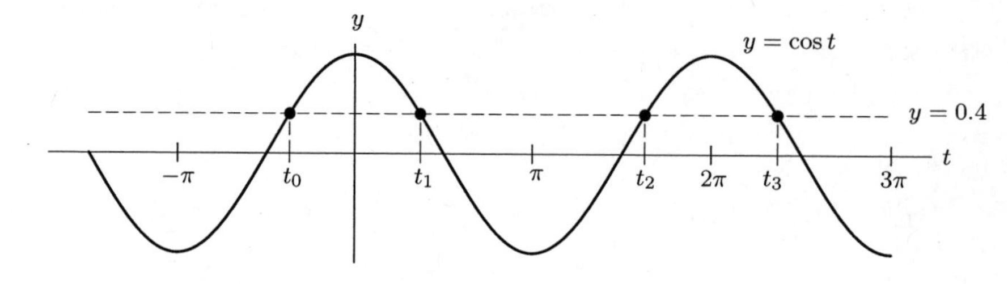

Figure 8.36: The points t_0, t_1, t_2, t_3 are solutions to the equation $\cos t = 0.4$

Problems

11. (a) Tracing along the graph in Figure 8.37, we see that the approximations for the two solutions are

$$t_1 \approx 1.88 \quad \text{and} \quad t_2 \approx 4.41.$$

Note that the first solution, $t_1 \approx 1.88$, is in the second quadrant and the second solution, $t_2 \approx 4.41$, is in the third quadrant. We know that the cosine function is negative in those two quadrants. You can check the two solutions by substituting them into the equation:

$$\cos 1.88 \approx -0.304 \quad \text{and} \quad \cos 4.41 \approx -0.298,$$

both of which are close to -0.3.

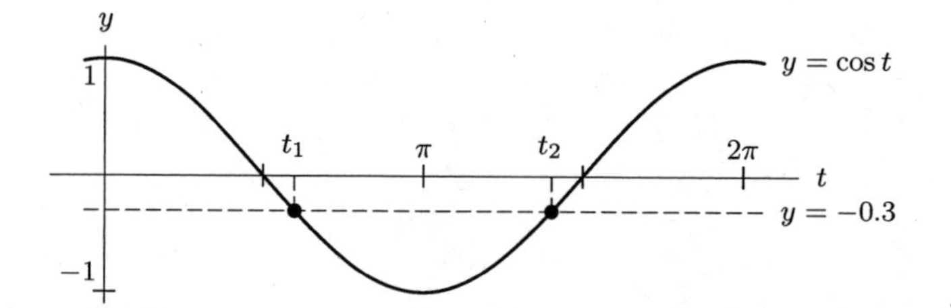

Figure 8.37: The angles t_1 and t_2 are the two solutions to $\cos t = -0.3$ for $0 \le t \le 2\pi$

(b) If your calculator is in radian mode, you should find

$$\cos^{-1}(-0.3) \approx 1.875,$$

which is one of the values we found in part (a) by using a graph. Using the $\boxed{\cos^{-1}}$ key gives only one of the solutions to a trigonometric equation. We find the other solutions by using the symmetry of the unit circle. Figure 8.38 shows that if $t_1 \approx 1.875$ is the first solution, then the second solution is

$$\begin{aligned} t_2 &= 2\pi - t_1 \\ &\approx 2\pi - 1.875 \approx 4.408. \end{aligned}$$

Thus, the two solutions are $t \approx 1.88$ and $t \approx 4.41$.

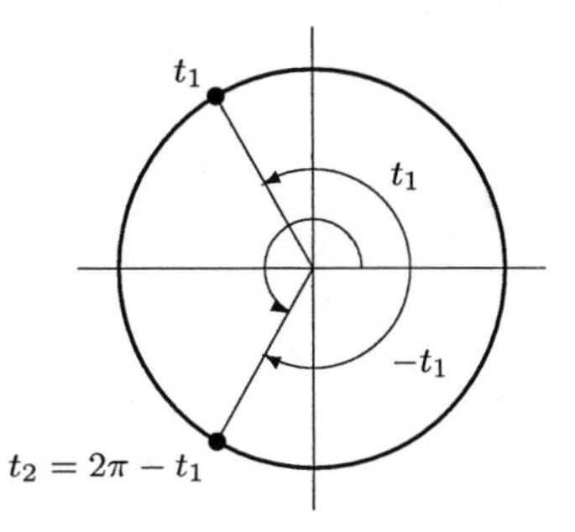

Figure 8.38: By the symmetry of the unit circle, $t_2 = 2\pi - t_1$

12. **(a)** We know that $\cos(\pi/3) = 1/2$. From the graph of $y = \cos t$ in Figure 8.39, we see that $t = \pi/3$, $t = 5\pi/3$, $t = -\pi/3$, and $t = -5\pi/3$ are all solutions, as are any values of t obtained by adding or subtracting multiples of 2π to these values.

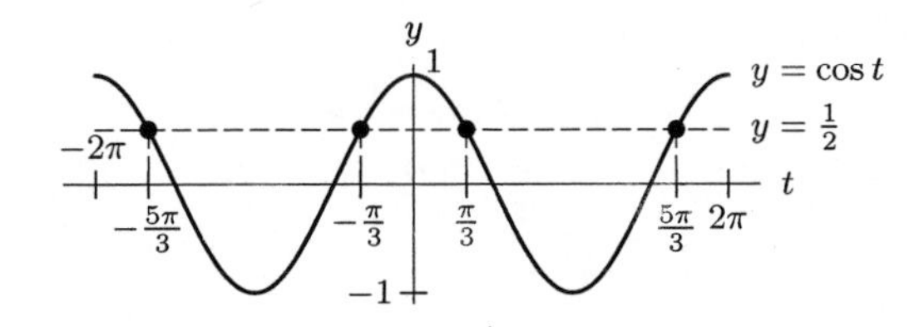

Figure 8.39: $y = \cos t$ has an infinite number of t-values for $y = 1/2$

(b) If we restrict our attention to the interval $0 \le t \le 2\pi$, we find two solutions, $t = \pi/3$ and $t = 5\pi/3$. To see why this is so, look at the unit circle in Figure 8.40. During one revolution around the circle, there are always two angles with the same cosine (or sine, or tangent), unless the cosine is 1 or -1. Therefore, we expect the equation $\cos t = 1/2$ to have two solutions in the interval $0 \le t \le 2\pi$.

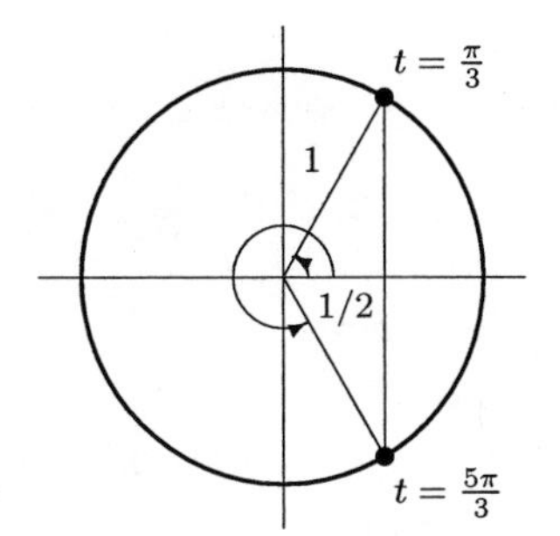

Figure 8.40: During one revolution around the unit circle, the two angles $\pi/3$ and $5\pi/3$ have the cosine value of $1/2$

13. We know that one solution to the equation $\cos t = 0.4$ is $t = \cos^{-1}(0.4) = 1.159$. From Figure 8.41, we see by the symmetry of the unit circle that another solution is $t = -1.159$. We know that additional solutions are given each time the angle t wraps around the circle in either direction. This means that

$$1.159 + 1 \cdot 2\pi = 7.442 \quad \text{wrap once around circle}$$

$$\begin{aligned} 1.159 + 2\cdot 2\pi &= 13.725 && \text{wrap twice around circle} \\ 1.159 + (-1)\cdot 2\pi &= -5.124 && \text{wrap once around the other way} \end{aligned}$$

and

$$\begin{aligned} -1.159 + 1\cdot 2\pi &= 5.124 && \text{wrap once around circle} \\ -1.159 + 2\cdot 2\pi &= 11.407 && \text{wrap once around circle} \\ -1.159 + (-1)\cdot 2\pi &= -7.442. && \text{wrap once around circle the other way} \end{aligned}$$

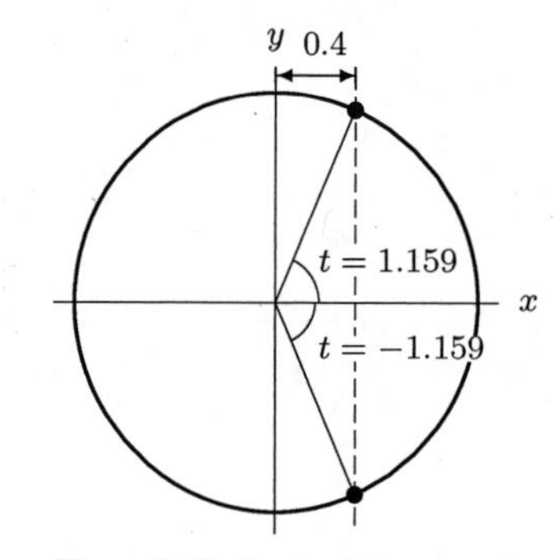

Figure 8.41: Two solutions to the equation $\cos t = 0.4$

14. Since $\sin t = -1$, we have $t = 3\pi/2$.

15. Since $\sin t = 1/2$, we have $t = \pi/6$ and $t = 5\pi/6$.

16. Since $\cos t = -1$, we have $t = \pi$.

17. Since $\cos t = 1/2$, we have $t = \pi/3$ and $t = 5\pi/3$.

18. Since $\tan t = 1$, we have $t = \pi/4$ and $t = 5\pi/4$.

19. Since $\tan t = -1$, we have $t = 3\pi/4$ and $t = 7\pi/4$.

20. Since $\tan t = \sqrt{3}$, we have $t = \pi/3$ and $t = 4\pi/3$.

21. Since $\tan t = 0$, we have $t = 0$, $t = \pi$, and $t = 2\pi$.

22. We have $x = \cos^{-1}(0.6) = 0.927$. A graph of $\cos x$ shows that the second solution is $x = 2\pi - 0.927 = 5.356$.

23. Collecting the $\sin x$ terms on the left gives

$$\begin{aligned} 2\sin x &= 1 - \sin x \\ 3\sin x &= 1 \\ \sin x &= 1/3 \\ x &= \sin^{-1}(1/3) = 0.340. \end{aligned}$$

A graph of $\sin x$ shows the second solutions is $x = \pi - 0.340 = 2.802$.

24. Multiplying both sides by $\cos x$ gives

$$\begin{aligned} 5\cos x &= 1/\cos x \\ 5\cos^2 x &= 1 \\ \cos^2 x &= 1/5 \\ \cos x &= \pm\sqrt{\frac{1}{5}}. \end{aligned}$$

Thus $x = \cos^{-1}\sqrt{1/5} = 1.107$ or $x = \cos^{-1}\left(-\sqrt{1/5}\right) = 2.034$. A graph of $\cos x$ shows that there are other solutions with $\cos x = \sqrt{1/5}$ given by $x = 2\pi - 1.107 = 5.176$ and with $\cos x = -\sqrt{1/5}$ given by $x = 2\pi - 2.034 = 4.249$. Thus the solutions are

$$1.107, 2.034, 4.249, 5.176.$$

25. Looking at the graph of $y = \sin(2x)$ in Figure 8.42, we see it crosses the line $y = 0.3$ four times between 0 and 2π, so there will be four solutions. Since $2x = \sin^{-1}(0.3) = 0.30469$, one solution is

$$x = \frac{0.30469}{2} = 0.1523.$$

The period of $y = \sin(2x)$ is π, so the other solutions in $0 \le x \le 2\pi$ are

$$\begin{aligned} x &= 0.1523 + \pi = 3.294 \\ x &= \pi/2 - 0.1523 = 1.418 \\ x &= 3\pi/2 - 0.1523 = 4.560. \end{aligned}$$

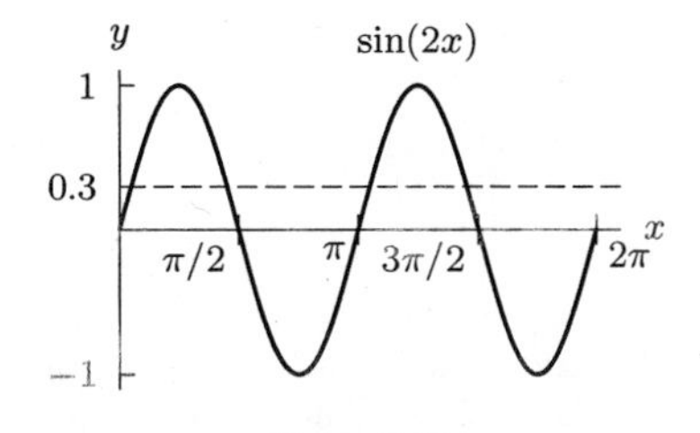

Figure 8.42

26. Since $\sin(x - 1) = 0.25$, we know

$$\begin{aligned} x - 1 &= \sin^{-1}(0.25) = 0.253. \\ x &= 1.253. \end{aligned}$$

Another solution for $x - 1$ is given by

$$\begin{aligned} x - 1 &= \pi - 0.253 = 2.889 \\ x &= 3.889. \end{aligned}$$

27. Since $5\cos(x + 3) = 1$, we have

$$\begin{aligned} \cos(x + 3) &= \frac{1}{5} = 0.2 \\ x + 3 &= \cos^{-1}(0.2) = 1.3694 \\ x &= 1.3694 - 3 = -1.6306. \end{aligned}$$

This value of x is not in the interval $0 \le x \le 2\pi$. To obtain values of x in this interval, we find values of $x + 3$ in the interval between 3 and $3 + 2\pi$, that is between 3 and 9.283. These values of $x + 3$ are

$$\begin{aligned} x + 3 &= 2\pi - 1.369 = 4.914 \\ x + 3 &= 2\pi + 1.369 = 7.653. \end{aligned}$$

Thus

$$x = 1.914, 4.653.$$

28. See Figure 8.43. The solutions to the equation $\cos\theta = -0.4226$ are the angles corresponding to the points P and Q on the unit circle with $x = -0.4226$.

Since $\cos^{-1}(-0.4226) = 115°$, point P corresponds to $115°$. Since P and Q both have reference angles of $65°$, Q corresponds to $180° + 65° = 245°$. Thus the solutions of the equation are

$$\theta = 115° \text{ and } \theta = 180° + 65° = 245°,$$
$$\theta = 360° + 115° = 475° \text{ and } \theta = 360° + 245° = 605°.$$

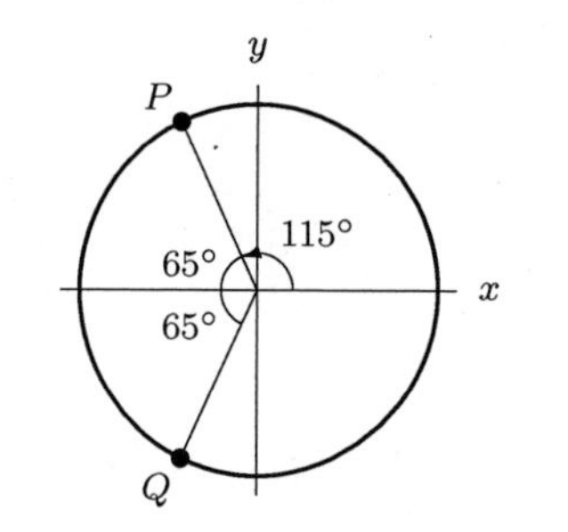

Figure 8.43: Points corresponding to angles with $\cos\theta = -0.4226$

29. **(a)** Since $\cos(65°) = 0.4226$, a calculator set in degrees gives $\cos^{-1}(0.4226) = 65°$. We see in Figure 8.44 that all the angles with a cosine of 0.4226 correspond either to the point P or to the point Q. We want solutions between $0°$ and $360°$, so Q is represented by $360° - 65° = 295°$. Thus, the solutions are

$$\theta = 65° \quad \text{and} \quad \theta = 295°.$$

(b) A calculator gives $\tan^{-1}(2.145) = 65°$. Since $\tan\theta$ is positive in the first and third quadrants, the angles with a tangent of 2.145 correspond either to the point P or the point R in Figure 8.45. Since we are interested in solutions between $0°$ and $720°$, the solutions are

$$\theta = 65°, \quad 245°, \quad 65° + 360°, \quad 245° + 360°.$$

That is

$$\theta = 65°, \quad 245°, \quad 425°, \quad 605°.$$

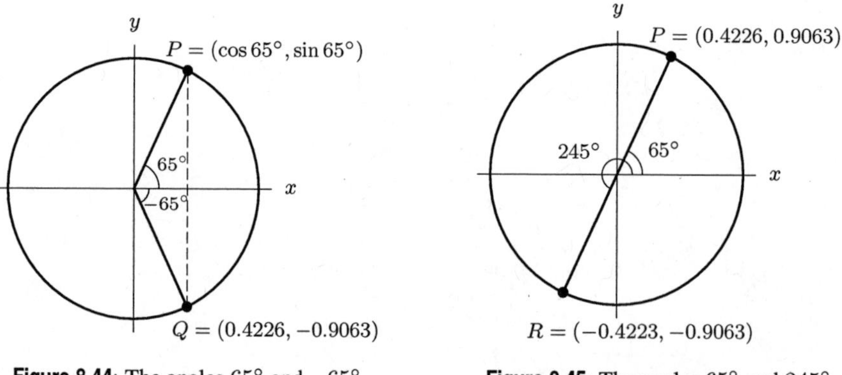

Figure 8.44: The angles $65°$ and $-65°$

Figure 8.45: The angles $65°$ and $245°$

30. By sketching a graph, we see that there are four solutions (see Figure 8.46). The first solution is given by $x = \cos^{-1}(0.6) = 0.927$, which is equivalent to the length labeled "b" in Figure 8.46. Next, note that, by the symmetry of the graph of the cosine function, we can obtain a second solution by subtracting the length b from 2π. Therefore, a second solution to the equation is given by $x = 2\pi - 0.927 = 5.356$. Similarly, our final two solutions are given by $x = 2\pi + 0.927 = 7.210$ and $x = 4\pi - 0.927 = 11.639$.

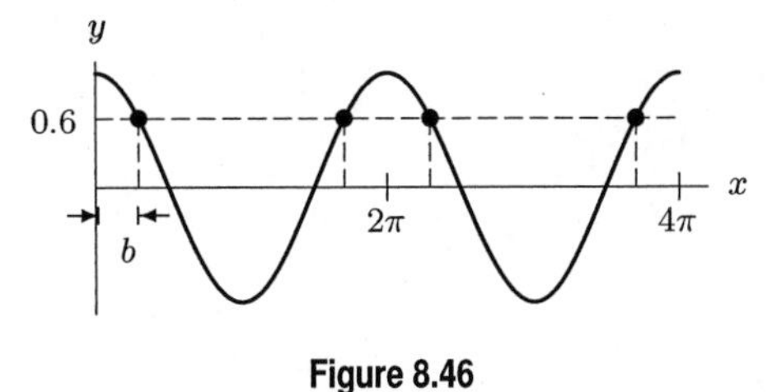

Figure 8.46

31. By sketching a graph, we see that there are two solutions (see Figure 8.47). The first solution is given by $x = \sin^{-1}(0.3) = 0.305$, which is equivalent to the length labeled "b" in Figure 8.47. Next, note that, by the symmetry of the graph of the sine function, we can obtain the second solution by subtracting the length b from π. Therefore, the other solution is given by $x = \pi - 0.305 = 2.837$.

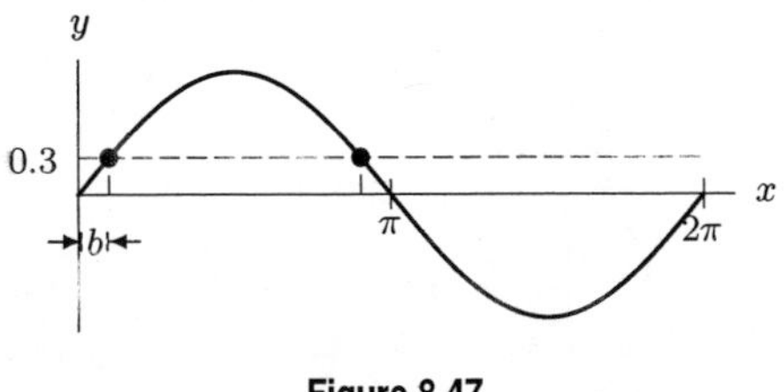

Figure 8.47

32. By sketching a graph, we see that there are two solutions (see Figure 8.48). The first solution is given by $x = \cos^{-1}(-0.7) = 2.346$, which corresponds to the leftmost point in Figure 8.48. To find the second solution, we first calculate the distance labeled "b" in Figure 8.48 to obtain $b = \pi - \cos^{-1}(-0.7) = 0.795$. Therefore, by the symmetry of the cosine function, the second solution is given by $x = \pi + b = 3.937$.

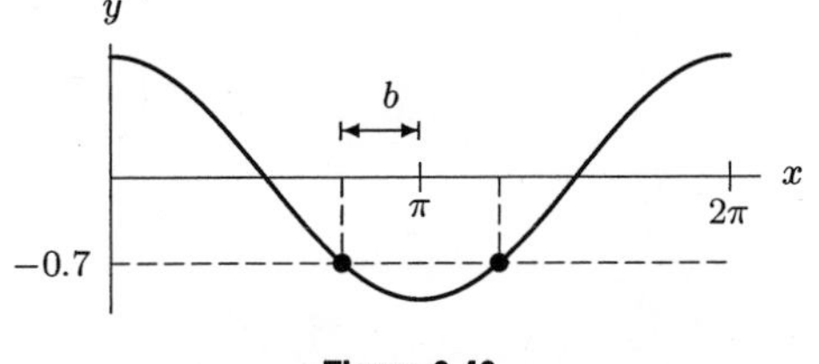

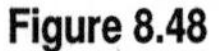
Figure 8.48

33. By sketching a graph, we see that there are four solutions (see Figure 8.49). To find the four solutions, we begin by calculating $\sin^{-1}(-0.8) = -0.927$. Therefore, the length labeled "b" in Figure 8.49 is given by $b = -\sin^{-1}(-0.8) = 0.927$. Now, using the symmetry of the graph of the sine function, we can see that the four solutions are given by $x = \pi + b = 4.069$, $x = 2\pi - b = 5.356$, $x = 3\pi + b = 10.352$, and $x = 4\pi - b = 11.639$.

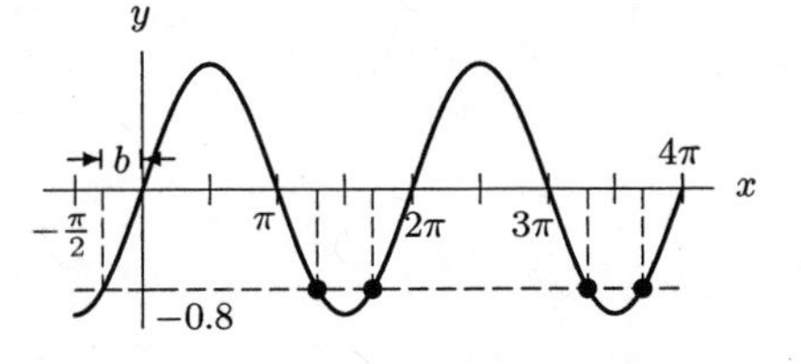

Figure 8.49

34. One solution is $\theta = \sin^{-1}(-\sqrt{2}/2) = -\pi/4$, and a second solution is $5\pi/4$, since $\sin(5\pi/4) = -\sqrt{2}/2$. All other solutions are found by adding integer multiples of 2π to these two solutions. See Figure 8.50.

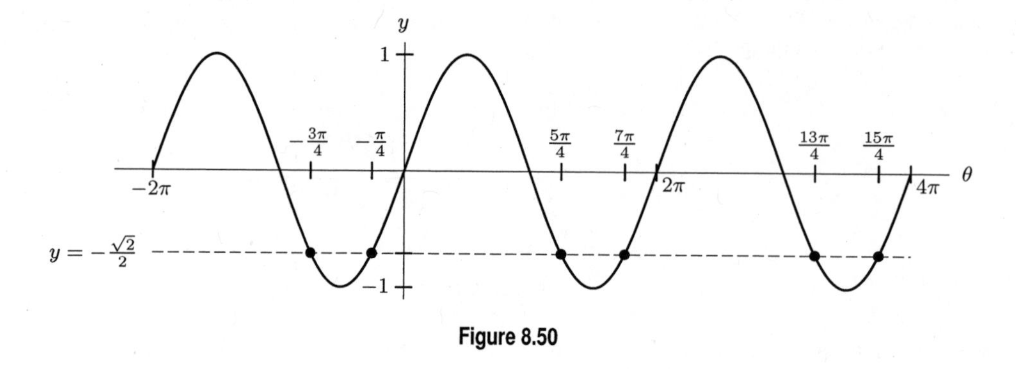

Figure 8.50

35. One solution is $\theta = \cos^{-1}(\sqrt{3}/2) = \pi/6$, and a second solution is $11\pi/6$, since $\cos(11\pi/6) = \sqrt{3}/2$. All other solutions are found by adding integer multiples of 2π to these two solutions. See Figure 8.51.

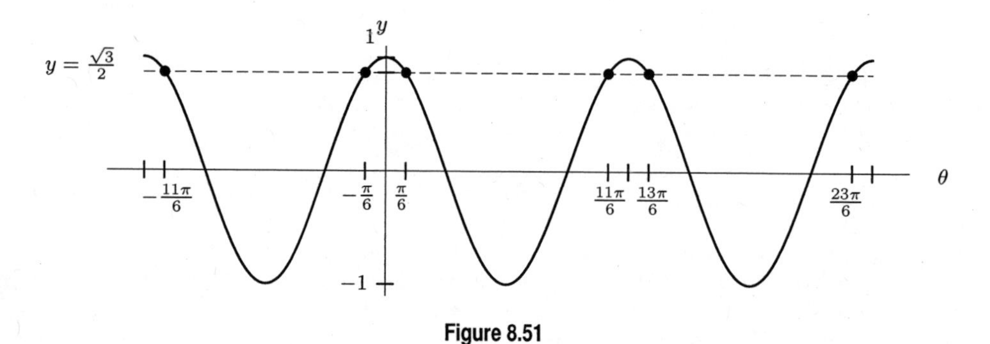

Figure 8.51

36. One solution is $\theta = \tan^{-1}(-\sqrt{3}/3) = -\pi/6$, All other solutions are found by adding integer multiples of π to $-\pi/6$. See Figure 8.52.

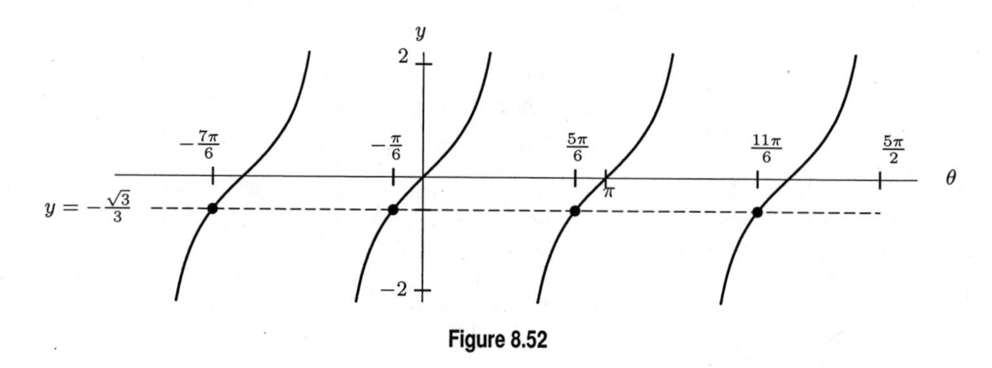

Figure 8.52

37. The angle between $-\pi/2$ and 0 whose tangent is -3 is $\tan^{-1}(-3)$. Thus, the angle in the second quadrant is

$$\theta = \pi + \tan^{-1}(-3) \approx 1.893$$

38.

$$\begin{aligned}\sec^2\alpha + 3\tan\alpha &= \tan\alpha\\ 1+\tan^2\alpha + 3\tan\alpha &= \tan\alpha\\ \tan^2\alpha + 2\tan\alpha + 1 &= 0\\ (\tan\alpha+1)^2 &= 0\\ \tan\alpha &= -1\\ \alpha &= \frac{3\pi}{4}, \frac{7\pi}{4}\end{aligned}$$

39. From Figure 8.53 we can see that the solutions lie on the intervals $\frac{\pi}{8} < t < \frac{\pi}{4}$, $\frac{3\pi}{4} < t < \frac{7\pi}{8}$, $\frac{9\pi}{8} < t < \frac{5\pi}{4}$ and $\frac{7\pi}{4} < t < \frac{15\pi}{8}$. Using the trace mode on a calculator, we can find approximate solutions $t = 0.52$, $t = 2.62$, $t = 3.67$ and $t = 5.76$.

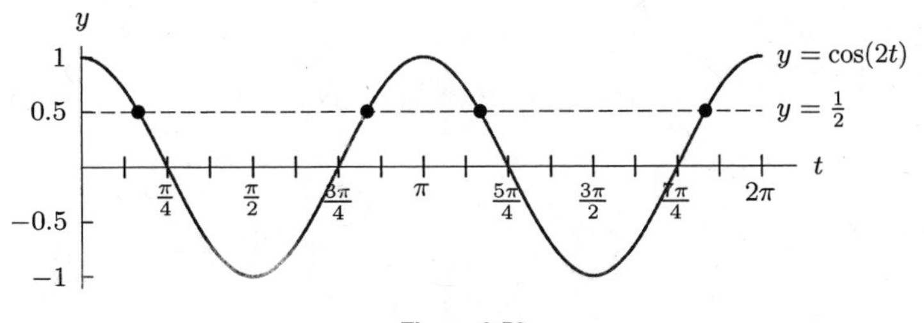

Figure 8.53

For a more precise answer we solve $\cos(2t) = \frac{1}{2}$ algebraically. To find $2t = \arccos(1/2)$. One solution is $2t = \pi/3$. But $2t = 5\pi/3$, $7\pi/3$, and $11\pi/3$ are also angles that have a cosine of $1/2$. Thus $t = \pi/6$, $5\pi/6$, $7\pi/6$, and $11\pi/6$ are the solutions between 0 and 2π.

40. To solve

$$\tan t = \frac{1}{\tan t}$$

we multiply both sides of the equation by $\tan t$. Multiplication gives us

$$\tan^2 t = 1 \qquad \text{or} \qquad \tan t = \pm 1.$$

From Figure 8.54, we see that there are two solutions for $\tan t = 1$, and two solutions for $\tan t = -1$, they are approximately $t = 0.79$, $t = 3.93$, and $t = 2.36$, $t = 5.50$.

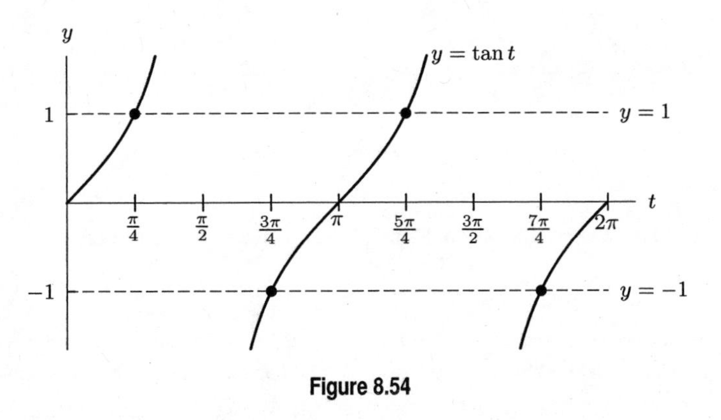

Figure 8.54

To find exact solutions, we have $t = \arctan(\pm 1) = \pm\pi/4$. There are other angles that have a tan of ± 1, namely $\pm 3\pi/4$. So $t = \pi/4$, $3\pi/4$, $5\pi/4$, and $7\pi/4$ are the solutions in the interval from 0 to 2π.

41. From Figure 8.55, we see that $2\sin t\cos t - \cos t = 0$ has four roots between 0 and 2π. They are approximately $t = 0.52$, $t = 1.57$, $t = 2.62$, and $t = 4.71$.

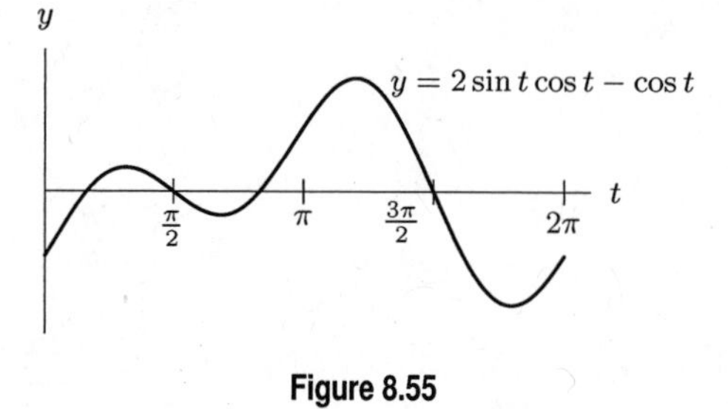

Figure 8.55

To solve the problem symbolically, we factor out $\cos t$:

$$2\sin t\cos t - \cos t = \cos t(2\sin t - 1) = 0.$$

So solutions occur either when $\cos t = 0$ or when $2\sin t - 1 = 0$. The equation $\cos t = 0$ has solutions $\pi/2$ and $3\pi/2$. The equation $2\sin t - 1 = 0$ has solution $t = \arcsin(1/2) = \pi/6$, and also $t = \pi - \pi/6 = 5\pi/6$. Thus the solutions to the original problem are

$$t = \frac{\pi}{2}, \frac{3\pi}{2}, \frac{\pi}{6} \text{ and } \frac{5\pi}{6}.$$

42. The solutions to the equation are at points where the graphs of $y = 3\cos^2 t$ and $y = \sin^2 t$ cross. From Figure 8.56, we see that $3\cos^2 t = \sin^2 t$ has four solutions between 0 and 2π, they are approximately $t = 1.05$, $t = 2.09$, $t = 4.19$, and $t = 5.24$.

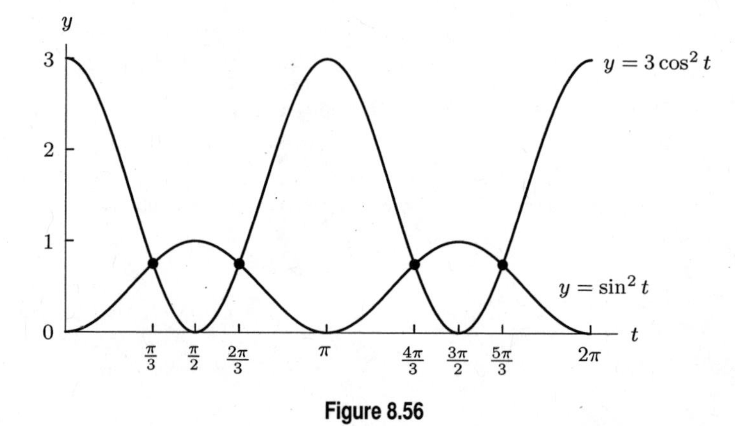

Figure 8.56

To solve $3\cos^2 t = \sin^2 t$ we divide both sides by $\cos^2 t$ and rewrite the equation as

$$3 = \tan^2 t \quad \text{or} \quad \tan t = \pm\sqrt{3}.$$

(Dividing by $\cos^2 t$ is valid only if $\cos t \neq 0$. Since $t = \pi/2$ and $t = 3\pi/2$ are not solutions, $\cos t \neq 0$.)

Using the inverse tangent and reference angles we find that the solutions occur at the points

$$t = \frac{\pi}{3}, \frac{4\pi}{3}, \frac{2\pi}{3} \text{ and } \frac{5\pi}{3}.$$

43. **(a)** The maximum is \$100,000 and the minimum is \$20,000. Thus in the function

$$f(t) = A\cos(B(t-h)) + k,$$

the midline is

$$k = \frac{100,000 + 20,000}{2} = \$60,000$$

and the amplitude is

$$A = \frac{100,000 - 20,000}{2} = \$40,000.$$

The period of this function is 12 since the sales are seasonal. Since

$$\text{period} = 12 = \frac{2\pi}{B},$$

we have

$$B = \frac{\pi}{6}.$$

The company makes its peak sales in mid-December, which is month -1 or month 11. Since the regular cosine curve hits its peak at $t = 0$ while ours does this at $t = -1$, we find that our curve is shifted horizontally 1 unit to the left. So we have

$$h = -1.$$

So the sales function is

$$f(t) = 40{,}000\cos\left(\frac{\pi}{6}(t+1)\right) + 60{,}000 = 40{,}000\cos\left(\frac{\pi}{6}t + \frac{\pi}{6}\right) + 60{,}000.$$

(b) Mid-April is month $t = 3$. Substituting this value into our function, we get

$$f(3) = \$40{,}000.$$

(c) To solve $f(t) = 60{,}000$ for t, we write

$$\begin{aligned} 60{,}000 &= 40{,}000\cos\left(\frac{\pi}{6}t + \frac{\pi}{6}\right) + 60{,}000 \\ 0 &= 40{,}000\cos\left(\frac{\pi}{6}t + \frac{\pi}{6}\right) \\ 0 &= \cos\left(\frac{\pi}{6}t + \frac{\pi}{6}\right). \end{aligned}$$

Therefore, $(\frac{\pi}{6}t + \frac{\pi}{6})$ equals $\frac{\pi}{2}$ or $\frac{3\pi}{2}$. Solving for t, we get $t = 2$ or $t = 8$. So in mid-March and mid-September the company has sales of \$60,000 (which is the average or midline sales value.)

44. **(a)** Let t be the time in hours since 12 noon. Let $d = f(t)$ be the depth in feet in Figure 8.57.

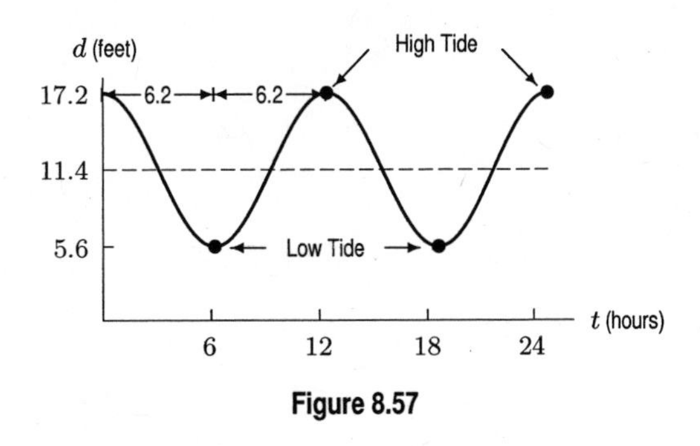

Figure 8.57

(b) The midline is $d = \dfrac{17.2 + 5.6}{2} = 11.4$ and the amplitude is $17.2 - 11.4 = 5.8$. The period is 12.4. Thus we get $d = f(t) = 11.4 + 5.8\cos\left(\dfrac{\pi}{6.2}t\right)$.

(c) We find the first t value when $d = f(t) = 8$:

$$8 = 11.4 + 5.8\cos\left(\frac{\pi}{6.2}t\right)$$

Using the $\cos^{-1}$ function

$$\frac{-3.4}{5.8} = \cos\left(\frac{\pi}{6.2}t\right)$$
$$\cos^{-1}\left(\frac{-3.4}{5.8}\right) = \frac{\pi}{6.2}t$$
$$t = \frac{6.2}{\pi}\cos^{-1}\left(\frac{-3.4}{5.8}\right) \approx 4.336 \text{ hours.}$$

Since $0.336(60) \approx 20$ minutes, the latest time the boat can set sail is 4:20 pm.

45. The curve is a sine curve with an amplitude of 5, a period of 8 and a vertical shift of -3. Thus the equation for the curve is $y = 5\sin\left(\frac{\pi}{4}x\right) - 3$. Solving for $y = 0$, we have

$$5\sin\left(\frac{\pi}{4}x\right) = 3$$
$$\sin\left(\frac{\pi}{4}x\right) = \frac{3}{5}$$
$$\frac{\pi}{4}x = \sin^{-1}\left(\frac{3}{5}\right)$$
$$x = \frac{4}{\pi}\sin^{-1}\left(\frac{3}{5}\right) \approx 0.819.$$

This is the x-coordinate of P. The x-coordinate of Q is to the left of 4 by the same distance P is to the right of O, by the symmetry of the sine curve. Therefore,

$$x \approx 4 - 0.819 = 3.181$$

is the x-coordinate of Q.

46. **(a)** $\sin^{-1} x$ is the angle whose sine is x. When $x = 0.5$, $\sin^{-1}(0.5) = \pi/6$.
(b) $\sin(x^{-1})$ is the sine of $1/x$. When $x = 0.5$, $\sin(0.5^{-1}) = \sin(2) \approx 0.909$.
(c) $(\sin x)^{-1} = \frac{1}{\sin x}$. When $x = 0.5$, $(\sin 0.5)^{-1} = \frac{1}{\sin(0.5)} \approx 2.086$.

47. **(a)** $\arccos(0.5) = \pi/3$
(b) $\arccos(-1) = \pi$
(c) $\arcsin(0.1) \approx 0.1$

48. Graph $y = 12 - 4\cos(3t)$ on $0 \le t \le 2\pi/3$ and locate the two points with y-coordinate 14. (See Figure 8.58.) These points have t-coordinates of approximately $t = 0.698$ and $t = 1.396$. There are six solutions in three cycles of the graph between 0 and 2π.

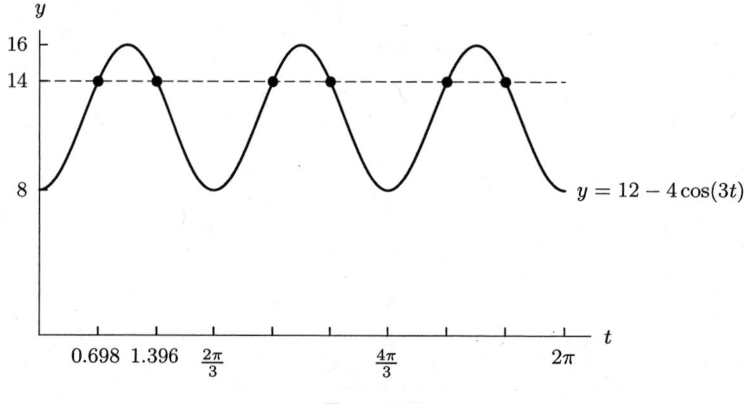

Figure 8.58

49. **(a)** Graph $y = 3 - 5\sin 4t$ on the interval $0 \le t \le \pi/2$, and locate values where the function crosses the t-axis. Alternatively, we can find the points where the graph $5\sin 4t$ and the line $y = 3$ intersect. By looking at the graphs of these two functions on the interval $0 \le t \le \pi/2$, we find that they intersect twice. By zooming in we can identify these points of intersection as roughly $t_1 \approx 0.16$ and $t_2 \approx 0.625$. See Figure 8.59.

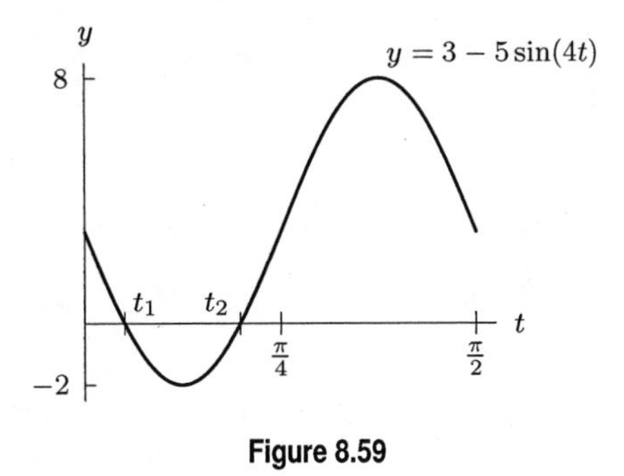

Figure 8.59

(b) Solve for $\sin(4t)$ and then use arcsine:

$$\begin{aligned} 5\sin(4t) &= 3 \\ \sin(4t) &= \frac{3}{5} \\ 4t &= \arcsin\left(\frac{3}{5}\right). \end{aligned}$$

So $t_1 = \dfrac{\arcsin(3/5)}{4} \approx 0.161$ is a solution. But the angle $\pi - \arcsin(3/5)$ has the same sine as $\arcsin(3/5)$. Solving $4t = \pi - \arcsin(3/5)$ gives $t_2 = \dfrac{\pi}{4} - \dfrac{\arcsin(3/5)}{4} \approx 0.625$ as a second solution.

50. **(a)** The domain of f is $-1 \le x \le 1$, so $\sin^{-1}(x)$ is defined only if x is between -1 and 1 inclusive. The range is between $-\pi/2$ and $\pi/2$ inclusive.

(b) The domain of g is $-1 \le x \le 1$, so $\cos^{-1}(x)$ is defined only if x is between -1 and 1 inclusive. The range is between 0 and π inclusive.

(c) The domain of h is all real numbers, so $\tan^{-1}(x)$ is defined for all values of x. The range is between but not including $-\pi/2$ and $\pi/2$.

51. **(a)** $\cos^{-1}\left(\frac{1}{2}\right)$ is the angle between 0 and π whose cosine is $1/2$. Since $\cos\left(\frac{\pi}{3}\right) = 1/2$, we have $\cos^{-1}\left(\frac{1}{2}\right) = \pi/3$.

(b) Similarly, $\cos^{-1}\left(-\frac{1}{2}\right)$ is the angle between 0 and π whose cosine is $-1/2$. Since $\cos\left(\frac{2\pi}{3}\right) = -1/2$, we have $\cos^{-1}\left(-\frac{1}{2}\right) = 2\pi/3$.

(c) From the first part, we saw that $\cos^{-1}\left(\frac{1}{2}\right) = \pi/3$. This gives

$$\cos\left(\cos^{-1}\left(\frac{1}{2}\right)\right) = \cos\left(\frac{\pi}{3}\right) = \frac{1}{2}.$$

This is not at all surprising; after all, what we are saying is that the cosine of the inverse cosine of a number is that number. However, the situation is not as straightforward as it may appear. The next part of this question illustrates the problem.

(d) The cosine of $5\pi/3$ is $1/2$. And since $\cos^{-1}\left(\frac{1}{2}\right) = \pi/3$, we have

$$\cos^{-1}\left(\cos\left(\frac{5\pi}{3}\right)\right) = \cos^{-1}\left(\frac{1}{2}\right) = \frac{\pi}{3}.$$

Thus, we see that the inverse cosine of the cosine of an angle does not necessarily equal that angle.

52. Statement II is always true, because $\arcsin x$ is an angle whose sine is x, and thus the sine of $\arcsin x$ will necessarily equal x. Statement I could be true or false. For example,

$$\arcsin\left(\sin\frac{\pi}{4}\right) = \arcsin\left(\frac{\sqrt{2}}{2}\right) = \frac{\pi}{4}.$$

On the other hand,

$$\arcsin(\sin\pi) = \arcsin(0) = 0,$$

which is not equal to π.

53. **(a)** The value of $\sin t$ will be between -1 and 1. This means that $k\sin t$ will be between $-k$ and k. Thus, $t^2 = k\sin t$ will be between 0 and k. So

$$-\sqrt{k} \le t \le \sqrt{k}.$$

(b) Plotting $2\sin t$ and t^2 on a calculator, we see that $t^2 = 2\sin t$ for $t = 0$ and $t \approx 1.40$.

(c) Compare the graphs of $k\sin t$, a sine wave, and t^2, a parabola. As k increases, the amplitude of the sine wave increases, and so the sine wave intersects the parabola in more points.

(d) Plotting $k\sin t$ and t^2 on a calculator for different values of k, we see that if $k \approx 20$, this equation will have a negative solution at $t \approx -4.3$, but that if k is any smaller, there will be no negative solution.

54. **(a)** In Figure 8.60, the earth's center is labeled O and two radii are extended, one through S, your ship's position, and one through H, the point on the horizon. Your line of sight to the horizon is tangent to the surface of the earth. A line tangent to a circle at a given point is perpendicular to the circle's radius at that point. Thus, since your line of sight is tangent to the earth's surface at H, it is also perpendicular to the earth's radius at H. This means that triangle OCH is a right triangle. Its hypotenuse is $r + x$ and its legs are r and d. From the Pythagorean theorem, we have

$$\begin{aligned} r^2 + d^2 &= (r+x)^2 \\ d^2 &= (r+x)^2 - r^2 \\ &= r^2 + 2rx + x^2 - r^2 = 2rx + x^2. \end{aligned}$$

Since d is positive, we have $d = \sqrt{2rx + x^2}$.

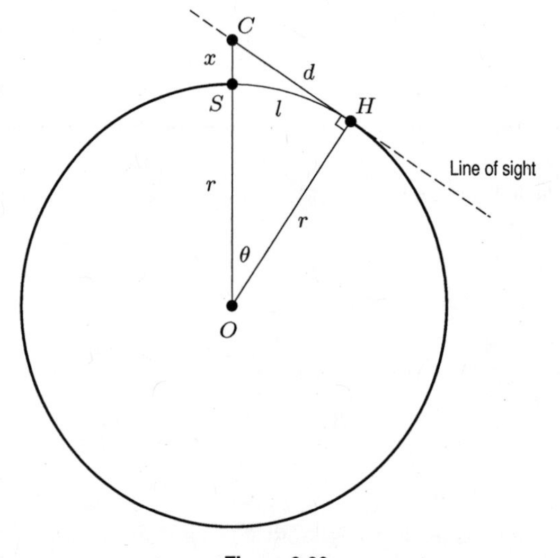

Figure 8.60

(b) We begin by using the formula obtained in part (a).

$$\begin{aligned} d &= \sqrt{2rx + x^2} \\ &= \sqrt{2(6{,}370{,}000)(50) + 50^2} \\ &\approx 25{,}238.908. \end{aligned}$$

Thus, you would be able to see a little over 25 kilometers from the crow's nest C.

Having found a formula for d, we will now try to find a formula for l, the distance along the earth's surface from the ship to the horizon H. In Figure 8.60, l is the arc length specified by the angle θ (in radians). The formula for arc length is

$$l = r\theta.$$

In this case, we must determine θ. From Figure 8.60 we see that

$$\cos\theta = \frac{\text{adjacent}}{\text{hypotenuse}} = \frac{r}{r+x}.$$

Thus,

$$\theta = \cos^{-1}\left(\frac{r}{r+x}\right)$$

since $0 \leq \theta \leq \pi/2$. This means that

$$\begin{aligned} l = r\theta &= r\cos^{-1}\left(\frac{r}{r+x}\right) \\ &= 6{,}370{,}000\cos^{-1}\left(\frac{6{,}370{,}000}{6{,}370{,}050}\right) \approx 25{,}238.776 \text{ meters.} \end{aligned}$$

There is very little difference—about 0.13 m or 13 cm—between the distance d that you can see and the distance l that the ship must travel to reach the horizon. If this is surprising, keep in mind that Figure 8.60 has not been drawn to scale. In reality, the mast height x is significantly smaller than the earth's radius r so that the point C in the crow's nest is very close to the ship's position at point S. Thus, the line segment d and the arc l are almost indistinguishable.

55. The angle between a line $y = b + mx$ and the x-axis is $\arctan m$. Thus, ℓ_1 makes an angle of $\arctan m_1$ with the x-axis, and ℓ_2 makes an angle of $\arctan m_2$ with the x-axis. The angle between the two lines is thus

$$\theta = \arctan(m_1) - \arctan(m_2).$$

Solutions for Section 8.5

Exercises

1. Quadrant IV.

2. Quadrant I.

3. Since $470° = 360° + 110°$, such a point is in Quadrant II.

4. Since $2.4\pi = 2\pi + 0.4\pi$, such a point is in Quadrant I.

5. Since $3.2\pi = 2\pi + 1.2\pi$, such a point is in Quadrant III.

6. Since $-2.9\pi = -2 \cdot 2\pi + 1.1\pi$, such a point is in Quadrant III.

7. Since $10.3\pi = 5 \cdot 2\pi + 0.3\pi$, such a point is in Quadrant I.

8. Since $-13.4\pi = -7 \cdot 2\pi + 0.6\pi$, such a point is in Quadrant II.

9. Since -7 is an angle of -7 radians, corresponding to a rotation of just over 2π, or one full revolution, in the clockwise direction, such a point is in Quadrant IV.

10. We have $0° < \theta < 90°$. See Figure 8.61.

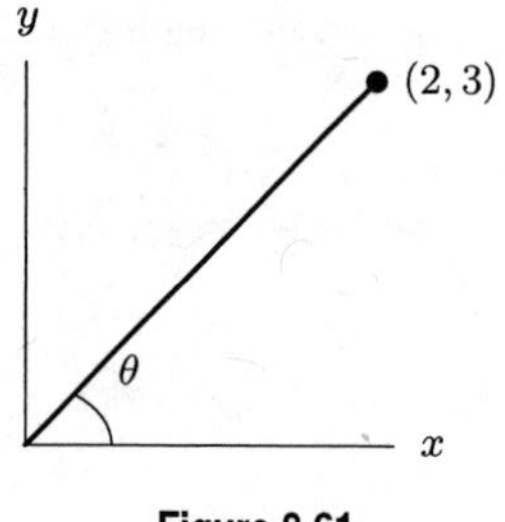

Figure 8.61

11. We have $90° < \theta < 180°$. See Figure 8.62.

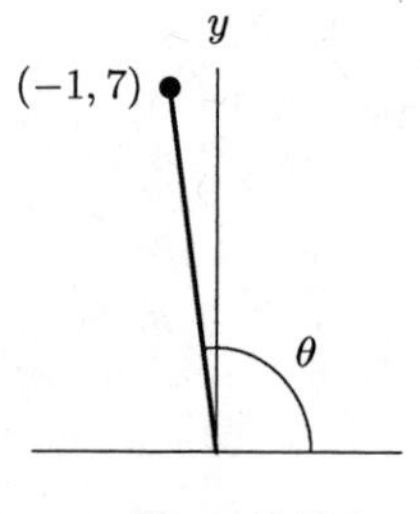

Figure 8.62

12. We have $270° < \theta < 360°$. See Figure 8.63.

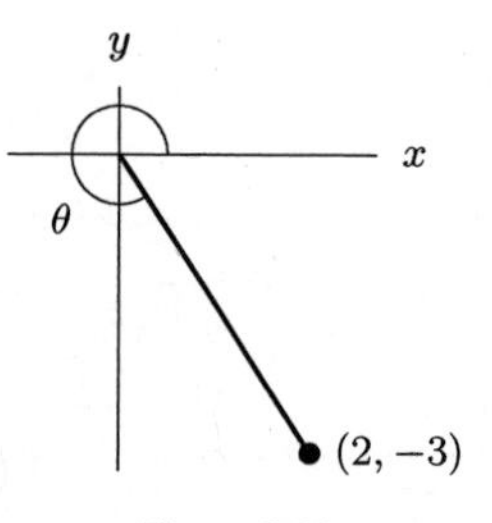

Figure 8.63

13. We have $180° < \theta < 270°$. See Figure 8.64.

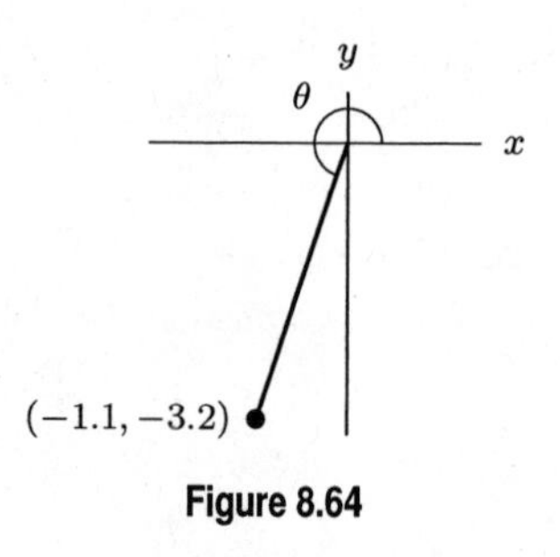

Figure 8.64

14. With $x = 1$ and $y = 1$, find r from $r = \sqrt{x^2 + y^2} = \sqrt{1^2 + 1^2} = \sqrt{2}$. Find θ from $\tan\theta = y/x = 1/1 = 1$. Thus, $\theta = \tan^{-1}(1) = \pi/4$. Since $(1, 1)$ is in the first quadrant this is a correct θ. The polar coordinates are $(\sqrt{2}, \pi/4)$.

15. With $x = -1$ and $y = 0$, find $r = \sqrt{x^2 + y^2} = \sqrt{(-1)^2 + 0^2} = 1$. Find θ from $\tan\theta = y/x = 0/(-1) = 0$. Thus, $\theta = \tan^{-1}(0) = 0$. Since $(-1, 0)$ is on the x-axis between the second and third quadrant, $\theta = \pi$. The polar coordinates are $(1, \pi)$.

16. With $x = \sqrt{6}$ and $y = -\sqrt{2}$, find $r = \sqrt{(\sqrt{6})^2 + (-\sqrt{2})^2} = \sqrt{8} = 2\sqrt{2}$. Find θ from $\tan\theta = y/x = -\sqrt{2}/\sqrt{6} = -1/\sqrt{3}$. Thus, $\theta = \tan^{-1}(-1/\sqrt{3}) = -\pi/6$. Since $(\sqrt{6}, -\sqrt{2})$ is in the fourth quadrant, this is the correct θ. The polar coordinates are $(2\sqrt{2}, -\pi/6)$.

17. With $x = -\sqrt{3}$ and $y = 1$, find $r = \sqrt{(-\sqrt{3})^2 + 1^2} = \sqrt{4} = 2$. Find θ from $\tan\theta = y/x = 1/(-\sqrt{3})$. Thus, $\theta = \tan^{-1}(-1/\sqrt{3}) = -\pi/6$. Since $(-\sqrt{3}, 1)$ is in the second quadrant, $\theta = -\pi/6 + \pi = 5\pi/6$. The polar coordinates are $(2, 5\pi/6)$.

18. With $r = 1$ and $\theta = 2\pi/3$, we find $x = r\cos\theta = 1\cdot\cos(2\pi/3) = -1/2$ and $y = r\sin\theta = 1\cdot\sin(2\pi/3) = \sqrt{3}/2$.
The rectangular coordinates are $(-1/2, \sqrt{3}/2)$.

19. With $r = \sqrt{3}$ and $\theta = -3\pi/4$, we find $x = r\cos\theta = \sqrt{3}\cos(-3\pi/4) = \sqrt{3}(-\sqrt{2}/2) = -\sqrt{6}/2$ and $y = r\sin\theta = \sqrt{3}\sin(-3\pi/4) = \sqrt{3}(-\sqrt{2}/2) = -\sqrt{6}/2$.
The rectangular coordinates are $(-\sqrt{6}/2, -\sqrt{6}/2)$.

20. With $r = 2\sqrt{3}$ and $\theta = -\pi/6$, we find $x = r\cos\theta = 2\sqrt{3}\cos(-\pi/6) = 2\sqrt{3}\cdot\sqrt{3}/2 = 3$ and $y = r\sin\theta = 2\sqrt{3}\sin(-\pi/6) = 2\sqrt{3}(-1/2) = -\sqrt{3}$.
The rectangular coordinates are $(3, -\sqrt{3})$.

21. With $r = 2$ and $\theta = 5\pi/6$, we find $x = r\cos\theta = 2\cos(5\pi/6) = 2(-\sqrt{3}/2) = -\sqrt{3}$ and $y = r\sin\theta = 2\sin(5\pi/6) = 2(1/2) = 1$.
The rectangular coordinates are $(-\sqrt{3}, 1)$.

Problems

22. Since $r = \sqrt{x^2 + y^2}$, write $\sqrt{x^2 + y^2} = 2$. Squaring both sides gives $x^2 + y^2 = 4$.

23. Multiply both sides by r to transform the right side of the equation into $6r\cos\theta$ so that we can substitute $x = r\cos\theta$. The left side is now $r^2 = x^2 + y^2$. In rectangular coordinates, the equation is $x^2 + y^2 = 6x$.

24. Since $\theta = \tan^{-1}(y/x)$, write $\tan^{-1}(y/x) = \frac{\pi}{4}$. Taking the tangent of both sides, we get $y/x = \tan(\pi/4) = 1$. In rectangular coordinates, the equation is $y = x$.

25. Rewrite the left side using $\tan\theta = \dfrac{\sin\theta}{\cos\theta}$:

$$\frac{\sin\theta}{\cos\theta} = r\cos\theta - 2$$

In order to substitute $x = r\cos\theta$ and $y = r\sin\theta$, multiply by the r in the numerator and denominator on the left

$$\frac{r\sin\theta}{r\cos\theta} = r\cos\theta - 2$$

and substitute:

$$\frac{y}{x} = x - 2.$$

So

$$y = x(x - 2)$$

or

$$y = x^2 - 2x.$$

26. By substitution the equation becomes $3r\cos\theta - 4r\sin\theta = 2$. This can be written as $r = \dfrac{2}{3\cos\theta - 4\sin\theta}$.

27. By substituting $r^2 = x^2 + y^2$, we have $r^2 = 5$. Since r is positive, this could also be written as $r = \sqrt{5}$.

28. By substituting $x = r\cos\theta$ and $y = r\sin\theta$, the equation becomes $r\sin\theta = (r\cos\theta)^2$. This could also be written as $r = \dfrac{\sin\theta}{\cos^2\theta}$.

29. By substituting $x = r\cos\theta$ and $y = r\sin\theta$, the equation becomes $2r\cos\theta \cdot r\sin\theta = 1$. This can be written as

$$r = \frac{1}{\sqrt{2\cos\theta\sin\theta}}.$$

30. Figure 8.65 shows that at 12 noon, we have:
In Cartesian coordinates, $H = (0,3)$. In polar coordinates, $H = (3,\pi/2)$; that is $r = 3, \theta = \pi/2$. In Cartesian coordinates, $M = (0,4)$. In polar coordinates, $M = (4,\pi/2)$, that is $r = 4, \theta = \pi/2$.

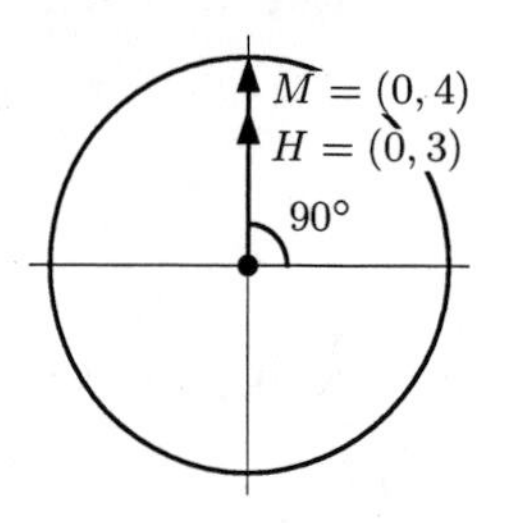

Figure 8.65

31. Figure 8.66 shows that at 3 pm, we have:
In Cartesian coordinates, $H = (3,0)$. In polar coordinates, $H = (3,0)$; that is $r = 3, \theta = 0$. In Cartesian coordinates, $M = (0,4)$. In polar coordinates, $M = (4,\pi/2)$, that is $r = 4, \theta = \pi/2$.

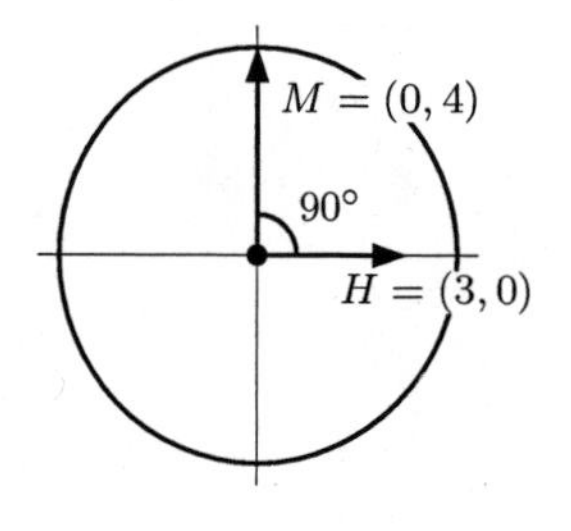

Figure 8.66

32. Figure 8.67 shows that at 9 am, we have:
In Cartesian coordinates, $H = (-3,0)$. In polar coordinates, $H = (3,\pi)$; that is $r = 3, \theta = \pi$. In Cartesian coordinates, $M = (0,4)$. In polar coordinates, $M = (4,\pi/2)$, that is $r = 4, \theta = \pi/2$.

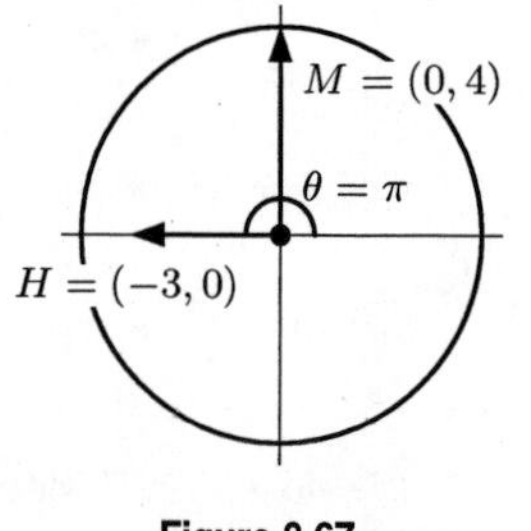

Figure 8.67

33. Figure 8.68 shows that at 1 pm, we have:
In Cartesian coordinates, $M = (0, 4)$. In polar coordinates, $M = (4, \pi/2)$; that is $r = 4, \theta = \pi/2$.
At 1 pm, the hour hand points toward 1, so for H, we have $r = 3$ and $\theta = \pi/3$. Thus, the Cartesian coordinates of H are

$$x = 3\cos\left(\frac{\pi}{3}\right) = 1.5, \qquad y = 3\sin\left(\frac{\pi}{3}\right) = \frac{3\sqrt{3}}{2} = 2.598.$$

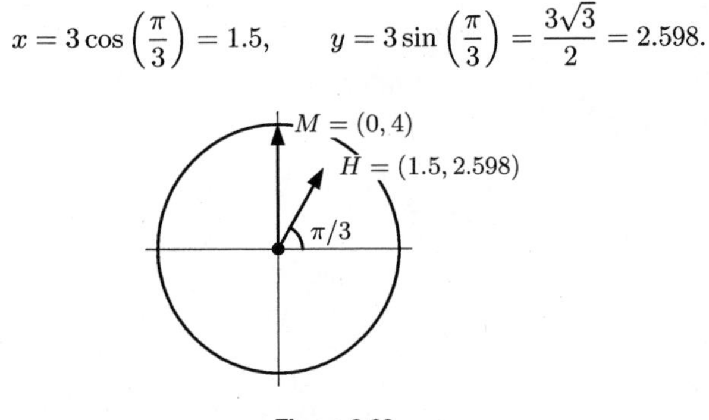

Figure 8.68

34. Figure 8.69 shows that at 1:30 pm, the polar coordinates of the point H (halfway between 1 and 2 on the clock face) are $r = 3$ and $\theta = 45° = \pi/4$. Thus, the Cartesian coordinates of H are given by

$$x = 3\cos\left(\frac{\pi}{4}\right) = \frac{3\sqrt{2}}{2} \approx 2.121, \quad y = 3\sin\left(\frac{\pi}{4}\right) = \frac{3\sqrt{2}}{2} \approx 2.121.$$

In Cartesian coordinates, $H \approx (2.121, 2.121)$. In polar coordinates, $H = (3, \pi/4)$. In Cartesian coordinates, $M = (0, -4)$. In polar coordinates, $M = (4, 3\pi/2)$.

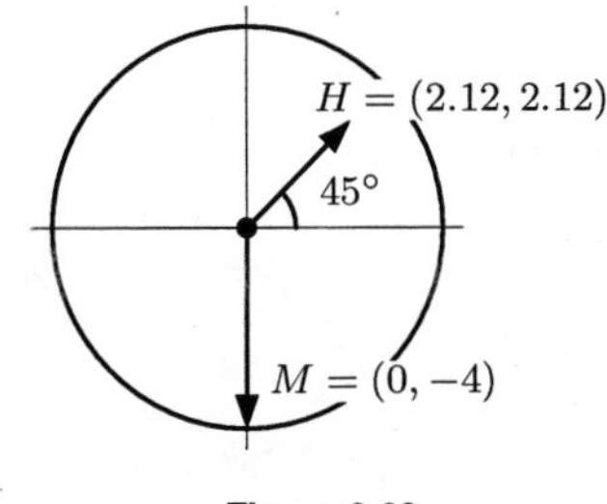

Figure 8.69

35. Figure 8.70 shows that at 7 am the polar coordinates of the point H are $r = 3$ and $\theta = 60° + 180° = 240° = 4\pi/3$. Thus, the Cartesian coordinates of H are given by

$$x = 3\cos\left(\frac{4\pi}{3}\right) = -\frac{3}{2} = -1.5, \quad y = 3\sin\left(\frac{4\pi}{3}\right) = -\frac{3\sqrt{3}}{2} \approx -2.598.$$

Thus, in Cartesian coordinates, $H = (-1.5, -2.598)$. In polar coordinates, $H = (3, 4\pi/3)$. In Cartesian coordinates, $M = (0, 4)$. In polar coordinates, $M = (4, \pi/2)$.

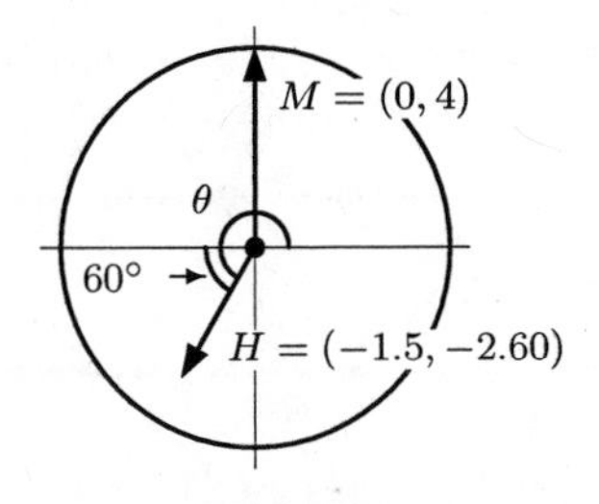

Figure 8.70

36. Figure 8.71 shows that at 3:30 pm, the polar coordinates of the point H (halfway between 3 and 4 on the clock face) are $r = 3$ and $\theta = 75° + 270° = 23\pi/12$. Thus, the Cartesian coordinates of H are given by

$$x = 3\cos\left(\frac{23\pi}{12}\right) \approx 2.898, \quad y = 3\sin\left(\frac{23\pi}{12}\right) \approx -0.776.$$

We have:

In Cartesian coordinates, $H \approx (2.898, -0.776)$; $M = (0, -4)$.
In polar coordinates, $H = (3, 23\pi/12)$; $M = (4, 3\pi/2)$.

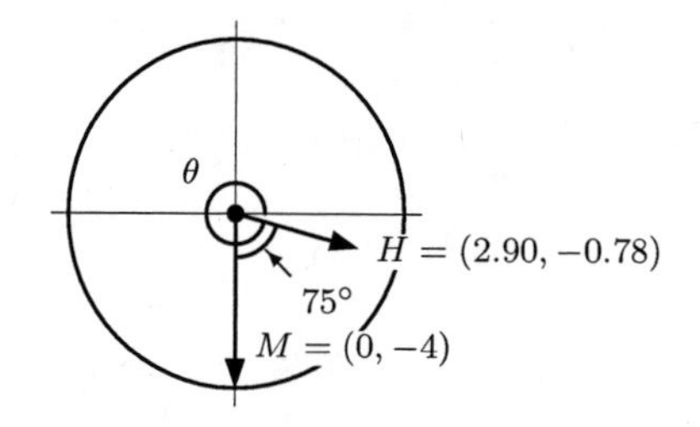

Figure 8.71

37. Figure 8.72 shows that at 9:15 am, the polar coordinates of the point H (half-way between 9 and 9:30 on the clock face) are $r = 3$ and $\theta = 82.5° + 90° = 172.5\pi/180$. Thus, the Cartesian coordinates of H are given by

$$x = 3\cos\left(\frac{172.5\pi}{180}\right) \approx -2.974, \quad y = 3\sin\left(\frac{172.5\pi}{180}\right) \approx 0.392.$$

In Cartesian coordinates, $H \approx (-2.974, 0.392)$. In polar coordinates, $H = (3, 172.5\pi/180)$. In Cartesian coordinates, $M = (4, 0)$. In polar coordinates, $M = (4, 0)$.

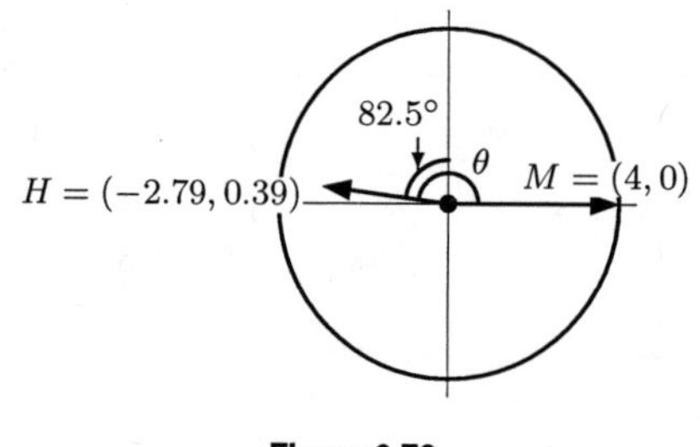

Figure 8.72

38. The region is given by $\sqrt{8} \le r \le \sqrt{18}$ and $\pi/4 \le \theta \le \pi/2$.

39. The region is given by $0 \le r \le 2$ and $-\pi/6 \le \theta \le \pi/6$.

40. The circular arc has equation $r = 1$, for $0 \le \theta \le \pi/2$. the vertical line $x = 2$ has polar equation $r\cos\theta = 2$, or $r = 2/\cos\theta$. So the region is described by $0 \le \theta \le \pi/2$ and $1 \le r \le 2/\cos\theta$.

41. **(a)** Table 8.3 contains values of $r = 1 - \sin\theta$, both exact and rounded to one decimal.

Table 8.3

θ	0	$\pi/3$	$\pi/2$	$2\pi/3$	π	$4\pi/3$	$3\pi/2$	$5\pi/3$	2π	$7\pi/3$	$5\pi/2$	$8\pi/3$
r	1	$1-\sqrt{3}/2$	0	$1-\sqrt{3}/2$	1	$1+\sqrt{3}/2$	2	$1+\sqrt{3}/2$	1	$1-\sqrt{3}/2$	0	$1-\sqrt{3}/2$
r	1	0.134	0	0.134	1	1.866	2	1.866	1	0.134	0	0.134

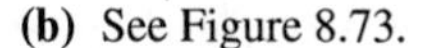

(b) See Figure 8.73.

Figure 8.73

Figure 8.74

(c) The circle has equation $r = 1/2$. The cardioid is $r = 1 - \sin\theta$. Solving these two simultaneously gives

$$1/2 = 1 - \sin\theta,$$

or

$$\sin\theta = 1/2.$$

Thus, $\theta = \pi/6$ or $5\pi/6$. This gives the points $(x, y) = ((1/2)\cos\pi/6, (1/2)\sin\pi/6) = (\sqrt{3}/4, 1/4)$ and $(x, y) = ((1/2)\cos 5\pi/6, (1/2)\sin 5\pi/6) = (-\sqrt{3}/4, 1/4)$ as the location of intersection.

(d) The curve $r = 1 - \sin 2\theta$, pictured in Figure 8.74, has two regions instead of the one region that $r = 1 - \sin\theta$ has. This is because $1 - \sin 2\theta$ will be 0 twice for every 2π cycle in θ, as opposed to once for every 2π cycle in θ for $1 - \sin\theta$.

42. There will be n loops. See Figures 8.75–8.78.

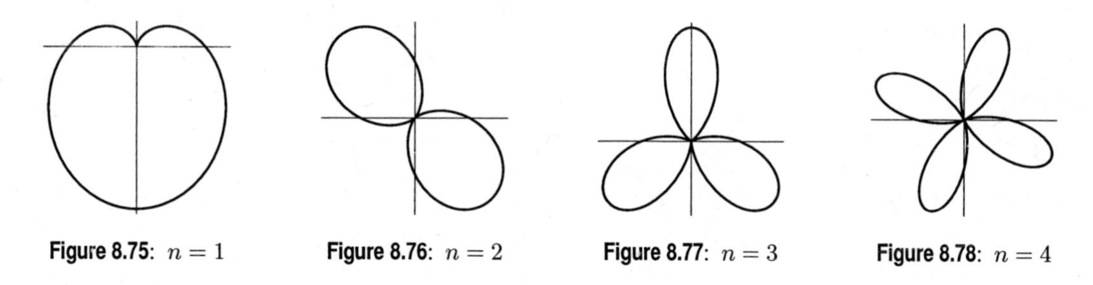

Figure 8.75: $n = 1$ Figure 8.76: $n = 2$ Figure 8.77: $n = 3$ Figure 8.78: $n = 4$

43. The graph will begin to draw over itself for any $\theta \geq 2\pi$ so the graph will look the same in all three cases. See Figure 8.79.

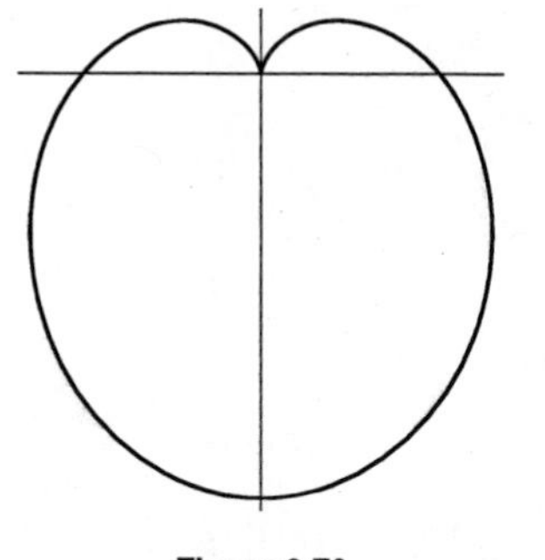

Figure 8.79

44. The curve will be a smaller loop inside a larger loop with an intersection point at the origin. Larger n values increase the size of the loops. See Figures 8.80-8.82.

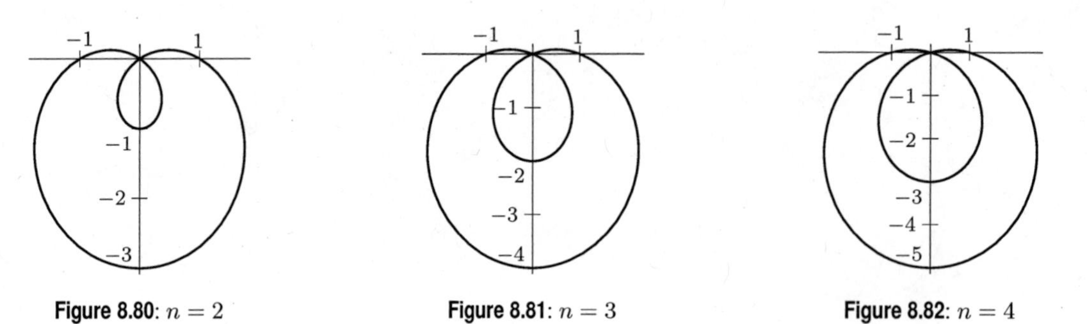

Figure 8.80: $n = 2$ **Figure 8.81**: $n = 3$ **Figure 8.82**: $n = 4$

45. See Figures 8.83 and 8.84. The first curve will be similar to the second curve, except the cardioid (heart) will be rotated clockwise by 90° ($\pi/2$ radians). This makes sense because of the identity $\sin\theta = \cos(\theta - \pi/2)$.

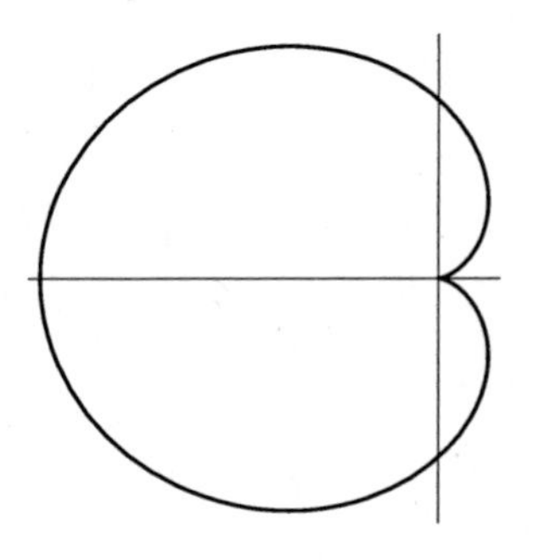

Figure 8.83: $r = 1 - \cos\theta$

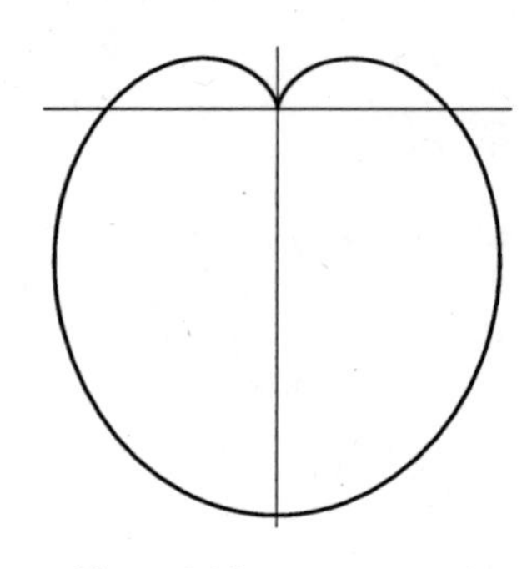

Figure 8.84: $r = 1 - \sin\theta$

46. Let $0 \le \theta \le 2\pi$ and $3/16 \le r \le 1/2$.

47. A loop starts and ends at the origin, that is, when $r = 0$. This happens first when $\theta = \pi/4$ and next when $\theta = 5\pi/4$. This can also be seen by using a trace mode on a calculator. Thus restricting θ so that $\pi/4 \le \theta \le 5\pi/4$ will graph the upper loop only. See Figure 8.85. To show only the other loop use $0 \le \theta \le \pi/4$ and $5\pi/4 \le \theta \le 2\pi$. See Figure 8.86.

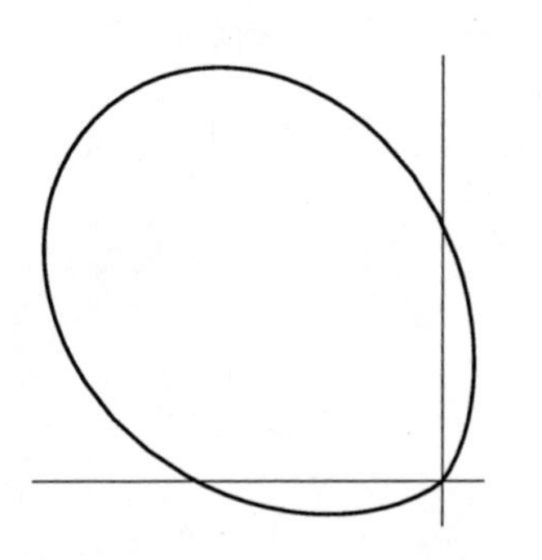

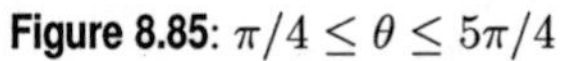

Figure 8.85: $\pi/4 \le \theta \le 5\pi/4$

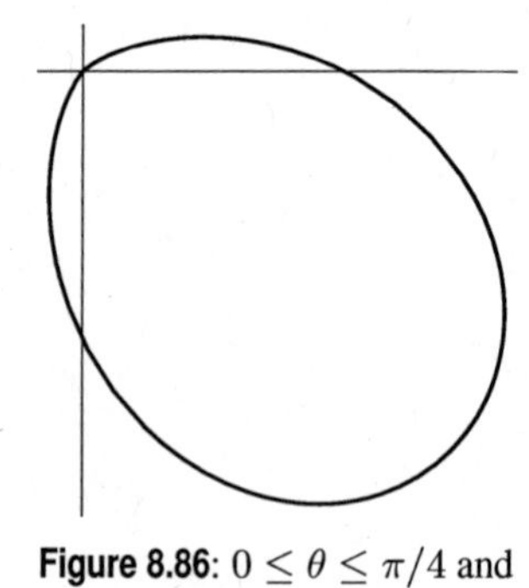

Figure 8.86: $0 \le \theta \le \pi/4$ and $5\pi/4 \le \theta \le 2\pi$

48. **(a)** Let $0 \le \theta \le \pi/4$ and $0 \le r \le 1$.
(b) Break the region into two pieces: one with $0 \le x \le \sqrt{2}/2$ and $0 \le y \le x$, the other with $\sqrt{2}/2 \le x \le 1$ and $0 \le y \le \sqrt{1 - x^2}$.

Solutions for Section 8.6

Exercises

1. $5e^{i\pi}$

2. $e^{\frac{i3\pi}{2}}$

3. $0e^{i\theta}$, for any θ.

4. $2e^{\frac{i\pi}{2}}$

5. We have $(-3)^2 + (-4)^2 = 25$, and $\arctan(4/3) \approx 4.069$. So the number is $5e^{i4.069}$

6. $\sqrt{10}e^{i\theta}$, where $\theta = \arctan(-3) \approx -1.249 + \pi = 1.893$ is an angle in the second quadrant.

7. $-5 + 12i$

8. $-11 + 29i$

9. $-3 - 4i$

10. $\frac{1}{4} - \frac{9i}{8}$

11. We have $(e^{i\pi/3})^2 = e^{i2\pi/3}$, thus $\cos\frac{2\pi}{3} + i\sin\frac{2\pi}{3} = -\frac{1}{2} + i\frac{\sqrt{3}}{2}$

12. $3 - 6i$

13. We have $\sqrt{e^{i\pi/3}} = e^{(i\pi/3)/2} = e^{i\pi/6}$, thus $\cos\frac{\pi}{6} + i\sin\frac{\pi}{6} = \frac{\sqrt{3}}{2} + \frac{i}{2}$.

14. $\sqrt[4]{10}\cos\frac{\pi}{8} + i\sqrt[4]{10}\sin\frac{\pi}{8}$ is one solution.

Problems

15. One value of $\sqrt{4i}$ is $\sqrt{4e^{i\pi/2}} = (4e^{i\pi/2})^{1/2} = 2e^{i\pi/4} = 2\cos\frac{\pi}{4} + i2\sin\frac{\pi}{4} = \sqrt{2} + i\sqrt{2}$

16. One value of $\sqrt{-i}$ is $\sqrt{e^{i\frac{3\pi}{2}}} = (e^{i\frac{3\pi}{2}})^{\frac{1}{2}} = e^{i\frac{3\pi}{4}} = \cos\frac{3\pi}{4} + i\sin\frac{3\pi}{4} = -\frac{\sqrt{2}}{2} + i\frac{\sqrt{2}}{2}$

17. One value of $\sqrt[3]{i}$ is $\sqrt[3]{e^{i\frac{\pi}{2}}} = (e^{i\frac{\pi}{2}})^{\frac{1}{3}} = e^{i\frac{\pi}{6}} = \cos\frac{\pi}{6} + i\sin\frac{\pi}{6} = \frac{\sqrt{3}}{2} + \frac{i}{2}$

18. One value of $\sqrt{7i}$ is $\sqrt{7e^{i\frac{\pi}{2}}} = (7e^{i\frac{\pi}{2}})^{\frac{1}{2}} = \sqrt{7}e^{i\frac{\pi}{4}} = \sqrt{7}\cos\frac{\pi}{4} + i\sqrt{7}\sin\frac{\pi}{4} = \frac{\sqrt{14}}{2} + i\frac{\sqrt{14}}{2}$

19. One value of $\sqrt[4]{-1}$ is $\sqrt[4]{e^{i\pi}} = (e^{i\pi})^{1/4} = e^{i\pi/4} = \cos\left(\frac{\pi}{4}\right) + i\sin(\frac{\pi}{4}) = \frac{\sqrt{2}}{2} + i\frac{\sqrt{2}}{2}$.

20. One value of $(1+i)^{2/3}$ is $(\sqrt{2}e^{i\frac{\pi}{4}})^{2/3} = (2^{\frac{1}{2}}e^{i\frac{\pi}{4}})^{\frac{2}{3}} = \sqrt[3]{2}e^{i\frac{\pi}{6}} = \sqrt[3]{2}\cos\frac{\pi}{6} + i\sqrt[3]{2}\sin\frac{\pi}{6} = \sqrt[3]{2}\cdot\frac{\sqrt{3}}{2} + i\sqrt[3]{2}\cdot\frac{1}{2}$

21. One value of $(\sqrt{3}+i)^{1/2}$ is
$(2e^{i\frac{\pi}{6}})^{1/2} = \sqrt{2}e^{i\frac{\pi}{12}} = \sqrt{2}\cos\frac{\pi}{12} + i\sqrt{2}\sin\frac{\pi}{12} \approx 1.366 + 0.366i$

22. One value of $(\sqrt{3}+i)^{-1/2}$ is
$(2e^{i\frac{\pi}{6}})^{-1/2} = \frac{1}{\sqrt{2}}e^{i(-\frac{\pi}{12})} = \frac{1}{\sqrt{2}}\cos(-\frac{\pi}{12}) + i\frac{1}{\sqrt{2}}\sin(-\frac{\pi}{12}) \approx 0.683 - 0.183i$

23. Since $\sqrt{5} + 2i = 3e^{i\theta}$, where $\theta = \arctan\frac{2}{\sqrt{5}} \approx 0.730$, one value of $(\sqrt{5}+2i)^{\sqrt{2}}$ is $(3e^{i\theta})^{\sqrt{2}} = 3^{\sqrt{2}}e^{i\sqrt{2}\theta} = 3^{\sqrt{2}}\cos\sqrt{2}\theta + i3^{\sqrt{2}}\sin\sqrt{2}\theta \approx 3^{\sqrt{2}}(0.513) + i3^{\sqrt{2}}(0.859) \approx 2.426 + 4.062i$

24. Substituting $A_1 = 2 - A_2$ into the second equation gives

$$(1-i)(2-A_2) + (1+i)A_2 = 0$$

so

$$\begin{aligned} 2iA_2 &= -2(1-i) \\ A_2 &= \frac{-(1-i)}{i} = \frac{-i(1-i)}{i^2} = i(1-i) = 1+i \end{aligned}$$

Therefore $A_1 = 2 - (1+i) = 1 - i$.

25. Substituting $A_2 = i - A_1$ into the second equation gives

$$iA_1 - (i - A_1) = 3,$$

so

$$\begin{aligned} iA_1 + A_1 &= 3 + i \\ A_1 &= \frac{3+i}{1+i} = \frac{3+i}{1+i}\cdot\frac{1-i}{1-i} = \frac{3-3i+i-i^2}{2} \\ &= 2 - i \end{aligned}$$

Therefore $A_2 = i - (2 - i) = -2 + 2i$.

26. If the roots are complex numbers, we must have $(2b)^2 - 4c < 0$ so $b^2 - c < 0$. Then the roots are

$$\begin{aligned} x = \frac{-2b \pm \sqrt{(2b)^2 - 4c}}{2} &= -b \pm \sqrt{b^2 - c} \\ &= -b \pm \sqrt{-1(c - b^2)} \\ &= -b \pm i\sqrt{c - b^2}. \end{aligned}$$

Thus, $p = -b$ and $q = \sqrt{c - b^2}$.

27. **(a)** The polar coordinates of i are $r = 1$ and $\theta = \pi/2$, as in Figure 8.87. Thus, the polar form of i is $i = re^{i\theta} = 1e^{i\pi/2} = e^{i\pi/2}$.

(b) Since $z = re^{i\theta}$ and $i = e^{i\pi/2}$, we have $iz = e^{i\pi/2} \cdot re^{i\theta} = re^{i(\theta+\pi/2)}$. The polar coordinates of z are (r, θ), while the polar coordinates of iz are $(r, \theta + \pi/2)$. The points z and iz are related as in Figure 8.88

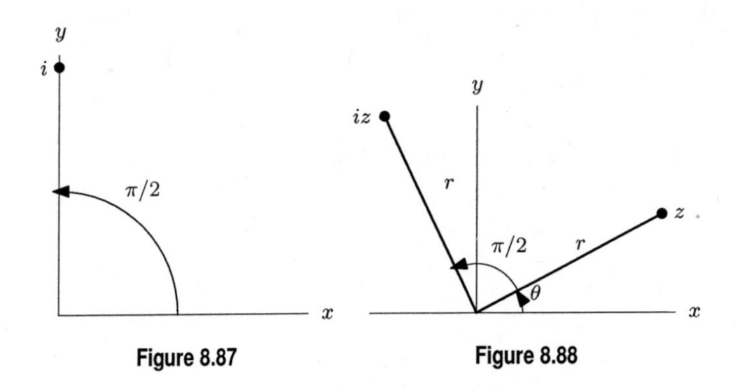

Figure 8.87 Figure 8.88

28. Using Euler's formula, we have:

$$e^{i(2\theta)} = \cos 2\theta + i \sin 2\theta$$

On the other hand,

$$e^{i(2\theta)} = \left(e^{i\theta}\right)^2 = (\cos\theta + i\sin\theta)^2 = (\cos^2\theta - \sin^2\theta) + i(2\cos\theta\sin\theta)$$

Equating imaginary parts, we find

$$\sin 2\theta = 2\sin\theta\cos\theta.$$

29. Using Euler's formula, we have:

$$e^{i(2\theta)} = \cos 2\theta + i \sin 2\theta$$

On the other hand,

$$e^{i(2\theta)} = \left(e^{i\theta}\right)^2 = (\cos\theta + i\sin\theta)^2 = (\cos^2\theta - \sin^2\theta) + i(2\cos\theta\sin\theta)$$

Equating real parts, we find

$$\cos 2\theta = \cos^2\theta - \sin^2\theta.$$

30. Since $e^{-i\theta} = \cos(-\theta) + i\sin(-\theta)$, we want to investigate $e^{-i\theta}$. By the definition of negative exponents,

$$e^{-i\theta} = \frac{1}{e^{i\theta}} = \frac{1}{\cos\theta + i\sin\theta}.$$

Multiplying by the conjugate $\cos\theta - i\sin\theta$ gives

$$e^{-i\theta} = \frac{1}{\cos\theta + i\sin\theta}\cdot\frac{\cos\theta - i\sin\theta}{\cos\theta - i\sin\theta} = \frac{\cos\theta - i\sin\theta}{\cos^2\theta + \sin^2\theta} = \cos\theta - i\sin\theta.$$

Thus

$$\cos(-\theta) + i\sin(-\theta) = \cos\theta - i\sin\theta,$$

so equating real parts, we see that cosine is even: $\cos(-\theta) = \cos\theta$.

31. Since $e^{-i\theta} = \cos(-\theta) + i\sin(-\theta)$, we want to investigate $e^{-i\theta}$. By the definition of negative exponents,

$$e^{-i\theta} = \frac{1}{e^{i\theta}} = \frac{1}{\cos\theta + i\sin\theta}.$$

Multiplying by the conjugate $\cos\theta - i\sin\theta$ gives

$$e^{-i\theta} = \frac{1}{\cos\theta + i\sin\theta}\cdot\frac{\cos\theta - i\sin\theta}{\cos\theta - i\sin\theta} = \frac{\cos\theta - i\sin\theta}{\cos^2\theta + \sin^2\theta} = \cos\theta - i\sin\theta.$$

Thus

$$\cos(-\theta) + i\sin(-\theta) = \cos\theta - i\sin\theta,$$

so equating imaginary parts, we see that sine is odd: $\sin(-\theta) = -\sin\theta$.

32. The number $z = e^{2i}$ is given in polar form with $(r, \theta) = (1, 2)$. Two more sets of polar coordinates for z are $(1, 2 + 2\pi)$, and $(1, 2 + 4\pi)$. Three cube roots of z are given by

$$\begin{aligned}
\left(1e^{2i}\right)^{1/3} = 1^{1/3}e^{1/3\cdot 2i} &= e^{2i/3} = \cos(2/3) + i\sin(2/3)\\
&= 0.786 + 0.618i.\\
\left(1e^{(2+2\pi)i}\right)^{1/3} = 1^{1/3}e^{1/3\cdot(2+2\pi)i} &= e^{(2+2\pi)i/3} = \cos((2+2\pi)/3) + i\sin((2+2\pi)/3)\\
&= -0.928 + 0.371i.\\
\left(1e^{(2+4i)i}\right)^{1/3} = 1^{1/3}e^{1/3\cdot(2+4\pi)i} &= e^{(2+4\pi)i/3} = \cos((2+4\pi)/3) + i\sin((2+4\pi)/3)i\\
&= 0.143 - 0.909i.
\end{aligned}$$

33. One polar form for $z=-8$ is $z=8e^{i\pi}$ with $(r,\theta)=(8,\pi)$. Two more sets of polar coordinates for z are $(8,3\pi)$, and $(8,5\pi)$. Three cube roots of z are given by

$$\begin{aligned}\left(8e^{\pi i}\right)^{1/3} &= 8^{1/3}e^{1/3\cdot\pi i} = 2e^{\pi i/3} = 2\cos(\pi/3)+i2\sin(\pi/3)\\ &= 1+1.732i.\\ \left(8e^{3\pi i}\right)^{1/3} &= 8^{1/3}e^{1/3\cdot 3\pi i} = 2e^{\pi i} = 2\cos\pi+i2\sin\pi\\ &= -2.\\ \left(8e^{5\pi i}\right)^{1/3} &= 8^{1/3}e^{1/3\cdot 5\pi i} = 2e^{5\pi i/3} = 2\cos(5\pi/3)+i2\sin(5\pi/3)\\ &= 1-1.732i.\end{aligned}$$

34. Three polar forms for $z=8$ have (r,θ) equal to $(8,0)$, $(8,2\pi)$, and $(8,4\pi)$. Three cube roots of z are given by

$$\begin{aligned}\left(8e^{0i}\right)^{1/3} &= 8^{1/3}e^{1/3\cdot 0i} = 2e^{0i} = 2\cos 0+i2\sin 0\\ &= 2.\\ \left(8e^{2\pi i}\right)^{1/3} &= 8^{1/3}e^{1/3\cdot 2\pi i} = 2e^{2\pi i/3} = 2\cos(2\pi/3)+i2\sin(2\pi/3)\\ &= -1+1.732i.\\ \left(8e^{4\pi i}\right)^{1/3} &= 8^{1/3}e^{1/3\cdot 4\pi i} = 2e^{4\pi i/3} = 2\cos(4\pi/3)+i2\sin(4\pi/3)\\ &= -1-1.732i.\end{aligned}$$

35. One polar form of the complex number $z=1+i$ has $(r,\theta)=(\sqrt{2},\pi/4)$. Two more sets of polar coordinates for z are $(\sqrt{2},\pi/4+2\pi)=(\sqrt{2},9\pi/4)$ and $(\sqrt{2},\pi/4+4\pi)=(\sqrt{2},17\pi/4)$. Three cube roots of z are given by

$$\begin{aligned}\left(\sqrt{2}e^{\pi i/4}\right)^{1/3} &= \left(2^{1/2}\right)^{1/3}e^{1/3\cdot\pi i/4} = 2^{1/6}e^{\pi i/12} = 2^{1/6}\cos(\pi/12)+i2^{1/6}\sin(\pi/12)\\ &= 1.084+0.291i.\\ \left(\sqrt{2}e^{9\pi i/4}\right)^{1/3} &= \left(2^{1/2}\right)^{1/3}e^{1/3\cdot 9\pi i/4} = 2^{1/6}e^{9\pi i/12} = 2^{1/6}\cos(9\pi/12)+i2^{1/6}\sin(9\pi/12)\\ &= -0.794+0.794i.\\ \left(\sqrt{2}e^{17\pi/i4}\right)^{1/3} &= \left(2^{1/2}\right)^{1/3}e^{1/3\cdot 17\pi i/4} = 2^{1/6}e^{17\pi i/12} = 2^{1/6}\cos(17\pi/12)+i2^{1/6}\sin(17\pi/12)\\ &= -0.291-1.084i.\end{aligned}$$

36. By de Moivre's formula we have

$$(\cos\pi/4+i\sin\pi/4)^4 = \cos(4\cdot\pi/4)+i\sin(4\cdot\pi/4) = -1+i0 = -1.$$

37. By de Moivre's formula we have

$$(\cos 2\pi/3+i\sin 2\pi/3)^3 = \cos(3\cdot 2\pi/3)+i\sin(3\cdot 2\pi/3) = 1+i0 = 1.$$

38. By de Moivre's formula we have

$$(\cos 2+i\sin 2)^{-1} = \cos(-2)+i\sin(-2) = \cos 2-i\sin 2.$$

39. By de Moivre's formula we have

$$(\cos\pi/4+i\sin\pi/4)^{-7} = \cos(-7\pi/4)+i\sin(-7\pi/4) = \frac{\sqrt{2}}{2}+i\frac{\sqrt{2}}{2}.$$

Solutions for Chapter 8 Review

Exercises

1. In order to change from degrees to radians, we multiply the number of degrees by $\pi/180$, so we have $330 \cdot \pi/180$, giving $\frac{11}{6}\pi$ radians.

2. In order to change from degrees to radians, we multiply the number of degrees by $\pi/180$, so we have $315 \cdot \pi/180$, giving $\frac{7}{4}\pi$ radians.

3. In order to change from degrees to radians, we multiply the number of degrees by $\pi/180$, so we have $-225 \cdot \pi/180$, giving $-\frac{5}{4}\pi$ radians.

4. In order to change from degrees to radians, we multiply the number of degrees by $\pi/180$, so we have $6\pi \cdot \pi/180$, giving $\pi^2/30 \approx 0.329$ radians.

5. In order to change from radians to degrees, we multiply the number of radians by $180/\pi$, so we have $\frac{3}{2}\pi \cdot 180/\pi$, giving 270 degrees.

6. In order to change from radians to degrees, we multiply the number of radians by $180/\pi$, so we have $180 \cdot 180/\pi$, giving $32{,}400/\pi \approx 10{,}313.240$ degrees.

7. In order to change from radians to degrees, we multiply the number of radians by $180/\pi$, so we have $(5\pi/\pi)(180/\pi)$, giving $900/\pi = 286.479$ degrees.

8. If we go around four times, we make four full circles, which is $2\pi \cdot 4 = 8\pi$ radians.

9. If we go around six times, we make six full circles, which is $2\pi \cdot 6 = 12\pi$ radians. Since we're going in the negative direction, we have -12π radians.

10. If we go around 16.4 times, we make 16.4 full circles, which is $2\pi \cdot 16.4 = 32.8\pi$ radians.

11. **(a)** Since $1.57 < 2 < 3.14$, you will be in the quadrant II.
(b) Since $3.14 < 4 < 4.71$, you will be in the quadrant III.
(c) Since $4.71 < 6 < 6.28$, you will be in the quadrant IV.
(d) Since $0 < 1.5 < 1.57$, you will be in the quadrant I.
(e) Since $3.14 < 3.2 < 4.71$, you will be in the quadrant III.

12. Writing $\cos^2 A = 1 - \sin^2 A$ and factoring, we have

$$\frac{\cos^2 A}{1+\sin A} = \frac{1-\sin^2 A}{1+\sin A} = \frac{(1-\sin A)(1+\sin A)}{1+\sin A} = 1 - \sin A.$$

13. Writing $\tan 2A = \sin 2A/\cos 2A$, we have

$$\frac{\sin 2A}{\tan 2A} + 2\cos 2A = \frac{\sin 2A}{\sin 2A/\cos 2A} + 2\cos 2A = 3\cos 2A.$$

14. The arc length, s, corresponding to an angle of θ radians in a circle of radius r is $s = r\theta$. In order to change from degrees to radians, we multiply the number of degrees by $\pi/180$, so we have $17 \cdot \pi/180$, giving $\frac{17}{180}\pi$ radians. Thus, our arc length is $6.2 \cdot 17\pi/180 \approx 1.840$.

15. The arc length, s, corresponding to an angle of θ radians in a circle of radius r is $s = r\theta$. In order to change from degrees to radians, we multiply the number of degrees by $\pi/180$, so we have $-585 \cdot \pi/180$, giving $-\frac{13}{4}\pi$ radians. Since length cannot be negative, we take the absolute value, giving us $13\pi/4$ radians. Thus, our arc length is $6.2 \cdot 13\pi/4 \approx 63.303$.

16. The arc length, s, corresponding to an angle of θ radians in a circle of radius r is $s = r\theta$. In order to change from degrees to radians, we multiply the number of degrees by $\pi/180$, so we have $-360/\pi \cdot \pi/180$, giving -2 radians. The negative sign indicates rotation in a clockwise, rather than counterclockwise, direction. Since length cannot be negative, we find the arc length corresponding to 2 radians. Thus, our arc length is $6.2 \cdot 2 = 12.4$.

17. **(a)** We are looking for the graph of a function with amplitude one but a period of π; only $C(t)$ qualifies.
(b) We are looking for the graph of a function with amplitude one and period 2π but which is shifted up by two units; only $D(t)$ qualifies.
(c) We are looking for the graph of a function with amplitude 2 and period 2π; only $A(t)$ qualifies.
(d) Only $B(t)$ is left and we are looking for the graph of a function with amplitude one and period 2π but which has been shifted to the left by two units. This checks with $B(t)$.

18. The midline is $y = 3$. The amplitude is 1. The period is 2π.

19. The midline is $y = 7$. The amplitude is 1. The period is 2π.

20. We first divide both sides of the equation by 6, giving

$$y = 2\sin(\pi t - 7) + 7.$$

The midline is $y = 7$. The amplitude is 2. The period is $2\pi/\pi = 2$.

21. Since $-299°$ is almost $300°$, clockwise, such a point is in Quadrant I.

22. Since $730° = 2 \cdot 360° + 10°$, such a point is in Quadrant I.

23. Since $-7.7\pi = -3 \cdot 2\pi - 1.7\pi$, such a point is in Quadrant I.

24. Since $14.4\pi = 7 \cdot 2\pi + 0.4\pi$, such a point is in Quadrant I.

25. With $r = \pi/2$ and $\theta = 0$, we find $x = r\cos\theta = (\pi/2)\cos 0 = 1.571$ and $y = r\sin\theta = (\pi/2)\sin 0 = 0$.
The rectangular coordinates are $(1.571, 0)$.

26. With $r = 2$ and $\theta = 2$, we find $x = r\cos\theta = 2\cos 2 = -0.832$ and $y = r\sin\theta = 2\sin 2 = 1.819$.
The rectangular coordinates are $(-0.832, 1.819)$.

27. With $r = 0$, the point specified is the origin, no matter what the angle measure. So $x = r\cos\theta = 0$ and $y = r\sin\theta = 0$.
The rectangular coordinates are $(0, 0)$.

28. With $r = 3$ and $\theta = 40°$, use a calculator in degree mode to find $x = r\cos\theta = 3\cos(40°) = 2.298$ and $y = r\sin\theta = 3\sin(40°) = 1.928$.
The rectangular coordinates are $(2.298, 1.928)$.

Problems

29. The amplitude is 4, the period is $\frac{2\pi}{1} = 2\pi$, the phase shift and horizontal shift are 0. See Figure 8.89.

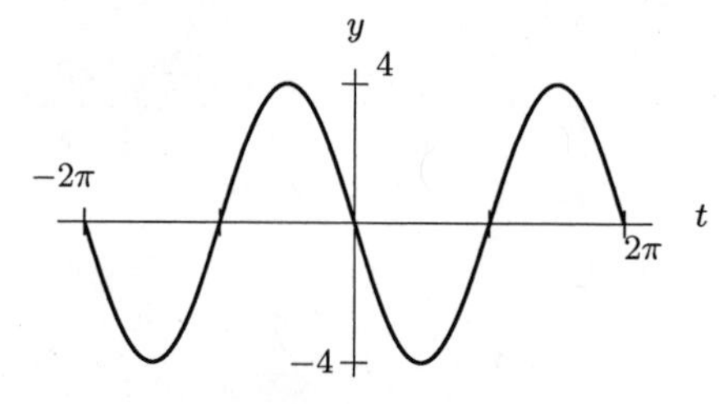

Figure 8.89: $y = -4\sin t$

30. The amplitude is 20, the period is $\frac{2\pi}{4\pi} = \frac{1}{2}$, the phase shift and horizontal shift are 0. See Figure 8.90.

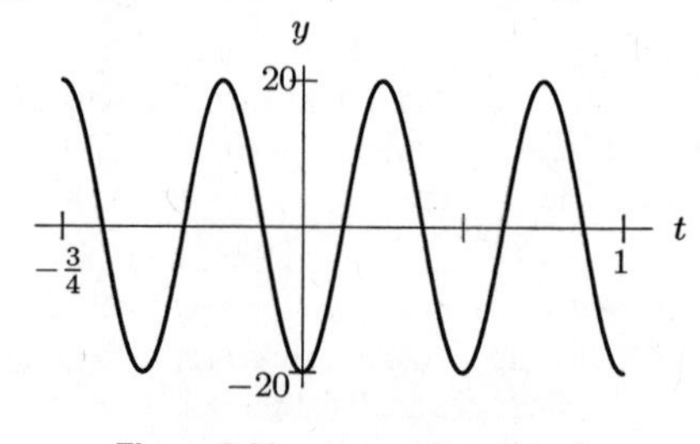

Figure 8.90: $y = -20\cos(4\pi t)$

31. The amplitude is 1, the period is $2\pi/2 = \pi$, the phase shift is $-\pi/2$, and

$$\text{Horizontal shift} = -\frac{\pi/2}{2} = -\frac{\pi}{4}.$$

Since the horizontal shift is negative, the graph of $y = \cos(2t)$ is shifted $\pi/4$ units to the left to give the graph in Figure 8.91.

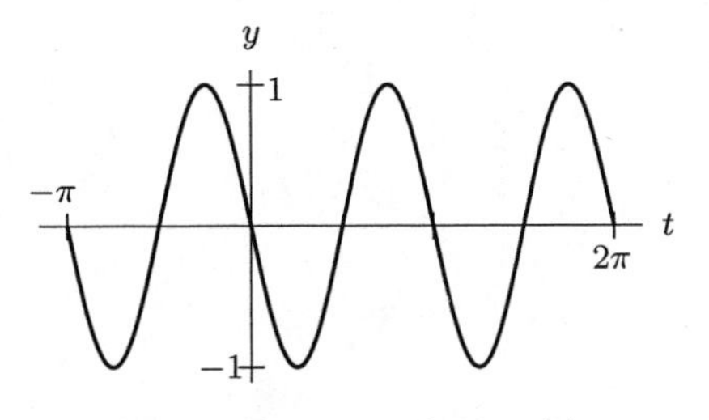

Figure 8.91: $y = \cos(2t + \pi/2)$

32. The amplitude is 3, the period is $\frac{2\pi}{4\pi} = \frac{1}{2}$, the phase shift is -6π, and

$$\text{Horizontal shift} = -\frac{6\pi}{4\pi} = -\frac{3}{2}.$$

Since the horizontal shift is negative, the graph of $y = 3\sin(4\pi t)$ is shifted $\frac{3}{2}$ units to the left to give the graph in Figure 8.92. Note that a shift of $\frac{3}{2}$ units produces the same graph as the unshifted graph. We expect this since $y = 3\sin(4\pi t + 6\pi) = 3\sin(4\pi t)$. See Figure 8.92.

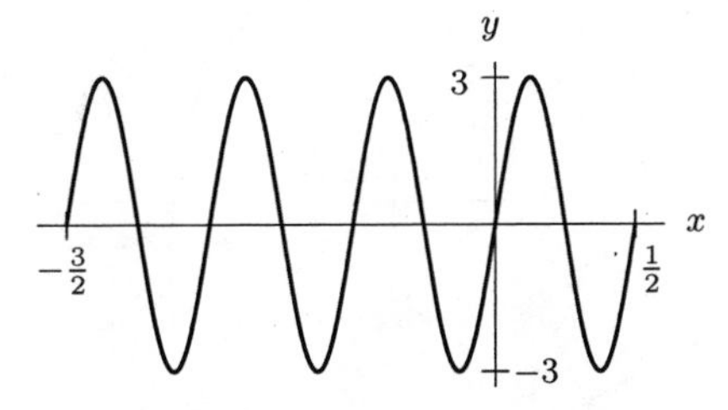

Figure 8.92: $y = 3\sin(4\pi t + 6\pi)$

33. The amplitude is 30, the midline is $y = 60$, and the period is 4.

34. The amplitude is 30; the midline is $y = 60$, and the period is 20.

35. Amplitude is 40; midline is $y = 50$; period is 16.

36. Amplitude is 50; midline is $y = 50$; period is 64.

37. We would use $y = \sin x$ for $f(x)$ and $k(x)$, as both these graphs cross the midline at $x = 0$. We would use $y = \cos x$ for $g(x)$ and $h(x)$, as both these graphs are as far as possible from the midline at $x = 0$.

38. See Figure 8.93.

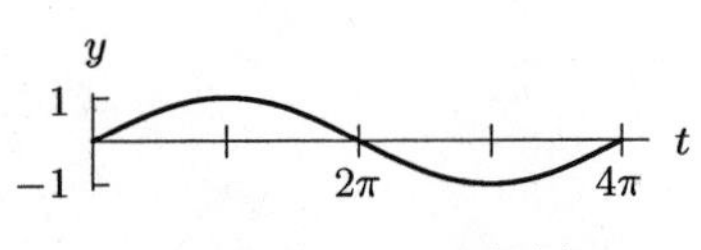

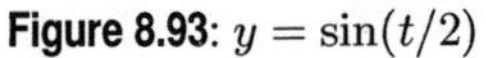

Figure 8.93: $y = \sin(t/2)$

39. See Figure 8.94.

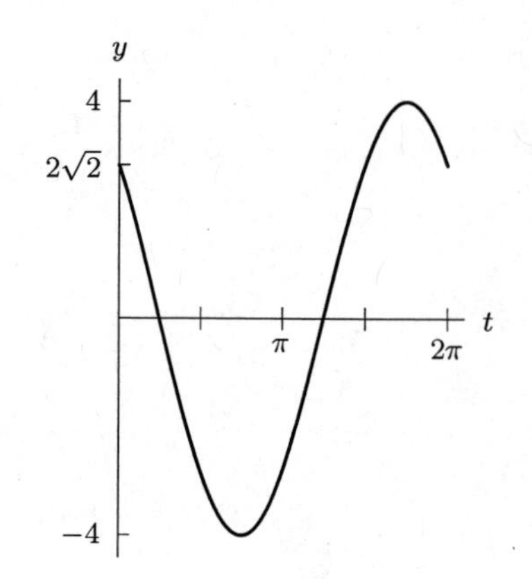

Figure 8.94: $y = 4\cos(t + \pi/4)$

40. See Figure 8.95.

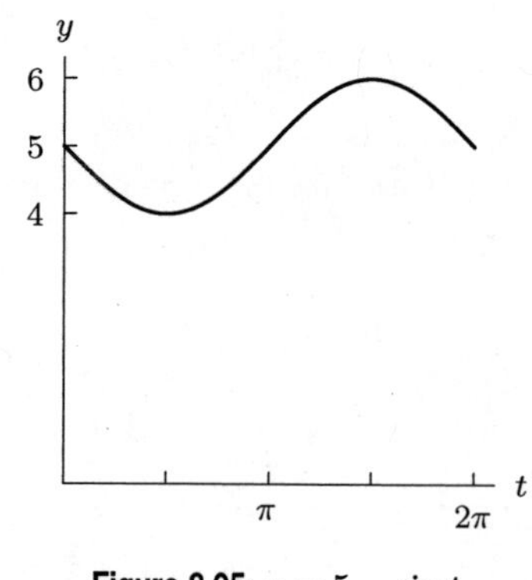

Figure 8.95: $y = 5 - \sin t$

41. See Figure 8.96.

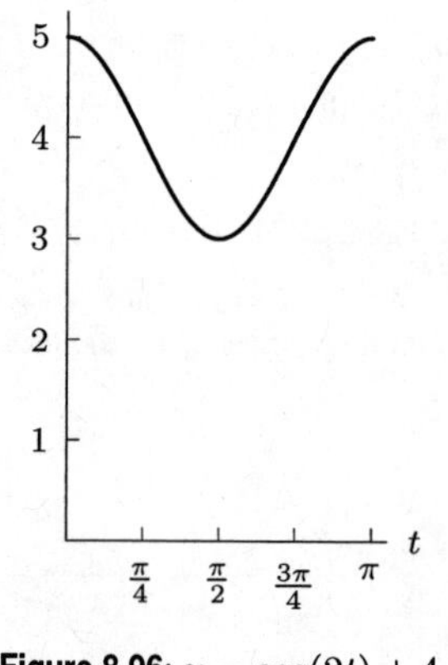

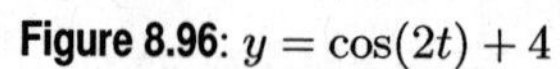

Figure 8.96: $y = \cos(2t) + 4$

42. **(a)** Since $\sin^2\theta + \cos^2\theta = 1$ and $\sin\theta = \frac{1}{7}$, we have

$$\left(\frac{1}{7}\right)^2 + \cos^2\theta = 1$$
$$\cos^2\theta = 1 - \frac{1}{49} = \frac{48}{49}$$
$$\cos\theta = \pm\frac{\sqrt{48}}{7}.$$

Now, θ is in the second quadrant ($\frac{\pi}{2} < \theta < \pi$), so $\cos\theta < 0$. Thus $\cos\theta = -\frac{\sqrt{48}}{7}$. Note this can also be written as $-\frac{4\sqrt{3}}{7}$.

(b) Use arccosine:

$$\theta = \arccos\left(-\frac{\sqrt{48}}{7}\right) \approx 2.998 \text{ radians.}$$

43. See Figure 8.97. They appear to be the same graph. This suggests the truth of the identity $\cos t = \sin(t + \frac{\pi}{2})$.

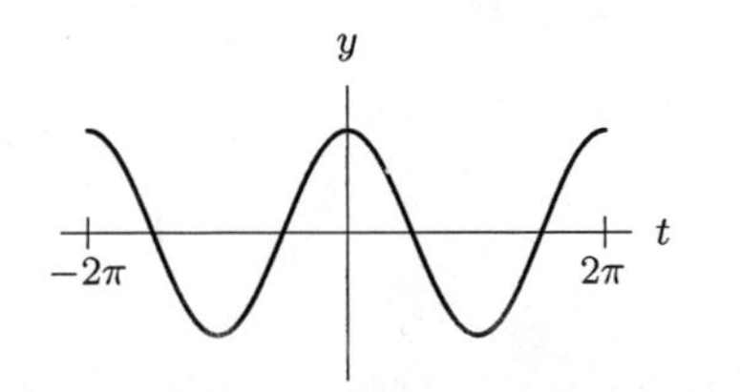

Figure 8.97: Graphs showing $\cos(t) = \sin\left(t + \frac{\pi}{2}\right)$

44. **(a)** Clearly $(x-1)^2+y^2 = 1$ is a circle with center $(1,0)$. To convert this to polar, use $x = r\cos\theta$ and $y = r\sin\theta$. Then $(r\cos\theta-1)^2+(r\sin\theta)^2 = 1$ or $r^2\cos^2\theta-2r\cos\theta+1+r^2\sin^2\theta = 1$. This means $r^2(\cos^2\theta+\sin^2\theta) = 2r\cos\theta$, or $r = 2\cos\theta$.

(b) 12 o'clock $\rightarrow (x,y) = (1,1)$ and $(r,\theta) = (\sqrt{2}, \pi/4)$,
3 o'clock $\rightarrow (x,y) = (2,0)$ and $(r,\theta) = (2,0)$,
6 o'clock $\rightarrow (x,y) = (1,-1)$ and $(r,\theta) = (\sqrt{2}, -\pi/4)$,
9 o'clock $\rightarrow (x,y) = (0,0)$ and $(r,\theta) = (0$, any angle $)$.

45. **(a)** Since B is the point $(1,0)$, the circle has radius 1. Thus, $\sin\theta =$ OE.
(b) From the definition of $\cos\theta$, we have $\cos\theta =$ OA.
(c) In ΔODB, we have $OB = 1$. Since $\tan\theta =$ Opp/Adj = DB/1, we have $\tan\theta =$ DB.
(d) Let P be the point of intersection of $\overline{FC}$ and $\overline{OD}$. Use the fact that ΔOFP and ΔOEP are similar, and form a proportion of the hypotenuse to the one unit side of the larger triangle.

$$\frac{\text{OF}}{1} = \frac{1}{\text{OE}} \quad \text{or} \quad \text{OF} = \frac{1}{\sin\theta}$$

(e) Use the fact that ΔOCP and ΔOAP are similar, and write

$$\frac{\text{OC}}{1} = \frac{1}{\text{OA}} \quad \text{or} \quad \text{OC} = \frac{1}{\cos\theta}$$

(f) Use the fact that ΔGOH and ΔDOB are similar, and write

$$\frac{\text{GH}}{1} = \frac{1}{\text{DB}} \quad \text{or} \quad \text{GH} = \frac{1}{\tan\theta}$$

46. $\cos 540^\circ = \cos(360^\circ + 180^\circ) = \cos 180^\circ = -1$.

47. Since we know that the y-coordinate on the unit circle at $7\pi/6$ is the y-coordinate at $\pi/6$ multiplied by -1, and since $\pi/6$ radians is the same as $30°$, we know that $\sin(7\pi/6) = -\sin(\pi/6) = -\sin 30° = -1/2$.

48. Since $\tan(-2\pi/3) = \sin(-2\pi/3)/\cos(-2\pi/3)$, we know that $\tan(-2\pi/3) = (-\sqrt{3}/2)/(-1/2) = \sqrt{3}$.

49. If we consider a triangle with opposite side of length 3 and hypotenuse 5, we can use the Pythagorean theorem to find the length of the adjacent side as

$$\sqrt{5^2 - 3^2} = 4.$$

This gives a triangle with sides 3 and 4, and hypotenuse 5. Since tangent is negative in the fourth quadrant, $\tan\theta = -\frac{3}{4}$. See Figure 8.98.

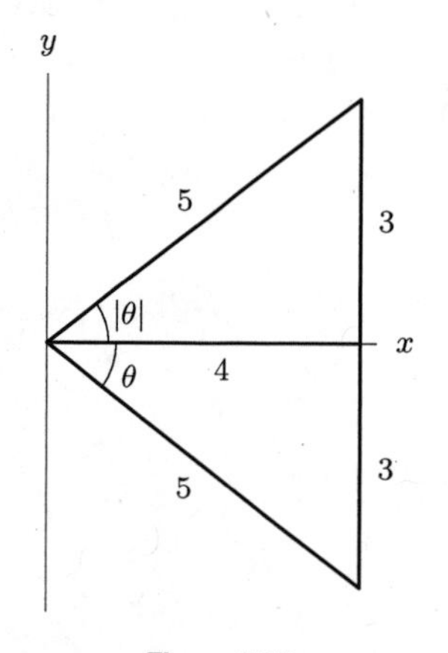

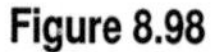
Figure 8.98

50. We use the inverse sine function on a calculator to get $\theta = 0.412$.

51. We use the inverse tangent function on a calculator to get $\theta - 1 = 0.168$, which gives us $\theta = 1.168$.

52. From the figure, we see that

$$\sin x = \frac{0.83}{1},$$

so

$$x = \sin^{-1}(0.83).$$

Using a calculator, we find that $x = \sin^{-1}(0.83) \approx 0.979$.

53. **(a)** The slope of the line is $m = \tan(5\pi/6) = -1/\sqrt{3}$, and the y-intercept is $b = 2$, so $y = (-1/\sqrt{3})x + 2$.
(b) Solve $0 = (-1/\sqrt{3})x + 2$ to get $x = 2\sqrt{3}$.

54. **(a)** $z_1 z_2 = (-3 - i\sqrt{3})(-1 + i\sqrt{3}) = 3 + (\sqrt{3})^2 + i(\sqrt{3} - 3\sqrt{3}) = 6 - i2\sqrt{3}$.

$$\frac{z_1}{z_2} = \frac{-3 - i\sqrt{3}}{-1 + i\sqrt{3}} \cdot \frac{-1 - i\sqrt{3}}{-1 - i\sqrt{3}} = \frac{3 - (\sqrt{3})^2 + i(\sqrt{3} + 3\sqrt{3})}{(-1)^2 + (\sqrt{3})^2} = \frac{i \cdot 4\sqrt{3}}{4} = i\sqrt{3}.$$

(b) We find (r_1, θ_1) corresponding to $z_1 = -3 - i\sqrt{3}$.

$r_1 = \sqrt{(-3)^2 + (\sqrt{3})^2} = \sqrt{12} = 2\sqrt{3}$.

$\tan\theta_1 = \frac{-\sqrt{3}}{-3} = \frac{\sqrt{3}}{3}$, so $\theta_1 = \frac{7\pi}{6}$.

Thus $-3 - i\sqrt{3} = r_1 e^{i\theta_1} = 2\sqrt{3}\, e^{i\frac{7\pi}{6}}$.

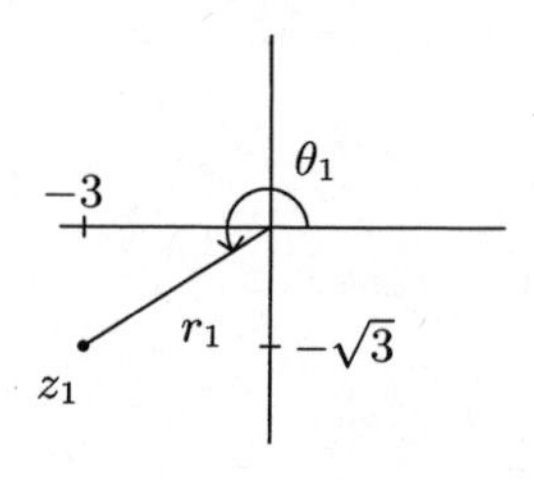

We find (r_2, θ_2) corresponding to $z_2 = -1 + i\sqrt{3}$.
$r_2 = \sqrt{(-1)^2 + (\sqrt{3})^2} = 2;$
$\tan\theta_2 = \dfrac{\sqrt{3}}{-1} = -\sqrt{3}$, so $\theta_2 = \dfrac{2\pi}{3}$.
Thus, $-1 + i\sqrt{3} = r_2 e^{i\theta_2} = 2e^{i\frac{2\pi}{3}}$.

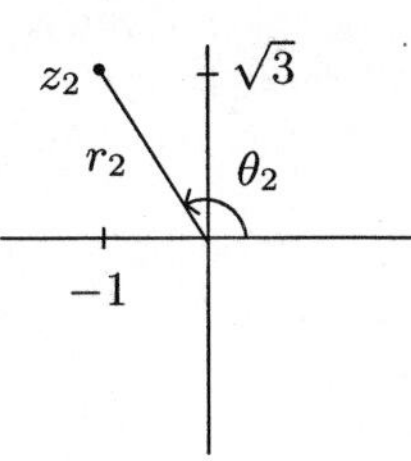

We now calculate $z_1 z_2$ and $\dfrac{z_1}{z_2}$.

$$z_1 z_2 = \left(2\sqrt{3}e^{i\frac{7\pi}{6}}\right)\left(2e^{i\frac{2\pi}{3}}\right) = 4\sqrt{3}e^{i\left(\frac{7\pi}{6}+\frac{2\pi}{3}\right)} = 4\sqrt{3}e^{i\frac{11\pi}{6}}$$
$$= 4\sqrt{3}\left[\cos\frac{11\pi}{6} + i\sin\frac{11\pi}{6}\right] = 4\sqrt{3}\left[\frac{\sqrt{3}}{2} - i\frac{1}{2}\right] = 6 - i2\sqrt{3}.$$

$$\frac{z_1}{z_2} = \frac{2\sqrt{3}e^{i\frac{7\pi}{6}}}{2e^{i\frac{2\pi}{3}}} = \sqrt{3}e^{i\left(\frac{7\pi}{6}-\frac{2\pi}{3}\right)} = \sqrt{3}e^{i\frac{\pi}{2}}$$
$$= \sqrt{3}\left(\cos\frac{\pi}{2} + i\sin\frac{\pi}{2}\right) = i\sqrt{3}.$$

These agree with the values found in (a).

55. **(a)** Since $z = 3 + 2i$, we have $iz = i(3 + 2i) = 3i + 2i^2 = -2 + 3i$. See Figure 8.99.

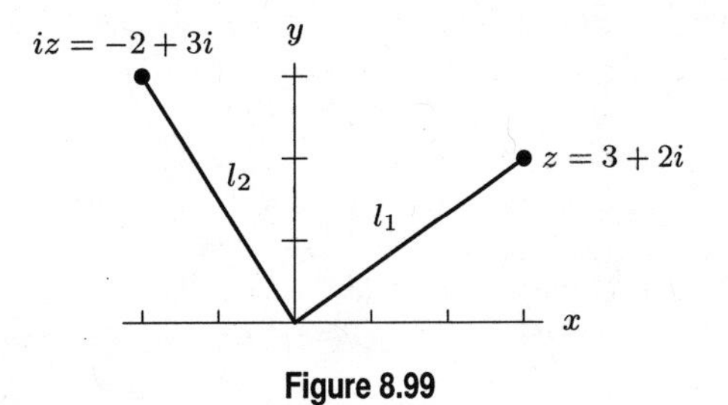

Figure 8.99

(b) The slope of the line l_1 in Figure 8.99 is $2/3$, and the slope of the line l_2 is $-3/2$. Since the product $(2/3)(-3/2) = -1$ of these two slopes is -1, the lines are perpendicular.

56. **(a)** The graph resembles a cosine function with midline $k = 8$, amplitude $A = 10$, and period $p = 60$, so

$$y = 10\cos\left(\frac{2\pi}{60}x\right) + 8.$$

(b) We find the zeros at x_1 and x_2 by setting $y = 0$. Solving gives

$$\begin{aligned} 10\cos\left(\frac{2\pi}{60}x\right) + 8 &= 0 \\ \cos\left(\frac{2\pi}{60}x\right) &= -\frac{8}{10} \\ \frac{2\pi}{60}x &= \cos^{-1}(-0.8) \\ x &= \frac{60}{2\pi}\cos^{-1}(-0.8) \\ &= 23.8550. \end{aligned}$$

Judging from the graph, this is the value of x_1. By symmetry, $x_2 = 60 - x_1 = 36.1445$.

57. **(a)** The period is $p = 80$, so $b = 2\pi/80$. The amplitude is $|a| = 300$, and the curve resembles an upside-down cosine, so $a = -300$. The midline is $k = 600$.

Putting this information together gives $y = 600 - 300\cos(2\pi x/80)$.

(b) These points fall on the line $y = 475$.

The period is $p = 80$, so $b = 2\pi/80$. The amplitude is $|a| = 300$, and the curve resembles an upside-down cosine, so $a = -300$. The midline is $k = 600$. Thus, a formula for the sinusoidal function is $y = 600 - 300\cos(2\pi x/80)$, so we obtain one possible solution as follows:

$$\begin{aligned}
600 - 300\cos\left(\frac{2\pi}{80}x\right) &= 475 \\
-300\cos\left(\frac{2\pi}{80}x\right) &= 475 - 600 = -125 \\
\cos\left(\frac{2\pi}{80}x\right) &= \frac{-125}{-300} = \frac{5}{12} \\
\frac{2\pi}{80}x &= \cos^{-1}\left(\frac{5}{12}\right) \\
x &= \frac{80}{2\pi}\cos^{-1}\left(\frac{5}{12}\right) \\
&= 14.5279.
\end{aligned}$$

We denote this solution by x_0, and it gives the x-coordinate of the left-most point. By symmetry, we see that

$$\begin{aligned}
x_1 &= 80 - x_0 = 65.4721 \\
x_2 &= 80 + x_0 = 94.5279.
\end{aligned}$$

58. First, by looking at the graph of f, we note that its amplitude is 5, and its midline is given by $y = 5$. Therefore, we have $A = \pm 5$ and $k = 5$. Also, since the graph of f completes one full cycle in 8 units, we see that the period of f is 8, so we have $B = 2\pi/8 = \pi/4$. Combining these observations, we see that we can take the four formulas to be horizontal translations of the functions $y_1 = 5\cos((\pi/4)x)+5, y_2 = -5\cos((\pi/4)x)+5, y_3 = 5\sin((\pi/4)x)+5$, and $y_4 = -5\sin((\pi/4)x)+5$, all of which have the same amplitude, period, and midline as f. After sketching y_1, we see that we can obtain the graph of f by shifting the graph of y_1 4 units to the right; therefore, $f_1(x) = 5\cos((\pi/4)(x-4))+5$. A sketch of the graph of y_2 reveals that the graphs of f and y_2 are identical, so $f_2(x) = -5\cos((\pi/4)x)+5$. A sketch of the graph of y_3 reveals that we can obtain the graph of f by shifting y_3 to the right 2 units, so $f_3(x) = 5\sin((\pi/4)(x-2))+5$. Similarly, a sketch of y_4 allows us to conclude that $f_4(x) = -5\sin((\pi/4)(x-6)) + 5$. Answers may vary.

59. First, by looking at the graph of f, we note that its amplitude is 6, and its midline is given by $y = 2$. Therefore, we have $A = \pm 6$ and $k = 2$. Also, since the graph of f completes one full cycle in 4π units, we see that the period of f is 4π, so we have $B = (2\pi)/(4\pi) = 1/2$. Combining these observations, we see that we can take the four formulas to be horizontal translations of the functions $y_1 = 6\cos((1/2)x)+2, y_2 = -6\cos((1/2)x)+2, y_3 = 6\sin((1/2)x)+2$, and $y_4 = -6\sin((1/2)x)+2$, all of which have the same amplitude, period, and midline as f. After sketching y_1, we see that we can obtain the graph of f by shifting the graph of y_1 to the right 3π units; therefore, $f_1(x) = 6\cos((1/2)(x-3\pi))+2$. Similarly, we can obtain the graph of f by shifting the graph of y_2 to the right π units, so $f_2(x) = -6\cos((1/2)(x - \pi)) + 2$, or by shifting the graph of y_3 to the right 2π units, so $f_3(x) = 6\sin((1/2)(x - 2\pi)) + 2$. Finally, a sketch of y_4 reveals that y_4 and f describe identical functions, so $f_4(x) = -6\sin((1/2)x) + 2$. Answers may vary.

60. First, by looking at the graph of f, we note that its amplitude is 3, and its midline is given by $y = -3$. Therefore, we have $A = \pm 3$ and $k = -3$. Also, since the graph of f completes one full cycle between $x = -3$ and $x = 5$, we see that the period of f is 8, so we have $B = 2\pi/8 = \pi/4$. Combining these observations, we see that we can take the four formulas to be horizontal translations of the functions $y_1 = 3\cos((\pi/4)x) - 3, y_2 = -3\cos((\pi/4)x) - 3, y_3 = 3\sin((\pi/4)x) - 3$, and $y_4 = -3\sin((\pi/4)x) - 3$, all of which have the same amplitude, period, and midline as f. After sketching y_1, we see that we can obtain the graph of f by shifting the graph of y_1 3 units to the right; therefore, $f_1(x) = 3\cos((\pi/4)(x-3)) - 3$. Similarly, we can obtain the graph of f by shifting the graph of y_2 to the left 1 unit, so $f_2(x) = -3\cos((\pi/4)(x+1)) - 3$, or by shifting the graph of y_3 to the right 1 unit, so $f_3(x) = 3\sin((\pi/4)(x-1)) - 3$. Finally, a sketch of y_4 reveals that we can obtain the graph of f by shifting the graph of y_4 to the right 5 units, so $f_4(x) = -3\sin((\pi/4)(x - 5)) - 3$. Answers may vary.

61. First, by looking at the graph of f, we see that the difference between the maximum and minimum output values is $8-(-2)=10$, so the amplitude is 5. Therefore, the midline of the graph is given by the line $y=-2+5=3$, so we have $A=\pm 5$ and $k=3$. We also note that the graph of f completes 1.5 cycles between $x=-2$ and $x=16$, so the period of the graph is 12, from which it follows that $B=2\pi/12=\pi/6$.

Combining these observations, we see that we can take the four formulas to be horizontal translations of the functions $y_1=5\cos((\pi/6)x)+3, y_2=-5\cos((\pi/6)x)+3, y_3=5\sin((\pi/6)x)+3$, and $y_4=-5\sin((\pi/6)x)+3$, all of which have the same amplitude, period, and midline as f. After sketching y_1, we see that we can obtain the graph of f by shifting the graph of y_1 2 units to the left; therefore, $f_1(x)=5\cos((\pi/6)(x+2))+3$. Similarly, we can obtain the graph of f by shifting the graph of y_2 to the right 4 units, so $f_2(x)=-5\cos((\pi/6)(x-4))+3$, or by shifting the graph of y_3 to the right 7 units, so $f_3(x)=5\sin((\pi/6)(x-7))+3$. Finally, a sketch of y_4 reveals that we can obtain the graph of f by shifting the graph of y_4 to the right 1 unit, so $f_4(x)=-5\sin((\pi/6)(x-1))+3$. Answers may vary.

62. We first solve for $\cos\alpha$,

$$\begin{aligned} 2\cos\alpha &= 1 \\ \cos\alpha &= \frac{1}{2} \\ \alpha &= \frac{\pi}{3}, \frac{5\pi}{3}. \end{aligned}$$

63. We first solve for $\tan\alpha$,

$$\begin{aligned} \tan\alpha &= \sqrt{3}-2\tan\alpha \\ 3\tan\alpha &= \sqrt{3} \\ \tan\alpha &= \frac{\sqrt{3}}{3} \\ \alpha &= \frac{\pi}{6}, \frac{7\pi}{6} \end{aligned}$$

64. We first solve for $\tan\alpha$,

$$\begin{aligned} 4\tan\alpha+3 &= 2 \\ 4\tan\alpha &= -1 \\ \tan\alpha &= -\frac{1}{4} \\ \alpha &= 2.897,\ 6.038. \end{aligned}$$

65. We first solve for $\sin\alpha$,

$$\begin{aligned} 3\sin^2\alpha+4 &= 5 \\ 3\sin^2\alpha &= 1 \\ \sin^2\alpha &= \frac{1}{3} \\ \sin\alpha &= \pm\sqrt{\frac{1}{3}} \end{aligned}$$

$$\sin\alpha=\sqrt{\tfrac{1}{3}} \qquad\qquad \sin\alpha=-\sqrt{\tfrac{1}{3}}$$
$$\alpha=0.616,\ 2.526 \qquad\qquad \alpha=3.757,\ 5.668$$

66. The arc length is equal to the radius times the radian measure, so

$$d=(2)\left(\frac{87}{60}\right)(2\pi)=5.8\pi\approx 18.221 \text{ inches.}$$

67. We know $r = 3960$ and $\theta = 1°$. Change θ to radian measure and use $s = r\theta$.

$$s = 3960(1)\left(\frac{\pi}{180}\right) \approx 69.115 \text{ miles.}$$

68. As the bob moves from one side to the other, as in Figure 8.100, the string moves through an angle on $10°$. We are therefore looking for the arc length on a circle of radius 3 feet cut off by an angle of $10°$. First we convert $10°$ to radians

$$10° = 10 \cdot \frac{\pi}{180} = \frac{\pi}{18} \text{ radians.}$$

Then we find

$$\begin{aligned} \text{arc length} &= \text{radius} \cdot \text{angle spanned in radians} \\ &= 3\left(\frac{\pi}{18}\right) \\ &= \frac{\pi}{6} \text{ feet.} \end{aligned}$$

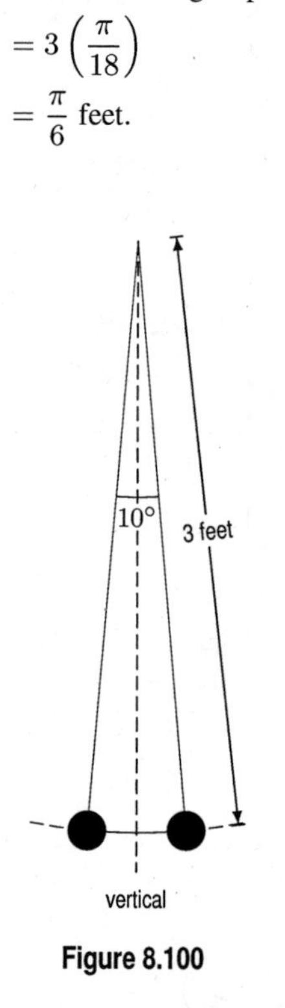

Figure 8.100

69. We can approximate this angle by using $s = r\theta$. The arc length is approximated by the moon diameter; and the radius is the distance to the moon. Therefore $\theta = s/r = 2160/238,860 \approx 0.009$ radians. Change this to degrees to get $\theta = 0.009(180/\pi) \approx 0.516°$. Note that we could also consider the radius to cut across the moon's center, in which case the radius would be $r = 238,860 + 2160/2 = 239,940$. The difference in the two answers is negligible.

70. The circumference of the outer edge is

$$6(2\pi) = 12\pi \text{ cm.}$$

A point on the outer edge travels 100 times this distance in one minute. Thus, a point on the outer edge must travel at the speed of 1200π cm/minute or roughly 3770 cm/min.

The circumference of the inner edge is

$$0.75(2\pi) = 1.5\pi \text{ cm.}$$

A point on the inner edge travels 100 times this distance in one minute. Thus, a point on the inner edge must travel at the speed of 150π cm/minute or roughly 471 cm/min.

71. Using $s = r\theta$, we know the arc length $s = 600$ and $r = 3960 + 500$. Therefore $\theta = 600/4460 \approx 0.1345$ radians.

72. **(a)** The midline, $D = 10$, the amplitude $A = 4$, and the period 1, so

$$1 = \frac{2\pi}{B} \quad \text{and} \quad B = 2\pi.$$

Therefore the formula is

$$f(t) = 4\sin(2\pi t) + 10.$$

(b) Solving $f(t) = 12$, we have

$$\begin{aligned} 12 &= 4\sin(2\pi t) + 10 \\ 2 &= 4\sin(2\pi t) \\ \sin 2\pi t &= \frac{1}{2} \\ 2\pi t &= \frac{\pi}{6}, \frac{5\pi}{6}, \frac{13\pi}{6}, \frac{17\pi}{6} \\ t &= \frac{1}{12}, \frac{5}{12}, \frac{13}{12}, \frac{17}{12} \end{aligned}$$

The spring is 12 centimeters from the ceiling at $1/12$ sec, $5/12$ sec, $13/12$ sec, $17/12$ sec.

73. The function has a maximum of 3000, a minimum of 1200 which means the upward shift is $\frac{3000+1200}{2} = 2100$. A period of eight years means the angular frequency is $\frac{\pi}{4}$. The amplitude is $|A| = 3000 - 2100 = 900$. Thus a function for the population would be an inverted cosine and $f(t) = -900\cos((\pi/4)t) + 2100$.

74. **(a)** The average monthly temperature in Fairbanks is shown in Figure 8.101.

(b) These data are best modeled by a vertically reflected cosine curve, which is reasonable for something like temperature that oscillates with a 12 month period.

(c) The midline temperature is $(61.3 + (-11.5))/2 = 24.9$. The amplitude is $61.3 - 24.9 = 36.4$ Since the period is 12, we have $B = 2\pi/12 = \pi/6$. There is no horizontal shift, so $C = 0$. Hence our function is

$$f(t) = 24.9 - 36.4\cos\left(\frac{\pi}{6}t\right).$$

(d) To solve $f(t) = 32$, we must solve the equation

$$\begin{aligned} 32 &= 24.9 - 36.4\cos\left(\frac{\pi}{6}t\right) \\ 7.1 &= -36.4\cos\left(\frac{\pi}{6}t\right) \\ \cos\left(\frac{\pi}{6}t\right) &= -0.195 \end{aligned}$$

$$\frac{\pi}{6}t = \cos^{-1}(-0.195) = 1.767 \qquad \frac{\pi}{6}t = 2\pi - 1.767 = 4.516$$

$$t = 3.375 \qquad\qquad t = 8.625$$

The temperature in Fairbanks reaches the freezing point in the middle of April and the middle of September.

(e) See Figure 8.102.

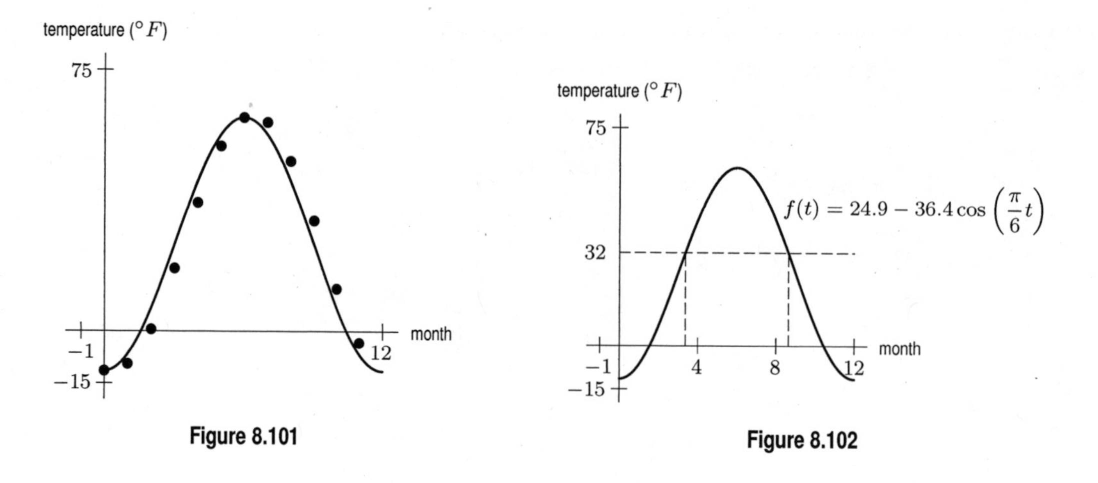

Figure 8.101

Figure 8.102

(f) The amplitude, period, and midline would be the same, but the function would be vertically reflected:

$$y = 24.9 + 36.4\cos\left(\frac{\pi}{6}t\right).$$

75. The data shows the period is about 0.6 sec. The angular frequency is $B = 2\pi/0.6 \approx 10.472$. The amplitude is (high − low)$/2 = (180 - 120)/2 = 60/2 = 30$ cm. The midline is (Minimum) + (Amplitude) $= 120 + 30 = 150$ cm. The weight starts in a low position so it is a quarter cycle behind the sine; that is, the phase shift is $\pi/2$:

$$y = 30\sin\left(10.5t - \frac{\pi}{2}\right) + 150.$$

CHECK YOUR UNDERSTANDING

1. False. One radian is about 57 degrees.
2. True. This is the definition of radian measure.
3. False. One radian is about 57 degrees, so three radians is more than 90° but less than 180°. Thus the point is in the second quadrant.
4. False. Multiply the angle by $\pi/180°$.
5. False. Use $s = r\theta$ to get, $s = 3 \cdot (\pi/3) = \pi$.
6. True. In radians, points in the second quadrant correspond to angles from $\pi/2$ to π.
7. True. Since $s = r\theta$, one complete revolution is 2π and $r = 1$, we have $s = 1 \cdot 2\pi \approx 6.28$.
8. True. We know $\cos 30° = \sqrt{3}/2 = \sin\frac{\pi}{3}$.
9. False. $\sin\frac{\pi}{6} = \frac{1}{2}$.
10. True. Since $\sin\frac{\pi}{4} = \sqrt{2}/2 = \cos\frac{\pi}{4}$.
11. True. Since sine is an odd function.
12. True. The period appears to be about 4.
13. False. The amplitude is (maximum − minimum)$/2 = (6 - 2)/2 = 2$.
14. True. Since $f(x+4) = f(x)$ and since $f(x+c) = f(c)$ is true for no smaller positive value of c, the period is 4.
15. True. The midline is $y =$ (maximum + minimum)$/2 = (6+2)/2 = 4$.
16. False. The period of g is only half as long as the period of f.

17. False. Amplitude is always positive. In this case it is equal to three.

18. False. The amplitude is 10.

19. True. The function is just a vertical stretch and an upward shift of the cosine function, so the period remains unchanged and is equal to 2π.

20. False. The maximum y-value is when $\cos x = 1$. It is 35.

21. True. The minimum y-value is when $\cos x = -1$. It is 15.

22. False. The midline equation is $y = 25$

23. True, since $\cos x = A \cos B(x - h) + k$ with $A = 1, B = 1, h = 0$, and $k = 0$.

24. True, since the values of $\sin x$ start to repeat after one rotation through 2π.

25. False, since $\sin(\pi x)$ has period $2\pi/\pi = 2$.

26. False. The amplitude is positive 2.

27. False. The cosine function has been vertically stretched and then shifted downward by 4. It has not been reflected.

28. True. The function $\sin(2t)$ is a sine function horizontally compressed by $\frac{1}{2}$ so it has half the period of $\sin t$.

29. False. The period is $\frac{1}{3}$ that of $y = \cos x$.

30. True. The period is $2\pi/B$.

31. False. A shift to the left by h units is given by $y = A \sin(2(x + h)) + k$.

32. True. The conditions for A, B and C in $y = A\cos(Bx) + C$ are perfectly met.

33. True. The amplitude is $\frac{1}{2}$, the period is π and the midline is $y = 1$ and its horizontal shift is correct.

34. False. Numerically, we could check the equation at $x = 0$ to find $y = -0.5\cos(2 \cdot 0 + \frac{\pi}{3}) + 1 = 0.75$, but the graph at $x = 0$ shows $y = 1$.

35. True. On the unit circle we define $x = \cos\theta$ and $y = \sin\theta$, so we have $\tan\theta = y/x = \sin\theta/\cos\theta$.

36. False. It is undefined at $\pi/2$ and $3\pi/2$.

37. True. Because the angles θ and $\theta + \pi$ determine the same line through the origin and hence have the same slope, which is the tangent.

38. False. The tangent is undefined at $\frac{\pi}{2}$. Note that $\tan x$ approaches $+\infty$ as x approaches $\pi/2$ from the left and approaches $-\infty$ as x approaches $\pi/2$ from the right.

39. True. Let $\theta = 5x$ in the identity $\sin^2\theta + \cos^2\theta = 1$.

40. False. In the second quadrant the tangent is negative.

41. True. We have $\sec\pi = 1/\cos\pi = 1/(-1) = -1$.

42. True. The value of $\csc\pi = 1/\sin\pi = 1/0$ is undefined.

43. False. The reciprocal of the sine function is the cosecant function.

44. True. This is true since $\cos\frac{\pi}{3} = 0.5$ and $0 < \pi/3 < \pi$.

45. True. Since $y = \arctan(-1) = -\pi/4$, we have $\sin(-\pi/4) = -\sqrt{2}/2$.

46. True. $\sin(\pi/3) = \sqrt{3}/2$.

47. False. The reference angle for 120° is 60°, since it takes 60° to reach the x-axis from 120°.

48. True. It takes 60° to reach the x-axis from 300°.

49. True, because $\sin x$ is an odd function.

50. False, because $\cos x$ is not an odd function.

51. False, because $\sin x$ is not an even function.

52. True, because $\cos x$ is an even function.

53. True. If $g(t) = f(t - \pi/2)$, the graph of g is related to the graph of f by a horizontal shift of $\pi/2$ units to the right.

54. True. When defined, $\cos(\cos^{-1}(x)) = x$.

55. False. $y = \sin^{-1} x = \arcsin x \neq 1/\sin x$.

56. False. The domain is the range of the cosine function, $-1 \le x \le 1$.

57. True. If $\cos t = 1$, then $\sin t = 0$ so $\tan t = 0/1 = 0$.

58. False. Find $x = \sin(\arcsin x) = \sin(0.5) \neq \pi/6$. The following statement is true. If $\arcsin 0.5 = x$ then $x = \pi/6$.

59. False. Any angle $\theta = \frac{\pi}{4} + n\pi$, or $\theta = -\frac{\pi}{4} + n\pi$ with n an integer, will have the same cosine value.

60. False; for instance, $\cos^{-1}(\cos(2\pi)) = 0$.

61. True

62. True; because $\tan A = \tan B$ means $A = B + k\pi$, for some integer k. Thus $\frac{A-B}{\pi} = k$ is an integer.

63. False; for example, $\cos\left(\frac{\pi}{4}\right) = \cos\left(\frac{7\pi}{4}\right)$ but $\sin\left(\frac{\pi}{4}\right) = -\sin\left(\frac{7\pi}{4}\right)$.

64. False. For example, $\cos(\pi/4) = \cos(-\pi/4)$.

65. False. The graph of $r = 1$ is the unit circle.

66. False. The graph is the ray from the origin forming an angle of 1 radian with the positive x-axis.

67. False. The point $(3, \pi)$ in polar coordinates is $(-3, 0)$ in Cartesian coordinates.

68. False. Cartesian coordinate values are unique, but polar coordinate values are not. For example, in polar coordinates, $(1, \pi)$ and $(1, 3\pi)$ represent the same point in the xy-plane.

69. True. The point is on the y axis three units down from the origin. Thus $r = 3$ and $\theta = 3\pi/2$. In polar coordinates this is $(3, 3\pi/2)$.

70. False. The region is a sector of a disk with radius two, centered at the origin. The angle θ sweeps out an angle of two radians, which is about 1/3 of a full circle.

71. True, since $\sqrt{a}$ is real for all $a \ge 0$.

72. True, since $(x - iy)(x + iy) = x^2 + y^2$ is real.

73. False, since $(1 + i)^2 = 2i$ is not real.

74. False. Let $f(x) = x$. Then $f(i) = i$ but $f(\bar{i}) = \bar{i} = -i$.

75. True. We can write any nonzero complex number z as $re^{i\beta}$, where r and β are real numbers with $r > 0$. Since $r > 0$, we can write $r = e^c$ for $c = \ln r$. Therefore, $z = re^{i\beta} = e^c e^{i\beta} = e^{c+i\beta} = e^w$ where $w = c + i\beta$ is a complex number.

76. False, since $(1 + 2i)^2 = -3 + 4i$.

77. True. This is Euler's formula, fundamental in higher mathematics.

78. True. $(1 + i)^2 = 1^2 + 2 \cdot 1 \cdot i + i^2 = 1 + 2i - 1 = 2i$.

79. True. Since $i^4 = 1$, we have $i^{101} = (i^4)^{25} i = 1 \cdot i = i$.

CHAPTER NINE

Solutions for Section 9.1

Exercises

1. We have:

$$\begin{aligned}\tan t\cos t-\frac{\sin t}{\tan t}&=\frac{\sin t}{\cos t}\cdot\cos t-\frac{\sin t}{\left(\frac{\sin t}{\cos t}\right)}\quad\text{because }\tan t=\tfrac{\sin t}{\cos t}\\&=\sin t-\sin t\cdot\frac{\cos t}{\sin t}\\&=\sin t-\cos t.\end{aligned}$$

2. We have:

$$\begin{aligned}2\cos t\,(3-7\tan t)&=6\cos t-14\cos t\cdot\tan t\\&=6\cos t-14\cos t\cdot\frac{\sin t}{\cos t}\\&=6\cos t-14\sin t.\end{aligned}$$

3. We have:

$$\begin{aligned}2\cos t-\cos t\,(1-3\tan t)&=2\cos t-\cos t+3\cos t\cdot\tan t\\&=\cos t+3\cos t\cdot\frac{\sin t}{\cos t}\\&=\cos t+3\sin t.\end{aligned}$$

4. We have:

$$\begin{aligned}2\cos t\,(3\sin t-4\tan t)&=6\sin t\cos t-8\tan t\cdot\cos t\\&=6\sin t\cos t-8\frac{\sin t}{\cos t}\cdot\cos t\\&=6\sin t\cos t-8\sin t.\end{aligned}$$

5. Writing $\sin 2\alpha=2\sin\alpha\cos\alpha$, we have

$$\frac{\sin 2\alpha}{\cos\alpha}=\frac{2\sin\alpha\cos\alpha}{\cos\alpha}=2\sin\alpha.$$

6. Writing $\cos^2\theta=1-\sin^2\theta$, we have

$$\frac{\cos^2\theta-1}{\sin\theta}=\frac{1-\sin^2\theta-1}{\sin\theta}=\frac{-\sin^2\theta}{\sin\theta}=-\sin\theta.$$

7. Writing $\cos 2t=\cos^2 t-\sin^2 t$ and factoring, we have

$$\frac{\cos 2t}{\cos t+\sin t}=\frac{\cos^2 t-\sin^2 t}{\cos t+\sin t}=\frac{(\cos t-\sin t)(\cos t+\sin t)}{\cos t+\sin t}=\cos t-\sin t.$$

8. Adding fractions gives

$$\frac{1}{1-\sin\theta}+\frac{1}{1+\sin\theta}=\frac{1+\sin\theta+1-\sin\theta}{(1-\sin\theta)(1+\sin\theta)}=\frac{2}{1-\sin^2\theta}=\frac{2}{\cos^2\theta}.$$

9. Combining terms and using $\cos^2\phi+\sin^2\phi=1$, we have

$$\frac{\cos\phi-1}{\sin\phi}+\frac{\sin\phi}{\cos\phi+1}=\frac{(\cos\phi-1)(\cos\phi+1)+\sin^2\phi}{\sin\phi(\cos\phi+1)}=\frac{\cos^2\phi-1+\sin^2\phi}{\sin\phi(\cos\phi+1)}=\frac{0}{\sin\phi(\cos\phi+1)}=0$$

10. Using $\tan t=\sin t/\cos t$, we have

$$\begin{aligned}\frac{1}{\sin t\cos t}-\frac{1}{\tan t}&=\frac{1}{\sin t\cos t}-\frac{1}{\sin t/\cos t}\\&=\frac{1}{\sin t\cos t}-\frac{\cos t}{\sin t}\\&=\frac{1-\cos^2 t}{\sin t\cos t}\\&=\frac{\sin^2 t}{\sin t\cos t}\\&=\frac{\sin t}{\cos t}\\&=\tan t.\end{aligned}$$

11. We have: $\dfrac{\sin\sqrt{\theta}}{\cos\sqrt{\theta}}=\tan\sqrt{\theta}.$

12. We have:

$$\begin{aligned}\frac{2\sin\frac{\alpha}{2}}{\cos\frac{\alpha}{2}}&=2\cdot\frac{\sin\frac{\alpha}{2}}{\cos\frac{\alpha}{2}}\\&=2\tan\frac{\alpha}{2}.\end{aligned}$$

13. We have:

$$\begin{aligned}\frac{3\sin(\phi+1)}{4\cos(\phi+1)}&=\frac{3}{4}\cdot\frac{\sin(\phi+1)}{\cos(\phi+1)}\\&=\frac{3}{4}\tan(\phi+1).\end{aligned}$$

14. We have:

$$\frac{1}{\left(\dfrac{\cos\left(r^2-s^2\right)}{\sin\left(r^2-s^2\right)}\right)}=\frac{\sin\left(r^2-s^2\right)}{\cos\left(r^2-s^2\right)}=\tan\left(r^2-s^2\right).$$

15.

$$\begin{aligned}2\sin\left(\frac{2}{k+3}\right)\cdot\frac{5}{3\cos\left(\frac{2}{k+3}\right)}&=\frac{10}{3}\cdot\frac{\sin\left(\frac{2}{k+3}\right)}{\cos\left(\frac{2}{k+3}\right)}\\&=\frac{10}{3}\tan\left(\frac{2}{k+3}\right).\end{aligned}$$

16. We have:

$$\frac{2}{\cos\left(1-\frac{1}{z}\right)} \cdot \frac{\sin\left(1-\frac{1}{z}\right)}{3} = \frac{2}{3}\frac{\sin\left(1-\frac{1}{z}\right)}{\cos\left(1-\frac{1}{z}\right)}$$
$$= \frac{2}{3}\tan\left(1-\frac{1}{z}\right).$$

17. The relevant identities are $\cos^2\theta+\sin^2\theta = 1$ and $\cos 2\theta = \cos^2\theta - \sin^2\theta = 2\cos^2\theta - 1 = 1-2\sin^2\theta$. See Table 9.1.

Table 9.1

θ in rad.	$\sin^2\theta$	$\cos^2\theta$	$\sin 2\theta$	$\cos 2\theta$
1	0.708	0.292	0.909	-0.416
$\pi/2$	1	0	0	-1
2	0.827	0.173	-0.757	-0.654
$5\pi/6$	$1/4$	$3/4$	$-\sqrt{3}/2$	$1/2$

18. We solve the Pythagorean identity for $\sin\theta$.

$$\sin^2\theta + \cos^2\theta = 1$$
$$\sin^2\theta = 1-\cos^2\theta$$
$$(\sin\theta)^2 = 1-\cos^2\theta.$$

If $\sin\theta \geq 0$,

$$\sin\theta = \sqrt{1-\cos^2\theta}.$$

If $\sin\theta < 0$,

$$\sin\theta = -\sqrt{1-\cos^2\theta}.$$

Problems

19. Divide both sides of $\cos^2\theta + \sin^2\theta = 1$ by $\sin^2\theta$. For $\sin\theta \neq 0$,

$$\frac{\cos^2\theta}{\sin^2\theta} + \frac{\sin^2\theta}{\sin^2\theta} = \frac{1}{\sin^2\theta}$$
$$\left(\frac{\cos\theta}{\sin\theta}\right)^2 + 1 = \left(\frac{1}{\sin\theta}\right)^2$$
$$\cot^2\theta + 1 = \csc^2\theta.$$

20. Since $\cos 2t$ has period π and $\sin t$ has period 2π, if the result we want holds for $0 \leq t \leq 2\pi$, it holds for all t. So let's concentrate on the interval $0 \leq t \leq 2\pi$.

Solving $\cos 2t = 0$ gives $t = \pi/4, 3\pi/4, 5\pi/4, 7\pi/4$.

From the graph in Figure 9.1, we see $\cos 2t > 0$ for $0 \leq t < \pi/4$, $3\pi/4 < t < 5\pi/4$, $7\pi/4 < t \leq 2\pi$.

Solving $1 - 2\sin^2 t = 0$ gives

$$\sin^2 t = \frac{1}{2}$$
$$\sin t = \pm\frac{1}{\sqrt{2}},$$

so $t = \pi/4, 3\pi/4, 5\pi/4, 7\pi/4$.

From the graph of $y = \sin t$ and the lines $y = 1/\sqrt{2}$ and $y = -1/\sqrt{2}$ in Figure 9.2, we see that $-1/\sqrt{2} < \sin t < 1/\sqrt{2}$ on the same intervals that $\cos 2t > 0$.

Now if $-1/\sqrt{2} < \sin t < 1/\sqrt{2}$, then

$$\sin^2 t < \frac{1}{2}$$
$$1 - 2\sin^2 t > 0.$$

Thus, $\cos 2t$ and $1 - 2\sin^2 t$ have the same sign for all t.

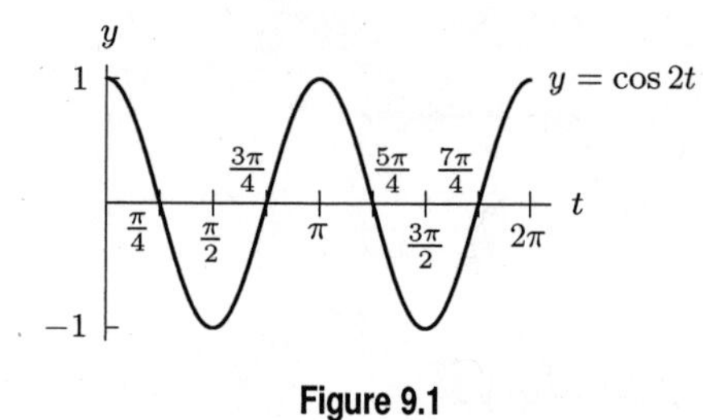

Figure 9.1

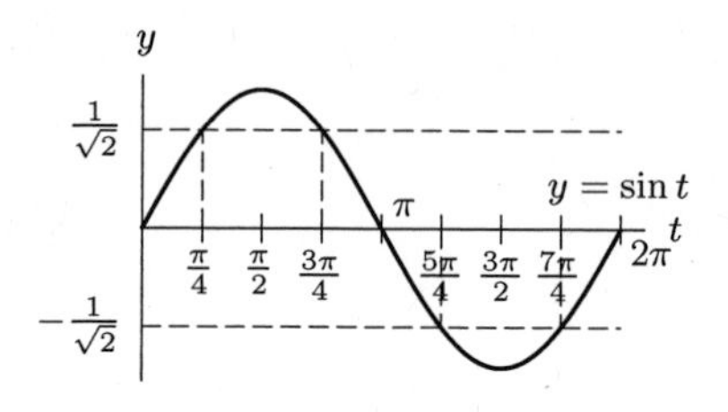

Figure 9.2

21. **(a)** $\cos 2\theta = 1 - 2\sin^2\theta = 1 - 2(1 - \cos^2\theta) = 1 - 2 + 2\cos^2\theta = 2\cos^2\theta - 1 = 2(\cos\theta)^2 - 1.$

(b) $\cos 2\theta = 1 - 2\sin^2\theta = (1 - \sin^2\theta) - \sin^2\theta = \cos^2\theta - \sin^2\theta = (\cos\theta)^2 - (\sin\theta)^2.$

22. We have

$$\tan 2\theta = \frac{\sin 2\theta}{\cos 2\theta} = \frac{2\sin\theta\cos\theta}{\cos^2\theta - \sin^2\theta}.$$

Dividing both top and bottom by $\cos^2\theta$ gives

$$\tan 2\theta = \frac{\dfrac{2\sin\theta\cos\theta}{\cos^2\theta}}{\dfrac{\cos^2\theta - \sin^2\theta}{\cos^2\theta}} = \frac{2\tan\theta}{1 - \tan^2\theta}.$$

23. Multiply the denominator by $1 + \cos t$ to get $\sin^2 t$:

$$\begin{aligned}\frac{\sin t}{1 - \cos t} &= \frac{\sin t}{1 - \cos t}\frac{(1 + \cos t)}{(1 + \cos t)}\\ &= \frac{\sin t(1 + \cos t)}{1 - \cos^2 t}\\ &= \frac{\sin t(1 + \cos t)}{\sin^2 t}\\ &= \frac{1 + \cos t}{\sin t}.\end{aligned}$$

24. Get a common denominator:

$$\begin{aligned}\frac{\cos x}{1 - \sin x} - \tan x &= \frac{\cos x}{1 - \sin x} - \frac{\sin x}{\cos x}\\ &= \frac{\cos^2 x - \sin x(1 - \sin x)}{(1 - \sin x)(\cos x)}\\ &= \frac{\cos^2 x - \sin x + \sin^2 x}{(1 - \sin x)(\cos x)}\\ &= \frac{1 - \sin x}{(1 - \sin x)\cos x} = \frac{1}{\cos x}.\end{aligned}$$

25. In order to get tan to appear, divide by $\cos x \cos y$:

$$\frac{\sin x\cos y+\cos x\sin y}{\cos x\cos y-\sin x\sin y}=\frac{\dfrac{\sin x\cos y}{\cos x\cos y}+\dfrac{\cos x\sin y}{\cos x\cos y}}{\dfrac{\cos x\cos y}{\cos x\cos y}-\dfrac{\sin x\sin y}{\cos x\cos y}}=\frac{\tan x+\tan y}{1-\tan x\tan y}$$

26. Using the trigonometric identity $\cos^2\theta = 1-\sin^2\theta$, we have

$$\begin{aligned}\sin^2\theta-\cos^2\theta&=\sin\theta\\ \sin^2\theta-(1-\sin^2\theta)&=\sin\theta\\ 2\sin^2\theta-\sin\theta-1&=0\\ (2\sin\theta+1)(\sin\theta-1)&=0\\ \sin\theta=-\frac{1}{2}\quad\text{or}&\quad\sin\theta=1.\end{aligned}$$

If $\sin\theta=1$, then $\theta=\pi/2$. On the other hand, if $\sin\theta=-1/2$, we first calculate the associated reference angle, which is $\sin^{-1}(1/2)=\pi/6$. Using a graph of the sine function on the interval $0\le\theta\le 2\pi$, we see that the two solutions to $\sin\theta=-1/2$ are given by $\theta=\pi+\pi/6=7\pi/6$ and $\theta=2\pi-\pi/6=11\pi/6$. Combining the above observations, we see that there are three solutions to the original equation: $\pi/2, 7\pi/6$, and $11\pi/6$.

27. Using the trigonometric identity $\sin(2\theta)=2\sin\theta\cos\theta$, we have

$$\begin{aligned}\sin(2\theta)-\cos\theta&=0\\ 2\sin\theta\cos\theta-\cos\theta&=0\\ \cos\theta(2\sin\theta-1)&=0\\ \cos\theta=0\quad\text{or}\quad\sin\theta&=\frac{1}{2}.\end{aligned}$$

If $\cos\theta=0$, then we have two solutions: $\theta=\pi/2$ and $\theta=3\pi/2$. On the other hand, if $\sin\theta=1/2$, we first calculate the associated reference angle, which is $\sin^{-1}(1/2)=\pi/6$. Using a graph of the sine function on the interval $0\le\theta\le2\pi$, we see that the two solutions to $\sin\theta=1/2$ are given by $\theta=\pi/6$ and $\theta=\pi-\pi/6=5\pi/6$. Combining the above observations, we see that there are four solutions to the original equation: $\pi/2, 3\pi/2, \pi/6$ and $5\pi/6$.

28. Using the trigonometric identity $\sec^2\theta=\tan^2\theta+1$, we have

$$\begin{aligned}\sec^2\theta&=1-\tan\theta\\ \tan^2\theta+1&=1-\tan\theta\\ \tan^2\theta+\tan\theta&=0\\ \tan\theta(\tan\theta+1)&=0\\ \tan\theta=0\quad\text{or}&\quad\tan\theta=-1.\end{aligned}$$

If $\tan\theta=0$, then we have three solutions: $\theta=0$ and $\theta=\pi$ and $\theta=2\pi$. On the other hand, if $\tan\theta=-1$, we first calculate the associated reference angle, which is $\tan^{-1}(1)=\pi/4$. Using a graph of the tangent function on the interval $0\le\theta\le2\pi$, we see that the two solutions to $\tan\theta=-1$ are given by $\theta=\pi-\pi/4=3\pi/4$ and $\theta=2\pi-\pi/4=7\pi/4$. Combining the above observations, we see that there are five solutions to the original equation: $0, \pi, 2\pi, 3\pi/4$, and $7\pi/4$.

29. Using the trigonometric identity $\tan(2\theta)=2\tan\theta/(1-\tan^2\theta)$, we have

$$\begin{aligned}\tan(2\theta)+\tan\theta&=0\\ \frac{2\tan\theta}{1-\tan^2\theta}&=-\tan\theta\\ 2\tan\theta&=-\tan\theta+\tan^3\theta\\ \tan^3\theta-3\tan\theta&=0\\ \tan\theta(\tan^2\theta-3)&=0\\ \tan\theta=0\quad\text{or}&\quad\tan\theta=\pm\sqrt{3}.\end{aligned}$$

If $\tan\theta = 0$, then we have three solutions: $\theta = 0$ and $\theta = \pi$ and $\theta = 2\pi$. On the other hand, if $\tan\theta = \sqrt{3}$, we first calculate the associated reference angle, which is $\tan^{-1}(\sqrt{3}) = \pi/3$. Using a graph of the tangent function on the interval $0 \le \theta \le 2\pi$, we see that the two solutions to $\tan\theta = \sqrt{3}$ are given by $\theta = \pi/3$ and $\theta = \pi + \pi/3 = 4\pi/3$. Finally, if $\tan\theta = -\sqrt{3}$, we again have a reference angle of $\pi/3$, and the two solutions to $\tan\theta = -\sqrt{3}$ are given by $\theta = \pi - \pi/3 = 2\pi/3$ and $\theta = 2\pi - \pi/3 = 5\pi/3$. Combining the above observations, we see that there are seven solutions to the original equation: $0, \pi, \pi/3, 4\pi/3, 2\pi/3, 5\pi/3$, and 2π.

30. By graphing we can see which expressions appear to be identically equal. The graphs all show the same window, $-2\pi \le x \le 2\pi$, $-4 \le y \le 4$. The following pairs of expressions look identical: a and i; b and l; c and d and f; e and g; h and j. Note that k and m are different functions. We can verify the identities algebraically. For example, a and i: $2\cos^2 t + \sin t + 1 = 2(1 - \sin^2 t) + \sin t + 1 = -2\sin^2 t + \sin t + 3$.

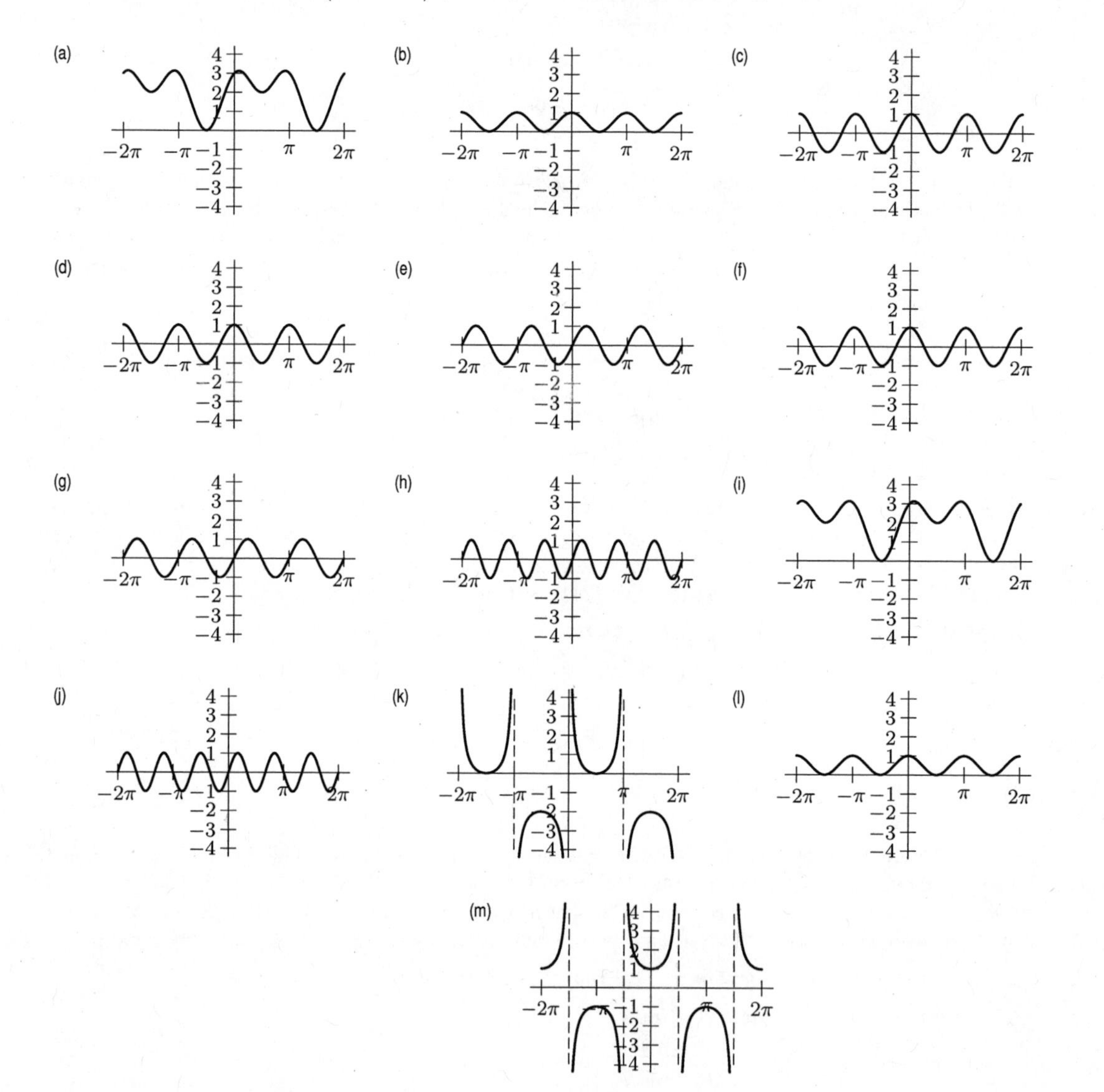

31. Not an identity. False for $x = 2$.

32. Not an identity. False for $x = 2$.

33. Not an identity. False for $x = 2$.

34. Not an identity. False for $x = \pi/2$.

35. Not an identity. False for $x = 2$.

36. If we let $x = 1$, then we have

$$\sin(x^2) = \sin(1^2) = 0.841 \neq 1.683 = 2\sin 1 = 2\sin x.$$

Therefore, since the equation is not true for $x = 1$, it is not an identity.

37. Identity. $\dfrac{\sin 2x}{1+\cos 2x} = \dfrac{2\sin x\cos x}{1+2\cos^2 x - 1} = \dfrac{2\sin x\cos x}{2\cos^2 x} = \dfrac{\sin x}{\cos x} = \tan x.$

38. If we let $A = 1$, then we have

$$\frac{\sin(2A)}{\cos(2A)} = \frac{\sin 2}{\cos 2} = -2.185 \neq 3.115 = 2\tan 1 = 2\tan A.$$

Therefore, since the equation is not true for $A = 1$, it is not an identity.

39. We have

$$\begin{aligned}\frac{\sin^2\theta - 1}{\cos\theta} &= \frac{-(1-\sin^2\theta)}{\cos\theta}\\ &= \frac{-\cos^2\theta}{\cos\theta}\\ &= -\cos\theta.\end{aligned}$$

Therefore, the equation is an identity.

40. Not an identity. False for $x = 0$.

41. Identity. $\sin x\tan x = \sin x \cdot \dfrac{\sin x}{\cos x} = \dfrac{\sin^2 x}{\cos x} = \dfrac{1-\cos^2 x}{\cos x}.$

42. Working on the left side, we have

$$\begin{aligned}\tan t + \frac{1}{\tan t} &= \frac{\sin t}{\cos t} + \frac{1}{\sin t/\cos t}\\ &= \frac{\sin t}{\cos t} + \frac{\cos t}{\sin t}\\ &= \frac{\sin^2 t + \cos^2 t}{\cos t\sin t}\\ &= \frac{1}{\sin t\cos t}.\end{aligned}$$

Therefore, the left side equals the right side and the equation is an identity.

43. If we let $x = 1$, then we have

$$\sin\left(\frac{1}{x}\right) = \sin 1 = 0.841 \neq 0 = \sin 1 - \sin 1 = \sin 1 - \sin x.$$

Therefore, since the equation is not true for $x = 1$, it is not an identity.

44. Identity. $\dfrac{2\tan x}{1+\tan^2 x}\cdot\dfrac{\cos^2 x}{\cos^2 x} = \dfrac{2\sin x\cos x}{\cos^2 x + \sin^2 x} = \dfrac{\sin 2x}{1} = \sin 2x.$

45. If we let $\theta = 1$, then we have

$$\frac{\sin\theta}{\cos\theta} - \frac{\cos\theta}{\sin\theta} = \frac{\sin 1}{\cos 1} - \frac{\cos 1}{\sin 1} = 0.915 \neq -0.458 = \frac{\cos 2}{\sin 2} = \frac{\cos(2\theta)}{\sin(2\theta)}.$$

Therefore, since the equation is not true for $\theta = 1$, it is not an identity.

46. Identity. $\dfrac{1-\tan^2 x}{1+\tan^2 x}\cdot\dfrac{\cos^2 x}{\cos^2 x} = \dfrac{\cos^2 x - \sin^2 x}{\cos^2 x + \sin^2 x} = \dfrac{\cos 2x}{1} = \cos 2x.$

47. **(a)** We can rewrite the equation as follows

$$0 = \cos 2\theta + \cos\theta = 2\cos^2\theta - 1 + \cos\theta.$$

Factoring we get

$$(2\cos\theta - 1)(\cos\theta + 1) = 0.$$

Thus the solutions occur when $\cos\theta = -1$ or $\cos\theta = \frac{1}{2}$. These are special values of cosine. If $\cos\theta = -1$ then we have $\theta = 180°$. If $\cos\theta = \frac{1}{2}$ we have $\theta = 60°$ or $300°$. Thus the solutions are

$$\theta = 60°,\ 180°,\ \text{and } 300°.$$

(b) Using the Pythagorean identity we can substitute $\cos^2\theta = 1 - \sin^2\theta$ and get

$$2(1 - \sin^2\theta) = 3\sin\theta + 3.$$

This gives

$$-2\sin^2\theta - 3\sin\theta - 1 = 0.$$

Factoring we get

$$-2\sin^2\theta - 3\sin\theta - 1 = -(2\sin\theta + 1)(\sin\theta + 1) = 0.$$

Thus the solutions occur when $\sin\theta = -\frac{1}{2}$ or when $\sin\theta = -1$. If $\sin\theta = -\frac{1}{2}$, we have

$$\theta = \frac{7\pi}{6} \quad \text{and} \quad \frac{11\pi}{6}.$$

If $\sin\theta = -1$ we have

$$\theta = \frac{3\pi}{2}.$$

48. Note the hypotenuse of the triangle is $\sqrt{1+y^2}$.

(a) $y = \frac{y}{1} = \tan\theta$.

(b) $\cos\phi = \sin(\pi/2 - \phi) = \sin\theta$.

(c) Since $\cos\theta = \dfrac{1}{\sqrt{1+y^2}}$, we have $\sqrt{1+y^2} = \dfrac{1}{\cos\theta}$, or $1+y^2 = \left(\dfrac{1}{\cos\theta}\right)^2$. (Alternatively, $1+y^2 = 1+\tan^2\theta$.)

(d) Triangle area $= \dfrac{1}{2}(\text{base})(\text{height}) = \dfrac{1}{2}(1)(y)$. But $y = \tan\theta$, so the area is $\frac{1}{2}\tan\theta$.

49. **(a)** By the Pythagorean theorem, the side adjacent to θ has length $\sqrt{1-y^2}$. So

$$\cos\theta = \sqrt{1-y^2}/1 = \sqrt{1-y^2}.$$

(b) Since $\sin\theta = y/1$, we have

$$\tan\theta = \frac{y}{\sqrt{1-y^2}}.$$

(c) Using the double angle formula,

$$\cos(2\theta) = 1 - 2\sin^2\theta = 1 - 2y^2.$$

(d) Supplementary angles have equal sines:

$$\sin(\pi - \theta) = \sin\theta = y.$$

(e) Since $\cos(\pi/2 - \theta) = y$, we have $\sin(\cos^{-1}(y)) = \sin(\pi/2 - \theta) = \sqrt{1-y^2}$. So

$$\sin^2(\cos^{-1}(y)) = 1 - y^2.$$

50. We know that $\cos^2\theta = 1 - \sin^2\theta = 1 - (3/5)^2 = 16/25$, and since θ is in the second quadrant, $\cos\theta = -\sqrt{16/25} = -4/5$. Thus $\sin 2\theta = 2\sin\theta\cos\theta = 2(3/5)(-4/5) = -24/25$. Furthermore, $\cos 2\theta = 1 - 2\sin^2\theta = 1 - 2(3/5)^2 = 7/25$, and $\tan 2\theta = \frac{\sin 2\theta}{\cos 2\theta} = \frac{-24}{25} \cdot \frac{25}{7} = \frac{-24}{7}$.

51. We have $\cos\theta = x/3$, so $\sin\theta = \sqrt{1-(x/3)^2} = \frac{\sqrt{9-x^2}}{3}$. Therefore,

$$\sin 2\theta = 2\sin\theta\cos\theta = 2\left(\frac{\sqrt{9-x^2}}{3}\right)\left(\frac{x}{3}\right) = \frac{2x}{9}\sqrt{9-x^2}.$$

52. We have $\sin\theta = \frac{x+1}{5}$, so $\cos 2\theta = 1-2\sin^2\theta = 1-2\left(\frac{x+1}{5}\right)^2 = 1-\frac{2(x+1)^2}{25}$.

53. **(a)** Let $\theta = \cos^{-1}x$, so $\cos\theta = x$. Then, since $0 \le \theta \le \pi$, $\sin\theta = \sqrt{1-x^2}$ and $\tan\theta = \frac{\sqrt{1-x^2}}{x}$, and $\tan(2\cos^{-1}x) = \tan 2\theta = \frac{2\tan\theta}{1-\tan^2\theta}$. Now $1-\tan^2\theta = 1-\frac{1-x^2}{x^2} = \frac{2x^2-1}{x^2}$, so $\tan 2\theta = \frac{2\sqrt{1-x^2}}{x}\cdot\frac{x^2}{2x^2-1} = \frac{2x\sqrt{1-x^2}}{2x^2-1}$.

(b) Let $\theta = \tan^{-1}x$, so $\tan\theta = x$. Then $\sin\theta = \frac{x}{\sqrt{1+x^2}}$ and $\cos\theta = \frac{1}{\sqrt{1+x^2}}$, so $\sin(2\tan^{-1}x) = \sin 2\theta = 2\sin\theta\cos\theta = 2\left(\frac{x}{\sqrt{1+x^2}}\right)\left(\frac{1}{\sqrt{1+x^2}}\right) = \frac{2x}{1+x^2}$.

54. We will use the identity $\cos(2x) = 2\cos^2 x - 1$, where x will be 2θ.

$$\begin{aligned}\cos 4\theta &= \cos(2x)\\ &= 2\cos^2 x - 1 \quad \text{(using the identity for } \cos(2x))\\ &= 2(2\cos^2\theta - 1)^2 - 1 \quad \text{(using the identity for } \cos(2\theta))\end{aligned}$$

55. First use $\sin(2x) = 2\sin x\cos x$, where $x = 2\theta$. Then

$$\sin(4\theta) = \sin(2x) = 2\sin(2\theta)\cos(2\theta).$$

Since $\sin(2\theta) = 2\sin\theta\cos\theta$ and $\cos(2\theta) = 2\cos^2\theta - 1$, we have

$$\sin 4\theta = 2(2\sin\theta\cos\theta)(2\cos^2\theta - 1).$$

56. **(a)** Since $\pi/2 < t \le \pi$ we have $0 \le \pi - t < \pi/2$ so the double angle formula for sine can be used for the angle $\theta = \pi - t$. Therefore $\sin 2\theta = 2\sin\theta\cos\theta$ tells us that

$$\sin 2(\pi - t) = 2\sin(\pi - t)\cos(\pi - t).$$

(b) By periodicity, $\sin 2(\pi - t) = \sin(2\pi - 2t) = \sin(-2t)$. By oddness, $\sin(-2t) = -\sin 2t$. Thus

$$\sin 2(\pi - t) = -\sin 2t.$$

(c) We have $\cos(\pi - t) = -\cos(-t) = -\cos t$ where the first equality is by what was given and the second by the evenness of cosine.

Similarly, we have $\sin(\pi - t) = -\sin(-t) = -(-\sin t) = \sin t$ where the first equality is by what was given and the second by the oddness of sine.

(d) Substitution of the results of parts (b) and (c) into part (a) shows that

$$-\sin 2t = 2\sin t(-\cos t).$$

Multiplication by -1 gives

$$\sin 2t = 2\sin t\cos t.$$

57. **(a)** Since $-\pi \le t < 0$ we have $0 < -t \le \pi$ so the double angle formula for sine can be used for the angle $\theta = -t$. Therefore $\sin 2\theta = 2\sin\theta\cos\theta$ tells us that

$$\sin(-2t) = 2\sin(-t)\cos(-t).$$

(b) Since sine is odd, we have $\sin(-2t) = -\sin 2t$. Since sine is odd and cosine is even, we have

$$2\sin(-t)\cos(-t) = 2(-\sin t)\cos t = -2\sin t\cos t.$$

Substitution of these results into the results of part (a) shows that

$$-\sin(2t) = -2\sin t\cos t.$$

Multiplication by -1 gives

$$\sin 2t = 2\sin t\cos t.$$

58. **(a)** Since $\pi/2 < t \leq \pi$ we have $0 \leq \pi - t < \pi/2$ so the double angle formula for cosine can be used for the angle $\theta = \pi - t$. Therefore $\cos 2\theta = 1 - 2\sin^2\theta$ tells us that

$$\cos 2(\pi - t) = 1 - 2\sin^2(\pi - t).$$

(b) By periodicity, $\cos 2(\pi - t) = \cos(2\pi - 2t) = \cos(-2t)$. By evenness, $\cos(-2t) = \cos 2t$. Thus

$$\cos 2(\pi - t) = \cos 2t.$$

(c) We have $1 - 2\sin^2(\pi - t) = 1 - 2(-\sin(-t))^2 = 1 - 2(\sin t)^2$ where the first equality is by what is given and the second by the oddness of sine.

(d) Substitution of the results of parts (b) and (c) into part (a) gives

$$\cos 2t = 1 - 2\sin^2 t.$$

59. **(a)** Since $-\pi \leq t < 0$ we have $0 < -t \leq \pi$ so the double angle formula for cosine can be used for the angle $\theta = -t$. Therefore $\cos 2\theta = 1 - 2\sin^2\theta$ tells us that

$$\cos(-2t) = 1 - 2\sin^2(-t).$$

(b) Since cosine is even we have $\cos(-2t) = \cos 2t$. Since sine is odd we have

$$-2\sin^2(-t) = 1 - 2(-\sin t)^2 = 1 - 2\sin^2 t.$$

Substitution of these results into the results of part (a) gives

$$\cos 2t = 1 - 2\sin^2 t.$$

60. From the exponent rules, we know

$$e^{i\theta} \cdot e^{-i\theta} = e^{i\theta - i\theta} = e^0 = 1.$$

Using Euler's formula, we rewrite this as

$$\underbrace{(\cos\theta + i\sin\theta)}_{e^{i\theta}}\underbrace{(\cos\theta - i\sin\theta)}_{e^{-i\theta}} = 1.$$

Multiplying out the left-hand side gives

$$(\cos\theta + i\sin\theta)(\cos\theta - i\sin\theta) = \cos^2\theta - i\sin\theta\cos\theta + i\sin\theta\cos\theta - i^2\sin^2\theta = 1.$$

Since $-i^2 = +1$, simplifying the left side gives the Pythagorean identity

$$\cos^2\theta + \sin^2\theta = 1.$$

61. We know that $\cos(\pi/3) = \cos 60^\circ = 1/2$ so $\cos^{-1}\left(\frac{1}{2}\right) = \pi/3$. This gives

$$\cos\left(\cos^{-1}\left(\frac{1}{2}\right)\right) = \cos\left(\frac{\pi}{3}\right) = \frac{1}{2}.$$

This is not at all surprising: after all, what we are saying is that the cosine of inverse cosine of a number is that number. However, the situation is not as straightforward as it may appear. For example, to evaluate the expression $\cos^{-1}\left(\cos\left(\frac{5\pi}{3}\right)\right)$, we write

$$\cos^{-1}\left(\cos\left(\frac{5\pi}{3}\right)\right) = \cos^{-1}\left(\frac{1}{2}\right),$$

because the cosine of $5\pi/3$ is $1/2$. And since again $\cos^{-1}\left(\frac{1}{2}\right) = \pi/3$, we have

$$\cos^{-1}\left(\cos\left(\frac{5\pi}{3}\right)\right) = \cos^{-1}\left(\frac{1}{2}\right) = \frac{\pi}{3}.$$

Thus, we see that the inverse cosine of the cosine of an angle does not necessarily equal that angle.

Solutions for Section 9.2

Exercises

1. We have $A = \sqrt{8^2 + (-6)^2} = \sqrt{100} = 10$. Since $\cos\phi = 8/10 = 0.8$ and $\sin\phi = -6/10 = -0.6$, we know that ϕ is in the fourth quadrant. Thus,

$$\tan\phi = -\frac{6}{8} = -0.75 \quad \text{and} \quad \phi = \tan^{-1}(-0.75) = -0.644,$$

so $8\sin t - 6\cos t = 10\sin(t - 0.644)$.

2. We have $A = \sqrt{8^2 + 6^2} = \sqrt{100} = 10$. Since $\cos\phi = 8/10 = 0.8$ and $\sin\phi = 6/10 = 0.6$ are both positive, ϕ is in the first quadrant. Thus,

$$\tan\phi = \frac{6}{8} = 0.75 \quad \text{and} \quad \phi = \tan^{-1}(0.75) = 0.644,$$

so $8\sin t + 6\cos t = 10\sin(t + 0.644)$.

3. Since $a_1 = -1$ and $a_2 = 1$, we have

$$A = \sqrt{(-1)^2 + 1^2} = \sqrt{2}$$

and

$$\cos\phi = -\frac{1}{\sqrt{2}} \quad \text{and} \quad \sin\phi = \frac{1}{\sqrt{2}} \quad \text{and} \quad \tan\phi = -1.$$

From the signs of $\cos\phi$ and $\sin\phi$, we see that ϕ is in the second quadrant. Since

$$\tan^{-1}(-1) = -\frac{\pi}{4},$$

and the tangent function has period π, we take

$$\phi = \pi - \frac{\pi}{4} = \frac{3\pi}{4}.$$

Thus

$$-\sin t + \cos t = \sqrt{2}\sin\left(t + \frac{3\pi}{4}\right).$$

4. Since $a_1 = -2$ and $a_2 = 5$, we have

$$A = \sqrt{(-2)^2 + 5^2} = \sqrt{29}$$

and

$$\cos\phi = \frac{-2}{\sqrt{29}} \quad \text{and } \sin\phi = \frac{5}{\sqrt{29}} \quad \text{and } \tan\phi = -\frac{5}{2}.$$

The signs of $\cos\phi$ and $\sin\phi$ show that ϕ must be in the second quadrant. Since

$$\tan^{-1}\left(-\frac{5}{2}\right) = -1.190$$

and the tangent function has period π, we take

$$\phi = \pi - 1.190 = 1.951.$$

Thus,

$$-2\sin 3t + 5\cos 3t = \sqrt{29}\sin(3t + 1.951).$$

5. Write $\sin 15^\circ = \sin(45^\circ - 30^\circ)$, and then apply the appropriate trigonometric identity.

$$\begin{aligned}\sin 15^\circ &= \sin(45^\circ - 30^\circ)\\ &= \sin 45^\circ \cos 30^\circ - \sin 30^\circ \cos 45^\circ\\ &= \frac{\sqrt{6}}{4} - \frac{\sqrt{2}}{4}\end{aligned}$$

Similarly, $\sin 75^\circ = \sin(45^\circ + 30^\circ)$.

$$\begin{aligned}\sin 75^\circ &= \sin(45^\circ + 30^\circ)\\ &= \sin 45^\circ \cos 30^\circ + \sin 30^\circ \cos 45^\circ\\ &= \frac{\sqrt{6}}{4} + \frac{\sqrt{2}}{4}\end{aligned}$$

Also, note that $\cos 75^\circ = \sin(90^\circ - 75^\circ) = \sin 15^\circ$, and $\cos 15^\circ = \sin(90^\circ - 15^\circ) = \sin 75^\circ$.

6. Though we could use the sum-of-angle and difference-of-angle formulas for cosine on each of the two parts, we can also use the formula for the sum of cosines:

$$\cos 165^\circ - \cos 75^\circ = -2\sin\frac{165+75}{2}\sin\frac{165-75}{2} = -2\sin 120 \sin 45 = -2\frac{\sqrt{3}}{2}\frac{\sqrt{2}}{2} = -\frac{\sqrt{6}}{2}.$$

7. Thus we could use the sum-of-angle and difference-of-angle formulas for cosine on each of the two parts, we can also use the formula for the sum of cosines:

$$\cos 75^\circ + \cos 15^\circ = 2\cos\frac{75+15}{2}\cos\frac{75-15}{2} = 2\cos 45 \cos 30 = 2\frac{\sqrt{2}}{2}\frac{\sqrt{3}}{2} = \frac{\sqrt{6}}{2}.$$

8. Since 345° is $300^\circ + 45^\circ$, we can use the sum-of-angle formula for sine and say that

$$\sin 345 = \sin 300 \cos 45 + \sin 45 \cos 300 = (-\sqrt{3}/2)(\sqrt{2}/2) + (\sqrt{2}/2)(1/2) = (-\sqrt{6} + \sqrt{2})/4.$$

9. Since 105° is $60^\circ + 45^\circ$, we can use the sum-of-angle formula for sine and say that

$$\sin 105 = \sin 60 \cos 45 + \sin 45 \cos 60 = (\sqrt{3}/2)(\sqrt{2}/2) + (\sqrt{2}/2)(1/2) = (\sqrt{6} + \sqrt{2})/4.$$

10. Since 285° is $240^\circ + 45^\circ$, we can use the sum-of-angle formula for cosine and say that

$$\cos 285 = \cos 240 \cos 45 - \sin 240 \sin 45 = (-1/2)(\sqrt{2}/2) - (-\sqrt{3}/2)(\sqrt{2}/2) = (-\sqrt{2} + \sqrt{6})/4.$$

11. **(a)** $\cos 35 + \cos 40 = 2\cos\left(\frac{35+40}{2}\right)\cos\left(\frac{35-40}{2}\right) = 1.585$. See Figure 9.3.

(b) $\cos 35 - \cos 40 = -2\sin\left(\frac{35+40}{2}\right)\sin\left(\frac{35-40}{2}\right) = 0.053$. See Figure 9.4.

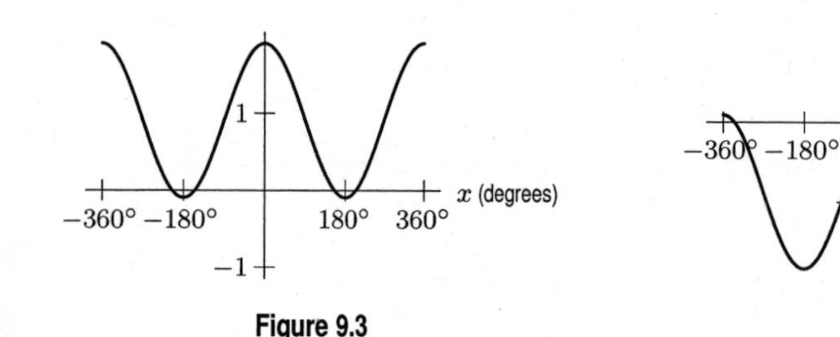

Figure 9.3

Figure 9.4

(c) $\sin 35 + \sin 40 = 2\sin\left(\frac{35+40}{2}\right)\cos\left(\frac{35-40}{2}\right) = 1.216$. See Figure 9.5.

(d) $\sin 35 - \sin 40 = 2\cos\left(\frac{35+40}{2}\right)\sin\left(\frac{35-40}{2}\right) = -0.069$. See Figure 9.6.

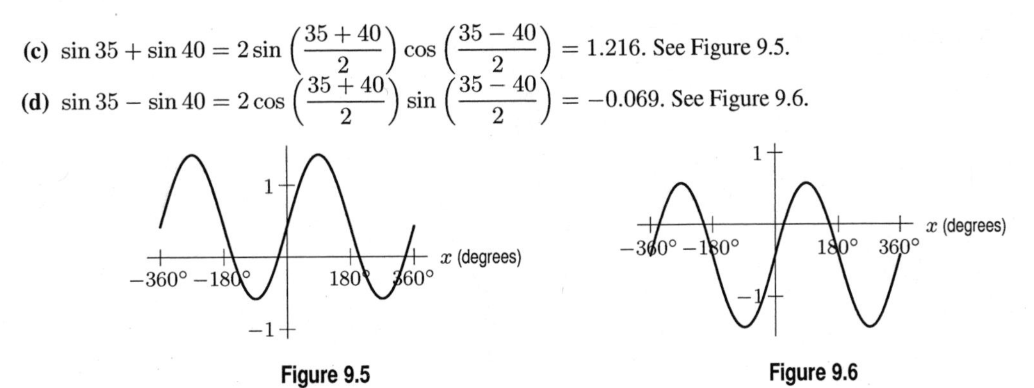

Figure 9.5 Figure 9.6

12. **(a)** $\sin(15+42) = \sin 15\cos 42 + \sin 42\cos 15 = 0.839$. See Figure 9.7.
(b) $\sin(15-42) = \sin 15\cos 42 - \sin 42\cos 15 = -0.454$. See Figure 9.8.

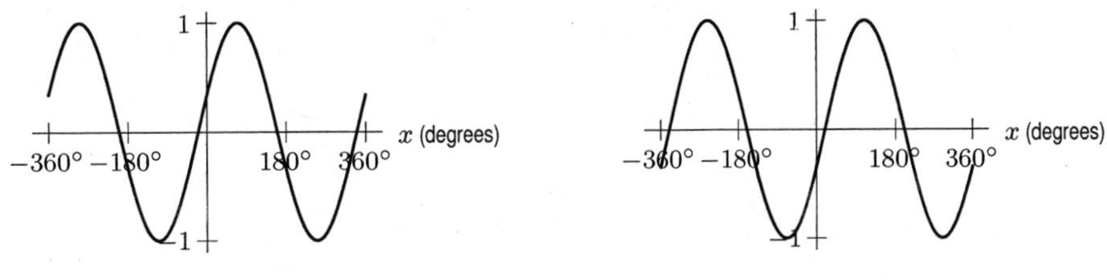

Figure 9.7 Figure 9.8

(c) $\cos(15+42) = \cos 15\cos 42 - \sin 15\sin 42 = 0.545$. See Figure 9.9.
(d) $\cos(15-42) = \cos 15\cos 42 + \sin 15\sin 42 = 0.891$. See Figure 9.10.

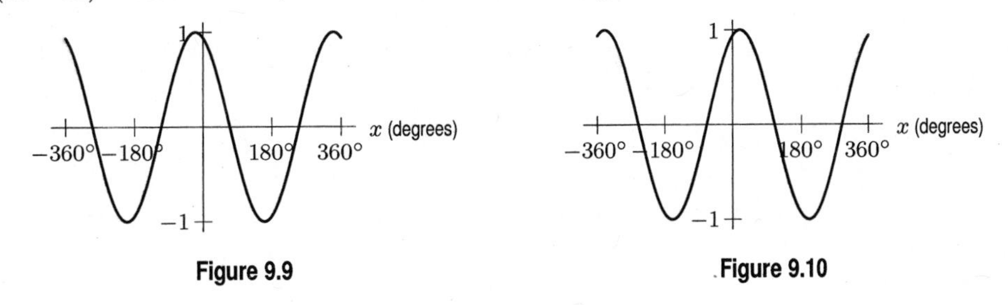

Figure 9.9 Figure 9.10

Problems

13. **(a)** $\cos(t-\pi/2) = \cos t\cos\pi/2 + \sin t\sin\pi/2 = \cos t\cdot 0 + \sin t\cdot 1 = \sin t$.
(b) $\sin(t+\pi/2) = \sin t\cos\pi/2 + \sin\pi/2\cos t = \sin t\cdot 0 + 1\cdot\cos t = \cos t$.

14. We can use the sum-of-angle identities

$$\cos(\theta+\phi) = \cos\theta\cos\phi - \sin\theta\sin\phi$$
$$\cos(\theta-\phi) = \cos\theta\cos\phi + \sin\theta\sin\phi$$

and subtract to obtain

$$\cos(\theta+\phi) - \cos(\theta-\phi) = -2\sin\theta\sin\phi.$$

We put $u = \theta+\phi$ and $v = \theta-\phi$ on the left side equation. Solving these simultaneous equations, we get $\theta = (u+v)/2$ and $\phi = (u-v)/2$. We put $\theta = (u+v)/2$ and $\phi = (u-v)/2$ on the right side to get

$$\cos u - \cos v = -2\sin\left(\frac{u+v}{2}\right)\sin\left(\frac{u-v}{2}\right).$$

15. We can use the sum-of-angle identities

$$\sin(\theta+\phi) = \sin\theta\cos\phi + \sin\phi\cos\theta$$
$$\sin(\theta-\phi) = \sin\theta\cos\phi - \sin\phi\cos\theta$$

and add to get

$$\sin(\theta+\phi) + \sin(\theta-\phi) = 2\sin\theta\cos\phi.$$

We put $u = \theta+\phi$ and $v = \theta-\phi$ on the left side of the equation. Solving these simultaneous equations, we get $\theta = (u+v)/2$ and $\phi = (u-v)/2$. We put $\theta = (u+v)/2$ and $\phi = (u-v)/2$ on the right side to get

$$\sin u + \sin v = 2\sin\left(\frac{u+v}{2}\right)\cos\left(\frac{u-v}{2}\right).$$

16. We start with

$$\sin u + \sin v = 2\sin\left(\frac{u+v}{2}\right)\cos\left(\frac{u-v}{2}\right).$$

Since $-\sin v = \sin(-v)$, we can write

$$\begin{aligned}
\sin u - \sin v &= \sin u + \sin(-v)\\
&= 2\sin\left(\frac{u+(-v)}{2}\right)\cos\left(\frac{u-(-v)}{2}\right)\\
&= 2\sin\left(\frac{u-v}{2}\right)\cos\left(\frac{u+v}{2}\right)\\
&= 2\cos\left(\frac{u+v}{2}\right)\sin\left(\frac{u-v}{2}\right).
\end{aligned}$$

17. For the sine, we have $\sin 2t = \sin(t+t) = \sin t\cos t + \sin t\cos t = 2\sin t\cos t$. This is the double angle formula for sine.

For cosine, we have $\cos 2t = \cos(t+t) = \cos t\cos t - \sin t\sin t = \cos^2 t - \sin^2 t$. This is the double angle formula for cosine.

18. We have

$$\begin{aligned}
\cos 3t &= \cos(2t+t)\\
&= \cos 2t\cos t - \sin 2t\sin t\\
&= (2\cos^2 t - 1)\cos t - (2\sin t\cos t)\sin t\\
&= \cos t((2\cos^2 t - 1) - 2\sin^2 t)\\
&= \cos t(2\cos^2 t - 1 - 2(1-\cos^2 t))\\
&= \cos t(2\cos^2 t - 1 - 2 + 2\cos^2 t)\\
&= \cos t(4\cos^2 t - 3)\\
&= 4\cos^3 t - 3\cos t,
\end{aligned}$$

as required.

19. We can use the identity $\sin u + \sin v = 2\sin((u+v)/2)\cos((u-v)/2)$. If we put $u = 4x$ and $v = x$ then our equation becomes

$$\begin{aligned}
0 &= 2\sin\left(\frac{4x+x}{2}\right)\cos\left(\frac{4x-x}{2}\right)\\
&= 2\sin\left(\frac{5}{2}x\right)\cos\left(\frac{3}{2}x\right)
\end{aligned}$$

The product on the right-hand side of this equation will be equal to zero precisely when $\sin(5x/2) = 0$ or when $\cos(3x/2) = 0$. We will have $\sin(5x/2) = 0$ when $5x/2 = n\pi$, for n an integer, or in other words for $x = (2\pi/5)n$.

This will occur in the stated interval for $x = 2\pi/5$, $x = 4\pi/5$, $x = 6\pi/5$, and $x = 8\pi/5$. We will have $\cos(3x/2) = 0$ when $3x/2 = n\pi + \pi/2$, that is, when $x = (2n+1)\pi/3$. This will occur in the stated interval for $x = \pi/3$, $x = \pi$, and $x = 5\pi/3$. So the given expression is solved by

$$x = \frac{2\pi}{5}, \frac{4\pi}{5}, \frac{6\pi}{5}, \frac{8\pi}{5}, \frac{\pi}{3}, \pi, \text{and} \frac{5\pi}{3}.$$

20. We are to prove that

$$\cos((n+1)\theta) = (2\cos\theta)(\cos(n\theta)) - \cos((n-1)\theta).$$

This holds if, and only if,

$$\cos((n+1)\theta) + \cos((n-1)\theta) = (2\cos\theta)(\cos(n\theta)).$$

Using the sum of cosines formula,

$$\cos u + \cos v = 2\cos\frac{u+v}{2}\cos\frac{u-v}{2},$$

with $u = (n+1)\theta$ and $v = (n-1)\theta$ gives the result:

$$\begin{aligned}\cos((n+1)\theta) + 2\cos((n-1)\theta) &= 2\cos\frac{(n+1)\theta + (n-1)\theta}{2}\cos\frac{(n+1)\theta - (n-1)\theta}{2}\\ &= 2\cos(n\theta)\cos\theta.\end{aligned}$$

21. We manipulate the equation for the average rate of change as follows:

$$\begin{aligned}\frac{\cos(x+h) - \cos x}{h} &= \frac{\cos x\cos h - \sin x\sin h - \cos x}{h}\\ &= \frac{\cos x\cos h - \cos x}{h} - \frac{\sin x\sin h}{h}\\ &= \cos x\left(\frac{\cos h - 1}{h}\right) - \sin x\left(\frac{\sin h}{h}\right).\end{aligned}$$

22. We manipulate the equation for the average rate of change as follows:

$$\begin{aligned}\frac{\sin(x+h) - \sin x}{h} &= \frac{\sin x\cos h + \sin h\cos x - \sin x}{h}\\ &= \frac{\sin x\cos h - \sin x}{h} + \frac{\sin h\cos x}{h}\\ &= \sin x\left(\frac{\cos h - 1}{h}\right) + \cos x\left(\frac{\sin h}{h}\right).\end{aligned}$$

23. We manipulate the equation for the average rate of change as follows:

$$\begin{aligned}\frac{\tan(x+h) - \tan x}{h} &= \frac{\dfrac{\tan x + \tan h}{1 - \tan x\tan h} - \tan x}{h}\\ &= \frac{(\tan x + \tan h - \tan x + \tan^2 x\tan h)/(1 - \tan x\tan h)}{h}\\ &= \frac{\tan h + \tan^2 x\tan h}{(1 - \tan x\tan h)\cdot h}\\ &= \frac{\dfrac{\sin h}{\cos h} + \tan^2 x\cdot\dfrac{\sin h}{\cos h}}{\left(1 - \tan x\cdot\dfrac{\sin h}{\cos h}\right)\cdot h}\end{aligned}$$

$$= \frac{(1+\tan^2 x)\dfrac{\sin h}{\cos h}}{\left(1-\tan x\dfrac{\sin h}{\cos h}\right)\cdot h}$$

$$= \frac{\left(\dfrac{1}{\cos^2 x}\right)\cdot\dfrac{\sin h}{\cos h}}{\left(1-\tan x\cdot\dfrac{\sin h}{\cos h}\right)\cdot h}$$

$$= \frac{\dfrac{1}{\cos^2 x}\cdot\sin h}{(\cos h-\tan x\sin h)\cdot h}$$

$$= \frac{\dfrac{1}{\cos^2 x}\cdot\sin h}{\cos h-\sin h\tan x}\cdot\left(\frac{1}{h}\right)$$

$$= \frac{1}{\cos^2 x}\frac{\sin h}{h}\cdot\frac{1}{\cos h-\sin h\tan x}.$$

24. **(a)** From $\cos 2u = 2\cos^2 u - 1$, we obtain $\cos u = \pm\sqrt{\dfrac{1+\cos 2u}{2}}$ and letting $u = \frac{v}{2}$, $\cos\dfrac{v}{2} = \pm\sqrt{\dfrac{1+\cos v}{2}}$.

(b) From $\tan\dfrac{1}{2}v = \dfrac{\sin\frac{1}{2}v}{\cos\frac{1}{2}v} = \dfrac{\pm\sqrt{\dfrac{1-\cos v}{2}}}{\pm\sqrt{\dfrac{1+\cos v}{2}}}$ we simplify to get $\tan\dfrac{1}{2}v = \pm\sqrt{\dfrac{1-\cos v}{1+\cos v}}$.

(c) The sign of $\sin\frac{1}{2}v$ is $+$, the sign of $\cos\frac{1}{2}v$ is $-$, and the sign of $\tan\frac{1}{2}v$ is $-$.

(d) The sign of $\sin\frac{1}{2}v$ is $-$, the sign of $\cos\frac{1}{2}v$ is $-$, and the sign of $\tan\frac{1}{2}v$ is $+$.

(e) The sign of $\sin\frac{1}{2}v$ is $-$, the sign of $\cos\frac{1}{2}v$ is $+$, and the sign of $\tan\frac{1}{2}v$ is $-$.

25. **(a)** Since ΔCAD and ΔCDB are both right triangles, it is easy to calculate the sine and cosine of their angles:

$$\sin\theta = \frac{c_1}{b}$$
$$\cos\theta = \frac{h}{b}$$
$$\sin\phi = \frac{c_2}{a}$$
$$\cos\phi = \frac{h}{a}.$$

(b) We can calculate the areas of the triangles using the formula Area $=$ Base $\cdot$ Height:

$$\text{Area } \Delta CAD = \frac{1}{2}c_1\cdot h = \frac{1}{2}(b\sin\theta)(a\cos\phi),$$
$$\text{Area } \Delta CDB = \frac{1}{2}c_2\cdot h = \frac{1}{2}(a\sin\phi)(b\cos\theta).$$

(c) We find the area of the whole triangle by summing the area of the two constituent triangles:

$$\text{Area } \Delta ABC = \text{Area } \Delta CAD + \text{Area } CDB = \frac{1}{2}(b\sin\theta)(a\cos\phi) + \frac{1}{2}(a\sin\phi)(b\cos\theta)$$

$$
\begin{aligned}
&= \frac{1}{2}ab(\sin\theta\cos\phi + \sin\phi\cos\theta) \\
&= \frac{1}{2}ab\sin(\theta+\phi) \\
&= \frac{1}{2}ab\sin C.
\end{aligned}
$$

26. **(a)** If B and C are acute angles, draw the altitude from A, dividing side a into two pieces, a_1 and a_2, as shown in Figure 9.11. Then $a_1 = b\cos C$ and $a_2 = c\cos B$, so $a = a_1 + a_2 = b\cos C + c\cos B$.

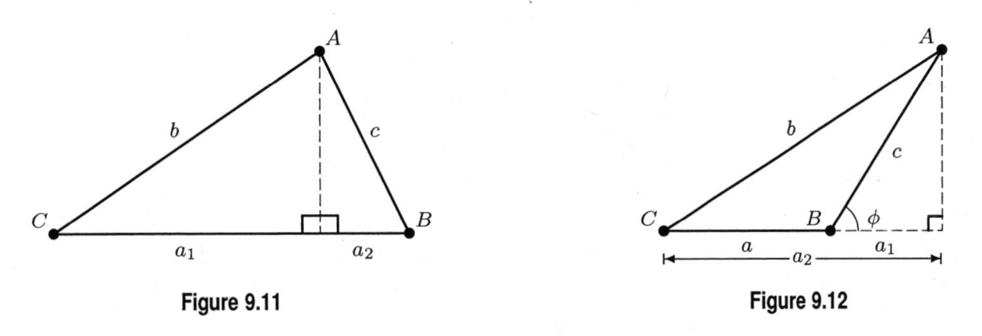

Figure 9.11 **Figure 9.12**

If one of the angles B or C (say B) is obtuse, draw the altitude from A as shown in Figure 9.12. Let a_1 be the extension of side a, so $a_2 = a_1 + a$, and let ϕ be the exterior angle at B, so that $\phi = 180^\circ - B$. Therefore $\cos\phi = -\cos B$. Then $a_2 = b\cos C$ and $a_1 = c\cos\phi = -c\cos B$. Thus $a = a_2 - a_1 = b\cos C - (-c\cos B) = b\cos C + c\cos B$.

(b) From the Law of Sines, $\frac{a}{\sin A} = \frac{b}{\sin B}$, so $b = \frac{a}{\sin A}\cdot\sin B$, and similarly $c = \frac{a}{\sin A}\cdot\sin C$. Substituting these expressions into the result of part (a), we have

$$a = b\cos C + c\cos B = \frac{a}{\sin A}\sin B\cos C + \frac{a}{\sin A}\sin C\cos B,$$

or

$$a = \frac{a}{\sin A}\left(\sin B\cos C + \sin C\cos B\right).$$

Thus $\sin A = \sin B\cos C + \sin C\cos B$. But $A + B + C = 180^\circ$, so $A = 180^\circ - (B+C)$, and hence $\sin A = \sin(B+C)$. Therefore,

$$\sin(B+C) = \sin B\cos C + \sin C\cos B.$$

27. **(a)** The coordinates of P_1 are $(\cos\theta, \sin\theta)$; for P_2 they are $(\cos(-\phi), \sin(-\phi)) = (\cos\phi, -\sin\phi)$; for P_3 they are $(\cos(\theta+\phi), \sin(\theta+\phi))$; and for P_4 they are $(1, 0)$.

(b) The triangles P_1OP_2 and P_3OP_4 are congruent by the side-angle-side property because $\angle P_1OP_2 = \theta + \phi = \angle P_3OP_4$. Therefore their corresponding sides P_1P_2 and P_3P_4 are equal.

(c) We have

$$
\begin{aligned}
(P_1P_2)^2 &= (\cos\theta - \cos\phi)^2 + (\sin\theta + \sin\phi)^2 \\
&= \cos^2\theta - 2\cos\theta\cos\phi + \cos^2\phi + \sin^2\theta + 2\sin\theta\sin\phi + \sin^2\phi \\
&= \cos^2\theta + \sin^2\theta + \cos^2\phi + \sin^2\phi - 2\cos\theta\cos\phi + 2\sin\theta\sin\phi \\
&= 2 - 2(\cos\theta\cos\phi - \sin\theta\sin\phi)
\end{aligned}
$$

We also have

$$
\begin{aligned}
(P_3P_4)^2 &= (\cos(\theta+\phi) - 1)^2 + (\sin(\theta+\phi) - 0)^2 \\
&= \cos^2(\theta+\phi) - 2\cos(\theta+\phi) + 1 + \sin^2(\theta+\phi) \\
&= 2 - 2\cos(\theta+\phi)
\end{aligned}
$$

The distances P_1P_2 and P_3P_4 are the square roots of these expressions (but we will use the squares of the distances).

(d) $(P_3P_4)^2 = (P_1P_2)^2$ by part (b), so

$$2 - 2\cos(\theta+\phi) = 2 - 2(\cos\theta\cos\phi - \sin\theta\sin\phi)$$
$$\cos(\theta+\phi) = \cos\theta\cos\phi - \sin\theta\sin\phi.$$

28. Using the exponent rules, we see from Euler's formula that

$$\begin{aligned} e^{i(\theta+\phi)} &= e^{i\theta}\cdot e^{i\phi} \\ &= (\cos\theta + i\sin\theta)(\cos\phi + i\sin\phi) \\ &= \cos\theta\cos\phi + \underbrace{i\cos\theta\sin\phi + i\sin\theta\cos\phi}_{i(\cos\theta\sin\phi + \sin\theta\cos\phi)} + \underbrace{i^2\sin\theta\sin\phi}_{-\sin\theta\sin\phi} \\ &= \underbrace{\cos\theta\cos\phi - \sin\theta\sin\phi}_{\text{Real part}} + i\underbrace{(\sin\theta\cos\phi + \cos\theta\sin\phi)}_{\text{Imaginary part}}. \end{aligned}$$

But Euler's formula also gives

$$e^{i(\theta+\phi)} = \underbrace{\cos(\theta+\phi)}_{\text{Real part}} + i\underbrace{\sin(\theta+\phi)}_{\text{Imaginary part}}.$$

Two complex numbers are equal only if their real and imaginary parts are equal. Setting real parts equal gives

$$\cos(\theta+\phi) = \cos\theta\cos\phi - \sin\theta\sin\phi.$$

Setting imaginary parts equal gives

$$\sin(\theta+\phi) = \sin\theta\cos\phi + \sin\phi\cos\theta.$$

Solutions for Section 9.3

Exercises

1.

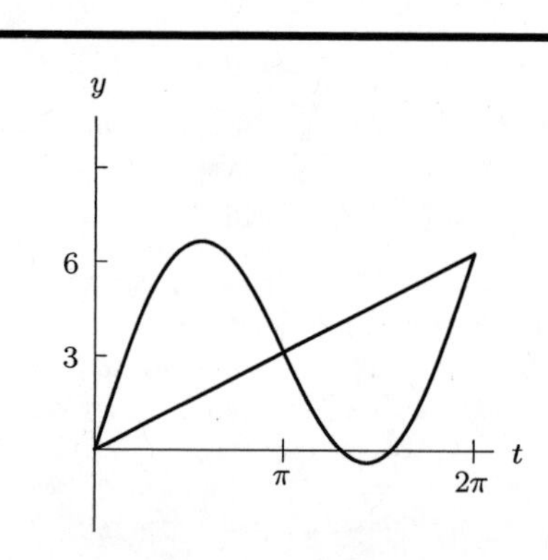

For the graphs to intersect, $t + 5\sin t = t$. So $\sin t = 0$, or $t =$ any integer multiple of π.

2.

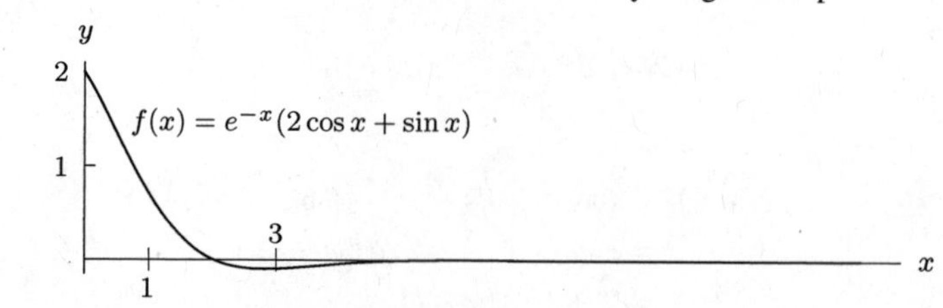

The maximum value of $f(x)$ is 2, which occurs when $x = 0$. The minimum appears to be $y \approx -0.094$, at $x \approx 2.820$.

3. **(a)** $P = 5000 + 300t$.
(b) $P = 3200(1.04)^t$.
(c) Since the population starts at a low value, it is a "reflected" cosine function of the form

$$P(t) = -A\cos(Bt) + k.$$

Low $= 1200$; High $= 3000$.
Midline $k = \frac{1}{2}(3000 + 1200) = 2100$.
Amplitude $A = 3000 - 2100 = 900$.
The period is 5, so $5 = 2\pi/B$, and we have $B = 2\pi/5$. Thus,

$$P(t) = -900\cos\left(\frac{2\pi}{5}t\right) + 2100.$$

4. **(a)** Substituting into f, we have

$$\begin{aligned} f(3) - f(2) &= \left(10{,}000 - 5000\cos\left(\frac{\pi}{6}\cdot 3\right)\right) - \left(10{,}000 - 5000\cos\left(\frac{\pi}{6}\cdot 2\right)\right) \\ &= -5000\cos\frac{\pi}{2} + 5000\cos\frac{\pi}{3} \\ &= 0 + 5000\left(\frac{1}{2}\right) = 2500. \end{aligned}$$

This means that the rabbit population increases by 2500 rabbits between March 1 (at $t = 2$) and April 1 (at $t = 3$).
(b) Using a graphing calculator to zoom in, or by using the inverse cosine function, we see that $f(t) = 12{,}000$ for $t = 3.786$ and $t = 8.214$. This means that the rabbit population reaches 12,000 sometime during late April (at $t = 3.786$) and falls back to 12,000 sometime during early September (at $t = 8.214$).

Problems

5. **(a)** We start the time count on Jan 1, so substituting $t = 0$ into $f(t)$ gives us the value of b, since both mt and $A\sin\frac{\pi t}{6}$ are equal to zero when $t = 0$. Thus, $b = f(0) = 20$. We see that in the 12 month period between Jan 1 and Jan 1 the value of the stock rose by $30.00. Therefore, the linear component grows at the rate of $30.00/year, or in terms of months, $30/12 =$ $2.50/month. So $m = 2.5$. Thus we have

$$P = f(t) = 2.5t + 20 + A\sin\frac{\pi t}{6}.$$

At an arbitrary data point, say $(Apr1, 37.50)$, we can solve for A. Since January 1 corresponds to $t = 0$, April 1 is $t = 3$. We have

$$37.50 = f(3) = 2.5(3) + 20 + A\sin\frac{3\pi}{6} = 7.5 + 20 + A\sin\frac{\pi}{2} = 27.5 + A.$$

Simplifying gives $A = 10$, and the function is

$$f(t) = 2.5t + 20 + 10\sin\frac{\pi t}{6}.$$

(b) The stock appreciates the most during the months when the sine function climbs the fastest. By looking at Figure 9.13 we see that this occurs roughly when $t = 0$ and $t = 11$, January and December.

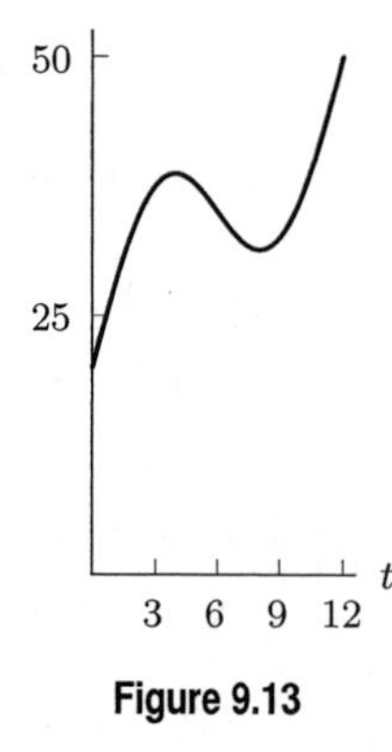

Figure 9.13

(c) Again, we look to Figure 9.13 to see when the graph actually decreases. It seems that the graph is decreasing roughly between the fourth and eighth months, that is, between May and September.

6. (a) We know that the minimum of $f(t)$ is 40 and that the maximum is 90. Thus the midline height is

$$\frac{90+40}{2} = 65$$

and the amplitude is

$$90 - 65 = 25.$$

We also know that $f(t)$ has a period of 24 hours and is at a minimum when $t = 0$. Thus a formula for $f(t)$ is

$$f(t) = 65 - 25\cos\left(\frac{\pi}{12}t\right).$$

(b) The amplitude is 30. The period is 24 hours, so the pattern repeats itself each day. The midline value is 80. So $g(t)$ goes up to a maximum of $80 + 30 = 110$ and down to a low of $80 - 30 = 50$ megawatts.

(c) From the graph in Figure 9.14, we can find that $t_1 \approx 4.160$ and $t_2 \approx 13.148$. Thus, the power required in both cities is the same at approximately 4 am and 1 pm.

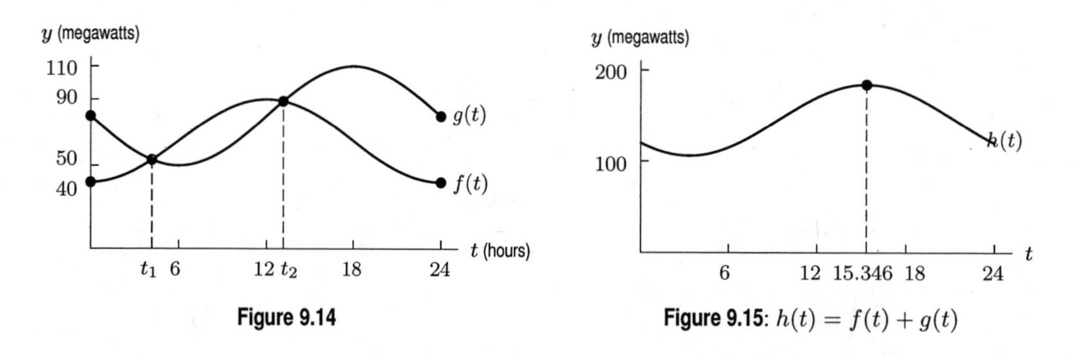

Figure 9.14

Figure 9.15: $h(t) = f(t) + g(t)$

(d) The function $h(t)$ in Figure 9.15 tells us the total amount of electricity required by both cities at a particular time of day. Using the trace key on a calculator, we find the maximum occurs at $t = 15.346$. So at 3:21 pm each day the most total power will be needed. The maximum power is 184.051 mw.

(e) Add the two functions to obtain

$$h(t) = 145 - \left(30\sin\left(\frac{\pi}{12}t\right) + 25\cos\left(\frac{\pi}{12}t\right)\right)$$

which can be written in the form

$$h(t) = 145 - A\sin\left(\frac{\pi}{12}t + \theta\right)$$

where $A = \sqrt{30^2 + 25^2} = \sqrt{1525} \approx 39.051$ and $\theta = \tan^{-1}(25/30) \approx 0.6947$. Thus

$$h(t) \approx 145 - 39.051 \sin\left(\frac{\pi}{12}t + 0.6947\right).$$

The function has an exact maximum of 145 + $\sqrt{1525}$ (about 184).

7. (a) As $t \to -\infty$, e^t gets very close to 0, which means that $\cos(e^t)$ gets very close to $\cos 0 = 1$. This means that the horizontal asymptote of f as $t \to -\infty$ is $y = 1$.

(b) As t increases, e^t increases at a faster and faster rate. We also know that as θ increases, $\cos\theta$ varies steadily between -1 and 1. But since e^t is increasing faster and faster as t gets large, $\cos(e^t)$ will vary between -1 and 1 at a faster and faster pace. So the graph of $f(t) = \cos(e^t)$ begins to wiggle back and forth between -1 and 1, faster and faster. Although f is oscillating between -1 and 1, f is not periodic, because the interval on which f completes a full cycle is not constant.

(c) The vertical axis is crossed when $t = 0$, so $f(0) = \cos(e^0) = \cos 1 \approx 0.540$ is the vertical intercept.

(d) Notice that the least positive zero of $\cos u$ is $u = \pi/2$. Thus the least zero t_1 of $f(t) = \cos(e^t)$ occurs where $e^{t_1} = \pi/2$ since e^t is always positive. So we have

$$e^{t_1} = \frac{\pi}{2}$$
$$t_1 = \ln\frac{\pi}{2}.$$

(e) We know that if $\cos u = 0$, then $\cos(u + \pi) = 0$. This means that if $\cos(e^{t_1}) = 0$, then $\cos\left(e^{t_1} + \pi\right) = 0$ will be the first zero of f coming after t_1. Therefore,

$$f(t_2) = \cos(e^{t_2}) = \cos\left(e^{t_1} + \pi\right) = 0.$$

This means that $e^{t_2} = e^{t_1} + \pi$. So

$$t_2 = \ln\left(e^{t_1} + \pi\right) = \ln(e^{\ln(\pi/2)} + \pi) = \ln\left(\frac{\pi}{2} + \pi\right) = \ln\left(\frac{3\pi}{2}\right).$$

Similar reasoning shows that the set of all zeros is $\{\ln(\frac{\pi}{2}), \ln(\frac{3\pi}{2}), \ln(\frac{5\pi}{2})...\}$.

8. (a) As $x \to \infty$, $\frac{1}{x} \to 0$ and we know $\sin 0 = 0$. Thus, $y = 0$ is the equation of the asymptote.

(b) As $x \to 0$ and $x > 0$, we have $\frac{1}{x} \to \infty$. This means that for small changes of x the change in $\frac{1}{x}$ is large. Since $\frac{1}{x}$ is a large number of radians, the function will oscillate more and more frequently as x becomes smaller.

(c) No, because the interval on which $f(x)$ completes a full cycle is not constant as x increases.

(d) $\sin\left(\frac{1}{x}\right) = 0$ means that $\frac{1}{x} = \sin^{-1}(0) + k\pi$ for k equal to some integer. Therefore, $x = \frac{1}{k\pi}$, and the greatest zero of $f(x) = \sin\frac{1}{x}$ corresponds to the smallest k, that is, $k = 1$. Thus, $z_1 = \frac{1}{\pi}$.

(e) There are an infinite number of zeros because $z = \frac{1}{k\pi}$ for all $k > 0$ are zeros.

(f) If $a = \frac{1}{k\pi}$ then the largest zero of $f(x)$ less then a would be $b = \frac{1}{(k+1)\pi}$.

9. Use the sum-of-angles identity for cosine on the second term to get

$$I\cos(\omega_c + \omega_d)t = I\cos\omega_c t\cos\omega_d t - I\sin\omega_c t\sin\omega_d t$$

and factor the term $\cos\omega_c t$ to show the equality.

10. (a) Types of video games are trendy for a length of time, during which they are extremely popular and sales are high, later followed by a cooling down period as the users become tired of that particular game type. The game players then become interested in a different game type — and so on.

(b) The sales graph does not fit the shape of the sine or cosine curve, and we would have to say that neither of those functions would give us a reasonable model. However from 1979–1989 the graph does have a basic negative cosine shape, but the amplitude varies.

(c) One way to modify the amplitude over time is to multiply the sine (or cosine) function by an exponential function, such as e^{kt}. So we choose a model of the form

$$s(t) = e^{kt}(-a\cos(Ct) + D),$$

where t is the number of years since 1979. Note the $-a$, which is due to the graph looking like an inverted cosine at 1979. The average value starts at about 1.6, and the period appears to be about 6 years. The amplitude is initially about 1.4, which is the distance between the average value of 1.6 and the first peak value of 3.0. This means

$$s(t) = e^{kt}\left(-1.4\cos\left(\frac{2\pi}{6}t\right) + 1.6\right).$$

By trial and error on your graphing calculator, you can arrive at a value for the parameter k. A reasonable choice is $k = 0.05$, which gives

$$s(t) = e^{0.05t}\left(-1.4\cos\left(\frac{2\pi}{6}t\right) + 1.6\right).$$

(d)

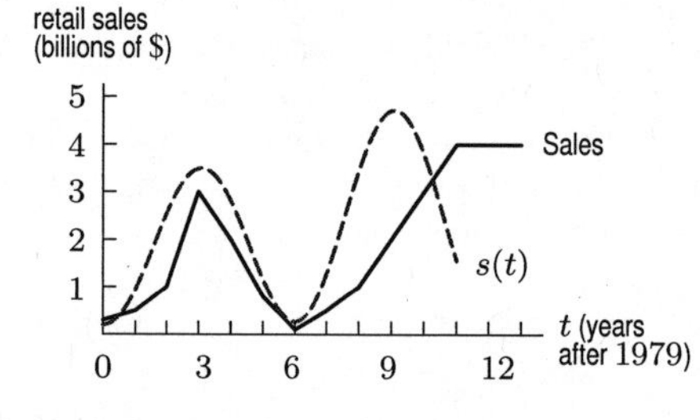

Notice that even though multiplying by the exponential function does increase the amplitude over time, it does not increase the period. Therefore, our model $s(t)$ does not fit the actual curve all that well.

(e) The predicted 1993 sales volume is $f(14) = 4.632$ billion dollars.

11. (a) First consider the height $h = f(t)$ of the hub of the wheel that you are on. This is similar to the basic Ferris wheel problem. Therefore $f_1(t) = 25 + 15\sin\left(\frac{\pi}{3}t\right)$, because the vertical shift is 25, the amplitude is 15, and the period is 6. Now the smaller wheel will also add or subtract height depending upon time. The difference in height between your position and the hub of the smaller wheel is given by $f_2(t) = 10\sin\left(\frac{\pi}{2}t\right)$ because the radius is 10 and the period is 4. Finally adding the two together we get:

$$f_1(t) + f_2(t) = f(t) = 25 + 15\sin(\frac{\pi}{3}t) + 10\sin(\frac{\pi}{2}t)$$

(b)

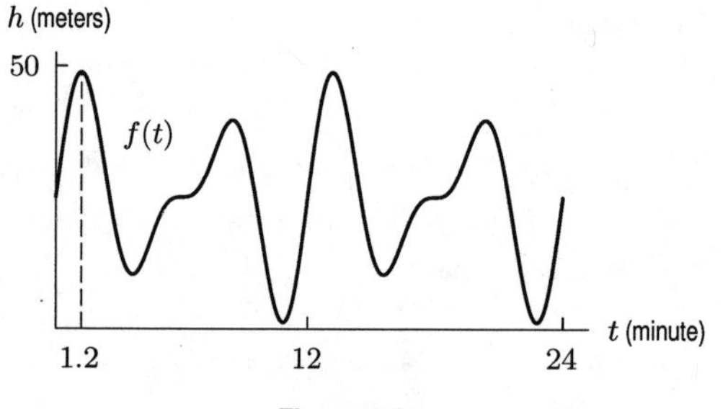

Figure 9.16

Looking at the graph shown in Figure 9.16, we see that $h = f(t)$ is periodic, with period 12. This can be verified by noting

$$\begin{aligned} f(t+12) &= 25 + 15\sin\left(\frac{\pi}{3}(t+12)\right) + 10\sin\left(\frac{\pi}{2}(t+12)\right) \\ &= 25 + 15\sin\left(\frac{\pi}{3}t + 4\pi\right) + 10\sin\left(\frac{\pi}{2}t + 6\pi\right) \\ &= 25 + 15\sin\left(\frac{\pi}{3}t\right) + 10\sin\left(\frac{\pi}{2}t\right) \\ &= f(t). \end{aligned}$$

(c) $h = f(1.2) = 48.776$ m.

12. **(a)** We have $\lambda = \frac{2\pi}{k} = \frac{2\pi}{2\pi} = 1$. Thus, the wavelength is 1 meter.

(b) The time for one wavelength to pass by is $\frac{2\pi}{\omega} = \frac{2\pi}{4\pi} = \frac{1}{2}$ of a second. Thus, two wavelengths pass by each second. The number of wavelengths which pass a point in a given unit of time is referred to as the frequency. It is sometimes written as 2 hertz (Hz), which equals 2 cycles per second.

(c) $y(x, 0) = 0.06\sin(2\pi x)$

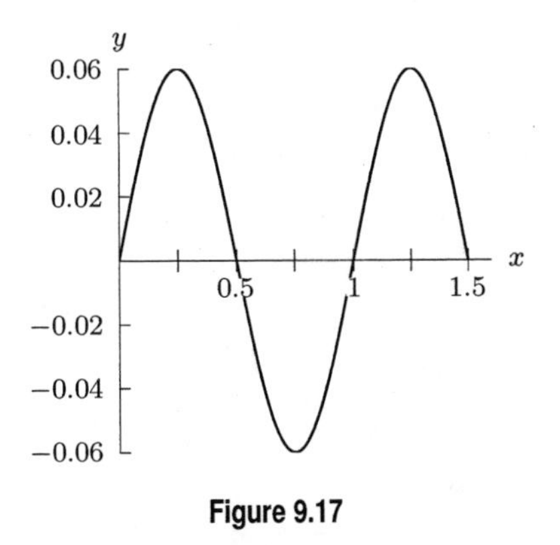

Figure 9.17

(d) Having the same graph means that $0.06\sin(2\pi x) = 0.06\sin(2\pi x - 4\pi t)$ for all x. In order for this to happen, $2\pi x$ and $2\pi x - 4\pi t$ must describe the same angle. This is the case any time $4\pi t$ is a multiple of 2π. Thus, $4\pi t = 2\pi k$ whenever $t = k/2$ for any integer k.

Solutions for Chapter 9 Review

Exercises

1. We have:

$$\begin{aligned}(1 - \sin t)(1 - \cos t) - \cos t \sin t &= \underbrace{1 - \cos t - \sin t + \sin t \cos t}_{(1-\sin t)(1-\cos t)} - \cos t \sin t \quad \text{multiply out}\\ &= 1 - \cos t - \sin t.\end{aligned}$$

2. We have:

$$\begin{aligned}2\cos t - 3\sin t - (2\sin t - 3\cos t) &= 2\cos t - 3\sin t - 2\sin t + 3\cos t\\ &= 5\cos t - 5\sin t.\end{aligned}$$

3. We have:

$$\begin{aligned}\sec t \sin t + 3\tan t &= \frac{1}{\cos t} \cdot \sin t + 3\tan t\\ &= \underbrace{\frac{\sin t}{\cos t}}_{\tan t} + 3\tan t\\ &= 4\tan t.\end{aligned}$$

4. We have:

$$\begin{aligned}\cot t\tan t\sin t &= \underbrace{\left(\frac{\cos t}{\sin t}\right)}_{\cot t}\cdot\underbrace{\left(\frac{\sin t}{\cos t}\right)}_{\tan t}\cdot\sin t\\ &= \sin t.\end{aligned}$$

5. We have:

$$\begin{aligned}\frac{\sec t}{\csc t} &= \frac{\left(\frac{1}{\cos t}\right)}{\left(\frac{1}{\sin t}\right)}\\ &= \frac{1}{\cos t}\cdot\sin t\\ &= \tan t.\end{aligned}$$

6. We have:

$$\begin{aligned}\frac{\cot t}{\csc t} &= \frac{\left(\frac{\cos t}{\sin t}\right)}{\left(\frac{1}{\sin t}\right)}\\ &= \frac{\cos t}{\sin t}\cdot\sin t\\ &= \cos t.\end{aligned}$$

7. We have:

$$\begin{aligned}(\sec t\cot t-\csc t\tan t)\sin t\cos t &= \sec t\cot t\sin t\cos t-\csc t\tan t\sin t\cos t \qquad \text{multiply out}\\ &= \underbrace{\left(\frac{1}{\cos t}\right)}_{\sec t}\underbrace{\left(\frac{\cos t}{\sin t}\right)}_{\cot t}\sin t\cos t-\underbrace{\left(\frac{1}{\sin t}\right)}_{\csc t}\underbrace{\left(\frac{\sin t}{\cos t}\right)}_{\tan t}\sin t\cos t\\ &= \cos t-\sin t.\end{aligned}$$

8. We have:

$$\begin{aligned}\frac{1}{2\csc t\cos t-3\cot t} &= \frac{1}{2\underbrace{\left(\frac{1}{\sin t}\right)}_{\csc t}\cos t-3\cot t}\\ &= \frac{1}{2\frac{\cos t}{\sin t}-3\underbrace{\left(\frac{\cos t}{\sin t}\right)}_{\cot t}}\\ &= \frac{1}{-\left(\frac{\cos t}{\sin t}\right)}=-\frac{\sin t}{\cos t}=-\tan t.\end{aligned}$$

9. Using $1-\cos^2\theta=\sin^2\theta$, we have

$$\frac{1-\cos^2\theta}{\sin\theta}=\frac{\sin^2\theta}{\sin\theta}=\sin\theta.$$

10. Writing $\cot x=\cos x/\sin x$ and $\csc x=1/\sin x$, we have

$$\frac{\cot x}{\csc x}=\frac{\cos x}{\sin x}\cdot\frac{1}{1/\sin x}=\cos x.$$

11. Writing $\cos 2\phi = 2\cos^2\phi - 1$, we have

$$\frac{\cos 2\phi + 1}{\cos\phi} = \frac{2\cos^2\phi - 1 + 1}{\cos\phi} = \frac{2\cos^2\phi}{\cos\phi} = 2\cos\phi.$$

12. Multiplying out and using the fact that $1 - \tan^2\theta = 1/\cos^2\theta$, we have

$$\cos^2\theta(1+\tan\theta)(1-\tan\theta) = \cos^2\theta(1-\tan^2\theta) = \cos^2\theta \cdot \frac{1}{\cos^2\theta} = 1.$$

Problems

13. They are both right. The first student meant that $\sin 2\theta = 2\sin\theta$ is not an identity meaning that it is not true for *all* θ. The second student had found one value for θ for which it was true.

14. Graphs of the four functions are in Figures 9.18 –9.21. The graphs in Figures9.19 and 9.20 suggest that $(\tan^2 x)(\sin^2 x)$ and $\tan^2 x - \sin^2 x$ may be identical.

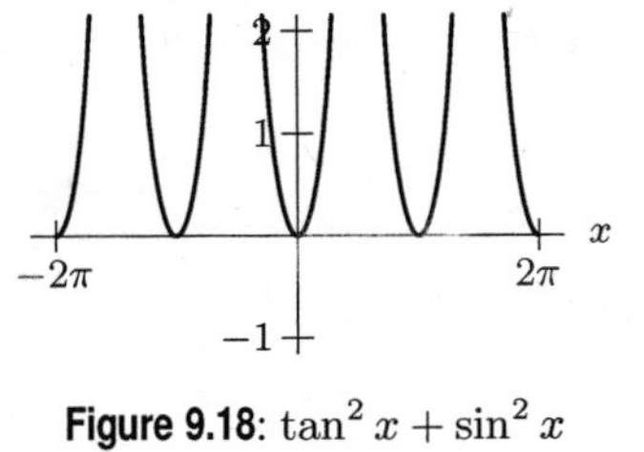

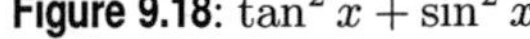
Figure 9.18: $\tan^2 x + \sin^2 x$

Figure 9.19: $(\tan^2 x)(\sin^2 x)$

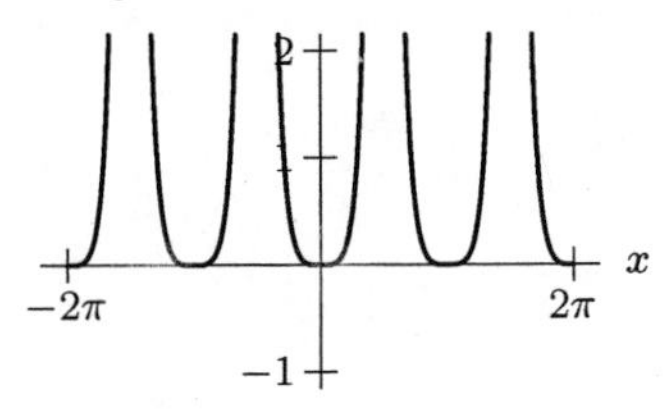

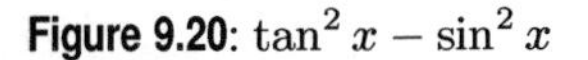
Figure 9.20: $\tan^2 x - \sin^2 x$

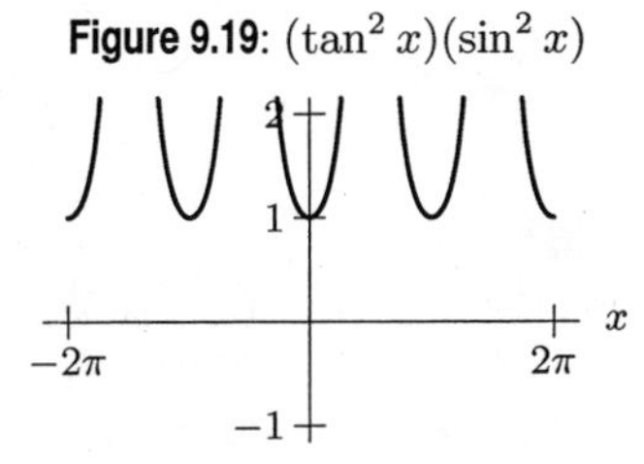

Figure 9.21: $\tan^2 x/\sin^2 x$

To prove the identity, use $\tan x = \sin x/\cos x$ to rewrite each side in terms of sine and cosine. We have

$$(\tan^2 x)(\sin^2 x) = \left(\frac{\sin x}{\cos x}\right)^2 (\sin^2 x) = \frac{\sin^4 x}{\cos^2 x}.$$

In addition,

$$\begin{aligned}
\tan^2 x - \sin^2 x &= \left(\frac{\sin x}{\cos x}\right)^2 - \sin^2 x \\
&= \frac{\sin^2 x}{\cos^2 x} - \sin^2 x \\
&= \sin^2 x\left(\frac{1}{\cos^2 x} - 1\right) \\
&= \frac{\sin^2 x(1-\cos^2 x)}{\cos^2 x} \\
&= \frac{\sin^2 x(\sin^2 x)}{\cos^2 x} \\
&= \frac{\sin^4 x}{\cos^2 x}.
\end{aligned}$$

15. **(a)** $\tan\theta = \dfrac{\text{opp}}{\text{adj}} = \dfrac{y}{1} = y.$

(b) $\cos\theta = \dfrac{\text{opp}}{\text{hyp}} = \dfrac{y}{\sqrt{1+y^2}}.$

(c) Since $\tan\theta = y$, we have $\theta = \tan^{-1} y$. Other answers are possible.

(d) $\cos(2\phi) = 2\cos\phi\sin\phi = 2\left(\dfrac{y}{\sqrt{1+y^2}}\right)\left(\dfrac{1}{\sqrt{1+y^2}}\right) = \dfrac{2y}{1+y^2}.$

16. Since $\csc\theta = 1/\sin\theta$, we have $\csc\theta = 1/(8/11) = 11/8$. Since $\cos^2\theta + \sin^2\theta = 1$,

$$\cos^2\theta + \left(\frac{8}{11}\right)^2 = 1$$
$$\cos^2\theta = 1 - \frac{64}{121}$$
$$\cos\theta = \pm\sqrt{\frac{57}{121}}.$$

Since $0 \le \theta \le \pi/2$, we know that $\cos\theta \ge 0$, so $\cos\theta = \sqrt{57/121}$. Since $\tan\theta = \sin\theta/\cos\theta$, we have $\tan\theta = (8/11)/\sqrt{57/121} = 8\sqrt{121}/11\sqrt{57} = 8/\sqrt{57}$.

17. Since $\sin\theta = 1/\csc\theta$, we have $\sin\theta = 1/94$. Using the Pythagorean Identity, $\cos^2\theta + \sin^2\theta = 1$, we have

$$\cos^2\theta + \left(\frac{1}{94}\right)^2 = 1$$
$$\cos^2\theta = 1 - \frac{1}{94^2}$$
$$\cos\theta = \pm\sqrt{\frac{8835}{8836}} = \pm\frac{\sqrt{8835}}{94}.$$

Since $0 \le \theta \le \pi/2$, we know that $\cos\theta \ge 0$, so $\cos\theta = \sqrt{8835}/94$.

Using the identity $\tan\theta = \sin\theta/\cos\theta$, we see that $\tan\theta = (1/94)/(\sqrt{8835}/94) = 1/\sqrt{8835}$.

18. By the Pythagorean identity, we know that $\cos^2\theta + \sin^2\theta = 1$, so

$$0.27^2 + \sin^2\theta = 1$$
$$\sin^2\theta = 1 - 0.27^2$$
$$\sin\theta = \pm\sqrt{0.927} \approx \pm 0.963.$$

However, we only need one value, so we take $\sin\theta = 0.963$, Since $\tan\theta = \sin\theta/\cos\theta$, we have $\tan\theta = \sqrt{0.927}/0.27 \approx 3.566$.

19. No. Since $\sin\theta$ cannot be greater than 1, $\sin\theta = 3$ is impossible. The problem is that $\dfrac{\sin\theta}{\cos\theta}$ is a ratio, so $\sin\theta$ could be 0.3 and $\cos\theta$ could be 0.4. We only know the ratio is $3/4$.

20. We have $2/7 = \cos 2\theta = 2\cos^2\theta - 1$. Solving for $\cos\theta$ gives

$$2\cos^2\theta = \frac{9}{7}$$
$$\cos^2\theta = \frac{9}{14}$$

Since θ is in the first quadrant, $\cos\theta = +\sqrt{9/14} = 3/\sqrt{14}$.

21. We will use one of the three double angle formulas for cosine to solve this equation algebraically. Since the equation involves $\sin\theta$, we will try the double angle formula for cosine which involves the sine:

$$\cos 2\theta = 1 - 2\sin^2\theta.$$

This gives

$$1 - 2\sin^2\theta = \sin\theta,$$

and we can rewrite this equation as follows:

$$2(\sin\theta)^2 + \sin\theta - 1 = 0.$$

Factoring, we have

$$(2\sin\theta - 1)(\sin\theta + 1) = 0.$$

The fact that the product $(2\sin\theta - 1)(\sin\theta + 1)$ equals zero implies that

$$2\sin\theta - 1 = 0 \quad \text{or} \quad \sin\theta + 1 = 0.$$

Solving $\sin\theta + 1 = 0$, we have $\sin\theta = -1$. This means $\theta = 3\pi/2$. Solving the other equation, we have

$$\begin{aligned} 2\sin\theta - 1 &= 0 \\ \sin\theta &= \frac{1}{2}. \end{aligned}$$

This gives

$$\theta = \frac{\pi}{6} \quad \text{or} \quad \theta = \pi - \frac{\pi}{6} = \frac{5\pi}{6}.$$

In summary, there are three solutions for $0 \le \theta < 2\pi : \theta = \pi/6, 5\pi/6$, and $3\pi/2$. Figure 9.22 illustrates these solutions graphically as the points where the graphs of $\cos 2\theta$ and $\sin\theta$ intersect.

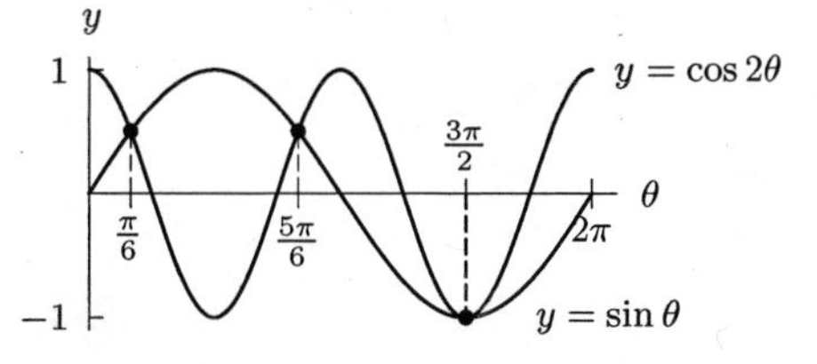

Figure 9.22: There are three solutions to the equation $\cos 2\theta = \sin\theta$, for $0 \le \theta < 2\pi$

22.

$$\begin{aligned} \cos(2\alpha) &= -\sin\alpha \\ 1 - 2\sin^2\alpha + \sin\alpha &= 0 \\ -2\sin^2\alpha + \sin\alpha + 1 &= 0 \\ 2\sin^2\alpha - \sin\alpha - 1 &= 0 \\ (2\sin\alpha + 1)(\sin\alpha - 1) &= 0 \end{aligned}$$

$$\begin{aligned} 2\sin\alpha + 1 &= 0 \\ 2\sin\alpha &= -1 \\ \sin\alpha &= -\frac{1}{2} \\ \alpha &= \frac{7\pi}{6}, \frac{11\pi}{6} \end{aligned} \qquad \begin{aligned} \sin\alpha - 1 &= 0 \\ \sin\alpha &= 1 \\ \alpha &= \frac{\pi}{2} \end{aligned}$$

23. Let $\theta = \cos^{-1}(\frac{5}{13})$. We can use the Pythagorean theorem and create a triangle with sides 5, 12, and 13 as in Figure 9.23. We see that $\sin\theta = \frac{12}{13}$. However we want $\sin(2\theta)$, so we use the double angle identity $\sin(2\theta) = 2\sin\theta\cos\theta = 2(\frac{12}{13})(\frac{5}{13}) = \frac{120}{169}$. You can check this on a calculator by finding that $\sin(2\cos^{-1}(5/13)) \approx 0.7100591716$, which is $120/169$ in decimal form.

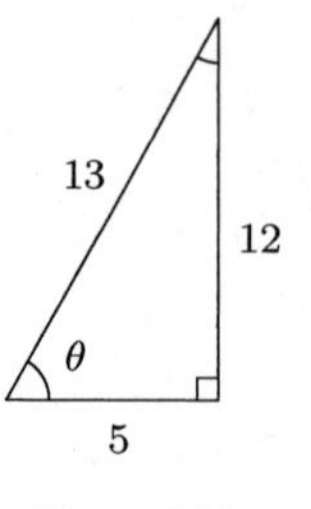

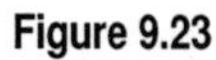
Figure 9.23

24. Start with the right side, which is more complex. Use a double angle identity, so both sides are expressed as trigonometric functions of θ.

$$\begin{aligned}\frac{1-\cos 2\theta}{2\cos\theta\sin\theta} &= \frac{1-(2\cos^2\theta-1)}{2\cos\theta\sin\theta}\\ &= \frac{2(1-\cos^2\theta)}{2\cos\theta\sin\theta}\\ &= \frac{2(\sin^2\theta)}{2\cos\theta\sin\theta}\\ &= \frac{\sin^2\theta}{\cos\theta\sin\theta}\\ &= \frac{\sin\theta}{\cos\theta}\\ &= \tan\theta.\end{aligned}$$

25. Start with the left side, which is more complex, and reduce it using the Pythagorean identity:

$$(\sin^2 2\pi t+\cos^2 2\pi t)^3 = (1)^3 = 1.$$

26. Start with the expression on the left and factor it as the difference of two squares, and then apply the Pythagorean identity to one factor.

$$\begin{aligned}\sin^4 x-\cos^4 x &= (\sin^2 x)^2-(\cos^2 x)^2\\ &= (\sin^2 x-\cos^2 x)(\sin^2 x+\cos^2 x)\\ &= (\sin^2 x-\cos^2 x)(1)\\ &= (\sin^2 x-\cos^2 x).\end{aligned}$$

27. Because both sides are equally complicated, many different approaches are possible. We multiply the right side by $(1+\sin\theta)/(1+\sin\theta)$ in order to introduce a factor of $1+\sin\theta$ to the numerator (the left side already has one) and to simplify the denominator. In the calculation which follows, we work only on the right side:

$$\begin{aligned}\frac{\cos\theta}{1-\sin\theta} &= \frac{\cos\theta}{1-\sin\theta}\cdot\frac{1+\sin\theta}{1+\sin\theta}\\ &= \frac{\cos\theta(1+\sin\theta)}{1-\sin^2\theta}\\ &= \frac{\cos\theta(1+\sin\theta)}{\cos^2\theta}\\ &= \frac{1+\sin\theta}{\cos\theta}.\end{aligned}$$

28. **(a)** $\cos\theta = x$.
(b) $\cos\left(\frac{\pi}{2} - \theta\right) = \sin\theta = \sqrt{1-\cos^2\theta} = \sqrt{1-x^2}$.
(c) $\tan^2\theta = \left(\frac{\sin\theta}{\cos\theta}\right)^2 = \left(\frac{\sqrt{1-x^2}}{x}\right)^2 = \frac{1-x^2}{x^2}$.
(d) $\sin(2\theta) = 2\sin\theta\cos\theta = 2\sqrt{1-x^2}(x)$.
(e) $\cos(4\theta) = 1 - 2\sin^2(2\theta)$. Now we use part (d):
$\cos(4\theta) = 1 - 2(2x\sqrt{1-x^2})^2 = 1 - 8x^2(1-x^2)$.
(f) $\sin(\cos^{-1}x) = \sin(\theta) = \sqrt{1-x^2}$.

29. We first solve for $\cos\alpha$,

$$\begin{aligned} 3\cos^2\alpha + 2 &= 3 - 2\cos\alpha \\ 3\cos^2\alpha + 2\cos\alpha - 1 &= 0 \\ (3\cos\alpha - 1)(\cos\alpha + 1) &= 0 \end{aligned}$$

$$\begin{aligned} 3\cos\alpha - 1 &= 0 \\ \cos\alpha &= \tfrac{1}{3}. \\ \alpha &= 1.231,\ 5.052 \end{aligned} \qquad\qquad \begin{aligned} \cos\alpha + 1 &= 0 \\ \cos\alpha &= -1 \\ \alpha &= \pi \end{aligned}$$

30. We first solve for $\sin\alpha$,

$$\begin{aligned} 3\sin^2\alpha + 3\sin\alpha + 4 &= 3 - 2\sin\alpha \\ 3\sin^2\alpha + 5\sin\alpha + 1 &= 0 \\ \sin\alpha &= \frac{-5\pm\sqrt{25-12}}{6} \\ \sin\alpha &= \frac{-5\pm\sqrt{13}}{6} \end{aligned}$$

$$\begin{aligned} \sin\alpha &= \tfrac{-5+\sqrt{13}}{6} \\ \alpha &= 3.376,\ 6.049 \end{aligned} \qquad\qquad \begin{aligned} \sin\alpha &= \tfrac{-5-\sqrt{13}}{6} \\ &\text{No solution } (-1\le\sin\alpha\le 1) \end{aligned}$$

31. Using the formula for $\cos(A+B)$ with $A = 2\theta$ and $B = \theta$:

$$\begin{aligned} \cos 3\theta = \cos(2\theta+\theta) &= \cos 2\theta\cos\theta - \sin 2\theta\sin\theta \\ &= (2\cos^2\theta - 1)\cos\theta - (2\sin\theta\cos\theta)\sin\theta \\ &= 2\cos^3\theta - \cos\theta - 2\cos\theta(\sin^2\theta) \\ &= 2\cos^3\theta - \cos\theta - 2\cos\theta(1-\cos^2\theta) \\ &= 4\cos^3\theta - 3\cos\theta \end{aligned}$$

32. Start with the double angle identity

$$\cos(2x) = 2\cos^2 x - 1.$$

Next, solve for $\cos x$:

$$\begin{aligned} 2\cos^2 x &= 1 + \cos(2x) \\ \cos^2 x &= \frac{1}{2}(1+\cos(2x)) \\ \cos x &= \sqrt{\frac{1}{2}(1+\cos(2x))}. \end{aligned}$$

Note that we chose the positive square root. We made this choice because we assumed that $0 \le \theta \le \pi/2$ which implies that $\cos x \ge 0$. Now we substitute $x = \theta/2$:

$$\cos\left(\frac{\theta}{2}\right) = \sqrt{\frac{1}{2}(1+\cos\theta)}.$$

33. Since $0 < \ln x < \frac{\pi}{2}$ and $0 < \ln y < \frac{\pi}{2}$, the angles represented by $\ln x$ and $\ln y$ are in the first quadrant. This means that both their sine and cosine values will be positive. Since $\ln(xy) = \ln x + \ln y$, we can write

$$\sin(\ln(xy)) = \sin\left(\ln x + \ln y\right).$$

By the sum-of-angle formula we have

$$\sin\left(\ln x + \ln y\right) = \sin(\ln x)\cos(\ln y) + \cos(\ln x)\sin(\ln y).$$

Since cosine is positive, we have

$$\cos(\ln x) = \sqrt{1 - \sin^2(\ln x)} = \sqrt{1 - \left(\frac{1}{3}\right)^2} = \frac{\sqrt{8}}{3}$$

and

$$\cos(\ln y) = \sqrt{1 - \sin^2(\ln y)} = \sqrt{1 - \left(\frac{1}{5}\right)^2} = \frac{\sqrt{24}}{5}.$$

Thus,

$$\begin{aligned}\sin\left(\ln x + \ln y\right) &= \sin(\ln x)\cos(\ln y) + \cos(\ln x)\sin(\ln y)\\ &= \left(\frac{1}{3}\right)\left(\frac{\sqrt{24}}{5}\right) + \left(\frac{\sqrt{8}}{3}\right)\left(\frac{1}{5}\right)\\ &= \frac{\sqrt{24}+\sqrt{8}}{15}\\ &\approx 0.515.\end{aligned}$$

34. **(a)** The graph of $g(\theta) = \sin\theta - \cos\theta$ is shown in Figure 9.24.

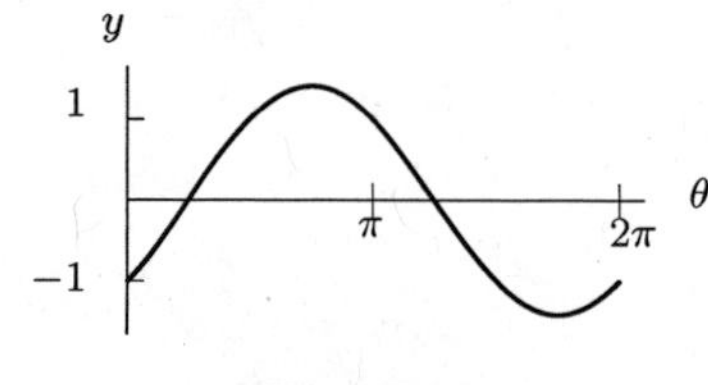

Figure 9.24

(b) We know $a_1 \sin t + a_2 \cos t = A\sin(t+\phi)$, where $A = \sqrt{a_1^2 + a_2^2}$ and $\tan\phi = a_2/a_1$. We let $a_1 = 1$ and $a_2 = -1$. This gives $A = \sqrt{2}$ and $\phi = \tan^{-1}(-1) = -\pi/4$, so

$$g(\theta) = \sqrt{2}\sin\left(\theta - \frac{\pi}{4}\right).$$

Note that we have chosen $B = 1$ and $k = 0$. We can check that this is correct by plotting the original function and $\sqrt{2}\sin(\theta - \pi/4)$ together.

We know that $\sin t = \cos(t - \pi/2)$, that is, the sine graph may be obtained by shifting the cosine graph $\pi/2$ units to the right. Thus,

$$g(\theta) = \sqrt{2}\cos\left(\theta - \frac{\pi}{4} - \frac{\pi}{2}\right) = \sqrt{2}\cos\left(\theta - \frac{3\pi}{4}\right).$$

35. **(a)** The population P_2 swings from a low of $3a - 2a = a$ to a high of $3a + 2a = 5a$, whereas P_1 swings from a low of $7a - a = 6a$ to a high of $7a + a = 8a$. Thus, P_2 has larger swings.
(b) The period of P_2 is $b + 2$, and the period of P_1 is b. Since P_2 has the longer period, it makes slower swings from low to high.
(c) At its smallest, $P_1 = 6a$, while at its largest, $P_2 = 5a$, so this statement describes P_1.
(d) The population P_2 reaches a low of a, where as P_1 reaches a low of $6a$, so P_2 appears more vulnerable to extinction.

CHECK YOUR UNDERSTANDING

1. True.

2. True. Substitute in $\tan\phi = \sin\phi/\cos\phi$ and simplify to see that this is an identity.

3. True. This is the definition of an identity.

4. False. An odd function would require $f(-\theta) = -f(\theta)$, but $f(-\frac{\pi}{4}) = f(\frac{\pi}{4}) = \frac{\sqrt{2}}{2}$.

5. True. This is an identity. Substitute using $\tan^2\theta = \sin^2\theta/\cos^2\theta$ and simplify to obtain $2 = 2\sin^2\theta + 2\cos^2\theta$. Divide by 2 to reach the Pythagorean Identity.

6. False. This is not the Pythagorean Identity, since the square is on the variable β. As a counterexample, let $\beta = 1$. Note that $\cos\beta^2 \neq (\cos\beta)^2$ and $\cos^2\beta = (\cos\beta)^2$.

7. True. First find $\sin\alpha$ and then use the double angle formula. Since $\cos^2\alpha + \sin^2\alpha = 1$, we have $\sin^2\alpha = 1 - \frac{4}{9} = \frac{5}{9}$. Because α is in the third quadrant, its sine is negative, so $\sin\alpha = -\frac{\sqrt{5}}{3}$. We find $\sin(2\alpha) = 2\sin\alpha\cos\alpha = 2(-\frac{\sqrt{5}}{3})(-\frac{2}{3}) = \frac{4\sqrt{5}}{9}$.

8. False. Although true for the value $\theta = 45°$, it is not true for all values of θ. A counterexample is $\sin 0° \neq \cos 0°$.

9. True. There are many ways to prove this identity. We use the identity $\cos 2\theta = \cos^2\theta - \sin^2\theta$ to substitute in the right side of the equation. This becomes $\frac{1}{2}(1 - (\cos^2\theta - \sin^2\theta))$. Now substitute using $1 - \cos^2\theta = \sin^2\theta$ (a form of the Pythagorean identity.) The right side then simplifies to $\sin^2\theta$ which is the left side.

10. False. A counterexample is $\cos(4 \cdot 0°) = 1 \neq -\cos(-4 \cdot 0°) = -1$.

11. False. For a counterexample, let $\theta = \phi = 90°$. Then $\sin(\theta + \phi) = \sin 180° = 0$, but $\sin\theta + \sin\phi = \sin 90° + \sin 90° = 1 + 1 = 2$.

12. False. For a counterexample let $\theta = 90°$. Then $\sin(2\theta) = \sin 180° = 0$, but $2\sin\theta = 2\sin 90° = 2$.

13. True. Start with the sine sum-of-angle identity:

$$\sin(\theta + \phi) = \sin\theta\cos\phi + \sin\phi\cos\theta$$

and let $\phi = \pi/2$, so

$$\sin(\theta + \pi/2) = \sin\theta\cos(\pi/2) + \sin(\pi/2)\cos\theta.$$

Simplify to

$$\sin(\theta + \pi/2) = \sin\theta \cdot 0 + 1 \cdot \cos\theta = \cos\theta.$$

14. False. We have $f(t) = \sin t\cos t = (1/2)\sin 2t$ which is periodic with period π.

15. True. We have $\sin(t - \pi) = \sin((t - \pi) + 2\pi) = \sin(t + \pi)$ where the first equality is by periodicity of the sine function.

16. False. For example, $t = \pi/2$ gives $\cos(t - \pi/2) = 1$ and $\cos(t + \pi/2) = -1$. The correct identity is $\cos(t - \pi/2) = -\cos(t + \pi/2)$.

17. True. Since $A\cos(Bt) = A\sin(Bt + \pi/2) = A\sin(B(t + \pi/(2B)))$, the graph of $A\cos(Bt)$ is a shift of $A\sin(Bt)$ to the left by $\pi/(2B)$.

18. True. We have

$$\begin{aligned} f(x + 2\pi) &= \sin 2(x + 2\pi) + \sin 3(x + 2\pi) \\ &= \sin(2x + 4\pi) + \sin(3x + 6\pi) \\ &= \sin 2x + \sin 3x = f(x) \end{aligned}$$

which shows that $f(x)$ is periodic. The period is 2π.

19. False. The sum of two nonzero sinusoidal functions with different periods is not a sinusoidal function. A graph of $f(x)$ on the interval $0 \le x \le 4\pi$ shows a more complicated shape that the graphs of a sinusoid.

20. False. If the periods are different the sum may not be sinusoidal. As a counterexample, graph $f(t) = \sin t + \sin 2t$, to see that $f(t)$ is not sinusoidal.

21. True. It is assumed that a_1 and a_2 are nonzero. The amplitude of the single sine function is $A = \sqrt{a_1^2 + a_2^2}$. Thus, A is greater than either a_1 or a_2.

22. True. We have $f(t) = A\sin(2\pi t + \phi)$ where $A = \sqrt{3^2 + 4^2} = 5$, $\cos\phi = \frac{3}{5}$ and $\sin\phi = \frac{4}{5}$.

23. False. The amplitude of g increases exponentially.

24. True. The maximum of g occurs when $\sin(\pi t/12) = -1$, so the maximum value of the function is $55 + 10 = 65$.

25. True. Hertz is a measure of cycles per second and so a single cycle will take 1/60th of a second.

26. True. The zeros occur when $\cos(\pi t) + 1 = 0$, so $\cos(\pi t) = -1$. Hence, $\pi t = n\pi$, where n is an odd integer. Thus, $t = n$ is an odd integer.

27. False. The variable B affects the period; the variable A affects the amplitude.

28. True; divide by e^t, which is never zero, and obtain $\tan t = 2/3$.

29. True. Since $0 \le \cos^{-1}(x) \le \pi$, we have $0 \le \sin(\cos^{-1} x) \le 1$. Thus, $\cos^{-1}(\sin(\cos^{-1} x))$ is an angle θ whose cosine is between 0 and 1. In addition, we have $0 \le \theta \le \pi$, as this is part of the definition of $\cos^{-1} x$. Hence $0 \le \theta \le \pi/2$.

30. True; we need $a + b\sin t$ to be positive for the natural logarithm to be defined. Since $\sin t \ge -1$, we have $a + b\sin t \ge a - b > 0$.

CHAPTER TEN

Solutions for Section 10.1

Exercises

1. We have
$$f(g(x)) = 2^{x/(x+1)}.$$

2. We have
$$g(f(x)) = \frac{2^x}{2^x + 1}.$$

3. Using substitution we have $f(g(x)) = \sin(4g(x)) = \sin(4\sqrt{x})$ and $g(f(x)) = \sqrt{f(x)} = \sqrt{\sin 4x}$.

4. Using substitution we have
$$m(n(x)) = 3 + (n(x))^2 = 3 + (\tan x)^2 = 3 + \tan^2 x$$
and
$$n(m(x)) = \tan(m(x)) = \tan(3 + x^2)$$

5. We have
$$\begin{aligned} w(x) &= p(p(x)) \\ &= p(\underbrace{2x+1}_{\text{input for } p}) && \text{because } p(x) = 2x + 1 \\ &= 2(2x+1) + 1 && \text{because } p(\text{input}) = 2 \cdot \text{input} + 1 \\ &= 4x + 3. \end{aligned}$$

6. To construct a table of values for r, we must evaluate $r(0), r(1), \ldots, r(5)$. Starting with $r(0)$, we have
$$r(0) = p(q(0)).$$
Therefore
$$r(0) = p(5) \qquad \text{(because } q(0) = 5\text{)}$$
Using the table given in the problem, we have
$$r(0) = 4.$$
We can repeat this process for $r(1)$:
$$r(1) = p(q(1)) = p(2) = 5.$$
Similarly,
$$\begin{aligned} r(2) &= p(q(2)) = p(3) = 2 \\ r(3) &= p(q(3)) = p(1) = 0 \\ r(4) &= p(q(4)) = p(4) = 3 \\ r(5) &= p(q(5)) = p(8) = \text{ undefined.} \end{aligned}$$

These results have been compiled in Table 10.1.

Table 10.1

x	0	1	2	3	4	5
$r(x)$	4	5	2	0	3	–

7.

x	0	1	2	3	4	5
$s(x)$	2	5	8	3	1	4

8. Since $g(x) = 9x - 2$, we substitute $9x - 2$ for x in $r(x)$, giving us $r(g(x)) = \sqrt{3(9x-2)}$, which simplifies to $r(g(x)) = \sqrt{27x-6}$.
9. Since $r(x) = \sqrt{3x}$, we substitute $\sqrt{3x}$ for x in $f(x)$, giving us $f(r(x)) = 3(\sqrt{3x})^2$, which simplifies to $f(r(x)) = 9x$.
10. Since $f(x) = 3x^2$, we substitute $3x^2$ for x in $r(x)$, giving us $r(f(x)) = \sqrt{3(3x^2)}$, which simplifies to $r(f(x)) = 3x$.
11. Since $f(x) = 3x^2$, we substitute $3x^2$ for x in $g(x)$, giving us $g(f(x)) = 9(3x^2) - 2$, which simplifies to $g(f(x)) = 27x^2 - 2$.
12. Since $f(x) = 3x^2$, we substitute $3x^2$ for x in $m(x)$, giving us $m(f(x)) = 4(3x^2)$, which simplifies to $m(f(x)) = 12x^2$, which we then substitute for x in $g(x)$, giving $g(m(f(x))) = 9(12x^2) - 2$, which simplifies to $g(m(f(x))) = 108x^2 - 2$.
13. Since $g(x) = 9x - 2$, we substitute $9x - 2$ for x in $m(x)$, giving us $m(g(x)) = 4(9x - 2)$, which simplifies to $m(g(x)) = 36x - 8$, which we then substitute for x in $f(x)$, giving $f(m(g(x))) = 3(36x - 8)^2$, which simplifies to $f(m(g(x))) = 3888x^2 - 1728x + 192$.
14. The inside function is $f(x) = \sin x$.
15. The inside function is $f(x) = \ln(x^2 + 4)$.
16. The inside function is $f(x) = x^3 + 3x + 1$.
17. The inside function is $f(x) = \cos 2x$.
18. The inside function is $f(x) = 5 + 1/x$.

Problems

19. The function $f(h(t))$ gives the area of the circle as a function of time, t.
20. The function $k(g(t))$ gives the length of the steel bar as a function of time, t.
21. The function $R(Y(q))$ gives revenue as a function of the quantity of fertilizer.
22. The function $t(f(H))$ gives the time of the trip as a function of temperature, H.
23. First, write:

$$v(u(x)) = \frac{1}{(x-1)^2} = \left(\frac{1}{x-1}\right)^2.$$

Then let $u(x) = \dfrac{1}{x-1}$ and $v(x) = x^2$. We can check that these work.

$$v(u(x)) = \left(\frac{1}{x-1}\right)^2 = \frac{1}{(x-1)^2}$$

$$u(v(x)) = \frac{1}{(x^2)-1} = \frac{1}{x^2-1}.$$

24. $g(x) = x^2$ and $h(x) = x + 3$
25. $g(x) = \sqrt{x}$ and $h(x) = 1 + \sqrt{x}$
26. $g(x) = x^2 + x$ and $h(x) = 3x$

27. $g(x) = \dfrac{1}{x^2}$ and $h(x) = x + 4$

28.

x	$f(x)$	$g(x)$	$h(x)$
0	1	2	5
1	9	0	1
2	5	1	9

According to the table, $h(0) = 5$. By definition, it is also true that $h(0) = f(g(0))$. Since $g(0) = 2$, $f(g(0)) = f(2)$. Put these pieces together:

$$h(0) = 5$$
$$h(0) = f(g(0)) = f(2)$$

So, $f(2) = 5$. We have $h(0) = 5 = f(g(0)) = f(2)$, so $f(2) = 5$. Also, $h(1) = f(g(1)) = f(0) = 1$. Finally, $h(2) = f(g(2)) = f(1) = 9$.

29. It is easiest to find values of h, because we can use the fact that $h(x) = g(f(x))$:

$$\begin{aligned} h(0) &= g(f(0)) \\ &= g(2) \qquad \text{because } f(0) = 2 \\ &= 3. \end{aligned}$$

Next, we will find values of f. To find $f(1)$, we know the output of $g(f(1))$ must be the same as $h(1)$, or 0. Since 0 is the output of g, we see from the table that its input must be 1. This means the value of $f(1)$ must be 1:

$$\begin{aligned} &g(f(1)) = h(1) = 0 \quad \text{because } h(1) = 0 \\ &g(\underbrace{f(1)}_{1}) = 0 \qquad \text{because } g(1) = 0 \\ &\text{so} \quad f(1) = 1. \end{aligned}$$

Likewise, for $f(2)$, we see that

$$\begin{aligned} &g(f(2)) = h(2) = 2 \quad \text{because } h(2) = 2 \\ &g(\underbrace{f(2)}_{4}) = 2 \qquad \text{because } g(4) = 2 \\ &\text{so} \quad f(2) = 4. \end{aligned}$$

Finally, we will find the values of g. To find $g(0)$, we know that the input of g is 0, so the output of f must be zero. This means that $x = 3$, because $f(3) = 0$. Thus, $g(0)$ is the same as $g(f(3))$, which equals $h(3)$ or 1:

$$\begin{aligned} g(0) &= g(\underbrace{f(3)}_{0}) \quad \text{because } f(3) = 0 \\ &= h(3) \qquad \text{because } h(3) = g(f(3)) \\ &= 1. \end{aligned}$$

Likewise, to find $g(3)$, we have

$$\begin{aligned} g(3) &= g(\underbrace{f(4)}_{3}) \quad \text{because } f(4) = 3 \\ &= h(4) \qquad \text{because } g(f(4)) = h(4) \\ &= 4. \end{aligned}$$

See Table 10.2.

Table 10.2

x	$f(x)$	$g(x)$	$h(x)$
0	2	1	3
1	1	0	0
2	4	3	2
3	0	4	1
4	3	2	4

30.

Table 10.3

t	u	v	w
0	2	3	1
1	3	4	2
2	1	1	4
3	4	2	0
4	0	0	3

31. Since $k(f(x)) = e^{f(x)} = e^{2x}$, we can let $f(x) = 2x$.

32. We have $f(k(x)) = f(e^x) = e^{2x} = (e^x)^2$. Thus, since $f(e^x) = (e^x)^2$, the formula for f must be $f(x) = x^2$.

33. Since $k(f(x)) = e^{f(x)}$, we have $e^{f(x)} = x$. Taking the natural log of both sides, we obtain the formula $f(x) = \ln x$.

34. First we find

$$f(x+h) - f(x) = (x+h)^2 + x + h - (x^2 + x) = (x^2 + 2xh + h^2 + x + h) - x^2 - x = 2xh + h^2 + h.$$

Then

$$\frac{f(x+h) - f(x)}{h} = \frac{2xh + h^2 + h}{h} = \frac{(2x + h + 1)h}{h} = 2x + h + 1.$$

35. For $f(x) = \sqrt{x}$

$$\frac{f(x+h) - f(x)}{h} = \frac{\sqrt{x+h} - \sqrt{x}}{h}.$$

36. First we calculate

$$\begin{aligned} f(x+h) - f(x) = \frac{1}{x+h} - \frac{1}{x} &= \frac{x}{x(x+h)} - \frac{x+h}{x(x+h)} \\ &= \frac{x - (x+h)}{x(x+h)} = \frac{-h}{x(x+h)} \end{aligned}$$

Then

$$\frac{f(x+h) - f(x)}{h} = \frac{\frac{-h}{x(x+h)}}{h} = \frac{-h}{x(x+h)} \cdot \frac{1}{h} = \frac{-1}{x(x+h)}$$

37. Since $f(x+h) = 2^{x+h}$,

$$\frac{f(x+h) - f(x)}{h} = \frac{2^{x+h} - 2^x}{h}.$$

38. **(a)** From the graphs, we see that $g(2) \approx 1.7$ and that $f(1.7) \approx 3.7$. Thus, $f(g(2)) \approx 3.7$.
(b) From the graphs, we see that $f(2) \approx 3.2$ and that $g(3.2) \approx 1.2$. Thus, $g(f(2)) \approx 1.2$.
(c) As above, $f(3) \approx 1.8$ and $f(1.8) \approx 3.6$, so $f(f(3)) \approx 3.6$.
(d) As above, $g(3) \approx 1.3$ and $g(1.3) \approx 2.2$, so $g(g(3)) \approx 2.2$.

39. (a) We have $f(1) = 2$, so $f(f(1)) = f(2) = 4$.
(b) We have $g(1) = 3$, so $g(g(1)) = g(3) = 1$.
(c) We have $g(2) = 2$, so $f(g(2)) = f(2) = 4$.
(d) We have $f(2) = 4$, so $g(f(2)) = g(4) = 0$.

40. (a) From the graph of $f(x)$, we see that $f(g(x)) = 0$ when $g(x) = 0$ and when $g(x) = 4$. Since the solution to $g(x) = 0$ is $x = 4$, and the solution to $g(x) = 4$ is $x = 0$, we see that $x = 0$ and $x = 4$ are the only solutions to the equation $f(g(x)) = 0$.
(b) From the graph of $g(x)$, we see that $g(f(x)) = 0$ only when $f(x) = 4$, which occurs only when $x = 2$. Thus, $x = 2$ is the only solution to $g(f(x)) = 0$.

41. (a) See Figures 10.1 and 10.2.

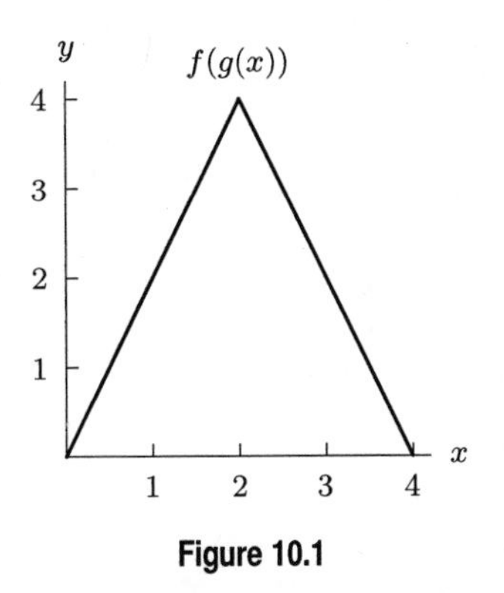

Figure 10.1

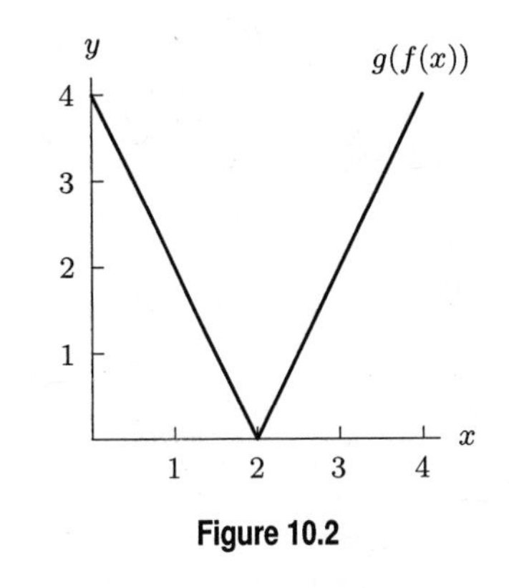

Figure 10.2

(b) From the graph, we see that $f(g(x))$ is increasing on the interval $0 < x < 2$.
(c) From the graph, we see that $g(f(x))$ is increasing on the interval $2 < x < 4$.

42. Reading values of the graph, we make an approximate table of values; we use these values to sketch Figure 10.3.

x	-2	-1	0	1	2
$f(x)$	-2	-0.3	0.7	1	0.7
$g(x)$	-0.7	-1	-0.7	0.3	2
$f(g(x))$	0.3	-0.3	0.3	0.9	0.7

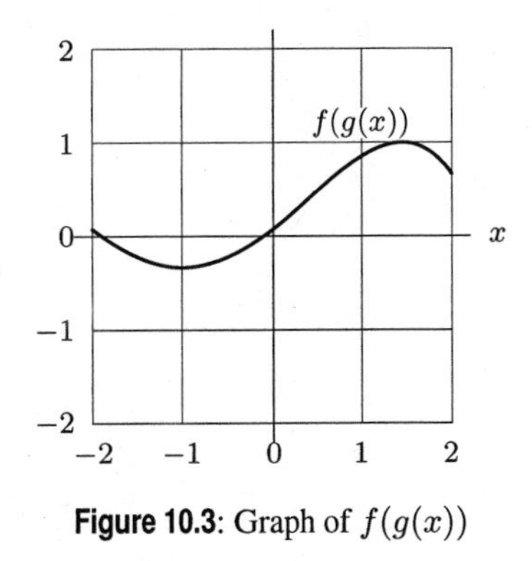

Figure 10.3: Graph of $f(g(x))$

43. Reading values of the graph, we make an approximate table of values; we use these values to sketch Figure 10.4.

x	-2	-1	0	1	2
$f(x)$	-2	-0.3	0.7	1	0.7
$g(x)$	-0.7	-1	-0.7	0.3	2
$g(f(x))$	-0.7	-0.9	-0.1	0.3	-0.1

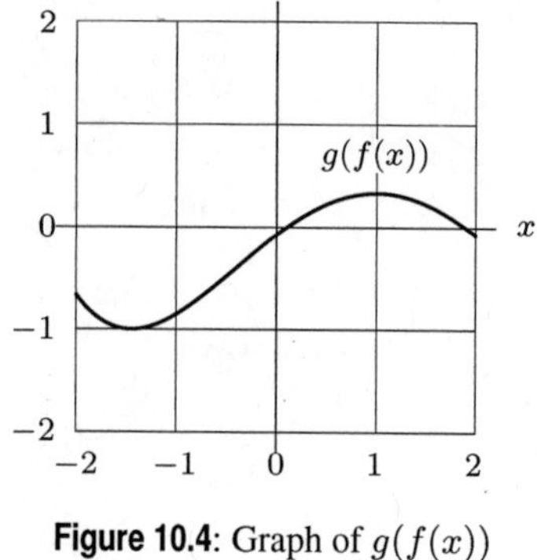

Figure 10.4: Graph of $g(f(x))$

44. Reading values of the graph, we make an approximate table of values; we use these values to sketch Figure 10.5.

x	-2	-1	0	1	2
$f(x)$	-2	-0.3	0.7	1	0.7
$f(f(x))$	-2	0.4	1	1	1

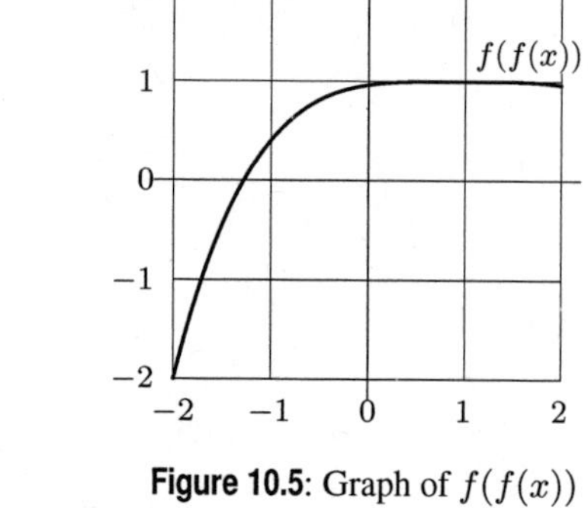

Figure 10.5: Graph of $f(f(x))$

45. Reading values of the graph, we make an approximate table of values; we use these values to sketch Figure 10.6.

x	-2	-1	0	1	2
$f(x)$	-2	-0.3	0.7	1	0.7
$g(g(x))$	-1	-1	-1	-0.4	2

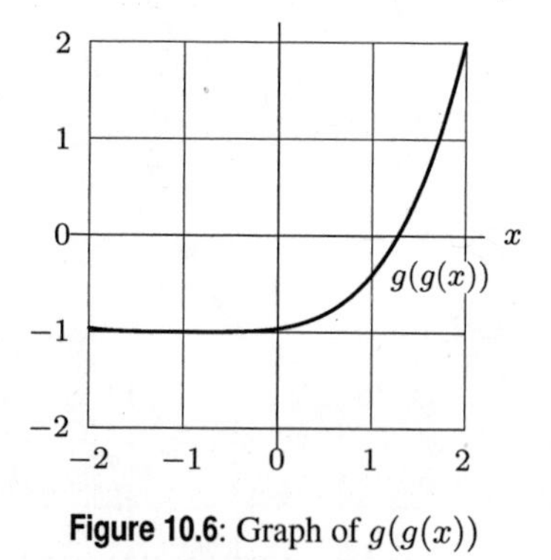

Figure 10.6: Graph of $g(g(x))$

46. We first need to find $f(1)$. In this case, $0 < x < 2$. During this interval, f is found using the formula $f(x) = 3x + 1$, so

$$f(1) = 3(1) + 1 = 4$$

We now know that

$$f(f(1)) = f(4).$$

In this case, $f(x) = x^2 - 3$ because $x = 4$, so

$$f(f(1)) = f(4) = 4^2 - 3 = 13$$

47. If $s(x) = 5 + \dfrac{1}{x+5} + x = x + 5 + \dfrac{1}{x+5}$ and $k(x) = x + 5$, then

$$s(x) = k(x) + \frac{1}{k(x)}.$$

However, $s(x) = v(k(x))$, so

$$v(k(x)) = k(x) + \frac{1}{k(x)}.$$

This is possible if

$$v(x) = x + \frac{1}{x}.$$

48. **(a)** From the definitions of the functions, we have:

$$u(x) = \frac{1}{12 - 4x}.$$

So the domain of u is all reals x, such that $x \neq 3$.

(b) From the definitions of the functions, we have:

$$u(x) = \sqrt{(12 - 4x) - 4} = \sqrt{8 - 4x}.$$

Since the quantity under the radical must be nonnegative, we require

$$\begin{aligned} 8 - 4x &\geq 0 \\ 8 &\geq 4x \\ 2 &\geq x. \end{aligned}$$

Thus the domain of u is all reals x such that $x \leq 2$.

49. **(a)** Since $v(x) = x^2$ and y can be written $\dfrac{1 + (x^2)}{2 + (x^2)}$, we take

$$u(x) = \frac{1 + x}{2 + x}.$$

(b) Since $v(x) = x^2 + 1$ and y can be written $\dfrac{1 + x^2}{1 + 1 + x^2}$, we take

$$u(x) = \frac{x}{1 + x}.$$

50. **(a)** Since $u(x) = e^x$ and $y = e^{-\sqrt{x}}$, we take $v(x) = -\sqrt{x}$.
(b) Since $v(x) = \sqrt{x}$ and $y = e^{-\sqrt{x}}$, we take $u(x) = e^{-x}$.

51. **(a)** Since $u(x) = \sqrt{1 + x^3}$ and y can be written as $y = \sqrt{1 + (-x)^3}$, we take $v(x) = -x$.
(b) Since $v(x) = x^3$ and y can be written as $y = \sqrt{1 - (x^3)}$, we take $u(x) = \sqrt{1 - x}$.

52. **(a)** Since $u(x) = 2x$ and y can be written as $y = 2 \cdot 2^x$, we take $v(x) = 2^x$.
(b) Since $v(x) = -x$ and y can be written $y = 2^{1-(-x)}$, we take $u(x) = 2^{1-x}$.

53. **(a)** Since $u(x) = x^2$ and y can be written as $y = (\sin x)^2$, we take $v(x) = \sin x$.
(b) Since $v(x) = x^2$ and y can be written $y = \sin^2(\sqrt{x})^2$, we take $u(x) = \sin^2(\sqrt{x})$.

54. **(a)** Since $u(x) = e^x$ and $y = e^{2\cos x}$, we take $v(x) = 2\cos x$.
(b) Since $v(x) = \cos x$ and $y = e^{2\cos x}$, we take $u(x) = e^{2x}$.

55. **(a)** (i) From the graph, we see that $g(4) = 2$, so from the table we get $f(g(4)) = f(2) = 3$.
(ii) From the table, we see that $f(4) = 1$, so from the graph we get $g(f(4)) = g(1) = 4$.
(iii) We have $f(0) = 2$, so $f(f(0)) = f(2) = 3$.
(iv) We have $g(0) = 1$, so $g(g(0)) = g(1) = 4$.

(b) First, by looking at the graph of $g(x)$, we can see that $g(0) = 1$ and $g(4.5) = 1$. Therefore, to solve the equation $g(g(x)) = 1$, we need to find all values of x for which $g(x) = 0$ and all values of x for which $g(x) = 4.5$. Since $g(x) = 4.5$ has no solution, and since $g(5) = 0$, we see that $x = 5$ is the only solution to $g(g(x)) = 1$.

56. To find the formula for $d(a(x))$ we start on the inside of the d function and replace $a(x)$ by its formula. Thus

$$d(\underbrace{a(x)}_{x+5}) = d(x+5).$$

Now apply the doubling to what was input, $(x+5)$, and we have

$$d(x+5) = 2(x+5) = 2x + 10.$$

To find the formula for $a(d(x))$, we again start on the inside and double the money.

$$a(\underbrace{d(x)}_{2x}) = a(2x).$$

Now apply the add 5 to what was input, $2x$, and we have

$$a(d(x)) = 2x + 5.$$

It should be clear that $d(a(x))$ is more profitable than $a(d(x))$ no matter what x is to begin with.

57. Substitute $q(p(t)) = \log(p(t))^2$ and solve:

$$\begin{aligned}
\log(p(t))^2 &= 0\\
\log(10(0.01)^t)^2 &= 0\\
\log(10^2 \cdot (0.01)^{2t}) &= 0\\
\log 10^2 + \log(0.01^{2t}) &= 0\\
2 + 2t\log(0.01) &= 0\\
2 + 2t(-2) &= 0\\
2 - 4t &= 0\\
t &= \frac{1}{2}
\end{aligned}$$

58. Substitute $f(g(t)) = \sin(3t - \frac{\pi}{4})$ and solve,

$$\begin{aligned}
\sin(3t - \frac{\pi}{4}) &= 1\\
\sin^{-1}(\sin(3t - \frac{\pi}{4})) &= \sin^{-1}(1)\\
3t - \frac{\pi}{4} &= \frac{\pi}{2}\\
3t &= \frac{\pi}{2} + \frac{\pi}{4}\\
3t &= \frac{3\pi}{4}\\
t &= \frac{\pi}{4}.
\end{aligned}$$

The solution $t = \frac{\pi}{4}$ is in the interval $0 \le t \le \frac{2\pi}{3}$. This is the only solution in this interval because $\sin(3t - \frac{\pi}{4})$ has period $\frac{2\pi}{3}$.

59. The function $f(g(x)) = \sqrt{x^2}$ is defined for all real numbers because x^2 is never negative, so the domain of $f(g(x))$ is all real numbers. However, $g(f(x)) = (\sqrt{x})^2$ is undefined for negative values of x, so the domain of $g(f(x))$ is all real numbers greater than or equal to 0.

60. **(a)** The function $y = f(x)$ has a y-intercept of 0, and a slope of $\frac{-10-0}{5-0} = -2$. So $f(x) = -2x$.
(b) The function is defined for the domain: $0 \le x \le 5$ and it takes values in the range: $-10 \le y \le 0$.
(c) The function $y = g(x)$ has a y-intercept of 1, and slope of $\frac{4-1}{1-0} = 3$. So $g(x) = 3x + 1$.
(d) The function $g(x)$ is defined on the domain: $0 \le x \le 1$ and takes values in the range: $1 \le y \le 4$.
(e) Since $f(x) = -2x$, and $g(x) = 3x + 1$,we know that $h(x) = f(g(x)) = -2(g(x)) = -2(3x + 1) = -6x - 2$.
(f) Since $g(x)$ is only defined for the domain: $0 \le x \le 1$, and the range of $g(x)$ is contained in the domain of $f(x)$, $h(x)$ has the same domain as $g(x)$. Since $h(x)$ is a linear function, we can find its range by evaluating $h(x)$ for the extreme values of its domain, e.g. at $x = 0$ and at $x = 1$. We find that $h(0) = -6(0) - 2 = -2$ and $h(1) = -6(1) - 2 = -8$, so the range of $h(x)$ is $-8 \le y \le -2$.
(g) Since $h(x) = -6x - 2$, we know that its y-intercept is -2 and its slope is -6. Given a domain of $0 \le x \le 1$, $h(x)$ goes from $(0, -2)$ to $(1, -8)$.

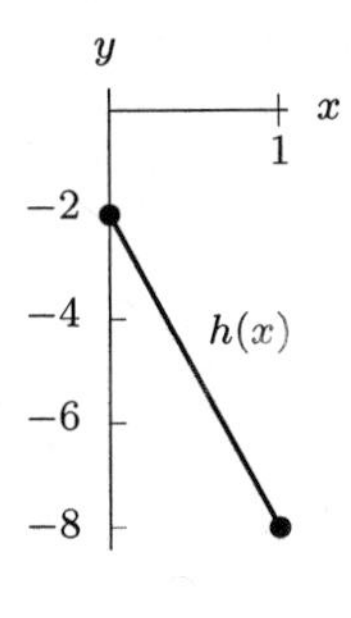

61. We have

$$\begin{aligned} q(r(x)) &= \frac{8^{x^3}}{16^{x^2}} \\ &= 8^{x^3} \cdot 16^{-x^2} && \text{exponent rule} \\ &= (2^3)^{x^3} \cdot (2^4)^{-x^2} \\ &= 2^{3x^3} \cdot 2^{-4x^2} && \text{exponent rule} \\ &= 2^{3x^3 - 4x^2} && \text{exponent rule} \\ &= 2^{r(x)}, \end{aligned}$$

so $q(x) = 2^x$.

62. We have

$$\begin{aligned} g(x) &= f(h(x)) \\ &= \frac{h(x)+1}{h(x)+3} && \text{because } f(x) = \tfrac{x+1}{x+3} \\ &= \frac{(3x^2-1)+1}{(3x^2-1)+3} && \text{because } h(x) = (3x^2-1) \\ &= \frac{3x^2}{3x^2+2}. \end{aligned}$$

63. We have

$$\begin{aligned} f(x) &= g(x) + \frac{2x+4}{x+3} \\ \underbrace{\frac{x+1}{x+3}}_{f(x)} &= g(x) + \frac{2x+4}{x+3} \end{aligned}$$

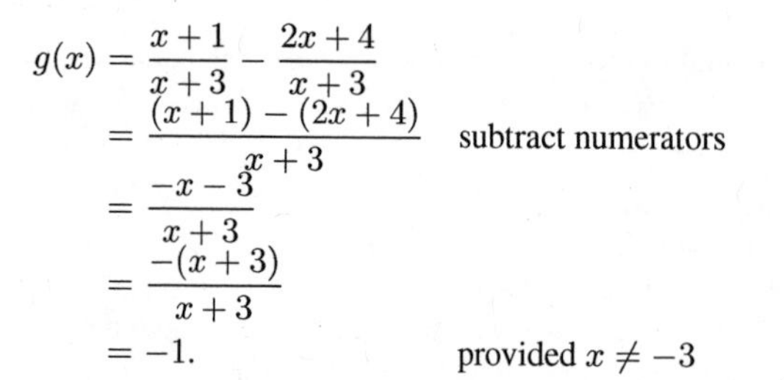

$$\begin{aligned} g(x) &= \frac{x+1}{x+3} - \frac{2x+4}{x+3} \\ &= \frac{(x+1)-(2x+4)}{x+3} && \text{subtract numerators} \\ &= \frac{-x-3}{x+3} \\ &= \frac{-(x+3)}{x+3} \\ &= -1. && \text{provided } x \neq -3 \end{aligned}$$

64. Statements (a) and (d) must be true.

(a) To see why this statement must be true, suppose that $f(2)$ is undefined. This means $h(2) = g(f(2))$ is undefined, so that the graph of h will not contain the point $(2, 5)$.

(b) To see why this statement need not be true, suppose $f(2) = 3$ and $g(3) = 5$ but that $g(2)$ is undefined. Then $h(2) = g(f(2)) = g(3) = 5$, so that the point $(2, 5)$ is on the graph of h, even though $g(2)$ is undefined.

(c) To see why this statement need not be true, suppose again that $f(2) = 3$ and $g(3) = 5$ but that 5 is not in the range of f. Then $h(2) = g(f(2)) = g(3) = 5$, so that the point $(2, 5)$ is on the graph of h, even though 5 is not in the range of f.

(d) To see why this statement must be true, note that if the point $(2, 5)$ is on the graph of h, then 5 must be a legal output value of g.

(e) To see why this statement need not be true, suppose $g(x) = 5$, so that the graph of g is a horizontal line. This function is not invertible. But provided $f(2)$ is defined, we see that $h(2) = g(f(2)) = 5$, so whether or not g is invertible is irrelevant.

65. Statements (a) and (e) must be true.

(a) To see why this statement must be true, suppose that f is undefined at a particular value of x, say $x = 6$. Then $h(6) = g(f(6))$ is undefined, because $f(6)$ is undefined. Thus, in order for h to be defined at all x, f must be too.

(b) To see why this statement need not be true, suppose that $g(x) = \sqrt{x}$, so that g is defined only for $x \geq 0$, and that $f(x) = x^2$. Then $h(x) = \sqrt{x^2}$. We see that h is defined for all x, even though g is not.

(c) To see why this statement need not be true, suppose that $f(x) = 3$, so that the graph of f is a horizontal line, and similarly that $g(x) = 4$. Then $h(x) = g(f(x)) = g(3) = 4$. We see that even though the range of f is the single value $y = 3$, the function h can be defined for all x.

(d) From part (c), we see that if $f(x) = 3$ and $g(x) = 4$, we have $h(x) = 4$. This means that even if the range of g is the single value $y = 4$, the function h can be defined for all x.

(e) To see why this statement must be true, suppose there that $f(2) = 3$ but that $g(x) = 1/(x-3)$. We see that the range of f includes 3, but that the domain of g does not, because $g(3) = 1/0$. This means that h is undefined at $x = 2$, because $h(2) = g(f(2)) = g(3)$, and $g(3)$ is undefined. The point is that: any legal output value of f must be a legal input value of g, or otherwise h will be undefined for some input values of f.

Solutions for Section 10.2

Exercises

1. It is not invertible.

2. It is not invertible.

3. A graph of this function on a window which contains both positive and negative values of x reveals that it fails the horizontal line test. Therefore, this function is not invertible.

4. A graph reveals that the output values of this function oscillate back and forth between $y = -1$ and $y = 1$. It is therefore not invertible.

5. It is not invertible.

6. A graph reveals that this function appears to be increasing everywhere. The function appears to pass the horizontal line test and is therefore invertible.

7. Since $f(x) = (x/4) - (3/2)$ and $g(t) = 4(t + 3/2)$, we have

$$f(g(t)) = \frac{4(t+\frac{3}{2})}{4} - \frac{3}{2} = t + \frac{3}{2} - \frac{3}{2} = t$$
$$g(f(x)) = 4\left(\frac{x}{4} - \frac{3}{2} + \frac{3}{2}\right) = 4\frac{x}{4} = x$$

8. One way to check that these functions are inverses is to make sure they satisfy the identities $f(f^{-1}(x)) = x$ and $f^{-1}(f(x)) = x$.

$$\begin{aligned} f(f^{-1}(x)) &= 1 + 7\left(\sqrt[3]{\frac{x-1}{7}}\right)^3 \\ &= 1 + 7\left(\frac{x-1}{7}\right) \\ &= 1 + (x-1) = x. \end{aligned}$$

Also,

$$\begin{aligned} f^{-1}(f(x)) &= \sqrt[3]{\frac{1+7x^3-1}{7}} \\ &= \sqrt[3]{x^3} = x. \end{aligned}$$

Thus, $f^{-1}(x) = \sqrt[3]{\dfrac{x-1}{7}}$.

9. One way to check that these functions are inverses is to make sure they satisfy the identities $g(g^{-1}(x)) = x$ and $g^{-1}(g(x)) = x$.

$$\begin{aligned} g(g^{-1}(x)) &= 1 - \frac{1}{\left(1 + \frac{1}{1-x}\right) - 1} \\ &= 1 - \frac{1}{\left(\frac{1}{1-x}\right)} \\ &= 1 - (1-x) \\ &= x. \end{aligned}$$

Also,

$$\begin{aligned} g^{-1}(g(x)) &= 1 + \frac{1}{1 - \left(1 - \frac{1}{x-1}\right)} \\ &= 1 + \frac{1}{\frac{1}{x-1}} \\ &= 1 + x - 1 = x. \end{aligned}$$

So the expression for g^{-1} is correct.

10. Since $h(x) = \sqrt{2x}$ and $k(t) = t^2/2$, we have

$$h(k(t)) = \sqrt{2\left(\frac{t^2}{2}\right)} = \sqrt{t^2} = t \qquad \text{provided } t \geq 0.$$

$$k(h(x)) = \frac{(\sqrt{2x})^2}{2} = \frac{2x}{2} = x \qquad \text{provided } x \geq 0.$$

11. Check using the two compositions

$$f(f^{-1}(x)) = e^{f^{-1}(x)+1} = e^{(\ln x - 1)+1} = e^{\ln x} = x$$

and

$$f^{-1}(f(x)) = \ln f(x) - 1 = \ln(e^{x+1}) - 1 = (x+1) - 1 = x.$$

They are inverses of one another.

12. Check using the two compositions

$$f(f^{-1}(x)) = e^{2f^{-1}(x)} = e^{2(\ln x)/2)} = e^{\ln x} = x$$

and

$$f^{-1}(f(x)) = \frac{\ln f(x)}{2} = \frac{\ln e^{2x}}{2} = \frac{2x}{2} = x.$$

They are inverses of one another.

13. Check using the two compositions

$$f(f^{-1}(x)) = e^{f^{-1}(x)/2} = e^{(2\ln x)/2} = e^{\ln x} = x$$

and

$$f^{-1}(f(x)) = 2\ln f(x) = 2\ln e^{x/2} = 2(x/2) = x.$$

They are inverses of one another.

14. Check by using the two compositions

$$f(f^{-1}(x)) = \ln \frac{f^{-1}(x)}{2} = \ln \frac{2e^x}{2} = \ln e^x = x$$

and

$$f^{-1}(f(x)) = 2e^{f(x)} = 2e^{\ln(x/2)} = 2(x/2) = x.$$

They are inverses of one another.

15. Let $y = x + 5$. Then $x = y - 5$, so $f^{-1}(x) = x - 5$.

16. Let $y = 1 - x$. Then $x = 1 - y$, so $g^{-1}(x) = 1 - x$.

17. Let $y = \sqrt{x}$. Then $x = y^2$, and $h^{-1}(x) = x^2$.

18. Let $y = \frac{1}{x}$. Then $x = \frac{1}{y}$, and $j^{-1}(x) = \frac{1}{x}$.

19. Start with $x = f(f^{-1}(x))$ and substitute $y = f^{-1}(x)$. We have

$$\begin{aligned} x &= f(y) \\ x &= 3y - 7 \\ x + 7 &= 3y \\ \frac{x+7}{3} &= y \end{aligned}$$

Therefore,

$$f^{-1}(x) = \frac{x+7}{3}.$$

20. Let $y = x/(x-1)$ Then

$$\begin{aligned} y(x-1) &= x \\ yx - y &= x \\ yx - x &= y \\ x(y-1) &= y \\ x &= \frac{y}{y-1}. \end{aligned}$$

Thus, $k^{-1}(x) = \frac{x}{x-1}$.

21. Let $y = \sqrt{1-2x^2}$. We have

$$\begin{aligned} y &= \sqrt{1-2x^2} \\ y^2 &= 1-2x^2 \\ 2x^2 &= 1-y^2 \\ x^2 &= \frac{1-y^2}{2} \\ x &= \sqrt{\frac{1-y^2}{2}}. \end{aligned}$$

Thus, $l^{-1}(x) = \sqrt{\frac{1-x^2}{2}}$.

22. Let $y = 1/x - x$. Assume $x > 0$. Then,

$$\begin{aligned} y &= \frac{1}{x} - x \\ y + x &= \frac{1}{x} \\ x^2 + yx - 1 &= 0. \end{aligned}$$

Using the quadratic formula, we have

$$x = \frac{-y + \sqrt{y^2+4}}{2}.$$

Note that we do not include the solution $x = \frac{-y-\sqrt{y^2+4}}{2}$, because x must be positive. This gives $m^{-1}(x) = \dfrac{-x+\sqrt{x^2+4}}{2}$.

23. Let $y = (1+x^2)^2$. Note that $y \geq 1$. Then $1 + x^2 = \sqrt{y}$, which means that $x^2 = \sqrt{y} - 1$. Thus, $x = \sqrt{\sqrt{y}-1}$, for $y \geq 1$. This means that $n^{-1}(x) = \sqrt{\sqrt{x}-1}$ for $x \geq 1$.

24. Let $y = 1/(1 + \dfrac{1}{x})$. Then

$$y = \frac{1}{\left(\frac{1+x}{x}\right)} = \frac{x}{x+1}.$$

This means that

$$\begin{aligned} y(x+1) &= x \\ yx + y &= x \\ x - yx &= y \\ x(1-y) &= y \\ x &= \frac{y}{1-y}. \end{aligned}$$

Thus, $o^{-1}(x) = x/(1-x)$.

25. Start with $x = j(j^{-1}(x))$ and substitute $y = j^{-1}(x)$. We have

$$\begin{aligned} x &= j(y) \\ x &= \sqrt{1+\sqrt{y}} \\ x^2 &= 1+\sqrt{y} \\ x^2 - 1 &= \sqrt{y} \\ (x^2-1)^2 &= y \end{aligned}$$

Therefore,

$$j^{-1}(x) = (x^2-1)^2.$$

26. Start with $x = h(h^{-1}(x))$ and substitute $y = h^{-1}(x)$. We have

$$\begin{aligned} x &= h(y) \\ x &= \frac{2y+1}{3y-2} \\ x(3y-2) &= 2y+1 \\ 3yx - 2x &= 2y+1 \\ 3yx - 2y &= 2x+1 \\ y(3x-2) &= 2x+1 \quad \text{(factor out a } y\text{)} \\ y &= \frac{2x+1}{3x-2} \end{aligned}$$

Therefore,

$$h^{-1}(x) = \frac{2x+1}{3x-2}.$$

27. Start with $x = k(k^{-1}(x))$ and substitute $y = k^{-1}(x)$. We have

$$\begin{aligned} x &= k(y) \\ x &= \frac{3-\sqrt{y}}{\sqrt{y}+2} \\ x(\sqrt{y}+2) &= 3-\sqrt{y} \\ x\sqrt{y}+2x &= 3-\sqrt{y} \\ x\sqrt{y}+\sqrt{y} &= 3-2x \\ \sqrt{y}(x+1) &= 3-2x \quad \text{(factor out a } \sqrt{y}\text{)} \\ \sqrt{y} &= \frac{3-2x}{x+1} \\ y &= \left(\frac{3-2x}{x+1}\right)^2 \end{aligned}$$

Therefore,

$$k^{-1}(x) = \left(\frac{3-2x}{x+1}\right)^2.$$

28. Start with $x = g(g^{-1}(x))$ and substitute $y = g^{-1}(x)$. We have

$$\begin{aligned} x &= g(y) \\ x &= \frac{\ln y - 5}{2\ln y + 7} \\ 2x\ln y + 7x &= \ln y - 5 \end{aligned}$$

$$\begin{aligned}
2x \ln y - \ln y &= -5 - 7x \\
\ln y(2x-1) &= -5 - 7x \\
\ln y &= \frac{-5-7x}{2x-1} \\
\ln y &= \frac{(-1)}{(-1)} \cdot \frac{(5+7x)}{(1-2x)} \\
\ln y &= \frac{5+7x}{1-2x} \\
y &= e^{\frac{5+7x}{1-2x}} \\
g^{-1}(x) &= e^{\frac{5+7x}{1-2x}}.
\end{aligned}$$

29. Start with $x = h(h^{-1}(x))$ and substitute $y = h^{-1}(x)$. We have

$$\begin{aligned}
x &= h(y) \\
x &= \log \frac{y+5}{y-4} \\
10^x &= \frac{y+5}{y-4} \\
10^x(y-4) &= y+5 \\
10^x y - 4 \cdot 10^x &= y+5 \\
10^x y - y &= 5 + 4 \cdot 10^x \\
y(10^x - 1) &= 5 + 4 \cdot 10^x \\
y &= \frac{5+4\cdot 10^x}{10^x - 1} \\
h^{-1}(x) &= \frac{5+4\cdot 10^x}{10^x - 1}.
\end{aligned}$$

30. We start with $f(f^{-1}(x)) = x$ and substitute $y = f^{-1}(x)$. We have

$$\begin{aligned}
f(y) &= x \\
\cos\sqrt{y} &= x \\
\sqrt{y} &= \arccos x \\
y &= (\arccos x)^2.
\end{aligned}$$

Thus

$$f^{-1}(x) = (\arccos x)^2.$$

31. We start with $g(g^{-1}(x)) = x$ and substitute $y = g^{-1}(x)$. We have

$$\begin{aligned}
g(y) &= x \\
2^{\sin y} &= x \\
\ln\left(2^{\sin y}\right) &= \ln x \\
\ln(2^{\sin y}) = (\sin y)\ln 2 &= \ln x \\
\sin y &= \frac{\ln x}{\ln 2} \\
y &= \arcsin\left(\frac{\ln x}{\ln 2}\right).
\end{aligned}$$

Thus

$$g^{-1}(x) = \arcsin\left(\frac{\ln x}{\ln 2}\right).$$

32. Both of these statements can be checked algebraically by composing the formulas for f and f^{-1}. To check that $f^{-1}(f(T)) = T$, evaluate

$$\begin{aligned} f^{-1}(f(T)) &= f^{-1}(150+5T) \\ &= \frac{1}{5}(150+5T) - 30 \\ &= 30 + T - 30 \\ &= T. \end{aligned}$$

To check that $f(f^{-1}(R)) = R$, evaluate

$$\begin{aligned} f(f^{-1}(R)) &= f\left(\frac{1}{5}R - 30\right) \\ &= 150 + 5\left(\frac{1}{5}R - 30\right) \\ &= 150 + R - 150 \\ &= R. \end{aligned}$$

Problems

33. The inverse function $f^{-1}(P)$ gives the time, t, in years at which the population is P thousand. Its units are years.

34. The inverse function, $f^{-1}(C)$, represents the quantity that can be produced at a cost of C. Its units are number of items.

35. **(a)** We have

$$R = 150 + 5T.$$

Solving for T we have

$$5T = R - 150,$$

so

$$T = f^{-1}(R) = \frac{1}{5}R - 30.$$

(b) Table 10.4 shows values of f, and Table 10.5 shows values of f^{-1}. Notice that the columns in the two tables are reversed. This is because input values of f are output values of f^{-1}, and input values of f^{-1} are output values of f.

Table 10.4 *Values of f*

T, temperature (°C)	$R = f(T)$, resistance (ohms)
−20	50
−10	100
0	150
10	200
20	250
30	300
40	350
50	400

Table 10.5 *Values of f^{-1}*

R, resistance (ohms)	$T = f^{-1}(R)$, temperature (°C)
50	−20
100	−10
150	0
200	10
250	20
300	30
350	40
400	50

36. **(a)** $f(125) = 2\sqrt{625} = 50$ mph

(b)

$$\begin{aligned} S &= 2\sqrt{5L} \\ S^2 &= (2\sqrt{5L})^2 \\ &= 4(5L) \\ &= 20L \\ L &= \frac{S^2}{20} \end{aligned}$$

Thus

$$f^{-1}(S) = \frac{S^2}{20}$$

(c) $f^{-1}(80) = \frac{80^2}{20} = \frac{6400}{20} = 320$ ft.

37. **(a)** $f(3) = 5^3 = 125$

$f^{-1}(\frac{1}{25}) = -2$ because $f(-2) = \frac{1}{25}$.

(b)

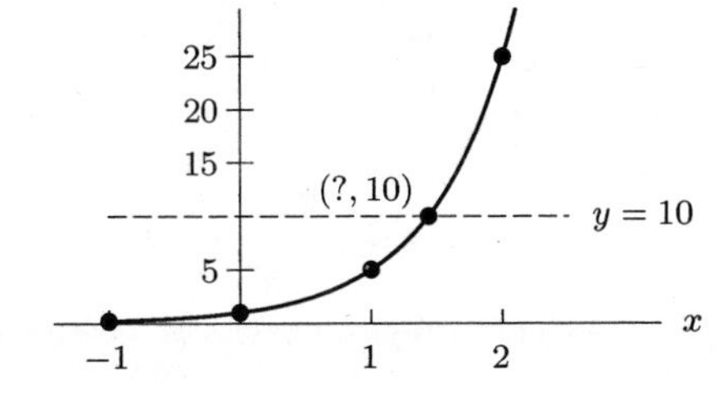

Using a calculator, $f^{-1}(10) \approx 1.43086$.

38. **(a)** When $t = 0$, $P = 37.8(1.044)^0 = 37.8(1) = 37.8$. This tells us that the population of the town when $t = 0$ is 37,800. The growth factor, 1.044, tells us that the population is 104.4% of what it had been the previous year, or that the town grows by 4.4% each year.

(b) Since $f(t) = 37.8(1.044)^t$, then $f(50) = 37.8(1.044)^{50} = 325.474$ This tells us that there will be 325,474 people after 50 years.

(c) To find $f^{-1}(P)$, which is the inverse function of $f(t)$, we need to solve

$$P = 37.8(1.044)^t$$

for t. Begin by dividing both sides by 37.8:

$$\frac{P}{37.8} = 1.044^t$$

Then, take the log of both sides, using the property $\log a^b = b \cdot \log a$.

$$\log\left(\frac{P}{37.8}\right) = \log 1.044^t = t \log 1.044.$$

So solving for t,

$$t = \frac{\log\left(\frac{P}{37.8}\right)}{\log 1.044}.$$

We can make this formula look a little simpler by recalling that $\log \frac{a}{b} = \log a - \log b$. The formula for our inverse function is now:

$$t = f^{-1}(P) = \frac{\log P - \log 37.8}{\log 1.044}.$$

(d) $f^{-1}(50) = \frac{\log 50 - \log 37.8}{\log 1.044} \approx 6.496$. It will take about 6.496 years for P to reach 50,000 people.

39. Reading the values from the graph, we get:

$$f(0) = 1.5, \quad f^{-1}(0) = 2.5, \quad f(3) = -0.5, \quad f^{-1}(3) = -5.$$

Ranking them in order from least to greatest, we get:

$$f^{-1}(3) < f(3) < 0 < f(0) < f^{-1}(0) < 3.$$

40. Solving for q gives

$$\begin{aligned} C &= 200 + 0.1q \\ q &= \frac{C - 200}{0.1} = 10C - 2000. \end{aligned}$$

The inverse function $q = f^{-1}(C)$ gives the number of kilograms of the chemical that can be manufactured for $\$C$.

41. Solving for t gives

$$\begin{aligned} P &= 10e^{0.02t} \\ \frac{P}{10} &= e^{0.02t} \\ 0.02t &= \ln\left(\frac{P}{10}\right) \\ t &= \frac{\ln(P/10)}{0.02} = 50\ln(P/10). \end{aligned}$$

The inverse function $t = f^{-1}(P) = 50\ln(P/10)$ gives the time, t, in years at which the population reaches P million.

42. Solving for I gives

$$\begin{aligned} N &= 10\log\left(\frac{I}{I_0}\right) \\ \frac{N}{10} &= \log\left(\frac{I}{I_0}\right) \\ \frac{I}{I_0} &= 10^{N/10} \\ I &= I_0 10^{N/10}. \end{aligned}$$

The inverse function $f^{-1}(N) = I_0 10^{N/10}$ gives the intensity of a sound with a decibel rating of N.

43. **(a)** Since each of 7 people uses 2 gallons a day, altogether 14 gallons a day are used. After t days, $14t$ gallons have been used, so since there were originally 800 gallons,

$$\text{Fresh water remaining} = f(t) = 800 - 14t \text{ gallons.}$$

(b) (i) $f(0) = 800$ gallons. This is the original amount of water brought to the island.

(ii) This represents the time when they will run out of water. To find $f^{-1}(0)$ we solve:

$$\begin{aligned} 800 - 14t &= 0 \\ 14t &= 800 \\ t &= \frac{800}{14} \approx 57.143 \quad \text{days} \end{aligned}$$

Since $f^{-1}(0) \approx 57.143$, they will run out of water after 57.143 days.

(iii) Want to find t such that $f(t) = \frac{1}{2}f(0)$:

$$\begin{aligned}\frac{1}{2}f(0) &= 400\\ 400 &= 800 - 14t\\ 14t &= 400\\ t &\approx 28.571 \text{ days.}\end{aligned}$$

This t value is the time when half of the original water is gone.

(iv) Substituting for $f(t)$ gives

$$800 - f(t) = 800 - (800 - 14t) = 14t.$$

This represents the total amount of water used in t days.

44. Solving for r gives

$$\begin{aligned}r &= \log\left(\frac{I}{I_0}\right)\\ r &= \log\left(\frac{I}{I_0}\right)\\ \frac{I}{I_0} &= 10^r\\ I &= I_0 10^r.\end{aligned}$$

The inverse function $f^{-1}(r) = I_0 10^r$ gives the intensity of an earthquake with Richter rating r.

45. **(a)** The compositions are

$$f(g(x)) = f(\ln x) = e^{\ln x} = x \quad \text{and} \quad g(f(x)) = g(e^x) = \ln e^x = x,$$

which shows that the two functions are inverses of one another.

(b) The graph of the two functions is symmetric about the line $y = x$. See Figure 10.7.

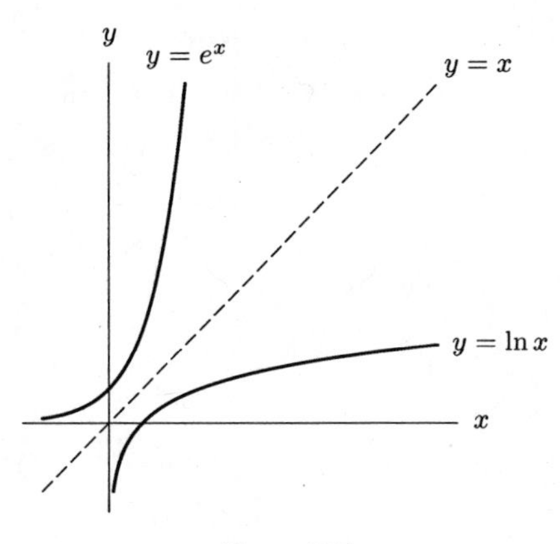

Figure 10.7

46. **(a)** The compositions are

$$p(q(t)) = p(\log t) = 10^{\log t} = t \quad \text{and} \quad q(p(t)) = q(10^t) = \log 10^t = t.$$

Thus, the two functions are inverses of one another.

(b) The graph of the two functions is symmetric about the line $y = t$. See Figure 10.8.

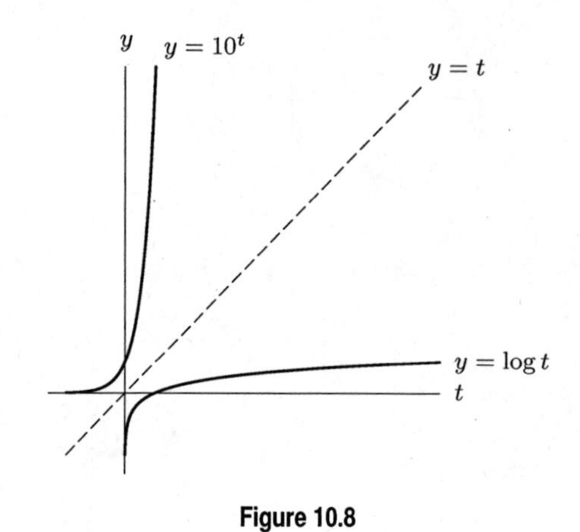

Figure 10.8

47. **(a)** We are told that the function is an exponential one, so we know that it must be of the form

$$P(t) = AB^t.$$

Since $P(0) = 150$,

$$\begin{aligned} P(0) &= AB^0 = 150 \\ &= A(1) = 150 \\ A &= 150. \end{aligned}$$

Thus

$$P(t) = 150B^t.$$

We know that $P(1) = 165$ so

$$\begin{aligned} P(1) &= 150B^1 = 165 \\ 150B &= 165 \\ B &= 1.1. \end{aligned}$$

Thus,

$$P(t) = 150(1.1)^t.$$

Checking our answer at $t = 2$ we indeed see that

$$\begin{aligned} P(2) &= AB^2 \\ &= 150(1.1)^2 \\ &= 150(1.21) \\ &\approx 182. \end{aligned}$$

(b) Letting $N = P(t)$ and solving for t we get

$$\begin{aligned} N &= 150(1.1)^t \\ \frac{N}{150} &= 1.1^t. \end{aligned}$$

Taking the log of both sides we get

$$\log\left(\frac{N}{150}\right) = \log(1.1^t) = t\log(1.1).$$

Dividing both sides by 1.1, we get

$$\frac{\log(\frac{N}{150})}{\log(1.1)} = t.$$

Recalling that

$$\log\left(\frac{a}{b}\right) = \log(a) - \log(b)$$

we get that

$$t = \frac{\log(N) - \log(150)}{\log(1.1)}.$$

Since this formula defines the inverse function of $P(t)$ and it is in terms of N, we can call this function $P^{-1}(N)$. This function tells us how many years it would take to have N cows.

$$P^{-1}(N) = \frac{\log(N) - \log(150)}{\log(1.1)}.$$

(c) Letting $N = 400$ in the function $P^{-1}(N)$ we get

$$\begin{aligned} P^{-1}(400) &= \frac{\log(400) - \log(150)}{\log(1.1)} \\ &\approx 10.3. \end{aligned}$$

Thus it would take roughly 10.3 years for the population of the cattle herd to reach 400. To check that this is indeed the correct answer we can let $t = 10.3$ in our original function $P(t)$

$$\begin{aligned} P(10.3) &= 150(1.1)^{10.3} \\ &\approx 400. \end{aligned}$$

48. (a) $N(20{,}000)$ is the number of units the company will sell if \$20,000 is spent on advertising. This amount is \$5,000 less than \$25,000; thus, the company will sell 20 fewer units than when it spends \$25,000, or 380 units total. Therefore, $N(20{,}000) = 380$.

(b) It costs \$5,000 to sell an additional 20 units, or \$250 to sell an additional unit. This means that $N(x)$ will increase by 1 when x increases by 250. In other words, the slope of $N(x)$ is $\frac{1}{250}$. We know that $N(25{,}000) = 400$. Thus, $(25{,}000, 400)$ is a point on the graph of $N(x)$. Since we know the slope of $N(x)$ and a point on its graph, we have

$$\begin{aligned} N(x) - 400 &= \frac{1}{250}(x - 25{,}000) \\ N(x) &= 400 + \frac{1}{250}x - 100 \\ N(x) &= \frac{1}{250}x + 300. \end{aligned}$$

(c) The slope of $N(x)$ is $1/250$, which means that an additional \$250 must be spent on advertising to sell an additional unit. The y-intercept of $N(x)$ is 300, which means that even if the company spends no money on advertising, it will still sell 300 units. The x-intercept is $-75{,}000$, which represents a negative amount of money spent on advertising. This has no obvious interpretation.

(d) $N^{-1}(500)$ is the advertising expenditure required to sell 500 units, or 100 units more than when it spends \$25,000. Since the company needs to spend an additional \$5,000 to sell an additional 20 units, it must spend an additional \$25,000 to sell 500 units, or \$50,000 total. Thus, $N^{-1}(500) = 50{,}000$.

(e) If only \$2000 in profits are made on the sale of ten units, then the per-unit profit, before advertising costs are accounted for, is \$200. Thus, the company makes an additional \$200 for each unit sold. However, it must spend \$250 on ads to sell an additional unit. Thus, the company must spend more on advertising to sell an additional unit than it makes on the sale of that unit. Therefore, it should discontinue its advertising campaign, or, at the very least, find an effective way to lower its per-unit advertising expenditure.

49. **(a)** Since $H(t)$ is a decreasing exponential function, we know that

$$H(t) = H_0 e^{-kt}$$

(Alternatively, we could have used $H(t) = H_0 a^t$, where $0 < a < 1$.) Since $H(0) = 200$, we have $H_0 = 200$. Thus, $H(t) = 200e^{-kt}$. Since $H(2) = 20$, we have

$$\begin{aligned} 200e^{-2k} &= 20 \\ e^{-2k} &= \frac{20}{200} = \frac{1}{10} \\ \ln e^{-2k} &= \ln \frac{1}{10} \\ -2k &= \ln \frac{1}{10} \\ k &= -\frac{\ln \frac{1}{10}}{2} \approx 1.15129. \end{aligned}$$

Thus,

$$H(t) = 200e^{-1.15129t}.$$

(b) After one quarter hour, $t = 0.25$ and

$$H(0.25) = 200e^{-1.15129(0.25)} = 149.979.$$

So the temperature dropped by about $200 - 149.979 = 50.021°$C in the first 15 minutes. After half an hour, $t = 0.5$ and

$$H(0.5) = 200e^{-1.15129(0.5)} \approx 112.468.$$

So the temperature dropped by about $150 - 112.468 = 37.532°$C in the next 15 minutes.

(c) If $t = H^{-1}(y)$, then t is the amount of time required for the brick's temperature to fall to $y°$C above room temperature. Letting $y = H(t)$, we have

$$\begin{aligned} y &= 200e^{-1.15129t} \\ e^{-1.15129t} &= \frac{y}{200} \\ \ln e^{-1.15129t} &= \ln\left(\frac{y}{200}\right) \\ -1.15129t &= \ln\left(\frac{y}{200}\right) \\ t &= \frac{\ln(y/200)}{-1.15129} \approx -0.86859 \ln\left(\frac{y}{200}\right) = H^{-1}(y). \end{aligned}$$

(d) We need to evaluate $H^{-1}(5)$:

$$H^{-1}(5) = -0.86859 \ln\left(\frac{5}{200}\right) = 3.204 \text{ hours},$$

or about 3 hours and 12 minutes.

(e) The function $H(t)$ has a horizontal asymptote at $y = 0$ (the t-axis). The temperature of the brick approaches room temperature. Since $H(t)$ is the number of degrees above room temperature, $y = 0$ must be room temperature.

50. We have

$$\begin{aligned} y &= 0.5(x^2 + A^2)^{0.5} \\ 2y &= (x^2 + A^2)^{0.5} \\ (2y)^2 &= x^2 + A^2 \\ x^2 &= 4y^2 - A^2 \\ x &= \left(4y^2 - A^2\right)^{0.5} \quad \text{because } x \geq 0 \\ \text{so} \quad f^{-1}(x) &= \left(4x^2 - A^2\right)^{0.5}. \end{aligned}$$

51. We have

$$\begin{aligned} y &= 0.5(x^{-1} + A^{-1})^{-1} \\ 2y &= (x^{-1} + A^{-1})^{-1} \\ (2y)^{-1} &= x^{-1} + A^{-1} \\ x^{-1} &= 0.5y^{-1} - A^{-1} \\ x &= \left(0.5y^{-1} - A^{-1}\right)^{-1} \\ \text{so}\quad f^{-1}(x) &= \left(0.5x^{-1} - A^{-1}\right)^{-1}. \end{aligned}$$

52. **(a)** Yes, given Table 10.6.

Table 10.6

x	-9	-8	-5	-4	6	7	9
$f^{-1}(x)$	3	2	1	0	-1	-2	-3

(b) No, because for example, $g(-3) = g(-1) = g(3) = 3$. Therefore, g^{-1} cannot exist, as we would be unable to determine whether $g^{-1}(3) = -1$ or $g^{-1}(3) = -3$ or $g^{-1}(3) = 3$.

(c)

Table 10.7

x	-3	-2	-1	0	1	2	3
$f(g(x))$	-9	-5	-9	-8	9	6	-9

(d) No element of the range of $f(x)$ is in the domain of $g(x)$. Therefore, $g(f(x))$ will be undefined for all values of x given by the above table.

53. We have

$$\begin{aligned} f(-1) &= -1 \cdot e^{-1} = -\frac{1}{e} && \text{so } \left(-1, -\frac{1}{e}\right) \text{ is on graph of } f \\ f(0) &= 0 \cdot e^{0} \quad\; = 0 && \text{so } (0,0) \text{ is on graph of } f \\ f(1) &= 1 \cdot e^{1} \quad\; = e. && \text{so } (1,e) \text{ is on graph of } f \end{aligned}$$

If (x, y) is a point on the graph of f, then (y, x) is a point on the graph of its inverse. This means three points on the graph of W are $(-1/e, -1)$, $(0, 0)$, and $(e, 1)$.

54. We have:

$$\begin{aligned} te^{2t} &= 5 \\ 2te^{2t} &= 10. \quad \text{multiply by 2} \end{aligned}$$

We know that $W(10)e^{W(10)} = 10$. This means

$$\underbrace{2t}_{W(10)} e^{\overbrace{2t}^{W(10)}} = 10$$

$$\begin{aligned} 2t &= W(10) \\ t &= 0.5W(10) \\ &= 0.5(1.746) \quad \text{since } W(10) = 1.746 \\ &= 0.873. \end{aligned}$$

Checking this answer in our original equation, we find

$$\begin{aligned} te^{2t} &= 0.873e^{2(0.873.)} \\ &= 5.004, \end{aligned}$$

which is correct within rounding.

55. **(a)** If $f(t)$ is exponential, then $f(t) = AB^t$, and

$$f(12) = AB^{12} = 20$$
$$\text{and} \quad f(7) = AB^7 = 13.$$

Taking the ratios we have
$$\frac{AB^{12}}{AB^7} = \frac{20}{13}$$
$$B^5 = \frac{20}{13}$$
$$B = \left(\frac{20}{13}\right)^{\frac{1}{5}} \approx 1.08998.$$

Substituting this value into $f(7)$, we get

$$f(7) = A(1.08998)^7 = 13$$
$$A = \frac{13}{(1.08998)^7} \approx 7.112.$$

Using these values of A and B, we have

$$f(t) = 7.112(1.08998)^t.$$

(b) Let $P = 7.112(1.08998)^t$, then solve for t:

$$P = 7.112(1.08998)^t$$
$$P/7.112 = 1.08998^t$$
$$\log 1.08998^t = \log\left(\frac{P}{7.112}\right)$$
$$t \log 1.08998 = \log\left(\frac{P}{7.112}\right)$$
$$t = \frac{\log(P/7.112)}{\log 1.08998}.$$

Thus

$$f^{-1}(P) = \frac{\log(P/7.112)}{\log 1.08998}.$$

(c)

$$f(25) = 7.112(1.08998)^{25} \approx 61.299.$$

This means that in year 25, the population is approximately 61,300 people.

$$f^{-1}(25) = \frac{\log(25/7.112)}{\log 1.08998} \approx 14.590.$$

This means that when the population is 25,000, the year is approximately 14.590.

56. **(a)** Her maximum height is approximately 36 m.

(b) She lands on the trampoline approximately 6 seconds later.

(c) The graph is a parabola, and hence is symmetric about the vertical line $x = 3$ through its vertex. We choose the right half of the parabola, whose domain is the interval $3 \leq t \leq 6$. See Figure 10.9.

(d) The gymnast is part of a complicated stunt involving several phases. At 3 seconds into the stunt, she steps off the platform for the high wire, 36 meters in the air. Three seconds later (at $t = 6$) she lands on a trampoline at ground level.

(e) See Figure 10.10. As the gymnast falls, her height decreases steadily from 36 m to 0 m. In other words, she occupies each height for one moment of time only, and does not return to that height at any other time. This means that each value of t corresponds to a single height, and therefore time is a function of height.

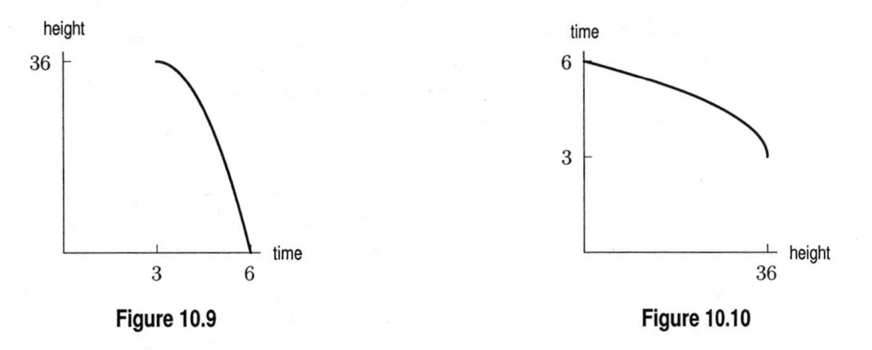

Figure 10.9

Figure 10.10

57. **(a)** $C(0)$ is the concentration of alcohol in the 100 ml solution after 0 ml of alcohol is removed. Thus, $C(0) = 99\%$.

(b) Note that there are initially 99 ml of alcohol and 1 ml of water.

$$C(x) = \begin{array}{c}\text{Concentration of alcohol}\\ \text{after removing } x \text{ ml}\end{array} = \frac{\text{Amount of alcohol remaining}}{\text{Amount of solution remaining}}$$

$$= \frac{\begin{array}{c}\text{Original amount}\\ \text{of alcohol}\end{array} - \begin{array}{c}\text{Amount of}\\ \text{alcohol removed}\end{array}}{\begin{array}{c}\text{Original amount}\\ \text{of solution}\end{array} - \begin{array}{c}\text{Amount of alcohol}\\ \text{removed}\end{array}} = \frac{99 - x}{100 - x}.$$

(c) If $y = C(x)$, then $x = C^{-1}(y)$. We have

$$\begin{aligned} y &= \frac{99 - x}{100 - x} \\ y(100 - x) &= 99 - x \\ 100y - xy &= 99 - x \\ x - xy &= 99 - 100y \\ x(1 - y) &= 99 - 100y \\ x &= \frac{99 - 100y}{1 - y}. \end{aligned}$$

Thus, $C^{-1}(y) = \dfrac{99 - 100y}{1 - y}$.

(d) The function $C^{-1}(y)$ tells us how much alcohol we need to remove in order to obtain a solution whose concentration is y.

58. **(a)** Since the amount of alcohol goes from 99% to 98%, you might expect that 1% of the alcohol should be removed. There are originally 99 ml of alcohol, and 1% of 99 = 0.99, and so this would mean that 0.99 ml needs to be removed. (As we shall see, this turns out to be completely wrong!)

(b) If $y = C(x)$, then y is the concentration of alcohol after x ml are removed. Since we want to remove an amount x of alcohol to yield a 98% solution, we have

$$C(x) = 0.98.$$

This means that

$$x = C^{-1}(0.98).$$

(c) We have

$$\begin{aligned} C^{-1}(0.98) &= \frac{99 - 100(0.98)}{1 - 0.98} \\ &= \frac{99 - 98}{0.02} \\ &= 50. \end{aligned}$$

Thus, we would need to remove 50 ml of alcohol from the 100 ml solution to obtain a 98% solution. This is much more than the 0.99 ml you might have expected. We can double check our answer to be sure. If we begin with 100 ml of solution containing 99 ml of alcohol, and then remove 50 ml of alcohol, we obtain a 50 ml solution containing 49 ml of alcohol. This gives a concentration of $\frac{49}{50} = 98\%$, which is what we wanted.

59. $f(2q_c)$ is a pulse: the predicted pulse of a person having 2 servings of coffee.

60. Since f is assumed to be an increasing function, its inverse is well-defined. This is an amount of caffeine: the amount predicted to give a pulse 20 bpm higher than r_c, that is, 20 bpm higher than the pulse of a person having 1 serving of coffee.

61. Since f is assumed to be an increasing function, its inverse is well-defined. This is an amount of caffeine. Notice that $r_c = f(q_c)$, and so $q_c = f^{-1}(r_c)$. Thus, $2f^{-1}(r_c) + 20 = 2q_c + 20$, making this 20 mg more caffeine than in 2 servings of coffee.

62. This is a pulse in bpm, or more specifically a *change* in pulse, since subtraction measures change. We know that $f(q_c)$ is the pulse of a person having 1 serving of coffee, and that $f(0)$ is the pulse of a person having no caffeine at all. Thus, this is the predicted *increase* in the pulse of a person after having 1 serving of coffee.

63. Since f is assumed to be an increasing function, its inverse is well-defined. This is an amount of caffeine, or more specifically a *change* in the amount of caffeine. We know from part (b) that $f^{-1}(r_c + 20)$ is the amount of caffeine resulting in a pulse 20 bpm higher than r_c. Thus, this is the amount of extra caffeine required after drinking a serving of coffee in order for the pulse to go up an extra 20 bpm.

64. Since f is assumed to be an increasing function, its inverse is well-defined. This is an amount of caffeine. We know that $1.1f(q_c) = 1.1r_c$ is 10% higher than the pulse of a person who has had 1 serving of coffee. This makes $f^{-1}(1.1f(q_c))$ is the amount of caffeine that will lead to a pulse 10% higher than will a serving of coffee.

Solutions for Section 10.3

Exercises

1. **(a)** We have $f(x) + g(x) = x + 1 + 3x^2 = 3x^2 + x + 1$.
(b) We have $f(x) - g(x) = x + 1 - 3x^2 = -3x^2 + x + 1$.
(c) We have $f(x)g(x) = (x + 1)(3x^2) = 3x^3 + 3x^2$.
(d) We have $f(x)/g(x) = (x + 1)/(3x^2)$.

2. **(a)** We have $f(x) + g(x) = x^2 + 4 + x + 2 = x^2 + x + 6$.
(b) We have $f(x) - g(x) = x^2 + 4 - (x + 2) = x^2 - x + 2$.
(c) We have $f(x)g(x) = (x^2 + 4)(x + 2) = x^3 + 2x^2 + 4x + 8$.
(d) We have $f(x)/g(x) = (x^2 + 4)/(x + 2) = ((x + 2)(x - 2))/(x + 2) = x - 2$.

3. **(a)** We have $f(x) + g(x) = x + 5 + x - 5 = 2x$.
(b) We have $f(x) - g(x) = x + 5 - (x - 5) = 10$.
(c) We have $f(x)g(x) = (x + 5)(x - 5) = x^2 - 25$.
(d) We have $f(x)/g(x) = (x + 5)/(x - 5)$.

4. **(a)** We have $f(x) + g(x) = x^2 + 4 + x^2 + 2 = 2x^2 + 6$.
(b) We have $f(x) - g(x) = x^2 + 4 - (x^2 + 2) = 2$.
(c) We have $f(x)g(x) = (x^2 + 4)(x^2 + 2) = x^4 + 6x^2 + 8$.
(d) We have $f(x)/g(x) = (x^2 + 4)/(x^2 + 2)$.

5. **(a)** We have $f(x) + g(x) = x^3 + x^2$.
(b) We have $f(x) - g(x) = x^3 - x^2$.
(c) We have $f(x)g(x) = (x^3)(x^2) = x^5$.
(d) We have $f(x)/g(x) = (x^3)/(x^2) = x$.

6. **(a)** We have $f(x) + g(x) = \sqrt{x} + x^2 + 2 = x^2 + \sqrt{x} + 2$.

(b) We have $f(x) - g(x) = \sqrt{x} - (x^2 + 2) = -x^2 + \sqrt{x} - 2$.
(c) We have $f(x)g(x) = \sqrt{x}(x^2 + 2) = x^{5/2} + 2\sqrt{x}$.
(d) We have $f(x)/g(x) = \sqrt{x}/(x^2 + 2)$.

7. $f(x) = (2x - 1) + (1 - x) = x$.

8. $g(x) = (1 - x) \cdot \frac{1}{x} = \frac{1}{x} - 1$.

9. $h(x) = 2(2x - 1) - 3(1 - x) = 4x - 2 - 3 + 3x = 7x - 5$.

10. $j(x) = \dfrac{2x - 1}{\frac{1}{x}} = x \cdot (2x - 1) = 2x^2 - x$.

11. $k(x) = (1 - x)^2 = (1 - x)(1 - x) = 1 - 2x + x^2$.

12. $l(x) = (2x - 1) - (1 - x) - \frac{1}{x} = 2x - 1 - 1 + x - \frac{1}{x} = 3x - \frac{1}{x} - 2$.

13. We have $f(x) = e^x(2x + 1) = 2xe^x + e^x$.

14. We have

$$\begin{aligned} g(x) &= (e^x)^2 + (2x + 1)^2 \\ &= e^{2x} + 4x^2 + 4x + 1. \end{aligned}$$

15. To obtain our formula, we must first calculate $v(u(x))$, and then square this result. We therefore have

$$h(x) = [v(u(x))]^2 = (2e^x + 1)^2 = 4e^{2x} + 4e^x + 1.$$

16. In this case, we must first square $u(x)$, and then plug this result into the function v. We therefore have

$$k(x) = v((e^x)^2) = v(e^{2x}) = 2e^{2x} + 1.$$

17. $f(x) + g(x) = \sin x + x^2$.

18. $g(x)f(x) = x^2 \sin x$.

19. $\dfrac{f(x)}{g(x)} = \dfrac{\sin x}{x^2}$.

20. $f(g(x)) = f(x^2) = \sin(x^2)$.

21. $g(f(x)) = g(\sin x) = \sin^2 x$.

22. $1 - (f(x))^2 = 1 - (\sin x)(\sin x) = 1 - \sin^2 x = \cos^2 x$.

Problems

23. Since $h(x) = f(x) + g(x)$, we know that $h(-1) = f(-1) + g(-1) = -4 + 4 = 0$. Similarly, $j(x) = 2f(x)$ tells us that $j(-1) = 2f(-1) = 2(-4) = -8$. Repeat this process for each entry in the table.

Table 10.8

x	$h(x)$	$j(x)$	$k(x)$	$m(x)$
-1	0	-8	16	-1
0	0	-2	1	-1
1	2	4	0	0
2	6	10	1	0.2
3	12	16	16	0.5
4	20	22	81	9/11

24. **(a)** A formula for $h(x)$ would be

$$h(x) = f(x) + g(x).$$

To evaluate $h(x)$ for $x = 3$, we use this equation:

$$h(3) = f(3) + g(3).$$

Since $f(x) = x + 1$, we know that

$$f(3) = 3 + 1 = 4.$$

Likewise, since $g(x) = x^2 - 1$, we know that

$$g(3) = 3^2 - 1 = 9 - 1 = 8.$$

Thus, we have

$$h(3) = 4 + 8 = 12.$$

To find a formula for $h(x)$ in terms of x, we substitute our formulas for $f(x)$ and $g(x)$ into the equation $h(x) = f(x) + g(x)$:

$$h(x) = \underbrace{f(x)}_{x+1} + \underbrace{g(x)}_{x^2-1}$$

$$h(x) = x + 1 + x^2 - 1 = x^2 + x.$$

To check this formula, we use it to evaluate $h(3)$, and see if it gives $h(3) = 12$, which is what we got before. The formula is $h(x) = x^2 + x$, so it gives

$$h(3) = 3^2 + 3 = 9 + 3 = 12.$$

This is the result that we expected.

(b) A formula for $j(x)$ would be

$$j(x) = g(x) - 2f(x).$$

To evaluate $j(x)$ for $x = 3$, we use this equation:

$$j(3) = g(3) - 2f(3).$$

We already know that $g(3) = 8$ and $f(3) = 4$. Thus,

$$j(3) = 8 - 2 \cdot 4 = 8 - 8 = 0.$$

To find a formula for $j(x)$ in terms of x, we again use the formulas for $f(x)$ and $g(x)$:

$$\begin{aligned} j(x) &= \underbrace{g(x)}_{x^2-1} - 2\underbrace{f(x)}_{x+1} \\ &= (x^2 - 1) - 2(x + 1) \\ &= x^2 - 1 - 2x - 2 \\ &= x^2 - 2x - 3. \end{aligned}$$

We check this formula using the fact that we already know $j(3) = 0$. Since we have $j(x) = x^2 - 2x - 3$,

$$j(3) = 3^2 - 2 \cdot 3 - 3 = 9 - 6 - 3 = 0.$$

This is the result that we expected.

(c) A formula for $k(x)$ would be

$$k(x) = f(x)g(x).$$

Evaluating $k(3)$, we have

$$k(3) = f(3)g(3) = 4 \cdot 8 = 32.$$

A formula in terms of x for $k(x)$ would be

$$\begin{aligned} k(x) &= \underbrace{f(x)}_{x+1} \cdot \underbrace{g(x)}_{x^2-1} \\ &= (x+1)(x^2-1) \\ &= x^3 - x + x^2 - 1 \\ &= x^3 + x^2 - x - 1. \end{aligned}$$

To check this formula,

$$k(3) = 3^3 + 3^2 - 3 - 1 = 27 + 9 - 3 - 1 = 32,$$

which agrees with what we already knew.

(d) A formula for $m(x)$ would be

$$m(x) = \frac{g(x)}{f(x)}.$$

Using this formula, we have

$$m(3) = \frac{g(3)}{f(3)} = \frac{8}{4} = 2.$$

To find a formula for $m(x)$ in terms of x, we write

$$\begin{aligned} m(x) = \frac{g(x)}{f(x)} &= \frac{x^2-1}{x+1} \\ &= \frac{(x+1)(x-1)}{(x+1)} \\ &= x - 1 \text{for } x \neq -1 \end{aligned}$$

We were able to simplify this formula by first factoring the numerator of the fraction $\frac{x^2-1}{x+1}$. To check this formula,

$$m(3) = 3 - 1 = 2,$$

which is what we were expecting.

(e) We have

$$n(x) = (f(x))^2 - g(x).$$

This means that

$$\begin{aligned} n(3) &= (f(3))^2 - g(3) \\ &= (4)^2 - 8 \\ &= 16 - 8 \\ &= 8. \end{aligned}$$

A formula for $n(x)$ in terms of x would be

$$\begin{aligned} n(x) &= (f(x))^2 - g(x) \\ &= (x+1)^2 - (x^2-1) \\ &= x^2 + 2x + 1 - x^2 + 1 \\ &= 2x + 2. \end{aligned}$$

To check this formula,

$$n(3) = 2 \cdot 3 + 2 = 8,$$

which is what we were expecting.

25. **(a)** Since the population consists only of men and women, the population size at any given time t will be the sum of the numbers of women and the number of men at that particular time. Thus

$$p(t) = f(t) + g(t).$$

(b) In any given year the total amount of money that women in Canada earn is equal to the average amount of money one woman makes in that year times the number of women. Thus

$$m(t) = g(t) \cdot h(t).$$

26.

$$\begin{aligned} f(g(65)) &= f(50) \quad \text{Because } g(65) = 50 \\ &= 65. \end{aligned}$$

27.

$$\begin{aligned} v(50) &= g(50)f(50) \\ &= 70 \cdot 65 = 4550. \end{aligned}$$

28. Using $f(x) = \dfrac{1}{x+1}$, we obtain

$$\begin{aligned} f\left(\frac{1}{x}\right) + \frac{1}{f(x)} &= \frac{1}{\frac{1}{x}+1} + \frac{1}{\frac{1}{x+1}} \\ &= \frac{1}{\frac{1+x}{x}} + x + 1 \\ &= \frac{x}{1+x} + x + 1 \end{aligned}$$

29. To compute this table, note that since $f(x) = r(x) + t(x)$, then $f(-2) = r(-2) + t(-2) = 4 + 8 = 12$ and since $g(x) = 4 - 2s(x)$, then $g(-2) = 4 - 2(-2) = 4 + 4 = 8$. Repeat this process for each entry in the table.

Table 10.9

x	$f(x)$	$g(x)$	$h(x)$	$j(x)$	$k(x)$	$l(x)$
-2	12	8	32	2	16	-12
-1	10	0	25	0	25	15
0	13	8	42	0.5	36	-8
1	4	0	-21	5	49	1
2	10	8	16	-3	64	4
3	22	0	117	-2	81	35

30. We have

$$\begin{aligned} H(x) &= F(G(x)) \\ &= F\left(\sqrt{x}\right) \\ &= \cos\left(\sqrt{x}\right) \\ h(x) &= f\left(G(x)\right) \cdot g(x) \\ &= f\left(\sqrt{x}\right) \cdot \frac{1}{2\sqrt{x}} \\ &= -\sin\left(\sqrt{x}\right) \cdot \frac{1}{2\sqrt{x}} \\ &= -\frac{\sin\left(\sqrt{x}\right)}{2\sqrt{x}}. \end{aligned}$$

31. Where $g(x) = 0$, we have $h(x) = f(x) \cdot 0 = 0$, and where $g(x) = 1$, $h(x) = f(x) \cdot 1 = f(x)$. Thus, h has an x-intercept wherever the graph of g does and h intersects the graph of f wherever $g(x) = 1$. Figure 10.11 shows the graph of h together with the graph of f dashed in.

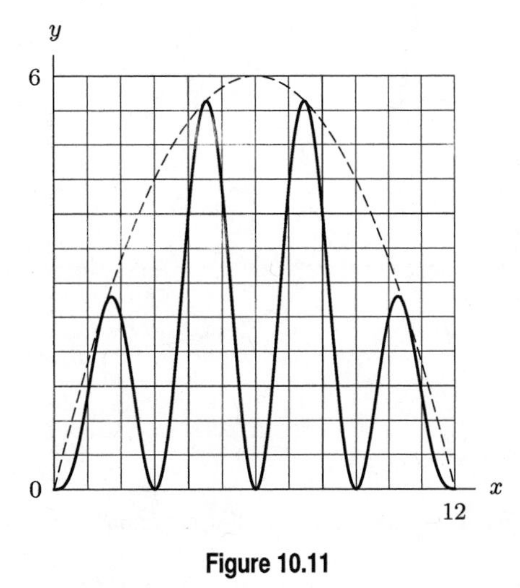

Figure 10.11

32. We have:

$$\begin{aligned} p(0) &= v(0)w(0) = 4 \cdot 2 = 8 \\ p(1) &= v(1)w(1) = 3 \cdot 1 = 3 \\ p(2) &= v(2)w(2) = 3 \cdot 3 = 9 \\ p(3) &= v(3)w(3) = 5 \cdot 4 = 20 \\ p(4) &= v(4)w(4) = 4 \cdot 0 = 0 \\ p(5) &= v(5)w(5) = 4 \cdot 5 = 20. \end{aligned}$$

See Table 10.10.

Table 10.10

x	0	1	2	3	4	5
$p(x)$	8	3	9	20	0	20

33. We have:

$$\begin{aligned}
q(0) &= w^{-1}(v(0)) = w^{-1}(4) = 3 \quad \text{because } v(0) = 4 \text{ and } w(3) = 4\\
q(1) &= w^{-1}(v(1)) = w^{-1}(3) = 2 \quad \text{because } v(1) = 3 \text{ and } w(2) = 3\\
q(2) &= w^{-1}(v(2)) = w^{-1}(3) = 2 \quad \text{because } v(2) = 3 \text{ and } w(2) = 3\\
q(3) &= w^{-1}(v(3)) = w^{-1}(5) = 5 \quad \text{because } v(3) = 5 \text{ and } w(5) = 5\\
q(4) &= w^{-1}(v(4)) = w^{-1}(4) = 3 \quad \text{because } v(4) = 4 \text{ and } w(3) = 4\\
q(5) &= w^{-1}(v(5)) = w^{-1}(4) = 3. \quad \text{because } v(5) = 4 \text{ and } w(3) = 4
\end{aligned}$$

See Table 10.11.

Table 10.11

x	0	1	2	3	4	5
$q(x)$	3	2	2	5	3	3

34. **(a)** See Figure 10.12.

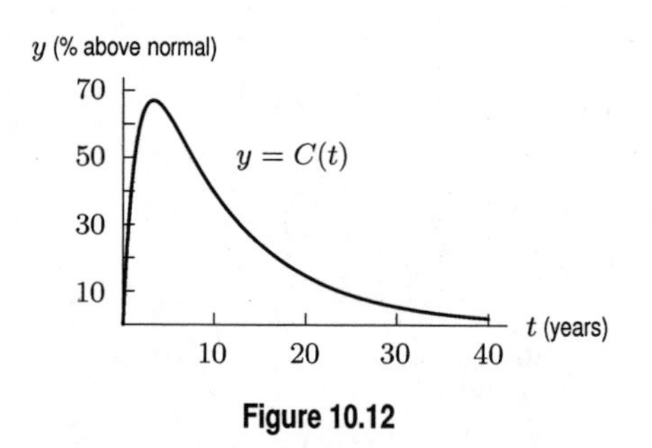

Figure 10.12

(b) From the graph we estimate that $t = 3.25$ years gives the maximum value of $C(t)$. The carbon-14 level is highest at that time.

(c) Since $C(t)$ tends toward 0, the CO_2 level eventually returns to normal.

35. We can find the revenue function as a product:

$$\text{Revenue} = (\#\text{ of customers}) \cdot (\text{price per customer}).$$

At the current price, 50,000 people attend every day. Since 2500 customers will be lost for each $1 increase in price, the function $n(i)$ giving the number of customers who will attend given i one-dollar price increases, is given by $n(i) = 50{,}000 - 2500i$. The price function $p(i)$ giving the price after i one-dollar price increases is given by $p(i) = 15 + i$. The revenue function $r(i)$ is given by

$$\begin{aligned}
r(i) &= n(i)p(i)\\
&= (50{,}000 - 2500i)(15 + i)\\
&= -2500i^2 + 12{,}500i + 750{,}000\\
&= -2500(i - 20)(i + 15).
\end{aligned}$$

The graph $r(i)$ is a downward-facing parabola with zeros at $i = -15$ and $i = 20$, so the maximum revenue occurs at $i = 2.5$ which is halfway between the zeros. Thus, to maximize profits the ideal price is $\$15 + 2.5(\$1.00) = \$15 + \$2.50 = \$17.50$.

36. See Figure 10.13.

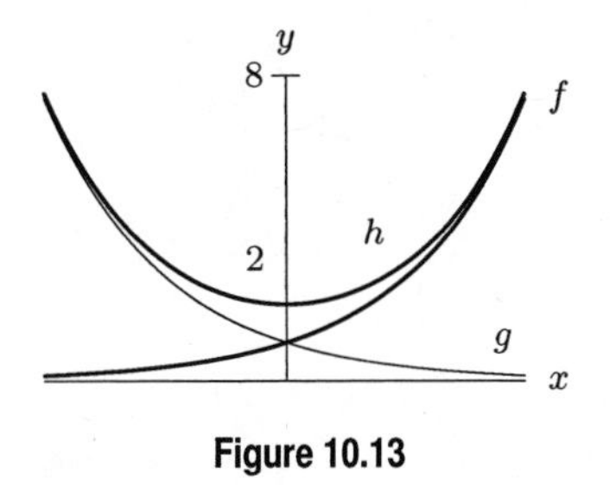

Figure 10.13

37. Since $f(a) = g(a)$, $h(a) = g(a) - f(a) = 0$. Similarly, $h(c) = 0$. On the interval $a < x < b$, $g(x) > f(x)$, so $h(x) = g(x) - f(x) > 0$. As x increases from a to b, the difference between $g(x)$ and $f(x)$ gets greater, becoming its greatest at $x = b$, then gets smaller until the difference is 0 at $x = c$. When $x < a$ or $x > b$, $g(x) < f(x)$ so $g(x) - f(x) < 0$. Subtract the length e from the length d to get the y-intercept. See Figure 10.14.

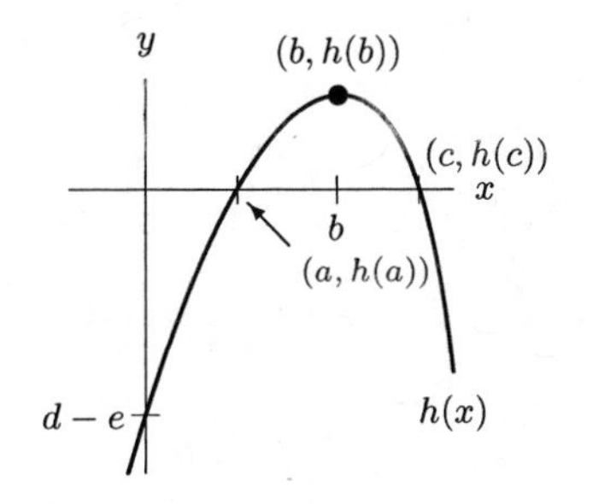

Figure 10.14

38. **(a)** See Table 10.12.

Table 10.12

t (yrs)	0	1	2	3	4	5
$f(t)$ ($)	127,300	132,479	132,086	136,136	138,666	146,822
$g(t)$	8.104	8.372	8.373	8.570	8.501	8.656

(b) f is the dollar difference between the cutoff income for a household in the 95thpercentile and a household in the 20thpercentile. In other words, f tells us how much more money a household at the bottom of the higher income bracket makes than a house at the top of the lower income bracket. In contrast, g is the ratio of these incomes. Notice that the dollar difference in incomes is rising for every year shown by the table (except for 2002 when it dropped slightly), ending (in 2005) with a difference of $146,822. In contrast, the ratio is much steadier, increasing from 8.104 to 8.656. This means that while the dollar gap between rich and poor households grew considerably during this time period (from about $127,300 to about $146,822), the rich households earned about 8.4 times the income of the poor households, with slight variation.

39. **(a)** Since the zeros of this quadratic function are 0 and 4, the formula for this function will be of the form $f(x) = kx(x-4)$, which is the product of two linear functions, $a(x) = kx$ and $b(x) = x - 4$. Figure 10.15 shows a possible graph of these two functions.

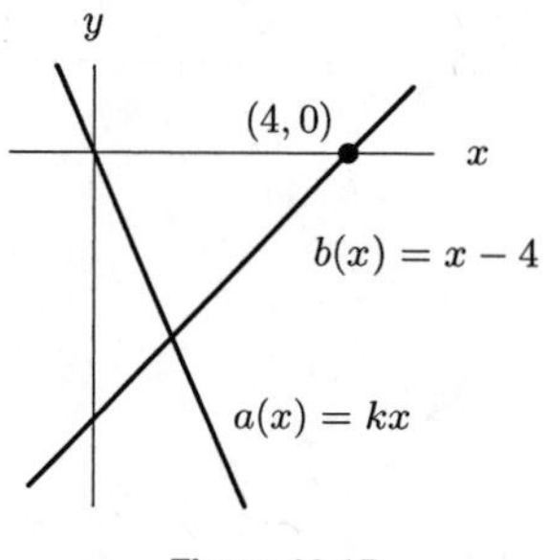

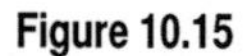

Figure 10.15

(b) If the formula of a quadratic function could be written as the product of two linear functions, $g(x) = (ax+b)(cx+d)$, then the function must have zeros at $-\frac{b}{a}$ and $-\frac{d}{c}$ (from $ax + b = 0$ and $cx + d = 0$). So any such quadratic function would have at least one zero (when $-\frac{b}{a}$ equals $-\frac{d}{c}$) and, more likely, two zeros. The function $y = q(x)$ has no zeros and therefore, cannot be the product of two linear functions.

40. Where $g(x) = f(x)$, we see that $h(x) = g(x) - f(x) = 0$. Thus, the graph of h has an x-intercept wherever the graphs of f and g cross. See Figure 10.16.

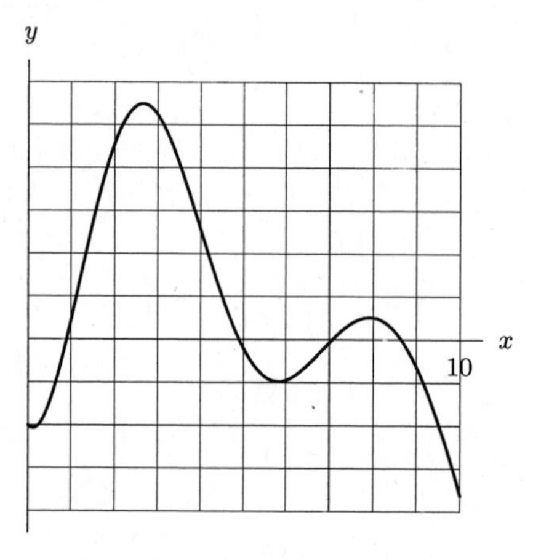

Figure 10.16

41. We have

$$
\begin{aligned}
H(x) &= \frac{F(x)}{G(x)} \\
&= \frac{e^{-x^2}}{x^4} = x^{-4} \cdot e^{-x^2} \\
h(x) &= \frac{f(x)G(x) - G(x)g(x)}{(G(x))^2}. \\
&= \frac{-2xe^{-x^2} \cdot x^4 - e^{-x^2} \cdot 4x^3}{(x^4)^2} \\
&= \frac{-2x^5e^{-x^2} - 4x^3e^{-x^2}}{x^8} = -2x^{-3}\left(1 + 2x^{-2}\right)e^{-x^2}.
\end{aligned}
$$

42. (a) Since the initial amount was 316.75 and the growth factor is 1.004, we have $A(t) = 316.75(1.004)^t$.

(b) Since the CO_2 level oscillates once per year, the period is 1 year. The amplitude is 3.25 pm, so one possible answer is $V(t) = 3.25\sin(2\pi t)$. Any sinusoidal function with the same amplitude and period could describe the variation.

(c) The graph of $y = 316.75(1.004)^t + 3.25\sin(2\pi t)$ is in Figure 10.17.

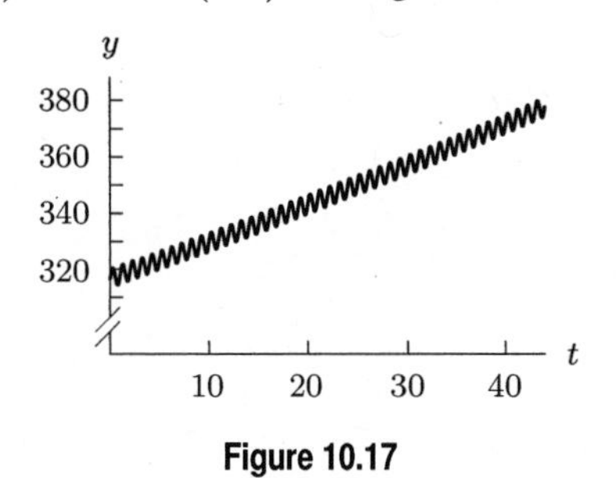

Figure 10.17

43. (a) The function $h_{\text{CA}}(t)$ gives the total number of pounds (in 1000s) of strawberries produced in California in year t. See Table 10.13.

Table 10.13 *Values for* $h_{CA}(t) = f_{CA}(t) \cdot g_{CA}(t)$

t	0	1	2	3	4
$h_{\text{CA}}(t)$	1,958,800	2,058,000	2,112,200	2,147,750	2,274,800

(b) A formula for $p(t)$ is given by

$$\begin{aligned} p(t) &= \frac{\text{Strawberries in CA and FL}}{\text{Strawberries in US total}} \\ &= \frac{\text{CA area} \times \text{CA yield} + \text{FL area} \times \text{FL yield}}{\text{US area} \times \text{US yield}} \\ &= \frac{f_{\text{CA}}(t) \cdot g_{\text{CA}}(t) + f_{\text{FL}}(t) \cdot g_{\text{FL}}(t)}{f_{\text{US}}(t) \cdot g_{\text{US}}(t)}. \end{aligned}$$

See Table 10.14.

Table 10.14 *Values for* $p(t)$*, the fraction of all US strawberries (by weight) grown in Florida and California in year t.*

t	0	1	2	3	4
$p(t)$	0.958	0.963	0.963	0.964	0.969

44. (a) Notice l_2 has slope -1 and has a y-intercept of 1. So the equation of l_2 is $y = -x + 1$. From Figure 10.18 we see $\tan\theta = y/x$. So $y = x\tan\theta$. Since points on the line l_2 satisfy both $y = -x + 1$ and $y = x\tan\theta$, we have

$$\underbrace{1 - y}_{x} = \underbrace{\frac{y}{\tan\theta}}_{x}.$$

We now solve for y:

$$\begin{aligned} y &= (1 - y)\tan\theta \\ y &= \tan\theta - y\tan\theta \\ y + y\tan\theta &= \tan\theta \\ y(1 + \tan\theta) &= \tan\theta \\ y &= \frac{\tan\theta}{1 + \tan\theta}. \end{aligned}$$

So

$$f(\theta) = \frac{\tan\theta}{1 + \tan\theta}.$$

(b) See Figure 10.19.

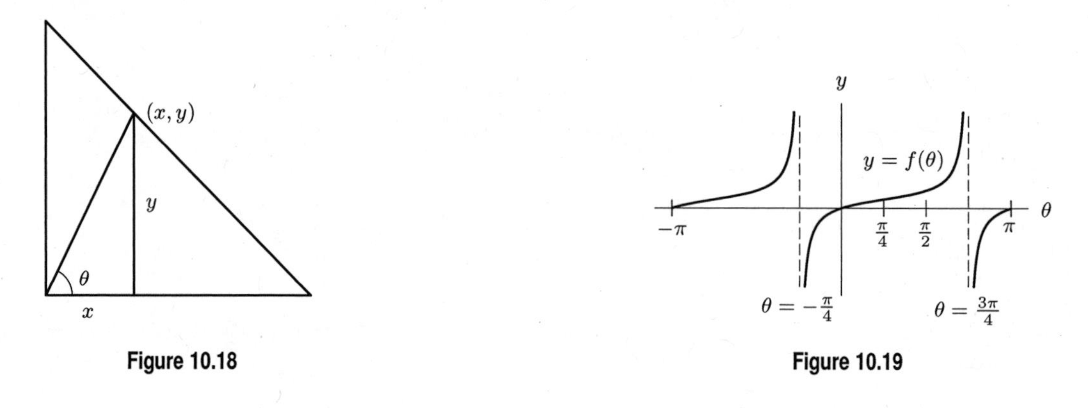

Figure 10.18

Figure 10.19

(c) Notice that y is undefined when $\theta = -\pi/4$ and when $\theta = 3\pi/4$. The reason is that line l_1 is parallel to l_2 for these angles, and no intersection occurs. See the figure in the problem.

As θ goes from $-\pi/4$ up to 0, the value of y increases from $-\infty$ to 0. This makes sense because for θ near $-\pi/4$, line l_1 intersects l_2 at very negative y-values; but when $\theta = 0$, the value of y equals zero. As θ increases from 0 to $3\pi/4$, the value of y increases to infinity, because l_1 intersects l_2 at very large y-values as θ approaches $3\pi/4$. Since $\tan\theta$ is periodic with period π, the function $f(\theta)$ repeats itself over and over again. Therefore, $f(\theta)$ is periodic.

45. In order to evaluate $h(3)$, we need to express the formula for $h(x)$ in terms of $f(x)$ and $g(x)$. Factoring gives

$$h(x) = C^{2x}(kx^2 + B + 1).$$

Since $g(x) = C^{2x}$ and $f(x) = kx^2 + B$, we can re-write the formula for $h(x)$ as

$$h(x) = g(x) \cdot (f(x) + 1).$$

Thus,

$$\begin{aligned} h(3) &= g(3) \cdot (f(3) + 1) \\ &= 5(7 + 1) \\ &= 40. \end{aligned}$$

46. The equation $f(2000) = 200{,}000$ tells us that Ace estimates that 2000 square feet of office space costs \$200,000.

47. Since $g(x) = \dfrac{f(x)}{x}$,

$$g(2000) = \frac{f(2000)}{2000} = \frac{200{,}000}{2000} = 100.$$

The value $g(2000)$ represents the dollar cost per square foot for building 2000 square feet of office space.

48. Since $2x$ represents twice as much office space as x, the cost of building twice as much space is $f(2x)$. The cost of building x amount of space is $f(x)$, so twice this cost is $2f(x)$. Thus, the contractors statement is expressed

$$f(2x) < 2f(x).$$

49. $g(q) < g(p)$ because the cost per square foot of building office space decreases as the total square footage increases. $g(p) < f(p)$ since the total cost of building more than one square foot is greater than the cost per square foot. $f(p) < f(q)$ since the total cost of building office space increases as the square footage increases. So

$$g(q) < g(p) < f(p) < f(q).$$

50. The equation tells us that Space estimates that 1500 square feet of office space costs $200,000.

51. We want to divide the cost of the office space by the number of square feet:

$$j(x) = \frac{x}{h(x)}.$$

52. The inequality $h(f(x)) < x$ tells us that Space can build fewer than x square feet of office space with the money Ace needs to build x square feet. You get more for your money with Ace.

Solutions for Chapter 10 Review

Exercises

1. Using substitution we have $h(k(x)) = 2^{k(x)} = 2^{x^2}$ and $k(h(x)) = (h(x))^2 = (2^x)^2 = 2^{2x} = 4^x$.

2. We want to replace each x in the formula for $f(x)$ with the value of $g(x)$, that is, $\frac{1}{x-3}$. The result is $\left(\frac{1}{x-3}\right)^2 + 1 = \frac{1}{x^2-6x+9} + 1 = \frac{1+x^2-6x+9}{x^2-6x+9} = \frac{x^2-6x+10}{x^2-6x+9}$.

3. We replace the x's that appear in the formula for $g(x)$ with x^2+1, the expression for $f(x)$. This gives $\frac{1}{(x^2+1)-3} = \frac{1}{x^2-2}$.

4. Replacing the x's in the formula for $f(x)$ with $\sqrt{x}$ gives $(\sqrt{x})^2 + 1 = x + 1$.

5. Substituting the expression x^2+1 for the x term in the formula for $h(x)$ gives $\sqrt{x^2+1}$.

6. We take the expression for $g(x)$, namely $\frac{1}{x-3}$, and substitute it back into the same expression wherever an x appears. The result is $\frac{1}{\frac{1}{x-3}-3}$. We need to simplify the denominator: $\frac{1}{x-3} - 3 = \frac{1}{x-3} - \frac{3(x-3)}{x-3} = \frac{1-(3x-9)}{x-3} = \frac{10-3x}{x-3}$.
So, $\frac{1}{\frac{1}{x-3}-3} = \frac{1}{\frac{10-3x}{x-3}} = \frac{x-3}{10-3x}$.

7. Two substitutions have to take place here. We first find $f(h(x))$, which is $x+1$. Next, we replace each x in the formula for $g(x)$ with an $x+1$. This gives the final result

$$g(f(h(x))) = \frac{1}{(x+1)-3} = \frac{1}{x-2}.$$

8. To complete this table, we need to first evaluate $f(x)$ for each value of x and then find $g(f(x))$. For example, if $x = \pi/6$, then $f(\pi/6) = 1/2$, so $g(f(\pi/6) = g(1/2) = \pi/3$. Similarly, if $x = \pi/2$, then $f(\pi/2) = 1$, so $g(f(\pi/2)) = g(1) = 0$. The results are in Table 10.15.

Table 10.15

x	$g(f(x))$
0	$\pi/2$
$\pi/6$	$\pi/3$
$\pi/4$	$\pi/4$
$\pi/3$	$\pi/6$
$\pi/2$	0

9. **(a)** A graph of this function on a window which contains both positive and negative values of x reveals that it fails the horizontal line test. Therefore, this function is not invertible.

(b) A graph reveals that the output values of this function oscillate back and forth between $y = -1$ and $y = 1$. It is therefore not invertible.

(c) A graph reveals that this function is always increasing. It passes the horizontal line test and is therefore invertible.

10. Solve for x in $y = h(x) = 12x^3$:

$$\begin{aligned} y &= 12x^3 \\ x^3 &= \frac{y}{12} \\ x &= h^{-1}(y) = \sqrt[3]{\frac{y}{12}}. \end{aligned}$$

Writing h^{-1} in terms of x gives $h^{-1}(x) = \sqrt[3]{\frac{x}{12}}$.

11. Start with our property of inverse functions

$$h(h^{-1}(x)) = x,$$

and substitute y for $h^{-1}(x)$ to get $h(y) = x$. Now, using the formula for h we get

$$h(y) = \frac{y}{2y+1} = x$$

and solving for y yields

$$\begin{aligned} \frac{y}{2y+1} &= x \\ y &= x(2y+1) \\ &= 2yx + x && \text{multiplying through by the denominator} \\ y - 2yx &= x && \text{collecting } y\text{-terms on the left} \\ y(1-2x) &= x && \text{factoring} \\ y &= \frac{x}{1-2x}. \end{aligned}$$

Now replacing y by $h^{-1}(x)$, we have our formula, $h^{-1}(x) = \frac{x}{1-2x}$. The graphs of both functions are shown in Figure 10.20. Note that the two graphs are symmetric about the line $y = x$.

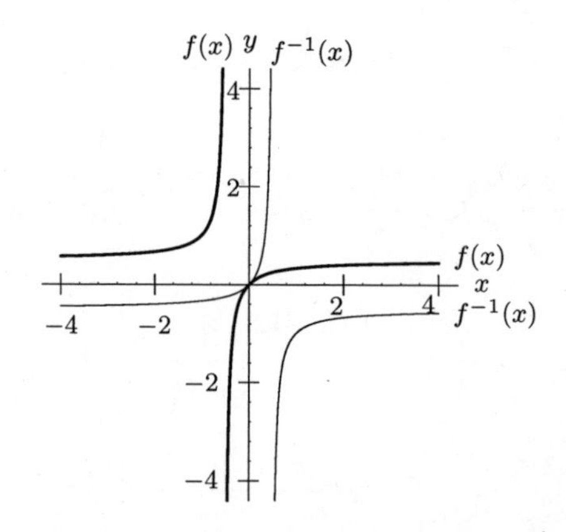

Figure 10.20: The graph of $h(x) = x/(2x+1)$ and the inverse $h^{-1}(x) = x/(1-2x)$

12. Start with $x = k(k^{-1}(x))$ and substitute $y = k^{-1}(x)$. We have

$$\begin{aligned} x &= k(y) \\ x &= 3e^{2y} \\ \frac{x}{3} &= e^{2y} \\ \ln\frac{x}{3} &= \ln e^{2y} = 2y \\ \frac{\ln\frac{x}{3}}{2} &= y \end{aligned}$$

So $y = k^{-1}(x) = \dfrac{\ln\frac{x}{3}}{2}$.

13. Start with $x = g(g^{-1}(x))$ and substitute $y = g^{-1}(x)$. We have

$$\begin{aligned} x &= g(y) \\ x &= e^{3y+1} \\ \ln x &= \ln e^{(3y+1)} \\ \ln x &= 3y + 1 \\ \ln x - 1 &= 3y \\ y &= \frac{1}{3}(\ln x - 1). \end{aligned}$$

Thus, $y = g^{-1}(x) = \dfrac{1}{3}(\ln x - 1)$.

14. Start with $x = n(n^{-1}(x))$ and substitute $y = n^{-1}(x)$. We have

$$\begin{aligned} x &= n(y) \\ x &= \log(y-3) \\ 10^x &= 10^{\log(y-3)} \\ 10^x &= y - 3 \\ y &= 10^x + 3 \end{aligned}$$

So $y = n^{-1}(x) = 10^x + 3$.

15. Start with $x = h(h^{-1}(x))$ and substitute $y = h^{-1}(x)$. We have

$$\begin{aligned} x &= h(y) \\ x &= \ln(1-2y) \\ e^x &= e^{\ln(1-2y)} \\ e^x &= 1 - 2y \\ y &= \frac{1}{2}(1 - e^x). \end{aligned}$$

Thus, $y = h^{-1}(x) = \dfrac{1}{2}(1 - e^x)$.

16. Solve for x in $y = h(x) = \sqrt{x}/(\sqrt{x}+1)$:

$$\begin{aligned} y &= \frac{\sqrt{x}}{\sqrt{x+1}} \\ \sqrt{x} &= y(\sqrt{x}+1) \\ \sqrt{x} &= y\sqrt{x} + y \\ \sqrt{x} - y\sqrt{x} &= y \end{aligned}$$

$$\begin{aligned}
\sqrt{x}(1-y) &= y \qquad \text{(factoring)}\\
\sqrt{x} &= \frac{y}{1-y}\\
x = h^{-1}(y) &= \left(\frac{y}{1-y}\right)^2.
\end{aligned}$$

Writing h^{-1} in terms of x gives

$$h^{-1}(x) = \left(\frac{x}{1-x}\right)^2.$$

17. Solving $y = g(x)$ for x gives

$$\begin{aligned}
y &= \frac{x-2}{2x+3}\\
(2x+3)y &= x-2\\
2xy+3y &= x-2\\
3y+2 &= x-2xy\\
x(1-2y) &= 3y+2\\
x &= \frac{3y+2}{1-2y},
\end{aligned}$$

so $g^{-1}(x) = \dfrac{3x+2}{1-2x}$.

18. Start with $x = f(f^{-1}(x))$ and let $y = f^{-1}(x)$. Then $x = f(y)$ means

$$\begin{aligned}
x &= \sqrt{\frac{4-7y}{4-y}}\\
x^2 &= \frac{4-7y}{4-y}\\
x^2(4-y) &= 4-7y\\
4x^2 - xy^2 &= 4-7y\\
4x^2-4 &= xy^2-7y\\
4x^2-4 &= y(x^2-7)\\
y &= \frac{4x^2-4}{x^2-7},
\end{aligned}$$

so $y = f^{-1}(x) = \dfrac{4x^2-4}{x^2-7}$.

19. Start with $x = f(f^{-1}(x))$ and let $y = f^{-1}(x)$. Then $x = f(y)$ means

$$\begin{aligned}
x &= \frac{\sqrt{y}+3}{11-\sqrt{y}}\\
x(11-\sqrt{y}) &= \sqrt{y}+3\\
11x - x\sqrt{y} &= \sqrt{y}+3\\
11x-3 &= \sqrt{y}+x\sqrt{y}\\
11x-3 &= \sqrt{y}(1+x)\\
\sqrt{y} &= \frac{11x-3}{1+x}\\
y &= \left(\frac{11x-3}{1+x}\right)^2,
\end{aligned}$$

so $y = f^{-1}(x) = \left(\dfrac{11x-3}{1+x}\right)^2$.

20. Solving $y = f(x)$ for x gives:

$$\begin{aligned} y &= \ln\left(1+\frac{1}{x}\right) \\ e^y &= 1+\frac{1}{x} \\ \frac{1}{x} &= e^y - 1 \\ x &= \frac{1}{e^y-1}, \end{aligned}$$

so $f^{-1}(x) = \dfrac{1}{e^x-1}$.

21. Start with $x = s(s^{-1}(x))$ and substitute $y = s^{-1}(x)$. We have

$$\begin{aligned} x &= s(y) \\ x &= \frac{3}{2+\log y} \\ \frac{x}{3} &= \frac{1}{2+\log y} \\ \frac{3}{x} &= 2+\log y \\ \frac{3}{x} - 2 &= \log y \\ 10^{\frac{3}{x}-2} &= 10^{\log y} \\ 10^{\frac{3}{x}-2} &= y \end{aligned}$$

So $s^{-1}(x) = 10^{(3/x)-2}$.

22. Start with $x = q(q^{-1}(x))$ and substitute $y = q^{-1}(x)$. We have

$$\begin{aligned} x &= q(y) \\ x &= \ln(y+3) - \ln(y-5) \\ x &= \ln\frac{y+3}{y-5} \\ e^x &= e^{\ln\left(\frac{y+3}{y-5}\right)} \\ e^x &= \frac{y+3}{y-5} \\ e^x(y-5) &= y+3 \\ ye^x - 5e^x &= y+3 \\ ye^x - y &= 3+5e^x \\ y(e^x-1) &= 3+5e^x \\ y &= \frac{3+5e^x}{e^x-1} \end{aligned}$$

So $q^{-1}(x) = \dfrac{3+5e^x}{e^x-1}$.

23. This function is not invertible. It does not pass the horizontal line test.

24. This function is invertible. To find the inverse, we say that $y = q(x) = 3x^3 - 2$ and solve for x:

$$\begin{aligned} y &= 3x^3 - 2 \\ \frac{y+2}{3} &= x^3 \end{aligned}$$

$$\sqrt[3]{\frac{y+2}{3}} = x.$$

Thus, $q^{-1}(y) = \sqrt[3]{\dfrac{y+2}{3}}$.

25. This function is not invertible. It does not pass the horizontal line test.

26. This function is invertible. To find the inverse, we say that $y = p(x) = 5x^5 + 4$ and solve for x:

$$\begin{aligned} y &= 5x^5 + 4 \\ \frac{y-4}{5} &= x^5 \\ \sqrt[5]{\frac{y-4}{5}} &= x. \end{aligned}$$

Thus, $p^{-1}(y) = \sqrt[5]{\dfrac{y-4}{5}}$.

27. This function is invertible. To find the inverse, we say that $y = r(x) = e^x - 7$ and solve for x:

$$\begin{aligned} y &= e^x - 7 \\ y + 7 &= e^x \\ \ln(y+7) &= x. \end{aligned}$$

Thus, $r^{-1}(y) = \ln(y+7)$.

28. This function is invertible. To find the inverse, we say that $y = b(x) = \sqrt[3]{x} - 2$ and solve for x:

$$\begin{aligned} y &= \sqrt[3]{x} - 2 \\ y + 2 &= \sqrt[3]{x} \\ (y+2)^3 &= x. \end{aligned}$$

Thus, $b^{-1}(y) = (y+2)^3$.

29. Since $h(x) = \sqrt{1-x}/x$ and $h^{-1}(x) = 1/(x^2+1)$, we have

$$\begin{aligned} h^{-1}(h(x)) &= h^{-1}\left(\sqrt{\frac{1-x}{x}}\right) \\ &= \frac{1}{\left(\sqrt{\frac{1-x}{x}}\right)^2 + 1} \\ &= \frac{1}{\frac{1-x}{x} + 1} \\ &= \frac{1}{\frac{1-x}{x} + \frac{x}{x}} \\ &= \frac{1}{\frac{1-x+x}{x}} = \frac{1}{\left(\frac{1}{x}\right)} = x. \end{aligned}$$

30. Solve for x in the equation

$$\begin{aligned} y &= \frac{x}{2x+1} \\ y(2x+1) &= x \\ 2yx + y &= x \\ x - 2yx &= y \\ x(1-2y) &= y \\ x &= \frac{y}{1-2y}. \end{aligned}$$

So

$$x = f^{-1}(y) = \frac{y}{1-2y}.$$

Thus

$$f^{-1}(x) = \frac{x}{1-2x}.$$

31. $g(f(x)) = g(e^x) = 2e^x - 1$

32. $g(x)f(x) = (2x-1)e^x$

33. $g(g(x)) = g(2x-1) = 2(2x-1) - 1 = 4x - 3$

34. Working from the inside out, we have

$$g(h(x)) = g(\sqrt{x}) = 2\sqrt{x} - 1.$$

This gives

$$f(g(h(x))) = f(2\sqrt{x} - 1) = e^{2\sqrt{x}-1}.$$

35. $f(g(x)) = f(2x-1) = e^{2x-1}$, so $f(g(x))h(x) = \sqrt{x}e^{2x-1}$.

36. $(f(h(x)))^2 = (f(\sqrt{x}))^2 = (e^{\sqrt{x}})^2 = e^{2\sqrt{x}}$

37. **(a)** $f(2x) = (2x)^2 + (2x) = 4x^2 + 2x$

(b) $g(x^2) = 2x^2 - 3$

(c) $h(1-x) = \dfrac{(1-x)}{1-(1-x)} = \dfrac{1-x}{x}$

(d) $(f(x))^2 = (x^2+x)^2$

(e) Since $g(g^{-1}(x)) = x$, we have

$$\begin{aligned} 2g^{-1}(x) - 3 &= x \\ 2g^{-1}(x) &= x + 3 \\ g^{-1}(x) &= \frac{x+3}{2}. \end{aligned}$$

(f) $(h(x))^{-1} = \left(\dfrac{x}{1-x}\right)^{-1} = \dfrac{1-x}{x}$

(g) $f(x)g(x) = (x^2+x)(2x-3)$

(h) $h(f(x)) = h(x^2+x) = \dfrac{x^2+x}{1-(x^2+x)} = \dfrac{x^2+x}{1-x^2-x}$

38. We have

$$v(x)/u(x) = \frac{e^x}{\frac{1}{1+x^2}} = e^x(1+x^2).$$

39. Find the two composite functions

$$u(v(x)) = u(e^x) = \frac{1}{1+(e^x)^2} = \frac{1}{1+e^{2x}}$$

and

$$w(v(x)) = w(e^x) = \ln(e^x) = x$$

to obtain

$$u(v(x)) \cdot w(v(x)) = \frac{1}{1+(e^x)^2} \cdot x = \frac{x}{1+(e^x)^2} = \frac{x}{1+e^{2x}}.$$

40. Evaluate $w(2+h) = \ln(2+h)$ and $w(2) = \ln 2$, so

$$\frac{w(2+h) - w(2)}{h} = \frac{\ln(2+h) - \ln 2}{h}.$$

41. To find $f(x)$, we add $m(x)$ and $n(x)$ and simplify: $m(x) + n(x) = 3x^2 - x + 2x = 3x^2 + x = f(x)$.

42. To find $g(x)$, we square $o(x)$, giving us $(o(x))^2 = (\sqrt{x+2})^2 = x + 2$.

43. To find $h(x)$, we multiply $n(x)$ and $o(x)$, giving $n(x)o(x) = 2x \cdot \sqrt{x+2}$.

44. To find $i(x)$, we first find $m(o(x)) = 3(\sqrt{x+2})^2 - \sqrt{x+2} = 3x+6-\sqrt{x+2}$. Then, $i(x) = (3x+6-\sqrt{x+2})(2x) = 6x^2 + 12x - 2x\sqrt{x+2}$.

45. To find $j(x)$, we divide $m(x)$by $n(x)$ and simplify: $(m(x))/n(x) = (3x^2 - x)/(2x) = 3x/2 - 1/2 = j(x)$.

46. To find $k(x)$, we subtract $n(x)$ and $o(x)$ from $m(x)$ and simplify: $m(x) - n(x) - o(x) = 3x^2 - x - 2x - \sqrt{x+2} = 3x^2 - 3x - \sqrt{x+2} = k(x)$.

47. Evaluate as $f(x)h(x) = x^{3/2} \tan 2x$.

48. Evaluate the denominator

$$f(g(x)) = \left(\frac{(3x-1)^2}{4}\right)^{3/2} = \frac{(3x-1)^3}{8}$$

then evaluate the fraction

$$\frac{h(x)}{f(g(x))} = \frac{\tan 2x}{\frac{(3x-1)^3}{8}} = \frac{8 \tan 2x}{(3x-1)^3}.$$

49. Evaluate the two parts of the subtraction

$$h(g(x)) = \tan\left(2\left(\frac{(3x-1)^2}{4}\right)\right) = \tan\frac{(3x-1)^2}{2} \quad \text{and} \quad f(9x) = (9x)^{3/2} = 9^{3/2} \cdot x^{3/2} = 27x^{3/2}$$

and subtract

$$h(g(x)) - f(9x) = \tan\left(\frac{(3x-1)^2}{2}\right) - 27x^{3/2}.$$

50. Evaluate $h(x/2) = \tan(2(x/2)) = \tan x$, so

$$h\left(\frac{x}{2}\right)\cos x = \tan x \cos x = \frac{\sin x}{\cos x}\cos x = \sin x.$$

Problems

51. First, we can find $r(0)$ and $r(4)$ by referring to the table:

$$\begin{aligned} r(0) &= q(p(0)) \\ &= q(4) \quad \text{Because } p(0) = 4 \end{aligned}$$

$$
\begin{aligned}
&= 5 \\
r(4) &= q(p(4)) \\
&= q(1) \quad \text{Because } p(4) = 1 \\
&= 2.
\end{aligned}
$$

Next, we find $p(1)$ and $p(2)$:

$$
\begin{aligned}
r(1) = q(p(1)) &= 1 \\
q(\underbrace{p(1)}_{5}) &= 1 \quad \text{From table, } q(5) = 1 \\
\text{so} \quad p(1) &= 5 \\
r(2) = q(p(2)) &= 0 \\
q(\underbrace{p(2)}_{3}) &= 0 \quad \text{From table, } q(3) = 0 \\
\text{so} \quad p(2) &= 3
\end{aligned}
$$

Finally, we find $q(0)$ and $q(2)$:

$$
\begin{aligned}
p(5) &= 0 \quad \text{From table} \\
\text{So} \quad q(0) &= q(p(5)) \\
&= r(5) = 3 \\
p(3) &= 2 \\
\text{So} \quad q(2) &= q(p(3)) \\
&= r(3) = 4.
\end{aligned}
$$

Putting this altogether, we have Table 10.16.

Table 10.16

t	$p(t)$	$q(t)$	$r(t)$
0	4	3	5
1	5	2	1
2	3	4	0
3	2	0	4
4	1	5	2
5	0	1	3

52. To find $h(-2)$, we need to use the definition $h(x) = g(f(x))$. So, $h(-2) = g(f(-2)) = g(4) = 0$. Similarly, $h(1) = g(f(1)) = g(5) = -1$.

To find $g(1)$, we need to find the right connection between $f(x)$, $g(x)$ and $h(x)$. Since $h(x) = g(f(x))$, we are looking for the value of x for which $f(x) = 1$. According to the table, $f(x) = 1$ at $x = 2$. Since $f(2) = 1$, we can express $g(1)$ as $g(f(2))$. However, $g(f(2)) = h(2) = -2$. If $g(f(2)) = -2$ and $f(2) = 1$, then by substitution, $g(1) = -2$.

To find $f(-1)$, let $f(-1) = k$. Then, $h(-1) = g(f(-1)) = 1$, or $g(k) = 1$. Looking at the second table, which gives the values of the function g, the only value of x for which $g(x) = 1$ is 2. So, k must equal 2. Since $f(-1)$ equals k, we can conclude that $f(-1) = 2$.

Similarly, we can let $f(0) = m$ and note that $h(0) = g(f(0)) = g(m) = 2$. Since $g(3) = 2$, $m = 3$, so $f(0) = 3$. So the completed tables look like this:

x	$f(x)$
-2	4
-1	2
0	3
1	5
2	1

x	$g(x)$
1	-2
2	1
3	2
4	0
5	-1

x	$h(x)$
-2	0
-1	1
0	2
1	-1
2	-2

53. If $f(x) = u(v(x))$, then one solution is $u(x) = \sqrt{x}$ and $v(x) = 3 - 5x$.

54. One possible solution is $g(x) = u(v(x))$ where $v(x) = x^2$ and $u(x) = \sin x$.

55. One possible solution is $h(x) = u(v(x))$ where $v(x) = \sin x$ and $u(x) = x^2$.

56. One possible solution is $k(x) = u(v(x))$ where $v(x) = \sin x$ and $u(x) = e^x + x$.

57. One possible solution is $F(x) = u(v(x))$ where $u(x) = x^3$ and $v(x) = 2x + 5$.

58. One possible solution is $G(x) = u(v(x))$ where $u(x) = \frac{2}{x}$ and $v(x) = 1 + \sqrt{x}$.

59. One possible solution is $H(x) = u(v(x))$ where $u(x) = 3^x$ and $v(x) = 2x - 1$.

60. One possible solution is $J(x) = u(v(x))$ where $u(x) = 8 - 2x$ and $v(x) = |x|$.

61. The troughs (where the graph is below the x-axis) are reflected about the horizontal axis to become humps. The humps (where the graph is above the x-axis) are unchanged.

62. (a)

-2π 2π x

Figure 10.21: $f(x) = \sin x$

(b)

-2π 2π x

Figure 10.22: $g(x) = |\sin x|$

(c)

-2π 2π x

Figure 10.23: $h(x) = \sin |x|$

(d)

-2π 2π x

Figure 10.24: $i(x) = |\sin |x||$

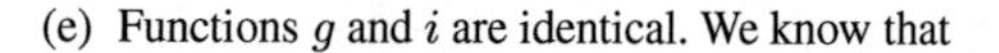

(e) Functions g and i are identical. We know that

$$\sin(-x) = -\sin x.$$

So if $x \geq 0$ we get

$$|\sin |x|| = |\sin x|$$

and if $x \leq 0$ we get

$$|\sin |x|| = |\sin(-x)| = |-\sin x| = |\sin x|.$$

63. **(a)** $r(x) = p(q(x)) = p(x-2) = \dfrac{1}{x-2} + 1 = \dfrac{1}{x-2} + \dfrac{x-2}{x-2} = \dfrac{1+x-2}{x-2} = \dfrac{x-1}{x-2}$.

(b) Let $s(x) = x + 1$ and $t(x) = \dfrac{1}{x}$. Then $s(t(x)) = \dfrac{1}{x} + 1 = p(x)$.

(c) Substituting $p(a)$ into p gives

$$\begin{aligned} p(p(a)) &= \frac{1}{p(a)} + 1 \\ &= \frac{1}{\frac{1}{a} + 1} + 1 \\ &= \frac{1}{\frac{1+a}{a}} + 1. \end{aligned}$$

Since

$$\frac{1}{\frac{1+a}{a}} = 1 \cdot \frac{a}{1+a} = \frac{a}{1+a}.$$

We can say that

$$p(p(a)) = \frac{a}{a+1} + 1 = \frac{a}{a+1} + \frac{a+1}{a+1} = \frac{2a+1}{a+1}$$

64. Since the domain of $g(x)$ is all real numbers, we need only avoid values of x where $f(g(x)) = 1/\sin x$ is undefined. Since $f(g(x))$ is undefined whenever $\sin x = 0$, the domain of $f(g(x))$ is given by all real numbers except integer multiples of π. The function $g(f(x)) = \sin(1/x)$ is defined everywhere except at $x = 0$, so the domain of $g(f(x))$ is all real numbers except 0.

65. First, we have $h(0) = f(g(0)) = f(1) = 0$, which completes the first row of the table. From the information in the second row of the table, we see that $h(1) = 1$. Therefore, since $h(1) = f(g(1))$, we conclude that $f(g(1)) = 1$, which is equivalent to $f(x) = 1$ if we let $x = g(1)$. Since f is invertible, our table indicates that $x = 2$ is the only solution to $f(x) = 1$. Therefore, $g(1) = x = 2$, which fills in the blank in the second row of the table. Finally, we have $h(2) = f(g(2)) = f(0) = 9$, which fills in the final entry in the table. See Table 10.17

Table 10.17

x	$f(x)$	$g(x)$	$h(x)$
0	9	1	0
1	0	2	1
2	1	0	9

66. **(a)** Substituting $t = 25$, we have

$$f(25) = 20 + 0.4 \cdot 25 = 30.$$

Thus, in 2010 (year $t = 25$), we have $P = 30$, so the population was 30,000 people.

(b) We have $t = f^{-1}(P)$. Thus, in $f^{-1}(25)$, the 25 is a population. So $f^{-1}(25)$ is the year in which the population reaches 25 thousand. We find t by solving the equation

$$20 + 0.4t = 25$$
$$0.4t = 5$$
$$t = 12.5.$$

Therefore, $f^{-1}(25) = 12.5$, which means that the population reached 25,000 people 12.5 years after 1985, or midway into 1997.

(c) We can estimate $f^{-1}(25)$ by reading the graph of $P = f(t)$ backward as shown in Figure 10.25.

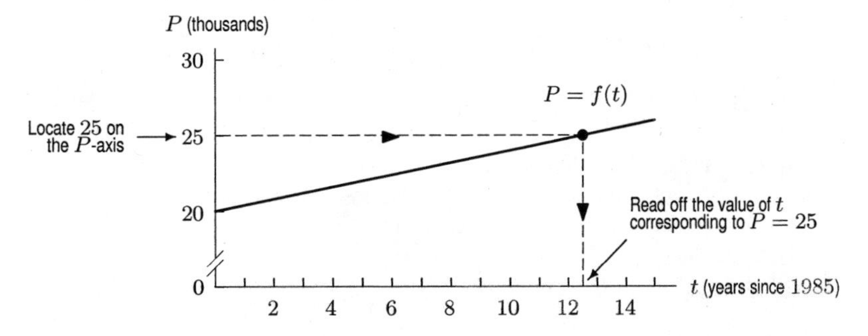

Figure 10.25: Using a graph of the function $P = f(t)$ to read off values of the inverse function $f^{-1}(P)$

67. **(a)** Since $P = 20 + 0.4t$, solving for t gives

$$0.4t = P - 20$$
$$t = \frac{P - 20}{0.4},$$

so

$$f^{-1}(P) = 2.5P - 50.$$

(b) The values in Table 10.18 were calculated using the formula $P = f(t) = 20 + 0.4t$; the values in Table 10.19 were calculated using the formula for $t = f^{-1}(P) = 2.5P - 50$. The table for f^{-1} can be obtained from the table for f by interchanging its columns, because the inverse function reverses the roles of inputs and outputs.

Table 10.18

t	$P = f(t)$
0	20
5	22
10	24
15	26
20	28

Table 10.19

P	$t = f^{-1}(P)$
20	0
22	5
24	10
26	15
28	20

68. **(a)** To find $t = f^{-1}(P)$, we solve for t:

$$P = 14 \cdot 2^{t/12}$$
$$2^{t/12} = \frac{P}{14} \qquad \text{divide}$$
$$\frac{t}{12} \cdot \ln 2 = \ln \frac{P}{14} \qquad \text{take logs}$$
$$t = 12 \frac{\ln\left(\frac{P}{14}\right)}{\ln 2}.$$

(b) We have

$$t = f^{-1}(24)$$
$$= 12 \frac{\ln\left(\frac{24}{14}\right)}{\ln 2}$$
$$= 9.331.$$

This tells us that the population will reach 24,000 after a bit more than 9 years. To verify our answer, we see that

$$P = f(9.331) = 14 \cdot 2^{9.331/12}$$
$$= 24,$$

as required.

69. The inverse function $g^{-1}(t)$ represents the velocity needed for a trip of t hours. Its units are mph.

70. **(a)** Since $(f(x))^{-1} = 1/f(x) = e^{-x}$, this equation is equivalent to $e^{-x} = 2$. Taking the natural logarithm of both sides, we obtain $-x = \ln 2$, so the solution is $x = -\ln 2$.

(b) Since the inverse function of e^x is given by $f^{-1}(x) = \ln x$, this equation is equivalent to $\ln x = 2$. Taking the exponential function of both sides, we obtain a solution of $x = e^2$.

(c) This equation is equivalent to $e^{1/x} = 2$. Taking the natural logarithm of both sides, we obtain $1/x = \ln 2$, so the final answer is given by $x = 1/\ln 2$.

71. If we let $z = \arcsin t$, then we want to simplify $\cos^2 z$. We know that

$$\sin^2 z + \cos^2 z = 1,$$

so we have

$$\cos^2 z = 1 - \sin^2 z.$$

Substituting for z gives

$$\cos^2(\arcsin t) = 1 - \sin^2(\arcsin t).$$

$$\cos^2(\arcsin t) = 1 - (\sin(\arcsin t))^2.$$

Since $\sin(\arcsin t) = t$, we have $\sin^2(\arcsin t) = (\sin(\arcsin t))^2 = t^2$, so

$$\cos^2(\arcsin t) = 1 - t^2.$$

72. Dividing by 7 gives $\sin(3x) = 2/7$. This has solutions

$$3x = \arcsin\left(\frac{2}{7}\right) + \quad \text{any multiple of} \quad 2\pi$$

and

$$3x = \pi - \arcsin\left(\frac{2}{7}\right) + \quad \text{any multiple of} \quad 2\pi.$$

So

$$x = \frac{1}{3}\arcsin\left(\frac{2}{7}\right) \pm k\left(\frac{2\pi}{3}\right) \quad \text{or} \quad \pi - \frac{1}{3}\arcsin\left(\frac{2}{7}\right) \pm k\left(\frac{2\pi}{3}\right),$$

where $k = 0, 1, 2, \ldots$

73. We take logarithms to help solve when x is in the exponent:

$$\begin{aligned} 2^{x+5} &= 3 \\ \ln(2^{x+5}) &= \ln 3 \\ (x+5)\ln 2 &= \ln 3 \\ x &= \frac{\ln 3}{\ln 2} - 5. \end{aligned}$$

74. We raise each side to the $1/(1.05)$ power:

$$\begin{aligned} x^{1.05} &= 1.09 \\ x &= 1.09^{1/1.05}. \end{aligned}$$

75. We take the exponential function to both sides since the exponential function is the inverse of logarithm:

$$\begin{aligned} \ln(x+3) &= 1.8 \\ x+3 &= e^{1.8} \\ x &= e^{1.8} - 3. \end{aligned}$$

76. Multiplying by the denominator gives:

$$\begin{aligned} \frac{2x+3}{x+3} &= 8 \\ 2x+3 &= 8x+24 \\ -21 &= 6x \\ x &= -\frac{21}{6} = -\frac{7}{2}. \end{aligned}$$

77. Squaring eliminates square roots:

$$\begin{aligned}\sqrt{x+\sqrt{x}} &= 3\\ x+\sqrt{x} &= 9\\ \sqrt{x} &= 9-x \quad (\text{so} \quad x\le 9)\\ x &= (9-x)^2 = 81-18x+x^2.\end{aligned}$$

So $x^2 - 19x + 81 = 0$. The quadratic formula gives the solutions

$$x = \frac{19 \pm \sqrt{37}}{2}.$$

The only solution is $x = \dfrac{19-\sqrt{37}}{2}$. The other solution is too large to satisfy the original equation.

78. **(a)** The amount of the radioactive substance present decreases by 4% every three years from a starting value of 20 grams.

(b) We have $f(8) \approx 17.9$, meaning that after 8 years, there are approximately 17.9 grams of the substance remaining. Thus, $f(8)$ represents the amount of radioactive substance remaining (in grams) after 8 years.

(c) We have

$$\begin{aligned}Q &= 20(0.96)^{t/3}\\ \frac{Q}{20} &= (0.96)^{t/3}\\ \ln\left(\frac{Q}{20}\right) &= (t/3)\ln(0.96)\\ t &= \frac{3\ln\left(\frac{Q}{20}\right)}{\ln(0.96)}.\end{aligned}$$

Thus, the inverse function is given by $f^{-1}(Q) = 3\ln(Q/20)/\ln(0.96)$.

(d) Substituting $Q = 8$ into the formula, we see that $f^{-1}(8) \approx 67.3$, meaning that it takes 67.3 years for the amount of substance remaining to decrease to 8 grams. Thus, $f^{-1}(8)$ is the amount of time (in years) that it takes for the amount of substance remaining to decline to 8 grams.

79. **(a)** $A = \pi r^2$

(b) The graph of the function in part (a) is in Figure 10.26.

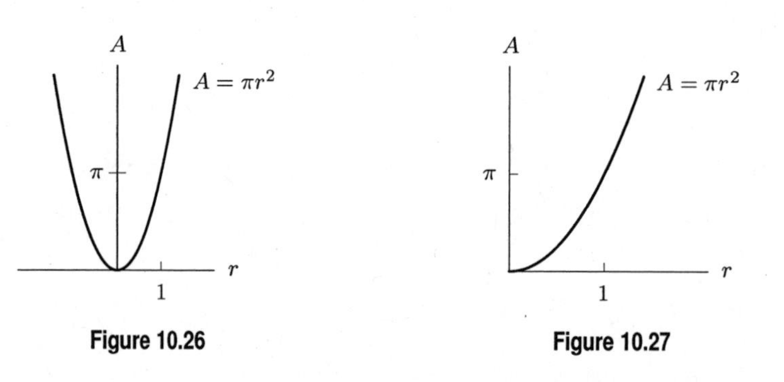

Figure 10.26

Figure 10.27

(c) Because a circle cannot have a negative radius, the domain is $r \ge 0$. See Figure 10.27.

(d) Solve the formula $A = f(r) = \pi r^2$ for r in terms of A:

$$\begin{aligned}r^2 &= \frac{A}{\pi}\\ r &= \pm\sqrt{\frac{A}{\pi}}\end{aligned}$$

The range of the inverse function is the same as the domain of f, namely non-negative real numbers. Thus, we choose the positive root, and $f^{-1}(A) = \sqrt{\dfrac{A}{\pi}}$.

(e) We rewrite both functions to be y in terms of x, and graph. See Figure 10.28.

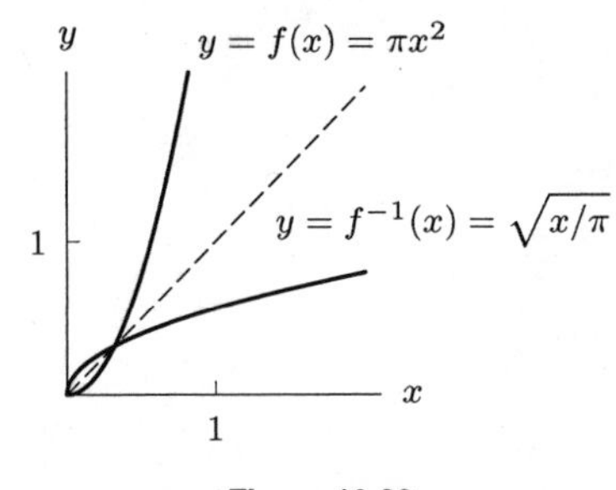

Figure 10.28

(f) Yes. If the function $A = \pi r^2$ refers to radius and area, its domain must be $r \geq 0$. On this domain the function is invertible, so radius is also a function of area.

80. **(a)** $D(5) = 100$. The demand at \$5 per unit would be 100 units per week.
(b) $D(p) = 500 - 200(p - 3) = 500 - 200p + 600 = 1100 - 200p$.
(c) Solve for p to get $D^{-1}(q)$. We have $p = \frac{1100-D(p)}{200}$. Rewriting we get $D^{-1}(q) = \dfrac{1100-q}{200}$. Thus $D^{-1}(5) = \dfrac{1100-5}{200} = \dfrac{1095}{200} = 5.475$. When 5 units are demanded the price per unit is \$5.48.
(d) The slope of $D(p)$ is -200, which means that the demand will go down by 200 when the unit price goes up by \$1.
(e) $p = D^{-1}(400) = 3.5$ so the price should be \$3.50.
(f) Revenue at 500 units per week: $500(\$3) = \$1500$. Revenue at 400 units per week: $400(\$3.50) = \$1400$. So it would go up by \$100.

81. **(a)** $f(g(a)) = f(a) = a$
(b) $g(f(c)) = g(c) = b$
(c) $f^{-1}(b) - g^{-1}(b) = 0 - c = -c$
(d) $0 < x \leq a$

82.

Table 10.20

x	$n(x)$	$p(x)$	$q(x)$
1	5	9	2/3
2	5	4	1/4
3	5	7	4
4	5	10	3/2

83.

$$\begin{aligned} p(q(x)) &= p(\sqrt{x} - 3) \\ &= 2(\sqrt{x} - 3) - 3 \\ &= 2\sqrt{x} - 9. \end{aligned}$$

84. Solving $y = r(x)$ for x, we have

$$y = \frac{2x - 1}{2x + 1}$$

$$\begin{aligned} y(2x+1) &= 2x-1 \\ 2xy+y &= 2x-1 \\ 2xy-2x &= -1-y \\ x(2y-2) &= -1-y \\ x &= \frac{-1-y}{2y-2} \\ &= \frac{y+1}{2-2y}. \end{aligned}$$

This means $r^{-1}(x) = (x+1)/(2-2x)$.

85.

$$\begin{aligned} p(x) &= r(x) \\ 2x-3 &= \frac{2x-1}{2x+1} \\ (2x-3)(2x+1) &= 2x-1 \\ 4x^2-4x-3 &= 2x-1 \\ 4x^2-6x-2 &= 0 \\ 2x^2-3x-1 &= 0. \end{aligned}$$

Using the quadratic formula, we have:

$$\begin{aligned} x &= \frac{-(-3) \pm \sqrt{(-3)^2 - 4 \cdot 2(-1)}}{2 \cdot 2} \\ &= \frac{3 \pm \sqrt{17}}{4}. \end{aligned}$$

86.

$$\begin{aligned} q(x) &= p(u(x)) \\ \underbrace{\sqrt{x}-3}_{q(x)} &= \underbrace{2u(x)-3}_{p(u(x))} \\ \sqrt{x} &= 2u(x) \\ u(x) &= 0.5\sqrt{x}. \end{aligned}$$

87. Notice that

$$\begin{aligned} p(s(x)) &= p((x-1)^2) \\ &= 2(x-1)^2 - 3. \end{aligned}$$

This is a quadratic function in vertex form. The vertex is $(h, k) = (1, -3)$. The y-intercept is $y = 2(2-1)^2 - 3 = -1$. The zeros are found by solving for $y = 0$:

$$\begin{aligned} 2(x-1)^2 - 3 &= 0 \\ 2(x-1)^2 &= 3 \\ (x-1)^2 &= \frac{3}{2} \\ x-1 &= \pm\sqrt{\frac{3}{2}} \\ x &= 1 \pm \sqrt{\frac{3}{2}}. \end{aligned}$$

See Figure 10.29.

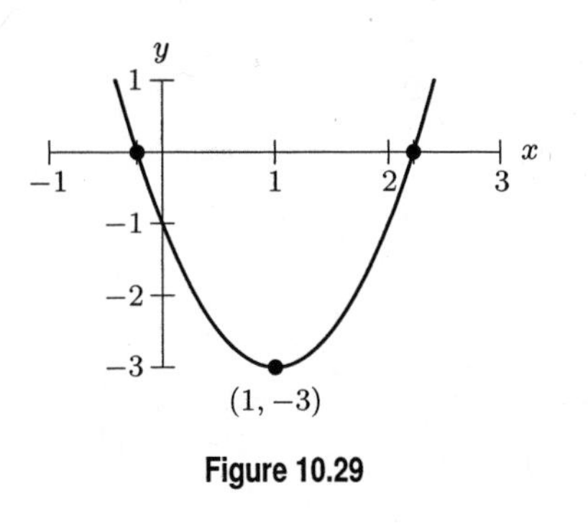

Figure 10.29

88. **(a)** This equation is equivalent to $2f(3x) = 1$, or $2e^{3x} = 1$. Dividing both sides by 2 and taking the natural log of both sides, we get $3x = \ln(1/2)$, giving an exact solution of $x = (1/3)\ln(1/2)$.

(b) This equation is equivalent to $e^x + e^{3x} = 1$. Since it is not possible to simplify the left hand side of this equation, we cannot use the rules of logarithms to isolate the variable x and find an exact solution. Using a graphing utility, we obtain an approximate solution of $x \approx -0.382$.

(c) This equation is equivalent to $e^{3x}e^{3x} = 2$, or $e^{6x} = 2$. Thus, the exact solution is given by $x = (1/6)\ln 2$.

(d) This equation is equivalent to $e^x e^{3x} = 2$, or $e^{4x} = 2$. Thus, the exact solution is given by $x = (1/4)\ln 2$.

89. **(a)** The function $f(x) = \sin^2 x$ is equal to $(u(x))^2$ but is not equal to $u(u(x))$. As an illustration of this, note that $f(\pi/2) = (\sin(\pi/2))^2 = 1$, but $u(u(\pi/2)) = \sin 1 \approx 0.84$. Since $f(1) \neq u(u(1))$, the functions $f(x)$ and $u(u(x))$ are not the same.

(b) First, we note that in the expression $p(x) = \sin(\cos^2 x)$, we are taking the composition of the sine function with $\cos^2 x$; we are not multiplying $\sin x$ by $\cos^2 x$. This tells us immediately that $u(x)(v(x))^2$ and $u(x)w(v(x))$ are not equal to $p(x)$. On the other hand, since $(v(x))^2 = w(v(x)) = \cos^2 x$, we see that $u((v(x))^2)$ and $u(w(v(x)))$ both equal $p(x)$, meaning that (ii) and (iii) are the only correct answers.

(c) (i) We have $(u(x)+v(x))^2 = \sin^2 x + 2\sin x\cos x + \cos^2 x$. Since $\sin^2 x + \cos^2 x = 1$ and $2\sin x\cos x = \sin 2x$, our answer simplifies to $1 + \sin 2x$.

(ii) We have $(u(x))^2 + (v(x))^2 = \sin^2 x + \cos^2 x = 1$.

(iii) We have $u(x^2) + v(x^2) = \cos(x^2) + \sin(x^2)$, which cannot be simplified.

90. See Figure 10.30.

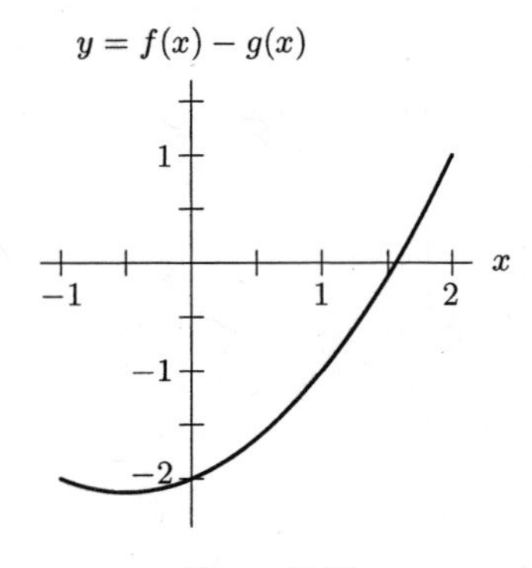

Figure 10.30

91. See Figure 10.31.

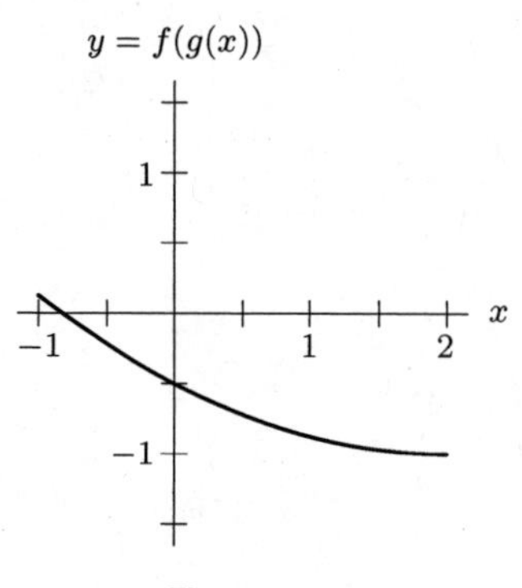

Figure 10.31

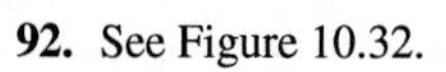
92. See Figure 10.32.

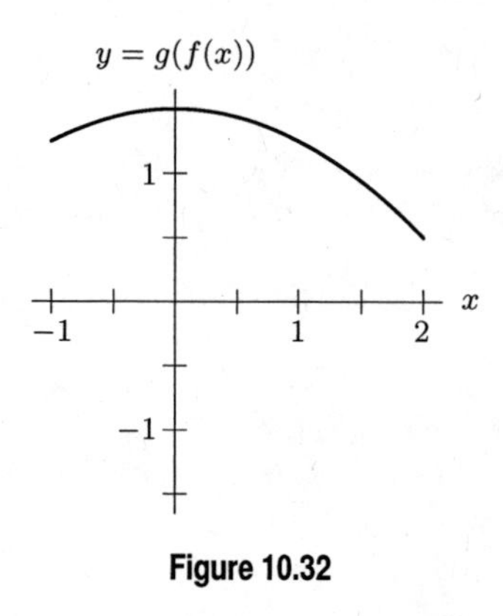

Figure 10.32

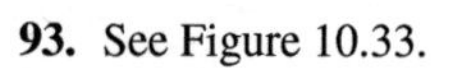
93. See Figure 10.33.

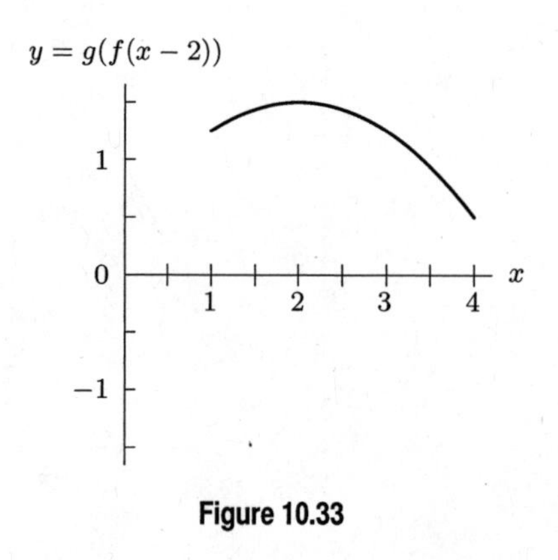

Figure 10.33

94. **(a)** Graph II, because the graph of $y = -f(x)$ is the reflection of $f(x)$ across the x-axis.
(b) Graph I, because the graph of $y = f(-x)$ reflects the graph across the y-axis.
(c) None. This graph would look like Graph I but with a y-intercept of -1.
(d) None. The graph of $f^{-1}(x)$ would be a reflection of the graph of $f(x)$ about the line $y = x$.
(e) Graph IV. This is the graph of $f(x)$ reflected about the line $y = x$ and then reflected about the x-axis.
(f) None. Adding 1 to the argument, we shift the graph of $f(x)$ to the left by 1.
(g) None. This is the graph of $f(x)$ reflected about the x-axis and raised by 2.

95. **(a)** Graphs of f and g are in Figure 10.34.

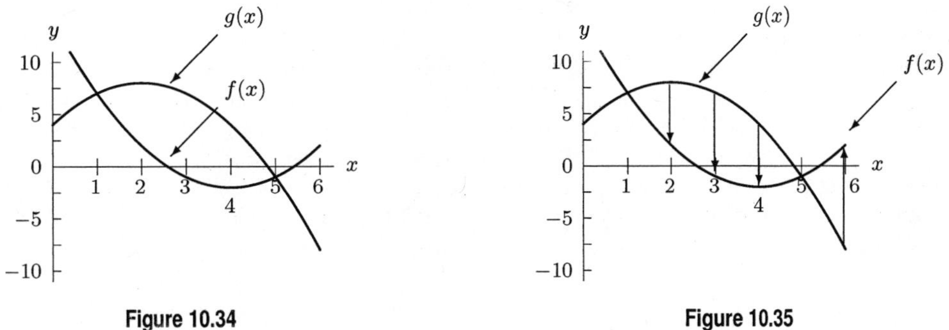

Figure 10.34

Figure 10.35

(b) Values of $f(x)$, $g(x)$, and $f(x) - g(x)$ are in Table 10.21.

Table 10.21

x	0	1	2	3	4	5	6
$f(x)$	14	7	2	-1	-2	-1	2
$g(x)$	4	7	8	7	4	-1	-8
$f(x) - g(x)$	10	0	-6	-8	-6	0	10

(c) See part (b).
(d) See Figure 10.35.
(e) See Figure 10.36.
(f) $f(x) = x^2 - 8x + 14$, $g(x) = -x^2 + 4x + 4$,
$f(x) - g(x) = 2x^2 - 12x + 10$.
(g) See Figure 10.36, where the graph of $f(x) - g(x)$ passes through the points plotted in part (e).

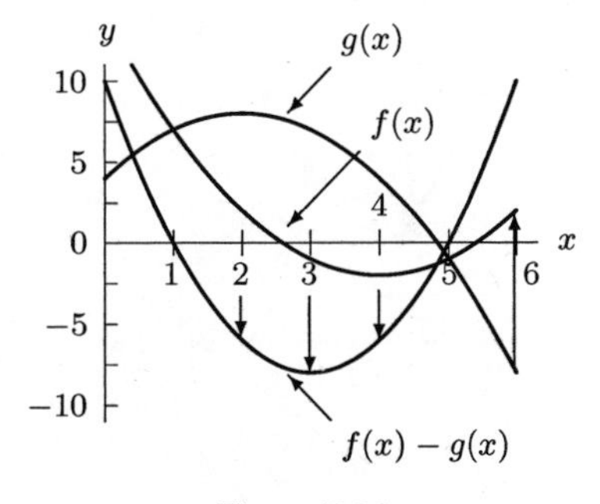

Figure 10.36

96. In all four of the functions, d oscillates. However, the functions $d = 2 + \cos t, d = 2 + \cos(e^t)$, and $d = 2 + e^{\cos t}$ do not accurately describe the motion of the weight because the magnitude of the oscillations does not decrease over time. The function $d = 2 + e^{-t}\cos t$ is therefore the only possible correct answer; the e^{-t} multiplied by $\cos t$ causes the magnitude of the vertical oscillations to decrease with time.

97. **(a)** Since $f(x)$ is a linear function, its formula can be written in the form $f(x) = mx + b$, where m represents the slope and b represents the y-intercept. According to the graph, the y-intercept is 4. Since $(-2, 0)$ and $(0, 4)$ both lie on the line, we know that

$$m = \frac{y_2 - y_1}{x_2 - x_1} = \frac{4 - 0}{0 - (-2)} = \frac{4}{2} = 2.$$

So we know that the formula is $f(x) = 2x + 4$. Similarly, we can find the slope of $g(x)$, $\frac{0-(-1)}{3-0} = \frac{1}{3}$, and the y-intercept, -1, so its formula is $g(x) = \frac{1}{3}x - 1$.

(b) To graph $h(x) = f(x) \cdot g(x)$, we first take note of where $f(x) = 0$ and $g(x) = 0$. At those places, $h(x) = 0$. Since the zero of $f(x)$ is -2 and the zero of $g(x)$ is 3, the zeros of $h(x)$ are -2 and 3. When $x < -2$, both $f(x)$ and $g(x)$ are negative, so we know that $h(x)$, their product, is positive. Similarly, when $x > 3$, both $f(x)$ and $g(x)$ are positive so $h(x)$ is positive. When $-2 < x < 3$, $f(x)$ is positive and $g(x)$ is negative, so $h(x)$ is negative. Also, since $h(x)$ is the product of two linear functions, we know that it is a quadratic function $h(x) = (2x+4)(\frac{1}{3}x - 1) = \frac{2}{3}x^2 - \frac{2}{3}x - 4$. Putting these pieces of information together, we know that the graph of $h(x)$ is a parabola with zeros at -2 and 3 (and, therefore, an axis of symmetry at $x = \frac{1}{2}$) and that it is positive when $x < -2$ or $x > 3$ and negative when $-2 < x < 3$. [Note: since you know the axis of symmetry is $x = \frac{1}{2}$, you know that the x-coordinate of the vertex is $\frac{1}{2}$. You could find the y-coordinates of its vertex by finding $h(\frac{1}{2}) = (2(\frac{1}{2}) + 4)(\frac{1}{3}(\frac{1}{2}) - 1) = (1 + 4)(\frac{1}{6} - 1) = 5(-\frac{5}{6}) = -\frac{25}{6} = -4\frac{1}{6}$.] See Figure 10.37.

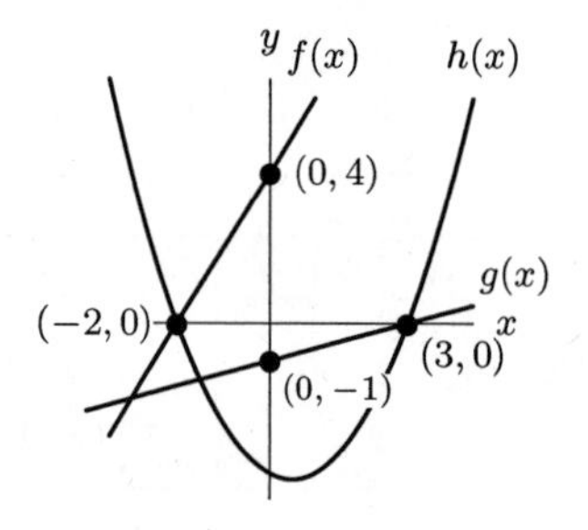

Figure 10.37

98. Since $c(x)$ is the product of $a(x)$ and $b(x)$, the zeros for $c(x)$ are the same as the zeros for $a(x)$ and $b(x)$, m and n. On the interval where both $a(x)$ and $b(x)$ are negative, $m < x < n$, their product, $c(x)$, is positive; similarly, $c(x)$ is negative when one of them is positive and one is negative. Since $c(x)$ is the product of two linear functions, it is a quadratic function with axis of symmetry halfway between the two zeros, at $x = \frac{m+n}{2}$. See Figure 10.38.

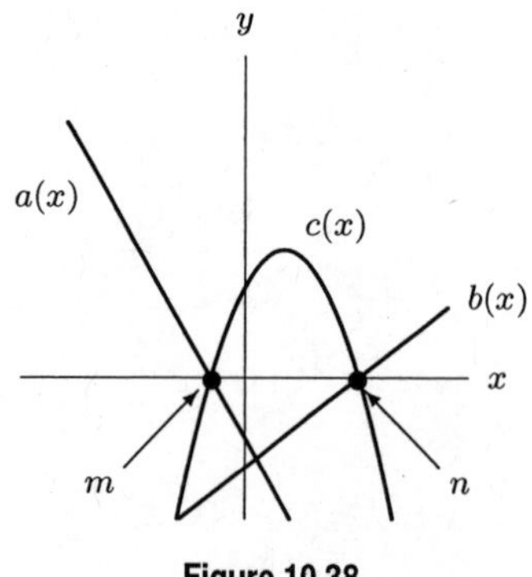

Figure 10.38

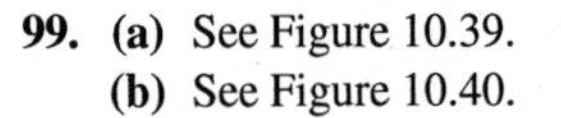

99. **(a)** See Figure 10.39.
(b) See Figure 10.40.

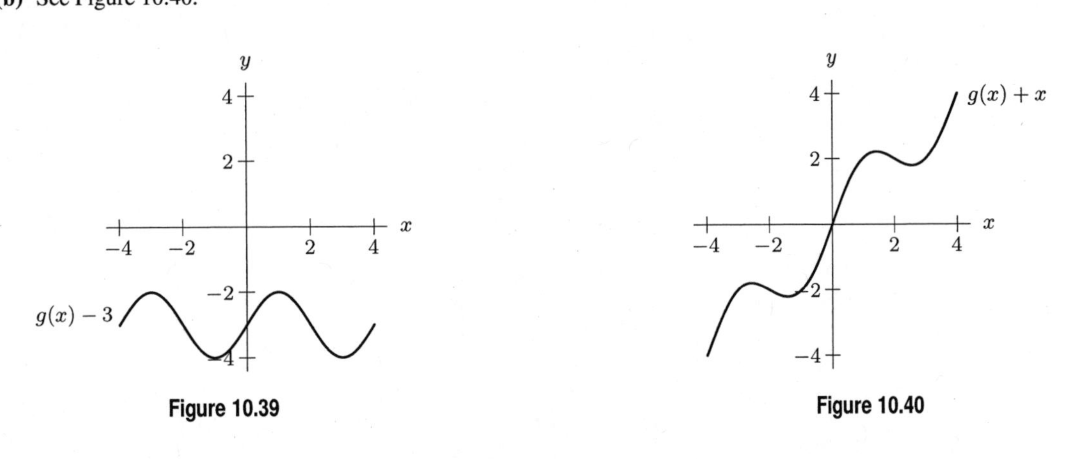

Figure 10.39

Figure 10.40

100. Both of these functions are odd functions which are undefined at $x = 0$. The cosecant function, $\csc x = \frac{1}{\sin x}$, is periodic while $\sin \frac{1}{x}$ is not. As $x \to 0$ the function $\sin \frac{1}{x}$ oscillates more and more rapidly. Also, $\sin \frac{1}{x}$ gets closer and closer to zero as x gets large while $\frac{1}{\sin x}$ continues to be periodic. Note also that

$$-1 \leq \sin \frac{1}{x} \leq 1,$$

while

$$\frac{1}{\sin x} \leq -1 \quad \text{or} \quad \frac{1}{\sin x} \geq 1$$

for all x.

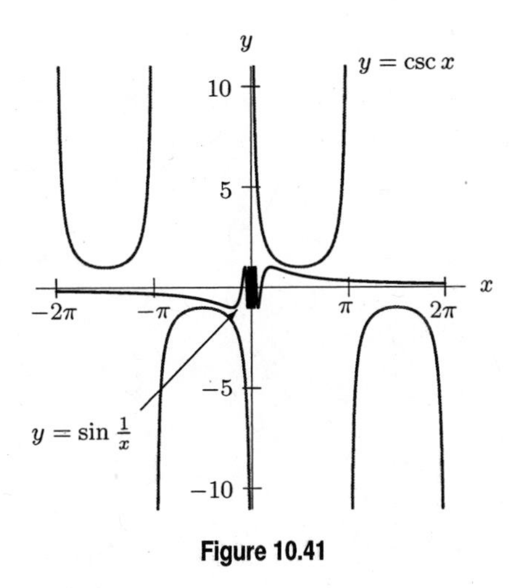

Figure 10.41

101. The statement is false. For example, if $f(x) = x$ and $g(x) = x^2$, then $f(x) \cdot g(x) = x^3$. In this case, $f(x) \cdot g(x)$ is an odd function, but $g(x)$ is an even function.

102. We have

$$\begin{aligned} g(f(x)) &= 64 \cdot 2^{x^2} \cdot 32^x \\ &= 2^6 \cdot 2^{x^2} \left(2^5\right)^x \end{aligned}$$

$$\begin{aligned} &= 2^{x^2}\cdot 2^{5x}\cdot 2^6 \\ &= 2^{x^2+5x+6} \\ &= 2^{f(x)} \qquad \text{because } f(x) = x^2+5x+6 \\ \text{so} \quad g(x) &= 2^x. \end{aligned}$$

103. We have

$$\begin{aligned} f(x) &= g(x)h(x) \\ \underbrace{x^2+5x+6}_{f(x)} &= g(x)\underbrace{(2x+6)}_{h(x)} \\ g(x) &= \frac{x^2+5x+6}{2x+6} \\ &= \frac{(x+2)(x+3)}{2(x+3)} \qquad \text{factor} \\ &= \frac{x+2}{2} = 0.5x+1. \end{aligned}$$

104. **(a)** We have

$$f(x) = h(g(x)) = 3\cdot 9^x.$$

Since $g(x) = 3^x$, we know that

$$\begin{aligned} h(g(x)) = h(3^x) &= 3\cdot 9^x \\ &= 3\cdot(3^2)^x \\ &= 3\cdot 3^{2x} \\ &= 3(3^x)^2. \end{aligned}$$

Since $h(3^x) = 3(3^x)^2$, we know that

$$h(x) = 3x^2.$$

(b) We have

$$f(x) = g(j(x)) = 3\cdot 9^x.$$

Since $g(x) = 3^x$, we know that

$$\begin{aligned} g(j(x)) = 3^{j(x)} &= 3\cdot 9^x \\ &= 3\cdot(3^2)^x \\ &= 3^1\cdot 3^{2x} \\ &= 3^{2x+1} \end{aligned}$$

Since $3^{j(x)} = 3^{2x+1}$, we know that

$$j(x) = 2x+1$$

105. **(a)** One possible answer is:

$$y = \underbrace{6x}_{g(x)}\cdot\underbrace{e^{\overbrace{3x^2}^{G(x)}}}_{f(G(x))}$$

$$\begin{aligned} \text{so} \quad f(x) &= e^x \\ g(x) &= 6x \\ G(x) &= 3x^2. \end{aligned}$$

(b) One possible answer is:

$$y = -\frac{\sin\left(\sqrt{x}\right)}{2\sqrt{x}} = \underbrace{-\frac{1}{2\sqrt{x}}}_{g(x)} \cdot \underbrace{\sin\left(\overbrace{\sqrt{x}}^{G(x)}\right)}_{f(G(x))}$$

$$\text{so} \quad f(x) = \sin x$$
$$g(x) = -\frac{1}{2\sqrt{x}}$$
$$G(x) = \sqrt{x}.$$

106. **(a)** The function f represents the exchange from dollars to yen, and 1 dollar buys 89.1823 yen. Thus, each of the x dollars buys 89.1823 yen, for a total of $89.1823x$ yen. Therefore,

$$f(x) = 89.1823x.$$

Referring to the table, we see that 1 dollar purchases 0.6686 European Union euros. If x dollars are invested, each of the x dollars will buy 0.6686 euros, for a total of $0.6686x$ euros. Thus,

$$g(x) = 0.6686x.$$

Finally, we see from the table that 1 yen buys 0.007496 euros. Each of the x yen invested buys 0.007496 euros, so $0.007496x$ euros can be purchased. Therefore,

$$h(x) = 0.007496x.$$

(b) We evaluate $h(f(1000))$ algebraically. Since $f(1000) = 89.1823(1000) = 89{,}182.3$, we have

$$\begin{aligned} h(f(1000)) &= h(89{,}182.3) \\ &= 0.007496(89{,}182.3) \\ &= 668.5105. \end{aligned}$$

To interpret this statement, we break the problem into steps. First, we see that $f(1000) = 89{,}182.3$ means 1000 dollars buy 89,182.3 yen. Second, we see that $h(89{,}182.3) = 668.5105$ means that 89,182.3 yen buys 668.5105 euros. In other words, $h(f(1000)) = 668.5105$ represents a trade of $1000 for 89,182.3 yen which is subsequently traded for 668.5105 euros (Of course, a direct trade of $1000 would yield 668.5105 euros).

107. **(a)** True. If a is in the domain of f and g, then $f(a)$ and $g(a)$ are defined, which means $h(a) = f(a) + g(a)$ is defined. On the other hand, if a is not in the domain of f and g, then neither $f(a)$ nor $g(a)$ is defined, so $h(a)$ is undefined.

(b) False. For instance, suppose $f(x) = 3$ and $g(x) = 3$ Then both functions have the same range, namely $y = 3$. However, $h(x) = f(x) + g(x) = 6$, which has as its range $y = 6$.

(c) False. Suppose $f(x) = 2x^2 + 1$ and $g(x) = -1 - x^2$. The graph of f lies entirely above the x-axis, and the graph of g below, so that neither function has a zero. However, $h(x) = f(x) + g(x) = 2x^2 + 1 - 1 - x^2 = x^2$, which has a zero at $x = 0$.

(d) True. On an interval from a to b,

$$\begin{aligned} \text{Average rate of change of } h &= \frac{h(b) - h(a)}{b - a} \\ &= \frac{f(b) + g(b) - (f(a) + g(a))}{b - a} && \text{since } h(x) = f(x) + g(x) \\ &= \frac{f(b) - f(a) + g(b) - g(a)}{b - a} && \text{regroup} \\ &= \underbrace{\frac{f(b) - f(a)}{b - a}}_{\frac{\Delta f}{\Delta x}} + \underbrace{\frac{g(b) - g(a)}{b - a}}_{\frac{\Delta g}{\Delta x}} && \text{split numerator} \\ &= \text{Average rate of change of } f + \text{Average rate of change of } g. \end{aligned}$$

108. **(a)** Let $f(x)$ and $g(x)$ be even functions defined for all x, and let $h(x)$ be the sum of $f(x)$ and $g(x)$.

Since $h(x) = f(x) + g(x)$, then

$$h(-x) = f(-x) + g(-x).$$

Since $f(x)$ and $g(x)$ are even functions, $f(-x) = f(x)$ and $g(-x) = g(x)$, so

$$h(-x) = f(x) + g(x).$$

But $f(x) + g(x) = h(x)$, so

$$h(-x) = h(x).$$

Since $h(-x) = h(x)$, we know that the sum of two even functions is even.

(b) Suppose both $f(x)$ and $g(x)$ are odd and $h(x) = f(x) + g(x)$. Then,

$$h(-x) = f(-x) + g(-x).$$

Since $f(-x)$ and $g(-x)$ are odd functions, $f(-x) = -f(x)$ and $g(-x) = -g(x)$. So

$$\begin{aligned} h(-x) &= -f(x) - g(x) \\ &= -(f(x) + g(x)) \\ &= -h(x). \end{aligned}$$

Thus, $h(-x) = -h(x)$, which means that the sum of two odd functions is odd.

(c) Suppose that $f(x)$ is even and $g(x)$ is odd. Then,

$$\begin{aligned} h(-x) &= f(-x) + g(-x) \\ &= f(x) - g(x). \end{aligned}$$

Thus, $h(-x) \neq h(x)$, and $h(-x) \neq -h(x)$, which means that h is neither even nor odd. The same argument holds if f is odd and g is even.

For example, consider $f(x) = x^2$ and $g(x) = x^3$. The function $h(x) = x^2 + x^3$ is neither even nor odd.

109. This is an increasing function, because as x increases, $f(x)$ increases, and as $f(x)$ increases, $f(f(x))$ increases.

110. This is a decreasing function, because as x increases, $f(x)$ increases, and as $f(x)$ increases, $g(x)$ decreases.

111. We can't tell. For example, suppose $f(x) = 2x$ and $g(x) = -x$. Then $f(x) + g(x) = x$, which is increasing. But if $f(x) = 2x$ and $g(x) = -3x$, then $f(x) + g(x) = -x$, which is decreasing.

112. This is an increasing function, because if $g(x)$ is a decreasing function, then $-g(x)$ will be an increasing function. Since $f(x) - g(x) = f(x) + [-g(x)]$, $f(x) - g(x)$ can be written as the sum of two increasing functions, and is thus increasing.

113. **(a)** $f(8) = 2$, because 8 divided by 3 equals 2 with a remainder of 2. Similarly, $f(17) = 2$, $f(29) = 2$, and $f(99) = 0$.

(b) $f(3x) = 0$ because, no matter what x is, $3x$ will be divisible by 3.

(c) No. Knowing, for example, that $f(x) = 0$ tells us that x is evenly divisible by 3, but gives us no other information regarding x.

(d) $f(f(x)) = f(x)$, because $f(x)$ equals either 0, 1, or 2, and $f(0) = 0$, $f(1) = 1$, and $f(2) = 2$.

(e) No. For example, $f(1) + f(2) = 1 + 2 = 3$, but $f(1 + 2) = f(3) = 0$.

114. Let $y = f(x)$. In order to find f^{-1}, we need to solve for x. But

$$y = g(h(x)), \text{so } g^{-1}(y) = g^{-1}(g(h(x))) = h(x).$$

Moreover,

$$h^{-1}(g^{-1}(y)) = h^{-1}(h(x)) = x,$$

hence $x = f^{-1}(y) = h^{-1}(g^{-1}(y))$. So $f^{-1}(x) = h^{-1}(g^{-1}(x))$.

115. Solving $L = L_\infty(1 - e^{-k(t+t_0)})$ for t, we have

$$\begin{aligned} 1 - e^{-k(t+t_0)} &= L/L_\infty \\ e^{-k(t+t_0)} &= 1 - L/L_\infty \\ -k(t+t_0) &= \ln(1 - L/L_\infty) \\ t &= -\frac{1}{k}\ln(1 - L/L_\infty), \end{aligned}$$

so

$$t = f^{-1}(L) = -\frac{1}{k}\ln(1 - L/L_\infty).$$

This function tells us the age t (in years) of fish whose mean length is L (in cm). To find the domain, note first that $L > 0$ (because a fish must have positive length) and that $1 - L/L_\infty > 0$ (otherwise the log function would be undefined). This gives $0 \leq L \leq L_\infty$, which means that L must be between 0 and the mean length of mature fish.

CHECK YOUR UNDERSTANDING

1. False, since $f(4) + g(4) = \frac{1}{4} + \sqrt{4}$ but $(f+g)(8) = \frac{1}{8} + \sqrt{8}$.

2. False, since $\dfrac{h(x)}{f(x)} = \dfrac{x-5}{1/x} = x(x-5)$.

3. True, since $f(4) + g(4) = \frac{1}{4} + \sqrt{4} = 2\frac{1}{4}$.

4. False. The function $f(g(x)) = \dfrac{1}{\sqrt{x}}$ is not defined for $x \leq 0$.

5. True, since $g(f(x)) = g\left(\dfrac{1}{x}\right) = \sqrt{\dfrac{1}{x}}$.

6. True. Since the two sides of the equation are defined only when $x > 0$, we have $f(x)g(x) = \dfrac{1}{x} \cdot \sqrt{x} = \dfrac{1}{\sqrt{x}}$, and $f(g(x)) = \dfrac{1}{\sqrt{x}}$.

7. True, since $2f(2) = 2 \cdot \frac{1}{2} = 1$ and $g(1) = \sqrt{1} = 1$.

8. True, since $f(1)g(1)h(1) = \frac{1}{1} \cdot \sqrt{1} \cdot (1-5) = -4$.

9. True. Evaluate $\dfrac{f(3)+g(3)}{h(3)} = \dfrac{\frac{1}{3}+\sqrt{3}}{3-5}$ and simplify.

10. False. $4h(2) = 4(2-5) = -12$, but $h(8) = 8 - 5 = 3$.

11. False. If $x < -3$ then $g(x)$ is not defined, thus $f(g(x))$ is not defined.

12. True. First, $g(6) = \sqrt{6+3} = 3$ then $f(3) = 3^2 = 9$.

13. False. As a counterexample, let $f(x) = x^2$ and $g(x) = x + 1$. Then $f(g(x)) = (x+1)^2 = x^2 + 2x + 1$, but $g(f(x)) = x^2 + 1$.

14. False. Solve the circumference formula for r, $r = C/(2\pi)$. Substitute for r in the area formula, $A = \pi(C/(2\pi))^2 = C^2/(4\pi)$.

15. True. First, $f(1) = 1^2 + 2 = 3$ then $f(3) = 3^2 + 2 = 11$.

16. True. First, $h(2) = f(g(2)) = f(\frac{1}{2})$. Now check to see if the formula $f(x) = x^2 + 1$ might work. It does, $f(\frac{1}{2}) = 1\frac{1}{4}$.

17. False. $f(x+h) = \dfrac{1}{x+h} \neq \dfrac{1}{x} + \dfrac{1}{h}$.

18. False. First $f(x+h) = (x+h)^2 + (x+h) = x^2 + 2xh + h^2 + x + h$. Now subtract $f(x) = x^2 + x$ to get $f(x+h) - f(x) = 2xh + h^2 + h$. Thus $(f(x+h) - f(x))/h = 2x + h + 1$.

19. False. $f(g(x)) = f(\sin x) = (\sin x)^2 = \sin^2 x$.

20. True. If $f(x) = b + mx$ and $g(x) = c + dx$, then

$$f(g(x)) = f(c + dx) = b + m(c + dx) = b + mc + mdx.$$

21. False. If $f(x) = ax^2 + bx + c$ and $g(x) = px^2 + qx + r$, then

$$f(g(x)) = f(px^2 + qx + r) = a(px^2 + qx + r)^2 + b(px^2 + qx + r) + c.$$

Expanding shows that $f(g(x))$ has an x^4 term.

22. True. For example, $h(x) = f(g(x))$ with $f(x) = 3x$ and $g(x) = (x^2 + 1)^3$, or $f(x) = 3x^3$ and $g(x) = x^2 + 1$

23. False. For example, if f and g are inverses, then $f(g(x)) = g(f(x)) = x$.

24. True. If $a < b$, then $g(a) < g(b)$ since g is increasing. Then $h(a) = f(g(a)) < f(g(b)) = h(b)$ since f is increasing. Therefore, h is increasing.

25. True, since $g(f(2)) = g(1) = 3$ and $f(g(3)) = f(1) = 3$.

26. False. For example, $f(x) = 2x$ is increasing and $f^{-1}(x) = \frac{1}{2}x$ is also increasing.

27. True. This is the vertical line test. It means that for some input value there is more than one output value, which means the graph does not represent a function.

28. False. For example $f(x) = x^2$ does not pass the horizontal line test, so it does not have an inverse.

29. True. The function g is not invertible if two different points in the domain have the same function value.

30. True. This is the horizontal line test. It means that for each output value there is exactly one input value. This ensures that reversing the input and output, the inverse will be a function.

31. False. None of them has an inverse.

32. True. Since $m \neq 0$, we can solve $y = f(x) = mx + b$ for x and get $x = \dfrac{1}{m}y - \dfrac{b}{m}$. Writing the inverse as a function of x we have $f^{-1}(x) = \dfrac{1}{m}x - \dfrac{b}{m}$.

33. True. Each x value has only one y value.

34. False. We see that for $y = 3$ there are two different values of x.

35. True. The graph passes the vertical line test.

36. False. A horizontal line through the y-intercept intersects the graph in several places, so the graph fails the horizontal line test.

37. True. The inverse of a function reverses the action of the function and returns the original value of the independent variable x.

CHAPTER ELEVEN

Solutions for Section 11.1

Skill Refresher

S1. $\sqrt{36t^2} = (36t^2)^{1/2} = 36^{1/2} \cdot (t^2)^{1/2} = 6|t^1| = 6|t|$

S2. Inside the parenthesis we write the radical as an exponent, which results in

$$\left(3x\sqrt{x^3}\right)^2 = \left(3x \cdot x^{3/2}\right)^2.$$

Then within the parenthesis we write

$$\left(3x^1 \cdot x^{3/2}\right)^2 = \left(3x^{5/2}\right)^2 = 3^2(x^{5/2})^2 = 9x^5.$$

S3. Raising (0.1) and $\left(4xy^2\right)$ to the second power yields $(0.1)^2 = (0.01)$ and $\left(4xy^2\right)^2 = 16x^2y^4$. Therefore $(0.1)^2\left(4xy^2\right)^2 = (0.01)\left(16x^2y^4\right) = 0.16x^2y^4$.

S4. In this example the same variable base w occurs in two separate factors: $w^{1/2}$ and $w^{1/3}$. Since we are multiplying these factors, we need to add the exponents, namely $1/2$ and $1/3$. This requires a common denominator of 6. Therefore

$$7\left(5w^{1/2}\right)\left(2w^{1/3}\right) = 70 \cdot w^{1/2} \cdot w^{1/3} = 70w^{3/6} \cdot w^{2/6} = 70w^{5/6}.$$

S5. We have

$$\begin{aligned} 10x^{5-2} &= 2 \\ 10x^3 &= 2 \\ x^3 &= 0.2 \\ x &= (0.2)^{1/3} = 0.585. \end{aligned}$$

S6. We have

$$\begin{aligned} \frac{5}{x^2} &= 500 \\ \frac{1}{x^2} &= 100 \\ x^2 &= 1/100 \\ x &= \pm(1/100)^{1/2} = \pm 0.1. \end{aligned}$$

S7. False

S8. False

S9. False

S10. False

Exercises

1. Yes. Writing the function as

$$g(x) = \frac{(-x^3)^3}{6} = \frac{(-1)^3(x^3)^3}{6} = \frac{-x^9}{6} = -\frac{1}{6}x^9,$$

we have $k = -1/6$ and $p = 9$.

2. Yes. Writing the function as

$$R(t) = \frac{4}{\sqrt{16t}} = \frac{4}{\sqrt{16}\cdot\sqrt{t}} = \frac{4}{4\sqrt{t}} = \frac{1}{\sqrt{t}} = \frac{1}{t^{1/2}} = t^{-1/2},$$

we have $k = 1$ and $p = -1/2$.

3. Although y is a power function of $(x+7)$, it is not a power function of x and cannot be written in the form $f(x) = kx^p$.

4. Yes. Writing the function as

$$T(s) = (6s^{-2})(es^{-3}) = 6es^{-2}s^{-3} = 6es^{-5},$$

we have $k = 6e$ and $p = -5$.

5. No. This function cannot be written in the form kx^p because the variable is in the exponent.

6. Yes. Writing the function as

$$K(w) = \frac{w^4}{4\sqrt{w^3}} = \frac{1}{4}\left(\frac{w^4}{\sqrt{w^3}}\right) = \frac{1}{4}\left(\frac{w^4}{w^{3/2}}\right) = \frac{1}{4}w^{4-(3/2)} = \frac{1}{4}w^{5/2},$$

we have $k = 1/4$ and $p = 5/2$.

7. This is a power function in the form $y = ax^p$:

$$y = \frac{48}{30625}\cdot x^{-2}, \qquad a = \frac{48}{30625}, \qquad p = -2.$$

We have

$$\begin{aligned} y &= 3\left(\frac{2}{5\sqrt{7x}}\right)^4 \\ &= 3\cdot\frac{2^4}{5^4\left(\sqrt{7}\sqrt{x}\right)^4} \\ &= \frac{48}{625\cdot 49x^2} \\ &= \frac{48}{30625}\cdot x^{-2}. \end{aligned}$$

8. This is a power function in the form $y = ax^p$:

$$y = \left(\sqrt{8\pi}\right)x^{1.5}, \qquad a = \sqrt{8\pi}, \qquad p = 1.5.$$

We see that

$$\begin{aligned} y &= \sqrt{\pi(2x)^3} \\ &= \sqrt{\pi\cdot 8x^3} \\ &= \sqrt{8\pi}\sqrt{x^3} \\ &= \left(\sqrt{8\pi}\right)x^{1.5}. \end{aligned}$$

9. Since the graph is symmetric about the y-axis, the power function is even.

10. Since the graph is symmetric about the origin, the power function is odd.

11. Since the graph is steeper near the origin and less steep away from the origin, the power function is fractional.

12. Since the graph is symmetric about the origin, the power function is odd.

13. We use the form $y = kx^p$ and solve for k and p. Using the point $(1,3)$, we have $3 = k1^p$. Since 1^p is 1 for any p, we know that $k = 3$. Using our other point, we see that

$$\begin{aligned} 13 &= 3 \cdot 4^p \\ \frac{13}{3} &= 4^p \\ \ln\left(\frac{13}{3}\right) &= p \ln 4 \\ \frac{\ln(13/3)}{\ln 4} &= p \\ 1.058 &\approx p. \end{aligned}$$

So $y = 3x^{1.058}$.

14. We use the form $y = kx^p$ and solve for k and p. Using the point $(1, 0.7)$, we have $0.7 = k1^p$. Since 1^p is 1 for any p, we know that $k = 0.7$. Using our other point, we see that

$$\begin{aligned} 8 &= 0.7 \cdot 7^p \\ \frac{8}{0.7} &= 7^p \\ \ln\left(\frac{8}{0.7}\right) &= p \ln 7 \\ \frac{\ln(8/0.7)}{\ln 7} &= p \\ 1.252 &\approx p. \end{aligned}$$

So $y = 0.7x^{1.252}$.

15. Since $f(1) = k \cdot 1^p = k$, we know $k = f(1) = \frac{3}{2}$

Since $f(2) = k \cdot 2^p = \frac{3}{8}$, and since $k = \frac{3}{2}$, we know

$$\left(\frac{3}{2}\right) \cdot 2^p = \frac{3}{8}$$

which implies

$$2^p = \frac{3}{8} \cdot \frac{2}{3} = \frac{1}{4}.$$

Thus $p = -2$, and $f(x) = \frac{3}{2} \cdot x^{-2}$.

16. Taking the ratio of $\frac{g(-\frac{1}{5})}{g(2)}$, we have

$$\frac{g(-\frac{1}{5})}{g(2)} = \frac{k(-\frac{1}{5})^p}{k(2)^p} = \frac{(-\frac{1}{5})^p}{(2)^p} = \left(\frac{-\frac{1}{5}}{2}\right)^p = \left(-\frac{1}{10}\right)^p.$$

Since

$$\frac{g(-\frac{1}{5})}{g(2)} = \frac{25}{-\frac{1}{40}} = 25(-40) = -1000,$$

we have

$$\left(-\frac{1}{10}\right)^p = -1000,$$

which implies that $(-10)^{-p} = (-10)^3$. So $p = -3$.

Then, using either point, for example, $g(2) = -\frac{1}{40}$, we have

$$g(2) = k(2)^{-3} = -\frac{1}{40},$$

so

$$k\left(\frac{1}{8}\right) = -\frac{1}{40},$$

so

$$k = -\frac{1}{40} \cdot 8 = -\frac{1}{5}.$$

Thus,

$$g(x) = -\frac{1}{5}x^{-3}.$$

17. Substituting into the general formula $c = kd^2$, we have $45 = k(3)^2$ or $k = 45/9 = 5$. So the formula for c is

$$c = 5d^2.$$

When $d = 5$, we get $c = 5(5)^2 = 125$.

18. Substituting into the general formula $c = k/d^2$, we have $45 = k/3^2$ or $k = 405$. So the formula for c is

$$c = \frac{405}{d^2}.$$

When $d = 5$, we get $c = 405/5^2 = 16.2$.

19. Substituting into the general formula $y = kx$, we have $6 = k(4)$ or $k = \frac{3}{2}$. So the formula for y is

$$y = \frac{3}{2}x.$$

When $y = 8$, we get $8 = \frac{3}{2}x$, so $x = \frac{2}{3} \cdot 8 = \frac{16}{3} = 5.33$.

20. Substituting into the general formula $y = k/x$, we have $6 = k/4$ or $k = 24$. So the formula for y is

$$y = \frac{24}{x}.$$

When $y = 8$, we get $8 = 24/x$, so $x = 24/8 = 3$.

21. We need to solve $f(x) = kx^p$ for p and k. To solve for p, take the ratio of any two values of $f(x)$, say $f(3)$ over $f(2)$:

$$\frac{f(3)}{f(2)} = \frac{27}{12} = \frac{9}{4}.$$

Since $f(3) = k \cdot 3^p$ and $f(2) = k \cdot 2^p$, we have

$$\frac{f(3)}{f(2)} = \frac{k \cdot 3^p}{k \cdot 2^p} = \frac{3^p}{2^p} = \left(\frac{3}{2}\right)^p = \frac{9}{4}.$$

Since $(\frac{3}{2})^p = \frac{9}{4}$, we know $p = 2$. Thus, $f(x) = kx^2$. To solve for k, use any point from the table. Note that $f(2) = k \cdot 2^2 = 4k = 12$, so $k = 3$. Thus, $f(x) = 3 \cdot x^2$.

22. Solve for $g(x)$ by taking the ratio of (say) $g(4)$ to $g(3)$:

$$\frac{g(4)}{g(3)} = \frac{-32/3}{-9/2} = \frac{-32}{3} \cdot \frac{-2}{9} = \frac{64}{27}.$$

We know $g(4) = k \cdot 4^p$ and $g(3) = k \cdot 3^p$. Thus,

$$\frac{g(4)}{g(3)} = \frac{k \cdot 4^p}{k \cdot 3^p} = \frac{4^p}{3^p} = \left(\frac{4}{3}\right)^p = \frac{64}{27}.$$

Thus $p = 3$. To solve for k, note that $g(3) = k \cdot 3^3 = 27k$. Thus, $27k = g(3) = -\frac{9}{2}$. Thus, $k = -\frac{9}{54} = -\frac{1}{6}$. This gives $g(x) = -\frac{1}{6}x^3$.

23. We need to solve $j(x) = kx^p$ for p and k. We know that $j(x) = 2$ when $x = 1$. Since $j(1) = k \cdot 1^p = k$, we have $k = 2$. To solve for p, use the fact that $j(2) = 16$ and also $j(2) = 2 \cdot 2^p$, so

$$2 \cdot 2^p = 16,$$

giving $2^p = 8$, so $p = 3$. Thus, $j(x) = 2x^3$.

24. Solve for $h(x)$ by taking the ratio of, say, $h(4)$ to $h(\frac{1}{4})$:

$$\frac{h(4)}{h(\frac{1}{4})} = \frac{-1/8}{-32} = \frac{-1}{8} \cdot \frac{-1}{32} = \frac{1}{256}.$$

We know $h(4) = k \cdot 4^p$ and $h(\frac{1}{4}) = k \cdot (\frac{1}{4})^p$. Thus,

$$\frac{h(4)}{h(\frac{1}{4})} = \frac{k \cdot 4^p}{k \cdot (\frac{1}{4})^p} = \frac{4^p}{(\frac{1}{4})^p} = 16^p = \frac{1}{256}.$$

Since $16^p = \frac{1}{256} = \frac{1}{16^2} = 16^{-2}$, $p = -2$. To solve for k, note that $h(4) = k \cdot 4^p = k \cdot 4^{-2} = \frac{k}{16}$. Since $h(4) = -\frac{1}{8}$, we have $\frac{k}{16} = -\frac{1}{8}$. Thus, $k = -2$,which gives $h(x) = -2x^{-2}$.

25. **(a)** $\lim_{x\to\infty} x^{-4} = \lim_{x\to\infty} (1/x^4) = 0$.

(b) $\lim_{x\to-\infty} 2x^{-1} = \lim_{x\to-\infty} (2/x) = 0$.

26. **(a)** $\lim_{t\to\infty} (t^{-3} + 2) = \lim_{t\to\infty} (1/t^3 + 2) = 0 + 2 = 2$.

(b) $\lim_{y\to-\infty} (5 - 7y^{-2}) = \lim_{y\to-\infty} (5 - 7/y^2) = 5 - 0 = 5$.

Problems

27. The graphs are shown in Figure 11.1.

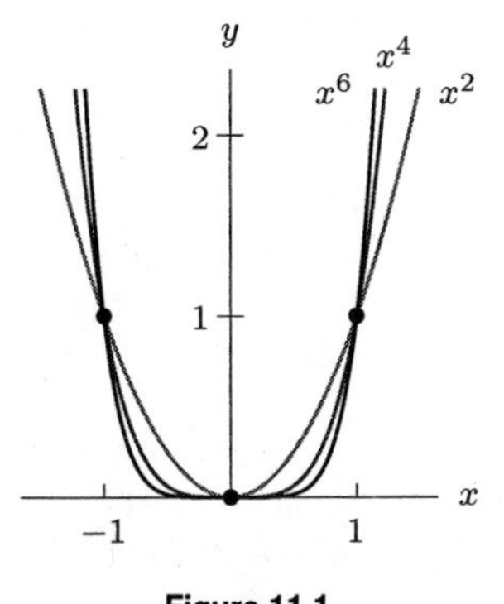

Figure 11.1

Some observations we can make are:

- All three curves pass through the points $(-1, 1)$, $(0, 0)$, and $(1, 1)$.
- All three curves have even symmetry and are more or less "U" shaped.
- x^4 and x^6 are much "flatter" near the origin than x^2, but climb much more steeply for $x < -1$ or $x > 1$.

28. **(a)** Note that $y = x^{-10} = (\frac{1}{x})^{10}$ is undefined at $x = 0$. Since $y = (\frac{1}{x})^{10}$ is raised to an even power, the graph "explodes" in the same direction as x approaches zero from the left and the right. Thus, as $x \longrightarrow 0$, $x^{-10} \longrightarrow +\infty$, $-x^{10} \longrightarrow 0$.

(b) As $x \longrightarrow \infty$, $x^{-10} \longrightarrow 0$, $-x^{10} \longrightarrow -\infty$.

(c) As $x \longrightarrow -\infty$, $x^{-10} \longrightarrow 0$, $-x^{10} \longrightarrow -\infty$.

29. **(a)** As $x \longrightarrow 0$ from the right, $x^{-3} \longrightarrow +\infty$, and $x^{1/3} \longrightarrow 0$.

(b) As $x \longrightarrow \infty$, $x^{-3} \longrightarrow 0$, and $x^{1/3} \longrightarrow \infty$.

30. (a) The power function will be of the form $g(x) = kx^p$, and from the graph we know p must be odd and k must be negative. Using $(-1, 3)$, we have

$$\begin{aligned} 3 &= k(-1)^p, \\ \text{so} \quad 3 &= -k \quad \text{(since } p \text{ is odd)} \\ \text{or} \quad k &= -3. \end{aligned}$$

We do not have enough information to solve for p, since any odd p will work. Therefore, we have $g(x) = -3x^p$, p odd.

(b) Since the function is of the form $g(x) = -3x^p$, with p odd, we know that the the graph of this function is symmetric about the origin. This implies that if (a, b) is a point on the graph, then $(-a, -b)$ is also a point on the graph. Thus the information that the point $(1, -3)$ is on the graph does not help us.

(c) Since we know that the function is symmetric about the origin it will follow that the points $(-2, 96)$ and $(0, 0)$ also lie on the graph. To get other points lying on the graph, we can find the formula for this function. We know that

$$g(x) = -3x^p$$

so plugging in the point $(2, -96)$ we get

$$-96 = -3(2)^p.$$

Solving for p

$$\begin{aligned} -96 &= -3(2)^p \\ 32 &= 2^p \\ p &= 5. \end{aligned}$$

Thus the formula for the function is given by

$$g(x) = -3x^5.$$

Any values satisfying this formula will describe points on the graph: e.g. $(3, -729)$ or $(-0.1, 0.00003)$ or $(\sqrt{7}, -147\sqrt{7})$ etc.

31. $c(t) = \frac{1}{t}$ is indeed one possible formula. It is not, however, the only one. Because the vertical and horizontal axes are asymptotes for this function, we know that the power p is a negative number and

$$c(t) = kt^p.$$

If $p = -3$ then $c(t) = kt^{-3}$. Since $(2, \frac{1}{2})$ lies on the curve, $\frac{1}{2} = k(2)^{-3}$ or $k = 4$. So, $c(t) = 4t^{-3}$ could describe this function. Similarly, so could $c(t) = 16x^{-5}$ or $c(t) = 64x^{-7}$...

32. (a) We have $A = x^3$, $B = x^2$, $C = x^{3/2}$, $D = x$, $E = x^{1/2}$, $F = x^{1/3}$. See Figure 11.2. For $x > 1$, the value of $y = x^3$ is greater than the value of $y = x^2$, which is greater than $y = x$, which is a line.

For large x (in fact, all $x > 1$), the graph of $y = x^{1/2}$ is below the graph of $y = x$, and $y = x^{1/3}$ is below $y = x^{1/2}$. This is reasonable since, for $x > 1$, squaring x makes it bigger and cubing it makes it bigger still; thus taking the square root of x makes it smaller and taking its cube root makes it smaller still. Between $x = 0$ and $x = 1$, the situation is reversed and $y = x^{1/3}$ is on top. (Why?) Not surprisingly, $y = x^{3/2}$ is between $y = x$ and $y = x^2$ for all x.

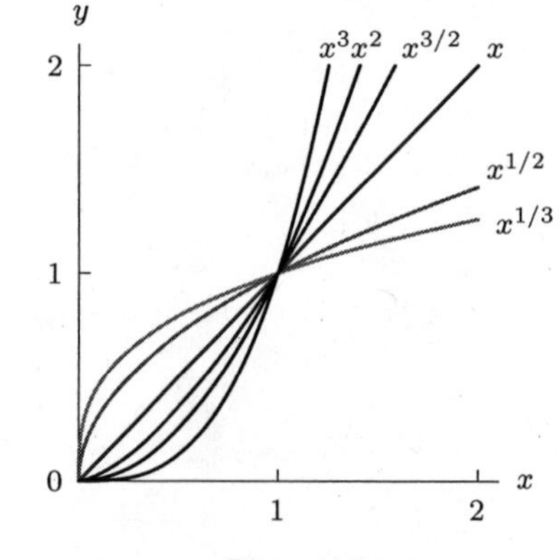

Figure 11.2

(b) We see that $y = x^2$ is concave up, whereas $y = x^{1/2}$ is concave down. For x positive, $y = x^3$ is concave up, whereas $y = x^{1/3}$ is concave down. This occurs because $y = x^2$ and $y = x^{1/2}$ are inverse functions (as are $y = x^3$ and $y = x^{1/3}$). The graphs of $y = x^2$ and $y = x^{1/2}$ are reflections in the line $y = x$, which changes the concavity.

33. All four graphs contain the point $(1, 1)$, so they all four have $k = 1$. Judging from their graphs, f and g appear to have positive values of p. The graph of g climbs faster as $x \to \infty$ so its value of p is larger. The graphs of v and w have asymptotes, suggesting their values of p are negative. The graph of v approaches its horizontal asymptote faster than the graph of w, so its value of p is "more negative."

Therefore, ranking these functions in order of their power, p, gives: v, w, f, g.

34. The graphs of g and w exhibit odd symmetry (symmetry across the origin), so these functions have an odd value of p.

35. We have

$$\begin{aligned} F(x) &= \frac{1}{\sqrt[3]{7x}} \\ &= \frac{1}{\sqrt[3]{7}\sqrt[3]{x}} \\ &= \underbrace{\frac{1}{\sqrt[3]{7}}}_{k} \cdot x^{\overbrace{-1/3}^{n}} \\ \text{so} \quad k &= \frac{1}{\sqrt[3]{7}} \\ n &= -\frac{1}{3} \\ \text{which means} \quad f(x) &= nkx^{n-1} \\ &= -\frac{1}{3} \cdot \frac{1}{\sqrt[3]{7}} \cdot x^{-\frac{1}{3}-1} \\ &= -\frac{1}{3\sqrt[3]{7}} \cdot x^{-\frac{4}{3}}. \end{aligned}$$

36. We have

$$\begin{aligned} f(x) &= \frac{\sqrt[5]{x^2}}{4} \\ &= \frac{1}{4} \cdot x^{\frac{2}{5}} \\ \text{so} \quad k &= \frac{1}{4} \\ n &= \frac{2}{5} \\ \text{which means} \quad F(x) &= \frac{kx^{n+1}}{n+1} \\ &= \frac{\frac{1}{4}x^{\frac{2}{5}+1}}{\frac{2}{5}+1} \\ &= \frac{\frac{1}{4}x^{\frac{7}{5}}}{\frac{7}{5}} \\ &= \frac{5}{28} \cdot x^{\frac{7}{5}}. \end{aligned}$$

37. (a) Since the cost of the fabric, $C(x)$, is directly proportional to the amount purchased, x, we know that the formula will be of the form

$$C(x) = kx.$$

(b) Since 3 yards cost \$28.50, we know that $C(3) = \$28.50$. Thus, we have

$$28.50 = 3k$$
$$k = 9.5$$

Our formula for the cost of x yards of fabric is

$$C(x) = 9.5x.$$

(c) Notice that the graph in Figure 11.3 goes through the origin.

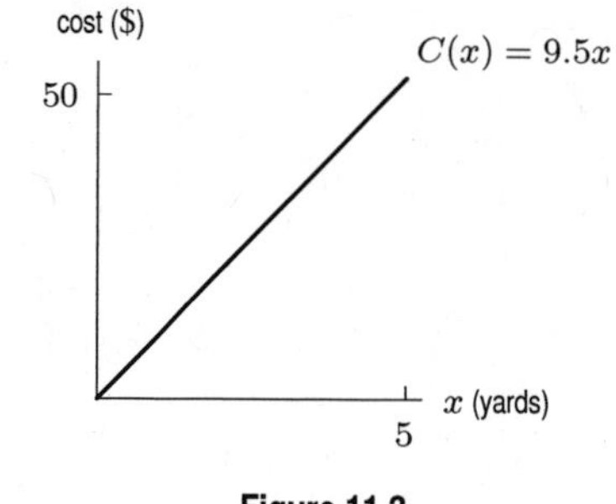

Figure 11.3

(d) To find the cost of 5.5 yards of fabric, we evaluate $C(x)$ for $x = 5.5$:

$$C(5.5) = 9.5(5.5) = \$52.25.$$

38. Calories are directly proportional to ounces because as the amount of beef, x, increases, the number of calories, c, also increases. Substituting into the general formula $c = kx$, we have $245 = k(3)$ or $k = 81.67$. So the formula is

$$c = 81.67x.$$

When $x = 4$, $c = 81.67(4) = 326.68$. Therefore, 4 ounces of hamburger contain 326.68 calories.

39. (a) Table 11.1 shows the circulation times in seconds for various mammals.

Table 11.1

Animal	Body mass (kg)	Circulation time (sec)
Blue whale	91000	302
African elephant	5450	150
White rhinoceros	3000	129
Hippopotamus	2520	123
Black rhinoceros	1170	102
Horse	700	90
Lion	180	64
Human	70	50

(b) If a mammal of mass m has a circulation time of T, then

$$T = 17.4m^{1/4}.$$

If a mammal of mass M has twice the circulation time, then

$$2T = 17.4M^{1/4}.$$

We want to find the relationship between m and M, so we divide these two equations, giving

$$\frac{2T}{T} = \frac{17.4M^{1/4}}{17.4m^{1/4}}.$$

Simplifying, we have

$$2 = \frac{M^{1/4}}{m^{1/4}}.$$

Taking the fourth power of both sides, we get

$$2^4 = \frac{M}{m},$$

and thus

$$16 = \frac{M}{m}.$$

The body mass of the animal with the larger circulation time is 16 times the body mass of the other animal.

40. **(a)** Since B is proportional to the three-quarters power of M, we have

$$B = kM^{\frac{3}{4}}.$$

(b) If M_E is the average mass of an elephant, and M_m that of a mouse then

$$\frac{\text{Metabolic rate of an elephant}}{\text{Metabolic rate of a mouse}} = \frac{kM_E^{\frac{3}{4}}}{kM_m^{\frac{3}{4}}} = \frac{(4.6 \cdot 1{,}000{,}000 \text{ grams})^{\frac{3}{4}}}{(20 \text{ grams})^{\frac{3}{4}}} = 230{,}000^{\frac{3}{4}} = 10{,}502.6,$$

so an elephant at rest uses over 10,000 times as much energy as a mouse!

41. Since the pitch is inversely proportional to the square root of the density, we have

$$P = k\frac{1}{\sqrt{\rho}} = k\rho^{-\frac{1}{2}},$$

for some constant k.

42. **(a)** We assume that the cost, c, in dollars, of a commercial is proportional to its time t, in seconds, so $c = k \cdot t$. The constant k is the cost per second that we want.

In Super Bowl XLIV in 2010, the cost was \$3 million $= \$3 \times 10^6$ for 30 seconds, so

$$3 \times 10^6 = k \cdot 30.$$

So

$$k = \frac{3{,}000{,}000}{30} = 100{,}000 \text{ dollars/sec.}$$

In Super Bowl I in 1967, for the same cost, \$3 million, advertisers could buy 34.439 minutes$= 34.439 \times 60$ seconds.

$$3 \times 10^6 = k \cdot 34.439 \times 60$$
$$k = \frac{3 \times 10^6}{34.439 \times 60} \approx 1451.84 \text{ dollars/sec.}$$

(b) The cost per second in 2010 was \$90,000 and in 1967, it was \$1451.84. Thus the cost per second has increased by a factor of

$$\frac{\$90{,}000}{\$1451.84} \approx 61.99.$$

So cost has increased by a factor of about 62.

43. The number of hours is inversely proportional to the speed, because as the speed, v, increases, the number of hours, h, decreases.

Substituting the given values into the general formula $h = k/v$, we get $3.5 = k/55$, so $k = 192.5$ and the formula is

$$h = \frac{192.5}{v}.$$

When $h = 3$, we get $3 = 192.5/v$, or $v = 64.167$. So getting to Albany in 3 hours would require the speed of 64.167 mph.

44. **(a)** With $u = 9$ and $l = 225$ we find $k = 9/\sqrt{225} = 3/5$. With $k = 0.6$ and $l = 4$, we find $u = (0.6)\sqrt{4} = 1.2$ meters/sec.

(b) Suppose the existing ship has speed u and length l, so

$$u = k\sqrt{l}.$$

The new ship has speed increased by 10%, so the new speed is $1.1u$. If the new length is L, since the constant remains the same, we have

$$1.1u = k\sqrt{L}.$$

Dividing these two equations we get

$$\frac{1.1u}{u} = \frac{k\sqrt{L}}{k\sqrt{l}}.$$

Simplifying and squaring we get

$$1.1 = \frac{\sqrt{L}}{\sqrt{l}}$$

$$(1.1)^2 = \frac{L}{l}$$

so

$$L = (1.1)^2 l = 1.21l.$$

Thus, the new hull length should be 21% longer than the hull length of the existing ship.

45. **(a)** Since the speed of sound is 340 meters/sec = 0.34 km/sec, if t is in seconds,

$$d = f(t) = 0.34t.$$

Thus, when $t = 5$ sec, $d = 0.34 \cdot 5 = 1.7$ km. Other values in the second row of Table 11.2 are calculated in a similar manner.

Table 11.2

Time, t	5 sec	10 sec	1 min	5 min
Distance, d (km)	1.7	3.4	20.4	102
Area, A (km^2)	9.1	36.3	1307	32685

(b) Using $d = 0.34t$, we want to calculate t when $d = 200$, so

$$200 = 0.34t$$

$$t = \frac{200}{0.34} = 588.24 \text{ sec} = 9.8 \text{ mins}.$$

(c) The values for A are listed in Table 11.2. These were calculated using the fact that the area of a circle of radius r is $A = \pi r^2$. At time t, the radius of the circle of people who have heard the explosion is $d = 0.34t$. Thus

$$A = \pi d^2 = \pi(0.34t)^2 = \pi(0.34)^2 t^2 = 0.363t^2.$$

This formula was used to calculate the values of A in Table 11.2.

(d) Since the population density is 31 people/km^2, the population, P, who have heard the explosion is given by

$$P = 31A.$$

Since $A = 0.363t^2$, we have

$$P = 31 \cdot 0.363t^2 = 11.25t^2.$$

So $P = f(t) = 11.25t^2$.

(e) The graph of $P = f(t) = 11.25t^2$ is in Figure 11.4.

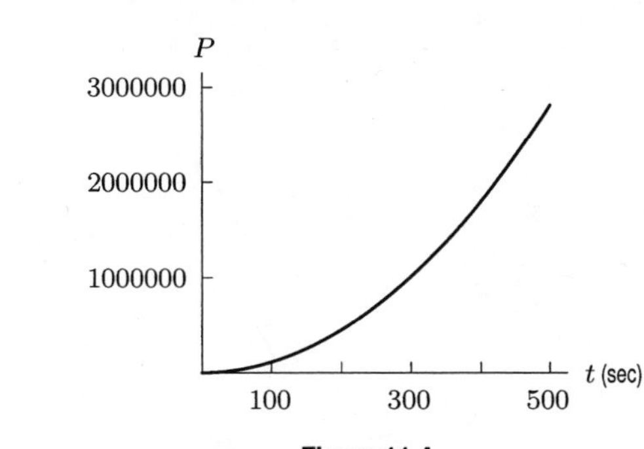

Figure 11.4

To find when 1 million people have heard the explosion, we find t when $P = 1{,}000{,}000$:

$$\begin{aligned} 1000000 &= 11.25t^2 \\ t^2 &= \frac{1000000}{11.25} \\ t &= \sqrt{\frac{1000000}{11.25}} = 298 \text{ sec} \approx 5 \text{ min.} \end{aligned}$$

46. (a) The radius, r, is 0 when $t = 0$ and increases by 200 meters per hour. Thus, after t hours, the radius in meters is given by

$$r = 200t.$$

(b) Since the spill is circular, its area, A, in square meters, is given by

$$A = \pi r^2.$$

Substituting for r from part (a), we have

$$A = \pi(r)^2 = \pi(200t)^2 = 40{,}000\pi t^2.$$

(c) When $t = 7$, the area (in square meters) is

$$A = 40{,}000 \cdot \pi \cdot 7^2 \approx 6{,}157{,}521.601.$$

47. (a) Since cooking time, t, is inversely proportional to power level, w, we know that for some constant k

$$t = \frac{k}{w}.$$

We know that $t = 6.5$ minutes when $w = 750$ watts, so we solve for k:

$$\begin{aligned} 6.5 &= \frac{k}{750} \\ k &= 6.5(750) = 4875. \end{aligned}$$

Thus, the function is

$$t = f(w) = \frac{4875}{w}.$$

(b) When $w = 250$ watts

$$t = \frac{4875}{250} = 19.5 \text{ minutes.}$$

Continuing in the same way gives the result in Table 11.3

Table 11.3

Power, w (watts)	250	300	500	650
Time, t (mins)	19.5	16.25	9.75	7.5

(c) The graph of $t = f(w) = 4875/w$ is shown in Figure 11.5

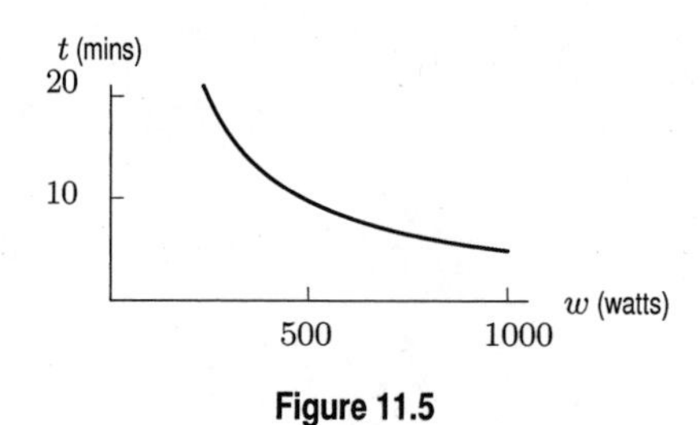

Figure 11.5

(d) For a new dish, there is a new value of k. Since $t = 2$ when $w = 250$

$$2 = \frac{k}{250}$$
$$k = 500,$$

Calculating t when $w = 500$,

$$t = \frac{500}{500} = 1 \text{ minute.}$$

48. **(a)** Since hailstones are approximately spherical, we expect their volume to be proportional to the cube of their radius. Since mass is proportional to volume, mass should be proportional to the cube of radius. To check, for each hailstone we calculate the ratio Mass/(Radius)^3.

$$\frac{\text{Mass}}{(\text{Radius})^3} = \frac{0.058}{(0.3)^3} = 2.148, \qquad \frac{1.835}{(0.95)^3} = 2.140, \qquad \frac{750}{(7.05)^3} = 2.140.$$

The fact that all three ratios are approximately 2.1 shows that m could be proportional to the cube of the radius.

(b) The constant of proportionality is about 2.1, so $m = 2.1r^3$.

(c) To find the radius, substitute $m = 3.4$ kg $= 3.4(1000$ gm$) = 3400$ gm and solve for r:

$$3400 = 2.1r^3$$
$$r^3 = \frac{3400}{2.1} \quad \text{so} \quad r = \sqrt[3]{\frac{3400}{2.1}} = 11.7 \text{ cm.}$$

(d) Pick one of the hailstones and calculate its volume. Using the formula for the volume of a sphere, for the average hailstone, we have

$$\text{Volume} = \frac{4}{3}\pi r^3 = \frac{4}{3}\pi(0.3)^3.$$
$$\text{Density} = \frac{\text{Mass}}{\text{Volume}} = \frac{0.058}{\frac{4}{3}\pi(0.3)^3} = 0.513 \text{ gm/cm}^3.$$

Thus, each cubic centimeter of ice weighs about 0.513 grams.

49. **(a)** The domain of $y = x^{-p}$ is all real numbers except $x = 0$. To find the range we must consider separately the cases when p is even and when p is odd. In the case that p is even, y is always positive. This is because if $p = 2k$ (where k is a positive integer) then

$$y = x^{-p} = \frac{1}{(x^2)^k}$$

and x^2 is never negative. Also there is no value of x for which y is equal to zero, because $y = \frac{1}{(x^2)^k}$ can never be zero. Since the function $(x^2)^k$ ranges over all positive numbers, so will the function $y = \frac{1}{(x^2)^k}$. Thus if p is even the range is all positive numbers.

If p is odd then we again note that zero is not in the range. We can rewrite

$$y = x^{-p} = \frac{1}{x^p}.$$

The range of the function x^p is all real numbers. Therefore, the range of $y = \frac{1}{x^p}$ will be the range of the function x^p excluding 0, or all real numbers except 0.

(b) If p is even we again write $p = 2k$ (where k is a positive integer) and

$$y = x^{-p} = \frac{1}{(x^2)^k}.$$

Now we note that y is symmetric with respect to the y-axis, since

$$\frac{1}{((-x)^2)^k} = \frac{1}{((-1)^2(x^2))^k} = \frac{1}{(x^2)^k}$$

If p is odd,

$$\frac{1}{(-x)^p} = \frac{1}{(-1)^p x^p} = -\frac{1}{x^p},$$

so $y(-x) = -y(x)$. Thus, when p is odd, $y = x^{-p}$ is symmetric with respect to the origin.

(c) When p is even,

$$x^{-p} \to \infty \quad \text{as} \quad x \to 0^+$$
$$x^{-p} \to \infty \quad \text{as} \quad x \to 0^-.$$

Thus, values of the function show the same pattern on each side of the y-axis – a property we would certainly expect for a function with even symmetry.

When p is odd, we found in part (b) that $y = x^{-p}$ is symmetric about the origin. Therefore, if $y \to +\infty$ as $x \to 0^+$ we would expect $y \to -\infty$ as $x \to 0^-$. This behavior is consistent with the behavior of $y = 1/x$ and $y = 1/x^3$.

(d) Again we will write

$$y = x^{-p} = \frac{1}{x^p}.$$

When x is a large positive number, x^p is a large positive number, so its reciprocal is a small positive number. If p is even (and thus the function is symmetric with respect to the y-axis) then y will be a small positive number for large negative values of x. If p is an odd number (and thus y is symmetric with respect to the origin) then y will be a small negative number for large negative values of x.

50. **(a)** We have

$$f(x) = g(h(x)) = 16x^4.$$

Since $g(x) = 4x^2$, we know that

$$g(h(x)) = 4(h(x))^2 = 16x^4$$
$$(h(x))^2 = 4x^4.$$
$$\text{Thus,} \quad h(x) = 2x^2 \text{ or } -2x^2.$$

Since $h(x) \leq 0$ for all x, we know that

$$h(x) = -2x^2.$$

(b) We have

$$f(x) = j\,(2g(x)) = 16x^4, \qquad j(x) \text{ a power function.}$$

Since $g(x) = 4x^2$, we know that

$$j\,(2g(x)) = j(8x^2) = 16x^4.$$

Since $j(x)$ is a power function, $j(x) = kx^p$. Thus,

$$\begin{aligned} j(8x^2) = k(8x^2)^p &= 16x^4 \\ k \cdot 8^p x^{2p} &= 16x^4. \end{aligned}$$

Since $x^{2p} = x^4$ if $p = 2$, letting $p = 2$, we have

$$\begin{aligned} k \cdot 64x^4 &= 16 \cdot x^4 \\ 64k &= 16 \\ k &= \frac{1}{4} \end{aligned}$$

Thus, $j(x) = \frac{1}{4}x^2$.

51. **(a)** If p is negative, the domain is all x except $x = 0$.
There are no domain restrictions if p is positive.
(b) If p is even and positive, the range is $y \geq 0$.
If p is even and negative, the range is $y > 0$.
If p is odd and positive, the range is all real numbers.
If p is odd and negative, the range is all real numbers except $y = 0$.
(c) If p is even, the graph is symmetric about the y-axis.
If p is odd, the graph is symmetric about the origin.

Solutions for Section 11.2

Exercises

1. Since 5^x is not a power function, this is not a polynomial.
2. This is a polynomial of degree one (since $x = x^1$).
3. This is a polynomial of degree two.
4. Yes, this is a polynomial (with variable t) of degree 6.
5. This is not a polynomial because $2e^x$ is not a power function.
6. Since $y = 4x^2 - 7x^{9/2} + 10$ and $9/2$ is not a nonnegative integer, this is not a polynomial.
7. $y = 2x^3 - 3x + 7$ is a third-degree polynomial with three terms. Its long-run behavior is that of $y = 2x^3$: as $x \to -\infty, y \to -\infty$, as $x \to +\infty, y \to +\infty$.
8. $y = 1 - 2x^4 + x^3$ is a fourth degree polynomial with three terms. Its long-run behavior is that of $y = -2x^4$: as $x \to \pm\infty, y \to -\infty$.
9. $y = (x+4)(2x-3)(5-x) = -2x^3 + 5x^2 + 37x - 60$ is a third-degree polynomial with four terms. Its long-run behavior is that of $y = -2x^3$: as $x \to -\infty, y \to +\infty$, as $x \to +\infty, y \to -\infty$.
10. **(a)** $\lim_{x\to\infty} (3x^2 - 5x + 7) = \lim_{x\to\infty} (3x^2) = \infty$.
(b) $\lim_{x\to-\infty} (7x^2 - 9x^3) = \lim_{x\to-\infty} (-9x^3) = \infty$.

Problems

11. See Figure 11.6. There are two real zeros at $x \approx 0.718$ and $x \approx 1.702$.

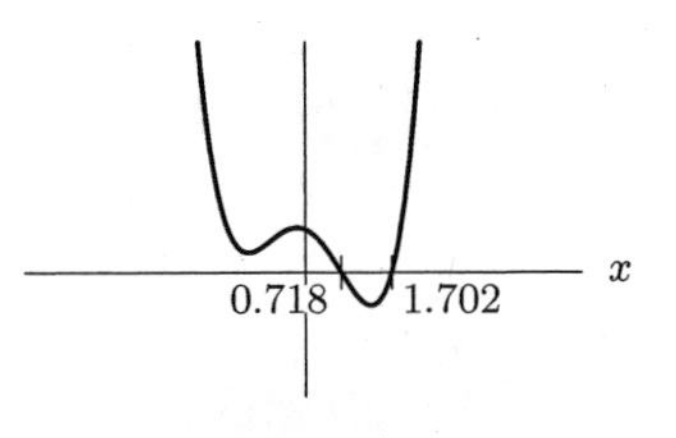

Figure 11.6

12. The graph of $y = g(x)$ is shown in Figure 11.7 on the window $-5 \leq x \leq 5$ by $-20 \leq y \leq 10$. The minimum value of g occurs at point B as shown in the figure. Using either a table feature or trace on a graphing calculator, we approximate the minimum value of g to be -16.543 (to three decimal places).

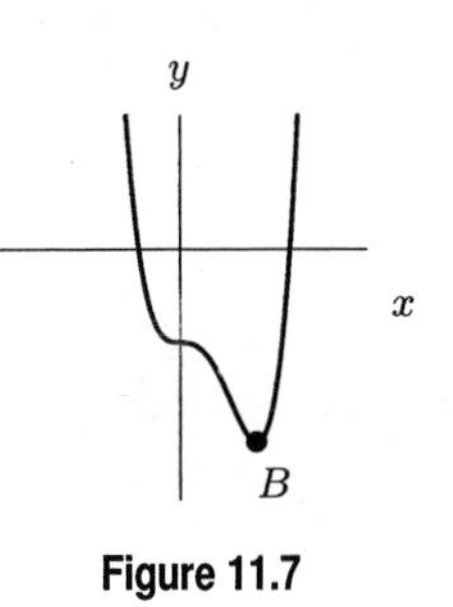

Figure 11.7

13. The window $-10 \leq x \leq 10$ by $-20 \leq y \leq 20$ gives a reasonable picture of both functions. See Figure 11.8. The functions cross the x-axis in the same places, which indicates the zeros of f and g are the same. The y-intercepts are different, since $f(0) = -5$ and $g(0) = 10$. In addition, the end behaviors of the functions differ. The function g has been flipped about the x-axis by the negative coefficient of x^3.

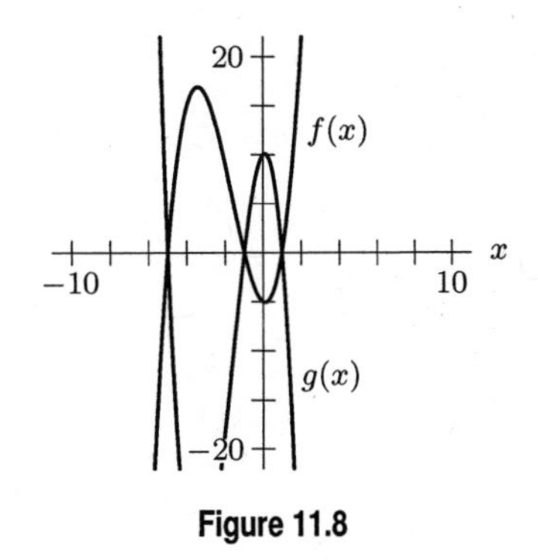

Figure 11.8

14. (a) The graphs of u and v have the same end behavior. As $x \to -\infty$, both $u(x)$ and $v(x) \to \infty$, and as $x \to \infty$, both $u(x)$ and $v(x) \to -\infty$.

The graphs have different y-intercepts, and u has three distinct zeros. The function v has a multiple zero at $x = 0$. See Figure 11.9.

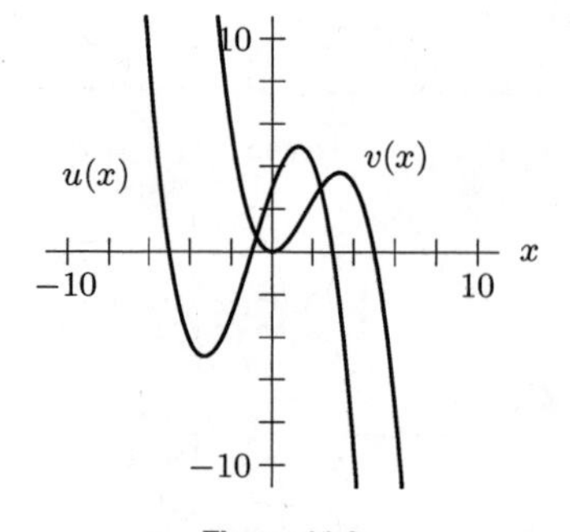

Figure 11.9

(b) On the window $-20 \le x \le 20$ by $-1600 \le y \le 1600$, the peaks and valleys of both functions are not distinguishable. Near the origin, the behavior of both functions looks the same. The functions are still distinguishable from one another on the ends.

On the window $-50 \le x \le 50$ by $-25{,}000 \le y \le 25{,}000$, the functions are still slightly distinct from one another on the ends—but barely.

On the last window the graphs of both functions appear identical. Both functions look like the function $y = -\frac{1}{5}x^3$.

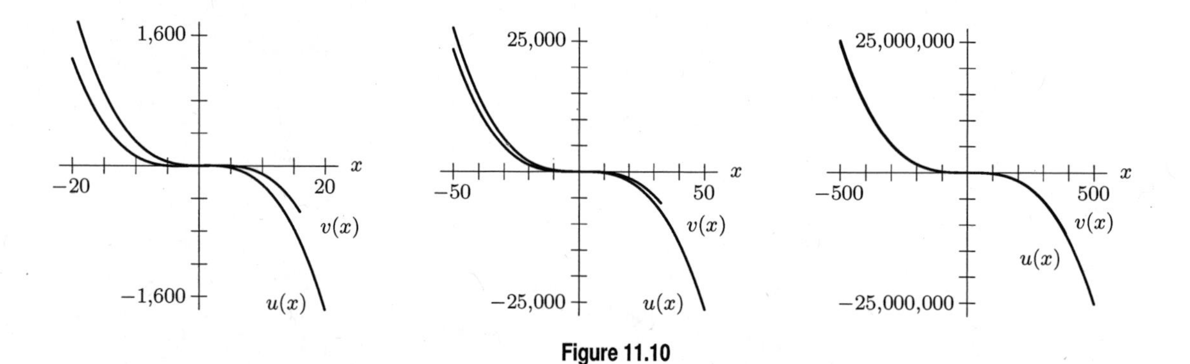

Figure 11.10

15. To find the x-intercept for $y = 2x - 4$ let $y = 0$. We have

$$0 = 2x - 4$$
$$2x = 4$$
$$x = 2.$$

When $x = 0$ on $y = x^4 - 3x^5 - 1 + x^2$, then $y = -1$. This gives the y-intercept for $y = x^4 - 3x^5 - 1 + x^2$. Thus, we have the points $(2, 0)$ and $(0, -1)$. The line through these points will have the same y-intercept, so the linear function is of the form

$$y = mx - 1.$$

The slope, m, is found by taking

$$\frac{0-(-1)}{2-0} = \frac{1}{2}.$$

Thus,

$$y = \frac{1}{2}x - 1$$

is the line through the required points.

16. **(a)** If $x = 10$, for example, we have :

$$f(x) = f(10) = \frac{1}{50{,}000}(10)^3 + \frac{1}{2}(10)$$
$$= \frac{1}{50} + 5 = 5.02$$

The most significant term is the linear term $\frac{1}{2}x$.

(b)

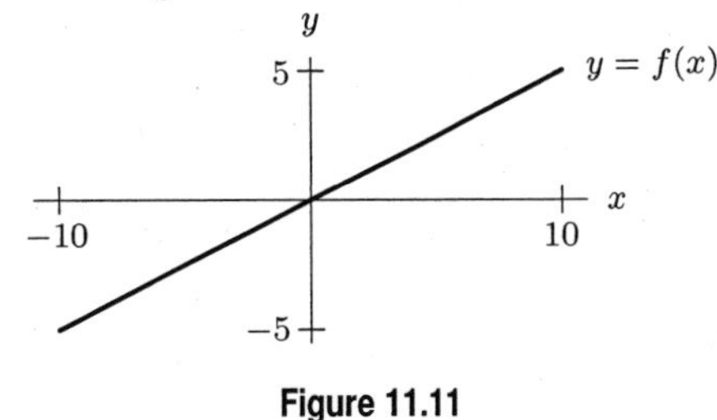

Figure 11.11

On this scale, the graph of $f(x)$ in Figure 11.11 is difficult to distinguish from the line $y = \frac{1}{2}x$; that is, the graph of the linear term alone. However, f itself is definitely not linear; it just looks this way near the origin.

(c) We would like to find the value of x for which the linear term is equal to the cubic term. That is, we want:

$$\frac{1}{50{,}000}x^3 = \frac{1}{2}x$$

which implies that:

$$x^3 = 25{,}000x$$
$$x^3 - 25{,}000x = 0$$
$$x(x^2 - 25{,}000) = 0$$

so

$$x = 0$$

or

$$x = \pm\sqrt{25{,}000} \approx \pm 158.114.$$

Thus, the cubic term becomes "important" for $x > 158.114$ or $x < -158.114$. (Neither term is "important" at $x = 0$, the origin.)

17. Here are the viewing windows used to create these figures. Your viewing windows may differ to a certain extent and still give similar-looking graphs.

(a) $-3 \le x \le -1, -5 \le y \le 5$
(b) $-3 \le x \le 4, -35 \le y \le 15$
(c) $1.25 \le x \le 2.35, -0 \le y \le 6$
(d) $-8 \le x \le 8, -50 \le y \le 2000$

18. We have $-0.1 \le x \le 0.1$, $0 \le y \le f(0.1)$ or $0 \le y \le 0.011$.

19. We have $-1.1 \le x \le -0.9$, $f(-1.1) \le y \le f(-0.9)$ or $-0.121 \le y \le 0.081$.

20. We have $-1.1 \le x \le 0.3$, $f(-1.1) \le y \le f(0.3)$ or $-0.121 \le y \le 0.117$.

21. We have $-20 \le x \le 20$, $f(-20) \le f(20)$ or $-7600 \le y \le 8400$.

22. **(a)** Using a computer or a graphing calculator, we can get a picture of $f(x)$ like the one in Figure 11.12. On this window f appears to be invertible because it passes the horizontal line test.

(b) Substituting gives

$$f(0.5) = (0.5)^3 + 0.5 + 1 = 1.625.$$

To find $f^{-1}(0.5)$, we solve $f(x) = 0.5$. With a computer or graphing calculator, we trace along the graph of f in Figure 11.13 to find

$$f^{-1}(0.5) \approx -0.424.$$

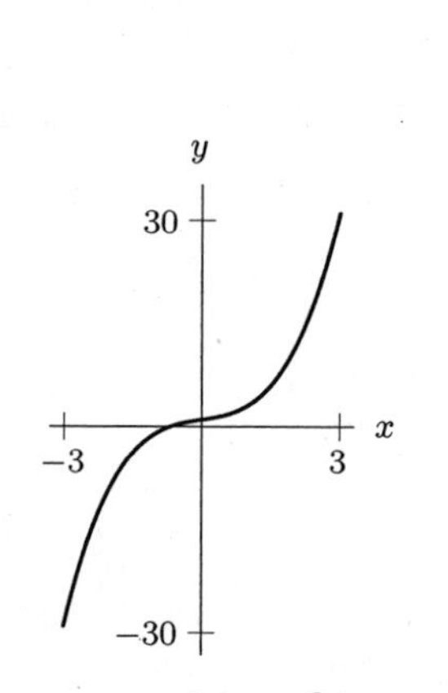

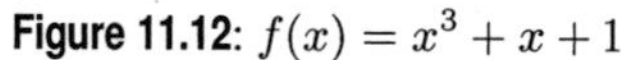

Figure 11.12: $f(x) = x^3 + x + 1$

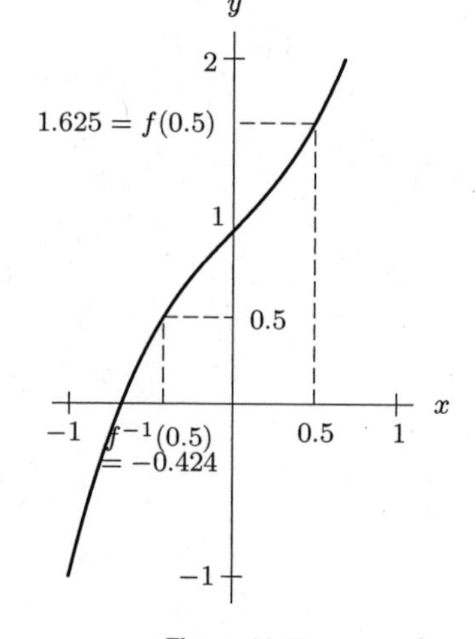

Figure 11.13

23. Use a graphing calculator or computer to approximate values where $f(x) = g(x)$ or to find the zeros for $f(x) - g(x)$. In either case, we find the points of intersection for f and g to be $x \approx -1.764$, $x \approx 0.875$ and $x \approx 3.889$. The values of x for which $f(x) < g(x)$ are on the interval $-1.764 < x < 0.875$ or $x > 3.889$.

24. We see that F is a degree $n = 4$ polynomial with: $a_4 = 3, a_3 = -4, a_2 = 0, a_1 = 5, a_0 = -4$. This means:

$$\begin{aligned}
f(x) &= \underbrace{n \cdot a_n x^{n-1}}_{4 \cdot 3x^3} && \text{because } n = 4 \text{ and } a_4 = 3 \\
&\quad + \underbrace{(n-1) \cdot a_{n-1} x^{n-2}}_{3(-4)x^5} && \text{because } n - 1 = 3 \text{ and } a_3 = -4 \\
&\quad + \underbrace{2a_2 x}_{2 \cdot 0 \cdot x} && \text{because } n - 2 = 2 \text{ and } a_2 = 0 \\
&\quad + \underbrace{a_1}_{5} && \text{because } a_1 = 5 \\
&= 4 \cdot 3x^3 + 3(-4)x^2 + 2 \cdot 0 \cdot x + 5 \\
&= 12x^3 - 12x^2 + 5.
\end{aligned}$$

25. **(a)** The graph of the function in the suggested window is shown in Figure 11.14.

(b) At $t = 0$ (when Liddleville was founded), the population was 1 hundred people.

(c) The t-intercept for $t > 0$ will show when the population was zero. This occurs at $t \approx 7.54$. Thus, Liddleville's population reached zero in July of 1897.

(d) The graph of y has a peak at $t \approx 3.12$. The population at that point is ≈ 10.1 hundred. So the maximum population was ≈ 1010 in February of 1893.

(e) Population predicted is -1.157 hundred. Since the actual population cannot be negative, we see the model does not predict well at $t = 8$.

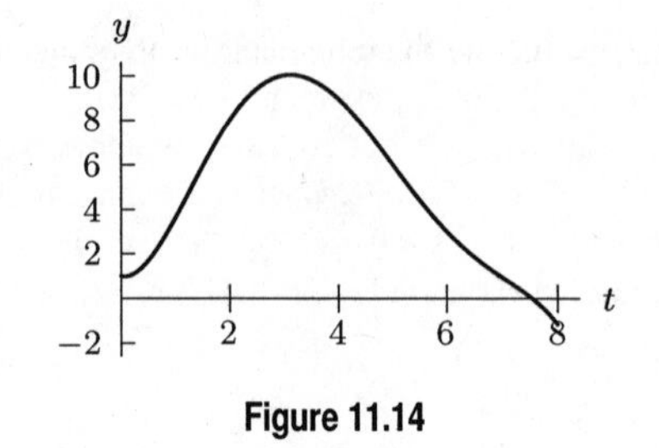

Figure 11.14

26. **(a)** The graph of $C(x) = 4(x-1)^2 + 4$ is the graph of $y = x^2$ shifted right one unit, a vertical stretch by a factor of 4, and then up by 4 units. The graph is shown in Figure 11.15.

(b) The price is \$10,000 per unit, since $R(1) = 10$ means selling 1000 units yields \$10,000,000.

(c) We have

$$\begin{aligned}\text{Profit} &= R(x) - C(x) \\ &= 10x - [4(x-1)^2 + 4] \\ &= 10x - 4(x-1)^2 - 4 \\ &= -8 + 18x - 4x^2\end{aligned}$$

The graph of $R(x) - C(x)$ is shown in Figure 11.16. Profit is negative for $x < 0.5$ and for $x > 4$. Profit $= 0$ at $x = 0.5$ and $x = 4$. Thus, the firm will break even with either 500 or 4000 units, make a profit for $500 < x < 4000$ units, and lose money for any number of units between 0 and 500 or greater than 4000.

Figure 11.15

Figure 11.16

27. **(a)** A graph of V is shown in Figure 11.17 for $0 \le t \le 5$, $0 \le V \le 1$.

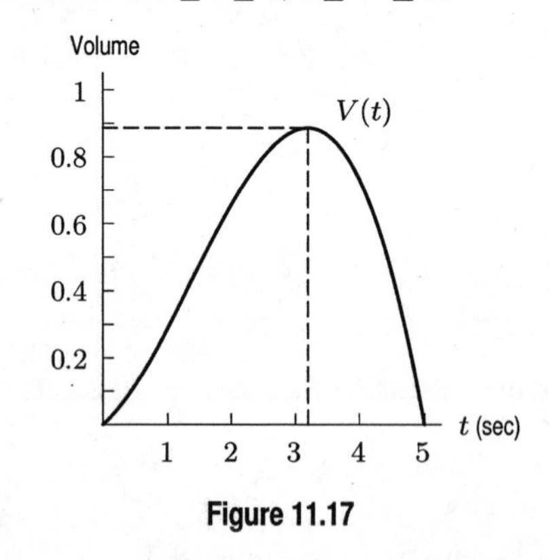

Figure 11.17

(b) The maximum value of V for $0 \le t \le 5$ occurs when $t \approx 3.195$, $V \approx .886$. Thus, at just over 3 seconds into the cycle, the lungs contain ≈ 0.88 liters of air.

(c) The volume is zero at $t = 0$ and again at $t \approx 5$. This indicates that at the beginning and end of the 5 second cycle the lungs are empty.

28. **(a)** We are interested in V for $0 \leq T \leq 30$, and the y-intercept of V occurs at (0, 999.87). If we look at the graph of V on the window $0 \leq x \leq 30$ by $0 \leq y \leq 1500$, the graph looks like a horizontal line. Since V is a cubic polynomial, we suspect more interesting behavior with a better choice of window. Note that $V(30) \approx 1003.77$, so we know V varies (at least) from $V = 999.87$ to $V \approx 1003.77$. Change the range to $998 \leq y \leq 1004$. On this window we see a more appropriate view of the behavior of V for $0 \leq T \leq 30$. See Figure 11.18. [Note: To view the function V as a cubic, a much larger window is needed. Try $-500 \leq x \leq 500$ by $-3000 \leq y \leq 5000$.]

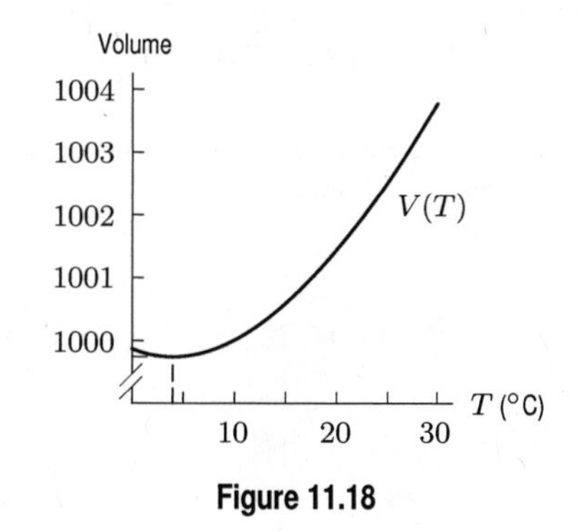

Figure 11.18

(b) The value of V decreases for $0 \leq T \leq 3.961$ and then increases for $3.961 < T < 30$. The graph is concave up on the interval $0 \leq T \leq 30$ (i.e., the graph bends upward). Thus, the volume of 1 kg of water decreases as T increases from 0° C to 3.961° C and increases thereafter. The volume increases at an increasing rate as the temperature increases.

(c) If density, d, is given by $d = m/V$ and m is constant, then the maximum density occurs when V is minimum. Thus, the maximum density occurs when $T \approx 3.961°C$. [Note: We have $m = 1$, but a graph of $y = 1/V$ is very difficult to distinguish from a horizontal line. One possible choice of window to view $y = 1/V$ is $0 \leq x \leq 30$, $0.000996 \leq y \leq 0.001001$.]

29. Yes. For the sake of illustration, suppose $f(x) = x^2 + x + 1$, a second-degree polynomial. Then

$$\begin{aligned} f(g(x)) &= (g(x))^2 + g(x) + 1 \\ &= g(x) \cdot g(x) + g(x) + 1. \end{aligned}$$

Since $f(g(x))$ is formed from products and sums involving the polynomial g, the composition $f(g(x))$ is also a polynomial. In general, $f(g(x))$ will be a sum of powers of $g(x)$, and thus $f(g(x))$ will be formed from sums and products involving the polynomial $g(x)$. A similar situation holds for $g(f(x))$, which will be formed from sums and products involving the polynomial $f(x)$. Thus, either expression will yield a polynomial.

30. **(a)** If f is even, then for all x

$$f(x) = f(-x).$$

Since $f(x) = ax^2 + bx + c$, this implies

$$\begin{aligned} ax^2 + bx + c &= a(-x)^2 + b(-x) + c \\ &= ax^2 - bx + c \end{aligned}$$

We can cancel the ax^2 term and the constant term c from both sides of this equation, giving

$$\begin{aligned} bx &= -bx \\ bx + bx &= 0 \\ 2bx &= 0. \end{aligned}$$

Since x is not necessarily zero, we conclude that b must equal zero, so that if f is even,

$$f(x) = ax^2 + c.$$

(b) If g is odd, then for all x

$$g(-x) = -g(x).$$

Since $g(x) = ax^3 + bx^2 + cx + d$, this implies

$$\begin{aligned} ax^3 + bx^2 + cx + d &= -(a(-x)^3 + b(-x)^2 + c(-x) + d) \\ &= -(-ax^3 + bx^2 - cx + d) \\ &= ax^3 - bx^2 + cx - d. \end{aligned}$$

The odd-powered terms cancel, leaving

$$\begin{aligned} bx^2 + d &= -bx^2 - d \\ 2bx^2 + 2d &= 0 \\ bx^2 + d &= 0. \end{aligned}$$

Since this must hold true for any value of x, we know that both b and d must equal zero. Therefore,

$$g(x) = ax^3 + cx.$$

31. (a) False. For example, $g(x) = x^3 + x^2$ is not odd.

(b) False. For example, $g(x) = (x-1)(x-2)(x-3)$ is not invertible, since there are three values ($x = 1, 2, 3$) where $g(x) = 0$.

(c) False. For example, $g(x) = -x^3$, where $\lim_{x\to\infty} g(x) = -\infty$.

(d) True. If $\lim_{x\to-\infty} g(x) = -\infty$, the leading coefficient is positive; that is, $a_n > 0$. Thus

$$\lim_{x\to\infty} g(x) = \lim_{x\to\infty} a_n x^n = \infty.$$

32. (a) See Figure 11.19.

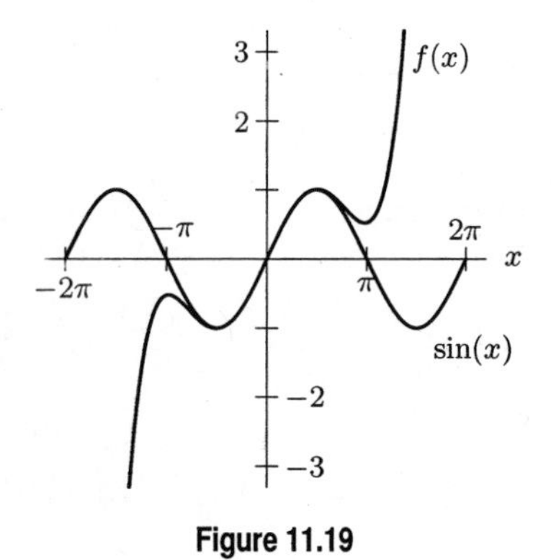

Figure 11.19

(b) The graphs are very similar on the interval

$$\frac{-\pi}{2} \le x \le \frac{\pi}{2},$$

and even slightly larger intervals show close similarity.

(c)

$$\begin{aligned} \sin\left(\frac{\pi}{8}\right) &= 0.382683432\cdots \\ f\left(\frac{\pi}{8}\right) &= \frac{\pi}{8} - \frac{1}{6}\left(\frac{\pi}{8}\right)^3 + \frac{1}{120}\left(\frac{\pi}{8}\right)^5 \\ &= 0.382683717\cdots. \end{aligned}$$

As you can see, $f\left(\frac{\pi}{8}\right)$ differs from $\sin\frac{\pi}{8}$ only in the 7th decimal place—that is, by less than 0.0001%.

(d) Since $\sin x$ is periodic with period 2π, we know that $\sin 18 = \sin(18-2\pi) = \sin(18-4\pi) = \sin(18-6\pi) = \cdots$. Notice that $18 - 6\pi = -0.8495\cdots$ is within the interval $-\frac{\pi}{2} \le x \le \frac{\pi}{2}$ on which f resembles $\sin x$. Thus, $f(18-6\pi) \approx \sin(18-6\pi)$, and since $\sin(18-6\pi) = \sin 18$, then

$$f(18-6\pi) \approx \sin 18.$$

Using a calculator, we find $f(18-6\pi) = -0.7510\cdots$, and $\sin(18) = -0.75098\cdots$. Thus, $f(18-6\pi)$ is an excellent approximation for sin 18. (In fact, your calculator evaluates trigonometric functions internally using a method similar to the one presented in this problem.)

33. (a) Substituting $x = 0.5$ into p, we have

$$p(0.5) = 1 - 0.5 + 0.5^2 - 0.5^3 + 0.5^4 - 0.5^5 \approx 0.65625.$$

Since $f(0.5) = 2/3 = 0.6666....$, the approximation is accurate to 2 decimal places.

(b) We have $p(1) = 1 - 1 + 1 - 1 + 1 - 1 = 0$, but $f(1) = 0.5$. Thus $p(1)$ is a poor approximation to $f(1)$.

(c) See Figure 11.20. The two graphs are difficult to tell apart for $-0.5 \le x \le 0.5$, but for x but outside this region the fit is not good.

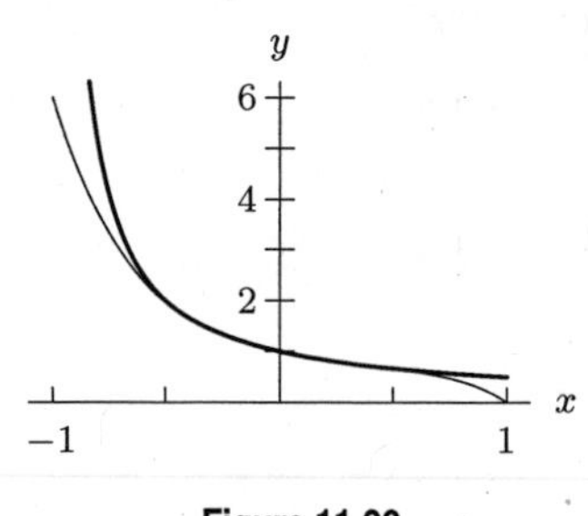

Figure 11.20

34. See Table 11.4.

Table 11.4 *Approximations for the speed of sound (in m/s) in water at temperature T (°C)*

T, °C	0	5	10	15	20	25	30
Linear	1402.4	1427.6	1452.8	1478.0	1503.2	1528.4	1553.5
Quadratic	1402.4	1426.1	1447.0	1464.9	1480.0	1492.1	1501.4
Cubic	1402.7	1426.5	1447.6	1466.3	1482.9	1497.5	1510.5

(a) From Table 11.4, we see that at 5°C, the linear approximation already differs from the actual temperature by more than 1°C. Thus, this approximation is valid only for $0 \le T < 5$.

(b) We see that the quadratic approximation is valid for $0 \le T < 15$.

(c) We see that the cubic approximation is valid for $0 \le T < 30$.

(d) At $T = 50$°C, the cubic approximation gives a velocity of 1551.0 m/sec, which is much higher than the actual value of 1542.6 m/sec. Thus, in order for the quartic term to give a better approximation at this temperature, it should be negative, so that it will lower the estimate. In fact, a better approximation can be given by adding the term $-1.398845 \cdot 10^{-6}T^4$. An even better approximation can be given by also adding the quintic (fifth-degree) term $2.787860 \cdot 10^{-9}T^5$. The resulting formula is known as the Marczak formula:

$$v = 1.402385 \cdot 10^3 + 5.038813T - 5.799136 \cdot 10^{-2}T^2 + 3.287156 \cdot 10^{-4}T^3 - 1.398845 \cdot 10^{-6}T^4 + 2.787860 \cdot 10^{-9}T^5.$$

Solutions for Section 11.3

Exercises

1. Zeros occur where $y = 0$, which we can find by factoring:

$$\begin{aligned} x^3 + 7x^2 + 12x &= 0 \\ x(x^2 + 7x + 12) &= 0 \\ x(x + 4)(x + 3) &= 0. \end{aligned}$$

Zeros are at $x = 0$, $x = -4$, and $x = -3$.

2. Zeros occur where $y = 0$. First we factor:

$$\begin{aligned} (x^2 + 2x - 7)(x^3 + 4x^2 - 21x) &= 0 \\ (x^2 + 2x - 7)x(x^2 + 4x - 21) &= 0 \\ x(x^2 + 2x - 7)(x + 7)(x - 3) &= 0. \end{aligned}$$

To find the points where $x^2 + 2x - 7 = 0$, we use the quadratic formula, so

$$\begin{aligned} x &= \frac{-2 \pm \sqrt{2^2 - 4 \cdot 1 \cdot -7}}{2 \cdot 1} \\ x &= \frac{-2 \pm \sqrt{32}}{2} \\ x &= \frac{-2 \pm 2\sqrt{8}}{2} \\ x &= -1 \pm \sqrt{8}. \end{aligned}$$

Thus, zeros are at $x = 0$, $x = -7$, $x = 3$, $x = -1 + \sqrt{8}$, and $x = -1 - \sqrt{8}$.

3. Zeros occur where $y = 0$, at $x = -3$, $x = 2$, and $x = -7$.

4. Zeros occur where $y = 0$, at $x = -2$ and $x = b$.

5. The graph of h shows zeros at $x = 0$, $x = 3$, and a multiple zero at $x = -2$. Thus

$$h(x) = x(x + 2)^2(x - 3).$$

Check by multiplying and gathering like terms.

6. The graph shows that $g(x)$ has zeros at $x = -2$, $x = 0$, $x = 2$, $x = 4$. Thus, $g(x)$ has factors of $(x + 2)$, x, $(x - 2)$, and $(x - 4)$, so

$$g(x) = k(x + 2)x(x - 2)(x - 4).$$

Since $g(x) = x^4 - 4x^3 - 4x^2 + 16x$, we see that $k = 1$, so

$$g(x) = x(x + 2)(x - 2)(x - 4).$$

7. The graph shows zeros at $x = -2$ and $x = 2$. The fact that f "lingers" at $x = 2$ before crossing the x-axis indicates a multiple zero at $x = 2$. Since the function changes sign at $x = 2$, the factor $(x - 2)$ is raised to an odd power. Thus, try

$$f(x) = (x + 2)(x - 2)^3$$

(Check this answer by expanding and gathering like terms.)

8. The function has a common factor of $4x$ which gives

$$f(x) = 4x(2x^2 - x - 15),$$

and the quadratic factor reduces further giving

$$f(x) = 4x(2x + 5)(x - 3).$$

Thus, the zeros of f are $x = 0$, $x = \frac{-5}{2}$, and $x = 3$.

9. This polynomial must be of fourth (or higher even-powered) degree, so either (but not both) of the zeros at $x = 2$ or $x = 5$ could be doubled. One possible formula is $y = k(x+2)(x-2)^2(x-5)$. Solving for k gives

$$\begin{aligned} k(0+2)(0-2)^2(0-5) &= 5 \\ -40k &= 5 \\ k &= -\frac{1}{8}, \end{aligned}$$

so

$$y = -\frac{1}{8}(x+2)(x-2)^2(x-5).$$

Another possible formula is $y = k(x+2)(x-2)(x-5)^2$. Solving for k gives

$$\begin{aligned} k(0+2)(0-2)(0-5)^2 &= 5 \\ -100k &= 5 \\ k &= -\frac{1}{20}, \end{aligned}$$

so

$$y = -\frac{1}{20}(x+2)(x-2)(x-5)^2.$$

There are other possible polynomials, but all are of degree higher than 4, so these are the simplest.

10. Factoring f gives $f(x) = -5(x+2)(x-2)(5-x)(5+x)$, so the x intercepts are at $x = -2, 2, 5, -5$.
The y intercept is at: $y = f(0) = -5(2)(-2)(5)(5) = 500$.

The polynomial is of fourth degree with the highest powered term $5x^4$. Thus, both ends point upward. A graph of $y = f(x)$ is shown in Figure 11.21.

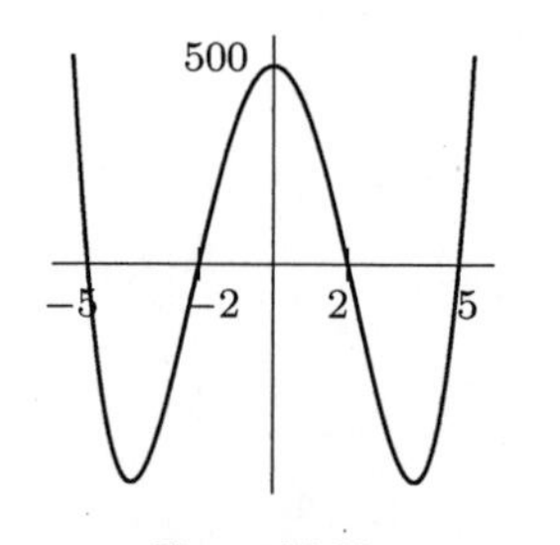

Figure 11.21

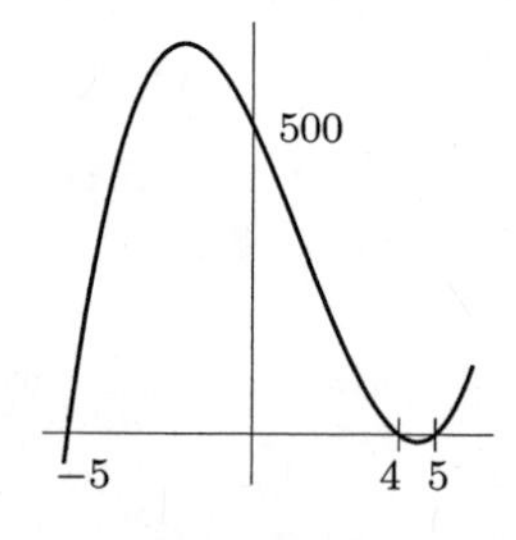

Figure 11.22

11. Factoring g gives $g(x) = 5(x-4)(x+5)(x-5)$, so the x intercepts are at $x = 4, -5, 5$.
The y intercept is at: $y = g(0) = 5(-4)5(-5) = 500$.

The polynomial is third degree with $5x^3$ the highest powered term. Thus, the end behavior is $g(x) \to \infty$ as $x \to \infty$ and $g(x) \to -\infty$ as $x \to -\infty$. A graph of $y = g(x)$ is shown in Figure 11.22.

Problems

12. **(a)** Setting each of the four factors equal to zero, we find that f has four zeros: $x, = 1/2 x = 1/3, x = 7$, and $x = 9$.

(b) It is not possible to find a single viewing window which shows all zeros and all turning points because of the fact that the turning point between $x = 1/2$ and $x = 7$ occurs at a very extreme y value (over 1000). The window required to see this turning point makes it difficult to see the zeros at $x = 1/2$ and $x = 1/3$ because they occur so close together.

(c) The viewing window $0 \leq x \leq 10, -500 \leq y \leq 1500$ clearly shows the turning points of f and the zeros $x = 7$ and $x = 9$, while the window $0 \leq x \leq 1, -2 \leq y \leq 2$ shows the zeros $x = 1/2$ and $x = 1/3$.

13. **(a)** The viewing window $-6 \leq x \leq 2, -10 \leq y \leq 10$ indicates that the zeros of f occur at approximately $x = -5, x = -1, x = 1/2$, and $x = 1$. We therefore guess that

$$f(x) = (x + 5)(x + 1)(2x - 1)(x - 1).$$

We check this by expanding the product to obtain $f(x)$.

(b) The window $-7 \leq x \leq 2, -150 \leq y \leq 10$ clearly shows all the turning points of f.

14. First, the viewing window $0 \leq x \leq 5, -25 \leq y \leq 25$ shows that p has zeros at approximately $x = 1$ and at $x = 3$. Also, since this view of p suggests that the graph of p "bounces off" the x-axis at $x = 1$, we guess that the factorization of $p(x)$ contains a positive even power of $(x - 1)$. Next, the viewing window $-20 \leq x \leq 10, -10000 \leq y \leq 10000$ indicates another zero at approximately $x = -15$. Putting all of this information together and noting that p is a fourth degree polynomial, we guess that the factorization of $p(x)$ must contain $(x - 3)$, $(x - 1)^2$, and $(x + 15)$. Thus, we have

$$p(x) = k(x + 15)(x - 1)^2(x - 3),$$

where k is some constant. Since the leading coefficient of p must equal 1, we take $k = 1$. We check this factorization by expanding the product to obtain $p(x)$.

15. The graph represents a polynomial of even degree, at least 4th. Zeros are shown at $x = -2, x = -1, x = 2$, and $x = 3$. Since $f(x) \to -\infty$ as $x \to \infty$ or $x \to -\infty$, the leading coefficient must be negative. Thus, of the choices in the table, only C and E are possibilities. When $x = 0$, function C gives

$$y = -\frac{1}{2}(2)(1)(-2)(-3) = -\frac{1}{2}(12) = -6,$$

and function E gives

$$y = -(2)(1)(-2)(-3) = -12.$$

Since the y-intercept appears to be $(0, -6)$ rather than $(0, -12)$, function C best fits the polynomial shown.

16. Clearly $f(x) = x$ works. However, the solution is not unique. If f is of the form $f(x) = ax^2 + bx + c$, then $f(0) = 0$ gives $c = 0$, and $f(1) = 1$ gives

$$a(1)^2 + b(1) + 0 = 1,$$

so

$$a + b = 1,$$

or

$$b = 1 - a.$$

Since these are the only conditions which must be satisfied, any polynomial of the form

$$f(x) = ax^2 + (1 - a)x$$

will work. If $a = 0$, we get $f(x) = x$.

17. Since all the three points fall on a horizontal line, the constant function $f(x) = 1$ (degree zero) is the only polynomial of degree ≤ 2 to satisfy the given conditions.

18. To pass through the given points, the polynomial must be of at least degree 2. Thus, let f be of the form

$$f(x) = ax^2 + bx + c.$$

Then using $f(0) = 0$ gives

$$a(0)^2 + b(0) + c = 0,$$

so $c = 0$. Then, with $f(2) = 0$, we have

$$\begin{aligned} a(2)^2 + b(2) + 0 &= 0 \\ 4a + 2b &= 0 \\ \text{so} \quad b &= -2a. \end{aligned}$$

Using $f(3) = 3$ and $b = -2a$ gives

$$a(3)^2 + (-2a)(3) + 0 = 3$$

so

$$\begin{aligned} 9a - 6a &= 3 \\ 3a &= 3 \\ a &= 1. \end{aligned}$$

Thus, $b = -2a$ gives $b = -2$. The unique polynomial of degree ≤ 2 which satisfies the given conditions is $f(x) = x^2 - 2x$.

19. The function f has zeros at $x = -3, 1, 4$. Thus, let $f(x) = k(x+3)(x-1)(x-4)$. Use $f(2) = 5$ to solve for k; $f(2) = k(2+3)(2-1)(2-4) = -10k$. Thus $-10k = 5$ and $k = -\frac{1}{2}$. This gives

$$f(x) = -\frac{1}{2}(x+3)(x-1)(x-4).$$

20. The function g has zeros at $x = -1$ and $x = 5$, and a double zero at $x = 3$. Thus, let $g(x) = k(x-3)^2(x-5)(x+1)$. Use $g(0) = 3$ to solve for k; $g(0) = k(-3)^2(-5)(1) = -45k$. Thus $-45k = 3$ and $k = -\frac{1}{15}$. So

$$g(x) = -\frac{1}{15}(x-3)^2(x-5)(x+1).$$

21. The points $(-3, 0)$ and $(1, 0)$ indicate two zeros for the polynomial. Thus, the polynomial must be of at least degree 2. We could let $p(x) = k(x+3)(x-1)$ as in the previous problems, and then use the point $(0, -3)$ to solve for k. An alternative method would be to let $p(x)$ be of the form

$$p(x) = ax^2 + bx + c$$

and solve for a, b, and c using the given points.

The point $(0, -3)$ gives

$$\begin{aligned} a \cdot 0 + b \cdot 0 + c &= -3, \\ \text{so} \quad c &= -3. \end{aligned}$$

Using $(1, 0)$, we have

$$\begin{aligned} a(1)^2 + b(1) - 3 &= 0 \\ \text{which gives} \quad a + b &= 3. \end{aligned}$$

The point $(-3, 0)$ gives

$$\begin{aligned} a(-3)^2 + b(-3) - 3 &= 0 \\ 9a - 3b &= 3 \\ \text{or} \quad 3a - b &= 1. \end{aligned}$$

From $a + b = 3$, substitute

$$a = 3 - b$$

into

$$3a - b = 1.$$

Then

$$\begin{aligned} 3(3-b) - b &= 1 \\ 9 - 3b - b &= 1 \\ -4b &= -8 \\ \text{so} \quad b &= 2. \end{aligned}$$

Then $a = 3 - 2 = 1$. Therefore,

$$p(x) = x^2 + 2x - 3$$

is the polynomial of least degree through the given points.

22. **(a)** We could think of $f(x) = (x-2)^3 + 4$ as $y = x^3$ shifted right 2 units and up 4. Thus, since $y = x^3$ is invertible, f should also be. Algebraically, we let

$$y = f(x) = (x-2)^3 + 4.$$

Thus,

$$\begin{aligned} y - 4 &= (x-2)^3 \\ \sqrt[3]{y-4} &= x - 2 \\ \sqrt[3]{y-4} + 2 &= x \end{aligned}$$

So $f(x)$ is invertible with an inverse

$$f^{-1}(x) = \sqrt[3]{x-4} + 2.$$

(b) Since g is not so obvious, we might begin by graphing $y = g(x)$. Figure 11.23 shows that the function $g(x) = x^3 - 4x^2 + 2$ does not satisfy the horizontal-line-test, so g is not invertible.

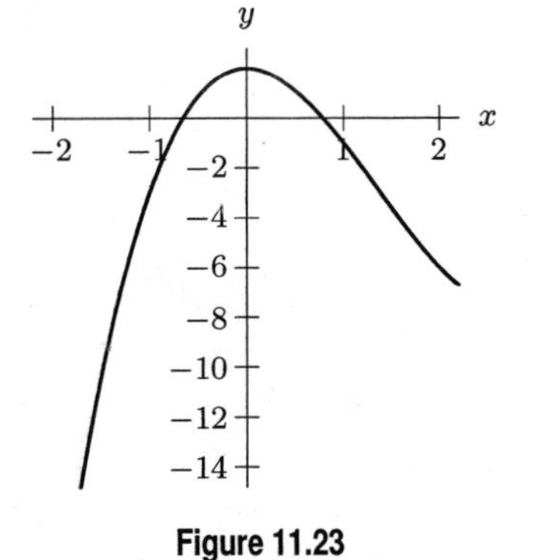

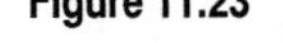

Figure 11.23

23. Try $f(x) = k(x+1)(x-1)^2$ because f has a zero at $x = -1$ and a double zero at $x = 1$. Since $f(0) = -1$, we have $f(0) = k(0+1)(0-1)^2 = k$; thus $k = -1$. So

$$f(x) = -(x+1)(x-1)^2.$$

24. Let $h(x) = k(x+2)(x+1)(x-1)^2(x-3)$, since h has zeros at $x = -2, -1, 3$ and a double zero at $x = 1$. To solve for k, use $h(2) = -1$. Since, $h(2) = k(2+2)(2+1)(2-1)^2(2-3) = k(4)(3)(1)(-1) = -12k$, then $-12k = -1$, or $k = \frac{1}{12}$. Thus

$$h(x) = \frac{1}{12}(x+2)(x+1)(x-1)^2(x-3)$$

is a possible choice.

25. To obtain the flattened effect of the graph near $x = 0$, let $x = 0$ be a multiple zero (of odd multiplicity). Thus, a possible choice would be $f(x) = kx^3(x+1)(x-2)$ for $k > 0$.

26. We know that $g(-2) = 0$, $g(-1) = -3$, $g(2) = 0$, and $g(3) = 0$. We also know that $x = -2$ is a multiple zero. Thus, let

$$g(x) = k(x+2)^2(x-2)(x-3).$$

Then, using $g(-1) = -3$, gives

$$g(-1) = k(-1+2)^2(-1-2)(-1-3) = k(1)^2(-3)(-4) = 12k,$$

so $12k = -3$, and $k = -\frac{1}{4}$. Thus,

$$g(x) = -\frac{1}{4}(x+2)^2(x-2)(x-3)$$

is a possible formula for g.

27. The function f has zeros at $x = -1$, $x = 0$, and at $x = 1$. The zero at $x = 1$ is at least double, so let $f(x) = kx(x+1)(x-1)^2$. To determine the value of k, use the fact that $f(-\frac{1}{2}) = -\frac{27}{16}$. Then,

$$k\left(-\frac{1}{2}\right)\left(-\frac{1}{2}+1\right)\left(-\frac{1}{2}-1\right)^2 = -\frac{27}{16},$$

$$k\left(-\frac{1}{4}\right)\left(-\frac{3}{2}\right)^2 = -\frac{27}{16},$$

$$k\left(-\frac{9}{16}\right) = -\frac{27}{16},$$

so $k = 3$. Thus,

$$f(x) = 3x(x+1)(x-1)^2$$

is a possible formula for f.

28. This polynomial has double zeros at $x = -4$ and at $x = 3$. The y-intercept is at $y = 4$. So

$$f(x) = k(x+4)^2(x-3)^2$$

where we want $f(0) = 4$. This implies that

$$k(0+4)^2(0-3)^2 = 4,$$

giving

$$k = \frac{4}{4^2 \cdot 3^2} = \frac{1}{36}.$$

So

$$f(x) = \frac{1}{36}(x+4)^2(x-3)^2.$$

29. We see that h has zeros at $x = -2$, $x = -1$ (a double zero), and $x = 1$. Thus, $h(x) = k(x+2)(x+1)^2(x-1)$. Then $h(0) = (2)(1)^2(-1)k = -2k$, and since $h(0) = -2$, $-2k = -2$ and $k = 1$. Thus,

$$h(x) = (x+2)(x+1)^2(x-1)$$

is a possible formula for h.

30. Notice that we can think of g as a vertically shifted polynomial. That is, if we let $g(x) = h(x) + 4$, then $h(x)$ is a polynomial with zeros at $x = -1$, $x = 2$, and $x = 4$; furthermore, since $g(-2) = 0$, $h(-2) = 0 - 4 = -4$. Thus,

$$h(x) = k(x+1)(x-2)(x-4).$$

To find k, note that $h(-2) = k(-2+1)(-2-2)(-2-4) = k(-1)(-4)(-6) = -24k$. Since $h(-2) = -24k = -4$, we have $k = \frac{1}{6}$, which gives

$$h(x) = \frac{1}{6}(x+1)(x-2)(x-4).$$

Thus since $g(x) = h(x) + 4$, we have

$$g(x) = \frac{1}{6}(x+1)(x-2)(x-4) + 4.$$

31. Let $g(x) = k(x+2)(x^2)(x-2)$, since g has zeros at $x = \pm 2$, and a double zero at $x = 0$. Since $g(1) = 1$, we have $k(1+2)(1^2)(1-2) = 1$; thus $-3k = 1$ and $k = -\frac{1}{3}$. So

$$g(x) = -\frac{1}{3}(x^2)(x+2)(x-2)$$

is a possible formula.

32. We see that g has zeros at $x = -2$, $x = 2$ and at $x = 0$. The zero at $x = 0$ is at least double, so let $g(x) = k(x+2)(x-2)x^2$. We have $g(1) = k(1+2)(1-2)(1)^2 = k \cdot 3(-1)(1)^2 = -3k$. Since $g(1) = 1$, $-3k = 1$, so $k = -\frac{1}{3}$ and

$$g(x) = -\frac{1}{3}(x+2)(x-2)x^2$$

is a possible formula for g.

33. $y = 4x^2 - 1 = (2x-1)(2x+1)$, which implies that $y = 0$ for $x = \pm\frac{1}{2}$.

34. Note that $y = x^4 + 6x^2 + 9 = (x^2+3)^2$. This implies that $y = 0$ if $x^2 = -3$, but $x^2 = -3$ has no real solutions. Thus, there are no zeros.

35. Zeros occur where $y = 0$, which we can find by factoring:

$$(x^2 - 8x + 12)(x-3) = y$$
$$(x-6)(x-2)(x-3) = y.$$

Zeros are at $x = 6$, $x = 2$, and $x = 3$.

36. Factoring $y = x^2 + 5x + 6$ gives $y = (x+2)(x+3)$. Thus $y = 0$ for $x = -2$ or $x = -3$.

37. $y = 4x^2 + 1 = 0$ implies that $x^2 = -\frac{1}{4}$, which has no solutions. There are no real zeros.

38. Zeros occur where $y = 0$, at $x = 0$ and $x = -3$. Since x^2 is never less than zero, $x^2 + 4$ is never less than 4, so $x^2 + 4$ has no zeros.

39. From its formula, we know that f has double zeros at $x = 5$ and $x = 3$ and a single zero at $x = 1$. Since it is an even function, we know the graph of f has even symmetry—that is, its graph is symmetrical across the y-axis. So we can use the unknowns r, s, and $g(x)$ to balance the zeros we do know. Here is one possibility:

- To balance the double zero at $x = 3$ with a double zero at $x = -3$, we can let $s = 2$.
- To balance the single zero at $x = 1$ with a single zero at $x = -1$, we can let $r = -1$.
- To balance the double zero at $x = 5$, we can let the second-degree polynomial $g(x) = k(x+5)^2$.

Putting this together gives

$$f(x) = (x-5)^2(x-3)^2(x-1)\underbrace{(x+1)}_{(x-r)}\underbrace{(x+3)^2}_{(x+3)^s}\cdot\underbrace{k(x+5)^2}_{g(x)}$$

Another possibility:

- To balance the double zero at $x = 3$ with a double zero at $x = -3$, we again let $s = 2$.
- If we let $g(x) = k(x+1)(x+5)$, then g is a second-degree polynomial that balances the single zero at $x = 1$ and one of the repeated zeros at $x = 5$.
- We can balance the other of the repeated zeros at $x = 5$ by letting $r = -5$.

Putting this together gives

$$f(x) = (x-5)^2(x-3)^2(x-1)\underbrace{(x+5)}_{(x-r)}\underbrace{(x+3)^2}_{(x+3)^s}\cdot\underbrace{k(x+5)(x+1)}_{g(x)}$$

Note that k is an arbitrary non-zero constant. In Figure 11.24, we assume $k = 1$.

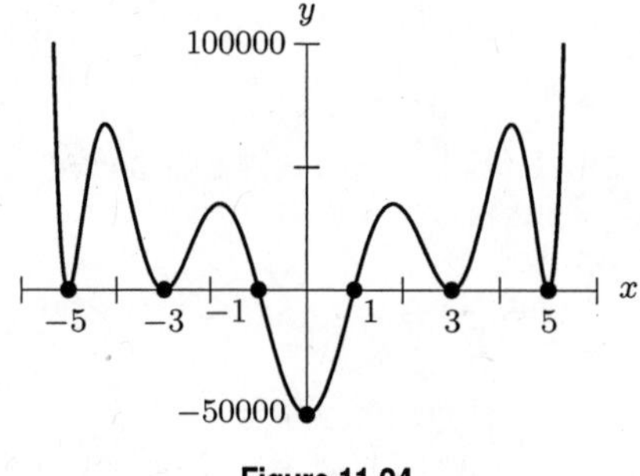

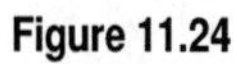

Figure 11.24

40. Two possibilities are

$$y = k_1(x+1)(x-2)^2$$
$$y = k_2(x+1)^2(x-2).$$

Given the y-intercept of $y = 3$, we can find the values of k_1 and k_2:

$$\begin{aligned} 3 &= k_1(0+1)(0-2)^2 \\ 4k_1 &= 3 \\ k_1 &= \frac{3}{4}, \\ 3 &= k_2(0+1)^2(0-2) \\ -2k_2 &= 3 \\ k_2 &= -\frac{3}{2}. \end{aligned}$$

This means our two possibilities are

$$y = \frac{3}{4}(x+1)(x-2)^2$$
$$y = -\frac{3}{2}(x+1)^2(x-2).$$

41. **(a)** $V(x) = x(6-2x)(8-2x)$

(b) Values of x for which $V(x)$ makes sense are $0 < x < 3$, since if $x < 0$ or $x > 3$ the volume is negative.

(c) See Figure 11.25.

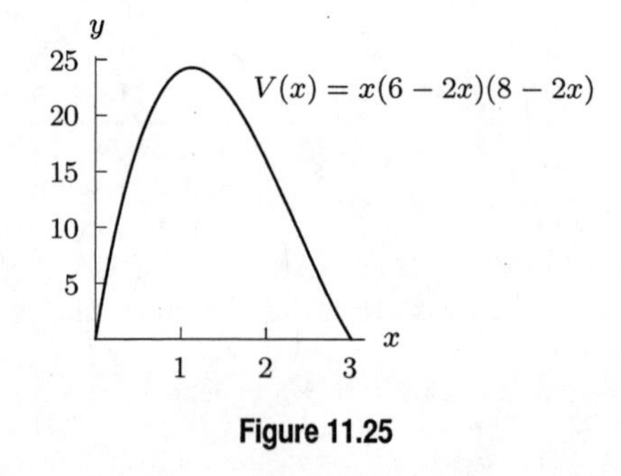

Figure 11.25

(d) Using a graphing calculator, we find the peak between $x = 0$ and $x = 3$ to occur at $x \approx 1.13$. The maximum volume is ≈ 24.26 in^3.

42. Let V be the amount of packing material you will need, then

$$\begin{aligned} V &= (\text{Volume of crate}) - (\text{Volume of box}) \\ &= V_c - V_b \end{aligned}$$

where V_c and V_b are the volumes of the crate and box, respectively. We have for the box's volume,

$$V_b = \underbrace{\text{length}}_{x} \cdot \overbrace{\text{width}}^{x+2} \cdot \underbrace{\text{depth}}_{x-1} = x(x+2)(x-1).$$

The wooden crate must be 1 ft longer than the cardboard box, so its length is $(x+1)$. This gives the required 0.5-ft clearance between the crate and the front and back of the box. Similarly, the crate's width must be 1 ft greater than the box's width of $(x+2)$, and its depth must be 2 ft greater than the box's depth of $(x-1)$. See Figure 11.26.

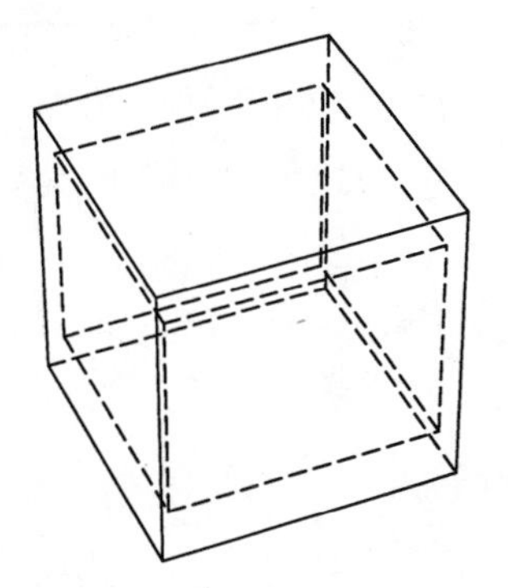

Figure 11.26: Packing a box inside a crate

We have for the crate's volume

$$V_c = \underbrace{\text{length}}_{x+1} \cdot \overbrace{\text{width}}^{(x+2)+1} \cdot \underbrace{\text{depth}}_{(x-1)+2} = (x+1)(x+3)(x+1).$$

Thus, the total amount of packing material will be

$$\begin{aligned} V &= V_c - V_b \\ &= (x+1)(x+3)(x+1) - x(x+2)(x-1). \end{aligned}$$

The formula for V is a difference of two third-degree polynomials. The format is not terribly convenient, so we simplify the formula by multiplying the factors for V_b and V_c and gathering like terms. Then for V we have

$$\begin{aligned} V &= (x+1)(x+3)(x+1) - x(x+2)(x-1) \\ &= (x^3+5x^2+7x+3) - (x^3+x^2-2x) \\ &= 4x^2+9x+3. \end{aligned}$$

The formula $V(x) = 4x^2+9x+3$ gives the necessary information for *appropriate values* of x. Note that the quadratic function $y = 4x^2+9x+3$ is defined for all values of x. However, since x represents the length of a box and $(x-1)$ is the depth of the box, the formula only makes sense as a model for $x > 1$. In this case, an understanding of the component polynomials representing V_b and V_c is necessary in order to determine the logical domain for $V(x)$.

43. We express the volume as a function of the length, x, of the square's side that is cut out in Figure 11.27.

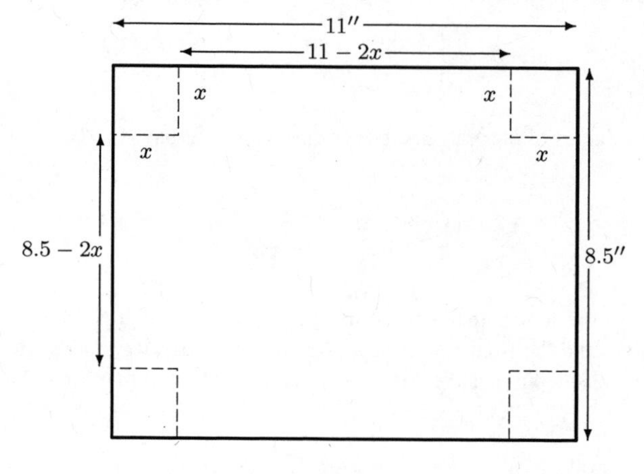

Figure 11.27

Since the sides of the base are $(11-2x)$ and $(8.5-2x)$ inches and the depth is x inches, the volume, $V(x)$, is given by

$$V(x) = x(11-2x)(8.5-2x).$$

The graph of $V(x)$ in Figure 11.28 suggest that the maximum volume occurs when $x \approx 1.585$ inches. (A good viewing window is $0 \le x \le 5$ and $0 \le y \le 70$.) So one side is $x = 1.585$, and therefore the others are $11 - 2(1.585) = 7.83$ and $8.5 - 2(1.585) = 5.33$.

The dimensions of the box are 7.83 by 5.33 by 1.585 inches.

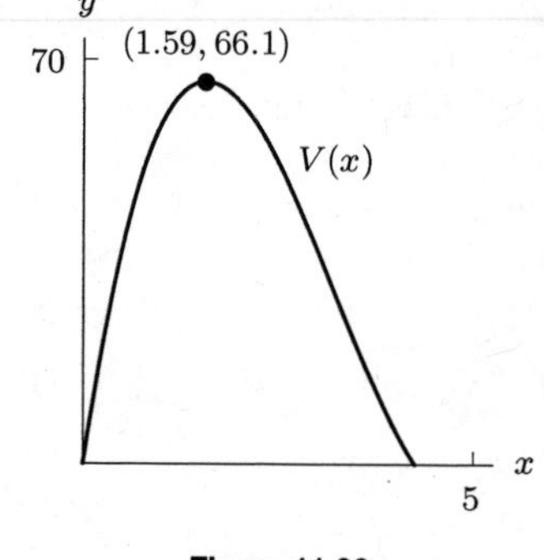

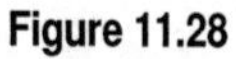

Figure 11.28

44. The ln function is defined only for positive inputs. Here, the input of the ln function is the polynomial $y = (x-3)^2(x+2)$. This polynomial has zeros at $x = 3$, where the graph bounces, and $x = -2$. Its long-run behavior is like $y = x^3$. From Figure 11.29, we see that for this polynomial, y is non-positive for $x \le -2$ and $x = 3$. Thus, the domain of g is all $x > -2$ except $x = 3$.

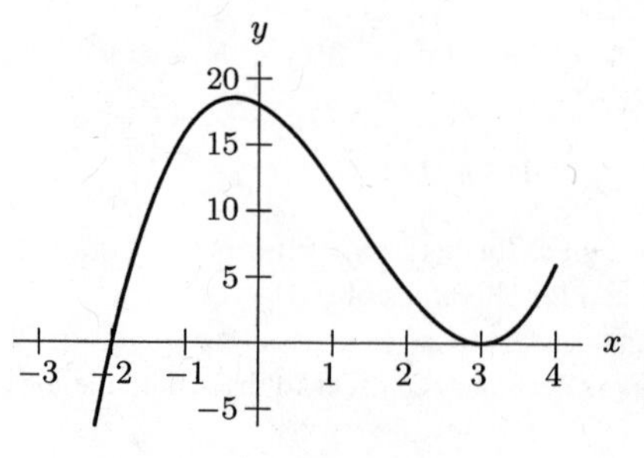

Figure 11.29

45. The domain is $x \geq c$ and $a \leq x \leq b$. Taking the hint, we see that the function $y = (x-a)(x-b)(x-c)$ has zeros at a, b, c and long-run behavior like $y = x^3$. Thus, the graph of $y = (x-a)(x-b)(x-c)$ looks something like the graph in the figure (though of course the zeros may be spaced differently). We see that $y < 0$ on the intervals $x < a$ and $b < x < c$. Thus, the function $y = \sqrt{(x-a)(x-b)(x-c)}$ is undefined on these intervals, but is defined everywhere else.

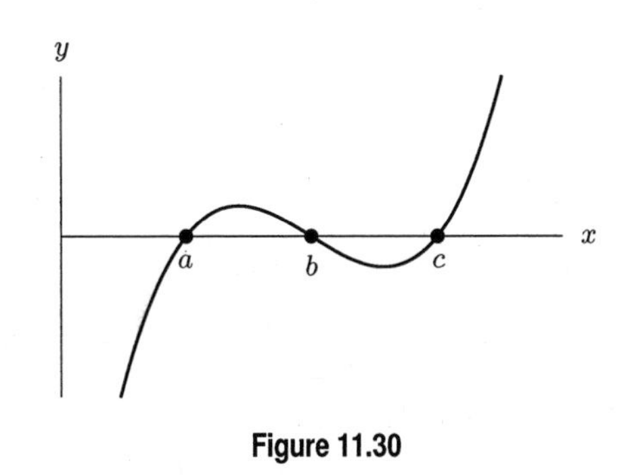

Figure 11.30

46. **(a)** Some things we know about the graph of a, are:

- As $x \to \infty, \quad a(x) \to \infty$. As $x \to -\infty, \quad a(x) \to -\infty$.
- a is an odd function, so it must be symmetric about the origin.
- There is a zero at $(0,0)$ on the graph of a.

(b) The zeros occur at $x = 0,\ 1.112$ and -1.112. Since the function is odd we already knew that for every positive zero there would be a corresponding negative zero.

(c) Since the function is symmetric about the origin, one only needs to concentrate on positive values of x. For values of x between zero and one, x^5 and $2x^3$ are very small, so $-4x$ dominates and $f(x) < 0$. But, for values of x larger than one, x^5 and $2x^3$ get large very quickly and $f(x) > 0$ soon after $x = 1$. Although $-4x$ becomes more and more negative, its magnitude is less and less important in relation to the other two terms. There is no chance that the graph is suddenly going to turn around and cross the x-axis once more.

We can also analyze a algebraically if we note that $a(x)$ can be rewritten as $a(x) = x(x^4 + 2x^2 - 4)$. Thus the zeros of $a(x)$ occur at zero and at the zeros of $(x^4 + 2x^2 - 4)$. Using the quadratic formula we get

$$\begin{aligned} x^2 &= \frac{-2 \pm \sqrt{2^2 - 4(-4)}}{2} \\ &= -1 \pm \frac{\sqrt{20}}{2} \\ \text{so} \quad x &= \pm\sqrt{-1 \pm \sqrt{5}}. \end{aligned}$$

Since we are only interested in the positive solutions we will look at

$$x = \sqrt{-1 \pm \sqrt{5}}.$$

Now

$$x = \sqrt{-1 - \sqrt{5}}$$

is not defined, so the only positive solution is

$$x = \sqrt{-1 + \sqrt{5}} \approx 1.112.$$

(d) The zeros of $b(x)$ occur at 0, 1.112 and -1.112. This should not be a surprise since $b(x) = 2a(x)$. To get the graph of $b(x)$, stretch the graph of $a(x)$ by a factor of two in a vertical direction. Note that the x-intercepts do not change.

47. **(a)** We could let $f(x) = k(x+2)(x-3)(x-5)$ to obtain the given zeros. For a y-intercept of 4, $f(0) = 4 = k(0+2)(0-3)(0-5) = 30k$. Thus $30k = 4$, so $k = \frac{2}{15}$. One possible formula is

$$f(x) = \frac{2}{15}(x+2)(x-3)(x-5).$$

(b) One possibility is that f looks like the function in Figure 11.31 and has a double zero at $x = 5$.

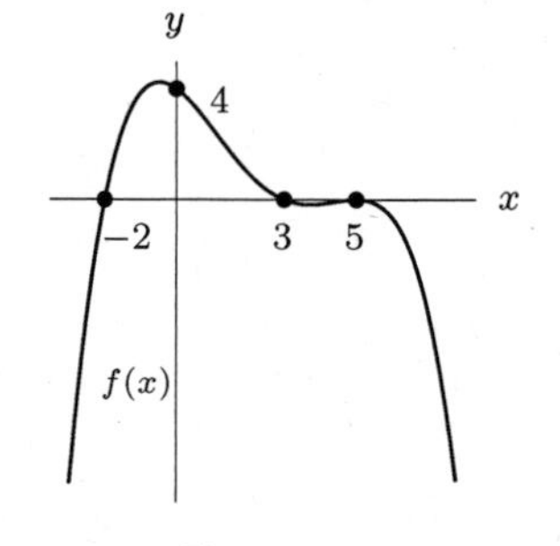

Figure 11.31

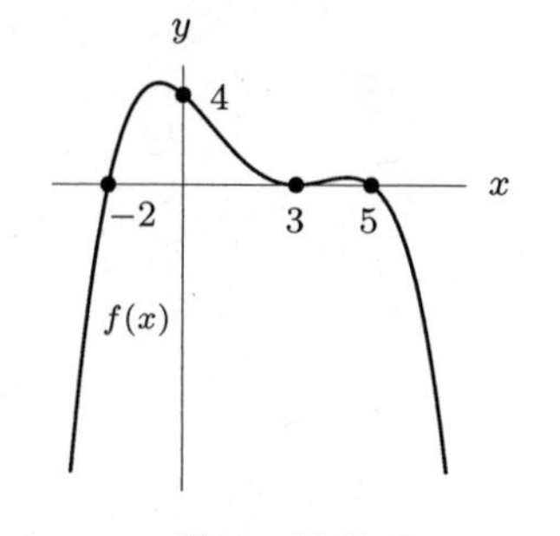

Figure 11.32

Then a formula for f is

$$f(x) = k(x+2)(x-3)(x-5)^2$$

and

$$f(0) = k(2)(-3)(-5)^2.$$

Thus,

$$-150k = 4$$

which gives us

$$k = -\frac{2}{75}.$$

Thus

$$f(x) = -\frac{2}{75}(x+2)(x-3)(x-5)^2.$$

Another possibility is that f has a double-zero at $x = 3$ instead of at $x = 5$. In this case f looks like the function in Figure 11.32. This gives the formula

$$f(x) = -\frac{2}{45}(x+2)(x-3)^2(x-5).$$

Note that if f had a double zero at $x = -2$, there must be another zero for $-2 < x < 0$ in order for f to satisfy $f(0) = 4$ and $y \to -\infty$ as $x \to \pm\infty$.

(c) One possibility is that f looks like the graph in Figure 11.33, with a double zero at $x = -2$ and single zeros at $x = 3$ and $x = 5$.

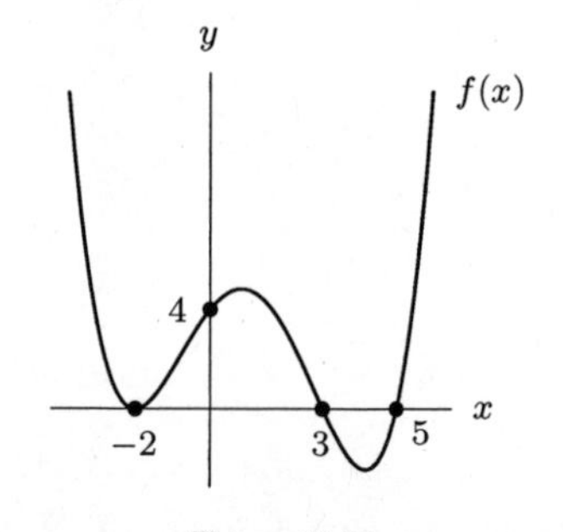

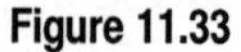

Figure 11.33

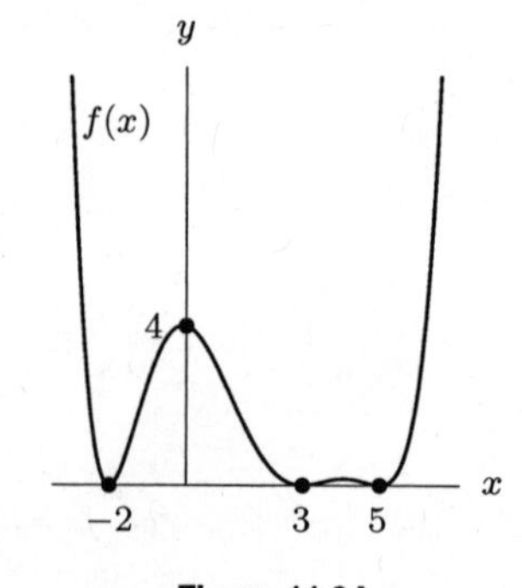

Figure 11.34

A formula for f is $f(x) = k(x+2)^2(x-3)(x-5)$, which gives us

$$k = \frac{1}{15}.$$

Thus,

$$f(x) = \frac{1}{15}(x+2)^2(x-3)(x-5).$$

It is also possible that f has 3 double-zeros at $x = -2$, $x = 3$ and $x = 5$. This leads to a 6th degree polynomial which looks like Figure 11.34. This gives the formula

$$f(x) = \frac{1}{225}(x+2)^2(x-3)^2(x-5)^2.$$

48. **(a)** Never true, because $f(x) \to -\infty$ as $x \to \pm\infty$, which means $f(x)$ must be of even degree.
(b) Sometimes true, since f could be an even-degree polynomial without being symmetric to the y-axis.
(c) Sometimes true, since f could have a multiple zero.
(d) Never true, because f must be of even degree.
(e) True, because, since f is of even degree, $f(-x)$ must have the same long-run behavior as $f(x)$.
(f) Never true, because, since $f(x) \to -\infty$ as $x \to \pm\infty$, f will fail the horizontal-line test.

Solutions for Section 11.4

Skill Refresher

S1. $\dfrac{6}{y} + \dfrac{7}{y^3} = \dfrac{6y^2+7}{y^3}$

S2. $\dfrac{13}{x-1} + \dfrac{14}{2x-2} = \dfrac{13}{x-1} + \dfrac{14}{2(x-1)} = \dfrac{13\cdot 2+14}{2(x-1)} = \dfrac{40}{2(x-1)} = \dfrac{20}{x-1}$

S3. $\dfrac{\frac{1}{x} - \frac{2}{x^2}}{\frac{2x-4}{x^5}} = \dfrac{x-2}{x^2}\cdot\dfrac{x^5}{2(x-2)} = \dfrac{x^3}{2}$

S4.

$$\begin{aligned}\frac{9}{x^2+5x+6} + \frac{12}{x+3} &= \frac{9}{(x+3)(x+2)} + \frac{12}{x+3}\\ &= \frac{9+12(x+2)}{(x+3)(x+2)}\\ &= \frac{12x+33}{(x+3)(x+2)}\\ &= \frac{3(4x+11)}{(x+3)(x+2)}\end{aligned}$$

S5.

$$\begin{aligned}\frac{5}{(x-2)^2(x+1)} - \frac{18}{(x-2)} &= \frac{5-18(x-2)(x+1)}{(x-2)^2(x+1)}\\ &= \frac{5-18x^2+18x+36}{(x-2)^2(x+1)0}\\ &= \frac{-18x^2+18x+41}{(x-2)^2(x+1)}\end{aligned}$$

S6. Dividing by $(x+y)$ is the same as multiplying by its reciprocal, $\frac{1}{x+y}$:

$$\frac{\frac{1}{x+y}}{x+y} = \frac{1}{x+y}\cdot\frac{1}{x+y} = \frac{1}{(x+y)^2}.$$

S7. We write this complex fraction as a multiplication problem. Therefore,

$$\frac{\frac{w+2}{2}}{w+2} = \frac{w+2}{2}\cdot\frac{1}{w+2} = \frac{1}{2}.$$

S8. In this example, the numerator and denominator have no common factor. Therefore the fraction cannot be simplified any further.

S9. $$\frac{x^{-1}+x^{-2}}{1-x^{-2}} = \frac{\frac{1}{x}+\frac{1}{x^2}}{1-\frac{1}{x^2}} = \frac{\frac{x+1}{x^2}}{\frac{x^2-1}{x^2}} = \frac{x+1}{x^2}\cdot\frac{x^2}{x^2-1} = \frac{x+1}{(x+1)(x-1)} = \frac{1}{x-1}.$$

Exercises

1. This is a rational function, and it is already in the form of one polynomial divided by another.
2. This is not a rational function, as we cannot put it in the form of one polynomial divided by another, since 4^x and 3^x are exponential, not power functions. Thus, $f(x)$ is not the ratio of polynomials.
3. This is a rational function, as we can put it in the form of one polynomial divided by another:
$$f(x) = \frac{x^2}{2}+\frac{1}{x} = \frac{x^3}{2x}+\frac{2}{2x} = \frac{x^3+2}{2x}.$$
4. This is not a rational function. We cannot put it in the form of one polynomial divided by another, since 3^x is an exponential, not a power function.
5. This is not a rational function, as we cannot put it in the form of one polynomial divided by another, since $\sqrt{x}+1$ is not a polynomial.
6. This is a rational function, as we can put it in the form of one polynomial divided by another:
$$f(x) = \frac{x^3}{2x^2}+\frac{1}{6} = \frac{x}{2}+\frac{1}{6} = \frac{3x}{6}+\frac{1}{6} = \frac{3x+1}{6}.$$
7. This is not a rational function, as we cannot put it in the form of one polynomial divided by another, since $4\sqrt{x}+7$ is not a polynomial.
8. We have
$$\lim_{x\to\infty}(2x^{-3}+4) = \lim_{x\to\infty}(2/x^3+4) = 0+4 = 4.$$
9. We have
$$\lim_{x\to\infty}(3x^{-2}+5x+7) = \lim_{x\to\infty}(3/x^2+5x+7) = \infty.$$
10. We have
$$\lim_{x\to\infty}\frac{4x+3x^2}{4x^2+3x} = \lim_{x\to\infty}\frac{3x^2}{4x^2} = \frac{3}{4}.$$
11. We have
$$\lim_{x\to-\infty}\frac{3x^2+x}{2x^2+5x^3} = \lim_{x\to-\infty}\frac{3x^2}{5x^3} = 0.$$
12. As $x\to\pm\infty$, $1/x\to 0$ and $x/(x+1)\to 1$, so $h(x)$ approaches $3-0+1=4$. Therefore $y=4$ is the horizontal asymptote.

13. As $x \to \pm\infty$, $1/x \to 0$, so $f(x) \to 1$. Therefore $y = 1$ is the horizontal asymptote.

14. $g(x) = \dfrac{-3x^2 + x + 2}{2x^2 + 1} = \dfrac{-3x^2}{2x^2} = \dfrac{-3}{2}$ as $x \to \pm\infty$.
Thus, $y = -\frac{3}{2}$ is the horizontal asymptote.

15. For the function f, $f(x) \to 1$ as $x \to \pm\infty$ since for large values of x, $f(x) \approx \frac{x^2}{x^2} = 1$.
The function $g(x) \approx \frac{x^3}{x^2} = x$ for large values of x. Thus, as $x \to \pm\infty$, $g(x)$ approaches the line $y = x$.
The function h will behave like $y = \frac{x}{x^2} = \frac{1}{x}$ for large values of x. Thus, $h(x) \to 0$ as $x \to \pm\infty$.

Problems

16. Let $y = f(x)$. Then $x = f^{-1}(y)$. Solving for x,

$$
\begin{aligned}
y &= \frac{4-3x}{5x-4} \\
y(5x-4) &= 4-3x \\
5xy - 4y &= 4-3x \\
5xy + 3x &= 4y+4 \\
x(5y+3) &= 4y+4 \quad \text{(factor out an } x\text{)} \\
x &= \frac{4y+4}{5y+3}
\end{aligned}
$$

Therefore,

$$f^{-1}(x) = \frac{4x+4}{5x+3}.$$

17. Note: There are many examples to fit these descriptions. Some choices are:

$$\text{Even: } f(x) = \frac{x^2}{x^2+1} \qquad \text{Odd: } f(x) = \frac{x^3}{x^2+1} \qquad \text{Neither: } f(x) = \frac{x+1}{x-1}.$$

If $f(x)$ is even, then $f(-x) = f(x)$. This will be true if and only if both $p(x)$ and $q(x)$ are even or both are odd. If one is even and one is odd, then and only then will $f(x)$ be odd. If one is neither then $f(x)$ is neither.

18. **(a)** If $\lim\limits_{x\to\infty} r(x) = 0$, then the degree of the denominator is greater than the degree of the numerator, so $n > m$.
(b) If $\lim\limits_{x\to\infty} r(x) = k$, with $k \neq 0$, the degree of the numerator and denominator must be equal, so $n = m$.

19. **(a)** A graph of f for $0 \le t \le 10$ and $0 \le y \le 1.5$ is shown in Figure 11.35.

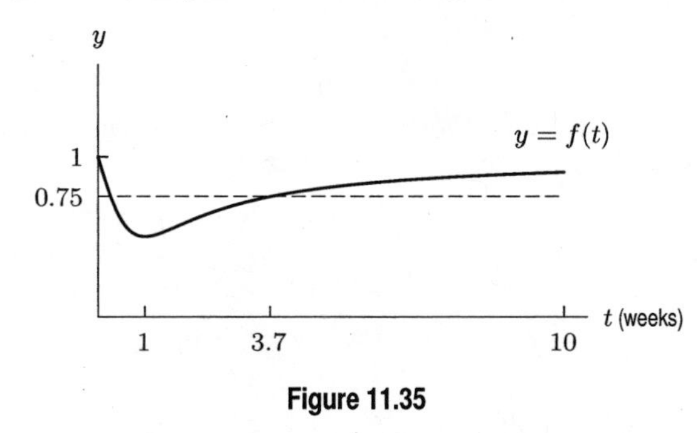

Figure 11.35

(b) At $t = 0$ the oxygen is at its normal level in the pond. The level decreases sharply for the first week after the waste is dumped into the pond. The oxygen level reaches its minimum of approximately one half the normal level by the end of the first week. Then the level begins to increase.
(c) Eventually, the oxygen level in the pond will once again approach the normal level of one.
(d) The line $y = 0.75$ is shown in Figure 11.35. After the level has reached its minimum, we can approximate the intersection of f and $y = 0.75$ at $t \approx 3.73$. Thus, it takes about 3.73 weeks for the level to return to 75% normal.

20. **(a)** To cover its costs of \$80,000 and make a profit of \$40,000, the printing house must take in \$120,000 from sales. Therefore, if 1000 copies are sold,

$$\text{Price per copy} = \frac{120{,}000}{1000} = \$120.$$

Calculating the price for other projected sales in the same way, we have the result in Table 11.5

Table 11.5

Number of copies sold	1000	2000	4000	6000
Price per copy, \$	120	60	30	20

(b) Generalizing the formula we used to calculate the costs in Table 11.5, we have

$$p = \frac{120{,}000}{s}.$$

(c) The graph of $p = 120000/s$ is shown in Figure 11.36

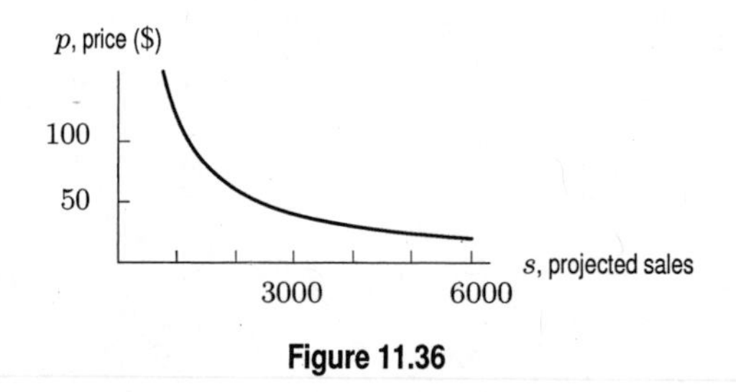

Figure 11.36

21. **(a)** Adding x kg of copper increases both the amount of copper and the total amount of alloy. Originally there are 3 kg of copper and 12 kg of alloy. Adding x kg of copper results in a total of $(3 + x)$ kg of copper and $(12 + x)$ kg of alloy. Thus, the new concentration is given by

$$f(x) = \frac{3 + x}{12 + x}.$$

(b) (i) $f(\frac{1}{2}) = \frac{3+\frac{1}{2}}{12+\frac{1}{2}} = \frac{\frac{7}{2}}{\frac{25}{2}} = \frac{7}{25} = 28\%$. Thus, adding one-half kilogram copper results in an alloy that is 28% copper.

(ii) $f(0) = \frac{3}{12} = \frac{1}{4} = 25\%$. This means that adding no copper results in the original alloy of 25% copper.

(iii) $f(-1) = \frac{2}{11} \approx 18.2\%$. This could be interpreted as meaning that the removal of 1 kg copper (corresponding to $x = -1$) results in an alloy that is about 18.2% copper.

(iv) Let $y = f(x) = \frac{3+x}{12+x}$. Then, multiplying both sides by the denominator we have

$$\begin{aligned}
(12 + x)y &= 3 + x \\
12y + xy &= 3 + x \\
xy - x &= 3 - 12y \\
x(y - 1) &= 3 - 12y \\
x &= \frac{3 - 12y}{y - 1}
\end{aligned}$$

and so

$$f^{-1}(x) = \frac{3 - 12x}{x - 1}$$

Using this formula, we have $f^{-1}(\frac{1}{2}) = \dfrac{-3}{-\frac{1}{2}} = 6$. This means that you must add 6 kg copper in order to obtain an alloy that is $\dfrac{1}{2}$, or 50%, copper. (You can check this by finding $f(6) = \frac{9}{18} = \frac{1}{2}$).

(v) $f^{-1}(0) = \dfrac{3}{-1} = -3$. Check: $f(-3) = \dfrac{0}{9} = 0$. This means that you must remove 3 kg copper to obtain an alloy that is 0% copper, or pure tin.

(c) The graph of $y = f(x)$ is in Figure 11.37. The axis intercepts are $(0, 0.25)$ and $(-3, 0)$. The y-intercept of $(0, 0.25)$ indicates that with no copper added the concentration is 0.25, or 25%, which is the original concentration of copper in the alloy. The x-intercept of $(-3, 0)$ indicates that to make the concentration of copper 0%, we would have to remove 3 kg of copper. This makes sense, as the alloy has only 3 kg of copper to begin with.

(d) The graph of $y = f(x)$ on a larger domain is in Figure 11.38.

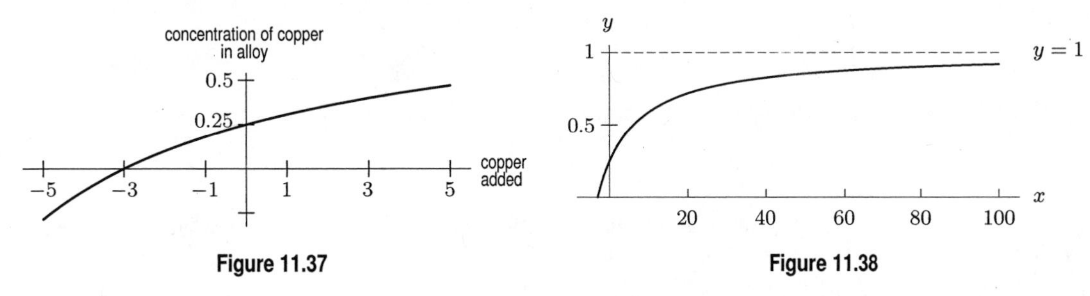

Figure 11.37 **Figure 11.38**

The concentration of copper in the alloy rises with the amount of copper added, x. However, the graph levels off for large values of x, never quite reaching $y = 1 = 100\%$. This is because as more and more copper is added, the concentration gets closer and closer to 100%, but the presence of the 9 kg of tin prevents the alloy from ever becoming 100% pure copper.

22. (a) Originally the total amount of the alloy is 2 kg, one half of which — or equivalently 1 kg — is tin. We have

$$\begin{aligned} C(x) &= \frac{\text{Total amount of tin}}{\text{Total amount of alloy}} \\ &= \frac{(\text{original amount of tin}) + (\text{added tin})}{(\text{original amount of alloy}) + (\text{added tin})} \\ &= \frac{1+x}{2+x} \end{aligned}$$

$C(x)$ is a rational function.

(b) Using our formula, we have

$$C(0.5) = \frac{1+0.5}{2+0.5} = \frac{1.5}{2.5} = 60\%.$$

This means that if 0.5 kg of tin is added, the concentration of tin in the resulting alloy will be 60%. As for $C(-0.5)$, we have

$$C(-0.5) = \frac{1-0.5}{2-0.5} = \frac{0.5}{1.5} \approx 33.333\%.$$

A negative x-value corresponds to the removal of tin from the original mixture, so the statement $C(-0.5) = 33.333\%$ would mean that removing 0.5 kg of tin results in an alloy that is 33.333% tin.

(c) To graph $y = C(x)$, let's see if we can represent the formula as a translation of a power function. We write

$$\begin{aligned} C(x) = \frac{x+1}{x+2} &= \frac{(x+2)-1}{x+2} \\ &= \frac{x+2}{x+2} - \frac{1}{x+2} \quad \text{(splitting the numerator)} \\ &= 1 - \frac{1}{x+2} \\ &= -\frac{1}{x+2} + 1. \end{aligned}$$

Thus the graph of C will resemble the graph of $f(x) = -\frac{1}{x}$ shifted two units to the left and then one unit up. Figure 11.39 shows this translation.

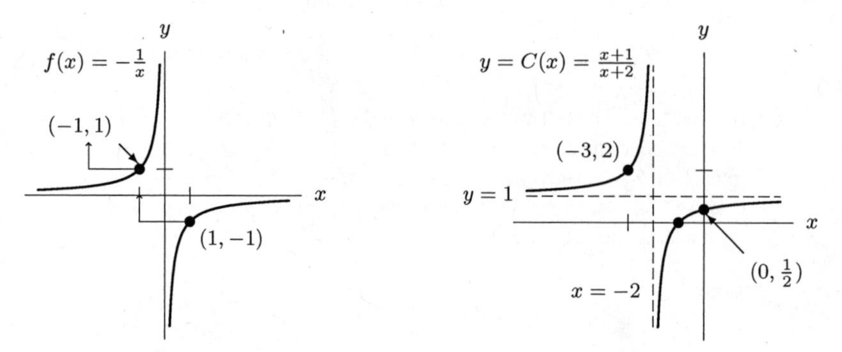

Figure 11.39: The graph of the rational function $y = C(x)$ is a translation of the graph of the power function $f(x) = -\frac{1}{x}$.

For "interesting features", we start with the intercepts and asymptotes. $C(x)$ has a y-intercept between $y = 0$ and $y = 1$, an x-intercept (or zero) at $x = -1$, a horizontal asymptote of $y = 1$, and a vertical asymptote of $x = -2$. What physical significance do these graphical features have?

First off, since

$$C(0) = \frac{0+1}{0+2} = \frac{1}{2} = 0.5$$

we see that the y-intercept is 0.5, or 50%. This means that if you add no tin (i.e. $x = 0$ kg), then the concentration is 50%, the original concentration of tin in the alloy.

Second, since

$$C(-1) = \frac{-1+1}{-1+2} = \frac{0}{1} = 0,$$

we see that the x-intercept is indeed at $x = -1$. This means that if you remove 1 kg of tin (i.e. $x = -1$ kg), then the concentration will be 0%, as there will be no tin remaining in the alloy.

This fact has a second implication: the graph of $C(x)$ is meaningless for $x < -1$, as it is impossible to remove more than 1 kg of tin. Thus, in the context of the problem at hand, the domain of $C(x)$ is $x \geq -1$. The graph on this domain is given by Figure 11.40. Notice that the vertical asymptote of the original graph (at $x = -2$) no longer appears, and it has no physical significance.

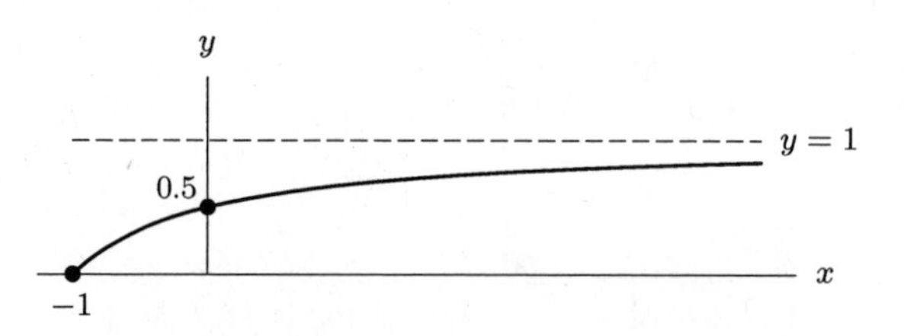

Figure 11.40: The domain of $C(x)$ is $x \geq -1$

The horizontal asymptote of $y = 1$ is, however, physically meaningful. As x grows large, we see that y approaches 1, or 100%. Since the amount of copper in the alloy is fixed at 1 kg, adding large amounts of tin results in an alloy that is nearly pure tin. For example, if we add 10 kg of tin then $x = 10$ and

$$C(x) = \frac{10+1}{10+2} = \frac{11}{12} = 0.916\ldots \approx 91.7\%.$$

Since the alloy now contains 11 kg of tin out of 12 kg total, it is relatively pure tin—at least, it is 91.7% pure. If instead we add 98 kg of tin, then $x = 98$ and

$$C(x) = \frac{98+1}{98+2} = \frac{99}{100} = 99\%.$$

Thus adding 98 kg of tin results in an alloy that is 99% pure. The 1 kg of copper is almost negligible. Therefore, the horizontal asymptote at $y = 1$ indicates that as the amount of added tin, x, grows large, the concentration of tin in the alloy approaches 1, or 100%.

23. The population of Mathville reached 20,000 when

$$\begin{aligned} 20\left(\frac{4t+3}{2t+5}\right) &= 20 \\ \frac{4t+3}{2t+5} &= 1 \\ 4t+3 &= 2t+5 \\ 2t &= 2 \\ t &= 1 \end{aligned}$$

so the population reached 20,000 after 1 year, in 2011.

The function $P(t)$ has a horizontal asymptote given by

$$\lim_{t\to\infty} P(t) = 20\left(\frac{4}{2}\right) = 40,$$

that is 40,000 people. So the population will never reach 50,000.

24. **(a)** The time to travel the first 10 miles is $\frac{10}{40} = 0.25$ hours. The time for the remaining 50 miles in $50/V$ hours so the total journey time is $T = 0.25 + 50/V$. Thus, the average speed is

$$\text{Average speed } = \frac{60}{T} = \frac{60}{\left(0.25 + \frac{50}{V}\right)} = \frac{240V}{V+200}.$$

(b) If you want to average 60 mph for the trip then you need

$$\frac{240V}{V+200} = 60.$$

Solving this equation gives $V = 200/3$ mph, nearly 70 mph.

25. **(a)** $f(x) = \dfrac{\text{Amount of Alcohol}}{\text{Amount of Liquid}} = \dfrac{x}{x+5}$

(b) $f(7) = \frac{7}{7+5} = \frac{7}{12} \approx 58.333\%$. Also, $f(7)$ is the concentration of alcohol in a solution consisting of 5 gallons of water and 7 gallons of alcohol.

(c) $f(x) = 0$ implies that $\dfrac{x}{x+5} = 0$ and so $x = 0$. The concentration of alcohol is 0% when there is no alcohol in the solution, that is, when $x = 0$.

(d) The horizontal asymptote is given by the ratio of the highest-power terms of the numerator and denominator:

$$y = \frac{x}{x} = 1 = 100\%$$

This means that as the amount of alcohol added, x, grows large, the concentration of alcohol in the solution approaches 100%.

26. **(a)** (i) $C(1) = 5050$ means the cost to make 1 unit is $5050.

(ii) $C(100) = 10{,}000$ means the cost to make 100 units is $10,000.

(iii) $C(1000) = 55{,}000$ means the cost to make 1000 units is $55,000.

(iv) $C(10000) = 505{,}000$ means the cost to make 10,000 units is $505,000.

(b) (i) $a(1) = C(1)/1 = 5050$ means that it costs $5050/unit to make 1 unit.

(ii) $a(100) = C(100)/100 = 100$ means that it costs \$100/unit to make 100 units.
(iii) $a(1000) = C(1000)/1000 = 55$ means that it costs \$55/unit to make 1000 units.
(iv) $a(10000) = C(10000)/10000 = 50.5$ means that it costs \$50.50/unit to make 10,000 units.

(c) As the number of units increases, the average cost per unit gets closer to \$50/unit, which is the unit (or marginal) cost. This makes sense because the fixed or initial \$5000 expenditure becomes increasingly insignificant as it is averaged over a large number of units.

27. (a) From Figure 11.41, we see that

$$\text{Slope of line } l = \frac{\Delta y}{\Delta x} = \frac{C(n_0) - 0}{n_0 - 0} = \frac{C(n_0)}{n_0}.$$

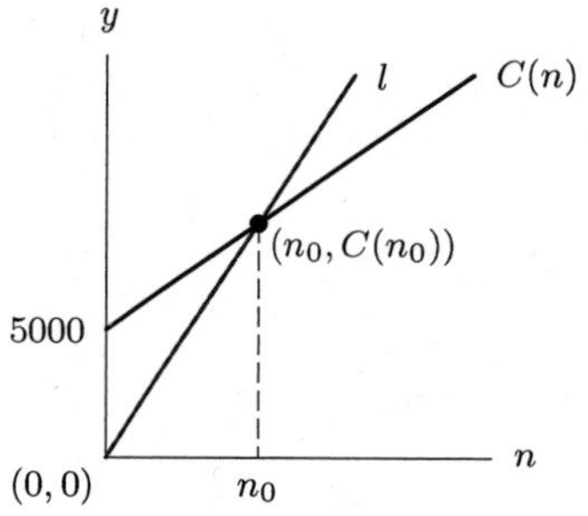

y l $C(n)$ $(n_0, C(n_0))$ 5000 n $(0,0)$ n_0

Figure 11.41

(b) $a(n_0) = C(n_0)/n_0$. Thus, the slope of line l is the same as the average cost of producing n_0 units.

28. Line l_1 has a smaller slope than line l_2. We know the slope of line l_1 represents the average cost of producing n_1 units, and the slope of l_2 represents the average cost of producing n_2 units. Thus, the average cost of producing n_2 units is more than that of producing n_1 units. For these goods, the average cost actually goes up between n_1 and n_2 units.

29. (a) $C(x) = 30000 + 3x$

(b) $a(x) = \dfrac{C(x)}{x} = \dfrac{30000 + 3x}{x} = 3 + \dfrac{30000}{x}$

(c) The graph of $y = a(x)$ is shown in Figure 11.42.

(d) The average cost, $a(x)$, approaches \$3 per unit as the number of units grows large. This is because the fixed cost of \$30,000 is averaged over a very large number of goods, so that each good costs only little more than \$3 to produce.

(e) The average cost, $a(x)$, grows very large as $x \to 0$, because the fixed cost of \$30000 is being divided among a small number of units.

(f) The value of $a^{-1}(y)$ tells us how many units the firm must produce to reach an average cost of \$$y$ per unit. To find a formula for $a^{-1}(y)$, let $y = a(x)$, and solve for x. Then

$$\begin{aligned} y &= \frac{30000 + 3x}{x} \\ yx &= 30000 + 3x \\ yx - 3x &= 30000 \\ x(y-3) &= 30000 \\ x &= \frac{30000}{y-3}. \end{aligned}$$

So, we have $a^{-1}(y) = \dfrac{30000}{y-3}$.

(g) We want to evaluate $a^{-1}(5)$, the total number of units required to yield an average cost of \$5 per unit.

$$a^{-1}(5) = \frac{30000}{5-3} = \frac{30000}{2} = 15000.$$

Thus, the firm must produce at least 15,000 units for the average cost per unit to be \$5. The firm must produce at least 15,000 units to make a profit.

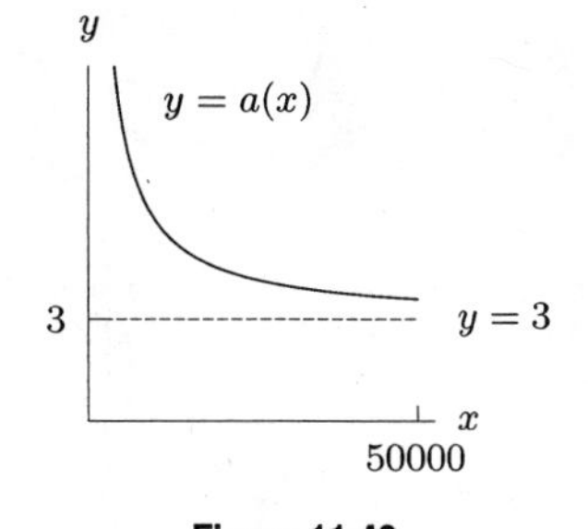

Figure 11.42

30. We need to find the coefficient c so that $R(x) = f(x)$ approximates $f(x)$. Notice that $R(0) = f(0) = 1$ for all values of c. We can find c by making sure that the function and approximation match at $x = 1$, so $R(1) = f(1) = e$. This gives

$$R(1) = \frac{1}{1 + c \cdot 1} = f(1) = e \quad \text{so} \quad c = e^{-1} - 1 = -0.6321,$$

so

$$R(x) = \frac{1}{1 - 0.6321x}.$$

Figure 11.43 shows $f(x)$ and $R(x)$ on the interval $(0, 1)$. There is reasonable agreement.

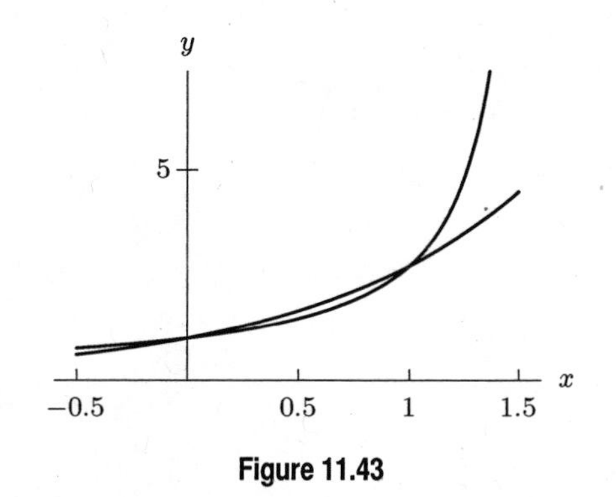

Figure 11.43

Solutions for Section 11.5

Exercises

1. The zero of this function is at $x = 4$. It has vertical asymptotes at $x = \pm 3$. Its long-run behavior is: $y \to 0$ as $x \to \pm\infty$. See Figure 11.44.

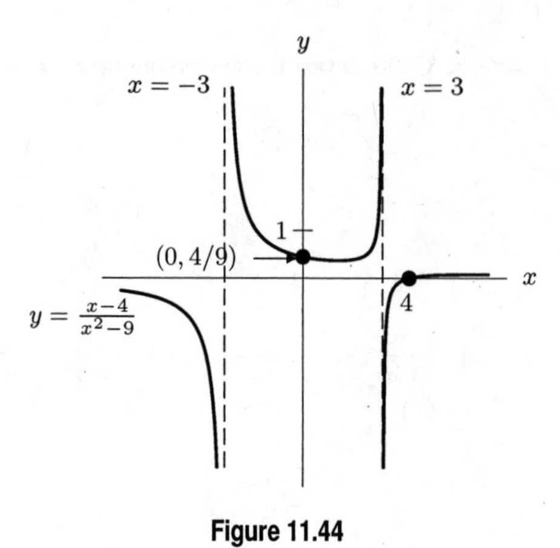

Figure 11.44

2. The zeros of this function are at $x = \pm 2$. It has a vertical asymptote at $x = 9$. Its long-run behavior is that it looks like the line $y = x$. See Figure 11.45.

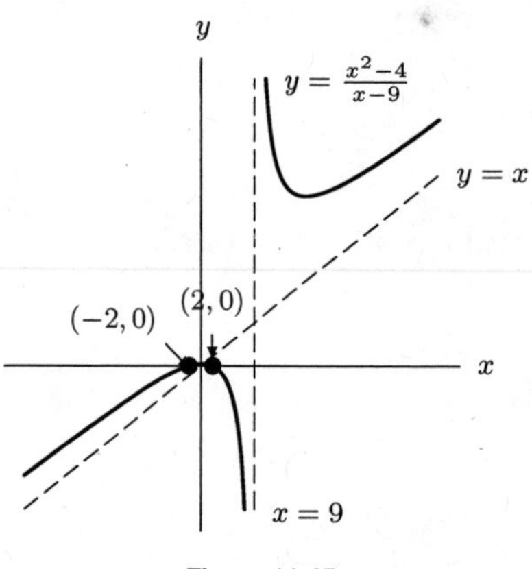

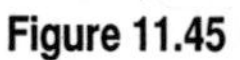
Figure 11.45

3. The zero of this function is at $x = -3$. It has a vertical asymptote at $x = -5$. Its long-run behavior is: $y \to 1$ as $x \to \pm\infty$. See Figure 11.46.

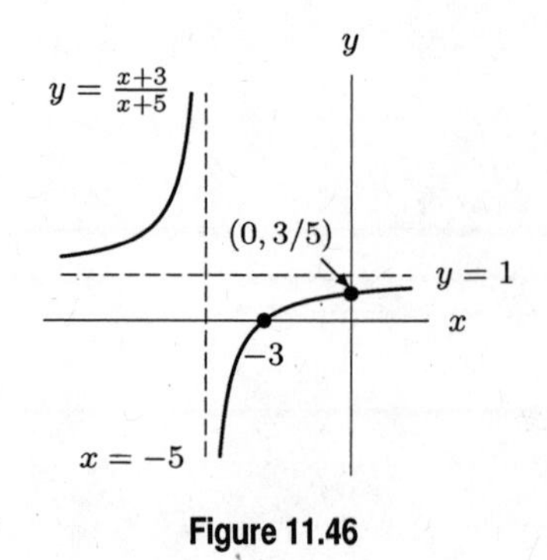

Figure 11.46

4. The zero of this function is at $x = -3$. It has a vertical asymptote at $x = -5$. Its long-run behavior is: $y \to 0$ as $x \to \pm\infty$. See Figure 11.47.

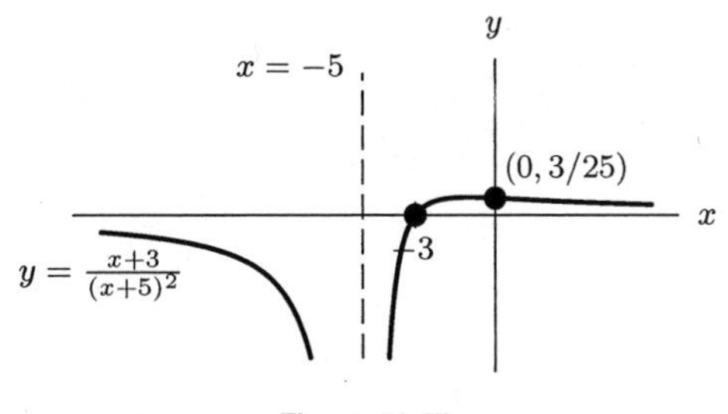

Figure 11.47

5. Since

$$g(x) = \frac{x^2 - 4}{x^3 + 4x^2} = \frac{(x-2)(x+2)}{x^2(x+4)},$$

the x-intercepts are $x = \pm 2$; there is no y-intercept; the horizontal asymptote is $y = 0$; the vertical asymptotes are $x = 0, x = -4$.

6. Since

$$k(x) = \frac{x(4-x)}{x^2 - 6x + 5} = \frac{x(4-x)}{(x-1)(x-5)},$$

the x-intercepts are $x = 0, x = 4$; the y-intercept is $y = 0$; the vertical asymptotes are $x = 1, x = 5$. Since we can also write

$$k(x) = \frac{4x - x^2}{x^2 - 6x + 5} = \frac{-x^2 + 4x}{x^2 - 6x + 5},$$

the horizontal asymptote is $y = -1$

7. The x-intercept is $x = 2$; the y-intercept is $y = -2/(-4) = 1/2$; the horizontal asymptote is $y = 1$; the vertical asymptote is $x = 4$.

8. Since

$$g(x) = \frac{x^2 - 9}{x^2 + 9} = \frac{(x-3)(x+3)}{x^2 + 9},$$

the x-intercepts are $x = \pm 3$; the y-intercept is $y = -9/9 = -1$; the horizontal asymptote is $y = 1$; there are no vertical asymptotes.

9. **(a)** See the following table.

x	-5	-4.1	-4.01	-4	-3.99	-3.9	-3
$G(x)$	10	82	802	Undef	-798	-78	-6

As x approaches -4 from the left the function takes on very large positive values. As x approaches -4 from the right the function takes on very large negative values.

(b)

x	5	10	100	1000
$G(x)$	1.111	1.429	1.923	1.992

x	-5	-10	-100	-1000
$G(x)$	10	3.333	2.083	2.008

For $x > -4$, as x increases, $f(x)$ approaches 2 from below. For $x < -4$, as x decreases, $f(x)$ approaches 2 from above.

(c) The horizontal asymptote is $y = 2$. The vertical asymptote is $x = -4$. See Figure 11.48.

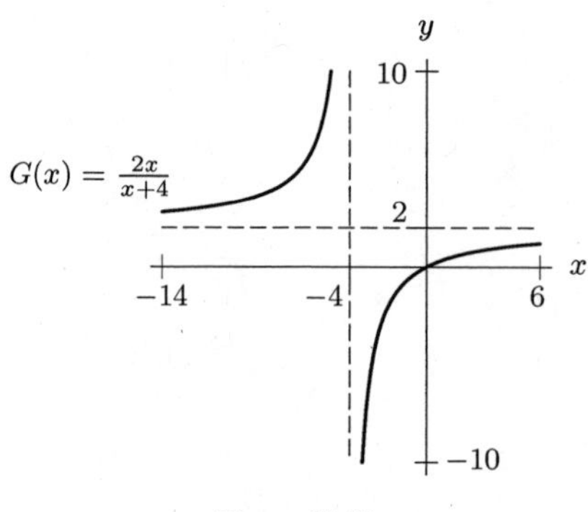

Figure 11.48

10. **(a)** See the following table.

x	−3	−2.1	−2.01	−2	−1.99	−1.9	−1
$g(x)$	1	100	10,000	Undef	10,000	100	1

As x approaches -2 from the left the function takes on very large positive values. As x approaches -2 from the right the function takes on very large positive values.

(b)

x	5	10	100	1000
$g(x)$	0.02	0.007	$9.6 \cdot 10^{-5}$	10^{-6}

x	−5	−10	−100	−1000
$g(x)$	0.111	0.016	10^{-4}	10^{-6}

For $x > -2$, as x increases, $f(x)$ approaches 0 from above. For $x < -2$ as x decreases, $f(x)$ approaches 0 from above.

(c) The horizontal asymptote is $y = 0$ (the x-axis). The vertical asymptote is $x = -2$. See Figure 11.49.

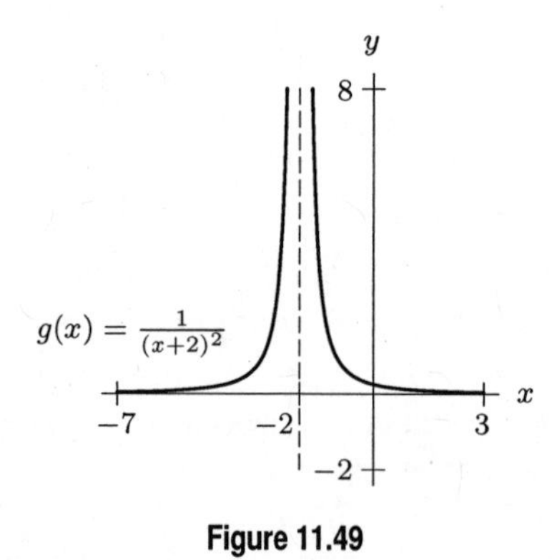

Figure 11.49

Problems

11. The graph is the graph of $y = 1/x$ moved up by 2. See Figure 11.50.

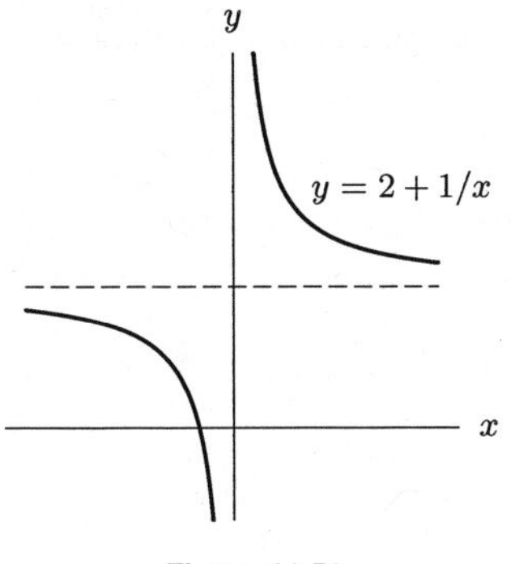

Figure 11.50

12. The graph will have vertical asymptotes at $x = \pm 4$ and zeros at $x = 3$ and $x = 2$. The y-intercept is $(0, -\frac{3}{4})$, and for large positive or negative values of x, we see that $y \to 2$—thus, there is a horizontal asymptote of $y = 2$. Note that the graph will intersect the horizontal asymptote if

$$\frac{2x^2 - 10x + 12}{x^2 - 16} = 2,$$

which implies

$$\begin{aligned} 2x^2 - 10x + 12 &= 2x^2 - 32 \\ -10x &= -44 \\ x &= 4.4 \end{aligned}$$

Putting all of this information together, we obtain a graph similar to that of Figure 11.51.

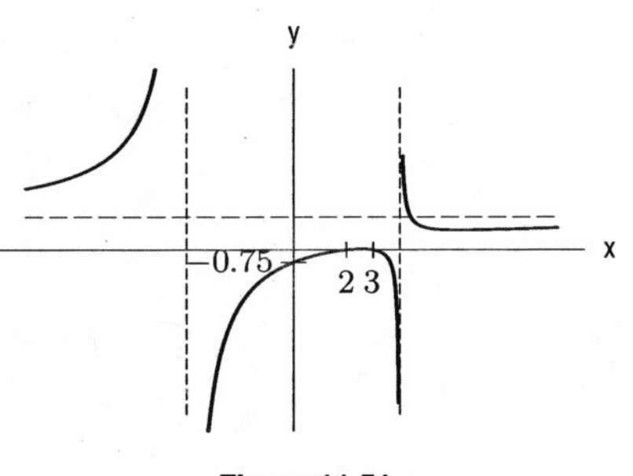

Figure 11.51

13. (a) To estimate

$$\lim_{x \to 5^+} \frac{x}{5 - x},$$

we consider what happens to the function when x is slightly larger than 5. The numerator is positive and the denominator is negative and is approaching 0 as x approaches 5. We suspect that $\frac{x}{5-x}$ gets more and more negative as x approaches 5 from the right. We can also use either a graph or a table of values as in Table 11.6 to estimate this limit. We see that

$$\lim_{x \to 5^+} \frac{x}{5 - x} = -\infty.$$

Table 11.6

x	5.1	5.01	5.001	5.0001
$f(x)$	−51	−501	−5001	−50001

(b) To estimate

$$\lim_{x\to 5^-} \frac{x}{5-x},$$

we consider what happens to the function when x is slightly smaller than 5. The numerator is positive and the denominator is positive and is approaching 0 as x approaches 5. We suspect that $\frac{x}{5-x}$ gets larger and larger as x approaches 5 from the left. We can also use either a graph or a table of values to estimate this limit. We see that

$$\lim_{x\to 5^-} \frac{x}{5-x} = +\infty.$$

14. (a) To estimate

$$\lim_{x\to 2^+} \frac{5-x}{(x-2)^2},$$

we consider what happens to the function when x is slightly larger than 2. The numerator is positive and the denominator is positive and is approaching 0 as x approaches 2. We suspect that $\frac{5-x}{(x-2)^2}$ gets larger and larger as x approaches 2 from the right. We can also use either a graph or a table of values as in Table 11.7 to estimate this limit. We see that

$$\lim_{x\to 2^+} \frac{5-x}{(x-2)^2} = +\infty.$$

Table 11.7

x	2.1	2.01	2.001	2.0001
$f(x)$	290	29900	2999000	299990000

(b) To estimate

$$\lim_{x\to 2^-} \frac{5-x}{(x-2)^2},$$

we consider what happens to the function when x is slightly smaller than 2. The numerator is positive and the denominator is positive and is approaching 0 as x approaches 2. We suspect that $\frac{5-x}{(x-2)^2}$ gets larger and larger as x approaches 2 from the left. We can also use either a graph or a table of values to estimate this limit. We see that

$$\lim_{x\to 2^-} \frac{5-x}{(x-2)^2} = +\infty.$$

15. (a) $y = -\frac{1}{(x-5)^2} - 1$ has a vertical asymptote at $x = 5$, no x intercept, horizontal asymptote $y = -1$: (iii)

(b) $y = \frac{x-2}{(x+1)(x-3)}$ has vertical asymptotes at $x = -1, 3$, x intercept at 2, horizontal asymptote $y = 0$: (i)

(c) $y = \frac{2x+4}{x-1}$ has a vertical asymptote at $x = 1$, x intercept at $x = -2$, horizontal asymptote $y = 2$: (ii)

(d) $y = \frac{x-3+x+1}{(x+1)(x-3)} = \frac{2x-2}{(x+1)(x-3)}$ has vertical asymptotes at $x = -1, 3$, x intercept at 1, horizontal asymptote at $y = 0$: (iv)

(e) $y = \frac{(1+x)(1-x)}{x-2}$ has vertical asymptote at $x = 2$, two x intercepts at ± 1: (vi)

(f) $y = \frac{1-4x}{2x+2}$ has a vertical asymptote at $x = -1$, x intercept at $x = \frac{1}{4}$, horizontal asymptote at $y = -2$: (v)

16. (a) Matches description (v). $y = \frac{f(x)}{g(x)} = \frac{x^2-4}{x^2+4}$: No vertical asymptotes. Horizontal asymptote at $y = 1$. x-intercepts at ± 2.

(b) Does not match any of the descriptions. $y = \frac{g(x)}{f(x)} = \frac{x^2+4}{x^2-4}$: Vertical asymptotes at $x = \pm 2$. Horizontal asymptote at $y = 1$. No zeros.

(c) Matches description (i). $y = \frac{h(x)}{f(x)} = \frac{x+5}{x^2-4}$: Vertical asymptotes at $x = \pm 2$. Horizontal asymptote at $y = 0$. x-intercept at -5.

(d) Matches description (vii). $y = f(\frac{1}{x}) = \frac{1}{x^2} - 4 = \frac{1-4x^2}{x^2}$: Vertical asymptote at $x = 0$. Horizontal asymptote at $y = -4$. x-intercepts at $\pm\frac{1}{2}$.

(e) Matches description (ii). $y = \frac{g(x)}{h(x)} = \frac{x^2+4}{x+5}$: Vertical asymptote at $x = -5$. No horizontal asymptotes. No x-intercepts.

(f) Matches description (ii). $y = \frac{h(x^2)}{h(x)} = \frac{x^2+5}{x+5}$: Vertical asymptote at $x = -5$. No horizontal asymptotes. No x-intercepts.

(g) Does not match any of the descriptions. $y = \frac{1}{g(x)} = \frac{1}{x^2+4}$: No vertical asymptotes. Horizontal asymptote at $y = 0$. No x-intercepts.

(h) Does not match any of the descriptions. $y = f(x) \cdot g(x) = (x^2 - 4)(x^2 + 4)$: No vertical asymptotes. No horizontal asymptotes. x-intercepts at ± 2.

17. (a) $\lim_{x\to\infty} f(x) = \lim_{x\to-\infty} f(x) = 0$.

(b) The vertical asymptote is $x = -2$ and we see

$$\lim_{x\to-2^+} f(x) = \infty \quad \text{and} \quad \lim_{x\to-2^-} f(x) = \infty.$$

18. (a) $\lim_{x\to\infty} f(x) = \lim_{x\to-\infty} f(x) = 2$.

(b) The vertical asymptote is $x = -4$ and we see

$$\lim_{x\to-4^+} f(x) = -\infty \quad \text{and} \quad \lim_{x\to-4^-} f(x) = \infty.$$

19. (a) If $f(n)$ is large, then $\frac{1}{f(n)}$ is small.

(b) If $f(n)$ is small, then $\frac{1}{f(n)}$ is large.

(c) If $f(n) = 0$, then $\frac{1}{f(n)}$ is undefined.

(d) If $f(n)$ is positive, then $\frac{1}{f(n)}$ is also positive.

(e) If $f(n)$ is negative, then $\frac{1}{f(n)}$ is negative.

20. (a) The graph of $y = \frac{1}{f(x)}$ will have vertical asymptotes at $x = 0$ and $x = 2$. As $x \to 0$ from the left, $\frac{1}{f(x)} \to -\infty$, and as $x \to 0$ from the right, $\frac{1}{f(x)} \to +\infty$. The reciprocal of 1 is 1, so $\frac{1}{f(x)}$ will also go through the point (1,1). As $x \to 2$ from the left, $\frac{1}{f(x)} \to +\infty$, and as $x \to 2$ from the right, $\frac{1}{f(x)} \to -\infty$. As $x \to \pm\infty$, $\frac{1}{f(x)} \to 0$ and is negative.

The graph of $y = \frac{1}{f(x)}$ is shown in Figure 11.52.

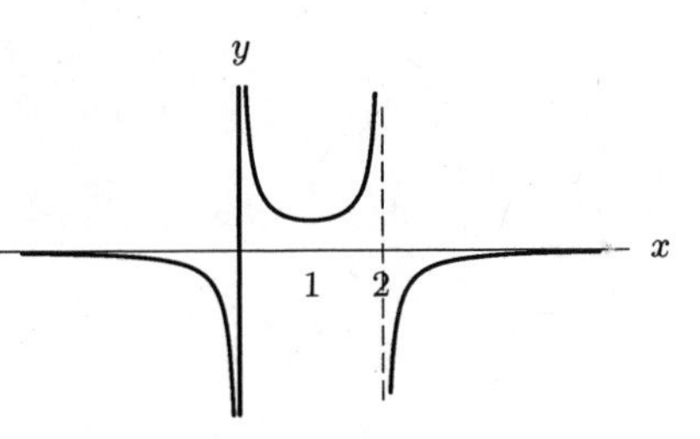

Figure 11.52

(b) A formula for f is of the form

$$f(x) = k(x-0)(x-2) \quad \text{and} \quad f(1) = 1.$$

Thus, $1 = k(1)(-1)$, so $k = -1$. Thus

$$f(x) = -x(x-2).$$

The reciprocal $\frac{1}{f(x)} = -\frac{1}{x(x-2)}$ is graphed as shown in Figure 11.52.

21. **(a)** The graph of $y = -f(-x) + 2$ will be the graph of f flipped about both the x-axis and the y-axis and shifted up 2 units . The graph is shown in Figure 11.53.

(b) The graph of $y = \frac{1}{f(x)}$ will have vertical asymptotes $x = -1$ and $x = 3$. As $x \to +\infty$, $\frac{1}{f(x)} \to -\frac{1}{2}$ and as $x \to -\infty$, $\frac{1}{f(x)} \to 0$. At $x = 0$, $\frac{1}{f(x)} = \frac{1}{2}$, and as $x \to -1$ from the left, $\frac{1}{f(x)} \to -\infty$; as $x \to -1$ from the right, $\frac{1}{f(x)} \to +\infty$; as $x \to 3$ from the left, $\frac{1}{f(x)} \to +\infty$; and as $x \to 3$ from the right, $\frac{1}{f(x)} \to -\infty$.

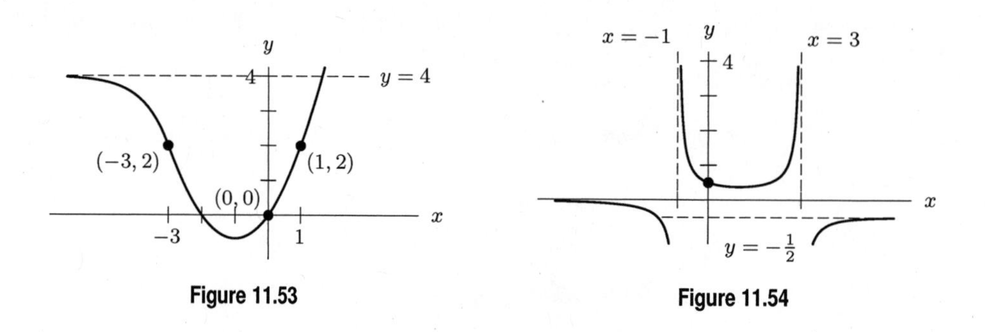

Figure 11.53

Figure 11.54

22. **(a)** The graph shows $y = 1/x$ shifted to the right one and up 2 units. Thus,

$$y = \frac{1}{x-1} + 2$$

is a choice for a formula.

(b) The equation $y = 1/(x-1) + 2$ can be written as

$$y = \frac{2x-1}{x-1}.$$

(c) We see that the graph has both an x-and y-intercept. When $x = 0$, $y = \frac{-1}{-1} = 1$, so the y-intercept is $(0, 1)$. If $y = 0$ then $2x - 1 = 0$, so $x = \frac{1}{2}$. The x-intercept is $(\frac{1}{2}, 0)$.

23. **(a)** The graph shows $y = 1/x$ flipped across the x-axis and shifted left 2 units. Therefore

$$y = -\frac{1}{x+2}$$

is a choice for a formula.

(b) The formula $y = -1/(x+2)$ is already written as a ratio of two linear functions.

(c) The graph has a y-intercept if $x = 0$. Thus, $y = -\frac{1}{2}$. Since y cannot be zero if $y = -1/(x+2)$, there is no x-intercept. The only intercept is $(0, -\frac{1}{2})$.

24. **(a)** The graph appears to be the graph of $y = 1/x$ shifted down 3 units. Thus,

$$y = \frac{1}{x} - 3$$

is a possible formula for the function.

(b) The formula $y = (1/x) - 3$ can be written as

$$y = \frac{-3x+1}{x}$$

after getting a common denominator and combining terms.

(c) The graph has no y-intercept since $x = 0$ is not in the domain of the function. Any x-intercept(s) will occur when the numerator of $y = (-3x+1)/x$ is zero. Then $-3x + 1 = 0$, so $x = \frac{1}{3}$. The only intercept is at $(\frac{1}{3}, 0)$.

25. The function f is the transformation of $y = \dfrac{1}{x}$, so $p = 1$. The graph of $y = \dfrac{1}{x}$ has been shifted three units to the right and four units up. To find the y-intercept, we need to evaluate $f(0)$:

$$f(0) = \frac{1}{-3} + 4 = \frac{11}{3}.$$

To find the x-intercepts, we need to solve $f(x) = 0$ for x.

$$\begin{aligned} \text{Thus,} \quad 0 &= \frac{1}{x-3} + 4, \\ -4 &= \frac{1}{x-3}, \\ -4(x-3) &= 1, \\ -4x + 12 &= 1, \\ -4x &= -11, \\ \text{so} \quad x &= \frac{11}{4} \quad \text{is the only } x\text{-intercept.} \end{aligned}$$

The graph of f is shown in Figure 11.55.

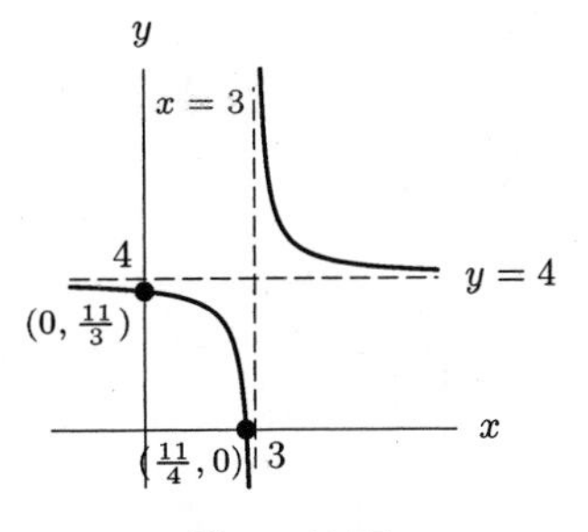

Figure 11.55

26. The function g is a transformation of $y = 1/x^2$, so $p = 2$. The graph of $y = 1/x^2$ has been shifted 2 units to the right, flipped over the x-axis and shifted 3 units down. To find the y-intercept, we need to evaluate $g(0)$:

$$g(0) = -\frac{1}{4} - 3 = -\frac{13}{4}.$$

Note that this function has no x-intercepts.

The graph of g is shown in Figure 11.56.

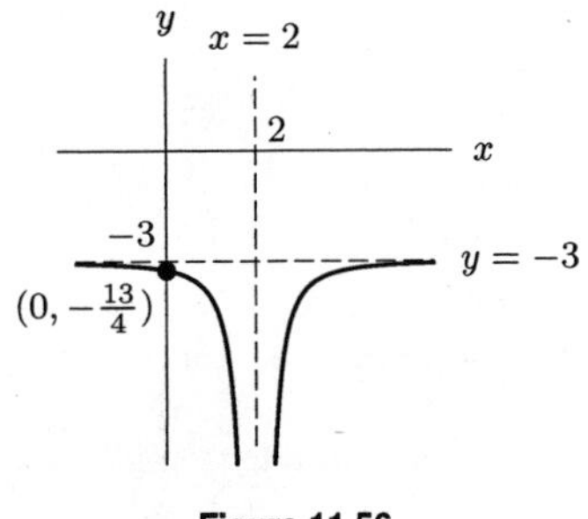

Figure 11.56

27. First, we can simplify the formula for $h(x)$:

$$\begin{aligned} h(x) &= \frac{1}{x-1} + \frac{2}{1-x} + 2 \\ &= \frac{1}{x-1} - \frac{2}{x-1} + 2 = -\frac{1}{x-1} + 2. \end{aligned}$$

Thus h is a transformation of $y = \frac{1}{x}$ with $p = 1$. The graph of $y = \frac{1}{x}$ has been shifted one unit to the right, flipped over the x-axis and shifted up 2 units. To find the y-intercept, we need to evaluate $h(0)$:

$$h(0) = -\frac{1}{-1} + 2 = 3.$$

To find the x-intercepts, we need to solve $h(x) = 0$ for x:

$$\begin{aligned} 0 &= -\frac{1}{x-1} + 2 \\ -2 &= -\frac{1}{x-1} \\ -2x + 2 &= -1 \\ -2x &= -3, \\ \text{so,} \quad x &= \frac{3}{2} \quad \text{is the only } x\text{-intercept.} \end{aligned}$$

The graph of h is shown in Figure 11.57.

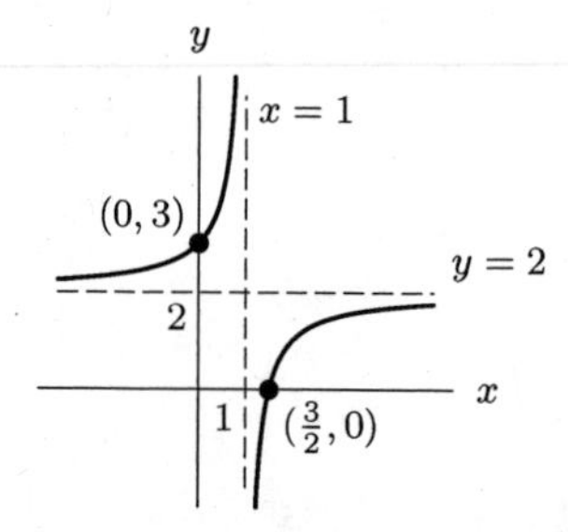

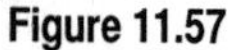

Figure 11.57

28. **(a)** The table indicates symmetry about the y-axis. The fact that the function values have the same sign on both sides of the vertical asymptotes indicates a transformation of $y = 1/x^2$ rather than $y = 1/x$.

(b) As x takes on large positive or negative values, $y \to 1$. Thus, we try

$$y = \frac{1}{x^2} + 1.$$

This formula works and can be expressed as

$$y = \frac{1 + x^2}{x^2}.$$

29. **(a)** The table indicates translation of $y = 1/x$ because the values of the function are headed in opposite directions near the vertical asymptote.

(b) The transformation of $y = 1/x$ given involves a shift to the right by 2 units, so we try

$$y = \frac{1}{x-2}.$$

A check of data from the table shows that we must add $\frac{1}{2}$ to each output of our guess. Thus, a formula would be

$$y = \frac{1}{x-2} + \frac{1}{2}.$$

As a ratio of polynomials, we have

$$y = \frac{1(2)}{(x-2)(2)} + \frac{1(x-2)}{2(x-2)}$$

so

$$y = \frac{x}{2x-4}.$$

30. **(a)** The table shows symmetry about the vertical asymptote $x = 3$. The fact that the function values have the same sign on both sides of the vertical asymptotes indicates a transformation of $y = 1/x^2$ rather than $y = 1/x$.

(b) In order to shift the vertical asymptote from $x = 0$ to $x = 3$ for the table, we try

$$y = \frac{1}{(x-3)^2}.$$

checking the x-values from the table in this formula gives y-values that are each 1 less than the y-values of the table. Therefore, we try

$$y = \frac{1}{(x-3)^2} + 1.$$

This formula works. To express the formula as a ratio of polynomials, we take

$$y = \frac{1}{(x-3)^2} + \frac{1(x-3)^2}{(x-3)^2},$$

so

$$y = \frac{x^2 - 6x + 10}{x^2 - 6x + 9}.$$

31. **(a)** The table indicates translation of $y = 1/x$ because the values of the function are headed in opposite directions near the vertical asymptote.

(b) The data points in the table indicate that $y \to \frac{1}{2}$ as $x \to \pm\infty$. The vertical asymptote does not appear to have been shifted. thus, we might try

$$y = \frac{1}{x} + \frac{1}{2}.$$

A check of x-values shows that this formula works. To express as a ratio of polynomials, we get a common denominator. Then

$$y = \frac{1(2)}{x(2)} + \frac{1(x)}{2(x)}$$

$$y = \frac{2+x}{2x}$$

32. We express the ratio of the volume to the surface area of the box. See Figure 11.58.

The sides of the base are $(11 - 2x)$ and $(8.5 - 2x)$ inches and the depth is x inches, so the volume $V(x) = x(11 - 2x)(8.5 - 2x)$. The surface area, $S(x)$, is the area of the complete sheet minus the area of the four squares, so $S(x) = 11 \cdot 8.5 - 4x^2$. Thus, the ratio we want to maximize is given by

$$R(x) = \frac{V(x)}{S(x)} = \frac{x(11-2x)(8.5-2x)}{11 \cdot 8.5 - 4x^2}.$$

The graph in Figure 11.59 suggests that the maximum of $R(x)$ occurs when $x \approx 1.939$ inches. (A good viewing window is $0 < x < 5$ and $0 < y < 1$.) So one side is $x = 1.939$ and therefore the others are $11 - 2(1.939) = 7.123$ and $8.5 - 2(1.939) = 4.623$ inches.

The dimensions of the box are 7.123 by 4.623 by 1.939 inches.

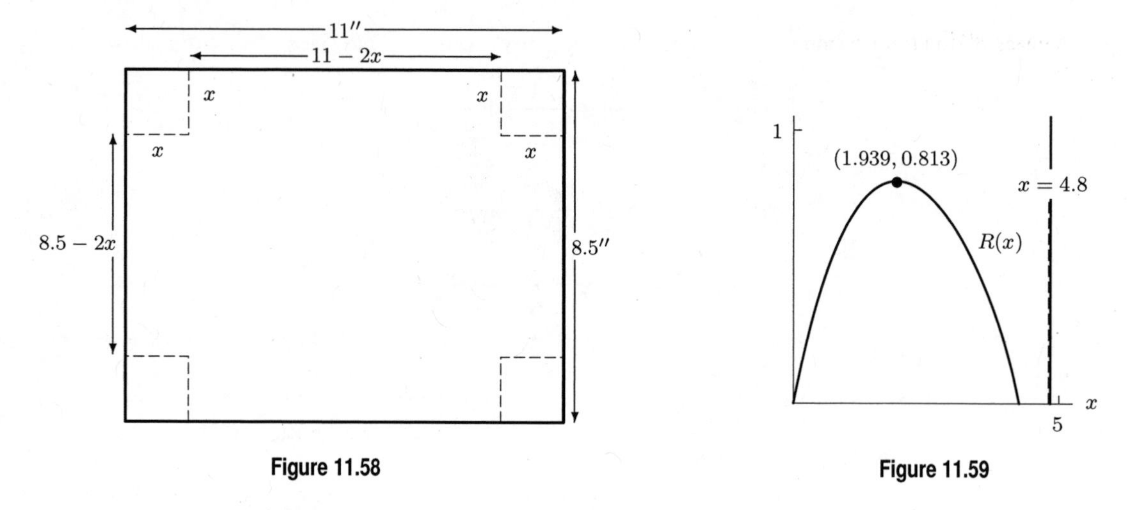

Figure 11.58

Figure 11.59

33.
- Since the graph has a vertical asymptote at $x = 2$, let the denominator be $(x - 2)$.
- Since the graph has a zero at $x = -1$, let the numerator be $(x + 1)$.
- Since the long–range behavior tends toward -1 as $x \to \pm\infty$, the ratio of the leading terms should be -1.

So a possible formula is $y = f(x) = -\left(\frac{x+1}{x-2}\right)$. You can check that the y–intercept is $y = \frac{1}{2}$, as it should be.

34. We try $(x+3)(x-1)$ in the numerator in order to get zeros at $x = -3$ and $x = 1$. There is only one vertical asymptote at $x = -2$, but in order to have the horizontal asymptote of $y = 1$, the numerator and denominator must be of same degree. Thus, try

$$y = \frac{(x+3)(x-1)}{(x+2)^2}.$$

Note that this answer gives the correct y-intercept of $(0, -\frac{3}{4})$ and $y \to 1$ as $x \to \pm\infty$.

35. A guess of $y = \frac{(x-3)(x+2)}{(x+1)(x-2)}$ fits the zeros and vertical asymptote of the graph. However, in order to satisfy the y-intercept at $(0, -3)$ and end behavior of $y \to -1$ as $x \to \pm\infty$, the graph should be "flipped" across the x-axis. Thus try $y = -\frac{(x-3)(x+2)}{(x+1)(x-2)}$.

36.
- Since the graph has asymptotes at $x = -1$ and $x = 2$, let the denominator be $(x-2)(x+1)$.
- Since the graph has zeros at $x = -2$ and $x = 3$, let the numerator be $(x+2)(x-3)$.
- Since the long–range behavior tends toward 1 as $x \to \pm\infty$, the ratio of the leading terms should be 1.

So a possible formula is $y = f(x) = \frac{(x+2)(x-3)}{(x-2)(x+1)}$. You can check that the y-intercept is $y = 3$, as it should be.

37. The graph has vertical asymptotes at $x = -1$ and $x = 1$. When $x = 0$, we have $y = 2$ and $y = 0$ at $x = 2$. The graph of $y = \frac{(x-2)}{(x+1)(x-1)}$ satisfies each of the requirements, including $y \to 0$ as $x \to \pm\infty$.

38. The graph of $y = \frac{x}{(x+2)(x-3)}$ fits.

39. Factoring the numerator, we have

$$f(x) = \frac{18 - 11x + x^2}{x-2} = \frac{(x-9)(x-2)}{x-2} = (x-9)\frac{(x-2)}{(x-2)}.$$

In this form we see that the graph of $y = f(x)$ is identical to that of $y = x - 9$, except that the graph of $y = f(x)$ has no y-value corresponding to $x = 2$. The line $y = x - 9$ goes through the point $(2, -7)$ so the graph of $y = f(x)$ will be the line $y = x - 9$ with a hole at $(2, -7)$.

40. Factoring the numerator, we have

$$g(x) = \frac{x^3+5x^2+x+5}{x+5} = \frac{x^2(x+5)+(x+5)}{x+5} = \frac{(x+5)(x^2+1)}{(x+5)} = (x^2+1)\frac{(x+5)}{(x+5)}.$$

In this form we see that the graph of $y = g(x)$ is identical to that of $y = x^2 + 1$, except that the graph of $y = g(x)$ has no y-value corresponding to $x = -5$. The parabola $y = x^2 + 1$ goes through the point $(-5, 26)$ so the graph of $y = g(x)$ will be the parabola $y = x^2 + 1$ with a hole at $(-5, 26)$.

41. In order to create a hole in the graph of $y = x^3$ at $(2, 8)$, we use the factor $\frac{x-2}{x-2}$. Multiplying by this factor is the same as multiplying by 1, except at $x = 2$, where the factor is undefined. So, our function is

$$h(x) = x^3\frac{(x-2)}{(x-2)} = \frac{x^4-2x^3}{x-2}.$$

42.

$$f(x) = \frac{p(x)}{q(x)} = \frac{-3(x-2)(x-3)}{(x-5)^2}$$

We need the factor of -3 in the numerator and the exponent of 2 in the denominator, because we have a horizontal asymptote of $y = -3$. The ratio of highest term of $p(x)$ to highest term of $q(x)$ will be $\frac{-3x^2}{x^2} = -3$.

43. The vertical asymptotes indicate a denominator of $(x + 2)(x - 3)$. The horizontal asymptote of $y = 0$ indicates that the degree of the numerator is less than the degree of the denominator. To get the point (5,0) we need $(x - 5)$ as a factor in the numerator. Therefore, try

$$g(x) = \frac{(x-5)}{(x+2)(x-3)}.$$

44. The description of h agrees with the description of g from part (b) except h has a horizontal asymptote of $y = 1$. Therefore, the degree of numerator and denominator must be the same. In fact, the highest-powered terms should be the same. Note that we can accomplish this without adding a zero by changing the function of (b) to

$$h(x) = \frac{(x-5)^2}{(x+2)(x-3)}.$$

Solutions for Section 11.6

Exercises

1. The function is exponential, because $p(x) = (5^x)^2 = 5^{2x} = (5^2)^x = 25^x$.

2. The function fits neither, because the variable in the exponent is squared.

3. The function fits neither form. If the expression in the parentheses expanded, then $m(x) = 3(9x^2 + 6x + 1) = 27x^2 + 18x + 3$.

4. The function is exponential, because $n(x) = 3 \cdot 2^{3x+1} = 3 \cdot 2^{3x} \cdot 2^1 = 6 \cdot 8^x$.

5. The function fits an exponential, because $r(x) = 2 \cdot 3^{-2x} = 2(3^{-2})^x = 2(\frac{1}{9})^x$.

6. The function is a power function, because $s(x) = \frac{4}{5x^{-3}} = \frac{4}{5}x^3$.

7. Since larger powers grow faster in the long-run,
A - (i)
B - (iv)
C - (ii)
D - (iii)

8. Larger powers of x give smaller values for $0 < x < 1$.
A - (iii)
B - (ii)
C - (iv)
D - (i)

9. **(a)**

Table 11.8

x	$f(x)$	$g(x)$
-3	1/27	-27
-2	1/9	-8
-1	1/3	-1
0	1	0
1	3	1
2	9	8
3	27	27

(b) As $x \to -\infty$, $f(x) \to 0$. For f, large negative values of x result in small $f(x)$ values because a large negative power of 3 is very close to zero. For g, large negative values of x result in large negative values of $g(x)$, because the cube of a large negative number is a larger negative number. Therefore, as $x \to -\infty$, $g(x) \to -\infty$.

As $x \to \infty$, $f(x) \to \infty$ and $g(x) \to \infty$. For $f(x)$, large x-values result in large powers of 3; for $g(x)$, large x values yield the cubes of large x-values. f and g both climb *fast*, but f climbs faster than g (for $x > 3$).

10. As $x \to \infty$, we know x^3 dominates x^2. Multiplying by a positive constant a or b, does not change the outcome, so $y = ax^3$ dominates.

11. Since 0.99^x is a decreasing exponential function, $y = 7 \cdot 0.99^x \to 0$ as $x \to \infty$, so $y = 6x^{35}$ dominates.

12. As $x \to \infty$, increasing exponential functions dominate power functions, so $y = 4e^x$ dominates.

13. As $x \to \infty$, the higher power dominates, so $x^{1.1}$ dominates $x^{1.08}$. The coefficients 1000 and 50 do not change this, so $y = 50x^{1.1}$ dominates.

Problems

14. Table 11.9 shows that 3^{-x} approaches zero faster than x^{-3} as $x \to \infty$.

Table 11.9

x	2	10	100
3^{-x}	1/9	0.000017	1.94×10^{-48}
x^{-3}	1/8	0.001	10^{-6}

15. The function $y = e^{-x}$ will approach zero faster. To see this, note that a doubling of x in the cubic function $y = x^{-3}$ will cause the y-value to decrease by a factor of $2^{-3} = \frac{1}{8}$; while a doubling of x in the exponential function $y = e^{-x}$ will cause the y-value to be squared. For small values of y, squaring decreases faster than multiplying by 2^{-3}. To see this numerically, look in Table 11.10.

Table 11.10

x	1	10
x^{-3}	1.0	0.001
e^{-x}	0.368	0.0000454

16. Neither. The formula for an exponential function is $y = b^x$; the formula for a power function is $y = x^p$, where b and p are constants. The function $y = x^x$ is a variable raised to a variable, and thus it fits neither description.

17. **(a)** Let $f(x) = ax + b$. Then $f(1) = a + b = 18$ and $f(3) = 3a + b = 1458$. Solving simultaneous equations gives us $a = 720, b = -702$. Thus $f(x) = 720x - 702$.

(b) Let $f(x) = a \cdot b^x$, then

$$\frac{f(3)}{f(1)} = \frac{ab^3}{ab} = b^2 = \frac{1458}{18} = 81.$$

Thus,

$$b^2 = 81$$
$$b = 9 \qquad \text{(since } b \text{ must be positive)}$$

Using $f(1) = 18$ gives

$$a(9)^1 = 18$$
$$a = 2.$$

Therefore, if f is an exponential function, a formula for f would be

$$f(x) = 2(9)^x.$$

(c) If f is a power function, let $f(x) = kx^p$, then

$$\frac{f(3)}{f(1)} = \frac{k(3)^p}{k(1)^p} = (3)^p$$

and

$$\frac{f(3)}{f(1)} = \frac{1458}{18} = 81.$$

Thus,

$$3^p = 81 \qquad \text{so} \qquad p = 4.$$

Solving for k, gives

$$18 = k(1^4) \qquad \text{so} \qquad k = 18.$$

Thus, a formula for f is

$$f(x) = 18x^4.$$

18. **(a)** If f is linear,

$$m = \frac{128 - 16}{2 - 1} = 112,$$

and

$$16 = 112(1) + b, \qquad \text{so} \qquad b = -96.$$

Thus,

$$f(x) = 112x - 96.$$

(b) If f is exponential, then

$$\frac{128}{16} = \frac{a(b)^2}{a(b)} = b, \qquad \text{so} \qquad b = 8$$

and

$$16 = a(8), \qquad \text{so} \qquad a = 2.$$

Therefore

$$f(x) = 2(8)^x.$$

(c) If f is a power function, $f(x) = k(x)^p$. Then

$$\frac{f(2)}{f(1)} = \frac{k(2)^p}{k(1)^p} = (2)^p = \frac{128}{16} = 8,$$

so $p = 3$. Using $f(1) = 16$ to solve for k, we have

$$16 = k(1^3), \qquad \text{so} \qquad k = 16.$$

Thus,

$$f(x) = 16x^3.$$

19. **(a)** If f is linear, then the formula for $f(x)$ is of the form

$$f(x) = mx + b,$$

where

$$m = \frac{\Delta y}{\Delta x} = \frac{48 - \frac{3}{4}}{2 - (-1)} = \frac{\frac{189}{4}}{3} = \frac{189}{12} = \frac{63}{4}$$

Thus, $f(x) = \frac{63}{4}x + b$. Since $f(2) = 48$, we have

$$48 = \frac{63}{4}(2) + b$$
$$48 - \frac{63}{2} = b$$
$$b = \frac{33}{2}$$

Thus, if f is a linear function,

$$f(x) = \frac{63}{4}x + \frac{33}{2}.$$

(b) If f is exponential, then the formula for $f(x)$ is of the form

$$f(x) = ab^x, \qquad b > 0,\ b \neq 1.$$

Taking the ratio of $f(2)$ to $f(-1)$, we have

$$\frac{f(2)}{f(-1)} = \frac{ab^2}{ab^{-1}} = b^3,$$

and

$$\frac{f(2)}{f(-1)} = \frac{48}{\frac{3}{4}} = 48 \cdot \frac{4}{3} = 64.$$

Thus

$$b^3 = 64,$$

and

$$b = 4.$$

To solve for a, note that

$$f(2) = a(4)^2 = 48,$$

which gives $a = 3$. Thus, an exponential model for f is $f(x) = 3 \cdot 4^x$.

(c) If f is a power function, then the formula for $f(x)$ is of the form

$$f(x) = kx^p, \qquad k \text{ and } p \text{ constants.}$$

Taking the ratio of $f(2)$ to $f(-1)$, we have

$$\frac{f(2)}{f(-1)} = \frac{k \cdot 2^p}{k \cdot (-1)^p} = \frac{2^p}{(-1)^p} = (-2)^p.$$

Since we know from part (b) that $\frac{f(2)}{f(-1)} = 64$, we have

$$(-2)^p = 64.$$

Thus, $p = 6$. To solve for k, note that

$$f(2) = k \cdot 2^6 = 48,$$

which gives

$$\begin{aligned} 64k &= 48 \\ k &= \frac{48}{64} = \frac{3}{4}. \end{aligned}$$

Thus, $f(x) = \frac{3}{4}x^6$ is a power function which satisfies the given data.

20. Explanations should include the recognition that Table 11.18 indicates a constant rate of change (linear behavior), and that Table 11.19 is the only candidate for the logarithmic function (since power functions do not have a zero at $x = 1$ and the change in function values is not linear). Table 11.20 must be the quadratic power function, since the function values are symmetric about the vertical axis, and Table 11.17 is left as the cubic power function.

Using the data, we find:

$$\begin{aligned} \text{Table 11.17} &: j(x) = .3x^3 \\ \text{Table 11.18} &: k(x) = 1.2x - 3 \\ \text{Table 11.19} &: m(x) = \log x \\ \text{Table 11.20} &: z(x) = .4x^2 \end{aligned}$$

21. Note: $\frac{5}{7} > \frac{9}{16} > \frac{3}{8} > \frac{3}{11}$, and we know that for $x > 1$, the higher the exponent, the more steeply the graph climbs, so

$$A \text{ is } kx^{5/7}, \quad B \text{ is } kx^{9/16}, \quad C \text{ is } kx^{3/8}, \quad D \text{ is } kx^{3/11}.$$

22. **(a)** For $0 < x < 1$, we know that $x^3 < x^2$ and we know that for $x > 0$, $x^2 < 2x^2$. Therefore, $f(x)$ is graph C, $g(x)$ is graph A, $h(x)$ is graph B.

(b) Yes, $B = g(x) = h(x)$ and $A = x^3 = 2x^2$, so

$$\begin{aligned} x^3 &= 2x^2 \\ x^3 - 2x^2 &= 0 \\ x^2(x-2) &= 0 \end{aligned}$$

So $x = 2$ is the only solution for $x > 0$.

(c) No, since $2x^2 > x^2$ for $x > 0$.

23. If $f(x) = mx^{1/3}$ goes through $(1, 2)$, then $m(1)^{1/3} = 2$, so $m = 2$ and $f(x) = 2x^{1/3}$. Using $x = 8$ in $f(x) = 2x^{1/3}$ gives $t = 4$. If $g(x) = kx^{4/3}$ goes through $(8, 4)$, then $k = \frac{1}{4}$. Thus, $m = 2, t = 4$, and $k = \frac{1}{4}$.

24. **(a)** We are given $t(v) = v^{-2} = 1/v^2$ and $r(v) = 40v^{-3} = 40/v^3$ therefore

$$\frac{1}{v^2} = \frac{40}{v^3}.$$

Multiplying by v^3 we get $v = 40$.

(b) We found in part (a) that $v = 40$. By graphing or substituting values of x between 0 and 40, we see that $r(x) > t(x)$ for $0 < x < 40$.

(c) For values of $x > 40$ we see by graphing or substituting values that $t(x) > r(x)$.

25. Since x^8 is much larger than $x^2 + 5$ for large x (either positive or negative), the ratio tends to zero. Thus, $y \to 0$ as $x \to \infty$ or $x \to -\infty$.

26. Since the denominator has highest power of t^6, which dominates t^2, the ratio tends to 0. Thus, $y \to 0$ as $t \to \infty$ or $t \to -\infty$.

27. For large positive t, the value of 5^t is much larger than the value of 2^t. Thus, $y \to 0$ as $t \to \infty$.
For large negative t, the values of 2^t and 5^t go to zero, so

$$y \to \frac{0+7}{0+9} = \frac{7}{9} \quad \text{as } t \to -\infty.$$

28. For large positive t, the value of $3^{-t} \to 0$ and $4^t \to \infty$. Thus, $y \to 0$ as $t \to \infty$.
For large negative t, the value of $3^{-t} \to \infty$ and $4^t \to 0$. Thus,

$$y \to \frac{\text{Very large positive number}}{0+7} \quad \text{as } t \to -\infty.$$

So $y \to \infty$ as $t \to -\infty$.

29. Multiplying out the numerator gives a polynomial with highest term x^3, which dominates the x^2 in the denominator. Thus, $y \to x$ as $x \to \infty$ or $x \to -\infty$. So $y \to \infty$ as $x \to \infty$, and $y \to -\infty$ as $x \to -\infty$.

30. For large positive x, the value of 2^x is much larger than the value of x^2. Thus, $y \to \infty$ as $x \to \infty$. For large negative x, the value of $2^x \to 0$, but $x^2 \to \infty$. Thus, $y \to 0$ as $x \to -\infty$.

31. Since $\sqrt{x}$ dominates $\ln x$, the value of y is very small for large positive x. Thus, $y \to 0$ as $x \to \infty$. Since negative x-values are not in the domain of this functions, we do not consider $x \to -\infty$.

32. Since e^t and t^2 both dominate $\ln|t|$, we have $y \to \infty$ as $t \to \infty$.
For large negative t, the value of $e^t \to 0$, but t^2 is large and dominates $\ln|t|$. Thus, $y \to \infty$ as $t \to -\infty$,

33. As $x \to \infty$, the value of $e^x \to \infty$ and $e^{-x} \to 0$. As $x \to -\infty$, the value of $e^{-x} \to \infty$ and $e^x \to 0$. Thus, $y \to \infty$ as $x \to \infty$ and $y \to -\infty$ as $x \to -\infty$.

34. As $x \to \infty$, the value of $e^x \to \infty$ and $e^{-x} \to 0$. As $x \to -\infty$, the value of $e^{-x} \to \infty$ and $e^x \to 0$. Thus,

$$\text{As } x \to \infty, \quad y = \frac{e^x - e^{-x}}{e^x + e^{-x}} \to \frac{e^x - 0}{e^x + 0} = 1.$$

$$\text{As } x \to -\infty, \quad y = \frac{e^x - e^{-x}}{e^x + e^{-x}} \to \frac{0 - e^{-x}}{0 + e^{-x}} = -1.$$

35. Since e^x dominates x^{100}, for large positive x, the value of y is very large. Thus, $y \to \infty$ as $x \to \infty$.
For large negative x, the value of e^x is very small, but x^{100} is very large. Thus, $y \to 0$ as $x \to -\infty$.

36. Since e^{3t} dominates e^{2t}, the value of y is very small for large positive t. Thus, $y \to 0$ as $t \to \infty$.
For large negative t, the value of $e^{2t} \to 0$, so $y \to 0$ as $t \to -\infty$.

37. The trigonometric function should oscillate, or in other words, the function values should move periodically back and forth between two extremes. It seems $f(x)$ best displays this behavior. The graph in Figure 11.60 shows the points for f from Table 11.21 in the problem. One possible curve has been dashed in. We can recognize the curve as having the same shape as the sine function. The amplitude is 2 (the curve only varies 2 units up or down from the central value of 4). It is raised 4 units from the x-axis. Also, the period is 4 because the curve makes one full cycle in the space of 4 units, which tells us that the frequency is $\frac{2\pi}{4} = \frac{\pi}{2}$. A formula for the curve in Figure 11.60 could be

$$f(x) = 2\sin\left(\frac{\pi}{2}x\right) + 4.$$

(Note: Answers are not unique!)

The exponential function should take the form $y = a \cdot b^x$. Since neither a nor b can be zero, the function cannot pass through the point $(0, 0)$. Therefore, $g(x)$ cannot be exponential. Try h as the exponential function. Figure 11.61 shows the points plotted from $h(x)$ with the curve dashed in.

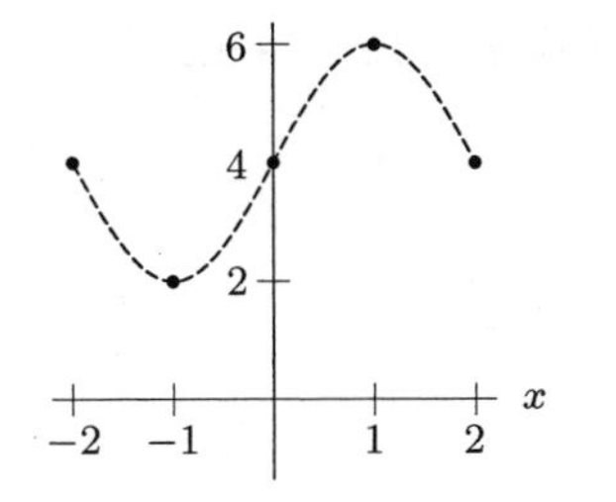

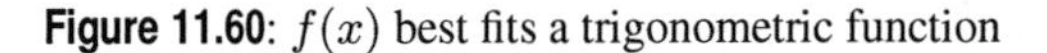
Figure 11.60: $f(x)$ best fits a trigonometric function

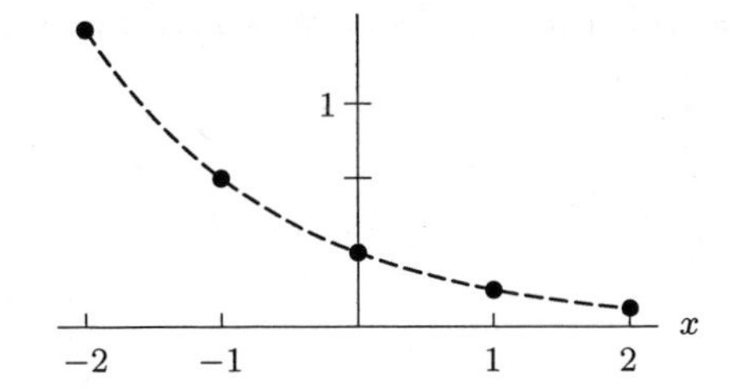

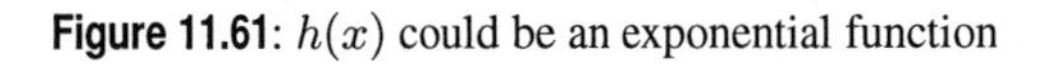
Figure 11.61: $h(x)$ could be an exponential function

Rewriting $h(0) = 0.33$ as $h(0) = \frac{1}{3}$ gives $a = \frac{1}{3}$. Thus, using any other point, say $(1, 0.17)$, we have

$$0.17 = \left(\frac{1}{3}\right) b$$

$$b \approx 0.50 = \frac{1}{2}.$$

We find a possible formula to be

$$h(x) = \frac{1}{3}\left(\frac{1}{2}\right)^x.$$

The power function is left for g. A power function takes the form $y = k \cdot x^p$. Solving for k and p, we find

$$g(x) = -\frac{5}{2}x^3.$$

We see that the data for g satisfy this formula.

38. (a)

No. of years elapsed	Start-of-year balance	End-of-year deposit	End-of-year interest
0	\$1000.00	\$1000	\$60.00
1	\$2060.00	\$1000	\$123.60
2	\$3183.60	\$1000	\$191.02
3	\$4374.62	\$1000	\$262.48
4	\$5637.10	\$1000	\$338.23
5	\$6975.33	\$1000	\$418.52

(b) The balance is not growing at a linear rate because the change (increase) in balance is increasing each year. If the growth were linear, the increase would be the same amount each year.

The balance is not growing exponentially, either, because the ratio of successive balance amounts is not constant. For example,

$$\frac{\text{Balance in year 2}}{\text{Balance in year 1}} = \frac{3183.60}{2060.00} \approx 1.55$$

and

$$\frac{\text{Balance in year 3}}{\text{Balance in year 2}} = \frac{4374.62}{3183.60} \approx 1.37.$$

Thus, neither a linear nor an exponential function represents the growth of the balance in this situation.

39. **(a)** If $p_n(r)$ represents the balance after n years where r is the rate of interest, we have the pattern in Table 11.11.

Table 11.11

In year $n =$	balance
0	1000
1	$1000 + 1000 \cdot r + 1000$ $=1000(1+r) + 1000$
2	$\underbrace{1000(1+r)+1000} + \underbrace{(1000(1+r)+1000)} \cdot r + 1000$ $=(1000(1+r)+1000)(1+r) + 1000$ $=\underbrace{1000(1+r)^2 + 1000(1+r) + 1000}$
3	$\underbrace{1000(1+r)^2 + 1000(1+r) + 1000} + [(1000)(1+r)^2 + 1000(1+r) + 1000] \cdot r + 1000$ $=(1000(1+r)^2 + 1000(1+r) + 1000)(1+r) + 1000$ $=1000(1+r)^3 + 1000(1+r)^2 + 1000(1+r) + 1000$
$\vdots$	
in year n	$1000(1+r)^n + 1000(1+r)^{n-1} + \cdots + 1000(1+r) + 1000$

Thus,

$$p_5(r) = 1000(1+r)^5 + 1000(1+r)^4 + 1000(1+r)^3 + 1000(1+r)^2 \\ +1000(1+r) + 1000$$

and

$$p_{10}(r) = 1000(1+r)^{10} + 1000(1+r)^9 + 1000(1+r)^8 + 1000(1+r)^7 \\ +1000(1+r)^6 + 1000(1+r)^5 + 1000(1+r)^4 + 1000(1+r)^3 + \\ 1000(1+r)^2 + 1000(1+r) + 1000.$$

(b) We can enter $p_5(r)$ as is on our calculator and solve for where $p_5(r) = 10{,}000$. Alternatively, let $x = (1+r)$, and solve

$$1000x^5 + 1000x^4 + 1000x^3 + 1000x^2 + 1000x + 1000 = 10,000$$

or equivalently solve for x such that

$$x^5 + x^4 + x^3 + x^2 + x - 9 = 0.$$

A graphical solution shows $x \approx 1.20279$. Since $x = 1 + r$, $r \approx 20.279\%$ interest.

40. The function $f(d) = b \cdot d^{p/q}$, with $p < q$, because $f(d)$ increases more and more slowly as d gets larger, and $g(d) = a \cdot d^{p/q}$, with $p > q$, because $g(d)$ increases more and more quickly as d gets larger.

Solutions for Section 11.7

Exercises

1. $f(x) = ax^p$ for some constants a and c. Since $f(1) = 1 = a(1)^p$, it follows that $a = 1$. Also, $f(2) = 2^p = c$. Solving for p we have $p = \ln c / \ln 2$. Thus, $f(x) = x^{\ln c/\ln 2}$

2. Regression on a calculator returns the exponential function $h(x) = 2.35(1.44)^x$.

3. Regression on a calculator returns the power function $g(x) = 2x^{1.2}$.

4. **(a)** (i) Calculator result: $y = 6.222t^{0.504}$. Answers may vary.

(ii) Calculator result: $y = 0.066t^2 - 1.515t + 25.571$. Answers may vary.

(b) For the time period 1970-2000, the quadratic function is the better fit. The power function is approximately the square root function so it is concave down, but the catch values increased rapidly toward the end of this period. Notice that the power function goes through $(0, 0)$, meaning that the predicted value of the 1965 catch is zero—not likely to be a realistic prediction. The quadratic function is shifted and stretched so is the better fit. See Figure 11.62. However, outside of the interval 1970-2000, there is no reason to suppose that either function is a good fit. Our results hold only for this time period.

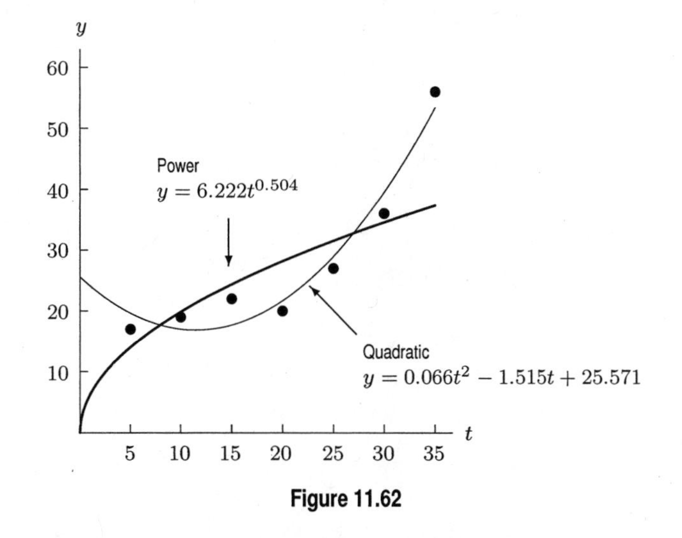

Figure 11.62

5. **(a)** Regression on a calculator returns the power function $f(x) = 201.353x^{2.111}$, where $f(x)$ represents the total dry weight (in grams) of a tree having an x cm diameter at breast height.

(b) Using our regression function, we obtain $f(20) = 201.535(20)^{2.111} = 112{,}313.62$ gm.

(c) Solving $f(x) = 201.353x^{2.111} = 100{,}000$ for x we get

$$x^{2.111} = \frac{100{,}000}{201.353}$$
$$x = \left(\frac{100{,}000}{201.353}\right)^{1/2.111} = 18.930 \text{ cm.}$$

6. **(a)** Performing regression with a power function on the calculator returns the function $Q(b) = 221.425b^{0.685}$.

(b) The estimated brain weight of the Erythrocebus is $Q(7800) = 221.425(7800)^{0.685} \approx 102{,}639$ mg.

7. The slope of this line is $\frac{6-0}{4-0} = \frac{3}{2}$. The vertical intercept is 0. Thus $\ln y = \frac{3}{2}\ln x$, and $y = e^{(3/2)\ln x} = e^{\ln(x^{3/2})} = x^{\frac{3}{2}}$.

8. The slope of this line is $\frac{2-0}{0-(-3)} = \frac{2}{3}$. The vertical intercept is 2. So $\ln y = 2 + \frac{2}{3}\ln x$, and $y = e^{2+\frac{2}{3}\ln x} = e^2 e^{\frac{2}{3}\ln x} = e^2 e^{\ln(x^{2/3})} = e^2 x^{\frac{2}{3}}$.

9. The slope of this line is $m = \frac{y_2 - y_1}{x_2 - x_1} = \frac{3}{2}$. The vertical intercept is 0, thus $y = \frac{3}{2}x$.

10. The slope of this line is $\frac{2-1}{0-(-1)} = 1$ and the vertical intercept is 2, thus $\ln y = x + 2$, so $y = e^{x+2}$.

11. The slope of this line is $\frac{2-0}{5-0} = 0.4$. The vertical intercept is 0. Thus $\ln y = 0.4x$, and $y = e^{0.4x}$.

12. The slope of this line is $\frac{1.7-0}{-1-0} = -1.7$. The vertical intercept is 0. Thus $\ln y = -1.7x$. So $y = e^{-1.7x}$.

Problems

13. **(a)** The function $y = -83.039 + 61.514x$ gives a superb fit, with correlation coefficient $r = 0.99997$.

(b) When the power function is plotted for $2 \leq x \leq 2.05$, it resembles a line. This is true for most of the functions we have studied. If you zoom in close enough on any given point, the function begins to resemble a line. However, for other values of x (say, $x = 3, 4, 5 \ldots$), the fit no longer holds.

14. **(a)** The FM band appears linear, because the FM frequency always increases by 4 as the distance increases by 10. See Figure 11.63.

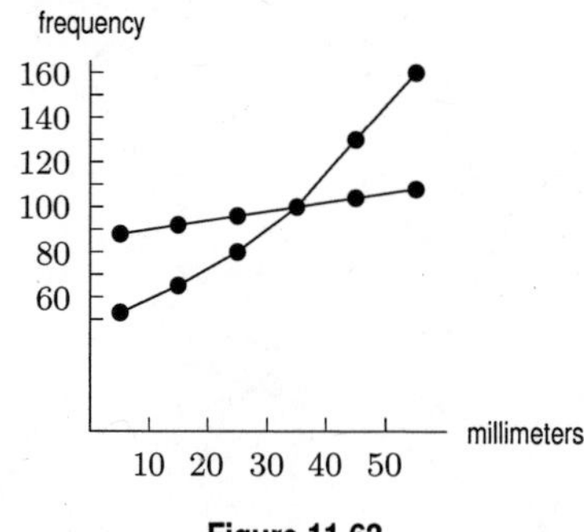

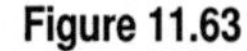
Figure 11.63

(b) The AM band is increasing at an increasing rate. These data could therefore represent an exponential relation.

(c) We recall that any linear function has a formula $f(x) = b + mx$. Since the rate of change, m, is the change in frequency, 4, compared to the change in length, 10, then $m = \frac{4}{10} = 0.4$. So

$$y = b + 0.4x.$$

But the table tells us that $f(5) = b + 0.4(5) = b + 2 = 88$. Therefore, $b = 86$, and

$$y = 86 + 0.4x.$$

We could have also used a calculator or computer to determine the coefficients for the linear regression.

(d) Since the data for the AM band appear exponential, we wish to plot the natural log of the frequency against the length. Table 11.12 gives the values of the AM station numbers, y, and the natural log, $\ln y$, of those station numbers as a function of their location on the dial, x.

Table 11.12

x	5	15	25	35	45	55
y	53	65	80	100	130	160
$\ln y$	3.97	4.17	4.38	4.61	4.87	5.08

The $(x, \ln y)$ data are very close to linear. Regression gives the coefficients for the linear equation $\ln y = b+mx$, yielding

$$\ln y = 3.839 + 0.023x.$$

Solving for y gives:

$$\begin{aligned} e^{\ln y} &= e^{3.839+0.023x} \\ y &= e^{3.839+0.023x} \quad \text{(since } e^{\ln y} = y\text{).} \\ y &= e^{3.839}e^{0.023x} \quad \text{(since } a^{x+y} = a^x a^y\text{)} \end{aligned}$$

Since $e^{3.839} \approx 46.5$,

$$y = 46.5e^{0.023x}.$$

15. By plotting x against $t^{0.4}$ we see a straight line with slope$\approx$ 3.50. Alternatively, by calculating $x/t^{0.4}$ for each of the data points, except $(0,0)$, we find a common value of approximately$\approx$ 3.50. These give an estimate of $a \approx 3.50$. Figure 11.64 shows the function $x = 3.50t^{0.4}$ and the data set from Table 11.32, which seems to model the situation well.

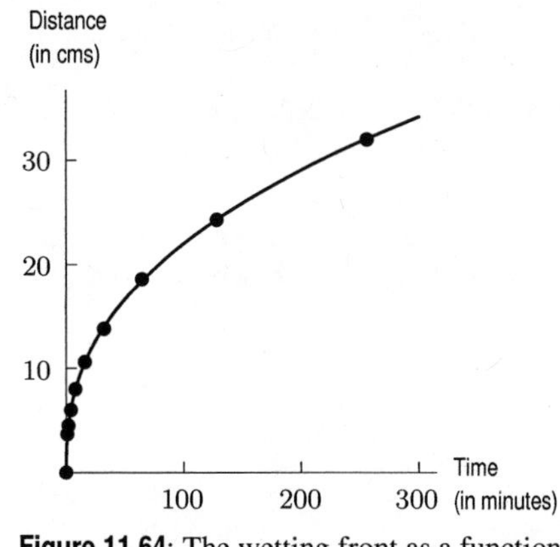

Figure 11.64: The wetting front as a function of time and the function $x = 3.49t^{0.4}$

16. **(a)** A computer or calculator gives

$$N = -14t^4 + 433t^3 - 2255t^2 + 5634t - 4397.$$

(b) The graph of the data and the quartic in Figure 11.65 shows a good fit between 1980 and 1996.

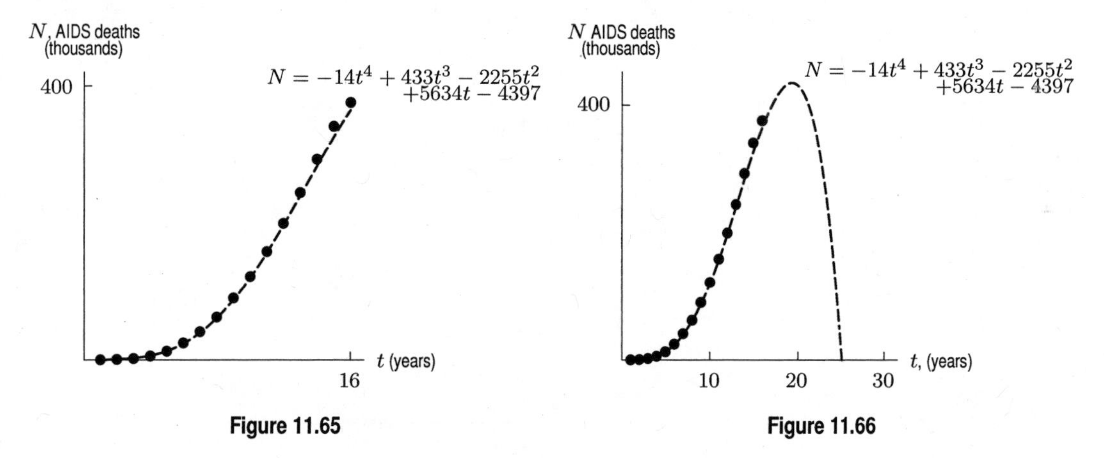

Figure 11.65 **Figure 11.66**

(c) Figure 11.65 shows that the quartic model fits the 1980-1996 data well. However, this model predicts that in 2000, the number of deaths decreases. See Figure 11.66. Since N is the total number of deaths since 1980, this is impossible. Therefore the quartic is definitely not a good model for $t > 20$.

17. **(a)**

Recliner price ($)	399	499	599	699	799
Demand (recliners)	62	55	47	40	34
Revenue ($)	24,738	27,445	28,153	27,960	27,166

(b) Using quadratic regression, we obtain the following formula for revenue, R, as a function of the selling price, p: $R(p) = -0.0565p^2 + 72.9981p + 4749.85$.

(c) By using a graphing calculator to zoom in, we see that the price which maximizes revenue is about $646. The revenue generated at this price is about $28,349.

18. **(a)** The function appears to be decreasing and concave up.

(b) Using a calculator and starting with $t = 1$ for 2006, a power regression function is obtained: $P(t) = 2.976t^{-0.002}$.

(c) Substituting $t = 7$, we get $P(7) = 2.976 \cdot 7^{-0.002} = 2.964$. In 2012, we extrapolate the Armenian population to be approximately 2,964,000.

19. **(a)** Use a calculator or computer to find an exponential regression function starting with $t = 0$ for 1985, $C(t) = 841.368(1.333)^t$. Answers may vary.

(b) From the equation, the growth factor is 1.333, so cellular subscriptions are increasing by 33.3% per year.

(c) Although the number of cell subscriptions may continue to increase, we eventually expect slower growth. The graph would become concave down.

20. **(a)** The curve does not fit well because the projected maximum, at the vertex, is about two years late.

(b) Use a calculator to find $v = -0.145h^3 + 1.850h^2 - 1.313h + 10.692$. Answers may vary. Yes, this equation is a better fit. See Figure 11.67.

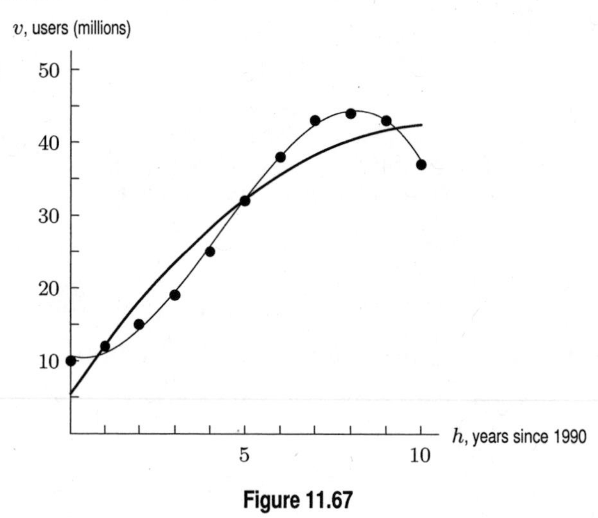

Figure 11.67

21. **(a)** See Figure 11.68.

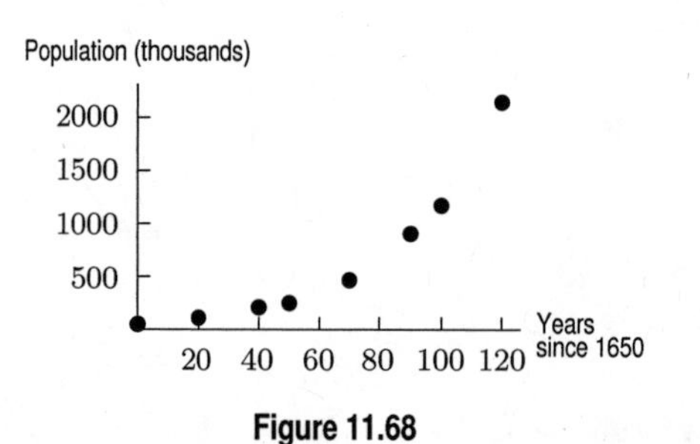

Figure 11.68

(b) Using a calculator or computer, we get $P(t) = 56.108(1.031)^t$. Answers may vary.

(c) The 56.108 represents a population of 56,108 people in 1650. Note that this is more than the actual population of 50,400. The growth factor of 1.031 means the rate of growth is approximately 3.1% per year.

(d) We find $P(100) = 1194.308$, which is slightly higher than the given data value of 1,170.8.

(e) The estimated population, $P(150) = 5510.118$, is higher than the given census population.

22. **(a)** $P(t) = 3.956(1.030)^t$. Answers may vary.

(b) The value, $P(10) = 3.956 \cdot 1.030^{10} = 5.314$ for 1800, is an interpolation, while in the other problem, 1800 was outside the data set of 1650-1790 and was a less accurate extrapolation.

(c) $P(210) = 3.956 \cdot 1.030^{220} = 2639.038$. This is more than 2.6 billion, and far from being correct, since the US population was about 310 million in 2010.

23. **(a)** The formula is $N = 1148.55e^{0.3617t}$. See Figure 11.69.

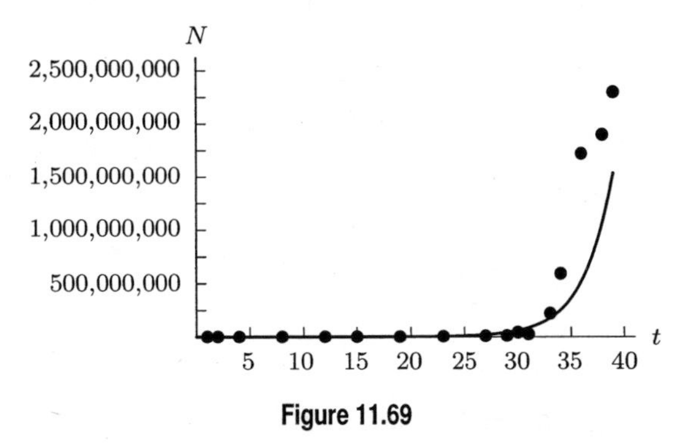

Figure 11.69

(b) The doubling time is given by $\ln 2/0.3617 \approx 1.916$. This is consistent with *Moore's Law*, which states that the number of transistors doubles about once every two years. Dr. Gordon E. Moore is Chairman Emeritus of Intel Corporation According to the Intel Corporation, "Gordon Moore made his famous observation in 1965, just four years after the first planar integrated circuit was discovered. The press called it 'Moore's Law' and the name has stuck. In his original paper, Moore observed an exponential growth in the number of transistors per integrated circuit and predicted that this trend would continue."

24. **(a)** Calculator result:$y = 2.183t^{1.271}$. Answers may vary.
(b) Calculator result: $y = 18.916 \cdot 1.060^t$. Answers may vary.
(c) Calculator result: $y = 0.553t^2 - 22.884t + 226.956$. Answers may vary.
(d) The power function is a poor fit because it does not rise fast enough to fit values from 1990 to 2000; its projection is likely too low for 2010. The exponential function fits the data better and although its values are low for the period 1990 to 2000; it probably gives the best projection for 2010. The quadratic function fits the data best of all by passing close to most points, but its 2010 projection is likely too high. See Figure 11.70.

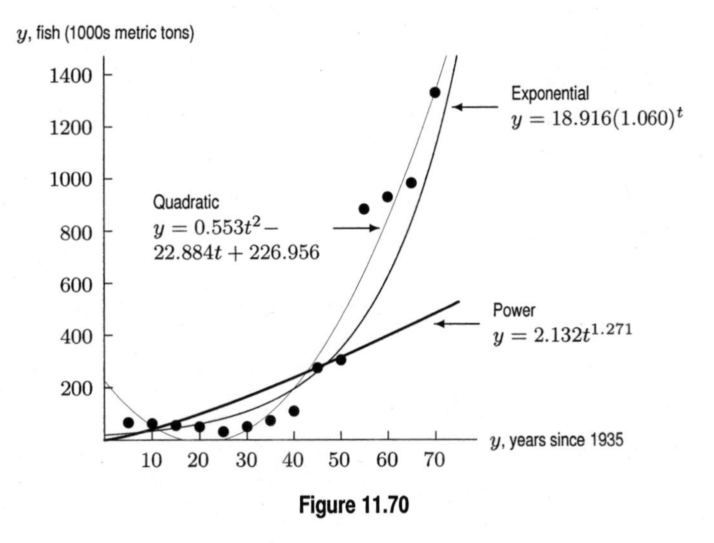

Figure 11.70

25. **(a)** Using $t = 5, 10, \ldots, 50$, yields $y = 0.310t^2 - 12.177t + 144.517$. Answers may vary.
(b) Using $t = 55, 60, 65, 70$, yields $y = 3.01t^2 - 348.43t + 10{,}955.75$. Answers may vary.
(c) $$f(t) = \begin{cases} y = 0.310t^2 - 12.177t + 144.517 & 5 \le t \le 50 \\ y = 3.01t^2 - 348.43t + 10{,}955.75 & 55 \le t \le 70 \end{cases}$$

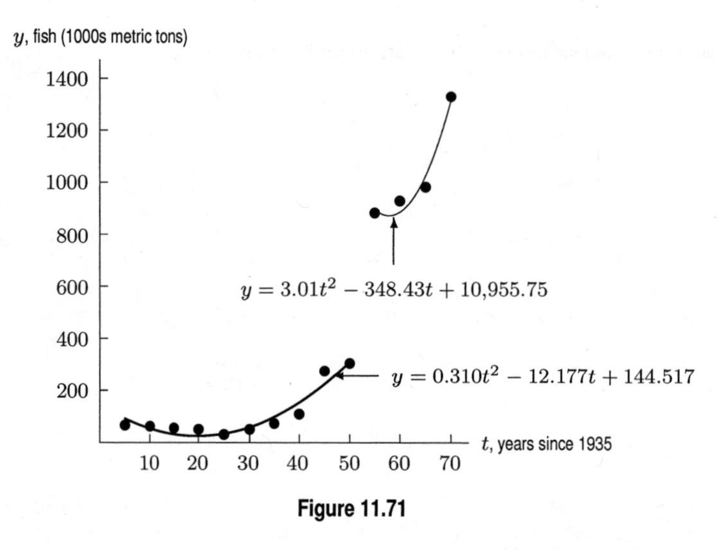

Figure 11.71

26. (a) Table 11.13 shows the transformed data, where $y = \ln N$. Figure 11.72 shows that the transformed data lie close to a straight line.

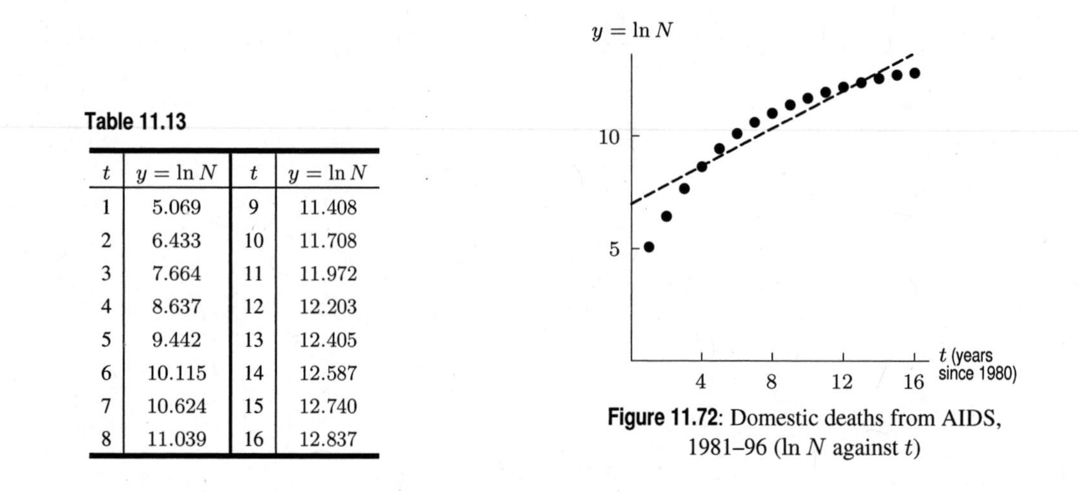

Table 11.13

t	$y = \ln N$	t	$y = \ln N$
1	5.069	9	11.408
2	6.433	10	11.708
3	7.664	11	11.972
4	8.637	12	12.203
5	9.442	13	12.405
6	10.115	14	12.587
7	10.624	15	12.740
8	11.039	16	12.837

Figure 11.72: Domestic deaths from AIDS, 1981–96 ($\ln N$ against t)

(b) We now use linear regression to estimate a line to fit the data points $(t, \ln N)$. The formula provided by a calculator (and rounded) is

$$y = 6.445 + 0.47t.$$

(c) To find the formula for N in terms of t, we substitute $\ln N$ for y and solve for N:

$$\begin{aligned} \ln N &= 6.445 + 0.47t \\ N &= e^{6.445+0.47t} \\ &= \left(e^{6.445}\right)\left(e^{0.47t}\right), \end{aligned}$$

and since $e^{6.445} \approx 630$, we have

$$N \approx 630e^{0.47t}.$$

27. (a) We take the log of both sides and make the substitution $y = \ln N$ to obtain

$$\begin{aligned} y = \ln N &= \ln(at^p) \\ &= \ln a + \ln t^p \\ &= \ln a + p \ln t. \end{aligned}$$

We make the substitution $b = \ln a$, so the equation becomes

$$y = b + p \ln t.$$

Notice that this substitution does not result in y being a linear function of t. However, if we make a second substitution, $x = \ln t$, we have

$$y = b + px.$$

So $y = \ln N$ *is* a linear function of $x = \ln t$.

(b) If a power function fits the original data, the points lie on a line. If the points do not lie on a line, a power function does not fit the data well.

28. (a) We transform the data using the substitutions $x = \ln t$ and $y = \ln N$. The transformed data in Table 11.14 are plotted in Figure 11.73.

Table 11.14

$x = \ln t$	$y = \ln N$	$x = \ln t$	$y = \ln N$
0	5.069	2.20	11.408
0.69	6.433	2.30	11.708
1.10	7.664	2.40	11.972
1.39	8.637	2.48	12.203
1.61	9.442	2.56	12.405
1.79	10.115	2.64	12.587
1.95	10.624	2.71	12.740
2.08	11.039	2.77	12.837

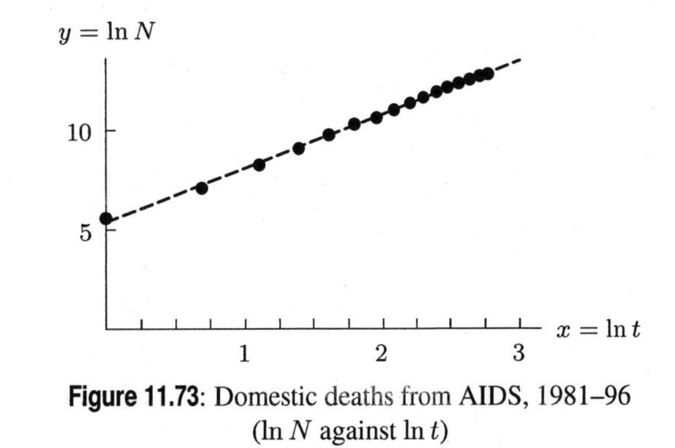

Figure 11.73: Domestic deaths from AIDS, 1981–96 ($\ln N$ against $\ln t$)

(b) Using regression to fit a line to these data gives

$$y = 4.670 + 3.005x.$$

(c) Now transform the equation back to the original variables, t and N. Since our original substitutions were $y = \ln N$ and $x = \ln t$, this equation becomes

$$\ln N = 4.670 + 3.005 \ln t.$$

Raise e to the power of both sides

$$\begin{aligned} e^{\ln N} &= e^{4.670+3.005 \ln t} \\ N &= (e^{4.670})(e^{3.005 \ln t}). \end{aligned}$$

Since $e^{4.670} \approx 107$ and $e^{3.005 \ln t} = \left(e^{\ln t}\right)^{3.005} = t^{3.005}$, we have

$$N \approx 107t^{3.005},$$

which is the formula of a power function.

29. (a) Quadratic is the only choice that increases and then decreases to match the data.

(b) Using ages of $x = 20, 30, \ldots, 80$, a quadratic function is $y = -34.136x^2 + 3497.733x - 39{,}949.714$. Answers may vary.

(c) The value of the function at 37 is $y = -34.136 \cdot 37^2 + 3497.733 \cdot 37 - 39{,}949.714 = \$42{,}734$.

(d) The value of the function for age 10 is $y = -34.136 \cdot 10^2 + 3497.733 \cdot 10 - 39{,}949.714 = -\$8386$. Answers may vary. Not reasonable, as income is positive. In addition, 10-year olds do not usually work.

30. **(a)** Formulas for the best fit curves are given by $h_e = 16.2e^{-0.0032t}$, $h_a = 12.3e^{-0.0076t}$, and $h_b = 14.7e^{-0.0067t}$. Since $1/0.0032 = 312.5$, we have $0.0032 = 1/312.5$. Thus the formula for h_e can be rewritten as

$$h_e = 16.2e^{-t/312.5}.$$

Similarly $0.0076 = 1/131.6$ and $0.0067 = 1/149.3$. Thus we have

$$h_a = 12.3e^{-t/131.6} \quad \text{and} \quad h_b = 14.7e^{-t/149.3}.$$

Figures 11.74–11.76 show the plots with the corresponding exponentials.

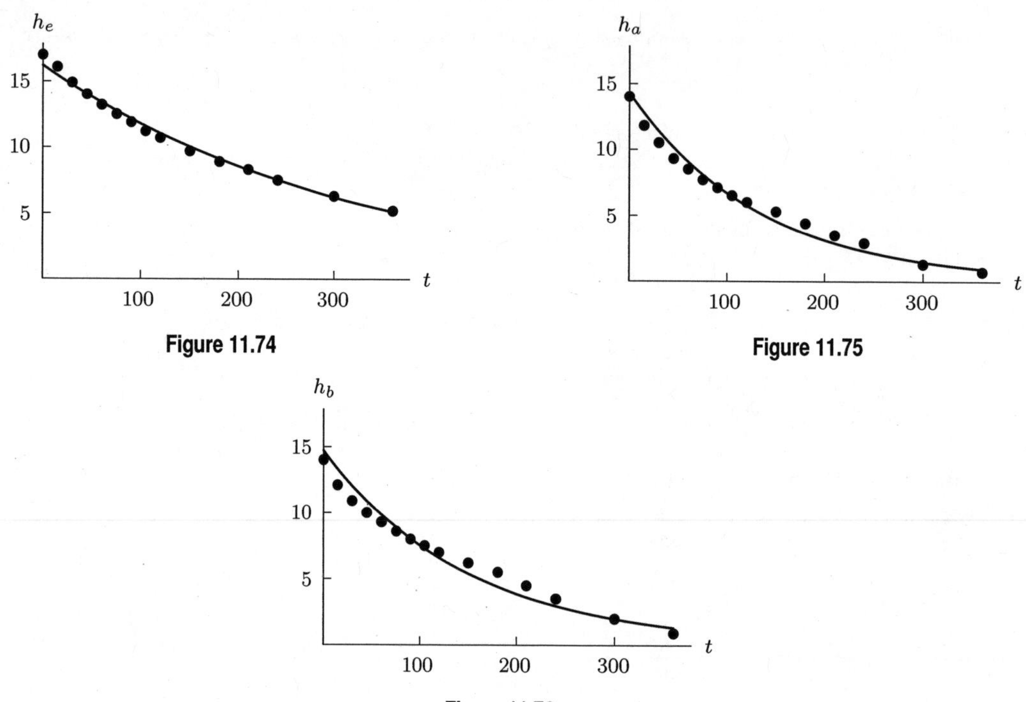

Figure 11.74

Figure 11.75

Figure 11.76

(b) The value of h_0 tells us the initial height of the foam, and the constant τ, called the time constant, tells us how long it takes the beer to drop by a factor of $1/e \approx 36.8\%$. In other words, every τ seconds 63.2% of the foam disappears. For Erdinger Weissbier, the foam is initially 16.2 cm high and goes down by 63.2% every 312.5 seconds. For Augustinerbräu München, the foam is initially 12.3 cm high and goes down by 63.2% every 131.6 seconds. For Budweiser Budvar, the foam is initially 14.7 cm high and goes down by 63.2% every 149.3 seconds.

31. **(a)** Using a computer or calculator, we find that $t = 8966.1H^{-2.3}$. See Figure 11.77.
(b) Using a computer or calculator, we find that $r = 0.0124H - 0.1248$. See Figure 11.78.

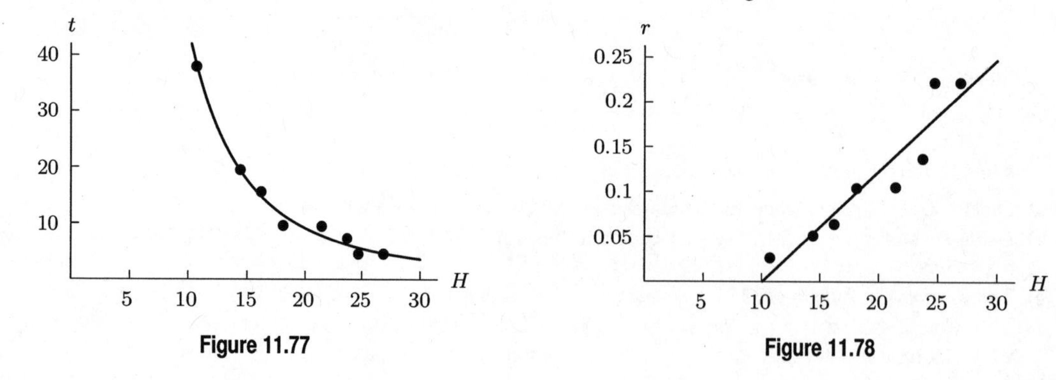

Figure 11.77

Figure 11.78

(c) From part (a), we see that $t \to \infty$ as $H \to 0$, and since $r = 1/t$, we have $r \to 0$ as $H \to 0$. From part (b), we can solve for the value of H making $r = 0$ as follows:

$$\begin{aligned} 0.0124H - 0.1248 &= 0 \\ 0.0124H &= 0.1248 \\ H &\approx 10.1. \end{aligned}$$

Thus, the first model predicts that the development rate will fall to $r = 0$ only at $H = 0°$C (the freezing point of water), whereas the second model predicts that r will reach 0 at around 10°C (or about 50°F). The latter prediction seems far more reasonable: certainly weevil eggs (or any other eggs) would not grow at temperatures near freezing.

32. (a) Because the data gives a decreasing function, we expect p to be negative.

(b) The point $(0, 244)$ must be omitted because a power function with a negative p is not defined for $c = 0$. A computer or calculator gives

$$l = 720c^{-0.722}.$$

(Different programs may give slightly different results.)

(c) If $l = kc^p$, we want a formula using natural logs such that

$$\begin{aligned} \ln l &= \ln(kc^p) \\ \ln l &= \ln k + \ln c^p \\ \ln l &= \ln k + p \ln c. \end{aligned}$$

Since $y = \ln l$ and $x = \ln c$,

$$y = \ln k + px.$$

Letting $b = \ln k$, we have

$$y = b + px.$$

(d) We must omit the point $(0, 244)$ because $\ln 0$ is undefined. The data is in Table 11.15.

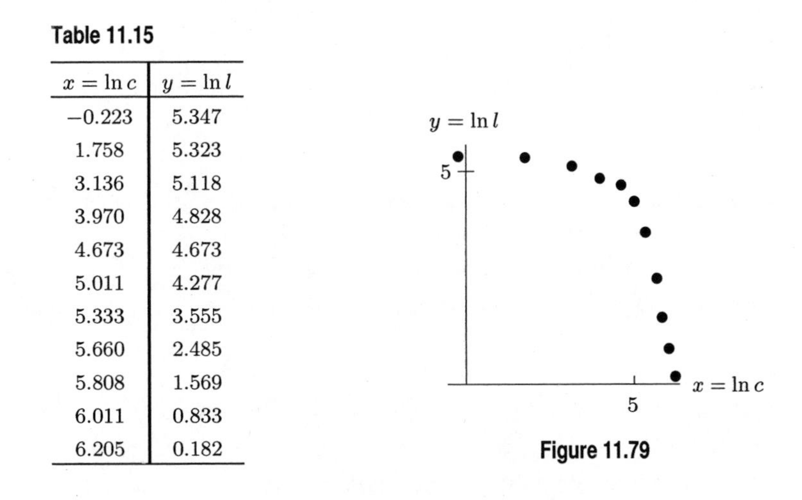

Table 11.15

$x = \ln c$	$y = \ln l$
−0.223	5.347
1.758	5.323
3.136	5.118
3.970	4.828
4.673	4.673
5.011	4.277
5.333	3.555
5.660	2.485
5.808	1.569
6.011	0.833
6.205	0.182

Figure 11.79

(e) The plotted points in Figure 11.79 don't look linear, so the power function won't give a good fit.

33. By plotting P against $D^{3/2}$ we see a straight line with slope$\approx$ 0.2. Alternatively, by calculating $P/D^{3/2}$ for each of the planets, we find a common value of approximately 0.2. We see that Kepler's model fits the data reasonably well, with $k \approx 0.2$. Kepler's equation is $P = 0.2d^{3/2}$. Figure 11.80 shows the function $P = 0.2D^{3/2}$ and the data set from Table 11.44, which models the situation well.

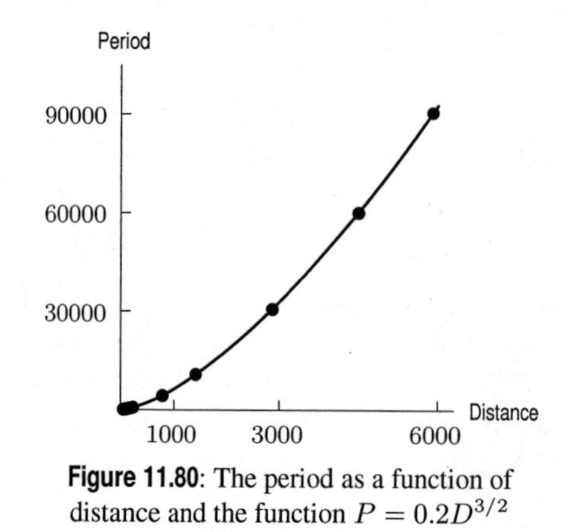

Figure 11.80: The period as a function of distance and the function $P = 0.2D^{3/2}$

Solutions for Chapter 11 Review

Exercises

1. This function represents proportionality to a power.

$$y = \frac{\frac{1}{3}}{2x^7} = \frac{1}{6x^7} = \left(\frac{1}{6}\right) x^{(-7)}.$$

Thus $k = 1/6$ and $p = -7$.

2. This function represents proportionality to a power.

$$y = \frac{6}{-2/x^5} = -3x^5 = (-3)x^{(5)}.$$

Thus $k = -3$ and $p = 5$.

3. This function does not represent proportionality to a power because we cannot get it into the form $y = kx^p$. Instead, we have a function of the form $y = kp^x$, with constant p.
4. This function does not represent proportionality to a power because we cannot get it into the form $y = kx^p$.
5. While $y = 6x^3$ is a power function, when we add two to it, it can no longer be written in the form $y = kx^p$, so this is not a power function.
6. Dividing both sides by three, we get $y = 3x^2$, so it is a power function.
7. Expanding the right side, we get $y - 9 = x^2 - 9$. Adding nine to both sides, we see that this is the power function, $y = x^2$.
8. Expanding the right side, we get $y = 4x^2 - 16 + 16 = 4x^2$. Thus, this is a power function, $y = 4x^2$.
9. Since the graph is symmetric about the y-axis, the power function is even.
10. Since the graph is steeper near the origin and less steep away from the origin, the power function has a power between 0 and 1.
11. Since the graph is symmetric about the origin, the power function has an odd power.
12. Since the graph is symmetric about the y-axis, the power function has an even power.
13. Since the graph is symmetric about the origin, the power function has an odd power.
14. Since the graph is steeper near the origin and less steep away from the origin, the power function is fractional.

15. The values are

$$k = 2\sqrt[3]{7}$$
$$p = \frac{11}{15}.$$

We have

$$\begin{aligned} r(x) = 2\sqrt[3]{7x}\sqrt[5]{x^2} &= 2\sqrt[3]{7}\sqrt[3]{x}\sqrt[5]{x^2} \\ &= 2\sqrt[3]{7}\cdot x^{1/3}\left(x^2\right)^{1/5} \\ &= 2\sqrt[3]{7}\cdot x^{1/3}\cdot x^{2/5} \\ &= 2\sqrt[3]{7}\cdot x^{\frac{1}{3}+\frac{2}{5}} \\ &= 2\sqrt[3]{7}\cdot x^{11/15}. \end{aligned}$$

16. By the ratio method,

$$\begin{aligned} \frac{f(5)}{f(3)} &= \frac{3}{5} \\ \frac{k(5)^p}{k(3)^p} &= \frac{3}{5} \\ \left(\frac{5}{3}\right)^p &= \frac{3}{5} \\ p &= -1. \end{aligned}$$

Solving for k, we have

$$\begin{aligned} k(3)^{-1} &= 5 \\ k &= 15. \end{aligned}$$

17. By multiplying out the expression $(x^2-4)(x^2-2x-3)$ and then simplifying the result, we see that

$$y = x^4 - 2x^3 - 7x^2 + 8x + 12$$

which is a fourth-degree polynomial.

18. Since the leading term of the polynomial is $16x^3$, the value of y goes to infinity as $x \to \infty$. The graph resembles $y = 16x^3$.

19. Since the leading term of the polynomial is $4x^4$, the value of y goes to infinity as $x \to \infty$. The graph resembles $y = 4x^4$.

20. Since $5x^2/x^{3/2} = 5x^{1/2} = 5\sqrt{x}$, this function can be written as $y = 5\sqrt{x} + 2$. Since the leading term is $5\sqrt{x}$, the value of y goes to infinity as $x \to \infty$. The graph resembles $y = 5\sqrt{x}$.

21. Since $2x^2/x^{-7} = 2x^9$, we can rewrite this polynomial as $y = 2x^9 - 7x^5 + 3x^3 + 2$. Since the leading term of the polynomial is $2x^9$, the value of y goes to infinity as $x \to \infty$. The graph resembles $y = 2x^9$.

22. This problem cannot be solved algebraically. Note that we cannot use the quadratic formula, as this is a 5th degree polynomial and not a 2nd degree polynomial. A graph of the function is shown in Figure 11.81 for $-1 \le x \le 1$, $-10 \le y \le 10$. From the graph, we approximate the zero to be at $x \approx -0.143$.

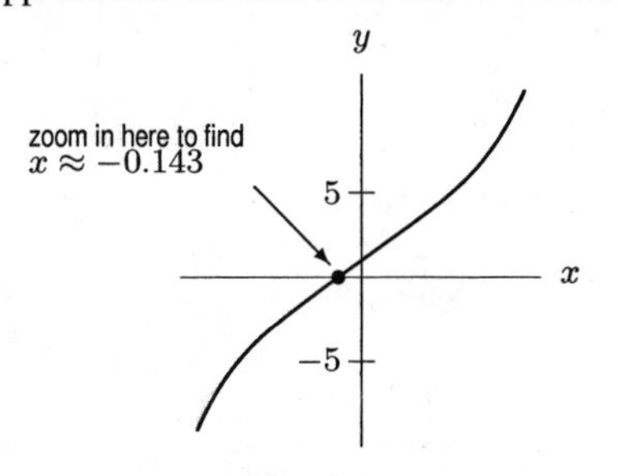

Figure 11.81

23. By using the quadratic formula, we find that $y = 0$ if

$$x = \frac{3 \pm \sqrt{9 - 4(2)(-3)}}{4} = \frac{3 \pm \sqrt{33}}{4}.$$

24. This is a rational function, as we can put it in the form of one polynomial divided by another:

$$f(x) = \frac{x^2}{x-3} - \frac{5}{x-3} = \frac{x^2-5}{x-3}.$$

25. This is not a rational function, as we cannot put it in the form of one polynomial divided by another, since e^x is an exponential function, not a polynomial.

26. Since $y = 7/x^{-4} = 7x^4$ and since x^4 dominates x^3, we see that $7/x^{-4}$ dominates.

27. Since $y = 4/e^{-x} = 4e^x$ and since increasing exponential functions dominate power functions, $y = 4/e^{-x}$ dominates.

28. **(a)** Since the long-run behavior of a rational function is given by the ratio of the leading terms, we have

$$\lim_{x\to\infty} \frac{x(x^2-4)}{5+5x^3} = \lim_{x\to\infty} \frac{x^3-4x}{5x^3+5} = \lim_{x\to\infty} \frac{x^3}{5x^3} = \frac{1}{5}.$$

(b) We have

$$\lim_{x\to-\infty} \frac{3x(x-1)(x-2)}{5-6x^4} = \lim_{x\to-\infty} \frac{3x^3-9x^2+6x}{-6x^4+5} = \lim_{x\to-\infty} \frac{3x^3}{-6x^4} = 0.$$

29. **(a)** Since the long-run behavior of $r(x) = p(x)/q(x)$ is given by the ratio of the leading terms of p and q, we have

$$\lim_{x\to\infty} \frac{2x+1}{x-5} = \lim_{x\to\infty} \frac{2x}{x} = 2.$$

(b) We have

$$\lim_{x\to-\infty} \frac{2+5x}{6x+3} = \lim_{x\to-\infty} \frac{5x}{6x} = \frac{5}{6}.$$

30. One way to approach this problem is to consider the graphical interpretations of even and odd functions. Recall that if a function is even, its graph is symmetric about the y-axis. If a function is odd, its graph is symmetric about the origin.

(a) The function $f(x) = x^2 + 3$ is $y = x^2$ shifted three units up. Note that $y = x^2$ is symmetric about the y-axis. An upward shift will not affect the symmetry. Therefore, f is even.

(b) We consider $g(x) = x^3 + 3$ as an upward shift (by three units) of $y = x^3$. However, although $y = x^3$ is symmetric about the origin, the upward-shifted function will not have that symmetry. Therefore, $g(x) = x^3 + 3$ is neither even nor odd.

(c) The function $y = \frac{1}{x}$ is symmetric about the origin (odd). The function $h(x) = \frac{5}{x}$ is merely a vertical stretch of $y = \frac{1}{x}$. This would not affect the symmetry of the function. Therefore, $h(x) = \frac{5}{x}$ is odd.

(d) If $y = |x|$, the graph is symmetric about the y-axis. However, a shift to the right by four units would make the resulting function symmetric to the line $x = 4$. The graph of $j(x) = |x - 4|$ is neither even nor odd.

(e) The function $k(x) = \log x$ is neither even nor odd. Since k is not defined for $x \le 0$, clearly neither type of symmetry would apply for $k(x) = \log x$.

(f) If we take $l(x) = \log(x^2)$, we have now included $x < 0$ into the domain of l. For $x < 0$, the graph looks similar to $y = \log x$ flipped about the y-axis. Thus, $l(x) = \log(x^2)$ is symmetric about the y-axis. —It is even.

(g) Clearly $y = 2^x$ is neither even nor odd. Likewise, $m(x) = 2^x + 2$ is neither even nor odd.

(h) We have already seen that $y = \cos x$ is even. Note that the graph of $n(x) = \cos x + 2$ is the cosine function shifted up two units. The graph will still be symmetric about the y-axis. Thus, $n(x) = \cos x + 2$ is even.

Problems

31. All power function go through the origin $(0, 0)$. The graph does not, so it does not represent a power function.

32. **(a)** $x^{1/n}$ is concave down: its values increase quickly at first and then more slowly as x gets larger. The function x^n, on the other hand, is concave up. Its values increase at an increasing rate as x gets larger. Thus

$$f(x) = x^{1/n}$$

and

$$g(x) = x^n.$$

(b) Since the point A is the intersection of $f(x)$ and $g(x)$, we want the solution of the equation $x^n = x^{1/n}$. Raising both sides to the power of n we get

$$(x^n)^{\cdot n} = (x^{1/n})^{\cdot n}$$

or in other words

$$x^{n^2} = x$$

Since $x \neq 0$ at the point A, we can divide both sides by x, giving

$$x^{n^2-1} = 1.$$

Since $n^2 - 1$ is just some integer we rewrite the equation as

$$x^p = 1$$

where $p = n^2 - 1$. If p is even, $x = \pm 1$, if p is odd $x = 1$. By looking at the graph we can tell that we are not interested in the situation when $x = -1$. When $x = 1$, the quantities x^n and $x^{1/n}$ both equal 1. Thus the coordinates of point A are $(1, 1)$.

33. Graph (i) looks periodic with amplitude of 2 and period of 2π, so it best corresponds to function J,

$$y = 2\sin(0.5x).$$

Graph (ii) appears to decrease exponentially with a y-intercept < 10, so it best corresponds to function L,

$$y = 2e^{-0.2x}.$$

Graph (iii) looks like a rational function with two vertical asymptotes, no zeros, a horizontal asymptote at $y = 0$ and a negative y-intercept, so it best corresponds to function O,

$$y = \frac{1}{x^2 - 4}.$$

Graph (iv) looks like a logarithmic function with a negative vertical asymptote and y-intercept at $(0, 0)$, so it best corresponds to function H,

$$y = \ln(x + 1).$$

34. Graph (i) looks trigonometric with period π and amplitude 2, so it is (B).
Graph (ii) looks like a logarithm which has been shifted right, so it is (L).
Graph (iii) looks cubic, shifted right and down, so it is (G).
Graph (iv) is a line with positive slope and negative y-intercept, so it is (H).
Graph (v) looks like a rational function with two vertical asymptotes and no zeros, so it is (U).
Graph (vi) looks like an exponential decay function flipped over the x-axis, so it is (S).
Graph (vii) looks trigonometric with amplitude less than 1 and period π, so it is (A).
Graph (viii) looks rational with two asymptotes and one positive zero, so it is (E).

35. The x intercepts are at $x = -4, -2, 2$, so let $y = k(x + 4)(x + 2)(x - 2)$. The y-intercept is at $(0, 24)$, so substituting $x = 0$, $y = 24$:

$$\begin{aligned} 24 &= k(4)(2)(-2) \\ 24 &= k(-16) \\ k &= \frac{24}{-16} = \frac{3}{-2}. \end{aligned}$$

Therefore, $y = -\frac{3}{2}(x + 4)(x + 2)(x - 2)$ is a possible formula.

36. Since the graph opens upward, we expect an even degree polynomial with a positive leading coefficient. Since the graph crosses the x-axis at -1 and -3, there are factors of $(x+1)$ and $(x+3)$, giving $y = a(x+1)(x+3)$. Since the graph crosses the y-axis at 12, we can find a by substituting $x = 0$:

$$\begin{aligned} 12 &= a(0+1)(0+3) \\ 12 &= 3a \\ 4 &= a. \end{aligned}$$

Thus, a possible polynomial is $y = 4(x+1)(x+3)$.

37. The graph appears to have x intercepts at $x = -\frac{1}{2}$, 3, 4, so let

$$y = k(x + \frac{1}{2})(x-3)(x-4).$$

The y-intercept is at $(0,3)$, so substituting $x = 0$, $y = 3$:

$$\begin{aligned} 3 &= k(\frac{1}{2})(-3)(-4), \\ \text{which gives} \quad 3 &= 6k, \\ \text{or} \quad k &= \frac{1}{2}. \end{aligned}$$

Therefore, $y = \frac{1}{2}(x + \frac{1}{2})(x-3)(x-4)$ is a possible formula for f.

38. The shape of the graph suggests an odd degree polynomial with a positive leading term. Since the graph crosses the x-axis at 0, -2, and 2, there are factors of x, $(x+2)$ and $(x-2)$, giving $y = ax(x+2)(x-2)$. To find a we use the fact that at $x = 1$, $y = -6$. Substituting:

$$\begin{aligned} -6 &= a(1)(1+2)(1-2) \\ -6 &= -3a \\ 2 &= a. \end{aligned}$$

Thus, a possible polynomial is $y = 2x(x+2)(x-2)$.

39. The graph has x intercepts at $x = -3$, 0, 2, so the equation is of the form

$$y = kx(x+3)(x-2).$$

The sign of k must be negative. To find its value, use the point $x = 1$, $y = 4$:

$$\begin{aligned} 4 &= k \cdot 1(1+3)(1-2) \\ 4 &= -4k \\ k &= -1. \end{aligned}$$

So the equation is $y = -x(x+3)(x-2)$.

40. The shape of the graph suggests an odd degree polynomial with a positive leading term. Since the graph crosses the x-axis at -2, there is a factor of $(x+2)$, and since it touches the x-axis at 3, there should be a factor of $(x-3)^2$, giving $y = a(x+2)(x-3)^2$. Since the graph crosses the y-axis at 126, we can find a by substituting $x = 0$:

$$\begin{aligned} 126 &= a(0+2)(0-3)^2 \\ 126 &= 18a \\ 7 &= a. \end{aligned}$$

Thus, a possible polynomial is $y = 7(x+2)(x-3)^2$.

41. We use the position of the "bounce" on the x-axis to indicate a multiple zero at that point. Since there is not a sign change at those points, the zero occurs an even number of times.

We have

$$y = k(x+3)x^2,$$

and using the point $(-1, 2)$ gives

$$2 = k(-1+3)(-1)^2 = 2k,$$

so

$$k = 1.$$

Thus, $y = x^2(x+3)$ is a possible formula.

42. We use the position of the "bounce" on the x-axis to indicate a multiple zero at that point. Since there is not a sign change at those points, the zero occurs an even number of times. Letting

$$y = k(x+2)^2(x)(x-2)^2$$

represent (c), we use the point $(1, -3)$ to get

$$-3 = f(1) = k(3)^2(1)(-1)^2,$$

so

$$-3 = 9k,$$
$$k = -\frac{1}{3}.$$

Thus, a possible is

$$y = -\frac{1}{3}(x+2)^2(x)(x-2)^2.$$

43. This one is tricky. However, we can view it as a translation of another function. Consider the graph in Figure 11.82. A formula for the graph in Figure 11.82 could be of the form $y = k(x+3)(x+2)(x+1)$. Since $y = 6$ if $x = 0$, $6 = k(0+3)(0+2)(0+1)$, therefore $6 = 6k$, which yields $k = 1$. Note that the graph of y is a vertical shift (by 4) of the graph in Figure 11.82, giving $y = (x+3)(x+2)(x+1) + 4$ as a possible formula for the function.

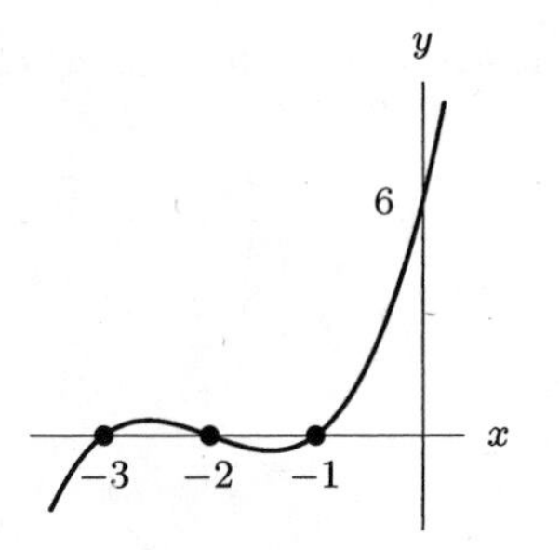

Figure 11.82

44. We know from the figure that $f(-3) = 0$, $f(0) = 0$, $f(1) = 2$, $f(3) = 0$. We also know that f does not change sign at $x = -3$ or at $x = 0$. Therefore, a possible formula for $f(x)$ is

$$f(x) = k(x+3)^2(x-3)x^2, k \neq 0.$$

Using $f(1) = 2$, we have

$$f(1) = k(1+3)^2(1-3)(1)^2 = -32k,$$

so $-32k = 2$. Thus, $k = -\frac{1}{16}$, and

$$f(x) = -\frac{1}{16}(x+3)^2(x-3)x^2$$

is a possible formula for f.

45. (a) The graph indicates the graph of $y = 1/x^2$ has been shifted to the right by 2 and down 1. Thus,

$$y = \frac{1}{(x-2)^2} - 1$$

is a possible formula for it.

(b) The equation $y = 1/(x-2)^2 - 1$ can be written as

$$y = \frac{-x^2 + 4x - 3}{x^2 - 4x + 4}$$

by obtaining a common denominator and combining terms.

(c) The graph has x-intercepts when $y = 0$ so the numerator of $y = (-x^2 + 4x - 3)/(x^2 - 4x + 4)$ must equal zero. Then

$$\begin{aligned}-x^2 + 4x - 3 &= 0\\ -(x^2 - 4x + 3) &= 0\\ -(x-3)(x-1) &= 0,\end{aligned}$$

so either $x = 3$ or $x = 1$. The x-intercepts are $(1, 0)$ and $(3, 0)$. Setting $x = 0$, we find $y = -\frac{3}{4}$, so $(0, -\frac{3}{4})$ is the y-intercept.

46. (a) The graph shows the graph of $y = 1/x^2$ shifted up 2 units. Therefore, a formula is

$$y = \frac{1}{x^2} + 2.$$

(b) The equation $y = (1/x^2) + 2$ can be written as

$$y = \frac{2x^2 + 1}{x^2}.$$

(c) We see that the graph has no intercepts on either axis. Algebraically this is seen by the fact that $x = 0$ is not in the domain of the function, and there are no real solutions to $2x^2 + 1 = 0$.

47. (a) The graph appears to be $y = 1/x^2$ shifted 3 units to the right and flipped across the x-axis. Thus,

$$y = -\frac{1}{(x-3)^2}$$

is a possible formula for it.

(b) The equation $y = -1/(x-3)^2$ can be written as

$$y = \frac{-1}{x^2 - 6x + 9}.$$

(c) Since y can not equal zero if $y = -1/(x^2 - 6x + 9)$, the graph has no x-intercept. The y-intercept occurs when $x = 0$, so $y = \frac{-1}{(-3)^2} = -\frac{1}{9}$. The y-intercept is at $(0, -\frac{1}{9})$.

48. (a) If g is a power function, $g(x) = kx^p$. Taking the ratio of $g(4)$ to $g(2)$, we have

$$\frac{g(4)}{g(2)} = \frac{k(4)^p}{k(2)^p} = 2^p \quad \text{and} \quad \frac{g(4)}{g(2)} = \frac{96}{24} = 4 = 2^2.$$

Therefore, $p = 2$. Then, using $g(2) = 24$ to find k, we have

$$\begin{aligned}g(2) = k \cdot 2^2 &= 24\\ k &= 6.\end{aligned}$$

Thus, if g is a power function, its formula is $g(x) = 6x^2$.

(b) If g is linear, then $g(x) = b + mx$. We use the two points $(2, 24)$ and $(4, 96)$ to solve for m:

$$m = \frac{\Delta y}{\Delta x} = \frac{96 - 24}{4 - 2} = \frac{72}{2} = 36.$$

Solving for b, we have

$$24 = b + 36(2)$$
$$b = -48.$$

Thus, if g is linear, its formula is $g(x) = -48 + 36x$.

(c) If g is exponential, then $g(x) = ab^x$. Taking ratios, we have

$$\frac{g(4)}{g(2)} = \frac{ab^4}{ab^2} = b^2.$$

We also know that

$$\frac{g(4)}{g(2)} = \frac{96}{24} = 4.$$

Putting these two statements together gives us

$$b^2 = 4$$

so

$$b = 2.$$

Having determined the value of b, we now have

$$g(x) = a(2^x).$$

To solve for a, we use the fact that $g(2) = 24$

$$g(2) = a(2)^2 = 24$$
$$a = 6.$$

This gives us the exponential formula $g(x) = 6(2^x)$.

49. (a) In factored form, $f(x) = x^2+5x+6 = (x+2)(x+3)$. Thus, f has zeros at $x = -2$ and $x = -3$. For $g(x) = x^2+1$ there are no real zeros.

(b) If $r(x) = \frac{f(x)}{g(x)}$, the zeros of r are where the numerator is zero (assuming g is not also zero at those points). Thus, r has zeros $x = -2$ and $x = -3$. There is no vertical asymptote since $g(x)$ is positive for all x. As $x \to \pm\infty$, r will behave like $y = \frac{x^2}{x^2} = 1$. Thus, $r(x) \to 1$ as $x \to \pm\infty$. The graph of $y = r(x)$ is shown in Figure 11.83.

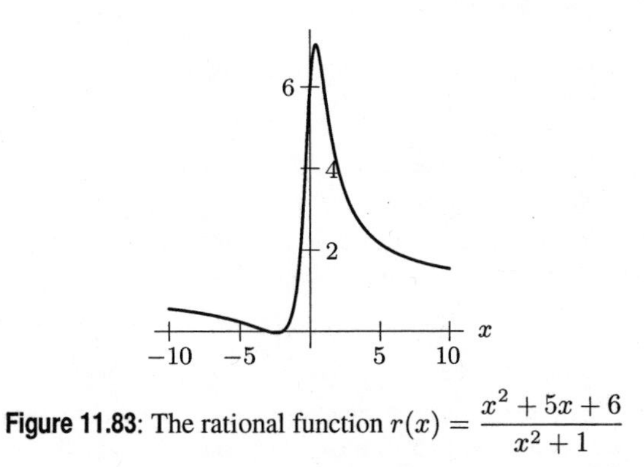

Figure 11.83: The rational function $r(x) = \dfrac{x^2 + 5x + 6}{x^2 + 1}$

(c) In fact s does not have a zero near the origin—it does not have a zero anywhere. If $s(x) = 0$ then $g(x) = 0$, which is never true. The function does have two vertical asymptotes, at the zeros of f, which are $x = -2$ and $x = 3$. As $x \to \pm\infty$, $s(x) \to 1$.

50. Notice that $f(x) = (x-3)^2$ has one zero (at $x = 3$), $g(x) = (x-2)(x+2)$ has 2 zeros (at $x = 2$ and $x = -2$), $h(x) = x+1$ has one zero (at $x = -1$), and $j(x)$ has no zeros.

(a) $s(x) = \dfrac{x^2-4}{x^2+1}$ has 2 zeros, no vertical asymptote, and a horizontal asymptote at $y = 1$.

(b) $r(x) = (x-3)^2(x+1)$ has 2 zeros, no vertical asymptote, and no horizontal asymptote.

(c) $h(x)/f(x)$ fits this description, but is not among the functions above.

(d) $p(x) = \dfrac{(x-3)^2}{x^2-4}$ and $q(x) = \dfrac{x+1}{x^2-4}$ each have 1 zero, 2 vertical asymptotes, and a horizontal asymptote. p has a horizontal asymptote at $y = 1$, and q has a horizontal asymptote at $y = 0$.

(e) $v(x) = \dfrac{x^2+1}{(x-3)^2}$ has no zeros, 1 vertical asymptote, and a horizontal asymptote at $y = 1$.

(f) $t(x) = \dfrac{1}{h(x)} = \dfrac{1}{x+1}$ has no zeros, 1 vertical asymptote, and a horizontal asymptote at $y = 0$.

51. **(a)** False. For example, $f(x) = x^2 + x$ is not even.

(b) False. For example, $f(x) = x^2$ is not invertible.

(c) True.

(d) False. For example, if $f(x) = x^2$, then

$$f(x) \to \infty \quad \text{as} \quad x \to \infty$$
$$f(x) \to \infty \quad \text{as} \quad x \to -\infty.$$

52. **(a)** On the standard viewing screen, the graph of f is shown in Figure 11.84:

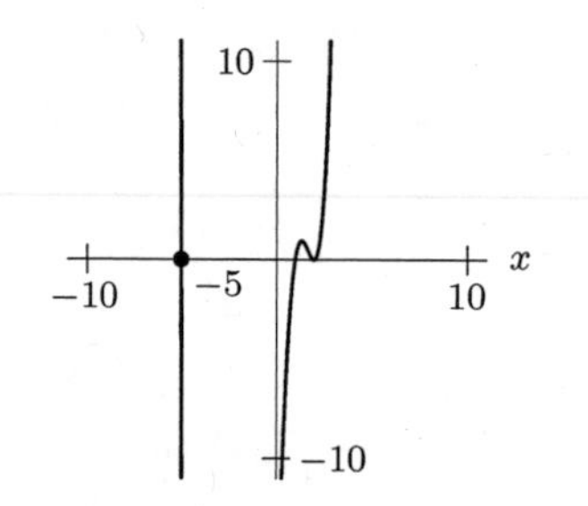

Figure 11.84: $f(x) = x^4 - 17x^2 + 36x - 20$

(b) No. The graph of f is very steep near $x = -5$, but that does not mean it has a vertical asymptote. Since f is a polynomial function, it is defined for all values of x.

(c) The function has 3 zeros. A good screen to see the zeros is $-6 \le x \le 3$, $-3 \le y \le 3$.

(d) Using a graphing calculator or a computer, we find that f has zeros at $x = -5$, $x = 1$, and a double zero at $x = 2$. Thus, $f(x) = (x+5)(x-1)(x-2)^2$.

(e) The function has 3 turning points, two of which are visible in the standard viewing window (see Figure 11.84), and one of which is in the third quadrant, but off the bottom of the screen. It is not possible to see all the turning points in the same window. To see the left-most turning point, a good window is $-6 \le x \le 6$, $-210 \le y \le 50$, but the other turning points are invisible on this scale. To see the other turning points, a good window is $0 \le x \le 3$, $-1 \le y \le 2$, but the left-most turning point is far too low to see on this window.

53.
$$f(x) = \frac{p(x)}{q(x)} = \frac{(x+3)(x-2)}{(x+5)(x-7)}$$

54. A denominator of $(x+1)$ will give the vertical asymptote at $x = -1$. The numerator will have a highest-powered term of $1 \cdot x^1$ to give a horizontal asymptote of $y = 1$. If there is a zero at $x = -3$, try

$$f(x) = \frac{(x+3)}{(x+1)}.$$

Note, this agrees with the y-intercept at $y = 3$.

55. The function f has a vertical asymptote at $x = 1$ and a zero at $x = -1$. A possible formula for $f(x)$ is

$$f(x) = \frac{x+1}{x-1}.$$

56. The function g has a double zero at $x = -1$ and a vertical asymptote at $x = 1$. A possible formula for g is

$$g(x) = \frac{(x+1)^2}{(x-1)^2}.$$

57. We have $f(x) = k(x+3)(x-2)(x-5)$. Solving for k,

$$\begin{aligned} k(0+3)(0-2)(0-5) &= -6 \\ 30k &= -6 \\ k &= -\frac{1}{5}. \end{aligned}$$

Thus, $f(x) = (-1/5)(x+3)(x-2)(x-5)$.

58.

$$\begin{aligned} p(x) &= k(x+3)(x-2)(x-5)(x-6)^2 \\ 7 = p(0) &= k(3)(-2)(-5)(-6)^2 \\ &= k(1080) \\ k &= \frac{7}{1080} \\ p(x) &= \frac{7}{1080}(x+3)(x-2)(x-5)(x-6)^2 \end{aligned}$$

59. Let $h(x) = j(x) + 7$ where $j(x)$ has zeros at $x = -5, -1, 4$. Since j has zeros at these x-values, h has "sevens" there—that is, the value of h equals 7 at these x-values. A formula for $j(x) = k(x+5)(x+1)(x-4)$, so $h(x) = k(x+5)(x+1)(x-4) + 7$. Solving for k, we have

$$\begin{aligned} h(0) &= 3 \\ k(5)(1)(-4) + 7 &= 3 \\ -20k &= -4 \\ k &= \frac{1}{5}. \end{aligned}$$

Thus, $h(x) = (1/5)(x+5)(x+1)(x-4) + 7$.

60. Using the hint, we have $w(x) = p(x) + v(x)$. If $w(x)$ equals $v(x)$ at $x = -4, 1, 3$, then $p(x) = 0$ at those x-values, so $p(x) = k(x+4)(x-1)(x-3)$. We have

$$\begin{aligned} w(x) &= \underbrace{k(x+4)(x-1)(x-3)}_{p(x)} + \underbrace{2x+5}_{v(x)} \\ w(0) &= k(4)(-1)(-3) + 2(0) + 5 \\ 2 &= 12k + 5 \\ k &= -\frac{1}{4}. \end{aligned}$$

Thus, $p(x) = (-1/4)(x+4)(x-1)(x-3)$ and $w(x) = (-1/4)(x+4)(x-1)(x-3) + 2x + 5$.

61. The map distance is directly proportional to the actual distance (mileage), because as the actual distance, x, increases, the map distance, d, also increases.

Substituting the values given into the general formula $d = kx$, we have $0.5 = k(5)$, so $k = 0.1$, and the formula is

$$d = 0.1x.$$

When $d = 3.25$, we have $3.25 = 0.1(x)$ so $x = 32.5$. Therefore, towns which are separated by 3.25 inches on the map are 32.5 miles apart.

62. Since the frequency, f, is inversely proportional to the length, L, we have

$$f = k\frac{1}{L},$$

so

$$L = k\frac{1}{f}.$$

Thus, the length of the string is inversely proportional to the frequency.

63. **(a)** We are given that if d is the radius of the earth,

$$180 = \frac{k}{d^2},$$

so

$$d^2 = \frac{k}{180}.$$

On a planet whose radius is three times the radius of the earth, the person's weight is

$$w = \frac{k}{(3d)^2} = \frac{k}{9d^2}.$$

Therefore,

$$w = \frac{k}{9(\frac{k}{180})} = \frac{180}{9} = 20 \text{ lbs.}$$

If the radius is one-third of the earth's,

$$\begin{aligned} w &= \frac{k}{(\frac{1}{3}d)^2} = \frac{k}{\frac{1}{9}d^2} = \frac{9k}{d^2} = \frac{9k}{\frac{k}{180}} \\ &= 9(180) = 1620 \text{ lbs.} \end{aligned}$$

(b) Let x be the fraction of the earth's radius, d. Since 1 ton $= 2000$ lb, we have

$$2000 = \frac{k}{(xd)^2} = \frac{k}{x^2d^2}.$$

We can use the information from part (a) to substitute

$$d^2 = \frac{k}{180}$$

so that

$$2000 = \frac{k}{x^2(\frac{k}{180})} = \frac{180}{x^2}.$$

Thus

$$x^2 = \frac{180}{2000} = 0.09$$

and

$$x = 0.3 \quad \text{(discarding negative } x\text{)}.$$

The radius is $\frac{3}{10}$ of the earth's radius.

64. **(a)** Kepler's Law states that the square of the period, P, is proportional to the cube of the distance, d. Thus, we have

$$P^2 = kd^3.$$

Solving for P gives

$$P = \sqrt{kd^3} = \sqrt{k}d^{3/2} = k_1d^{3/2}.$$

For the earth, $P = 365$ and $d = 93{,}000{,}000$. Thus,

$$365 = k_1(93{,}000{,}000)^{3/2},$$

so

$$k_1 = \frac{365}{(93{,}000{,}000)^{3/2}} = 4.1 \cdot 10^{-10}.$$

This gives

$$P = \frac{365}{(93{,}000{,}000)^{3/2}} \cdot d^{3/2} = 365 \frac{d^{3/2}}{(93{,}000{,}000)^{3/2}} = 365 \left(\frac{d}{93{,}000{,}000} \right)^{3/2},$$

or

$$P = 4.1 \cdot 10^{-10} d^{3/2}.$$

(b) For Jupiter, $d = 483{,}000{,}000$, so we have

$$P = 365 \left(\frac{483{,}000{,}000}{93{,}000{,}000} \right)^{3/2}$$

which gives

$$P \approx 4320 \text{ earth days},$$

or almost 12 earth years.

65. (a) The graph of the function on the suggested window is shown in Figure 11.85. At $x = 0$ (when Smallsville was founded), the population was 5 hundred people.

(b) The x-intercept for $x > 0$ will show when the population was zero. This occurs at $x \approx 8.44$. Thus, Smallsville became a ghost town in June of 1908.

(c) There are two peaks on the graph on $0 \le x \le 10$, but the first occurs before $x = 5$ (i.e.,before 1905). The second peak occurs at $x \approx 7.18$. The population at that point is ≈ 7.9 hundred. So the maximum population was ≈ 790 in February of 1907.

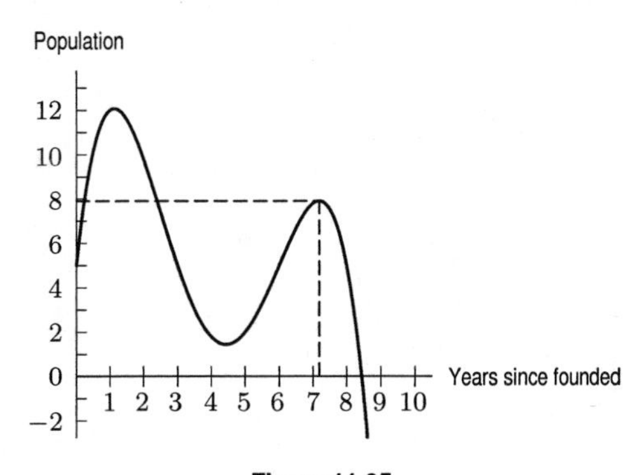

Figure 11.85

66. (a) The graph of $C(x) = (x-1)^3 + 1$ is the graph of $y = x^3$ shifted right one unit and up one unit. The graph is shown in Figure 11.86.

(b) The price is \$1000 per unit, since $R(1) = 1$ means selling 1000 units yields \$1,000,000.

(c)

$$\begin{aligned} \text{Profit } &= R(x) - C(x) \\ &= x - [(x-1)^3 + 1] \\ &= x - (x-1)^3 - 1. \end{aligned}$$

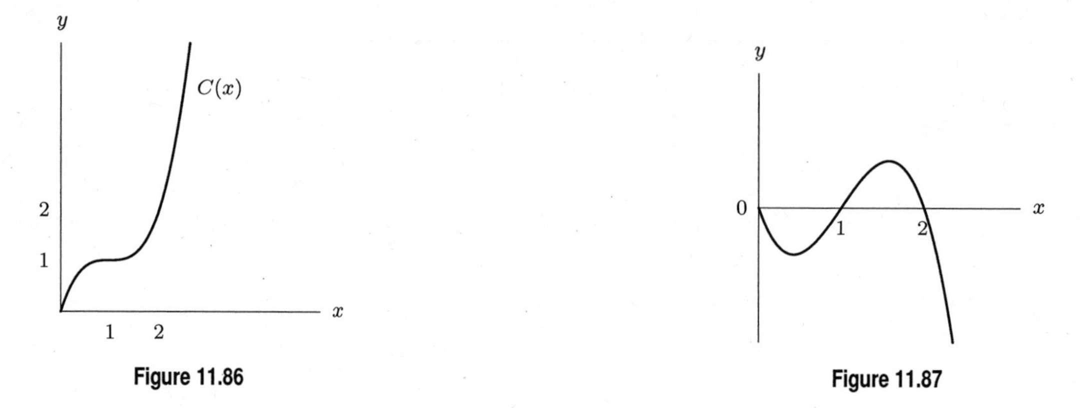

Figure 11.86 **Figure 11.87**

The graph of $R(x) - C(x)$ is shown in Figure 11.87. Profit is negative for $x < 1$ and for $x > 2$. Profit $= 0$ at $x = 1$ and $x = 2$. Thus, the firm will break even with either 1000 or 2000 units, make a profit for $1000 < x < 2000$ units, and lose money for any number of units between 0 and 1000 or greater than 2000.

67. **(a)** The length x of a fish proportional to L implies that $x = aL$, where a is a positive constant. The weight y proportional to its volume, and therefore to L^3, implies $y = bL^3$, where b is a positive constant. Eliminating L between these two equations gives $y = (b/a^3)x^3 = kx^3$ where $k = b/a^3$.

(b) The ratio x^3/y should be approximately constant. The successive values of x^3/y are 113.2, 113.1, 114.4, 116.1, 115.9, 114.1, 114.6, 115.7, 114.7, 113.9, and 113.7, with an average of 114.5. Since $y = kx^3$, the ratio $x^3/y = 1/k = 114.5$. Thus, $k \approx 1/114.5 = 0.0087$.

(c) Figure 11.88 shows the function $y = 0.0087x^3$ and the data in Table 11.47. This function is a reasonable model.

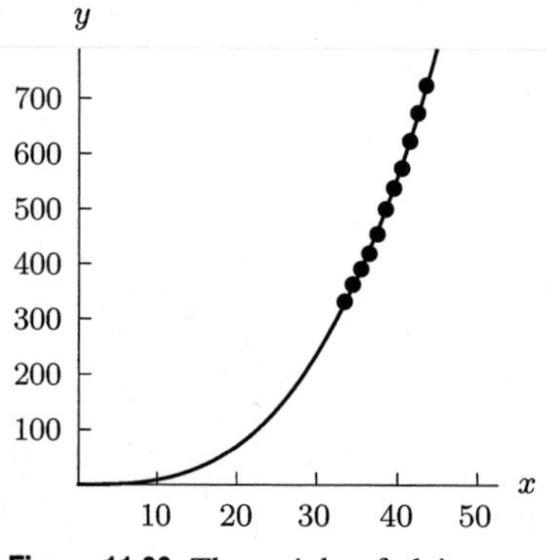

Figure 11.88: The weight of plaice as a function of its length and the function $y = 0.0087x^3$

68. **(a)** We have $T = kR^2D^4$.

(b) If R doubles, the thrust increases by a factor of 4. To see why, suppose

$$T_1 = kR_1^2D^4 \qquad \text{and} \qquad T_2 = kR_2^2D^4.$$

Now suppose $R_2 = 2R_1$. Then

$$T_2 = k(2R_1)^2D^4 = k2^2R_1^2D^4 = 4kR_1^2D^4.$$

So

$$\frac{T_2}{T_1} = \frac{4kR_1^2D^4}{kR_1^2D^4} = 4.$$

Thus, the thrust has been multiplied by 4.

(c) If D doubles, the thrust increases by a factor of $2^4 = 16$. To see why, suppose

$$T_1 = kR^2D_1^4 \quad \text{and} \quad T_2 = kR^2D_2^4$$

and let $D_2 = 2D_1$. Then

$$T_2 = kR^2(2D_1)^4 = kR^2 2^4 D_1^4 = 16kR^2D_1^4,$$

so

$$\frac{T_2}{T_1} = \frac{16kR^2D_1^4}{kR^2D_1^4} = 16.$$

Thus, the thrust has been multiplied by 16.

(d) Suppose R_1 and D_1 are the original values of the speed and diameter, and R_2 and D_2 are the new values. We are told that $D_2 = 1.5D_1$, and we want to find the relation between R_1 and R_2. Since T remains constant,

$$T = kR_1^2D_1^4 = kR_2^2D_2^4.$$

Substituting $D_2 = 1.5D_1$

$$kR_1^2D_1^4 = kR_2^2(1.5D_1)^4$$
$$kR_1^2D_1^4 = kR_2^2(1.5)^4D_1^4$$

Canceling k and D_1^4 gives

$$R_1^2 = (1.5)^4R_2^2.$$

We are interested in the ratio R_2/R_1, so we solve::

$$\frac{R_2^2}{R_1^2} = \frac{1}{(1.5)^4}$$
$$\left(\frac{R_2}{R_1}\right)^2 = \frac{1}{(1.5)^4}$$
$$\frac{R_2}{R_1} = \frac{1}{(1.5)^2} = 0.444 = 44.4\%.$$

Thus, the speed should be reduced to 44.4% of its previous value.

69. (a)

$$p(1) = 1 + 1 + \frac{1^2}{2} + \frac{1^3}{6} + \frac{1^4}{24} + \frac{1^5}{120} \approx 2.71666\ldots.$$

This is accurate to 2 decimal places, since $e \approx 2.718$.

(b) $p(5) \approx 91.417$. This is not at all close to $e^5 \approx 148.4$.

(c) See Figure 11.89. The two graphs are difficult to tell apart for $-2 \le x \le 2$, but for x much less than -2 or much greater than 2, the fit gets worse and worse.

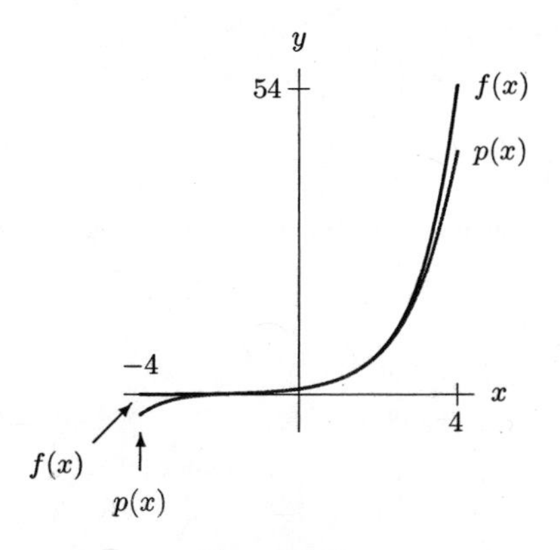

Figure 11.89

70. **(a)** See Figure 11.90.

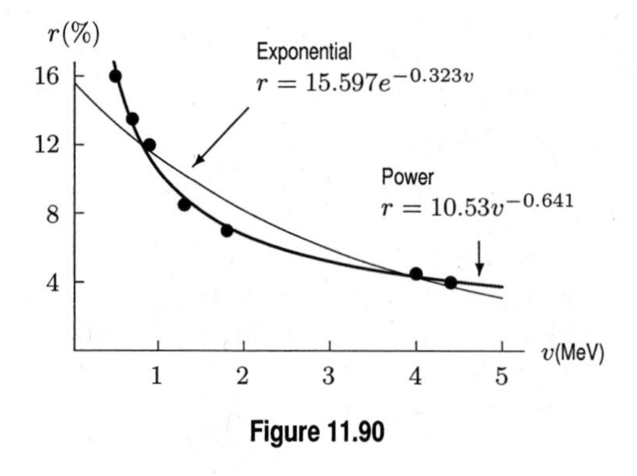

Figure 11.90

(b) The value of r tends to go down as v increases. This means that the telescope is better able to distinguish between high-energy gamma ray photons than low-energy ones.

(c) The first curve is a power function given by $r = 10.53v^{-0.641}$. The second is an exponential function given by $r = 15.597e^{-0.323v}$. The power function appears to give a better fit.

(d) The power function predicts that $r \to \infty$ as $v \to 0$, and so is most consistent with the prediction that the telescope gets rapidly worse and worse at low energies. In contrast, exponential function predicts that r gets close to 15.6% as E gets close to 0.

CHECK YOUR UNDERSTANDING

1. False. The quadratic function $y = 3x^2 + 5$ is not of the form $y = kx^n$, so it is not a power function.
2. False. This is an exponential function.
3. True. Evaluate $g(-1) = (-1)^p$. Since p is even, $g(-1) = 1$, so the point $(-1, 1)$ is on the graph.
4. True. Evaluate $g(-x) = (-x)^p$. Since p is even, $g(-x) = g(x)$, so the graph of g is symmetric about the y-axis.
5. True. All positive even power functions have an upward opening U shape.
6. False. Since $f(0) = 1/0$, we see that $f(x)$ is undefined at $x = 0$.
7. False. The y-axis is also an asymptote.
8. False. The function values of $f(x) = x^{-1}$ approach $+\infty$ as x approaches zero from the right side, but the function values approach $-\infty$ as x approaches zero from the left side.
9. True. The x-axis is an asymptote for $f(x) = x^{-1}$, so the values approach zero.
10. True. In the long run, an increasing exponential function grows faster than any power function.
11. True. The function $g(x) = \ln x$ grows slower than $h(x) = x^a$, for every a greater than 0.
12. False. We have $2^x < x^2$ at $x = 3$.
13. False. As x grows very large the exponential decay function g approaches the x-axis faster than any power function with a negative power.
14. False. This is an exponential function.
15. True. It is of the form $y = kx^n$, where $k = 3$ and $n = 1$.
16. True. A quadratic function is a polynomial function of degree 2.
17. False. For example, the polynomial $x^2 + x^3$ has degree 3 because the degree is the highest power, not the first power, in the formula for the polynomial.
18. True. In the long run the highest-power term eventually dominates.
19. False. A zero of a polynomial is any x value for which $p(x) = 0$.

20. True. The graph of a polynomial p intersects the x-axis where $p(x) = 0$, and these values of x are the zeros of p.

21. True. The graph crosses the y-axis at the point $(0, p(0))$.

22. True. Since g is a polynomial of degree 4 with positive leading coefficient, it will eventually dominate any polynomial of degree 3.

23. False. Any polynomial of positive even degree has a graph which in the long run looks like a U, so it does not pass the horizontal line test.

24. False. Some odd degree polynomials have an inverse, but not all of them do. As a counterexample, consider $p(x) = x(x-1)(x+1)$, a polynomial of degree 3 that has no inverse because $p(x) = 0$ has more than one solution.

25. True. We can write $p(x) = (x-a) \cdot C(x)$. Evaluating at $x = a$ we get $p(a) = (a-a) \cdot C(a) = 0 \cdot C(a) = 0$.

26. True. If a polynomial $p(x)$ has $n+1$ zeros, $r_1, r_2, \ldots, r_{n+1}$, then the product $(x-r_1)(x-r_2)\cdots(x-r_{n+1})$ is a factor of $p(x)$ of degree $n+1$, so $p(x)$ must have degree $n+1$ or higher. This is impossible if $p(x)$ has degree n.

27. True. There is a multiple zero at $x = -2$ because the graph is tangent to the x-axis at $x = -2$.

28. False. There is a single zero at $x = 0$ because the graph crosses the x-axis at $x = 0$. The fact that the graph does not appear to be tangent to the x-axis at $x = 0$ leads us to conclude that there is not a multiple zero at $x = 0$.

29. True. This is the definition of a rational function.

30. True. This is the quotient of two polynomials, $f(x) = 1$ and $g(x) = x$.

31. True. The highest-power term in a polynomial determines its long-run behavior, so by considering the ratio of the highest-power terms in the numerator and denominator, we are determining the long-run behavior of the rational function.

32. False. The ratio of the highest degree terms in the numerator and denominator is $x/x = 1$ so for large positive x-values, y approaches 1.

33. True. The ratio of the highest degree terms in the numerator and denominator is $2x/x^2 = 2/x$, so for large positive x-values, y approaches 0.

34. False. The ratio of the highest degree terms in the numerator and denominator is $x^3/4x^3 = 1/4$ so the asymptote is $y = 1/4$.

35. False. The ratio of the highest degree terms in the numerator and denominator is $-4x^2/x^2 = -4$. So for large positive x-values, y approaches -4.

36. False. The ratio of the highest terms in the numerator and denominator is $5x/x = 5$. So for large positive x-values, y approaches 5.

37. False. The ratio of the highest degree terms in the numerator and denominator is $3x^4/x^2 = 3x^2$. So for large positive x-values, y behaves like $y = 3x^2$.

38. True. The ratio of the highest terms in the numerator and denominator is $x^3/(-x^2) = -x$. Since the x-values are negative, y approaches positive infinity.

39. True. A zero numerator forces the fraction value to be zero when the fraction is defined. It is undefined when the denominator is zero.

40. True. This is the definition of the zero of a function.

41. True. At $x = -4$, we have $f(-4) = (-4+4)/(-4-3) = 0/(-7) = 0$, so $x = -4$ is a zero.

42. False. At $x = -2$, we have $f(-2) = (-2+2)/((-2)^2 - 4) = 0/0$, so the function is not defined at $x = -2$, and $x = -2$ is not a zero of the function.

43. False. The numerator is never zero so g has no zeros.

44. False. The rational function r has a zero, not an asymptote, at each of the zeros of $p(x)$.

45. False. If $p(x)$ has no zeros, then $r(x)$ has no zeros. For example, if $p(x)$ is a nonzero constant or $p(x) = x^2 + 1$, then $r(x)$ has no zeros.

46. False. If the asymptote is horizontal it is possible. For example, the graph of $f(x) = (x-1)(x+4)/((x+2)(x-2))$ crosses the horizontal asymptote $y = 1$ where $x = 0$.

47. False. It has a zero at $w = 1$, since $g(1) = 0/(-55) = 0$.

Solutions to Skills for Chapter 11

1. $\frac{3}{5}+\frac{4}{7}=\frac{3\cdot 7+4\cdot 5}{35}=\frac{21+20}{35}=\frac{41}{35}$
2. $\frac{7}{10}-\frac{2}{15}=\frac{7\cdot 15-2\cdot 10}{150}=\frac{105-20}{150}=\frac{85}{150}=\frac{17}{30}$
3. $\frac{1}{2x}-\frac{2}{3}=\frac{1\cdot 3-2(2x)}{6x}=\frac{3-4x}{6x}$
4. $\frac{6}{7y}+\frac{9}{y}=\frac{6\cdot y+9\cdot 7y}{7y^2}=\frac{6y+63y}{7y^2}=\frac{69y}{7y^2}=\frac{69}{7y}$
5. $\frac{-2}{yz}+\frac{4}{z}=\frac{-2z+4yz}{yz^2}=\frac{-2+4y}{yz}=\frac{-2(1-2y)}{yz}$
6. $\frac{-2z}{y}+\frac{4}{y}=\frac{-2z+4}{y}=\frac{-2(z-2)}{y}$
7. $\frac{2}{x^2}-\frac{3}{x}=\frac{2-3x}{x^2}$
8. $\frac{\frac{3}{4}}{\frac{7}{20}}=\frac{3}{4}\cdot\frac{20}{7}=\frac{60}{28}=\frac{15}{7}$
9. $\frac{\frac{5}{6}}{15}=\frac{5}{6}\cdot\frac{1}{15}=\frac{1}{18}$
10. $\frac{\frac{3}{x}}{\frac{x^2}{6}}=\frac{3}{x}\cdot\frac{6}{x^2}=\frac{18}{x^3}$
11. $\frac{\frac{3}{x}}{\frac{6}{x^2}}=\frac{3}{x}\cdot\frac{x^2}{6}=\frac{x}{2}$
12. $\frac{14}{x-1}+\frac{13}{2x-2}=\frac{14}{x-1}+\frac{13}{2(x-1)}=\frac{2\cdot 14+13}{2(x-1)}=\frac{41}{2(x-1)}$
13. $\frac{4z}{x^2y}-\frac{3w}{xy^4}=\frac{4zxy^4-3wx^2y}{x^3y^5}=\frac{xy(4zy^3-3wx)}{x^3y^5}=\frac{4y3z-3wx}{x^2y^4}$
14. $\frac{10}{y-2}+\frac{3}{2-y}=\frac{10}{y-2}-\frac{3}{y-2}=\frac{7}{y-2}$
15. $\frac{8y}{y-4}+\frac{32}{y-4}=\frac{8y+32}{y-4}=\frac{8(y+4)}{y-4}$
16. $\frac{8y}{y-4}+\frac{32}{4-y}=\frac{8y}{y-4}-\frac{32}{y-4}=\frac{8y-32}{y-4}=\frac{8(y-4)}{y-4}=8$
17. $$\begin{aligned}\frac{8}{3x^2-x-4}-\frac{9}{x+1}&=\frac{8}{(x+1)(3x-4)}-\frac{9}{x+1}\\&=\frac{8-9(3x-4)}{(x+1)(3x-4)}\\&=\frac{-27x+44}{(x+1)(3x-4)}\end{aligned}$$
18. $$\begin{aligned}\frac{15}{(x-3)^2(x+5)}+\frac{7}{(x-3)(x+5)^2}&=\frac{15(x+5)+7(x-3)}{(x-3)^2(x+5)^2}\\&=\frac{15x+75+7x-21}{(x-3)^2(x+5)^2}\\&=\frac{22x+54}{(x-3)^2(x+5)^2}\\&=\frac{2(11x+27)}{(x-3)^2(x+5)^2}\end{aligned}$$

19. The common denominator is $(x-4)(x+4)=x^2-16$. Therefore,

$$\frac{3}{x-4}-\frac{2}{x+4}=\frac{3(x+4)}{(x-4)(x+4)}-\frac{2(x-4)}{(x+4)(x-4)}$$
$$\frac{3(x+4)-2(x-4)}{x^2-16}=\frac{3x+12-2x+8}{x^2-16}$$
$$=\frac{x+20}{x^2-16}.$$

20. If we rewrite the second fraction $-\frac{1}{1-x}$ as $\frac{1}{x-1}$, the common denominator becomes $x-1$. Therefore,

$$\frac{x^2}{x-1}-\frac{1}{1-x}=\frac{x^2}{x-1}+\frac{1}{x-1}=\frac{x^2+1}{x-1}.$$

21. The second denominator $4r^2+6r=2r(2r+3)$, while the first denominator is $2r+3$. Therefore the common denominator is $2r(2r+3)$. We have:

$$\frac{1}{2r+3}+\frac{3}{4r^2+6r}=\frac{1}{2r+3}+\frac{3}{2r(2r+3)}$$
$$=\frac{1\cdot 2r}{2r(2r+3)}+\frac{3}{2r(2r+3)}$$
$$=\frac{2r+3}{2r(2r+3)}=\frac{1}{2r}.$$

22. The common denominator is $u+a$. Therefore,

$$u+a+\frac{u}{u+a}=\frac{(u+a)(u+a)}{u+a}+\frac{u}{u+a}=\frac{(u+a)^2+u}{u+a}.$$

23. The common denominator is $(\sqrt{x})^3$.

$$\frac{1}{\sqrt{x}}-\frac{1}{(\sqrt{x})^3}=\frac{(\sqrt{x})^2}{(\sqrt{x})^3}-\frac{1}{(\sqrt{x})^3}=\frac{x-1}{(\sqrt{x})^3}$$

It is fine to leave the answer in the form $\frac{x-1}{(\sqrt{x})^3}$, or we can rationalize the denominator:

$$\frac{x-1}{(\sqrt{x})^3}=\frac{x-1}{x\sqrt{x}}=\frac{\sqrt{x}(x-1)}{x\sqrt{x}\sqrt{x}}=\frac{x\sqrt{x}-\sqrt{x}}{x^2}.$$

24. The common denominator is e^{2x}. Thus,

$$\frac{1}{e^{2x}}+\frac{1}{e^x}=\frac{1}{e^{2x}}+\frac{e^x}{e^{2x}}=\frac{1+e^x}{e^{2x}}.$$

25. If we factor the number and denominator of the second fraction, we can cancel some terms with the first,

$$\frac{a+b}{2}\cdot\frac{8x+2}{b^2-a^2}=\frac{a+b}{2}\cdot\frac{2(4x+1)}{(b+a)(b-a)}=\frac{4x+1}{b-a}.$$

26. The common denominator is $4M$. Therefore,

$$\frac{0.07}{M}+\frac{3}{4}M^2=\frac{(0.07)(4)}{4M}+\frac{(3M^2)M}{4M}=\frac{.28+3M^3}{4M}.$$

27. Each of the denominators are different and therefore the common denominator is $r_1r_2r_3$. Accordingly,

$$\frac{1}{r_1}+\frac{1}{r_2}+\frac{1}{r_3}=\frac{r_2r_3+r_1r_3+r_1r_2}{r_1r_2r_3}.$$

28. $\dfrac{8y}{y-4}-\dfrac{32}{y-4}=\dfrac{8y-32}{y-4}=\dfrac{8(y-4)}{y-4}=8$

29. $\dfrac{2a+3}{(a+3)(a-3)}$

30. We change this division example to a multiplication problem by writing the reciprocal of the second faction. Therefore,

$$\frac{x^3}{x-4}\Big/\frac{x^2}{x^2-2x-8}=\frac{x^3}{x-4}\cdot\frac{x^2-2x-8}{x^2}=\frac{x(x-4)(x+2)}{x-4}=x(x+2).$$

31. First we find a common denominator for the two fractions in the numerator. Thus,

$$\begin{aligned}\frac{\frac{1}{(x+h)^2}-\frac{1}{x^2}}{h}&=\frac{x^2-(x+h)^2}{x^2(x+h)^2}\cdot\frac{1}{h}\\&=\frac{x^2-x^2-2xh-h^2}{x^2(x+h)^2}\cdot\frac{1}{h}=\frac{h(-2x-h)}{x^2(x+h)^2}\cdot\frac{1}{h}\\&=\frac{-2x-h}{x^2(x+h)^2}.\end{aligned}$$

32. Recall that the terms a^{-2} and b^{-2} can be written as $\dfrac{1}{a^2}$ and $\dfrac{1}{b^2}$ respectively. Therefore,

$$\frac{a^{-2}+b^{-2}}{a^2+b^2}=\frac{\frac{1}{a^2}+\frac{1}{b^2}}{a^2+b^2}=\frac{\frac{b^2+a^2}{a^2b^2}}{a^2+b^2}=\frac{b^2+a^2}{a^2b^2}\cdot\frac{1}{a^2+b^2}=\frac{1}{a^2b^2}.$$

33. We expand within the first brackets first. Therefore,

$$\begin{aligned}\frac{[4-(x+h)^2]-[4-x^2]}{h}&=\frac{[4-(x^2+2xh+h^2)]-[4-x^2]}{h}\\&=\frac{[4-x^2-2xh-h^2]-4+x^2}{h}=\frac{-2xh-h^2}{h}\\&=-2x-h.\end{aligned}$$

34.

$$\frac{b^{-1}(b-b^{-1})}{b+1}=\frac{\frac{1}{b}\left(b-\frac{1}{b}\right)}{b+1}=\frac{1}{b^2}\frac{(b^2-1)}{b+1}=\frac{(b+1)(b-1)}{b^2(b+1)}=\frac{b-1}{b^2}.$$

35.

$$\frac{1-a^{-2}}{1+a^{-1}}=\frac{1-\frac{1}{a^2}}{1+\frac{1}{a}}=\frac{a^2-1}{a^2}\cdot\frac{a}{a+1}=\frac{(a-1)(a+1)}{a(a+1)}=\frac{a-1}{a}=1-\frac{1}{a}.$$

36. We simplify the second complex fraction first. Thus,

$$\begin{aligned}p-\frac{q}{\frac{p}{q}+\frac{q}{p}}&=p-\frac{q}{\frac{p^2+q^2}{qp}}=p-q\cdot\frac{qp}{p^2+q^2}\\&=\frac{p(p^2+q^2)-q^2p}{p^2+q^2}=\frac{p^3}{p^2+q^2}.\end{aligned}$$

37. $$\frac{\frac{3}{xy}-\frac{5}{x^2y}}{\frac{6x^2-7x-5}{x^4y^2}} = \frac{\frac{3x-5}{x^2y}}{\frac{(3x-5)(2x+1)}{y^4y^2}} = \frac{3x-5}{x^2y}\cdot\frac{x^4y^2}{(3x-5)(2x+1)} = \frac{x^2y}{2x+1}$$

38. Cancellation is employed here to simplify. Therefore,

$$\frac{\frac{1}{x}\left(3x^2\right)-(\ln x)(6x)}{(3x^2)^2} = \frac{3x-(\ln x)(6x)}{9x^4} = \frac{1-(\ln x)(2)}{3x^3} = \frac{1-2\ln x}{3x^3}.$$

39. We cancel the common factor x^3+1 in both numerator and denominator. Therefore,

$$\frac{2x(x^3+1)^2-x^2(2)(x^3+1)(3x^2)}{[(x^3+1)^2]^2} = \frac{2x(x^3+1)-x^2(2)(3x^2)}{(x^3+1)^3} = \frac{2x^4+2x-6x^4}{(x^3+1)^3} = \frac{2x-4x^4}{(x^3+1)^3}.$$

40. Write

$$\frac{\frac{1}{2}(2x-1)^{-1/2}(2)-(2x-1)^{1/2}(2x)}{(x^2)^2} = \frac{\frac{1}{(2x-1)^{1/2}}-\frac{2x(2x-1)^{1/2}}{1}}{(x^2)^2}.$$

Next a common denominator for the top two fractions is $(2x-1)^{1/2}$. Therefore we obtain,

$$\frac{\frac{1}{(2x-1)^{1/2}}-\frac{2x(2x-1)}{(2x-1)^{1/2}}}{x^4} = \frac{1-4x^2+2x}{(2x-1)^{1/2}}\cdot\frac{1}{x^4} = \frac{-4x^2+2x+1}{x^4\sqrt{2x-1}}.$$

41. Dividing $2x^3$ into each term in the numerator yields:

$$\frac{26x+1}{2x^3} = \frac{26x}{2x^3}+\frac{1}{2x^3} = \frac{13}{x^2}+\frac{1}{2x^3}.$$

42. Dividing $3\sqrt{x}$ into both terms in the numerator yields:

$$\frac{\sqrt{x}+3}{3\sqrt{x}} = \frac{\sqrt{x}}{3\sqrt{x}}+\frac{3}{3\sqrt{x}} = \frac{1}{3}+\frac{1}{\sqrt{x}}.$$

43.

$$\frac{6l^2+3l-4}{3l^4} = \frac{6l^2}{3l^4}+\frac{3l}{3l^4}-\frac{4}{3l^4} = \frac{2}{l^2}+\frac{1}{l^3}-\frac{4}{3l^4}$$

44. The denominator p^2+11 is divided into each of the two terms of the numerator. Thus,

$$\frac{7+p}{p^2+11} = \frac{7}{p^2+11}+\frac{p}{p^2+11}.$$

45.

$$\frac{\frac{1}{3}x-\frac{1}{2}}{2x} = \frac{\frac{x}{3}}{2x}-\frac{\frac{1}{2}}{2x} = \frac{x}{3}\cdot\frac{1}{2x}-\frac{1}{2}\cdot\frac{1}{2x} = \frac{1}{6}-\frac{1}{4x}$$

46. In this example, dividing the denominator into each term of the numerator involves the same base t. Therefore we subtract exponents.

$$\frac{t^{-1/2}+t^{1/2}}{t^2} = \frac{t^{-1/2}}{t^2}+\frac{t^{1/2}}{t^2} = t^{-1/2-2}+t^{1/2-2} = t^{-5/2}+t^{-3/2} = \frac{1}{t^{5/2}}+\frac{1}{t^{3/2}}$$

47. We write the numerator $x - 2$ as $x + 5 - 7$. Therefore,

$$\frac{x-2}{x+5} = \frac{(x+5)-7}{x+5} = 1 - \frac{7}{x+5}.$$

48. The numerator $q - 1 = q - 4 + 3$. Thus,

$$\frac{q-1}{q-4} = \frac{(q-4)+3}{q-4} = 1 + \frac{3}{q-4}.$$

49. Dividing the denominator R into each term in the numerator yields,

$$\frac{R+1}{R} = \frac{R}{R} + \frac{1}{R} = 1 + \frac{1}{R}.$$

50. Rewrite $3 + 2u = 2u + 3 = (2u + 1) + 2$. Thus,

$$\frac{3+2u}{2u+1} = \frac{2u+3}{2u+1} = \frac{(2u+1)+2}{2u+1} = 1 + \frac{2}{2u+1}.$$

51. Dividing by $\cos x$ yields:

$$\frac{\cos x + \sin x}{\cos x} = \frac{\cos x}{\cos x} + \frac{\sin x}{\cos x} = 1 + \frac{\sin x}{\cos x}.$$

52.

$$\frac{1+e^x}{e^x} = \frac{1}{e^x} + \frac{e^x}{e^x} = \frac{1}{e^x} + 1 = 1 + \frac{1}{e^x}$$

53. False

54. True

55. False

56. False

57. True

58. True, factor $x^{-1/3}$ on the left and rewrite it with positive exponent in the denominator.

CHAPTER TWELVE

Solutions for Section 12.1

Exercises

1. We can describe an elevation with one number, so this is a scalar.
2. Temperature is measured by a single number, and so is a scalar.
3. The wind velocity is a vector because it has both a magnitude (the speed of the wind) and a direction (the direction of the wind).
4. Scalar.
5. Scalar.
6. We can describe a population with one number, so this is a scalar.
7. We need three numbers—for instance, height, longitude, and latitude—to describe this location, so this is a vector.
8. We need two numbers, one for men and one for women, so this is a vector.
9. See Figure 12.1.

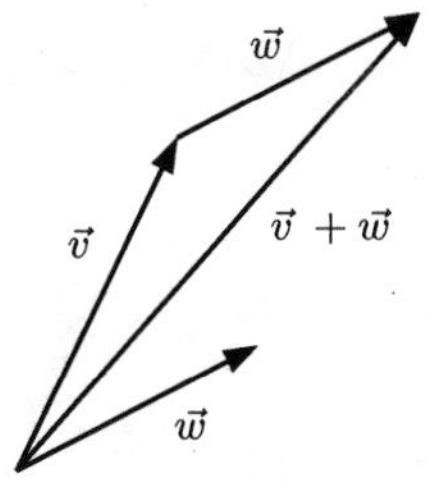

$\vec{w}$

$\vec{v}$ $\vec{v} + \vec{w}$

$\vec{w}$

Figure 12.1

10. See Figures 12.2 and 12.3.

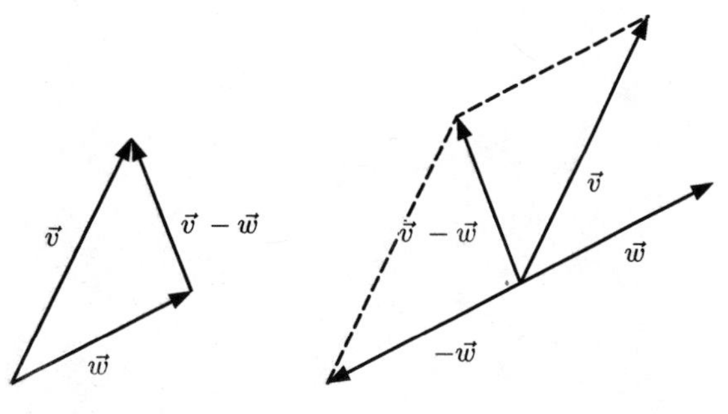

Figure 12.2 Figure 12.3

11. See Figure 12.4.

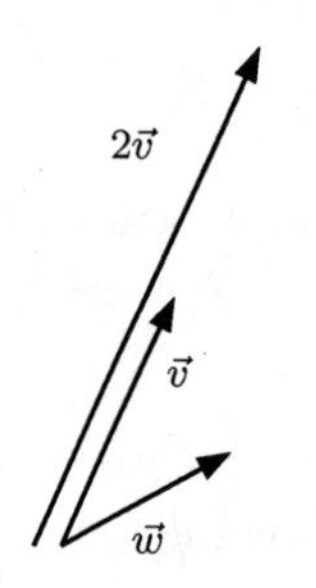

Figure 12.4

12. See Figure 12.5.

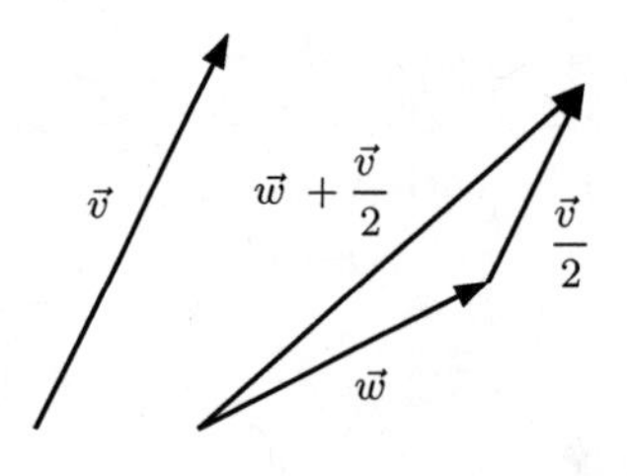

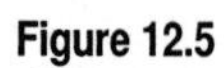

Figure 12.5

13. See Figure 12.6.

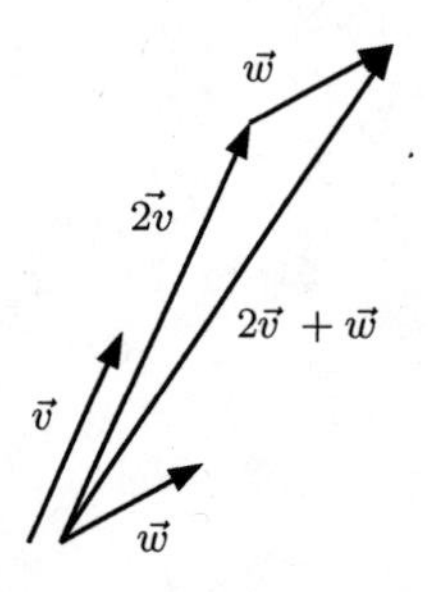

Figure 12.6

14.

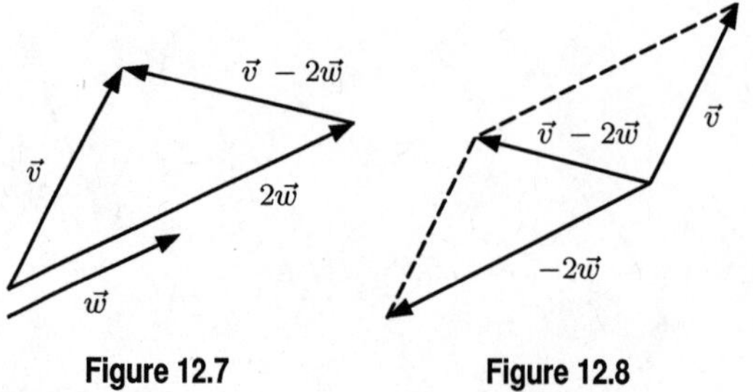

Figure 12.7 Figure 12.8

15.

$$\vec{p} = 2\vec{w}, \quad \vec{q} = -\vec{u}, \quad \vec{r} = \vec{w} + \vec{u} = \vec{u} + \vec{w},$$
$$\vec{s} = \vec{p} + \vec{q} = 2\vec{w} - \vec{u}, \quad \vec{t} = \vec{u} - \vec{w}$$

Problems

16. **(a)** The kite's position is shown in Figure 12.9. The vector $\vec{v}$ from the kite to the ground has length

$$\|\vec{v}\| = 50 \sin 20^\circ = 17.101 \text{ feet}$$

and points directly downward.

(b) The distance between the two positions is given by the Law of Cosines

$$d^2 = 50^2 + 50^2 - 2 \cdot 50 \cdot 50 \cdot \cos 20^\circ = 301.5369$$
$$d = 17.365 \text{ feet.}$$

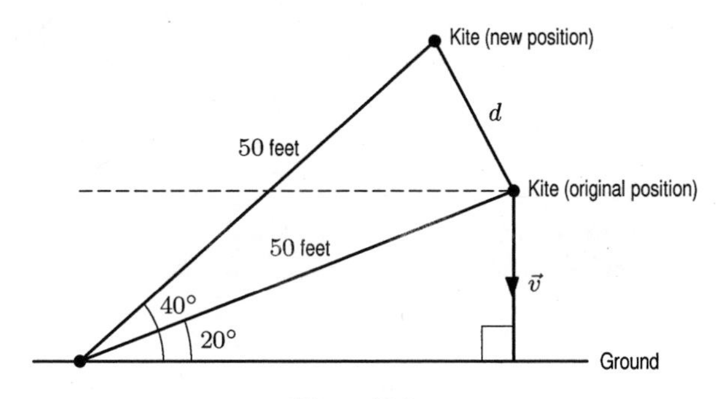

Figure 12.9

17. **(a)** From Figure 12.10,

$$\text{Distance from Oracle Road} = 5 \sin 20^\circ = 1.710 \text{ miles.}$$

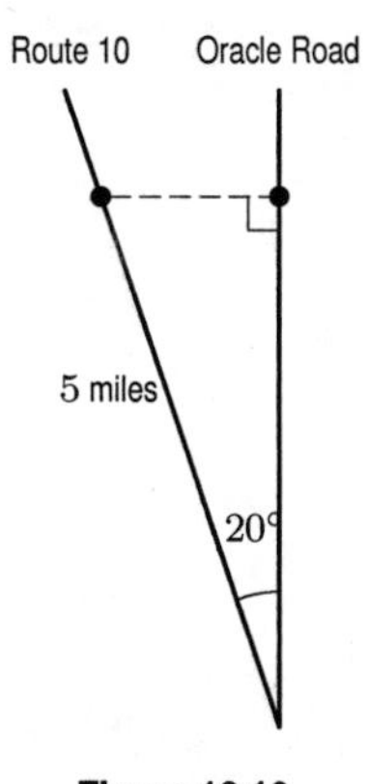

Figure 12.10

(b) If the distance along Route 10 is x miles, we have

$$x \sin 20^\circ = 2 \text{ miles}$$
$$x = \frac{2}{\sin 20^\circ} = 5.848 \text{ miles.}$$

18. In Figure 12.11 we choose north as the positive y-axis, east as the positive x-axis and home as the origin, O.

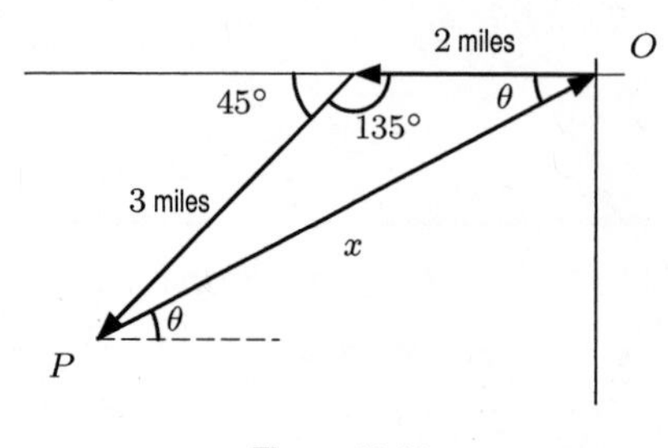

Figure 12.11

We know that the first vector is 2 units long, the second vector is 3 units and is at an angle of 45° from the first. Joining the tail of the first vector and the head of the second vector forms a triangle.

The length of the third side, x, can be found by applying the Law of Cosines:

$$x^2 = 2^2 + 3^2 - 2 \cdot 2 \cdot 3 \cdot \cos(135^\circ) = 13 - 12\left(-\frac{\sqrt{2}}{2}\right)$$
$$x^2 = 13 + 6\sqrt{2} = 21.4853$$
$$x = 4.635.$$

To obtain the angle θ, we apply the Law of Sines:

$$\frac{\sin\theta}{3} = \frac{\sin 135^\circ}{x}$$
$$\sin\theta = 3 \cdot \frac{\sqrt{2}/2}{4.635} = 0.458.$$

So $\theta = 27.236^\circ$.

Therefore, the person should walk 27.236° north of east for 4.635 miles to go directly back home. See Figure 12.11.

19. From Figure 12.12 we see that the vector $\overrightarrow{PQ}$ is 4 units long and makes an angle of $\theta + 45^\circ = 72.236^\circ$ with vector $\overrightarrow{OP}$ (from Problem 18). Joining Q to O makes a triangle. We now can apply the Law of Cosines to find y:

$$\begin{aligned} y^2 &= x^2 + 4^2 - 2 \cdot 4 \cdot x \cdot \cos(\theta + 45^\circ) \\ &= 21.485 + 16 - 37.082 \cdot \cos(72.236^\circ) \\ &= 26.172 \\ y &= 5.116. \end{aligned}$$

The angle made by $\overrightarrow{QO}$ with the y-axis is $\phi - 45^\circ$, where ϕ is found by applying the Law of Sines:

$$\frac{\sin\phi}{4.635} = \frac{\sin 72.236^\circ}{y}$$
$$\sin\phi = 4.635\frac{\sin 72.236^\circ}{5.116} = 0.863$$
$$\phi = 59.639^\circ.$$

Thus, the person must walk 5.116 miles at an angle of 14.639° east of north.

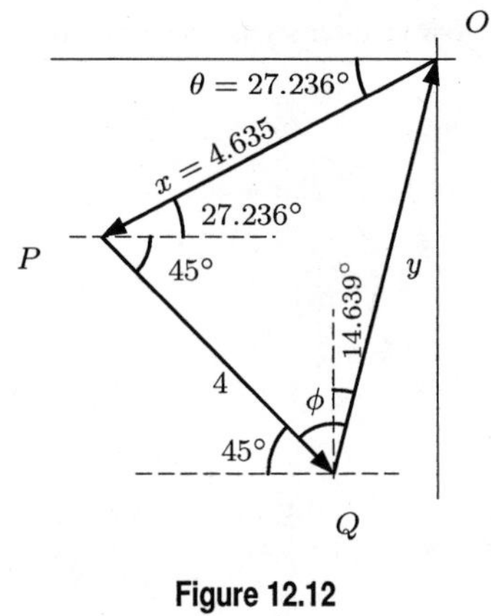

Figure 12.12

20. After 2 seconds,

$$\text{Distance moved by puck } = 7 \cdot 2 = 14 \text{ feet.}$$

See Figure 12.13. Then

$$d = \text{Distance from edge } = 14 \sin 35^\circ = 8.030 \text{ feet.}$$

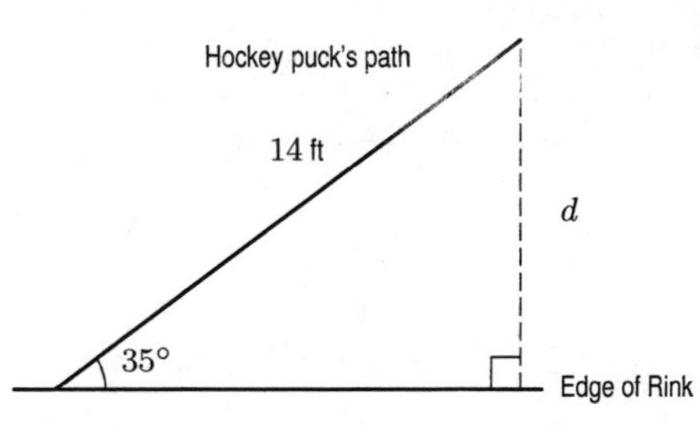

Figure 12.13

21.

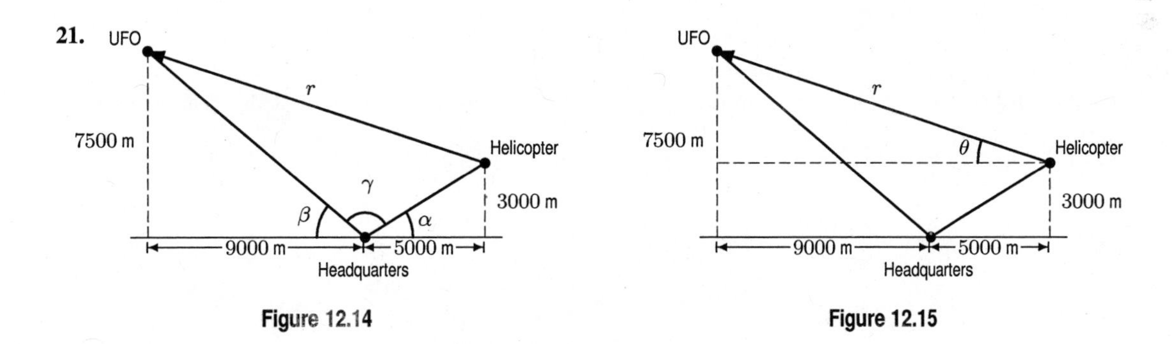

Figure 12.14

Figure 12.15

Figure 12.14 shows the headquarters at the origin, and a positive y-value as up, and a positive x-value as east. To solve for r, we must first find γ:

$$\begin{aligned}\gamma &= 180^\circ - \alpha - \beta \\ &= 180^\circ - \arctan\frac{3000}{5000} - \arctan\frac{7500}{9000} \\ &= 109.231^\circ.\end{aligned}$$

We now can find r using the Law of Cosines in the triangle formed by the position of the headquarters, the helicopter and the UFO.

In kilometers:

$$\begin{aligned} r^2 &= 34 + 137.250 - 2 \cdot \sqrt{34} \cdot \sqrt{137.250} \cdot \cos\gamma \\ r^2 &= 216.250 \\ r &= 14.705 \text{ km} \\ &= 14{,}705 \text{ m.} \end{aligned}$$

From Figure 12.15 we see:

$$\begin{aligned} \tan\theta &= \frac{4500}{14{,}000} \\ \theta &= 17.819^\circ. \end{aligned}$$

Therefore, the helicopter must fly 14,705 meters with an angle of 17.819° from the horizontal.

22. To find l, the distance the helicopter must travel to intercept the UFO, we must first find x as shown in Figure 12.16. From the Pythagorean theorem (in kilometers):

$$\begin{aligned} x^2 &= 5^2 + 12^2 \\ x &= 13. \end{aligned}$$

We can now find l using the Pythagorean theorem:

$$\begin{aligned} l^2 &= 13^2 + 4.5^2 \\ l &= 13.757. \end{aligned}$$

To find the angle θ, we say:

$$\begin{aligned} \sin\theta &= \frac{12}{13} \\ \theta &= 67.380^\circ. \end{aligned}$$

To find the angle ϕ, we say:

$$\begin{aligned} \sin\phi &= \frac{4.5}{13.757} \\ \phi &= 19.093^\circ. \end{aligned}$$

Therefore, the helicopter must travel 13,757 meters at an angle of 67.380° north of west and 19.093° from the horizontal.

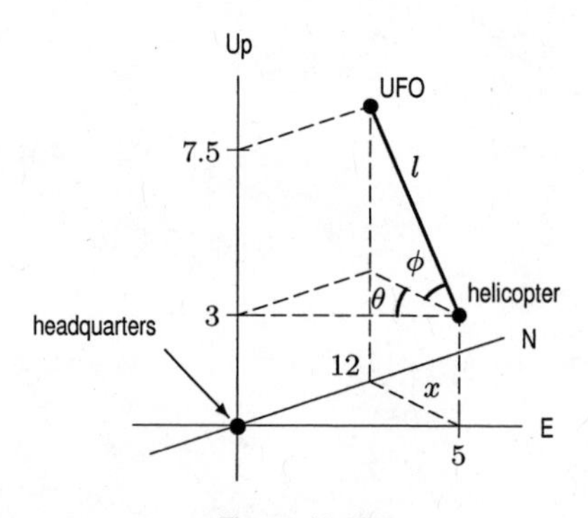

Figure 12.16

23. **(a)** We begin by using the Law of Cosines to find the magnitude of the net force on the spacecraft due to Saturn and Titan. Since $\vec{F}_{ST} = \vec{F}_S + \vec{F}_T$, the force due to Saturn and Titan, $\vec{F}_{ST}$, is obtained by joining the tail of $\vec{F}_T$ to the head of $\vec{F}_S$. We use the Law of Cosines to find the length of $\vec{F}_{ST}$:

$$\begin{aligned}
||\vec{F}_{ST}||^2 &= 7^2 + 10^2 - 2 \cdot 7 \cdot 10 \cos 140^\circ \\
&= 149 - 140 \cos 140^\circ \\
||\vec{F}_{ST}|| &= 16.00769.
\end{aligned}$$

Now we use the Law of Sines to find the direction of $\vec{F}_{ST}$. If θ is the angle between $\vec{F}_{ST}$ and the line joining the spacecraft to Saturn, measured to the right:

$$\begin{aligned}
\frac{\sin\theta}{10} &= \frac{\sin 140^\circ}{||\vec{F}_{ST}||} \\
\sin\theta &= 10\left(\frac{\sin 140^\circ}{||\vec{F}_{ST}||}\right) = 0.40155 \\
\theta &= \arcsin(0.40155) = 23.675^\circ.
\end{aligned}$$

To find $||\vec{F}_{STR}||$, we use the Law of Cosines on the triangle formed by joining the head of $\vec{F}_{ST}$ with the tail of $\vec{F}_R$. We have

$$\begin{aligned}
||\vec{F}_{STR}||^2 &= (16.00769)^2 + 5^2 - 2(16.00769)5 \cos 76.325^\circ \\
&= 243.4017 \\
||\vec{F}_{STR}|| &= 15.6013.
\end{aligned}$$

(b) To determine the direction of the spacecraft, we must determine the direction of $\vec{F}_{STR}$. We use the Law of Sines (as we did in part (a) for the direction of $\vec{F}_{ST}$), but with ψ measured to the left of the line joining the spacecraft to Saturn:

$$\begin{aligned}
\frac{\sin\psi}{5} &= \frac{\sin 76.325^\circ}{||\vec{F}_{STR}||} \\
\sin\psi &= 5\left(\frac{\sin 76.325^\circ}{15.6013}\right) = 0.31140 \\
\psi &= \arcsin(0.31140) = 18.144^\circ.
\end{aligned}$$

So we know the spacecraft does not stay on course. Since $18.144 < 23.675$, we know that the spacecraft veers slightly right, 5.531°, toward Titan.

(c) We want the sum of the forces acting on the spacecraft to have direction 0°, so the engines must (if possible) be directed such that they eliminate the 5.531° to the right. The triangle has angle 5.531° facing the length of 1. We use the Law of Sines to try and find the angle facing 15.6013, and then find the angle of thrust from it. We have

$$\begin{aligned}
\frac{\sin\phi}{15.6013} &= \frac{\sin 5.531^\circ}{1} \\
\sin\phi &= 15.6013 \sin 5.531^\circ \\
\phi &= \arcsin(15.6013 \sin 5.531^\circ) \\
&= \arcsin(1.5037).
\end{aligned}$$

We see that ϕ is undefined, since no real angle has a sin of 1.5037. Thus, the engines cannot be used to correct the spacecraft's course.

24. Let $\vec{v}$ be the velocity of the ball relative to the ground. Figure 12.17 shows the wind in the same direction as the ball; Figure 12.18 Shows the wind opposing the ball.

In Figure 12.17,

$$\begin{aligned}
||\vec{v}\,||^2 &= 5^2 + 3^2 - 2 \cdot 3 \cdot 5 \cdot \cos 135^\circ = 55.213 \\
||\vec{v}\,|| &= \sqrt{55.2132} = 7.4306 = 7.431 \text{ ft/sec}.
\end{aligned}$$

The vector $\vec{v}$ makes an angle θ with the ball's path. By the Law of Sines

$$\frac{\sin\theta}{3} = \frac{\sin 135^\circ}{7.4306}$$
$$\sin\theta = \frac{3\sin 135^\circ}{7.4306} = 0.28548$$
$$\theta = \sin^{-1}(0.28548) = 16.588^\circ.$$

In Figure 12.18

$$\|\vec{v}\|^2 = 5^2 + 3^2 - 2\cdot 3\cdot 5\cdot \cos 45^\circ = 12.7868$$
$$\|\vec{v}\| = \sqrt{12.7868} = 3.5759 = 3.576 \text{ ft/sec}.$$

The vector $\vec{v}$ makes an angle θ with the ball's path. By the Law of Sines

$$\frac{\sin\theta}{3} = \frac{\sin 45^\circ}{3.5759}$$
$$\sin\theta = \frac{3\sin 45^\circ}{3.5759} = 0.59323$$
$$\theta = \sin^{-1}(0.59323) = 36.387^\circ.$$

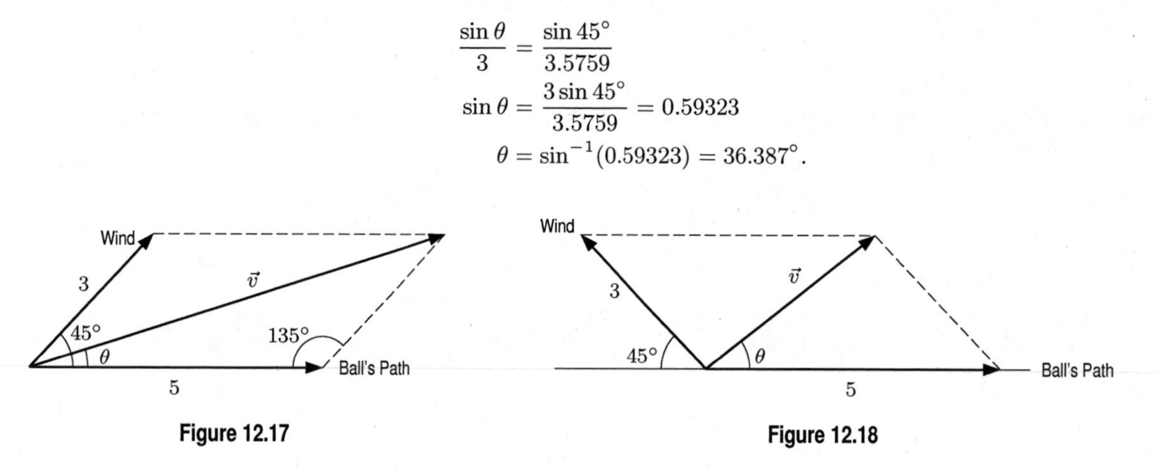

Figure 12.17

Figure 12.18

25. The vector $\vec{v} + \vec{w}$ is equivalent to putting the vectors $\overrightarrow{OA}$ and $\overrightarrow{AB}$ end-to-end as shown in Figure 12.19; the vector $\vec{w} + \vec{v}$ is equivalent to putting the vectors $\overrightarrow{OC}$ and $\overrightarrow{CB}$ end-to-end. Since they form a parallelogram, $\vec{v} + \vec{w}$ and $\vec{w} + \vec{v}$ are both equal to the vector $\overrightarrow{OB}$, we have $\vec{v} + \vec{w} = \vec{w} + \vec{v}$.

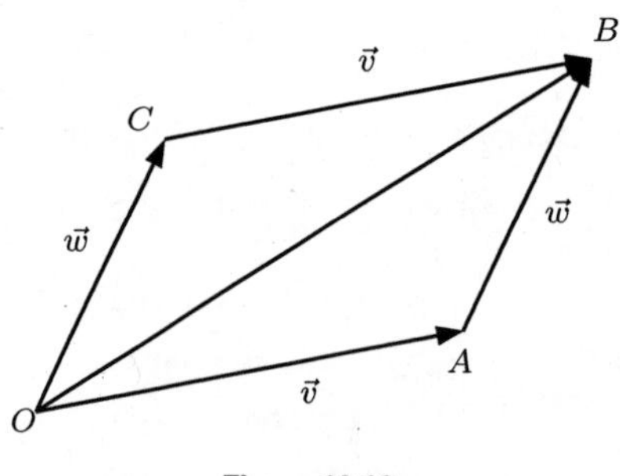

Figure 12.19

26. The vectors $\vec{v}$, $a\vec{v}$ and $b\vec{v}$ are all parallel. Figure 12.20 shows them with $a, b > 1$, so all the vectors are in the same direction. Notice that $a\vec{v}$ is a vector a times as long as $\vec{v}$ and $b\vec{v}$ is b times as long as $\vec{v}$. Therefore $a\vec{v} + b\vec{v}$ is a vector $(a + b)$ times as long as $\vec{v}$, and in the same direction. Thus,

$$a\vec{v} + b\vec{v} = (a + b)\vec{v}.$$

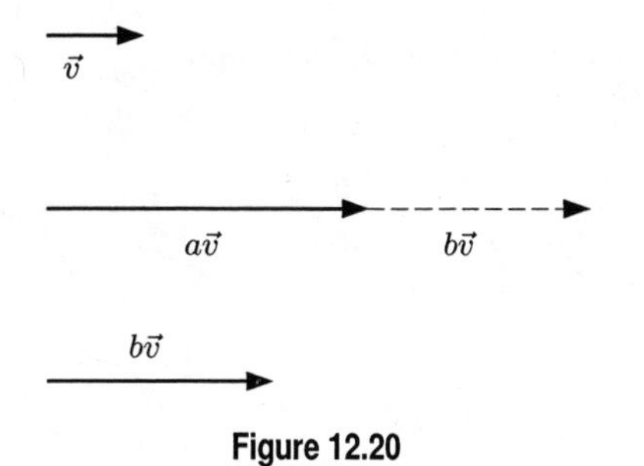

Figure 12.20

27. The effect of scaling the left-hand picture in Figure 12.21 is to stretch each vector by a factor of a (shown with $a > 1$). Since, after scaling up, the three vectors $a\vec{v}$, $a\vec{w}$, and $a(\vec{v} + \vec{w})$ form a similar triangle, we know that $a(\vec{v} + \vec{w})$ is the sum of the other two: that is

$$a(\vec{v} + \vec{w}) = a\vec{v} + a\vec{w}.$$

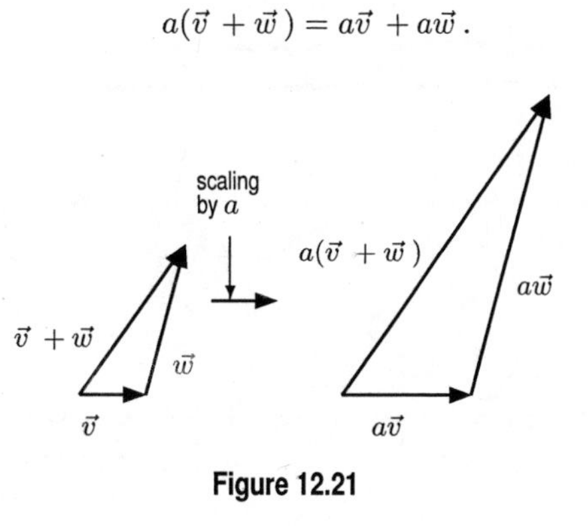

Figure 12.21

28. Since the zero vector has zero length, adding it to $\vec{v}$ has no effect. Thinking of $\vec{v}$ as a displacement, adding a zero displacement leaves us with the same displacement.

29. According to the definition of scalar multiplication, $1 \cdot \vec{v}$ has the same direction and magnitude as $\vec{v}$, so it is the same as $\vec{v}$.

30. By Figure 12.22, the vectors $\vec{v} + (-1)\vec{w}$ and $\vec{v} - \vec{w}$ are equal.

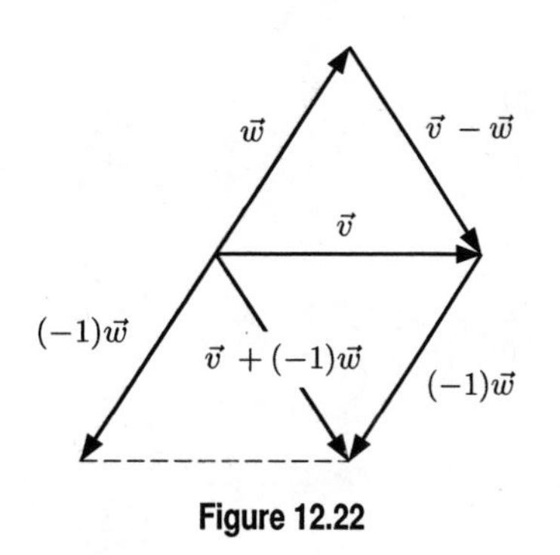

Figure 12.22

31. See Figure 12.23. The vector $\vec{u} + \vec{v}$ is represented by $\overrightarrow{OB}$. The vector $(\vec{u} + \vec{v}) + \vec{w}$ is represented by $\overrightarrow{OB}$ followed by $\overrightarrow{BC}$, which is therefore $\overrightarrow{OC}$. Now $\vec{v} + \vec{w}$ is represented by $\overrightarrow{AC}$. So $\vec{u} + (\vec{v} + \vec{w})$ is $\overrightarrow{OA}$ followed by $\overrightarrow{AC}$, which is $\overrightarrow{OC}$. Since we get the vector $\overrightarrow{OC}$ by both methods, we know

$$(\vec{u} + \vec{v}) + \vec{w} = \vec{u} + (\vec{v} + \vec{w}).$$

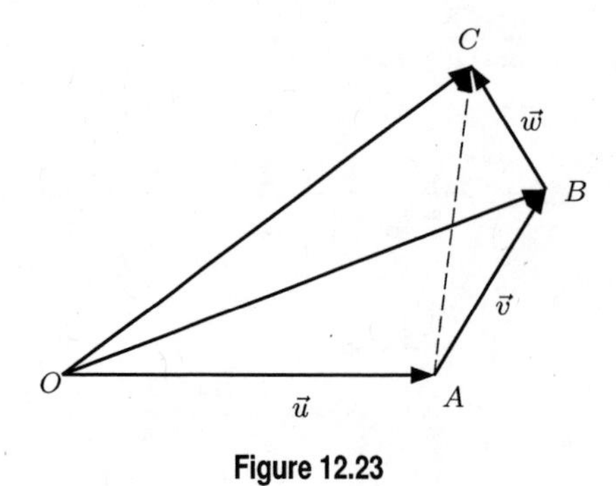

Figure 12.23

Solutions for Section 12.2

Exercises

1. The vector we want is the displacement from Q to P, which is given by

$$\overrightarrow{QP} = (1-4)\vec{i} + (2-6)\vec{j} = -3\vec{i} - 4\vec{j}.$$

2. The vector we want is the displacement from P to Q, which is given by

$$\overrightarrow{PQ} = (4-1)\vec{i} + (6-2)\vec{j} = 3\vec{i} + 4\vec{j}.$$

3. The vector $\vec{w}$ appears to consist of $9.2 - 6.3$ units to the left on the x-axis, and $4.5 - 0.7$ units down on the y-axis. Multiplying by 0.25 to convert to inches gives,

$$\vec{v} \approx -0.725\vec{i} - 0.95\vec{j}.$$

4. $5(2\vec{i} - \vec{j}) + \vec{j} = 10\vec{i} - 5\vec{j} + \vec{j} = 10\vec{i} - 4\vec{j}.$

5. $4\vec{i} + 2\vec{j} - 3\vec{i} + \vec{j} = \vec{i} + 3\vec{j}.$

6. $-(\vec{i} + \vec{j}) + 2(2\vec{i} - 3\vec{j}) = -\vec{i} - \vec{j} + 4\vec{i} - 6\vec{j} = 3\vec{i} - 7\vec{j}.$

7. $(0.9\vec{i} - 1.8\vec{j} - 0.02\vec{k}) - (0.6\vec{i} - 0.05\vec{k}) = 0.3\vec{i} - 1.8\vec{j} + 0.03\vec{k}$

8.

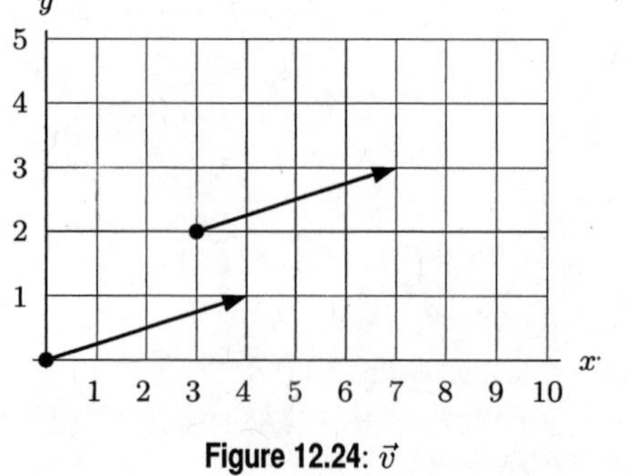

Figure 12.24: $\vec{v}$

9. $\|\vec{v}\| = \sqrt{1^2 + (-1)^2 + 3^2} = \sqrt{11} \approx 3.317$

10. $\|\vec{v}\| = \sqrt{1^2 + (-1)^2 + 2^2} = \sqrt{6} \approx 2.449$

11. $\|\vec{v}\| = \sqrt{7.2^2 + (-1.5)^2 + 2.1^2} = \sqrt{58.5} \approx 7.649$

12. $\|\vec{v}\| = \sqrt{1.2^2 + (-3.6)^2 + 4.1^2} = \sqrt{31.21} \approx 5.587$

Problems

13. The velocity of the ship in still water is $10\vec{j}$ knots and the velocity of the current is $-5\vec{i}$ since the current is east to west. The velocity of the ship is $-5\vec{i} + 10\vec{j}$ knots.

14. **(a)** $\vec{v} = 2\vec{i} + \vec{j}$
(b) Since $\vec{w} = \vec{i} - \vec{j}$, we have $2\vec{w} = 2\vec{i} - 2\vec{j}$.
(c) Since $\vec{v} = 2\vec{i} + \vec{j}$ and $\vec{w} = \vec{i} - \vec{j}$, we have $\vec{v} + \vec{w} = (2\vec{i} + \vec{j}) + (\vec{i} - \vec{j}) = 3\vec{i}$.
(d) Since $\vec{w} = \vec{i} - \vec{j}$ and $\vec{v} = 2\vec{i} + \vec{j}$, we have $\vec{w} - \vec{v} = (\vec{i} - \vec{j}) - (2\vec{i} + \vec{j}) = -\vec{i} - 2\vec{j}$.
(e) $\overrightarrow{PQ} = \vec{i} + \vec{j}$
(f) Since P is at the point $(1, -2)$, the vector we want is $(2-1)\vec{i} + (0-(-2))\vec{j} = \vec{i} + 2\vec{j}$.
(g) The vector must be horizontal, so $\vec{i}$ will work.
(h) The vector must be vertical, so $\vec{j}$ will work.

15. The angle is $45°$, or $\pi/4$.

16. Figure 12.25 shows the vector $\vec{w}$ redrawn with its head at the tail of $\vec{v}$.

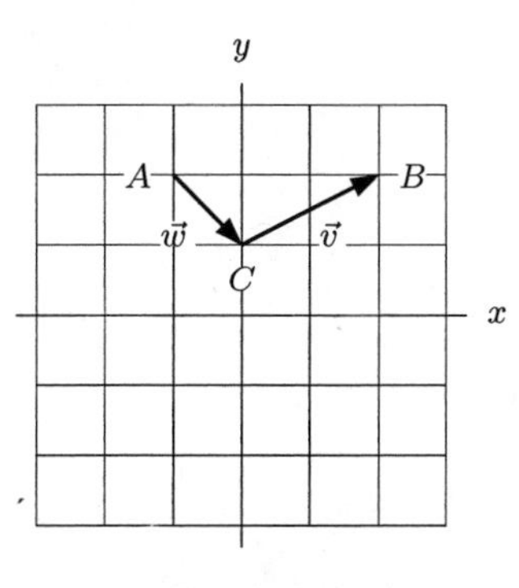

Figure 12.25

By the Pythagorean theorem, the length of the line AC is $\sqrt{2}$ and the length of CB is$\sqrt{2^2 + 1^2} = \sqrt{5}$. Finally, the length of AB is 3, so by the Law of Cosines

$$
\begin{aligned}
(\sqrt{5})^2 + (\sqrt{2})^2 - 2\sqrt{5}\sqrt{2}\cos(\theta) &= 3^2 \\
-2\sqrt{10}\cos(\theta) &= 2 \\
\cos(\theta) &= -\frac{1}{\sqrt{10}} \\
\theta &= \cos^{-1}(-\frac{1}{\sqrt{10}}) \\
\theta &= 108.435°
\end{aligned}
$$

Comparing this with Figure 12.25 this looks about right.

17. Figure 12.26 shows the vector $\vec{w}$ redrawn to show that it is perpendicular to the displacement vector $\overrightarrow{PQ}$, which lies along the dotted line. Thus, the angle is $90°$ or $\pi/2$.

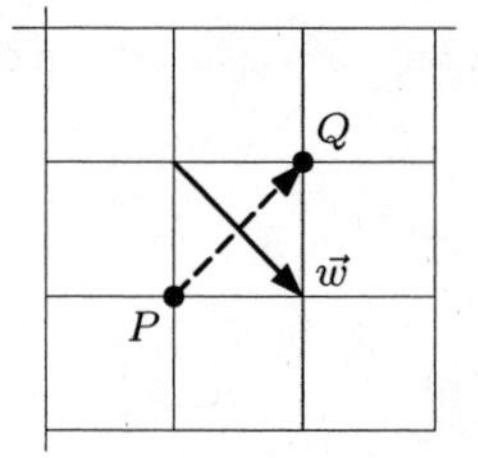

Figure 12.26

18. **(a)** If the car is going east, it is going solely in the positive x-direction, so its velocity vector is $50\vec{i}$. The $\vec{j}$-component is $\vec{0}$.

(b) If the car is going south, it is going solely in the negative y-direction, so its velocity vector is $-50\vec{j}$. The $\vec{i}$-component is $\vec{0}$.

(c) If the car is going southeast, the angle between the x-axis and the velocity vector is $-45°$. Therefore

$$\begin{aligned}\text{Velocity vector} &= 50\cos(-45°)\vec{i} + 50\sin(-45°)\vec{j} \\ &= (25\sqrt{2})\vec{i} - (25\sqrt{2})\vec{j}.\end{aligned}$$

(d) If the car is going northwest, the velocity vector is at a $45°$ angle to the y-axis, which is $135°$ from the x-axis. Therefore:

$$\text{Velocity vector} = 50(\cos 135°)\vec{i} + 50(\sin 135°)\vec{j} = -(25\sqrt{2})\vec{i} + (25\sqrt{2})\vec{j}.$$

19. Let the velocity vector of the airplane be $\vec{V} = x\vec{i} + y\vec{j} + z\vec{k}$ in km/hr. We know that $x = -y$ because the plane is traveling northwest. Also, $\|\vec{V}\| = \sqrt{x^2+y^2+z^2} = 200$ km/hr and $z = 300$ m/min $= 18$ km/hr. We have $\sqrt{x^2+y^2+z^2} = \sqrt{x^2+x^2+18^2} = 200$, so $x = -140.847, y = 140.847, z = 18$. (The value of x is negative and y is positive because the plane is heading northwest.) Thus,

$$\vec{v} = -140.847\vec{i} + 140.847\vec{j} + 18\vec{k}.$$

20. **(a)** The displacement from P to Q is given by

$$\overrightarrow{PQ} = (4\vec{i} + 6\vec{j}) - (\vec{i} + 2\vec{j}) = 3\vec{i} + 4\vec{j}.$$

Since

$$\|\overrightarrow{PQ}\| = \sqrt{3^2+4^2} = 5,$$

a unit vector $\vec{u}$ in the direction of $\overrightarrow{PQ}$ is given by

$$\vec{u} = \frac{1}{5}\overrightarrow{PQ} = \frac{1}{5}(3\vec{i} + 4\vec{j}) = \frac{3}{5}\vec{i} + \frac{4}{5}\vec{j}.$$

(b) A vector of length 10 pointing in the same direction is given by

$$10\vec{u} = 10(\frac{3}{5}\vec{i} + \frac{4}{5}\vec{j}) = 6\vec{i} + 8\vec{j}.$$

21. We need to calculate the length of each vector.

$$\begin{aligned}\|21\vec{i} + 35\vec{j}\| &= \sqrt{21^2+35^2} = \sqrt{1666} \approx 40.8, \\ \|40\vec{i}\| &= \sqrt{40^2} = 40.\end{aligned}$$

So the first car is faster.

22. Since both vehicles reach the crossroad in exactly one hour, at the present the truck is at O in Figure 12.27; the police car is at P and the crossroads is at C. If $\vec{r}$ is the vector representing the displacement of the truck with respect to the police car:

$$\vec{r} = -40\vec{i} - 30\vec{j}.$$

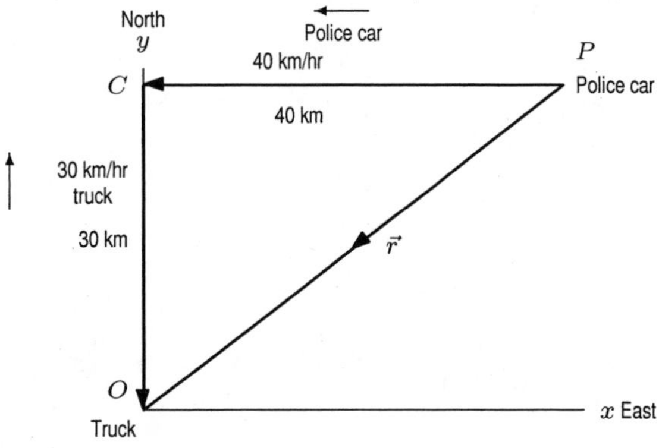

Figure 12.27

23. **(a)** The shot is moving at 50 km/hr, which is the magnitude of the velocity vector.

(b) Thinking of the velocity vector as the hypotenuse of a right triangle making a 30° angle with the ground, the horizontal component is the leg adjacent to the 30° angle, and the vertical component is the leg opposite. This means:

$$\text{Horizontal component} = 50\cos 30^\circ = 43.301$$
$$\text{Vertical component} = 50\sin 30^\circ = 25.$$

24. To determine if two vectors are parallel, we need to see if one vector is a scalar multiple of the other one. Since $\vec{u} = -2\vec{w}$, and $\vec{v} = \frac{1}{4}\vec{q}$ and no other pairs have this property, only $\vec{u}$ and $\vec{w}$, and $\vec{v}$ and $\vec{q}$ are parallel.

25. **(a)** The velocity, $\vec{v}$, is represented by a vector of length 5 in a northeasterly direction. The vector $\vec{i} + \vec{j}$ points northeast, but has length $\sqrt{1^2 + 1^2} = \sqrt{2}$. Thus,

$$\vec{v} = \frac{5}{\sqrt{2}}(\vec{i} + \vec{j}) = 3.536(\vec{i} + \vec{j})$$

(b) The current flows northward, so it is represented by $\vec{c} = 1.2\vec{j}$. The swimmer's velocity relative to the riverbed is

$$\vec{s} = \vec{c} + \vec{v} = 1.2\vec{j} + 3.536(\vec{i} + \vec{j}) = 3.536\vec{i} + \vec{4}.736\vec{j}.$$

26. See Figure 12.28. The speed of the boat is

$$\|\vec{v}\| = \sqrt{2.5^2 + 3.1^2} = 3.982 \text{ mph}.$$

The angle with the coastline is

$$\theta = \tan^{-1}\left(\frac{3.1}{2.5}\right) = 51.116^\circ.$$

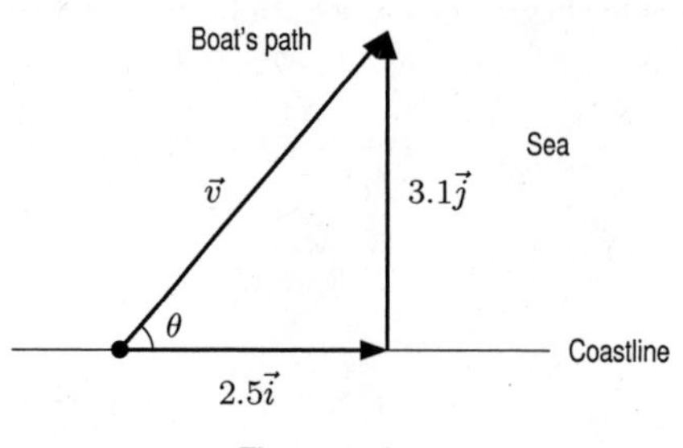

Figure 12.28

27. **(a)** (See Figure 12.29.)

(i) $\vec{m} = 3\vec{j}, \vec{h} = 2\vec{j}$.

(ii) $\vec{m} = 3\vec{j}, \vec{h} = 2\vec{i}$.

(iii) $\vec{m} = 3\vec{j}, \vec{h} = (2\cos 60°)\vec{i} + (2\sin 60°)\vec{j} = \vec{i} + \sqrt{3}\vec{j} = \vec{i} + 1.732\vec{j}$.

(iv) $\vec{m} = -3\vec{j}, \vec{h} = (2\cos 45°)\vec{i} + (2\sin 45°)\vec{j} = \sqrt{2}\vec{i} + \sqrt{2}\vec{j} = 1.414\vec{i} + 1.414\vec{j}$.

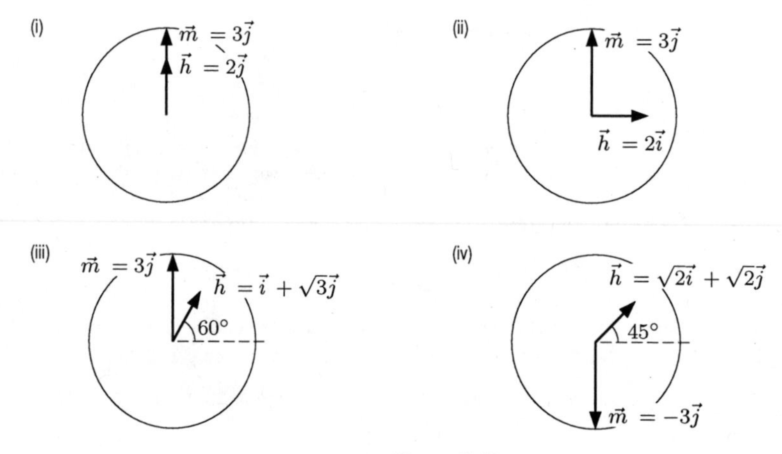

Figure 12.29

(b) From Figure 12.30, we see the displacement vector, $\vec{D}$, is $\vec{D} = 3\vec{j} - 2\vec{i}$.

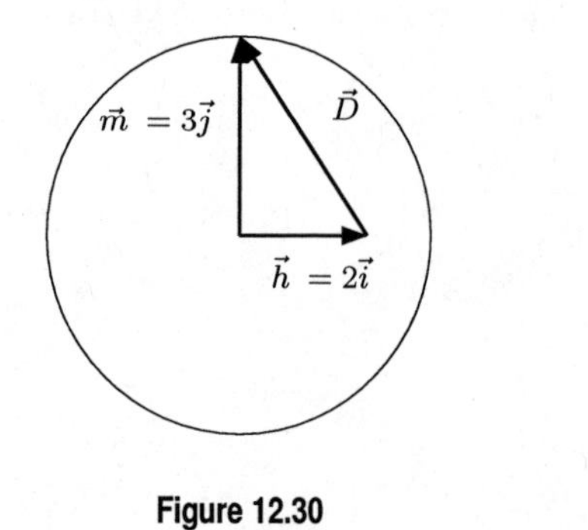

Figure 12.30

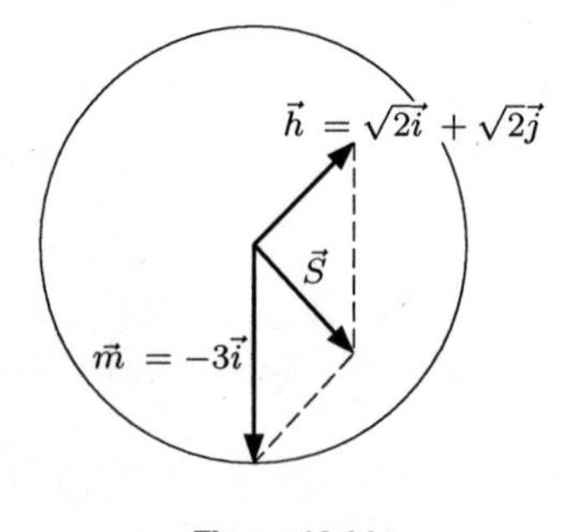

Figure 12.31

(c) From Figure 12.31, we see that the sum, $\vec{S}$, is given by

$$\vec{S} = \sqrt{2}\vec{i} + \sqrt{2}\vec{j} - 3\vec{j} = \sqrt{2}\vec{i} + (\sqrt{2} - 3)\vec{j} = 1.414\vec{i} - 1.586\vec{j}.$$

28. We get displacement by subtracting the coordinates of the origin $(0,0,0)$ from the coordinates of the cat $(1,4,0)$, giving

$$\text{Displacement} = (1-0)\vec{i} + (4-0)\vec{j} + (0-0)\vec{k} = \vec{i} + 4\vec{j}.$$

29. We get displacement by subtracting the coordinates of the bottom of the tree, $(2,4,0)$, from the coordinates of the squirrel, $(2,4,1)$, giving:

$$\text{Displacement} = (2-2)\vec{i} + (4-4)\vec{j} + (1-0)\vec{k} = \vec{k}.$$

30.

$$\begin{aligned}\text{Displacement} &= \text{Cat's coordinates} - \text{Bottom of the tree's coordinates}\\ &= (1-2)\vec{i} + (4-4)\vec{j} + (0-0)\vec{k} = -\vec{i}.\end{aligned}$$

31.

$$\begin{aligned}\text{Displacement} &= \text{Squirrel's coordinates} - \text{Cat's coordinates}\\ &= (2-1)\vec{i} + (4-4)\vec{j} + (1-0)\vec{k} = \vec{i} + \vec{k}.\end{aligned}$$

Solutions for Section 12.3

Exercises

1. $\vec{B} = 2\vec{M} = 2(1,1,2,3,5,8) = (2,2,4,6,10,16)$.

2. $\vec{G} = \vec{N} + \vec{M} = (5,6,7,8,9,10) + (1,1,2,3,5,8) = (6,7,9,11,14,18)$.

3. $\vec{A} = \vec{M} - \vec{N} = (1,1,2,3,5,8) - (5,6,7,8,9,10) = (-4,-5,-5,-5,-4,-2)$.

4. $\vec{\epsilon} = 2\vec{N} - 7\vec{M} = 2(5,6,7,8,9,10) - 7(1,1,2,3,5,8) = (10,12,14,16,18,20) - (7,7,14,21,35,56)$
$= (3,5,0,-5,-17,-36)$.

5. $\vec{K} = \dfrac{\vec{N}}{3} + \dfrac{2\vec{N}}{3} = \dfrac{3\vec{N}}{3} = \vec{N} = (5,6,7,8,9,10)$.

6. $\vec{p} = 1.067\vec{M} + 2.361\vec{N} = 1.067(1,1,2,3,5,8) + 2.361(5,6,7,8,9,10) = (1.067, 1.067, 2.134, 3.201, 5.335, 8.536) +$ $(11.805, 14.166, 16.527, 18.888, 21.249, 23.61) = (12.872, 15.233, 18.661, 22.089, 26.584, 32.146)$.

7. $\vec{Z} = \dfrac{\vec{N}}{3} + \dfrac{\vec{M}}{2} = \dfrac{1}{3}(5,6,7,8,9,10) + \dfrac{1}{2}(1,1,2,3,5,8) = \left(\dfrac{5}{3}, 2, \dfrac{7}{3}, \dfrac{8}{3}, 3, \dfrac{10}{3}\right) + \left(\dfrac{1}{2}, \dfrac{1}{2}, 1, \dfrac{3}{2}, \dfrac{5}{2}, 4\right)$
$= \left(\frac{13}{6}, \frac{5}{2}, \frac{10}{3}, \frac{25}{6}, \frac{11}{2}, \frac{22}{3}\right)$.

8. We have

$$\begin{aligned}\vec{R} = \sqrt{3}\vec{M} + 4\sqrt{3}\vec{N} &= \sqrt{3}(1,1,2,3,5,8) + 4\sqrt{3}(5,6,7,8,9,10)\\ &= (\sqrt{3}, \sqrt{3}, 2\sqrt{3}, 3\sqrt{3}, 5\sqrt{3}, 8\sqrt{3}) + (20\sqrt{3}, 24\sqrt{3}, 28\sqrt{3}, 32\sqrt{3}, 36\sqrt{3}, 40\sqrt{3})\\ &= (21\sqrt{3}, 25\sqrt{3}, 30\sqrt{3}, 35\sqrt{3}, 41\sqrt{3}, 48\sqrt{3}) = \sqrt{3}(21,25,30,35,41,48).\end{aligned}$$

9. Since the components of $\vec{Q}$ represent millions of people, an increase of 120,000 people will increase each component by 0.12. Therefore,

$$\begin{aligned}\vec{S} &= \vec{Q} + (0.12, 0.12, 0.12, 0.12, 0.12, 0.12)\\ &= (3.51, 1.32, 6.40, 1.31, 1.08, 0.62) + (0.12, 0.12, 0.12, 0.12, 0.12, 0.12)\\ &= (3.63, 1.44, 6.52, 1.43, 1.20, 0.74).\end{aligned}$$

10. Since each population increases by 2%, we multiply $\vec{Q}$ by 1.02 to find $\vec{R}$:

$$\vec{R} = 1.02\vec{Q} = 1.02(3.51, 1.32, 6.40, 1.31, 1.08, 0.62) = (3.580, 1.346, 6.528, 1.336, 1.102, 0.632).$$

11. Since the components of $\vec{Q}$ represent millions of people, a decrease of 43,000 people decreases each component by 0.043. Therefore,

$$\begin{aligned}\vec{S} &= \vec{Q} - (0.043, 0.043, 0.043, 0.043, 0.043, 0.043)\\ &= (3.51, 1.32, 6.40, 1.31, 1.08, 0.62) - (0.043, 0.043, 0.043, 0.043, 0.043, 0.043)\\ &= (3.467, 1.277, 6.357, 1.267, 1.037, 0.577).\end{aligned}$$

12. Since each population decreases by 22%, we multiply $\vec{Q}$ by $1 - 0.22 = 0.78$ to find $\vec{T}$:

$$\vec{T} = 0.78\vec{Q} = 0.78(3.51, 1.32, 6.40, 1.31, 1.08, 0.62) = (2.738, 1.030, 4.992, 1.022, 0.842, 0.484).$$

Problems

13. The total scores are out of 300 and are given by the total score vector $\vec{v} + 2\vec{w}$:

$$\begin{aligned}\vec{v} + 2\vec{w} &= (73, 80, 91, 65, 84) + 2(82, 79, 88, 70, 92)\\ &= (73, 80, 91, 65, 84) + (164, 158, 176, 140, 184)\\ &= (237, 238, 267, 205, 268).\end{aligned}$$

To get the scores as a percentage, we divide by 3, giving

$$\frac{1}{3}(237, 238, 267, 205, 268) \approx (79.000, 79.333, 89.000, 68.333, 89.333).$$

14. Suppose $\vec{u}$ represents the velocity of the plane relative to the air and $\vec{w}$ represents the velocity of the wind. We can add these two vectors by adding their components. Suppose north is in the y-direction and east is the x-direction. The vector representing the airplane's velocity makes an angle of 45° with north; the components of $\vec{u}$ are

$$\vec{u} = 700\sin 45°\vec{i} + 700\cos 45°\vec{j} \approx 494.975\vec{i} + 494.975\vec{j}.$$

Since the wind is blowing from the west, $\vec{w} = 60\vec{i}$. By adding these we get a resultant vector $\vec{v} = 554.975\vec{i} + 494.975\vec{j}$. The direction relative to the north is the angle θ shown in Figure 12.32 given by

$$\begin{aligned}\theta = \tan^{-1}\frac{x}{y} &= \tan^{-1}\frac{554.975}{494.975}\\ &\approx 48.271°.\end{aligned}$$

The magnitude of the velocity is

$$\|\vec{v}\| = \sqrt{494.975^2 + 554.975^2} = \sqrt{552{,}996.970} = 743.638 \text{ km/hr}.$$

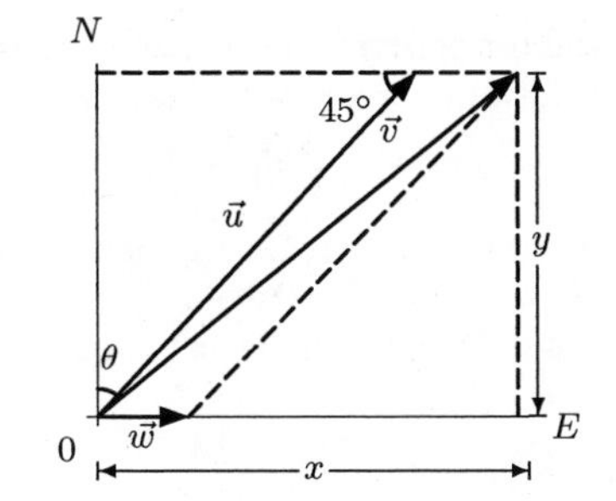

Figure 12.32: Note that θ is the angle between north and the vector $\vec{v}$

15. Let the x-axis point east and the y-axis point north. Since the wind is blowing from the northeast at a speed of 50 km/hr, the velocity of the wind is

$$\vec{w} = -50\cos 45°\vec{i} - 50\sin 45°\vec{j} \approx -35.355\vec{i} - 35.355\vec{j}.$$

Let $\vec{a}$ be the velocity of the airplane, relative to the air, and let ϕ be the angle from the x-axis to $\vec{a}$; since $\|\vec{a}\| = 600$ km/hr, we have $\vec{a} = 600\cos\phi\vec{i} + 600\sin\phi\vec{j}$. (See Figure 12.33.)

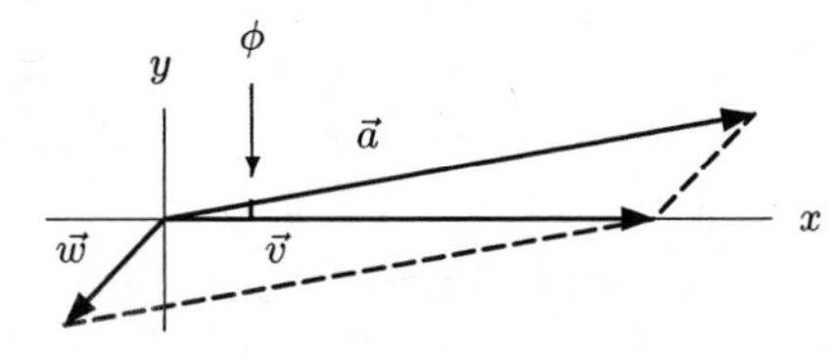

Figure 12.33

Now the resultant velocity, $\vec{v}$, is given by

$$\begin{aligned}\vec{v} &= \vec{a} + \vec{w} = (600\cos\phi\vec{i} + 600\sin\phi\vec{j}) + (-35.355\vec{i} - 35.355\vec{j})\\ &= (600\cos\phi - 35.355)\vec{i} + (600\sin\phi - 35.355)\vec{j}.\end{aligned}$$

Since the airplane is to fly due east, i.e., in the x direction, then the y-component of the velocity must be 0, so we must have

$$\begin{aligned}600\sin\phi - 35.355 &= 0\\ \sin\phi &= \frac{35.355}{600}.\end{aligned}$$

Thus $\phi = \arcsin(35.355/600) \approx 3.378°$.

16. (a)

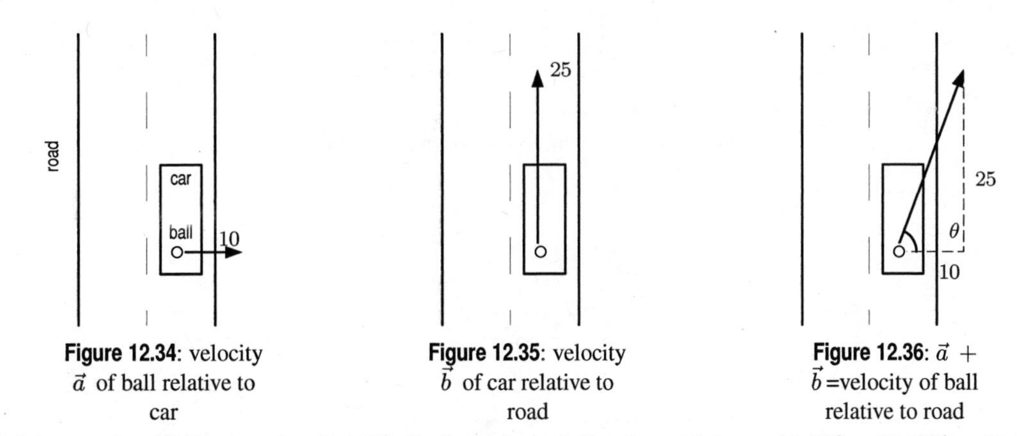

Figure 12.34: velocity $\vec{a}$ of ball relative to car

Figure 12.35: velocity $\vec{b}$ of car relative to road

Figure 12.36: $\vec{a} + \vec{b}$ =velocity of ball relative to road

(b) Solving $\tan\theta = 25/10$ gives $\theta = 68.199°$. So the angle the ball makes with the road is $90° - 68.199° = 21.801°$.

17. **(a)** See the sketch in Figure 12.37, where $\vec{v}$ represents the first part of the man's walk, and $\vec{w}$ represents the second part. Since the man first walks 5 miles, we know $\|\vec{v}\| = 5$. Since he walks 30° north of east, resolving gives

$$\vec{v} = 5\cos 30^\circ \vec{i} + 5\sin 30^\circ \vec{j} = 4.330\vec{i} + 2.500\vec{j}.$$

For the second leg of his journey, the man walks a distance x miles due east, so $\vec{w} = x\vec{i}$.

(b) The vector from finish to start is $-(\vec{v} + \vec{w}) = -(4.330 + x)\vec{i} - 2.500\vec{j}$. This vector is at an angle of 10° south of west. So, using the magnitudes of the sides in the triangle in Figure 12.38:

$$\begin{aligned}\frac{2.500}{4.330 + x} &= \tan(10^\circ) = 0.176\\ 2.500 &= 0.176(4.330 + x)\\ x &= \frac{2.500 - 0.176 \cdot 4.330}{0.176} = 9.848.\end{aligned}$$

This means that $x = 9.848$.

Figure 12.37

Figure 12.38

(c) The distance from the starting point is $\| - (4.330 + 9.848)\vec{i} - (2.500)\vec{j}\| = \sqrt{14.178^2 + 2.500^2} = 14.397$ miles.

18. **(a)** The only components that have the same value (in the same position in the vectors) have a value of 40 in the fourth position, so outfits $\vec{a}$ and $\vec{c}$ must have the same belt, which costs \$40.

(b) (i) Their wishes sum to $3\vec{a} + 2\vec{c} + \vec{d}$.

(ii) Writing out the sum, we get:

$$\begin{aligned}3\vec{a} + 2\vec{c} + \vec{d} &= 3(75, 30, 120, 40, 200) + 2(145, 50, 100, 40, 300) + (60, 45, 200, 35, 150)\\ &= (225, 90, 360, 120, 600) + (290, 100, 200, 80, 600) + (60, 45, 200, 35, 150)\\ &= (575, 235, 760, 235, 1350)\end{aligned}$$

(c) (i) As expected, we know that the second component in each vector must represent the price of shirts, because if David buys 5 outfits and spends only \$145 on shirts, he must spend on average less than \$30 on each shirt. Only the first and second components have items selling for less than \$30, and only the second allows us to get to an exact sum of \$145, since in the first component, $5 \cdot 25 = 125 \neq 145$ and $4 \cdot 25 + 1 \cdot 60 = 160 > 145$, which are the the two lowest possible sums of 5 items. We solve the equations $30w + 20x + 50y + 45z = 145$ and $w + x + y + z = 5$, $0 \leq w, x, y, z \leq 5$. It is clear that we must have either 1 or 3 of outfit $\vec{d}$ in order to have a 5 in the units' digit of the sum. Since $3 \cdot 45 = 135$, that is not possible, as adding two more of any outfit will exceed \$145. So we know that $z = 1$. This give us

$$30w + 20x + 50y = 100 \quad \text{and} \quad w + x + y = 4.$$

We see that if $y = 1$, the only way for the sum $30w + 20x + 50y$ to equal 100 is if $w = x = 1$, but then $w + x + y = 3$, and we still have an outfit unaccounted for. Thus $y = 0$ and $w = x = 2$. Thus, altogether we have 2 outfits $\vec{a}$ and 2 outfits $\vec{b}$ and 1 outfit $\vec{d}$:

$$2\vec{a} + 2\vec{b} + \vec{d}.$$

(ii) We have

$$\begin{aligned}\vec{e} &= 2\vec{a} + 2\vec{b} + \vec{d}\\ &= 2(75, 30, 120, 40, 200) + 2(25, 20, 75, 30, 90) + (60, 45, 200, 35, 150)\\ &= (150 + 60 + 240 + 80 + 400) + (50 + 40 + 150 + 60 + 180) + (60 + 45 + 200 + 35 + 150)\\ &= (260 + 145 + 590 + 175 + 730).\end{aligned}$$

19. **(a)** We have

$$\begin{aligned}\vec{F}_{\text{net}} &= \vec{F}_1 + \vec{F}_2 + \vec{F}_3 \\ &= (3,6) + (-2,5) + (7,-4) \\ &= (3-2+7, 6+5-4) = (8,7).\end{aligned}$$

(b) We want $\vec{F}_{\text{net}} + \vec{F}_4 = \vec{0}$, and so $\vec{F}_4 = \vec{0} - \vec{F}_{\text{net}} = (0,0) - (8,7) = (-8,-7)$.

20. **(a)** We have $||\vec{r}\,|| = \sqrt{3^2+5^2} = \sqrt{34}$.

(b) We have $||\vec{F}\,|| = qQ/||r||^2 = 20 \cdot 30/34 = 3/17 = 17.6$.

(c) The angle that $\vec{r}$ makes with the x-axis has $\tan\theta = r_y/r_x = 5/3$, so $\theta = \tan^{-1}(5/3) = 59.0^\circ$. Thus, $F_x = F\cos\theta = 17.6\cos(59.0^\circ) = 9.065$ and $F_y = F\sin\theta = 17.6\sin(59.0^\circ) = 15.086$, so $\vec{F} = (9.065, 15.086)$.

21. In an actual video game, our rectangle would be replaced with a more sophisticated graphic (perhaps an airplane or an animated figure). But the principles involved in rotation about the origin are the same, and it will be easier to think about them using rectangles instead of fancy graphics.

We can represent the four corners of the rectangle (before rotation) using the position vectors $\vec{p}_a$, $\vec{p}_b$, $\vec{p}_c$, and $\vec{p}_d$. For instance, the components of $\vec{p}_a$ are $\vec{p}_a = 2\vec{i} + \vec{j}$.

After the rectangle has been rotated, its four corners are given by the position vectors $\vec{q}_a$, $\vec{q}_b$, $\vec{q}_c$, and $\vec{q}_d$. Notice that the lengths of these vectors have not changed; in other words,

$$||\vec{p}_a|| = ||\vec{q}_a||, \qquad ||\vec{p}_b|| = ||\vec{q}_b||, \qquad ||\vec{p}_c|| = ||\vec{q}_c||, \qquad \text{and} \qquad ||\vec{p}_d|| = ||\vec{q}_d||.$$

This is because in a rotation the only thing that changes is orientation, not length.

When the rectangle is rotated through a 35° angle, the angle made by corner a increases by 35°. So do the angles made by the other three corners. Letting θ be the angle made by corner a, we have

$$\begin{aligned}\tan\theta &= \frac{1}{2} \\ \theta &= \arctan 0.5 = 26.565^\circ.\end{aligned}$$

This is the direction of the position vector $\vec{p}_a$. After rotation, the angle θ is given by

$$\theta = 26.565^\circ + 35^\circ = 61.565^\circ.$$

This is the direction of the new position vector $\vec{q}_a$. The length of $\vec{q}_a$ is the same as the length of $\vec{p}_a$ and is given by

$$||\vec{q}_a||^2 = ||\vec{p}_a||^2 = 2^2 + 1^2 = 5,$$

and so $||\vec{q}_a|| = \sqrt{5}$. Thus, the components of $\vec{q}_a$ are given by

$$\begin{aligned}\vec{q}_a &= (\sqrt{5}\cos 61.565^\circ)\vec{i} + (\sqrt{5}\sin 61.565^\circ)\vec{j} \\ &= 1.065\vec{i} + 1.966\vec{j}.\end{aligned}$$

This process can be repeated for the other three corners. You can see for yourself that the angles made with the origin by the corners a, b, and c, respectively, are 14.036°, 26.565°, and 45°. After rotation, these angles are 49.036°, 61.565°, and 80°. Similarly, the lengths of the position vectors for these three points (both before and after rotation) are $\sqrt{17}$, $\sqrt{20}$, and $\sqrt{8}$. Thus, the final positions of these three points are

$$\begin{aligned}\vec{q}_b &= 2.703\vec{i} + 3.113\vec{j}, \\ \vec{q}_c &= 2.129\vec{i} + 3.933\vec{j}, \\ \vec{q}_d &= 0.491\vec{i} + 2.785\vec{j}.\end{aligned}$$

22. **(a)** See Figure 12.39.

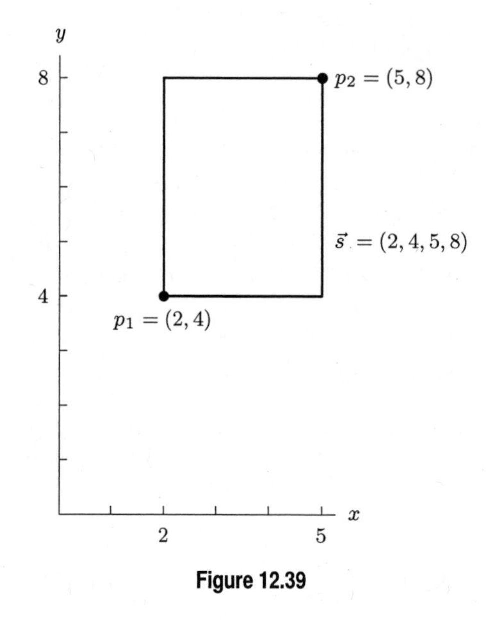

Figure 12.39

(b) Let $\vec{t} = (x_1, y_1, x_2, y_2)$, so that one corner of the rectangle is $p_1 = (x_1, y_1)$ and the opposite corner is $p_2 = (x_2, y_2)$.

(i) We have

$$\vec{t} + \vec{u} = (x_1, y_1, x_2, y_2) + (1, 0, 1, 0) = (x_1 + 1, y_1, x_2 + 1, y_2).$$

One corner of the new rectangle is $(x_1 + 1, y_1)$, and the opposite corner is $(x_2 + 1, y_2)$. These corners are each one unit farther to the right than the old corners, and so the whole rectangle is one unit farther to the right.

(ii) We have

$$\vec{t} + \vec{v} = (x_1, y_1, x_2, y_2) + (0, 1, 0, 1) = (x_1, y_1 + 1, x_2, y_2 + 1).$$

One corner of the new rectangle is $(x_1, y_1 + 1)$, and the opposite corner is $(x_2, y_2 + 1)$. These corners are each one unit farther up than the old corners, and so the whole rectangle is one unit farther up.

(iii) By a similar argument to part (i), the new rectangle is one unit to the left of the old one.

(iv) The new rectangle is 2 units to the right of the old rectangle and 3 units up.

(v) The new rectangle is 1 unit to the right and 1 unit up.

(vi) The new rectangle is k units to the right and k units up.

Solutions for Section 12.4

Exercises

1. $\vec{z} \cdot \vec{a} = (\vec{i} - 3\vec{j} - \vec{k}) \cdot (2\vec{j} + \vec{k}) = 1 \cdot 0 + (-3)2 + (-1)1 = 0 - 6 - 1 = -7.$

2. $\vec{a} \cdot \vec{z} = (2\vec{j} + \vec{k}) \cdot (\vec{i} - 3\vec{j} - \vec{k}) = 0 \cdot 1 + 2(-3) + 1(-1) = 0 - 6 - 1 = -7.$

3. $\vec{c} \cdot \vec{y} = (\vec{i} + 6\vec{j}) \cdot (4\vec{i} - 7\vec{j}) = 1 \cdot 4 + 6(-7) = 4 - 42 = -38.$

4. $\vec{a} \cdot \vec{y} = (2\vec{j} + \vec{k}) \cdot (4\vec{i} - 7\vec{j}) = -14.$

5. $\vec{a} \cdot \vec{b} = (2\vec{j} + \vec{k}) \cdot (-3\vec{i} + 5\vec{j} + 4\vec{k}) = 0(-3) + 2 \cdot 5 + 1 \cdot 4 = 0 + 10 + 4 = 14.$

6. $\vec{b} \cdot \vec{z} = (-3\vec{i} + 5\vec{j} + 4\vec{k}) \cdot (\vec{i} - 3\vec{j} - \vec{k}) = -22.$

7. $\vec{c} \cdot \vec{a} + \vec{a} \cdot \vec{y} = (\vec{i} + 6\vec{j}) \cdot (2\vec{j} + \vec{k}) + (2\vec{j} + \vec{k}) \cdot (4\vec{i} - 7\vec{j}) = 12 - 14 = -2.$

8. $\vec{c} + \vec{y} = (\vec{i} + 6\vec{j}) + (4\vec{i} - 7\vec{j}) = 5\vec{i} - \vec{j}$, so

$$\vec{a} \cdot (\vec{c} + \vec{y}) = (2\vec{j} + \vec{k}) \cdot (5\vec{i} - \vec{j}) = -2.$$

9. Since $\vec{a} \cdot \vec{b}$ is a scalar and $\vec{a}$ is a vector, the expression is a vector parallel to $\vec{a}$. We have

$$\vec{a} \cdot \vec{b} = (2\vec{j} + \vec{k}) \cdot (-3\vec{i} + 5\vec{j} + 4\vec{k}) = 0(-3) + 2(5) + 1(4) = 14.$$

Thus,

$$(\vec{a} \cdot \vec{b}) \cdot \vec{a} = 14\vec{a} = 14(2\vec{j} + \vec{k}) = 28\vec{j} + 14\vec{k}.$$

10. Since $\vec{c} \cdot \vec{c}$ is a scalar and $(\vec{c} \cdot \vec{c})\vec{a}$ is a vector, the expression is another scalar. We could calculate $\vec{c} \cdot \vec{c}$, then $(\vec{c} \cdot \vec{c})\vec{a}$, and then take the dot product $((\vec{c} \cdot \vec{c})\vec{a}) \cdot \vec{a}$. Alternatively, we can use the fact that

$$((\vec{c} \cdot \vec{c})\vec{a}) \cdot \vec{a} = (\vec{c} \cdot \vec{c})(\vec{a} \cdot \vec{a}).$$

Since

$$\vec{c} \cdot \vec{c} = (\vec{i} + 6\vec{j}) \cdot (\vec{i} + 6\vec{j}) = 1^2 + 6^2 = 37$$
$$\vec{a} \cdot \vec{a} = (2\vec{j} + \vec{k}) \cdot (2\vec{j} + \vec{k}) = 2^2 + 1^2 = 5,$$

we have,

$$(\vec{c} \cdot \vec{c})(\vec{a} \cdot \vec{a}) = 37(5) = 185.$$

11. Since $\vec{a} \cdot \vec{y}$ and $\vec{c} \cdot \vec{z}$ are both scalars, the expression is the product of two numbers and therefore a number. We have

$$\vec{a} \cdot \vec{y} = (2\vec{j} + \vec{k}) \cdot (4\vec{i} - 7\vec{j}) = 0(4) + 2(-7) + 1(0) = -14$$
$$\vec{c} \cdot \vec{z} = (\vec{i} + 6\vec{j}) \cdot (\vec{i} - 3\vec{j} - \vec{k}) = 1(1) + 6(-3) + 0(-1) = -17.$$

Thus,

$$(\vec{a} \cdot \vec{y})(\vec{c} \cdot \vec{z}) = 238.$$

12. Since $\vec{z} \cdot \vec{c}$ and $\vec{y} \cdot \vec{a}$ are both scalars, the expression is the product of two numbers and therefore a number. We have

$$\vec{z} \cdot \vec{c} = (\vec{i} - 3\vec{j} - \vec{k}) \cdot (\vec{i} + 6\vec{j}) = 1(1) + (-3)6 + (-1)0 = -17.$$
$$\vec{y} \cdot \vec{a} = (4\vec{i} - 7\vec{j}) \cdot (2\vec{j} + \vec{k}) = (4)0 + (-7)2 + (0)1 = -14$$

Thus,

$$(\vec{z} \cdot \vec{c})(\vec{y} \cdot \vec{a}) = 238.$$

Problems

13. We use the dot product to find this angle.
We have $\vec{w} = \vec{i} - \vec{j}$ and $\vec{v} = \vec{i} + 2\vec{j}$ so

$$\vec{w} \cdot \vec{v} = (\vec{i} - \vec{j}) \cdot (\vec{i} + 2\vec{j}) = -1,$$

therefore

$$\vec{w} \cdot \vec{v} = ||\vec{w}|| \cdot ||\vec{v}|| \cos\theta.$$

Since $\|\vec{w}\| = \sqrt{5}$ and $\|\vec{v}\| = \sqrt{2}$, and $\vec{w} \cdot \vec{v} = -1$, we have

$$-1 = \sqrt{2}\sqrt{5}\cos\theta$$
$$\cos\theta = -\frac{1}{\sqrt{10}}$$
$$\theta = \arccos -\frac{1}{\sqrt{10}} = 108.435^\circ.$$

14. **(a)** The force has magnitude 2 downward and the displacement has magnitude 3 downward. Therefore $\theta = 0^\circ$, so

$$\text{Work done} = 2 \cdot 3\cos 0^\circ = 2 \cdot 3 \cdot 1 = 6 \text{ ft-lb.}$$

(b) The force is the same, the displacement has magnitude 5 feet and the angle between the force and the displacement is 180°. Thus,

$$\text{Work done} = 2 \cdot 5\cos 180^\circ = 2 \cdot 5(-1) = -10 \text{ ft-lb.}$$

15. Intuitively, we know that we have to do more work to push the refrigerator up this ramp than the one in the text because it is steeper. In this case

$$\text{Work} = -\|\vec{F}\|\|\vec{d}\|\cos 120^\circ = -350 \cdot 12\cos 120^\circ = 2100.0 \text{ ft-lbs,}$$

so nearly three times as much work is needed.

The reason that more work is needed is that the refrigerator ends up farther from the ground.

16. Since $3\vec{i} + \sqrt{3}\vec{j} = \sqrt{3}(\sqrt{3}\vec{i} + \vec{j})$, we know that $3\vec{i} + \sqrt{3}\vec{j}$ and $\sqrt{3}\vec{i} + \vec{j}$ are scalar multiples of one another, and therefore parallel.

Since $(\sqrt{3}\vec{i} + \vec{j}) \cdot (\vec{i} - \sqrt{3}\vec{j}) = \sqrt{3} - \sqrt{3} = 0$, we know that $\sqrt{3}\vec{i} + \vec{j}$ and $\vec{i} - \sqrt{3}\vec{j}$ are perpendicular.

Since $3\vec{i} + \sqrt{3}\vec{j}$ and $\sqrt{3}\vec{i} + \vec{j}$ are parallel, $3\vec{i} + \sqrt{3}\vec{j}$ and $\vec{i} - \sqrt{3}\vec{j}$ are perpendicular, too.

17. Using the dot product, the angle is given by

$$\cos\theta = \frac{(\vec{i} + \vec{j} + \vec{k}) \cdot (\vec{i} - \vec{j} - \vec{k})}{\|\vec{i} + \vec{j} + \vec{k}\|\|\vec{i} - \vec{j} - \vec{k}\|} = \frac{1 \cdot 1 + 1(-1) + 1(-1)}{\sqrt{1^1 + 1^2 + 1^2}\sqrt{1^2 + (-1)^2 + (-1)^2}} = -\frac{1}{3}.$$

So, $\theta = \arccos(-\frac{1}{3}) \approx 1.911$ radians, or $\approx 109.471^\circ$.

18. In general, $\vec{u}$ and $\vec{v}$ are perpendicular when $\vec{u} \cdot \vec{v} = 0$.

In this case, $\vec{u} \cdot \vec{v} = (t\vec{i} - \vec{j} + \vec{k}) \cdot (t\vec{i} + t\vec{j} - 2\vec{k}) = t^2 - t - 2$.

This is zero when $t^2 - t - 2 = 0$, i.e. when $(t-2)(t+1) = 0$, so $t = 2$ or -1.

In general, $\vec{u}$ and $\vec{v}$ are parallel if and only if $\vec{v} = \alpha\vec{u}$ for some real number α.

Thus we need $\alpha t\vec{i} - \alpha\vec{j} + \alpha\vec{k} = t\vec{i} + t\vec{j} - 2\vec{k}$, so we need $\alpha t = t$, and $-\alpha = t$, and $\alpha = -2$. But if $\alpha = -2$, we can't have $\alpha t = t$ unless $t = 0$, and if $t = 0$, we can't have $-\alpha = t$, so there are no values of t for which $\vec{u}$ and $\vec{v}$ are parallel.

19. The maximum value of $\|\vec{a} + \vec{b}\|$ is $7 + 4 = 11$, which occurs when $\vec{a}$ and $\vec{b}$ are in the same direction. The minimum value of $\|\vec{a} + \vec{b}\|$ is $7 - 4 = 3$, which occurs when they are in opposite directions.

The maximum value of $\|\vec{a} - \vec{b}\|$ is $7+4 = 11$, which occurs when $\vec{a}$ and $\vec{b}$ are in opposite directions. The minimum value of $\|\vec{a} - \vec{b}\|$ is $7 - 4 = 3$, which occurs when $\vec{a}$ and $\vec{b}$ are in the same direction.

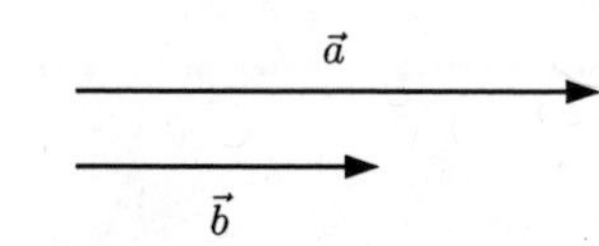

Figure 12.40: Vectors $\vec{a}$ and $\vec{b}$ in same direction: Max $\|\vec{a} + \vec{b}\|$, min $\|\vec{a} - \vec{b}\|$

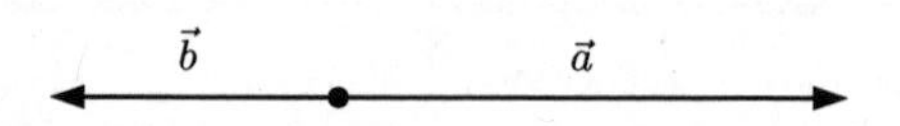

Figure 12.41: Vectors $\vec{a}$ and $\vec{b}$ in opposite direction: max $\|\vec{a} - \vec{b}\|$, min $\|\vec{a} + \vec{b}\|$

20. We need to find the speed of the wind in the direction of the track. Looking at Figure 12.42, we see that we want the component of $\vec{w}$ in the direction of $\vec{v}$. We calculate

$$\|\vec{w}_{\text{parallel to } \vec{v}}\| = \|\vec{w}\| \cos\theta = \frac{\vec{w} \cdot \vec{v}}{\|\vec{v}\|} = \frac{(5\vec{i} + \vec{j}) \cdot (2\vec{i} + 6\vec{j})}{\|2\vec{i} + 6\vec{j}\|}$$
$$= \frac{16}{\sqrt{40}} \approx 2.53 < 5.$$

Therefore, the race results will not be disqualified.

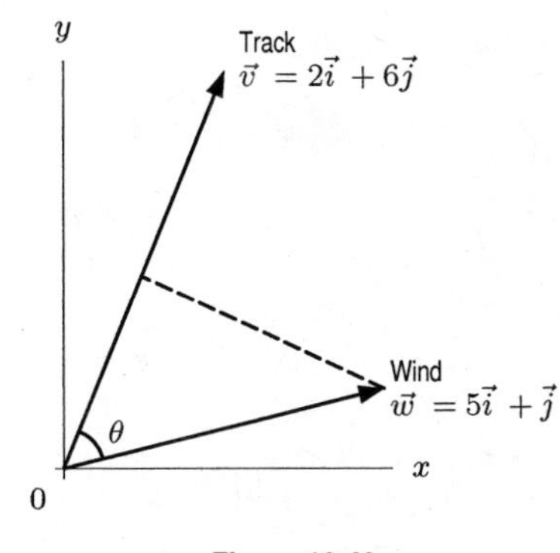

Figure 12.42

21. It is clear from the Figure 12.43 that only angle $\angle CAB$ could possibly be a right angle. Subtraction of x, y values for the points gives $\overrightarrow{AB} = 3\vec{i} - \vec{j}$ and $\overrightarrow{AC} = 1\vec{i} + 2\vec{j}$. Taking the dot product yields $\overrightarrow{AB} \cdot \overrightarrow{AC} = (3)(1) + (-1)(2) = 1$. Since this is non-zero, the angle can not be a right angle.

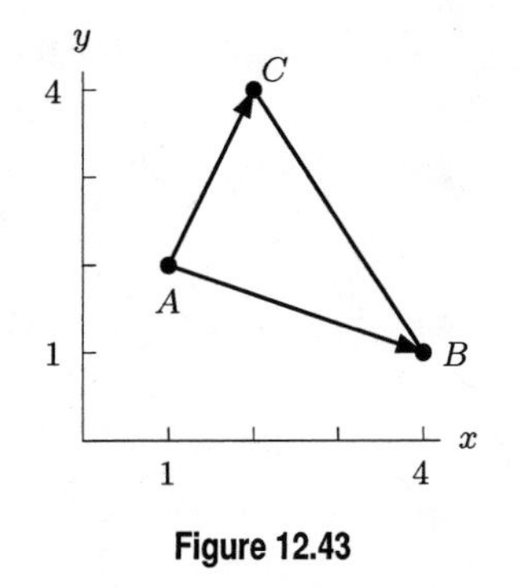

Figure 12.43

22. We want to show that $(\vec{b} \cdot \vec{c})\vec{a} - (\vec{a} \cdot \vec{c})\vec{b}$ and $\vec{c}$ are perpendicular. We do this by taking their dot product:

$$((\vec{b} \cdot \vec{c})\vec{a} - (\vec{a} \cdot \vec{c})\vec{b}) \cdot \vec{c} = (\vec{b} \cdot \vec{c})(\vec{a} \cdot \vec{c}) - (\vec{a} \cdot \vec{c})(\vec{b} \cdot \vec{c}) = 0.$$

Since the dot product is 0, the vectors $(\vec{b} \cdot \vec{c})\vec{a} - (\vec{a} \cdot \vec{c})\vec{b}$ and $\vec{c}$ are perpendicular.

23. We have:

$$\vec{u} \cdot (\vec{v} + \vec{w}) = (u_1, u_2) \cdot (v_1 + w_1, v_2 + w_2)$$
$$= u_1(v_1 + w_1) + u_2(v_2 + w_2) = u_1v_1 + u_1w_1 + u_2v_2 + u_2w_2$$
$$= (u_1v_1 + u_2v_2) + (u_1w_1 + u_2w_2) = \vec{u} \cdot \vec{v} + \vec{u} \cdot \vec{w}.$$

24. By the distributive law,

$$\begin{aligned}\vec{u}\cdot\vec{v} &= (u_1\vec{i}+u_2\vec{j})\cdot(v_1\vec{i}+v_2\vec{j})\\ &= u_1v_1\vec{i}\cdot\vec{i}+u_1v_2\vec{i}\cdot\vec{j}+u_2v_1\vec{j}\cdot\vec{i}+u_2v_2\vec{j}\cdot\vec{j}.\end{aligned}$$

Since $\|\vec{i}\| = \|\vec{j}\| = 1$ and the angle between $\vec{i}$ and $\vec{j}$ is $90°$,

$$\vec{i}\cdot\vec{i}=1,\quad \vec{j}\cdot\vec{j}=1,\quad \vec{i}\cdot\vec{j}=\vec{j}\cdot\vec{i}=0.$$

Thus,

$$\vec{u}\cdot\vec{v}=u_1v_1+u_2v_2.$$

25. **(a)** We have the price vector $\vec{a}=(3,2,4)$. Let the consumption vector $\vec{c}=(c_b,c_e,c_m)$, then $3c_b+2c_e+4c_m=40$ or $\vec{a}\cdot\vec{c}=40$.

(b) Note $\vec{a}\cdot\vec{c}$ is the cost of consuming $\vec{c}$ groceries at Acme Store, so $\vec{b}\cdot\vec{c}$ is the cost of consuming $\vec{c}$ groceries at Beta Mart. Thus $\vec{b}\cdot\vec{c}-\vec{a}\cdot\vec{c}=(\vec{b}-\vec{a})\cdot\vec{c}$ is the difference in costs between Beta and Acme for the same $\vec{c}$ groceries.

For $\vec{b}-\vec{a}$ to be perpendicular to $\vec{c}$, we must have $(\vec{b}-\vec{a})\cdot\vec{c}=0$. Since $\vec{b}-\vec{a}=(0.20,-0.20,0.50)$, the vector $\vec{b}-\vec{a}$ is perpendicular to $\vec{c}$ if $0.20c_b-0.20c_e+0.50c_m=0$. For example, this occurs when we consume the same number of loaves of bread as dozens of eggs, but no milk.

(c) Since $\vec{b}\cdot\vec{c}$ is the cost of groceries at Beta, you might think of $(1/1.1)\vec{b}\cdot\vec{c}$ as the "freshness-adjusted" cost at Beta. Then $(1/1.1)\vec{b}\cdot\vec{c}<\vec{a}\cdot\vec{c}$ means the "freshness-adjusted" cost is cheaper at Beta.

26. We have

$$\begin{aligned}\|\vec{a}_2\| &= \sqrt{0.10^2+0.08^2+0.12^2+0.69^2}=0.7120\\ \|\vec{a}_3\| &= \sqrt{0.20^2+0.06^2+0.06^2+0.66^2}=0.6948\\ \|\vec{a}_4\| &= \sqrt{0.22^2+0.00^2+0.20^2+0.57^2}=0.6429\end{aligned}$$

$$\begin{aligned}\vec{a}_2\cdot\vec{a}_3 &= 0.10\cdot0.20+0.08\cdot0.06+0.12\cdot0.06+0.69\cdot0.66=0.4874\\ \vec{a}_3\cdot\vec{a}_4 &= 0.20\cdot0.22+0.06\cdot0.00+0.06\cdot0.20+0.66\cdot0.57=0.4322.\end{aligned}$$

The distance between the English and the Bantus is given by θ where

$$\cos\theta=\frac{\vec{a}_2\cdot\vec{a}_3}{\|\vec{a}_2\|\|\vec{a}_3\|}=\frac{0.4874}{(0.7120)(0.6948)}\approx0.9852$$

so $\theta\approx9.9°$.
The distance between the English and the Koreans is given by ϕ where

$$\cos\phi=\frac{\vec{a}_3\cdot\vec{a}_4}{\|\vec{a}_3\|\|\vec{a}_4\|}=\frac{0.4322}{(0.6948)(0.6429)}\approx0.9676$$

so $\phi\approx14.6°$. Hence the English are genetically closer to the Bantus than to the Koreans.

27. Let the room be put in the coordinate system as shown in Figure 12.44.

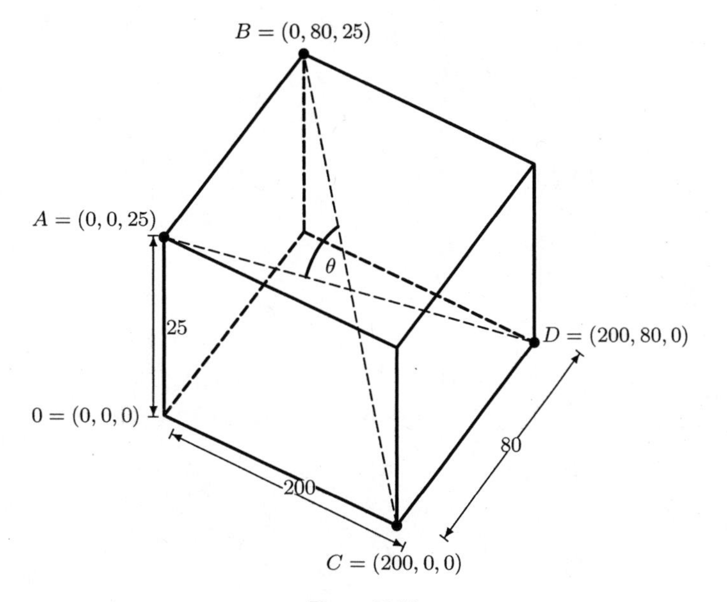

Figure 12.44

Then the vectors of the two strings are given by:
$\overrightarrow{AD} = (200\vec{i} + 80\vec{j} + 0\vec{k}) - (0\vec{i} + 0\vec{j} + 25\vec{k}) = 200\vec{i} + 80\vec{j} - 25\vec{k}$
$\overrightarrow{BC} = (200\vec{i} + 0\vec{j} + 0\vec{k}) - (0\vec{i} + 80\vec{j} + 25\vec{k}) = 200\vec{i} - 80\vec{j} - 25\vec{k}$.
Let the angle between $\overrightarrow{AD}$ and $\overrightarrow{BC}$ be θ. Then we have

$$\begin{aligned}
\cos\theta &= \frac{\overrightarrow{AD} \cdot \overrightarrow{BC}}{\|\overrightarrow{AD}\|\,\|\overrightarrow{BC}\|} \\
&= \frac{200(200) + (80)(-80) + (-25)(-25)}{\sqrt{200^2 + 80^2 + (-25)^2}\sqrt{(200)^2 + (-80)^2 + (-25)^2}} \\
&= \frac{34225}{47025} \\
&= 0.727804 \\
\theta &= \arccos 0.727804 \\
&= 43.297^\circ.
\end{aligned}$$

28. **(a)** The points A, B and C are shown in Figure 12.45.

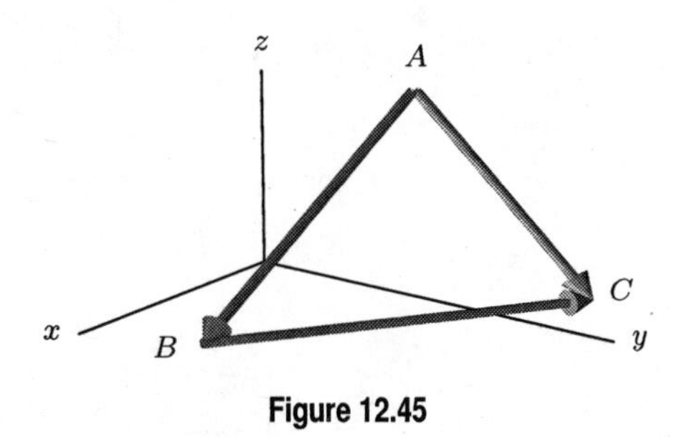

Figure 12.45

First, we calculate the vectors which form the sides of this triangle:
$\overrightarrow{AB} = (4\vec{i} + 2\vec{j} + \vec{k}) - (2\vec{i} + 2\vec{j} + 2\vec{k}) = 2\vec{i} - \vec{k}$

$\overrightarrow{BC} = (2\vec{i} + 3\vec{j} + \vec{k}) - (4\vec{i} + 2\vec{j} + \vec{k}) = -2\vec{i} + \vec{j}$
$\overrightarrow{AC} = (2\vec{i} + 3\vec{j} + \vec{k}) - (2\vec{i} + 2\vec{j} + 2\vec{k}) = \vec{j} - \vec{k}$.
Now we calculate the lengths of each of the sides of the triangles:
$\|\overrightarrow{AB}\| = \sqrt{2^2 + (-1)^2} = \sqrt{5}$
$\|\overrightarrow{BC}\| = \sqrt{(-2)^2 + 1^2} = \sqrt{5}$
$\|\overrightarrow{AC}\| = \sqrt{1^2 + (-1)^2} = \sqrt{2}$.
Thus the length of the shortest side of S is $\sqrt{2}$.

(b) $\cos \angle BAC = \frac{\overrightarrow{AB} \cdot \overrightarrow{AC}}{\|\overrightarrow{AB}\| \cdot \|\overrightarrow{AC}\|} = \frac{2 \cdot 0 + 0 \cdot 1 + (-1) \cdot (-1)}{\sqrt{5} \cdot \sqrt{2}} \approx 0.32.$

29. (a) We have

$$\begin{aligned} \vec{r} \cdot \vec{w} &= (x_1, y_1, x_2, y_2) \cdot (-1, 0, 1, 0) \\ &= -x_1 + x_2 = x_2 - x_1, \end{aligned}$$

so, since $x_2 > x_1$, this quantity represents the width w.

(b) We have

$$\begin{aligned} \vec{r} \cdot \vec{h} &= (x_1, y_1, x_2, y_2) \cdot (0, -1, 0, 1) \\ &= -y_1 + y_2 = y_2 - y_1, \end{aligned}$$

so, since $y_2 > y_1$, this quantity represents the height h.

(c) We have

$$\begin{aligned} 2\vec{r} \cdot (\vec{w} + \vec{h}) &= 2\vec{r} \cdot \vec{w} + 2\vec{r} \cdot \vec{h} \\ &= 2w + 2h, \end{aligned}$$

where from part (a) w is the width and from part (b) h is the height. Thus, this quantity represents the perimeter p.

Solutions for Section 12.5

Exercises

1. (a) We have

$$5\mathbf{R} = 5\begin{pmatrix} 3 & 7 \\ 2 & -1 \end{pmatrix} = \begin{pmatrix} 5\cdot 3 & 5\cdot 7 \\ 5\cdot 2 & 5\cdot -1 \end{pmatrix} = \begin{pmatrix} 15 & 35 \\ 10 & -5 \end{pmatrix}.$$

(b) We have

$$-2\mathbf{S} = -2\begin{pmatrix} 1 & -5 \\ 0 & 8 \end{pmatrix} = \begin{pmatrix} -2\cdot 1 & -2\cdot -5 \\ -2\cdot 0 & -2\cdot 8 \end{pmatrix} = \begin{pmatrix} -2 & 10 \\ 0 & -16 \end{pmatrix}.$$

(c) We have

$$\begin{aligned} \mathbf{R} + \mathbf{S} &= \begin{pmatrix} 3 & 7 \\ 2 & -1 \end{pmatrix} + \begin{pmatrix} 1 & -5 \\ 0 & 8 \end{pmatrix} \\ &= \begin{pmatrix} 3+1 & 7-5 \\ 2+0 & -1+8 \end{pmatrix} = \begin{pmatrix} 4 & 2 \\ 2 & 7 \end{pmatrix}. \end{aligned}$$

(d) Writing $\mathbf{S} - 3\mathbf{R} = \mathbf{S} + (-3)\mathbf{R}$, we first find $-3\mathbf{R}$:

$$-3\mathbf{R} = -3\begin{pmatrix} 3 & 7 \\ 2 & -1 \end{pmatrix} = \begin{pmatrix} -3\cdot 3 & -3\cdot 7 \\ -3\cdot 2 & -3\cdot -1 \end{pmatrix} = \begin{pmatrix} -9 & -21 \\ -6 & 3 \end{pmatrix}.$$

This gives

$$\begin{aligned} \mathbf{S} + (-3)\mathbf{R} &= \begin{pmatrix} 1 & -5 \\ 0 & 8 \end{pmatrix} + \begin{pmatrix} -9 & -21 \\ -6 & 3 \end{pmatrix} \\ &= \begin{pmatrix} 1-9 & -5-21 \\ 0-6 & 8+3 \end{pmatrix} = \begin{pmatrix} -8 & -26 \\ -6 & 11 \end{pmatrix}. \end{aligned}$$

(e) Writing $\mathbf{R} + 2\mathbf{R} + 2(\mathbf{R} - \mathbf{S}) = 5\mathbf{R} + (-2)\mathbf{S}$, we use our answers to parts (a) and (b):

$$\begin{aligned} 5\mathbf{R} + (-2)\mathbf{S} &= \begin{pmatrix} 15 & 35 \\ 10 & -5 \end{pmatrix} + \begin{pmatrix} -2 & 10 \\ 0 & -16 \end{pmatrix} \\ &= \begin{pmatrix} 15-2 & 35+10 \\ 10+0 & -5-16 \end{pmatrix} = \begin{pmatrix} 13 & 45 \\ 10 & -21 \end{pmatrix}. \end{aligned}$$

(f) We have

$$k\mathbf{S} = \begin{pmatrix} k\cdot 1 & k\cdot -5 \\ k\cdot 0 & k\cdot 8 \end{pmatrix} = \begin{pmatrix} k & -5k \\ 0 & 8k \end{pmatrix}.$$

2. **(a)** We have

$$\begin{aligned} 2\mathbf{A} &= 2\begin{pmatrix} 2 & 5 & 7 \\ 4 & -6 & 3 \\ 16 & -5 & 0 \end{pmatrix} \\ &= \begin{pmatrix} 2\cdot 2 & 2\cdot 5 & 2\cdot 7 \\ 2\cdot 4 & 2\cdot -6 & 2\cdot 3 \\ 2\cdot 16 & 2\cdot -5 & 2\cdot 0 \end{pmatrix} = \begin{pmatrix} 4 & 10 & 14 \\ 8 & -12 & 6 \\ 32 & -10 & 0 \end{pmatrix}. \end{aligned}$$

(b) We have

$$\begin{aligned} -3\mathbf{B} &= -3\begin{pmatrix} 8 & -6 & 0 \\ 5 & 3 & -2 \\ 3 & 7 & 12 \end{pmatrix} \\ &= \begin{pmatrix} -3\cdot 8 & -3\cdot -6 & -3\cdot 0 \\ -3\cdot 5 & -3\cdot 3 & -3\cdot -2 \\ -3\cdot 3 & -3\cdot 7 & -3\cdot 12 \end{pmatrix} = \begin{pmatrix} -24 & 18 & 0 \\ -15 & -9 & 6 \\ -9 & -21 & -36 \end{pmatrix}. \end{aligned}$$

(c) We have

$$\mathbf{A} + \mathbf{B} = \begin{pmatrix} 2 & 5 & 7 \\ 4 & -6 & 3 \\ 16 & -5 & 0 \end{pmatrix} + \begin{pmatrix} 8 & -6 & 0 \\ 5 & 3 & -2 \\ 3 & 7 & 12 \end{pmatrix}$$

$$= \begin{pmatrix} 2+8 & 5-6 & 7+0 \\ 4+5 & -6+3 & 3-2 \\ 16+3 & -5+7 & 0+12 \end{pmatrix} = \begin{pmatrix} 10 & -1 & 7 \\ 9 & -3 & 1 \\ 19 & 2 & 12 \end{pmatrix}.$$

(d) Notice that we can write $2\mathbf{A} - 3\mathbf{B}$ as $2\mathbf{A} + (-3\mathbf{B})$. We found $2\mathbf{A}$ in part (a) and $-3\mathbf{B}$ in part (b). We have:

$$2\mathbf{A} + (-3\mathbf{B}) = \begin{pmatrix} 4 & 10 & 14 \\ 8 & -12 & 6 \\ 32 & -10 & 0 \end{pmatrix} + \begin{pmatrix} -24 & 18 & 0 \\ -15 & -9 & 6 \\ -9 & -21 & -36 \end{pmatrix}$$

$$= \begin{pmatrix} 4-24 & 10+18 & 14+0 \\ 8-15 & -12-9 & 6+6 \\ 32-9 & -10-21 & 0-36 \end{pmatrix} = \begin{pmatrix} -20 & 28 & 14 \\ -7 & -21 & 12 \\ 23 & -31 & -36 \end{pmatrix}.$$

(e) We found $(\mathbf{A} + \mathbf{B})$ in part (c). We have:

$$5(\mathbf{A} + \mathbf{B}) = 5\begin{pmatrix} 10 & -1 & 7 \\ 9 & -3 & 1 \\ 19 & 2 & 12 \end{pmatrix}$$

$$= \begin{pmatrix} 5\cdot 10 & 5\cdot -1 & 5\cdot 7 \\ 5\cdot 9 & 5\cdot -3 & 5\cdot 1 \\ 5\cdot 19 & 5\cdot 2 & 5\cdot 12 \end{pmatrix} = \begin{pmatrix} 50 & -5 & 35 \\ 45 & -15 & 5 \\ 95 & 10 & 60 \end{pmatrix}.$$

(f) There are many ways to work this problem. One way is to write $\mathbf{A} + (\mathbf{A} + (\mathbf{A} + \mathbf{B})) = 3\mathbf{A} + \mathbf{B}$. This gives:

$$3\mathbf{A} + \mathbf{B} = 3\begin{pmatrix} 2 & 5 & 7 \\ 4 & -6 & 3 \\ 16 & -5 & 0 \end{pmatrix} + \begin{pmatrix} 8 & -6 & 0 \\ 5 & 3 & -2 \\ 3 & 7 & 12 \end{pmatrix}$$

$$= \begin{pmatrix} 3\cdot 2 & 3\cdot 5 & 3\cdot 7 \\ 3\cdot 4 & 3\cdot -6 & 3\cdot 3 \\ 3\cdot 16 & 3\cdot -5 & 3\cdot 0 \end{pmatrix} + \begin{pmatrix} 8 & -6 & 0 \\ 5 & 3 & -2 \\ 3 & 7 & 12 \end{pmatrix}$$

$$= \begin{pmatrix} 6+8 & 15-6 & 21+0 \\ 12+5 & -18+3 & 9-2 \\ 48+3 & -15+7 & 0+12 \end{pmatrix} = \begin{pmatrix} 14 & 9 & 21 \\ 17 & -15 & 7 \\ 51 & -8 & 12 \end{pmatrix}.$$

3. (a) We have

$$4\mathbf{U} = 4\begin{pmatrix} 3 & 2 & 5 & 1 \\ 4 & 6 & 7 & 3 \\ 1 & 9 & 5 & 8 \\ 0 & -2 & 4 & 6 \end{pmatrix}$$

$$= \begin{pmatrix} 4\cdot 3 & 4\cdot 2 & 4\cdot 5 & 4\cdot 1 \\ 4\cdot 4 & 4\cdot 6 & 4\cdot 7 & 4\cdot 3 \\ 4\cdot 1 & 4\cdot 9 & 4\cdot 5 & 4\cdot 8 \\ 4\cdot 0 & 4\cdot -2 & 4\cdot 4 & 4\cdot 6 \end{pmatrix} = \begin{pmatrix} 12 & 8 & 20 & 4 \\ 16 & 24 & 28 & 12 \\ 4 & 36 & 20 & 32 \\ 0 & -8 & 16 & 24 \end{pmatrix}.$$

(b) We have

$$-2\mathbf{V} = -2\begin{pmatrix} 1 & 6 & 4 & 2 \\ 3 & 5 & -1 & 7 \\ 9 & 4 & 7 & 3 \\ 2 & 8 & 4 & 5 \end{pmatrix}$$

$$= \begin{pmatrix} -2\cdot 1 & -2\cdot 6 & -2\cdot 4 & -2\cdot 2 \\ -2\cdot 3 & -2\cdot 5 & -2\cdot -1 & -2\cdot 7 \\ -2\cdot 9 & -2\cdot 4 & -2\cdot 7 & -2\cdot 3 \\ -2\cdot 2 & -2\cdot 8 & -2\cdot 4 & -2\cdot 5 \end{pmatrix} = \begin{pmatrix} -2 & -12 & -8 & -4 \\ -6 & -10 & 2 & -14 \\ -18 & -8 & -14 & -6 \\ -4 & -16 & -8 & -10 \end{pmatrix}.$$

(c) We have

$$\mathbf{U} - \mathbf{V} = \begin{pmatrix} 3 & 2 & 5 & 1 \\ 4 & 6 & 7 & 3 \\ 1 & 9 & 5 & 8 \\ 0 & -2 & 4 & 6 \end{pmatrix} - \begin{pmatrix} 1 & 6 & 4 & 2 \\ 3 & 5 & -1 & 7 \\ 9 & 4 & 7 & 3 \\ 2 & 8 & 4 & 5 \end{pmatrix}$$

$$= \begin{pmatrix} 3-1 & 2-6 & 5-4 & 1-2 \\ 4-3 & 6-5 & 7-(-1) & 3-7 \\ 1-9 & 9-4 & 5-7 & 8-3 \\ 0-2 & -2-8 & 4-4 & 6-5 \end{pmatrix} = \begin{pmatrix} 2 & -4 & 1 & -1 \\ 1 & 1 & 8 & -4 \\ -8 & 5 & -2 & 5 \\ -2 & -10 & 0 & 1 \end{pmatrix}.$$

(d) Notice that we can write $3\mathbf{U} - 3\mathbf{V}$ as $3(\mathbf{U} - \mathbf{V})$. We found $\mathbf{U} - \mathbf{V}$ in part (c). We have:

$$3(\mathbf{U} - \mathbf{V}) = 3\begin{pmatrix} 2 & -4 & 1 & -1 \\ 1 & 1 & 8 & -4 \\ -8 & 5 & -2 & 5 \\ -2 & -10 & 0 & 1 \end{pmatrix}$$

$$= \begin{pmatrix} 3\cdot 2 & 3\cdot -4 & 3\cdot 1 & 3\cdot -1 \\ 3\cdot 1 & 3\cdot 1 & 3\cdot 8 & 3\cdot -4 \\ 3\cdot -8 & 3\cdot 5 & 3\cdot -2 & 3\cdot 5 \\ 3\cdot -2 & 3\cdot -10 & 3\cdot 0 & 3\cdot 1 \end{pmatrix} = \begin{pmatrix} 6 & -12 & 3 & -3 \\ 3 & 3 & 24 & -12 \\ -24 & 15 & -6 & 15 \\ -6 & -30 & 0 & 3 \end{pmatrix}.$$

(e) Notice that we can write $\mathbf{U} + \mathbf{U} - (\mathbf{U} - \mathbf{V})$ as $\mathbf{U} + \mathbf{V}$. We have

$$\mathbf{U} + \mathbf{V} = \begin{pmatrix} 3 & 2 & 5 & 1 \\ 4 & 6 & 7 & 3 \\ 1 & 9 & 5 & 8 \\ 0 & -2 & 4 & 6 \end{pmatrix} + \begin{pmatrix} 1 & 6 & 4 & 2 \\ 3 & 5 & -1 & 7 \\ 9 & 4 & 7 & 3 \\ 2 & 8 & 4 & 5 \end{pmatrix}$$

$$= \begin{pmatrix} 3+1 & 2+6 & 5+4 & 1+2 \\ 4+3 & 6+5 & 7-1 & 3+7 \\ 1+9 & 9+4 & 5+7 & 8+3 \\ 0+2 & -2+8 & 4+4 & 6+5 \end{pmatrix} = \begin{pmatrix} 4 & 8 & 9 & 3 \\ 7 & 11 & 6 & 10 \\ 10 & 13 & 12 & 11 \\ 2 & 6 & 8 & 11 \end{pmatrix}.$$

(f) Notice that we can write $2(2\mathbf{U}-\mathbf{V})$ as $4\mathbf{U}+(-2\mathbf{V})$. We found $4\mathbf{U}$ in part (a) and $-2\mathbf{V}$ in part (b). We have

$$4\mathbf{U}+(-2\mathbf{V}) = \begin{pmatrix} 12 & 8 & 20 & 4 \\ 16 & 24 & 28 & 12 \\ 4 & 36 & 20 & 32 \\ 0 & -8 & 16 & 24 \end{pmatrix} + \begin{pmatrix} -2 & -12 & -8 & -4 \\ -6 & -10 & 2 & -14 \\ -18 & -8 & -14 & -6 \\ -4 & -16 & -8 & -10 \end{pmatrix}$$

$$= \begin{pmatrix} 12-2 & 8-12 & 20-8 & 4-4 \\ 16-6 & 24-10 & 28+2 & 12-14 \\ 4-18 & 36-8 & 20-14 & 32-6 \\ 0-4 & -8-16 & 16-8 & 24-10 \end{pmatrix}$$

$$= \begin{pmatrix} 10 & -4 & 12 & 0 \\ 10 & 14 & 30 & -2 \\ -14 & 28 & 6 & 26 \\ -4 & -24 & 8 & 14 \end{pmatrix}.$$

4. (a) We have

$$\mathbf{R}\vec{p} = \begin{pmatrix} 3 & 7 \\ 2 & -1 \end{pmatrix}\begin{pmatrix} 3 \\ 1 \end{pmatrix} = \begin{pmatrix} 3\cdot 3+7\cdot 1 \\ 2\cdot 3-1\cdot 1 \end{pmatrix} = \begin{pmatrix} 16 \\ 5 \end{pmatrix}.$$

(b) We have

$$\mathbf{S}\vec{q} = \begin{pmatrix} 1 & -5 \\ 0 & 8 \end{pmatrix}\begin{pmatrix} -1 \\ 5 \end{pmatrix} = \begin{pmatrix} 1\cdot -1-5\cdot 5 \\ 0\cdot -1+8\cdot 5 \end{pmatrix} = \begin{pmatrix} -26 \\ 40 \end{pmatrix}.$$

(c) We have $\vec{q}+\vec{p} = (3,1)+(-1,5) = (2,6)$, and so

$$\mathbf{S}(\vec{q}+\vec{p}) = \begin{pmatrix} 1 & -5 \\ 0 & 8 \end{pmatrix}\begin{pmatrix} 2 \\ 6 \end{pmatrix} = \begin{pmatrix} 1\cdot 2-5\cdot 6 \\ 0\cdot 2+8\cdot 6 \end{pmatrix} = \begin{pmatrix} -28 \\ 48 \end{pmatrix}.$$

(d) We have

$$\mathbf{R}+\mathbf{S} = \begin{pmatrix} 3 & 7 \\ 2 & -1 \end{pmatrix} + \begin{pmatrix} 1 & -5 \\ 0 & 8 \end{pmatrix}$$

$$= \begin{pmatrix} 3+1 & 7-5 \\ 2+0 & -1+8 \end{pmatrix} = \begin{pmatrix} 4 & 2 \\ 2 & 7 \end{pmatrix}$$

and so

$$(\mathbf{R}+\mathbf{S})\vec{p} = \begin{pmatrix} 4 & 2 \\ 2 & 7 \end{pmatrix}\begin{pmatrix} 3 \\ 1 \end{pmatrix} = \begin{pmatrix} 4\cdot 3+2\cdot 1 \\ 2\cdot 3+7\cdot 1 \end{pmatrix} = \begin{pmatrix} 14 \\ 13 \end{pmatrix}.$$

(e) From parts (a) and (b), we have

$$\mathbf{R}\vec{p}\cdot\mathbf{S}\vec{q} = (16,5)\cdot(-26,40) = -216.$$

(f) We have $\vec{p}\cdot\vec{q} = (3,1)(-1,5) = 2$ and so

$$(\vec{p}\cdot\vec{q})\mathbf{S} = 2\mathbf{S} = 2\begin{pmatrix} 1 & -5 \\ 0 & 8 \end{pmatrix} = \begin{pmatrix} 2\cdot 1 & 2\cdot -5 \\ 2\cdot 0 & 2\cdot 8 \end{pmatrix} = \begin{pmatrix} 2 & -10 \\ 0 & 16 \end{pmatrix}.$$

5. (a) We have

$$\mathbf{A}\vec{u} = \begin{pmatrix} 2 & 5 & 7 \\ 4 & -6 & 3 \\ 16 & -5 & 0 \end{pmatrix}\begin{pmatrix} 3 \\ 2 \\ 5 \end{pmatrix}$$

$$= \begin{pmatrix} 2\cdot 3+5\cdot 2+7\cdot 5 \\ 4\cdot 3-6\cdot 2+3\cdot 5 \\ 16\cdot 3-5\cdot 2+0\cdot 5 \end{pmatrix} = \begin{pmatrix} 51 \\ 15 \\ 38 \end{pmatrix}.$$

(b) We have

$$\mathbf{B}\vec{v} = \begin{pmatrix} 8 & -6 & 0 \\ 5 & 3 & -2 \\ 3 & 7 & 12 \end{pmatrix}\begin{pmatrix} -1 \\ 0 \\ 3 \end{pmatrix}$$

$$= \begin{pmatrix} 8\cdot -1-6\cdot 0+0\cdot 3 \\ 5\cdot -1+3\cdot 0-2\cdot 3 \\ 3\cdot -1+7\cdot 0+12\cdot 3 \end{pmatrix} = \begin{pmatrix} -8 \\ -11 \\ 33 \end{pmatrix}.$$

(c) Letting $\vec{w} = \vec{u} + \vec{v} = (2,2,8)$, we have:

$$\mathbf{A}\vec{w} = \begin{pmatrix} 2 & 5 & 7 \\ 4 & -6 & 3 \\ 16 & -5 & 0 \end{pmatrix}\begin{pmatrix} 2 \\ 2 \\ 8 \end{pmatrix}$$

$$= \begin{pmatrix} 2\cdot 2+5\cdot 2+7\cdot 8 \\ 4\cdot 2-6\cdot 2+3\cdot 8 \\ 16\cdot 2-5\cdot 2+0\cdot 8 \end{pmatrix} = \begin{pmatrix} 70 \\ 20 \\ 22 \end{pmatrix}.$$

Another to work this problem would be to write $\mathbf{A}(\vec{u} + \vec{v})$ as $\mathbf{A}\vec{u} + \mathbf{A}\vec{v}$ and proceed accordingly.

(d) Letting $\mathbf{C} = \mathbf{A} + \mathbf{B}$, we have

$$\mathbf{C} = \begin{pmatrix} 2 & 5 & 7 \\ 4 & -6 & 3 \\ 16 & -5 & 0 \end{pmatrix} + \begin{pmatrix} 8 & -6 & 0 \\ 5 & 3 & -2 \\ 3 & 7 & 12 \end{pmatrix} = \begin{pmatrix} 10 & -1 & 7 \\ 9 & -3 & 1 \\ 19 & 2 & 12 \end{pmatrix}.$$

We can now write $(\mathbf{A} + \mathbf{B})\vec{v}$ as $\mathbf{C}\vec{v}$, and so:

$$\mathbf{C}\vec{v} = \begin{pmatrix} 10 & -1 & 7 \\ 9 & -3 & 1 \\ 19 & 2 & 12 \end{pmatrix}\begin{pmatrix} -1 \\ 0 \\ 3 \end{pmatrix}$$

$$= \begin{pmatrix} 10\cdot -1-1\cdot 0+7\cdot 3 \\ 9\cdot -1-3\cdot 0+1\cdot 3 \\ 19\cdot -1+2\cdot 0+12\cdot 3 \end{pmatrix} = \begin{pmatrix} 11 \\ -6 \\ 17 \end{pmatrix}.$$

(e) From part (a) we have $\mathbf{A}\vec{u} = (51, 15, 38)$, and from part (b) we have $\mathbf{B}\vec{v} = (-8, -11, 33)$. This gives

$$\begin{aligned}\mathbf{A}\vec{u} \cdot \mathbf{B}\vec{v} &= (51, 15, 38) \cdot (-8, -11, 33) \\ &= 51 \cdot -8 + 15 \cdot -11 + 38 \cdot 33 = 681.\end{aligned}$$

(f) We have $\vec{u} \cdot \vec{v} = 3 \cdot -1 + 2 \cdot 0 + 5 \cdot 3 = 12$, and so

$$(\vec{u} \cdot \vec{v})\mathbf{A} = 12\mathbf{A} = 12\begin{pmatrix} 2 & 5 & 7 \\ 4 & -6 & 3 \\ 16 & -5 & 0 \end{pmatrix} = \begin{pmatrix} 24 & 60 & 84 \\ 48 & -72 & 36 \\ 192 & -60 & 0 \end{pmatrix}.$$

6. (a) We have

$$\begin{aligned}\mathbf{U}\vec{t} &= \begin{pmatrix} 3 & 2 & 5 & 1 \\ 4 & 6 & 7 & 3 \\ 1 & 9 & 5 & 8 \\ 0 & -2 & 4 & 6 \end{pmatrix}\begin{pmatrix} 4 \\ 5 \\ 1 \\ -1 \end{pmatrix} \\ &= \begin{pmatrix} 3\cdot 4+2\cdot 5+5\cdot 1+1\cdot -1 \\ 4\cdot 4+6\cdot 5+7\cdot 1+3\cdot -1 \\ 1\cdot 4+9\cdot 5+5\cdot 1+8\cdot -1 \\ 0\cdot 4-2\cdot 5+4\cdot 1+6\cdot -1 \end{pmatrix} = \begin{pmatrix} 26 \\ 50 \\ 46 \\ -12 \end{pmatrix}.\end{aligned}$$

(b) We have

$$\begin{aligned}\mathbf{V}\vec{s} &= \begin{pmatrix} 1 & 6 & 4 & 2 \\ 3 & 5 & -1 & 7 \\ 9 & 4 & 7 & 3 \\ 2 & 8 & 4 & 5 \end{pmatrix}\begin{pmatrix} 2 \\ 0 \\ -1 \\ 7 \end{pmatrix} \\ &= \begin{pmatrix} 1\cdot 2+6\cdot 0+4\cdot -1+2\cdot 7 \\ 3\cdot 2+5\cdot 0-1\cdot -1+7\cdot 7 \\ 9\cdot 2+4\cdot 0+7\cdot -1+3\cdot 7 \\ 2\cdot 2+8\cdot 0+4\cdot -1+5\cdot 7 \end{pmatrix} = \begin{pmatrix} 12 \\ 56 \\ 32 \\ 35 \end{pmatrix}.\end{aligned}$$

(c) From part (a) we know that $\mathbf{U}\vec{t} = (26, 50, 46, -12)$. We can find $\mathbf{U}\vec{s}$ as follows:

$$\begin{aligned}\mathbf{U}\vec{s} &= \begin{pmatrix} 3 & 2 & 5 & 1 \\ 4 & 6 & 7 & 3 \\ 1 & 9 & 5 & 8 \\ 0 & -2 & 4 & 6 \end{pmatrix}\begin{pmatrix} 2 \\ 0 \\ -1 \\ 7 \end{pmatrix} \\ &= \begin{pmatrix} 3\cdot 2+2\cdot 0+5\cdot -1+1\cdot 7 \\ 4\cdot 2+6\cdot 0+7\cdot -1+3\cdot 7 \\ 1\cdot 2+9\cdot 0+5\cdot -1+8\cdot 7 \\ 0\cdot 2-2\cdot 0+4\cdot -1+6\cdot 7 \end{pmatrix} = \begin{pmatrix} 8 \\ 22 \\ 53 \\ 38 \end{pmatrix}.\end{aligned}$$

and so

$$\begin{aligned}\mathbf{U}\vec{t}\cdot\mathbf{U}\vec{s} &= (26,50,46,-12)\cdot(8,22,53,38)\\ &= 26\cdot 8+50\cdot 22+46\cdot 53-12\cdot 38 = 3290.\end{aligned}$$

(d) We can write this as $\mathbf{U}\vec{s}-\mathbf{U}\vec{t}$. From parts (a) and (c) we know that $\mathbf{U}\vec{s} = (8,22,53,38)$ and $\mathbf{U}\vec{t} = (26,50,46,-12)$. This gives

$$\mathbf{U}\vec{s}-\mathbf{U}\vec{t} = (8,22,53,38)-(26,50,46,-12) = (-18,-28,7,50).$$

(e) We first find $\mathbf{U}+\mathbf{V}$:

$$\begin{aligned}\mathbf{U}+\mathbf{V} &= \begin{pmatrix}3&2&5&1\\4&6&7&3\\1&9&5&8\\0&-2&4&6\end{pmatrix}+\begin{pmatrix}1&6&4&2\\3&5&-1&7\\9&4&7&3\\2&8&4&5\end{pmatrix}\\ &= \begin{pmatrix}3+1&2+6&5+4&1+2\\4+3&6+5&7-1&3+7\\1+9&9+4&5+7&8+3\\0+2&-2+8&4+4&6+5\end{pmatrix} = \begin{pmatrix}4&8&9&3\\7&11&6&10\\10&13&12&11\\2&6&8&11\end{pmatrix}.\end{aligned}$$

We next find $\vec{s}+\vec{t}$:

$$\vec{s}+\vec{t} = (2,0,-1,7)+(4,5,1,-1) = (6,5,0,6).$$

This gives:

$$\begin{aligned}(\mathbf{U}+\mathbf{V})(\vec{s}+\vec{t}) &= \begin{pmatrix}4&8&9&3\\7&11&6&10\\10&13&12&11\\2&6&8&11\end{pmatrix}\begin{pmatrix}6\\5\\0\\6\end{pmatrix}\\ &= \begin{pmatrix}4\cdot 6+8\cdot 5+9\cdot 0+3\cdot 6\\7\cdot 6+11\cdot 5+6\cdot 0+10\cdot 6\\10\cdot 6+13\cdot 5+12\cdot 0+11\cdot 6\\2\cdot 6+6\cdot 5+8\cdot 0+11\cdot 6\end{pmatrix} = \begin{pmatrix}82\\157\\191\\108\end{pmatrix}.\end{aligned}$$

(f) We found $\mathbf{U}+\mathbf{V}$ in part (e). Evaluating $\vec{s}\cdot\vec{t}$, we have:

$$\begin{aligned}\vec{s}\cdot\vec{t} &= (2,0,-1,7)\cdot(4,5,1,-1)\\ &= 2\cdot 4+0\cdot 5-1\cdot 1+7\cdot -1 = 0.\end{aligned}$$

Thus,

$$(\vec{s}\cdot\vec{t})(\mathbf{U}+\mathbf{V}) = 0(\mathbf{U}+\mathbf{V}) = \begin{pmatrix}0&0&0&0\\0&0&0&0\\0&0&0&0\\0&0&0&0\end{pmatrix}.$$

Problems

7. **(a)** $\mathbf{A}$ is 2×2 and $\vec{u}$ has dimension 2 so this is defined. We have

$$\begin{aligned} \mathbf{A}\vec{u} &= \begin{pmatrix} 3 & 7 \\ 2 & -1 \end{pmatrix}\begin{pmatrix} 1 \\ 1 \end{pmatrix} \\ &= \begin{pmatrix} 3 \cdot 1 + 7 \cdot 1 \\ 2 \cdot 1 - 1 \cdot 1 \end{pmatrix} = \begin{pmatrix} 10 \\ 1 \end{pmatrix}. \end{aligned}$$

(b) $\mathbf{B}$ is 3×3 but $\vec{u}$ has dimension 2 so this is not defined.
(c) The vector $\vec{u}$ has dimension 2 but $\vec{v}$ has dimension 3 so $(\vec{u} + \vec{v})$ is not defined.
(d) $\mathbf{A}$ is matrix and $\vec{u}$ is a vector so $\mathbf{A} + \vec{u}$ is not defined.
(e) $\vec{v} \cdot \vec{v} = 1 \cdot 1 + 3 \cdot 3 = 10$ so from part (a) we have

$$\frac{\mathbf{A}\vec{u}}{\vec{v} \cdot \vec{v}} = \frac{1}{10}\begin{pmatrix} 10 \\ 1 \end{pmatrix} = \begin{pmatrix} 1 \\ 0.1 \end{pmatrix}.$$

(f) The vector $\vec{u}$ has dimension 2 but $\vec{v}$ has dimension 3 so $\vec{u} \cdot \vec{v}$ is not defined.

8. **(a)** We have

$$\begin{aligned} \vec{w} &= \mathbf{A}\vec{u} \\ &= \begin{pmatrix} 1 & 2 \\ 2 & -1 \end{pmatrix}\begin{pmatrix} 5 \\ -5 \end{pmatrix} \\ &= \begin{pmatrix} 1 \cdot 5 + 2 \cdot (-5) \\ 2 \cdot 5 + (-1) \cdot (-5) \end{pmatrix} = \begin{pmatrix} -5 \\ 15 \end{pmatrix}. \end{aligned}$$

(b) Let $\vec{w} = (x, y)$ then

$$\begin{aligned} \mathbf{A}\vec{w} &= \begin{pmatrix} 1 & 2 \\ 2 & -1 \end{pmatrix}\begin{pmatrix} x \\ y \end{pmatrix} \\ &= \begin{pmatrix} 1 \cdot x + 2 \cdot y \\ 2 \cdot x + (-1) \cdot y \end{pmatrix} \\ &= \begin{pmatrix} x + 2y \\ 2x - y \end{pmatrix} \end{aligned}$$

so $\mathbf{A}\vec{w} = \vec{u}$ is equivalent to

$$\begin{pmatrix} x + 2y \\ 2x - y \end{pmatrix} = \begin{pmatrix} 5 \\ -5 \end{pmatrix}.$$

Comparing the components, we obtain the simultaneous equations

$$\begin{aligned} x + 2y &= 5 \\ 2x - y &= -5 \end{aligned}$$

which have the solution $x = -1$, $y = 3$, so $\vec{w} = (-1, 3)$.

9. **(a)** We have

$$
\begin{aligned}
s_{\text{new}} &= s_{\text{old}} - \underbrace{0.10 s_{\text{old}}}_{\text{10\% infected}} \\
&= 0.90 s_{\text{old}} \\
i_{\text{new}} &= i_{\text{old}} + \underbrace{0.10 s_{\text{old}}}_{\text{10\% infected}} - \underbrace{0.50 i_{\text{old}}}_{\text{50\% recover}} + \underbrace{0.02 r_{\text{old}}}_{\text{2\% reinfected}} \\
&= 0.10 s_{\text{old}} + 0.50 i_{\text{old}} + 0.02 r_{\text{old}} \\
r_{\text{new}} &= r_{\text{old}} + \underbrace{0.50 i_{\text{old}}}_{\text{50\% recover}} - \underbrace{0.02 r_{\text{old}}}_{\text{2\% reinfected}} \\
&= 0.50 i_{\text{old}} + 0.98 r_{\text{old}}.
\end{aligned}
$$

Using matrix multiplication, we can rewrite these 3 equations as

$$
\begin{pmatrix} s_{\text{new}} \\ i_{\text{new}} \\ r_{\text{new}} \end{pmatrix} = \begin{pmatrix} 0.90 & 0 & 0 \\ 0.10 & 0.50 & 0.02 \\ 0 & 0.50 & 0.98 \end{pmatrix} \begin{pmatrix} s_{\text{old}} \\ i_{\text{old}} \\ r_{\text{old}} \end{pmatrix},
$$

and so $\vec{p}_{\text{new}} = \mathbf{T}\vec{p}_{\text{old}}$ where $\mathbf{T} = \begin{pmatrix} 0.90 & 0 & 0 \\ 0.10 & 0.50 & 0.02 \\ 0 & 0.50 & 0.98 \end{pmatrix}$.

(b) We have

$$
\begin{aligned}
\vec{p_1} = \mathbf{T}\vec{p_0} &= \begin{pmatrix} 0.90 & 0 & 0 \\ 0.10 & 0.50 & 0.02 \\ 0 & 0.50 & 0.98 \end{pmatrix} \begin{pmatrix} 2.0 \\ 0.0 \\ 0.0 \end{pmatrix} \\
&= \begin{pmatrix} 0.9(2) + 0(0) + 0(0) \\ 0.1(2) + 0.5(0) + 0.02(0) \\ 0(2) + 0.5(0) + 0.98(0) \end{pmatrix} = \begin{pmatrix} 1.8 \\ 0.2 \\ 0.0 \end{pmatrix} \\
\vec{p_2} = \mathbf{T}\vec{p_1} &= \begin{pmatrix} 0.90 & 0 & 0 \\ 0.10 & 0.50 & 0.02 \\ 0 & 0.50 & 0.98 \end{pmatrix} \begin{pmatrix} 1.8 \\ 0.2 \\ 0.0 \end{pmatrix} \\
&= \begin{pmatrix} 0.9(1.8) + 0(0.2) + 0(0) \\ 0.1(1.8) + 0.5(0.2) + 0.02(0) \\ 0(1.8) + 0.5(0.2) + 0.98(0) \end{pmatrix} = \begin{pmatrix} 1.62 \\ 0.28 \\ 0.10 \end{pmatrix} \\
\vec{p_3} = \mathbf{T}\vec{p_2} &= \begin{pmatrix} 0.90 & 0 & 0 \\ 0.10 & 0.50 & 0.02 \\ 0 & 0.50 & 0.98 \end{pmatrix} \begin{pmatrix} 1.62 \\ 0.28 \\ 0.10 \end{pmatrix} \\
&= \begin{pmatrix} 0.9(1.62) + 0(0.28) + 0(0.1) \\ 0.1(1.62) + 0.5(0.28) + 0.02(0.1) \\ 0(1.62) + 0.5(0.28) + 0.98(0.1) \end{pmatrix} = \begin{pmatrix} 1.458 \\ 0.304 \\ 0.238 \end{pmatrix}.
\end{aligned}
$$

10. **(a)** We have

$$
\begin{pmatrix} f_{\text{new}} \\ s_{\text{new}} \\ t_{\text{new}} \end{pmatrix} = \begin{pmatrix} 0.3 & 0.6 & 0.5 \\ 0.7 & 0 & 0 \\ 0 & 0.4 & 0 \end{pmatrix} \begin{pmatrix} f_{\text{old}} \\ s_{\text{old}} \\ t_{\text{old}} \end{pmatrix},
$$

and so

$$\begin{aligned} f_{\text{new}} &= 0.3f_{\text{old}} + 0.6s_{\text{old}} + 0.5t_{\text{old}} \\ s_{\text{new}} &= 0.7f_{\text{old}} \\ t_{\text{new}} &= 0.4s_{\text{old}}. \end{aligned}$$

From the first equation we can conclude that in a given year, 30% of the first-year insects lay eggs, as do 60% of the second-year insects and 50% of the third-year insects. From the second equation, we see that 70% of the first-year insects survive into their second year. From the third equation, we see that 40% of the second-year insects survive into their third year.

(b) We have

$$\begin{aligned} \vec{p_1} = \mathbf{T}\vec{p_0} &= \begin{pmatrix} 0.3 & 0.6 & 0.5 \\ 0.7 & 0 & 0 \\ 0 & 0.4 & 0 \end{pmatrix}\begin{pmatrix} 2000 \\ 0 \\ 0 \end{pmatrix} \\ &= \begin{pmatrix} 0.3(2000) + 0.6(0) + 0.5(0) \\ 0.7(2000) + 0(0) + 0(0) \\ 0(2000) + 0.4(0) + 0(0) \end{pmatrix} = \begin{pmatrix} 600 \\ 1400 \\ 0 \end{pmatrix} \\ \vec{p_2} = \mathbf{T}\vec{p_1} &= \begin{pmatrix} 0.3 & 0.6 & 0.5 \\ 0.7 & 0 & 0 \\ 0 & 0.4 & 0 \end{pmatrix}\begin{pmatrix} 600 \\ 1400 \\ 0 \end{pmatrix} \\ &= \begin{pmatrix} 0.3(600) + 0.6(1400) + 0.5(0) \\ 0.7(600) + 0(1400) + 0(0) \\ 0(600) + 0.4(1400) + 0(0) \end{pmatrix} = \begin{pmatrix} 1020 \\ 420 \\ 560 \end{pmatrix} \\ \vec{p_3} = \mathbf{T}\vec{p_2} &= \begin{pmatrix} 0.3 & 0.6 & 0.5 \\ 0.7 & 0 & 0 \\ 0 & 0.4 & 0 \end{pmatrix}\begin{pmatrix} 1020 \\ 420 \\ 560 \end{pmatrix} \\ &= \begin{pmatrix} 0.3(1020) + 0.6(420) + 0.5(560) \\ 0.7(1020) + 0(420) + 0(560) \\ 0(1020) + 0.4(420) + 0(560) \end{pmatrix} = \begin{pmatrix} 838 \\ 714 \\ 168 \end{pmatrix}. \end{aligned}$$

11. **(a)** Let $\vec{p}_{\text{old}} = (A_{\text{old}}, B_{\text{old}})$ and $\vec{p}_{\text{new}} = (A_{\text{new}}, B_{\text{new}})$. We have

$$\begin{aligned} A_{\text{new}} &= A_{\text{old}} - \underbrace{0.03A_{\text{old}}}_{3\%\text{ leave }A} + \underbrace{0.05B_{\text{old}}}_{5\%\text{ leave }B} \\ &= 0.97A_{\text{old}} + 0.05B_{\text{old}} \\ B_{\text{new}} &= B_{\text{old}} - \underbrace{0.05B_{\text{old}}}_{5\%\text{ leave }B} + \underbrace{0.03A_{\text{old}}}_{3\%\text{ leave }A} \\ &= 0.03A_{\text{old}} + 0.95B_{\text{old}}. \end{aligned}$$

Using matrix multiplication, we can rewrite these two equations as

$$\begin{pmatrix} A_{\text{new}} \\ B_{\text{new}} \end{pmatrix} = \begin{pmatrix} 0.97 & 0.05 \\ 0.03 & 0.95 \end{pmatrix}\begin{pmatrix} A_{\text{old}} \\ B_{\text{old}} \end{pmatrix},$$

and so $\vec{p}_{\text{new}} = \mathbf{T}\vec{p}_{\text{old}}$ where $\mathbf{T} = \begin{pmatrix} 0.97 & 0.05 \\ 0.03 & 0.95 \end{pmatrix}$.

(b) We have

$$\vec{p}_{2006} = \mathbf{T}\vec{p}_{2005} = \begin{pmatrix} 0.97 & 0.05 \\ 0.03 & 0.95 \end{pmatrix} \begin{pmatrix} 200 \\ 400 \end{pmatrix}$$

$$= \begin{pmatrix} 0.97(200) + 0.05(400) \\ 0.03(200) + 0.95(400) \end{pmatrix} = \begin{pmatrix} 214 \\ 386 \end{pmatrix}$$

$$\vec{p}_{2007} = \mathbf{T}\vec{p}_{2006} = \begin{pmatrix} 0.97 & 0.05 \\ 0.03 & 0.95 \end{pmatrix} \begin{pmatrix} 214 \\ 386 \end{pmatrix}$$

$$= \begin{pmatrix} 0.97(214) + 0.05(386) \\ 0.03(214) + 0.95(386) \end{pmatrix} = \begin{pmatrix} 226.88 \\ 373.12 \end{pmatrix}.$$

12. **(a)** For part (i) we have

$$\mathbf{I_2}\vec{u} = \begin{pmatrix} 1 & 0 \\ 0 & 1 \end{pmatrix} \begin{pmatrix} 3 \\ 2 \end{pmatrix} = \begin{pmatrix} 1\cdot 3 + 0\cdot 2 \\ 0\cdot 3 + 1\cdot 2 \end{pmatrix} = \begin{pmatrix} 3 \\ 2 \end{pmatrix}.$$

For part (ii) we have

$$\mathbf{I_2}\vec{u} = \begin{pmatrix} 1 & 0 \\ 0 & 1 \end{pmatrix} \begin{pmatrix} 0 \\ 7 \end{pmatrix} = \begin{pmatrix} 1\cdot 0 + 0\cdot 7 \\ 0\cdot 0 + 1\cdot 7 \end{pmatrix} = \begin{pmatrix} 0 \\ 7 \end{pmatrix}.$$

For part (iii) we have

$$\mathbf{I_2}\vec{u} = \begin{pmatrix} 1 & 0 \\ 0 & 1 \end{pmatrix} \begin{pmatrix} a \\ b \end{pmatrix} = \begin{pmatrix} 1\cdot a + 0\cdot b \\ 0\cdot a + 1\cdot b \end{pmatrix} = \begin{pmatrix} a \\ b \end{pmatrix}.$$

Notice that in each case, $\mathbf{I_2}\vec{u} = \vec{u}$. In other words, $\mathbf{I_2}$ has no effect on $\vec{u}$.

(b) For part (i) we have

$$\mathbf{I_3}\vec{u} = \begin{pmatrix} 1 & 0 & 0 \\ 0 & 1 & 0 \\ 0 & 0 & 1 \end{pmatrix} \begin{pmatrix} -1 \\ 5 \\ 7 \end{pmatrix}$$

$$= \begin{pmatrix} 1\cdot -1 + 0\cdot 5 + 0\cdot 7 \\ 0\cdot -1 + 1\cdot 5 + 0\cdot 7 \\ 0\cdot -1 + 0\cdot 5 + 1\cdot 7 \end{pmatrix} = \begin{pmatrix} -1 \\ 5 \\ 7 \end{pmatrix}.$$

For part (ii) we have

$$\mathbf{I_3}\vec{u} = \begin{pmatrix} 1 & 0 & 0 \\ 0 & 1 & 0 \\ 0 & 0 & 1 \end{pmatrix} \begin{pmatrix} 3 \\ 8 \\ 1 \end{pmatrix}$$

$$= \begin{pmatrix} 1\cdot 3 + 0\cdot 8 + 0\cdot 1 \\ 0\cdot 3 + 1\cdot 8 + 0\cdot 1 \\ 0\cdot 3 + 0\cdot 8 + 1\cdot 1 \end{pmatrix} = \begin{pmatrix} 3 \\ 8 \\ 1 \end{pmatrix}.$$

For part (iii) we have

$$\mathbf{I_3}\vec{u} = \begin{pmatrix} 1 & 0 & 0 \\ 0 & 1 & 0 \\ 0 & 0 & 1 \end{pmatrix} \begin{pmatrix} a \\ b \\ c \end{pmatrix} = \begin{pmatrix} 1\cdot a + 0\cdot b + 0\cdot c \\ 0\cdot a + 1\cdot b + 0\cdot c \\ 0\cdot a + 0\cdot b + 1\cdot c \end{pmatrix} = \begin{pmatrix} a \\ b \\ c \end{pmatrix}.$$

Notice that in each case, $\mathbf{I_3}\vec{u} = \vec{u}$. In other words, $\mathbf{I_3}$ has no effect on $\vec{u}$.

(c) Following the pattern, we might expect $\mathbf{I_4}$ to be a 4-by-4 matrix with 1s along the diagonal and 0s everywhere else. Letting $\vec{w} = (a, b, c, d)$, we see that:

$$\mathbf{I_4}\vec{w} = \begin{pmatrix} 1 & 0 & 0 & 0 \\ 0 & 1 & 0 & 0 \\ 0 & 0 & 1 & 0 \\ 0 & 0 & 0 & 1 \end{pmatrix}\begin{pmatrix} a \\ b \\ c \\ d \end{pmatrix}$$

$$= \begin{pmatrix} 1\cdot a+0\cdot b+0\cdot c+0\cdot d \\ 0\cdot a+1\cdot b+0\cdot c+0\cdot d \\ 0\cdot a+0\cdot b+1\cdot c+0\cdot d \\ 0\cdot a+0\cdot b+0\cdot c+1\cdot d \end{pmatrix} = \begin{pmatrix} a \\ b \\ c \\ d \end{pmatrix}.$$

Once again, we see that $\mathbf{I_4}$ has no effect on $\vec{w}$.

Matrices like $\mathbf{I_2}, \mathbf{I_3}$, and $\mathbf{I_4}$ are referred to as *identity matrices*. Multiplying a vector (or another matrix) by an identity matrix has no effect. The notion is similar to multiplying a number by 1, which is why the number 1 is sometimes referred to as the *multiplicative identity*.

13. **(a)** We first find $\vec{v}$:

$$\vec{v} = \mathbf{A}\vec{u} = \begin{pmatrix} 2 & 1 \\ 3 & 2 \end{pmatrix}\begin{pmatrix} 3 \\ 5 \end{pmatrix} = \begin{pmatrix} 11 \\ 19 \end{pmatrix}.$$

Now, we show that $\vec{u} = \mathbf{A}^{-1}\vec{v}$:

$$\mathbf{A}^{-1}\vec{v} = \begin{pmatrix} 2 & -1 \\ -3 & 2 \end{pmatrix}\begin{pmatrix} 11 \\ 19 \end{pmatrix} = \begin{pmatrix} 3 \\ 5 \end{pmatrix} = \vec{u}.$$

(b) We first find $\vec{v}$:

$$\vec{v} = \mathbf{A}\vec{u} = \begin{pmatrix} 2 & 1 \\ 3 & 2 \end{pmatrix}\begin{pmatrix} -1 \\ 7 \end{pmatrix} = \begin{pmatrix} 5 \\ 11 \end{pmatrix}.$$

Now, we show that $\vec{u} = \mathbf{A}^{-1}\vec{v}$:

$$\mathbf{A}^{-1}\vec{v} = \begin{pmatrix} 2 & -1 \\ -3 & 2 \end{pmatrix}\begin{pmatrix} 5 \\ 11 \end{pmatrix} = \begin{pmatrix} -1 \\ 7 \end{pmatrix} = \vec{u}.$$

(c) We first find $\vec{v}$:

$$\vec{v} = \mathbf{A}\vec{u} = \begin{pmatrix} 2 & 1 \\ 3 & 2 \end{pmatrix}\begin{pmatrix} a \\ b \end{pmatrix} = \begin{pmatrix} 2a+b \\ 3a+2b \end{pmatrix}.$$

Now, we show that $\vec{u} = \mathbf{A}^{-1}\vec{v}$:

$$\mathbf{A}^{-1}\vec{v} = \begin{pmatrix} 2 & -1 \\ -3 & 2 \end{pmatrix}\begin{pmatrix} 2a+b \\ 3a+2b \end{pmatrix}$$

$$= \begin{pmatrix} 2(2a+b)-(3a+2b) \\ -3(2a+b)+2(3a+2b) \end{pmatrix}$$

$$= \begin{pmatrix} 4a+2b-3a-2b \\ -6a-3b+6a+4b \end{pmatrix}$$

$$= \begin{pmatrix} a \\ b \end{pmatrix} = \vec{u}.$$

14. **(a)** From Problem 13, we know that $\mathbf{A} = \begin{pmatrix} 2 & 1 \\ 3 & 2 \end{pmatrix}$ and so $a = 2, b = 1, c = 3, d = 2$. This gives $D = ad - bc = 2(2) - 1(3) = 1$. We have

$$\mathbf{A}^{-1} = \frac{1}{D}\begin{pmatrix} d & -b \\ -c & a \end{pmatrix} = (1)\begin{pmatrix} 2 & -1 \\ -3 & 2 \end{pmatrix} = \begin{pmatrix} 2 & -1 \\ -3 & 2 \end{pmatrix},$$

as required.

(b) For $\mathbf{B} = \begin{pmatrix} 3 & 11 \\ 1 & 7 \end{pmatrix}$, we have $a = 3, b = 11, c = 1, d = 7$. This gives $D = ad - bc = 3(7) - 11(1) = 10$. We have

$$\mathbf{B}^{-1} = \frac{1}{D}\begin{pmatrix} d & -b \\ -c & a \end{pmatrix} = \frac{1}{10}\begin{pmatrix} 7 & -11 \\ -1 & 3 \end{pmatrix} = \begin{pmatrix} 0.7 & -1.1 \\ -0.1 & 0.3 \end{pmatrix}.$$

To verify this, we first find $\vec{v} = \mathbf{B}\vec{u}$:

$$\vec{u} = \mathbf{B}\vec{u} = \begin{pmatrix} 3 & 11 \\ 1 & 7 \end{pmatrix}\begin{pmatrix} a \\ b \end{pmatrix} = \begin{pmatrix} 3a + 11b \\ a + 7b \end{pmatrix}.$$

We now show that $\vec{v} = \mathbf{B}^{-1}\vec{u}$:

$$\begin{aligned} \mathbf{B}^{-1}\vec{u} &= \begin{pmatrix} 0.7 & -1.1 \\ -0.1 & 0.3 \end{pmatrix}\begin{pmatrix} 3a + 11b \\ a + 7b \end{pmatrix} \\ &= \begin{pmatrix} 0.7(3a + 11b) - 1.1(a + 7b) \\ -0.1(3a + 11b) + 0.3(a + 7b) \end{pmatrix} \\ &= \begin{pmatrix} 2.1a + 7.7b - 1.1a - 7.7b \\ -0.3a - 1.1b + 0.3a + 2.1b \end{pmatrix} \\ &= \begin{pmatrix} a \\ b \end{pmatrix} = \vec{u}. \end{aligned}$$

(c) For $\mathbf{C} = \begin{pmatrix} 2 & 8 \\ 3 & 12 \end{pmatrix}$ we have $a = 2, b = 8, c = 3, d = 12$. This gives $D = ad - bc = 2(12) - 8(3) = 0$. Since $1/D$ is undefined, $\mathbf{C}^{-1} = (1/D)\begin{pmatrix} d & -b \\ -c & a \end{pmatrix}$ is undefined.

15. **(a)** We have

$$\mathbf{A}\vec{v_2} = \begin{pmatrix} -2 & -1 \\ 8 & 7 \end{pmatrix}\begin{pmatrix} 1 \\ -1 \end{pmatrix} = \begin{pmatrix} -1 \\ 1 \end{pmatrix} = -1\begin{pmatrix} -1 \\ 1 \end{pmatrix},$$

so $\lambda_2 = -1$.

(b) We have

$$\mathbf{A}\vec{v_3} = \begin{pmatrix} -2 & -1 \\ 8 & 7 \end{pmatrix}\begin{pmatrix} -3 \\ 3 \end{pmatrix} = \begin{pmatrix} 3 \\ -3 \end{pmatrix} = -1\begin{pmatrix} -3 \\ 3 \end{pmatrix},$$

so $\lambda_3 = -1$, the same as λ_2.

The reason $\lambda_2 = \lambda_3$ is because $\vec{v_3} = -3\vec{v_2}$, that is, because $\vec{v_3}$ is a multiple of $\vec{v_2}$. Since $\mathbf{A}$ multiplies $\vec{v_2}$ by -1, it also multiples $\vec{v_3}$ by -1.

(c) Since $\vec{v}$ is an eigenvector of $\mathbf{A}$, we have $\mathbf{A}\vec{v} = \lambda\vec{v}$, $\lambda \neq 0$. We know that multiplying a vector $\vec{v}$ by a non-zero scalar λ produces a vector whose direction is the same as $\vec{v}$ and whose length is λ times the length of $\vec{v}$. This means that $\mathbf{A}\vec{v}$ is parallel to $\vec{v}$.

16. **(a)** We have

$$\vec{v} = 3\vec{r_1} + 5\vec{r_2} = 3\underbrace{(3,2)}_{\vec{r_1}} + 5\underbrace{(0,1)}_{\vec{r_2}}$$

$$= \underbrace{\begin{pmatrix} \boxed{\begin{matrix} 3 \\ 2 \end{matrix}} & \boxed{\begin{matrix} 0 \\ 1 \end{matrix}} \end{pmatrix}}_{\vec{r_1}\ \ \vec{r_2}} \underbrace{\begin{pmatrix} 3 \\ 2 \end{pmatrix}}_{\vec{u}}$$

$$= (9, 11),$$

so $\vec{v} = (9,11)$, $\vec{u} = (3,2)$, and $\mathbf{R} = \begin{pmatrix} 3 & 0 \\ 2 & 1 \end{pmatrix}$.

(b) We have

$$\vec{q} = \underbrace{\begin{pmatrix} \boxed{\begin{matrix} 3 \\ 2 \end{matrix}} & \boxed{\begin{matrix} 4 \\ 3 \end{matrix}} \end{pmatrix}}_{\vec{s_1}\ \ \vec{s_2}} \underbrace{\begin{pmatrix} 3 \\ -2 \end{pmatrix}}_{\vec{p}}$$

$$= 3\underbrace{(3,2)}_{\vec{s_1}} - 2\underbrace{(4,3)}_{\vec{s_2}}$$

$$= 3\vec{s_1} - 2\vec{s_2} = (1, 0).$$

17. **(a)** We have $\mathbf{C} = \begin{pmatrix} 3 & 5 \\ 2 & 4 \end{pmatrix}$ and so

$$\begin{aligned}
\mathbf{C}\vec{u} &= \begin{pmatrix} 3 & 5 \\ 2 & 4 \end{pmatrix}\begin{pmatrix} a \\ b \end{pmatrix} \\
&= \begin{pmatrix} 3a + 5b \\ 2a + 4b \end{pmatrix} \\
&= \begin{pmatrix} 3a \\ 2a \end{pmatrix} + \begin{pmatrix} 5b \\ 4b \end{pmatrix} \\
&= a\begin{pmatrix} 3 \\ 2 \end{pmatrix} + b\begin{pmatrix} 5 \\ 4 \end{pmatrix} \\
&= a\vec{c_1} + b\vec{c_2}.
\end{aligned}$$

(b) Provided $\mathbf{C}^{-1}$ exists, we can write $\vec{u} = \mathbf{C}^{-1}v$. We can find $\mathbf{C}^{-1}$ using the approach from Problem 14: we have $a = 3, b = 5, c = 2, d = 4$, and so $D = ad - bc = 3(4) - 5(2) = 2$. We have

$$\mathbf{C}^{-1} = \frac{1}{D}\begin{pmatrix} d & -b \\ -c & a \end{pmatrix} = \frac{1}{2}\begin{pmatrix} 4 & -5 \\ -2 & 3 \end{pmatrix} = \begin{pmatrix} 2 & -2.5 \\ -1 & 1.5 \end{pmatrix}.$$

This means that

$$\vec{u} = \mathbf{C}^{-1}\vec{v}$$

$$= \begin{pmatrix} 2 & -2.5 \\ -1 & 1.5 \end{pmatrix} \begin{pmatrix} 2 \\ 5 \end{pmatrix}$$

$$= \begin{pmatrix} 2(2) - 2.5(5) \\ -1(2) + 1.5(5) \end{pmatrix}$$

$$= \begin{pmatrix} -8.5 \\ 5.5 \end{pmatrix}.$$

(c) From parts (a) and (b), we see that

$$\vec{v} = \begin{pmatrix} 2 \\ 5 \end{pmatrix} = \mathbf{C}\vec{u} = \begin{pmatrix} 3 & 2 \\ 5 & 4 \end{pmatrix} \begin{pmatrix} -8.5 \\ 5.5 \end{pmatrix}.$$

Referring to Problem 16, we see that

$$\vec{v} = \begin{pmatrix} 2 \\ 5 \end{pmatrix} = \Bigg(\underbrace{\boxed{\begin{matrix} 3 \\ 2 \end{matrix}}}_{\vec{c_1}} \underbrace{\boxed{\begin{matrix} 5 \\ 4 \end{matrix}}}_{\vec{c_2}} \Bigg) \underbrace{\begin{pmatrix} -8.5 \\ 5.5 \end{pmatrix}}_{\vec{u}}$$

$$= -8.5 \underbrace{(3,2)}_{\vec{c_1}} + 5.5 \underbrace{(5,4)}_{\vec{c_2}}$$

$$= -8.5\vec{c_1} + 5.5\vec{c_2},$$

which gives $\vec{v}$ as a combination of $\vec{c_1}$ and $\vec{c_2}$.

18. Let $\mathbf{C} = \begin{pmatrix} 6 & 7 \\ 1 & 2 \end{pmatrix}$ be the matrix with columns $\vec{c_1}$ and $\vec{c_2}$, and let $\vec{u} = (a, b)$. Then $\vec{v} = \mathbf{C}\vec{u}$. Solving for $\vec{u}$, we first find $\mathbf{C}^{-1}$: we have $a = 6$, $b = 7$, $c = 1$, $d = 2$, and so $D = ad - bc = 6(2) - 7(1) = 5$. This gives

$$\mathbf{C}^{-1} = \frac{1}{D} \begin{pmatrix} d & -b \\ -c & a \end{pmatrix} = \frac{1}{5} \begin{pmatrix} 2 & -7 \\ -1 & 6 \end{pmatrix} = \begin{pmatrix} 0.4 & -1.4 \\ -0.2 & 1.2 \end{pmatrix}.$$

This means that

$$\vec{u} = \mathbf{C}^{-1}\vec{v} = \begin{pmatrix} 0.4 & -1.4 \\ -0.2 & 1.2 \end{pmatrix} \begin{pmatrix} 5 \\ 8 \end{pmatrix} = \begin{pmatrix} 0.4(5) - 1.4(8) \\ -0.2(5) + 1.2(8) \end{pmatrix} = \begin{pmatrix} -9.2 \\ 8.6 \end{pmatrix}.$$

We can now write

$$\vec{v} = \begin{pmatrix} 5 \\ 8 \end{pmatrix} = \Bigg(\underbrace{\boxed{\begin{matrix} 6 \\ 1 \end{matrix}}}_{\vec{c_1}} \underbrace{\boxed{\begin{matrix} 7 \\ 2 \end{matrix}}}_{\vec{c_2}} \Bigg) \underbrace{\begin{pmatrix} -9.2 \\ 8.6 \end{pmatrix}}_{\vec{u}}$$

$$= -9.2 \underbrace{(6,1)}_{\vec{c_1}} + 8.6 \underbrace{(7,2)}_{\vec{c_2}}$$

$$= -9.2\vec{c_1} + 8.6\vec{c_2},$$

which gives $\vec{v}$ as a combination of $\vec{c_1}$ and $\vec{c_2}$.

Solutions for Chapter 12 Review

Exercises

1. $3\vec{c} = 3(1,1,2) = (3,3,6)$.

2. $\vec{a} + \vec{b} = (5,1,0) + (2,-1,9) = (7,0,9)$

3. $\vec{b} - \vec{a} = (2,-1,9) - (5,1,0) = (-3,-2,9)$.

4. $\vec{a} + 2(\vec{b} + \vec{c}) = (5,1,0) + 2((2,-1,9) + (1,1,2)) = (5,1,0) + 2(3,0,11) = (11,1,22)$.

5. $2\vec{a} - 3(\vec{b} - \vec{c}) = 2(5,1,0) - 3((2,-1,9) - (1,1,2)) = (10,2,0) - 3(1,-2,7) = (7,8,-21)$.

6. $2(\vec{b} + 4(\vec{a} + \vec{c})) = 2((2,-1,9) + 4((5,1,0) + (1,1,2)) = 2((2,-1,9) + 4(6,2,2)) = 2(26,7,17) = (52,14,34)$.

7. $\vec{a} + \vec{b} - (\vec{a} - \vec{b}) = \vec{a} + \vec{b} - \vec{a} + \vec{b} = 2\vec{b} = 2(2,-1,9) = (4,-2,18)$.

8. $4(\vec{c} + 2(\vec{a} - \vec{c}) - 2\vec{a}) = 4(-\vec{c}) = -4(1,1,2) = (-4,-4,-8)$.

9. $-4\vec{i} + 8\vec{j} - 0.5\vec{i} + 0.5\vec{k} = -4.5\vec{i} + 8\vec{j} + 0.5\vec{k}$

10. $\vec{i} + 2\vec{j} - 6\vec{i} - 3\vec{j} = -5\vec{i} - \vec{j}$.

11. $(3\vec{i} + \vec{j}) \cdot (5\vec{i} - 2\vec{j}) = 15 - 2 = 13$.

12. $(3\vec{j} - 2\vec{k} + \vec{i}) \cdot (4\vec{k} - 2\vec{i} + 3\vec{j}) = (\vec{i} + 2\vec{j} - 3\vec{k}) \cdot (-2\vec{i} + 3\vec{j} + 4\vec{k}) = -2 + 6 - 12 = -8$.

13. $(5\vec{i} - \vec{j} - 3\vec{k}) \cdot (2\vec{i} + \vec{j} + \vec{k}) = 5 \cdot 2 - 1 \cdot 1 - 3 \cdot 1 = 6$.

14. Since $2\vec{i} \cdot 5\vec{j} = 0$,

$$(2\vec{i} \cdot 5\vec{j})(\vec{i} + \vec{j} + \vec{k}) = \vec{0}.$$

15. Since the dot product is a scalar, $(2\vec{i} + 5\vec{j}) \cdot 3\vec{i} = 6$, we have $\vec{i} + \vec{j} + \vec{k}$ multiplied by 6, that is

$$(2\vec{i} + 5\vec{j}) \cdot 3\vec{i}(\vec{i} + \vec{j} + \vec{k}) = 6(\vec{i} + \vec{j} + \vec{k}) = 6\vec{i} + 6\vec{j} + 6\vec{k}.$$

16. Since $(2\vec{i} + 3\vec{j} + \vec{k}) \cdot (3\vec{i} + \vec{j} + 4\vec{k}) = 6 + 3 + 4 = 13$, we have

$$(\vec{i} + \vec{j} + \vec{k})(2\vec{i} + 3\vec{j} + \vec{k}) \cdot (3\vec{i} + \vec{j} + 4\vec{k}) = (\vec{i} + \vec{j} + \vec{k})13 = 13\vec{i} + 13\vec{j} + 13\vec{k}.$$

17. $\vec{a} = \vec{b} = \vec{c} = 3\vec{k}, \quad \vec{d} = 2\vec{i} + 3\vec{k}, \quad \vec{e} = \vec{j}, \quad \vec{f} = -2\vec{i}$.

18. $\vec{u} = \vec{i} + \vec{j} + 2\vec{k}$ and $\vec{v} = -\vec{i} + 2\vec{k}$.

19. $\|\vec{u}\| = \sqrt{1^2 + 1^2 + 2^2} = \sqrt{6}, \|\vec{v}\| = \sqrt{(-1)^2 + 2^2} = \sqrt{5}$.

20. The length of the vector $\vec{i} - \vec{j} + 2\vec{k}$ is $\sqrt{1^2 + (-1)^2 + 2^2} = \sqrt{6}$. We can scale the vector down to length 2 by multiplying it by $\frac{2}{\sqrt{6}}$. So the answer is $\frac{2}{\sqrt{6}}\vec{i} - \frac{2}{\sqrt{6}}\vec{j} + \frac{4}{\sqrt{6}}\vec{k}$.

Problems

21. **(a)** To be parallel, vectors must be scalar multiples. The $\vec{k}$ component of the first vector is 2 times the $\vec{k}$ component of the second vector. So the $\vec{i}$ components of the two vectors must be in a 2:1 ratio, and the same is true for the $\vec{j}$ components. Thus, $4 = 2a$ and $a = 2(a-1)$. These equations have the solution $a = 2$, and for that value, the vectors are parallel.

(b) Perpendicular means a zero dot product. So $4a + a(a-1) + 18 = 0$, or $a^2 + 3a + 18 = 0$. Since $b^2 - 4ac = 9 - 4 \cdot 1 \cdot 18 = -63 < 0$, there are no real solutions. This means the vectors are never perpendicular.

22. **(a)** Let $\vec{a}$ be the velocity of P relative to the ground. Since the wheel is spinning at $6\pi/2\pi(1) = 3$ revolutions/second, the magnitude $\|\vec{a}\|$ is 3 rev/sec $\cdot 2\pi(1)$ ft/rev $= 6\pi$ ft/sec. See Figure 12.46.

(b) Let $\vec{b}$ be the velocity of the axle relative to the ground. Then we are given that $\|\vec{b}\| = 6\pi$ ft/sec. See Figure 12.47.

(c) Let $\vec{c}$ be the velocity of P relative to the ground. If $\vec{c}$ is the velocity of P relative to the ground, then we know that $\vec{c} = \vec{a} + \vec{b}$. We can construct the diagram in Figure 12.48 by adding $\vec{a}$ and $\vec{b}$, which are both of magnitude 6π. See Figure 12.48.

(d) The point P stops when it touches the ground. There $\vec{a} = -\vec{b}$ and so $\vec{c} = \vec{a} + \vec{b} = 0$ at that point. The fastest speed is at the top, when $\vec{a} = \vec{b}$ and so $\|\vec{c}\| = \|\vec{a}\| + \|\vec{b}\| = 12\pi$ ft/sec.

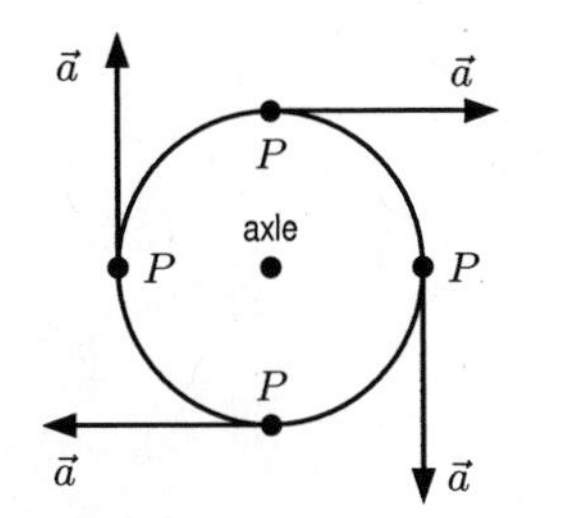

Figure 12.46: $\vec{a}$ is the velocity of P relative to the axle

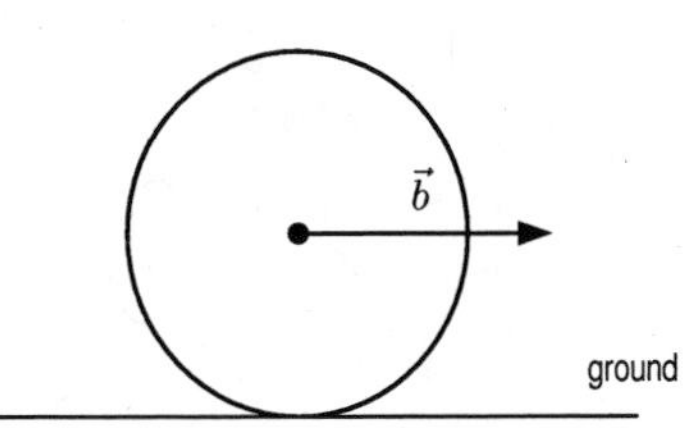

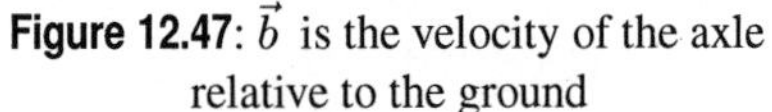

Figure 12.47: $\vec{b}$ is the velocity of the axle relative to the ground

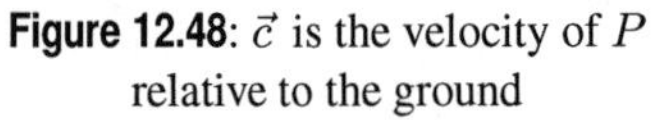

Figure 12.48: $\vec{c}$ is the velocity of P relative to the ground

23. **(a)** The wait list is given by

$$\vec{L} = (51, 47, 41, 22, 23) - (40, 40, 30, 15, 10) = (11, 7, 11, 7, 13).$$

(b) We have

$$\vec{F} = (40, 40, 30, 15, 10) - (8, 4, 9, 7, 6) = (32, 36, 21, 8, 4)$$
$$\vec{G} = (11, 7, 11, 7, 13) - (8, 4, 9, 7, 6) = (3, 3, 2, 0, 7).$$

$\vec{F}$ tells us the enrollment at each level after the dropouts have occurred. $\vec{G}$ tells us the number remaining on the wait list after the opened spots have been filled.

24. **(a)** We have

$$\begin{aligned}\vec{E}_{\text{max}} \cdot \vec{T} &= (40, 40, 30, 15, 10) \cdot (30, 30, 40, 80, 120)\\ &= 40 \cdot 30 + 40 \cdot 30 + 30 \cdot 40 + 15 \cdot 80 + 10 \cdot 120 = 6000,\end{aligned}$$

so the total weekly tuition for all students is \$6000.

(b) We have

$$\vec{T}_{\text{new}} = (30, 30, 40, 80, 120) + (5, 5, 10, 20, 30) = (35, 35, 50, 100, 150).$$

This is the new tuition vector. We see that

$$\begin{aligned}\vec{E}_{\text{max}} \cdot \vec{R} &= (40, 40, 30, 15, 10) \cdot (5, 5, 10, 20, 30)\\ &= 40 \cdot 5 + 40 \cdot 5 + 30 \cdot 10 + 15 \cdot 20 + 10 \cdot 30 = 1300,\end{aligned}$$

which tells us that the total change in weekly tuition is \$1300. Also,

$$\begin{aligned}\vec{E}_{\text{max}} \cdot \vec{T}_{\text{new}} &= (40, 40, 30, 15, 10) \cdot (35, 35, 50, 100, 150)\\ &= 40 \cdot 35 + 40 \cdot 35 + 30 \cdot 50 + 15 \cdot 100 + 10 \cdot 150 = 7300,\end{aligned}$$

which is the new tuition.

Notice that $\vec{E}_{\text{max}} \cdot (\vec{T} + \vec{R}) = \vec{E}_{\text{max}} \cdot \vec{T} + \vec{E}_{\text{max}} \cdot \vec{R}$, as required by the distributive law for the dot product.

25. See Figure 12.49, where $\vec{g}$ is the acceleration due to gravity, and $g = \|\vec{g}\,\|$.
If $\theta = 0$ (the plank is at ground level), the sliding force is $F = 0$.
If $\theta = \pi/2$ (the plank is vertical), the sliding force equals g, the force due to gravity.
Therefore, we can guess that F is proportional with $\sin\theta$:

$$F = g\sin\theta.$$

This agrees with the bounds at $\theta = 0$ and $\theta = \pi/2$, and with the fact that the sliding force is smaller than g between 0 and $\pi/2$.

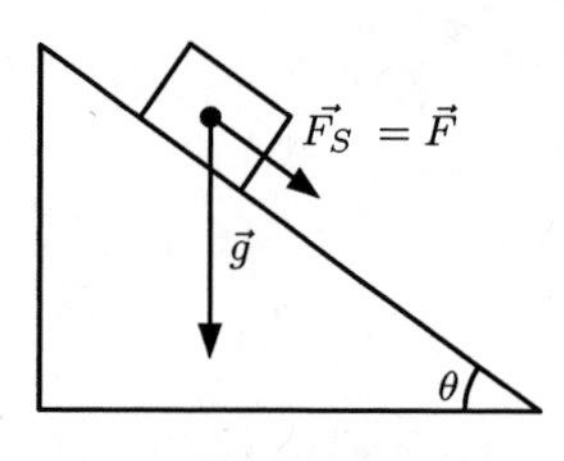

Figure 12.49

26. **(a)** See Figure 12.50, where $\vec{g}$ is the acceleration due to gravity, and $g = \|\vec{g}\,\|$. As in Problem 25, $F = g\sin\theta$. If we let l be the length of the plank, then

$$\sin\theta = \frac{2t}{l},$$

because the plank's height increases by 2 ft/sec. Therefore,

$$F = h(t) = g\frac{2t}{l} = 2\frac{gt}{l}.$$

(b) The block starts sliding when time t satisfies

$$F = 3 \text{ lbs}$$
$$\frac{2gt}{l} = 3$$
$$t = \frac{3l}{2g}.$$

Therefore the block will begin to slide at $t = 3l/2g$.

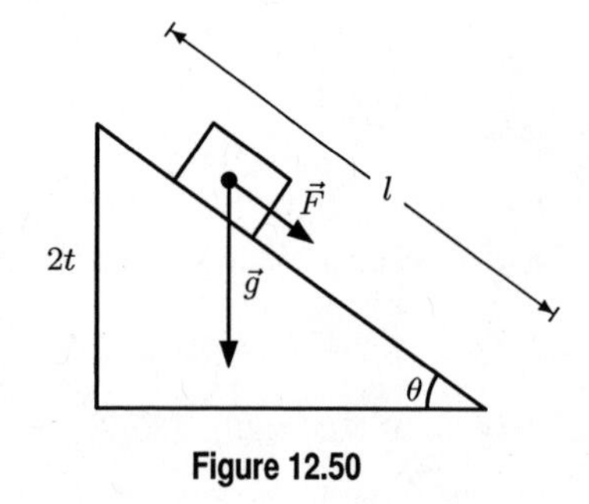

Figure 12.50

27. See Figure 12.51, where $\vec{g}$ is the acceleration due to gravity, and $g = \|\vec{g}\,\|$. Writing $F_S = \|\vec{F_S}\,\|$, we see that

$$\sin\theta = \frac{F_S}{g}.$$

Therefore $F_S = g\sin\theta$, which agrees with the answer for Problem 25.

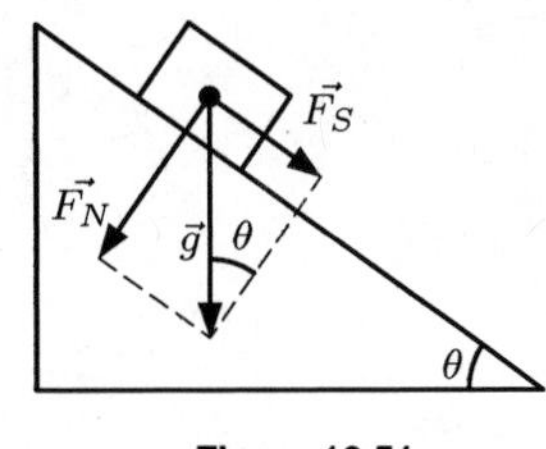

Figure 12.51

28. The velocity vector of the plane with respect to the calm air has the form

$$\vec{v} = a\vec{i} + 80\vec{k} \text{ where } \|\vec{v}\| = 480.$$

(See Figure 12.52.) Therefore $\sqrt{a^2 + 80^2} = 480$ so $a = \sqrt{480^2 - 80^2} \approx 473.286$ km/hr. We conclude that $\vec{v} \approx 473.286\vec{i} + 80\vec{k}$.

The wind vector is

$$\begin{aligned}\vec{w} &= 100(\cos 45°)\vec{i} + 100(\sin 45°)\vec{j} \\ &\approx 70.711\vec{i} + 70.711\vec{j}.\end{aligned}$$

The velocity vector of the plane with respect to the ground is then

$$\begin{aligned}\vec{v} + \vec{w} &= (473.286\vec{i} + 80\vec{k}) + (70.711\vec{i} + 70.711\vec{j}) \\ &= 544\vec{i} + 70.711\vec{j} + 80\vec{k}.\end{aligned}$$

From Figure 12.53, we see that the velocity relative to the ground is

$$543.997\vec{i} + 70.7\vec{j}.$$

The ground speed is therefore $\sqrt{543.997^2 + 70.7^2} \approx 548.573$ km/hr.

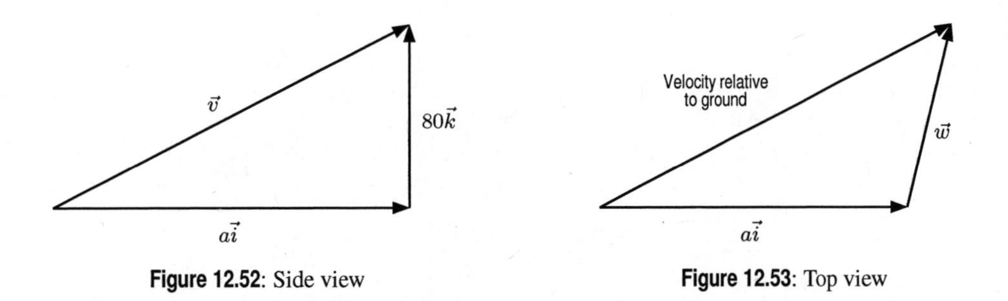

Figure 12.52: Side view

Figure 12.53: Top view

29. The speed of the particle before impact is v, so the speed after impact is $0.8v$. If we consider the barrier as being along the x-axis (see Figure 12.54), then the $\vec{i}$-component is $0.8v\cos 60° = 0.8v(0.5) = 0.4v$.

Similarly, the $\vec{j}$-component is $0.8v\sin 60° = 0.8v(0.8660) \approx 0.693v$. Thus

$$\vec{v}_{\text{after}} = 0.4v\vec{i} + 0.693v\vec{j}.$$

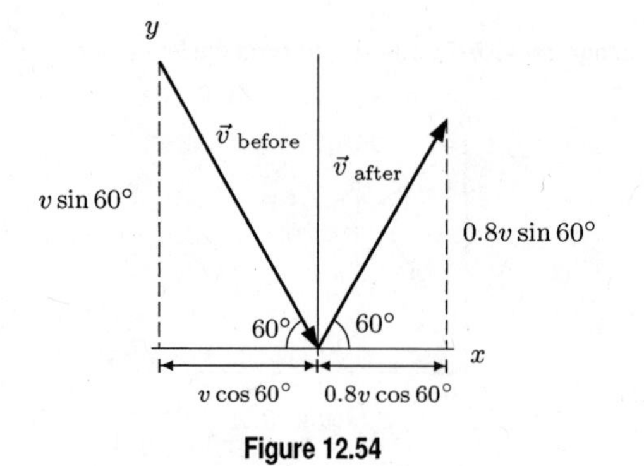

Figure 12.54

30. We must check that all the points are the same distance apart, i.e., the magnitude of the displacement vectors $\overrightarrow{OA}$, $\overrightarrow{OB}$, $\overrightarrow{OC}$, $\overrightarrow{BA}$, $\overrightarrow{CB}$ and $\overrightarrow{CA}$ is the same. Here goes:

$$\begin{aligned}
\|\overrightarrow{OA}\| &= \|(2\vec i + 0\vec j + 0\vec k) - (0\vec i + 0\vec j + 0\vec k)\| = \sqrt{2^2+0^2+0^2} = 2\\
\|\overrightarrow{OB}\| &= \|(1\vec i + \sqrt{3}\vec j + 0\vec k) - (0\vec i + 0\vec j + 0\vec k)\| = \sqrt{1^2+(\sqrt{3})^2+0^2} = 2\\
\|\overrightarrow{OC}\| &= \|(1\vec i + 1/\sqrt{3}\vec j + 2\sqrt{2/3}\vec k) - (0\vec i + 0\vec j + 0\vec k)\| = \sqrt{1+1/3+4(2/3)} = 2\\
\|\overrightarrow{BA}\| &= \|(2\vec i + 0\vec j + 0\vec k) - (1\vec i + \sqrt{3}\vec j + 0\vec k)\| = \sqrt{1+3+0} = 2\\
\|\overrightarrow{CB}\| &= \|(1\vec i + \sqrt{3}\vec j + 0\vec k) - (1\vec i + 1/\sqrt{3}\vec j + 2\sqrt{2/3}\vec k)\|\\
&= \sqrt{0^2+(\sqrt{3}-1/\sqrt{3})^2+4(2/3)} = \sqrt{3-2+1/3+8/3} = 2\\
\|\overrightarrow{CA}\| &= \|(2\vec i + 0\vec j + 0\vec k) - (1\vec i + 1/\sqrt{3}\vec j + 2\sqrt{2/3}\vec k)\| = \sqrt{1+1/3+4(2/3)} = 2.
\end{aligned}$$

31. **(a)** To get from A to B, you must go down 7, to the left 2, and forward 2. So $\overrightarrow{AB} = 2\vec i - 2\vec j - 7\vec k$. Similarly, $\overrightarrow{AC} = -2\vec i + 2\vec j - 7\vec k$.

(b) Remember

$$\cos\theta = \frac{\overrightarrow{AB}\cdot\overrightarrow{AC}}{\|\overrightarrow{AB}\|\|\overrightarrow{AC}\|} = \frac{(2)(-2)+(-2)(2)+(-7)(-7)}{\sqrt{57}\sqrt{57}} = \frac{41}{57}.$$

So $\theta = 44.003°$.

32. Suppose θ is the angle between $\vec u$ and $\vec v$.

(a) By the definition of scalar multiplication, we know that $-\vec v$ is in the opposite direction of $\vec v$, so the angle between $\vec u$ and $-\vec v$ is $\pi - \theta$. (See Figure 12.55.) Hence,

$$\begin{aligned}
\vec u \cdot (-\vec v) &= \|\vec u\|\|-\vec v\|\cos(\pi-\theta)\\
&= \|\vec u\|\|\vec v\|(-\cos\theta)\\
&= -(\vec u\cdot\vec v).
\end{aligned}$$

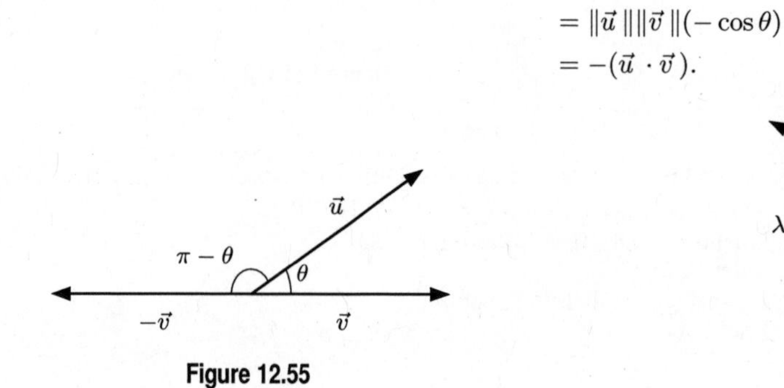

Figure 12.55

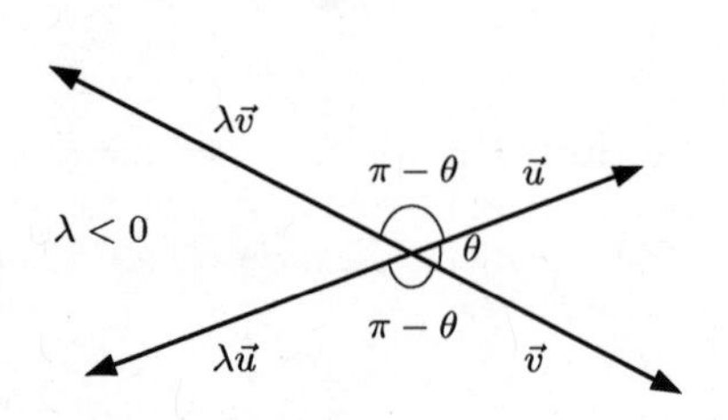

Figure 12.56

(b) If $\lambda < 0$, the angle between $\vec{u}$ and $\lambda\vec{v}$ is $\pi - \theta$, and so is the angle between $\lambda\vec{u}$ and $\vec{v}$. (See Figure 12.56.) So we have,

$$\begin{aligned}\vec{u} \cdot (\lambda\vec{v}) &= \|\vec{u}\|\|\lambda\vec{v}\| \cos(\pi - \theta) \\ &= |\lambda|\|\vec{u}\|\|\vec{v}\|(-\cos\theta) \\ &= -\lambda\|\vec{u}\|\|\vec{v}\|(-\cos\theta) \quad \text{since } |\lambda| = -\lambda \\ &= \lambda\|\vec{u}\|\|\vec{v}\|\cos\theta \\ &= \lambda(\vec{u} \cdot \vec{v}).\end{aligned}$$

By a similar argument, we have

$$\begin{aligned}(\lambda\vec{u}) \cdot \vec{v} &= \|\lambda\vec{u}\|\|\vec{v}\| \cos(\pi - \theta) \\ &= -\lambda\|\vec{u}\|\|\vec{v}\|(-\cos\theta) \\ &= \lambda(\vec{u} \cdot \vec{v}).\end{aligned}$$

33. If $\vec{x}$ and $\vec{y}$ are two consumption vectors corresponding to points satisfying the same budget constraint, then

$$\vec{p} \cdot \vec{x} = k = \vec{p} \cdot \vec{y}.$$

Therefore we have

$$\vec{p} \cdot (\vec{x} - \vec{y}) = \vec{p} \cdot \vec{x} - \vec{p} \cdot \vec{y} = 0.$$

Thus $\vec{p}$ and $\vec{x} - \vec{y}$ are perpendicular; that is, the difference between two consumption vectors on the same budget constraint is perpendicular to the price vector.

34. $\overrightarrow{AB} = 3\vec{i}$ and $\overrightarrow{CD} = -3\vec{i} - 4\vec{j}$.

35. $\overrightarrow{AB} = -\vec{u}$; $\overrightarrow{BC} = 3\vec{v}$; $\overrightarrow{AC} = \overrightarrow{AB} + \overrightarrow{BC} = -\vec{u} + 3\vec{v}$; $\overrightarrow{AD} = 3\vec{v}$.

36. The six sides are given in Figure 12.57.

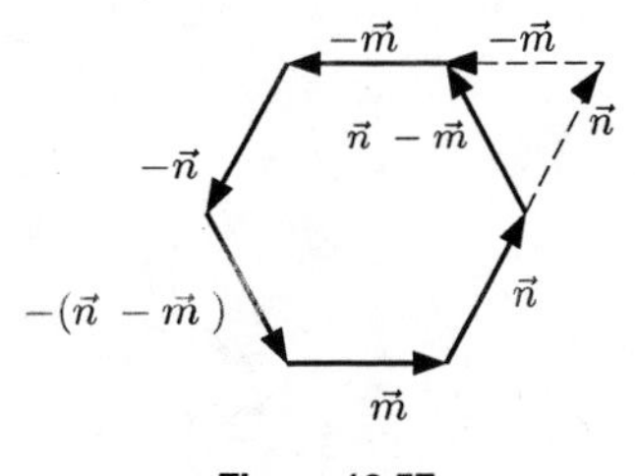

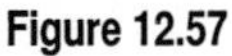
Figure 12.57

The three diagonals, $\vec{u}$, $\vec{v}$, $\vec{w}$, are given in Figures 12.58-12.60. We have

$$\vec{u} = 2\vec{m}, \quad \vec{v} = 2\vec{n}, \quad \vec{w} = 2\vec{n} - 2\vec{m}.$$

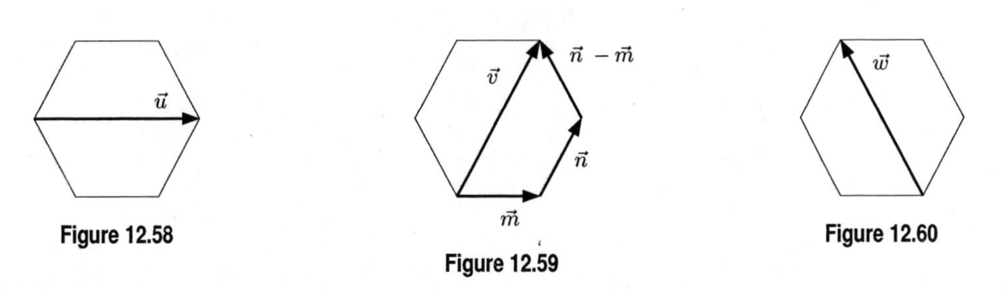

Figure 12.58

Figure 12.59

Figure 12.60

37. Using the result of Problem 36, we have $\overrightarrow{AC} = \vec{w} + \vec{n} - \vec{m} = 3\vec{n} - 3\vec{m}$; $\overrightarrow{AB} = \vec{v} + \vec{m} + \vec{n} = 3\vec{m} + \vec{n}$; $\overrightarrow{AD} = \vec{v} + \vec{m} - (\vec{n} - \vec{m}) = 4\vec{m} - \vec{n}$; $\overrightarrow{BD} = (-\vec{n}) - (\vec{n} - \vec{m}) = \vec{m} - 2\vec{n}$.

CHECK YOUR UNDERSTANDING

1. False. The length $||0.5\vec{i} + 0.5\vec{j}\,|| = \sqrt{(0.5)^2 + (0.5)^2} = \sqrt{0.5} \neq 1$.
2. False. It is longer only if $|k| > 1$.
3. False. We have $\| -\vec{u}\,\| = \|\vec{u}\,\| = 5$. Multiplying a vector by -1 does not change its length. The length of a vector is never negative.
4. True. Since $6\vec{i} - 2\vec{j} + 2\vec{k} = 2(3\vec{i} - \vec{j} + \vec{k}\,)$, the second vector is a multiple of the first and thus parallel to it.
5. False. The dot product $(2\vec{i} + \vec{j}\,) \cdot (2\vec{i} - \vec{j}\,) = 3$ is not zero.
6. False. Vectors in the direction opposite to $2\vec{i} + 3\vec{j}$ are all of the form $-2k\vec{i} - 3k\vec{j}$ where $k > 0$.
7. False. The angle θ between $\vec{u}$ and $\vec{v}$ satisfies $\vec{u} \cdot \vec{v} = \|\vec{u}\,\| \cdot \|\vec{u}\,\| \cos\theta < 0$, so $\cos\theta < 0$ and therefore $90° < \theta < 180°$.
8. False. The dot product of two vectors is a scalar.
9. True. If $\vec{u} = (u_1, u_2)$ and $\vec{v} = (v_1, v_2)$, then $\vec{u} + \vec{v}$ and $\vec{v} + \vec{u}$ both equal $(u_1 + v_1, u_2 + v_2)$.
10. False. For example, if $\vec{u} = \vec{i}$ and $\vec{v} = \vec{j}$ then $\|\vec{u} + \vec{v}\,\| = \sqrt{2}$ but $\|\vec{u}\,\| + \|\vec{v}\,\| = 2$.
11. True. If $\vec{u} = (u_1, u_2)$ and $\vec{v} = (v_1, v_2)$, then $\vec{u} \cdot \vec{v}$ and $\vec{v} \cdot \vec{u}$ both equal $u_1v_1 + u_2v_2$.
12. True. A dot product of zero ensures perpendicularity.
13. False. Distance is not a vector. The vector $\vec{i} + \vec{j}$ is the displacement vector from P to Q. The distance between the points is the length of the displacement vector: $\|\vec{i} + \vec{j}\,\| = \sqrt{2}$.
14. False. The vector $-4\vec{i} + 5\vec{j}$ is found by subtracting corresponding coordinates, not by adding them.
15. True. This is one definition of vector addition.
16. False. The length of $\vec{v}$ is the square root of $\vec{v} \cdot \vec{v}$.
17. True. Both vectors have length $\sqrt{13}$.
18. True. The first number, m, gives the number of rows and the second number, n, gives the number of columns.
19. False. We have $a_{23} = 6$, the entry in the second row, third column of the matrix $\mathbf{A}$.
20. False. The matrix $\mathbf{A}$ is 2×3 because it has 2 rows (horizontal) and three columns (vertical).
21. True. In the subscript, the first number gives the row and the second number gives the column.
22. False. The correct rotation matrix is $R = \begin{pmatrix} \cos\phi & -\sin\phi \\ \sin\phi & \cos\phi \end{pmatrix}$.
23. False. Multiplication is not commutative. For example, if $A = \begin{pmatrix} 0 & 1 \\ 1 & 0 \end{pmatrix}$ and $B = \begin{pmatrix} 2 & 0 \\ 0 & 3 \end{pmatrix}$, then $AB = \begin{pmatrix} 0 & 3 \\ 2 & 0 \end{pmatrix}$ and $BA = \begin{pmatrix} 0 & 2 \\ 3 & 0 \end{pmatrix}$.
24. True. The sum $\mathbf{A} + \mathbf{B}$ of two matrices is defined only when $\mathbf{A}$ and $\mathbf{B}$ have the same dimensions.
25. True. We have

$$A\vec{v} = \begin{pmatrix} 2 & 3 \\ -1 & 5 \end{pmatrix}\begin{pmatrix} 10 \\ 5 \end{pmatrix} = \begin{pmatrix} 2 \cdot 10 + 3 \cdot 5 \\ -10 + 5 \cdot 5 \end{pmatrix} = \begin{pmatrix} 35 \\ 15 \end{pmatrix}.$$

CHAPTER THIRTEEN

Solutions for Section 13.1

Exercises

1. Not arithmetic. The differences are 5, 4, 3.

2. Arithmetic. The differences are all -7.

3. Arithmetic. The differences are all 5.

4. Not arithmetic. The differences are $-7, -6, -5$.

5. Arithmetic, with $a = 6$, $d = 3$, so $a_n = 6 + (n-1)3 = 3 + 3n$.

6. Not arithmetic. The differences are $-2, 3, -4, \ldots$.

7. Arithmetic, with $a = -1$, $d = -0.1$, so $a_n = -1 + (n-1)(-0.1) = -0.9 - 0.1n$.

8. Geometric. The ratios of successive terms are all 3.

9. Not geometric. The ratios of successive terms are $2, 2, \frac{3}{2}$.

10. Geometric. The ratios of successive terms are all $-\frac{1}{2}$.

11. Geometric. The ratios of successive terms are all 0.1.

12. Geometric, since the ratios of successive terms are all 3. Thus, $a = 4$ and $r = 3$, so $a_n = 4 \cdot 3^{n-1}$.

13. Not geometric, since the ratios of successive terms are $1/4, 1/4, 1/2$.

14. Geometric, since the ratios of successive terms are all -2. Thus, $a = 2$ and $r = -2$, so $a_n = 2(-2)^{n-1}$.

15. Geometric, since the ratios of successive terms are all $1/2$. Thus, $a = 4$ and $r = 1/2$, so $a_n = 4(\frac{1}{2})^{n-1}$.

16. Geometric, since the ratios of successive terms are all 0.1. Thus, $a = 4$ and $r = 0.1$, so $a_n = 4(0.1)^{n-1}$.

17. Geometric, since the ratios of successive terms are all $1/1.2$. Thus, $a = 1$ and $r = 1/1.2$, so $a_n = 1(1/1.2)^{n-1} = 1/(1.2)^{n-1}$.

Problems

18. $a_1 = 2$, $a_2 = 2^2 = 4$, $a_3 = 2^3 = 8$, $a_4 = 2^4 = 16$. This is geometric.

19. $a_1 = 3/3 = 1$, $a_2 = 5/4$, $a_3 = 7/5$, $a_4 = 9/6 = 3/2$. This is not a geometric sequence.

20. $a_1 = (-1/3)^1 = -1/3$, $a_2 = (-1/3)^2 = 1/9$, $a_3 = (-1/3)^3 = -1/27$, $a_4 = (-1/3)^4 = 1/81$. This is a geometric sequence.

21. $a_1 = \cos(\pi) = -1$, $a_2 = \cos(2\pi) = 1$, $a_3 = \cos(3\pi) = -1$, $a_4 = \cos(4\pi) = 1$. This is a geometric sequence.

22. $a_1 = 1^2 - 1 = 0$, $a_2 = 2^2 - 2 = 2$, $a_3 = 3^2 - 3 = 6$, $a_4 = 4^2 - 4 = 12$. This is not a geometric sequence.

23. $a_1 = 1/\sqrt{1} = 1$, $a_2 = 1/\sqrt{2}$, $a_3 = 1/\sqrt{3}$, $a_4 = 1/\sqrt{4} = 1/2$. This is not a geometric sequence.

24. Since the first term is 10 and the difference is 5, the arithmetic sequence is given by

$$a_n = 10 + (n-1)5,$$

so $a_{10} = 10 + (10-1) \cdot 5 = 55$.

25. Since the first term is 5 and the difference is 10, the arithmetic sequence is given by

$$a_n = 5 + (n-1)10.$$

For $a_n = 5 + (n-1) \cdot 10 > 1000$, we must have

$$\begin{aligned} 10(n-1) &> 995 \\ n-1 &> 99.5 \\ n &> 100.5. \end{aligned}$$

The terms of the sequence exceed 1000 when $n \geq 101$.

26. Since $a_1 = 3$ and $d = 5 - 3 = 2$, we have

$$\begin{aligned} a_5 &= 3 + (5-1)2 = 11 \\ a_{50} &= 3 + (50-1)2 = 101 \\ a_n &= 3 + (n-1)2 = 1 + 2n. \end{aligned}$$

27. Since $a_1 = 6$ and $d = 7.2 - 6 = 1.2$, we have

$$\begin{aligned} a_5 &= 6 + (5-1)1.2 = 10.8 \\ a_{50} &= 6 + (50-1)1.2 = 64.8 \\ a_n &= 6 + (n-1)1.2 = 4.8 + 1.2n. \end{aligned}$$

28. Since $a_3 - a_1 = 2d = 4.7 - 2.1 = 2.6$, we have $d = 1.3$, so

$$\begin{aligned} a_5 &= 2.1 + (5-1)1.3 = 7.3 \\ a_{50} &= 2.1 + (50-1)1.3 = 65.8 \\ a_n &= 2.1 + (n-1)1.3 = 0.8 + 1.3n. \end{aligned}$$

29. Since $a_6 - a_3 = 3d = 9 - 5.7 = 3.3$, we have $d = 1.1$, so $a_1 = a_3 - 2d = 5.7 - 2 \cdot 1.1 = 3.5$. Thus

$$\begin{aligned} a_5 &= 3.5 + (5-1)1.1 = 7.9 \\ a_{50} &= 3.5 + (50-1)1.1 = 57.4 \\ a_n &= 3.5 + (n-1)1.1 = 2.4 + 1.1n. \end{aligned}$$

30. Since $a = 1$ and $r = 2$, we have

$$\begin{aligned} a_6 &= ar^5 = 1 \cdot 2^5 = 32 \\ a_n &= ar^{n-1} = 2^{n-1}. \end{aligned}$$

31. Since $a = 7$ and $r = 5.25/7 = 0.75$, we have

$$\begin{aligned} a_6 &= ar^5 = 7 \cdot (0.75)^5 = 1.661 \\ a_n &= ar^{n-1} = 7(0.75)^{n-1}. \end{aligned}$$

32. Since $a_1 = a = 3$ and $a_3 = ar^2 = 48$, we have

$$\frac{a_3}{a_1} = \frac{ar^2}{a} = r^2 = \frac{48}{3} = 16,$$

so

$$r = 4.$$

$$\begin{aligned} a_6 &= ar^5 = 3 \cdot (4)^5 = 3072 \\ a_n &= ar^{n-1} = 3 \cdot 4^{n-1}. \end{aligned}$$

33. Since $a_2 = ar = 6$ and $a_4 = ar^3 = 54$, we have

$$\frac{a_4}{a_2} = \frac{ar^3}{ar} = r^2 = \frac{54}{6} = 9,$$

so

$$r = 3.$$

We have $a = 6/r = 6/3 = 2$. Thus

$$a_6 = ar^5 = 2 \cdot (3)^5 = 486$$
$$a_n = ar^{n-1} = 2 \cdot 3^{n-1}.$$

34. **(a)** We are told that , in millions of barrels, $a_1 = 81$. In the same units, we have

$$a_1 = 81,$$
$$a_2 = 81(1.012) = 81.972,$$
$$a_3 = 81(1.012)^2 = 82.956,$$
$$a_4 = 81(1.012)^3 = 83.951,$$

and

$$a_n = 81(1.012)^{n-1}.$$

(b) Consumption reaches 100 million barrels per day when

$$a_n = 81(1.012)^{n-1} = 100.$$

Using logs, we have

$$(1.012)^{n-1} = \frac{100}{81}$$
$$(n-1)\log(1.012) = \log\left(\frac{100}{81}\right)$$
$$n - 1 = \frac{\log(100/81)}{\log(1.012)} = 17.665$$
$$n = 18.665.$$

Thus, consumption exceeds 100 million barrels per day about 18.7 years after 2003, that is, first in 2022.

35. **(a)** Since $a_1 = 646.7$ bn m^3 and $r = 1.006$, we have, in bn m^3,

$$a_1 = 646.7$$
$$a_2 = 646.7(1.006) = 650.580$$
$$a_3 = 646.7(1.006)^2 = 654.484$$
$$a_4 = 646.7(1.006)^3 = 658.411.$$

(b) Since $b_1 = 367.7$ bn m^3 and $r = 1.079$, we have, in bn m^3,

$$b_1 = 367.7$$
$$b_2 = 367.7(1.079) = 396.748$$
$$b_3 = 367.7(1.079)^2 = 428.091$$
$$b_4 = 367.7(1.079)^3 = 461.911.$$

(c) Consumption is equal when $a_n = b_n$. Since $a_n = 646.7(1.006)^{n-1}$ and $b_n = 367.7(1.079)^{n-1}$, we have

$$646.7(1.006)^{n-1} = 367.7(1.079)^{n-1},$$

so

$$\begin{aligned}\ln 646.7 + (n-1)\ln(1.006) &= \ln 367.7 + (n-1)\ln(1.079)\\ (n-1)(\ln(1.079) - \ln(1.006)) &= \ln 646.7 - \ln 367.7\\ n-1 &= \frac{\ln(646.7/367.7)}{\ln(1.079/1.006)} = 8.060.\end{aligned}$$

Thus, $n = 9.060$; That is, Asian consumption will overtake US consumption in 2012.

36. (a) If the yearly growth factor is r, then we know

$$2.4 = 2.2r^3,$$

so

$$\begin{aligned}r^3 &= \frac{2.4}{2.2}\\ r &= \left(\frac{2.4}{2.2}\right)^{1/3} = 1.0294285.\end{aligned}$$

Thus, n years after 2005, the population is

$$a_n = 2.4(1.0294285)^n \approx 2.4(1.029)^n \text{ million.}$$

(b) The population reaches 10 million when

$$\begin{aligned}2.4(1.0294285)^n &= 10\\ (1.0294285)^n &= \frac{10}{2.4}\\ n\ln(1.0294285) &= \ln\left(\frac{10}{2.4}\right)\\ n &= \frac{\ln(10/2.4)}{\ln(1.0294285)} = 49.204.\end{aligned}$$

Thus the population is predicted to reach 10 million just over 49 years after 2005; that is in the year 2054.

37. (a) The growth factor of the population is

$$r = \frac{17.960}{17.613} = 1.0197.$$

Thus, one and two years after 2005, the population is

$$\begin{aligned}a_1 &= 17.960(1.0197) = 18.314\\ a_2 &= 17.960(1.0197)^2 = 18.675.\end{aligned}$$

(b) Using the growth factor in part (a), n years after 2005, the population is

$$a_n = 17.960(1.0197)^n.$$

(c) The doubling time is the value of n for which

$$\begin{aligned}a_n &= 2\cdot 17.960\\ 17.960(1.0197)^n &= 2\cdot 17.960\\ (1.0197)^n &= 2\\ n\ln(1.0197) &= \ln 2\\ n &= \frac{\ln 2}{\ln(1.0197)} = 35.531.\end{aligned}$$

Thus the doubling time is about 36.5 years

38. Not arithmetic, because points do not lie on a line, so geometric. The sequence is increasing, so $r > 1$.

39. Arithmetic, because points lie on a line. The sequence is increasing, so $d > 0$.

40. Not arithmetic, because points do not lie on a line, so geometric. The sequence is decreasing, so $r < 1$.

41. Arithmetic, because points lie on a line. The sequence is decreasing, so $d < 0$.

42. Since $a_1 = 3$ and $a_n = 2a_{n-1}$, we have $a_2 = 2a_1 = 6$, $a_3 = 2a_2 = 12$, $a_4 = 2a_3 = 24$.
If we do not multiply out, we can see the general pattern more easily:

$$a_2 = 2 \cdot 3, \quad a_3 = 2 \cdot 2 \cdot 3 = 2^2 \cdot 3, \quad a_4 = 2 \cdot 2^2 \cdot 3 = 2^3 \cdot 3.$$

Thus a_n is a geometric sequence, $a_n = 2^{n-1} \cdot 3$.

43. Since $a_1 = 2$ and $a_n = a_{n-1} + 5$, we have $a_2 = 7$, $a_3 = a_2 + 5 = 12$, $a_4 = a_3 + 5 = 17$. We have added 5 each time, so this is an arithmetic sequence with $a_n = 2 + (n-1)5 = -3 + 5n$.

44. Since $a_1 = 1$ and $a_n = -a_{n-1}$, we have $a_2 = -1$, $a_3 = 1$, $a_4 = -1$. In general, $a_n = (-1)^{n-1}$.

45. Since $a_1 = 3$ and $a_n = 2a_{n-1} + 1$, we have $a_2 = 2a_1 + 1 = 7$, $a_3 = 2a_2 + 1 = 15$, $a_4 = 2a_3 + 1 = 31$. Writing out the terms without simplification to try and guess the pattern, we have

$$\begin{aligned} a_1 &= 3 \\ a_2 &= 2 \cdot 3 + 1 \\ a_3 &= 2(2 \cdot 3 + 1) + 1 = 2^2 \cdot 3 + 2 + 1 \\ a_4 &= 2(2^2 \cdot 3 + 2 + 1) + 1 = 2^3 \cdot 3 + 2^2 + 2 + 1. \end{aligned}$$

Thus

$$a_n = 2^{n-1} \cdot 3 + 2^{n-2} + 2^{n-3} + \cdots + 2 + 1.$$

46. Since the first term is 10 and the ratio is -0.2, the geometric sequence is given by

$$a_n = 10(-0.2)^n.$$

We need $|10(-0.2)^n| < 10^{-7}$, which simplifies to $|10(0.2)^n| < 10^{-7}$ using the properties of the absolute value function. Taking logarithms, base 10, we have

$$\begin{aligned} \log(10(0.2)^n) = \log 10 + n\log(0.2) &< \log 10^{-7} \\ 1 + n\log 0.2 &< -7 \\ n &> \frac{-8}{\log 0.2} = 11.445. \end{aligned}$$

The sequence will be smaller than 10^{-7} for $n \geq 12$.

47. **(a)** $p_0 = 150$, $p_1 = 187.5$, $p_2 = 199.219$, $p_3 = 199.997$, $p_4 \approx 200$, $p_5 \approx 200$ and $p_5 > p_4$. This sequence appears to converge.

(b) We write $p_{n+1} = 2p_n - \frac{p_n^2}{200}$ as

$$p_{n+1} - p_n = p_n - \frac{p_n^2}{200} = \frac{p_n}{200}(200 - p_n).$$

Thus if $0 < p_n < 200$, the right-hand side is positive, so the sequence is increasing.
Now we write $p_{n+1} = 2p_n - \frac{p_n^2}{200}$ as

$$p_{n+1} = p_n\left(\frac{400 - p_n}{200}\right).$$

We need to find the maximum value of the right hand side for $0 < p_n < 200$. Notice that it is a parabola with zeros at $p_n = 0$ and $p_n = 400$, so it has a maximum of 200 at the midpoint where $p_n = 200$, and so $p_{n+1} < 200$.
This value is called the carrying capacity.

48. **(a)** We have $a_1 = a_2 = 1$, $a_3 = 2$, $a_4 = 3$, $a_5 = 5$, so $a_6 = 3 + 5 = 8$, $a_7 = 5 + 8 = 13$, $a_8 = 8 + 13 = 21$.

(b) Since a_n is the sum of the two previous terms, $a_n = a_{n-1} + a_{n-2}$.

(c) We have $r_n = a_n/a_{n-1}$ and $r_{n-1} = a_{n-1}/a_{n-2}$. Using our answers to part (b),

$$a_n = a_{n-1} + a_{n-2},$$

we have

$$\frac{a_n}{a_{n-2}} = \frac{a_{n-1}}{a_{n-2}} + 1,$$

so, multiplying by a_{n-1} in the numerator and denominator on the left, we have

$$\begin{aligned} \frac{a_{n-1}}{a_{n-1}} \cdot \frac{a_n}{a_{n-2}} = \frac{a_n}{a_{n-1}} \cdot \frac{a_{n-1}}{a_{n-2}} &= \frac{a_{n-1}}{a_{n-2}} + 1 \\ r_n \cdot r_{n-1} &= r_{n-1} + 1 \\ r_n &= \frac{r_{n-1} + 1}{r_{n-1}}. \end{aligned}$$

(d) If $r_n \to r$ as $n \to \infty$, then $r_{n-1} \to r$ as $n \to \infty$. From part (c), we know that $r_n \cdot r_{n-1} = r_{n-1} + 1$. Thus, $r_n \cdot r_{n-1} \to r^2$ and $r_{n-1} + 1 \to r + 1$ as $n \to \infty$. Putting this together, we have

$$\begin{aligned} r^2 &= r + 1 \\ r^2 - r - 1 &= 0. \end{aligned}$$

This equation has solution

$$r = \frac{1 \pm \sqrt{(-1)^2 - 4 \cdot 1 \cdot (-1)}}{2} = \frac{1 \pm \sqrt{5}}{2}.$$

Since the ratios are all positive, we expect r to be positive, so $r = (1 + \sqrt{5})/2$. This is the golden ratio.

49. **(a)** You send the letter to 4 friends; at that stage your name is on the bottom of the list. The 4 friends send it to 4 each, so 4^2 people have the letter when you are third on the list. These 4^2 people send it to 4 each, so 4^3 people have the letter when you are second on the list. By similar reasoning, 4^4 people have the letter when you are at the top of the list. Thus, you should receive $\$4^4 = \$256$. (The catch is, of course, that someone usually breaks the chain.)

(b) By the same reasoning as in part (a), we see that $d_n = 4^n$.

50. **(a)** Since there are more people with incomes less than or equal to \$50,000 than there are people with incomes less than or equal to \$40,000 (because all those with incomes less than or equal to \$40,000 also have incomes less than or equal to \$50,000), we have $a_{40} < a_{50}$.

(b) The quantity $a_{50} - a_{40}$ represents the fraction of the US population who have incomes between \$40,000 and \$50,000.

(c) No, because there are some people (for example, children) whose income is zero. This fraction of the population is included in a_n for all n. Thus, a_n is never 0.

(d) The value of a_n increases as n increases. If n is larger than all incomes in the US, then $a_n = 1$.

Solutions for Section 13.2

Exercises

1. This series is not arithmetic, as each term is twice the previous one.

2. This series is arithmetic, since each term is 2 less than the previous term.

3. This series is not arithmetic, since the difference between successive terms is sometimes 1 and sometimes 2.

4. This series is arithmetic, since each term is $1 = 3/3$ greater than the previous term.

5. $\sum_{i=-1}^{5} i^2 = (-1)^2 + 0^2 + 1^2 + 2^2 + 3^2 + 4^2 + 5^2.$

6. $\sum_{i=10}^{20}(i+1)^2 = 11^2 + 12^2 + 13^2 + \cdots + 21^2.$

7. $(2\cdot 0+1)+(2\cdot 1+1)+(2\cdot 2+1)+(2\cdot 3+1)+(2\cdot 4+1)+(2\cdot 5+1) = 1+3+5+7+9+11$

8. $3(1-3)+3(2-3)+3(3-3)+3(4-3)+3(5-3)+3(6-3) = -6-3+0+3+6+9$

9. $\sum_{j=2}^{10}(-1)^j = (-1)^2 + (-1)^3 + (-1)^4 + \cdots + (-1)^{10}.$

10. $\sum_{n=1}^{7}(-1)^{n-1}2^n = (-1)^0 2^1 + (-1)^1 2^2 + (-1)^2 2^3 + \cdots (-1)^6 2^7.$

11. The first term is $3\cdot 1$, the second term is $3\cdot 2$, up to $3\cdot 7$. Thus, one possible answer is

$$\sum_{n=1}^{7} 3n.$$

12. The first term is 10 more than $3\cdot 0$, followed by 10 more than $3\cdot 1$, etc. One possible answer is

$$\sum_{n=0}^{4}(10+3n).$$

13. The pattern is $1/2, 2/2, 3/2, \ldots, 8/2$. Thus, a possible solution is

$$\sum_{n=1}^{8}\frac{1}{2}n.$$

14. The first term is 30 and then each term is five less than the preceding term until we reach $5 = 30 - 5\cdot 5$. So, a possible solution is

$$\sum_{n=0}^{5}(30-5n).$$

15. **(a)** $\sum_{i=1}^{10} 2i$

(b) This is an arithmetic series with $a_1 = 2$, $n = 10$, and $d = 2$. Thus, the sum of the first 10 terms of the series is

$$S_{10} = \frac{1}{2}\cdot 10(2\cdot 2 + 9\cdot 2) = 110.$$

16. One way to work this problem is to complete the first few values of a_n, and then continue the pattern across the row. Then, the values of S_n can be found by summing the values of a_n from left to right. See Table 13.1. We see that $a_1 = 2$ and that $d = a_2 - a_1 = 7 - 2 = 5$.

Table 13.1

n	1	2	3	4	5	6	7	8
a_n	2	7	12	17	22	27	32	37
S_n	2	9	21	38	60	87	119	156

17. One way to work this problem is to complete the first few values of a_n, and then continue the pattern across the row. Then, the values of S_n can be found by summing the values of a_n from left to right. See Table 13.2. We see that $a_1 = 3$ and $d = a_2 - a_1 = 7 - 3 = 4$.

Table 13.2

n	1	2	3	4	5	6	7	8
a_n	3	7	11	15	19	23	27	31
S_n	3	10	21	36	55	78	105	136

18. One way to work this problem is to complete the first few values of a_n, and then continue the pattern across the row. Then, the values of S_n can be found by summing the values of a_n from left to right. See Table 13.3. We see that $a_1 = 7$ and that $a_4 = 16$. To find the value of d, we can use the equation $a_n = a_1 + (n-1)d$:

$$a_4 = a_1 + (4-1)d$$
$$16 = 7 + 3d$$
$$d = 3.$$

Table 13.3

n	1	2	3	4	5	6	7	8
a_n	7	10	13	16	19	22	25	28
S_n	7	17	30	46	65	87	112	140

19. One way to work this problem is to complete the first few values of a_n, and then continue the pattern across the row. Then, the values of S_n can be found by summing the values of a_n from left to right. See Table 13.4. We see that $a_1 = 2$. We can figure out the value of a_2 as follows:

$$S_2 = a_1 + a_2$$
$$13 = 2 + a_2$$
$$a_2 = 11.$$

This means that $d = a_2 - a_1 = 11 - 2 = 9$.

Table 13.4

n	1	2	3	4	5	6	7	8
a_n	2	11	20	29	38	47	56	65
S_n	2	13	33	62	100	147	203	268

20. One way to work this problem is to complete the first few values of a_n, and then continue the pattern across the row. Then, the values of S_n can be found by summing the values of a_n from left to right. See Table 13.5. We knew that $S_6 = 201$, $S_7 = 273$, and $S_8 = 356$. Note that $S_7 = S_6 + a_7$, and so $273 = 201 + a_7$, which means that $a_7 = 72$. Similarly , we know that $S_8 = S_7 + a_8$, and so $356 = 273 + a_8$, which means that $a_8 = 83$. Now we know two consecutive values of a_n, and so we can find d: we have $d = a_8 - a_7 = 83 - 72 = 11$. Finally, we can find a_1 as follows:

$$a_7 = a_1 + (7-1)d$$
$$72 = a_1 + 6 \cdot 11$$
$$a_1 = 6.$$

Table 13.5

n	1	2	3	4	5	6	7	8
a_n	6	17	28	39	50	61	72	83
S_n	6	23	51	90	140	201	273	356

21. We use the formula $S_n = 1 + 2 + 3 + \cdots + n = \frac{1}{2}n(n+1)$ with $n = 1000$:

$$S_{1000} = \frac{1}{2} \cdot 1000 \cdot 1001 = 500{,}500.$$

22. $\sum_{i=1}^{50} 3i = 3 + 6 + 9 + \cdots$. This is an arithmetic series, so we can use the formula

$$S_{50} = \frac{1}{2}50(2 \cdot 3 + 49 \cdot 3) = 3825.$$

23. $\sum_{i=1}^{30} (5i + 10) = 15 + 20 + 25 + \cdots$. We use the formula to find the sum of this arithmetic series

$$S_{30} = \frac{1}{2} \cdot 30(2 \cdot 15 + 29 \cdot 5) = 2625.$$

24. $\sum_{n=0}^{15} (2 + \frac{1}{2}n) = 2 + \frac{5}{2} + 3 + \frac{7}{2} + 4 + \cdots$. This is an arithmetic series with 16 terms (we start at $n = 0$ and continue to $n = 15$). Thus,

$$S_{16} = \frac{1}{2} \cdot 16(2 \cdot 2 + 15 \cdot \frac{1}{2}) = 92.$$

25. $\sum_{n=0}^{10} (8 - 4n) = 8 + 4 + 0 + (-4) + \cdots$. This is an arithmetic series with 11 terms and $d = -4$.

$$S_{11} = \frac{1}{2} \cdot 11(2 \cdot 8 + 10(-4)) = -132.$$

26. This is an arithmetic series with $a_1 = \sqrt{2}$, $n = 25$, and $d = \sqrt{2}$. Thus,

$$S_{25} = \frac{1}{2} \cdot 25(2\sqrt{2} + 24\sqrt{2}) = 325\sqrt{2}.$$

27. This is an arithmetic series with $a_1 = -101$, $n = 11$, and $d = 10$. Thus

$$S_{11} = \frac{1}{2} \cdot 11(2(-101) + 10 \cdot 10) = -561.$$

28. This is an arithmetic series with $a_1 = 26.5$, $n = 14$, and $d = -2$. Thus

$$S_{14} = \frac{1}{2} \cdot 14(2 \cdot 26.5 + 13(-2)) = 189.$$

29. This is an arithmetic series with $a_1 = -3.01$, $n = 35$, and $d = -0.01$. Thus

$$S_{35} = \frac{1}{2} \cdot 35(2(-3.01) + 34(-0.01)) = -111.3.$$

Problems

30. To raise \$400 Jenny needs 1600 quarters. If the triangle has n rows then the total number of quarters is

$$S_n = 1 + 2 + 3 + \ldots + n = \frac{1}{2}n(n+1).$$

We solve $n(n+1) = 3200$ to get $n = -57.071$ or $n = 56.071$. We can discard the negative solution, so Jenny needs 57 rows of quarters in her triangle.

31. This is an arithmetic series. We have $n = 30$, $a_1 = 5$, $d = 5$. To find the 30th multiple of 5 we use the formula $a_n = a_1 + (n-1)d$. Substituting, we get

$$a_{30} = 5 + 29 \cdot 5 = 150.$$

To find the sum of the arithmetic series we use the formula $S_n = (1/2)n(2a_1 + (n-1)d)$, substituting, we get

$$S_{30} = \frac{30}{2}(2 \cdot 5 + 29 \cdot 5) = 2325.$$

32. **(a)** (i) The text gives $S_5 = 12{,}607$, the total number of AIDS deaths (since 1980) in 1985. Then adding one more term each year gives

$$\begin{aligned} S_6 &= S_5 + 12{,}110 = 24{,}717 = \text{Total number of deaths up to 1986.} \\ S_7 &= S_6 + 16{,}412 = 41{,}129 = \text{Total number of deaths up to 1987.} \\ S_8 &= S_7 + 21{,}119 = 62{,}248 = \text{Total number of deaths up to 1988.} \end{aligned}$$

(ii) The differences $S_6 - S_5$, etc., give the yearly death, so

$$\begin{aligned} S_6 - S_5 &= 24{,}717 - 12{,}607 = 12{,}110 = \text{Deaths in 1986.} \\ S_7 - S_6 &= 41{,}129 - 24{,}717 = 16{,}412 = \text{Deaths in 1987.} \\ S_8 - S_7 &= 62{,}248 - 41{,}129 = 21{,}119 = \text{Deaths in 1988.} \end{aligned}$$

(b) In general, $S_{n+1} - S_n = a_{n+1}$, the number of deaths in year $n+1$, or $n+1$ years after 1980.

33. **(a)** (i) From the table, we have $S_4 = 226.6$, the population of the US in millions 4 decades after 1940, that is, in 1980. Similarly, $S_5 = 248.7$, the population in millions in 1990, and $S_6 = 281.4$, the population in 2000.

(ii) We have $a_2 = S_2 - S_1 = 179.3 - 150.7 = 28.6$; that is, the increase in the US population in millions in the 1950s.

Similarly, $a_5 = S_5 - S_4 = 248.7 - 226.6 = 22.1$, the population increase in millions during the 1980s. In the same way, $a_6 = S_6 - S_5 = 281.4 - 248.7 = 32.7$, the population increase during the 1990s.

(iii) Using the answer to (ii), we have $a_6/10 = 32.7/10 = 3.27$, the average yearly population growth during the 1990s.

(iv) We have

$S_n =$ US population, in millions, n decades after 1940.
$a_n = S_n - S_{n-1} =$ growth in US population in millions, during the nth decade after 1940.
$a_n/10 =$ Average yearly growth, in millions, during the nth decade after 1940.

34. These two series give the same terms so their difference is 0:

$$\begin{aligned} \sum_{i=1}^{5} i^2 - \sum_{j=0}^{4} (j+1)^2 &= (1^2 + 2^2 + 3^2 + 4^2 + 5^2) - ((0+1)^2 + (1+1)^2 + (2+1)^2 + (3+1)^2 + (4+1)^2) \\ &= 1^2 + 2^2 + 3^2 + 4^2 + 5^2 - (1^2 + 2^2 + 3^2 + 4^2 + 5^2) = 0. \end{aligned}$$

35. Expanding, we have

$$\begin{aligned} \sum_{i=4}^{20} i - \sum_{j=4}^{20} (-2j) &= (4 + 5 + \cdots + 20) - (-2 \cdot 4 - 2 \cdot 5 - \cdots - 2 \cdot 20) \\ &= (4 + 5 + \cdots + 20) + 2(4 + 5 + \cdots + 20) \\ &= 3(4 + 5 + \cdots + 20). \end{aligned}$$

The sum in parentheses is an arithmetic series with $a = 4$, $d = 1$, $n = 17$, so, using the formula for the sum, we have

$$\sum_{i=4}^{20} i - \sum_{j=4}^{20} (-2j) = 3 \cdot \frac{17}{2}(2 \cdot 4 + 16 \cdot 1) = 612.$$

36. The expression $\sum_{i=1}^{20} 1$ is asking us to add a 1 for every number between 1 and 20, so $\sum_{i=1}^{20} 1 = 20$.

37. Expanding both sums, we see

$$\begin{aligned}\sum_{i=1}^{15} i^3 - \sum_{j=3}^{15} j^3 &= (1^3 + 2^3 + 3^3 + 4^3 + \cdots + 15^3) - (3^3 + 4^3 + \cdots + 15^3)\\ &= 1^3 + 2^3 = 9.\end{aligned}$$

38. We know that $S_7 = 16 + 48 + 80 + \cdots + 208$. Here, $a_1 = 16$, $n = 7$, and $d = 32$. This gives

$$S_7 = \frac{1}{2}n\left(2a_1 + (n-1)d\right) = \frac{1}{2} \cdot 7\left(2 \cdot 16 + (7-1)32\right) = 784.$$

39. **(a)** Using the formula for the sum of an arithmetic series with $a_1 = 16$ and $d = 32$, we have

$$\begin{aligned}S_4 &= \frac{1}{2} \cdot 4\left(2 \cdot 16 + (3)32\right) = 256 \text{ feet},\\ S_5 &= \frac{1}{2} \cdot 5\left(2 \cdot 16 + (4)32\right) = 400 \text{ feet},\\ S_6 &= \frac{1}{2} \cdot 6\left(2 \cdot 16 + (5)32\right) = 576 \text{ feet}.\end{aligned}$$

(b) At $t = 4$, the height is $h = 1000 - 256 = 744$ feet.
At $t = 5$, the height is $h = 1000 - 400 = 600$ feet.
At $t = 6$, the height is $h = 1000 - 576 = 424$ feet.
The values of h we have calculated are those shown in Figure 13.2 in the text, namely $h = 744$, $h = 600$, and $h = 424$ feet.

40. We have $a_1 = 16$ and $d = 32$. The distance fallen by the object in n seconds is the sum

$$S_n = \frac{1}{2}n\left(2 \cdot 16 + (n-1)32\right) = \frac{1}{2}n(32 + 32n - 32) = \frac{1}{2}n(32n) = 16n^2.$$

So

$$f(n) = 16n^2.$$

41. We have $a_1 = 16$ and $d = 32$. At the end of n seconds, the object has fallen

$$S_n = \frac{1}{2}n\left(2 \cdot 16 + (n-1)32\right) = \frac{1}{2}n(32 + 32n - 32) = 16n^2.$$

The height of the object at the end of n seconds is

$$h = 1000 - S_n = 1000 - 16n^2.$$

When the object hits the ground, $h = 0$, so

$$0 = 1000 - 16n^2$$

$$n^2 = \frac{1000}{16} \quad \text{so} \quad n = \pm\sqrt{\frac{1000}{16}} = \pm 7.906 \text{ sec.}$$

Since n is positive, it takes 7.906 seconds, that is, nearly 8 seconds, for the object to hit the ground.

42. **(a)** On the nth round, the boy gives his sister 1 M&M and takes n for himself.
(b) After n rounds, the sister has n M&Ms. The number of M&Ms that the boy has is given by the arithmetic series

$$\text{Number of M\&Ms} = 1 + 2 + 3 + \cdots + n = \sum_{i=1}^{n} i = \frac{n(n+1)}{2}.$$

43. The first row has $a_1 = 30$ seats, the second row has $a_2 = 30 + 4 = 34$ seats, $a_3 = 30 + 2 \cdot 4$ seats. The number of seats in each row is a arithmetic sequence. To find the number of seats in the last row, we use the formula $a_n = a_1 + (n-1)d$. Thus,

$$a_{20} = 30 + 19(4) = 106.$$

To find the total number of seats in the auditorium, we have to find the sum of the series $30 + 34 + 38 + \cdots$ and use the formula $S = (1/2)n(2a_1 + (n-1)d)$. Therefore,

$$S_{20} = \frac{1}{2} \cdot 20(2 \cdot 30 + 19 \cdot 4) = 1360.$$

44. **(a)** $n^3 - (n-1)^3 = n^3 - (n^3 - 3n^2 + 3n - 1) = 3n^2 - 3n + 1.$

(b) We have

$$\begin{aligned} n^3 &= n^3 - (n-1)^3 + (n-1)^3 - (n-2)^3 + (n-2)^3 - \ldots - 2^3 + 2^3 - 1^3 + 1^3 - 0^3 \\ &= (3n^2 - 3n + 1) + (3(n-1)^2 - 3(n-1) + 1) + \ldots + (3 \cdot (2)^2 - 3 \cdot (2) + 1) + (3 \cdot 1^2 - 3 \cdot 1 + 1) \\ &= \sum_{j=1}^{j=n} (3j^2 - 3j + 1). \end{aligned}$$

(c) We see that

$$\begin{aligned} n^3 &= \sum_{j=1}^{j=n} (3j^2 - 3j + 1) \\ &= 3\sum_{j=1}^{j=n} j^2 - 3\sum_{j=1}^{j=n} j + \sum_{j=1}^{j=n} 1 \\ &= 3\sum_{j=1}^{j=n} j^2 - \frac{3}{2}n(n+1) + n. \end{aligned}$$

Rearranging gives

$$\begin{aligned} 3\sum_{j=1}^{j=n} j^2 &= n^3 + \frac{3}{2}n(n+1) - n \\ &= \frac{n}{2}(2n^2 + 3n + 1) \end{aligned}$$

so

$$\sum_{j=1}^{j=n} j^2 = \frac{n}{6}(n+1)(2n+1).$$

45. **(a)** Since there are 9 terms, we can group them into 4 pairs each totaling 66. The middle or fifth term, $a_5 = 33$, remains unpaired. This means that

$$\text{Sum of series} = 4(66) + 33 = 297.$$

This is the same answer we got by adding directly.

(b) We can use our formula derived for the sum of an even number of terms to add the first 8 terms. We have $a_1 = 5$, $n = 8$, and $d = 7$. This gives

$$\text{Sum of first 8 terms} = \frac{1}{2}(8)(2(5) + (8-1)(7)) = 236.$$

Adding the ninth term gives

$$\text{Sum of series} = 236 + 61 = 297.$$

This is the same answer we got by adding directly.

(c) Using the method from part (a), we add the first and last terms and obtain $a_1 + a_n$. Notice that $a_n = a_1 + (n-1)d$ and so this expression can be rewritten as $a_1 + a_1 + (n-1)d = 2a_1 + (n-1)d$. We then add the second and next to last terms and obtain $a_2 + a_{n-1}$. We know that $a_2 = a_1 + d$ and that $a_{n-1} = a_1 + (n-2)d$, and so this expression can be rewritten as $a_1 + d + a_1(n-2)d = 2a_1 + (n-1)d$. Continuing in this manner, we see that the sum of each pair is $2a_1 + (n-1)d$. The total number of such terms is given by $\frac{1}{2}(n-1)$. For instance, if there are 9 terms, the total number of pairs is $\frac{1}{2}(9-1) = 4$. Thus, the subtotal of these pairs is given by

$$\text{Subtotal of pairs} = \frac{1}{2}(n-1)(2a_1 + (n-1)d).$$

As you can check for yourself, the unpaired (middle) term is given by

$$\text{Unpaired (middle) term} = a_1 + \frac{1}{2}(n-1)d.$$

For instance, in the case of the series given in the question, the unpaired term is given by

$$5 + \frac{1}{2}(9-1)(7) = 5 + 4(7) = 33.$$

Therefore, the sum of the arithmetic series is given by

$$\begin{aligned}\text{Sum} &= \text{Subtotal of pairs} + \text{Unpaired term}\\ &= \underbrace{\frac{1}{2}(n-1)(2a_1 + (n-1)d)}_{\text{Subtotal of pairs}} + \underbrace{a_1 + \frac{1}{2}(n-1)d}_{\text{Unpaired term}}.\end{aligned}$$

One way to simplify this expression is to first factor out 1/2:

$$\text{Sum} = \frac{1}{2}\left[(n-1)(2a_1 + (n-1)d) + 2a_1 + (n-1)d\right].$$

The bracketed part of this expression involves $(n-1)$ terms equaling $2a_1 + (n-1)d$, plus one more such term, for a total number of n such terms. We have

$$\text{Sum} = \frac{1}{2}n(2a_1 + (n-1)d),$$

which is the same as the formula derived in the text.

Using the method from part (b), we see that since n is odd, $(n-1)$ is even, so we can use the formula derived for an even number of terms to add the first $(n-1)$ terms. Substituting $(n-1)$ for n in our formula, we have

$$\text{Sum of first } (n-1) \text{ terms} = \frac{1}{2}(n-1)(2a_1 + (n-2)d).$$

The total sum is given by

$$\text{Sum} = \text{Sum of first } (n-1) \text{ terms} + n^{\text{th}} \text{ term}.$$

Since the nth term is given by $a_1 + (n-1)d$, we have

$$\text{Sum} = \underbrace{\frac{1}{2}(n-1)(2a_1 + (n-2)d)}_{\text{Sum of first } (n-1) \text{ terms}} + \underbrace{a_1 + (n-1)d}_{n^{\text{th}} \text{ term}}.$$

To simplify this expression, we first factor out 1/2, as before:

$$\text{Sum} = \frac{1}{2}\left[(n-1)(2a_1 + (n-2)d) + 2a_1 + 2(n-1)d\right].$$

We can rewrite $2a_1 + 2(n-1)d$ as $2a_1 + 2nd - 2d$, and then as $2a_1 + nd - 2d + nd$, and finally as $2a_1 + (n-2)d + nd$. This gives

$$\text{Sum} = \frac{1}{2}\left[(n-1)(2a_1 + (n-2)d) + 2a_1 + (n-2)d + nd\right].$$

We have $n-1$ terms each equaling $2a_1+(n-2)d$ plus one more such term plus a term equaling nd. This gives a total of n terms equaling $2a_1+(n-2)d$ plus the nd term:

$$\begin{aligned}\text{Sum} &= \frac{1}{2}\left[n(2a_1+(n-2)d)+nd\right]\\ &= \frac{1}{2}n\left[2a_1+(n-2)d+d\right] \qquad \text{factoring out } n\\ &= \frac{1}{2}n(2a_1+(n-1)d),\end{aligned}$$

which is the same answer as before.

Solutions for Section 13.3

Exercises

1. Since

$$\sum_{j=5}^{18} = 3\cdot 2^5+3\cdot 2^6+3\cdot 2^7+\cdots+3\cdot 2^{18},$$

there are $18-4=14$ terms in the series. The first term is $a=3\cdot 2^5$ and the ratio is $r=2$, so

$$\text{Sum} = \frac{3\cdot 2^5(1-2^{14})}{1-2} = 1{,}572{,}768.$$

2. Since

$$\sum_{k=2}^{20}(-1)^k 5(0.9)^k = 5(0.9)^2-5(0.9)^3+5(0.9)^4-\cdots+5(0.9)^{20},$$

the series has $20-1=19$ terms. The first term is $a=5(0.9)^2$ and the ratio is $r=-0.9$, so

$$\text{Sum} = \frac{5(0.9)^2(1-(-0.9)^{19})}{1+0.9} = 2.420.$$

3. $3+\frac{3}{2}+\frac{3}{4}+\frac{3}{8}\cdots+\frac{3}{2^{10}} = 3\left(1+\frac{1}{2}+\cdots+\frac{1}{2^{10}}\right) = \frac{3\left(1-\frac{1}{2^{11}}\right)}{1-\frac{1}{2}} = 5.997.$

4. This is a geometric series whose first term is 5 and whose ratio is 3. We are looking for the sum of the first thirteen terms. Note that the first term is $5(3^0)$. Thus,

$$S_{13} = \frac{5(1-3^{13})}{1-3} = 3{,}985{,}805.$$

5. This is a geometric series with a first term is $a=1/125$ and a ratio of $r=5$. To determine the number of terms in the series, use the formula, $a_n=ar^{n-1}$. Calculating successive terms of the series $(1/125)5^{n-1}$, we get

$$1/125, 1/25, 1/5, 1, 5, 25, 125, 625.$$

Thus we sum the first eight terms of the series:

$$S_8 = \frac{(1/125)(1-5^8)}{1-5} = \frac{97656}{125} = 781.248.$$

6. We will use the formula, $S_{10} = \dfrac{8(1-2^{10})}{1-2} = 8184.$

7. Note that we are seeking the sum of the first eight terms. Thus,

$$S_8 = \frac{2\left(1-\left(\frac{3}{4}\right)^8\right)}{1-\frac{3}{4}} = 7.199.$$

8. Yes, $a = 2$, ratio $= 1/2$.

9. Yes, $a = 1$, ratio $= -1/2$.

10. No. Ratio between successive terms is not constant: $\dfrac{1/3}{1/2} = \dfrac{2}{3}$, while $\dfrac{1/4}{1/3} = \dfrac{3}{4}$.

11. Yes, $a = 5$, ratio $= -2$.

12. Notice that these numbers are all powers of 4, starting with 4^0 and ending with 4^4. Thus a possible answer is: $\sum_{n=0}^{4} 4^n$.

13. These numbers are all power of 3, with signs alternating from positive to negative. We need to change the signs on an alternating basis. By raising -1 to various powers, we can create the pattern shown. One possible answer is: $\sum_{n=1}^{6}(-1)^{n+1}(3^n)$.

14. After the initial term, each of these terms comes from multiplying 5 by the preceding number. So we know that we need powers of five to produce the correct series. One possibility is: $\sum_{n=0}^{6} 2(5^n)$.

15. We again need powers of -1 to create the alternating pattern. The first term is 32, and each term is $1/2$ of the preceding term, so we could use powers of $1/2$ to create the series. One possible answer is: $\sum_{n=0}^{5}(-1)^n 32\left(\frac{1}{2}\right)^n$.

16. $\sum_{n=0}^{n=4}(0.1)^n = 1 + 0.1 + 0.1^2 + 0.1^3 + 0.1^4 = (1-0.1^5)/(1-0.1) = 1.1111.$

17. $\sum_{n=0}^{n=5} 3/(2^n) = 3 + 3/2 + 3/4 + 3/8 + 3/16 + 3/32 = 3(1-1/2^6)/(1-1/2)) = 189/32$

18.

$$\begin{aligned}\sum_{j=0}^{j=n-1} e^{jx} &= 1 + e^x + e^{2x} + \ldots + e^{(n-1)x}\\ &= 1 + e^x + (e^x)^2 + \ldots + (e^x)^{n-1}\\ &= (1-e^{nx})/(1-e^x)\end{aligned}$$

19. $\sum_{n=0}^{n=N}(-1)^n = 1 + (-1) + (-1)^2 + (-1)^3 + \ldots + (-1)^N$, so $\sum_{n=0}^{n=N}(-1)^n = 1$ if N is even and 0 is N is odd.

Problems

20. We write

$$\begin{aligned}ar^3 + ar^5 + ar^7 + ar^9 + ar^{11} &= ar^3(1 + r^2 + r^4 + r^6 + r^8)\\ &= ar^3(1 + r^2 + (r^2)^2 + (r^2)^3 + (r^2)^4)\\ &= ar^3\sum_{j=0}^{j=4}(r^2)^j.\end{aligned}$$

21. **(a)** Let a_n be worldwide oil consumption n years after 2003. Then, $a_1 = 81$, $a_2 = 81(1.012)$, and $a_n = 81(1.012)^{n-1}$. Thus, between 2004 and 2028,

$$\text{Total oil consumption } = \sum_{n=1}^{25} 81(1.012)^{n-1} \text{ billion barrels.}$$

(b) Using the formula for the sum of a finite geometric series, we have

$$\text{Total oil consumption } = \frac{81\left(1-(1.012)^{25}\right)}{1-1.012} = 2345.291 \text{ billion barrels.}$$

22. **(a)** The amount of atenolol in the blood for a period of one day is given by $Q = Q_0e^{-kt}$, where Q_0 is the initial quantity, k is a constant, and t is in hours. Since the half-life is 6.3 hours,

$$\frac{1}{2} = e^{-6.3k}, \quad k = -\frac{1}{6.3}\ln\frac{1}{2} \approx 0.110.$$

After 24 hours

$$Q = Q_0e^{-k(24)} \approx Q_0e^{-0.110(24)} \approx Q_0(0.071).$$

Thus, the percentage of the atenolol that remains after 24 hours is about 7.1%.

(b)

$$\begin{aligned}
Q_0 &= 50\\
Q_1 &= 50 + 50(0.071)\\
Q_2 &= 50 + 50(0.071) + 50(0.071)^2\\
Q_3 &= 50 + 50(0.071) + 50(0.071)^2 + 50(0.071)^3\\
&\cdots\\
Q_n &= 50 + 50(0.071) + 50(0.071)^2 + \cdots + 50(0.071)^n\\
&= \frac{50(1-(0.071)^{n+1})}{1-0.071}
\end{aligned}$$

(c)

$$\begin{aligned}
P_1 &= 50(0.071)\\
P_2 &= 50(0.071) + 50(0.071)^2\\
P_3 &= 50(0.071) + 50(0.071)^2 + 50(0.071)^3\\
P_4 &= 50(0.071) + 50(0.071)^2 + 50(0.071)^3 + 50(0.071)^4\\
&\cdots\\
P_n &= 50(0.071) + 50(0.071)^2 + 50(0.071)^3 + \cdots + 50(0.071)^n\\
&= 50(0.071)\left(1 + (0.071) + (0.071)^2 + \cdots + (0.071)^{n-1}\right)\\
&= \frac{0.071(50)(1-(0.071)^n)}{1-0.071}.
\end{aligned}$$

23. **(a)** Let B_n be the balance in dollars in the account right after the nth deposit. Then

$$\begin{aligned}
B_1 &= 3000\\
B_2 &= 3000(1.05) + 3000\\
B_3 &= 3000(1.05)^2 + 3000(1.05) + 3000\\
&\vdots\\
B_{15} &= 3000(1.05)^{14} + 3000(1.05)^{13} + \cdots + 3000.
\end{aligned}$$

This is a finite geometric series whose sum is given by

$$B_{15} = \frac{3000(1-(1.05)^{15})}{1-1.05} = 64{,}735.69 \text{ dollars.}$$

(b) For continuous compounding, the ratio is $e^{0.05}$ instead of 1.05, so the series becomes

$$B_{15} = 3000(e^{0.05})^{14} + 3000(e^{0.05})^{13} + \cdots + 3000$$
$$B_{15} = \frac{3000(1 - (e^{0.05})^{15})}{1 - e^{0.05}} = 65{,}358.46 \text{ dollars.}$$

Notice that, as expected, continuous compounding leads to a larger balance than annual compounding.

24. Let B_n be the balance in dollars right after the nth deposit. Then

$$\begin{aligned}
B_1 &= 1000\\
B_2 &= 1000(1.03) + 1000\\
B_3 &= 1000(1.03)^2 + 1000(1.03) + 1000\\
&\vdots\\
B_{20} &= 1000((1.03)^{19} + (1.03)^{18} + \cdots + 1).
\end{aligned}$$

Using the formula for the sum of a finite geometric series,

$$B_{20} = \frac{1000(1 - (1.03)^{20})}{1 - 1.03} = 26{,}870.37 \text{ dollars.}$$

25. The answer to Problem 24 is given by

$$B_{20} = \frac{1000(1 - (1.03)^{20})}{1 - 1.03} = 26{,}870.37.$$

(a) Replacing \$1000 by \$2000 doubles the answer, giving \$53,740.75.
(b) Doubling the interest rate to 6% by replacing 1.03 by 1.06 less than doubles the answer, giving \$36,785.60.
(c) Doubling the number of deposits to 40 by replacing $(1.03)^{20}$ by $(1.03)^{40}$ more than doubles the answer, giving \$75,401.26.

26. **(a)** The interest rate is 3% per year $= 3/12 = 0.25\%$ per month. Let B_n be the balance in the account right after the nth deposit. Then

$$\begin{aligned}
B_1 &= 500\\
B_2 &= B_1e^{0.0025} + 500 = 500e^{0.0025} + 500\\
B_3 &= B_2e^{0.0025} + 500 = 500(e^{0.0025})^2 + 500e^{0.0025} + 500\\
B_4 &= B_3e^{0.0025} + 500 = 500((e^{0.0025})^3 + (e^{0.0025})^2 + e^{0.0025} + 1).
\end{aligned}$$

In 2 years, there are 24 months, so

$$B_{24} = 500((e^{0.0025})^{23} + \cdots + (e^{0.0025})^2 + e^{0.0025} + 1).$$

This is a finite geometric series with sum

$$B_{24} = \frac{500(1 - (e^{0.0025})^{24})}{1 - e^{0.0025}} = 12{,}351.86 \text{ dollars.}$$

(b) The balance goes over \$10,000 when

$$\begin{aligned}
B_n = \frac{500(1 - (e^{0.0025})^n)}{1 - e^{0.0025}} &= 10{,}000\\
1 - e^{0.0025n} &= \frac{10{,}000}{500}(1 - e^{0.0025})\\
e^{0.0025n} &= 1 + 20(e^{0.02} - 1)\\
n &= \frac{\ln(1 + 20(e^{0.0025} - 1))}{0.0025} = 19.54 \text{ months.}
\end{aligned}$$

Thus the balance first goes over \$10,000 before the 20th deposit, which takes place at the start of the 20th month.

27. (a) Let B_n be the balance in the account right after the nth payment is made. Then

$$\begin{aligned}
B_1 &= 50{,}000 - 1000 = 49{,}000\\
B_2 &= B_1(1.04) - 1000 = 49{,}000(1.04) - 1000\\
B_3 &= B_2(1.04) - 1000 = 49{,}000(1.04)^2 - 1000(1.04) - 1000\\
B_4 &= B_3(1.04) - 1000 = 49{,}000(1.04)^3 - 1000((1.04)^2 + 1.04 + 1)\\
B_5 &= 49{,}000(1.04)^4 - 1000((1.04)^3 + (1.04)^2 + 1.04 + 1)\\
&\vdots\\
B_{10} &= 49{,}000(1.04)^9 - 1000((1.04)^8 + (1.04)^7 + \cdots + 1).
\end{aligned}$$

Using the formula for the sum of a finite geometric series, we have

$$B_{10} = 49{,}000(1.04)^9 - \frac{1000(1-(1.04)^9)}{1-1.04} = 59{,}159.48 \text{ dollars.}$$

(b) If the payment is $\$P$ instead of \$1000, we solve for $\$P$ making $B_{10} = 0$. We have

$$\begin{aligned}
B_{10} = (50{,}000 - P)(1.04)^9 - \frac{P(1-(1.04)^9)}{1-1.04} &= 0,\\
P\left(\frac{1-(1.04)^9}{1-1.04} + (1.04)^9\right) &= 50{,}000(1.04)^9\\
P &= \frac{50{,}000(1.04)^9}{((1-(1.04)^9)/(1-1.04) + (1.04)^9)} = 5927.45 \text{ dollars.}
\end{aligned}$$

Solutions for Section 13.4

Exercises

1. Yes, $a = 1$, ratio $= -x$.

2. Yes, $a = y^2$, ratio $= y$.

3. No. Ratio between successive terms is not constant: $\dfrac{2x^2}{x} = 2x$, while $\dfrac{3x^3}{2x^2} = \dfrac{3}{2}x$.

4. No. Ratio between successive terms is not constant: $\dfrac{6z^2}{3z} = 2z$, while $\dfrac{9z^3}{6z^2} = \dfrac{3}{2}z$.

5. Yes, $a = e^x$, ratio $= e^x$.

6. Yes, $a = 1$, ratio $= -e^{-3x}$.

7. Yes, $a = 1$, ratio $= \sqrt{2}$.

8. Sum $= \dfrac{y^2}{1-y}$, $|y| < 1$.

9. Sum $= \dfrac{1}{1-(-x)} = \dfrac{1}{1+x}$, $|x| < 1$.

10. $-2 + 1 - \frac{1}{2} + \frac{1}{4} - \frac{1}{8} + \frac{1}{16} - \cdots = \sum_{i=0}^{\infty} (-2)\left(-\frac{1}{2}\right)^i$, a geometric series.

Let $a = -2$ and $r = -\frac{1}{2}$. Then

$$\sum_{i=0}^{\infty} (-2)\left(-\frac{1}{2}\right)^i = \frac{a}{1-r} = \frac{-2}{1-(-\frac{1}{2})} = -\frac{4}{3}.$$

11. Since the first term is $a = 11$ and $r = -0.1$,

$$\text{Sum } = 11 - 11(0.1) + 11(0.1)^2 - \cdots = \frac{a}{1-r} = \frac{11}{1-(-0.1)} = \frac{11}{1.1} = 10.$$

12. Since

$$\sum_{i=2}^{\infty}(0.1)^i = (0.1)^2 + (0.1)^2 + (0.1)^3 \cdots,$$

the first term is $a = (0.1)^2$ and the ratio is $r = 0.1$, so

$$\text{Sum } = \sum_{i=2}^{\infty}(0.1)^i = \frac{a}{1-r} = \frac{(0.1)^2}{1-0.1} = 0.011.$$

13.
$$\sum_{i=4}^{\infty}\left(\frac{1}{3}\right)^i = \left(\frac{1}{3}\right)^4 + \left(\frac{1}{3}\right)^5 + \cdots = \left(\frac{1}{3}\right)^4\left(1 + \frac{1}{3} + \left(\frac{1}{3}\right)^2 + \cdots\right) = \frac{(\frac{1}{3})^4}{1-\frac{1}{3}} = \frac{1}{54}.$$

14. $\displaystyle\sum_{i=0}^{\infty}\frac{3^i+5}{4^i} = \sum_{i=0}^{\infty}\left(\frac{3}{4}\right)^i + \sum_{i=0}^{\infty}\frac{5}{4^i}$, a sum of two geometric series.

$$\sum_{i=0}^{\infty}\left(\frac{3}{4}\right)^i = \frac{1}{1-\frac{3}{4}} = 4,$$

$$\sum_{i=0}^{\infty}\frac{5}{4^i} = \frac{5}{1-\frac{1}{4}} = \frac{20}{3},$$

$$\text{so } \sum_{i=0}^{\infty}\frac{3^i+5}{4^i} = 4 + \frac{20}{3} = \frac{32}{3}.$$

15. Since

$$\sum_{i=0}^{\infty}\frac{3^i}{2^{2i}} = \sum_{i=0}^{\infty}\frac{3^i}{4^i} = \sum_{i=0}^{\infty}\left(\frac{3}{4}\right)^i = 1 + 3/4 + (3/4)^2 + \cdots,$$

the first term is $a = 1$ and the ratio is $r = 3/4$, so

$$\text{Sum } = \sum_{i=0}^{\infty}\frac{3^i}{2^{2i}} = \frac{a}{1-r} = \frac{1}{1-3/4} = 4.$$

16. Expanding gives

$$\begin{aligned}\sum_{j=1}^{\infty}7((0.1)^j + (0.2)^{j+2}) &= \sum_{j=1}^{\infty}7(0.1)^j + \sum_{j=1}^{\infty}7(0.2)^{j+2} \\ &= (7(0.1) + 7(0.1)^2 + 7(0.1)^3 + \cdots) + (7(0.2)^3 + 7(0.2)^4 + 7(0.2)^5 + \cdots).\end{aligned}$$

We have the sum of two infinite geometric series, the first with first term $a = 7(0.1)$ and ratio $r = 0.1$, and the second with first term $a = 7(0.2)^3$ and ratio $r = 0.2$. Thus

$$\text{Sum } = \sum_{j=1}^{\infty}7((0.1)^j + (0.2)^{j+2}) = \frac{7(0.1)}{1-0.1} + \frac{7(0.2)^3}{1-0.2} = 0.848.$$

17. Since

$$\sum_{i=1}^{\infty} x^{2i} = x^2 + x^4 + x^6 \cdots,$$

the first term is $a = x^2$ and the ratio is $r = x^2$. Since $|x| < 1$,

$$\text{Sum} = \sum_{i=1}^{\infty} x^{2i} = \frac{a}{1-r} = \frac{x^2}{1-x^2}.$$

Problems

18. **(a)** $0.232323\ldots = 0.23 + 0.23(0.01) + 0.23(0.01)^2 + \ldots$ which is a geometric series with $a = 0.23$ and $x = 0.01$.

(b) The sum is $\dfrac{0.23}{1-0.01} = \dfrac{0.23}{0.99} = \dfrac{23}{99}$.

19. $0.235235235\ldots = \dfrac{235}{1{,}000} + \dfrac{235}{1{,}000{,}000} + \dfrac{235}{1{,}000{,}000{,}000} + \cdots$.

This is an infinite geometric series whose ratio is $1/1000$. Thus,

$$S = \frac{\frac{235}{1000}}{1-\frac{1}{1000}} = \frac{235}{999}.$$

20. $6.19191919\ldots = 6 + \dfrac{19}{100} + \dfrac{19}{10{,}000} + \dfrac{19}{1{,}000{,}000} + \cdots$. Thus,

$$S = 6 + \frac{\frac{19}{100}}{1-\frac{1}{100}} = 6 + \frac{19}{99} = \frac{613}{99}.$$

21. $0.122222\ldots = 0.1 + \dfrac{2}{100} + \dfrac{2}{1000} + \dfrac{2}{10000} + \dfrac{2}{100000} + \cdots$. Thus,

$$S = 0.1 + \frac{\frac{2}{100}}{1-\frac{1}{10}} = \frac{1}{10} + \frac{2}{90} = \frac{11}{90}.$$

22. $0.4788888\ldots = 0.47 + \dfrac{8}{1{,}000} + \dfrac{8}{10{,}000} + \dfrac{8}{100{,}000} + \cdots$. Thus,

$$S = 0.47 + \frac{\frac{8}{1000}}{1-\frac{1}{10}} = \frac{47}{100} + \frac{8}{900} = \frac{431}{900}.$$

23. $0.7638383838\ldots = 0.76 + \dfrac{38}{10{,}000} + \dfrac{38}{1{,}000{,}000} + \dfrac{38}{100{,}000{,}000} + \cdots$. Thus,

$$S = 0.76 + \frac{\frac{38}{10000}}{1-\frac{1}{100}} = \frac{76}{100} + \frac{38}{9900} = \frac{7562}{9900} = \frac{3781}{4950}.$$

24. Let Q_n represent the quantity, in milligrams, of ampicillin in the blood right after the nth tablet. Then

$$\begin{aligned}
Q_1 &= 250 \\
Q_2 &= 250 + 250(0.04) \\
Q_3 &= 250 + 250(0.04) + 250(0.04)^2 \\
&\vdots \\
Q_n &= 250 + 250(1.04) + 250(1.04)^2 + \cdots + 250(0.04)^{n-1}.
\end{aligned}$$

This is a geometric series. Its sum is given by

$$Q_n = \frac{250(1-(0.04)^n)}{1-0.04}.$$

Thus,

$$Q_3 = \frac{250(1-(0.04)^3)}{1-0.04} = 260.400$$

and

$$Q_{40} = \frac{250(1-(0.04)^{40})}{1-0.04} = 260.417.$$

In the long run, as $n \to \infty$, we know that $(0.04)^n \to 0$, and so

$$Q_n = \frac{250(1-(0.04)^n)}{1-0.04} \to \frac{250(1-0)}{1-0.04} = 260.417.$$

In the long run, the drug level approaches 260.417 mg.

25. (a)

$$\begin{aligned} P_1 &= 0 \\ P_2 &= 250(0.04) \\ P_3 &= 250(0.04) + 250(0.04)^2 \\ P_4 &= 250(0.04) + 250(0.04)^2 + 250(0.04)^3 \\ &\vdots \\ P_n &= 250(0.04) + 250(0.04)^2 + \cdots + 250(0.04)^{n-1} \end{aligned}$$

(b) Factoring our formula for P_n, we see that it involves a geometric series of $n-2$ terms:

$$P_n = 250(0.04)\underbrace{\left[1 + 0.04 + (0.04)^2 + \cdots + (0.04)^{n-2}\right]}_{n-2 \text{ terms}}.$$

The sum of this series is given by

$$1 + 0.04 + (0.04)^2 + \cdots + (0.04)^{n-2} = \frac{1-(0.04)^{n-1}}{1-0.04}.$$

Thus,

$$\begin{aligned} P_n &= 250(0.04)\left(\frac{1-(0.04)^{n-1}}{1-0.04}\right) \\ &= 10\left(\frac{1-(0.04)^{n-1}}{1-0.04}\right). \end{aligned}$$

(c) In the long run, that is, as $n \to \infty$, we know that $(0.04)^{n-1} \to 0$, and so

$$P_n = 10\left(\frac{1-(0.04)^{n-1}}{1-0.04}\right) \to 10\left(\frac{1-0}{1-0.04}\right) = 10.417.$$

Thus, P_n gets closer to 10.417 and Q_n gets closer to 260.42. We'd expect these limits to differ because one is right before taking a tablet and one is right after. We'd expect the difference between them to be exactly 250 mg, the amount of ampicillin in one tablet.

26.

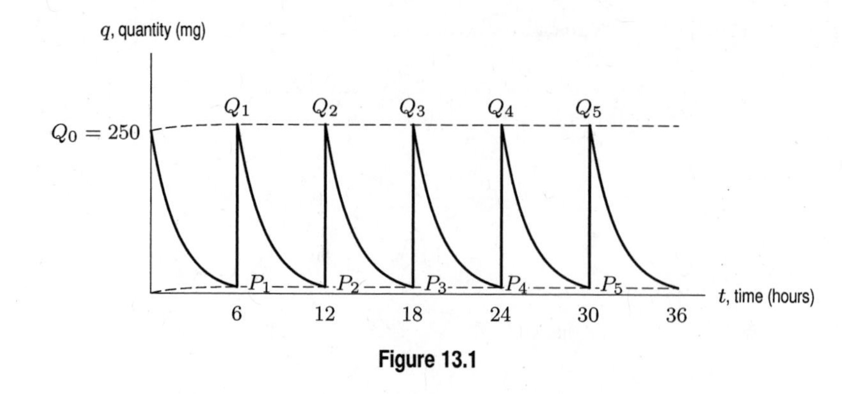

Figure 13.1

27. On the date of the first payment,

$$\text{Present value of first payment, in millions of dollars} = 3.$$

Since the second payment is made a year in the future, with continuous compounding,

$$\text{Present value of second payment, in millions of dollars} = 3e^{-0.07}.$$

Since the next payment is two years in the future,

$$\text{Present value of third payment, in millions of dollars} = 3e^{-0.07(2)}.$$

Similarly,

$$\text{Present value of tenth payment, in millions of dollars} = 3e^{-0.07(9)}.$$

Thus, in millions of dollars,

$$\text{Total present value} = 3 + 3e^{-0.07} + 3e^{-0.07(2)} + 3e^{-0.07(3)} + \cdots + 3e^{-0.07(9)}.$$

Since $e^{-0.07(n)} = \left(e^{-0.07}\right)^n$ for any n, we can write

$$\text{Total present value} = 3 + 3e^{-0.07} + 3\left(e^{-0.07}\right)^2 + 3\left(e^{-0.07}\right)^3 + 3\left(e^{-0.07}\right)^9.$$

This is a finite geometric series with $a = 3$ and $x = e^{-0.07}$; the sum will be

$$\text{Total present value of contract, in millions of dollars} = \frac{3\left(1 - (e^{-0.07})^{10}\right)}{1 - e^{-0.07}} \approx 22.3.$$

28.

$$\begin{aligned}\text{Present value of first coupon} &= \frac{50}{1.06}\\ \text{Present value of second coupon} &= \frac{50}{(1.06)^2}, \text{ etc.}\end{aligned}$$

$$\begin{aligned}\text{Total present value} &= \underbrace{\frac{50}{1.06} + \frac{50}{(1.06)^2} + \cdots + \frac{50}{(1.06)^{10}}}_{\text{coupons}} + \underbrace{\frac{1000}{(1.06)^{10}}}_{\text{principal}}\\ &= \frac{50}{1.06}\left(1 + \frac{1}{1.06} + \cdots + \frac{1}{(1.06)^9}\right) + \frac{1000}{(1.06)^{10}}\\ &= \frac{50}{1.06}\left(\frac{1 - \left(\frac{1}{1.06}\right)^{10}}{1 - \frac{1}{1.06}}\right) + \frac{1000}{(1.06)^{10}}\\ &= 368.004 + 558.395\\ &= \$926.40\end{aligned}$$

29.

$$\text{Present value of first coupon} = \frac{50}{1.04}$$
$$\text{Present value of second coupon} = \frac{50}{(1.04)^2}, \text{etc.}$$

$$\begin{aligned}
\text{Total present value} &= \underbrace{\frac{50}{1.04} + \frac{50}{(1.04)^2} + \cdots + \frac{50}{(1.04)^{10}}}_{\text{coupons}} + \underbrace{\frac{1000}{(1.04)^{10}}}_{\text{principal}} \\
&= \frac{50}{1.04}\left(1 + \frac{1}{1.04} + \cdots + \frac{1}{(1.04)^9}\right) + \frac{1000}{(1.04)^{10}} \\
&= \frac{50}{1.04}\left(\frac{1 - \left(\frac{1}{1.04}\right)^{10}}{1 - \frac{1}{1.04}}\right) + \frac{1000}{(1.04)^{10}} \\
&= 405.545 + 675.564 \\
&= \$1081.11.
\end{aligned}$$

30. (a)

$$\text{Present value of first coupon} = \frac{50}{1.05}$$
$$\text{Present value of second coupon} = \frac{50}{(1.05)^2}, \text{etc.}$$

$$\begin{aligned}
\text{Total present value} &= \underbrace{\frac{50}{1.05} + \frac{50}{(1.05)^2} + \cdots + \frac{50}{(1.05)^{10}}}_{\text{coupons}} + \underbrace{\frac{1000}{(1.05)^{10}}}_{\text{principal}} \\
&= \frac{50}{1.05}\left(1 + \frac{1}{1.05} + \cdots + \frac{1}{(1.05)^9}\right) + \frac{1000}{(1.05)^{10}} \\
&= \frac{50}{1.05}\left(\frac{1 - \left(\frac{1}{1.05}\right)^{10}}{1 - \frac{1}{1.05}}\right) + \frac{1000}{(1.05)^{10}} \\
&= 386.087 + 613.913 \\
&= \$1000.
\end{aligned}$$

(b) When the interest rate is 5%, the present value equals the principal.

(c) When the interest rate is more than 5%, the present value is smaller than it is when interest is 5% and must therefore be less than the principal. Since the bond will sell for around its present value, it will sell for less than the principal; hence the description *trading at discount.*

(d) When the interest rate is less than 5%, the present value is more than the principal. Hence the bound will be selling for more than the principal, and is described as *trading at a premium.*

Solutions for Chapter 13 Review

Exercises

1. We have $n = 18$, $a_1 = 8$, $d = 3$. To find the sum of the series we use the formula $S = (1/2)n(2a_1 + (n-1)d)$. So

$$S_{18} = \frac{18}{2}(2(8) + 17(3)) = 603.$$

To find the 18th term, we use the formula $a_n = a_1 + (n-1)d$. Thus,

$$a_{18} = 8 + 17(3) = 59.$$

2. We start with 100, decrease by tens until we reach 0, which is $100 - 10(10)$. A possible answer is

$$\sum_{n=0}^{10}(100 - 10n).$$

3. (a) $\sum_{n=1}^{5}(4n-3) = (4\cdot 1 - 3) + (4\cdot 2 - 3) + (4\cdot 3 - 3) + (4\cdot 4 - 3) + (4\cdot 5 - 3) = 1 + 5 + 9 + 13 + 17.$

(b) The sum is an arithmetic series with $a_1 = 1$ and $d = 4$. Thus,

$$S_5 = \frac{1}{2}5(2\cdot 1 + (5-1)4) = \frac{5}{2}(2+16) = 45.$$

4. Yes, $a = 1$, ratio $= 2z$.

5. No. Ratio between successive terms is not constant: $\dfrac{x^2/3}{x/2} = \dfrac{2x}{3}$, while $\dfrac{x^3/4}{x^2/3} = \dfrac{3x}{4}$.

6. Yes, $a = 1$, ratio $= -y^2$.

7. Sum $= \dfrac{1}{1-2z}, |z| < 1/2$.

8. Sum $= \dfrac{1}{1-(-y^2)} = \dfrac{1}{1+y^2}, |y| < 1$.

9. This is an arithmetic series. We have $n = 9$, $a_1 = 7$, $d = 7$. Using our formula, $S_n = (1/2)n(2a_1 + (n-1)d)$, we have

$$S_9 = \frac{9}{2}(2\cdot 7 + (9-1)7) = 315.$$

Problems

10. (a) The sequence is 1, 4, 9, 16. The nth grid has n dots per side, that is, n rows of n dots each, for a total of n^2 dots.

(b) The number of black dots in each grid is 1, 3, 5, 7, ... This appears to be the sequence of odd numbers. To see that this is true, notice (in Figure 13.2) that starting with a grid with 2 dots on each side, we add 2 dots to the right, 2 dots to the top, and 1 dot to the corner. Likewise, starting with a grid with 3 dots on each side, we add 3 dots to the right, 3 dots to the top, and 1 dot to the corner:

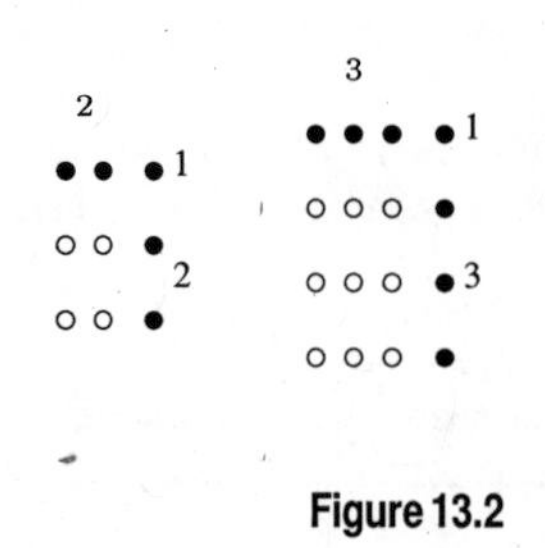

Figure 13.2

In general, starting with a grid with $n-1$ dots on each side, we add $n-1$ dots to the right, $n-1$ dots to the top, and 1 dot to the corner to obtain a grid with n dots on each side:

$$\text{Number of black dots in } n^{\text{th}} \text{ grid} = (n-1) + (n-1) + 1 = 2n - 1.$$

The nth grid has $2n - 1$ black dots.

(c) Notice that the total number of dots in the second grid equals the total number of *black* dots in the first two grids. Likewise, the total number of dots in the third grid equals the total number of *black* dots in the first three grids, and so on. Letting a_n be the number of black dots in the nth grid, and S_n be the total number of dots in the nth grid, we see that

$$S_n = a_1 + a_2 + \cdots + a_n.$$

From part (b), we have $a_n = 2n - 1$. If we rewrite this as $a_n = 2(n-1) + 1$, we have $a_n = a_1 + (n-1)d$ where $a_1 = 1$ and $d = 2$. Using our formula for the sum of an arithmetic series, this gives

$$\begin{aligned} S_n &= \frac{1}{2}n(2a_1 + (n-1)d) \\ &= \frac{1}{2}n(2 + (n-1)(2)) \\ &= n + n(n-1) \\ &= n^2, \end{aligned}$$

which agrees with our formula in part (a), thus verifying the relationship.

11. The numbers in the sequence 1, 4, 9, 16, ... from Judging from the diagram, the first 4 triangular numbers are 1, 3, 6, 10. The numbers of black dots in successive triangles are 1, 2, 3, 4. This is an arithmetic sequence given by $a_n = n$. If we rewrite this as $a_n = (n-1) + 1$, then we have $a_n = a_1 + (n-1)d$ where $a_1 = 1$ and $d = 1$. As with square numbers (see Problem 10), the number of dots in the nth triangle equals the sum of the number of *black* dots in the first n triangles. Letting S_n be the number of dots in the nth triangle, we have

$$\begin{aligned} S_n &= a_1 + a_2 + \cdots + a_n \\ &= \frac{1}{2}n(2a_1 + (n-1)d) \\ &= \frac{1}{2}n(2 + (n-1)(1)) \\ &= n + n(n-1)/2 \\ &= n(n+1)/2. \end{aligned}$$

12. (a) See Table 13.6.

Table 13.6 *Marginal cost for producing furniture*

n	c	Δc	n	c	Δc
1	300	−20	7	225	−2
2	280	−17	8	223	1
3	263	−14	9	224	4
4	249	−11	10	228	7
5	238	−8	11	235	10
6	230	−5	12	245	13

(b) We see that

$$\begin{aligned} c_2 &= 300 + (-20) \\ c_3 &= 300 + (-20) + (-17) \\ c_4 &= 300 + (-20) + (-17) + (-14) \\ c_n &= 300 + \underbrace{(-20) + (-17) + (-14) + \cdots}_{n-1 \text{ terms}} \\ &= 300 + \sum_{i=1}^{n-1} a_i, \end{aligned}$$

where a_i is the ith term in the arithmetic sequence $-20, -17, -14, \ldots$, where $a_1 = -20$ and where $d = 3$. Using our formula for the sum of an arithmetic sequence, we know that

$$\begin{aligned}\sum_{i=1}^{n-1} a_i &= \frac{1}{2}(n-1)\left[2a_1 + (n-2)d\right] \qquad a_1 = -20 \text{ and } d = 3\\ &= \frac{1}{2}(n-1)\left[2(-20) + (n-2)(3)\right]\\ &= \frac{1}{2}(n-1)(3n-46).\end{aligned}$$

Thus, $c_n = 300 + \frac{1}{2}(n-1)(3n-46)$. To check this formula, we let $n = 12$:

$$c_{12} = 300 + \frac{1}{2}(12-1)(3 \cdot 12 - 46) = 245,$$

which is the answer we got in part (a). To find the cost of producing the 50th piece of furniture, we have

$$c_{50} = 300 + \frac{1}{2}(50-1)(3 \cdot 50 - 46) = 2848.$$

(c) From part (a), we see that producing the 8th piece costs only \$223, but that each subsequent piece costs more. From part (b), we see that producing the 50th piece costs \$2848—far more than is profitable. Checking numbers, we see that the 19th and 20th pieces cost

$$c_{19} = 300 + \frac{1}{2}(19-1)(3 \cdot 19 - 46) = \$399$$

$$c_{20} = 300 + \frac{1}{2}(20-1)(3 \cdot 20 - 46) = \$433.$$

We see that pieces 1 through 19 can be produced at a profit, although the profit for the 19th piece is only \$1. Pieces 20 and on are produced at a loss. Thus, the workshop should produce 19 pieces of furniture a day.

13. We know $S_8 = 108$, $a_1 = 24$ and $n = 8$. We need to find d. Substituting in the formula $S = (1/2)n(2a_1 + (n-1)d)$, we get

$$\begin{aligned}108 &= \frac{1}{2}(8)(2(24) + 7d)\\ 108 &= 4(48 + 7d)\\ 108 &= 192 + 28d\\ d &= -3.\end{aligned}$$

He can make 8 rows of cans, starting with 24 cans on the bottom. Each row of cans will have 3 fewer cans than the row underneath it.

14. In 2008 the enrollment will be $8000(1.02) = 8160$. In 2009 the enrollment will be $8160(1.02) = 8323.3$; that is 8323 students. In 2010 the enrollment will be $8323.2(1.02) = 8489.664$; that is 8490 students. In 2011 the enrollment will be $8489.664(1.035) = 8786.802$; that is 8787 students. Each year after this the enrollment increases by 3.5 percent. See Table 13.7.

Table 13.7 *University enrollment by year*

Year	2007	2008	2009	2010	2011	2012	2013	2014	2015	2016	2017
Enrollment	8000	8160	8323	8490	8787	9094	9413	9742	10,083	10,436	10,801

15. We can picture the 30 people lined up in a row. The first person goes down the row and shakes hands with 29 people. The second person goes down the row and shakes hands with 28 people, and so on. The second to last person has only one person left to shake hands with. Thus

$$\text{Total number of handshakes } = 29 + 28 + \cdots + 1 = \frac{29(29+1)}{2} = 435.$$

A second way to solve the problem is the following: Each of the 30 people shakes hands with 29 people. Since we count each handshake only once, i.e., person A shakes hands with person B also counts as person B shaking hands with person A, so

$$\text{Total number of handshakes } = \frac{30 \cdot 29}{2}.$$

16. **(a)** Let B_n be the balance in dollars in the account right after the nth payment is made. Then, if B_0 is the initial balance,

$$\begin{aligned}
B_0 &= 75{,}000 \\
B_1 &= B_0(1.02) - 1000 = 75{,}000(1.02) - 1000 \\
B_2 &= B_1(1.02) - 1000 = 75{,}000(1.02)^2 - 1000(1.02) - 1000 \\
B_3 &= B_2(1.02) - 1000 = 75{,}000(1.02)^3 - 1000((1.02)^2 + 1.02 + 1) \qquad \text{factored 1000} \\
&\vdots \\
B_{24} &= 75{,}000(1.02)^{24} - 1000((1.02)^{23} + \cdots + 1.02 + 1).
\end{aligned}$$

Using the formula for the sum of a finite geometric series, we have

$$B_{24} = 75{,}000(1.02)^{24} - \frac{1000(1 - (1.02)^{24})}{1 - 1.02} = 90{,}210.93 \text{ dollars.}$$

(b) If the payments are \$3000 instead of \$1000,

$$B_{24} = 75{,}000(1.02)^{24} - \frac{3000(1 - (1.02)^{24})}{1 - 1.02} = 29{,}367.21 \text{ dollars.}$$

17.

$$\begin{aligned}
\text{Total present value, in dollars} &= 1000 + 1000e^{-0.04} + 1000e^{-0.04(2)} + 1000e^{-0.04(3)} + \cdots \\
&= 1000 + 1000(e^{-0.04}) + 1000(e^{-0.04})^2 + 1000(e^{-0.04})^3 + \cdots
\end{aligned}$$

This is an infinite geometric series with $a = 1000$ and $x = e^{(-0.04)}$, and sum

$$\text{Total present value, in dollars} = \frac{1000}{1 - e^{-0.04}} = 25{,}503.33.$$

18. **(a)** Since \$100,000 earns $0.03 \cdot 100{,}000 = \$3000$ interest in one year, and \$98,000 earns only slightly less (\$2940), the withdrawal is smaller than the interest earned, so we expect the balance to increase with time. Thus, we expect the balance to be higher after the second withdrawal.

(b) Let B_n be the balance in the account in dollars right after the nth withdrawal of \$2000. Then

$$\begin{aligned}
B_1 &= 100{,}000 - 2000 = 98{,}000 \\
B_2 &= B_1(1.03) - 2000 = 98{,}000(1.03) - 2000 \\
B_3 &= B_2(1.03) - 2000 = 98{,}000(1.03)^2 - 2000(1.03) - 2000 \\
B_4 &= B_3(1.03) - 2000 = 98{,}000(1.03)^3 - 2000((1.03)^2 + 1.03 + 1) \\
B_5 &= B_4(1.03) - 2000 = 98{,}000(1.03)^4 - 2000((1.03)^3 + (1.03)^2 + 1.03 + 1) \\
&\vdots \\
B_{20} &= 98{,}000(1.03)^{19} - 2000((1.03)^{18} + (1.03)^{17} + \cdots + 1.03 + 1).
\end{aligned}$$

Using the formula for the sum of a finite geometric series, we have

$$B_{20} = 98{,}000(1.03)^{19} - \frac{2000(1 - (1.03)^{19})}{1 - 1.03} = 121{,}609.86 \text{ dollars.}$$

(c) With withdrawals of \$2000 a year, the balance increases. The balance remains constant if the withdrawals exactly balance the interest earned; larger withdrawals cause the balance to decrease. If the largest withdrawal is $\$x$, then $B_1 = 100{,}000 - x$. The interest earned on this equals x, so

$$0.03(100{,}000 - x) = x.$$

Thus

$$\begin{aligned} 0.03 \cdot 100{,}000 - 0.03x &= x \\ x &= \frac{0.03 \cdot 100{,}000}{1.03} = 2912.62 \text{ dollars.} \end{aligned}$$

19. **(a)** Let a_n by the number of yards walked on the nth day. The first day he walks 300 yards, thus $a_1 = 300$, the second day he walks 50 additional yards, so $a_2 = 350$; the next day he walks 50 additional yards, so $a_3 = 400$, and so on. The sequence for the first week is: 300, 350, 400, 450, 500, 550, 600.

(b) Note that a_n is an arithmetic sequence with $a_1 = 300$ and $d = 50$. Therefore, $a_n = 300 + (n-1)50$. After two weeks, so for $n = 14$ days, the patient walks $a_{14} = 300 + (14-1)50 = 950$ yards.

(c) A mile is equal to 1760 yards. Setting $a_n = 1760$, we have

$$300 + (n-1)50 = 1760.$$

Solving the equation for n:

$$\begin{aligned} (n-1)50 &= 1460 \\ n - 1 &= 29.2 \\ n &= 30.2. \end{aligned}$$

Since n denotes the nth day of the patient's recovery, he first walks at least one mile on the 31st day.

20. **(a)** The first person calls 4 people, so there are $4 + 1 = 5$ total by the end of stage 1. In stage two, each of the 4 people calls 4 more, so

$$\text{Total number with news at second stage } = 1 + 4 + 16.$$

In the third stage, each of the 16 people call 4 more people so there are $16 \cdot 4$ more calls and

$$\text{Total number with news at third stage } = 1 + 4 + 16 + 64.$$

Similarly

$$\begin{aligned} \text{Total number with news at fourth stage } &= 1 + 4 + 16 + 64 + 256 \\ \text{Total number with news at fifth stage } &= 1 + 4 + 16 + 64 + 256 + 1024 = 1365. \end{aligned}$$

So there are 1365 people with the news at the fifth stage.

(b) As in part (a) at each stage of the tree, we multiply the number of people in the previous stage by 4 to find the number of new people contacted at the next stage. We then add up the number of people in each stage. Therefore

$$\text{Total number of members in a tree that contains 10 stages } = \sum_{i=0}^{10} 4^i.$$

This is a finite geometric series with $a = 1$ and $r = 4$ so

$$\sum_{i=0}^{10} 4^i = \frac{1(1-4^{11})}{1-4} = 1{,}398{,}101.$$

(c) In part (a) we saw that a tree with 5 stages reaches 1365 people. Thus, we only need one more stage to reach $1365 + 4 \cdot 1024 = 5461$ people, which is above the required number of 5000.

21. **(a)** Let h_n be the height of the nth bounce after the ball hits the floor for the nth time. Then from Figure 13.3,

$$\begin{aligned} h_0 &= \text{height before first bounce } = 10 \text{ feet}, \\ h_1 &= \text{height after first bounce } = 10\left(\frac{3}{4}\right) \text{ feet}, \\ h_2 &= \text{height after second bounce } = 10\left(\frac{3}{4}\right)^2 \text{ feet}. \end{aligned}$$

Generalizing this gives

$$h_n = 10\left(\frac{3}{4}\right)^n.$$

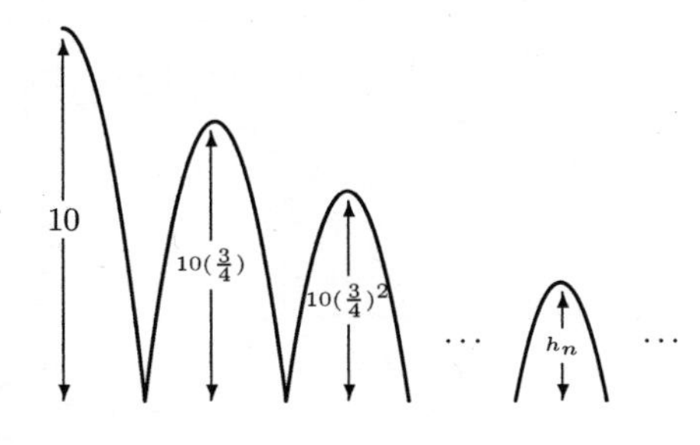

Figure 13.3

(b) When the ball hits the floor for the first time, the total distance it has traveled is just $D_1 = 10$ feet. (Notice that this is the same as $h_0 = 10$.) Then the ball bounces back to a height of $h_1 = 10\left(\frac{3}{4}\right)$, comes down and hits the floor for the second time. The total distance it has traveled is

$$D_2 = h_0 + 2h_1 = 10 + 2 \cdot 10\left(\frac{3}{4}\right) = 25 \text{ feet}.$$

Then the ball bounces back to a height of $h_2 = 10\left(\frac{3}{4}\right)^2$, comes down and hits the floor for the third time. It has traveled

$$D_3 = h_0 + 2h_1 + 2h_2 = 10 + 2 \cdot 10\left(\frac{3}{4}\right) + 2 \cdot 10\left(\frac{3}{4}\right)^2 = 25 + 2 \cdot 10\left(\frac{3}{4}\right)^2 = 36.25 \text{ feet}.$$

Similarly,

$$\begin{aligned} D_4 &= h_0 + 2h_1 + 2h_2 + 2h_3 \\ &= 10 + 2 \cdot 10\left(\frac{3}{4}\right) + 2 \cdot 10\left(\frac{3}{4}\right)^2 + 2 \cdot 10\left(\frac{3}{4}\right)^3 \\ &= 36.25 + 2 \cdot 10\left(\frac{3}{4}\right)^3 \\ &\approx 44.688 \text{ feet}. \end{aligned}$$

(c) When the ball hits the floor for the nth time, its last bounce was of height h_{n-1}. Thus, by the method used in part (b), we get

$$\begin{aligned} D_n &= h_0 + 2h_1 + 2h_2 + 2h_3 + \cdots + 2h_{n-1} \\ &= 10 + \underbrace{2 \cdot 10\left(\frac{3}{4}\right) + 2 \cdot 10\left(\frac{3}{4}\right)^2 + 2 \cdot 10\left(\frac{3}{4}\right)^3 + \cdots + 2 \cdot 10\left(\frac{3}{4}\right)^{n-1}}_{\text{finite geometric series}} \end{aligned}$$

$$= 10 + 2 \cdot 10 \cdot \left(\frac{3}{4}\right)\left(1 + \left(\frac{3}{4}\right) + \left(\frac{3}{4}\right)^2 + \cdots + \left(\frac{3}{4}\right)^{n-2}\right)$$

$$= 10 + 15\left(\frac{1 - \left(\frac{3}{4}\right)^{n-1}}{1 - \left(\frac{3}{4}\right)}\right)$$

$$= 10 + 60\left(1 - \left(\frac{3}{4}\right)^{n-1}\right).$$

22. The first drop from 10 feet takes $\frac{1}{4}\sqrt{10}$ seconds. The first full bounce (to $10 \cdot (\frac{3}{4})$ feet) takes $\frac{1}{4}\sqrt{10 \cdot (\frac{3}{4})}$ seconds to rise, therefore the same time to come down. Thus, the full bounce, up and down, takes $2(\frac{1}{4})\sqrt{10 \cdot (\frac{3}{4})}$ seconds. The next full bounce takes $2(\frac{1}{4})\sqrt{10 \cdot (\frac{3}{4})^2} = 2(\frac{1}{4})\sqrt{10}\left(\sqrt{\frac{3}{4}}\right)^2$ seconds. The nth bounce takes $2(\frac{1}{4})\sqrt{10}\left(\sqrt{\frac{3}{4}}\right)^n$ seconds. Therefore:

$$\text{Total amount of time} = \frac{1}{4}\sqrt{10} + \underbrace{\frac{2}{4}\sqrt{10}\sqrt{\frac{3}{4}} + \frac{2}{4}\sqrt{10}\left(\sqrt{\frac{3}{4}}\right)^2 + \frac{2}{4}\sqrt{10}\left(\sqrt{\frac{3}{4}}\right)^3}_{\text{Geometric series with } a = \frac{2}{4}\sqrt{10}\sqrt{\frac{3}{4}} = \frac{1}{2}\sqrt{10}\sqrt{\frac{3}{4}} \text{ and ratio } = \sqrt{\frac{3}{4}}} + \cdots$$

$$= \frac{1}{4}\sqrt{10} + \frac{1}{2}\sqrt{10}\sqrt{\frac{3}{4}}\left(\frac{1}{1 - \sqrt{3/4}}\right) \text{ seconds.}$$

23. **(a)**

$$\begin{aligned}\text{Total amount of money deposited} &= 100 + 92 + 84.64 + \cdots \\ &= 100 + 100(0.92) + 100(0.92)^2 + \cdots \\ &= \frac{100}{1 - 0.92} = 1250 \quad \text{dollars.}\end{aligned}$$

(b) Credit multiplier $= 1250/100 = 12.50$
The 12.50 is the factor by which the bank has increased its deposits, from \$100 to \$1250.

24. **(a)** See Figure 13.4.

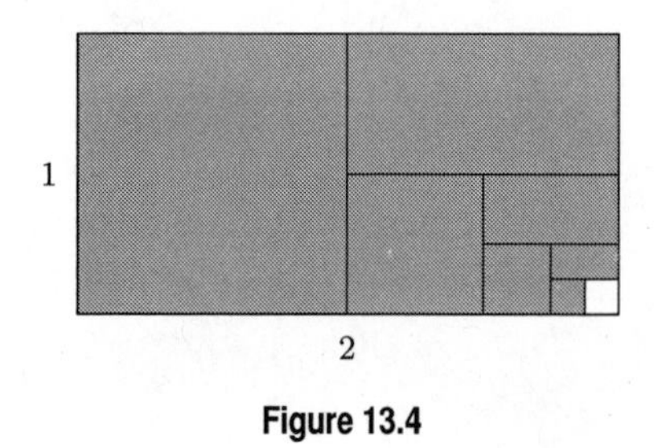

Figure 13.4

(b) The area of the first shaded piece is 1 by 1. The area of the second shaded piece is $1(\frac{1}{2}) = \frac{1}{2}$. The third shaded piece is half of the previous piece, so its area is $(\frac{1}{2})(\frac{1}{2}) = \frac{1}{4}$, and so forth. After n divisions

$$\text{Total shaded area} = 1 + \frac{1}{2} + \frac{1}{4} + \cdots + \frac{1}{2^{n-1}}.$$

(c) If we could continue the described process indefinitely, we would eventually shade all of the area, without ever shading anything outside of the original rectangle; therefore, the total shaded area is 2.

Using part (b), note that we have an infinite geometric series with $a = 1$ and $r = \frac{1}{2}$. Therefore

$$\text{Total shaded area} = \sum_{i=0}^{\infty} \frac{1}{2^i} = \frac{1}{1 - 1/2} = 2.$$

CHECK YOUR UNDERSTANDING

1. True. $a_1 = (1)^2 + 1 = 2$.
2. False. The first term can be any real number.
3. True. The differences between successive terms are all zero, so the sequence is arithmetic. The ratios of successive terms are all one, so it is also geometric.
4. True. The differences between successive terms are all a.
5. True. The differences between successive terms are all 1.
6. True. Factor d to obtain the usual formula: $a_n = a_1 + d(n - 1)$.
7. True. The ratios of successive terms are all x^2.
8. False. A sequence of positive numbers is not alternating. A sequence is alternating only if the terms alternate between positive and negative.
9. True. The first partial sum is just the first term of the sequence.
10. True. The total college credits are an accumulating sum of the credits completed each year.
11. False. That formula applies only to the series $1 + 2 + 3 + \cdots n$.
12. False. There are five.
13. True. The sum is n terms of 3. That is, $3 + 3 + \cdots + 3 = 3n$.
14. True. The formula $S_n = \frac{1}{2}n(a_1 + a_n)$ can be rewritten as $S_n = \frac{(a_1+a_n)}{2} \cdot n$ to correspond to the given statement.
15. False. The credits are cumulative, so this is your total credit earned during the first four years.
16. False. $C_3 - C_2$ is the credit earned in year 3.
17. False. The terms of the series can be negative so partial sums can decrease.
18. True. The terms alternate between $+1$ and -1.
19. False. For example, in case $n = 2$,
$$S_2 = \frac{a(1 - r^2)}{1 - r} = a \cdot \frac{(1 - r)(1 + r)}{1 - r} = a(1 + r).$$
20. False. The balance after 10 years is greater since the early money will earn interest for a longer time. This can be confirmed by calculating the two finite sums, $\sum_{i=1}^{5} 2000(1 + r)^i$ and $\sum_{i=1}^{10} 1000(1 + r)^i$ for some interest rate r.
21. True. If $a = 1$ and $r = -\frac{1}{2}$, it can be written $\sum_{i=0}^{5}(-\frac{1}{2})^i$.
22. True. Each term is a constant multiple of the preceding term. It can be written as $\sum_{i=0}^{7} ar^i$ with $a = 2000(1.06)^3$ and $r = 1.06$.
23. False. To be a geometric series would require the first term to be 2000. Note that the last ten terms are a geometric series.
24. True. The result is obtained by writing $5000 = 3000 + 2000$ and using the finite geometric sum formula $\frac{a(1-r^n)}{1-r}$, with $a = 2000, r = 1.06$.

25. False. If payments are made at the end of each year, after 20 years, the balance at 5% is about \$66,000, while at 10% it would be about \$115,000. If payments are made at the start of each year, the corresponding figures are \$69,000 and \$126,000.

26. True. This is the definition.

27. True. For each payment, B, we know that if r is the effective annual rate, the present value is $P = \frac{B}{(1+r)^n}$, so $2P = \frac{2B}{(1+r)^n}$.

28. False. Checking the first three terms $1 + \frac{1}{4} + \frac{1}{9}$, we see that successive terms do not have the same ratio.

29. False. The series does not converge since the odd terms (Q_1, Q_3, etc.) are all -1.

30. True. Using the infinite sum formula we find,

$$\sum_{i=1}^{\infty} 100(0.95)^i = \frac{100}{1 - 0.95} = 2000 \text{ mg}.$$

31. False. We need $|r| < 1$ for convergence.

32. True. A geometric series with positive a and $0 < r < 1$ has an infinite number of positive terms and a finite value.

33. False. An arithmetic series with $d \neq 0$ diverges.

34. False. Infinite geometric series converge only if the absolute value of the ratio of successive terms is less than one.

CHAPTER FOURTEEN

Solutions for Section 14.1

Exercises

1. We use a parameter t so that when $t = 0$ we have $x = 1$ and $y = 3$. One possible parameterization is $x = 1 + 2t$, $y = 3 + t, 0 \leq t \leq 1$.
2. We use a parameter t so that when $t = 0$, we have $x = 1$ and $y = 3$. One possible parameterization is $x = 1+t(-3-1) = 1 - 4t, y = 3 + t(5 - 3) = 3 + 2t, 0 \leq t \leq 1$.
3. We can parameterize the line from $(0, 0)$ to $(1, 1)$ by $x = t, y = t, 0 \leq t \leq 1$. The line from $(1, 1)$ to $(2, 0)$ can then be parameterized by $x = 1 + (t - 1)(2 - 1) = t, y = 1 + (t - 1)(0 - 1) = 2 - t$, for $1 \leq t \leq 2$. To check this note that when $t = 1$, we have $x = 1, y = 2 - 1 = 1$, and when $t = 2$ we have $x = 2, y = 2 - 2 = 0$, as required.
4. The unit circle centered at the origin, traversed anti-clockwise, can be parameterized by $x = \cos t, y = \sin t, 0 \leq t \leq 2\pi$, so a the quarter circle of radius 3 can be parametrized by $x = 3\cos t, y = 3\sin t, 0 \leq t \leq \frac{\pi}{2}$.
5. True. Eliminating t gives $x = 3(y - 3)$. Thus we have the straight line $y = 3 + \frac{x}{3}$. We must also check the end points $t = 0$ and $t = 1$. When $t = 0$ we have $x = 0$ and $y = 3$, and when $t = 1$ we have $x = 3$ and $y = 3 + 1 = 4$.
6. False. Eliminating t gives $x + y = 2$. Thus we have the straight line $y = 2 - x$ We must also check the end points $t = 0$ and $t = 1$. When $t = 0$ we have $x = 1$ and $y = 1$, and when $t = 1$ we have $x = -3$ and $y = 5$, so the line is traversed in the wrong direction.
7. False. Eliminating t gives $x^2 + y^2 = 4\cos^2 t + 4\sin^2 t = 4$, which is the circle of radius 2 centered at the origin. Checking the end points $t = -\frac{\pi}{2}$ and $t = \frac{\pi}{2}$, we have $x = 0$ and $y = 2$, and $x = 0$ and $y = -2$, as required but the curved traced out is in the wrong half plane because $x < 0$, for $-\frac{\pi}{2} < t < \frac{\pi}{2}$.
8. True. Eliminating t gives $(x-3)^2 + (y-4)^2 = 4\cos^2 t + 4\sin^2 t = 4$, which is the circle of radius 2 centered at $(3, 4)$.
9. The graph of the parametric equations is in Figure 14.1.

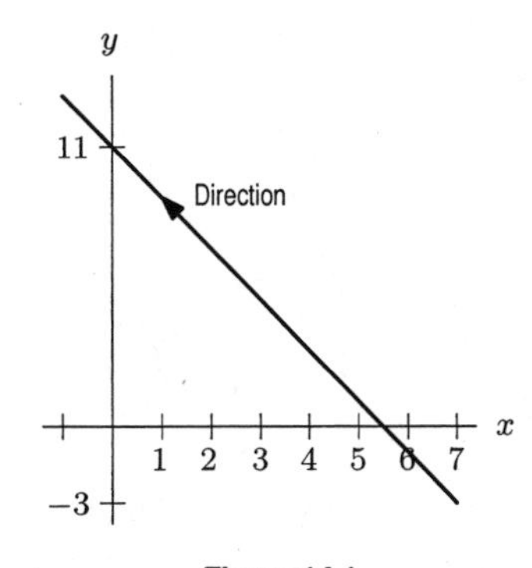

Figure 14.1

It is given that $x = 5 - 2t$, thus $t = (5 - x)/2$. Substitute this into the second equation:

$$y = 1 + 4t$$
$$y = 1 + 4\frac{(5 - x)}{2}$$
$$y = 11 - 2x.$$

10. The graph of the parametric equations is in Figure 14.2.

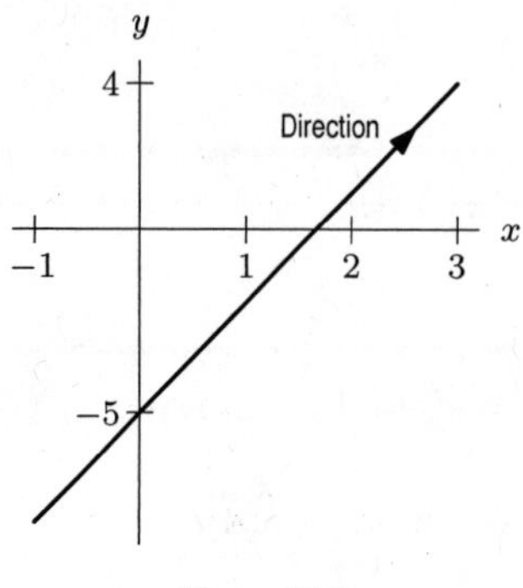

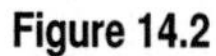
Figure 14.2

Since $x = t + 1$, we have $t = x - 1$. Substitute this into the second equation:

$$y = 3t - 2$$
$$y = 3(x - 1) - 2$$
$$y = 3x - 5.$$

11. The graph of the parametric equations is in Figure 14.3.

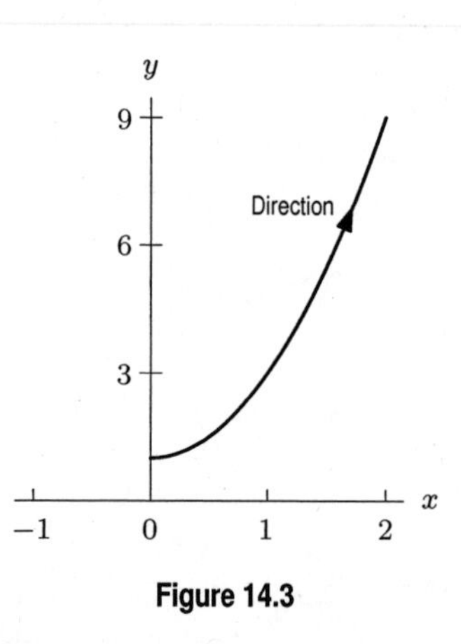

Figure 14.3

To solve $x = \sqrt{t}$ for t, we square both sides to get $t = x^2$. Substituting this into the second equation, we have:

$$y = 2t + 1$$
$$y = 2x^2 + 1.$$

12. The graph of the parametric equations is in Figure 14.4.

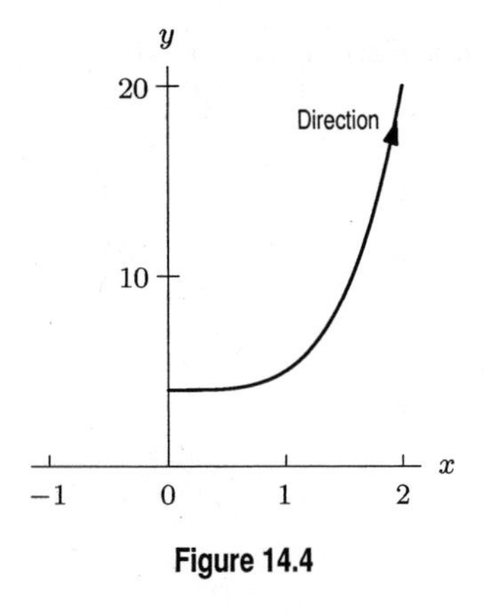

Figure 14.4

To solve $x = \sqrt{t}$ for t, we square both sides to get $t = x^2$. Substituting this into the second equation, we have:

$$y = t^2 + 4$$
$$y = (x^2)^2 + 4 = x^4 + 4.$$

13. The graph of the parametric equations is in Figure 14.5.

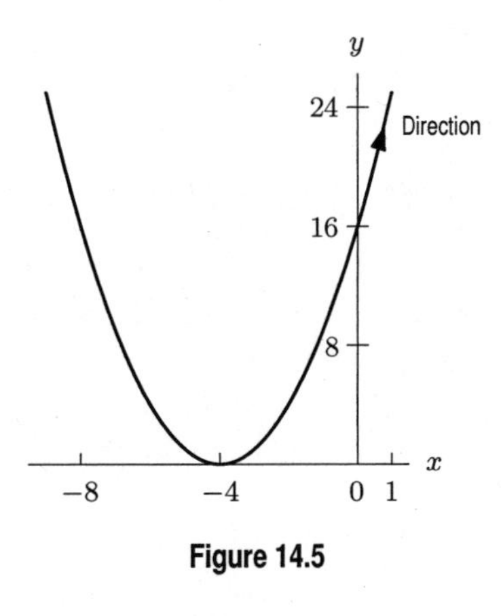

Figure 14.5

Since $x = t - 3$, we have $t = x + 3$. Substitute this into the second equation:

$$\begin{aligned} y &= t^2 + 2t + 1 \\ y &= (x+3)^2 + 2(x+3) + 1 \\ &= x^2 + 8x + 16 = (x+4)^2. \end{aligned}$$

14. The graph of the parametric equations is in Figure 14.6.

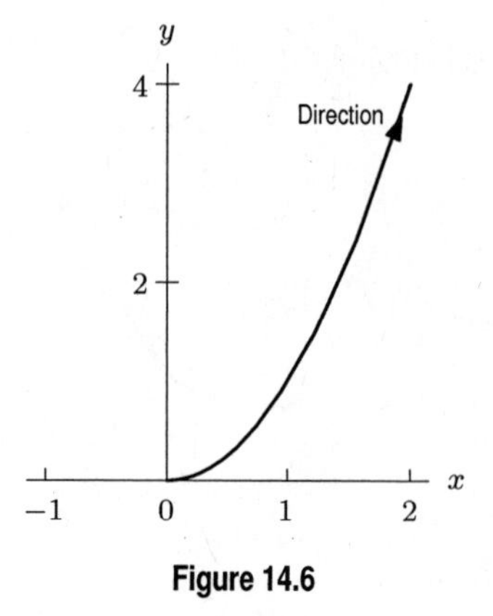

Figure 14.6

We are given that $x = e^t$ and $y = e^{2t}$ which is $(e^t)^2$. Thus, we can replace e^t with x and get $y = x^2$.

15. The graph of the parametric equations is in Figure 14.7.

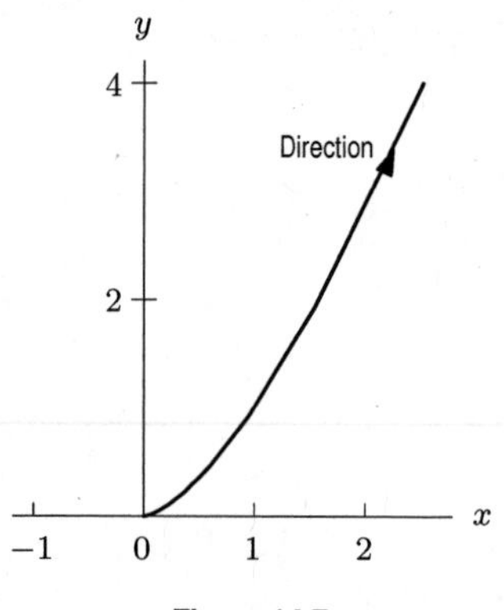

Figure 14.7

Since $x = e^{2t}$, we take the square root of both sides and get $\sqrt{x} = e^t$. Now, since $y = e^{3t}$ which is $(e^t)^3$, we can replace e^t with $\sqrt{x}$ and get

$$y = (\sqrt{x})^3 \quad \text{or} \quad y = x^{3/2}.$$

16. The graph of the parametric equations is in Figure 14.8.

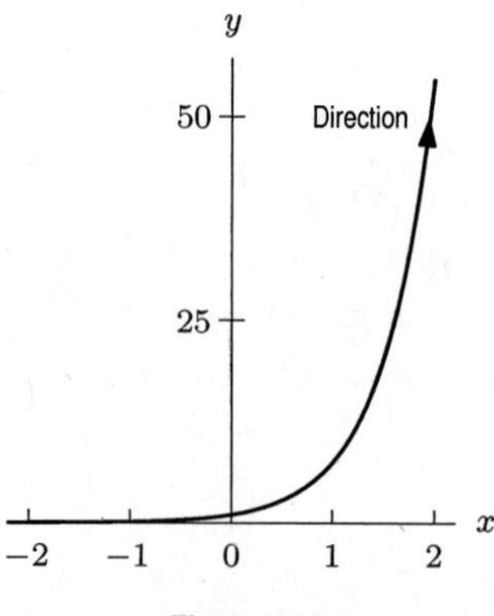

Figure 14.8

Since $x = \ln t$, we have $t = e^x$. Now, $y = t^2 = (e^x)^2 = e^{2x}$.

17. The graph of the parametric equations is in Figure 14.9.

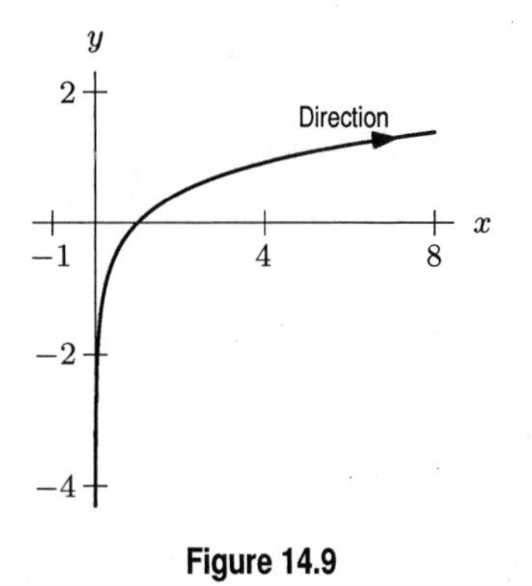

Figure 14.9

Since $x = t^3$, we take the natural log of both sides and get $\ln x = 3\ln t$ or $\ln t = 1/3 \ln x$. We are given that $y = 2\ln t$, thus,

$$y = 2\left(\frac{1}{3}\ln x\right) = \frac{2}{3}\ln x.$$

18. The graph of the parametric equations is in Figure 14.10.

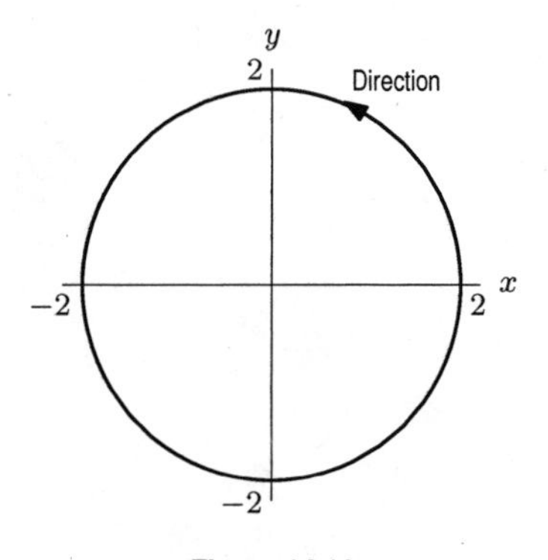

Figure 14.10

We are given that

$$x = 2\cos t$$
$$y = 2\sin t,$$

by squaring both sides, we get

$$x^2 = 4\cos^2 t$$
$$y^2 = 4\sin^2 t.$$

By adding the two equations and factoring out a 4, we get $x^2 + y^2 = 4(\cos^2 t + \sin^2 t)$. Since $\cos^2 t + \sin^2 t = 1$, we get

$$x^2 + y^2 = 4.$$

19. The graph of the parametric equations is in Figure 14.11.

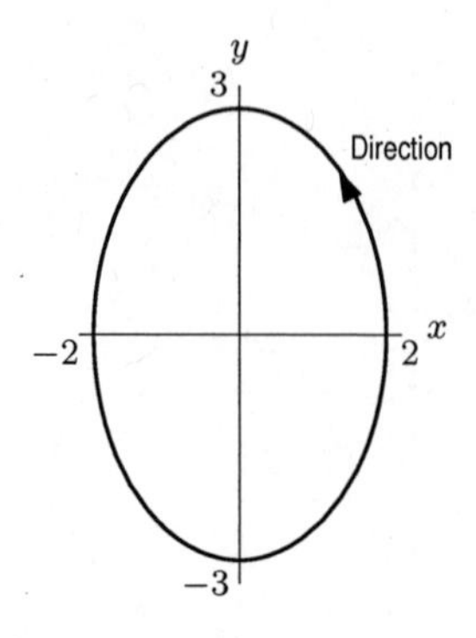

Figure 14.11

We are given that

$$x = 2\cos t$$
$$y = 3\sin t.$$

Isolating the trig functions, we get

$$\frac{x}{2} = \cos t$$
$$\frac{y}{3} = \sin t.$$

Squaring both sides gives

$$\left(\frac{x}{2}\right)^2 = \cos^2 t$$
$$\left(\frac{y}{3}\right)^2 = \sin^2 t.$$

By adding the two equations, we get $\frac{x^2}{4} + \frac{y^2}{9} = \cos^2 t + \sin^2 t$. Since $\cos^2 t + \sin^2 t = 1$, we get

$$\frac{x^2}{4} + \frac{y^2}{9} = 1.$$

20. The graph of the parametric equations is in Figure 14.12.

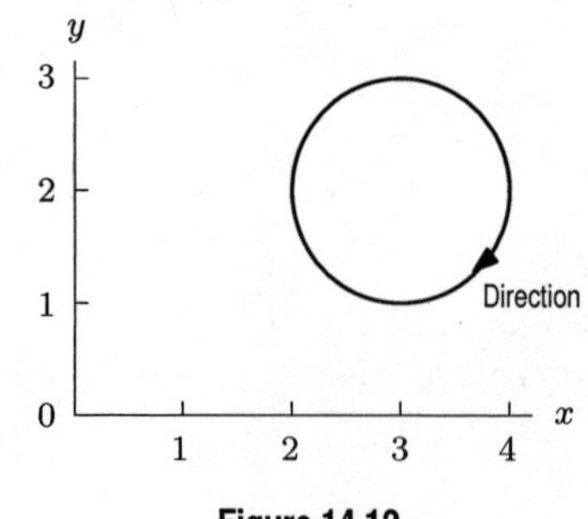

Figure 14.12

We are given that

$$x = 3 + \sin t$$
$$y = 2 + \cos t.$$

Isolating the trig functions, we get

$$x - 3 = \sin t$$
$$y - 2 = \cos t.$$

By squaring both sides, we get

$$(x-3)^2 = \sin^2 t$$
$$(y-2)^2 = \cos^2 t.$$

Adding the two equations gives us $(x-3)^2 + (y-2)^2 = (\sin^2 t + \cos^2 t)$.
Since $\sin^2 t + \cos^2 t = 1$, we get

$$(x-3)^2 + (y-2)^2 = 1.$$

21. This is like Example 5 on page 573 of the text, except that the x-coordinate goes all the way to 2 and back. So the particle traces out the rectangle shown in Figure 14.13.

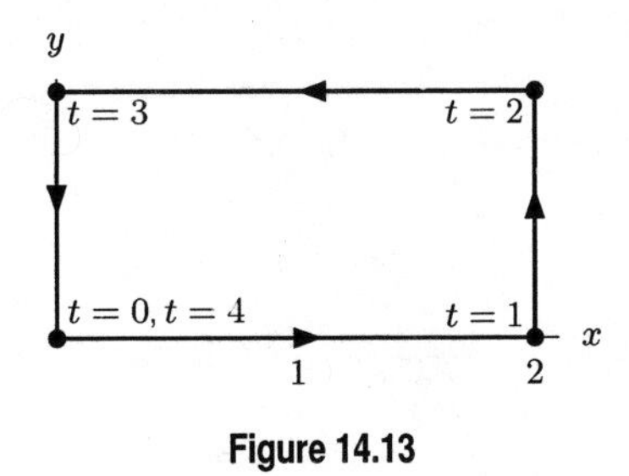

Figure 14.13

22. Between times $t = 0$ and $t = 1$, x goes at a constant rate from 0 to 1 and y goes at a constant rate from 1 to 0. So the particle moves in a straight line from $(0, 1)$ to $(1, 0)$. Similarly, between times $t = 1$ and $t = 2$, it goes in a straight line to $(0, -1)$, then to $(-1, 0)$, then back to $(0, 1)$. So it traces out the diamond shown in Figure 14.14.

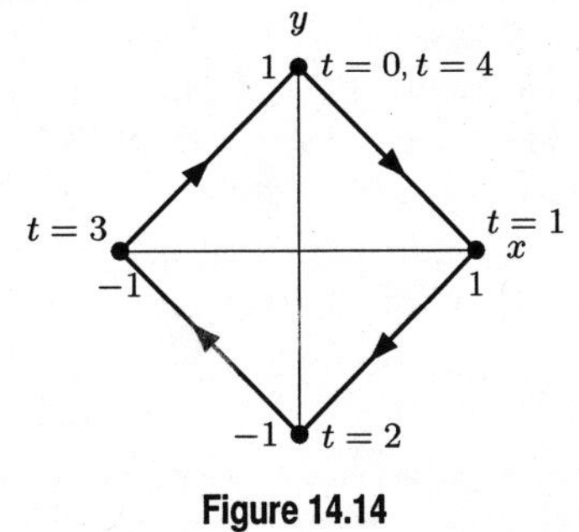

Figure 14.14

23. As the x-coordinate goes at a constant rate from 2 to 0, the y-coordinate goes from 0 to 1, then down to -1, then back to 0. So the particle zigs and zags from $(2, 0)$ to $(1.5, 1)$ to $(1, 0)$ to $(0.5, -1)$ to $(0, 0)$. Then it zigs and zags back again, forming the shape in Figure 14.15.

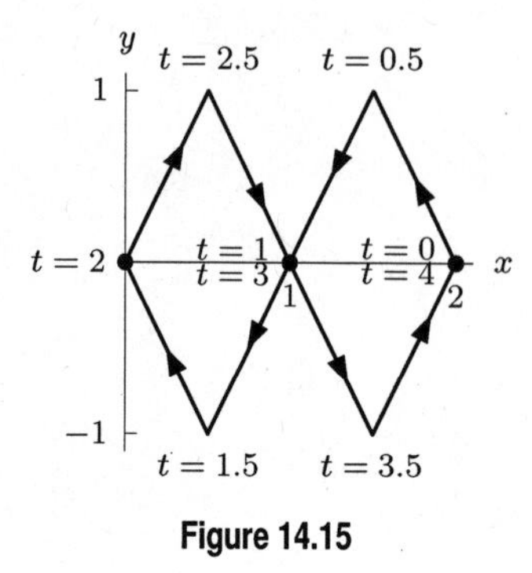

Figure 14.15

24. Between times $t = 0$ and $t = 1$, x goes from -1 to 1, while y stays fixed at 1. So the particle goes in a straight line from $(-1, 1)$ to $(1, 1)$. Then both the x- and y-coordinates decrease at a constant rate from 1 to -1. So the particle goes in a straight line from $(1, 1)$ to $(-1, -1)$. Then it moves across to $(1, -1)$, then back diagonally to $(-1, 1)$. See Figure 14.16.

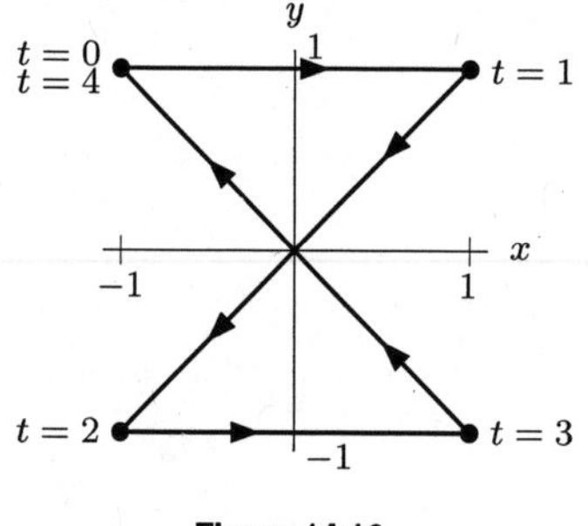

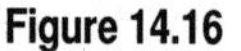

Figure 14.16

Problems

25. The particle moves clockwise: For $0 \leq t \leq \frac{\pi}{2}$, starting at $(1, 0)$ as t increases, we have $x = \cos t$ decreasing and $y = -\sin t$ decreasing. Similarly, for the time intervals $\frac{\pi}{2} \leq t \leq \pi, \pi \leq t \leq \frac{3\pi}{2}$, and $\frac{3\pi}{2} \leq t \leq 2\pi$, we see that the particle moves clockwise. The same is true for all $-\infty < t < +\infty$.

26. For $0 \leq t \leq \frac{\pi}{2}$, starting at $(0, 1)$ as t increases, we have $x = \sin t$ increasing and $y = \cos t$ decreasing, so the motion is clockwise for $0 \leq t \leq \frac{\pi}{2}$. Similarly, we see that the motion is clockwise for the time intervals $\frac{\pi}{2} \leq t \leq \pi, \pi \leq t \leq \frac{3\pi}{2}$, and $\frac{3\pi}{2} \leq t \leq 2\pi$. The same is true for all $-\infty < t < +\infty$.

27. Let $f(t) = t^2$. The particle is moving clockwise when $f(t)$ is decreasing, so when $t < 0$. The particle is moving counterclockwise when $f(t)$ is increasing, so when $t > 0$.

28. Let $f(t) = \ln t$. The particle is moving counterclockwise when $t > 0$. Any other time, when $t \leq 0$, the position is not defined.

29. In all three cases, $y = x^2$, so that the motion takes place on the parabola $y = x^2$.

In case (a), the x-coordinate always increases at a constant rate of one unit distance per unit time, so the equations describe a particle moving to the right on the parabola at constant horizontal speed.

In case (b), the x-coordinate is never negative, so the particle is confined to the right half of the parabola. As t moves from $-\infty$ to $+\infty$, $x = t^2$ goes from ∞ to 0 to ∞. Thus the particle first comes down the right half of the parabola, reaching the origin $(0, 0)$ at time $t = 0$, where it reverses direction and goes back up the right half of the parabola.

In case (c), as in case (a), the particle traces out the entire parabola $y = x^2$ from left to right. The difference is that the horizontal speed is not constant. This is because a unit change in t causes larger and larger changes in $x = t^3$ as t approaches $-\infty$ or ∞. The horizontal motion of the particle is faster when it is farther from the origin.

30. **(a)** If $t \geq 0$, we have $x \geq 2, y \geq 4$, so we get the part of the line to the right of and above the point $(2, 4)$.
(b) When $t = 0, (x, y) = (2, 4)$. When $t = -1, (x, y) = (-1, -3)$. Restricting t to the interval $-1 \leq t \leq 0$ gives the part of the line between these two points.
(c) If $x < 0$ then $2 + 3t < 0$, so $t < -2/3$. Thus $t < -2/3$ gives the points on the line to the left of the y-axis.

31. **(a)** We can replace x with t and $t + 1$ to get two parameterizations:

$$x = t, \quad y = t^2 \qquad \text{and} \qquad x = t + 1, \quad y = (t + 1)^2.$$

Alternatively, the second parameterization could look totally different:

$$x = t^3, \quad y = t^6.$$

(b) Altering the answers to part (a) gives:

$$x = t, \quad y = (t + 2)^2 + 1 \qquad \text{and} \qquad x = t + 1, \quad y = (t + 3)^2 + 1.$$

32. For $0 \leq t \leq 2\pi$, the graph is in Figure 14.17.

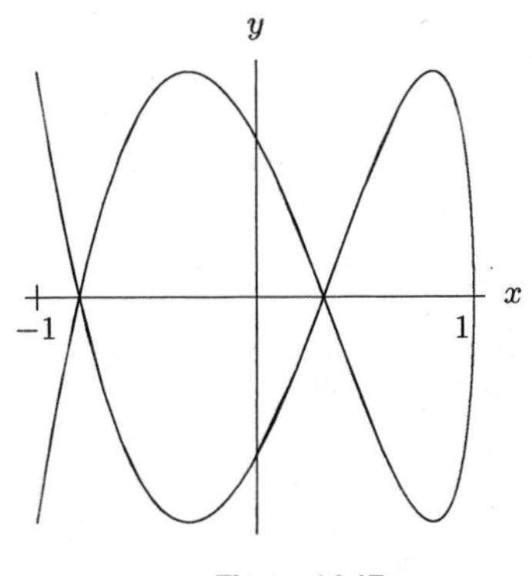

Figure 14.17

33. For $0 \leq t \leq 2\pi$, the graph is in Figure 14.18.

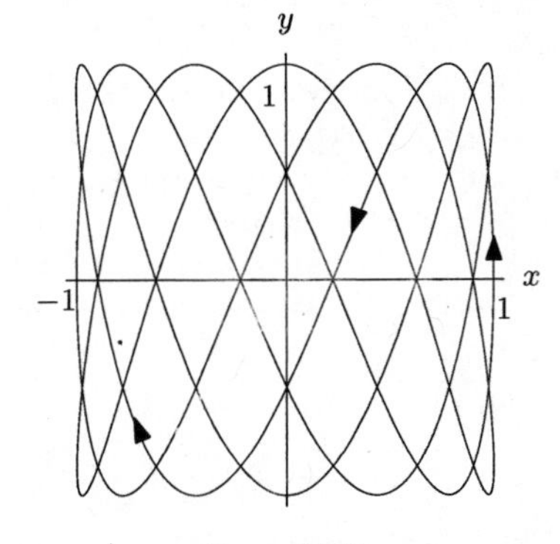

Figure 14.18

34. For $0 \le t \le 2\pi$, the graph is in Figure 14.19.

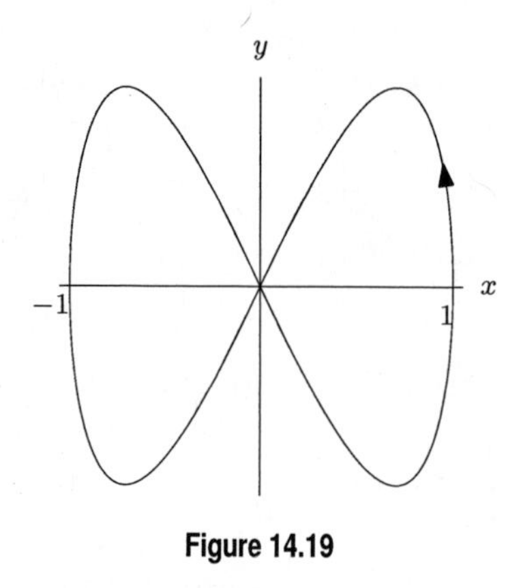

Figure 14.19

35. This curve never closes on itself. Figure 14.20 shows how it starts out.
The plot for $0 \le t \le 8\pi$ is in Figure 14.20.

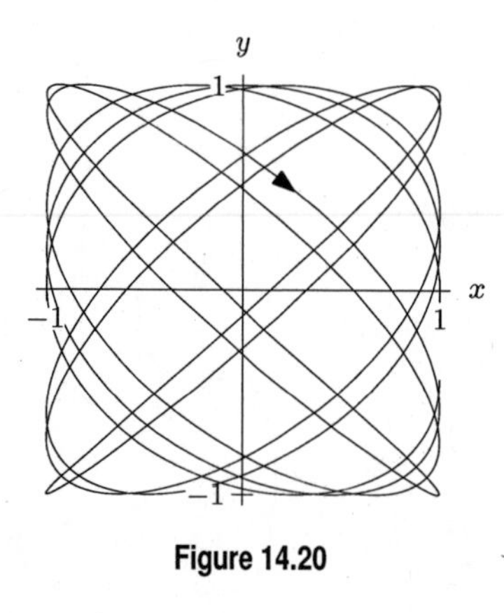

Figure 14.20

36. (I) has a positive slope and so must be l_1 or l_2. Since its y-intercept is negative, these equations must describe l_2. (II) has a negative slope and positive x-intercept, so these equations must describe l_3.

37. **(a)** Since the x-coordinate and the y-coordinate are always the same (they both equal t), the bug follows the path $y = x$.
(b) The bug starts at $(1, 0)$ because $\cos 0 = 1$ and $\sin 0 = 0$. Since the x-coordinate is $\cos x$, and the y-coordinate is $\sin x$, the bug follows the path of a unit circle, traveling counterclockwise. It reaches the starting point of $(1, 0)$ when $t = 2\pi$, because $\sin t$ and $\cos t$ are periodic with period 2π.
(c) Now the x-coordinate varies from 1 to -1, while the y-coordinate varies from 2 to -2; otherwise, this is much like part (b) above. If we plot several points, the path looks like an ellipse, which is a circle stretched out in one direction.

38. One possible answer is $x = -2, y = t$.

39. The slope of the line is

$$m = \frac{3 - (-1)}{1 - 2} = -4.$$

The equation of the line with slope -4 through the point $(2, -1)$ is $y - (-1) = (-4)(x - 2)$, so one possible parameterization is $x = t$ and $y = -4t + 8 - 1 = -4t + 7$.

40. The particle starts moving from left to right, then reverses its direction for a short time, then continues motion left to right. See Figure 14.21.

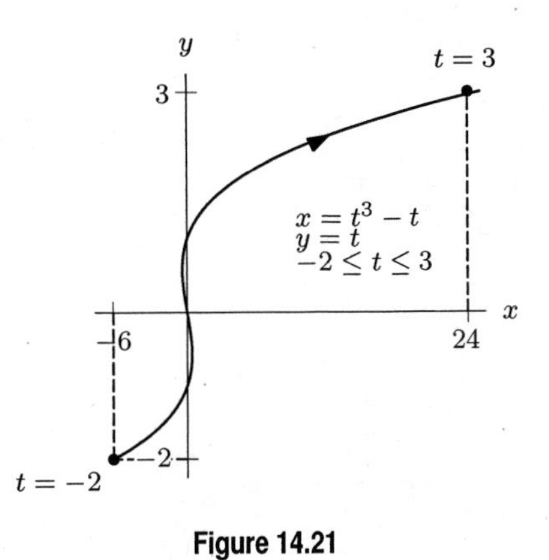

Figure 14.21

41. The particle moves back and forth between -1 and 1. See Figure 14.22.

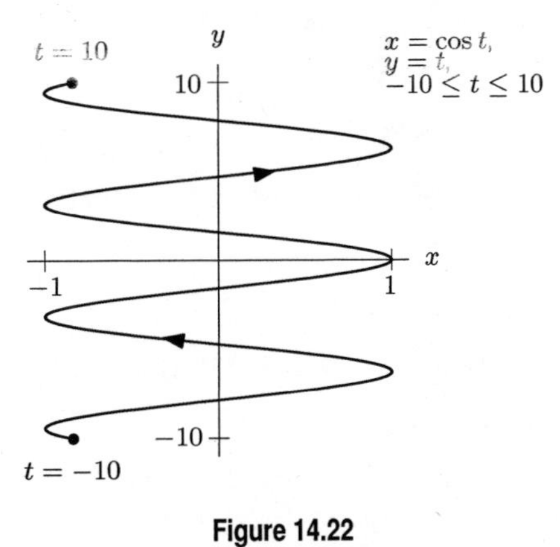

Figure 14.22

42. The particle starts moving to the left, reverses direction three times, then ends up moving to the right. See Figure 14.23.

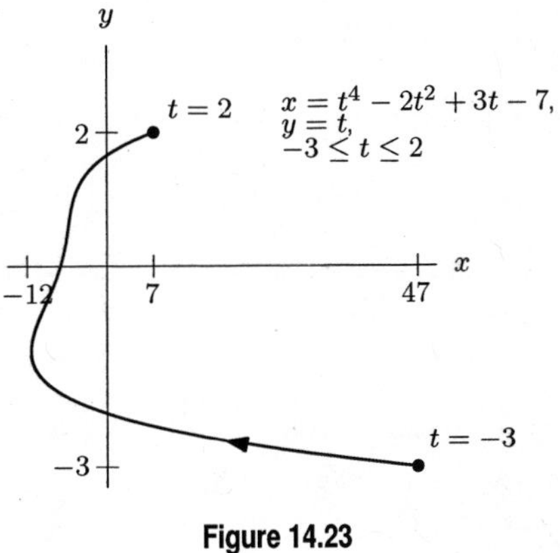

Figure 14.23

43. **(a)** A parameterization for the curve $d(t)$ is

$$x = t$$
$$y = -16t^2 + 48t + 6.$$

(b) The graph of $d(t)$ is in Figure 14.24.

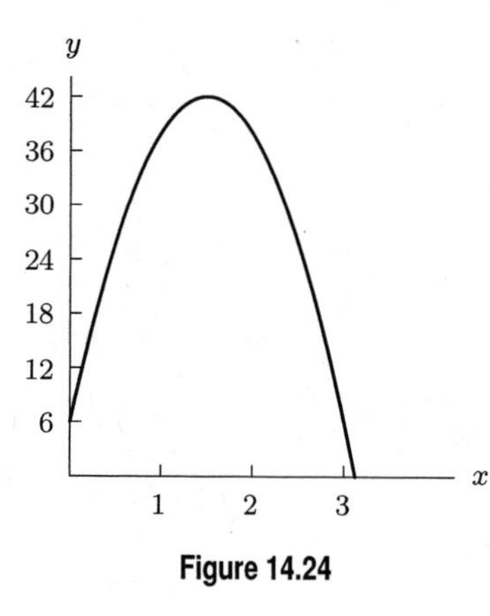

Figure 14.24

(c) Substituting 0 for t, we get $y = 6$. So the height is 6 feet; perhaps the hand of a person throwing the ball is 6 feet off the ground.

(d) Substituting 6 for y, we get

$$\begin{aligned} 6 &= -16t^2 + 48t + 6 \\ 0 &= -16t^2 + 48t \\ 0 &= -16t(t-3) \\ t = 0 &\text{ and } t = 3. \end{aligned}$$

So at $t = 3$, the ball will reach 6 feet again as it falls to the ground.

(e) Either graph the function and see where it reaches its maximum value or, solve algebraically. Complete the square:

$$\begin{aligned} y &= -16t^2 + 48t + 6 \\ y &= -16(t^2 - 3t + \frac{9}{4}) + 6 + 36 \\ &= -16\left(t - \frac{3}{2}\right)^2 + 42. \end{aligned}$$

At 1.5 seconds the ball reaches its maximum height of 42 feet above the ground.

Solutions for Section 14.2

Exercises

1. Explicit. For each x we can write down the corresponding values of y.

2. Explicit. For each t we can write down the values of x and y.

3. Implicit. Solving for y we obtain $y = \frac{a}{b}\sqrt{b^2 - x^2}$ and $y = -\frac{a}{b}\sqrt{b^2 - x^2}$.

4. Explicit. Solving for y we obtain $y = (x+1)^2$.

5. Implicit. This is a quadratic in y so solving for y by completing the square, we obtain $y = 1+\sqrt{2+x}$ and $y = 1-\sqrt{2+x}$.

6. Dividing by 4 and rewriting the equation as

$$x^2 + y^2 = \frac{9}{4},$$

we see that the center is $(0,0)$, and the radius is $3/2$.

7. Dividing by 3 and rewriting the equation as

$$x^2 + y^2 = \frac{10}{3},$$

we see that the center is $(0,0)$, and the radius is $\sqrt{10/3}$.

8. Moving the $-(x+1)^2$ term to the left and dividing by 2 gives

$$\begin{aligned} 2(x+1)^2 + 2(y+3)^2 &= 32 \\ (x+1)^2 + (y+3)^2 &= 16. \end{aligned}$$

Thus, the center is $(-1,-3)$, and the radius is 4.

9. Multiplying by $(y-4)^2$ gives

$$\begin{aligned} 4 - (x-4)^2 - (y-4)^2 &= 0 \\ (x-4)^2 + (y-4)^2 &= 4. \end{aligned}$$

Thus, the center is $(4,4)$, and the radius is 2.

10. We can use $x = 3\sin t$, $y = 3\cos t$ for $0 \le t \le 2\pi$.

11. We consider $x = 4\cos t, y = 4\sin t$ for $0 \le t \le 2\pi$. However, since y increases from 0 as t increases from 0, this goes counterclockwise. We can use

$$x = 4\cos t,\quad y = -4\sin t \quad \text{for } 0 \le t \le 2\pi.$$

12. We consider $x = -7\cos t$, $y = 7\sin t$ for $0 \le t \le 2\pi$. However, as t increases from 0, the value of y increases, so this starts at the correct point but goes clockwise. We can use

$$x = -7\cos t,\quad y = -7\sin t,\quad \text{for } 0 \le t \le 2\pi.$$

13. We can use $x = 5\sin t$, $y = -5\cos t$ for $0 \le t \le 2\pi$.

14. We can use $x = 3 + 4\sin t$, $y = 1 + 4\cos t$ for $0 \le t \le 2\pi$.

15. We can use $x = 3 + 5\cos t$, $y = 4 + 5\sin t$ for $0 \le t \le 2\pi$.

16. We consider $x = -1 - 3\cos t$, $y = -2 + 3\sin t$ for $0 \le t \le 2\pi$. However, as t increases from 0, the value of y increases, so this starts at the correct point but goes clockwise. We can use

$$x = -1 - 3\cos t,\quad y = -2 - 3\sin t,\quad \text{for } 0 \le t \le 2\pi.$$

17. We consider $x = -2 + \sqrt{5}\sin t$, $y = 1 + \sqrt{5}\cos t$ for $0 \le t \le 2\pi$. However, as t increases from 0, the value of x increases, so this starts at the correct point but goes clockwise. We can use

$$x = -2 - \sqrt{5}\sin t,\quad y = 1 + \sqrt{5}\cos t,\quad \text{for } 0 \le t \le 2\pi.$$

Problems

18. False. The center is at $(2,-2)$ but the radius is $\sqrt{2}$

19. True. Since $x^2 + 2x + y^2 - 2y + 2 = 0$ can be rewritten as $(x+1)^2 + (y-1)^2 = 4$, this is a circle with center at $(-1,1)$ with radius 2.

20. True. The center is at $(2, 0)$ and the radius is 2.

21. **(a)** Center is $(2, -4)$ and radius is $\sqrt{20}$.

(b) Rewriting the original equation and completing the square, we have

$$\begin{aligned}
2x^2 + 2y^2 + 4x - 8y &= 12 \\
x^2 + y^2 + 2x - 4y &= 6 \\
(x^2 + 2x + 1) + (y^2 - 4y + 4) - 5 &= 6 \\
(x + 1)^2 + (y - 2)^2 &= 11.
\end{aligned}$$

So the center is $(-1, 2)$, and the radius is $\sqrt{11}$.

22. The circle $(x - 2)^2 + (y - 2)^2 = 1$.

23. Since $x - 2 = \cos t$ and $y = \cos^2 t$, this parameterization traces out the parabola $y = (x - 2)^2$, for $1 \le x \le 3$.

24. The line segment $y + x = 4$, for $1 \le x \le 3$.

25. Since $y - 3 = \sin t$ and $x = 4\sin^2 t$, this parameterization traces out the parabola $x = 4(y - 3)^2$ for $2 \le y \le 4$.

26. **(a)** Separate the ant's path into three parts: from $(0, 0)$ to $(1, 0)$ along the x-axis; from $(1, 0)$ to $(0, 1)$ via the circle; and from $(0, 1)$ to $(0, 0)$ along the y-axis. (See Figure 14.25.) The lengths of the paths are 1, $\frac{2\pi}{4} = \frac{\pi}{2}$, and 1 respectively. Thus, the time it takes for the ant to travel the three paths are (using the formula $t = \frac{d}{v}$) $\frac{1}{2}$, $\frac{1}{3}$, and $\frac{1}{2}$ seconds.

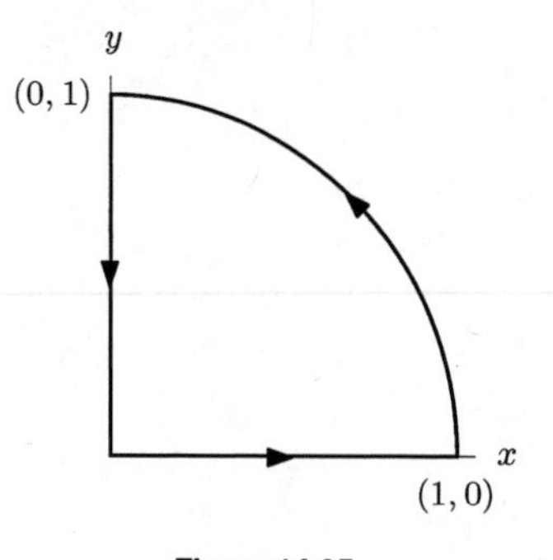

Figure 14.25

From $t = 0$ to $t = \frac{1}{2}$, the ant is heading toward $(1, 0)$ so its coordinate is $(2t, 0)$. From $t = \frac{1}{2}$ to $t = \frac{1}{2} + \frac{1}{3} = \frac{5}{6}$, the ant is veering to the left and heading toward $(0, 1)$. At $t = \frac{1}{2}$, it is at $(1, 0)$ and at $t = \frac{5}{6}$, it is at $(0, 1)$. Thus its position is $(\cos[\frac{3\pi}{2}(t - \frac{1}{2})], \sin[\frac{3\pi}{2}(t - \frac{1}{2})])$. Finally, from $t = \frac{5}{6}$ to $t = \frac{5}{6} + \frac{1}{2} = \frac{4}{3}$, the ant is headed home. Its coordinates are $(0, -2(t - \frac{4}{3}))$.

In summary, the function expressing the ant's coordinates is

$$(x(t), y(t)) = \begin{cases} (2t, 0) & \text{when } 0 \le t \le \frac{1}{2} \\ \left(\cos(\frac{3\pi}{2}(t - \frac{1}{2})), \sin(\frac{3\pi}{2}(t - \frac{1}{2}))\right) & \text{when } \frac{1}{2} < t \le \frac{5}{6} \\ (0, -2(t - \frac{4}{3})) & \text{when } \frac{5}{6} \le t \le \frac{4}{3}. \end{cases}$$

(b) To do the reverse path, observe that we can reverse the ant's path by interchanging the x and y coordinates (flipping it with respect to the line $y = x$), so the function is

$$(x(t), y(t)) = \begin{cases} (0, 2t) & \text{when } 0 \le t \le \frac{1}{2} \\ \left(\sin(\frac{3\pi}{2}(t - \frac{1}{2})), \cos(\frac{3\pi}{2}(t - \frac{1}{2}))\right) & \text{when } \frac{1}{2} < t \le \frac{5}{6} \\ (-2(t - \frac{4}{3}), 0) & \text{when } \frac{5}{6} < t \le \frac{4}{3}. \end{cases}$$

27. Implicit: $xy = 1$, $x > 0$
Explicit: $y = 1/x$, $x > 0$
Parametric: $x = t$, $y = 1/t$, $t > 0$

28. Parametric: $x = e^t$, $y = e^{2t}$ for all t. Explicit: $y = x^2$, for $x > 0$. Implicit: $x^2 - y = 0$, for $x > 0$.

29. Explicit: $y = \sqrt{4 - x^2}$
Implicit: $y^2 = 4 - x^2$ or $x^2 + y^2 = 4$, $y > 0$
Parametric: $x = 4\cos t$, $y = 4\sin t$, with $0 \le t \le \pi$.

30. Implicit: $x^2 - 2x + y^2 = 0$, $y < 0$. Explicit: $y = -\sqrt{-x^2 + 2x}$, $0 \le x \le 2$. Parametric: The curve is the lower half of a circle centered at $(1, 0)$ with radius 1, so $x = 1 + \cos t$, $y = \sin t$, for $\pi \le t \le 2\pi$.

31. **(a)** The center of the wheel moves horizontally, so its y-coordinate does not change; it equals 1 at all times. In one second, the wheel rotates 1 radian, which corresponds to 1 meter on the rim of a wheel of radius 1 meter, so the rolling wheel advances at a rate of 1 meter/sec. Thus the x-coordinate of the center, which equals 0 at $t = 0$, equals t at time t. At time t the center is at the point $x = t$, $y = 1$.

(b) By time t the spot on the rim has rotated t radians clockwise, putting it at angle $-t$ as in Figure 14.26. The coordinates of the spot with respect to the center of the wheel are $x = \cos(-t)$, $y = \sin(-t)$. Adding these to the coordinates $x = t$, $y = 1$ of the center gives the location of the spot as $x = t + \cos t$, $y = 1 - \sin t$. See Figure 14.27.

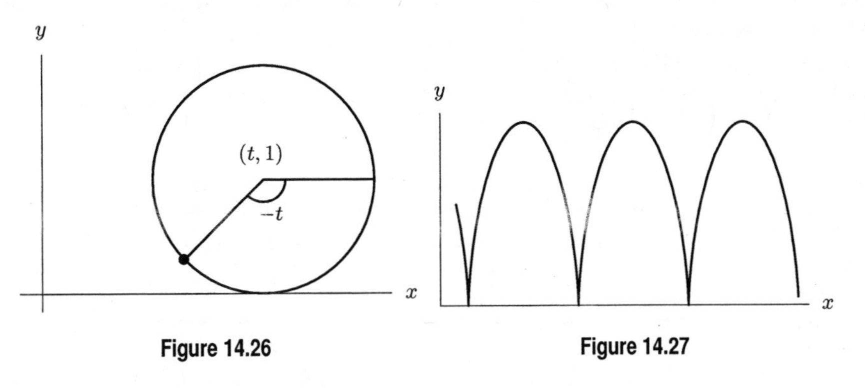

Figure 14.26

Figure 14.27

32. **(a)** C_1 has center at the origin and radius 5, so $a = b = 0, k = 5$ or -5.
(b) C_2 has center at $(0, 5)$ and radius 5, so $a = 0, b = 5, k = 5$ or -5.
(c) C_3 has center at $(10, -10)$, so $a = 10, b = -10$. The radius of C_3 is $\sqrt{10^2 + (-10)^2} = \sqrt{200}$, so $k = \sqrt{200}$ or $k = -\sqrt{200}$.

Solutions for Section 14.3

Exercises

1. **(a)** The center is at the origin. The diameter in the x-direction is 4 and the diameter in the y-direction is $2\sqrt{5}$.
(b) The equation of the ellipse is

$$\frac{x^2}{2^2} + \frac{y^2}{(\sqrt{5})^2} = 1 \quad \text{or} \quad \frac{x^2}{4} + \frac{y^2}{5} = 1.$$

2. **(a)** The center is at the origin. The diameter in the x-direction is 24 and the diameter in the y-direction is 10.
(b) The equation of the ellipse is

$$\frac{x^2}{12^2} + \frac{y^2}{5^2} = 1 \quad \text{or} \quad \frac{x^2}{144} + \frac{y^2}{25} = 1.$$

3. **(a)** The center is $(1, 0)$. The diameter in the x-direction is 16 and the diameter in the y-direction is 12.
(b) The equation of the ellipse is

$$\frac{(x-1)^2}{8^2} + \frac{y^2}{6^2} = 1 \quad \text{or} \quad \frac{(x-1)^2}{64} + \frac{y^2}{36} = 1.$$

4. **(a)** The center is $(-2,-1)$. The diameter in the x-direction is 6 and the diameter in the y-direction is 10.
(b) The equation of the ellipse is

$$\frac{(x+2)^2}{3^2}+\frac{(y+1)^2}{5^2}=1 \quad \text{or} \quad \frac{(x+2)^2}{9}+\frac{(y+1)^2}{25}=1.$$

5. We consider $x=-2\cos t$, $y=5\sin t$, for $0\le t\le 2\pi$. However, in this parameterization, y increases as t increases from 0, so it traces clockwise. Thus, we take $x=-2\cos t$, $y=-5\sin t$, for $0\le t\le 2\pi$.

6. We can use $x=12\cos t$, $y=5\sin t$ for $0\le t\le 2\pi$.

7. The center is $(1,0)$, and $a=8$, $b=6$, so we consider $x=1+8\cos t$, $y=0+6\sin t$, for $0\le t\le 2\pi$. However, this parameterization traces out the ellipse counterclockwise, so we use $x=1+8\cos t$, $y=-6\sin t$, for $0\le t\le 2\pi$.

8. The center is $(-2,-1)$, and $a=3$, $b=5$, so we can use $x=-2-3\cos t$, $y=-1+5\sin t$, for $0\le t\le 2\pi$.

9. The fact that the parameter is called s, not t, makes no difference. The minus sign means that the ellipse is traced out in the opposite direction. The graph of the ellipse in the xy-plane is the same as the ellipse in the example, and it is traced out once as s increases from 0 to 2π.

10. This ellipse has a graph which is the same curve in the xy plane, but it is traced out at twice the speed. At $t=\pi$, this parameterization has already returned to its starting point $(9,4)$. This parameterization traces out the ellipse twice during $0\le t\le 2\pi$.

Problems

11. **(a)** The center is $(-1,3)$, major axis $a=\sqrt{6}$, minor axis $b=2$. Figure 14.28 shows the graph.

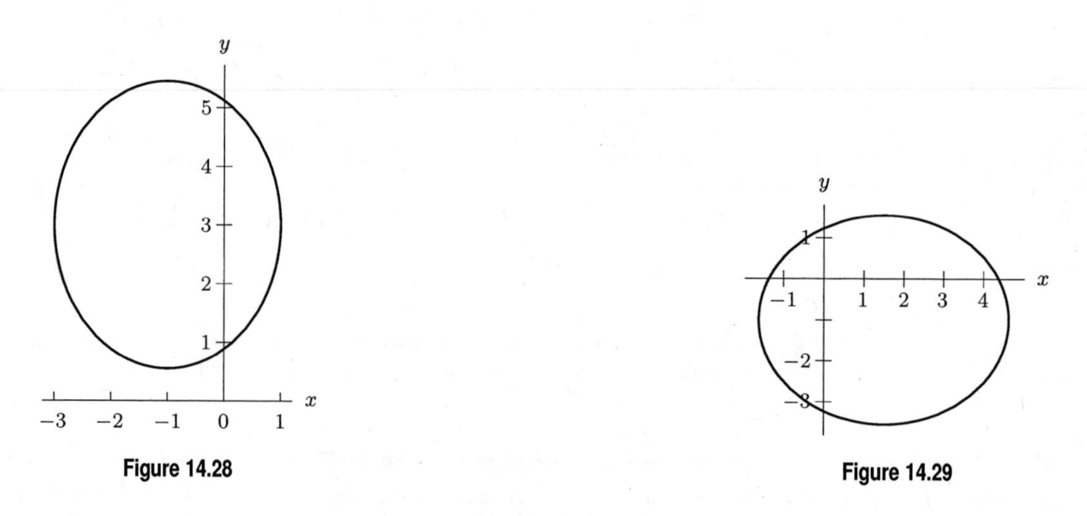

Figure 14.28

Figure 14.29

(b) Rewriting the equation and completing the square, we have

$$\begin{aligned}
2x^2+3y^2-6x+6y&=12\\
2(x^2-3x)+3(y^2+2y)&=12\\
2(x^2-3x+\frac{9}{4})+3(y^2+2y+1)-\frac{9}{2}-3&=12\\
2(x-\frac{3}{2})^2+3(y+1)^2&=\frac{39}{2}\\
\frac{(x-\frac{3}{2})^2}{(39/4)}+\frac{(y+1)^2}{(13/2)}&=1.
\end{aligned}$$

So the center is $(3/2,-1)$, major axis is $a=\sqrt{39}/2$, minor axis is $b=\sqrt{13/2}$. Figure 14.29 shows the graph.

12. Completing the square on $x^2 + 4x$ and $y^2 + 10y$:

$$\begin{aligned}
\frac{1}{9}(x^2+4x) + \frac{1}{25}(y^2+10y) &= -\frac{4}{9} \\
\frac{1}{9}((x+2)^2-4) + \frac{1}{25}((y+5)^2-25) &= -\frac{4}{9} \\
\frac{1}{9}(x+2)^2 - \frac{4}{9} + \frac{1}{25}(y+5)^2 - 1 + \frac{13}{4} &= -\frac{4}{9} \\
\frac{(x+2)^2}{9} + \frac{(y+5)^2}{25} &= 1.
\end{aligned}$$

The center is $(-2,-5)$, and $a = 3$, $b = 5$.

13. Completing the square on $x^2 - 2x$ and $y^2 + 4y$:

$$\begin{aligned}
\frac{1}{4}(x^2-2x) + y^2 + 4y + \frac{13}{4} &= 0 \\
\frac{1}{4}((x-1)^2-1) + (y+2)^2 - 4 + \frac{13}{4} &= 0 \\
\frac{1}{4}(x-1)^2 - \frac{1}{4} + (y+2)^2 - 4 + \frac{13}{4} &= 0 \\
\frac{(x-1)^2}{4} + (y+2)^2 &= 1.
\end{aligned}$$

The center is $(1,-2)$, and $a = 2$, $b = 1$.

14. Factoring out the 4 from $4x^2 - 4x = 4(x^2 - x)$ and completing the square on $x^2 - x$ and $y^2 + 2y$ gives

$$\begin{aligned}
4(x^2-x) + y^2 + 2y &= 2 \\
4\left(\left(x-\frac{1}{2}\right)^2 - \frac{1}{4}\right) + (y+1)^2 - 1 &= 2 \\
4\left(x-\frac{1}{2}\right)^2 - 1 + (y+1)^2 - 1 &= 2 \\
4\left(x-\frac{1}{2}\right)^2 + (y+1)^2 &= 4.
\end{aligned}$$

Dividing by the 4 to get 1 on the right:

$$\left(x-\frac{1}{2}\right)^2 + \frac{(y+1)^2}{4} = 1.$$

The center is $(\frac{1}{2},-1)$, and $a = 1$, $b = 2$.

15. Factoring out the 4 from $4x^2 + 16 = 4(x^2 + 4x)$ and completing the square on $x^2 + 4x$ and $y^2 + 2y$:

$$\begin{aligned}
4(x^2+4x) + y^2 + 2y + 13 &= 0 \\
4((x+2)^2-4) + (y+1)^2 - 1 + 13 &= 0 \\
4(x+2)^2 - 16 + (y+1)^2 - 1 + 13 &= 0 \\
4(x+2)^2 + (y+1)^2 - 4 &= 0.
\end{aligned}$$

Moving the -4 to the right and dividing by 4 to get 1 on the right side:

$$\begin{aligned}
4(x+2)^2 + (y+1)^2 &= 4 \\
(x+2)^2 + \frac{(y+1)^2}{4} &= 1.
\end{aligned}$$

The center is $(-2,-1)$, and $a = 1$, $b = 2$.

16. Factoring out the 9 from $9x^2 - 54x = 9(x^2 - 6x)$ and the 4 from $4y^2 - 16y = 4(y^2 - 4)$ and completing the square on $x^2 - 6x$ and $y^2 - 4y$:

$$\begin{aligned}
9(x^2 - 6x) + 4(y^2 - 4y) - 61 &= 0 \\
9((x-3)^2 - 9) + 4((y-2)^2 - 4) + 61 &= 0 \\
9(x-3)^2 - 81 + 4(y-2)^2 - 16 + 61 &= 0 \\
9(x-3)^2 + 4(y-2)^2 - 36 &= 0.
\end{aligned}$$

Moving the -36 to the right and dividing by 36 to get 1 on the right side,

$$\begin{aligned}
\frac{9(x-3)^2}{36} + \frac{4(y-2)^2}{36} &= \frac{36}{36} \\
\frac{(x-3)^2}{4} + \frac{(y-2)^2}{9} &= 1.
\end{aligned}$$

The center is $(3, 2)$, and $a = 2$, $b = 3$.

17. Factoring out the 9 from $9x^2 + 9x = 9(x^2 + x)$ and the 4 from $4y^2 - 4y = 4(y^2 - y)$ and completing the square on $x^2 + x$ and $y^2 - y$:

$$\begin{aligned}
9(x^2 + x) + 4(y^2 - y) &= \frac{131}{4} \\
9\left(\left(x + \frac{1}{2}\right)^2 - \frac{1}{4}\right) + 4\left(\left(y - \frac{1}{2}\right)^2 - \frac{1}{4}\right) &= \frac{131}{4} \\
9\left(x + \frac{1}{2}\right)^2 - \frac{9}{4} + 4\left(y - \frac{1}{2}\right)^2 - 1 &= \frac{131}{4} \\
9\left(x + \frac{1}{2}\right)^2 + 4\left(y - \frac{1}{2}\right)^2 &= \frac{144}{4} = 36.
\end{aligned}$$

Dividing by 36 to get 1 on the right:

$$\begin{aligned}
\frac{9\left(x + \frac{1}{2}\right)^2}{36} + \frac{4\left(y - \frac{1}{2}\right)^2}{36} &= \frac{36}{36} \\
\frac{\left(x + \frac{1}{2}\right)^2}{4} + \frac{\left(y - \frac{1}{2}\right)^2}{9} &= 1.
\end{aligned}$$

The center is $\left(-\frac{1}{2}, \frac{1}{2}\right)$, and $a = 2$, $b = 3$.

18. The center of the ellipse is at (h, k); both coordinates are negative here. Figure 14.30 shows a, b, $|h|$, $|k|$ on the ellipse. We see that

$$h < k < 0 < a < b.$$

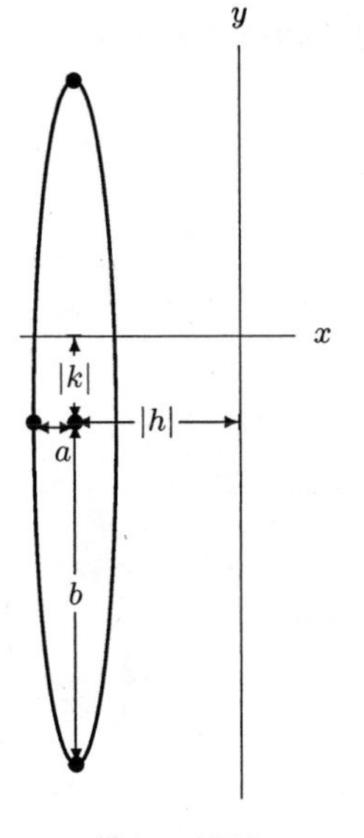

Figure 14.30

19. The center of the ellipse is at (h, k); both coordinates are positive here. Figure 14.31 shows a, b, h, k on the ellipse. We see that

$$0 < k < b < h < a.$$

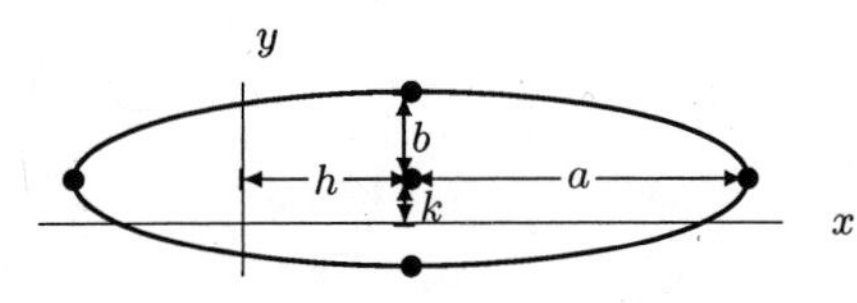

Figure 14.31

20. We have

$$\begin{aligned}
Ax^2 - Bx + y^2 &= r_0^2 \\
A\left(x^2 - \frac{B}{A}x\right) + y^2 &= r_0^2 \\
A\left(x^2 - \frac{B}{A}x + \left(\frac{B}{2A}\right)^2\right) + y^2 &= r_0^2 + \frac{B^2}{4A} \qquad \text{completing the square} \\
A\left(x - \frac{B}{2A}\right)^2 + y^2 &= \frac{4Ar_0^2 + B^2}{4A}.
\end{aligned}$$

Dividing both sides by $(4Ar_0^2 + B^2)/(4A)$, we obtain

$$\frac{(x - B/2A)^2}{a^2} + \frac{y^2}{b^2} = 1,$$

where $a^2 = (4Ar_0^2 + B^2)/(4A^2)$ and $b^2 = (4Ar_0^2 + B^2)/(4A)$.

21. **(a)** Clearing the denominator, we have

$$r(1 - \epsilon\cos\theta) = r_0$$

$$\begin{aligned}
r - r\epsilon\cos\theta &= r_0 \\
r - \epsilon x &= r_0 \qquad \text{because } x = r\cos\theta \\
r &= \epsilon x + r_0 \\
r^2 &= (\epsilon x + r_0)^2 \qquad \text{squaring both sides} \\
x^2 + y^2 &= \epsilon^2 x^2 + 2\epsilon r_0 x + r_0^2 \qquad \text{because } r^2 = x^2 + y^2 \\
x^2 - \epsilon^2 x^2 - 2\epsilon r_0 x + y^2 &= r_0^2 \qquad \text{regrouping} \\
(1-\epsilon^2)x^2 - 2\epsilon r_0 x + y^2 &= r_0^2 \qquad \text{factoring} \\
Ax^2 - Bx + y^2 &= r_0^2,
\end{aligned}$$

where $A = 1 - \epsilon^2$ and $B = 2\epsilon r_0$. From Question 20, we see that this is the formula of an ellipse.

(b) In the fraction $r_0/(1-\epsilon\cos\theta)$, the numerator is a constant. Thus, the fraction's value is largest when the denominator is smallest, and smallest when the denominator is largest. The largest value of the denominator occurs at $\cos\theta = -1$, that is, at $\theta = \pi$, and so the minimum value of r is

$$r = \frac{r_0}{1-\epsilon\cos\pi} = \frac{r_0}{1+\epsilon}.$$

The smallest value of the denominator occurs at $\cos\theta = 1$, that is, at $\theta = 0$, and so the minimum value of r is

$$r = \frac{r_0}{1-\epsilon\cos 0} = \frac{r_0}{1-\epsilon}.$$

(c) See Figure 14.32. We know from part (b) that at $\theta = 0$, $r = r_0/(1-\epsilon) = 6/(1-1/2) = 12$, and that at $\theta = \pi$, $r = r_0/(1+\epsilon) = 6/(1+1/2) = 4$. In addition, at $\theta = \pi/2$ and $\theta = 3\pi/2$, we see that $\cos\theta = 0$ and so $r = r_0 = 6$. This gives us the four points labeled in the figure. By symmetry, the center of the ellipse must lie between the points $(4,\pi)$ and $(12,0)$, or, in Cartesian coordinates, the points $(-4,0)$ and $(12,0)$. Thus, the center is at $(8,0)$.

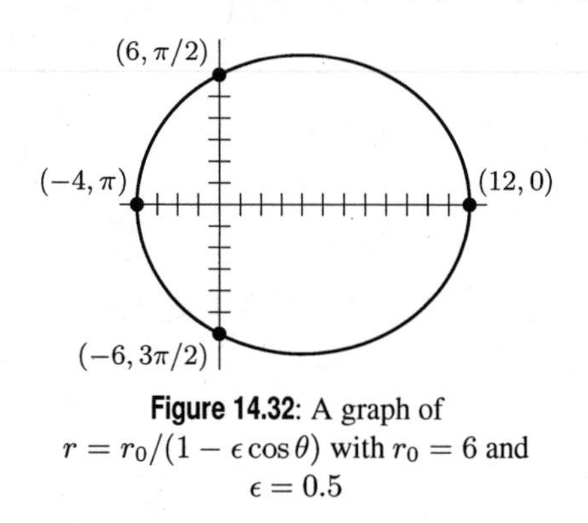

Figure 14.32: A graph of $r = r_0/(1-\epsilon\cos\theta)$ with $r_0 = 6$ and $\epsilon = 0.5$

(d) The length of the horizontal axis is given by the minimum value of r (at $\theta = \pi$) plus the maximum value of r (at $\theta = 0$):

$$\begin{aligned}
\text{Horizontal axis length} &= r_{\text{min}} + r_{\text{max}} \\
&= \frac{r_0}{1+\epsilon} + \frac{r_0}{1-\epsilon} \\
&= \frac{r_0(1-\epsilon) + r_0(1+\epsilon)}{(1-\epsilon)(1+\epsilon)} \\
&= \frac{2r_0}{1-\epsilon^2}.
\end{aligned}$$

(e) At $\epsilon = 0$, we have $r = r_0/(1-0) = r_0$, which is the equation (in polar coordinates) of a circle of radius r_0. In our formula in part (d), we see that as ϵ gets closer to 1, the denominator gets closer to 0, and so the length of the horizontal axis increases rapidly. However, the y-intercepts at $(r_0, \pi/2)$ and $(r_0, 3\pi/2)$ do not change. For this to be true, the ellipse must be getting longer and longer along the horizontal axis as the eccentricity gets close to 1, but not be changing very much along the vertical axis.

22. **(a)** Looking at the side view in the diagram, we see that when tilted through an angle θ, the diameter of the disk forms the hypotenuse of a right triangle, and the length ℓ is given by leg opposite θ. See Figure 14.33. This means that

$$\ell = 2r \sin\theta.$$

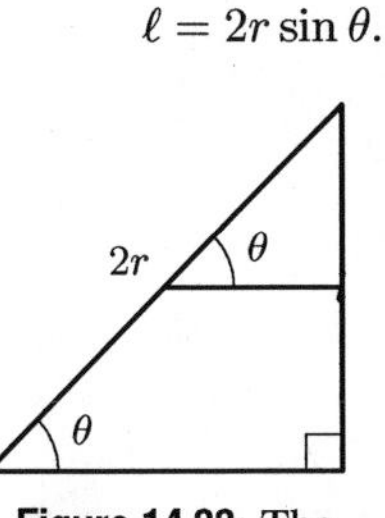

Figure 14.33: The length ℓ is the leg in a right triangle

(b) At $\theta = 0°$, we have $\ell = 2r\sin 0° = 0$. The same is true for $\theta = 180°$. This makes sense, because both a $0°$ tilt and a $180°$ tilt correspond to tilting the disk onto its side, so that the diameter is not visible. At $\theta = 90°$, we have $\ell = 2r\sin 90° = 2r$. This makes sense, because a $90°$ tilt corresponds to looking at the disk head on, and the full diameter of the disk is visible.

(c) The disk is centered at the origin, and so its formula is given by $x^2/a^2 + y^2/b^2 = 1$. Here, the length of the horizontal axis is $2a$, which must equal $2r$, the diameter of the disk, and so $a = r$. The length of the vertical axis is given by $2b = \ell = 2r\sin\theta$, and so $b = r\sin\theta$. Thus, a formula for the ellipse is given by

$$\frac{x^2}{r^2} + \frac{y^2}{(r\sin\theta)^2} = 1.$$

(d) From part (c), we have $r^2 = 16$ and $(r\sin\theta)^2 = 7$. Taking the positive solutions, we have $r = 4$ from the first equation, and so from the second equation we see that $(4\sin\theta)^2 = 7$. This gives $4\sin\theta = \sqrt{7}$, and so $\theta = \arcsin(\sqrt{7}/4)$, or about $41.4°$.

23. **(a)** From Figure 14.34, we see that the orbital ellipse is centered at $(h, 0)$, and so we can write its formula as $(x - h)^2/a^2 + y^2/b^2 = 1$. We also see that the ellipse is bounded by a rectangle of width $2a = r_e + r_m$, and so $a = (r_e + r_m)/2$. Notice that $(h, 0)$ is midway between $(-r_e, 0)$ and $(+r_m, 0)$, and so using the midpoint formula we have $h = (r_m + (-r_e))/2$ which gives $h = (r_m - r_e)/2$. The equation of the ellipse is

$$\left(\frac{x - (r_m - r_e)/2}{(r_e + r_m)/2}\right)^2 + \frac{y^2}{b^2} = 1$$

$$\left(\frac{2x - r_m + r_e}{r_e + r_m}\right)^2 + \frac{y^2}{b^2} = 1.$$

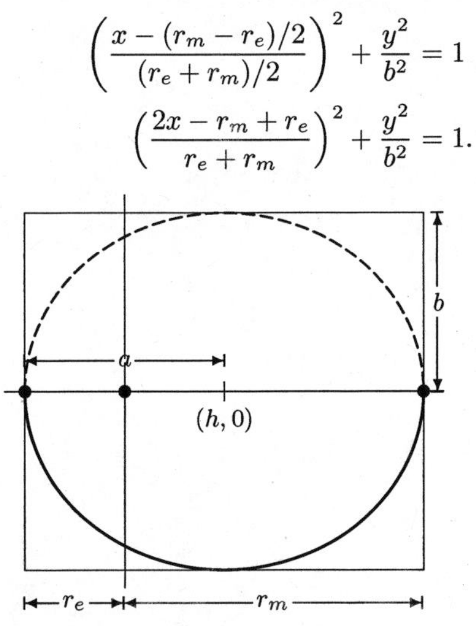

Figure 14.34: The center is $(h, 0)$. The length of the horizontal axis is $2a = r_e + r_m$

(b) We have

$$\begin{aligned} b^2 &= 2ar_e - r_e^2 \\ &= (r_e + r_m)r_e - r_e^2 \qquad \text{from part (a) we have } 2a = r_e + r_m \\ &= r_e^2 + r_m r_e - r_e^2 \\ &= r_e r_m, \end{aligned}$$

and so $b = \sqrt{r_e r_m}$. We say that b is the the *geometric mean* of r_e and r_m.

Solutions for Section 14.4

Exercises

1. **(a)** The vertices are at $(0, 7)$ and $(0, -7)$. The center is at the origin.

(b) The asymptotes have slopes $7/2$ and $-7/2$. The equations of the asymptotes are

$$y = \frac{7}{2}x \quad \text{and} \quad y = -\frac{7}{2}x.$$

(c) The equation of the hyperbola is

$$\frac{y^2}{7^2} - \frac{x^2}{2^2} = 1 \quad \text{or} \quad \frac{y^2}{49} - \frac{x^2}{4} = 1.$$

2. **(a)** The vertices are at $(5, 0)$ and $(-5, 0)$. The center is at the origin.

(b) The asymptotes have slopes $1/5$ and $-1/5$. The equations of the asymptotes are

$$y = \frac{1}{5}x \quad \text{and} \quad y = -\frac{1}{5}x.$$

(c) The equation of the hyperbola is

$$\frac{x^2}{5^2} - \frac{y^2}{1^2} = 1 \quad \text{or} \quad \frac{x^2}{25} - y^2 = 1.$$

3. **(a)** The vertices are at $(4, 4)$ and $(2, 4)$. The center is at $(3, 4)$.

(b) The asymptotes have slopes $(7 - 1)/(4 - 2) = 3$ and -3. The equations of the asymptotes are

$$y - 4 = 3(x - 3) \quad \text{and} \quad y - 4 = -3(x - 3)$$

or

$$y = 3x - 5 \quad \text{and} \quad y = -3x + 13.$$

(c) The equation of the hyperbola is

$$\frac{(x-3)^2}{1^2} - \frac{(y-4)^2}{3^2} = 1 \quad \text{or} \quad (x-3)^2 - \frac{(y-4)^2}{9} = 1.$$

4. **(a)** The vertices are at $(2, 7)$ and $(2, -1)$. The center is at $(2, 3)$.

(b) The asymptotes have slopes $(7 - (-1))/(5 - (-1)) = 8/6 = 4/3$ and $-4/3$. The equations of the asymptotes are

$$y - 3 = \frac{4}{3}(x - 2) \quad \text{and} \quad y - 3 = -\frac{4}{3}(x - 2)$$

or

$$y = \frac{4}{3}x + \frac{1}{3} \quad \text{and} \quad y = -\frac{4}{3}x + \frac{17}{3}.$$

(c) The equation of the hyperbola is

$$\frac{(y-3)^2}{4^2} - \frac{(x-2)^2}{3^2} = 1 \quad \text{or} \quad \frac{(y-3)^2}{16} - \frac{(x-2)^2}{9} = 1.$$

5. The hyperbola is centered at the origin, and $a = 2$, $b = 7$. We can use $x = 2\tan t$, $y = 7\sec t = 7/\cos t$.
If $0 < t < \pi/2$, then $x > 0$, $y > 0$, so we have Quadrant I.
If $\pi/2 < t < \pi$, then $x < 0$, $y < 0$, so we have Quadrant III.
If $\pi < t < 3\pi/2$, then $x > 0$, $y < 0$, so we have Quadrant IV.
If $3\pi/2 < t < 2\pi$, then $x < 0$, $y > 0$, so we have Quadrant II.
So the upper half is given by $0 \le t < \pi/2$ together with $3\pi/2 < t < 2\pi$.

6. The hyperbola is centered at the origin, and $a = 5$, $b = 1$. We can use $x = 5\sec t = 5/\cos t$, $y = \tan t$.
If $0 < t < \pi/2$, then $x > 0$, $y > 0$, so we have Quadrant I.
If $\pi/2 < t < \pi$, then $x < 0$, $y < 0$, so we have Quadrant III.
If $\pi < t < 3\pi/2$, then $x < 0$, $y > 0$, so we have Quadrant II.
If $3\pi/2 < t < 2\pi$, then $x > 0$, $y < 0$, so we have Quadrant IV.
So the right half is given by $0 \le t < \pi/2$ together with $3\pi/2 < t < 2\pi$.

7. The hyperbola is centered at $(3, 4)$, and $a = 1$, $b = 3$. We can use $x = 3 + \sec t = 3 + 1/\cos t$, $y = 4 + 3\tan t$.
If $0 < t < \pi/2$, then $x > 0$, $y > 0$, so we have Quadrant I.
If $\pi/2 < t < \pi$, then $x < 0$, $y < 0$, so we have Quadrant III.
If $\pi < t < 3\pi/2$, then $x < 0$, $y > 0$, so we have Quadrant II.
If $3\pi/2 < t < 2\pi$, then $x > 0$, $y < 0$, so we have Quadrant IV.
So the left half is given by $\pi/2 < t \le \pi$ together with $\pi \le t < 3\pi/2$; that is, $\pi/2 < t < 3\pi/2$.

8. The hyperbola is centered at $(2, 3)$, and $a = 4$, $b = 3$. We can use $x = 2 + 4\tan t$, $y = 3 + 3\sec t = 3 + 3/\cos t$.
If $0 < t < \pi/2$, then $x > 0$, $y > 0$, so we have Quadrant I.
If $\pi/2 < t < \pi$, then $x < 0$, $y < 0$, so we have Quadrant III.
If $\pi < t < 3\pi/2$, then $x > 0$, $y < 0$, so we have Quadrant IV.
If $3\pi/2 < t < 2\pi$, then $x < 0$, $y > 0$, so we have Quadrant II.
So the lower half is given by $\pi/2 < t \le \pi$ together with $\pi \le t < 3\pi/2$; that is, $\pi/2 < t < 3\pi/2$.

Problems

9. Factoring out -1 from $-y^2 + 4y = -(y^2 - 4y)$ and completing the square on $x^2 - 2x$ and $y^2 - 4y$ gives

$$\begin{aligned}
\frac{1}{4}(x^2 - 2x) - (y^2 - 4y) &= \frac{19}{4} \\
\frac{1}{4}((x-1)^2 - 1) - ((y-2)^2 - 4) &= \frac{19}{4} \\
\frac{(x-1)^2}{4} - \frac{1}{4} - (y-2)^2 + 4 &= \frac{19}{4} \\
\frac{(x-1)^2}{4} - (y-2)^2 &= 1.
\end{aligned}$$

The center is $(1, 2)$, the hyperbola opens right-left, and $a = 2$, $b = 1$.

10. Completing the square on $y^2 + 2y$ and $x^2 - 4x$ gives

$$\begin{aligned}
\frac{1}{4}(y^2 + 2y) - \frac{1}{9}(x^2 - 4x) &= \frac{43}{36} \\
\frac{1}{4}((y+1)^2 - 1) - \frac{1}{9}((x-2)^2 - 4) &= \frac{43}{36} \\
\frac{(y+1)^2}{4} - \frac{1}{4} - \frac{(x-2)^2}{9} + \frac{4}{9} &= \frac{43}{36} \\
\frac{(y+1)^2}{4} - \frac{(x-2)^2}{9} &= 1.
\end{aligned}$$

The center is $(2, -1)$, the hyperbola opens up-down, and $a = 3$, $b = 2$.

11. Factoring out -4 from $-4y^2 - 24y = -4(y^2 + 6y)$ and completing the square on $x^2 + 2x$ and $y^2 + 6y$ gives

$$x^2 + 2x - 4(y^2 + 6y) = 39$$

$$\begin{aligned}(x+1)^2-1-4((y+3)^2-9)&=39\\(x+1)^2-1-4(y+3)^2+36&=39\\(x+1)^2-4(y+3)^2&=4.\end{aligned}$$

Dividing by 4 to get a 1 on the right,

$$\frac{(x+1)^2}{4}-(y+3)^2=1.$$

The center is $(-1,-3)$, the hyperbola opens right-left, and $a=2$, $b=1$.

12. Factoring out 9 from $9x^2-36x=9(x^2-4x)$ and -4 from $-4y^2+8y=-4(y^2-2y)$ and completing the square on x^2-4x and y^2-2y gives

$$\begin{aligned}9(x^2-4x)-4(y^2-2y)&=4\\9((x-2)^2-4)-4((y-1)^2-1)&=4\\9(x-2)^2-36-4(y-1)^2+4&=4\\9(x-2)^2-4(y-1)^2&=36.\end{aligned}$$

Dividing by 36 to get a 1 on the right,

$$\begin{aligned}\frac{9(x-2)^2}{36}-\frac{4(y-1)^2}{36}&=\frac{36}{36}\\\frac{(x-2)^2}{4}-\frac{(y-1)^2}{9}&=1.\end{aligned}$$

The center is $(2,1)$, the hyperbola opens right-left, and $a=2$, $b=3$.

13. Factoring out 4 from $4x^2-8x=4(x^2-2x)$ and 36 from $36y^2-36y=36(y^2-y)$ and completing the square on x^2-2x and y^2-y gives

$$\begin{aligned}4(x^2-2x)&=36(y^2-y)-31\\4((x-1)^2-1)&=36\left(\left(y-\frac{1}{2}\right)^2-\frac{1}{4}\right)-31\\4(x-1)^2-4&=36\left(y-\frac{1}{2}\right)^2-9-31\\4(x-1)^2&=36\left(y-\frac{1}{2}\right)^2-36.\end{aligned}$$

Moving $36(y-\frac{1}{2})^2$ to the left and dividing by -36 to get 1 on the right:

$$\begin{aligned}\frac{4(x-1)^2}{-36}-\frac{36\left(y-\frac{1}{2}\right)^2}{-36}&=-\frac{36}{-36}\\-\frac{(x-1)^2}{9}+\left(y-\frac{1}{2}\right)^2&=1\\\left(y-\frac{1}{2}\right)^2-\frac{(x-1)^2}{9}&=1.\end{aligned}$$

The center is $(1,\frac{1}{2})$, the hyperbola opens up-down, and $a=3$, $b=1$.

14. Factoring out 9 from $9y^2+6y=9(y^2+\frac{2}{3}y)$ and 8 from $8x^2+24x=8(x^2+3x)$ and completing the square on $y^2+\frac{2}{3}y$ and x^2+3x gives

$$\begin{aligned}9\left(y^2+\frac{2}{3}y\right)&=89+8(x^2+3x)\\9\left(\left(y+\frac{1}{3}\right)^2-\frac{1}{9}\right)&=89+8\left(\left(x+\frac{3}{2}\right)^2-\frac{9}{4}\right)\end{aligned}$$

$$9\left(y+\frac{1}{3}\right)^2 - 1 = 89 + 8\left(x+\frac{3}{2}\right)^2 - 18$$
$$9\left(y+\frac{1}{3}\right)^2 = 72 + 8\left(x+\frac{3}{2}\right)^2.$$

Moving $8(x+\frac{3}{2})^2$ to the left and dividing by 72 to get 1 on the right:

$$\frac{9(y+\frac{1}{3})^2}{72} - \frac{8\left(x+\frac{3}{2}\right)^2}{72} = \frac{72}{72}$$
$$\frac{(y+\frac{1}{3})^2}{8} - \frac{\left(x+\frac{3}{2}\right)^2}{9} = 1.$$

The center is $(-\frac{3}{2}, -\frac{1}{3})$, the hyperbola opens up-down, and $a = 3$, $b = \sqrt{8} = 2\sqrt{2}$.

15. Form II of the equation applies because the hyperbola opens up-down, rather than right-left.

The center is at (h, k); both coordinates are negative here. Figure 14.35 shows a, b, h, k on the hyperbola. We see $k < h < 0 < a < b$.

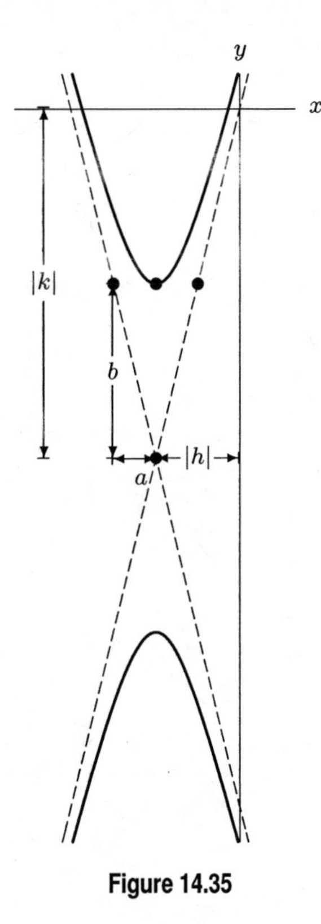

Figure 14.35

16. Form I of the equation applies because the hyperbola opens right-left, rather than up-down.

The center is at (h, k); both coordinates are positive here. Figure 14.36 shows a, b, h, k on the hyperbola. We see $0 < b < k < h < a$.

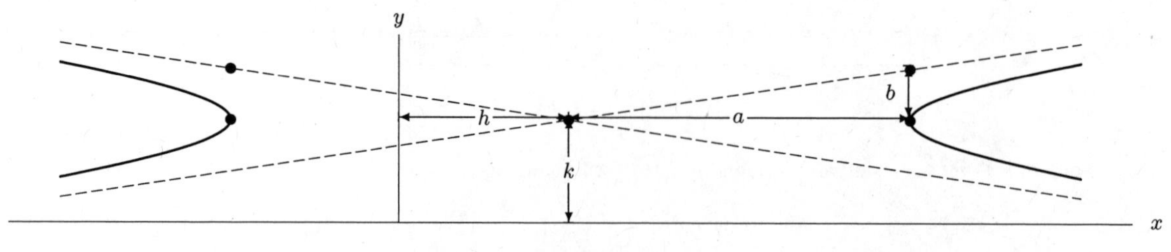

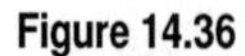

Figure 14.36

17. **(a)** The center is $(-5, 2)$, vertices are $(-5+\sqrt{6}, 2)$ and $(-5-\sqrt{6}, 2)$. The asymptotes are $y = \pm\frac{2}{\sqrt{6}}(x+5)+2$. Figure 14.37 shows the hyperbola.

(b) Rewriting the equation and competing the square, we have

$$\begin{aligned} x^2 - y^2 + 2x &= 4y + 17 \\ x^2 - y^2 + 2x - 4y &= 17 \\ (x^2 + 2x + 1) - (y^2 + 4y + 4) + 3 &= 17 \\ (x+1)^2 - (y+2)^2 &= 14 \\ \frac{(x+1)^2}{14} - \frac{(y+2)^2}{14} &= 1. \end{aligned}$$

Thus the center is $(-1, -2)$; vertices are $(-1-\sqrt{14}, -2)$ and $(-1+\sqrt{14}, -2)$; asymptotes are $y = \pm(x+1) - 2$, that is, $y = x - 1$ and $y = -x - 3$. Figure 14.38 shows the hyperbola.

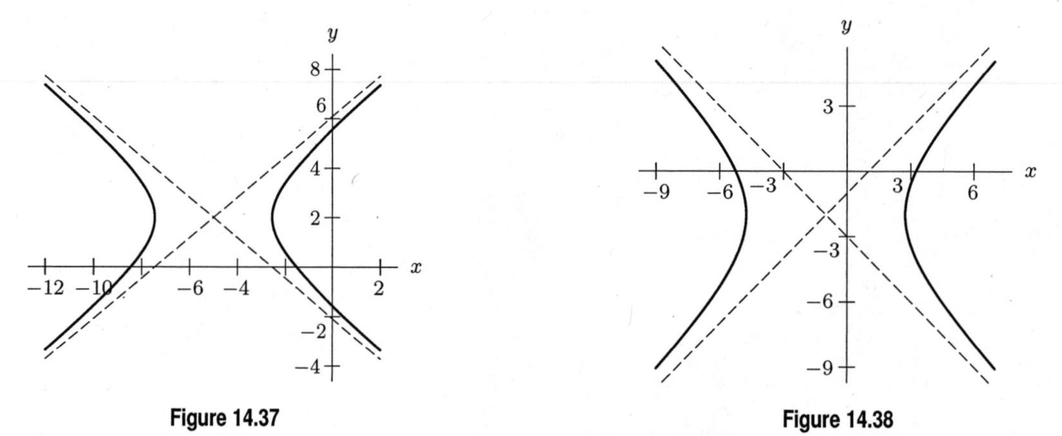

Figure 14.37

Figure 14.38

18. **(a)** We see from Figure 14.39 that $\tan\theta = b/a$ and so $b = a\tan\theta$. This means that $x^2/a^2 - y^2/(a\tan\theta)^2 = 1$.

(b) From the figure, we see that if the trajectory corresponds to a repulsive force, then $a > d$.

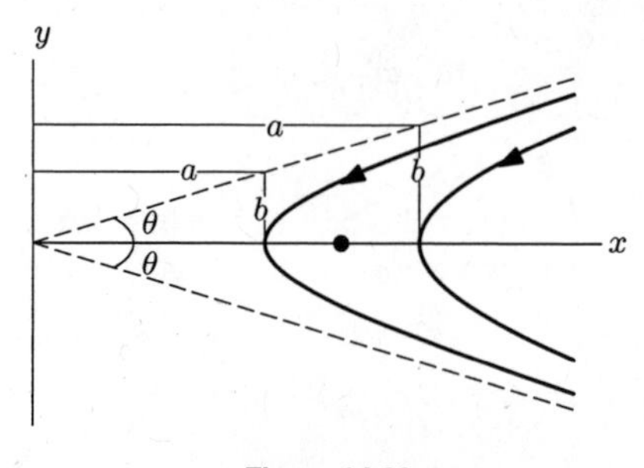

Figure 14.39

Solutions for Section 14.5

Exercises

1. This is the standard form of an ellipse. Since $a > b$ the major axis is the x-axis.
2. This is the standard form of a hyperbola. Since the x^2 term is positive, the vertices are on the x-axis.
3. Divide both sides by -1 to get the standard form of a hyperbola. Since the y^2 term is positive, the vertices are on the y-axis.
4. Multiply both sides by 2 and subtract the y^2 term from both sides to get the standard form of a parabola. Since the squared term involves y, the axis of symmetry is horizontal.
5. Subtract the x^2 term from both sides to get the standard form of a hyperbola. Since the y^2 term is positive, the vertices are on the y-axis.
6. The radius is $r = \sqrt{(-1)^2 + 3^2} = \sqrt{10}$.
7. The radius is $r = \sqrt{1^2 + 2^2} = \sqrt{5}$. This is the distance from the center to any point on the circle, so when the circle intercepts the y-axis, $x = 0$ and $y = \pm\sqrt{5}$. The points are $(0, \sqrt{5})$ and $(0, -\sqrt{5})$.
8. The sum of the distances to the two focal points on an ellipse is constant. Since the ellipse is centered at the origin, it is symmetric about both the vertical and horizontal axes and the sum of the distances to the two focal points is preserved for (a), (b), and (c). This is not always the case for (d).
9. The two focal points lie on the major axis, and are equidistant from the center. Therefore, the major axis is vertical and the other focal point is at $(0, -2)$.
10. In the form
$$\frac{x^2}{a^2} + \frac{y^2}{b^2} = 1$$
we see that $b > a$. Therefore the focal points are at $(0, \pm c)$, where $c = \sqrt{b^2 - a^2} = \sqrt{16 - 9} = \sqrt{7}$. The two focal points are $(0, \sqrt{7})$ and $(0, -\sqrt{7})$.
11. The difference of the distances to the two focal points on a hyperbola is constant. Since the hyperbola is centered at the origin, it is symmetric about both the vertical and horizontal axes and the difference of the distances to the two focal points is preserved for (a), (b), and (c). This is not always the case for (d).
12. The two focal points lie on the same axis as the vertices, and they are equally spaced from the center. Therefore, the axis through the vertices is vertical and the other focal point is at $(0, 3)$.
13. In the form
$$\frac{x^2}{a^2} - \frac{y^2}{b^2} = 1$$
we see that $a^2 = b^2 = 1$. The focal points lie on the same axis as the positive squared term. Therefore the focal points are at $(\pm c, 0)$, where $c = \sqrt{a^2 + b^2} = \sqrt{1 + 1} = \sqrt{2}$. The two focal points are $(\sqrt{2}, 0)$ and $(-\sqrt{2}, 0)$.
14. Since the parabola is centered at the origin, and opens downward, it is symmetric about the vertical axis. Only (b) is symmetric with $(-1, -2)$ about the y-axis.
15. Being centered at the origin and having a focal point on the x-axis, the equation must have the form $x = ay^2$. Using the relationship $c = 1/(4a)$, we have$-1 = 1/(4a)$, so $a = -1/4$. The equation is $x = -(1/4)y^2$.
16. Being centered at the origin and having a focal point on the y-axis, the equation must have the form $y = ax^2$. Using the relationship $c = 1/(4a)$, we have$-3 = 1/(4a)$, so $a = -1/12$. The equation is $y = -(1/12)x^2$.

Problems

17. The reflective properties of a parabola are somewhat helpful, because the transmission can be focused to go in only one direction. However, the location of the transmission will be perfectly clear to anyone noticing the transmission. A series of elliptical mirrors would be much more complicated and expensive to build, and by following the signal, someone could figure out where the transmission was coming from without your being aware, because surveying a series of mirrors is

complicated. A hyperbolic reflection allows you to survey a small mirror near your camp while at the same time making it appear that your camp is somewhere else (at the other focus of the hyperbola). So you should follow the mess sergeant's advice.

18. Solve the equation $(x+2)^2 = 5(y-1)$ for y to obtain

$$y = \frac{1}{5}(x+2)^2 + 1.$$

From this form of the equation we read the vertex coordinates to be $(-2, 1)$, with the parabola opening upward. The distance from the vertex to the focus is $c = 1/(4a)$ with $a = 1/5$, so

$$c = \frac{1}{4(1/5)} = \frac{5}{4}.$$

This distance is added to 1, the vertex y value, to obtain the focal point coordinates $(-2, 9/4)$.

19. Solve the equation $y^2 = \frac{1}{4}(x+2)$ for x to obtain

$$x = 4y^2 - 2.$$

From this form of the equation we read the vertex coordinates to be $(-2, 0)$, with the parabola opening to the right. The distance from the vertex to the focus is $c = 1/(4a)$ with $a = 4$, so

$$c = \frac{1}{4(4)} = \frac{1}{16}.$$

This distance is added to -2, the vertex x value, to obtain the focal point coordinates $(-31/16, 0)$.

20. We put the equation in vertex form by completing the square, where $b = -1$, that is, we add and subtract $(-1/2)^2 = 1/4$ to the right side:

$$y = x^2 - x + \frac{1}{4} + 1 - \frac{1}{4} = \left(x - \frac{1}{2}\right)^2 + \frac{3}{4}.$$

From this form of the equation we read the vertex coordinates to be $(1/2, 3/4)$, with the parabola opening upward. The distance from the vertex to the focus is $c = 1/(4a)$ with $a = 1$, so $c = 1/4$. This distance is added to $3/4$, the vertex y value, to obtain the focal point coordinates $(1/2, 1)$.

21. From the figure, we see the vertex has coordinates $(1, 0)$ and the equation must be of the form

$$x = ay^2 + 1$$

The distance, c from the vertex to the focus is 2. Using $c = 1/(4a)$, we solve for a in $2 = 1/(4a)$ and find $a = 1/8$. The equation is $x = (1/8)y^2 + 1$. The directrix is the line $x = -1$, since it is 2 units to the left of the vertex.

22. From the figure, we see the vertex has coordinates $(2, -3)$ and the parabola passes through the origin. Using the vertex form of the parabolic equation we have,

$$y = a(x-2)^2 - 3.$$

By substituting the origin values, $(0, 0)$, in the equation

$$0 = a(0-2)^2 - 3$$

we find $a = \frac{3}{4}$. The equation is

$$y = \frac{3}{4}(x-2)^2 - 3.$$

Find the distance from the vertex to the focus as $c = 1/(4a)$, so

$$c = \frac{1}{4(3/4)} = \frac{1}{3}.$$

The focus is at the point $(2, -3 + 1/3) = (2, -8/3)$. The directrix is a horizontal line at a distance of c below the vertex. The line is $y = -3 - (1/3) = -10/3$.

23. This is a circle with center at $(0,0)$ and focal point at $(0,0)$.

24. Rewrite the formula as

$$\frac{x^2}{1^2}+\frac{(y-3)^2}{5^2}=1$$

to see that this is an ellipse with center at $(0,3)$, and with $b=5$ and $a=1$. Since $b>a$ the distance, c, from the center to a focus point is found from

$$c=\sqrt{b^2-a^2}=\sqrt{5^2-1^2}=\sqrt{24}.$$

With $c=\sqrt{24}$ the focal points are at $(0,3+\sqrt{24})$ and $(0,3-\sqrt{24})$.

25. Rewrite the formula as

$$(x^2+4x)+2(y^2-6y)=3.$$

Complete the square of both the x and y expressions

$$(x^2+4x+4)+2(y^2-6y+9)=3+4+2(9),$$

and simplify to

$$(x+2)^2+2(y-3)^2=25$$

or

$$\frac{(x+2)^2}{25}+\frac{(y-3)^2}{(25/2)}=1$$

This is an ellipse with center at $(-2,3)$, and $a=5$ and $b=\sqrt{25/2}$. Since $a>b$, the distance c, from the center to a focus point is found from

$$c=\sqrt{a^2-b^2}=\sqrt{25-25/2}=\sqrt{25/2}=5/\sqrt{2}.$$

With $c=5/\sqrt{2}$ the focal points are at $(-2+5/\sqrt{2},3)$ and $(-2-5/\sqrt{2},3)$.

26. The ellipse is centered at the origin so its equation is

$$\frac{x^2}{a^2}+\frac{y^2}{b^2}=1.$$

The point P shows that $a=5$. To find b we use the coordinates of point Q in the equation to get

$$\frac{1^2}{5^2}+\frac{4^2}{b^2}=1.$$

This simplifies to $b^2=50/3$ and the final equation is

$$\frac{x^2}{25}+\frac{y^2}{50/3}=1.$$

Since $a>b$, the distance c from the vertex to the focus is

$$c=\sqrt{a^2-b^2}=\sqrt{25-50/3}=\sqrt{25/3}=5/\sqrt{3}.$$

The focal points are $(\pm 5/\sqrt{3},0)$.

27. The ellipse is centered at $(0,2)$ so the equation is of the form

$$\frac{x^2}{a^2}+\frac{(y-2)^2}{b^2}=1.$$

The foci $(0,1)$ and $(0,3)$ are on the vertical axis, and we see from the graph that $c=1$ and $b=2$. When the major axis is vertical, we use the equation $c=\sqrt{b^2-a^2}$. So, $1=\sqrt{2^2-a^2}$ or $a=3$. Thus our equation is

$$\frac{x^2}{3}+\frac{(y-2)^2}{4}=1.$$

28. Because the y^2 term is positive and the x^2 term is negative this hyperbola has vertices on a vertical line through its center. The center can be read from the form of the equation and is $(2, -2)$. The length from the center to the vertex is the distance $a = 5$ which is read from the equation since $a^2 = 25$. Thus the vertices are at the points $(2, -2 \pm 5)$ or $(2, 3)$ and $(2, -7)$. Since $b^2 = 4$ and $a^2 = 25$, we find $c^2 = a^2 + b^2 = 29$, and $c = \sqrt{29}$. The focal points are $(2, -2 + \sqrt{29})$ and $(2, -2 - \sqrt{29})$.

29. First rewrite the equation in the form,

$$\frac{x^2}{8} - \frac{y^2}{16} = 1.$$

Because the x^2 term is positive and the y^2 term is negative this hyperbola has vertices on a horizontal line through its center. The center is at the origin. The length from the center to the vertex is the distance $a = 2\sqrt{2}$ which is read from the equation since $a^2 = 8$. Thus the vertices are at the points $(2\sqrt{2}, 0)$ and $(-2\sqrt{2}, 0)$. Since $b^2 = 16$ and $a^2 = 8$, we find $c^2 = a^2 + b^2 = 24$, and $c = \sqrt{24}$. The focal points are $(\sqrt{24}, 0)$ and $(-\sqrt{24}, 0)$.

30. Rewrite the equation by preparing to complete the square,

$$4(y^2 + 14y) - (x^2 + 8x) = -100$$

complete the square of both the x and y expression

$$4(y^2 + 14y + 7^2) - (x^2 + 8x + 4^2) = -100 + 4(7^2) - 4^2$$

and simplify to

$$4(y + 7)^2 - (x + 4)^2 = 80$$

or

$$\frac{(y + 7)^2}{20} - \frac{(x + 4)^2}{80} = 1.$$

Because the y^2 term is positive and the x^2 term is negative this hyperbola has vertices on a vertical line through its center. The center is at the point $(-4, -7)$. The vertical distance from the center to the vertex is $b = \sqrt{20}$ which is read from the equation since $b^2 = 20$. Thus the vertices are at the points $(-4, -7 \pm \sqrt{20})$. Since $a^2 = 80$ and $b^2 = 20$, we find $c^2 = a^2 + b^2 = 100$, and $c = 10$. The focal points are $(-4, -7 \pm 10)$.

31. The hyperbola is centered at the origin and opens upward so the equation is of the form

$$\frac{y^2}{b^2} - \frac{x^2}{a^2} = 1.$$

The dotted box shows that $a = 2$ and $b = 7$,

$$\frac{y^2}{49} - \frac{x^2}{4} = 1.$$

Using $c^2 = a^2 + b^2 = 4 + 49 = 53$, we find $c = \sqrt{53}$. The focal points are at $(0, \pm\sqrt{53})$.

32. The hyperbola is centered at $(3, 4)$ and opens left and right, so the equation is of the form

$$\frac{(x - 3)^2}{a^2} - \frac{(y - 4)^2}{b^2} = 1.$$

The dotted box shows that $a = 1$ and $b = 3$,

$$(x - 3)^2 - \frac{(y - 4)^2}{9} = 1.$$

Using $c^2 = a^2 + b^2 = 10$, we find $c = \sqrt{10}$. The foci are at $(3 \pm \sqrt{10}, 4)$.

33. After passing through the second focal point it is reflected back to the original focal point.

34. The ellipse is centered at the origin and has the equation

$$\frac{x^2}{a^2} + \frac{y^2}{b^2} = 1.$$

When the pencil reaches the positive x-axis, it is at $(5, 0)$, giving us $a = 5$. We know $a > b$, since the focal points lie on the horizontal axis, and we use the formula $c = \sqrt{a^2 - b^2}$ to find b, with $c = 2$.

$$2 = \sqrt{5^2 - b^2} \text{ and } b = \sqrt{21}.$$

The ellipse has the equation

$$\frac{x^2}{25} + \frac{y^2}{21} = 1.$$

35. No, any reflective ray that passes through the near focal point must be aimed at the other focal point below the mirror. Since the beam is aimed straight down but off center, it is never aimed at the focal point below the mirror.

36. Yes, any reflective ray aimed straight down into the mirror will be reflected back through the focal point.

37. For a cross-section of the dish centered at the origin, the equation of the parabola is $y = ax^2$. We know that the rim of the dish is the point $(12, 4)$, thus $4 = a(12)^2$ and solving for a we find $a = 1/36$. Using the equation $c = 1/(4a)$, we find

$$c = \frac{1}{4(1/36)} = 9.$$

The end of the arm is placed 9 inches above the center of the dish to be the focus of the incoming signal.

38. For a cross-section of the dish centered at the origin, the equation of the parabola is $y = ax^2$. We know that the graph of the rim of the dish passes through the coordinate point (w, d), where w is the radius and d is the depth. We find a from $d = a(w)^2$ or $a = d/w^2$. Using the equation $c = 1/(4a)$, we find

$$c = \frac{1}{4(d/w^2)} = \frac{w^2}{4d}.$$

From this equation we see that when d is halved, the distance of the focal point is twice as far from the center of the dish.

39. Consider the search light centered at the origin and pointed vertically upward. The equation of a cross-section of the parabolic mirror is $y = ax^2$. With a depth of 3/4 ft and a radius of 3, we find a by substitution, $3/4 = a(3)^2$, so $a = 1/12$. The focal point, $(0, c)$ is found from $c = 1/(4a)$, so $c = 1/(4(1/12)) = 3$. The lamp should be 3 ft above the center.

40. **(a)** The major axis has length $146 + 152 = 298$ million km.

(b) Since $2a = 298$, we find $a = 149$. Let $(\pm c, 0)$ be the focal points. Since $a = 149 = c + 146$, we find $c = 3$. To find b we use the formula $c^2 = a^2 - b^2$ and find $b^2 = 149^2 - 3^2 = 22{,}192$, so $b = 148.970$. Shifting the axis so the sun is at the origin we have,

$$\frac{(x-3)^2}{149^2} + \frac{y^2}{148.970^2} = 1.$$

41. **(a)** The major axis has length $88 + 5250 = 5338$ million km.

(b) Since $2a = 5338$, we find $a = 2669$. Let $(\pm c, 0)$ be the two focal points. Since $a = 2669 = c + 88$, we find $c = 2581$. To find b we use the formula $c^2 = a^2 - b^2$ and find $b^2 = 2669^2 - 2581^2 = 462{,}000$, so $b = 680$. Shifting the axis so the sun is at the origin we have,

$$\frac{(x-2581)^2}{2669^2} + \frac{y^2}{680^2} = 1.$$

(c) Use the parametric equations of an ellipse centered at (h, k),

$$x = h + a\cos t, \quad y = k + b\sin t, \quad 0 \le t \le 2\pi$$

to obtain

$$x = 2581 + 2669\cos t, \quad y = 680\sin t, \quad 0 \le t \le 2\pi.$$

42. **(a)** Hyperbola.

(b) The radius values are half of the given diameter values and $A_t = A_b = 14/2 = 7$. Substituting, we find the upper shadow equation,

$$\frac{y^2}{((8 \cdot 7)/4)^2} - \frac{x^2}{8^2} = 1$$

or

$$\frac{y^2}{196} - \frac{x^2}{64} = 1.$$

The lower shadow equation is

$$\frac{y^2}{7^2} - \frac{x^2}{8^2} = 1$$

or

$$\frac{y^2}{49} - \frac{x^2}{64} = 1.$$

(c) Isolate y as function of x for the upper shadow curve,

$$\begin{aligned}
\frac{y^2}{196} - \frac{x^2}{64} &= 1 \\
\frac{y^2}{196} &= 1 + \frac{x^2}{64} \\
\sqrt{\frac{y^2}{196}} &= \sqrt{1 + \frac{x^2}{64}} \\
y &= \pm 14\sqrt{1 + \frac{x^2}{64}}.
\end{aligned}$$

Since the shadow is above the shade, we use the positive branch,

$$y = 14\sqrt{1 + \frac{x^2}{64}}.$$

For the lower shadow,

$$\begin{aligned}
\frac{y^2}{49} - \frac{x^2}{64} &= 1 \\
\frac{y^2}{49} &= 1 + \frac{x^2}{64} \\
\sqrt{\frac{y^2}{49}} &= \sqrt{1 + \frac{x^2}{64}} \\
y &= \pm 7\sqrt{1 + \frac{x^2}{64}}.
\end{aligned}$$

Since the shadow is below the shade, we use the negative branch,

$$y = -7\sqrt{1 + \frac{x^2}{64}}.$$

43. **(a)** The difference is $5 - \sqrt{13}$, so $2a = 5 - \sqrt{13}$ and $a = (5 - \sqrt{13})/2$. Center the hyperbola at $(0, 4)$. The distance from the center to the focus is $c = 1$. Thus

$$b^2 = c^2 - a^2 = 1 - (\frac{5 - \sqrt{13}}{2})^2.$$

Shifting the center results in the equation,

$$\frac{(y-4)^2}{(\frac{5-\sqrt{13}}{2})^2} - \frac{x^2}{1 - (\frac{5-\sqrt{13}}{2})^2} = 1$$

or

$$\frac{(y-4)^2}{0.486} - \frac{x^2}{0.514} = 1.$$

(b) The difference is $5-\sqrt{10}$ so $2a = 5-\sqrt{10}$ and $a = (5-\sqrt{10})/2$. Center the hyperbola at $(0, 2.5)$. The distance from the center to a focal point is $c = 2.5$. Thus,

$$b^2 = c^2 - a^2 = 6.25 - (\frac{5-\sqrt{10}}{2})^2.$$

Shifting the center results in the equation,

$$\frac{(y-2.5)^2}{(\frac{5-\sqrt{10}}{2})^2} - \frac{x^2}{6.25 - (\frac{5-\sqrt{10}}{2})^2} = 1$$

or

$$\frac{(y-2.5)^2}{0.844} - \frac{x^2}{5.406} = 1.$$

(c) Isolate y as function of x,

$$\frac{(y-4)^2}{0.486} - \frac{x^2}{0.514} = 1$$

$$\frac{(y-4)^2}{0.486} = 1 + \frac{x^2}{0.514}$$

$$\sqrt{\frac{(y-4)^2}{0.486}} = \sqrt{1 + \frac{x^2}{0.514}}$$

$$y - 4 = \pm 0.697\sqrt{1 + \frac{x^2}{0.514}}.$$

Since the position is below the center, we use the negative branch,

$$y = -0.697\sqrt{1 + \frac{x^2}{0.514}} + 4.$$

Similarly, we obtain

$$y = -0.919\sqrt{1 + \frac{x^2}{5.406}} + 2.5.$$

(d)

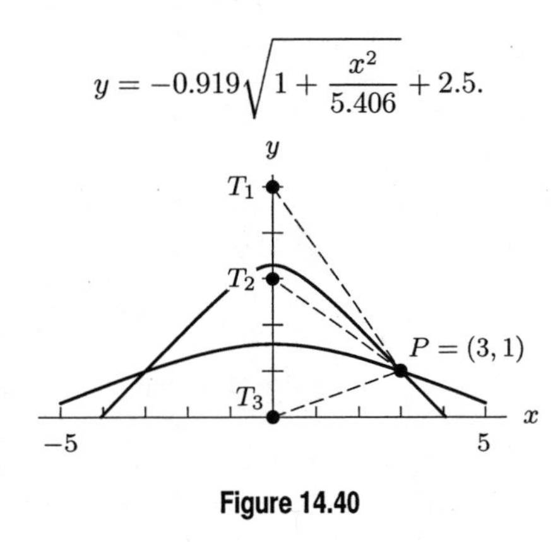

Figure 14.40

44. Consider the set of points (x, y) so that the sum of the distances to two focal points $(\pm c, 0)$ is constant. Note that the intercepts $(\pm a, 0)$ satisfy this condition.

Using the distance formula, we see that the distance from (x, y) to $(c, 0)$ is $\sqrt{(x-c)^2 + y^2}$ and the distance from (x, y) to $(-c, 0)$ is $\sqrt{(x+c)^2 + y^2}$. The distance from $(a, 0)$ to $(c, 0)$ is $a - c$ and the distance from $(a, 0)$ to $(-c, 0)$ is

$a + c$. We have:

$$\begin{aligned}
\text{Sum of distances from } (x,y) \text{ to focal points} &= \text{Sum of distances from } (a,0) \text{ to focal points}\\
\sqrt{(x-c)^2+y^2} + \sqrt{(x+c)^2+y^2} &= (a-c) + (a+c)\\
\sqrt{(x-c)^2+y^2} + \sqrt{(x+c)^2+y^2} &= 2a\\
\sqrt{(x-c)^2+y^2} &= 2a - \sqrt{(x+c)^2+y^2}.
\end{aligned}$$

We square both sides and simplify:

$$\begin{aligned}
(x-c)^2 + y^2 &= 4a^2 - 4a\sqrt{(x+c)^2+y^2} + ((x+c)^2 + y^2)\\
x^2 - 2cx + c^2 + y^2 &= 4a^2 - 4a\sqrt{(x+c)^2+y^2} + x^2 + 2cx + c^2 + y^2.
\end{aligned}$$

We cancel x^2, c^2, and y^2 from both sides, and subtract $2cx$ from both sides to obtain:

$$-4cx = 4a^2 - 4a\sqrt{(x+c)^2+y^2}.$$

We divide through by 4:

$$-cx = a^2 - a\sqrt{(x+c)^2+y^2},$$

and then isolate the square root:

$$a\sqrt{(x+c)^2+y^2} = a^2 + cx.$$

We square both sides again to obtain:

$$\begin{aligned}
a^2((x+c)^2 + y^2) &= a^4 + 2cxa^2 + c^2x^2\\
a^2(x^2 + 2cx + c^2 + y^2) &= a^4 + 2cxa^2 + c^2x^2\\
a^2x^2 + 2cxa^2 + a^2c^2 + a^2y^2 &= a^4 + 2cxa^2 + c^2x^2\\
a^2x^2 + a^2c^2 + a^2y^2 &= a^4 + c^2x^2.
\end{aligned}$$

Since $c < a$, there is a positive number b such that $c = \sqrt{a^2 - b^2}$. Using this fact, we have:

$$\begin{aligned}
a^2x^2 + a^2(a^2 - b^2) + a^2y^2 &= a^4 + (a^2 - b^2)x^2\\
a^2x^2 + a^4 - a^2b^2 + a^2y^2 &= a^4 + a^2x^2 - b^2x^2\\
-a^2b^2 + a^2y^2 &= -b^2x^2\\
b^2x^2 + a^2y^2 &= a^2b^2.
\end{aligned}$$

Dividing through by a^2b^2, we arrive at the equation for an ellipse given in Section 14.3:

$$\begin{aligned}
\frac{b^2x^2}{a^2b^2} + \frac{a^2y^2}{a^2b^2} &= \frac{a^2b^2}{a^2b^2}\\
\frac{x^2}{a^2} + \frac{y^2}{b^2} &= 1.
\end{aligned}$$

This is the equation for an ellipse with x-intercepts $\pm a$ and y-intercepts $\pm b$.

45. Consider the set of points (x, y) so that the difference of the distances to two focal points $(\pm c, 0)$ is constant. Note that the x-intercepts $(\pm a, 0)$ satisfy this condition.

Using the distance formula, we see that the distance from (x, y) to $(c, 0)$ is $\sqrt{(x-c)^2+y^2}$ and the distance from (x, y) to $(-c, 0)$ is $\sqrt{(x+c)^2+y^2}$. The distance from $(a, 0)$ to $(-c, 0)$ is $c + a$ and the distance from $(a, 0)$ to $(c, 0)$ is $c - a$ since $c > a$. We have:

$$\begin{aligned}
\text{Difference of distances from } (x,y) \text{ to focal points} &= \text{Difference of distances from } (a,0) \text{ to focal points}\\
\sqrt{(x-c)^2+y^2} - \sqrt{(x+c)^2+y^2} &= (c+a) - (c-a)\\
\sqrt{(x-c)^2+y^2} - \sqrt{(x+c)^2+y^2} &= 2a\\
\sqrt{(x-c)^2+y^2} &= 2a + \sqrt{(x+c)^2+y^2}.
\end{aligned}$$

We square both sides and simplify:

$$\begin{aligned}(x-c)^2+y^2 &= 4a^2+4a\sqrt{(x+c)^2+y^2}+((x+c)^2+y^2)\\ x^2-2cx+c^2+y^2 &= 4a^2+4a\sqrt{(x+c)^2+y^2}+x^2+2cx+c^2+y^2.\end{aligned}$$

We cancel x^2, c^2, and y^2 from both sides, and subtract $2cx$ from both sides to obtain:

$$-4cx = 4a^2+4a\sqrt{(x+c)^2+y^2}.$$

We divide through by 4:

$$-cx = a^2+a\sqrt{(x+c)^2+y^2},$$

and then isolate the square root:

$$-a\sqrt{(x+c)^2+y^2} = a^2+cx.$$

We square both sides again to obtain:

$$\begin{aligned}a^2((x+c)^2+y^2) &= a^4+2cxa^2+c^2x^2\\ a^2(x^2+2cx+c^2+y^2) &= a^4+2cxa^2+c^2x^2\\ a^2x^2+2cxa^2+a^2c^2+a^2y^2 &= a^4+2cxa^2+c^2x^2\\ a^2x^2+a^2c^2+a^2y^2 &= a^4+c^2x^2.\end{aligned}$$

Since $c > a$, there is a positive number b such that $c^2 = a^2+b^2$:

$$\begin{aligned}a^2x^2+a^2(a^2+b^2)+a^2y^2 &= a^4+(a^2+b^2)x^2\\ a^2x^2+a^4+a^2b^2+a^2y^2 &= a^4+a^2x^2+b^2x^2\\ a^2b^2+a^2y^2 &= b^2x^2\\ b^2x^2-a^2y^2 &= a^2b^2.\end{aligned}$$

Dividing through by a^2b^2, we arrive at the equation for a hyperbola given in Section 14.3:

$$\begin{aligned}\frac{b^2x^2}{a^2b^2}-\frac{a^2y^2}{a^2b^2} &= \frac{a^2b^2}{a^2b^2}\\ \frac{x^2}{a^2}-\frac{y^2}{b^2} &= 1.\end{aligned}$$

This is the equation for a hyperbola.

Solutions for Section 14.6

Exercises

1. We use $x = \sinh t$, $y = \cosh t$, for $-\infty < t < \infty$.

2. We use $x = \sinh t$, $y = -\cosh t$, for $-\infty < t < \infty$.

3. We use $x = -\cosh t$, $y = \sinh t$, for $-\infty < t < \infty$.

4. We use $x = 2+\cosh t$, $y = 3+\sinh t$, for $-\infty < t < \infty$.

5. We use $x = 1+2\sinh t$, $y = -1-3\cosh t$, for $-\infty < t < \infty$.

6. We use $x = 1+2\cosh t$, $y = -1+3\sinh t$, for $-\infty < t < \infty$.

7. Rewrite the equation as

$$\frac{(y+3)^2}{1/9} - \frac{(x+1)^2}{1/4} = 1, \quad y < -3,$$

and use $x = -1 + \frac{1}{2}\sinh t$, $y = -3 - \frac{1}{3}\cosh t$, for $-\infty < t < \infty$.

8. Rewrite the equation as

$$\frac{(x+3)^2}{1/4} - (y-1)^2 = 1, \quad x > -3,$$

and use $x = -3 + \frac{1}{2}\cosh t$, $y = 1 + \sinh t$, for $-\infty < t < \infty$.

9. Divide by 36 to rewrite the equation as

$$\frac{(y+3)^2}{4} - \frac{(x+1)^2}{9} = 1, \quad y > -3,$$

and use $x = -1 + 3\sinh t$, $y = -3 + 2\cosh t$, for $-\infty < t < \infty$.

10. Divide by 6 to rewrite the equation as

$$\frac{12(x-1)^2}{6} - \frac{3(y+2)^2}{6} = 1, \quad x > 1$$
$$\frac{(x-1)^2}{1/2} - \frac{(y+2)^2}{2} = 1, \quad x > 1,$$

so use $x = 1 + (1/\sqrt{2})\cosh t$, $y = -2 + \sqrt{2}\sinh t$, for $-\infty < t < \infty$.

11. Substituting $-x$ for x in the formula for $\sinh x$ gives

$$\sinh(-x) = \frac{e^{-x} - e^{-(-x)}}{2} = \frac{e^{-x} - e^{x}}{2} = -\frac{e^{x} - e^{-x}}{2} = -\sinh x.$$

12. Substitute $x = 0$ into the formula for $\sinh x$. This yields

$$\sinh 0 = \frac{e^0 - e^{-0}}{2} = \frac{1-1}{2} = 0.$$

13. The graph of $\sinh x$ in the text suggests that

$$\text{As } x \to \infty, \quad \sinh x \to \tfrac{1}{2}e^x$$
$$\text{As } x \to -\infty, \quad \sinh x \to -\tfrac{1}{2}e^{-x}.$$

Using the facts that

$$\text{As } x \to \infty, \quad e^{-x} \to 0,$$
$$\text{As } x \to -\infty, \quad e^{x} \to 0,$$

we can predict the same results algebraically:

$$\text{As } x \to \infty, \quad \sinh x = \tfrac{e^x - e^{-x}}{2} \to \tfrac{1}{2}e^x$$
$$\text{As } x \to -\infty, \quad \sinh x = \tfrac{e^x - e^{-x}}{2} \to -\tfrac{1}{2}e^{-x}.$$

Problems

14. Completing the square on $x^2 - 2x$ and recognizing $y^2 - 4y + 4$ as a perfect square gives

$$x^2 - 2x = y^2 - 4y + 4$$
$$(x-1)^2 - 1 = (y-2)^2$$
$$(x-1)^2 - (y-2)^2 = 1.$$

Thus, we use $x = 1 + \cosh t$, $y = 2 + \sinh t$ for $-\infty < t < \infty$.

15. Factoring out -4 from $-4y^2+8y=-4(y^2-2y)$ and completing the square on x^2+2x and y^2-2y gives

$$\begin{aligned} x^2+2x-4(y^2-2y)&=7, \quad x<-1\\ (x+1)^2-1-4((y-1)^2-1)&=7, \quad x<-1\\ (x+1)^2-1-4(y-1)^2+4&=7, \quad x<-1\\ (x+1)^2-4(y-1)^2&=4, \quad x<-1. \end{aligned}$$

Dividing by 4 to get 1 on the right,

$$\frac{(x+1)^2}{4}-(y-1)^2=1, \quad x<-1,$$

so we use $x=-1-2\cosh t, y=1+\sinh t$ for $-\infty<t<\infty$.

16. Factoring out -4 from $-4x^2+8x=-4(x^2-2x)$ and completing the square on x^2-2x gives

$$\begin{aligned} y^2-4(x^2-2x)&=12, \quad y<0\\ y^2-4((x-1)^2-1)&=12, \quad y<0\\ y^2-4(x-1)^2+4&=12, \quad y<0\\ y^2-4(x-1)^2&=8, \quad y<0. \end{aligned}$$

Dividing by 8 to get 1 on the right,

$$\frac{y^2}{8}-\frac{(x-1)^2}{2}=1, \quad y<0,$$

so we use $x=1+\sqrt{2}\sinh t, y=-\sqrt{8}\cosh t=-2\sqrt{2}\cosh t$ for $-\infty<t<\infty$.

17. Factoring out 2 from $2x^2-12x=2(x^2-6x)$ and 4 from $4y^2+4y=4(y^2+y)$ and completing the square on x^2-6x and y^2+y gives

$$\begin{aligned} 25+2(x^2-6x)&=4(y^2+y), \quad y>-\frac{1}{2}\\ 25+2((x-3)^2-9)&=4\left(\left(y+\frac{1}{2}\right)^2-\frac{1}{4}\right), \quad y>-\frac{1}{2}\\ 25+2(x-3)^2-18&=4\left(y+\frac{1}{2}\right)^2-1, \quad y>-\frac{1}{2}\\ 2(x-3)^2&=4\left(y+\frac{1}{2}\right)^2-8, \quad y>-\frac{1}{2}. \end{aligned}$$

Then, moving $4(y+\frac{1}{2})^2$ to the left and dividing by -8 to get 1 on the right,

$$\begin{aligned} \frac{2(x-3)^2}{-8}-\frac{4\left(y+\frac{1}{2}\right)^2}{-8}&=1, \quad y>-\frac{1}{2}\\ \frac{\left(y+\frac{1}{2}\right)^2}{2}-\frac{(x-3)^2}{4}&=1, \quad y>-\frac{1}{2}, \end{aligned}$$

so we use $x=3+2\sinh t, y=-\frac{1}{2}+\sqrt{2}\cosh t$ for $-\infty<t<\infty$.

18. **(a)** See Figure 14.41.
(b) See Figure 14.42.

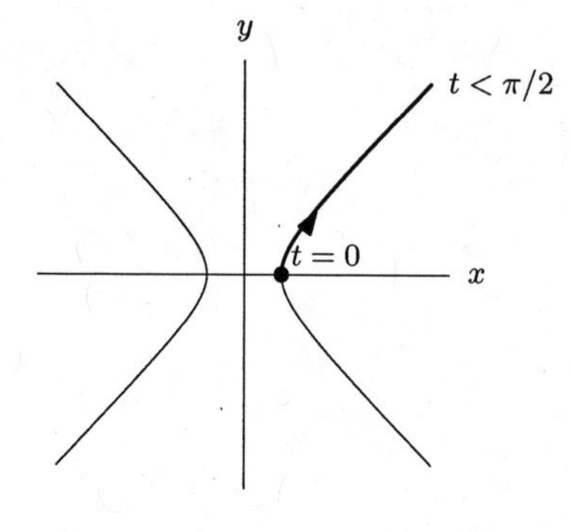

Figure 14.41

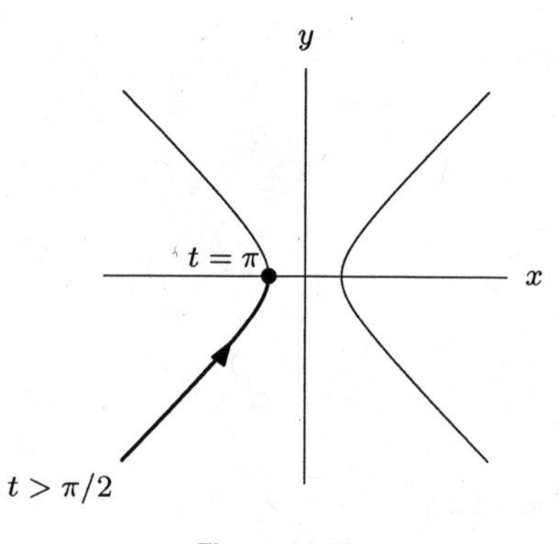

Figure 14.42

(c) See Figure 14.43.
(d) See Figure 14.44.

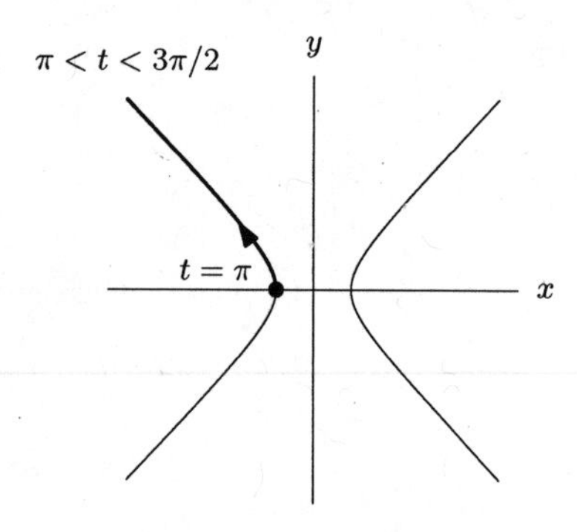

Figure 14.43

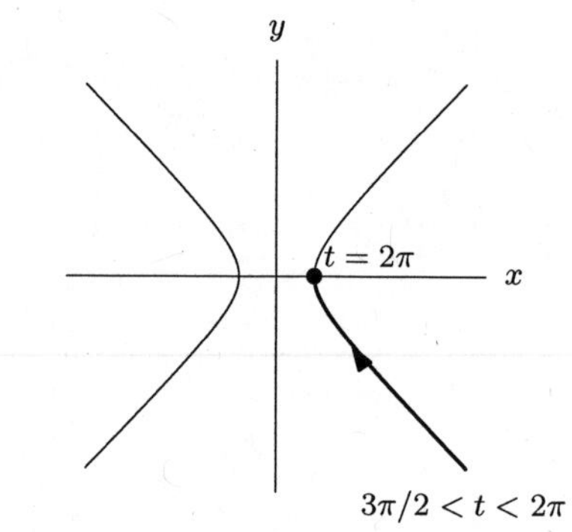

Figure 14.44

19. Comparison of the hyperbola equation

$$\frac{(x-h)^2}{a^2} - \frac{(y-k)^2}{b^2} = 1$$

with the identity

$$\cosh^2 t - \sinh^2 t = 1$$

suggests the equations

$$\frac{x-h}{a} = \cosh t \quad \text{and} \quad \frac{y-k}{b} = \sinh t.$$

This gives the parameterization:

$$x = h + a\cosh t \quad \text{and} \quad y = k + b\sinh t.$$

20. Yes. First we observe that

$$\sinh 2x = \frac{e^{2x} - e^{-2x}}{2}.$$

Now let's calculate

$$(\sinh x)(\cosh x) = \left(\frac{e^x - e^{-x}}{2}\right)\left(\frac{e^x + e^{-x}}{2}\right)$$

$$= \frac{(e^x)^2 - (e^{-x})^2}{4}$$
$$= \frac{e^{2x} - e^{-2x}}{4}$$
$$= \frac{1}{2} \sinh 2x.$$

Thus, we see that

$$\sinh 2x = 2 \sinh x \cosh x.$$

21. Yes. First, we observe that

$$\cosh 2x = \frac{e^{2x} + e^{-2x}}{2}.$$

Now, using the fact that $e^x \cdot e^{-x} = 1$, we calculate

$$\begin{aligned}
\cosh^2 x &= \left(\frac{e^x + e^{-x}}{2}\right)^2 \\
&= \frac{(e^x)^2 + 2e^x \cdot e^{-x} + (e^{-x})^2}{4} \\
&= \frac{e^{2x} + 2 + e^{-2x}}{4}.
\end{aligned}$$

Similarly, we have

$$\begin{aligned}
\sinh^2 x &= \left(\frac{e^x - e^{-x}}{2}\right)^2 \\
&= \frac{(e^x)^2 - 2e^x \cdot e^{-x} + (e^{-x})^2}{4} \\
&= \frac{e^{2x} - 2 + e^{-2x}}{4}.
\end{aligned}$$

Thus, to obtain $\cosh 2x$, we need to add (rather than subtract) $\cosh^2 x$ and $\sinh^2 x$, giving

$$\begin{aligned}
\cosh^2 x + \sinh^2 x &= \frac{e^{2x} + 2 + e^{-2x} + e^{2x} - 2 + e^{-2x}}{4} \\
&= \frac{2e^{2x} + 2e^{-2x}}{4} \\
&= \frac{e^{2x} + e^{-2x}}{2} \\
&= \cosh 2x.
\end{aligned}$$

Thus, we see that the identity relating $\cosh 2x$ to $\cosh x$ and $\sinh x$ is

$$\cosh 2x = \cosh^2 x + \sinh^2 x.$$

22. For $-5 \leq x \leq 5$, we have the graphs of $y = a\cosh(x/a)$ shown in Figure 14.45.

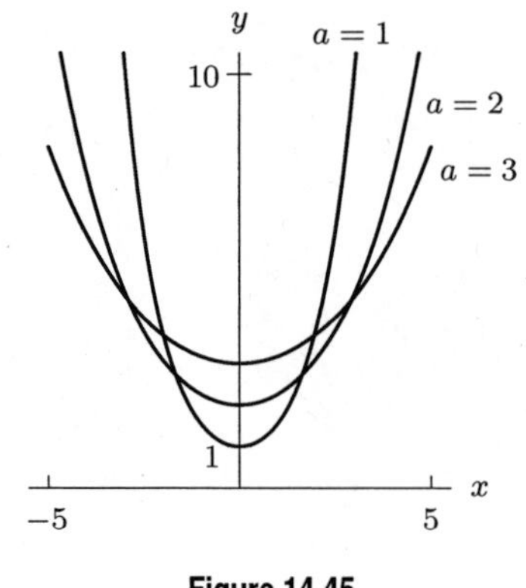

Figure 14.45

Increasing the value of a makes the graph flatten out and raises the minimum value. The minimum value of y occurs at $x = 0$ and is given by

$$y = a\cosh\left(\frac{0}{a}\right) = a\left(\frac{e^{0/a} + e^{-0/a}}{2}\right) = a.$$

23. Using the formula for $\sinh x$ with imaginary inputs, we have

$$\sinh(ix) = \frac{e^{ix} - e^{-ix}}{2}.$$

Substituting $e^{ix} = \cos x + i\sin x$ and $e^{-ix} = \cos x - i\sin x$, we have

$$\begin{aligned}\sinh(ix) &= \frac{(\cos x + i\sin x) - (\cos x - i\sin x)}{2}\\ &= i\sin x.\end{aligned}$$

24. We know that $\cosh(iz) = \cos z$, where z is real. Substituting $z = ix$, where x is real so z is imaginary, we have

$$\begin{aligned}\cosh(iz) &= \cos z\\ \cosh(i\cdot ix) &= \cos(ix) \qquad \text{substitute } z = ix\\ \cosh(-x) &= \cos(ix).\end{aligned}$$

But $\cosh(-x) = \cosh(x)$, thus

$$\cosh x = \cos(ix).$$

25. We know that $\sinh(iz) = i\sin z$, where z is real. Substituting $z = ix$, where x is real so z is imaginary, we have

$$\begin{aligned}\sinh(iz) &= i\sin z\\ \sinh(i\cdot ix) &= i\sin(ix) \qquad \text{substituting } z = ix\\ \sinh(-x) &= i\sin(ix).\end{aligned}$$

But $\sinh(-x) = -\sinh(x)$, thus we have

$$-\sinh x = i\sin(ix).$$

Multiplying both sides by i gives

$$-i\sinh x = -1\sin(ix).$$

Thus,

$$i\sinh x = \sin(ix).$$

Solutions for Chapter 14 Review

Exercises

1. The coefficients of x^2 and y^2 are equal, so this is a circle with center $(0, 3)$ and radius $\sqrt{5}$.

2. Rearranging the terms gives

$$2(x+1)^2 + 2(y-4)^2 = 7,$$

so, dividing by 2,

$$(x+1)^2 + (y-4)^2 = \frac{7}{2}.$$

The coefficients of x^2 and y^2 are equal, so this is a circle with center $(-1, 4)$ and radius $\sqrt{7/2}$.

3. Hyperbola, centered at $(0, 1)$, with $a = 2$, $b = 3$ and opening left-right.

4. Rewriting as

$$(x+1)^2 + \frac{(y+2)^2}{8} = 1$$

shows that this is an ellipse centered at $(-1, -2)$, with $a = 1$, $b = \sqrt{8} = 2\sqrt{2}$.

5. Dividing by 36 to get 1 on the right gives

$$\frac{9(x-5)^2}{36} + \frac{4y^2}{36} = \frac{36}{36}$$
$$\frac{(x-5)^2}{4} + \frac{y^2}{9} = 1.$$

This is an ellipse centered at $(5, 0)$, with $a = 2$, $b = 3$.

6. Rewriting as

$$7(y-1)^2 - 9x^2 = 63$$

and dividing by 63 to get 1 on the right gives

$$\frac{7(y-1)^2}{63} - \frac{9x^2}{63} = 63$$
$$\frac{(y-1)^2}{9} - \frac{x^2}{7} = 1.$$

This is a hyperbola centered at $(0, 1)$, with $a = \sqrt{7}$, $b = 3$ and opening up-down.

7. Rewriting as

$$2\left(x + \frac{1}{3}\right) - 3\left(y - \frac{1}{2}\right)^2 = -6$$

and dividing by -6 to get 1 on the right gives

$$\frac{2\left(x+\frac{1}{3}\right)^2}{-6} - \frac{3\left(y-\frac{1}{2}\right)^2}{-6} = 1$$
$$\frac{\left(y-\frac{1}{2}\right)^2}{2} - \frac{\left(x+\frac{1}{3}\right)^2}{3} = 1.$$

This is a hyperbola centered at $(-\frac{1}{3}, \frac{1}{2})$, with $a = \sqrt{3}$, $b = \sqrt{2}$ and opening up-down.

8. Multiplying both sides by $(y+2)^2$ gives

$$4 - 2(x-1)^2 = (y+2)^2.$$

Rearranging terms gives

$$2(x-1)^2 + (y+2)^2 = 4.$$

Dividing by 4 to get 1 on the right gives

$$\frac{(x-1)^2}{2} + \frac{(y+2)^2}{4} = 1.$$

This is an ellipse centered at $(1, -2)$, with $a = \sqrt{2}, b = 2$.

9. One possible answer is $x = 3\cos t, y = -3\sin t, 0 \le t \le 2\pi$.

10. One possible answer is $x = 2 + 5\cos t, y = 1 + 5\sin t, 0 \le t \le 2\pi$.

11. The parameterization $x = 2\cos t,\ y = 2\sin t,\ 0 \le t \le 2\pi$, is a circle of radius 2 traced out counterclockwise starting at the point $(2, 0)$. To start at $(-2, 0)$, put a negative in front of the first coordinate

$$x = -2\cos t \quad y = 2\sin t, \qquad 0 \le t \le 2\pi.$$

Now we must check whether this parameterization traces out the circle clockwise or counterclockwise. Since when t increases from 0, $\sin t$ is positive, the point (x, y) moves from $(-2, 0)$ into the second quadrant. Thus, the circle is traced out clockwise, so this is one possible parameterization.

12. The parameterization $(x, y) = (4 + 4\cos t, 4 + 4\sin t)$ gives the correct circle, but starts at $(8, 4)$. To start on the x-axis we need $y = 0$ at $t = 0$, thus one possible parameterization is

$$x = 4 + 4\cos\left(t - \frac{\pi}{2}\right), \quad y = 4 + 4\sin\left(t - \frac{\pi}{2}\right).$$

13. The ellipse $x^2/25 + y^2/49 = 1$ can be parameterized by $x = 5\cos t,\ y = 7\sin t, 0 \le t \le 2\pi$.

14. We have $a = 1$ and $b = 2$, so the implicit equation for the hyperbola is

$$x^2 - \frac{y^2}{4} = 1,$$

so we can use the parameterization $x = \sec t = 1/\cos t, y = 2\tan t, 0 \le t \le 2\pi$.

15. The parameterization $x = -3\cos t,\ y = 7\sin t,\ 0 \le t \le 2\pi$, starts at the right point but sweeps out the ellipse in the wrong direction (the y-coordinate becomes positive as t increases). Thus, a possible parameterization is $x = -3\cos(-t) = -3\cos t, y = 7\sin(-t) = -7\sin t, 0 \le t \le 2\pi$.

16. The center of the hyperbola is $(0, 5)$, so $b = 7 - 5 = 2$. Since the slope of the asymptote is $b/a = 4$, we have $2/a = 4$ and $a = 1/2$. The implicit equation for the hyperbola is

$$\frac{(y-5)^2}{2} - \frac{x^2}{1/2} = 1.$$

Thus, one possible parameterization is $x = \frac{1}{2}\sec t = 1/(2\cos t), y = 5 + 2\tan t$, for $0 \le t \le 2\pi$,

17. In the form

$$\frac{x^2}{a^2} + \frac{y^2}{b^2} = 1$$

we see that $a > b$. Therefore the focal points are at $(\pm c, 0)$, where $c = \sqrt{a^2 - b^2} = \sqrt{25 - 4} = \sqrt{21}$. The two focal points are $(\sqrt{21}, 0)$ and $(-\sqrt{21}, 0)$.

18. Divide by 5 we get the form

$$\frac{y^2}{b^2} - \frac{x^2}{a^2} = 1$$

with $a^2 = b^2 = 5$. The focal points lie on the same axis as the positive squared term. Therefore the focal points are at $(0, \pm c)$, where $c = \sqrt{a^2 + b^2} = \sqrt{5 + 5} = \sqrt{10}$. The two focal points are $(0, \sqrt{10})$ and $(0, -\sqrt{10})$.

19. In the parabola form $y = ax^2$ we see that $a = 5$. In this form the focal point lies on the vertical axis and the distance c, from center to the focal point, is $c = 1/(4a)$. We find

$$c = \frac{1}{4(5)} = \frac{1}{20}.$$

The focal point is at $(0, 1/20)$. The directrix line is perpendicular to the y-axis and a distance of $c = 1/20$ below the vertex. The equation of that line is $y = -1/20$.

Problems

20. $x = t, y = 5$.

21. A parametric equation for the circle is

$$x = \cos t, y = \sin t.$$

As t increases from 0, we have x increasing and y decreasing, which is a counterclockwise movement, so this parameterization is correct.

22. The parameterization $x = 2\cos t$, $y = 2\sin t$ has the right radius but starts at the point $(2, 0)$. To start at $(0, 2)$, we can use $x = 2\cos(t + \pi/2)$, $y = 2\sin(t + \pi/2)$.

23. Completing the square on $x^2 + 2x$ gives

$$\begin{aligned} x^2 + 2x + y^2 &= 0 \\ (x+1)^2 - 1 + y^2 &= 0 \\ (x+1)^2 + y^2 &= 1. \end{aligned}$$

The coefficients of x^2 and y^2 are equal, so this is a circle with center $(-1, 0)$ and radius 1.

24. Rewriting as

$$2x^2 - 4x - y^2 + 2y,$$

factoring out 2 from $2x^2 - 4x = 2(x^2 - 2x)$ and -1 from $-y^2 + 2y = -(y^2 - 2y)$, and completing the square on $x^2 - 2x$ and $y^2 - 2y$ gives

$$\begin{aligned} 2(x^2 - 2x) - (y^2 - 2y) &= 0 \\ 2((x-1)^2 - 1) - ((y-1)^2 - 1) &= 0 \\ 2(x-1)^2 - 2 - (y-1)^2 + 1 &= 0 \\ 2(x-1)^2 - (y-1)^2 &= 1. \end{aligned}$$

This is a hyperbola centered at $(1, 1)$ with $2 = 1/a^2$, so $a = 1/\sqrt{2}$, $b = 1$, and opening left-right.

25. Factoring out 6 from $6x^2 - 12x = 6(x^2 - 2x)$ and 9 from $9y^2 + 6y = 9(y^2 + \frac{2}{3}y)$, and completing the square on $x^2 - 2x$ and $y^2 + \frac{2}{3}y$ gives

$$\begin{aligned} 6(x^2 - 2x) - 9\left(y^2 + \frac{2}{3}y\right) + 1 &= 0 \\ 6((x-1)^2 - 1) + 9\left(\left(y + \frac{1}{3}\right)^2 - \frac{1}{9}\right) + 1 &= 0 \\ 6(x-1)^2 - 6 + 9\left(y + \frac{1}{3}\right)^2 - 1 + 1 &= 0 \\ 6(x-1)^2 + 9\left(y + \frac{1}{3}\right)^2 &= 6. \end{aligned}$$

(Alternatively, you may recognize $9y^2 + 6y + 1$ as the perfect square $(3y + 1)^2$). Dividing by 6 to get 1 on the right

$$\begin{aligned} (x-1)^2 + \frac{9\left(y + \frac{1}{3}\right)^2}{6} &= 1 \\ (x-1)^2 + \frac{3\left(y + \frac{1}{3}\right)^2}{2} &= 1. \end{aligned}$$

This is an ellipse centered at $(1, -\frac{1}{3})$ with $a = 1$, $b^2 = 2/3$, so $b = \sqrt{2/3}$.

26. Multiplying by $2y - y^2$ gives

$$\begin{aligned} x^2 - 4x - (2y - y^2) &= 0 \\ x^2 - 4x + y^2 - 2y &= 0. \end{aligned}$$

Completing the square leads to

$$\begin{aligned} (x-2)^2 - 4 + (y-1)^2 - 1 &= 0 \\ (x-2)^2 + (y-1)^2 &= 5. \end{aligned}$$

The coefficients of x^2 and y^2 are equal, so this is a circle with center $(2, 1)$ and radius $\sqrt{5}$.

27. The graph of these equations with $-50 \le t \le 50$ is shown in Figure 14.46:

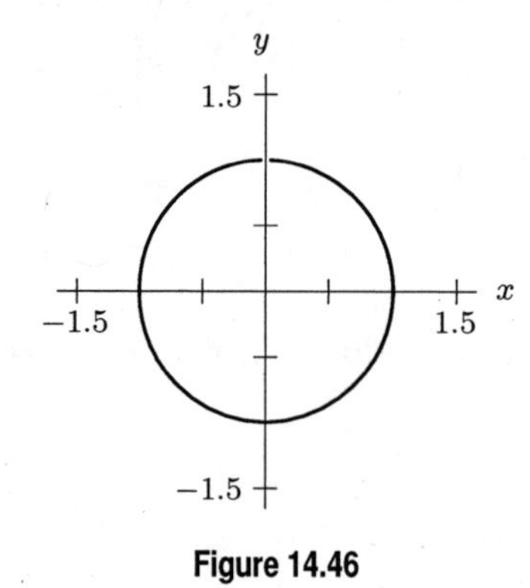

Figure 14.46

All of the points lie on the unit circle. (You can check this since $x^2 + y^2 = 1$.) While your graph may appear to be a complete circle, there is no value of t that gives the point $x = 0$, $y = 1$. This is because the equation

$$y = \frac{t^2 - 1}{t^2 + 1} = 1$$

has no real solution. Only when t approaches positive or negative infinity does the point get close to $(0, 1)$. Technically, it is not a circle.

28. The plot looks like Figure 14.47.

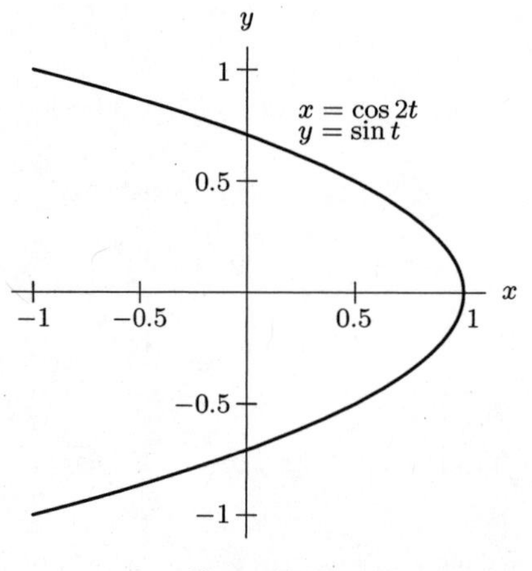

Figure 14.47

which does appear to be part of a parabola. To prove that it is, we note that we have

$$x = \cos 2t$$

$$y = \sin t$$

and must somehow find a relationship between x and y. Recall the trigonometric identity

$$\cos 2t = 1 - 2\sin^2 t.$$

Thus we have $x = 1 - 2y^2$, which is a parabola lying along the x-axis, for $-1 \le y \le 1$.

29. **(a)** Since P moves in a circle we have

$$x = 10\cos t$$

$$y = 10\sin t.$$

This completes a revolution in time 2π.

(b) First, consider the planet as stationary at (x_0, y_0). Then the equations for M are

$$x = x_0 + 3\cos 8t$$

$$y = y_0 + 3\sin 8t.$$

The factor of 8 is inserted because for every $2\pi/8$ units of time, $8t$ covers 2π, which is one orbit. But since P moves, we must replace (x_0, y_0) by the position of P. So we have

$$x = 10\cos t + 3\cos 8t$$

$$y = 10\sin t + 3\sin 8t.$$

(c) See Figure 14.48.

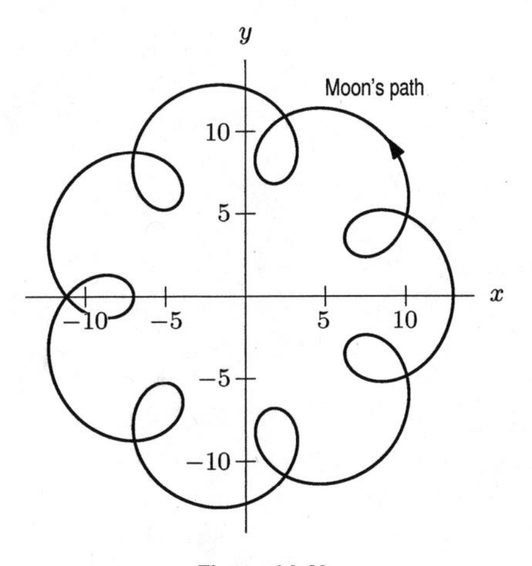

Figure 14.48

30. **(a)** The hyperbola is centered at the origin and opens upward so the standard equation is of the form

$$\frac{y^2}{1} - \frac{x^2}{3} = 1.$$

Thus $a^2 = 3$ and $b^2 = 1$. Using $c^2 = a^2 + b^2 = 3 + 1 = 4$, we find $c = 2$. The focal points are at $(0, \pm 2)$. Note that $Q = (0, 2)$ is a focus.

(b) A reflective beam from P to Q is possible since Q is a focus of the hyperbola. We need to aim our beam toward the other focal point $(0, -2)$. We find the equation of the line from $(8, 8)$ to $(0, -2)$. The slope is $m = 10/8 = 5/4$ and the y-intercept is -2, thus the desired linear equation is $y = \frac{5}{4}x - 2$.

CHECK YOUR UNDERSTANDING

1. True. The path is on the line $3x - 2y = 0$.
2. True. The path is on the line $3x - 2y = 0$.
3. True. Both paths are given by the line $y = x$.
4. True. Since $x = \sin(\pi/2) = 1$ and $y = \cos(\pi/2) = 0$, the object is at $(1, 0)$.
5. False. Starting at $t = 0$ the object is at $(0, 1)$. For $0 \le t \le \pi$, $x = \sin(t/2)$ increases and $y = \cos(t/2)$ decreases, thus the motion is clockwise.
6. True. At $t = 5$ we have $x = 10$, $y = 16$.
7. False. It is the graph of two straight lines, $y = x$ and $y = -x$.
8. False. The center is correct but the radius is 2.
9. False. The standard form of the circle is $(x + 4)^2 + (y + 5)^2 = 141$. The center is $(-4, -5)$.
10. False. The standard form of the circle is $(x + 4)^2 + (y + 5)^2 = 141$. The radius is $\sqrt{141}$.
11. True. The equations are a parametric description of the graph of $y = x^2$.
12. False. The graph is two lines through the origin, $y = 2x$ and $y = -2x$. The graph of $y^2 - 4x^2 = a$ is a hyperbola if $a \neq 0$.
13. False. There are many parameterizations; $x = \cos t, y = \sin t$ and $x = \sin t, y = \cos t$ are two of them.
14. True. Solving $x = 2t + 1$ for t, we get $t = \frac{1}{2}(x - 1)$. Substitute this expression for t in $y = t - 1$ and simplify to obtain $y = \frac{1}{2}(x - 3)$.
15. True. The given equation is the standard form of the equation of a circle with center $(1, -1)$ and radius $\sqrt{5}$.
16. False. Since the coefficients of x^2 and y^2 are different, and the equation can be written as
$$\frac{x^2}{9} + \frac{y^2}{4} = 1,$$
this is an ellipse.
17. True. Since
$$\frac{x^2}{4} + y^2 = \frac{(2\cos t)^2}{4} + (\sin t)^2 = \cos^2 t + \sin^2 t = 1,$$
these equations parameterize the ellipse $(x^2/4) + y^2 = 1$.
18. False. The horizontal axis extends from $(-1/3, 0)$ to $(1/3, 0)$ and has length $2/3$. The vertical axis extends from $(0, -1/2)$ to $(0, 1/2)$ and has length 1. The vertical axis is longer.
19. True. Since the positive term is $\dfrac{(y-k)^2}{b^2}$, the hyperbola opens up and down.
20. True. The asymptotes are the diagonal lines through the corners of the box of width 4 and height 6 centered at the origin. The box has vertices $(\pm 2, \pm 3)$ and its diagonals have slope $\pm 3/2$. The two asymptotes are $y = (3/2)x$ and $y = -(3/2)x$. On the other hand, the equation $x^2/4 - y^2/9 = 0$ can be written as $(x/2 - y/3)(x/2 + y/3) = 0$ so its graph is the two lines $x/2 - y/3 = 0$ and $x/2 + y/3 = 0$, which are indeed the asymptotes.
21. False. It is defined as $\sinh x = \dfrac{e^x - e^{-x}}{2}$.
22. True. As x approaches infinity, the value of e^{-x} approaches zero in the definition $\cosh x = \dfrac{e^x + e^{-x}}{2}$. The remaining value is $\frac{1}{2}e^x$.
23. False. The correct identity is $\cosh^2 x - \sinh^2 x = 1$.
24. False. The hyperbolic functions are not periodic.
25. True. We have $\cosh(-x) = (e^{-x} + e^{-(-x)})/2 = (e^x + e^{-x})/2 = \cosh x$, so the hyperbolic cosine is an even function.
26. True. We have $\cosh x > e^x/2$ and increasing exponential functions all grow faster than any power of x as $x \to \infty$.

27. False. $\cosh 0 = 1$.

28. True. We have

$$\sinh x = \frac{e^x}{2} - \frac{e^{-x}}{2} < \frac{e^x}{2} + \frac{e^{-x}}{2} = \cosh x.$$

29. False. We have

$$\sinh \pi = \frac{e^\pi}{2} - \frac{e^{-\pi}}{2} = 11.549 \neq 0.$$

We do have $\sin \pi = 0$.

30. True. We have

$$\cosh(\ln 2) = \frac{e^{\ln 2} + e^{-\ln 2}}{2} = \frac{2 + 1/2}{2} = \frac{5}{4}.$$

31. True. If the lengths of the major and minor axes are a and b, then the distance between the two foci is $2c = 2\sqrt{a^2 - b^2}$.

32. True.

33. False. The two asymptotes are entirely in the region between the two branches of the hyperbola. The two focal points are outside this region. To move from an asymptote to a focal point you must cross the hyperbola.